Personalized Management in Orthopedics and Traumatology

Personalized Management in Orthopedics and Traumatology

Guest Editors

Markus Gesslein
Eduard Kurz

MDPI

Basel • Beijing • Wuhan • Barcelona • Belgrade • Novi Sad • Cluj • Manchester

Guest Editors
Markus Gesslein
Sport Orthopedics
Paracelsus Medical University
Nuremberg
Germany

Eduard Kurz
Department of Orthopedic
and Trauma Surgery
Martin-Luther-University
Halle-Wittenberg
Halle (Saale)
Germany

Editorial Office
MDPI AG
Grosspeteranlage 5
4052 Basel, Switzerland

This is a reprint of the Special Issue, published open access by the journal *Journal of Personalized Medicine* (ISSN 2075-4426), freely accessible at: www.mdpi.com/journal/jpm/special_issues/94N117Q23A.

For citation purposes, cite each article independently as indicated on the article page online and using the guide below:

Lastname, A.A.; Lastname, B.B. Article Title. *Journal Name* **Year**, *Volume Number*, Page Range.

ISBN 978-3-7258-3284-2 (Hbk)
ISBN 978-3-7258-3283-5 (PDF)
https://doi.org/10.3390/books978-3-7258-3283-5

Contents

About the Editors

Markus Gesslein

Prof. Dr. Markus Gesslein works at the Paracelsus Medical University Nuernberg and has a special interest in research concerning arthroscopy, sports injuries, knee surgery, shoulder fracture, septic arthritis, ACL surgery, rotator cuff injuries, and concussion.

Eduard Kurz

Dr. Eduard Kurz, PhD, works as a researcher at the Department of Orthopedic and Trauma Surgery, Martin-Luther-University Halle-Wittenberg and has a special interest in research concerning rehabilitation, neuromuscular control, biomechanics, and return to sport.

Preface

Dear Colleagues,

There is an ongoing discussion about optimizing the treatment of acute and chronic conditions in orthopedics and after trauma to fit the needs of an aging population as well as those of the growing crowd of highly active patients. It becomes increasingly clear that the "one treatment fits all" concept has been shifting towards individualized treatment, tailored to the individual patient needs and activity level.

Individual treatment, for example, after joint injury in high-level or even professional sports, now includes an injury-specific "return to competition" rehabilitation which has been proven successful for returning to the same level of activity.

On the other hand, we are also facing enormous challenges concerning elderly patients with traumatic injuries or degenerative joint diseases. Treatment is often complicated by a vast amount of individual risk factors like osteoporosis, circulatory disorder, or enhanced risk of falling following immobilization.

This Special Issue aims to promote the integration of individual patients' needs, outcomes, and risk factors into clinical routine in orthopedics and after trauma.

Markus Gesslein and Eduard Kurz
Guest Editors

Journal of
Personalized Medicine

MDPI

Article

Investigating Painkiller Use in Amateur Football: A Coach's Perspective

Andreas Kopf [1], Werner Krutsch [2,3,4], Dominik Szymski [2,3], Johannes Weber [2,3], Volker Alt [2,3], Hermann Josef Bail [1], Niklas Engel [1], Johannes Rüther [1] and Lorenz Huber [2,3,*]

1 Department of Orthopaedics and Traumatology, Paracelsus Medical University, Breslauer Straße 201, 90471 Nuremberg, Germany; andreas.kopf@klinikum-nuernberg.de (A.K.); niklas.engel@klinikum-nuernberg.de (N.E.)

2 Department of Trauma Surgery, University Medical Centre Regensburg, Franz-Josef-Strauss Allee 11, 93053 Regensburg, Germany

3 FIFA Medical Centre of Excellence, University Medical Centre Regensburg, 93053 Regensburg, Germany

4 SportDocsFranken, 90473 Nuernberg, Germany

* Correspondence: lorenz1.huber@ukr.de

Abstract: Background/Objectives: Painkiller use in amateur sports and, especially, in football is increasingly being discussed, but the scientific data on this field are very limited. Therefore, the aim of this study was to investigate the prevalence of painkiller use in amateur football from the perspective of coaches, as well as to determine how and to which extent coaches can influence their teams in terms of painkiller use and prevention. Methods: Using an online questionnaire, a cross-sectional analysis of painkiller use in German amateur football from the 4th league to the lowest amateur classes was carried out from the perspective of team coaches. A total of 628 participants were contacted, and 400 (63.7%) completed the questionnaire completely and were therefore included in the evaluation. Results: Of the 400 participating team coaches in amateur football, 369 (92.3%) were male and 31 (7.7%) were female. The coaches reported that 36.2% (SD = 29.1) of their players have used painkillers at some point due to football-related pain in their career. The majority of coaches believed that the use of painkillers is not compatible with competition (74%), and even more believed that it is not compatible with football training (90.8%). Furthermore, 56.2% of the coaches themselves had already taken painkillers in their own football career for football-related pain, and 64% had already bought over-the-counter painkillers without a prescription. The use of painkillers increased in the higher playing levels. The availability of painkillers in first aid kits was reported by around 60%, but they were reported as freely accessible in the dressing room by only 10% of the coaches. Conclusions: This is the first study to describe the painkiller use in amateur football from the perspective of coaches. The prevalence of painkiller use in this study was found to be significantly lower than what is indicated in the data from the existing literature. The majority of coaches see the use of painkillers during games and training as incompatible, even though there is a large proportion of coaches who have already bought over-the-counter painkillers for football-related pain. As the first scientific analysis of team coaches, this study provides fundamental data for the prevention of excessive painkiller use in amateur football.

Keywords: football; soccer; amateur sports; medication; recreational; painkiller use

Citation: Kopf, A.; Krutsch, W.; Szymski, D.; Weber, J.; Alt, V.; Bail, H.J.; Engel, N.; Rüther, J.; Huber, L. Investigating Painkiller Use in Amateur Football: A Coach's Perspective. *J. Pers. Med.* **2024**, *14*, 1003. https://doi.org/10.3390/jpm14091003

Academic Editor: Jih-Yang Ko

Received: 15 August 2024
Revised: 6 September 2024
Accepted: 14 September 2024
Published: 20 September 2024

1. Introduction

The use of painkillers in football has received great public and scientific interest in recent years. Although the majority of the scientific data come from professional sports, there are also studies on the prevalence of painkiller use in amateur sports [1,2]. In professional sports, the use of painkillers has been described as having a higher prevalence than in the general population, regardless of the type of sport [3]. Most studies on painkiller use in sports focus on the major events with many participants. However, individual events like marathons often show high rates of painkiller use among non-professional athletes, as shown by quantitative studies [4,5].

In football, previous research has systematically collected data that depicts the use of medication at the FIFA World Cups from 2002 to 2018. These studies showed that more than half (55%) of the players took NSAIDs over the course of the tournament, with a enduringly constant prevalence of use over the years. They also showed that around 30% of players take NSAIDs directly before a game [6–9]. It is not known whether the use of painkillers was medically indicated or was due to dependence for treating pain or injuries. In addition, it was not described whether the medication was self-administered or prescribed by team doctors. Individual studies in professional football that have investigated the prevalence of painkiller use over an entire season have shown an even higher prevalence of 66% to over 90% [10,11]. However, the results of a meta-analysis involving around 40,000 data sets suggest that these findings cannot simply be transferred to amateur sports as it was shown that non-competitive athletes have a significantly lower use of painkillers compared to professional athletes [12]. The high intensity in professional football, especially at major tournaments such as the World Cup, raises doubts about whether the data collected applies to amateur football. Leyk et al. (2023) highlighted this research deficit in their review and called for the collection of extensive data on amateur athletes, especially footballers (who have come under particular public criticism) [3].

The goal of this study was, therefore, to analyse the painkiller use in national amateur football from the perspective of responsible coaches. This includes both the painkiller prevalence from the coaches' perspective and the role of coaches in relation to painkiller use. The coaches' attitude towards the compatibility of painkiller use in football, as well as their knowledge of the effects and side effects of painkillers, were also investigated. The hypothesis of this study was to understand whether the prevalence and patterns of painkiller use in amateur football are comparable to those found in other amateur sports or professional football cohorts investigated in previously published studies. Additionally, this study aimed to identify potential preventive strategies to reduce painkiller use in amateur football.

2. Methods

In this prospective cohort study, an online questionnaire was specifically designed for team coaches of German amateur football teams. The analysed football teams belong to the regional Football Association in Germany, the Bavarian FA, which includes more than 1.6 million football players and more than 4500 football clubs. The study population includes amateur teams from the highest amateur playing level (4th league) to the lowest playing levels and veteran football teams. All of the invited team coaches of the attending football clubs were able to complete the questionnaire anonymously from January 2022 to July 2022. The coaches were contacted via the Bavarian Football Association (BFV) and its regional offices, as well as by the Association of Football Coaches (GFT) from each of Bavaria's administrative districts. The participating team coaches were informed about the methods of this study and the questionnaire. Coaches who did not complete the questionnaire in full were excluded from this study.

The online questionnaire, which analysed the prevalence of pain, the use of painkillers in amateur football players, the availability of painkillers, and the indication of painkiller use in detail, consisted of 45 items on different topics. Since coach recommendations can influence player painkiller use, it was analysed how much the coaches knew about the painkiller use in their teams and how often advice was given. Another important point, which was analysed, was the access to painkillers in the locker room. The questionnaire was based on the consensus statement on injury definitions and data collection procedures of Fuller et al. [13], as well as on the most common types of injuries in football [14], on the use of nutritional supplements, and the use of pain medication in football, as detailed in previous studies [15]. The most relevant items were summarized and adapted specifically for amateur football.

IBM SPSS Statistics Version 28.0 was used for the statistical analysis and graphical presentation of the tables and figures. For the descriptive presentation of the results, frequencies were used in a percentage in addition to absolute numbers.

The Ethics Committee of the University of Regensburg (No. 22-2781-101) approved this study.

3. Results

A total of 628 coaches from amateur football accepted the invitation to take part in this study, and 400 of them (63.7%) completed and submitted the questionnaire. In total, 369 male coaches and 31 female coaches were included in the survey in the end. Almost 95% of the participating team coaches were former football players. In terms of coaching licenses, more than 85% of the participants owned the official UEFA-B or C-license. Moreover, 12.7% of the coaches worked in salaried amateur football from the fourth to the seventh playing level in Germany, while 87.3% of them worked in real recreational amateur football (Table 1).

Table 1. The anthropometric and football-specific data of the population.

	Mean ± SD
Age in years	45.6 ± 11.9
Height in cm	183.6 ± 81.6
Weight in kg	85.7 ± 13.6
Sex = male, *n* (%)	369 (92.3%)
Coaching career in years	14.5 ± 10.3
Former football player, *n* (%)	369 (92.3%)
Paid amateur football, *n* (%)	42 (12.7%)
Pure amateur football, *n* (%)	278 (87.3%)
Coaching license, *n* (%)	
- UEFA A-license	9 (2.4%)
- UEFA B-license	137 (36.5%)
- UEFA C-license	198 (52.8%)
- Elite youth license	10 (2.7%)
- License precursors	21 (5.6%)

Overall, the team coaches stated that 36.2% of the players in their current teams already used painkillers for football-associated pain (SD = 29.2). The majority (84.7%) of them reported about football-associated pain in their teams over the entire season, and 60% of them reported football-associated pain occurring every month. The fact that football playing may be painful was shown by the coaches' own previous injury histories, where just 0.5% of team coaches had no injury in their own career. The team coaches reported that 56.2% of them had used painkillers for football-related pain during their own playing careers (Table 2).

Table 2. The frequency of football-associated pain or injuries and the painkiller use in the team coaches' careers.

Football-Associated Pain, *n* (%)	
Every match and training	46 (11.6%)
Once a week	95 (24.3%)
Once a month	96 (24.6%)
Once every six months	94 (24.0%)
Not at all	60 (15.3%)

Table 2. *Cont.*

Football-associated painkiller use in their own career, *n* (%)	
Already used once	218 (56.2%)
Never used	170 (43.8%)
Football-associated injuries during their career, *n* (%)	
Contusion	305 (77.4%)
Ligament injury	278 (70.6%)
Muscle strain	249 (63.2%)
Skin injury (abrasion/laceration)	224 (56.9%)
Muscle fibre tear	205 (52.0%)
Back pain	186 (47.2%)
Cartilage/meniscus injury	170 (43.1%)
Capsule injury	157 (39.8%)
Groin/hip pain	157 (39.8%)
Foot/heel pain	130 (33.0%)
Tendon pain (patella/Achilles tendon)	121 (30.7%)
Fracture	107 (27.2%)
Disc/nerve pain	67 (17.0%)
Muscle bundle rupture	64 (16.2%)
Concussion	60 (15.2%)
Other	41 (10.4%)
None	2 (0.5%)

The availability of painkillers in amateur football showed that 85% of the coaches had received painkillers prescribed by a doctor for medically justifiable reasons. Additionally, 64% had bought over-the-counter painkillers at the pharmacy without a prescription. In contrast to this high availability, only around 10% of coaches left freely available painkillers for their players in the dressing room (Table 3).

Table 3. The availability of painkillers for amateur football players.

Medically Prescribed Painkillers, *n* (%)	
Have ever received	337 (85.5%)
Never received	57 (14.5%)
Painkillers purchased over-the-counter, *n* (%)	
Have ever acquired	251 (64.0%)
Never acquired	141 (36.0%)
Availability in the dressing room, *n* (%)	
Freely available	41 (10.6%)
Not available	347 (89.4%)

Around 75% of amateur team coaches saw no compatibility between painkillers and football competition, while a minority (around 25%) saw compatibility. Similarly, only a minority of around 7% of coaches would recommend the use of painkillers to maintain readiness to play, while 45% would never do so. According to the coaches, most pain medication was only taken after acute injuries or complaints in football (71.6%). However, 46% of coaches also stated that painkillers were sometimes taken preventively before a game (Table 4).

There were 80% of male team coaches who were in favour of the free availability of painkillers, while only 47.6% of their female colleagues said the same. Furthermore, 61.9% of the female coaches reported occasional or more frequent sport-independent pain, while only 47.6% of male coaches did so.

Significant age-related differences showed that team coaches over the age of 50 were more critical of the compatibility of painkillers and training compared to younger team coaches (over 50 years: 50%, under 32 years: 30%). It was also shown that coaches were rarely involved in the decision to take painkillers, regardless of age.

Around two thirds of the team coaches in the higher amateur football leagues (fourth division) saw a strong compatibility of painkillers with competition, while only around 20% of coaches in the lowest leagues saw this as the case. Painkillers were also available in around 80% of the first aid kits in the fourth division and around only 40% in the lowest leagues (Table 5). Due to the small number of participants from the fourth league, the differences in the league comparison were not statistically significant.

Table 4. The team coaches' views on the compatibility of painkillers and football.

Compatibility of Painkillers and Competition, *n* (%)	
Fully compatible	4 (1.0%)
Very compatible	16 (4.1%)
Predominantly compatible	82 (20.9%)
Predominantly incompatible	194 (49.5%)
Not compatible	96 (24.5%)
Use of painkillers to maintain readiness to play, *n* (%)	
Many cases	7 (1.8%)
Occasionally	21 (5.4%)
In exceptional cases	185 (47.4%)
Never	177 (45.4%)
Reasons for painkiller use, *n* (%)	
Injuries	156 (71.6%)
Preventive before match	102 (46.8%)
Preventive before training	28 (12.8%)
General everyday life	36 (16.5%)
Not specified	2 (0.9%)

Table 5. The differences in the occurrence of pain and painkiller use between age groups and divisions.

Compatibility of Painkillers and Training, *n* (%)	Age under 32	Age over 50
Predominantly compatible	5 (9.8%)	10 (6.4%)
Predominantly incompatible	29 (56.9%)	68 (43.3%)
Not compatible	17 (33.3%)	79 * (50.3%) $p = 0.047$
Request for advice on painkillers, *n* (%)		
Always	0 (0.0%)	1 (0.6%)
Often	1 (2.0%)	16 (10.3%)
Occasionally	8 (15.7%)	33 (21.2%)
Rare	22 (43.1%)	57 (36.5%)
Never	20 (39.2%)	49 (31.4%)
Compatibility of painkillers and competition, *n* (%)	4th league	Lowest 5 leagues
Very compatible	1 (11.1%)	9 (3.8%)
Predominantly compatible	5 (55.6%)	46 (19.4%)
Predominantly incompatible	1 (11.1%)	126 (52.9%)
Not compatible	2 (22.2%)	57 (23.9%)

Table 5. *Cont.*

Availability of painkillers in the first aid kit, *n* (%)		
Available	7 (77.8%)	89 (37.4%)
Not available	2 (22.2%)	149 (62.6%)

* $p < 0.05$.

4. Discussion

The primary objective of this study, i.e., to analyse the use of painkillers in amateur football from coach perspectives, can be answered, for the first time, with a lifetime prevalence over the coaches' own football careers of 36.2%, which is a lower rate when compared to comparable studies of professional football and other sports [3,15–17]. Some previous studies have described prevalences of painkiller use being over 90%; however, these numbers were recorded using different methods like doping control forms and, therefore, a direct comparison cannot be made in a sustainable manner [3]. Additionally, since the coaches, not the players, were surveyed in this study, it may have caused selection bias and led to a lower prevalence.

In total, 75% of the coaches in this study considered painkillers incompatible or rarely compatible with competition. This makes clear that the majority of the coaches in this study did not urge athletes to take painkillers and subordinate their health to the team's sporting performance goals, as is often criticized in the literature [18,19]. Thus, a minority of team coaches reported that there were freely available painkillers in the dressing rooms; in particular, in the higher leagues, there was also a high prevalence of painkillers in the first aid kits. It was also found that 25% of all the coaches considered painkillers and competition to be at least predominantly compatible. This study is the first to describe this fact, which is the key point that will help to support future preventive strategies.

The data from this study also show that playing football is highly associated with pain during or after sports performance. Around 10% of the team coaches reported football-associated pain every match day or training session, and just under a quarter reported it once a week. These results are consistent with the data from previous literature. In a recent study from Sweden, in which over 500 amateur footballers were examined by means of a questionnaire regarding their football-related pain, a football-associated pain prevalence of around 20% per week was also reported [20]. The fact that acute injuries and overuse complaints are the most common reason for taking painkillers is again shown by the data from this study. Just over 70% of the team coaches stated that acute injuries are a reason for taking painkillers, and it was by far the largest group of occasions for taking painkillers in this study population. More problematic, however, was the fact that around 45% of coaches also stated that painkillers are occasionally taken prophylactically before matches. This has already been shown in other sports [21], and the results of this study make it clear that there is still a need for education regarding painkillers use that is not medically indicated, meaning there is potential for improvement.

It is also known that the role of coaches is very important for players and can influence player behaviour with regard to painkiller use [22]. However, coaches are usually only asked about this topic sporadically or not at all in this population. Only 8% of coaches are regularly involved with their players in their painkiller use. Regarding the coaching staff surveyed, only few coaches were found to directly influence the painkiller use of their players. Furthermore, the management style of the team and the attitude towards painkillers can also have an influence on the team [22]. It was a minority, but around 5% of the coaches approved the use of painkillers during or before important games. A study of comparable parameters could not be found in the literature, but this seems reason enough to promote more education and information in this regard such that as many coaches as possible can be reached. The data show that responsible use of painkillers and a solid understanding of potential side effects should be part of player and coach educations.

The availability of painkillers in the club was also recorded as easy access could increase use. Painkiller availability in first aid kits was reported by around 60% of coaches.

The access also varied with playing level as there was a much higher availability in the highest analysed league (77.8%). It can also be claimed that the coaches were very responsible with the access to painkillers for their team as it was stated that, in over 90% of cases, medication was not left unattended in the changing room. The consideration of these environmental conditions, as in the present study, could not be found in the literature. Nevertheless, we saw that almost two thirds of the team coaches had already purchased over-the-counter painkillers for football-related pain without receiving a prescription.

When analysing the use of painkillers from the coaches' point of view, this study also revealed differences in the subgroups that had not been described at all in the previous scientific literature. A major difference between the sexes was also evident in the opinion on the free sale of common painkillers. In this study, only 47.6% of the female coaches were in favour of continued free sale, while this same sentiment was significantly more common among the male coaches (80%). Female coaches in this study reported non-sport-related pain more frequently (61.9% vs. 47.6%). Similar sex-specific differences in pain frequency and painkiller use have also been noted in the literature. For instance, Hager et al. (2021) found a higher frequency of pain among female volleyball players compared to males, and Wezenberg et al. (2023) also demonstrated this for football-related pain [20,23]. The higher use of painkillers by women is not limited to sports as studies of the general population also show this trend, and comorbidities such as dysmenorrhea, psychosocial circumstances, and stereotypical gender roles can contribute to differences in pain expression [24,25].

Furthermore, the older coaches were much more critical of the use of painkillers than their younger colleagues. Accordingly, 50% of older coaches also considered the use of medication during their players' training to be wrong, and 36% saw no occasion justifying the use of painkillers in football. Among younger coaches, this figure was only 33%. The players asked the older coaches, who were more critical of painkillers, for advice on painkiller use more frequently than the younger coaches. In total, 32% of the older coaches stated that they were occasionally or frequently asked for advice regarding painkillers and their use, while only 17% of the younger coaches group stated this. Data showing such age-related differences in coach attitudes to medication use have not been previously found in the literature and they were addressed for the first time in this study.

The differences in the painkiller analysis with regard to different league levels showed that coaches in the lower leagues considered the compatibility of painkiller use and football to be significantly less compatible than coaches in the fourth league (22% vs. 66%). This is consistent with the data from the literature, where the higher performance levels also showed a higher use of painkillers in football as well as in other sports [8,12,26,27]. However, the available data suggest that a higher level of performance generally correlates with an increased use of painkillers or a higher availability of painkillers. Increased sporting intensity and a higher level of training effort, as well as the resulting increase in pain and injuries, can explain this. The increased pressure, such as the players' own ambitions and the expectations of the sporting environment, especially in the transition to the semi-professional sector, can also be a reason for higher painkiller use. It is particularly important to educate all the players involved about the use of painkillers, as well as their possible negative consequences for the players and also, therefore, for the teams. This can further improve the prevention of unindicated painkiller use [28].

Due to the online questionnaire-based prospective cohort study design, there were some limitations. First, investigating topics like painkiller use, which may be influenced by social desirability, poses challenges in verifying the authenticity of the responding coaches' answers. There was also a possible selection bias in the separate analysis of the coaches. Furthermore, the participants were recruited exclusively from the federal state of Bavaria as the intensive cooperation between the Bavarian Football Association and its coaching associations made such a study possible However, even though there are some methodological weaknesses of this study, there are no apparent reasons for the results not being transferable to the whole of German amateur football. The comparability of the subgroups may also be limited due to the fact that the population studied differed greatly in terms of sex and playing levels. The results of the current status of the painkiller prevalence in German amateur football from the perspective of coaches are the only ones in the literature to date, which makes it very difficult to classify them in the context of the literature. However, the fact that these initial results were presented with a large study population of 400 coaches provides a solid basis of data for further study projects on this topic. The large number of test subjects is a major strength of this study, particularly given the lack of data in amateur sports and the often poor response rates in the studies in this area.

5. Conclusions

This is the first study with information about painkiller use in amateur football from the perspective of team coaches. It can be shown that the prevalence of painkiller use in this study was significantly lower than was found in the data from the existing literature. The majority of coaches see the use of painkillers during games and training as incompatible, and only 10% indicated that painkillers were freely accessible in the dressing room. This study also shows that football is painful and that a large proportion of team coaches are already taking over-the-counter painkillers, which are freely available without a prescription.

For future educational work, the information of our study is essential. The importance of well-founded advice for players and coaches regarding injury breaks and the avoidance of painkillers in training or matches must be ensured, and a lively exchange between players and coaches in this regard must be promoted. Future research should also approach amateur football players directly in order to analyse their intrinsic urge to use painkillers in football.

Author Contributions: Conceptualization: A.K. and L.H.; Methodology, A.K. and W.K.; Investigation, A.K.; Writing—Original Draft Preparation, A.K. and L.H.; Writing—Review and Editing, W.K., D.S., N.E., J.W. and J.R.; Visualization, L.H.; Supervision, H.J.B. and V.A. All authors have read and agreed to the published version of the manuscript.

Funding: This research received no external funding.

Institutional Review Board Statement: The study was conducted in accordance with the Declaration of Helsinki, and approved by the Ethics Committee of the University of Regensburg (protocol code No. 22-2781-101, date of approval 23 February 2022).

Informed Consent Statement: Informed consent was obtained from all the subjects involved in this study.

Data Availability Statement: The data generated in this study are included in the results of the published article.

Conflicts of Interest: W.K. is member of the Medical Committee of the German FA (DFB) and the Bavarian Football Association (BFV). All other authors state that they have no conflicts of interest.

References

1. Pöppel, K.; Kreutz, G.; Müller, M.; Büsch, D. Painkiller intake and problematic health literacy in sport and music students—A cross-sectional study. *Sci. Rep.* **2024**, *14*, 12517. [CrossRef] [PubMed]
2. Sachse, J.; Steinberg, A.; Mayr, I.; Schubert, B.; Argüeso, O.; Daniels, J.v. PILLENKICK—Schmerzmittelmissbrauch im Fußball. Available online: https://correctiv.org/en/top-stories/2020/06/08/kickonpills/ (accessed on 7 April 2024).
3. Leyk, D.; Rüther, T.; Hartmann, N.; Vits, E.; Staudt, M.; Hoffmann, M.A. Analgesic Use in Sports. *Dtsch. Arztebl. Int.* **2023**, *120*, 155–161. [CrossRef] [PubMed]
4. Almond, C.S.D.; Shin, A.Y.; Fortescue, E.B.; Mannix, R.C.; Wypij, D.; Binstadt, B.A.; Duncan, C.N.; Olson, D.P.; Salerno, A.E.; Newburger, J.W.; et al. Hyponatremia among runners in the Boston Marathon. *N. Engl. J. Med.* **2005**, *352*, 1550–1556. [CrossRef] [PubMed]
5. Küster, M.; Renner, B.; Oppel, P.; Niederweis, U.; Brune, K. Consumption of analgesics before a marathon and the incidence of cardiovascular, gastrointestinal and renal problems: A cohort study. *BMJ Open* **2013**, *3*, e002090. [CrossRef]
6. Tscholl, P.M.; Junge, A.; Dvorak, J. The use of medication and nutritional supplements during FIFA World Cups 2002 and 2006. *Br. J. Sports Med.* **2008**, *42*, 725–730. [CrossRef]

7. Tscholl, P.M.; Dvorak, J. Abuse of medication during international football competition in 2010—Lesson not learned. *Br. J. Sports Med.* **2012**, *46*, 1140–1141. [CrossRef]
8. Vaso, M.; Weber, A.; Tscholl, P.M.; Junge, A.; Dvorak, J. Use and abuse of medication during 2014 FIFA World Cup Brazil: A retrospective survey. *BMJ Open* **2015**, *5*, e007608. [CrossRef]
9. Oester, C.; Weber, A.; Vaso, M. Retrospective study of the use of medication and supplements during the 2018 FIFA World Cup Russia. *BMJ Open Sport Exerc. Med.* **2019**, *5*, e000609. [CrossRef]
10. Kavukcu, E.; Burgazlı, K.M. Preventive Health Perspective in Sports Medicine: The Trend at the Use of Medications and Nutritional Supplements during 5 Years Period between 2003 and 2008 in Football. *Balkan Med. J.* **2013**, *30*, 74–79. [CrossRef]
11. Taioli, E. Use of permitted drugs in Italian professional soccer players. *Br. J. Sports Med.* **2007**, *41*, 439–441. [CrossRef]
12. Pedersen, J.R.; Andreucci, A.; Thorlund, J.B.; Koes, B.; Møller, M.; Storm, L.K.; Bricca, A. Prevalence, frequency, adverse events, and reasons for analgesic use in youth athletes: A systematic review and meta-analysis of 44,381 athletes. *J. Sci. Med. Sport* **2022**, *25*, 810–819. [CrossRef] [PubMed]
13. Fuller, C.W.; Ekstrand, J.; Junge, A.; Andersen, T.E.; Bahr, R.; Dvorak, J.; Hägglund, M.; McCrory, P.; Meeuwisse, W.H. Consensus statement on injury definitions and data collection procedures in studies of football (soccer) injuries. *Br. J. Sports Med.* **2006**, *40*, 193–201. [CrossRef] [PubMed]
14. Ekstrand, J.; Krutsch, W.; Spreco, A.; van Zoest, W.; Roberts, C.; Meyer, T.; Bengtsson, H. Time before return to play for the most common injuries in professional football: A 16-year follow-up of the UEFA Elite Club Injury Study. *Br. J. Sports Med.* **2020**, *54*, 421–426. [CrossRef]
15. Tscholl, P.M.; Vaso, M.; Weber, A.; Dvorak, J. High prevalence of medication use in professional football tournaments including the World Cups between 2002 and 2014: A narrative review with a focus on NSAIDs. *Br. J. Sports Med.* **2015**, *49*, 580–582. [CrossRef]
16. Overbye, M. Walking the line? An investigation into elite athletes' sport-related use of painkillers and their willingness to use analgesics to train or compete when injured. *Int. Rev. Sociol. Sport* **2021**, *56*, 1091–1115. [CrossRef]
17. Pedrinelli, A.; Ejnisman, L.; Fagotti, L.; Dvorak, J.; Tscholl, P.M. Medications and Nutritional Supplements in Athletes during the 2000, 2004, 2008, and 2012 FIFA Futsal World Cups. *Biomed Res. Int.* **2015**, *2015*, 870308. [CrossRef]
18. Thiel, A. Spitzensport und Gesundheit—Eine Risikokultur im Fokus. *DZSM* **2014**, *6*, 147–148. [CrossRef]
19. Mayer, J.; Thiel, A. Verletzungen im Leistungssport aus soziologischer Perspektive. *Sportwiss* **2011**, *41*, 124–136. [CrossRef]
20. Wezenberg, D.; Lindblom, H.; Sonesson, S.; Hägglund, M. Prevalence and intensity of pain in male and female amateur football players: A prospective cohort study. *J. Sci. Med. Sport* **2024**, *27*, 222–227. [CrossRef] [PubMed]
21. Brune, V.K.; Niederweis, U.; Kaufmann, A.; Küster-Kaufmann, M. Analgetikamissbrauch bei Marathonläufern: Jeder Zweite nimmt vor dem Start ein Schmerzmittel. *MMW Fortschr. Med.* **2009**, *40*, 39–41.
22. Mayer, J.; Giel, K.E.; Malcolm, D.; Schneider, S.; Diehl, K.; Zipfel, S.; Thiel, A. Compete or rest? Willingness to compete hurt among adolescent elite athletes. *Psychol. Sport Exerc.* **2018**, *35*, 143–150. [CrossRef]
23. Hager, L.; Averbeck, B.; Voelcker-Rehage, C.; Kutz, D.F. Sex differences in the consumption of over-the-counter analgesics among amateur volleyball players. *BMC Sports Sci. Med. Rehabil.* **2021**, *13*, 45. [CrossRef] [PubMed]
24. Bartley, E.J.; Fillingim, R.B. Sex differences in pain: A brief review of clinical and experimental findings. *Br. J. Anaesth.* **2013**, *111*, 52–58. [CrossRef] [PubMed]
25. Stuart, M.; Farooq, M.; Thomas, T.; Mohamed-Ali, N.; Al-Maadheed, M.; Mohamed-Ali, V. The Essential and Optimal Analgesic and Anti-Inflammatory Medicines for Athletes at the Olympic Games. *Sports Med. Open* **2024**, *10*, 80. [CrossRef]
26. Gjelstad, A.; Herlofsen, T.M.; Bjerke, A.-L.; Lauritzen, F.; Björnsdottir, I. Use of pharmaceuticals amongst athletes tested by Anti-Doping Norway in a five-year period. *Front. Sports Act. Living* **2023**, *5*, 1260806. [CrossRef]
27. Holgado, D.; Hopker, J.; Sanabria, D.; Zabala, M. Analgesics and Sport Performance: Beyond the Pain-Modulating Effects. *PM&R* **2018**, *10*, 72–82. [CrossRef]
28. Tscholl, P.M.; Gard, S.; Schindler, M. A sensible approach to the use of NSAIDs in sports medicine. *SSEM* **2017**, *65*, 15–20. [CrossRef]

Journal of Personalized Medicine

MDPI

Article

Sports-Related Concussion Is a Personalized Issue—Evaluation of Medical Assessment and Subjective Feeling of the Athlete in a German Level 1 Trauma Center

Johannes Weber [1,2,*], Lorenz Huber [1,2], Borys Frankewycz [1,2], Werner Krutsch [1,2,3], Volker Alt [1,2] and Dominik Szymski [1,2]

1 Department of Trauma Surgery, University Medical Centre Regensburg, Franz-Josef-Strauss Allee 11, 93053 Regensburg, Germany
2 FIFA Medical Centre of Excellence, University Medical Centre Regensburg, 93053 Regensburg, Germany
3 SportDocsFranken, 90473 Nuernberg, Germany
* Correspondence: johannes1.weber@ukr.de

Abstract: Sports-related concussions (SRC) have developed into a highly discussed topic in sports medicine over the last few years and demonstrate a severe issue in the personalized treatment of patients. This retrospective cohort study investigated 86 patients with sports-related concussions in a level 1 trauma center, relating to the mechanism, symptoms, medical history, acute therapy including first assessment and the return to sport. The research is based on medical records as well as questionnaires six months after hospitalization. Loss of consciousness for under 30 min (41.2%), headache (36.5%) and amnesia (29.4%) were the most frequent symptoms when presenting in the emergency room. During the hospitalization, mainly headache and vertigo were documented. Most concussions occurred after incidents in equitation and cycling sports; the most common mechanism was falling to the ground with a subsequent impact (59.3%). At the time of discharge from hospital, in 13.4% of all cases, concussion symptoms were still documented in medical records, in contrast to 39.5% of the concerned athletes who reported symptoms for longer than 24 h, and 41.0% who reported ongoing post-concussion symptoms after six months. Concussions are difficult-to-treat disorders with a challenging diagnostic process and many symptoms in various values and levels of persistence. Therefore, a patient-involving treatment with a complaint-dependent return to sport process should be applied to concerned athletes.

Keywords: athletic injury; sports medicine; emergency medicine; head trauma; concussion

Citation: Weber, J.; Huber, L.; Frankewycz, B.; Krutsch, W.; Alt, V.; Szymski, D. Sports-Related Concussion Is a Personalized Issue—Evaluation of Medical Assessment and Subjective Feeling of the Athlete in a German Level 1 Trauma Center. *J. Pers. Med.* **2022**, *12*, 1596. https://doi.org/10.3390/jpm12101596

Academic Editors: Eduard Kurz and Markus Gesslein

Received: 29 August 2022
Accepted: 24 September 2022
Published: 28 September 2022

1. Introduction

In recent years, sports-related concussions (SRC) have become one of the most thrilling topics in sports medicine [1–3]. Historically speaking, the investigation of concussions in sports had already started back in 1928, with the description of neurological deficits in boxers ("punch drunk") [4,5]. Since at least 2005, with the investigation of the cohesion of chronic traumatic encephalopathy (CTE) in NFL players after repetitive concussions, public interest has grown rapidly [6,7]. Nowadays, and especially in regard to football, heading and the impact on the head are heavily discussed. Prohibitions of headers in youths, which were recently introduced, for example in the USA, have also raised awareness for this subject [8,9]. In addition to soccer, the most common sports in Germany are team sports such as basketball and handball, or individual sports such as running, cycling and weight training. Concussions can thereby be caused by a direct impact on the head, or a transmitted force induced by an impact somewhere else on the body. Neurological impairment, as the main symptom, occurs rapidly and mostly disappears spontaneously after a short period of time. A loss of consciousness might also be part of the symptoms. Structural changes are detected rarely in conventional imaging, but functional MRI (fMRI) has shown relevant changes in brain activity in affected areas. Pathophysiologically, a complex cascade leads to an energy discrepancy with reduced functionality of the brain cells [1,10,11]. After experiencing a traumatic brain injury (TBI), a period of time where athletes are still vulnerable and at a high risk of permanent damage due to repeated concussions was reported [12].

Even though the main part of the evaluation of concussions is obtained in emergency departments, there is still a lack of knowledge and practical skills, and an underdiagnosis of SRC is expected [13,14]. Due to the various number of symptoms and the time of their occurrence, a personalized approach to concussion treatment is required.

Therefore, this study investigates first-time sports-related concussions in a German Level 1 trauma center, focusing on initial and long-term symptoms. The comparison of the first examination and a follow-up questionnaire forms the basis of this research project.

2. Materials and Methods

2.1. Study Population

The present study retrospectively analyzes sport-associated head concussions in a level 1 trauma department within the year 2018. All athletes aged over 18 and younger than 50 years, who were treated with the diagnosis of a sports-related concussion in our emergency department, were selected for inclusion in the study population. In Germany, patients with a loss of consciousness or neurological issues are generally admitted for inpatient monitoring. After completing a surveillance period in hospital, the athletes were informed about this study and a written consent was obtained to participate in the study. Six months after the incident, a written standardized follow-up questionnaire was sent to the patients to evaluate persistent symptoms, the return to training and sports after the end of hospitalization and to evaluate possible further injuries after discharge. Further medical surveillance was performed by local general physicians directly after discharge from the hospital. Further follow-ups were carried out in personalized time frames by the general physicians. A follow-up investigation in our hospital was only performed in severe cases, 4 to 6 weeks after discharge. Missing consent or missing injury report data were exclusion criteria. The study was approved by the Ethics Committee of the University of Regensburg (no. 15-101-0134).

2.2. Data collection

The data collection was inspired by the Berlin Consensus statement of concussions and the multicenter study of Winkler et al. (2016) [2,15]. A standardized questionnaire on basis of the Consensus statement was used for data collection. The first part of the research analyzed the doctor's letter and the clinical record data of the patient in a standardized pattern. The focus was set on the cause of the injury and the type of sport being practiced during the injury. In addition, occurring symptoms and secondary injuries during the clinical surveillance, as well as diagnostic procedures, were included in the statistics. Six months after the initial trauma, the follow-up questionnaire was sent to the participating athletes. Here, previous concussion history, former injuries to the head and the detailed trauma mechanism were evaluated. Furthermore, the diagnosis placement, interruption of the activity, persistence of complaints, return to activity and knowledge about concussion were evaluated.

2.3. Statistical Analysis

Continuous data were expressed as means ± standard deviation (SD) and categorical data as frequency counts (percentages). Odds ratios are reported as effect estimates. The significance level was set to $p < 0.05$. All analyses were performed with IBM SPSS Statistics, version 25.0 (Armonk, NY, USA).

3. Results

3.1. Analysis of Medical Record

Eighty-six patients, who were treated for sports-related concussion in a trauma level 1 emergency department, were included in this study. During the accident, the injured patients were mainly practicing equitation (24/85, 28.2%), football (22/85, 25.8%) and cycling or mountain biking (MTX, 18/85, 21.2%).

In total, most athletes sustained their SRC after a fall, with the subsequent impact occurring without any physical contact (45/85, 52.9%; $p < 0.01$). Impact caused by direct physical contact (19/85, 22.4%) and impact caused by a ball or another object (8/85, 9.4%) were further frequent mechanisms. Observing the different types of sports, an accumulation of concussions caused by direct physical contact in ball sports (18/19, 94.7%; $p < 0.01$) attracted attention. Whereas in equitation, the type of sport with the most SRCs, 91.7% of SRCs were reported from a fall with a subsequent impact without physical contact (22/24, 91.7%; $p < 0.01$). Cycling and mountain biking (MTX) were also mainly characterized by falls (12/18, 66.7%; $p < 0.05$) (Table 1).

Table 1. Distribution of sports-related concussions (SRC) by type of sport and injury mechanism, $n = 85$.

Type of Sport	Impact through Direct Physical Contact	Impact through Ball or Other Object	Fall with Subsequent Impact without Contact	Fall with Subsequent Impact with Contact	Other
Soccer	13 *	5	4	-	-
Cycling sport	-	-	9 *	4	-
Equitation	-	-	22 *	-	2
Mountain Biking	-	1	3	-	1
Winter Sports	-	-	3	-	1
Climbing	-	-	5	-	-
Swimming	1	-	-	-	-
Martial arts	-	1	-	-	-
Paragliding	-	-	1	-	-
Running	-	-	1	-	-
Other ball sports	5 *	1	2	-	-

* $p < 0.05$.

Relating to the doctor's letter at the end of the treatment, 13.1% (11/84) of the athletes showed remaining symptoms at the end of the monitoring period. Only one patient needed an extension of observation (1.2%), due to persistent severe symptoms which faded after 48 h surveillance. Radiological imaging, which was made after 24 h, showed no intracerebral bleeding in this case. The mean duration of hospitalization was 38.4 h (Table 2), including patients with applied surgical therapy due to secondary diagnoses, with 76.5% (65/85; $p < 0.05$) of all patients being released within the first day after experiencing an SRC (Table 3). A total of 5.9% (5/85) refused to be monitored in hospital. The level of surveillance ranged between no inpatient monitoring necessary and 5 days. During the treatment, 41.2% (35/85) received radiological head imaging by cranial computer tomography (CCT) or magnetic resonance imaging (cMRI). In particular, in patients with SRC sustained in equitation and cycling, radiological imaging was applied more often than in ball sports (odds ratio: 2.3, 2.7, respectively). Isolated concussions without any additional injuries occurred only in half of the treated patients. Further injuries were mainly contusions of the trunk and the extremities (40%), followed by skin lesions (21.2%). Severe musculoskeletal trauma such as fractures or ligament injuries occurred in 16.5% of the cases. Directly after concussion, athletes mainly reported unconsciousness ($n = 35$), headache ($n = 31$) and retrograde amnesia ($n = 25$) (Table 2).

Table 2. Medical record information on sports-related concussion.

Hospitalization in Days	% (n = 85)
no hospitalization	5.9 (5)
6 h	1.2 (1)
24 h	75.3 (64)
48 h	8.2 (7)
72 h	2.4 (2)
4 days or longer	7.0 (6)
Main symptoms after concussion	% (n = 85)
unconsciousness < 30 min	41.2 (35)
headache	36.5 (31)
retrograde amnesia	29.4 (25)
nausea	23.5 (20)
vertigo	21.2 (18)
vomiting	11.8 (10)
Symptom duration	% (n = 38)
1 to 4 h	42.1 (16)
4 to 12 h	13.2 (5)
12 to 24 h	5.3 (2)
24 to 72 h	10.5 (4)
3 to 7 days	21.1 (8)
more than 7 days	7.9 (3)
Pretrauma existing neuropsychiatric disorders	% (n = 39)
chronic headache or migraine	33.3 (13)
ADHD, dyslexia	7.7 (3)
depression or any other psychiatric disease	12.8 (5)
one of these disorders in family	25.6 (10)
First assessment after trauma	% (n = 38)
medical personnel	18.4 (7)
physiotherapist	7.9 (3)
coach	10.5 (4)
patient himself/herself	15.8 (6)
teammate	31.6 (12)
others	15.8 (6)

ADHD: attention deficit hyperactivity disorder.

Table 3. Return to sports information of the athletes with a sports-related concussion.

Schedule for Return to Sports (RTS)	% (n = 39)
yes	48.7 (19)
no	48.7 (19)
not known	2.6 (1)
Person who set the schedule	% (n = 15)
doctor in hospital	73.3 (11)
general doctor	13.3 (2)
athlete himself/herself	13.3 (2)
Adherence to deadline	% (n = 21)
yes	71.4 (15)
no	28.6 (4)
Successful return to sport	% (n = 39)
yes	89.7 (35)
no	10.3 (4)

3.2. *Medical History and Behavior after Incident*

Overall, 43.6% of the athletes indicated having had a previous concussion in their medical history. Of these, 58.8% had already reported more than one concussion. Most of the previous concussions (88.2%) also occurred in sports. Only 59.0% of the athletes who had already experienced an SRC were treated in hospital due to the previous insult. In addition, the medical history of neuropsychological disorders of these patients was also collected (Table 3). Here, only 38.5% declared having sufficient personal knowledge about concussions. Moreover, athletes with a lack of knowledge more often classified concussions as a severe injury (70.8%; $p < 0.05$).

After the incident, 89.7% of the patients with an SRC interrupted activity, but only 69.2% terminated it for the rest of the day. First assessment of trauma sequelae was made by the athletes or teammates in nearly 50% of cases. In 41.1%, concussion symptoms disappeared within the first four hours after the incident, whereas nearly one third (11/38) reported symptoms lasting for more than three days (Table 3).

3.3. Remaining Symptoms after 6 Months and Return to Sports

Six months after the SRC, 41.0% of the athletes still reported remaining symptoms caused by the SRC after their discharge from hospital. Women were significantly more likely to report being affected by long-term symptoms ($p = 0.02$) compared to men. The most reported symptoms were headache (75.0%) and nausea (37.5%) (Figure 1). A return to sports at the same level was possible in nearly 90% of all cases, but 24 (61.5%) of the athletes complained about lower activity levels at the six-month follow-up (Table 3).

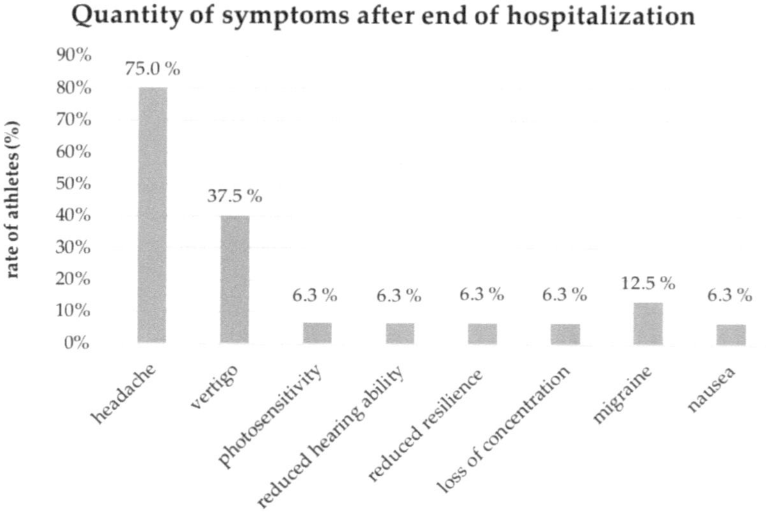

Figure 1. Symptoms after discharge from hospital.

4. Discussion

The present study retrospectively investigated sports-related concussions in a level 1 trauma department in Germany, with its focus on the persistence of symptoms, injury patterns, diagnostics and the return to sport, and thereby turned attention to individual differences and personal perceptions.

The most important result of the examination is the significant difference between the symptoms indicated at the time of discharge from medical treatment in the doctor's letter and the patient's questionnaire performed 6 months after discharge. In the evaluation of the medical records, 13.1% of the athletes had indicated symptoms after completing the medical surveillance period, whilst in the questionnaire performed 6 months after an SRC, 41.0% still indicated remaining symptoms in the self-report. Symptoms of a concussion sometimes can be difficult to discover because of their diversity and late onset, up to 48 h after the impact [16], and therefore, are overlooked in clinical settings [13,17–19]. Directly after the accident, the most common reported symptoms were the loss of consciousness (41.2%) and headache (36.5%). However, in the patient-reported follow-up investigation after 6 months, the most common described symptoms were headache (75%) and vertigo (37.5%). Often, re-evaluating SRCs' symptoms is insufficient during the surveillance period by medical personnel. In many cases, no follow-up treatment is applied to the patients, although they need a detailed second or ongoing evaluation with strict guidelines [14,20,21]. In other situations, patients themselves are sometimes not aware of the symptoms due to a lack of knowledge, while others want to enforce a fast discharge and return to work or sport. In particular, successful athletes may take a higher risk to conceal symptoms in order to avoid a longer break from training [1]. Long-term signs of concussions are a major cause of prolonged observation and longer hospitalization [1,14,20,21]. Depending on the type of sport, rates of persistent symptoms between 10 and 30% have been described in the literature. In particular, contact sports and younger athletes showed an increased risk [22,23]. The persistence of symptoms after a concussion is known as post-concussion syndrome (PCS). The World Health Organization did not define a specific time frame in the International Classification of Disease-10 (ICD-10) coding, but the persistence of three or more of the following symptoms—headache, fatigue, dizziness, irritability, insomnia and concentration or memory difficulty—is obligatory [24]. Immediately after the incident, the majority of the athletes in this study reported losing consciousness (41.2%) and headache (36.5%). Although many symptoms faded after hours, nearly one third of the patients had to deal with symptoms for more than 3 days. In detail, headache and neck pain as well as fatigue and drowsiness are described in the literature as the main symptoms of patients with mTBI [25]. Persistent headache is a common symptom after concussion and was noticed in some populations in up to 37% of the cases, 45 days after the injury [25]. The variance of cerebral blood flow is discussed as one of the main mechanisms of PCS. During rest, a reduced blood flow was noticed, whilst during exercise, an excessive increase with an accompanying headache proceeded. Furthermore, abnormal cerebral blood flow was recorded for up to a month after the impact [22]. The high rate of patients with a loss of consciousness as a symptom may be explained by the structure of the study center as university hospital serving as a level 1 trauma center. In addition, a reduced rate of admission of patients when not experiencing a loss of consciousness is possible.

The analyzed patients experienced an SRC mainly after a fall with a subsequent impact, but without any physical contact. This mostly occurred in equitation and cycling sports, whereas direct contact and an impact with a ball or any other object was the key injury mechanism in ball and contact sports. This distribution of injury pattern is caused by the different sports' biomechanics. Team sports with a higher proportion of physical contact (e.g., ice hockey, American football) and equitation have already been shown to account for a large proportion of SRCs in the past [2,26–28]. The mechanisms of concussion vary by type of sport and position, but do not necessarily require an impact to the head. For example, 5% of all concussions in American football, soccer and ice hockey are related to a trauma below the neck [29,30]. A longitudinal study determined a growth in the total number of concussions over the last decades. The evolution of professional athletes, who are better trained than in former times, and the development of sports to involve more speed, contact and risk has led to a higher risk of injuries to the head [31]. Moreover, recently developed protective gear can potentially lead to an increase in concussion rates by making athletes trust too much in the material and therefore take more risks during sports [32]. An important point of this investigation is the report of German data, which according to the epidemiology and types of sports is comparable to other European countries. Due to cultural specifics in treatment, the injury mechanism and epidemiology, the results have to be checked carefully before drawing worldwide conclusions.

Another interesting finding of this study was that women reported significantly more long-term symptoms than men after 6 months. It is known that women report more symptoms in acute trauma situations [33,34] and that these symptoms can last up to a month, especially in young patients [34].

In addition, the application of radiological imaging of the head with cCT and cMRI was investigated, according to the various types of sports and injury mechanisms. Overall,

41.1% of the patients had according radiological imaging. Odds ratio analysis showed at least a doubling of rates in cycling and equitation compared to ball sports. This was mainly conditioned by the injury mechanism, as in cycling and equitation, trauma was mostly caused by high-speed incidents and falls from a height. Due to trauma guidelines, radiological imaging should be processed in these cases [35,36]. The high rate of head imaging was also caused by the hierarchy of the study center. More severe cases are sent to the level 1 trauma departments, while milder injuries are more often treated in local hospitals [26,37]. However, in comparison to other studies, similar rates of sustained radiological imaging of the head have been observed [19].

When analyzing the behavior patterns and the first medical assessment directly after the injury, a lack of knowledge of how to act and behave was indicated by more than every second patient. In particular, in persons with—in their opinion—sufficient knowledge of concussions, as well as in athletes with insufficient knowledge, concussions were regarded as severe injuries. Nevertheless, 10.3% did not interrupt their training or match and 30.8% did not terminate physical activity directly after the incident. Although guidelines for red flag symptoms of concussions and different sideline screening tools exist, an integration of daily preclinical routine is rarely observed. Guidelines for sport-related concussions, for example, demand an immediate stop to the match or training and an interruption of the physical activity. In the next step, the affected athlete should be referred to a healthcare professional in order to be screened and treated [1,14,21,38,39]. However, many cases of concussions are still undetected, due to avoided medical assessment, and ignored or misunderstood symptoms. The introduction of a concussion law revealed this problem by noticing significantly more patients with concussion in emergency departments [40]. A good understanding of concussion, especially its symptoms and initial assessment, is a key factor in treatment and outcome. A lack of knowledge has been described among coaches, match officials and athletes and thus forms a basis for future prevention [41]. Bad management of SRC was not only found amongst athletes, but also in emergency departments. Rowe et al. found that one out of six patients in emergency departments were misdiagnosed despite, having clear symptoms of concussion [19]. Other investigations postulated bad management and lack of guidelines in hospitals, as well as insufficient realization [13,18,42]. Recently discovered correlation between biomarkers and concussions could help to reduce the number of misdiagnosed cases [21,43].

The importance of fundamental skills and information is reflected in the investigation of the first medical contact. In almost half of the incidents, other athletes (31.6%) or the patient himself (15.8%) were responsible for the initial assessment after trauma, as only a small number of patients had a medical professional as the initial contact. This turns out to be problematic, since interpreting the unspecific symptoms could result in symptoms being misunderstood or dismissed by nonprofessional consultation [19].

The perfect time to return to sport (RTS) after having an SRC is a highly discussed topic [15,29,42,44–46]. Each patient has to be treated individually by a step-by-step return to sport protocol. The Zurich statement, for example, recommends a return to moderate physical activity when the athlete is free of any symptoms, after an initial rest of 2 days without somatic or mental burden, and a return-to-competition after one week at the earliest [15]. In our study, the date of sport resumption was mainly set by the treating doctor in the hospital (73.3%), since there was no possibility for a secondary assessment of the patient after being released from the hospital. The median time of training absence was 14 days, but 28.6% of the patients did not comply with the determined timing of the return to sport. In a systematic review, Buettner et al. (2020) postulated a typical recovery time within 7 to 10 days after SRC [44]. In most recommendations, adequate rest is the main principle of treatment. Taking a break until concussion signs fade seems to be the best way to avoid the prolongation of symptoms so far [31,45], even if recent studies showed a converse thesis and demonstrated a reduced rate of PCS and a faster rehabilitation process after a shorter break, as well as an early beginning of moderate training [47,48]. Light aerobic exercise with cervical, vestibular, cognitive behavioral or vision therapy is also discussed as a potentially positive influence on returning to sport [46,49]. As the fading of symptoms can last longer or symptoms sometimes appear with a delay, the return to sport process should be surveilled by a physician [16,49]. Although coaches are more likely to refer to the physician's advice, the majority of athletes want to decide on their own about the timing of the RTS. In particular, more ambitious and advanced athletes are known to force a faster return to training and competition activities, even if they take higher personal risks [17,21].

In our study, more than every second athlete already had experience of concussions in their past. The larger part had already been hospitalized for a concussion, and the majority had also been classified as exercise-related concussions. In some cases, the personal risk behavior of the concerned athletes seems to be associated with the appearance of SRC [31,50,51]. Another thesis for the repetitive cumulation of concussions among certain patients is based on a lack of regeneration after a previous SRC [31,52]. The pathophysiological correlate of this vulnerable phase is reduced cerebral blood flow, which was registered in athletes up to one month after concussion [22].

For future prevention, knowledge about sport-associated concussions should be increased [3,41], especially in athletes. For example, sideline evaluation tests (e.g., SCAT-5) are certified assessment tools for a fast evaluation of symptoms and should be more integrated in comprehensive usage [39,53–56]. As biomarker testing in the emergency department is still in the developmental process [43], medical personnel should be sensitized to SRC as well.

5. Conclusions

This retrospective investigation analyzed sports-related concussions in a level 1 emergency department in the course of a year. The study findings, such as the difference in the symptoms athletes complained to doctors about at the time of discharge from hospital in comparison to those they indicated in a questionnaire which was performed later on, or that long-term symptoms of SRC seem to significantly affect women more than men after 6 months, show that SRC is a difficult-to-treat disorder which requires a detailed, individually planned and patient-involving treatment process.

Author Contributions: Conceptualization, J.W. and W.K.; methodology, L.H.; software, J.W. and D.S.; validation, B.F., W.K. and J.W.; formal analysis, L.H and D.S.; investigation, J.W.; resources, V.A.; data curation, J.W. and B.F.; writing—original draft preparation, D.S, L.H. and J.W..; writing—review and editing, B.F., W.K. and V.A.; visualization, D.S.; supervision, V.A. and W.K.; project administration, J.W. All authors have read and agreed to the published version of the manuscript.

Funding: This research received no external funding.

Institutional Review Board Statement: The study was conducted in accordance with the Declaration of Helsinki and approved by the Institutional Review Board of the University of Regensburg, Germany (no: 15-101-0134).

Informed Consent Statement: Informed consent was obtained from all the subjects involved in the study.

Data Availability Statement: The dataset is available on request.

Conflicts of Interest: The authors declare no conflict of interest.

References

1. Norton, C. Concussion in sport: The importance of accurate and reliable discharge advice. *Emerg. Nurse J. RCN Accid. Emerg. Nurs. Assoc.* **2018**, *25*, 35–42. [CrossRef]
2. Winkler, E.A.; Yue, J.K.; Burke, J.F.; Chan, A.K.; Dhall, S.S.; Berger, M.S.; Manley, G.T.; Tarapore, P.E. Adult sports-related traumatic brain injury in United States trauma centers. *Neurosurg. Focus* **2016**, *40*, E4. [CrossRef]
3. Yue, J.K.; Winkler, E.A.; Burke, J.F.; Chan, A.K.; Dhall, S.S.; Berger, M.S.; Manley, G.T.; Tarapore, P.E. Pediatric sports-related traumatic brain injury in United States trauma centers. *Neurosurg. Focus* **2016**, *40*, E3. [CrossRef]
4. Martland, H.S. Punch Drunk. *J. Am. Med. Assoc.* **1928**, *91*, 1103–1107. [CrossRef]

5. Parker, H.L. Traumatic Encephalopathy ('Punch Drunk') of Professional Pugilists. *J. Neurol. Psychopathol.* **1934**, *15*, 20–28. [CrossRef]
6. Lindsley, C.W. Chronic Traumatic Encephalopathy (CTE): A Brief Historical Overview and Recent Focus on NFL Players. *ACS Chem. Neurosci.* **2017**, *8*, 1629–1631. [CrossRef]
7. Omalu, M.B.I.; DeKosky, M.S.T.; Minster, M.R.L.; Kamboh, M.I.; Hamilton, M.R.L.; Wecht, M.C.H. Chronic Traumatic Encephalopathy in a National Football League Player. *Neurosurgery* **2005**, *57*, 128–134; discussion 128–134. [CrossRef]
8. Sandmo, S.B.; Andersen, T.E.; Koerte, I.K.; Bahr, R. Head impact exposure in youth football—Are current interventions hitting the target? *Scand. J. Med. Sci. Sports* **2020**, *30*, 193–198. [CrossRef]
9. Siva, N. Scotland to ban heading in children's football. *Lancet* **2020**, *395*, 258. [CrossRef]
10. Gänsslen, A.; Krutsch, W.; Schmehl, I.; Rickels, E. Chronisch Traumatische Enzephalopathie: Wie Sportverletzungen das Gehirn schädigen können. *Dtsch. Ärztebl.* **2016**, *113*, 13. [CrossRef]
11. Harmon, K.G.; Clugston, J.; Dec, K.; Hainline, B.; Herring, S.A.; Kane, S.; Kontos, A.P.; Leddy, J.J.; McCrea, M.A.; Poddar, S.K.; et al. American Medical Society for Sports Medicine Position Statement on Concussion in Sport. *Clin. J. Sport Med. Off. J. Can. Acad. Sport. Med.* **2019**, *29*, 87–100. [CrossRef]
12. Cantu, R.C. Second-impact syndrome. *Clin. Sports Med.* **1998**, *17*, 37–44. [CrossRef]
13. Carson, J.D.; Rendely, A.; Garel, A.; Meaney, C.; Stoller, J.; Kaicker, J.; Hayden, L.; Moineddin, R.; Frémont, P. Are Canadian clinicians providing consistent sport-related concussion management advice? *Can Fam Physician* **2016**, *62*, 494–500.
14. Wright, D.W.; Bazarian, J.J. Emergency department evaluation of the concussed athlete. *Handb. Clin. Neurol.* **2018**, *158*, 81–90. [CrossRef]
15. McCrory, P.; Meeuwisse, W.; Dvorak, J.; Aubry, M.; Bailes, J.; Broglio, S.; Cantu, R.C.; Cassidy, D.; Echemendia, R.J.; Castellani, R.J.; et al. Consensus statement on concussion in sport—The 5th international conference on concussion in sport held in Berlin, October Br. *J. Sports Med.* **2017**, *51*, 838–847. [CrossRef]
16. Fuller, C.W.; Fuller, G.W.; Kemp, S.P.T.; Raftery, M. Evaluation of World Rugby's concussion management process: Results from Rugby World Cup 2015. *Br. J. Sports Med.* **2017**, *51*, 64–69. [CrossRef]
17. Loose, O.; Achenbach, L.; Fellner, B.; Lehmann, J.; Jansen, P.; Nerlich, M.; Angele, P.; Krutsch, W. Injury prevention and return to play strategies in elite football: No consent between players and team coaches. *Arch. Orthop. Trauma. Surg.* **2018**, *138*, 985–992. [CrossRef]
18. Deichman, J.J.; Graves, J.M.; A Klein, T.; Mackelprang, J.L. Characteristics of youth who leave the emergency department without being seen following sports-related concussion. *Concussion Lond. Engl.* **2019**, *4*, CNC68. [CrossRef]
19. Rowe, B.H.; Eliyahu, L.; Lowes, J.; Gaudet, L.A.; Beach, J.; Mrazik, M.; Cummings, G.; Voaklander, D. Concussion diagnoses among adults presenting to three Canadian emergency departments: Missed opportunities. *Am. J. Emerg. Med.* **2018**, *36*, 2144–2151. [CrossRef]
20. Conder, A.; Conder, R.; Friesen, C. Neurorehabilitation of Persistent Sport-Related Post-Concussion Syndrome. *Neuro Rehabilitation* **2020**, *46*, 167–180. [CrossRef]
21. Hubertus, V.; Marklund, N.; Vajkoczy, P. Management of concussion in soccer. *Acta Neurochir.* **2019**, *161*, 425–433. [CrossRef] [PubMed]
22. Leddy, J.; Baker, J.G.; Haider, M.N.; Hinds, A.; Willer, B. A Physiological Approach to Prolonged Recovery From Sport-Related Concussion. *J. Athl. Train.* **2017**, *52*, 299–308. [CrossRef] [PubMed]
23. Makdissi, M.; Cantu, R.C.; Johnston, K.M.; McCrory, P.; Meeuwisse, W.H. The difficult concussion patient: What is the best approach to investigation and management of persistent (>10 days) postconcussive symptoms? *Br. J. Sports Med.* **2013**, *47*, 308–313. [CrossRef]
24. ICD-10 Version:2019. Available online: https://icd.who.int/browse10/2019/en (accessed on 9 May 2022).
25. King, J.A.; McCrea, M.A.; Nelson, L.D. Frequency of Primary Neck Pain in Mild Traumatic Brain Injury/Concussion Patients. *Arch. Phys. Med. Rehabil.* **2020**, *101*, 89–94. [CrossRef]
26. Krutsch, W.; Krutsch, V.; Hilber, F.; Pfeifer, C.; Baumann, F.; Weber, J.; Schmitz, P.; Kerschbaum, M.; Nerlich, M.; Angele, P. 11.361 sports injuries in a 15-year survey of a Level I emergency trauma department reveal different severe injury types in the 6 most common team sports. *Sport. Sportschaden. Organ. Ges. Orthop.-Traumatol. Sport.* **2018**, *32*, 111–119. [CrossRef]
27. Lyons, T.W.; Miller, K.A.; Miller, A.F.; Mannix, R. Racial and Ethnic Differences in Emergency Department Utilization and Diagnosis for Sports-Related Head Injuries. *Front. Neurol.* **2019**, *10*, 690. [CrossRef]
28. Nguyen, J.V.K.; Brennan, J.H.; Mitra, B.; Willmott, C. Frequency and Magnitude of Game-Related Head Impacts in Male Contact Sports Athletes: A Systematic Review and Meta-Analysis. *Sports Med.* **2019**, *49*, 1575–1583. [CrossRef]
29. Refakis, C.A.; Turner, C.D.; Cahill, P.J. Sports-related Concussion in Children and Adolescents. *Clin. Spine Surgery A Spine Publ.* **2017**, *30*, 191–196. [CrossRef]
30. Delaney, J.S.; Puni, V.; Rouah, F. Mechanisms of Injury for Concussions in University Football, Ice Hockey, and Soccer: A pilot study. *Clin. J. Sport Med. Off. J. Can. Acad. Sport Med.* **2006**, *16*, 162–165. [CrossRef]
31. Pauelsen, M.; Nyberg, G.; Tegner, C.; Tegner, Y. Concussion in Ice Hockey—A Cohort Study Across 29 Seasons. *Clin. J. Sport Med.* **2017**, *27*, 283–287. [CrossRef]
32. Torg, J.S.; Truex, R.; Quedenfeld, T.C.; Burstein, A.; Spealman, A.; Nichols, C. The National Football Head and Neck Injury Registry. Report and conclusions 1978. *JAMA* **1979**, *241*, 1477–1479. [CrossRef] [PubMed]

33. Covassin, T.; Schatz, P.; Swanik, C.B. Sex differences in neuropsychological function and post-concussion symptoms of concussed collegiate athletes. *Neurosurgery* **2007**, *61*, 345–350; discussion 350–351. [CrossRef] [PubMed]
34. Resch, J.E.; Rach, A.; Walton, S.; Broshek, D.K. Sport Concussion and the Female Athlete. *Clin. Sports Med.* **2017**, *36*, 717–739. [CrossRef] [PubMed]
35. Ernstberger, A.; Schreyer, A.; Schleder, S.; Baumer, S.; Angerpointner, K.; Diepold, E.; Nerlich, M. Computertomographie bei Polytrauma. *Trauma und Berufskrankh.* **2017**, *19*, 57–63. [CrossRef]
36. Lendemans, S.; Ruchholtz, S. S3-Leitlinie Polytrauma/Schwerverletzten-Behandlung. *Der Unfallchirurg* **2012**, *115*, 14–21. [CrossRef] [PubMed]
37. Padegimas, E.M.; Stepan, J.; Stoker, G.E.; Polites, G.M.; Brophy, R.H. Epidemiology and Severity of Sports and Recreation Injuries Presenting to a Tertiary Adult Emergency Department. *Phys. Sportsmed.* **2016**, *44*, 263–268. [CrossRef]
38. McKee, A.C.; Daneshvar, D.; Alvarez, V.E.; Stein, T. The neuropathology of sport. *Acta Neuropathol.* **2014**, *127*, 29–51. [CrossRef]
39. Putukian, M.; Raftery, M.; Guskiewicz, K.; Herring, S.; Aubry, M.; Cantu, R.C.; Molloy, M. Onfield assessment of concussion in the adult athlete. *Br. J. Sports Med.* **2013**, *47*, 285–288. [CrossRef]
40. Trojian, T.; Violano, P.; Hall, M.; Duncan, C.C. The effects of a state concussion law on the frequency of sport-related concussions as seen in two emergency departments. *Inj. Epidemiol.* **2015**, *2*, 2. [CrossRef]
41. Yeo, P.C.; Yeo, E.Q.Y.; Probert, J.; Sim, S.H.S.; Sirisena, D. A Systematic Review and Qualitative Analysis of Concussion Knowledge amongst Sports Coaches and Match Officials. *J. Sports Sci. Med.* **2020**, *19*, 65–77.
42. McLeod, T.C.V.; Kostishak, N.; Anderson, B.E.; Bacon, C.E.W.; Lam, K.C. Patient, Injury, Assessment, and Treatment Characteristics and Return-to-Play Timelines After Sport-Related Concussion: An Investigation from the Athletic Training Practice-Based Research Network. *Clin. J. Sport Med. Off. J. Can. Acad. Sport Med.* **2019**, *29*, 298–305. [CrossRef]
43. Meier, T.B.; Huber, D.L.; Bohorquez-Montoya, L.; Nitta, M.E.; Savitz, J.; Teague, T.K.; Bazarian, J.J.; Hayes, R.L.; Nelson, L.D.; McCrea, M.A. A Prospective Study of Acute Blood-Based Biomarkers for Sport-Related Concussion. *Ann. Neurol.* **2020**, *87*, 907–920. [CrossRef] [PubMed]
44. Büttner, F.; Howell, D.R.; Ardern, C.L.; Doherty, C.; Blake, C.; Ryan, J.; Catena, R.; Chou, L.-S.; Fino, P.; Rochefort, C.; et al. Concussed athletes walk slower than non-concussed athletes during cognitive-motor dual-task assessments but not during single-task assessments 2 months after sports concussion: A systematic review and meta-analysis using individual participant data. *Br. J. Sports Med.* **2020**, *54*, 94–101. [CrossRef] [PubMed]
45. Gupta, A.; Summerville, G.; Senter, C. Treatment of Acute Sports-Related Concussion. *Curr. Rev. Musculoskelet. Med.* **2019**, *12*, 117–123. [CrossRef]
46. Leddy, J.J.; Baker, J.G.; Willer, B. Active Rehabilitation of Concussion and Post-concussion Syndrome. *Phys. Med. Rehabilitation Clin. North Am.* **2016**, *27*, 437–454. [CrossRef]
47. Grool, A.M.; Aglipay, M.; Momoli, F.; Meehan, W.P.; Freedman, S.; Yeates, K.; Gravel, J.; Gagnon, I.; Boutis, K.; Meeuwisse, W.; et al. Association Between Early Participation in Physical Activity Following Acute Concussion and Persistent Postconcussive Symptoms in Children and Adolescents. *JAMA* **2016**, *316*, 2504–2514. [CrossRef]
48. Howell, D.R.; Mannix, R.C.; Quinn, B.; Taylor, J.A.; Tan, C.O.; Meehan, I.W.P. Physical Activity Level and Symptom Duration Are Not Associated After Concussion. *Am. J. Sports Med.* **2016**, *44*, 1040–1046. [CrossRef]
49. Sufrinko, A.M.; Kontos, A.; Apps, J.N.; McCrea, M.; Hickey, R.W.; Collins, M.W.; Thomas, D.G. The Effectiveness of Prescribed Rest Depends on Initial Presentation After Concussion. *J. Pediatr.* **2017**, *185*, 167–172. [CrossRef]
50. Burman, E.; Lysholm, J.; Shahim, P.; Malm, C.; Tegner, Y. Concussed athletes are more prone to injury both before and after their index concussion: A data base analysis of 699 concussed contact sports athletes. *BMJ Open Sport Exerc. Med.* **2016**, *2*, e000092. [CrossRef]
51. Manley, G.; Gardner, A.J.; Schneider, K.; Guskiewicz, K.M.; Bailes, J.; Cantu, R.C.; Castellani, R.J.; Turner, M.; Jordan, B.D.; Randolph, C.; et al. A systematic review of potential long-term effects of sport-related concussion. *Br. J. Sports Med.* **2017**, *51*, 969–977. [CrossRef]
52. Giza, C.C.; Kutcher, J.S.; Ashwal, S.; Barth, J.; Getchius, T.S.; Gioia, G.A.; Gronseth, G.S.; Guskiewicz, K.; Mandel, S.; Manley, G.; et al. Summary of evidence-based guideline update: Evaluation and management of concussion in sports: Report of the Guideline Development Subcommittee of the American Academy of Neurology. *Neurology* **2013**, *80*, 2250–2257. [CrossRef] [PubMed]
53. Kruse, A.J.; Nugent, A.S.; Peterson, A.R. Using sideline concussion tests in the emergency department. *Open Access Emerg. Med. OAEM* **2018**, *10*, 113–121. [CrossRef] [PubMed]
54. Putukian, M. Clinical Evaluation of the Concussed Athlete: A View From the Sideline. *J. Athl. Train.* **2017**, *52*, 236–244. [CrossRef] [PubMed]
55. Schepart, Z.; Putukian, M. Sideline assessment of concussion. *Handb. Clin. Neurol.* **2018**, *158*, 75–80. [CrossRef] [PubMed]
56. Bin Zahid, A.; Hubbard, M.E.; Dammavalam, V.M.; Balser, D.Y.; Pierre, G.; Kim, A.; Kolecki, R.; Mehmood, T.; Wall, S.P.; Frangos, S.G.; et al. Assessment of acute head injury in an emergency department population using sport concussion assessment tool—3rd edition. *Appl. Neuropsychol. Adult* **2018**, *25*, 110–119. [CrossRef] [PubMed]

Journal of *Personalized Medicine*

MDPI

Article

Knee Orthotics Do Not Influence Coordinative Skills—A Randomized Controlled Crossover Pilot Trial

Robert Prill [1,2,*], Caren Cruysen [1,2], Aleksandra Królikowska [3], Sebastian Kopf [1,2] and Roland Becker [1,2]

1 Center of Orthopaedics and Traumatology, University Hospital Brandenburg/Havel, Brandenburg Medical School Theodor Fontane, 14770 Brandenburg an der Havel, Germany
2 Faculty of Health Sciences Brandenburg, Brandenburg Medical School Theodor Fontane, 14770 Brandenburg an der Havel, Germany
3 Ergonomics and Biomedical Monitoring Laboratory, Department of Physiotherapy, Faculty of Health Sciences, Wroclaw Medical University, 50-367 Wroclaw, Poland
* Correspondence: robert.prill@mhb-fontane.de

Abstract: Objective: This single-blind randomized controlled crossover pilot trial investigated whether hard or soft knee orthotics affect the back in action (BIA) test battery performance. Methods: Twenty-four healthy participants (13 males, 11 females) were randomly assigned into three equal groups differentiated through the order of device use. The data were collected in a laboratory setting. BIA test battery (balance tests, vertical jumps, and parkour hop tests) was run with a rigid orthotic device, a soft brace, or no aid in a crossover order. Analysis of Variance repeated measures and Friedman Test were used to calculate depended-group differences. Results: No significant or clinically relevant effect or differences was observed between running the BIA with a soft brace, rigid orthosis, or no aid (p = 0.53–0.97) for all included tests. No adverse events have been observed. Conclusion: Soft and rigid knee braces do not affect performance in healthy participants. Missing experience with the devices might explain a few influences on feedback mechanisms. There is no disadvantage to be expected regarding healthy participants running back to sports.

Keywords: anterior cruciate ligament; biomedical monitoring; physiotherapy; rehabilitation; return-to-sports; sports

Citation: Prill, R.; Cruysen, C.; Królikowska, A.; Kopf, S.; Becker, R. Knee Orthotics Do Not Influence Coordinative Skills—A Randomized Controlled Crossover Pilot Trial. *J. Pers. Med.* **2022**, *12*, 1509. https://doi.org/10.3390/jpm12091509

Academic Editors: Markus Gesslein and Eduard Kurz

Received: 26 August 2022
Accepted: 11 September 2022
Published: 14 September 2022

1. Introduction

According to the International Organization for Standardization (ISO), orthoses, called an orthotic device, is defined as an externally applied device used to compensate for impairments of the structure and function of the neuro-muscular and skeletal systems [1]. Consecutively, knee braces have been classified by the American Academy of Orthopedic Surgeons into prophylactic, functional, rehabilitation, and patellofemoral ones [2]. Even though they are widely used, controversy exists about whether braces should be used for protection or treatment purposes, especially for the knee during and after rehabilitation. For instance, braces and orthoses are well-established for rehabilitation after anterior cruciate ligament (ACL) injury, but their usefulness is still discussed with controversy [3]. Prophylactic knee bracing does not seem to reduce the incidence of knee injuries in uninjured patients [4].

There is also conflicting evidence in terms of the impact of braces on knee proprioception. It has been shown that orthoses did not seem to increase kinesthetic awareness [5]. The role of proprioception in motor control and functional joint stability seems to play an important role. More attention should be paid to the role of feedforward and feedback strategies [6].

Another common question when returning to sports is if braces or orthoses may impact athletes' performance. A review comparing functional bracing with non-bracing after Anterior Cruciate Ligament Reconstruction (ACLR) showed significant improvement in knee kinematics and gait, but decreased quadriceps activation without affecting function tests, range of motion, and proprioception [3]. However, reducing quadriceps activation may cause the impact of specific tests such as hop tests.

The purpose of this crossover trial was to evaluate the impact of hard and soft orthoses on the performance of healthy subjects. It was hypothesized that braces or orthosis might reduce performance when completing the back in the action (BIA) test battery. To prove this hypothesis, we developed a crossover design with healthy adults, running a test battery without any knee stabilizers aids, a brace, and an orthosis.

2. Materials and Methods

A single-blind controlled crossover design was used and the study was conducted in accordance with CONSORT checklist of information to include when reporting randomized crossover trials [7]. Current recommendations for reporting sports medicine and orthopedic clinical trials have been respected [8]. The study was approved by the appropriate ethical committee related to the institution in which it was performed. All the participants were voluntarily recruited from medical and physical therapy students. Inclusion criteria were the age of 20 to 30 years, healthy well-being, moderate sports activity, and the absence of current pain in the limbs or trunk. Subjects with a previous trunk, hip, knee, or ankle surgery were excluded. Subjects were selected by the person who also instructed the BIA test battery. This person was not involved in the statistical analysis. After written informed consent was obtained, subjects were randomly allocated to run 1, 2, or 3 using the Microsoft Excel (© 2022 Microsoft Corporation) function with the formula '= INT(RAND()*(4 − 1) + 1)' by a not further involved person, resulting in a random allocation of 1:1:1 for all three types of runs. They either started with "soft brace," "hard Orthotic device," or "no aid," followed by the other two in a different order (Table 1).

Table 1. Test protocol.

Studied Group	Group A (n = 8)	Group B (n = 8)	Group C (n = 8)
Test run preceded by a warm-up			
Run 1	No aid	Brace	Orthosis
Run 2	Brace	Orthosis	No aid
Run 3	Orthosis	No aid	Brace

n, number of individuals.

All three groups, being actually differentiated through the order of used devices they perform the test battery with, ran the described parkour three times with a full rest of about 40 min in between: once without an external stabilizer, once with a functional brace (*Genumedi® E+motion, Medi, Bayreuth, Germany*) as presented in Figure 1, and once with an orthosis (*M.4s® comfort, Medi, Bayreuth, Germany*) as presented in Figure 2.

Randomizing was essential to reduce the interaction effects of the three runs of the test battery. Participant blinding was not possible due to the nature of the study.

A simple power-based effect size calculation to detect a large effect of not wearing external knee stabilizers was performed for three groups using G*Power 3.1 (Duesseldorf, Germany). with an alpha error of five percent and a power 1-β of 90 percent, suggesting a sample size of 24 to be included.

The BIA consists of stability tests, two-leg and one-leg countermovement jumps (CMJ), standardized hop parkour, and a speed test [9]. The BIA showed excellent test-retest reliability for two-leg CMJ, moderate reliability for the two-leg stability test, and good reliability for all further included test items [10]. Individuals did warm up with a standardized protocol, including 10 min (min) of ergometer cycling and one test run through all tests of BIA. The BIA test battery was already described in detail [9,10]. In brief, it includes seven tests. First, individuals must balance with both legs for 30 s (s) on a balance board (*TST Trendsport, Grosshöflein, Austria*), measuring the center of pressure and giving instant biofeedback to participants about the position of the disc (Figure 1). A higher score will be reached when few movements and centered balance are measured. The second task is similar, but patients stand on one leg only, first on the dominant and second on the non-dominant leg. The third task is a two-leg countermovement jump (CMJ) wearing a movement sensor (*Myotest S.A., Sion, Switzerland*) in a belt for capturing height differences. A one-leg CMJ follows a two-leg CMJ, performed in the same manner just with a single leg. This is followed by a three-repetition plyometric jump test. The sixth task is a speed jump test. Hereby, a standardized single-leg hop parkour test (*TST Trendsport, Grosshöflein, Austria*) (Figure 3) with 16 forward-backward and sideway jumps must be completed with both dominant and dominant non-dominant legs.

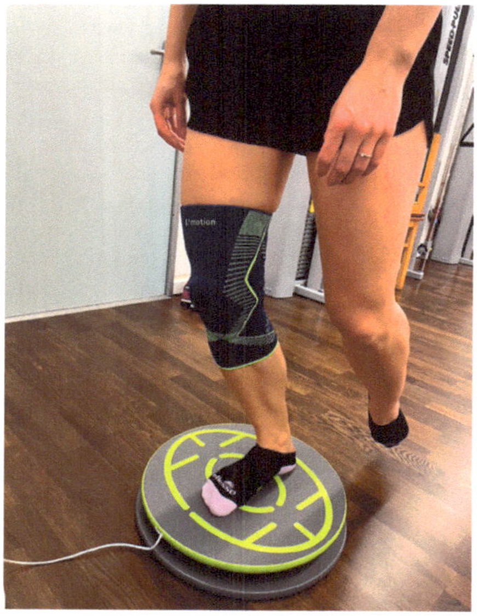

Figure 1. Single Leg Balance Test.

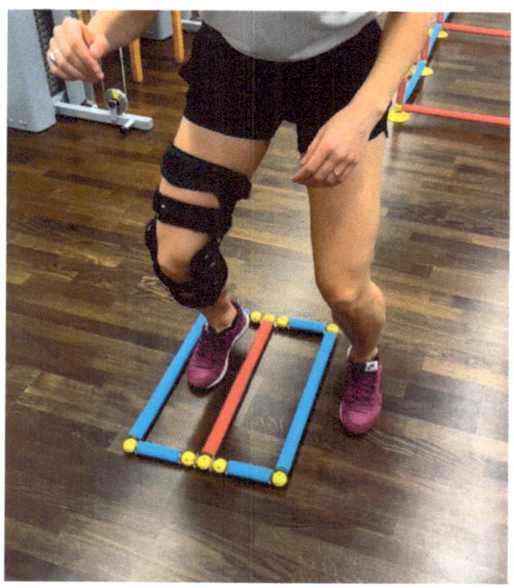

Figure 2. Quick feet test.

Figure 3. Speedy Jump Test.

The seventh test is a quick feet test using the Speedy Basic Jump Set (*TST Trendsport, Grosshöflein, Austria, Figure 2*). Time for 15 repetitions was measured. When unexpected events occurred, like participants stepping down from the challenge disc or jumping in the wrong order, they were asked to start the attempt again. The tests were performed according to a previously published quality criteria study [10]. After the 24 participants ran the test, battery results were blinded and transferred to an examiner. Without the knowledge of group assignment, an initial statistical analysis was performed.

IBM SPSS Statistics V28.0.0.0 was used for statistical calculations. Descriptive statistics were used to show data distribution. Data were presented as median and interquartile ranges. Analysis of Variance repeated measures (ANOVA) and, due to non parametric data in the jump test being ordinal scaled, the Friedman Test was used to calculate dependend between-group differences. For additional blinding, a group assignment was added to the results after statistical calculation and discussion of results in the author team.

3. Results

Twenty-four subjects (13 female, 11 male) with an age of 24 ± 2 years, a height of 1.75 ± 0.08 m, and a weight of 65 ± 8.5 kg were included. The median score for the two-leg stability test was 2.3 (1.72 to 2.90,) and for the one-leg stability test, 2.3 (1.90 to 2.88) (Figure 4).

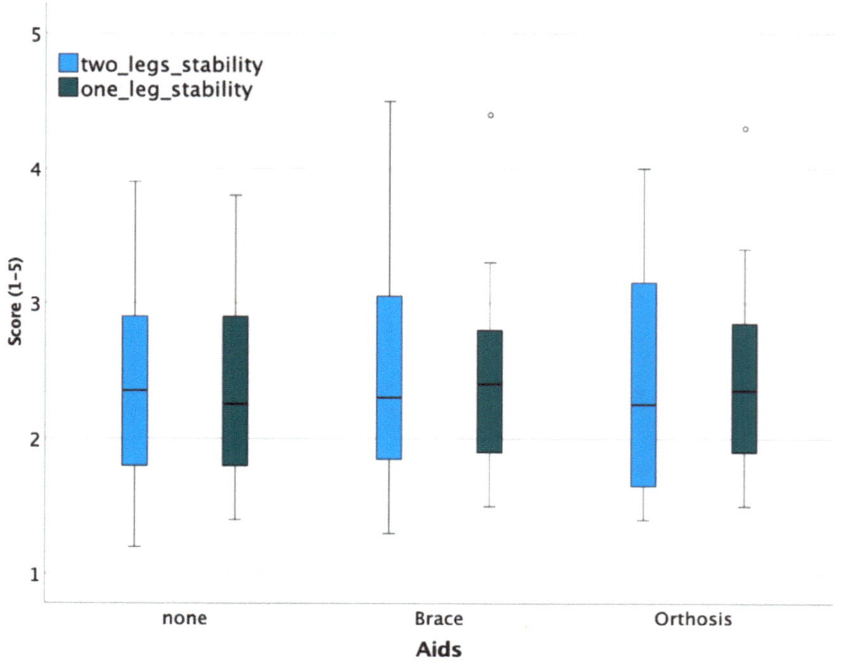

Figure 4. Stability tests show no clinically relevant difference between devices (° are outliers).

A statistically significant difference was not observed between groups (Table 2).

Table 2. Test statistics for between-group differences.

		SQUARE SUM	Chi²/F *	Sig.	Effect Size (ETA Square)
Two-legged stability	Friedmann		2.396	0.302	0.031
One-legged stability	Friedmann		1.542	0.463	0.026
Plyometric jump test	ANOVA		1.203	0.303	0.053
Two-legged CMJ	ANOVA		0.598	0.528	0.025
one-leg CMJ	ANOVA		0.536	0.589	0.023
Parkour test	ANOVA		0.643	0.484	0.027
Quick feet test	ANOVA		0.765	0.471	0.032

* F for ANOVA (Analysis of Variance repeated measures) repeated measures, Chi-Square for Friedman Test.

Participants performed the plyometric jump test without any knee stabilizer with a mean jump height of 27.6 ± 8.4 cm, with a brace of 25.9 ± 7.0 cm, and with an orthosis of 25.9 ± 6.4. For two-leg CMJ, the range of means between the three groups was 36.6 to 37.3 cm and 21.0 to 21.7 cm for one-leg CMJ. Median and IQRs are displayed in Figure 5.

None of the between-group mean differences were significant (Table 2). Subjects completed the parkour on average in 7.8 ± 3.6 s, with a group range of means between 7.68–7.88 s. The quick feet test was finished with a mean time of 7.68 ± 1.06 s without aids, 7.72 ± 1.58 s with a brace, and 7.88 ± 1.15 s with an orthosis. None of the between-group mean differences were significant (Table 2).

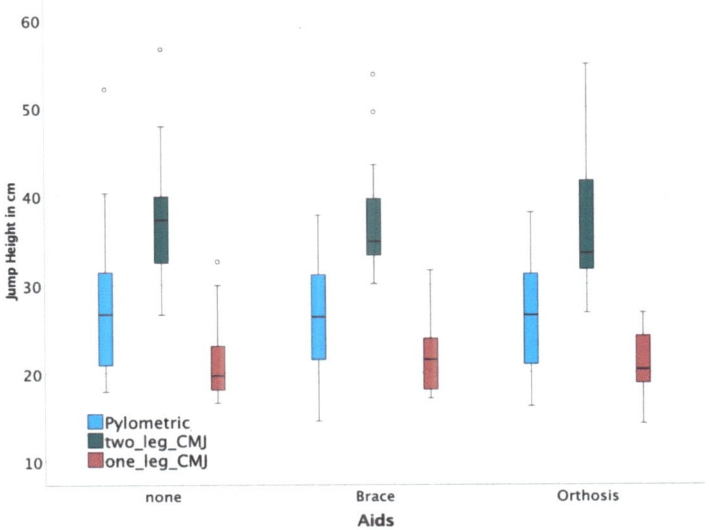

Figure 5. Jump tests show no clinically relevant difference between devices. (° are outliers).

4. Discussion

Our main findings in the present study were that no significant differences in performance of the back in action test battery were observed, regardless of if participants performed the test battery with a hard orthotic device, a soft brace, or no aid. Therefore, those devices did show a significant or clinical relevant effect on the performance of the participants.

The BIA consists of stability tests, two-leg and one-leg CMJ, standardized hop parkour, and a speed test. The BIA showed excellent test–retest reliability for two-leg CMJ, moderate reliability for the two-leg stability test, and good reliability for all further included test items [9]. The current test battery was used to test more than one skill and allow a better overview of coordinative abilities and the possibility to draw conclusions on the physiology.

Initiation of feedback control is based on actions predominantly influenced by the experience with the detected stimulus, which could be somatosensory, visual, or vestibular. The answer of the organism is based on different regulatory systems like reflexes or experience-based modulation. In contrast, feedforward control is based on anticipatory actions before homeostatic disruption is detected with one of the before-mentioned stimuli and is immediately used until feedback mechanisms help better react to a specific task [11,12]. Again, feedforward control is based on experiences with the expected task.

Feedforward strategies are predominantly used when motor learning for a certain task is finished and a person is familiar with the task. When external factors are likely to influence the success during the process of answering a task, like in Judo, feedback mechanisms become more important [13]. The functional joint stability is accomplished through a complementary relationship between static and dynamic components [12].

A potential explanation for similar results with or without orthotic devices might be few kinesthetic benefits for healthy participants due to the rather fixed and feedforward-based neuromuscular motor program they use for answering the tasks. One potential explanation is that lacking the experience for benefiting from feedback advantages, they could take additional external proprioceptive information through braces or orthosis.

For example, the combination of laterally wedged insoles and knee braces might influence balance in medial knee-osteoarthritis patients [14]. In contrast to those findings in our study, no influence was observed on a single leg or two leg balance tests performance, which is predominantly feedback control-based, which could be explained through the few benefits participants can take out of the additional sensory information. There was no clinically relevant difference between groups observed in the Counter Movement Jump Tests, parkour, and quick feed tests. This is an indicator of not being handicapped in developing speed and power by the used orthotic devices.

The same test design might show different results in patients after ACLR. The feedforward and the feedback systems are likely to be negatively influenced through not properly working active contributors, as to be seen in the onset latency of muscle activation. Neuromuscular impairments are present in patients early and late after ACLR [15,16]. The Sensorimotor system is also not sufficiently working through the absence of passive contributors like the ACL and misinforming structures like the swollen and inflamed joint capsular after surgery [17]. This results in high-risk knee positions. For example, larger knee abduction during landing seems to have a predictive value for ACL ruptures [18]. There probably is a mechanical, passive function of braces, as some impact of braces on physical performance was shown. The usage of braces or sleeves reduced the maximal flexion angle, abduction angle, and adduction moment of the knee [19]. Correct aligned knee brace might also slightly reduce ligament strain [20]. Therefore, feedforward and feedback strategies are likely to be negatively influenced by injury and reconstruction of the ACL, and those systems might benefit from braces or orthosis if athletes gather a certain amount of experience. This means that the longer the brace was used, the more benefits the participant could perhaps take out for more advantages in terms of feedback.

Strategies for test batteries are strongly discussed, and approaches like indexes, sensor-based testing, or rehabilitation strategies become more popular [21,22]. While significantly higher values in healthy athletes and patients after ACLR performing a drop jump with a knee brace [23] or for symmetry in increased hop distance during the single and crossover hop tests have been observed [24], no differences in performance for patients after ACLR in jump height or a modified agility t test have been seen at the timepoint of return to sport (RTS) and three months after [25]. In an RCT, including 150 patients post-ACLR, no significant difference was observed in the short or long term for the outcomes ACL-QOL Questionnaire, KT-1000 arthrometer side-to-side difference, for hop limb symmetry index, or Tegner Activity Scale between patients wearing neoprene sleeve or hard orthotic device. In a study with 19 male alpine skiers, neither rigid knee braces nor a sleeve influenced negative drop jumps. A musculoskeletal modeling analysis showed reduced flexion and frontal plane movement when wearing one of the devices, but they did not help reduce any shear force in the knee [19].

With respect to these controversies, the usage of hard or soft orthotic knee braces generally seems to result in some benefits with regard to of safety, but not in disadvantages in terms of performance.

In a recent study, athletes with ACLR eleven months after surgery were evaluated. They still showed a side-to-side difference (LSI) in the BIA test items. This was in contrast to the four-hop test and the measured isokinetic strength, which both did not show side-to-side differences anymore. This indicates the test might be superior compared to a single hop or isokinetic test for detecting limb symmetry index (LSI) differences after surgery, and thus is a better guide to allowing an athlete [9]. Using the BIA test battery, it was shown that patients eight months after ACLR were not ready for a safe RTS in terms of neuromuscular abilities [26]. These findings align with another study showing that nearly half of highly active patients present strength and functional fitness deficits six months after ACLR, no matter which graft type was used [27]. Other factors such as the sensorimotor system may be considered at the time of return to sports. Even athletes with higher strength and function results were at similar risk of an ACL graft tear with a higher risk for a contralateral ACL tear. High-performance patients are also at increased risk for an ACL graft tear [28]. Therefore, thinking about evaluating expressions of the Sensorimotor System must not only be done in terms of the system; it must be done in terms of the demands as well. Kyritsis et al. showed predictive value for isokinetic strength testing, a running *t*-test, triple-hop, and triple crossover hop test, and also found a decreased hamstrings to quadriceps ratio of the involved leg to be of higher risk for graft rupture [29]. Besides the additional use of established questionnaires, like the Lysholm Score [30] and generally the inclusion of psychological factors for RTS decision making [31], new issues should be considered when carrying on defining back to sport criteria. Activities around the hip joint, gluteal muscles, and lateral trunk obliquity seem to be of high interest for evaluating the RTS performance [32–34]. Additionally, fatigue should be more highlighted in ACLR risk evaluations [35]. More activity of the hamstrings seems to be promising for less ACL loading and, therefore, less risk of re-rupture [36]. This should be considered for improving rehabilitation programs, perhaps brace designing, and RTS battery development.

The following limitations of the study should be mentioned. Despite the possibility of a few feedback advantages, the results of the balance tests show a widespread distribution of performance. Therefore, the measurement error might be larger than the influence of orthotic devices on performance due to a lack of discriminatory power of the test in the studied population. The used test battery was developed for athletes and might require a higher level of performance to produce more reliable results. The test design was only applied to healthy subjects, and therefore caution must be taken when using results for injured athletes.

5. Conclusions

Soft and rigid knee braces do not affect performance in healthy participants. Missing experience with the devices might explain a few influences on feedback mechanisms. There is no disadvantage to be expected regarding healthy participants running back to sports test batteries with a brace or orthosis. Further studies after knee injuries are required.

Author Contributions: Conceptualization, R.P. and R.B.; methodology, R.P.; software, S.K.; validation, C.C., R.B. and S.K.; formal analysis, R.P.; investigation C.C.; resources, R.B. and S.K.; writing—original draft preparation, R.B.; writing—review and editing, A.K., R.B. and S.K.; visualization, R.P.; supervision, R.B.; project administration, R.P. and C.C.; funding acquisition R.B. and R.P. All authors have read and agreed to the published version of the manuscript.

Funding: The study gained external funding from medi GmbH & Co. KG. The source of funding had no role in the study design, in the collection, analysis, and interpretation of data, in the writing of the report, and in the decision to submit the article for publication.Funded by the MHB publication fund supported by DFG.

Institutional Review Board Statement: The study was approved by the appropriate ethical committee from the local medical chamber (AS 107(dB)/2018). All the participants signed informed consent to the work. The first author obtained written permission from participants to include their images in the manuscript.

Informed Consent Statement: Informed consent was obtained from all subjects involved in the study.

Data Availability Statement: Data is available from the authors on request from the corresponding author.

Conflicts of Interest: The authors have no conflict of interest to declare.

References

1. *ISO 8549-1:2020*; Prosthetics and Orthotics—Vocabulary—Part 1: General Terms for External Limb Prostheses and External Orthoses. Available online: https://www.iso.org/obp/ui/#iso:std:iso:8549:-3:ed-2:v1:en (accessed on 5 September 2022).
2. Paluska, S.A.; McKeag, D.B. Knee braces: Current evidence and clinical recommendations for their use. *Am. Fam. Phys.* **2000**, *61*, 411–418, 423–424.
3. Sugimoto, D.; LeBlanc, J.C.; Wooley, S.E.; Micheli, L.J.; Kramer, D.E. The Effectiveness of a Functional Knee Brace on Joint-Position Sense in Anterior Cruciate Ligament-Reconstructed Individuals. *J. Sport Rehabil.* **2016**, *25*, 190–194. [CrossRef] [PubMed]
4. Salata, M.J.; Gibbs, A.E.; Sekiya, J.K. The effectiveness of prophylactic knee bracing in american football: A systematic review. *Sports Health* **2010**, *2*, 375–379. [CrossRef] [PubMed]
5. Ramstrand, N.; Gjøvaag, T.; Starholm, I.M.; Rusaw, D.F. Effects of knee orthoses on kinesthetic awareness and balance in healthy individuals. *J. Rehabil. Assist. Technol. Eng.* **2019**, *6*, 2055668319852537. [CrossRef]
6. Riemann, B.L.; Lephart, S.M. The Sensorimotor System, Part II: The Role of Proprioception in Motor Control and Functional Joint Stability. *J. Athl. Train.* **2002**, *37*, 80–84.

7. Dwan, K.; Li, T.; Altman, D.G.; Elbourne, D. CONSORT 2010 statement: Extension to randomised crossover trials. *BMJ* **2019**, *366*, l4378. [CrossRef]
8. Schulz, R.; Langen, G.; Prill, R.; Cassel, M.; Weissgerber, T.L. Reporting and transparent research practices in sports medicine and orthopaedic clinical trials: A meta-research study. *BMJ Open* **2022**, *12*, e059347. [CrossRef]
9. Ebert, J.R.; Edwards, P.; Currie, J.; Smith, A.; Joss, B.; Ackland, T.; Buelow, J.-U.; Hewitt, B. Comparison of the 'Back in Action' Test Battery to Standard Hop Tests and Isokinetic Knee Dynamometry in Patients Following Anterior Cruciate Ligament Reconstruction. *Int. J. Sports Phys. Ther.* **2018**, *13*, 389–400. [CrossRef]
10. Hildebrandt, C.; Muller, L.; Zisch, B.; Huber, R.; Fink, C.; Raschner, C. Functional assessments for decision-making regarding return to sports following ACL reconstruction. Part I: Development of a new test battery. *Knee Surg. Sports Traumatol. Arthrosc.* **2015**, *23*, 1273–1281. [CrossRef]
11. Collins, J.J.; De Luca, C.J. Open-loop and closed-loop control of posture: A random-walk analysis of center-of-pressure trajectories. *Exp. Brain Res.* **1993**, *95*, 308–318. [CrossRef]
12. Riemann, B.L.; Lephart, S.M. The sensorimotor system, part I: The physiologic basis of functional joint stability. *J. Athl. Train.* **2002**, *37*, 71–79.
13. Prill, R.; Appell Coriolano, H.J.; Michel, S.; Alfuth, M. The Influence of the Special Throwing Technique on the Prevalence of Knee Joint Injuries in Judo. *Arch. Budo* **2014**, *10*, 211–216.
14. Khan, S.J.; Khan, S.S.; Usman, J.; Mokhtar, A.H.; Abu Osman, N.A. Combined effects of knee brace, laterally wedged insoles and toe-in gait on knee adduction moment and balance in moderate medial knee osteoarthritis patients. *Gait Posture* **2018**, *61*, 243–249. [CrossRef]
15. Norte, G.E.; Hertel, J.; Saliba, S.A.; Diduch, D.R.; Hart, J.M. Quadriceps Neuromuscular Function in Patients With Anterior Cruciate Ligament Reconstruction With or Without Knee Osteoarthritis: A Cross-Sectional Study. *J. Athl. Train.* **2018**, *53*, 475–485. [CrossRef]
16. Prill, R.; Michel, S.; Schulz, R.; Coriolano, H.A. Body Composition and Strength Parameters in Elite Judo Athletes 5 Years after Anterior Cruciate Ligament Reconstruction. *Int. J. Sports Med.* **2019**, *40*, 38–42. [CrossRef]
17. Relph, N.; Herrington, L.; Tyson, S. The effects of ACL injury on knee proprioception: A meta-analysis. *Physiotherapy* **2014**, *100*, 187–195. [CrossRef]
18. Bates, N.A.; Myer, G.D.; Hale, R.F.; Schilaty, N.D.; Hewett, T.E. Prospective Frontal Plane Angles Used to Predict ACL Strain and Identify Those at High Risk for Sports-Related ACL Injury. *Orthop. J. Sports Med.* **2020**, *8*, 2325967120957646. [CrossRef]
19. Moon, J.; Kim, H.; Lee, J.; Panday, S.B. Effect of wearing a knee brace or sleeve on the knee joint and anterior cruciate ligament force during drop jumps: A clinical intervention study. *Knee* **2018**, *25*, 1009–1015. [CrossRef]
20. Hacker, S.P.; Schall, F.; Ignatius, A.; Dürselen, L. The effect of knee brace misalignment on the anterior cruciate ligament: An experimental study. *Prosthet. Orthot. Int.* **2019**, *43*, 309–315. [CrossRef]
21. Oleksy, Ł.; Królikowska, A.; Mika, A.; Kuchciak, M.; Szymczyk, D.; Rzepko, M.; Bril, G.; Prill, R.; Stolarczyk, A.; Reichert, P. A Compound Hop Index for Assessing Soccer Players' Performance. *J. Clin. Med.* **2022**, *11*, 255. [CrossRef]
22. Prill, R.; Walter, M.; Królikowska, A.; Becker, R. A Systematic Review of Diagnostic Accuracy and Clinical Applications of Wearable Movement Sensors for Knee Joint Rehabilitation. *Sensors* **2021**, *21*, 8221. [CrossRef]
23. Rebel, M.; Paessler, H.H. The effect of knee brace on coordination and neuronal leg muscle control: An early postoperative functional study in anterior cruciate ligament reconstructed patients. *Knee Surg. Sports Traumatol. Arthrosc.* **2001**, *9*, 272–281. [CrossRef]
24. Peebles, A.T.; Miller, T.K.; Moskal, J.T.; Queen, R.M. Hop testing symmetry improves with time and while wearing a functional knee brace in anterior cruciate ligament reconstructed athletes. *Clin. Biomech.* **2019**, *70*, 66–71. [CrossRef]
25. Dickerson, L.C.; Peebles, A.T.; Moskal, J.T.; Miller, T.K.; Queen, R.M. Physical Performance Improves With Time and a Functional Knee Brace in Athletes After ACL Reconstruction. *Orthop. J. Sports Med.* **2020**, *8*, 2325967120944255. [CrossRef]
26. Herbst, E.; Hoser, C.; Hildebrandt, C.; Raschner, C.; Hepperger, C.; Pointner, H.; Fink, C. Functional assessments for decision-making regarding return to sports following ACL reconstruction. Part II: Clinical application of a new test battery. *Knee Surg. Sports Traumatol. Arthrosc.* **2015**, *23*, 1283–1291. [CrossRef]
27. Csapo, R.; Pointner, H.; Hoser, C.; Gfoller, P.; Raschner, C.; Fink, C. Physical Fitness after Anterior Cruciate Ligament Reconstruction: Influence of Graft, Age, and Sex. *Sports* **2020**, *8*, 30. [CrossRef]
28. Marigi, E.M.; Bernard, C.D.; Hale, R.F.; Stuart, M.J.; Levy, B.A.; Dahm, D.L.; Hewett, T.E.; Krych, A.J. Return to Sport: Does 6-month Functional Testing Predict Second ACL Injuries at Long-term Follow-Up? *Orthop. J. Sports Med.* **2019**. [CrossRef]
29. Kyritsis, P.; Bahr, R.; Landreau, P.; Miladi, R.; Witvrouw, E. Likelihood of ACL graft rupture: Not meeting six clinical discharge criteria before return to sport is associated with a four times greater risk of rupture. *Br. J. Sports Med.* **2016**, *50*, 946–951. [CrossRef]
30. Briggs, K.K.; Lysholm, J.; Tegner, Y.; Rodkey, W.G.; Kocher, M.S.; Steadman, J.R. The reliability, validity, and responsiveness of the Lysholm score and Tegner activity scale for anterior cruciate ligament injuries of the knee: 25 years later. *Am. J. Sports Med.* **2009**, *37*, 890–897. [CrossRef]
31. Kaplan, Y.; Witvrouw, E. When Is It Safe to Return to Sport After ACL Reconstruction? Reviewing the Criteria. *Sports Health* **2019**, *11*, 301–305. [CrossRef]

32. Nagelli, C.; Di Stasi, S.; Tatarski, R.; Chen, A.; Wordeman, S.; Hoffman, J.; Hewett, T.E. Neuromuscular Training Improves Self-Reported Function and Single-Leg Landing Hip Biomechanics in Athletes After Anterior Cruciate Ligament Reconstruction. *Orthop. J. Sports Med.* **2020**, *8*, 2325967120959347. [CrossRef] [PubMed]
33. Rinaldi, V.G.; Prill, R.; Jahnke, S.; Zaffagnini, S.; Becker, R. The influence of gluteal muscle strength deficits on dynamic knee valgus: A scoping review. *J. Exp. Orthop.* **2022**, *9*, 81. [CrossRef] [PubMed]
34. Taniguchi, S.; Ishida, T.; Yamanaka, M.; Ueno, R.; Ikuta, R.; Chijimatsu, M.; Samukawa, M.; Koshino, Y.; Kasahara, S.; Tohyama, H. Landing Biomechanics During a Single-leg Landing With Lateral Trunk Obliquity in Female Athletes. *Res. Square* **2021**. [CrossRef]
35. Bourne, M.N.; Webster, K.E.; Hewett, T.E. Is Fatigue a Risk Factor for Anterior Cruciate Ligament Rupture? *Sports Med.* **2019**, *49*, 1629–1635. [CrossRef]
36. Ueno, R.; Navacchia, A.; Schilaty, N.D.; Myer, G.D.; Hewett, T.E.; Bates, N.A. Hamstrings Contraction Regulates the Magnitude and Timing of the Peak ACL Loading During the Drop Vertical Jump in Female Athletes. *Orthop. J. Sports Med.* **2021**, *9*, 23259671211034487. [CrossRef]

Journal of *Personalized Medicine*

MDPI

Study Protocol

IRO/Shift Test Is Comparable to the Jobe Test for Detection of Supraspinatus Lesions

George Fieseler [1], René Schwesig [2], Julia Sendler [1,2], Jakob Cornelius [2], Stephan Schulze [2], Wolfgang Lehmann [3], Souhail Hermassi [4], Karl-Stefan Delank [2] and Kevin Laudner [5,*]

1 Clinic for Orthopedic and Trauma Surgery, Sports Medicine, Clinic Hannoversch Münden, 34346 Hannoversch Münden, Germany
2 Department of Orthopedic and Trauma Surgery, Martin-Luther-University Halle-Wittenberg, 06120 Halle, Germany
3 Clinic for Orthopedic, Trauma and Reconstructive Surgery, Georg August University Göttingen, 37075 Göttingen, Germany
4 Physical Education Department, College of Education, Qatar University, Doha 2713, Qatar
5 Department of Health Sciences, University of Colorado Colorado Springs, Colorado Springs, CO 80918, USA
* Correspondence: klaudner@uccs.edu

Abstract: To determine how the internal rotation and shift (IRO/shift) test compares to the gold standard of clinical tests (Jobe test) for diagnosing supraspinatus lesions and to confirm these clinical results with surgical findings, 100 symptomatic patients were clinically examined between October 2018 and November 2019. All 100 patients were evaluated using both the IRO/shift test and Jobe test. A total of 48 of these patients received surgical intervention. Based on these data, sensitivity, specificity, positive predictive value (PPV), negative predictive value (NPV), and accuracy for both the IRO/shift test and Jobe test were calculated. The IRO/shift test had a sensitivity of 96% (95% CI: 82–100%), specificity of 50% (95% CI: 27–73%), PPV of 73% (95% CI: 56–86%), NPV of 91% (95% CI: 59–100%), and an accuracy of 77% (95% CI: 63–88%). The Jobe test had a sensitivity of 89% (95% CI: 72–98%), specificity of 60% (95% CI: 36–81%), PPV of 76% (95% CI: 58–89%), NPV of 80% (95% CI: 52–96%), and an accuracy of 77% (95% CI: 54–81%). These results suggest that the IRO/shift test is comparable to the Jobe test, which is often viewed as the gold standard clinical examination for assessing supraspinatus lesions. This study was approved by the Ethics Commission of the Martin Luther University Halle-Wittenberg (reference number: 2018-05).

Keywords: clinical; examination; rotator cuff; special test; sensitivity; specificity

Citation: Fieseler, G.; Schwesig, R.; Sendler, J.; Cornelius, J.; Schulze, S.; Lehmann, W.; Hermassi, S.; Delank, K.-S.; Laudner, K. IRO/Shift Test Is Comparable to the Jobe Test for Detection of Supraspinatus Lesions. *J. Pers. Med.* **2022**, *12*, 1422. https://doi.org/10.3390/jpm12091422

Academic Editors: Markus Gesslein and Eduard Kurz

Received: 11 August 2022
Accepted: 27 August 2022
Published: 31 August 2022

1. Introduction

The excessive mobility of the glenohumeral joint often comes at the expense of stability. The rotator cuff muscles are key stabilizers of the shoulder and are routinely stressed during activities of daily living, as well as during overhead sporting motions [1]. It is therefore not surprising that rotator cuff injuries account for the majority of all shoulder-related pathologies [2]. Research reported that 34% of the general population had symptomatic tears, while 65% had asymptomatic tears, and the prevalence of tears significantly increased with age beginning in the 50s [3]. This pathology was also noted as a common cause of morbidity among the elderly [1]. Not only is rotator cuff pathology common but it can also be very debilitating, with 40% of patients with a rotator cuff disorder complaining of persistent symptoms lasting longer than 12 months [4]. Common symptoms include pain, lost range of motion, weakness, and sleep disturbances [5,6]. This combination of symptoms contributes to the debilitating nature of this pathology and negatively impacts activities of daily living, performance in sport, and leisure activities, as well as an individual's ability to adequately perform their job [5].

Due to the common occurrence and debilitating nature of rotator cuff injuries, it is imperative that such pathologies are accurately diagnosed to ensure appropriate treatment. There has been a plethora of studies investigating the efficiency of different clinical tests for rotator cuff pathology. Common highly favorable tests for supraspinatus lesions include the drop arm test, full can test, and Hawkin's sign, just to name a few [7–9]. Although special tests such as the drop arm test have been shown to have good specificity (87–99%) and variable sensitivity (6–71%) [9,10], the Jobe's test commonly outperforms the others with more consistent specificity (62–75%) and sensitivity (72–88%) [7,9,11] and is considered the gold standard of clinical testing [7,9].

Due to the wide ranges of specificity and sensitivity between tests, several reports have suggested that clinicians should not rely on just one test but instead use multiple tests to increase diagnostic accuracy [9,12–14]. As such, the lead author of this study recently developed the internal rotation and shift (IRO/shift) test to add to the existing special tests in an effort to enhance the diagnostic efficiency of supraspinatus tears.

The objective of our study was to determine how the IRO/shift test compares to the gold standard of clinical tests (Jobe test) and confirm with surgical findings. More specifically, this study compared the specificity, sensitivity, positive predictive value (PPV), and negative predictive value (NPV) of the IRO/shift test and Jobe test compared to surgical findings. We hypothesized that the IRO/shift test would provide similar findings compared to those of the Jobe test and surgical findings and would be considered a beneficial clinical test to include in supraspinatus evaluation protocols.

2. Materials and Methods

2.1. Subjects

A sample of convenience was used among 100 patients who initially volunteered to participate in this study between October 2018 and November 2019. Subjects were 18 years or older and were patients at a single surgery and rehabilitation clinic that specializes in shoulder pathologies. All subjects provided informed consent prior to any data collection as approved by the Ethics Commission of the Martin Luther University Halle-Wittenberg (reference number: 2018-05). Patients were excluded from the study if they had a status of post instability, arthrosis, fracture, rheumatoid arthritis, or adhesive capsulitis or had upper extremity range of motion (ROM) restrictions. Among the 100 patients evaluated clinically, 48 were scheduled for arthroscopic surgery (31 male, 17 female, age 55.4 ± 11.4 years, height 1.76 ± 0.08 m, weight 85.2 ± 18.6 kg, and BMI 27.3 ± 4.5 kg/m^2) between 2018 and 2019. Additional demographic information can be viewed in Table 1. Both partial and full-thickness tears identified during surgery were considered positive for a supraspinatus lesion and data analysis. Reasons for not receiving surgery among the remaining 52 patients included pre-operative and anesthetic contraindications, the patient not wanting to pursue surgical treatment, major and retractive supraspinatus lesion, being a declared candidate for primary reverse shoulder arthroplasty, or conservative treatment having been recommended.

Table 1. Patient anthropometric and demographic data (n = 48).

Age (years)	55.4 ± 11.4
Height (m)	1.76 ± 0.08
Weight (kg)	85.2 ± 18.6
Body mass index (kg/m^2)	27.3 ± 4.53
Sex (male: female)	31:17
Duration of pain (<6 months: >6 months)	2:46
accident anamnesis (yes: no)	11:37
pain localization (left: right)	21:27
Cause of the lesion (traumatic: after shoulder dislocation: degenerative/impingement)	9:2:37

Reported as mean ± standard deviation.

2.2. Procedures

A blinded, prospective research design was used for this study. Each of the 48 surgical patients was clinically evaluated using both the IRO/shift test and the Jobe test and subsequently received surgery. All pre-operative examinations were conducted by an experienced orthopedic physician who has been conducting shoulder surgeries since 2001 and who performs approximately 200 shoulder surgeries annually.

For the IRO/shift test [15], each patient stood in a relaxed position with their feet shoulder-width apart. The patient was instructed to actively move their involved side arm behind their back and then internally rotate so that the dorsal aspect of their hand moved superiorly along their spine. At the end range of pain-free motion, the examiner then applied an additional passive movement of the arm into greater internal rotation and therefore elevated the hand further superiorly along the spine (Figure 1). Passively moving the shoulder into end-range adduction and internal rotation causes further anterior translation of the humeral head and subsequent increased pressure on the supraspinatus and any potential muscle defect, which results in increased pain (positive test). The intention of the IRO/shift test is to determine the presence of supraspinatus pathology; however, pain during this motion may also be caused by damage to the long head of the biceps. Therefore, biceps involvement must subsequently be ruled out. For this, the examiner performed subsequent special tests, such as the O'Brien's, Speed's, and/or Yergason's tests. If pain was present during these additional special tests, indicating biceps involvement, then the IRO/shift test was considered negative; however, if the biceps special tests were negative, then the IRO/shift test was considered positive for supraspinatus pathology.

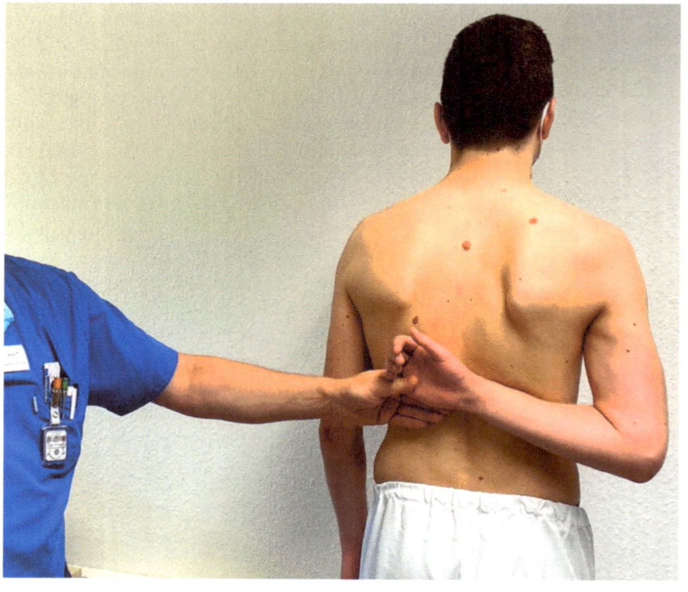

Figure 1. Passively applying elevation overpressure (shoulder adduction and internal rotation) to the patient's end range of motion.

2.3. Statistical Analysis

SPSS version 28.0 for Windows (IBM, Armonk, NY, USA) was used for all statistical analyses. Surgical findings were used to determine the sensitivity, specificity, positive predictive value (PPV), negative predictive value (NPV), and accuracy for both the Jobe test and IRO/shift test. Sensitivity of these diagnostic tests was defined as the test's ability to demonstrate a positive test when the condition was actually present (i.e., true positive). Conversely, specificity was the test's ability to produce a negative test when the pathology was actually absent (i.e., true negative). In addition to sensitivity and specificity, values for prediction were calculated to provide perspectives into the feasibility of the tests. The PPV determines the likelihood that a patient who presents with a positive clinical test has the pathology, while NPV determines the likelihood that a patient who presents with a negative test is actually free from that pathology. The accuracy of a test considers both sensitivity and specificity and thereby describes how correct the test is in identifying and excluding a pathology.

A sample size of 70 was calculated based on the prevalence of surgical intervention of 50% (48/100) and an expected sensitivity of 90% (95% CI, width of 10%) [16].

3. Results

The IRO/shift test has shown good–excellent reliability (intrarater (ICC = 0.73, 95% CI: 60–82), interrater (ICC = 0.89, 95% CI: 81–94)), as well as 92% sensitivity (95% CI: 86–99%) and 67% specificity (95% CI: 50–84%) when confirmed with MRI. This test also demonstrated 86% positive predictive value and 80% negative predictive value [15].

3.1. Surgical Findings

Among the 48 patients who underwent surgery, 28 were found to have a supraspinatus tear (partial or full). Various pathologies were found among the 20 patients who did not have a supraspinatus tear including biceps tendon pathology, SLAP lesions, osteophytosis, bursitis, acromioclavicular joint arthrosis, and calcific tendinitis. Table 2 depicts the comparison of the IRO/shift test and the Jobe test with the surgical findings.

Table 2. IRO/shift and Jobe Tests compared to surgical findings.

Test	IRO/Shift Test % (95% CI)	Jobe Test % (95% CI)
Sensitivity	96 (82–100)	89 (72–98)
Specificity	50 (27–73)	60 (36–81)
PPV	73 (56–86)	76 (58–89)
NPV	91 (59–100)	80 (52–96)
Accuracy	77 (63–88)	77 (54–81)

CI: confidence interval; PPV: positive predictive value; NPV: negative predictive value.

3.2. Comparison of IRO/Shift Test and Jobe Test

Among the 28 patients with a supraspinatus lesion identified upon surgery, 26 (93%) presented with a positive IRO/shift test and 25 (89%) had a positive Jobe test. Among the 20 patients who had no evidence of supraspinatus lesion found during surgery, 10 (50%) had a negative IRO/shift test, while 12 (60%) had a negative Jobe test. For the IRO/shift test and Jobe test, a similar performance was calculated (Table 2).

4. Discussion

Rotator cuff pathology is one of the leading orthopedic injuries with an increased prevalence of both partial and full-thickness tears as individuals age [17–19]. What makes this pathology even more difficult is that reliance on imaging (MRI, diagnostic ultrasound) alone can be detrimental due to the increased risk of inaccurate diagnoses [20–22]. As such, the purpose of this study was to compare the accuracy of the IRO/shift and Jobe tests with surgical intervention as opposed to MRI. Therefore, multiple clinical tests should be used during evaluation for enhanced diagnostic accuracy. Although numerous special tests have been investigated in the literature, there is still a paucity of information on which are most effective. The results of this study indicated that the IRO/shift test is a valid clinical test that is comparable to the Jobe test and should be considered when examining patients with potential supraspinatus pathology.

We chose to compare the IRO/shift test with the Jobe test because the Jobe test is arguably the gold standard of clinical exams for evaluating superior supraspinatus pathologies. Jain et al. [23] used a predictive model to determine the probability of supraspinatus tears without using imaging. These predictors included sex, the involved-to-uninvolved ratio of external rotation strength, and results of the Jobe test and lift-off test. The authors reported that this model could accurately recognize a supraspinatus tear with satisfactory discrimination. However, this is not the first study to recognize the efficiency of the Jobe test. Numerous investigations have noted the Jobe test as being highly sensitive and specific for supraspinatus tears; however, there has been some disagreement on whether the test is more sensitive or specific [7,9,24,25]. Other studies have used surgical findings to determine the accuracy of other special tests compared to the Jobe test for diagnosing supraspinatus tears. Franca et al. [26] reported that compared to the Jobe test, the drop arm test was less sensitive but more specific. Similarly, the drop arm test had higher NPV but lower PPV. Zou et al. [27] found that the Jobe test and the hug-up test had similar accuracy, while the drop arm had less sensitivity, specificity, PPV, and NPV and the Hawkins test had similar specificity and PPV but less sensitivity and NPV than Jobe. Our study found that when confirmed with surgical findings, the Jobe test demonstrated good sensitivity (89%) and less specificity (60%); however, both PPV (76%) and NPV (80%) were laudable.

We believe that these results demonstrate the value of the IRO/shift test. However, past investigations have consistently shown that multiple clinical tests should be used because no single clinical test can sufficiently diagnose supraspinatus tears [9,12–14]. Furthermore, the use of multiple reliable and valid clinical tests may also decrease the clinicians need to rely on expensive imaging [7,9]. As such, we suggest clinicians incorporate the IRO/shift test into their existing protocol for clinical examination of supraspinatus pathology.

There are a few limitations to this study worth noting. Patients must also have ample active and passive glenohumeral internal rotation and adduction ROM to perform the IRO/shift test. As such, this special test would not be appropriate for patients with severe restrictions or pain in these motions. Additionally, we chose to compare the IRO/shift test to the Jobe test because the Jobe test is predominantly considered to be the gold standard for clinical examinations of superior rotator cuff pathology; however, there are numerous other special tests that should be considered during evaluation, such as Hawkin's sign, full can test, and drop arm test, to name but a few [7–9]. Future investigations are necessary to compare the clinical usefulness of the IRO/shift test with these other exams. Lastly, the involved number of patients did not meet our calculated sample size (48 vs. 70). Therefore, the results should not be overinterpreted.

5. Conclusions

The IRO/shift test had similar sensitivity and specificity to that of the Jobe test and demonstrated good validity for assessing supraspinatus pathology. As such, we would suggest clinicians consider using the IRO/shift test along with other clinical tests and advanced imaging preferred by the respective clinician.

Author Contributions: Conceptualization, G.F., R.S., J.S., J.C., S.S., W.L., S.H., K.-S.D. and K.L.; methodology, G.F., R.S., J.S., J.C., S.S., W.L., S.H., K.-S.D. and K.L.; software, G.F., R.S., J.S., J.C., S.S., W.L., S.H., K.-S.D. and K.L.; validation, G.F., R.S., J.S., J.C., S.S., W.L., S.H., K.-S.D. and K.L.; formal analysis, G.F., R.S., J.S., J.C., S.S., W.L., S.H., K.-S.D. and K.L.; investigation, G.F., R.S., J.S., J.C., S.S., W.L., S.H., K.-S.D. and K.L.; resources, G.F., R.S., J.S., J.C., S.S., W.L., S.H., K.-S.D. and K.L.; data curation, G.F., R.S., J.S., J.C., S.S., W.L., S.H., K.-S.D. and K.L.; writing original draft preparation, G.F., R.S., J.S., J.C., S.S., W.L., S.H., K.-S.D. and K.L.; writing review and editing, G.F., R.S., J.S., J.C., S.S., W.L., S.H., K.-S.D. and K.L.; visualization, G.F., R.S., J.S., J.C., S.S., W.L., S.H., K.-S.D. and K.L.; supervision, G.F., R.S., J.S., J.C., S.S., W.L., S.H., K.-S.D. and K.L.; project administration, G.F., R.S., J.S., J.C., S.S., W.L., S.H., K.-S.D. and K.L. All authors have read and agreed to the published version of the manuscript.

Funding: This research received no external funding.

Institutional Review Board Statement: This study was approved by the Ethics Commission of the Martin Luther University Halle-Wittenberg (reference number: 2018-05).

Informed Consent Statement: All subjects provided informed consent.

Data Availability Statement: Please contact the corresponding author for inquiries regarding study data.

Conflicts of Interest: The authors declare no conflict of interest.

References

1. Abdelwahab, A.; Ahuja, N.; Iyengar, K.P.; Jain, V.K.; Bakti, N.; Singh, B. Traumatic rotator cuff tears—Current concepts in diagnosis and management. *J. Clin. Orthop. Trauma* **2021**, *18*, 51–55. [CrossRef] [PubMed]
2. Zaid, M.B.; Young, N.M.; Pedoia, V.; Feeley, B.T.; Ma, C.B.; Lansdown, D.A. Anatomic shoulder parameters and their relationship to the presence of degenerative rotator cuff tears and glenohumeral osteoarthritis: A systematic review and meta-analysis. *J. Shoulder Elb. Surg.* **2019**, *28*, 2457–2466. [CrossRef]
3. Minagawa, H.; Yamamoto, N.; Abe, H.; Fukuda, M.; Seki, N.; Kikuchi, K.; Kijima, H.; Itoi, E. Prevalence of symptomatic and asymptomatic rotator cuff tears in the general population: From mass-screening in one village. *J. Orthop.* **2013**, *10*, 8–12. [CrossRef] [PubMed]
4. van der Windt, D.A.; Koes, B.W.; Boeke, A.J.; Devillé, W.; De Jong, B.A.; Bouter, L.M. Shoulder disorders in general practice: Prognostic indicators of outcome. *Br. J. Gen. Pract.* **1996**, *46*, 519–523.
5. Minns Lowe, C.J.; Moser, J.; Barker, K. Living with a symptomatic rotator cuff tear 'bad days, bad nights': A qualitative study. *BMC Musculoskelet. Disord.* **2014**, *15*, 228. [CrossRef] [PubMed]
6. Nyman, P.; Palenius, K.; Panula, H.; Mälkiä, E. Patients' experiences of shoulder problems prior to and following intervention. *Physiother. Theory Pract.* **2012**, *28*, 221–231. [CrossRef] [PubMed]
7. Jain, N.B.; Luz, J.; Higgins, L.D.; Dong, Y.; Warner, J.J.; Matzkin, E.; Katz, J.N. The Diagnostic Accuracy of Special Tests for Rotator Cuff Tear: The ROW Cohort Study. *Am. J. Phys. Med. Rehabil.* **2017**, *96*, 176–183. [CrossRef]
8. Junior, J.A.Y.; Nicolao, F.A.; Matsunaga, F.T.; Netto, N.A.; Belloti, J.C.; Tamaoki, M.J.S. Supraspinatus tears: Predictability of magnetic resonance imaging findings based on clinical examination. *J. Shoulder Elb. Surg.* **2021**, *30*, 1834–1843. [CrossRef]
9. Lädermann, A.; Meynard, T.; Denard, P.J.; Ibrahim, M.; Saffarini, M.; Collin, P. Reliable diagnosis of posterosuperior rotator cuff tears requires a combination of clinical tests. *Knee Surg. Sports Traumatol. Arthrosc.* **2021**, *29*, 2118–2133. [CrossRef]
10. van Kampen, D.A.; van den Berg, T.; van der Woude, H.J.; Castelein, R.M.; Scholtes, V.A.; Terwee, C.B.; Willems, W.J. The diagnostic value of the combination of patient characteristics, history, and clinical shoulder tests for the diagnosis of rotator cuff tear. *J. Orthop. Surg. Res.* **2014**, *9*, 70. [CrossRef]
11. Águila-Ledesma, I.R.; Córdova-Fonseca, J.L.; Medina-Pontaza, O.; A Núñez-Gómez, D.; Calvache-García, C.; Pérez-Atanasio, J.M.; Torres-González, R. Diagnostic value of a predictive model for complete ruptures of the rotator cuff associated to subacromial impingement. *Acta Ortop. Mex.* **2017**, *31*, 108–112.
12. Cadogan, A.; McNair, P.J.; Laslett, M.; Hing, W. Diagnostic Accuracy of Clinical Examination and Imaging Findings for Identifying Subacromial Pain. *PLoS ONE.* **2016**, *11*, e0167738. [CrossRef]
13. Sgroi, M.; Loitsch, T.; Reichel, H.; Kappe, T. Diagnostic Value of Clinical Tests for Supraspinatus Tendon Tears. *Arthroscopy* **2018**, *34*, 2326–2333. [CrossRef]
14. Somerville, L.E.; Willits, K.; Johnson, A.M.; Litchfield, R.; LeBel, M.E.; Moro, J.; Bryant, D. Clinical Assessment of Physical Examination Maneuvers for Rotator Cuff Lesions. *Am. J. Sports Med.* **2014**, *42*, 1911–1919. [CrossRef]
15. Fieseler, G.; Laudner, K.; Sendler, J.; Cornelius, J.; Schulze, S.; Lehmann, W.; Hermassi, S.; Delank, K.-S.; Schwesig, R. The internal rotation and shift-test for the detection of superior lesions of the rotator cuff: Reliability and clinical performance. *JSES Int.* **2022**, *6*, 495–499. [CrossRef]
16. Buderer, N.M. Statistical methodology: I. Incorporating the prevalence of disease into the sample size calculation for sensitivity and specificity. *Acad. Emerg. Med.* **1996**, *3*, 895–900. [CrossRef]
17. Milgrom, C.; Schaffler, M.; Gilbert, S.; Van Holsbeeck, M. Rotator-cuff changes in asymptomatic adults. The effect of age, hand dominance and gender. *J. Bone Jt. Surg. Br.* **1995**, *77*, 296–298. [CrossRef]
18. Sher, J.S.; Uribe, J.W.; Posada, A.; Murphy, B.J.; Zlatkin, M.B. Abnormal findings on magnetic resonance images of asymptomatic shoulders. *J. Bone Jt. Surg. Am.* **1995**, *77*, 10–15. [CrossRef]
19. Frost, P.; Andersen, J.H.; Lundorf, E. Is supraspinatus pathology as defined by magnetic resonance imaging associated with clinical sign of shoulder impingement? *J. Shoulder Elb. Surg.* **1999**, *8*, 565–568. [CrossRef]
20. Bordage, G. Where are the history and the physical? *CMAJ* **1995**, *152*, 1595–1598.
21. Cook, C. The lost art of the clinical examination: An overemphasis on clinical special tests. *J. Man Manip. Ther.* **2010**, *18*, 3–4. [CrossRef]
22. Roquelaure, Y.; Ha, C.; Leclerc, A.; Touranchet, A.; Sauteron, M.; Melchior, M.; Imbernon, E.; Goldberg, M. Epidemiologic surveillance of upper-extremity musculoskeletal disorders in the working population. *Arthritis. Rheum.* **2006**, *55*, 765–778. [CrossRef] [PubMed]

23. Jain, N.B.; Fan, R.; Higgins, L.D.; Kuhn, J.E.; Ayers, G.D. Does My Patient with Shoulder Pain Have a Rotator Cuff Tear? A Predictive Model from the ROW Cohort. *Orthop. J. Sports Med.* **2018**, *6*, 2325967118784897. [CrossRef] [PubMed]
24. Beaudreuil, J.; Nizard, R.; Thomas, T.; Peyre, M.; Liotard, J.P.; Boileau, P.; Marc, T.; Dromard, C.; Steyer, E.; Bardin, T.; et al. Contribution of clinical tests to the diagnosis of rotator cuff disease: A systematic literature review. *Jt. Bone Spine* **2009**, *76*, 15–19. [CrossRef]
25. Cotter, E.J.; Hannon, C.P.; Christian, D.; Frank, R.M.; Bach, B.R. Comprehensive Examination of the Athlete's Shoulder. *Sports Health* **2018**, *10*, 366–375. [CrossRef] [PubMed]
26. Franca, F.O.; Godinho, G.G.; Freitas, J.M.A.; Lang, A.S.; Ammar, C.D.; Martinelli, F. Correlation of Physical Examination with Arthroscopic Findings in the Treatment of Rotator Cuff Tear. *Rev. Bras. Ortop.* **2022**, *57*, 467–471.
27. Zou, A.P.; An, F.M.; Xin, Y.Q. Diagnostic value of physical examination tests for supraspinatus tendon tears. *Zhongguo Gu Shang* **2022**, *35*, 220–224.

Journal of *Personalized Medicine*

MDPI

Article

An Innovative Non-Invasive Method for Early Detection and Monitoring of Acute Compartment Syndrome

Razvan Tudor Tepordei [1,2], Carmen Lacramioara Zamfir [1], Alin Horatiu Nedelcu [1,*], Ovidiu Gabriel Avadanei [3], Tudor Cozma [2,4], Ovidiu Alexa [2,4], Manuela Ursaru [5], Lacramioara Perianu [1], Iuliana Magdalena Starcea [6], Ileana Ioniuc [6], Vasile Valeriu Lupu [6] and Ancuta Lupu [6]

1 Department of Morpho-Functional Science I, Discipline of Anatomy, "Grigore T. Popa" University of Medicine and Pharmacy, 16 Universitatii Street, 700115 Iasi, Romania; razvan.tepordei@umfiasi.ro (R.T.T.); carmen.zamfir@umfiasi.ro (C.L.Z.); lacramioara.perianu@umfiasi.ro (L.P.)
2 Department of Orthopaedics and Traumatology, "Sf. Spiridon" Emergency Universitary Hospital, 700115 Iasi, Romania; tudor.cozma@umfiasi.ro (T.C.); ovidiu.alexa@umfiasi.ro (O.A.)
3 Department of Physics, Al. I. Cuza University, 11 Carol I Bv, 700506 Iasi, Romania; minu@uaic.ro
4 Department of Surgical Sciences II, "Grigore T. Popa" University of Medicine and Pharmacy, 16 Universitatii Street, 700115 Iasi, Romania
5 Department of Surgical Sciences I, "Grigore T. Popa" University of Medicine and Pharmacy, 16 Universitatii Street, 700115 Iasi, Romania; manuela.ursaru@umfiasi.ro
6 Department of Mother and Child, "Grigore T. Popa" University of Medicine and Pharmacy, 16 Universitatii Street, 700115 Iasi, Romania; iuliana.starcea@umfiasi.ro (I.M.S.); ileana.ioniuc@umfiasi.ro (I.I.); vasile.lupu@umfiasi.ro (V.V.L.); ancuta.ignat1@umfiasi.ro (A.L.)
* Correspondence: alin.nedelcu@umfiasi.ro

Abstract: Background: Acute compartment syndrome is a major surgical emergency with complex pathophysiology and a highly unpredictable pattern of evolution. We hypothesized that the onset of acute compartment syndrome of the leg or forearm is associated with variations in the surface temperature of the distal segment (foot or hand) with a distinct pattern, which acts as an early warning sign. **Materials and Methods:** We developed a monitoring device that consists of two thermic sensors attached to a modular limb splint, which continuously measure the temperature difference between the proximal and distal regions of the limb (i.e., arm–hand, thigh–foot). Firstly, we investigated both the arm–hand and thigh–foot temperature gradients of hospitalized patients' healthy limbs (43 patients, 56 upper limbs, 64 lower limbs) in order to establish a baseline. Secondly, we examined the correlation between the thermic gradients and intracompartmental pressure values in compartment syndrome limbs (20 patients, 6 upper limbs, 14 lower limbs). **Results:** For the control group, the mean values for the normal limb thermic gradients were −0.17 °C for the upper limbs, and 0.03 °C for the lower limbs. In the impending compartment syndrome group (defined by intracompartmental pressure values), the mean index was −0.38 °C. In the fully developed compartment syndrome group, the mean value was 4.11 °C. **Discussions:** Analysis was performed using the ANOVA one-way statistical method. This showed significant differences between the compartment syndrome group and the impending and control groups. A decreasing trend in the thermic gradient in patients with impending compartment syndrome compared with the control group was noted. **Conclusions:** The thermic gradient of limbs presenting signs of impending compartment syndrome decreases as a result of the increased temperature of the distal segment. This pattern can be used as an early diagnostic method for acute compartment syndrome. This technique is non-invasive and bears no risk to the patient, allowing facile continuous monitoring during immobilization.

Keywords: trauma; acute compartment syndrome; intracompartmental pressure; thermic gradient; temperature sensor

check for updates

Citation: Tepordei, R.T.; Zamfir, C.L.; Nedelcu, A.H.; Avadanei, O.G.; Cozma, T.; Alexa, O.; Ursaru, M.; Perianu, L.; Starcea, I.M.; Ioniuc, I.; et al. An Innovative Non-Invasive Method for Early Detection and Monitoring of Acute Compartment Syndrome. *J. Pers. Med.* **2024**, *14*, 477. https://doi.org/10.3390/jpm14050477

Academic Editor: Günther Maderbacher

Received: 5 April 2024
Revised: 22 April 2024
Accepted: 28 April 2024
Published: 29 April 2024

1. Introduction

Acute compartment syndrome (ACS), regardless of its localization, is a severe complication, with significantly debilitating potential. Muscle and nerve ischemia results from an imbalance between the perfusion pressure and the metabolic demands of a tissue. Its early diagnosis is difficult, especially in children or unresponsive patients, and its evolution is highly unpredictable. In this context, the methods for early recognition of ACS are still a source of debate, leading to treatment delays and both local and systemic complications. The prognosis can be only belatedly expressed. Because of its heterogeneous etiology and its complex pathology, which is still incompletely understood, this syndrome represents a redoubtable condition that even with adequate treatment may have a fatal outcome [1]. In stark contrast with these multiple considerations, the sole recommended treatment is decompressive fasciotomy [2,3].

Therefore, the problem is not in choosing an appropriate method of treatment, but in the recognition of acute compartment syndrome at its onset. Outcomes are closely related both to the promptness of the diagnosis and to the period between injury and treatment. The orthopedic clinician must have a high index of suspicion in all trauma cases, especially in comatose or pediatric patients. The gold standard for diagnosis is the direct measurement of intracompartmental pressure. This is achieved through invasive techniques that offer only a brief look into the evolution of the disease. The risk of permanent muscular and nervous damage versus an almost complete recovery is firmly correlated with the promptness of therapeutic decisions.

Our diagnosis method provides a solution to the problem of identifying the onset of local microcirculation alteration characteristics associated with ACS. The purpose of our research was to investigate the hypothesis that an increased temperature difference between the proximal and the distal regions of a limb has real diagnostic significance. To measure this temperature difference, we used a modular splint with thermic sensors attached to the proximal and distal segments. The device we propose here has a double function: to immobilize a fractured limb, and to realize continuous, non-invasive monitoring of the thermic gradient of the limb. The results produced can be useful for clinicians, offering an alternative to invasive pressure determination.

2. Materials and Methods

To investigate our hypothesis, we first established a baseline by determining the normal values of the thermic gradient of both the upper limb (temperature difference between the arm and hand, upper limb index) and lower limb (temperature difference between the thigh and the foot, lower limb index) in the healthy, uninjured limbs of hospitalized patients who had been admitted for surgical treatment of various musculoskeletal lesions. Consequently, we examined the correlations between intracompartmental pressure values and the value of the thermic index reported in the limbs affected by acute compartment syndrome.

In order to identify the normal values of the thermic gradients in hospitalized patients, we examined 43 patients from the Orthopedics and Traumatology Clinic at "Sf. Spiridon" Emergency University Hospital, Iasi. The demographic details of these patients were as follows: 26 men and 17 women; 20 pre-operative and 23 post-operative; aged between 19 and 57 years; 56 healthy upper limbs, and 64 healthy lower limbs. These patients formed the control group. They were hospitalized for acute injuries of the locomotor system, excluding polytrauma and coma. Patients with associated pathologies that may have caused variations in skin temperature in the extremities were excluded. Another reason for excluding patients with other pathologies was their use of medication that may have modified the recorded surface thermal values of the extremities.

The patients in the post-operative period did not receive drugs likely to impair blood circulation in the limbs. All of the patients had temperatures measured at an interval of at least 72 h post-operatively in order to minimize the effects of anesthetic drugs administered during surgical intervention.

For each patient, the local cutaneous temperature was measured from a contralateral healthy and uninjured limb with a Thermoval Clinical Thermometer (Paul Hartmann AG, Heidenheim, Germany). In the upper limbs, we measured the temperatures between two points located as follows: on the anterior face of the arm midpoint between the acromion and the cubital fossa crease, and on the dorsum of the hand at the second interosseous space. The two measurement points corresponding to the lower limbs were: on the anterior face of the thigh midpoint between the anterior superior iliac spine and the base of the patella, and on the dorsum of the foot at the second interosseous space. Using these benchmarks, we avoided the regions superposed to important blood vessels, and those isolated from the environment by the dorsal decubitus of the patients. The limbs were exposed to the standard environmental conditions at least 10 min before the measurements were taken.

To continuously monitor the temperature and to easily calculate the proximal–distal thermal gradient, we used an Arduino Kit µC ATMEGA328 (BCMI Labs, Monza, Italy), which is an open-source programmable electronic platform, and two low-cost and low-consumption thermal sensors, LM335 (STMicroelectronics, Plan-les-Ouates, Switzerland) (Figure 1). The Arduino microcontroller presented the registered thermal values and the difference between them on an RT162 LCD (Nortech Engineering Inc., Port Charlotte, FL, USA) display in real time, using the output voltages of the two sensors.

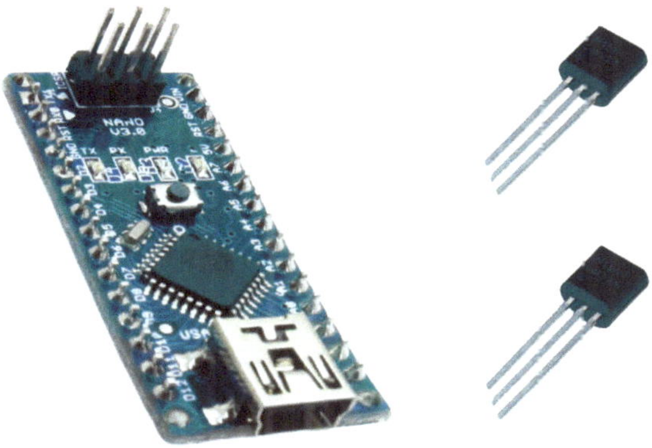

Figure 1. The Arduino microcontroller and the sensors.

If the thermal gradient was more than 2 °C (a value that can be easily adjusted), the microcontroller sent a 1 KHz signal to a 5V Buzzer via a 2N7000 transistor (STMicroelectronics, Plan-les-Ouates, Switzerland), which provided the necessary current required by the Buzzer. We also used a 5V 1A LM7805 stabilizer (STMicroelectronics, Plan-les-Ouates, Switzerland) in order to ensure that the entire device would work properly, even if the voltage supply was not constant.

The device was attached to a modular splint for limb immobilization (Figure 2). The sensors were placed in the following manner: one on the proximal component of the splint, and the other on the distal component (Figure 3). By applying the splint on the affected limb, the thermic sensors contacted the tegument of the proximal and distal regions, thus detecting the temperature difference between the two segments.

As an alternative, the sensor pair was able to be utilized independently, separate from the modular splint, incorporated in a plaster cast. In this simplified version, the device lacked the digital display, but gained a high degree of autonomy, operating in a continuous manner with a single 9V battery for almost 14 days. The complete version, because of its higher power demands, used a 12 V wall adapter.

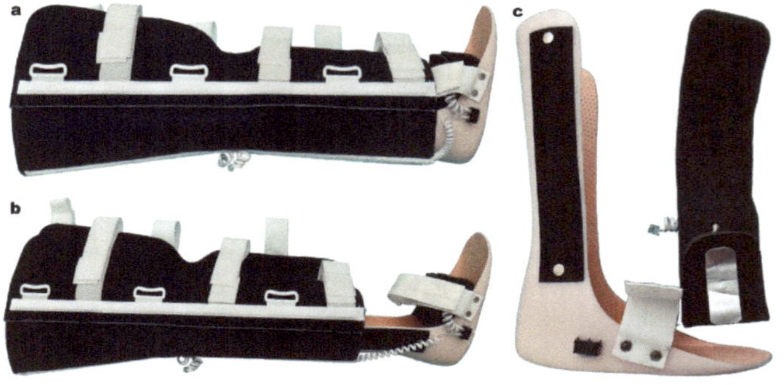

Figure 2. Lower limb modular splint. (**a**,**b**) length control; (**c**) distal thermic sensors.

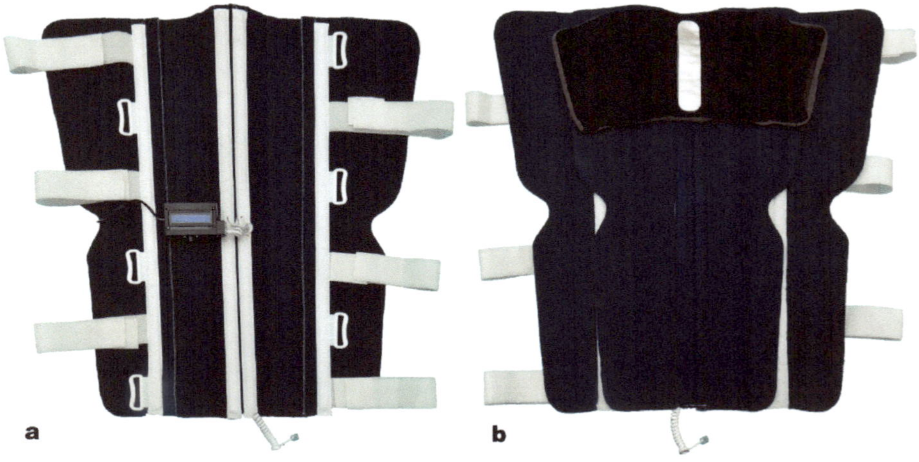

Figure 3. Lower limb modular splint. Proximal component. (**a**) Outer aspect and the electronic device. (**b**) Internal surface with thermic sensor.

The splint contained two components. The proximal component was made of reinforced neoprene with aluminum bars (Figure 3). The distal component was made of a thermoform material, using a universal left-right mold, with fixed external Velcro bands, which allowed connection with the proximal component, and length control (Figure 2). RJ11 connectors were attached to both components.

We fixed the thermic sensors to a flexible metal plate using a thermoconductive paste, similar to the type used in electrosurgery, which is well tolerated on skin contact (Figure 4).

To correlate the intracompartmental pressure values with the thermic gradient values, we included 20 patients, 15 males and 5 females, aged between 19 and 62, with either impending or fully developed ACS. Of this group, 11 patients had leg fractures (6 ACS and 5 impending ACS), 3 had tibia plateau fractures (1 ACS and 2 impending ACS), 3 had distal radius fractures (2 ACS and 1 impending ACS), 1 had an olecranon fracture and ACS, and 2 had grade III elbow sprains (1 ACS and 1 impending ACS). We excluded patients with skin lesions on the proximal and distal segments. The inclusion and exclusion criteria are presented in Table 1.

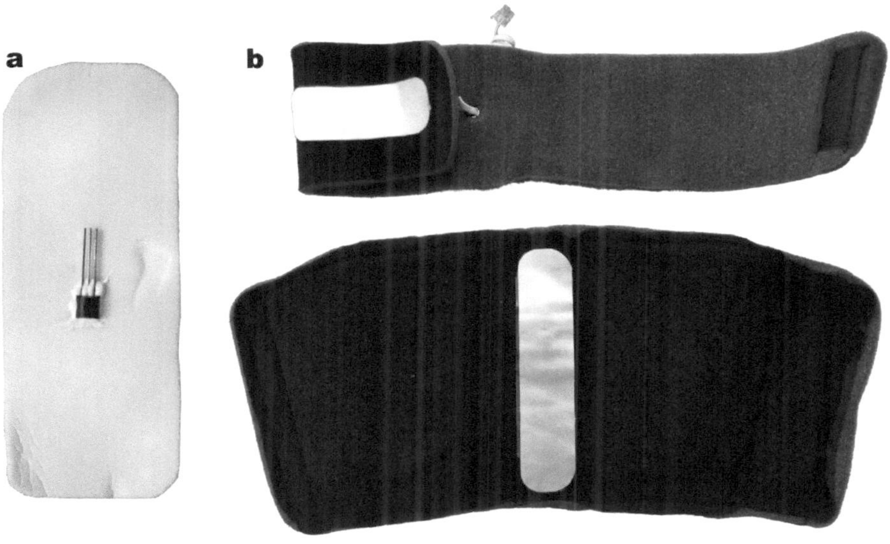

Figure 4. (**a**) Thermic sensor attached to the metallic foil. (**b**) Thermic sensors for the distal and proximal component.

Table 1. Inclusion and exclusion criteria.

Inclusion Criteria	Exclusion Criteria
1. Age over 18 years	1. Age under 18 years
2. Acute injuries of the locomotor system	2. Medication that can modify the thermal values
3. Only one limb affected	3. Skin lesions on proximal and distal segments.

We used a compartment pressure measuring system (CPMS—Mammendorf Institute for Physics and Medicine, Mammendorf, Germany) in continuous mode to assess intracompartmental pressure and to identify ACS cases, following the principle of the pressure gradient ΔP (the difference between diastolic systemic pressure and the intracompartmental pressure).

Each patient with leg/forearm musculoskeletal lesions was assessed clinically for signs of ACS (pain disproportionate to the lesion, pain with passive mobilization, and tension oedema of the affected segment). For those cases which raised reasonable suspicion, we continuously (24/24) measured the intracompartmental pressure of the affected limb, and applied the thermic monitoring splint (Figure 5).

All patients with values of $\Delta P < 30$ mmHg were considered to have ACS, and those with ΔP between 31 and 40 mmHg were considered to have impending ACS. All lower limb fractures with increased intracompartmental pressures were immobilized with the continuous monitoring splint, both in the pre- and post-operative periods (Figure 6). All ACS cases were surgically confirmed.

Statistical analysis was performed with SPSS 17 software, using an independent samples T-test, paired samples T-test, Pearson bivariate correlations, Spearman correlations, and ANOVA one-way.

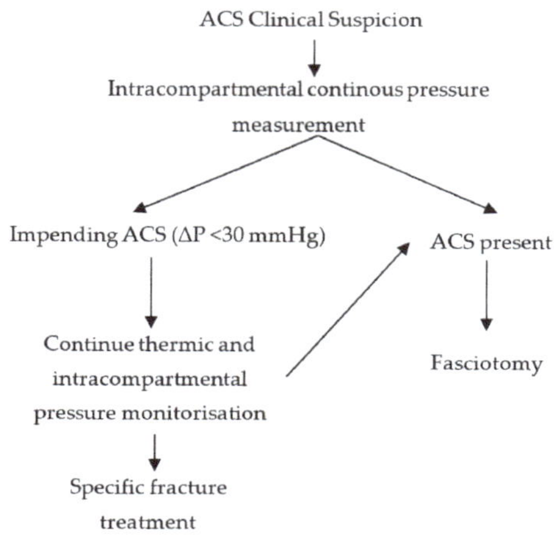

Figure 5. Therapeutic protocol diagram.

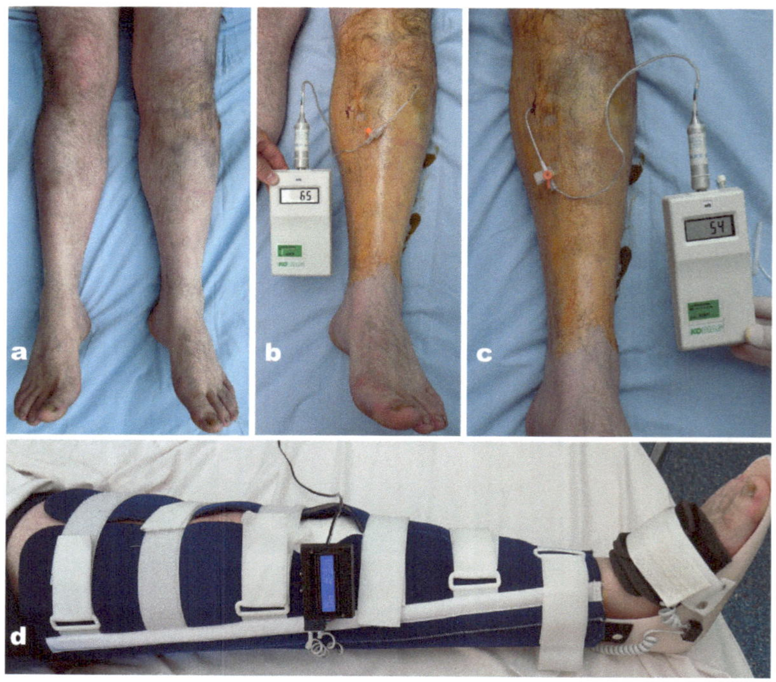

Figure 6. (**a**) Fracture of the left tibia plateau; ACS suspected. (**b**,**c**) Increased values of intracompartmental pressure in the anterior and posterior compartments (BP = 151/101 mmHg, $\Delta P > 30$); impending ACS. (**d**) Immobilization with the thermic monitoring splint ($\Delta T = 0.38$ °C).

3. Results

The main characteristics of the control and ACS groups were studied (Table 2). Results showed that the control group had an average hand temperature (M = 36.65) significantly higher than the average arm temperature (M = 36.50; $t(63) = -3.53$, $p < 0.01$), and that this difference was present both in pre-operative ($M_{arm} = 36.50$, $M_{hand} = 36.65$, $t(17) = -2.38$,

$p < 0.05$) and post-operative patients ($M_{arm} = 36.44$, $M_{hand} = 36.64$, $t(38) = -3.71$, $p < 0.01$). There were no significant differences between the temperatures measured in the lower limb. In the general sample, the average upper and lower limb temperature gradients were not significantly different, but in the post-operative condition, the results showed a lower limb average gradient ($M = 0.03$) significantly higher than the upper limb gradient ($M = -0.17$, $t(21) = 2.43$, $p < 0.05$).

Table 2. Main characteristics of the explored groups.

Characteristics	Control Group (n = 43)	Explored Group (n = 20)
Mean age	41.84 years	40.45 years
Male/Female	1.5/1	3/1
Upper limb	64	6
Lower limb	56	14
Diastolic blood pressure	86.4 mmHg	81.1 mmHg
Environmental temperature	24 ± 0.5 °C	
Patients' mean temperature	36.4 (36–37.5)	36.9 (36.6–37.6)
Time from fracture	48–96 h	6–72 h
Associated pathology	6	2
Associated fractures	44	4
Open fractures	0	1
Craniocerebral trauma	0	1
Coma	0	1
Pre-surgery	20	20
Post-surgery	23	0

Gender significantly influenced the temperatures of the upper limb, with male patients having higher arm ($M = 36.63$) and hand ($M = 36.76$) temperatures than female patients ($M_{arm} = 36.31$, $t(62) = 4.06$, $p < 0.01$ and $M_{hand} = 36.50$, $t(62) = 2.88$, $p < 0.01$). Similar gender differences were found in average foot temperatures, with significantly higher values in men ($M = 36.47$) than women ($M = 36.27$, $t(53) = 2.03$, $p < 0.05$) (Figure 7).

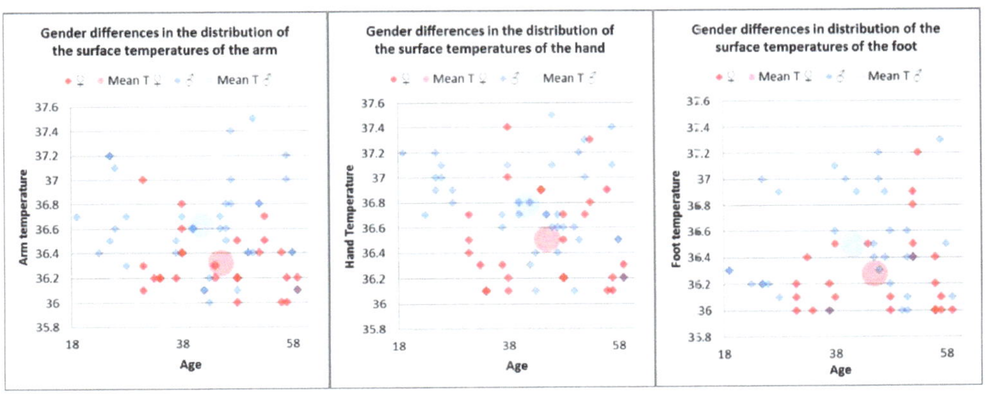

Figure 7. Gender differences in distribution of the surface temperatures of the arm, hand, and foot in the control group.

The results also indicated within-gender differences between temperatures in the upper limb, with the arm temperature being significantly lower than the hand temperature in both male (M_{arm} = 36.63, M_{hand} = 36.76, t(27) = −2.44, $p < 0.05$) and female patients (M_{arm} = 36.31, M_{hand} = 36.50, t(25) = −2.52, $p < 0.05$). Moreover, in female patients, the lower limb average temperature index (M = 0.05) was significantly higher than the upper limb index (M = −0.15, t(16) = 2.24, $p < 0.05$).

We also found a marginal positive correlation (r = 0.29, $p = 0.06$) between the temperature indexes of the upper and the lower limb, meaning that an increase in one index tends to be associated with an increase in the other index. A closer examination of the results shows that this correlation was statistically significant and stronger in the post-operative patients (r = 0.47, $p < 0.05$), while in the pre-operative condition, it remained weak and nonsignificant (r = 0.21, $p > 0.05$). For the pre-operative condition only, age was significantly negatively correlated with both arm (r = −0.75, $p < 0.01$) and hand (r = −0.81, $p < 0.01$) temperatures, meaning that with increasing age, their values tended to decrease.

Regarding the ACS group, we observed considerably increased thermic index values in patients with manifested ACS. In the impending ACS group, we detected thermic indexes close to zero and negative, with the distal segment temperature being higher than the proximal (Figure 8).

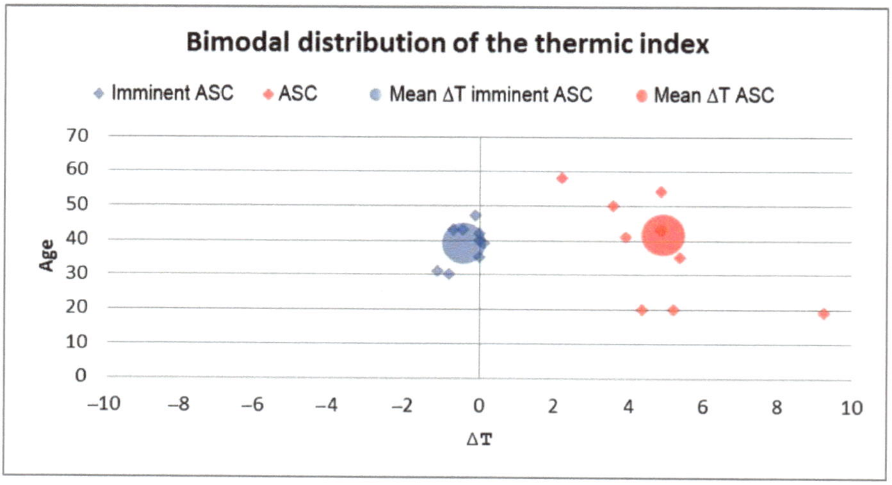

Figure 8. Bimodal distribution of the thermic index in patients with impending ACS and present ACS. The impending ACS group had a negative thermic index mean compared to the ACS group.

Using the ANOVA one-way statistical method, we observed that there were significant differences regarding the thermic indexes of the patients, depending on the absence, imminence, or presence of ACS (F(2, 137) = 114.32, $p < 0.01$). Means, standard deviations, and the number of patients for each group are presented in Table 3.

Table 3. Mean thermic index distribution in the control group, impending, and fully developed ASC groups.

ASC/Impending/Control	n	Thermic Index	
		M	SD
ASC	11	4.11	3.02
Impending	9	−0.35	0.51
Control	120	−0.11	0.37

To verify the presence of significant differences between groups of patients, we applied the post-hoc Bonferroni test. Results showed that the thermic index values in patients with ACS were significantly increased compared with those of the patients with impending ACS (Bonferroni = 11.15, $p < 0.01$), and with those of the control group, respectively (Bonferroni = 15.07, $p < 0.01$). Although not statistically significant, we found a tendency towards a decreased thermic index in patients with impending ACS compared with the control group (Figure 9).

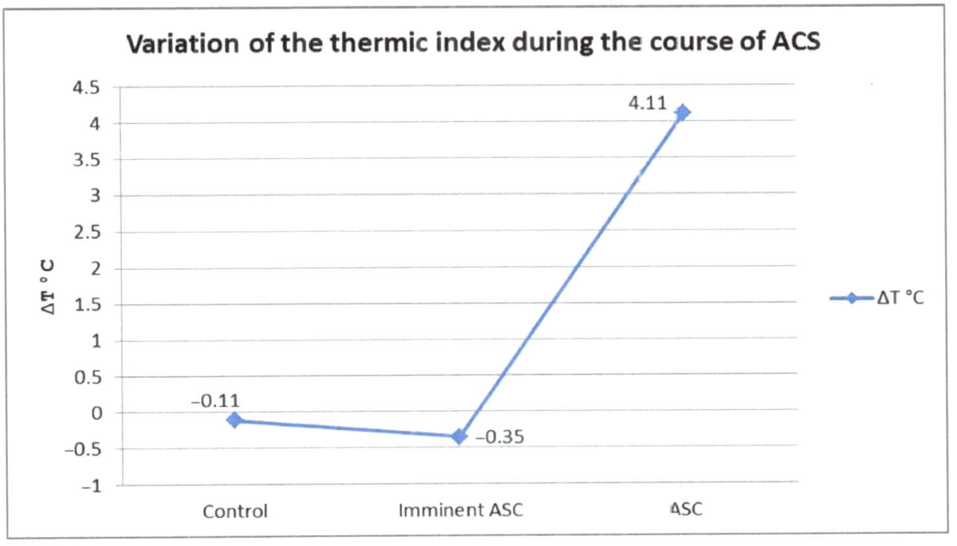

Figure 9. The dynamic of ACS evolution. Slight drop of thermic index in the impending ACS group as a result of distal segment surface temperature increase, followed by a marked increase in thermic index in the ACS group.

Another element of our study was to assess any statistically significant differences between the thermic indexes of the patients according to the number of affected compartments. The results indicated an important effect of this variable on thermic index variability ($F(4, 127) = 228.34$, $p < 0.01$). Means, standard deviations, and the number of patients for each group are presented in Table 4. The Bonferroni post-hoc test revealed the following differences between groups: the thermic index of the control group was significantly smaller compared with the patients with two (Bonferroni = 9.60, $p < 0.01$), three (Bonferroni = 18.83, $p < 0.01$), and four (Bonferroni = 22.59, $p < 0.01$) affected compartments, respectively. Those with one affected compartment had a significantly lower thermic index than those with two (Bonferroni = 7.85, $p < 0.01$), three (Bonferroni = 12.34, $p < 0.01$), and four (Bonferroni = 17.84, $p < 0.01$), respectively.

The values of the thermic index were also lower in patients with two affected compartments compared with the patients with three (Bonferroni = 4.31, $p < 0.01$) and four (Bonferroni = 11.41, $p < 0.01$) affected compartments, with similar differences being present between the last two groups (Bonferroni = 8.72). As Figure 10 shows, when compared to the control group, the thermic index tended to decrease in patients with only one affected compartment (though this was not statistically significant). After that, the thermic index values followed an ascendant trend in direct relation with an increasing number of affected compartments.

Table 4. Subject distribution by the number of the affected compartments.

Number of Affected Compartments	n	Thermic Index	
		Mean	SD
Control	120	−0.11	0.37
1 compartment	2	−1.02	0.85
2 compartments	3	2.77	1.01
3 compartments	5	4.41	1.86
4 compartments	2	8.25	1.44

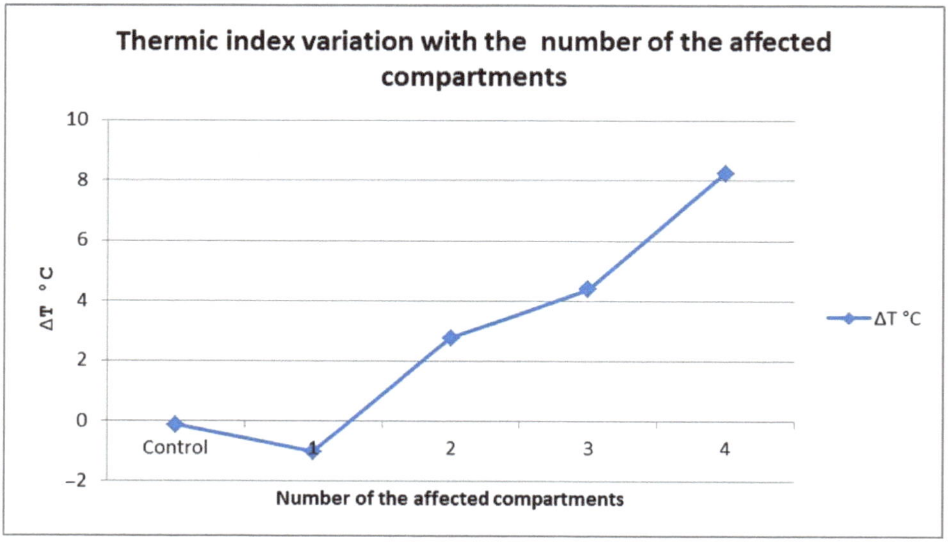

Figure 10. ΔT variation with the number of affected compartments.

In order to observe the relationship between the thermic index and the pressure gradient in all four compartments of the upper limb, we calculated the ρ Spearman correlation coefficient. The results showed that in the anterior ($\rho = -0.77$, $p > 0.05$), lateral ($\rho = -0.71$, $p > 0.05$), and deep ($\rho = -0.77$, $p > 0.05$) compartments, respectively, there were strong negative correlations between the two factors, meaning that as the thermic index became higher, the pressure gradient tended to decrease. However, there was an exception; in the pronator quadratus compartment, the correlation was positive ($\rho = 0.48$, $p > 0.05$), with high thermic gradient values being associated with high values of the pressure gradient. Although these correlations did not meet the statistical significance criteria, most probably because of our reduced sample, they were high enough to have practical significance in diagnosis of the syndrome, with the value of the R2 determining coefficient being around 0.45 (indicating that almost 45% of the pressure gradient variance can be explained by the variability of the thermic index).

Similar analysis conducted for the four compartments of the lower limb revealed the same negative, strong, and this time statistically significant, correlational pattern between the values of the thermic index and those of the pressure gradient in all of the compartments, meaning that as the thermic index became higher, the pressure gradient tended to decrease. The strongest correlation appeared in the lateral compartment ($\rho = -0.82$, $p < 0.01$), followed the anterior ($\rho = -0.77$, $p < 0.01$), deep posterior ($\rho = -0.73$, $p < 0.01$) and superficial posterior ($\rho = -0.67$, $p < 0.01$) compartments, respectively.

4. Discussion

In healthy, non-hospitalized subjects, normal temperatures of the foot and hand vary widely with the environmental temperature (from 15.9 °C during winter to 37.5 °C in summer) [4] and subject activity level [5–7]. Nardin et al. [4] examined variations in the surface temperature of the foot in 39 healthy subjects, aged between 18 and 65, during daytime and nighttime. They reported a mean temperature of the foot of 30.6 °C during the daytime and 34 °C at night, but this varied widely in relation to the ambient temperature, from 15.9 °C in winter to 37.5 °C during summer. Ceron et al. [5] investigated variations in hand temperature in workers in cold (13 °C) and warm (23.9 °C) environments and obtained a wide distribution of surface temperatures which were directly correlated to environmental conditions. Our data indicated a quasi-constant temperature of around 36.5 °C. This discrepancy can be explained by the fact that our subjects were in-patients with reduced activity levels; the majority of them were confined to bed, thus eliminating activity level from equation. In addition, the hospital ambient temperature is constant (24 ± 0.5 °C); hence, variations in body temperature cannot be linked to variations in the ambient temperature.

To our knowledge, there are no descriptions in the literature of the investigated thermic gradients (thigh–foot and arm–hand) in healthy limbs, which prompted us to investigate further to obtain a baseline. During our study, we obtained gradient values close to zero (−0.17 °C arm–hand thermic gradient and −0.05 °C thigh–foot thermic gradient). Hence, we considered these values as baseline.

The value of 30 mmHg for the intracompartmental pressure was considered an absolute indication for decompressive fasciotomy in order to prevent the evolution of compartment syndrome [8], although many authors admit that this value is heavily influenced by the hemodynamic status of the patient [9–11]. Tissue tolerance to increased pressure values, such as those present in compartment syndrome, varies widely depending on three factors: the specific effect on the local blood flow, the metabolic needs of the tissues, and the duration of the pressure. The effects of these factors are nuanced by the clinical characteristics of the patient, including the presence of hypotension, shock, arterial spasm, and prone position of the limb. Thus, a single absolute intracompartmental pressure value cannot be applied indiscriminately as a clear indication for fasciotomy. For this reason, Whiteside proposed the use of the difference between the blood diastolic pressure and the intracompartmental pressure as a reference [12,13]. The value of this pressure gradient for the indication of decompressive fasciotomy is 30 mmHg, and this correlates much better with clinical practice.

In simple trauma fractures (without ACS), there is a local hyperemia explained by the nonspecific inflammatory reaction, providing local defense. This hyperemia caused by local vasodilation is clinically expressed by an increase in the local temperature, without alterations of the distal segment [14–16]. In ACS, regardless of the involved mechanism, the common factor is a decrease in tissue perfusion through a decrease in the local blood flow, as a consequence of increased intracompartmental pressure [17,18]. Arterial pressure, venous pressure, vascular resistance, and interstitial pressure represent essential factors in blood pressure [19,20]. The arterio-venous gradient is directly involved in the maintenance of normal blood pressure values. This gradient is defined as the difference between arterial and venous pressure. Conditions that decrease arterial blood pressure (hypovolemia, shock, and elevated position of the anatomic segment) or raise the venous pressure will reduce the arterio-venous gradient and the perfusion rate [21]. This theory of the arterio-venous gradient explains, but in an incomplete manner, the appearance and evolution of ACS [22]. At the same time, it can explain the increase in the distal segment temperature detected by us in the early stages of ACS.

Katz et al. [23] hypothesized that ACS is associated with a decrease in the cutaneous temperature of the distal segment from the injury. They investigated this hypothesis in the lower limb, using an infrared video camera. The results showed statistically significant differences between the temperatures of the thigh and the foot (thigh–foot index) in ACS legs (8 °C), demonstrating the usefulness of this method in the detection of this syndrome.

Distal thermic cutaneous alterations and the evolution of ACS have an interdependent dynamic, which allowed us to identify two distinct groups, one with impending ACS, and one with present ACS.

We have identified a pattern in the dynamics of ACS onset: the average of the thermic gradient in patients with impending ACS decreases and becomes negative (mean $\Delta t = -0.38$ °C) because of the increasing temperature of the distal segment. After that, along with the evolution of ACS, this gradient increases and becomes positive (mean $\Delta t = 4.11$ °C) in accordance with a drop in the cutaneous temperature of the foot or hand. This initial increase in the distal segment temperature acts as an early warning signal and can be used to identify cases of impending ACS.

The number of affected compartments also controls the distal segment temperature. When only one compartment was affected, the distal segment temperature was increased, but we were not able to find a statistic significant difference between this group and the control group. During the successive recruitment of all of the neighboring compartments, the distal segment temperature decreased; in those cases, the measurement of the thermic gradient became useful to define the evolutive stages of ACS.

The triggering factor is represented by tissue edema, in the context of two main physiopathological mechanisms that reduce the transmural pressure gradient: the decrease in arterial-capillary flow and the presence of a non-specific local inflammatory reaction, secondary to the trauma. A vicious cycle of edema-ischemia-pressure is created, with progressive evolution, which can successively affect all of the osteofascial compartments of an anatomical segment. The different weight of these two mechanisms in a given case will give the clinical picture distinct characters of different intensities and durations, making it difficult to identify risk factors and stage the disease. None of the theories that have tried to explain the pathophysiological mechanisms of compartment syndrome have provided a definitive verdict. Rather, the mechanisms are intricate and complement each other, making a characterization with predictive value on the evolution and prognosis of this disease impossible.

There has been growing interest in the use of NIR spectroscopy (near-infrared spectroscopy), which is based on specific changes in the transmission and absorption of the near-infrared electromagnetic spectrum by the oxygen saturation level of hemoglobin. Variations in tissue hemoglobin saturation percentage lead to variations in the absorption of infrared radiation. One of the limitations of this method is that its tissue penetration does not exceed 3 cm [24,25], making it useful only for measuring superficial compartments. Rostosky et al. [26] measured subcutaneous adipose tissue thickness at the ankle in 50 patients with severe calf injuries by ultrasonography. They did not notice differences between injured and healthy limbs (6.98 mm vs. 7.06 mm) and did not obtain thicknesses of subcutaneous adipose tissue greater than 2 cm in any patient, validating the usefulness of this method.

It should be noted that in uncomplicated leg trauma with acute compartment syndrome, local hyperemia due to the non-specific inflammatory process will increase the percentage of oxygenated hemoglobin, a phenomenon objectified by Shuler and colleagues [24], who compared 26 patients with unilateral tibial fractures with 25 healthy subjects, spectroscopically measuring the percentage of oxygenated hemoglobin in all four compartments of the calf. They obtained values that showed that the average hemoglobin oxygen saturation increased by 15.4% in the compartments of the anatomical segment with fracture injuries as compared to the contralateral limb and healthy subjects. More recent studies [14,27,28] have aimed to objectify the perfusion deficit by using NIR spectroscopy in acute compartment syndrome located at the level of the calf. In all of these patients, invasive determination of the intracompartmental pressure as well as spectroscopic measurement of the contralateral intact leg were performed. At the level of the affected compartments (with high pressure values and differences between diastolic and intracompartmental pressure < 10 mmHg), the percentage of oxygenated hemoglobin was shown to have decreased by approximately 10% compared to the contralateral limb, thus demonstrating a positive correlation between the differences in the spectroscopic values of the unaffected and affected compartments and the pressure gradient ΔP.

Thus, a phenomenon similar to the one we identified in our thermal and pressure correlation study appears: initially, local hyperemia increases the temperature of the affected segment, and then the perfusion deficit leads to a decrease in temperature. In this light, continuous temperature monitoring of the distal segment is of real use for the detection of impending acute compartment syndrome, and represents a much cheaper and more accessible technique than NIR spectroscopy.

The design and execution of a universal splint that correctly immobilizes a fracture is difficult, having to take into account the differences in the dimensions of the limbs, both in length and in diameter. We opted for the principle of modularity. because even if this increases the complexity, it provides a better adaptation to size variations in the anatomical segments involved. The contact of the splint with the integument must be as intimate as possible, especially at the level of the components containing the temperature sensors, in order not to introduce errors in the measurement of thermal values. This was achieved through a compromise between universality and stability. Any form of immobilization, to be perfectly stable, must be personalized to the patient, and any method of universal immobilization will have to give up customization in favor of adaptability. Thus, the principle of modularity was applied, combining the two points of view located at opposite ends.

The presence of acute or chronic skin conditions can prevent the device from working optimally, given that the contact point of the sensors with the skin cannot be changed. In the case of open fractures, this impediment does not exist, because the injuries are at the level of the intervening segment (leg or forearm), and not at the level of the proximal (thigh or arm) or distal (leg or hand) segments where the temperature sensors are placed. Chronic vascular trophic skin conditions can lead to erroneous temperature values by changing the values of the distal segment. Usually, these conditions are found in the elderly (over 60 years old), in whom compartment syndrome has a very low incidence rate. In young people, the presence of these chronic conditions is exceptional, so this associated pathology is not an impediment. Skin lesions of any nature, acute or chronic, located at the sensor contact areas (second interosseous space on the dorsal side of the foot, second interosseous space on the dorsal side of the hand, anterior face of the arm, and anterior face of the thigh) is a contraindication for temperature measurement at these levels. In this case, the sensors of the splint can be disassembled, and applied separately to regions with intact skin.

The limitations of our study come from the fact that acute compartment syndrome is a rare complication, and consequently, the number of selected cases was relatively small. Another limitation is the fact that we used a control group consisting of healthy, uninjured limbs (the contralateral uninjured limb of a patient with an uncomplicated fracture). The literature agrees that in response to injury, the temperature of the involved segment rises. To our knowledge, in a trauma uncomplicated by ACS, local temperature increase (ROI) does not involve the temperature of the distal or proximal limb segment. Moreover, our study did not determine the absolute thermal values, but the variation in the thermal gradient during continuous monitoring between the extremities of the affected limb (thigh–foot and arm–hand). A higher number of cases, perhaps using a multicentric study, should be employed to achieve better statistical confidence.

5. Conclusions

The initial stages of ACS can be detected using a dynamic measurement of thermic gradient, which signals the onset of ACS by the recruitment of the first compartment. Our device acts as an early warning alarm, identifying the beginning of ACS.

Despite the abovementioned limitations, we believe our method to be valid and useful in clinical practice. Moreover, with the development of artificial intelligence, we will be able to create a connection between AI and our medical device made. Future research should extend the investigation, testing the scientific reliability of both our proposed alternative diagnostic method and our monitoring device.

Author Contributions: Conceptualization, R.T.T. and A.H.N.; methodology, R.T.T., O.A. and C.L.Z.; formal analysis, T.C., M.U. and L.P.; investigation, A.L., V.V.L. and I.I.; electronic devices, O.G.A. and R.T.T.; data curation, R.T.T., T.C. and O.A.; software, I.M.S. and V.V.L.; writing—original draft preparation, R.T.T., O.G.A., T.C., O.A., M.U. and L.P.; writing—review and editing, A.H.N., C.L Z., I.M.S., I.I., V.V.L. and A.L.; project administration, A.H.N.; C.L.Z., A.H.N., O.G.A., T.C., O.A., M.U., L.P., I.M.S., I.I., V.V.L. and A.L. contributed equally with R.T.T. to this article. All authors have read and agreed to the published version of the manuscript.

Funding: This research received no external funding.

Institutional Review Board Statement: The study was conducted according to the guidelines of the Declaration of Helsinki and was approved by the Ethics Committee of the "Sf. Spiridon" Emergency University Hospital Iasi, (no. 29/27.03.2024).

Informed Consent Statement: Informed consent was obtained from the subjects involved in the study.

Data Availability Statement: The data presented in this article are available on request from the corresponding author.

Acknowledgments: We thank the orthopedists at the "Sf. Spiridon" Emergency University Hospital Iasi who provided us with the data relating to the functional explorations of the patients.

Conflicts of Interest: The authors declare no conflict of interest.

References

1. Zhu, B.L.; Zhao, R.; Li, R.B.; Wu, X.; Guan, D.W.; Zhang, G.H. Fatal acute compartment syndrome in patients after surgical treatment: 2 cases report. *Leg. Med.* **2009**, *11*, 544–545. [CrossRef] [PubMed]
2. Janakiram, N.B.; Motherwell, J.M.; Goldman, S.M.; Dearth, C.L. Efficacy of non-surgical interventions for promoting improved functional outcomes following acute compartment syndrome: A systematic review. *PLoS ONE* **2022**, *17*, e0274132. [CrossRef] [PubMed]
3. Etemad-Rezaie, A.; Yang, S.; Kirklys, M.; Higginbotham, D.O.; Zalikha, A.K.; Nasr, K. Single incision fasciotomy for acute compartment syndrome of the leg: A systematic review of the literature. *J. Orthop.* **2022**, *31*, 134–139. [CrossRef]
4. Nardin, R.A.; Fogerson, P.M.; Nie, R.; Rutkove, S.B. Foot temperature in healthy individuals: Effects of ambient temperature and age. *J. Am. Podiatr. Med. Assoc.* **2010**, *100*, 258–264.
5. Ceron, R.J.; Radwin, R.G.; Henderson, C.J. Hand skin temperature variations for work in moderately cold environments and the effectiveness of periodic rewarming. *Am. Ind. Hyg. Assoc. J.* **1995**, *56*, 558–567. [CrossRef]
6. Zontak, A.; Sideman, S.; Verbitsky, O.; Beyar, R. Dynamic Thermography: Analysis of Hand Temperature During Exercise. *Ann. Biomed. Eng.* **1998**, *26*, 988–993. [CrossRef] [PubMed]
7. Balci, G.A.; Avci, A.B.; Colakoglu, M.; Basaran, T. Estimation of Heat Production Rate using Thermal Data During Exercise in Indoor Environments: A Study of Heat Storage Rate in Male Athletes. *Int. J. Biometeorol.* **2024**. [CrossRef]
8. Janzing, H.M.; Broos, P.L. Routine monitoring of compartment pressure in patients with tibial fractures: Beware of overtreatment! *Injury* **2001**, *32*, 415–421. [CrossRef]
9. Dean, R.S.; Farley, K.X.; Waterman, B.R.; Guettler, J.; Bicos, J. Chronic exertional compartment syndrome is frequently diagnosed through static compartment pressure measurements and managed with fasciotomy: A systematic review. *J. ISAKOS* **2024**, *9*, 71–78. [CrossRef]
10. Lorange, J.P.; Laverdière, C.; Corban, J.; Montreuil, J.; Harvey, E.J. Diagnosis Accuracy for Compartment Syndrome: A Systematic Review and Meta-Analysis. *J. Orthop. Trauma* **2023**, *37*, e319–e325. [CrossRef]
11. Aweid, O.; Del Buono, A.; Malliaras, P.; Iqbal, H.; Morrissey, D.; Maffulli, N.; Padhiar, N. Systematic review and recommendations for intracompartmental pressure monitoring in diagnosing chronic exertional compartment syndrome of the leg. *Clin. J. Sport Med.* **2012**, *22*, 356–370. [CrossRef]
12. Willy, C.; Sterk, J.; Gerngroß, H. *Methods of Measurement of Tissue Pressure. Das Kompartment-Syndrom*; Springer: Berlin, Germany, 1998.
13. Mortensen, S.J.; Vora, M.M.; Mohamadi, A.; Wright, C.L.; Hanna, P.; Lechtig, A.; Egan, J.; Williamson, P.M.; Wixted, J.J.; Rutkove, S.B.; et al. Diagnostic Modalities for Acute Compartment Syndrome of the Extremities: A Systematic Review. *JAMA Surg.* **2019**, *154*, 655–665. [CrossRef]

14. Shuler, M.S.; Reisman, W.M.; Kinsey, T.L.; Whitesides, T.E., Jr.; Hammerberg, E.M.; Davila, M.G.; Moore, T.J. Correlation between muscle oxygenation and compartment pressures in acute compartment syndrome of the leg. *J. Bone Jt. Surg. Am.* **2010**, *92*, 863–870. [CrossRef]
15. Haluzan, D.; Davila, S.; Antabak, A.; Dobric, I.; Stipic, J.; Augustin, G.; Ehrenfreund, T.; Prlic, I. Thermal changes during healing of distal radius fractures-Preliminary findings. *Injury* **2015**, *46* (Suppl. S6), S103–S106. [CrossRef] [PubMed]
16. der Strasse, W.A.; Campos, D.P.; Mendonça, C.J.A.; Soni, J.F.; Tuon, F.; Mendes, J.; Nohama, P. Evaluating physiological progression of chronic tibial osteomyelitis using infrared thermography. *Res. Biomed. Eng.* **2022**, *38*, 941–953. [CrossRef]
17. Bihari, A.; McClure, J.A.; Inculet, C.; Del Balso, C.; Vinden, C.; Schemitsch, E.; Sanders, D.; Lawendy, A.R. Fasciotomy and rate of amputation after tibial fracture in adults: A population-based cohort study. *OTA Int.* **2024**, *7*, e333. [CrossRef] [PubMed]
18. Rockwood, C.A.; Bucholz, R.W.; Court-Brown, C.M.; Heckman, J.D.; Tornetta, P. *Rockwood and Green's Fractures in Adults*, 9th ed.; Wolters Kluwer: Philadelphia, PA, USA, 2020.
19. Khoshhal, K.I.; Alsaygh, E.F.; Alsaedi, O.F.; Alshahir, A.A.; Alzahim, A.F.; Al Fehaid, M.S. Etiology of trauma-related acute compartment syndrome of the forearm: A systematic review. *J. Orthop. Surg. Res.* **2022**, *17*, 342. [CrossRef]
20. Mortensen, S.J.; Orman, S.; Serino, J.; Mohamadi, A.; Nazarian, A.; von Keudell, A. Factors Associated with Development of Traumatic Acute Compartment Syndrome: A Systematic Review and Meta-analysis. *Arch. Bone Jt. Surg.* **2021**, *9*, 263–271. [CrossRef] [PubMed]
21. Morello, V.; Gamulin, A. Clinical and radiological risk factors associated with the occurrence of acute compartment syndrome in tibial fractures: A systematic review of the literature. *EFORT Open Rev.* **2023**, *8*, 926–935. [CrossRef]
22. Korrapati, A.; Ta, C.N.; Mitchell, B.C.; Wall, P.V.; Gurusamy, P.; Dwight, K.; Girard, P.J.; Schwartz, A.K.; Kent, W.T. Fracture blisters: Predictors for time to definitive fixation in pilon fractures. *Eur. J. Orthop. Surg. Traumatol.* **2024**, *34*, 161–166. [CrossRef]
23. Katz, L.M.; Varidhi Nauriyal, V.; Nagaraj, S.; Finch, A.; Pearlstein, K.; Szymanowski, A.; Sproule, C.; Rich, P.B.; Guenther, B.D.; Pearlstein, R.D. Infrared imaging of trauma patients for detection of acute compartment syndrome of the leg. *Crit. Care Med.* **2008**, *36*, 1756–1761. [CrossRef] [PubMed]
24. Shuler, M.S.; Reisman, W.M.; Whitesides, T.E., Jr.; Kinsey, T.L.; Hammerberg, E.M.; Davila, M.G.; Moore, T.J. Near-infrared spectroscopy in lower extremity trauma. *J. Bone Jt. Surg. Am.* **2009**, *91*, 1360–1368. [CrossRef] [PubMed]
25. Kim, Y.; Han, I.; Jung, J.; Yang, S.; Lee, S.; Koo, B.; Ahn, S.; Nam, Y.; Song, S.H. Measurements of Electrodermal Activity, Tissue Oxygen Saturation, and Visual Analog Scale for Different Cuff Pressures. *Sensors* **2024**, *24*, 917. [CrossRef] [PubMed]
26. Roskosky, M.; Robinson, G.; Reisman, W.; Ziran, B.; Shuler, M.S.; Freedman, B. Subcutaneous depth in a traumatized lower extremity. *J. Trauma Acute Care Surg.* **2014**, *77*, S190–S193. [CrossRef] [PubMed]
27. Davidson, A.L.; Sutherland, M.A.; Siska, R.C.; Janis, J.E. Practical Review on the Contemporary Diagnosis and Management of Compartment Syndrome. *Plast. Reconstr. Surg. Glob Open.* **2024**, *12*, e5637. [CrossRef]
28. Novak, M.; Penhaker, M.; Raska, P.; Pleva, L.; Schmidt, M. Extremity compartment syndrome: A review with a focus on non-invasive methods of diagnosis. *Front. Bioeng. Biotechnol.* **2022**, *10*, 801586. [CrossRef]

Journal of *Personalized Medicine*

MDPI

Article

Does Treatment of Adolescent Fractures Differ between Specialties? A Survey among Pediatric and Trauma Surgeons

Alexander Hanke [1,2], Eva Scheerer-Harbauer [1], Christian Wulbrand [2] and Clemens Memmel [1,3,*]

[1] Department of Pediatric Surgery and Pediatric Orthopedics, Clinic St. Hedwig, Barmherzige Brueder Regensburg, KUNO Pediatric University Medical Center, 93049 Regensburg, Germany; alexander.hanke@barmherzige-regensburg.de (A.H.)
[2] Department of Trauma Surgery, Orthopedics and Sports Medicine, Barmherzige Brueder Regensburg, 93049 Regensburg, Germany
[3] Department of Trauma Surgery, University Medical Center Regensburg, 93053 Regensburg, Germany
* Correspondence: clemens.memmel@ukr.de

Abstract: From a traumatological point of view, adolescents (12–18 years) represent a special group of patients. This is due to their biomechanical characteristics being between pediatric and adult fracture types. In Germany, they are treated by both pediatric and trauma surgeons. For this survey, seven cases of adolescent fractures were evaluated by both pediatric and trauma surgeons and their preferred treatment options were raised. The questionnaires were completed anonymously. Additionally, information on the specialty and years of experience were asked. In total, 126 valid questionnaires were obtained (from 78 pediatric and 48 trauma surgeons). The respondents' mean clinical experience was high (71.5% stated more than 10 years of surgical experience). For every single exemplary case, a significant difference in therapy decisions between the groups could be found. For the demonstrated seven cases, a tendency toward more operative and more invasive treatments was found with trauma surgeons compared to pediatric surgeons. On the other hand, there was a risk of underestimating the severity of fracture entities similar to adult fractures in pediatric surgeons. Overall, a continuous interdisciplinary exchange between both surgical specialties is necessary to ensure optimal treatment for adolescent fractures and to develop guidelines in the future.

Keywords: adolescent fracture; pediatric traumatology; trauma surgery; standardized treatment

Citation: Hanke, A.; Scheerer-Harbauer, E.; Wulbrand, C.; Memmel, C. Does Treatment of Adolescent Fractures Differ between Specialties? A Survey among Pediatric and Trauma Surgeons. *J. Pers. Med.* **2024**, *14*, 842. https://doi.org/10.3390/jpm14080842

Academic Editors: Jih-Yang Ko and Norio Yamamoto

Received: 3 July 2024
Revised: 15 July 2024
Accepted: 8 August 2024
Published: 9 August 2024

1. Introduction

In many countries, e.g., Germany, fractures in adolescents (12–18 years) are treated by either pediatric or trauma surgeons, depending on regional or hospital policy. This is not only due to the rescue service system but also to often significant differences between the nominal age of the patient and the skeletal maturity of the patient. In adolescents, fracture types can be found that belong to adults due to the beginning or completion epiphyseal closure.

For pediatric fractures, reference literature [1,2] and guidelines [3–5] exist. Furthermore, extensive literature and guidelines for adult and geriatric trauma can be found. However, except for femoral fractures [6], there are no guidelines specifically for adolescent fractures in Germany. These fractures are extremely challenging to treat as age, physeal status, corrective potential, height, and weight have to be considered and often do not correlate [7].

A recent survey showed that the application of X-ray and aftercare regimens of pediatric and adolescent fractures differ significantly [8,9]. In line with this finding, the following survey was conducted to evaluate whether the treatment strategy of adolescent fractures also differs depending on the specialty of the treating physician. The overall aim of this research project is to develop standardized treatment concepts for adolescent fractures, irrespective of the specialty of pediatric or trauma surgery. This pilot study represents the first step in this direction, as it is intends to provide a descriptive representation of current differences in the treatment of adolescent fractures.

2. Materials and Methods

This study presents the results of a survey in the form of an online questionnaire, which was sent to German departments of pediatric and trauma surgery. For the questionnaire, X-ray images of adolescent fractures (age 12–18) in past years were screened. Seven cases were chosen by the research team, which were deemed fitting for the questionnaire. These cases were as follows, by order of appearance in the questionnaire:

1. Displaced subcapital humeral Salter–Harris Type 2 fracture (growth plate open, f, 13 y, 150 cm, 48 kg; Figure 1);
2. Distal tibia Salter–Harris Type 2 fracture with concomitant distal fibula shaft fracture (growth plate still visible, m, 15 y, 170 cm, 60 kg; Figure 2);
3. Midshaft clavicle fracture (proximal humeral growth plate open, m, 13, 160 cm, 40 kg; Figure 3);
4. Mulifragmentary femoral shaft fracture in the proximal third (growth plates closed, f, 17 y, 165 cm, 55 kg; Figure 4);
5. Transverse forearm shaft fracture (growth plates almost closed, f, 14 y, 155 cm, 45 kg; Figure 5);
6. Dorsally displaced Salter–Harris Type 2 distal radius fracture (growth plates still visible, m, 17 y, 185 cm, 85 kg; Figure 6);
7. Bimalleolar fracture with additional Tillaux fragment and loose joint body (growth plates open, m, 14, 175 cm, 55 kg; Figure 7).

In addition, information on age, weight, and height was provided, as this must be included in the decision-making process for the preferred therapy. Different treatment options were given as choices for every fracture; furthermore, a field for free input was provided if the treatment of choice was not listed. An option for respondents' comments existed for every case. The questionnaire finished with general questions about the primarily responsible department for adolescent fractures and the frequency of pediatric, adolescent, adult, and geriatric fractures in daily practice. Furthermore, the respondents' specialty and their years of experience were evaluated. The questionnaire was completely anonymous for the shown X-rays and the respondents' data. The materials and methods were checked and approved by the institutional review board and the local ethics committee of the University of Regensburg before launching the survey (protocol code: 24-3804-180; date of approval: 22 December 2023).

The survey was conducted with a free online survey tool (SoSci Survey GmbH, Munich, Germany). The link to the survey was open and sent to all pediatric surgical departments in Germany (85 clinics) and the corresponding departments for trauma surgery in the same clinic with the bid to share the questionnaire with the whole team. Furthermore, the link was shared with the members of the German association of pediatric surgery (DGKCH) and the German young forum for orthopedics and trauma surgery (JFOU). Reminders were sent two weeks later. The survey was accessible for two months (April and May 2024). In addition to fully completed questionnaires, a further inclusion criterion was the respondent's membership of a pediatric surgery or trauma surgery department with a certain level of clinical experience in the treatment of adolescent fractures. Incomplete questionnaires, especially with no information on the specialty, were excluded.

The evaluation and statistical analysis were conducted with SPSS software package (Version 24.0, IBM SPSS Inc., Chicago, IL, USA). Differences between the groups were tested by Pearson's chi-squared test. The level of significance was set to $p = 0.05$.

3. Results

Two clinics for pediatric surgery declined participation, because they stated they were not involved in pediatric or adolescent trauma. The link to the questionnaire was clicked 339 times. A total of 144 participated in the survey, although 18 did not finish it. This resulted in 126 questionnaires suitable for analysis. Of those, 71 were pediatric surgeons ("ped") and 48 trauma surgeons ("trauma"). A total of seven surgeons who were specialized in both were exclusively treating children and adolescents, thus their results were added to the "ped" group.

A total of 31.0% (39/126) reported more than 20 years of surgical experience, 40.5% (51/126) between 10 and 20 years, 15.9% (20/126) between 5 and 10 years, and 12.7% (16/126) less than 5 years. This was similar in both groups ($p = 0.56$).

The cut-off value for the treatment of adolescent trauma by pediatric surgeons was primarily the patient's 18th birthday (95/126, 75.4%). Furthermore, the 16th (10/126, 7.9%) and 15th (5/126, 4.0%) birthday and, in two cases, the 14th and 10th birthday were named. A total of 11.1% (14/126) reported a general interdisciplinary approach to adolescent trauma. No significant difference between either group regarding cut-off values was found ($p = 0.08$).

As expected, the reported frequency of pediatric trauma (0–12 years) more than once a week was significantly higher for pediatric surgeons ($p < 0.01$). On the other hand, adult and geriatric trauma was treated almost exclusively by trauma surgeons ($p < 0.01$). Adolescent trauma (12–18 years) was treated more often on a daily to weekly basis by pediatric surgeons (ped: 61/78, 78.2%; trauma: 23/48, 47.9%; $p < 0.01$) (Table 1).

Table 1. Reported frequency in treatment of different age groups, sorted by specialty; ped = pediatric surgeons ($n = 78$); trauma = trauma surgeons ($n = 48$). Field coloring marks percentage of number of participants: blank: <12.5%, shaded light green: 12.5% to 25.0%, medium: 25.0% to 50%, dark green: >50% of the group.

		Never	1x/mon.	1x/week	>1x/week	Daily
children	ped (n)	3	0	2	21	52
(0–12 y)	trauma (n)	8	9	9	7	15
adolescents	ped (n)	2	3	12	27	34
(12–18 y)	trauma (n)	1	10	14	11	12
adults	ped (n)	67	7	0	0	4
(>18 y)	trauma (n)	6	3	2	8	29
geriatrics	ped (n)	74	1	1	0	2
(>75 y)	trauma (n)	10	0	1	5	32

A statistical effect between the frequency of treating adolescent fractures and the treatment strategy could only be found for the forearm fracture in case 5 (radius: $p = 0.03$, ulna: $p < 0.01$) and the fibula fracture in case 7 ($p = 0.03$). In these named cases, the treatment option of elastic stable intramedullary nailing (ESIN) was preferred to plate osteosynthesis the more often adolescent trauma occurred in the daily practice. For all other cases and questions, no significant effect was found ($p > 0.05$, respectively).

3.1. *Case 1—Proximal Humeral Fracture*

Closed reduction and ESIN osteosynthesis were the favored options for both pediatric (58/78, 74.4%) and trauma surgeons (23/48, 47.9%). Closed reduction and percutaneous K-Wires were the second most frequent options (ped: 11/78, 14.1%; trauma: 9/48, 18.8%). Non-operative treatment was named 11 times (ped: 7/78, 9.0%; trauma: 4/48, 8.3%). Primary open reduction was named mainly by trauma surgeons (ped: 2/78, 2.6%; trauma: 11/48, 22.9%). For osteosynthesis after open reduction, besides ESIN and K-wires, locking plates (twice) and cannulated screws (once) were named by trauma surgeons. Finally, one trauma surgeon proposed closed reduction and then conservative treatment. The differences in the treatment strategies between both specialties were significant ($p = 0.01$, see Figure 1).

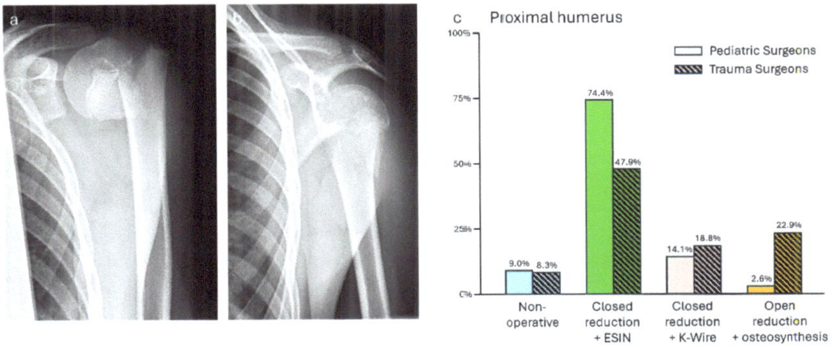

Figure 1. (**a**) Anterior–posterior and (**b**) lateral X-rays of a proximal subcapital humeral fracture (f, 13 y, 150 cm, 48 kg). (**c**) Descriptive data of the survey results, indicated in percentages.

3.2. Case 2—Distal Lower Leg Fracture

The three frequently named options for treatment of the distal tibia fracture were similar in approach (ped: 67/78, 85.9%; trauma: 30/48, 62.5%): closed reduction and retention either with overstaying (ped: 27/78, 34.6%: trauma: 9/48, 18.8%), buried K-wires (ped: 21/78, 26.7%; trauma: 12/48, 25.0%), or percutaneous screws (ped: 19/78, 24.4%; trauma: 9/48, 18.8%). Both closed reduction and no further osteosynthesis (ped: 3/78, 3.8%; trauma: 7/48, 14.6%) or osteosynthesis with plates (ped: 2/78, 2.6%; trauma: 9/48, 18.8%) were more often suggested by trauma surgeons. Among individual approaches (ped: 6/78, 7.7%; trauma: 2/48, 4.2%), three pediatric surgeons opted for open reduction to remove interfragmentary periost followed by K-wire fixation. An additional preoperative computed tomography (CT) was requested by one pediatric and one trauma surgeon.

No additional treatment of the fibula and immobilization by cast or walker was suggested by most of the surgeons (ped: 65/78, 83.3%, trauma: 23/48, 47.9%). As a rationale, indirect reduction via the interosseous membrane was additionally given by some participants. Nevertheless, osteosynthesis of the fibula by either tubular plates (ped: 3/78, 3.8%; trauma 14/48, 29.2%) or ESIN (ped: 5/78, 6.4%; trauma: 11/48, 22.9%) was more often named by trauma surgeons. Four pediatric surgeons (4/78, 5.1%) opted for an individual approach due to intraoperative stability testing and one opted for open reduction and locking plate osteosynthesis (1/78, 1.3%).

The treatment of the distal tibia and the distal fibula showed significant differences between the two groups ($p < 0.01$, respectively, see Figure 2).

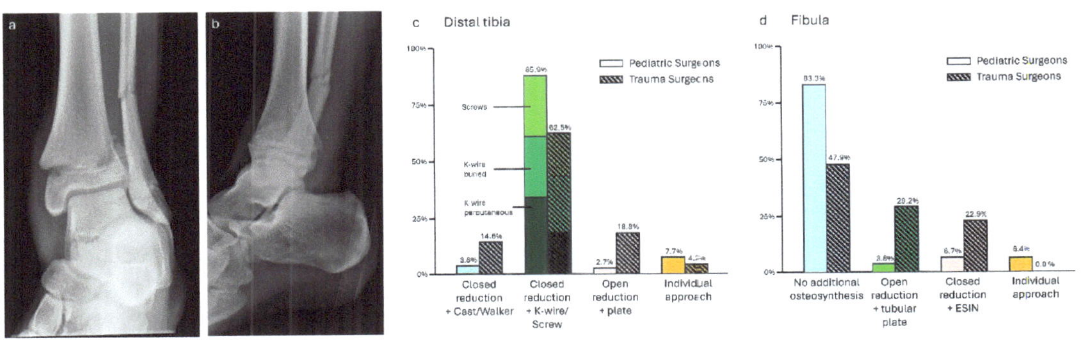

Figure 2. (**a**) Anterior–posterior and (**b**) lateral X-rays of a distal lower leg fracture (m, 15 y, 170 cm, 60 kg). Descriptive data of the survey results for the (**c**) distal tibia and (**d**) fibula, indicated in percentages.

3.3. Case 3—Clavicle Fracture

Non-operative treatment (ped: 46/78, 59.0%; trauma: 22/48, 45.8%), by either arm sling (ped: 23/78, 29.5%; trauma: 12/48, 25%) or backpack bandage (ped: 23/78, 29.5%; trauma: 10/48, 20.1%), was favored by both groups. Nevertheless, operative treatment by closed reduction and ESIN was chosen more frequently by trauma surgeons (ped: 10/78, 12.8%; trauma: 18/48, 37.5%). Less often named was open reduction and ESIN (ped: 6/78, 7.7%; trauma: 2/48, 2.6%) or plating, which was only chosen by trauma surgeons (3/48, 6.3%). It has to be mentioned that, in particular, pediatric surgeons opted to promote both operative and non-operative as equal options to the parents. Furthermore, it was added by the participants in the free text box that the decision should be made based on individual factors, especially the patient's sport demand (ped: 16/78, 20.5%; trauma 3/48, 6.3%). Although conservative treatment was the most common choice for both specialties, more trauma surgeons tended to favor surgical treatment ($p < 0.01$, see Figure 3).

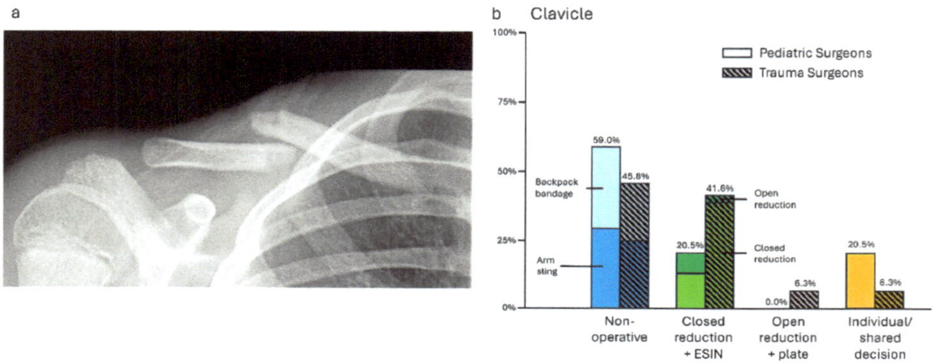

Figure 3. (**a**) Anterior–posterior X-ray of a dislocated clavicle shaft fracture (m, 13 y, 160 cm, 40 kg). (**b**) Descriptive data of the survey results, indicated in percentages.

3.4. Case 4—Multifragmentary Femoral Fracture

All surgeons saw the need for operative treatment. For patient positioning and reduction of the fracture, both pediatric and trauma surgeons preferred a fracture table (ped: 62/73, 84.9%; trauma: 37/48, 77.1%) over a surgical table (ped: 11/73, 15.1%; trauma: 11/48, 22.9%). No significant difference in the patient positioning was found ($p = 0.08$).

Osteosynthesis antegrade nailing, with or without the femoral neck component, was the treatment of choice, although trauma surgeons were much more in favor (ped: 38/77, 49.4%; trauma: 44/48, 91.6%). Except for locking plates (ped: 15/77, 19.5%; trauma: 4/48, 8.3%), trauma surgeons did not choose any other option. Pediatric surgeons also opted for ESIN without endcaps (11/77, 14.3%) and with endcaps (2/77, 2.6%). Other options were external fixation as the definitive treatment (4/77, 5.2%) and retrograde nailing (3/77, 3.9%). Four pediatric surgeons (4/77, 5.2%) noted that they would refer such a fracture type to the trauma department. This difference in surgical approaches was significant ($p < 0.01$, see Figure 4). One pediatric surgeon did not give an answer to this case.

3.5. Case 5—Forearm Shaft Fracture

All surgeons saw the need for surgical treatment. Both specialties preferred osteosynthesis with ESIN. To address the fracture of the radius, a retrograde ESIN with entry point proximal to the growth plate was the treatment of choice (ped: 65/78, 83.3%; trauma: 23/48, 47.9%). Retrograde ESIN with an entry point distal to the growth plate (ped: 4/78, 5.1%; trauma: 7/48, 14.6%) or ESIN in antegrade fashion (ped: 6/78, 7.7%; trauma: 3/48, 6.3%) were also named. The need for locking plating was seen more by trauma surgeons (ped:

3/78, 3.8%; trauma: 15/48, 31.3%). The treatment of the ulna, in analogy to the radius, occurred mainly with ESIN in antegrade fashion (ped: 70/78, 89.7%; trauma: 30/48, 62.5%) and sometimes in retrograde fashion (ped: 5/78, 6.4%; trauma: 3/48, 6.3%). Locking plates were preferred by the same surgeons that opted for locking plating of the radius (ped: 3/78, 3.8%; trauma: 15/48, 31.3%). No hybrid option (i.e., combination of ESIN and plates) was promoted. Trauma surgeons were more likely to favor locking plates than pediatric surgeons. ($p < 0.01$, respectively, see Figure 5). It must be noted that three trauma surgeons added the information that a non-locking plate would also be sufficient (3/15, 20.0%).

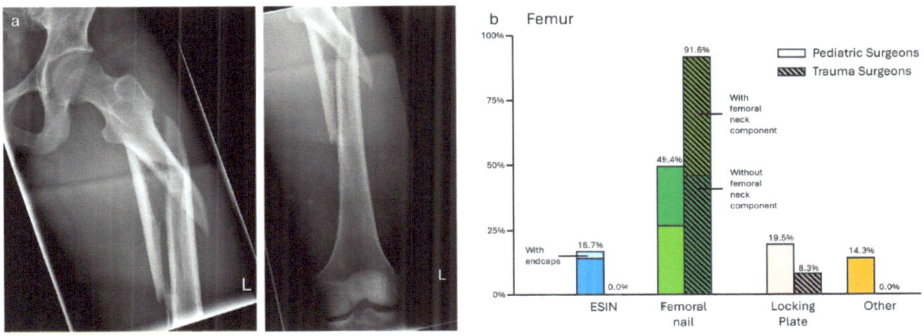

Figure 4. (**a**) X-ray of a multifragmentary femoral fracture (f, 17 y, 165 cm, 55 kg). (**b**) Descriptive data of the survey results, indicated in percentage.

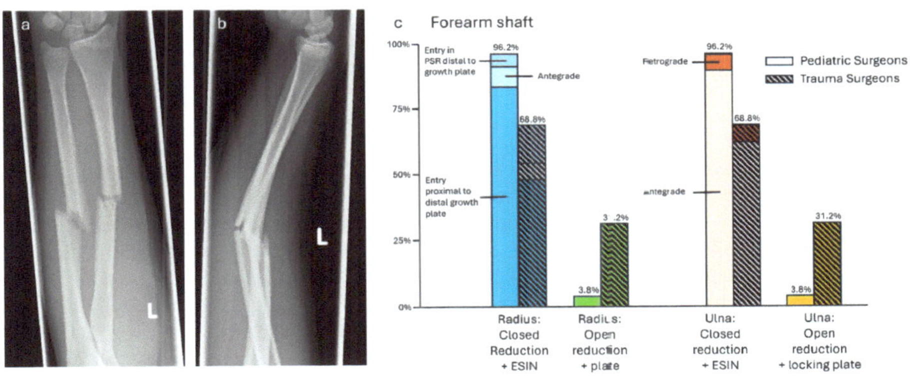

Figure 5. (**a**) Anterior–posterior and (**b**) lateral X-rays of a dislocated forearm shaft fracture (f, 14 y, 155 cm, 45 kg). (**c**) Descriptive data of the survey results, indicated in percentages.

3.6. Case 6—Distal Radius Fracture

Closed reduction and percutaneous K-wiring was the treatment of choice (ped: 65/78, 83.3; trauma: 33/48, 68.8%). Leaving the K-wires overstay was mentioned slightly more often (ped: 35/78, 44.9%; 20/48, 41.7%) than burying them (ped: 30/78, 38.5%; 13/48, 27.1%). While closed reduction and casting was the second choice of pediatric surgeons (ped: 9/78, 11.5%; trauma: 4/48, 8.3%), open reduction and plating was the second choice for trauma surgeons (ped: 3/78, 3.8%; trauma: 11/48, 22.9%). Complete non-operative treatment was only named by one pediatric surgeon. The more frequent choice of plating among trauma surgeons was significant ($p = 0.02$, see Figure 6).

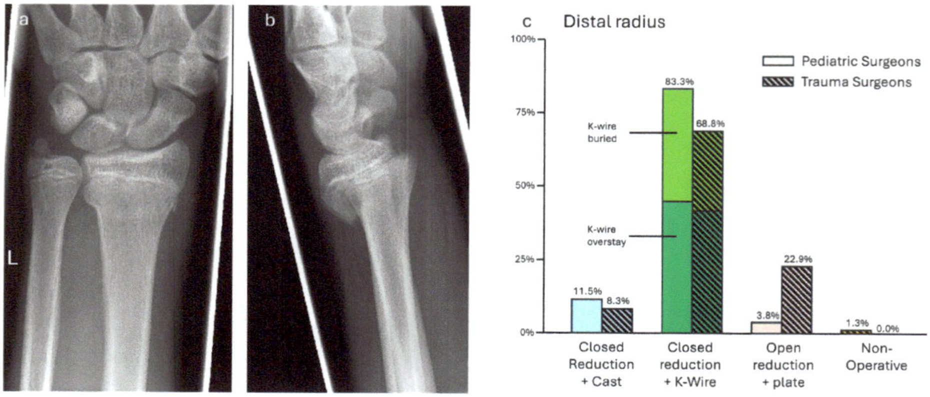

Figure 6. (**a**) Anterior–posterior and (**b**) lateral X-rays of a distal radial fracture with concomitant non-dislocated fracture of the styloid process of the distal ulna (m, 17 y, 185 cm, 85 kg). (**c**) Descriptive data of the survey results, indicated in percentages.

3.7. *Case 7—Bimalleolar Fracture*

For the treatment of the medial malleolus, screws were the treatment of choice for both specialties (ped: 68/77, 88.3%; trauma: 42/48, 87.5%). K-wires were proposed by some participants (ped: 7/77, 9.1%; trauma: 6/48, 12.5%). One pediatric surgeon opted for non-operative treatment and one for plating. Overall, no significant difference was found ($p = 0.81$, see Figure 7).

In contrast to this, the fibula was mainly treated non-operative by pediatric surgeons (ped: 43/77, 55.8%, trauma: 9/48, 18.8%), while trauma surgeons preferred tubular plating (ped: 11/77, 14.3%; trauma: 24/48, 50.0%). ESIN (ped: 14/77, 18.2%; trauma: 4/48, 8.3%), intramedullary K-wire (ped: 4/77, 5.2%; trauma: 1/48, 2.1%), and locking plates (ped: 3/77, 3.9%; trauma: 10/48, 20.8%) were also named. Two pediatric surgeons opted for intraoperative stability testing of the fibula before making a treatment decision. This variance between the treatment options had a significant difference ($p < 0.01$, see Figure 7).

Screw fixation of the Tillaux fragment was mainly chosen by both groups (ped: 51/76, 67.1%; trauma: 40/48, 83.3%). Although non-operative treatment was higher with pediatric surgeons (ped: 17/76, 22.4%; trauma: 4/48, 8.3%), there was no significant difference in this treatment option ($p = 0.17$).

Trauma surgeons mainly recommended removing the free joint body through an arthrotomy of existing approaches on the medial or laterals side (ped: 23/76, 30.3%, trauma: 27/48, 56.3%). Pediatric surgeons mainly favored leaving it (ped: 28/76, 36.8%; trauma: 9/48, 18.8%). Other options like arthroscopic removal (ped: 7/76, 9.2%; trauma: 5/48, 10.4%) or refixation (ped: 9/76, 11.8%; trauma: 2/48, 4.2%) were chosen. An individual approach with intraoperative decision was promoted by the free answer option (ped: 9/76, 11.8%; trauma: 5/48, 10.4%). The differences in the approach for the free joint body were significant ($p = 0.04$, see Figure 7).

While pediatric surgeons mainly did not choose special treatment of the syndesmosis (ped: 48/76, 63.1%; trauma: 14/48, 29.2%), trauma surgeons preferred a syndesmotic screw transfixation (ped: 22/76, 28.9%; trauma: 27/48, 56.3%). It has to be noted that, besides a few votes for dynamic syndesmotic systems (ped: 2/76, 2.6%; trauma: 3/48, 6.3%), some pediatric and trauma surgeons recommended intraoperative testing after refixation of the Tillaux fragment (ped: 4/76, 5.3%; trauma: 4/48, 8.3%). Nevertheless, the difference between both groups was significant ($p < 0.01$, see Figure 7).

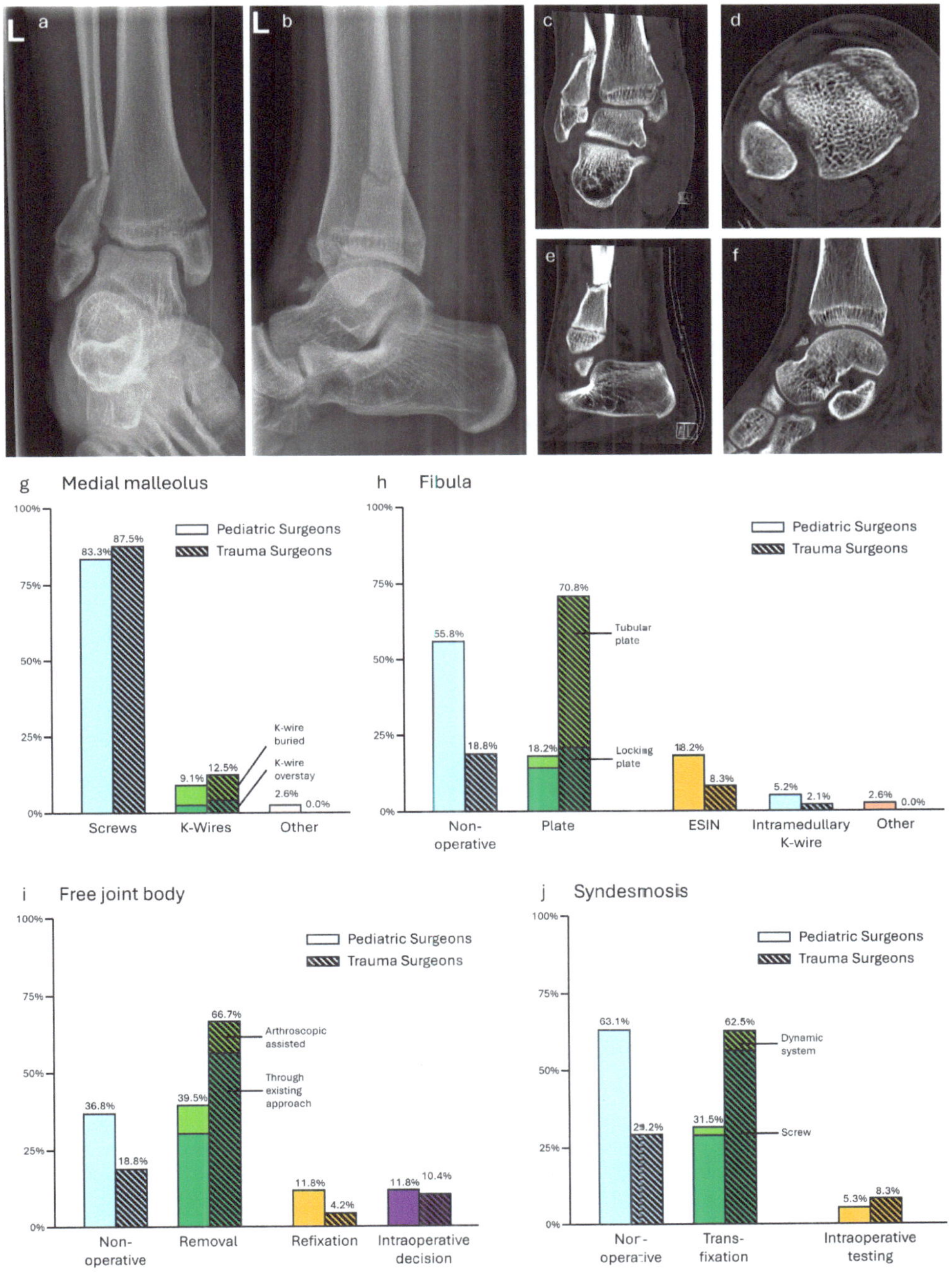

Figure 7. (**a**,**b**) X-rays and (**c**–**f**) representative CT layers of a bimalleolar fracture with concomitant a Tillaux fracture and free joint body (m, 14 y, 175 cm, 55 kg). (**g**–**j**) Descriptive data of the survey results, indicated in percentages.

Finally, most trauma surgeons proposed the usage of an intraoperative CT scan (ped: 11/77, 14.3%; trauma: 31/48, 64.6%), while pediatric surgeons did not (ped: 30/77, 39.0%; trauma: 10/48, 20.8) or did not have the equipment to do so (ped: 25/77, 32.5%; trauma: 6/48, 12.5). Some pediatric surgeons would obtain a postoperative CT scan (ped: 11/77, 14.3%; trauma: 1/48, 2.1%). This difference in the approach to intra- and postoperative CT imaging was significant ($p < 0.01$).

Two pediatric surgeons did not give answers to all questions associated with this case.

4. Discussion

This study reports therapy decisions made by pediatric and trauma surgeons in an online survey regarding seven different adolescent fractures. As the main result, the survey clearly states that in every single case, the given therapy decisions significantly depend on the physician's specialty. For the demonstrated seven cases, a tendency toward more operative and more invasive treatments was found for trauma surgeons compared to pediatric surgeons. On the other hand, there was a risk of underestimating the severity of fracture entities similar to adult fractures in pediatric surgeons. In the following, we discuss the results for each underlying case.

Proximal humeral fracture—Proximal humeral fractures in adolescents are known to be mainly Salter–Harris Type 2. Surgery is recommended if there is more than >20° angulation and more than one-third shaft width displacement because of the reduced remodeling potential with >13-year-olds [10,11]. With an angulation of more than 45° and full displacement of the shaft in the shown X-ray, operative treatment would be advisable. This was recommended in consensus by more than 90% of both pediatric and trauma surgeons. Although closed reduction and retrograde ESINs are recommended [11,12], more trauma surgeons tended to use more invasive procedures. Nevertheless, independent of the treatment, subcapital fractures in adolescents are described to have a good outcome [13].

Distal lower leg fracture—A survey among 37 pediatric orthopedic surgeons by Swarup et al. [14] showed significant differences in the treatment of Salter–Harris Type 2 distal tibia fractures in children and adolescents. Although closed reduction and casting can be successful, it has a known risk of failure [15] and was not the treatment of choice for both pediatric and trauma surgeons in this survey. Closed reduction and K-wires or screws were the most suggested. Therefore, trauma surgeons were more likely to use plate osteosynthesis than pediatric surgeons. Greater differences were seen in the treatment of the concomitant fibula fracture. A total of 52.1% of trauma surgeons proposed osteosynthesis in accordance with the literature for adult fractures [16]. On the other hand, Wang et al. demonstrated that additional fixation of the fibula in children (mean age 7.8 years) was not beneficial compared to conservative management [17]. This is in line with the majority opinion (83.3%) of pediatric surgeons. Therefore, in our opinion, factors like the patient's weight have to be taken into consideration, to see if additional stability is needed.

Clavicle fracture—Luo et al. [18] found that adult trauma surgeons were more likely to operate on adolescent clavicle fractures than pediatric orthopedics (32.6% vs. 10.3%) in their institution. This is in accordance with this study's results (trauma: 41.6% vs. ped: 20.5%). Furthermore, the existing literature does not offer a guideline for the indication for non-operative or operative treatment [19,20]. Even the choice between ESIN and plate osteosynthesis is not conclusively discussed, although there is evidence that even multi-fragmentary clavicle fractures can be treated well with intramedullary osteosynthesis [21]. All in all, not only the fracture pattern but also individual risk factors and activity levels must be considered for making treatment decisions [22].

Multifragmentary femoral fracture—This case showed a scenario of an adolescent with adult bone status and adult fracture type. While both specialties had the same approach to reduction, there were significant differences in the mode of fixation. Almost all trauma surgeons opted for antegrade femoral nailing according to the AO Principles [23] and national guidelines [6]. It should be noted that if the physis is still open, special nail designs with a lateral entry point should be used [24]. Pediatric surgeons gave a broad range of treatment options, which were in analogy to the treatment of pediatric trauma. Although ESINs are a valid treatment option for adolescent femoral fractures, the fracture pattern and weight have to be taken into consideration [25,26]. Locking plates, external fixation, and retrograde nails might also be options, but offer less postoperative weight bearing compared to intramedullary nailing. It has to be noted that some pediatric surgeons explicitly stated that they would refer such an adult fracture pattern to trauma surgeons.

Forearm shaft fracture—ESINs are the recommended treatment for forearm shaft fractures in children [27,28]. In adolescence, the decision between ESIN and plating is difficult, as Ho et al. described for radius shaft fractures, as no predictive parameters exist for the need for stability and remodeling potential exists to some extent [29]. This can be transferred to forearm shaft fractures. In this survey, both specialties preferred ESINs. Nevertheless, the results show that trauma surgeons tend more towards plating than pediatric surgeons. It is noteworthy that although hybrid fixation with ESIN and plate [30] is a described and known technique for adolescent forearm shaft fractures, neither pediatric nor trauma surgeons opted for such treatment.

Distal radius fracture—Salter–Harris Type 2 fractures at the distal radius are very common in adolescents [31]. Instead of the initial degree of displacement, age was found as the main factor for loss of reduction after closed reduction and casting, which occurred in about 30% [32]. A review by Greig et al. [33] found that up to 15° of displacement is tolerable in adolescents and recommended K-wire fixation and a cast as the operative treatment, over plating. Nevertheless, it has to mentioned that plating offers the option of direct anatomical reduction and cast-free aftercare

Bimalleolar fracture—The Tillaux fracture is described as a typical adolescent fracture. Because of its intra-articular component, surgical treatment is recommended when displacement is >2 mm [34,35]. It is still under discussion if closed or open reduction is superior [36]. The shown example had a concomitant bimalleolar fracture making it a rare but well-described type of adolescent fracture [37–39]. Pediatric and trauma surgeons showed similar survey results: both the Tillaux fracture and the medial malleolus needed screw fixation. More pediatric surgeons addressed the syndesmosis with refixation of the Tillaux fragment and saw no need for osteosynthesis of the fibula, according to pediatric trauma [40]. On the other hand, trauma surgeons opted for additional fixation of the fibula, in line with adult trauma [23]. Therefore, mainly syndesmotic screw fixation was named, although new dynamic techniques might be superior [41]. Removal of the free joint body was more often proposed by trauma surgeons. As they opted for more invasive osteosynthesis and, thus, bigger surgical approaches, it can be assumed that envisioning the ankle joint is easier for them and they have more familiarity with open joint surgery. The anatomical reduction of the syndesmosis is vital for a good clinical outcome. The advantages of intraoperative (cone beam) CT imaging in adult ankle fractures with syndesmotic injuries provides an anatomical reduction in the syndesmosis [41,42]. Nevertheless, no data for this type of intraoperative diagnostic exist for pediatric and adolescent ankle fractures. The survey showed not only significant differences in the availability of intraoperative (cone beam) CT imaging between pediatric and trauma surgeons but also in their application. It can be assumed that pediatric surgeons keep more strictly to the 'as low as reasonably achievable' principle (ALARA) for X-ray [9].

The cases of the survey show clearly that the recommended treatment is highly dependent on the physicians' affiliation to the respective specialty—in this case, pediatric or trauma surgery. First, one reason for the different treatment decisions may be the simple fact that preference is given to procedures that are familiar and frequently used. This bias, of course, applies for both specialties, as it could be seen for the forearm fracture (case 5) and the distal fibula fracture of the last case (case 7). In addition, the lack of clear guidelines for most adolescent fractures leads to disagreement in their care in everyday clinical practice. Even though the patient's outcome of the hypothetical treatments, unfortunately, cannot be recorded via this study, it must be assumed that such a broad range of treatment recommendations cannot be beneficial for the outcome in general. It is important to note that each treatment decision for adolescent fractures should be made on an individual and multifactorial basis but while following standardized recommendations. Validated guidelines have proven to be the best way to ensure an ideal outcome for each patient. The result of this survey, therefore, shows the need for an interdisciplinary approach for establishing guidelines for adolescent fracture types.

Despite the high response rate of experienced surgeons, the return rate of resident surgeons was rather low. The reason for this effect remains unclear. Nevertheless, in clinical practice, it is the experienced specialist physician who is entitled to make the indication for conservative surgical treatment and who chooses the type of osteosynthesis. The distribution of the respondents, therefore, reflects the everyday clinical practice more closely. Nevertheless, the survey was sent to all clinics with a pediatric surgery department and the corresponding trauma surgery departments. Usually, these are major clinics. It can be assumed that no or, by chance, only a few results were obtained from smaller trauma surgery departments that usually treat both pediatric and adolescent trauma. Although the seven cases were chosen with care, many other adolescent fracture types exist, and the results cannot be generalized. More cases would have made the survey longer and less practicable. The most common treatment options from non-operative to maximally invasive were given to be chosen from. Although a free input field for other treatment options was offered, the bias presented by these pre-given answers cannot be fully excluded. This survey does not provide information about the outcome of the hypothetic therapy decision of the respondents. This marks a limitation within the discussion of our results, as there might be more than one adequate decision. The lack of further information on comorbidities, the patients' activity level, or similar factors means that the surgeon's decision is hypothetical and not all factors influencing the therapy decision were named. Nevertheless, the questionnaire was suitable for answering the research question.

5. Conclusions

The treatment of adolescent fractures depends, to some extent, on the specialty of the treating surgeon. For the demonstrated seven cases, a tendency towards more operative and more invasive treatment was found for trauma surgeons compared to pediatric surgeons. There was a risk of underestimating the severity of fracture entities similar to adult fractures in pediatric surgeons. In conclusion, this survey calls for continuous interdisciplinary exchange between both surgical specialties to ensure optimal treatment of adolescent fractures and to develop guidelines for the future.

Author Contributions: Conceptualization, A.H.; methodology, A.H.; software, A.H.; validation, C.W., E.S.-H. and C.M.; formal analysis, C.M.; investigation, A.H.; resources, E.S.-H.; data curation, A.H.; writing—original draft preparation, A.H. and C.M.; writing—review and editing, E.S.-H. and C.W.; visualization, A.H.; supervision, C.M.; project administration, A.H. All authors have read and agreed to the published version of the manuscript.

Funding: This research received no external funding.

Institutional Review Board Statement: The study was conducted in accordance with the Declaration of Helsinki and approved by the Institutional Ethics Committee of the University of Regensburg (protocol code: 24-3804-180; date of approval: 22 December 2023).

Informed Consent Statement: Not applicable.

Data Availability Statement: The data that support the findings of this study are available on request from the corresponding author (C.M.). The data are not publicly available due to restrictions, as the information they contain could compromise the privacy of the research participants.

Conflicts of Interest: The authors declare no conflicts of interest.

References

1. Von Laer, L.; Schneidmüller, D.; Hell, A.K. *Frakturen und Luxationen im Wachstumsalter*, 7th ed.; Thieme: Stuttgart, Germany, 2020.
2. Marzi, I. *Kindertraumatologie*, 3rd ed.; Springer: Berlin/Heidelberg, Germany, 2016.

3. Deutsche Gesellschaft für Unfallchirurgie. S1 Leitlinie Suprakondyläre Humerusfraktur beim Kind. AWMF Nr. 012-014. 12/2014. Available online: https://register.awmf.org/de/leitlinien/detail/187-018 (accessed on 23 June 2024).
4. Deutsche Gesellschaft für Unfallchirurgie. S2k Leitlinie Frakturen des distalen Unterschenkels im Kindesalter, AWMF Nr. 012-026. 10/2017. Available online: https://register.awmf.org/de/leitlinien/detail/187-029 (accessed on 23 June 2024).
5. Deutsche Gesellschaft für Unfallchirurgie. S2e Leitlinie Proximale Femurfrakturen des Kindes, AWMF Nr. 012-020. 10/2021. Available online: https://register.awmf.org/de/leitlinien/detail/187-024 (accessed on 23 June 2024).
6. Deutsche Gesellschaft für Kinderchirurgie. S1 Leitlinie Femurschaftfraktur im Kindes- und Jugendalter, AWMF Nr. 006-016. 04/2023. Available online: https://register.awmf.org/de/leitlinien/detail/006-016 (accessed on 23 June 2024).
7. Parikh, S.N.; Wells, L.; Mehlman, C.T.; Scherl, S.A. Management of fractures in adolescents. *J Bone Jt. Surg. Am.* **2010**, *92*, 2947–2958. [CrossRef]
8. Dresing, K.; Kraus, R.; Fernandez, F.; Schmittenbecher, P.; Dresing, K.; Strohm, P.; Spering, C. Imaging after trauma in clinics and practice for children and adolescents: Part 1 of the results of a nationwide online survey of the Pediatric Traumatology Section of the German Trauma Society. *Unfallchirurgie* **2023**, *126*, 34–41. [CrossRef]
9. Dresing, K.; Fernandez, F.; Schmittenbecher, P.; Dresing, K.; Strohm, P.; Spering, C.; Kraus, R. Intraoperative imaging of children and adolescents, for selected fractures and in follow-up after conservative and operative treatment: Part 2 of the results of a nationwide online survey of the Pediatric Traumatology Section of the German Trauma Society. *Unfallchirurgie* **2023**, *126*, 42–54. [CrossRef]
10. Desai, K.B.; Waghmare, G.; Sathe, A.; Chinraj, R. Bilateral Traumatic Proximal Humeral Physeal Fracture in an Adolescent Child–A Rare Case Report and Review of Literature. *J. Orthop. Case Rep.* **2020**, *10*, 75–79. [CrossRef]
11. Hohloch, L.; Eberbach, H.; Wagner, F.C.; Strohm, P.C.; Reising, K.; Südkamp, N.P.; Zwingmann, J. Age- and severity-adjusted treatment of proximal humerus fractures in children and adolescents-A systematical review and meta-analysis. *PLoS ONE* **2017**, *12*, e0183157. [CrossRef]
12. Schmittenbecher, P.P. Frakturen der oberen Extremität im Kindes-/Wachstumsalter. Fractures of the upper limbs in childhood and adolescence. *Chirurg* **2017**, *88*, 451–466. [CrossRef] [PubMed]
13. Lähdeoja, T.; Pakkasjärvi, N.; Aronen, P.; Willamo, P.; Aronniemi, J.; Sintonen, H.; Nietosvaara, Y.; Paavola, M. Outcomes at a Mean of 13 Years After Proximal Humeral Fracture During Adolescence. *J. Bone Jt. Surg. Am.* **2023**, *105*, 839–848. [CrossRef] [PubMed]
14. Swarup, I.; Pearce, R.; Sanborn, R.; Shore, B.J. Variations in the Management of Closed Salter-Harris II Distal Tibia Fractures. *J. Pediatr. Orthop.* **2023**, *43*, e742–e746. [CrossRef] [PubMed]
15. Ramalingam, W.G.; Carry, P.; Brazell, C.; Calkins, R.; Linza-Moscati, S.; Stonebacк, J.; Miller, N.H. Outcomes of displaced adolescent distal third tibia fractures: Can we do better? *J. Pediatr. Orthop. B* **2022**, *31*, e147–e153. [CrossRef]
16. Bonnevialle, P.; Lafosse, J.-M.; Pidhorz, L.; Poichotte, A.; Asencio, G.; Dujardin, F. Distal leg fractures: How critical is the fibular fracture and its fixation? *Orthop. Traumatol. Surg. Res.* **2010**, *96*, 667–673. [CrossRef]
17. Wang, W.; Canavese, F.; Lin, R.; Pan, Y.; Huang, D.; Xiong, Z.; Chen, S. Clinical, radiographic and pedobarographic analysis of skeletally immature patients with surgically treated distal metaphyseal fractures of the tibia: Is concomitant fixation of the fibula necessary? *J. Pediatr. Orthop. B* **2020**, *29*, 490–498. [CrossRef]
18. Luo, T.D.; Ashraf, A.; Larson, A.N.; Stans, A.A.; Shaughnessy, W.J.; McIntosh, A.L. Complications in the treatment of adolescent clavicle fractures. *Orthopedics* **2015**, *38*, e287–e291. [CrossRef]
19. Scott, M.L.; Baldwin, K.D.; Mistovich, R.J. Operative Versus Nonoperative Treatment of Pediatric and Adolescent Clavicular Fractures: A Systematic Review and Critical Analysis. *JBJS Rev.* **2019**, *7*, e5. [CrossRef]
20. Yang, S.; Andras, L. Clavicle Shaft Fractures in Adolescents. *Orthop. Clin. N. Am.* **2017**, *48*, 47–58. [CrossRef]
21. Ode, G.E.; Larson, T.B.; Connor, P.M.; Fleischli, J.E.; D'Alessandro, D.F. Intramedullary Fixation for Displaced Clavicle Fractures in the Adolescent Athlete. *J. Surg. Orthop. Adv.* **2020**, *29*, 81–87.
22. Moverley, R.; Little, N.; Gulihar, A.; Singh, B. Current concepts in the management of clavicle fractures. *J Clin Orthop Trauma* **2020**, *11* (Suppl. S1), 25–30. [CrossRef]
23. Buckley, R.; Moran, C.G.; Apivatthakakul, T. *AO Principles of Fracture Management: Vol. 1: Principles, Vol. 2: Specific Fractures*, 3rd ed.; Thieme: Stuttgart, Germany, 2017.
24. Angadi, D.S.; Shepherd, D.E.T.; Vadivelu, R.; Barrett, T. Rigid intramedullary nail fixation of femoral fractures in adolescents: What evidence is available? *J. Orthop. Traumatol.* **2014**, *15*, 147–153. [CrossRef]
25. Andreacchio, A.; Alberghina, F.; Marengo, L.; Canavese, F. Pediatric tibia and femur fractures in patients weighing more than 50 kg (110 lb): Mini-review on current treatment options and outcome. *Musculoskelet Surg.* **2019**, *103*, 23–30. [CrossRef]
26. Frei, B.; Mayr, J.; de Bernardis, G.; Camathias, C.; Holland-Cunz, S.; Rutz, E. Elastic stabile intramedullary nailing (ESIN) of diaphyseal femur fractures in children and adolescents: A strobe-compliant study. *Medicine* **2019**, *98*, e15085. [CrossRef]
27. Schmittenbecher, P.P. State-of-the-art treatment of forearm shaft fractures. *Injury* **2005**, *36* (Suppl. S1), A25–A34. [CrossRef]
28. Poutoglidou, F.; Metaxiotis, D.; Kazas, C.; Alvanos, D.; Mpeletsiotis, A. Flexible intramedullary nailing in the treatment of forearm fractures in children and adolescents, a systematic review. *J. Orthop.* **2020**, *20*, 125–130. [CrossRef]
29. Ho, C.A. Radius Shaft Fractures-What Alignment is Acceptable at What Age? Plates or Flexible Nails? *J. Pediatr. Orthop.* **2021**, *41*, S14–S19. [CrossRef]

30. Feng, Y.; Shui, X.; Wang, J.; Cai, L.; Wang, G.; Hong, J. Comparison of hybrid fixation versus dual intramedullary nailing fixation for forearm fractures in older children: Case-control study. *Int. J. Surg.* **2016**, *30*, 7–12. [CrossRef]
31. Deng, H.; Zhao, Z.; Xiong, Z.; Gao, F.; Tang, S.; Li, Y.; Li, W.; Huang, J.; Cui, S.; Chen, X.; et al. Clinical characteristics of 1124 children with epiphyseal fractures. *BMC Musculoskelet. Disord.* **2023**, *24*, 598. [CrossRef]
32. Rivera, A.E.; Welbeck, A.; Cummings, J.L.; Goldfarb, C.A.; Hosseinzadeh, P. Distal Radius Salter-Harris II Fracture Loss of Reduction: The Importance of Coronal Plane Angulation in Older Children. *J. Orthop. Trauma* **2023**, *37*, 417–422. [CrossRef]
33. Greig, D.; Silva, M. Management of Distal Radius Fractures in Adolescent Patients. *J. Pediatr. Orthop.* **2021**, *41*, S1–S5. [CrossRef]
34. Tiefenboeck, T.M.; Binder, H.; Joestl, J.; Tiefenboeck, M.M.; Boesmueller, S.; Krestan, C.; Schurz, M. Displaced juvenile Tillaux fractures: Surgical treatment and outcome. *Wien. Klin. Wochenschr.* **2017**, *129*, 169–175. [CrossRef]
35. Tak, S.; Qureshi, M.K.; Ackland, J.A.; Arshad, R.; Salim, J. Adolescent Tillaux Fractures: A Systematic Review of the Literature. *Cureus* **2021**, *13*, e12860. [CrossRef]
36. Ayas, M.S.; Köse, A.; Terzi, E.; Dincer, R.; Topal, M.; Uymur, E.Y.; Şahin, A. Surgical management of displaced adolescent Tillaux Fractures with the mini-open technique. *Turk. J. Trauma Emerg. Surg.* **2021**, *27*, 109–114. [CrossRef]
37. Wood, M.B.; Bier, A.D.; Otsuka, N.Y. Bimalleolar ankle fracture in an adolescent with tillaux fracture. *J. Surg. Orthop. Adv.* **2009**, *18*, 103–105.
38. Helmig, K.C.; Choi, T.J.; Silva, S.R. Combined Tillaux and Bimalleolar Ankle Fracture in a Pediatric Patient: A Case Report. *Cureus* **2022**, *14*, e21648. [CrossRef]
39. Yuan, Q.; Guo, Z.; Wang, X.; Dai, J.; Zhang, F.; Fang, J.; Yin, C.; Yu, W.; Zhen, Y. Concurrent ipsilateral Tillaux fracture and medial malleolar fracture in adolescents: Management and outcome. *J. Orthop. Surg. Res.* **2020**, *15*, 423. [CrossRef]
40. Shore, B.J.; Kramer, D.E. Management of Syndesmotic Ankle Injuries in Children and Adolescents. *J. Pediatr. Orthop.* **2016**, *36* (Suppl. S1), S11–S14. [CrossRef] [PubMed]
41. Van Heest, T.J.; Lafferty, P.M. Injuries to the ankle syndesmosis. *J. Bone Jt. Surg. Am.* **2014**, *96*, 603–613. [CrossRef]
42. Vetter, S.Y.; Euler, J.; Beisemann, N.; Swartman, B.; Keil, H.; Grützner, P.A.; Franke, J. Validation of radiological reduction criteria with intraoperative cone beam CT in unstable syndesmotic injuries. *Eur. J. Trauma Emerg. Surg.* **2021**, *47*, 897–903. [CrossRef]

Journal of *Personalized Medicine*

MDPI

Article

The Clinical Efficacy of Contouring Periarticular Plates on a 3D Printed Bone Model

Seung-yeob Sakong [1], Jae-Woo Cho [2], Beom-Soo Kim [3], Sung-Jun Park [4], Eic-Ju Lim [5,*] and Jong-Keon Oh [2,*]

1 Department of Orthopaedic SurgeryAjou University Hospital, Ajou University College of Medicine, Suwon 16499, Republic of Korea; sgsy4040@gmail.com
2 Department of Orthopedic Surgery, Korea University Guro Hospital, Seoul 08308, Republic of Korea; jeus1103@gmail.com
3 Department of Orthopedic Surgery, Keimyung University Dongsan Hospital, Keimyung University Medicine, Daegu 41931, Republic of Korea; kbs090216@gmail.com
4 Department of Mechanical Engineering, Korea National University of Transportation, Chungbuk 27469, Republic of Korea; park@ut.ac.kr
5 Department of Orthopaedic Surgery, Chungbuk National University Hospital, Chungbuk National University College of Medicine, Cheongju 28644, Republic of Korea
* Correspondence: limeicju@gmail.com (E.-J.L.); jongkeon.oh@gmail.com (J.-K.O.); Tel.: +82-43-269-6077 (E.J.L.); +82-2-2626-1869 (J.-K.O.)

Abstract: We report our experience of preoperative plate contouring for periarticular fractures using three-dimensional printing (3DP) technology and describe its benefits. We enrolled 34 patients, including 11 with humerus midshaft fractures, 12 with tibia plateau fractures, 2 with pilon fractures, and 9 with acetabulum fractures. The entire process of plate contouring over the 3DP model was videotaped and retrospectively analyzed. The total time and number of trials for the intraoperative positioning of precontoured plates and any further intraoperative contouring events were prospectively recorded. The mismatch between the planned and postoperative plate positions was evaluated. The average plate contouring time was 9.2 min for humerus shaft, 13.8 min for tibia plateau fractures, 8.8 min for pilon fractures, and 11.6 min for acetabular fractures. Most precontoured plates (88%, 30/34) could sit on the planned position without mismatch. In addition, only one patient with humerus shaft fracture required additional intraoperative contouring. Preoperative patient specific periarticular plate contouring using a 3DP model is a simple and efficient method that may alleviate the surgical challenges involved in plate contouring and positioning.

Keywords: preoperative plate contouring; three-dimensional printing; periarticular fracture; plate fixation

Citation: Sakong, S.-y.; Cho, J.-W.; Kim, B.-S.; Park, S.-J.; Lim, E.-J.; Oh, J.-K. The Clinical Efficacy of Contouring Periarticular Plates on a 3D Printed Bone Model. *J. Pers. Med.* **2023**, *13*, 1145. https://doi.org/10.3390/jpm13071145

Academic Editors: Markus Gesslein and Eduard Kurz

Received: 14 June 2023
Revised: 12 July 2023
Accepted: 14 July 2023
Published: 17 July 2023

1. Introduction

Periarticular fracture is one of the most challenging fractures to treat due to their complex anatomy, morphological variation among individuals, and limited surgical access to the fracture site. The main treatment goal in these fractures is the achievement of anatomical reduction of articular fragments. If not treated properly, various complications, including delayed union, nonunion, malunion, implant failure, joint stiffness, and post-traumatic arthritis may easily arise [1,2]. Therefore, open reduction and internal fixation for periarticular fractures need to be accurate and individualized.

We generally use anatomical preshaped conventional plates for periarticular fractures. However, these plates are designed as per the average size and shape of a human body. Therefore, given the individual anatomical variations in patients, it is not possible to prepare a single plate that fits every patient. Thus, several studies have suggested the need for a distinct plate design to accommodate the differences arising due to sex and race [3–5]. Each fracture complexity would necessitate a different contour and screw trajectory, and the anatomical preshaped plates may require fine contouring before use. In complex cases that involve the acetabulum, other periarticular area contouring of the implant in the three-dimensional plane is usually necessary [6].

As of present, plate contouring is usually performed during surgery. However, it is impossible to expose the whole 3D bony structure clearly, and it is challenging to repeat plate contouring and positioning on the bony surface to confirm whether it fits well, owing to the anatomical obstacles and reduction tools. Therefore, plate contouring is usually regarded as is an imprecise and time-consuming process.

Surgeons usually contour plates preoperatively with commercialized saw bone models when they perform surgical intervention for complex anatomical structures, such as the pelvis. However, as the commercialized saw bone model is manufactured as per the average human pelvis size, it cannot accurately reproduce the anatomy of each patient.

Technical advances have considerably improved medical treatment practices, and 3D printing (3DP) techniques have been applied to the field of orthopedics [7–12]. Recently, several studies have shown that 3DP, the rapid prototyping technology, can be used for efficient preoperative plate contouring as per the exact surface of the individualized 3DP bone model [13–16].

However, most of the previous studies were related to precontoured plates, including acetabular fractures [17–23]. Furthermore, these studies did not provide detailed description of the process of plate precontouring using 3DP bone model, leaving out practical information regarding the time required and the number of trials necessary for precontouring of the plates. This study aimed to report our experience of contouring plates preoperatively for periarticular fractures using 3DP technology and describe its benefits by estimating the surgeon's contouring time, numbers of trials for plate contouring in between trial positioning, number of trials for plate positioning, and reporting events of intraoperative further contouring and mismatch between the planned and postoperative plate position. Furthermore, we retrospectively compared 3DP techniques with conventional surgical treatment for acetabular fractures to evaluate the effectiveness of precontoured plate using the 3DP technology.

2. Materials and Methods

2.1. Patient Selection

In this retrospective study, we enrolled 41 patients with periarticular fracture for whom fixation was attempted using precontoured plates with 3DP models in our hospital from January 2019 to January 2020. The following inclusion criteria were applied: (1) age > 18 years (2) fresh closed fractures, (3) no fracture, deformity, or surgical history on the contralateral normal side, (4) periarticular fracture or shaft fracture that extended to the periarticular area, and (5) complex articular fracture with severe comminution. The exclusion criteria were as follows: (1) open fracture, (2) peri implant fracture, (3) old and pathologic fracture, and (4) contralateral side fractures and dislocation.

2.2. Three-Dimensional Printing

The site of the fracture and the opposite site of the uninjured bone were examined using computed tomography (CT) scan (slice thickness, 2 mm). The medical image data were saved as digital imaging and communication in medicine formats and converted to standard triangulation language (STL) file formats to use 3DP specialized software (MIMICS; Materialise, Belgium).

When we used the image of the fractured site, each fragment was segmented and categorized. A comparison with the mirrored image of the uninjured site as a reference showed that the segmented fragments had virtually reduced. The reduced fragments were merged and wrapped thereafter. Two expert trauma orthopedic surgeons confirmed the quality of virtual reduction; then, an STL file of the virtually reduced bone model was sent to a 3D printer.

If the affected site had a comminuted fracture with small fragments for which virtual reduction was difficult to perform, we printed the 3D bone model using a mirror image of the unaffected site. After segmentation of the whole unaffected site, it was produced as the replica of the fractured site using a mirroring function of MIMICS. We drew the fracture line using a pen on the 3D printed bone model referring to the 3D reconstructive image of the fracture site to improve our understanding of the fracture pattern.

2.3. Precontouring of the Plates

Two skilled surgeons performed precontouring of the plates with the 3DP model. Precontouring of the plates could be accomplished by repeating the procedures of plate contouring using bending instruments and positioning the contoured plates on the 3DP model to evaluate whether they fit well until the desired plate was created by the surgeon. All the processes were recorded as video clips and saved. Two orthopedic trauma fellows who did not participate in the precontouring checked the surgeon's contouring time, number of trials for plate contouring in between trial positionings, and number of trials performed for plate positioning by reviewing the video clips. After the precontoured plates were taped to the 3DP model, we took the clinical photos and saved fluoroscopic images to help decide the proper position of the plate during the surgery (Figure 1). Completed precontoured plates were sterilized preoperatively.

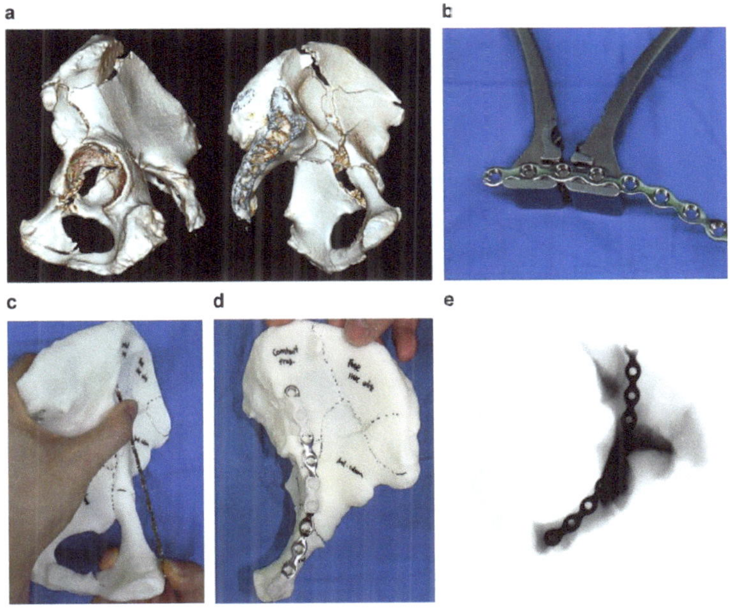

Figure 1. A representative case of a precontoured plate using a three-dimensional printing (3DP) model for acetabular fracture. (**a**) A 61-year-old man with associated both-column acetabular fracture. (**b**) A 3.5-mm low-profile reconstruction plate was contoured. (**c**) Positioning of the contoured plate using the 3DP model. (**d**,**e**) Clinical and fluoroscopic images of the precontoured pelvic brim plate taped to the 3DP model.

2.4. Intraoperative Precontoured Plate Positioning

The total time and number of trials for intraoperative precontoured plate positioning were retrospectively checked and recorded from C-arm images.

The total time of intraoperative plate positioning was defined as the time from the C-arm image with the first appearance of the precontoured plate to the time of the C-arm image right before the first screw appears on the precontoured plate. The total number of trials for intraoperative precontoured plate positioning was defined as the number of trials in which the plate position was changed until the final position.

Any events of intraoperative further contouring caused by an anatomical mismatch of the precontoured plate were described. Further, we reported the mismatch between the planned and postoperative plate position by superimposing the preoperative C-arm or clinical images to the corresponding postoperative radiographs or 3D-reconstructed CT views after adjustment of transparency.

2.5. Comparative Study for Acetabular Fracture

We retrospectively assessed 11 patients who underwent open reduction for acetabular fracture and internal fixation with intraoperatively contoured locking plates from January 2016 to December 2018 before using the 3DP technique. Nine acetabular fractures were fixed with precontoured plates using 3DP models. These 20 patients were divided into the following 2 groups: the conventional surgery group (Group 1) and the 3D printing group (Group 2).

Total operation time, instrumentation time, and blood loss amounts were compared between the groups. The operation time was recorded from the initial skin incision to complete wound closure based on medical records. The instrumentation time was defined as the time from the C-arm image of the first appearance of the plate to that of the completed whole screw fixation. Intraoperative blood loss was quantified by measuring the amount of irrigation fluid and weighing the surgical sponges used for blood and fluid collection during the surgery. Any events, including complications, were also documented.

The Mann–Whitney test was used to compare the total operation time, instrumentation time, and blood loss amounts between the two groups.

2.6. Saw Bone Oriented Contoured Plates

Furthermore, we contoured the plates using commercial pelvis and tibia saw bones as templates and used them in the 3DP models of the 12 tibia plateau and 9 acetabular fracture cases included in this study to assess anatomical mismatches.

The lateral rim plate for the lateral tibia plateau using a 2.7 variable-angle (VA) foot plate as well as 7 cases of pelvic brim plates and 2 cases of posterior wall plate using 3.5-mm low profile curved reconstruction plate were contoured with the saw bone for the tibia plateau and acetabular fracture cases, respectively. After the saw bone oriented contoured plates were applied to each 3DP model, the need for additional contouring due to anatomical mismatch was evaluated. In case of additional contouring, we estimated the total additional time for contouring, number of plate contouring, and number of plate positioning.

3. Results

3.1. Clinical Data of the 3DP Technique

Overall, 41 periarticular fracture patients were treated using plate contouring with 3DP. However, 7 patients, including 2 with distal humerus fractures, 2 with tibia plateau fractures, and 3 with acetabular fractures, did not require plate contouring because the plates fit well on the 3D printed bone models; thus, these patients were excluded.

A total of 16 mirrored images from the unaffected side and 18 virtually reduced images were used for 3DP (Table 1).

Printing of the 3D models took within 12 h in all cases. There are some differences based on the complexity, numbers, and size of the fracture fragment; approximately 30 min to 1 h was required, on an average, for virtual reduction (Table 2).

Table 1. General data of the 3DP models.

Location of Fracture	Numbers of Cases	Mirrored Image	Virtually Reduced Image
Fractures of humerus shaft	11	2	9
Fractures of tibia plateau	12	7	5
Fractures of pilon	2	0	2
Fractures of Acetabulum	9	7	2
Total	34	16	18

Table 2. Required time for 3DP.

Location of Fracture	3D Printing (hours)	Virtual Reduction (hours)
Fractures of humerus shaft	10	0.5
Fractures of tibia plateau	5	1
Fractures of pilon	4.5	0.7
Fractures of acetabulum	12	1

3.2. *Data for the Precontoured Plate*

On an average, 9.2 min were needed for fractures of the humerus shaft, 13.8 min for fractures of the tibia plateau, 8.8 min for fractures of pilon, and 11.6 min for fractures of the acetabulum for the plate contouring performed by the surgeon (Table 3).

Table 3. Data for precontouring.

Location of Fracture	Number of Contoured Plates	Contouring Time of the Surgeon (min)	Numbers of Plate Contouring in between Trial Positioning	Numbers of Trial for Plate Positioning
Fractures of humerus shaft	1	9.2	7	8.3
Fractures of tibia plateau	2	13.8	15.1	13.5
Fractures of pilon	1.5	8.8	6.5	6
Fractures of acetabulum	1.3	11.6	10	8.5

3.3. *Data for Intraoperative Precontoured Plate Positioning*

Most of the cases did not require additional intraoperative contouring because of a mismatch—except for one case (Table 4).

Table 4. Intraoperative contoured plate positioning.

Location of Fracture	Numbers of Plates	Total Time for Plate Positioning (min)	Trial Numbers of Plate Positioning	Numbers of Cases Need for Further Contouring
Fractures of humerus shaft	1	3.7	1.3	1
Fractures of tibia plateau	2	7.7	3.2	0
Fractures of pilon	1.5	2.6	1.5	0
Fractures of acetabulum	1.3	5.4	1.6	0

3.4. *Precontoured Plate Mismatch and Complications*

Mismatch between the precontoured plate and the initially planned plate position was reported in three cases of the tibia plateau fractures, and one case of acetabulum fracture, and significant complications were reported in two cases. Lateral tibia plateau resorption after open reduction and plate fixation for tibia plateau fractures was observed at 6 and

4 months postoperatively, respectively. One of them, which is associated with progressive valgus deformity, is planned for distal femur osteotomy after union. Implant removal was performed at 9 months after the procedure because of discomfort during ambulation for the other case.

3.5. Comparative Data of Acetabular Fracture Cases

The instrumentation time in Group 2 was significantly shorter than in Group 1, with a mean duration of 40.6 and 91 min, respectively (Table 5).

Table 5. Comparative data of acetabular fracture.

	Group 1 (n = 11)	Group 2 (n = 9)	*p* Value
Instrumentation time (min)	91	40.6	<0.05
Total operation time (min)	332.5	294.3	0.211
Blood loss (mL)	1040	866.6	0.611
Complications	L5 nerve root irritation	None	

One case in Group 1 developed a major complication. An intraoperatively contoured suprapectineal plate was placed beyond the SI (Sacroiliac) joint, which resulted in a suspected iatrogenic L5 nerve root injury. Foot drop symptom was observed immediately postoperatively, but the patient showed a gradual improvement after 3 days.

3.6. Data of Saw Bone Oriented Contoured Plates

The saw bone oriented plates of all 12 fractures of tibia plateau and 7 fractures of acetabulum were necessary for additional contouring due to anatomical mismatch (Figure 2, Table 6). In one tibia plateau fracture case, we should have contoured another plate because the complication occurred as the plate was broken during the additional contouring.

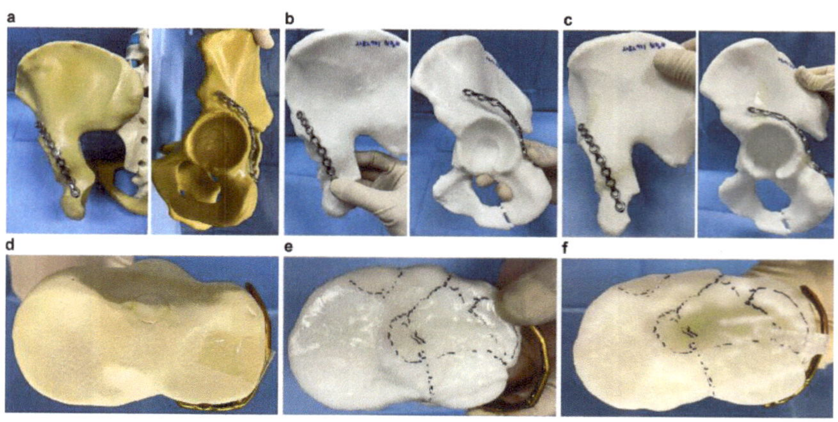

Figure 2. Saw bone oriented contoured plates on (**a**–**c**) pelvis and (**e**–**f**) proximal tibia models. (**a**) Acetabular posterior wall plate was contoured on a pelvis saw bone model. (**b**) The contoured plate was positioned on a three-dimensional printing (3DP) model, but anatomical mismatch was indicated. (**c**) The additional contoured plate fits well with the 3DP model; additional contouring required 5 min and 15 s. (**d**) A lateral tibia plateau rim plate was contoured on a tibia saw bone model. (**e**) Mismatch between the 3DP model and contoured plate. (**f**) Additional contoured plate fits well with the 3DP model; additional contouring required 8 min and 30 s.

Table 6. Data of saw bone oriented contoured plates.

Location of Fracture	Numbers of Cases Needed for Additional Contouring	Surgeon's Additional Contouring Time (min)	Number of Additional Contouring in between Trial Positioning	Number of Additional Trial for Plate Positioning
Tibia plateau facture	12 (12/12)	8.4	8.2	7.8
Acetabulum fracture	7 (7/9)	5.6	6.2	6.5

4. Discussion

The most significant contribution of this study is that we have described the process of precontouring of plates for periarticular fracture using 3DP technique in detail, including information about the surgeon's contouring time and the number of trials for plate contouring and plate positioning using 3DP models. Moreover, we evaluated the effectiveness of this method by reporting the time required for intraoperative plate positioning, number of trials for intraoperative plate positioning, any events of intraoperative further contouring, and mismatch between the planned and postoperative plate position.

Most importantly, we were able to perform plate contouring in a simple manner within a relatively short time, using the 3DP model.

However, even though this study was performed by experienced surgeons in vitro, free from any structural obstacles, several numbers of trials were required for plate contouring and positioning. The following aspects should be considered during plate precontouring: (1) optimal plate positioning; (2) screw trajectory after plate positioning; (3) selection of the portion of the plates that is to be contoured; and (4) degree of contouring, tending to under or over contour. In case of intraoperative plate contouring after considering all these factors, as contouring and trial of plate positioning, including repetitive temporary positioning pinning and confirmation by C-arm back and forth, would be performed on real bony structure surrounded by anatomical barriers and reduction tools, a considerably longer time and more trials would be required. Moreover, accurate contouring, as per the surgeons' satisfaction, is challenging to achieve because of these reasons (Figure 3). Hung et al. [18] and Miani et al. [20] demonstrated that using the 3D technique for the precontouring process could shorten the operative time and help reduce intraoperative blood loss in the treatment of anterior pelvic ring and acetabulum fractures, respectively. These results were similar to our comparative study for acetabular fracture cases, indicating that the plate precontouring could relatively and substantially reduce the total operation time and instrumentation time. Zheng et al. [24] also reported a substantially shorter operation time, less blood loss, and shorter fluoroscopy times in the treatment of pilon fractures using the 3DP models. Thus, preoperative contouring using 3DP models can reduce the surgical time for laborious intraoperative trials for plate contouring and instrumentation time.

In our evaluation of commercialized saw bone oriented contoured plate, the considerable time and numbers of trial and plate positionings for additional contouring due to anatomical mismatch were wasted in most cases. Besides, we could observe one case with a broken plate due to additional repeated contouring back and forth. Thus, our study shows that the saw bone, which is manufactured as per the average human anatomy, may not perfectly reproduce the anatomy of individual patients compared with customized 3DP models. Furthermore, accurate plate contouring using 3DP models can prevent plate weakening or breakage after repetitive contouring by reducing the unnecessary numbers of contouring and plate positioning.

The precontoured plates could be positioned at the reduced bone surrounded by anatomical barriers and reduction tools in a short time with an adequate number of attempts during the surgery. Moreover, 3DP models contribute to the contouring of well-fitted plates in the desired fashion, eliminating the need for additional contouring due to mismatch in most cases (97%, 33 of 34) (Figures 4 and 5). Moreover, we found that most precontoured plates (88%, 30 of 34) could sit on the optimal position as planned, without derailment (Figure 6). By referring to the fluoroscopic and clinical images of the precontoured plate attached to the 3DP model during the surgery, we could prevent significant derailment of the plate.

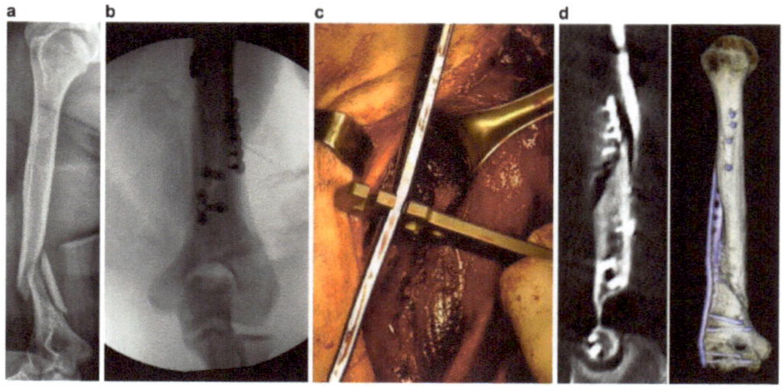

Figure 3. A case of intraoperatively contoured plate mismatch in humerus shaft fracture. (**a**) A 36-year-old woman with right humerus shaft fracture extending to the periarticular lesion. (**b**) Wedge fragment and spiral fracture were reduced with mini plates. (**c**) Anatomical precontoured humerus lateral column plate was contoured intraoperatively in 25 min. (**d**) Immediate postoperative CT shows loss of reduction owing to significant mismatch between the intraoperative contoured plate and bone.

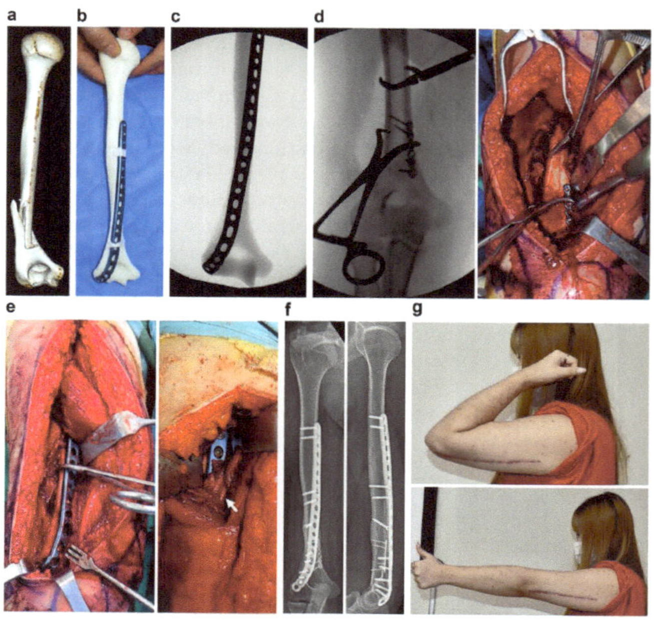

Figure 4. A case demonstrating a precontoured plate using the 3DP model in a humerus fracture. (**a**) A 40-year-old woman with right humerus fracture that extends to the periarticular lesion. (**b**) Anatomical precontoured humerus lateral column plate was contoured using the 3DP model in 7 min and 30 s. (**c**) C-arm image of precontoured plate taped on the 3DP model was saved. (**d**) Spiral fracture was reduced with a mini plate and was prepared for the main posterolateral position. (**e**) Precontoured humerus lateral column plate could sit well under the radial nerve (white arrow) without additional contouring. (**f**,**g**) Complete union and full range of motion could be achieved at postoperative 5 months.

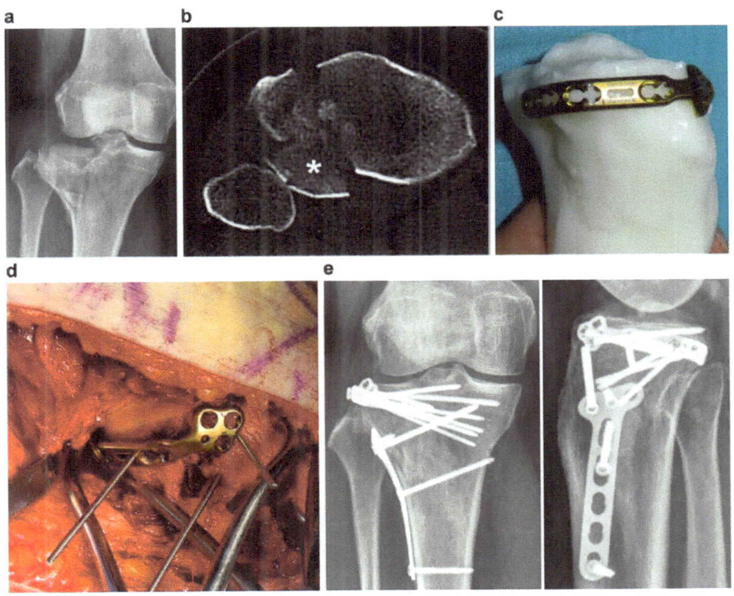

Figure 5. A case demonstrating a precontoured plate using the 3DP model in tibia plateau fracture. (**a**,**b**) A 40-year-old woman with right tibia plateau fracture after 3-m fall from height. Radiographs show anterolateral split depression fracture with a posterolateral fragment involvement. (**c**) A 2.7-mm variable-angle locking compression plate (VA LCP) plate was contoured using the 3DP model in 9 min. (**d**) The precontoured lateral rim plate could be positioned as it required no additional contouring. (**e**) Complete union and full range of motion without implant irritation could be achieved 3 months after the surgery.

Although lateral plateau resorption was observed in two cases, as the initial joint depression was too severe (21.63 mm and 9.27 mm), we were unable to restore the level of the joint to the degree that we desired, using the 3D-printed bone models. Therefore, a precontoured lateral plateau rim plate and screws were placed closer to the joint than we had expected.

All 3DP models could be manufactured within 12 h. Even when virtual reduction was attempted, it did not exceed 2 h, irrespective of the fracture complexity. There is an operative delay in most cases of periarticular fractures after stabilization of soft tissue; therefore, the time for 3DP would not affect the appropriate time for surgery.

The 3DP models cost between USD 50$between USD 100$ depending on size and location. The use of 3DP models for orthopedic surgeries is not covered under public insurance in South Korea; therefore, this study was sponsored by the government. We believe that technical advancements, increasing use, and validation of the use of 3DP in orthopedic surgery would lower the cost and time duration required for 3DP in the near future.

There are certain limitations of this study. Our trial was retrospective in nature, involved a relatively small numbers of cases, and we did not employ a control group. Furthermore, precontouring was performed by two experts only. A larger case series with a control group and the participation of beginners or trainees would be more beneficial to validate the usefulness of precontoured plates using the 3DP technique. Moreover, we focused on the process of plate precontouring using a 3DP model itself, rather than the clinical data or functional outcomes. Therefore, the clinical results of plate precontouring using 3DP model would be indicated in a future study.

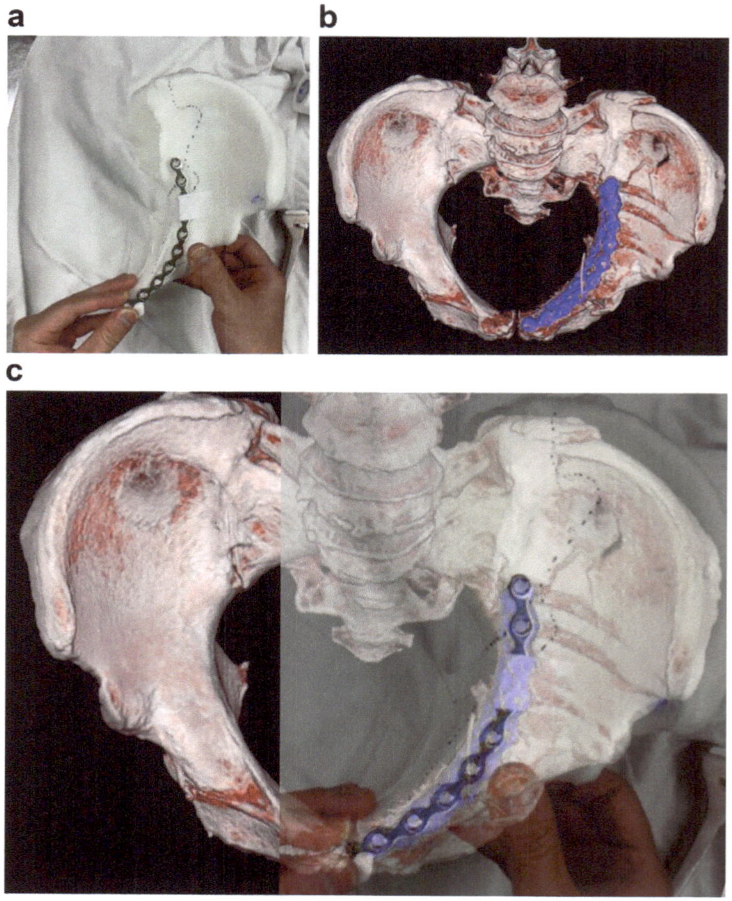

Figure 6. Evaluation for anatomical mismatch of the precontoured plate. (**a**) Clinical photograph of the precontoured pelvic brim plate taped on the 3DP model. (**b**) Corresponding postoperative 3D reconstructed view. (**c**) Superimposed photograph of (**a**,**b**) demonstrate no evidence of a mismatch.

A real-time 3DP model is an effective technique for preoperative contouring of the patient specific plates, and can promote the reduction of the operative time and laborious trials for plate contouring during surgery.

5. Conclusions

In patients with periarticular fractures, 3DP-assisted precontoured plates are useful because they enable the surgeon to accurately contour the plate in the desired fashion before the surgery. This helps in shortening the operation time and alleviating the surgical challenges involved in plate contouring and positioning.

Author Contributions: Methodology, data curation, writing—original draft, formal analysis, S.-y.S.; visualization, data curation, E.-J.L., B.-S.K.; investigation and resources, validation, J.-W.C., S.-J.P.; data curation, supervision, J.-K.O.; project administration, E.-J.L.; conceptualization, investigation and resources, supervision, writing—review and editing J.-K.O. All authors have read and agreed to the published version of the manuscript.

Funding: This research was supported by a material parts special accounting development Fund of the ministry of science and ICT (MSIT), Republic of Korea. (Grant number: S1222-20-1001).

Institutional Review Board Statement: Institutional Review Board has been approved (No. 2020 GRO300).

Informed Consent Statement: Not applicable.

Data Availability Statement: Not applicable.

Conflicts of Interest: The authors declare no conflict of interest.

References

1. Dillin, L.; Slabaugh, P. Delayed wound healing, infection, and nonunion following open reduction and internal fixation of tibial plafond fractures. *J. Trauma* **1986**, *26*, 1116–1119. [CrossRef] [PubMed]
2. McFerran, M.A.; Smith, S.W.; Boulas, H.J.; Schwartz, H.S. Complications encountered in the treatment of pilon fractures. *J. Orthop. Trauma* **1992**, *6*, 195–200. [CrossRef] [PubMed]
3. Lin, K.J.; Wei, H.W.; Lin, K.P.; Tsai, C.L.; Lee, P.Y. Proximal femoral morphology and the relevance to design of anatomically precontoured plates: A study of the Chinese population. *Sci. World J.* **2014**, *2014*, 106941. [CrossRef] [PubMed]
4. Oppermann, J.; Bredow, J.; Beyer, F.; Neiss, W.F.; Spies, C.K.; Eysel, P.; Dargel, J.; Wacker, M. Distal radius: Anatomical morphometric gender characteristics. Do anatomical pre-shaped plates pay attention on it? *Arch. Orthop. Trauma Surg.* **2015**, *135*, 133–139. [CrossRef]
5. Hwang, J.H.; Oh, J.K.; Oh, C.W.; Yoon, Y.C.; Choi, H.W. Mismatch of anatomically pre-shaped locking plate on Asian femurs could lead to malalignment in the minimally invasive plating of distal femoral fractures: A cadaveric study. *Arch. Orthop. Trauma Surg.* **2012**, *132*, 51–56. [CrossRef]
6. Bagaria, V.; Deshpande, S.; Rasalkar, D.D.; Kuthe, A.; Paunipagar, B.K. Use of rapid prototyping and three-dimensional reconstruction modeling in the management of complex fractures. *Eur. J. Radiol.* **2011**, *80*, 814–820. [CrossRef]
7. Assink, N.; Reininga, I.H.F.; Ten Duis, K.; Doornberg, J.N.; Hoekstra, H.; Kraeima, J.; Witjes, M.J.H.; de Vries, J.P.M.; FFA, I.J. Does 3D-assisted surgery of tibial plateau fractures improve surgical and patient outcome? A systematic review of 1074 patients. *Eur. J. Trauma Emerg. Surg.* **2022**, *48*, 1737–1749. [CrossRef]
8. Dust, T.; Hartel, M.J.; Henneberg, J.E.; Korthaus, A.; Ballhause, T.M.; Keller, J.; Ohlmeier, M.; Maas, K.J.; Frosch, K.H.; Krause, M. The influence of 3D printing on inter- and intrarater reliability on the classification of tibial plateau fractures. *Eur. J. Trauma Emerg. Surg.* **2023**, *49*, 189–199. [CrossRef]
9. Li, K.; Liu, Z.; Li, X.; Wang, J. 3D printing-assisted surgery for proximal humerus fractures: A systematic review and meta-analysis. *Eur. J. Trauma Emerg. Surg.* **2022**, *48*, 3493–3503. [CrossRef]
10. Yammine, K.; Karbala, J.; Maalouf, A.; Daher, J.; Assi, C. Clinical outcomes of the use of 3D printing models in fracture management: A meta-analysis of randomized studies. *Eur. J. Trauma Emerg. Surg.* **2022**, *48*, 3479–3491. [CrossRef]
11. Huitema, J.M.; van der Gaast, N.; Brouwers, L.; Jaarsma, R.L.; Doornberg, J.N.; Edwards, M.J.R.; Hermans, E. Are 3D-printed Models of Tibial Plateau Fractures a Useful Addition to Understanding Fractures for Junior Surgeons? *Clin. Orthop. Relat. Res.* **2022**, *480*, 1170–1177. [CrossRef]
12. Fidanza, A.; Perinetti, T.; Logroscino, G.; Saracco, M. 3D Printing Applications in Orthopaedic Surgery: Clinical Experience and Opportunities. *Appl. Sci.* **2022**, *12*, 3245. [CrossRef]
13. Shon, H.C.; Bang, J.Y.; Lee, Y.; Koh, K.H.; Kim, J.W. Optimal plate position in minimally invasive plate osteosynthesis for mid-shaft clavicle fractures: Simulation using 3D-printed models of actual clinical cases. *Eur. J. Trauma Emerg. Surg.* **2021**, *47*, 1411–1416. [CrossRef] [PubMed]
14. Kim, J.W.; Lee, Y.; Seo, J.; Park, J.H.; Seo, Y.M.; Kim, S.S.; Shon, H.C. Clinical experience with three-dimensional printing techniques in orthopedic trauma. *J. Orthop. Sci.* **2018**, *23*, 383–388. [CrossRef] [PubMed]
15. Yang, L.; Grottkau, B.; He, Z.; Ye, C. Three dimensional printing technology and materials for treatment of elbow fractures. *Int. Orthop.* **2017**, *41*, 2381–2387. [CrossRef]
16. Chung, K.J.; Hong, D.Y.; Kim, Y.T.; Yang, I.; Park, Y.W.; Kim, H.N. Preshaping plates for minimally invasive fixation of calcaneal fractures using a real-size 3D-printed model as a preoperative and intraoperative tool. *Foot Ankle Int.* **2014**, *35*, 1231–1236. [CrossRef] [PubMed]
17. Hsu, C.L.; Chou, Y.C.; Li, Y.T.; Chen, J.E.; Hung, C.C.; Wu, C.C.; Shen, H.C.; Yeh, T.T. Pre-operative virtual simulation and three-dimensional printing techniques for the surgical management of acetabular fractures. *Int. Orthop.* **2019**, *43*, 1969–1976. [CrossRef] [PubMed]
18. Hung, C.C.; Li, Y.T.; Chou, Y.C.; Chen, J.E.; Wu, C.C.; Shen, H.C.; Yeh, T.T. Conventional plate fixation method versus pre-operative virtual simulation and three-dimensional printing-assisted contoured plate fixation method in the treatment of anterior pelvic ring fracture. *Int. Orthop.* **2019**, *43*, 425–431. [CrossRef]
19. Upex, P.; Jouffroy, P.; Riouallon, G. Application of 3D printing for treating fractures of both columns of the acetabulum: Benefit of pre-contouring plates on the mirrored healthy pelvis. *Orthop. Traumatol. Surg. Res.* **2017**, *103*, 331–334. [CrossRef] [PubMed]

20. Maini, L.; Sharma, A.; Jha, S.; Sharma, A.; Tiwari, A. Three-dimensional printing and patient-specific pre-contoured plate: Future of acetabulum fracture fixation? *Eur. J. Trauma Emerg. Surg.* **2018**, *44*, 215–224. [CrossRef] [PubMed]
21. Maini, L.; Verma, T.; Sharma, A.; Sharma, A.; Mishra, A.; Jha, S. Evaluation of accuracy of virtual surgical planning for patient-specific pre-contoured plate in acetabular fracture fixation. *Arch. Orthop. Trauma Surg.* **2018**, *138*, 495–504. [CrossRef] [PubMed]
22. Chana-Rodríguez, F.; Mañanes, R.P.; Rojo-Manaute, J.; Gil, P.; Martínez-Gómiz, J.M.; Vaquero-Martín, J. 3D surgical printing and pre contoured plates for acetabular fractures. *Injury* **2016**, *47*, 2507–2511. [CrossRef] [PubMed]
23. Papotto, G.; Testa, G.; Mobilia, G.; Perez, S.; Dimartino, S.; Giardina, S.M.C.; Sessa, G.; Pavone, V. Use of 3D printing and pre-contouring plate in the surgical planning of acetabular fractures: A systematic review. *Orthop. Traumatol. Surg. Res.* **2022**, *108*, 103111. [CrossRef]
24. Zheng, W.; Chen, C.; Zhang, C.; Tao, Z.; Cai, L. The Feasibility of 3D Printing Technology on the Treatment of Pilon Fracture and Its Effect on Doctor-Patient Communication. *Biomed Res. Int.* **2018**, *2018*, 8054698. [CrossRef] [PubMed]

Journal of Personalized Medicine

MDPI

Review

Advancements in Diagnosis and Management of Distal Radioulnar Joint Instability: A Comprehensive Review Including a New Classification for DRUJ Injuries

Awad Dmour [1], Stefan-Dragos Tirnovanu [1,2,*], Dragos-Cristian Popescu [1,2], Norin Forna [1,3], Tudor Pinteala [1,3], Bianca-Ana Dmour [4,5], Liliana Savin [1,3], Bogdan Veliceasa [1,2,*], Alexandru Filip [1,2], Adrian Claudiu Carp [1,2], Paul Dan Sirbu [1,3] and Ovidiu Alexa [1,2]

1 Department of Orthopedics and Traumatology, Faculty of Medicine, "Grigore T. Popa" University of Medicine and Pharmacy, 700115 Iasi, Romania; awad.dmour@d.umfiasi.ro (A.D.); dragos.popescu@umfiasi.ro (D.-C.P.); norin.forna@umfiasi.ro (N.F.); tudor_pinteala@umfiasi.ro (T.P.); liliana.savin@umfiasi.ro (L.S.); alexandru-filip@umfiasi.ro (A.F.); adrian-claudiu.carp@umfiasi.ro (A.C.C.); paul.sirbu@umfiasi.ro (P.D.S.); ovdiu.alexa@umfiasi.ro (O.A.)
2 Department of Orthopaedics and Traumatology, "Sf. Spiridon" Emergency Universitary Hospital, 700115 Iasi, Romania
3 Department Orthopedics and Traumatology, Clinical Rehabilitation Hospital, 700661 Iasi, Romania
4 Department of Internal Medicine, "Grigore T. Popa" University of Medicine and Pharmacy, 700115 Iasi, Romania; bianca-ana.gherasim-dmour@d.umfiasi.ro
5 Department of III Internal Medicine Clinic, "St. Spiridon" County Clinical Emergency Hospital, 700111 Iasi, Romania
* Correspondence: stefan-dragos.tirnovanu@d.umfiasi.ro (S.-D.T.); bogdan.veliceasa@umfiasi.ro (B.V.)

Abstract: Distal radioulnar joint (DRUJ) instability is a complex condition that can severely affect forearm function, causing pain, limited range of motion, and reduced strength. This review aims to consolidate current knowledge on the diagnosis and management of DRUJ instability, emphasizing a new classification system that we propose. The review synthesizes anatomical and biomechanical factors essential for DRUJ stability, focusing on the interrelationship between the bones and surrounding soft tissues. Our methodology involved a thorough examination of recent studies, incorporating clinical assessments and advanced imaging techniques such as MRI, ultrasound, and dynamic CT. This approach allowed us to develop a classification system that categorizes DRUJ injuries into three distinct grades. This system is intended to be practical for both clinical and radiological evaluations, offering clear guidance for treatment based on injury severity. The review discusses a range of treatment options, from conservative measures like splinting and physiotherapy to surgical procedures, including arthroscopy and DRUJ arthroplasty. The proposed classification system enhances the accuracy of diagnosis and supports more effective decision making in clinical practice. In summary, our findings suggest that the integration of advanced imaging techniques with minimally invasive surgical interventions can lead to better outcomes for patients. This review serves as a valuable resource for clinicians, providing a structured approach to managing DRUJ instability and improving patient care through the implementation of our new classification system.

Keywords: distal radioulnar joint (DRUJ) instability; distal radioulnar joint (DRUJ) classification; triangular fibrocartilage complex (TFCC) injuries; imaging modalities; DRUJ arthroscopy

Citation: Dmour, A.; Tirnovanu, S.-D.; Popescu, D.-C.; Forna, N.; Pinteala, T.; Dmour, B.-A.; Savin, L.; Veliceasa, B.; Filip, A.; Carp, A.C.; et al. Advancements in Diagnosis and Management of Distal Radioulnar Joint Instability: A Comprehensive Review Including a New Classification for DRUJ Injuries. *J. Pers. Med.* **2024**, *14*, 943. https://doi.org/10.3390/jpm14090943

Academic Editors: Markus Gesslein and Eduard Kurz

Received: 8 August 2024
Revised: 28 August 2024
Accepted: 31 August 2024
Published: 5 September 2024

1. Introduction

The distal radioulnar joint (DRUJ) is a critical component of the forearm, enabling pronation and supination actions necessary for functional movements. It consists of the radius, which serves as the mobile portion of the distal part of the forearm, rotating around the ulna. The DRUJ, in conjunction with the proximal radioulnar joint, acts as a unified mechanism responsible for forearm rotation [1].

Ensuring the stability of the DRUJ is essential for maintaining optimal forearm function. This stability is maintained by the integrity of both the bones and the surrounding soft tissues, including muscles and ligaments that attach to the joint [2]. Disruptions in either the bony structures or the soft tissues can lead to joint instability, resulting in pain, restricted range of motion, and functional impairments.

Joint instability and arthritis are common pathologies that can affect the DRUJ. These conditions may occur independently or in combination, significantly impacting patient well-being [3]. Arthritis involves inflammation and degeneration of the joint, leading to pain, stiffness, and reduced joint function. Joint instability, on the other hand, refers to excessive joint movement or laxity, causing instability and functional limitations.

In this review article, we aim to provide a comprehensive analysis of the DRUJ, encompassing its anatomy, function, and various pathologies. Drawing from the latest research and clinical insights, we explore the mechanisms of DRUJ stability, highlighting the intricate interplay between bones and soft tissues. Furthermore, we delve into the etiology, clinical presentation, diagnostic approaches, and treatment modalities for DRUJ pathology.

By enhancing our understanding of the DRUJ and its associated pathologies, we can improve diagnosis, treatment, and patient outcomes. This review article aims to serve as a valuable resource for healthcare professionals, researchers, and individuals interested in the field of upper-limb orthopedics.

2. Anatomy of the Distal Radioulnar Joint

Understanding the anatomy of the DRUJ is vital for evaluating and diagnosing conditions like DRUJ instability, triangular fibrocartilage complex (TFCC) tears, and fractures. The DRUJ consists of the distal radius (specifically, the ulnar notch) and the ulnar head, the radius of curvature of which is smaller by approximately 4–7 mm [3,4]. Achieving stability in this joint requires a delicate balance between the bones, contributing to around 20% of stability, and the soft tissues, which provide passive and active stability [4,5].

The ulnar notch covers 130°, and the ulnar head covers 200° of joint surfaces with cartilage, allowing for necessary translational and rotational movements for pronation and supination [6]. The fovea is a small depression on the ulnar head where the foveal ligament attaches; it is an important stabilizer of the DRUJ and is often injured in TFCC tears [7]. The ulnar styloid, a bony projection continuing distally from the dome of the ulna, features a shallow concavity known as the fovea, which serves as a critical area for soft tissue attachment [8].

The vascular supply to the TFC (triangular fibrocartilage) is primarily derived from the ulnar artery through its palmar and dorsal radiocarpal branches, as well as the dorsal branch of the anterior interosseous artery and the palmar branch of the anterior interosseous artery. These vessels are arranged radially in relation to the fibrocartilage. Histological examinations of the TFC demonstrate vascularity in the outer 15% to 20% of the disc, while the rest of the structure is avascular. The branches originating from the ulnar artery play a significant role in supplying blood to the TFC region of the complex [9].

The interosseous membrane (IOM) serves as a delicate, fibrous structure that acts as a barrier between the radius and ulna bones in the forearm. Injuries to the IOM can arise from various factors, such as trauma, infections, or inflammatory conditions. These lesions can give rise to symptoms such as discomfort, diminished strength, and instability within the forearm [10].

The three-locker system classifies forearm joint injuries by focusing on the proximal radioulnar joint (PRUJ), mid-radioulnar joint (MRUJ), and distal radioulnar joint (DRUJ) (Figure 1). It categorizes injuries into three types based on the extent of damage to key structures like the interosseous membrane, the TFCC, and the forearm bones. This system helps in understanding the anatomical impact across the entire radioulnar joint complex and guides appropriate treatment strategies [11].

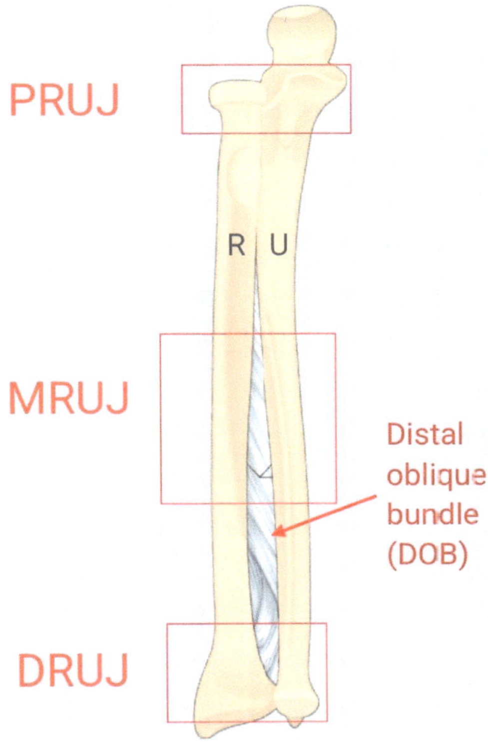

Figure 1. Anatomy of the radius (R) and ulna (U) showing the DOB and illustrating of the three-locker system concept.

In conclusion, a thorough understanding of DRUJ anatomy is fundamental for clinicians managing wrist and forearm conditions, enabling accurate diagnoses and effective treatment plans tailored to each patient's requirements.

3. Biomechanics of the Distal Radioulnar Joint (DRUJ)

Motion at the DRUJ is not purely axial but combines anterior-to-posterior translation with proximal-to-distal translation [12]. The joint allows pronation–supination movements of approximately 150° and ensures stability during joint translation [4,5]. Axial loading with wrist hyperextension and forearm pronation can cause separation between the ulna and radius at the DRUJ, leading to changes in ulnar translation and rotation [13]

Positive ulnar variance, indicating a longer distal ulna, can result in increased load transmission at the ulnocarpal joint and increase the likelihood of ulnar impaction or ulnocarpal abutment syndrome [14]. The relative length index between the distal ulna and the distal radius, known as ulnar variance or Hulten's variance, determines whether the ulna is distally longer (positive ulnar variance, when the level of the ulna is >2.5 mm beyond the radius margin at the distal radioulnar joint) or shorter (negative ulnar variance, when the ulna is ≤2.5 mm than the radius at the DRUJ) [4,9]. This variance in ulnar length relative to the radius can have implications for the stability and biomechanics of the wrist joint.

Soft tissues (both static and dynamic) contribute to DRUJ stability. Static structures such as the TFCC, the dorsal and volar radio ulnar ligaments, the collateral ulnar ligament, and the joint capsule provide stability. Dynamic muscle stabilizers include the extensor carpi ulnaris tendon and the pronator quadratus muscle [4]. The TFCC and IOM collectively provide 70% to 80% of the total stability during forearm rotation [3].

The distal oblique bundle (DOB), part of the distal portion of the IOM, plays a crucial role in providing stability to the entire forearm [15]. Additionally, the distal IOM serves as an important secondary stabilizer of the DRUJ [16]. In the context of a distal radius fracture, where the TFCC is frequently injured, the usually intact DIOM plays a central role in maintaining DRUJ stability [16].

The three-locker system also helps in assessing the biomechanical effects of DRUJ injuries by linking specific anatomical disruptions to changes in joint stability and function. By categorizing injuries, it offers a clearer understanding of their biomechanical impact, assisting in both surgical and non-surgical treatment decisions [11].

Understanding the intricate anatomy and biomechanics of the DRUJ is crucial for diagnosing and managing various pathologies affecting this joint. It helps in selecting the best therapeutic options for forearm injuries that often involve bone fractures with joint instability [17].

4. Clinical Evaluation

The diagnosis of DRUJ instability requires a thorough clinical examination and specific tests. Wrist examination is crucial for assessing DRUJ instability, particularly in patients with ligament laxity [1,3]. Careful inspection of both fists can reveal signs such as protrusion of the ulnar head, indicating dorsal displacement relative to the radius, or a dimple instead of the ulnar head, suggesting volar displacement [1,18].

In the acute setting, assessing the stability of the DRUJ involves various maneuvers. With the patient's forearm in a neutral position and the elbow stabilized, the amount of volar and dorsal translation of the distal ulna is evaluated while stabilizing the radius. Pronation and supination are also tested, with the ulna pushed toward the patient to assess the firmness of the end point and elicit any pain. Comparing these findings to those on the contralateral side is important, and audible or palpable clinking may occur during active and passive pronation and supination. Obvious subluxation of the ulnar head indicates DRUJ instability [18,19].

The DRUJ ballottement test is essential in clinical practice for evaluating the stability of the joint. It involves applying pressure to the ulnar head while stabilizing the radius to assess the mobility of the ulnar head relative to the radius. Typically performed with the forearm in a neutral position, this maneuver evaluates both active and passive pronation and supination to further gauge DRUJ stability. Recent studies utilizing advanced techniques like three-dimensional electromagnetic tracking devices have underscored the test's reliability and validity by demonstrating a significant correlation between measured DRUJ movement and clinical instability [20]. Despite variations in predictive values and sensitivity noted in studies involving specialist hand surgeons [21], the DRUJ ballottement test remains indispensable for the diagnosis of DRUJ instability and in guiding treatment decisions in clinical settings. Ongoing research and standardization efforts aim to improve the test's reliability and validity, potentially leading to more accurate diagnoses and better patient outcomes [20–22].

Chronic cases benefit from specific diagnostic tests, such as the ulnar fovea sign, which elicits pain between the ulnar styloid and the flexor tendon of the carpi ulnaris, indicative of ligament involvement [3,19]. The dorso-palmar stress test assesses instability based on the degree of ulnar head translation during firm radius stabilization [23]. Additionally, the press test, where ulno-carpal pain upon rising from an armchair suggests possible TFCC lesions, is also valuable [23,24].

The Beighton score serves as a screening tool for joint hypermobility, evaluating various joints to assess hypermobility-related issues alongside DRUJ-specific tests [13]. Integration of these tests provides a thorough evaluation of DRUJ function, facilitating accurate diagnosis and guiding appropriate treatment decisions.

5. Imaging of the DRUJ

Assessing DRUJ pathology and determining suitable treatment options may require a multimodal imaging approach due to the complex anatomy and small structures involved. Obtaining clear images of the DRUJ can be challenging due to its intricate nature [25]. A comprehensive evaluation should encompass static, dynamic, and functional relationships, as well as abnormalities affecting the bones, ligaments, and muscles within the joint [26]. The integration of automated methods addresses the limitations of manual two-dimensional measurements, leading to reduced inter-observer variability and improved accuracy in diagnosing and planning treatment for patients with DRUJ pathologies [27].

5.1. Conventional Radiography

Conventional radiography is valuable for evaluating joint instability and excluding other potential causes of wrist pain, such as fractures or arthritis [12]. However, it primarily focuses on bone imaging and may not fully assess soft tissue structures, although sometimes, DRUJ injuries can be seen via simple X-ray, as illustrated in Figure 1, which shows a forearm fracture with an associated DRUJ injury (there is an increased radioulnar distance, as well as a dorsal dislocation of the ulnar head) (Figure 2). Dynamic joint motion cannot be evaluated with static radiographic images. In cases of suspected ligamentous injuries or more complex DRUJ pathology, additional imaging modalities like MRI or CT may be necessary for a comprehensive assessment [25,26].

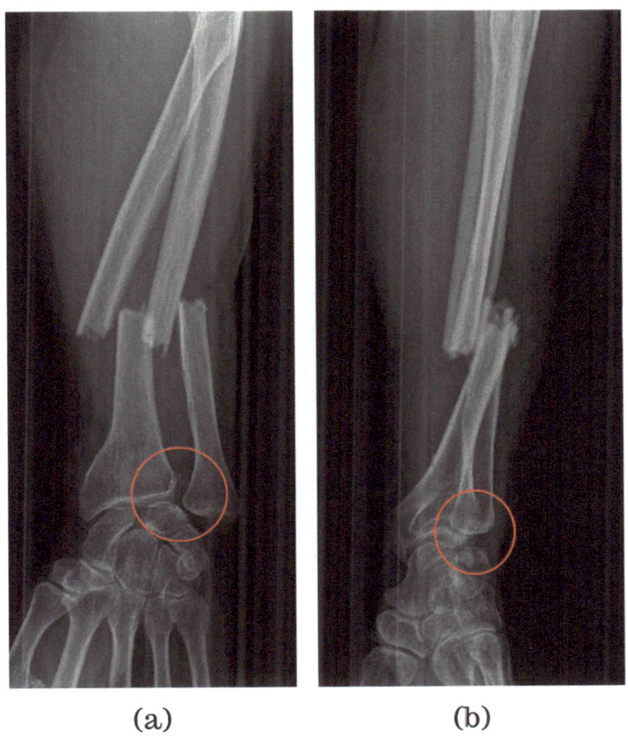

Figure 2. Forearm fracture with radiographic evidence indicating distal radioulnar joint (DRUJ) injury: (**a**) antero-posterior view (increased radioulnar space); (**b**) lateral view (dorsal dislocation of the ulnar head).

5.2. Magnetic Resonance Imaging (MRI)

Magnetic Resonance Imaging (MRI) is the most sensitive imaging modality for diagnosing TFCC tears and associated abnormalities. Its excellent soft tissue resolution allows for accurate visualization and characterization of TFCC tears, and it is invaluable for evaluating structures involved in DRUJ stability, such as the dorsal and volar radioulnar ligaments, especially after fractures (Figure 3). MRI also helps identify bone contusions and soft tissue injuries around the joint. However, MRI has limitations for patients with contraindications like metal implants or claustrophobia and may not be as effective as radiography or CT in assessing bone-related pathologies (Figure 4). Expertise in musculoskeletal imaging is crucial for proper interpretation of MRI findings [12,25,26].

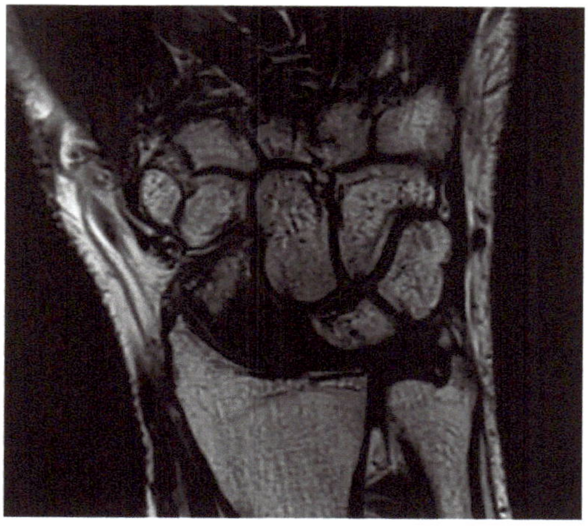

Figure 3. Wrist MRI evaluation of the soft tissues (intact TFCC) following a scaphoid nonunion fracture.

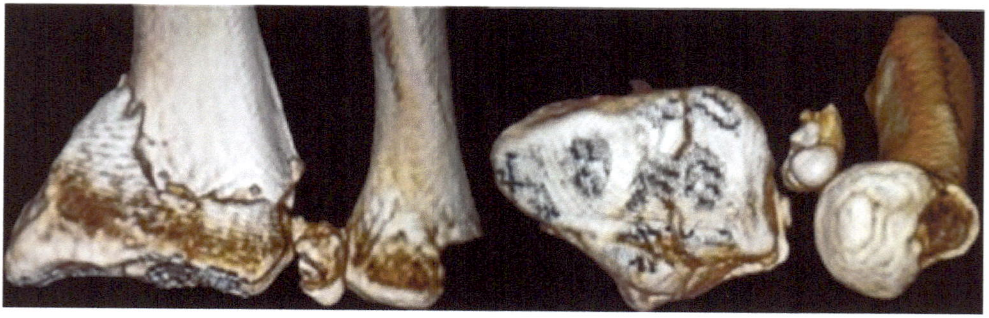

Figure 4. 3D reconstruction of an articular distal radius fracture with associated ulnar styloid fracture displaced inside the DRUJ.

5.3. Ultrasound Imaging (USI)

Ultrasound Imaging (USI) is a promising, noninvasive modality for detecting translational movement in the DRUJ among healthy individuals, diagnosing interosseous membrane (IOM) tears [10], and assessing tendon disorders like tenosynovitis. USI offers real-time imaging during active range-of-motion exercises, enabling earlier detection of DRUJ instability and facilitating more effective treatment options [12,28]. USI also demonstrates high sensitivity in detecting screw protrusion following volar plating of the distal radius and aids in diagnosing associated tendon disorders [29,30].

5.4. Dynamic Imaging Techniques (Stress Radiography and 4D-CT)

Dynamic imaging techniques (stress radiography and 4D-CT) offer valuable insights for the evaluation of DRUJ instability during active range-of-motion exercises. Dynamic CT (4D-CT) captures multiple images of the joint during motion, generating a dynamic 3D representation that provides detailed imaging of the carpal bones and ligaments, identifying subtle abnormalities not visible on static images and playing a crucial role in surgical planning and treatment decisions. However, dynamic CT exposes patients to higher radiation levels than other imaging modalities and should be used with appropriate precautions [25,26,31].

5.5. Dual Fluoroscopic and Computed Tomography (DFIS-CT)

Dual Fluoroscopic and Computed Tomography (DFIS-CT) involves capturing dynamic fluoroscopic images and simultaneously acquiring high-resolution computed tomography scans during active wrist motion. This approach allows for real-time evaluation of carpal kinematics, providing detailed 3D imaging of the carpal bones and ligaments. Future studies with larger sample sizes and comparisons with established gold-standard imaging techniques are needed to establish the accuracy and reliability of DFIS-CT [26,32].

A summary of different imaging techniques and their applications is provided in Table 1.

Table 1. Comparison of imaging techniques for DRUJ instability evaluation.

Imaging Technique	Key Advantages	Applications
Conventional radiography	High-quality bone imaging; Cost-effective and widely available; Quick acquisition for emergency situations [12].	Evaluation of fractures, dislocations, and degenerative changes in the DRUJ [12,25].
Magnetic resonance imaging (MRI)	Excellent soft tissue resolution; Differentiation of acute from chronic injuries; Non-invasive and preferred for soft tissue assessment [24].	Diagnosis of TFCC tears and associated abnormalities; Assessment of structures involved in DRUJ stability; Identification of bone contusions and soft tissue injuries [12,24].
Ultrasound imaging (USI)	Real-time imaging during active movements [12]; Detection of TFCC tears with high sensitivity [27]; Ability to detect screw protrusion post volar plating [30].	Detection of translational movement in the DRUJ among healthy individuals [27]; Diagnosis of interosseous membrane (IOM) tears [10]; Assessment of tendon disorders, such as tenosynovitis [29].
Dynamic imaging techniques (stress radiography and 4D-CT)	Provides insights during active range-of-motion exercises [24]; Offers real-time assessment of carpal kinematics; Enables comprehensive evaluation of carpal bones and ligaments [25].	Evaluation of DRUJ instability during active range-of-motion exercises; Identification of subtle abnormalities not visible in static images; Surgical planning and treatment decisions [24,25].
DFIS-CT (dual fluoroscopic and computed tomography)	Captures dynamic fluoroscopic and CT images; Real-time evaluation of carpal kinematics [25,31].	In vivo assessment of 3D carpal kinematics; Evaluation of SLIL repair failure and other carpal abnormalities; Potential low-dose alternative for wrist kinematics assessment [25,31].

While MRI remains the most useful diagnostic tool in diagnosing DRUJ instability and TFCC injuries [33,34], wrist arthroscopy stands out as the "gold standard" for identifying injuries to the TFCC (which is the main stabilizer of the DRUJ). It is the sole diagnostic tool capable of dynamically assessing the degree of instability and the healing potential of the injury, serving dual roles in both diagnosis and repair of the affected TFCC [35,36]. Wrist arthroscopy consistently demonstrates higher diagnostic accuracy than MRI [31].

6. Diverse Injuries Associated with the Distal Radioulnar Joint (DRUJ)

6.1. Isolated DRUJ Dislocation

Isolated acute dislocation of the distal radioulnar joint (DRUJ) is rare but noteworthy, with cases reported in the medical literature [19]. Unlike dislocations with fractures of the radius or distal ulna, isolated DRUJ dislocations are less common [37]. Dorsal dislocations present with a dorsal ulnar prominence and restrict supination, while volar dislocations manifest with a volar ulnar prominence and hinder pronation [19]. The DRUJ's stability is maintained by various ligamentous structures, including the distal oblique bundle (DOB) [5]. These injuries can result from a fall on an outstretched hand or a direct blow to the wrist [7]. Athletes in sports requiring repetitive forearm movements, like tennis and golf, are at increased risk [38]. Treatment may involve closed reduction techniques under anesthesia, followed by immobilization with a splint or cast, and physical therapy [19]. In cases of instability, Kirschner wires may be inserted, or arthroscopic repair of the TFCC can be performed [39]. Due to its rarity, further research is needed for better understanding and management.

6.2. TFCC Injuries

The TFCC is essential for wrist stability and movement, consisting of the dorsal radioulnar ligament (DRIL), palmar radioulnar ligament (PRIL), and meniscus homologue (MH). It stabilizes the DRUJ and acts as a shock absorber [17,40].

TFCC tears can occur due to traumatic events (e.g., twisting or falls), degenerative changes, avulsion (forceful pulls), or ischemic conditions (reduced blood flow) [40]. These tears are relatively common, affecting 1–2 people per 100,000 annually, and may require surgical treatment if not healed spontaneously [19].

Foveal TFCC tears resulting from recurrent stress can lead to perforation and erosion of the ulnar carpus [41,42]. Schmitt et al. introduced the "CUP" classification for TFCC lesions in 2022, although it does not consider adjacent structures like the DRUJ and ulnar carpus [43]. The "iceberg concept" proposed by Atzei and Luchetti describes the TFCC's anatomy, highlighting its distal and proximal components, and notes other structures contributing to DRUJ stability, including the ulnar collateral ligaments, extensor carpi ulnaris tendon, pronator quadratus muscle, and IOM [2,44].

Diagnostic maneuvers for foveal TFCC tears include the ulnocarpal stress test, fovea sign, DRUJ instability test, and piano-key sign [7]. Early diagnosis and treatment of TFCC injuries are crucial to prevent chronic pain, instability, and osteoarthritis [22]. Non-surgical treatment, which is often effective for patients without DRUJ instability, includes rest, immobilization, physical therapy, and NSAIDs, with a 6-month conservative period before considering surgery [22,45].

Timely management of TFCC injuries is vital for alleviating pain, restoring stability, and improving wrist function, enhancing patients' quality of life. Healthcare professionals should prioritize early recognition and treatment, along with patient education on prevention and early reporting of wrist issues [22].

6.3. Ulnar Styloid Fractures

Ulnar styloid fractures, often associated with distal radius fractures (DRFs), impact forearm rotation and stability [46]. These fractures are classified as tip and base fractures, aiding in their management [47,48]. The necessity of surgical fixation remains debated, with studies showing mixed results in terms of outcomes (Figure 5) [43,49,50]. Early detection and treatment are crucial, with nonunion repair or fragment excision sometimes being necessary [51,52].

While it was previously believed that ulnar styloid fractures might contribute to DRUJ instability, recent research suggests otherwise. The union status of the ulnar styloid, whether at the tip or base, does not affect the wrist's range of motion or functional status. However, temporary immobilization of the DRUJ may facilitate the recovery of the TFCC [53,54].

When considering treatment options, improvements in function and symptoms can be achieved through nonunion repair or fragment excision, possibly combined with TFCC repair, tailored to the size and location of the ulnar styloid fracture [53,55]. Delayed treatment can lead to complications, such as resorption of the ulnar styloid [51]. Further research is needed to optimize management strategies [43,46,49,52].

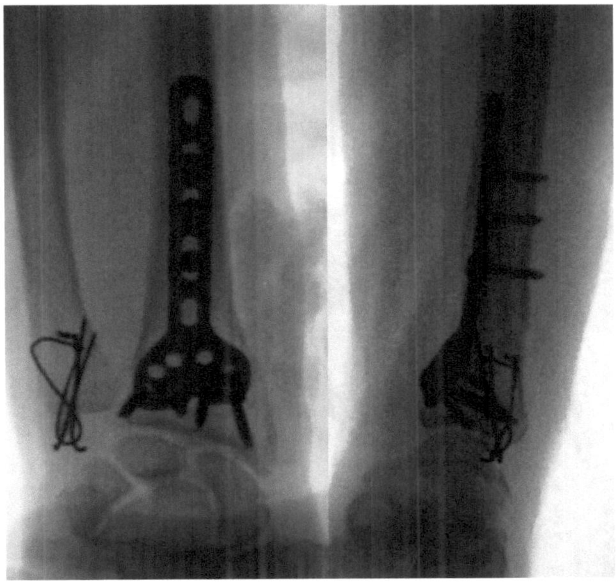

Figure 5. ORIF of a distal radius fracture associated with an ulnar styloid fracture.

6.4. Ulnar Impaction Syndrome

Ulnar impaction syndrome results from repetitive wrist loading, causing excessive pressure on the ulnar side and leading to changes primarily affecting the ulnar head and the lunate's proximal ulnar corner [54]. This increased load can cause symptoms like ulnar abutment syndrome, proximal migration of the radius, and longitudinal instability, as often seen in Essex–Lopresti fractures [38].

Diagnosis is crucial for management and can be made using imaging techniques like X-rays or MRI scans, which identify structural abnormalities and assess severity [7]. Bone scans are also valuable when combined with radiographic assessments [56].

Treatment depends on the condition's extent and impact on wrist function. Conservative management may involve rest, immobilization, and anti-inflammatory medication [7]. Severe cases with persistent symptoms may require surgical intervention. The main surgical options are ulnar-shortening osteotomy and the wafer procedure, performed either openly or arthroscopically [56]. These procedures reduce ulnar side load and restore wrist stability [7]. Early and accurate imaging is essential in planning treatment and preventing long-term joint complications.

6.5. Distal Radius Fractures

Distal radius fractures (DRFs) are common, with a significant percentage leading to DRUJ instability and symptomatic wrist pain [41]. Proper reduction and positioning of hardware during initial treatment are crucial, with immobilization or surgical intervention sometimes being necessary [8]. Intra-articular DRFs are often associated with ligamentous injuries, contributing to DRUJ instability [57]. Comprehensive evaluation and management of these fractures are essential to prevent complications and ensure optimal recovery.

6.6. Essex–Lopresti Injuries

Essex–Lopresti injuries involve a fracture of the radial head, dislocation of DRUJ, and rupture of IOM, leading to forearm instability [7,38]. These injuries often result from high-energy impacts, such as falls on an outstretched hand, and affect the proximal, middle, and distal radioulnar joints [38,58].

Estimates suggest the Essex–Lopresti Lesions (ELLs) occur in about 1% of radial head fractures, although recent evidence indicates that the prevalence might be closer to 5% [59]. Diagnosis is crucial for prompt management and can be made using X-rays or MRI scans, which reveal fractures, dislocations, and ligament ruptures [7,58]. Clinical signs, such as DRUJ widening or radial head migration, also aid in diagnosis [58].

Treatment typically involves surgery to restore stability and function. Procedures may include radial head replacement, DRUJ stabilization, and IOM repair or reconstruction [7,59]. In chronic cases of longitudinal radioulnar dissociation (LRUD), central band reconstruction of the IOM may be necessary [38].

Early and accurate diagnosis of Essex–Lopresti injuries is essential due to their complexity and potential long-term consequences. Prompt management improves outcomes and prevents disability. Healthcare professionals should maintain high suspicion in radial head fracture cases and use a multidisciplinary approach for optimal care [38,58,60].

6.7. Galeazzi Fractures

Galeazzi fractures involve a fracture of the radius and dislocation of DRUJ [7,61], comprising about 7% of forearm fractures and often resulting from high-energy trauma [11]. Symptoms include forearm swelling and deformity [62]. Various classification systems categorize these fractures based on their location relative to the DRUJ and radial styloid involvement [63–65]. In adults, radial shaft fractures with distal ulna fractures are termed Galeazzi equivalents [62].

Accurate diagnosis relies on X-rays or CT scans to evaluate the DRUJ and detect neurovascular damage, although radiographic predictors of instability may not always be reliable (Figure 6) [7,62].

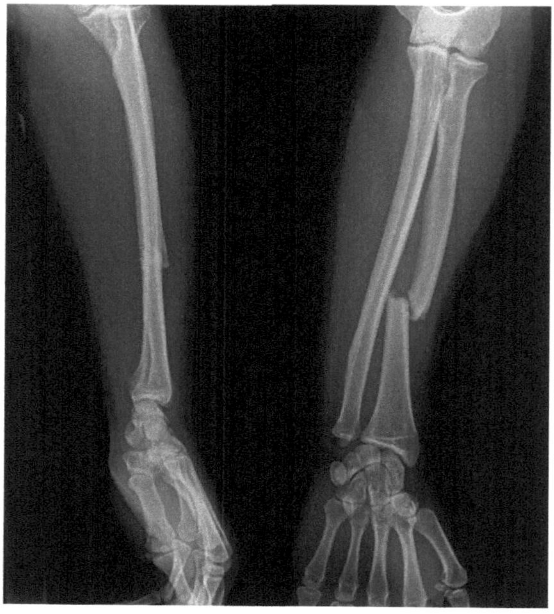

Figure 6. AP and lateral views of a Galeazzi fracture.

Treatment typically involves surgical intervention, such as open reduction and internal fixation (ORIF) and DRUJ stabilization (Figure 7) [7,65,66]. Postoperative care includes immobilization and rehabilitation, depending on DRUJ stability [62,65,67]. Complications can include pain, stiffness, instability, malunion, nonunion, and rare neurovascular damage [58,61,62].

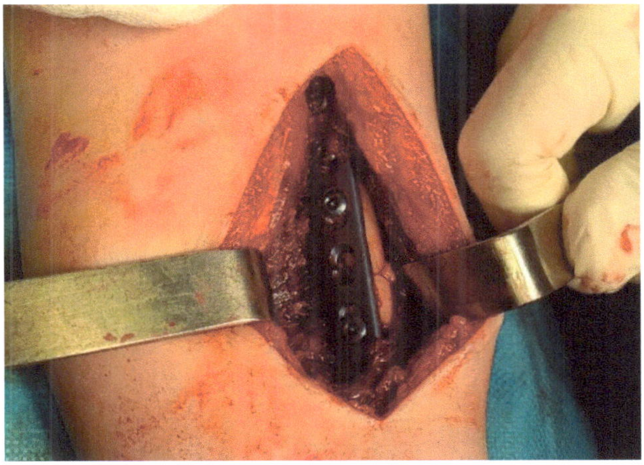

Figure 7. ORIF of a Galeazzi fracture via the Henry approach using a titanium plate.

Prompt and appropriate treatment, often involving surgery, is crucial for optimal outcomes and preventing long-term complications. A multidisciplinary approach involving orthopedic specialists and radiologists is essential for comprehensive care and recovery [7,61,62,68].

6.8. Criss-Cross Injury

Classically described as simultaneous dislocation of the radial head and DRUJ without fracture [69], criss-cross injury may also involve various fractures and/or convergent elbow dislocation [70,71]. Named for the distinctive criss-cross appearance on lateral radiographs due to ulna and radius displacement, it was first described by Leung et al. in 2002 [63]. This rare injury [69] is classified by Leung et al. into the following two types: type I (anterior displacement of radial and ulnar heads at the PRUJ and DRUJ) and type II (posterior displacement of radial and ulnar heads at the PRUJ and DRUJ) [68].

Treatment involves reducing both the PRUJ and DRUJ by closed reduction or surgery, followed by early rehabilitation. Most patients regain good forearm function with either conservative or surgical treatment [70].

6.9. Chronic Instabilities and DRUJ Arthritis

Chronic instability of the DRUJ can lead to joint degeneration and arthritis, often triggered by trauma, rheumatoid arthritis, or congenital abnormalities [17,72]. DRUJ stability is crucial for grip strength and forearm movement and function, relying on joint surface alignment and soft tissue integrity within the TFCC, DRUJ capsule, and IOM [12].

Instability can cause ulnar impaction syndrome due to repetitive wrist loading, leading to ulnar lengthening and joint wear [7,43]. Improperly managed Galeazzi fractures and Essex–Lopresti injuries can also result in chronic DRUJ instability, potentially leading to arthritis [61,62,66].

Diagnosis involves imaging techniques like MRI or arthroscopy (Figure 8) [7,66]. Treatment options include conservative management or surgical intervention, such as

DRUJ arthroplasty or arthroscopy. There is no single universally accepted treatment algorithm, so individualized plans are necessary [64,65,72].

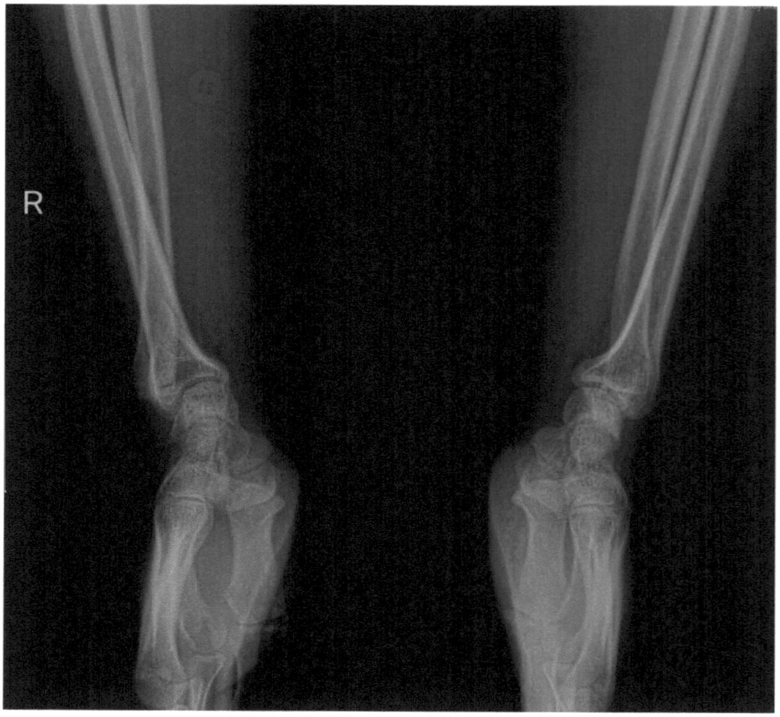

Figure 8. Dorsal dislocation of the right ulnar head following a distal radius fracture managed with closed reduction and immobilization.

In conclusion, chronic DRUJ instability can lead to joint degeneration and arthritis, highlighting the importance of early and proper management. A comprehensive, multidisciplinary approach is essential for effective treatment and joint preservation [61,62,72].

An overview of these injuries, including their mechanisms, diagnostic approaches, treatment options, and prevalence/incidence rates is shown in Table 2.

Table 2. Main injuries associated with DRUJ Instability.

Type of Injury	Mechanism	Diagnosis and Imaging	Treatment	Prevalence/Incidence
Isolated DRUJ dislocation	Fall on outstretched hand or direct blow to the wrist	Clinical presentation: dorso-, or volar–ulnar prominence, restricted supination or pronation. Imaging: X-rays, MRI [18,36].	Closed reduction techniques, immobilization with splint or cast, physical therapy. Surgical options: Kirschner wire insertion, arthroscopic repair of TFCC [18,38].	Rare; less common than dislocations with fractures [36].
TFCC injuries	Traumatic, degenerative avulsion, ischemic tears	Clinical presentation: wrist pain and dysfunction. Imaging: X-rays, MRI, bone scans. Physical testing maneuvers: ulnocarpal stress test, fovea sign, etc. [7,39].	Conservative management: rest, immobilization, NSAIDs, physical therapy [21]. Surgical options: arthroscopic or open repair of TFCC [44].	Incidence: 1–2 per 100,000 people per year [18].

Table 2. *Cont.*

Type of Injury	Mechanism	Diagnosis and Imaging	Treatment	Prevalence/Incidence
Ulnar styloid fractures	High-energy trauma, falls on outstretched hand	Clinical presentation: forearm rotation and stability affected. Imaging: X-rays, CT scans [45].	Classification and diagnosis based on location and pattern type [47]. Treatment options: conservative management, surgical fixation, ulnar shortening osteotomy [45,51].	21–61% of DRFs have ulnar styloid fractures [45].
Ulnar impaction syndrome	Repetitive wrist joint loading	Imaging: X-rays, MRI. Bone scans may also be used [7,54].	Conservative management: rest, immobilization, anti-inflammatory medication [7]. Surgical intervention: ulnar shortening osteotomy, wafer procedure [73].	Common cause of ulnar sided wrist pain [73].
Distal radius fractures	Falls, high-energy trauma	Clinical presentation: wrist pain, reduced range of motion. Imaging: X-rays, CT scans [8].	Management: proper reduction, immobilization, surgical intervention when necessary [8]. Complications: associated with DRUJ instability and ligamentous injuries [40,56].	Common injuries: DRUJ instability in 2–37% of cases [40].
Essex–Lopresti injuries	High-energy loads applied axially to the forearm	Imaging: X-rays, MRI scans. Clinical signs: DRUJ widening, radial head migration, subluxation of ulna [7].	Surgical procedures: radial head replacement, DRUJ stabilization, interosseous ligament repair or reconstruction [7,59], multidisciplinary approach for optimal care [37].	Estimated prevalence: 1% to possibly as high as 5% [58].
Galeazzi fractures	High-energy trauma	Clinical presentation: forearm swelling, deformity [60]. Imaging: X-rays, CT scans [7].	Surgical intervention: ORIF, DRUJ stabilization [7,65,66], postoperative care and rehabilitation. Complications may include persistent pain, stiffness, instability [60].	Comprises approximately 7% of all forearm fractures [61].
Criss-cross injury	Simultaneous dislocation of radial head and DRUJ	Diagnosis: imaging (X-rays, MRI) [68,70]. Classification: types I and II based on displacement direction [67].	Treatment: closed reduction or surgery, early rehabilitation. Good prognosis with either conservative or surgical treatment [69].	Extremely rare injury [68].
Chronic DRUJ instabilities	Traumatic injuries, rheumatoid arthritis	Diagnosis: imaging (MRI, arthroscopy). Treatment: conservative management or surgical intervention (arthroplasty, arthroscopy), multidisciplinary approach [17,71].	Chronic instability can lead to joint degeneration and ulnar impaction syndrome. Treatment aims to preserve joint health and function [71].	Associated with various conditions [17,71].

7. Treatment

In addressing DRUJ instability, it is crucial to understand the anatomical foundations of DRUJ stability and utilize clinical and paraclinical methods for evaluation [5].

The severity of the injury guides treatment. For mild to moderate DRUJ injuries, non-surgical approaches are common. These include activity modification, immobilization in a cast or brace, and physical therapy to reduce symptoms and improve joint stability [5,17,74]. External wristband braces are effective in stabilizing the DRUJ during various forearm rotations, reducing volar and dorsal translations [74]. Non-surgical methods are also the primary treatment for all types of TFCC injuries [75,76].

Severe cases may require surgical interventions to restore stability and function. Surgical options include repair (open or arthroscopic) or reconstruction of the TFCC or DOB, ulnar shortening osteotomy, and arthrodesis [5,22,75,76]. The choice of surgical treatment depends on the injury's specifics and the patient's overall health.

When it comes to DRUJ instability associated with fractures, after the surgical fixation of the main fracture, DRUJ pinning is an option for severe or chronic injuries, although it carries a higher risk of complications (Figure 9) [60]. Immobilization is suitable for milder injuries or patients not fit for surgery. The choice depends on factors like injury severity, chronicity, patient health, surgeon experience, and patient preferences. An experienced orthopedic surgeon should evaluate to determine the most appropriate approach for Essex–Lopresti injuries [59,60].

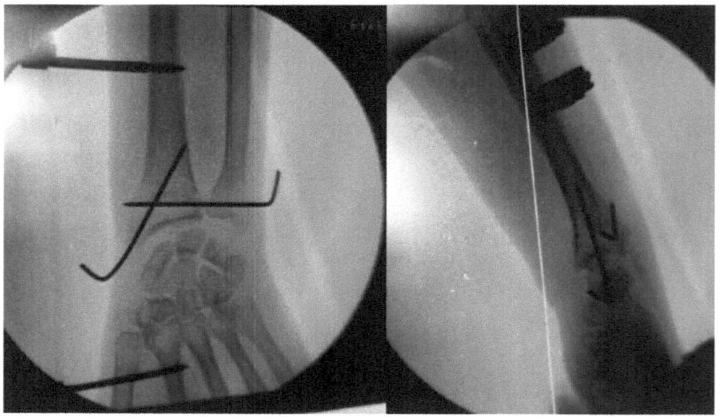

(a) (b)

Figure 9. Intraoperative images of DRUJ pinning after distal radius fracture fixation with K wires and an external fixator: (**a**) antero-posterior view; (**b**) lateral view.

According to recent research, trans fixation of the DRUJ should be conducted in a position where optimal reduction is attainable. Based on the findings of a recent study, this position typically corresponds to supination [67].

Gaspar et al. proposed a treatment involving the reconstruction of the IOM with the Mini TightRope Device® (Arthrex, Inc., Naples, FL, USA), ulnar shortening osteotomy, and simultaneous arthroscopic TFCC repair, reporting improvements in motion, grip strength, and QuickDASH scores [77]. Reconstructing the central band (CB) in Essex–Lopresti injuries restores stability, reduces distal ulnar impaction forces, and alleviates stress on the radio-capitellar joint, although complications can occur [38,58].

Both open and arthroscopic TFCC repair is used with success, achieving similar outcomes in term of DRUJ stability [78]. Most techniques utilize micro-sutures anchors or trans-osseous sutures, as well as rein-type capsular sutures, with similar functional results achieved across all techniques [28,76,78,79].

Arthroscopy is a minimally invasive surgical procedure that can be used to directly visualize the TFCC [40,78,80]. Recent anatomical studies have highlighted the proximity of the transverse branch of the dorsal branch of the ulnar nerve to the portals used in DRUJ arthroscopy, raising concerns about safety [81,82]. However, the advantage of DRUJ arthroscopy is that it allows for the visualization of the deep parts of the TFCC and its insertion area in the fovea, which cannot be adequately visualized with other methods [76,82]. This expanded visualization capability has increased the potential applicability of DRUJ arthroscopy.

Arthroscopically assisted reduction of dislocated or subluxated DRUJ has shown promising results, with minimal complications compared to traditional open reduction techniques [17]. Arthroscopy also affords the opportunity to identify and subsequently address associated soft tissue injuries, particularly to the scapholunate interosseous ligament (SLIL), lunotriquetral interosseous ligament (LTIL), and TFCC, which are common in intra-articular distal radius fractures [57].

Despite initial concerns regarding the proximity of the ulnar nerve during DRUJ arthroscopy, recent anatomical studies underscore its advantages in visualizing the TFCC and its foveal insertion. However, outcomes of arthroscopic debridement for TFCC tears may vary based on the tear type and ulnar-plus variance, potentially affecting postoperative satisfaction [25]. A recent study comparing the outcomes of suture anchor and rein-type capsular suture TFCC repair showed that the two techniques achieve similar results with satisfactory outcomes [28].

DRUJ arthroplasty involves replacing the joint with a prosthesis and is indicated for patients with severe DRUJ instability who have not responded to other treatments [41,83]. There are two main types of DRUJ prosthesis, namely total DRUJ replacement and hemi-arthroplasty. The choice of prosthesis depends on the patient's individual anatomy and needs. Total DRUJ arthroplasty distinguishes itself from other resection arthroplasties and ulnar head replacements by replacing all elements of the DRUJ, including the ulnar head, TFCC, and sigmoid notch [83]. While DRUJ arthroplasty can provide relief from pain and improve the range of motion, patients should be aware of potential complications, such as infection, nerve damage, and prosthesis loosening [41]. The use of a DRUJ arthroplasty prosthesis is a viable alternative to other DRUJ salvage procedures [83].

Several "palliative" surgical procedures are available to address ulnar head pathology, including the Darrach procedure, hemiresection procedures, the Sauvé–Kapandji procedure, and total ulnar head arthroplasty. However, these procedures may have drawbacks, such as altering the relationship with the DRUJ or leading to an unstable ulnar stump despite soft tissue stabilization procedures [83,84] (Table 3).

Table 3. Palliative surgical procedures for DRUJ.

Bower's Procedure	This method includes the removal of a significant part of the ulnar head while keeping the ulnar styloid, TFCC, and ulnar cortical column intact. It ensures continuity between the ulnar column and radius, providing support from the ulnar styloid and TFCC.
Sauvé–Kapandji Procedure	This technique preserves stability on the ulnar aspect of the joint by keeping the ulnar head intact and securing it with screws to the ulnar notch of the radius. To restore forearm rotation, a segment of the ulna is removed at the ulnar neck.
Darrach's Procedure	Darrach's procedure involves removing the distal portion of the ulna. It was initially described by William Darrach in 1912 for the treatment of distal radius pseudarthrosis. The recommended excision site is at the upper end of the ulnar notch.

Rehabilitation after surgical treatment of DRUJ injuries typically involves a period of immobilization followed by gradual range-of-motion exercises and strengthening exercises to restore function and prevent recurrence [17,22,45,84].

Treating DRUJ instability requires a comprehensive approach considering injury severity, patient factors, and available non-surgical and surgical options. Conservative treatments are effective for mild to moderate cases, while severe cases may necessitate surgical interventions. Recent advances in arthroscopic techniques offer promising outcomes with less postoperative pain and faster recovery. IOM and CB reconstructions in Essex–Lopresti injuries provide significant benefits, but potential complications must be considered. Ultimately, the combination of conservative measures, surgical interventions, and rehabilitation is vital for the restoration of joint stability and function.

8. Suggested New Classification of DRUJ Instability by the Authors

The management of distal radioulnar joint (DRUJ) injuries is complicated by the joint's complex anatomy and the involvement of multiple stabilizing structures. Existing treatment algorithms, like the "Four-Leaf Clover" approach [85], focus on addressing all relevant components but do not provide a straightforward classification system for clinical use. Current classifications, such as those proposed by Palmer and Atzei, primarily address specific injuries, like TFCC lesions, and lack comprehensive coverage of the full spectrum of DRUJ pathologies, particularly those involving extrinsic stabilizers like the interosseous membrane and pronator quadratus [44,86]. This limitation can lead to inconsistencies in diagnosis and treatment, especially in complex cases. Moreover, the variability in osseous anatomy and the intricate nature of soft-tissue stabilizers pose additional challenges in applying these classifications uniformly across patient populations [87]. Therefore, there is a clear need for a more practical classification system that integrates the latest understanding of DRUJ biomechanics and pathology, facilitating better diagnostic and treatment outcomes [86,87].

Based on our review and existing literature, we devised this classification system to improve the diagnostic approach for DRUJ injuries, integrating clinical evaluations and imaging methods. This system aims to streamline treatment decisions, as detailed in Table 4 and demonstrated in Figure 10. This classification can be applied using basic clinical and X-ray images, along with advanced imaging techniques, to accurately determine the severity of the injury.

Table 4. DRUJ injury classification.

Grade	Imaging Findings	Clinical Presentation	Treatment
Grade 1	- X-ray/CT: normal joint congruency - US/MRI: ligamentous integrity may appear intact	- Occasional discomfort or mild limitation of wrist movement	- Conservative management: - Activity modification - Immobilization - Physical therapy
Grade 2	- X-ray/CT: joint subluxation or incongruity may be present in stress tests or 4D-CT scans - US/MRI: Partial ligamentous tears or attenuations (TFCC/DOB/volar and dorsal radioulnar ligaments)	- Persistent pain, occasional clicking or catching sensations, mild to moderate limitation of wrist movement	- Conservative management: - Immobilization - Physical therapy - Possible surgical intervention (DRUJ pinning, TFCC repair) depending on severity and chronicity of symptoms.
Grade 3	- X-ray/CT: severe joint incongruity/dislocation - US/MRI: complete ligamentous tears (complex TFCC, DOB, and pronator quadratus tears) and joint dislocation	- Persistent pain, significant functional impairment, instability during daily activities - Possible signs of nerve compression or vascular compromise.	- Surgical intervention: Repair or reconstruction of ligaments (TFCC and DOB), ulnar shortening osteotomy, arthrodesis, or DRUJ arthroplasty depending on injury and patient's health.

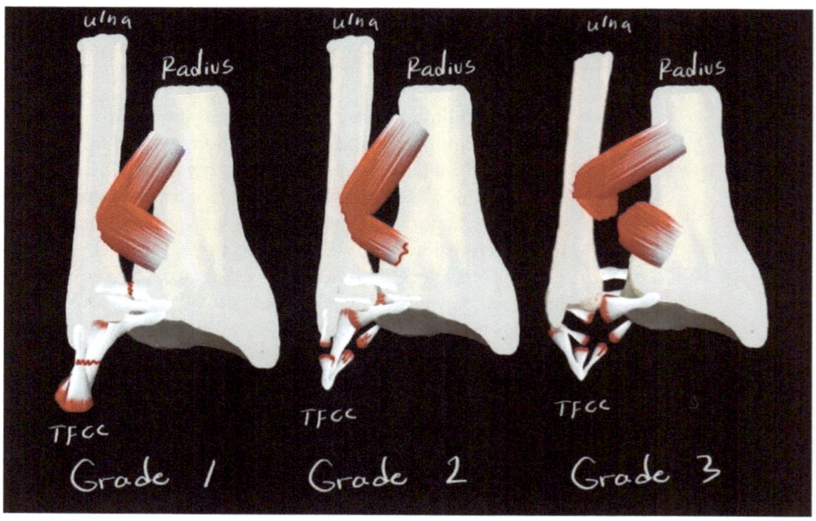

Figure 10. Illustration of DRUJ injury classification. Grade 1 shows normal joint alignment, with intact ligaments. Grade 2 displays subluxation and partial ligament tears. Grade 3 depicts severe dislocation with complete tears of the TFCC, as well as the DOB.

9. Discussion and Future Stances

DRUJ instability and TFCC pathologies pose significant challenges in the field of orthopedic medicine. Over the years, efforts have been made to better understand the anatomical foundations of DRUJ stability and develop effective treatment approaches. In this academic discussion, we delve into the latest developments in the management of DRUJ instability, with a particular focus on emerging technologies and innovative approaches that promise to revolutionize the diagnosis, treatment, and overall outcomes of patients with DRUJ instability and TFCC injuries.

One notable advancement in the management of DRUJ instability is alternative classification systems for forearm fracture–dislocations, such as the three-locker system classification, which aims to differentiate between uncomplicated dislocations and severe dislocations that involve the forearm joint [11,61,85]. By focusing on dislocations or fracture–dislocations affecting two or more lockers of the forearm joint, this classification system provides a more intricate categorization of forearm fracture–dislocations, enhancing the understanding of various injury patterns and enabling tailored treatment approaches. This advancement is crucial in guiding clinicians to adopt a more nuanced and personalized approach when managing DRUJ instability cases [11].

In the context of evaluating the TFCC using imaging techniques, ultrasonography (US) has emerged as a promising modality [12]. US complements other imaging modalities such as radiography and MRI, offering high inter-observer reliability in detecting translational movement of the DRUJ in healthy individuals. This suggests the potential application of US in future musculoskeletal assessments and highlights its utility in accurately diagnosing and assessing DRUJ instability.

Moreover, the integration of automated 3D methods for quantifying DRUJ morphology has shown promise in enhancing surgical planning and treatment outcomes for various joints in the body [27]. By leveraging these advanced imaging technologies, clinicians can gain a deeper understanding of the complex anatomical structures involved in DRUJ instability and TFCC injuries, leading to more precise and effective treatment decisions.

Innovative imaging techniques have also addressed the challenges encountered when evaluating the TFCC using grayscale ultrasound. Power Doppler, a specialized ultrasound technique that detects blood flow, has proven valuable in improving the evaluation and diagnosis of TFCC injuries [88]. By assessing the presence of increased vascularity at the deep border of the ulnocarpal ligament and the outer margin of the TFCC, along with other indications, such as joint widening and visible gaps, clinicians can enhance diagnostic accuracy. Additionally, magnetic resonance arthrography (MRA) has emerged as another valuable imaging tool in assessing TFCC injuries [89,90]. MRA provides detailed images of the TFCC's structure and its relationship with surrounding tissues, aiding in identifying subtle tears and occult injuries that may not be apparent in conventional MRI.

The incorporation of these innovative imaging techniques and technologies in the assessment of DRUJ and TFCC conditions holds great promise for advancing the field of musculoskeletal diagnostics and treatment planning. The ability to accurately visualize and assess these complex structures will undoubtedly lead to better-informed treatment decisions and improved patient outcomes.

In addition to advancements in imaging, research into minimally invasive techniques for DRUJ conditions is gaining momentum. Emerging methods such as arthroscopy-assisted procedures and percutaneous fixation techniques are being explored to offer less invasive and efficient treatment options. Arthroscopic surgery continues to gain popularity as an effective treatment option for DRUJ instability, offering advantages like minimal invasiveness, reduced soft tissue damage, and quicker recovery. Anatomical reconstruction of the distal radioulnar ligaments (DRULs) through a tunnel created between the ulnar head and the sigmoid notch of the radius, along with arthroscopic repair of the TFCC, has shown promising outcomes, with the potential to enhance patient recovery and overall treatment success [91,92]. These advancements pave the way for better management of DRUJ conditions and improved patient outcomes.

Furthermore, wrist arthroscopy performed under local anesthesia without the need for a tourniquet and sedation presents a promising strategy for addressing DRUJ issues [89]. This approach offers several advantages, including reduced risks and costs associated with the procedure. Patients can remain conscious throughout the surgery, avoiding potential complications related to general anesthesia or regional blocks. The absence of a tourniquet enhances safety by minimizing the risk of nerve injury, skin necrosis, and postoperative discomfort. Moreover, wrist arthroscopy performed under local anesthesia allows patients to provide real-time feedback to the surgeon, aiding in accurate diagnosis and treatment of DRUJ problems. With positive outcomes and high patient satisfaction, this technique shows potential as an effective and cost-efficient option for managing DRUJ conditions, leading to its potentially wider adoption in clinical settings.

In conclusion, the field of orthopedics has witnessed remarkable advancements in the management of DRUJ instability and TFCC pathologies. These developments, ranging from novel classification systems to innovative imaging techniques and minimally invasive approaches, are transforming the approach to diagnosis and treatment. As emerging technologies continue to shape the field, the future of managing these challenging wrist injuries looks brighter than ever before.

It is essential to acknowledge that while these advancements hold immense potential, further research and validation are necessary to establish their widespread clinical applicability. Collaborative efforts among researchers, clinicians, and industry partners are critical to continue pushing the boundaries of orthopedic care and bringing these innovations to patients globally. Additionally, the cost-effectiveness and accessibility of these advanced technologies must be carefully considered to ensure equitable distribution and utilization in healthcare settings worldwide.

As the field of orthopedics continues to evolve, the combined efforts of clinicians, researchers, and technology developers will drive forward the frontiers of orthopedic care, ultimately benefiting patients and transforming the landscape of wrist injury management. With a comprehensive understanding of the anatomical foundations, combined with cutting-edge technologies and less invasive approaches, the future holds great promise for improved outcomes and enhanced quality of life for patients facing these complex wrist injuries. The evolving landscape of orthopedic medicine presents an exciting era of innovation and progress in the management of DRUJ instability and TFCC pathologies. As ongoing research and technological advancements continue to shape the field, the future of managing these challenging wrist injuries looks brighter than ever before.

10. Conclusions

The management of DRUJ instability remains complex, involving both conservative and surgical treatments. This review provides a thorough overview of DRUJ instability, examining various treatment approaches and future directions in the field.

Accurate evaluation of DRUJ instability requires understanding of its anatomical basis and the employment of precise diagnostic methods. Advanced imaging techniques such as high-resolution MRI and 3D-CT scanning have enhanced diagnostic accuracy and surgical planning [12,17,27]. The revised classification system offers a more comprehensive approach to complex forearm fracture–dislocations, moving beyond traditional eponyms [17,61].

Conservative treatments, including activity modification, immobilization, and physical therapy, are effective for mild to moderate cases [5,17]. Severe cases may necessitate surgical interventions, such as TFCC repair, ulnar shortening osteotomy, or arthrodesis, tailored to the individual patient's needs [5,23]. Recent advancements in arthroscopy provide benefits like reduced invasiveness and faster recovery [79,89]. Additionally, wrist arthroscopy under local anesthesia offers a promising approach, improving patient safety and allowing for real-time feedback [91].

Procedures like IOM and CB reconstruction for Essex–Lopresti injuries aim to restore stability but carry risks such as iatrogenic fractures and nerve injury. Future research should focus on optimizing these techniques and exploring new strategies [38].

Overall, integrating advanced imaging and arthroscopic techniques, along with ongoing technological innovations, holds promise for better management of DRUJ instability. Continued advancements are expected to improve patient outcomes and care quality.

Author Contributions: A.D., S.-D.T., D.-C.P., N.F., T.P., B.-A.D., L.S., B.V., A.F., A.C.C., P.D.S. and O.A. contributed equally to this article. All authors have read and agreed to the published version of the manuscript.

Funding: This research received no external funding.

Institutional Review Board Statement: Not applicable.

Data Availability Statement: Not applicable.

Conflicts of Interest: The authors declare no conflicts of interest.

References

1. Poppler, L.H.; Moran, S.L. Acute Distal Radioulnar Joint Instability: Evaluation and Treatment. *Hand Clin.* **2020**, *36*, 429–441. [CrossRef]
2. Haugstvedt, J.R.; Langer, M.F.; Berger, R.A. Distal Radioulnar Joint: Functional Anatomy, Including Pathomechanics. *J. Hand Surg. Eur. Vol.* **2017**, *42*, 338–345. [CrossRef] [PubMed]
3. Faucher, G.K.; Zimmerman, R.M.; Zimmerman, N.B. Instability and Arthritis of the Distal Radioulnar Joint: A Critical Analysis Review. *JBJS Rev.* **2016**, *4*, e3. [CrossRef]
4. Huang, J.I.; Hanel, D.P. Anatomy and Biomechanics of the Distal Radioulnar Joint. *Hand Clin.* **2012**, *28*, 157–163. [CrossRef] [PubMed]
5. Marès, O. Distal Radioulnar Joint Instability. *Hand Surg. Rehabil.* **2017**, *36*, 305–313. [CrossRef] [PubMed]
6. Tolat, A.R.; Stanley, J.K.; Trail, I.A. A Cadaveric Study of the Anatomy and Stability of the Distal Radioulnar Joint in the Coronal and Transverse Planes. *J. Hand Surg. Br.* **1996**, *21*, 587–594. [CrossRef] [PubMed]
7. Nakamura, T.; Sato, K.; Okazaki, M.; Toyama, Y.; Ikegami, H. Repair of Foveal Detachment of the Triangular Fibrocartilage Complex: Open and Arthroscopic Transosseous Techniques. *Hand Clin.* **2011**, *27*, 281–290. [CrossRef]
8. Nypaver, C.; Bozentka, D.J. Distal Radius Fracture and the Distal Radioulnar Joint. *Hand Clin.* **2021**, *37*, 293–307. [CrossRef]
9. Thiru, R.G.; Ferlic, D.C.; Clayton, M.L.; McClure, D.C. Arterial Anatomy of the Triangular Fibrocartilage of the Wrist and Its Surgical Significance. *J. Hand Surg. Am.* **1986**, *11*, 258–263. [CrossRef]
10. Soubeyrand, M.; Lafont, C.; Oberlin, C.; France, W.; Maulat, I.; Degeorges, R. The "Muscular Hernia Sign": An Original Ultrasonographic Sign to Detect Lesions of the Forearm's Interosseous Membrane. *Surg. Radiol. Anat.* **2006**, *28*, 372–378. [CrossRef]
11. Artiaco, S.; Fusini, F.; Sard, A.; Dutto, E.; Massè, A.; Battiston, B. Fracture-dislocations of the forearm joint: A systematic review of the literature and a comprehensive locker-based classification system. *J. Orthop. Traumatol.* **2020**, *21*, 21. [CrossRef] [PubMed] [PubMed Central]
12. Oldfield, C.E.; Boland, M.R.; Greybe, D.; Hing, W. Ultrasound Imaging of the Distal Radioulnar Joint: A New Method to Assess Ulnar Radial Translation in Forearm Rotation. *J. Hand Surg. Eur. Vol.* **2017**, *42*, 389–394. [CrossRef] [PubMed]
13. Hojo, J.; Omokawa, S.; Iida, A.; Ono, H.; Moritomo, H.; Tanaka, Y. Three-Dimensional Kinematic Analysis of the Distal Radioulnar Joint in the Axial-Loaded Extended Wrist Position. *J. Hand Surg. Am.* **2019**, *44*, 336.e1–336.e6. [CrossRef] [PubMed]
14. Schuind, F.; An, K.N.; Berglund, L.; Rey, R.; Cooney, W.P., 3rd; Linscheid, R.L.; Chao, E.Y. The Distal Radioulnar Ligaments: A Biomechanical Study. *J. Hand Surg. Am.* **1991**, *16*, 1106–1114. [CrossRef]
15. Hohenberger, G.M.; Schwarz, A.M.; Weiglein, A.H.; Krassnig, R.; Kuchling, S.; Plecko, M. Prevalence of the Distal Oblique Bundle of the Interosseous Membrane of the Forearm: An Anatomical Study. *J. Hand Surg. Eur. Vol.* **2018**, *43*, 426–430. [CrossRef]
16. Moritomo, H. The Distal Interosseous Membrane: Current Concepts in Wrist Anatomy and Biomechanics. *J. Hand Surg. Am.* **2012**, *37*, 1501–1507. [CrossRef]
17. Orbay, J.L.; Cambo, R.A. Biomechanical Factors in Stability of the Forearm. *Hand Clin.* **2020**, *36*, 407–415. [CrossRef]
18. Pang, E.Q.; Yao, J. Ulnar-Sided Wrist Pain in the Athlete (TFCC/DRUJ/ECU). *Curr. Rev. Musculoskelet. Med.* **2017**, *10*, 53–61. [CrossRef]
19. Qazi, S.; Graham, D.; Regal, S.; Tang, P.; Hammarstedt, J.E. Distal Radioulnar Joint Instability and Associated Injuries: A Literature Review. *J. Hand Microsurg.* **2021**, *13*, 123–131. [CrossRef]
20. Nagashima, M.; Omokawa, S.; Hasegawa, H.; Nakanishi, Y.; Kawamura, K.; Tanaka, Y. Reliability and Validity Analysis of the Distal Radioulnar Joint Ballottement Test. *J. Hand Surg. Am.* **2024**, *49*, 15–22. [CrossRef] [PubMed]
21. Pickering, G.T.; Fine, N.F.; Knapper, T.D.; Giddins, G.E.B. The Reliability of Clinical Assessment of Distal Radioulnar Joint Instability. *J. Hand Surg. Eur. Vol.* **2022**, *47*, 375–378. [CrossRef]
22. Jawed, A.; Ansari, M.T.; Gupta, V. TFCC Injuries: How We Treat? *J. Clin. Orthop. Trauma* **2020**, *11*, 570–579. [CrossRef] [PubMed]
23. Spies, C.K.; Langer, M.; Müller, L.P.; Oppermann, J.; Unglaub, F. Distal Radioulnar Joint Instability: Current Concepts of Treatment. *Arch. Orthop. Trauma. Surg.* **2020**, *140*, 639–650. [CrossRef] [PubMed]
24. Quadlbauer, S.; Pezzei, C.; Hintringer, W.; Hausner, T.; Leixnering, M. Clinical examination of the distal radioulnar joint. *Orthopade* **2018**, *47*, 628–636. [CrossRef] [PubMed]
25. Ng, A.W.H.; Tong, C.S.L.; Hung, E.H.Y.; Griffith, J.F.; Tse, W.L.; Wong, C.W.Y.; Mak, M.C.K.; Ho, P.C. Top-Ten Tips for Imaging the Triangular Fibrocartilaginous Complex. *Semin. Musculoskelet. Radiol.* **2019**, *23*, 436–443. [CrossRef] [PubMed]

26. Amrami, K.K.; Moran, S.L.; Berger, R.A.; Ehman, E.C.; Felmlee, J.P. Imaging the Distal Radioulnar Joint. *Hand Clin.* **2010**, *26*, 467–475. [CrossRef] [PubMed]
27. Roner, S.; Fürnstahl, P.; Scheibler, A.-G.; Sutter, R.; Nagy, L.; Carrillo, F. Three-Dimensional Automated Assessment of the Distal Radioulnar Joint Morphology According to Sigmoid Notch Surface Orientation. *J. Hand Surg. Am.* **2020**, *45*, 1083.e1–1083.e11. [CrossRef]
28. Lo, I.-N.; Chen, K.-J.; Yin, C.-Y.; Huang, H.-K.; Wang, J.-P.; Huang, Y.-C. Comparing the Outcomes of Suture Anchor Repair and Rein-Type Capsular Suture for Triangular Fibrocartilage Complex Foveal Tears with a Minimum 2-Year Follow-Up. *J. Hand Surg. Am.* **2024**, *49*, 321–328. [CrossRef]
29. Williams, D.; Singh, J.; Heidari, N.; Ahmad, M.; Noorani, A.; Di Mascio, L. Assessment of Penetration of Dorsal Screws after Fixation of the Distal Radius Using Ultrasound: Cadaveric Study. *Ann. R. Coll. Surg. Engl.* **2016**, *98*, 138–142. [CrossRef] [PubMed]
30. Sabzghabaei, A.; Shojaee, M.; Dolatabadi, A.A.; Manouchehrifar, M.; and Asadi, M. Ultrasound-Guided Reduction of Distal Radius Fractures. *Emergency* **2016**, *4*, 132–135. [PubMed]
31. Oonk, J.G.M.; Dobbe, J.G.G.; Strackee, S.D.; Strijkers, G.J.; Streekstra, G.J. Quantification of the Methodological Error in Kinematic Evaluation of the DRUJ Using Dynamic CT. *Sci. Rep.* **2023**, *13*, 3159. [CrossRef] [PubMed]
32. Harris, E.; Atthakomol, P.; Korteweg, J.; Li, G.; Hosseini, A.; Kachooei, A.; Chen, N.C. Non-invasive 3D fluoroscopic/CT imaging of the Wrist. *Orthop. J. Harv. Med. Sch.* **2021**, *21*, 40–48. Available online: https://musculoskeletalkey.com/the-kinematics-and-clinical-implications-of-the-dart-throwing-motion/ (accessed on 6 August 2024).
33. Zhao, X.; Yu, A.; Zhao, H.; Qiu, Y. Diagnostic Value of MRI in Traumatic Triangular Fibrocartilage Complex Injuries: A Retrospective Study. *BMC Musculoskelet. Disord.* **2024**, *25*, 63. [CrossRef] [PubMed]
34. Squires, J.H.; England, E.; Mehta, K.; Wissman, R.D. The Role of Imaging in Diagnosing Diseases of the Distal Radioulnar Joint, Triangular Fibrocartilage Complex, and Distal Ulna. *AJR Am. J. Roentgenol.* **2014**, *203*, 146–153. [CrossRef]
35. Petsatodis, E.; Pilavaki, M.; Kalogera, A.; Drevelegas, A.; Agathangelidis, F.; Ditsios, K. Comparison between Conventional MRI and MR Arthrography in the Diagnosis of Triangular Fibrocartilage Tears and Correlation with Arthroscopic Findings. *Injury* **2019**, *50*, 1464–1469. [CrossRef] [PubMed]
36. Phillips, R.; Choo, S.; Fletcher, B.; Nuelle, J.A.V. Dry Wrist Arthroscopy: Technique and Rationale. *Arthroscopy* **2023**, *39*, 1779–1780. [CrossRef]
37. Carlsen, B.T.; Dennison, D.G.; Moran, S.L. Acute Dislocations of the Distal Radioulnar Joint and Distal Ulna Fractures. *Hand Clin.* **2010**, *26*, 503–516. [CrossRef] [PubMed]
38. Masouros, P.T.; Apergis, E.P.; Babis, G.C.; Pernientakis, S.S.; Igoumenou, V.G.; Mavrogenis, A.F.; Nikolaou, V.S. Essex-Lopresti Injuries: An Update. *EFORT Open Rev.* **2019**, *4*, 143–150. [CrossRef]
39. Wolfe, S.W.; Pederson, W.C.; Kozin, S.H.; Cohen, M.S. *Green's Operative Hand Surgery E-Book: 2-Volume Set*; Elsevier Health Sciences: New York, NY, USA, 2021; ISBN 9780323697941.
40. Eschweiler, J.; Li, J.; Quack, V.; Rath, B.; Baroncini, A.; Hildebrand, F.; Migliorini, F. Anatomy, Biomechanics, and Loads of the Wrist Joint. *Life* **2022**, *12*, 188. [CrossRef]
41. Moradi, A.; Binava, R.; Vahedi, E.; Ebrahimzadeh, M.H.; Jupiter, J.B. Distal Radioulnar Oint Prosthesis. *Arch. Bone Jt. Surg.* **2021**, *9*, 22–32.
42. Makita, A.; Nakamura, T.; Takayama, S.; Toyama, Y. The Shape of the Triangular Fibrocartilage during Pronation-Supination. *J. Hand Surg. Br.* **2003**, *28*, 537–545. [CrossRef]
43. Zenke, Y.; Sakai, A.; Oshige, T.; Moritani, S.; Nakamura, T. The Effect of an Associated Ulnar Styloid Fracture on the Outcome after Fixation of a Fracture of the Distal Radius. *J. Bone Jt. Surg. Br.* **2009**, *91*, 102–107. [CrossRef]
44. Atzei, A.; Luchetti, R. Foveal TFCC Tear Classification and Treatment. *Hand Clin.* **2011**, *27*, 263–272. [CrossRef] [PubMed]
45. Im, J.; Kang, S.J.; Lee, S.J. A Comparative Study between Conservative and Surgical Treatments of Triangular Fibrocartilage Complex Injury of the Wrist with Distal Radius Fractures. *Clin. Orthop. Surg.* **2021**, *13*, 105–109. [CrossRef] [PubMed]
46. Wijffels, M.; Ring, D. The Influence of Non-Union of the Ulnar Styloid on Pain, Wrist Function and Instability after Distal Radius Fracture. *J. Hand Microsurg.* **2011**, *3*, 11–14. [CrossRef]
47. Logan, A.J.; Lindau, T.R. The Management of Distal Ulnar Fractures in Adults: A Review of the Literature and Recommendations for Treatment. *Strateg. Trauma Limb Reconstr.* **2008**, *3*, 49–56. [CrossRef] [PubMed]
48. Biyani, A.; Simison, A.J.; Klenerman, L. Fractures of the Distal Radius and Ulna. *J. Hand Surg. Br.* **1995**, *20*, 357–364. [CrossRef] [PubMed]
49. Almedghio, S.; Arshad, M.S.; Almari, F.; Chakrabarti, I. Effects of Ulnar Styloid Fractures on Unstable Distal Radius Fracture Outcomes: A Systematic Review of Comparative Studies. *J. Wrist Surg.* **2018**, *7*, 172–181. [CrossRef]
50. Sammer, D.M.; Shah, H.M.; Shauver, M.J.; Chung, K.C. The Effect of Ulnar Styloid Fractures on Patient-Rated Outcomes after Volar Locking Plating of Distal Radius Fractures. *J. Hand Surg. Am.* **2009**, *34*, 1595–1602. [CrossRef]
51. Shi, X.; Pan, T.; Wu, D.; Chen, R.; Lin, Z.; Pan, J. The Impact of Varus Angulation on Proximal Fractures of the Ulna. *BMC Musculoskelet. Disord.* **2018**, *19*, 103. [CrossRef]
52. Chen, A.C.-Y.; Chiu, C.-H.; Weng, C.-J.; Chang, S.-S.; Cheng, C.-Y. Early and Late Fixation of Ulnar Styloid Base Fractures Yields Different Outcomes. *J. Orthop. Surg. Res.* **2018**, *13*, 193. [CrossRef]
53. Meyer, M.A.; Leversedge, F.J. Management of Ulnar Styloid Nonunions. *Hand Clin.* **2024**, *40*, 97–103. [CrossRef] [PubMed]

54. Ozcelik, I.B.; Jusoh, M.H.; Cavit, A. Reverse Wafer Procedure for Ulnar Impaction Syndrome. *Tech. Hand Up. Extrem. Surg.* **2024**, *28*, 16–18. [CrossRef] [PubMed]
55. Ozben, H.; Uludağ, S.; Ataker, Y.; Comert Ece, S.; Gudemez, E. Distal Radius Fractures: What Are the Effects of Ulnar Styloid Fracture and Face-Toface Rehabilitation on the Functional Outcome? *Acta Orthop. Traumatol. Turc.* **2023**, *57*, 361–365. [CrossRef] [PubMed]
56. Mirza, A.; Mirza, J.B.; Zappia, L.C.; Thomas, T.L. Ulnar-Sided Wrist Pain: A Diagnostic Evaluation Guide From 30-Plus Years of Experience. *Cureus* **2024**, *16*, e53332. [CrossRef]
57. Yao, J.; Fogel, N. Arthroscopy in Distal Radius Fractures: Indications and When to Do It. *Hand Clin.* **2021**, *37*, 279–291. [CrossRef]
58. Artiaco, S.; Fusini, F.; Colzani, G.; Massè, A.; Battiston, B. Chronic Essex-Lopresti Injury: A Systematic Review of Current Treatment Options. *Int. Orthop.* **2019**, *43*, 1413–1420. [CrossRef]
59. Ahlfors, R.; Eriksen, M. Essex-Lopresti-Laesion er Hyppigt Overset. Available online: https://ugeskriftet.dk/videnskab/essex-lopresti-laesion-er-hyppigt-overset (accessed on 6 August 2024).
60. Verhiel, S.H.W.L.; Özkan, S.; Langhammer, C.G.; Chen, N.C. The Serially-Operated Essex-Lopresti Injury: Long-Term Outcomes in a Retrospective Cohort. *J. Hand Microsurg.* **2020**, *12*, 47–55. [CrossRef] [PubMed]
61. Dabas, V.; Bhatia, N.; Agrawal, G. Both Bone Forearm Fracture with Complex DRUJ Dislocation: What Are the Odds? *J. Wrist Surg.* **2023**, *12*, 81–85. [CrossRef]
62. Garg, R.; Mudgal, C. Galeazzi Injuries. *Hand Clin.* **2020**, *36*, 455–462. [CrossRef]
63. Walsh, H.P.; McLaren, C.A.; Owen, R. Galeazzi Fractures in Children. *J. Bone Jt. Surg. Br.* **1987**, *69*, 730–733. [CrossRef]
64. Rettig, M.E.; Raskin, K.B. Galeazzi Fracture-Dislocation: A New Treatment-Oriented Classification. *J. Hand Surg. Am.* **2001**, *26*, 228–235. [CrossRef]
65. Maculé Beneyto, F.; Arandes Renú, J.M.; Ferreres Claramunt, A.; Ramón Soler, R. Treatment of Galeazzi Fracture-Dislocations. *J. Trauma* **1994**, *36*, 352–355. [CrossRef] [PubMed]
66. Alajmi, T. Galeazzi Fracture Dislocations: An Illustrated Review. *Cureus* **2020**, *12*, e9367. [CrossRef] [PubMed]
67. El Barbari, J.S.; Kohlhas, L.; Franke, J.; Grützner, P.A.; Schnetzke, M.; Swartman, B J. Analyzing the Reduction Quality of the Distal Radioulnar Joint after Closed K-Wire Transfixation in a Cadaver Model: Is Supination or Neutral Position Superior? *Arch. Orthop. Trauma. Surg.* **2024**, *144*, 1603–1609. [CrossRef] [PubMed]
68. Leung, Y.F.; Ip, S.P.S.; Ip, W.Y.; Kam, W.L.; Wai, Y.L. The Crisscross Injury Mechanism in Forearm Injuries. *Arch. Orthop. Trauma. Surg.* **2005**, *125*, 298–303. [CrossRef] [PubMed]
69. Jin, X.-Y.; Zhao, W.-B.; Dong, Y.-Q.; Huang, Y.-G. Simultaneous Dislocation of the Radial Head and Distal Radio-Ulnar Joint without Fracture in an Adult Patient: A Case Report and Review of Literature. *BMC Surg.* **2020**, *20*, 71. [CrossRef]
70. Tan, J.; Xiong, J.; Zhang, X.; Yang, M.; Li, T.; Gong, M. The Diagnosis and Treatment of Criss-Cross Injury of the Forearm: A Retrospective Analysis. *Injury* **2024**, *55*, 111295. [CrossRef]
71. Fontana, M.; Rotini, M.; Battiston, B.; Artiaco, S.; Dutto, E.; Sard, A.; Colozza, A.; Vicenti, G.; Cavallo, M.; Rotini, R. Unstable Lesions of the Forearm: Terminology, Evaluative Score and Synoptic Table. *Injury* **2023**, *54* (Suppl. S1), S85–S95. [CrossRef] [PubMed]
72. Carr, L.W.; Adams, B. Chronic Distal Radioulnar Joint Instability. *Hand Clin.* **2020**, *36*, 443–453. [CrossRef]
73. Ajit Singh, V.; Jia, T.Y.; Devi Santharalinggam, R.; Gunasagaran, J. Relationship of Ulna Styloid Fracture to the Distal Radio-Ulnar Joint Stability. A Clinical, Functional, and Radiographic Outcome Study. *PLoS ONE* **2023**, *18*, e0279210. [CrossRef]
74. Shin, S.-H.; Park, T.; Hong, E.; Kwak, D.-S.; Chung, Y.-G. Stabilization of the Distal Radioulnar Joint with or without Triangular Fibrocartilage Complex Tear by an External Wrist Band Brace: A Cadaveric Study. *Healthcare* **2022**, *10*, 828. [CrossRef]
75. Yamamoto, M.; Koh, S.; Tatebe, M.; Shinohara, T.; Shionoya, K.; Hirata, H.; Nakamura, R. Importance of Distal Radioulnar Joint Arthroscopy for Evaluating the Triangular Fibrocartilage Complex. *J. Orthop. Sci.* **2010**, *15*, 210–215. [CrossRef]
76. Khair, Y.; Mustafa, A.; Mestrihi, S.; Azzam, E.; Al-Qasaimeh, M.; Awad, D.; Ovidiu, A. Outcome in TFCC Repair Using Micro Anchor and Trans-osseous Technique. *Exp. Ther. Med.* **2023**, *26*, 557. [CrossRef]
77. Gaspar, M.P.; Kane, P.M.; Pflug, E.M.; Jacoby, S.M.; Osterman, A.L.; Culp, R.W. Interosseous Membrane Reconstruction with a Suture-Button Construct for Treatment of Chronic Forearm Instability. *J. Shoulder Elb. Surg.* **2016**, *25*, 1491–1500. [CrossRef] [PubMed]
78. Andersson, J.K.; Åhlén, M.; Andernord, D. Open versus Arthroscopic Repair of the Triangular Fibrocartilage Complex: A Systematic Review. *J. Exp. Orthop.* **2018**, *5*, 6. [CrossRef]
79. Liu, E.H.; Suen, K.; Tham, S.K.; Ek, E.T. Surgical Repair of Triangular Fibrocartilage Complex Tears: A Systematic Review. *J. Wrist Surg.* **2021**, *10*, 70–83. [CrossRef] [PubMed]
80. Kim, B.; Yoon, H.-K.; Nho, J.-H.; Park, K.H.; Park, S.-Y.; Yoon, J.-H.; Song, H.S. Arthroscopically Assisted Reconstruction of Triangular Fibrocartilage Complex Foveal Avulsion in the Ulnar Variance-Positive Patient. *Arthroscopy* **2013**, *29*, 1762–1768. [CrossRef]
81. Abrams, R.A.; Petersen, M.; Botte, M.J. Arthroscopic Portals of the Wrist: An Anatomic Study. *J. Hand Surg. Am.* **1994**, *19*. 940–944. [CrossRef] [PubMed]
82. Pillukat, T.; Mühldorfer-Fodor, M.; Windolf, J.; van Schoonhoven, J. Arthroscopy of the distal radioulnar joint. *Orthopade* **2018**, *47*, 647–654. [CrossRef] [PubMed]

83. Levina, Y.; Mesa, L.; Hannon, P.J.; Coutelle, N.A.; Hess, A.V.; Garcia, M.J. Retrospective and Prospective Outcomes of Distal Radioulnar Joint Prosthesis Arthroplasty at a Single Center. *J. Hand Surg. Glob. Online* **2023**, *5*, 620–623. [CrossRef] [PubMed]
84. Rockwood, C.A.; Green, D.; Bucholz, R. *Rockwood and Green's Fractures in Adults*; Lippincott Williams & Wilkins: Philadelphia, PA, USA, 1991.
85. Kakar, S.; Garcia-Elias, M. The "Four-Leaf Clover" Treatment Algorithm: A Practical Approach to Manage Disorders of the Distal Radioulnar Joint. *J. Hand Surg. Am.* **2016**, *41*, 551–564. [CrossRef] [PubMed]
86. Flores, D.V.; Umpire, D.F.; Rakhra, K.S.; Jibri, Z.; Belmar, G.A.S. Distal Radioulnar Joint: Normal Anatomy, Imaging of Common Disorders, and Injury Classification. *Radiographics* **2023**, *43*, e220109. [CrossRef] [PubMed]
87. Hsiao, Y.N.; Chien, C.S.; Wu, T.M. Treatment of acute distal radioulnar joint instability with distal oblique bundle augmentation of the interosseous membrane by suture-button suspension: A case series. *Jt. Dis. Relat. Surg.* **2022**, *33*, 658–665. [CrossRef] [PubMed] [PubMed Central]
88. Hung, C.-Y.; Chang, K.-V.; Özçakar, L. Dynamic and Doppler Ultrasound Imaging for the Diagnosis of Triangular Fibrocartilage Complex Injury and Ulnocarpal Wrist Instability. *Am. J. Phys. Med. Rehabil.* **2016**, *95*, e111–e112. [CrossRef] [PubMed]
89. Pavić, R.; Margetić, P.; Benšić, M.; Hnatešen, D.; Kopljar, M.; Margetić, K. Comparative radiological diagnostics of acute triangular fibrocartilage complex injuries. *Acta Clin. Croat.* **2022**, *61*, 129–137. [CrossRef]
90. Ng, A.W.H.; Griffith, J.F.; Fung, C.S.Y.; Lee, R.K.L.; Tong, C.S.L.; Wong, C.W.Y.; Tse, W.L.; Ho, P.C. MR Imaging of the Traumatic Triangular Fibrocartilaginous Complex Tear. *Quant. Imaging Med. Surg.* **2017**, *7*, 443–460. [CrossRef]
91. Ho, P.-C. Arthroscopic Partial Wrist Fusion. *Tech. Hand Up. Extrem. Surg.* **2008**, *12*, 242–265. [CrossRef]
92. Bain, G.I.; Baker, A.; Whipple, T.L.; Poehling, G.G.; Mathoulin, C.; Ho, P.C. History of Wrist Arthroscopy. *J. Wrist Surg.* **2022**, *11*, 96–119. [CrossRef] [PubMed] [PubMed Central]

Journal of
Personalized Medicine

MDPI

Article

Outcome Analysis of Distal Radius Fracture with Orthosis Versus Cast Immobilization after Palmar Plate Osteosynthesis: A Randomized Controlled Study

Tim Klopfer [1,2,*], Philipp Hemmann [1], Verena Rupprecht [3,4], Fabian Stuby [1,5], Ulrich Stöckle [1,6] and Adrian Meder [1,7]

1 Department of Traumatology and Reconstructive Surgery, BG Trauma Center, Tuebingen, Eberhard Karls University Tuebingen, Schnarrenbergstraße 95, 72076 Tuebingen, Germany
2 OC Bayreuth, Parsifalstrasse 5, 95445 Bayreuth, Germany
3 Orthopedic Department, University of Tuebingen, Hoppe-Seyler-Strasse 3, 72076 Tuebingen, Germany
4 Trauma Center Klinik am Eichert, Eichertstraße 3, 73035 Goeppingen, Germany
5 BG Trauma Center Murnau, Professor-Kuentscher-Strasse 8, 82418 Murnau, Germany
6 Charité Berlin, Augustenburger Platz 1, 13353 Berlin, Germany
7 Chirurgische und Orthopädische Gemeinschaftspraxis Reutlingen, In Laisen 11, 72766 Reutlingen, Germany
* Correspondence: tim.klopfer@oc-bayreuth.de

Abstract: Although the benefits of hand orthoses were shown in previous studies, they have not been able to establish themselves in clinical routines. With a focus on patient satisfaction, this study aimed to evaluate the latest generation of hand orthoses after palmar plate osteosynthesis for isolated distal radius fractures in comparison with circular plaster casts. 50 patients (16% dropout rate) were randomly assigned to an orthotic group (immobilization by orthosis, OG) or a control group (immobilization by a plaster cast, CG). Intra-articular fractures were present in 74% of the cases, and unstable AO C3 fractures in 26%. Questionnaires on patient satisfaction, documentation of the time required, clinical scores (DASH, SF-36), range of motion, grip measurements and radiographs were used for evaluation. The OG proved to be equivalent to the plaster treatment in terms of patient satisfaction, and stability of the reduction, as well as clinical scores DASH and SF-36. The OG was even superior in terms of personal hygiene ($p = 0.011$), handling ($p = 0.008$) and better adaptability ($p = 0.013$). Significantly less time was required to apply the orthosis ($p < 0.001$). In addition to the good results achieved so far, the study showed that the latest generation of orthoses has several advantages over plaster cast therapy, and could therefore become established in everyday clinical practice.

Keywords: distal radius fracture; palmar plate osteosynthesis; hand immobilization; orthosis; circular plaster cast; patient satisfaction; Disabilities of Arm, Shoulder and Hand (DASH-Score)

Citation: Klopfer, T.; Hemmann, P.; Rupprecht, V.; Stuby, F.; Stöckle, U.; Meder, A. Outcome Analysis of Distal Radius Fracture with Orthosis Versus Cast Immobilization after Palmar Plate Osteosynthesis: A Randomized Controlled Study. *J. Pers. Med.* **2023**, *13*, 130. https://doi.org/10.3390/jpm13010130

Academic Editors: Markus Gesslein and Eduard Kurz

Received: 13 December 2022
Revised: 27 December 2022
Accepted: 27 December 2022
Published: 9 January 2023

1. Introduction

Fracture of the distal radius (DRF) is generally one of the most common fractures, representing approximately 240 per 100,000 cases, and is therefore of major importance in everyday clinical practice. There are peak rates of fracture at 340–500 cases per 100,000 persons, in both younger (5–14 years) and older people (>50 years) [1,2]. In older patients, fractures are usually associated with osteoporosis, which complicates both the surgical treatment and follow-up [3]. Surgical therapy is the established treatment for dislocated fractures, with good results, and it has been performed increasingly often in recent years [4–6]. Immobilization, at least for a short time, is usually carried out both preoperatively and postoperatively.

Complications in the treatment of DRF are not uncommon, with estimated rates of 3–36% [7–9]. They include wound healing disorders and infections (<1% in closed fractures [9], <44% in open fractures [10]), compartment syndrome (<0.3% [11]), nerve lesions (<10% [12]), complex regional pain syndrome (CRPS; <1.4% [9]), secondary dislocation/misalignment (<1.4% [9]), tendon ruptures (<1.4% [9,13]), and pseudarthrosis (<0.2% [14]). Adequate reduction and immobilization are required for decongestion and pain reduction. Immobilization in a plaster cast or splint is initially carried out after closed reduction, as well as after surgery. The period of postoperative immobilization is determined on an individual basis depending on the severity of the fracture, bone quality, and quality of care, and is usually 0–6 weeks [15–18]. Little changed in this approach for many years. Quadlbauer et al., showed that early mobilization leads to improved functional results [19]. Occasional attempts to switch to more advanced orthoses were generally not successful despite good research results—probably due to cost differences and, above all, for reasons of manageability and practicability in everyday use [20,21]. In some cases, research studies on orthoses have even been discontinued due to problems such as secondary dislocation [22].

However, there are also limitations in classic plaster cast treatment for medical staff as well as for patients. For the medical staff, the application of a circular plaster cast is time-consuming, personnel-intensive and sometimes challenging—especially on days with high patient volume and extreme strains on staff. From a clinical point of view, a modern orthosis should therefore be easy to apply, allow uncomplicated radiographic control, and still be comparable to the gold standard cast in terms of the quality of immobilization. From the patient's point of view, a modern orthosis should also be easy to use, allow mobilization of the free joints, and provide a better sense of hygiene. Therefore, the aim of the present study is to analyze the results of a modern orthosis compared with plaster immobilization after palmar plate osteosynthesis in distal radius fractures. For this, patient satisfaction is defined as the primary end point while retention of reduction, regaining functionality and the time required for application are defined as secondary end points.

2. Materials and Methods

The study was conducted in accordance with the Declaration of Helsinki and approved in the framework of "good clinical practice" by the Ethics Committee of the University of Tübingen (Project Number 173/2017BO1, 24.05.2017). The study is registered in the German Registry for Clinical Studies (Deutsches Register für klinische Studien, DRKS), number DRKS00012933. Furthermore, the study was designed according to the Consolidated Standards of Reporting Trials (CONSORT). In a monocenter, parallel group design, fifty participants were included between December 2017 and August 2019 at BG Trauma Center Tuebingen, Germany. The sample size was calculated a priori assuming an effect size of 0.9. The sample size calculation was supported by the center for clinical studies at the university of Tuebingen. Included were male and female patients between the ages of 18 and 80 years who had suffered a DRF with an AO classification 23-A2 to C3 and were scheduled for open reduction and isolated palmar plate osteosynthesis. Patients were excluded when they met one of the following criteria: (1) planned conservative therapy, (2) closed reduction not possible, (3) open fractures, (4) additional fractures in other locations and multiple injuries, (5) higher-grade soft-tissue trauma (G2 and G3 in Tscherne–Oestern classification), (6) cancer and pathological fractures, (7) initial loss of sensitivity or reduced sensation, (8) neurodegenerative diseases, including dementia, (9) wrist deformity, (10) planned surgical treatment with external fixator and/or dorsal plate osteosynthesis and, (11) lack of compliance.

Randomization was performed with a computer using a random number generator, labeling 50 participation certificates before the trial started. These were folded and sealed in numbered envelopes from 1–50. These envelopes were stored sequentially and numbered in a folder in the emergency department. Opening occurred after completing patient education and obtaining written consent to the study from the doctor in the emergency department. The patients were randomly assigned either to the orthosis group (OG) or the control group (CG) and were given a corresponding identification number and anonymized for the study. Since the type of splint—orthosis or plaster cast—is visible, double blinding was only possible until the time of immobilization. However, until the time of reduction and subsequent fixation, neither the staff nor the patient knew the path of immobilization. The study protocol is shown in the CONSORT flow diagram (Figure 1).

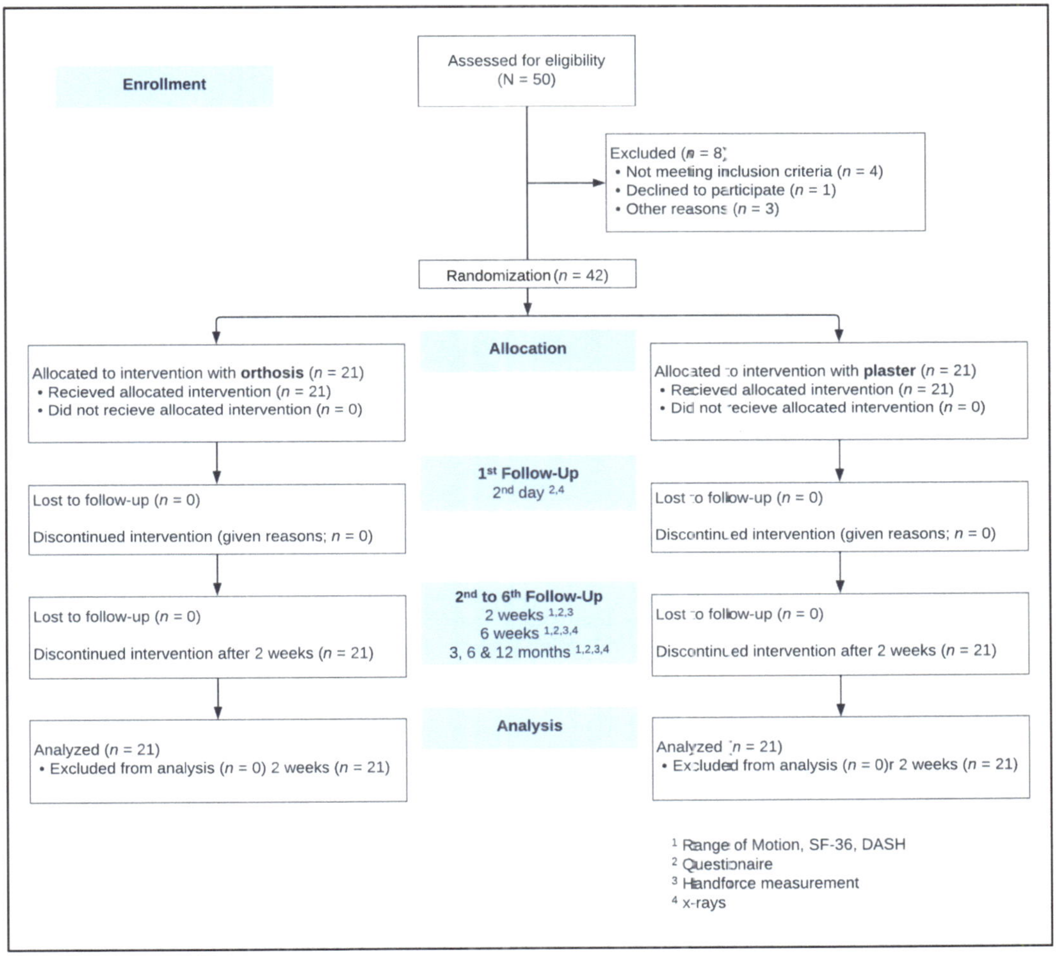

Figure 1. The CONSORT 2010 flow diagram.

In accordance with the hospital's standard procedure, first, the fractured forearm was lifted by finger-trap traction and was then reduced. The reduced fracture was then splinted with a circular cast or an orthosis (OPTIVOhand, OPED GmbH, Valley, Germany). After the traction was lifted, another radiograph was obtained. If an intra-articular fracture was suspected, a computed tomography (CT) scan was performed. Palmar plate osteosynthesis was performed in all patients after the swelling had subsided (VA-LCP, Synthes GmbH, Oberdorf, Switzerland; or Aptus, Medartis AG, Basel, Switzerland, Figure 2).

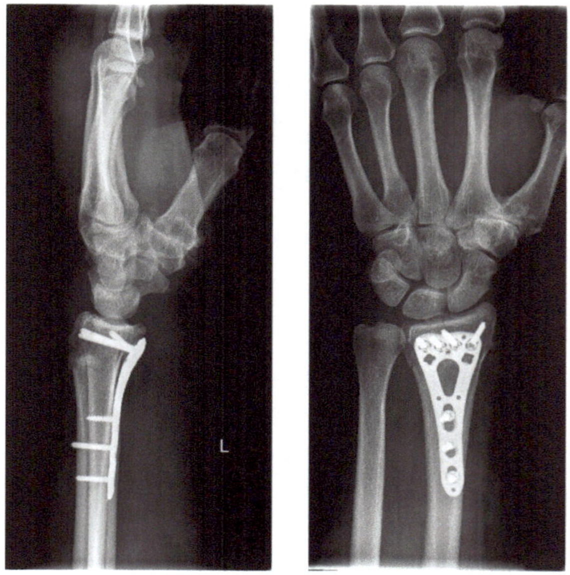

Figure 2. Standard operative treatment with palmar plate.

Postoperatively, immobilization was carried out with a dorsal plaster splint versus the OPTIVOhand orthosis for 2 weeks, with mobilization of the adjacent joints. The used orthosis consists of a flexible polyester textile fabric with flexible adjustable Velcro fasteners and is reinforced by aluminum splints. It weighs 600–800 g. It is washable and can be used with an interchangeable splint for either the left or right wrist. The two available sizes were selected depending on the circumference of the patient's wrist. In both plaster casting and orthotic therapy, a thin layer of cotton wool padding was applied to avoid pressure points caused by swelling (Figure 3A–E).

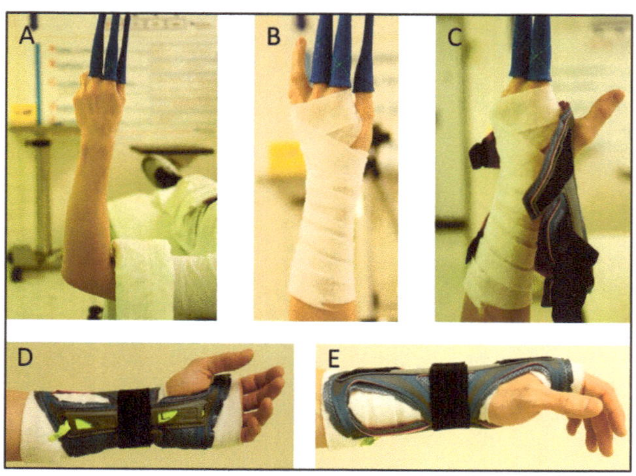

Figure 3. Orthosis application procedure: (**A**) Hanging in extension, (**B**) Cotton-wool padding, (**C**) Placement of the orthosis in suspension, (**D**,**E**) Correct placement of the orthosis.

The time required to apply the orthosis and the cast was measured during the initial application. In view of the several variables capable of affecting the time measurements, only the placement of the circular white plaster after finger-trap traction was measured using a stopwatch. The time required for hardening, splitting, and checking was excluded from the measuring. Casts and orthoses were both applied in the emergency room for primary care as well as in the operating room after surgery under the supervision of the medical staff; any problems were recorded using a questionnaire. Follow-up examinations were performed in accordance with the scheme after 2 days, 2 weeks, 6 weeks, and 3, 6, and 12 months (Figure 1). The sutures and immobilization devices were removed 14 days postoperatively. For the follow-up, patients completed the disabilities of the arm, shoulder and hand questionnaire (DASH), and the Short Form-36 questionnaire (SF-36) including physical (PCS) and mental component summary (MCS). A questionnaire on patient satisfaction with the splint used was also given to both groups. This questionnaire included the items handling, hygiene, the accuracy of fit, pressure complaints, aesthetics, and dressing on a 5-point Likert scale. Additional to the ROM and grip strength assessment, surgeons completed a questionnaire about fit accuracy, signs of secondary dislocation, ease of application and pressure points. Furthermore, radiographic imaging was carried out on two planes.

Statistical analysis was performed using IBM SPSS Statistics for Windows, version 25 (IBM Corporation, Armonk, New York, NY, USA). The Kolmogorov–Smirnov test was used to test the distribution of the groups, with a value of $p < 0.05$ indicating no significant differences in the distribution. Chi-square tests, for categorical variables, and t-tests, for interval-scaled variables, were used to test if the group characteristics differed initially. Furthermore, a repeated measures analysis of variance (ANOVA) was used to examine differences between the study groups over the entire course of treatment and follow up. If the assumption of sphericity was violated, tested by Mauchly's Test, the results were corrected by the Greenhouse–Geisser method. To avoid the error accumulation by multiple testing, the Bonferroni-Holm method was used in post hoc analysis. The significance level was set at $p < 0.05$.

3. Results

A total of 50 patients with isolated DRFs were enrolled in the study between December 2017 and August 2019. The dropout rate was eight patients (16%). Follow-up data were completed for 21 patients in each study group. The most common reason for dropouts was a change in the treatment regimen after inclusion in the study, such as a decision to use an external fixator due to increased instability ($n = 3$) or additional dorsal plate osteosynthesis ($n = 1$). Table 1 shows the patients' age and sex distribution. The right arm was affected more often in both groups, at 62% in the CG and 57% in the OG ($p = 0.753$). In both groups, there was a predominance of severe C-type fractures in the AO classification, affecting a total of 31 patients (74%). There were no significant differences in the distribution of AO classifications ($p = 0.520$).

3.1. Study-Specific Questionnaire

There were no significant differences in levels of patient satisfaction regarding general physical condition, except for physical capacity 6 weeks postoperatively. Patients in the OG felt significantly more restricted than those in the CG ($p = 0.042$). However, the OG patients were significantly more satisfied with the splint's accuracy of fit ($p = 0.007$). There were no significant differences between the two groups in relation to wrist pain during the postoperative follow-up examinations. There were no significant differences between the groups in relation to exercise capacity at 2 weeks postoperatively ($p = 0.469$). Patients in the CG could do slightly more exercise after 6 weeks ($p = 0.042$). However, this difference was no longer evident 3 months postoperatively ($p = 0.458$). The handling of the splint was rated as significantly easier ($p = 0.008$). The patients in the OG felt significantly less restricted in personal hygiene ($p = 0.011$). There was a significant difference in the patients' assessment of subjective adaptability in favor of the OG ($p = 0.013$). No significant differences were noted in relation to pressure symptoms, disturbance of appearance, or difficulties in dressing.

Table 1. Subject characteristics. Independent *t*-test (age) and Chi-square test (sex, operated side, type of fractures) were used to test if both groups—OG and CG—differed initially.

	Total (n = 42)	**Orthosis Group (n = 21)**	**Control Group (n = 21)**	p
Age in years (± SD)	57 (14)	58 (12)	56 (17)	0.793
Female sex (n, %)	32 (76 %)	17 (81 %)	15 (71 %)	0.469
Operated side right (n, %)	25 (60 %)	12 (57 %)	13 (62 %)	0.753
Type A fractures				
A2	2 (5%)	0 (0%)	2 (10%)	
A3	7 (17%)	5 (24%)	2 (10%)	
Type B fractures				
B3	2 (5%)	1 (5%)	1 (5%)	0.520
Type C fractures				
C1	8 (19%)	3 (14%)	5 (24%)	
C2	12 (29%)	7 (33%)	5 (24%)	
C3	11 (26%)	5 (24%)	6 (29%)	

3.2. SF-36

Statistical analysis of the PCS score of the SF-36 revealed a significant effect on time after surgery (F (3.680, 139.842) = 71.481, $p < 0.001$). However, there was no significant difference between both groups during the whole follow-up (F (1, 38) = 0.547, $p = 0.464$). In both groups, the PCS scores significantly declined between the second day after surgery to a minimum at 2 weeks follow-up and recovered until 6 months follow-up. Further, there was no statistically significant interaction effect between time after surgery and the study groups (F (3.680, 139.842), $p = 0.522$; Figure 4A).

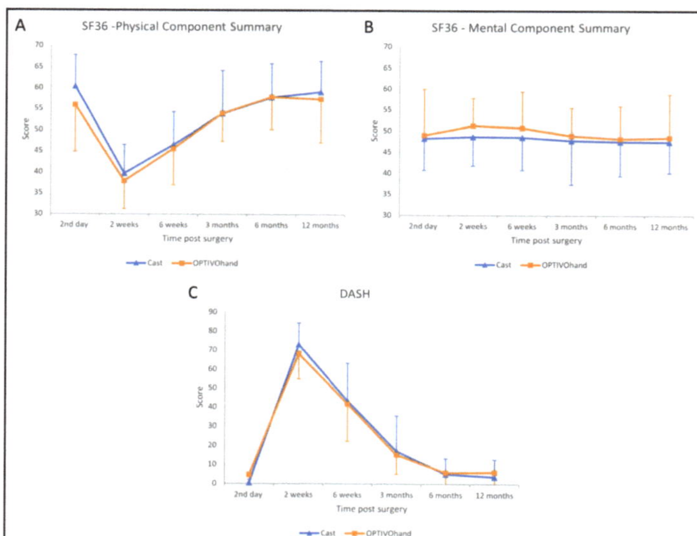

Figure 4. Patient outcome scores during follow-up ((**A**) SF36—Physical Component Summary; (**B**) SF36—Mental Component Summary, (**C**) DASH). Blue triangles represent immobilization by circular cast, orange squares represent immobilization by OPTIVOhand orthosis.

Similar to the PCS score, the MCS score differs significantly over time (F (3.299, 125.354) = 45.070, $p < 0.001$). Hereby, the MCS score was significantly increased over 2 to 6 weeks post-surgery. The significant effect between both groups (F (1, 38) = 5.345, $p = 0.026$) was not confirmed in the post hoc t-test due to Bonferroni-Holm correction. There was no significant interaction between the group and time after surgery (Figure 4B).

3.3. DASH

The significant effect over time was also confirmed on the DASH score (F (3.204, 124.940) = 255.468, $p < 0.001$). Both groups significantly improved 2 weeks after surgery and recovered after 6 months (Figure 4C). There was neither a significant effect on the DASH score between both groups over the whole follow-up (F (1, 39) = 0.004, $p = 0.948$) nor an interaction effect between the type of orthosis and follow-up (F (3.204, 124.940) = 0.906, $p = 0.445$).

3.4. Maintenance of Reduction

To determine the loss of reduction, the reduction was controlled during finger-trap tracking before immobilization and after the application of cast or orthosis application without traction. Thus, the preservation of reduction was verified twice consecutively by two radiographs or one radiograph and one CT scan. Loss of reduction was defined as a loss of palmar inclination of >10°. This occurred seven times in the OG (14%) and 10 times in the CG (20%). There were no statistically significant correlations between the groups and the occurrence of loss of reduction ($p = 0.412$).

3.5. Application Time

The OG clearly showed a significant advantage in terms of the time required for the application: the plaster application took an average of 7 min 30 s (SD 2 min 32 s), while the orthosis only took 2 min 5 s (SD 1 min 38 s), $p < 0.001$. Not included was the time for hardening and splitting of the plaster.

3.6. Clinical Examination

The functional outcome 2 weeks postoperatively after removal of the splints was still limited in comparison with the healthy hand. In both groups, patients were able to significantly improve their ROM (with extension and flexion added, F (2.844, 79.623) = 96.840, $p < 0.001$; Figure 5A) and grip strength (as a percentage of the strength on the healthy side, F (2.679, 61.628) = 112.577, $p < 0.001$; Figure 5B) over the entire study period. Independently of the type of splint (ROM: F (1, 28) = 0.011, $p = 0.918$), patients achieved almost comparable mobility and strength of the healthy side after 1 year. The total ROM 2 weeks postoperatively was limited by a mean of 66° (SD 6°), whereas after 1 year the deficit was only 9° (SD 3°). With no sig. differences between both groups (F (1, 23) = 0.044, $p = 0.836$), all patients increased the relative grip strength from initially 17% 2 weeks after surgery to 96% after 1 year on average. There were no significant differences between the OG and CG in relation to mean circumference differences, as a measure of swelling.

3.7. Surgeon Questionnaire

The surgeons did not observe any significant differences in the assessment of fit accuracy, signs of secondary dislocation, or ease of application and pressure points.

3.8. Complications

A total of four complications occurred that required surgical treatment, three in the CG (14%; (1) secondary dislocation of the dorsal fracture zone after palmar plate osteosynthesis, (2) Secondary dislocation with consecutive shortening of the distal radius (3) B3 fracture "Reversed-Barton-Fracture") and one in the OG (5%; rupture of the extensor policis longus tendon). None of these complications were caused by the use of the orthosis or the plaster cast. All of the complications were successfully treated with an appropriate revision operation.

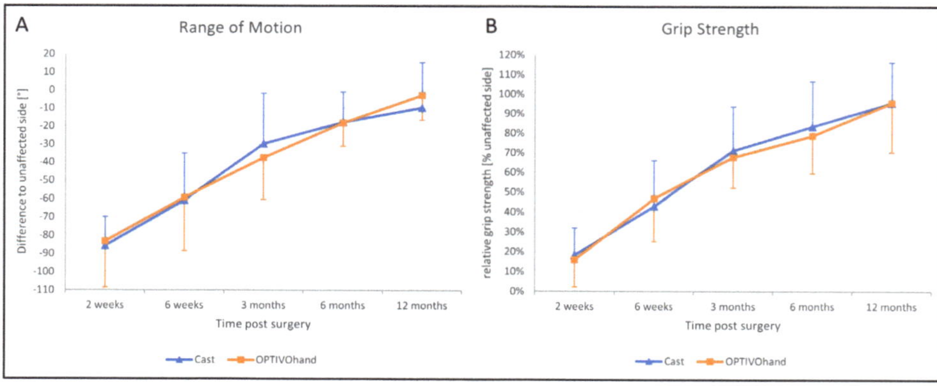

Figure 5. Functional outcome ((**A**) range of motion; (**B**) grip strength) after 2 weeks to 12 months follow-up.

4. Discussion

This study shows that using a prefabricated orthosis for the immobilization of DRFs before and after surgery is equivalent to the standard treatment with plaster casting. Aspects such as perceived personal hygiene, handling, and adaptability were rated better by patients of the OG. One reason why orthoses might be integrated more into everyday clinical work is the fairly small amount of time and expertise that is required to apply and adjust it to the patient. The DRF is still one of the most common types of fracture in general. Even today, the largest age peak is seen in 75–84-year-old patients [2], and a significant increase in these numbers is expected because of demographic trends [23]. As they are usually a result of osteoporosis, these fractures are challenging both for surgical procedures and for follow-up treatment [3]. In younger individuals, early or even immediate mobilization may seem reasonable and may be associated with better functional outcomes [24,25]. However, other studies have shown that initial immobilization also provides benefits. For example, it has been shown that the need for analgesics, especially opioids, can be reduced [26]. The previous gold standard, plaster casting, has not yet been replaced in everyday clinical work by any of the manufactured orthosis models, that have received some good research results. Stuby et al., were already able to demonstrate very good results with a similar orthosis in relation to ROM, SF-36, and DASH in 2015 [20]. One reason for orthoses being rarely used in everyday clinical work may have been their size and the more difficult adjustment to the limb when there is a lack of practice. Switching from plaster casting to orthoses postoperatively may then be impracticable for various reasons. The aim of the present study was therefore to further eliminate the above-mentioned problems to demonstrate that by using a suitable orthosis improved care can be achieved.

The standard treatment approach in our Level I trauma center was not altered by the study protocol, except from the placement of the orthosis. In our approach, we were able to adjust the orthosis used immediately during finger-trap traction and show that there was no increase in secondary dislocations on the radiographic and CT control imaging and, moreover, that diagnostic procedures were in no way restricted when the orthosis was in place. The quality of detail was not limited in any way in the lateral projection, very less in the posteroanterior view (Figure 6A–D). Additionally, the use of the orthosis did not affect the CT scan. As shown in Figure 7, there are no metal artifacts induced by the aluminum splints. Hence, modern orthoses enable comprehensive radiographic diagnostics.

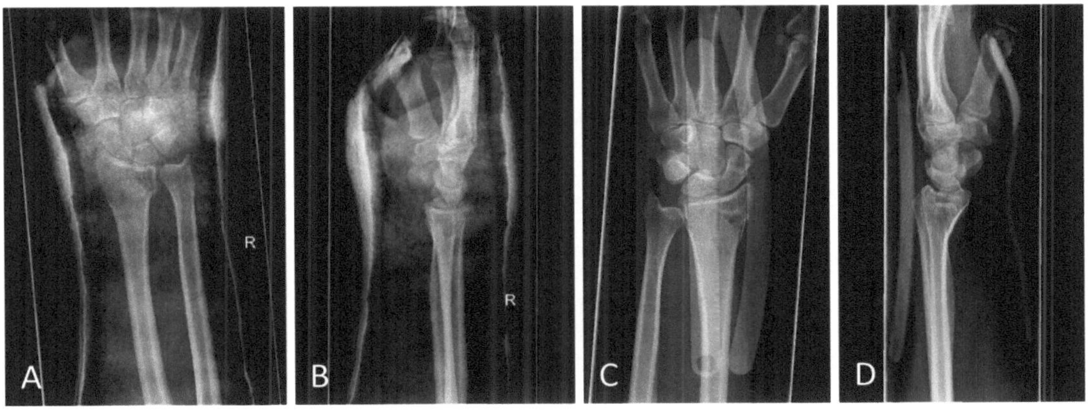

Figure 6. (**A**) Posteroanterior radiograph with cast. (**B**) Mediolateral radiograph with cast. (**C**) Posteroanterior radiograph with orthosis. (**D**) Mediolateral radiograph with orthosis. All radiographs were taken after reduction before surgery.

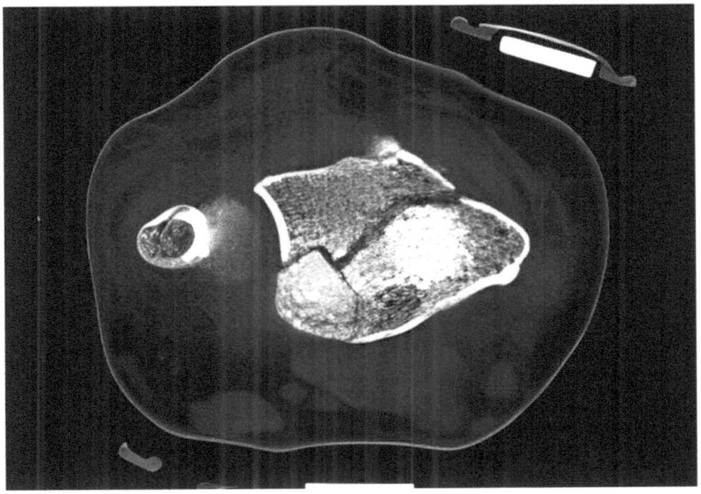

Figure 7. CT scan of a DRF in transversal view. The use of aluminum-based orthosis does not induce metal artifacts.

Even in unstable AO type C3 fractures, the orthosis achieved very good stability and acceptance. No increased dislocation in the orthosis was observed. A significant time saving was also noted, as well as easy handling during placement. A clear advantage in comparison with previous studies is the fact that the orthosis remained with the patient from the first day of trauma care and could be used and adjusted at any time without problems. As the orthoses have a slim and lightweight design, significantly better personal hygiene, adjustability, and handling were observed. In the regular postoperative follow-up visits, no other significant differences were observed in relation to loss of reduction (>10%), DASH score, SF-36 scores, or grip strength, apart from better mental health component responses.

After the removal of the orthoses and plaster splints 2 weeks postoperatively, both study groups received accompanying physiotherapy to maintain or regain mobility and strength. Although various studies have shown that early mobilization can also achieve early mobility, the authors of this study believe that immobilization for 2 weeks is advisable, especially in the case of initially very unstable fractures and concomitant osteoporosis. This does not entail any significantly increased risk of persistent movement restriction [25,27,28]. If mobilization is permitted during a physiotherapy session, the orthosis can be removed and reapplied in an easy and fast manner.

A meta-analysis by van Delft et al. recommended in 2019 that immobilization should not last longer than 3 weeks, if possible [16]. In exceptional cases, prolonged immobilization may also have to be accepted. This is possible without any problems with the orthosis used in this study, as it can be easily adjusted and washed to maintain hygiene. On very hot and humid days, the orthosis has clear advantages over the standard plaster cast in terms of stability and hygiene of the material. In contrast to a circular cast, orthoses allow preoperative and postoperative wound checking due to their integrated openings. It can also be adjusted in case of increasing or decreasing swelling. Particularly in busy emergency rooms or when there are staff shortages, the simple and quick application of orthosis can save a lot of time and stress. These orthoses are not recommended in patients with dementia or lack of cooperation, as they can be removed too easily in comparison with a circular cast. The problems reported in previous studies and orthoses have been analyzed and consistently improved throughout the latest device generations. The clinical results in the present study are sufficiently persuasive, so further studies on conservative therapy, especially in geriatric patients, would be desirable.

The study has a few limitations: the number of patients is small and dropouts occurred. These were mainly due to the rigorous inclusion and exclusion criteria. Patients were excluded if procedures were changed during the decision-making process; this occurred more often than expected. Only palmar plate osteosyntheses were included, in order to ensure sufficient comparability.

5. Conclusions

The results of this study indicate that both preoperative and postoperative reduction can be maintained with an orthosis and that there are no disadvantages in comparison with the current gold standard of plaster casting. The orthosis allows quick and easy application, opportunities for adjustment during therapy, better personal hygiene and can reduce the workload of hospital staff.

Author Contributions: Conceptualization, T.K. and P.H.; Data curation, T.K., V.R. and A.M.; Formal analysis, T.K. and V.R.; Funding acquisition, T.K. and A.M.; Investigation, T.K., P.H., V.R. and A.M.; Methodology, T.K. and P.H.; Project administration, T.K.; Resources, T.K.; Software, T.K. and V.R.; Supervision, F.S., U.S. and A.M.; Validation, T.K., P.H., V.R. and A.M.; Visualization, T.K.; Writing—original draft, T.K.; Writing—review and editing, T.K. All authors have read and agreed to the published version of the manuscript.

Funding: The study was financed by a grant from OPED GmbH Medizinpark 1, D-83626 Valley, Germany.

Institutional Review Board Statement: The study was conducted in accordance with the Declaration of Helsinki, and approved by the Ethics Committee of University of Tübingen and approved in the framework of "good clinical practice" (Project Number 173/2017BO1, 24.05.2017).

Informed Consent Statement: Informed consent was obtained from all subjects involved in the study.

Data Availability Statement: All data, tables and figures presented in this manuscript are original. Further inquiries can be directed to the corresponding author.

Conflicts of Interest: The funders had no role in study design, data collection and analysis, interpretation of the data, decision to publish, or preparation of the manuscript. There are no financial or non-financial competing interests. This includes: employment, consultancy, patents, products in development, marketed products, etc. The authors declare no further conflict of interests.

References

1. De Putter, C.E.; Selles, R.W.; Polinder, S.; Hartholt, K.A.; Looman, C.W.; Panneman, M.J.M.; Verhaar, J.A.N.; Hovius, S.E.R.; van Beeck, E.F. Epidemiology and Health-Care Utilisation of Wrist Fractures in Older Adults in The Netherlands, 1997–2009. *Injury* **2013**, *44*, 421–426. [CrossRef] [PubMed]
2. Chung, K.C.; Spilson, S.V. The Frequency and Epidemiology of Hand and Forearm Fractures in the United States. *J. Hand Surg.* **2001**, *26*, 908–915. [CrossRef] [PubMed]
3. Diaz-Garcia, R.J.; Oda, T.; Shauver, M.J.; Chung, K.C. A Systematic Review of Outcomes and Complications of Treating Unstable Distal Radius Fractures in the Elderly. *J. Hand Surg.* **2011**, *36*, 824–835. [CrossRef]
4. Mattila, V.M.; Huttunen, T.T.; Sillanpää, P.; Niemi, S.; Pihlajamäki, H.; Kannus, P. Significant Change in the Surgical Treatment of Distal Radius Fractures: A Nationwide Study between 1998 and 2008 in Finland *J. Trauma* **2011**, *71*, 939–942. [CrossRef]
5. Azad, A.; Kang, H.P.; Alluri, R.K.; Vakhshori, V.; Kay, H.F.; Ghiassi, A. Epidemiological and Treatment Trends of Distal Radius Fractures across Multiple Age Groups. *J. Wrist Surg.* **2019**, *8*, 305–311. [CrossRef]
6. Fowler, J.R.; Ilyas, A.M. Prospective Evaluation of Distal Radius Fractures Treated with Variable-Angle Volar Locking Plates. *J. Hand Surg.* **2013**, *38*, 2198–2203. [CrossRef]
7. Thorninger, R.; Madsen, M.L.; Wæver, D.; Borris, L.C.; Rölfing, J.H.D. Complications of Volar Locking Plating of Distal Radius Fractures in 576 Patients with 3.2 Years Follow-Up. *Injury* **2017**, *48*, 1104–1109. [CrossRef] [PubMed]
8. Wichlas, F.; Haas, N.P.; Disch, A.; Machó, D.; Tsitsilonis, S. Complication Rates and Reduction Potential of Palmar versus Dorsal Locking Plate Osteosynthesis for the Treatment of Distal Radius Fractures. *J. Orthop. Traumatol. Off. J. Ital. Soc. Orthop. Traumatol.* **2014**, *15*, 259–264. [CrossRef]
9. Esenwein, P.; Sonderegger, J.; Gruenert, J.; Ellenrieder, B.; Tawfik, J.; Jakubietz, M. Complications Following Palmar Plate Fixation of Distal Radius Fractures: A Review of 665 Cases. *Arch. Orthop. Trauma Surg.* **2013**, *133*, 1155–1162. [CrossRef]
10. Rozental, T.D.; Blazar, P.E. Functional Outcome and Complications after Volar Plating for Dorsally Displaced, Unstable Fractures of the Distal Radius. *J. Hand Surg.* **2006**, *31*, 359–365. [CrossRef] [PubMed]
11. Hwang, R.W.; de Witte, P.B.; Ring, D. Compartment Syndrome Associated with Distal Radial Fracture and Ipsilateral Elbow Injury. *J. Bone Jt. Surg. Am. Vol.* **2009**, *91*, 642–645. [CrossRef] [PubMed]
12. Patel, V.P.; Paksima, N. Complications of Distal Radius Fracture Fixation. *Bull. NYU Hosp. Jt. Dis.* **2010**, *68*, 112–118. [PubMed]
13. Soong, M.; van Leerdam, R.; Guitton, T.G.; Got, C.; Katarincic, J.; Ring, D. Fracture of the Distal Radius: Risk Factors for Complications after Locked Volar Plate Fixation. *J. Hand Surg.* **2011**, *36*, 3–9. [CrossRef] [PubMed]
14. Bacorn, R.W.; Kurtzke, J.F. Colles' Fracture: A Study of Two Thousand Cases from the New York State Workmen's Compensation Board. *J. Bone Jt. Surg. Am. Vol.* **1953**, *35-A*, 643–658. [CrossRef]
15. Handoll, H.H.G.; Elliott, J. Rehabilitation for Distal Radial Fractures in Adults. *Cochrane Database Syst. Rev.* **2015**, CD003324. [CrossRef] [PubMed]
16. Van Delft, E.A.K.; van Gelder, T.G.; de Vries, R.; Vermeulen, J.; Bloemers, F.W. Duration of Cast Immobilization in Distal Radial Fractures: A Systematic Review. *J. Wrist Surg.* **2019**, *8*, 430–438. [CrossRef]
17. Bhan, K.; Hasan, K.; Pawar, A.S.; Patel, R. Rehabilitation Following Surgically Treated Distal Radius Fractures: Do Immobilization and Physiotherapy Affect the Outcome? *Cureus* **2021**, *13*, e16230. [CrossRef]
18. Watson, N.; Haines, T.; Tran, P.; Keating, J.L. A Comparison of the Effect of One, Three, or Six Weeks of Immobilization on Function and Pain After Open Reduction and Internal Fixation of Distal Radial Fractures in Adults: A Randomized Controlled Trial. *J. Bone Jt. Surg. Am. Vol.* **2018**, *100*, 1118–1125. [CrossRef]
19. Quadlbauer, S.; Pezzei, C.; Jurkowitsch, J.; Kolmayr, B.; Keuchel, T.; Simon, D.; Hausner, T.; Leixnering, M. Early Rehabilitation of Distal Radius Fractures Stabilized by Volar Locking Plate: A Prospective Randomized Pilot Study. *J. Wrist Surg.* **2017**, *6*, 102–112. [CrossRef]
20. Stuby, F.M.; Döbele, S.; Schäffer, S.-D.; Mueller, S.; Ateschrang, A.; Baumann, M.; Zieker, D. Early Functional Postoperative Therapy of Distal Radius Fracture with a Dynamic Orthosis: Results of a Prospective Randomized Cross-over Comparative Study. *PLoS ONE* **2015**, *10*, e0117720. [CrossRef]
21. Khudairy, A.A.; Hirpara, K.M.; Kelly, I.P.; Quinlan, J.F. Conservative Treatment of the Distal Radius Fracture Using Thermoplastic Splint: Pilot Study Results. *Eur. J. Orthop. Surg. Traumatol. Orthop. Traumatol.* **2013**, *23*, 647–650. [CrossRef] [PubMed]
22. Janzing, H.M.J.; Bessems, S.A.M.; Ligthart, M.A.P.; van Lieshout, E.M.M.; Theeuwes, H.P.; Barten, D.G.; Verhofstad, M.H.J. Treatment of Dorsally Dislocated Distal Radius Fractures with Individualized 3D Printed Bracing: An Exploratory Study. *3D Print. Med.* **2020**, *6*, 22. [CrossRef] [PubMed]
23. Stirling, E.R.B.; Johnson, N.A.; Dias, J.J. Epidemiology of Distal Radius Fractures in a Geographically Defined Adult Population. *J. Hand Surg. Eur. Vol.* **2018**, *43*, 974–982. [CrossRef] [PubMed]
24. Karantana, A.; Downing, N.D.; Forward, D.P.; Hatton, M.; Taylor, A.M.; Scammell, B.E.; Moran, C.G.; Davis, T.R.C. Surgical Treatment of Distal Radial Fractures with a Volar Locking Plate versus Conventional Percutaneous Methods: A Randomized Controlled Trial. *J. Bone Jt. Surg. Am. Vol.* **2013**, *95*, 1737–1744. [CrossRef] [PubMed]
25. Berner, S.H.; Willis, F.B. Dynamic Splinting in Wrist Extension Following Distal Radius Fractures. *J. Orthop. Surg. Res.* **2010**, *5*, 53. [CrossRef]

26. Andrade-Silva, F.B.; Rocha, J.P.; Carvalho, A.; Kojima, K.E.; Silva, J.S. Influence of Postoperative Immobilization on Pain Control of Patients with Distal Radius Fracture Treated with Volar Locked Plating: A Prospective, Randomized Clinical Trial. *Injury* **2019**, *50*, 386–391. [CrossRef]
27. Jongs, R.A.; Harvey, L.A.; Gwinn, T.; Lucas, B.R. Dynamic Splints Do Not Reduce Contracture Following Distal Radial Fracture: A Randomised Controlled Trial. *J. Physiother.* **2012**, *58*, 173–180. [CrossRef]
28. Zeckey, C.; Späth, A.; Kieslich, S.; Kammerlander, C.; Böcker, W.; Weigert, M.; Neuerburg, C. Early Mobilization Versus Splinting After Surgical Management of Distal Radius Fractures. *Dtsch. Arztebl. Int.* **2020**, *117*, 445–451. [CrossRef]

Journal of Personalized Medicine

MDPI

Systematic Review

Bilateral Scaphoid Fractures: A Systematic Literature Review

Lorenzo D'Itri , Maria Serena Gattuso , Claudio Domenico Cobisi, Massimo Ferruzza, Ludovico Lucenti * and Lawrence Camarda

Department of Orthopaedics and Traumatology, University of Palermo, 90133 Palermo, Italy; lorenzo.ditri@unipa.it (L.D.); mariaserena.gattuso@you.unipa.it (M.S.G.); claudiodomenico.cobisi@community.unipa.it (C.D.C.); massimoferruzza@libero.it (M.F.); lawrence.camarda@unipa.it (L.C.)

* Correspondence: ludovico.lucenti@gmail.com

Abstract: Bilateral scaphoid fractures are rare lesions, warranting a review to synthesize current knowledge, identify gaps, and suggest research directions. Two authors, adhering to PRISMA guidelines, in January 2024 identified 16 case reports (1976–2023). Data extraction included demographics, injury mechanisms, associated injuries, fracture sites, treatments, and outcomes. Among 121 initial outcomes, 16 articles met the criteria, predominantly affecting young people (93.75% males, mean age 22 years). High-energy traumas (75%) often caused associated wrist injuries (68.75%). Most fractures required surgical intervention (68.75%), primarily headless compression screws. Bilateral scaphoid fractures, which are rare but associated with high-energy traumas, commonly involve wrist injuries. Surgical management is often necessary, yielding better outcomes with fewer complications. Further research is essential to understand the epidemiology, optimal management, and long-term results. Early diagnosis and appropriate treatment are crucial for preventing complications and ensuring favorable patient outcomes.

Keywords: scaphoid; carpal bones; trauma; surgery; headless compression screws; distal radius fractures

Citation: D'Itri, L.; Gattuso, M.S.; Cobisi, C.D.; Ferruzza, M.; Lucenti, L.; Camarda, L. Bilateral Scaphoid Fractures: A Systematic Literature Review. *J. Pers. Med.* **2024**, *14*, 424. https://doi.org/10.3390/jpm14040424

Academic Editors: Markus Gesslein and Eduard Kurz

Received: 27 March 2024
Revised: 9 April 2024
Accepted: 15 April 2024
Published: 16 April 2024

1. Introduction

The scaphoid, the largest bone in the proximal carpal row [1], plays a pivotal role in transferring compressive loads from the hand to the forearm and is crucial for maintaining carpal stability [2]. Fractures of the scaphoid are the most frequently encountered fractures among the carpal bones [2], accounting for approximately 2.4% of all wrist fractures in the United States, with an incidence rate of 1.47 per 100,000 person–years [3]. These fractures are more frequent among younger males, particularly athletes and military personnel. They are less common in children, constituting only 3% of all pediatric hand and wrist fractures [4].

Fractures of the scaphoid can be categorized into two primary mechanisms: traumatic and stress-induced. The typical cause is a fall onto an outstretched hand, resulting in forceful hyperextension of the wrist [5,6].

The diagnosis of scaphoid fractures is sometimes challenging. X-rays may miss fractures in up to 25% of cases, leading to delays and unnecessary immobilization. Techniques like CT, MRI, bone scans, and ultrasound offer higher accuracy and quicker results. Ultrasound shows promise due to its availability and accuracy. MRI is considered the best for hidden fractures and associated injuries. MRI or CT scans can be useful for patients with positive clinical findings, alongside repeat X-rays if needed. CT scans and the use of 3D printing can be helpful for diagnosis and especially for surgical treatment planning [7].

Bilateral scaphoid fractures, involving both wrists simultaneously, are exceptionally rare [8] and have been infrequently reported in the literature. The occurrence of bilateral scaphoid fractures can significantly diminish patients' quality of life, impair functionality, and impose limitations on basic activities of daily living, such as dressing, grooming, and household tasks, but also sports activities and hobbies, thereby highlighting the importance of prompt diagnosis and comprehensive treatment to optimize recovery and minimize disability.

This paper presents a comprehensive review of the literature pertaining to bilateral scaphoid fractures. The objectives of this review were to consolidate the existing knowledge on this uncommon condition, identify literature gaps, and propose directions for future research.

2. Materials and Methods

2.1. Study Selection

This review of the existing literature adhered to the guidelines outlined in the Preferred Reporting Items for Systematic Reviews and Meta-Analyses (PRISMA) [9].

In January 2024, we conducted searches of PubMed and Web of Science databases using the following research string: "(Bilateral) AND (Scaphoid) AND (Fractures)". The searches were carried out by two authors (L.D. and M.S.G.).

A standardized data entry form was employed to extract information, including the year of the study, number of patients, sex, age, associated injuries, mechanism of injury, site of fractures, treatment, complications, and outcomes.

Study quality evaluation was performed by two independent reviewers (L.D. and M.S.G.) using the tool provided by Murad et al. [10].

Conflicts about data were resolved by consultation with two senior surgeons (L.C. and L.L.).

2.2. Selection Criteria

The selection of studies was based on various inclusion criteria: (1) studies written in the English language; (2) studies of any level of evidence; (3) studies concerning patients with bilateral scaphoid fractures including clinical and radiological diagnosis, type of treatment, outcome, and follow-up.

Several parameters were used as exclusion criteria: (1) inaccessible studies; (2) studies focused on different topics; (3) studies without an accessible abstract; (4) studies with poor scientific methodology. Any duplicates were also excluded.

2.3. Data Extraction and Criteria Assessment

Two researchers independently retrieved all information from the text, tables, and figures in the articles. The data were methodically arranged in a specific order: (1) year of publication; (2) first author; (3) paper title; (4) number of patients; (5) sex of patients; (6) age at the injury time; (7) associated lesions; (8) comorbidities; (9) mechanism of injury; (10) CT scan performed; (11) site of fractures; (12) type and timing of treatment; (13) complications; (14) hardware removal; (15) outcomes; (16) follow-up. Conflicts between the two reviewers were resolved by two senior surgeons. The PRISMA flow chart utilized for the selection and screening of studies is shown in Figure 1.

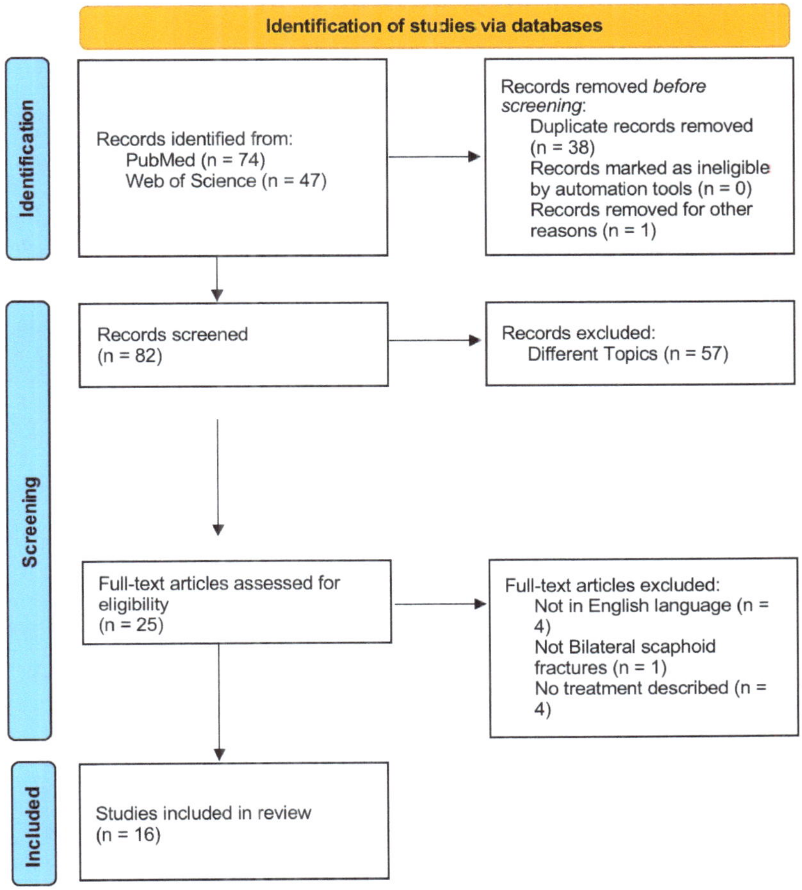

Figure 1. The Preferred Reporting Items for Systematic Reviews and Meta-Analyses (PRISMA) flow chart used for study selection and screening.

3. Results

In the search engine query, a total of 121 outcomes were identified (74 from PubMed and 47 from Web of Science). Subsequently, 38 duplicates and one letter to the author were excluded.

Following the initial screening, a total of 25 articles were selected for a thorough full-text examination. In the final analysis, after an in-depth review of the full texts and scrutiny of the reference lists, a total of 16 articles were chosen following the pre-established criteria.

A flowchart illustrating the selection and screening methodology, following PRISMA guidelines [9], is presented (Figure 1).

Data from the studies were collected. The main parameters are shown in Table 1.

Table 1. Main parameters of each study.

Year	Lead Author	Article	No. of Patients	Sex	Age
1976	Stother I.G. [11]	A report of 3 cases of simultaneous Colles' and scaphoid fractures	1	Male	26
1999	Kay R.M [12]	Bilateral proximal radial and scaphoid fractures in a child	1	Male	13
2000	Kaneko K. [13]	Bilateral transcapholunate dislocation	1	Male	35
2008	Ozkan K. [14]	Fractures of the bilateral distal radius and scaphoid: a case report	1	Male	28
2010	Yinusa W. [15]	Bilateral Simultaneous Fracture of the Carpal Scaphoid Successfully Treated with Conservative Cast Immobilisation: A Case Report	1	Male	28
2012	Pidemunt G. [12]	Bilateral Stress Fracture of the Carpal Scaphoid: Report in a Child and Review of the Literature	1	Male	13
2014	Yildirim C. [16]	Bilateral dorsal trans-scaphoid perilunate fracture-dislocation: A case report	1	Male	21
2014	Saglam F. [17]	Chronic wrist pain in a goalkeeper; bilateral scaphoid stress fracture: A case report	1	Male	19
2014	Mohamed Haflah N.H. [18]	Bilateral scaphoid stress fracture in a platform diver presenting with unilateral symptoms	1	Male	16
2016	Virani S.R. [19]	A unique case of bilateral trans-scaphoid perilunate dislocation with dislocation of lunate into the forearm	1	Male	35
2019	Afshar A. [20]	Bilateral Scaphocapitate Fracture Syndrome: A Case Report with Long-Term Follow up	1	Male	25
2020	Ghargozloo D. [1]	Traumatic bilateral scaphoid fractures	1	Male	17
2022	Meraghni N. [21]	Bilateral Combined Fractures of the Scaphoid and Distal Radius: A Case Report	1	Female	16
2023	Johnson N.A. [22]	Recurrence of Scaphoid Stress Fracture: A Case Report	1	Male	20
2023	Korkoman A.J [23].	Bilateral greater arc Perilunate injury: A case report	1	Male	25
2023	Reigstad O. [24]	Bilateral carpal pilon-type fractures due to clenched fist trauma: a case report	1	Male	17

3.1. Subjects

All studies, performed between 1976 and 2023, were case reports, each including a single patient with a bilateral scaphoid fracture.

Hence, as shown in Table 1, the cohort comprised 16 participants, including 15 males (93.75%) and 1 female (6.25%). The mean age was 22.13 years.

3.2. Mechanism of Injury

In 75% of the cases (n = 12), the injury mechanism involved high-energy trauma. Specifically, 37.5% of the cases (n = 6) resulted from falls from a height [11–14,20,23], 6.25% (n = 1) were related to a traffic accident [24], 18.75% (n = 3) were associated with falls onto both outstretched hands [15,16,19], and 12.5% (n = 2) were attributed to sports-related trauma (Table 2) [1,21].

Table 2. Main characteristics of the trauma and fracture.

Lead Author	Associated Injuries	Mechanism of Fracture	Herbert's Classification	CT Scan Performed
Stother I.G. [11]	Bilateral distal radius epiphysis fractures; D12 fracture	Traumatic: fall from a height	B2	No
Kay R.M. [12]	Bilateral fractures of the proximal radius	Traumatic: fall from a height	A1	No
Kaneko K. [13]	Bilateral carpal bone dislocations	Traumatic: fall from a height	B4	No
Ozkan K. [14]	Bilateral distal radius epiphysis fractures; L1 fracture	Traumatic: fall from a height	B2	No
Yinusa W. [15]	Bilateral subluxation of the scapho-capito-trapezoid joints	Traumatic: FOOSH *	Right: B2 Left: B3	No
Pidemunt G. [12]	Not described	Stress	B2	Yes
Yildirim C. [16]	Bilateral carpal bone dislocations	Traumatic: FOOSH	B4	No
Saglam F. [17]	Not described	Stress: sports-related	B2	No MRI performed
Mohamed Haflah N.H. [18]	Not described	Stress: sports-related	Right: D2 Left: B2	No
Virani S.R. [19]	Bilateral carpal bone dislocations	Traumatic: FOOSH	B4	Yes
Afshar A. [20]	Bilateral carpal bone dislocations; bilateral trans-capitate fractures	Traumatic: fall from a height	B4	No
Ghargozloo D. [1]	Not described	Traumatic: sports-related	Right: D2 Left: B2	Yes (after 8 months)
Meraghni N. [21]	Bilateral distal radius epiphysis fractures	Traumatic: sports-related	A2	No
Johnson N.A. [22]	Not described	Stress: sports-related	B2	Yes
Korkoman A.J. [23]	Bilateral distal radius epiphysis fractures; bilateral carpal bone dislocations	Traumatic: fall from a height	B4	Yes
Reigstad O. [24]	Capitate and hamate fractures on the left wrist; capitate fracture on the right wrist	Traumatic: road traffic accident	Right: A2 Left: B1	Yes

* FOOSH = fall on outstretched hands.

Four patients (n = 4; 25%) did not disclose specific traumatic events; instead, they reported multiple minor traumas occurring repeatedly over time, categorizing these fractures as stress fractures [17,18,22,25]. Among these cases, 18.75% (n = 3) could be characterized as sports-related [17,18,22], while, in 5.88% (n = 1), there was no clarification regarding the patient's engagement in any sports activity [25].

3.3. *Associated Injuries*

In 68.75% of cases (n = 11), associated wrist injuries were identified; of these, 36.36% (n = 4) presented with bilateral distal radius fractures [11,14,21,23], 45.45% (n = 5) had bilateral dislocations of carpal bones [13,16,19,20,23], and 18.18% (n = 2) had bilateral non-scaphoid carpal bone fractures [20,24].

In 6.25% of cases (n = 1), a bilateral proximal radius fracture was described [12], while, in 12.5% of cases (n = 2), injuries to the vertebral column were reported [11,14].

3.4. Fracture Site

According to Herbert's classification for scaphoid fractures, 31.25% of cases (n = 5) exhibited bilateral B2-type fractures [11,14,17,22,25], 29.41% of cases (n = 5) bilateral B4-type fractures [13,16,19,20,23], 6.25% of cases (n = 1) bilateral A2-type fractures [21], and 6.25% (n = 1) bilateral A1-type fractures [12].

Fractures exhibiting discordance bilaterally based on Herbert's classification were documented in 25% of cases (n = 4). In one case (n = 1; 6.25%), a B2 fracture on the right and a B3 fracture on the left were reported [15]. In another case, an A2 fracture on the right and a B1 fracture on the left were observed [24]. Additionally, two cases presented a B2 fracture on the left and a D2 fracture on the right [1,18].

3.5. Treatment

As shown in Table 3, in 31.25% of cases (n = 5), non-surgical management with cast immobilization proved adequate [11,12,16,17,22].

Table 3. Main characteristics of treatments, complications, and outcomes.

Lead Author	Treatment	Complications	Outcome	Follow-Up (in Months)
Stother I.G. [11]	Conservative	Not described	Union	11
Kay R.M. [12]	Conservative	Not described	No pain	1
Kaneko K. [13]	Bilateral headless compression screws	Not described	Union	24
Ozkan K. [14]	Bilateral headless compression screws	Not described	Union	9
Yinusa W. [15]	Conservative	Not described	Union	3
Pidemunt G. [12]	Two-stage (right first, left after 6 months) curettage, iliac crest bone grafting, and fixation with headless compression screws	Not described	No pain	15
Yildirim C. [16]	Bilateral headless compression screws	Non-union of the right scaphoid	Left: union Right: non-union without symptoms	24
Saglam F. [17]	Conservative	Not described	Union	28
Mohamed Haflah N.H. [18]	Bilateral headless compression screws	Not described	Union	4
Virani S.R. [19]	Bilateral headless compression screws	Not described	Union	12
Afshar A. [20]	Bilateral headless compression screws	Not described	Union	60
Ghargozloo D. [1]	Bilateral headless compression screws (right: with non-vascularized bone graft)	Not described	Union	11
Meraghni N. [21]	K-wires percutaneous fixation	Not described	Union	12
Johnson N.A. [22]	Conservative	Recurrent stress fracture of the right scaphoid after 9 months treated with headless compression screws	Union	12
Korkoman A.J. [23]	Bilateral headless compression screws	Not described	Union	12
Reigstad O. [24]	Bilateral headless compression screws	Not described	Union	6

The remaining 68.75% (n = 11) underwent surgery. Among these, 72.73% (n = 8) were treated with headless compression screws [13,14,16,18–20,23,24], 9.09% (n = 1) were treated with headless compression screws and bone grafting [1], 9.09% (n = 1) were treated with percutaneous fixation with k-wires [21], and 9.09% (n = 1) underwent a two-stage procedure that included fracture curettage, bone graft from iliac crest, and fixation with headless compression screws, first the right side and then the left side after 6 months [25].

3.6. Follow-Up and Outcome

The average follow-up was 15.25 months. Two patients (12.50%) underwent follow-up at 6 months after the trauma; in 50% of cases (n = 8), follow-up was between 6 months and 1 year; in 25% of cases (n = 4), follow-up continued beyond one year.

Complications were absent in 80% of cases (n = 4) treated non-operatively [11,12,15,17]. In the remaining 90.91% of surgically managed cases (n = 10), no complications were reported.

4. Discussion

Scaphoid fractures represent the most prevalent type of carpal fracture, accounting for 60% of all carpal bone fractures [26]. They mainly occur in young adults and constitute around 3% of all fractures [27]. The scaphoid bone is notably the most frequently fractured carpal bone, with bilateral involvement occurring in about 1% of cases [28,29].

According to a recent review on scaphoid fractures published in 2023 by Almigdad A. et al. [30], it was found the majority of patients affected by bilateral scaphoid fractures were men (15 patients, 93.75%), and the site of the fracture was predominantly in the waist of the scaphoid (84.38%).

In contrast to the findings described by Almigdad et al. [30] for scaphoid fractures, in the present review, it was found that only 43.75% (n = 7) of patients with bilateral scaphoid fractures were between the ages of 25 and 40. Indeed, more than half of the patients with bilateral scaphoid fractures (56.25%, n = 9) fall within the age group under 25 years.

Our analysis revealed a variety of traumatic mechanisms leading to bilateral scaphoid fractures, with falls from height being the most common cause (37.50%). Only 12.50% were due to acute injuries during sports activities. This contrasts with the pattern seen in unilateral scaphoid fractures, where sports-related injuries are more typical [31]. Even when considering stress fractures associated with sports, falls from height remain the predominant cause of bilateral injuries.

As reported for unilateral scaphoid fractures by Wells M.E. et al. [32], the most associated injury with a scaphoid fracture is a distal radius fracture. In our study, we observed that bilateral scaphoid fractures are associated with bilateral distal radius fractures in 25% of cases (n = 4). However, in contrast to the findings for unilateral fractures, the most frequently associated injury, reported in 31.25% of patients (n = 5), is bilateral dislocation of the carpal bones, likely attributable to the increased intensity of trauma, often caused by falls from a height. In two cases described by Afshar A. et al. [20] and Reigstad O. et al. [24], accounting for 12.5% of cases, associated fractures of other carpal bones were identified: in the first case, a bilateral trans-capitate fracture was associated with perilunate dislocation while, in the second case, fractures of the capitate and hamate on the left and only the capitate on the right were observed. Kay R.M. et al. [12] described a single case (6.25%) of a 13-year-old patient with an associated fracture of the proximal radius after a fall from a height of 2 m. Stother I.G. et al. [11] and Ozkan K. et al. [14] delineated the instances of two patients (12.5%) in whom two vertebral injuries were concomitant following a fall from a height. From our results, based on the Herbert classification, the most frequent fracture pattern in bilateral scaphoid fractures was type B2, accounting for 31.25% (n = 5). This finding closely aligns with the study by Duckworth A.D. et al. [33] concerning unilateral scaphoid fractures, where they identified a prevalence of B2 fractures in 36.4% of cases. In 29.41% of cases (n = 5), characterized by the occurrence of bilateral trans-scaphoid perilunate dislocation, a manifestation of type-B4 fractures in accordance with Herbert's classification was observed in both wrists [13,16,19,20,23].

Meraghni et al. [21] reported a single case (6.25%) involving bilateral scaphoid fractures classified as A2 according to Herbert's classification, whereas Kay et al. [12] documented a single case (6.25%) featuring bilateral fractures classified as A1. Fractures exhibiting discordance bilaterally based on Herbert's classification were documented in 25% of cases (n = 4) as follows:

- Yinusa W. et al. [19] described one case (6.25%) with a B2 fracture of the right wrist and a B3 fracture of the left wrist;
- Reigstad et al. [24] described one case (6.25%) with an A2 fracture of the right wrist and a B1 fracture of the left wrist;
- Mohamed Haflah N. H. [18] et al. and Ghargozloo D. [1] et al. each reported two cases (12.5%) of patients presenting with a B2 fracture on the left and a D2 fracture on the right. It was emphasized that the right-sided fracture was asymptomatic in the first case and went undetected for a period of 8 months in the second case.

In our study, we observed that while non-surgical treatment was sufficient for five patients (31.25%), most patients, accounting for 68.75% (n = 11), underwent surgical intervention. All patients treated conservatively did not exhibit dislocations of the carpal bones. In the studies examined, carpal bone dislocation was one of the primary indications for surgical intervention. In the performed surgery, the most frequently employed treatment (72.73%) was fixation using headless compression screws, as also reported in the study by Suh N. et al. [30]. Due to the rotational forces observed at the level of the first row of the carpus, Gray R.R.L et al. [34] considered the possibility of using a double headless compression screw in highly unstable fracture patterns, citing satisfactory results reported in the studies by Garcia et al. [35] and DiPrinzio et al. [36]. In two patients, the use of bone grafts was necessary: in the case described by Ghargozloo et al. [1], this was only performed for the right wrist as the fracture had gone unnoticed, while, in the case reported by Pidemunt G. et al. [25], a bone graft was used bilaterally. The use of autologous bone graft often involves harvesting from a different site, requiring two surgical procedures. In a 2022 study, Vigni G.E. et al. [37] demonstrated improved bone regeneration using an artificial polybutylene succinate scaffold in a rabbit model. This scaffold has already been employed for nerve regeneration in a murine model [38,39]. This application could potentially be considered in the future for cases of bilateral scaphoid fractures in cases not suitable for performing an autologous graft harvesting procedure [40].

As indicated, the mean follow-up duration was 15 months, with 50% of the studies having a follow-up period ranging from 6 months to one year. Follow-up for up to 5 years was described in only one case by Afshar A. et al. [20]. Complications were absent in 80% of cases (n = 4) treated non-operatively [11,12,17,19]. Johnson N.A. et al. [22] reported a singular non-operatively treated case, wherein a subsequent stress fracture of the right scaphoid occurred. This was managed successfully with a single compression headless screw, ultimately achieving resolution. Yildirim C. et al. [16] documented a solitary case within the surgically treated cohort (9.09%) exhibiting a non-painful pseudoarthrosis on the right side. In the remaining 90.91% of surgically managed cases (n = 10), no complications were reported. This highlights that surgical treatment tends to have a lower complication rate compared to conservative treatment, especially because most surgically treated fractures exhibit a pattern of greater instability and/or associated injuries more frequently.

Given the limited sample size, particularly in the conservative treatment group comprising only five patients, it is crucial to approach conclusions with caution. Although the results may appear less favorable, with one complication representing 20% of the patients, extrapolating broader implications remains limited without a larger cohort. While it is plausible that surgical patients may present with more severe injuries, the occurrence of complications should not be presumed solely based on the treatment type.

For non-surgically treated fractures, Almigdad A. et al. [30] indeed recommend a more aggressive management approach and clinical follow-up with a CT scan at 3 months to early detect fractures that are progressing towards nonunion. This allows for the timely referral of these patients to a hand surgery center for optimal case management.

Our study revealed that six patients were under the age of 18 (37.5%); two patients were under 14 years old (12.5%). Pediatric scaphoid fractures are indeed the most common carpal bone fractures, with an annual incidence rate of 11 per 100,000 [41]. On average, they predominantly affect males at around 12.2 years old, while females typically experience them at around 10.3 years old. Scaphoid fractures in boys under 11 years old and girls under 9 years old are indeed rare. Typically, these fractures occur at the distal pole of the scaphoid in children and often heal without complications [41–43]. Pediatric scaphoid fractures can sometimes be diagnosed late (in 8% to 29% of cases), leading to uncertainties in treatment approaches. It is not entirely clear when casting is the best option, but it seems preferable over surgery in this group of patients. In our study, one pediatric patient was treated surgically for a bilateral stress fracture of the scaphoid [25].

5. Conclusions

Bilateral scaphoid fractures are not very common in the literature and have been documented in only a few patients. This type of injury is often the result of high-energy trauma, typically falls from a height. Given the severity of the trauma, wrist injuries, such as fractures and dislocations, are frequently associated, while elbow injuries are less commonly observed. Patients most prone to experiencing this type of bilateral injury are males aged between 13 and 35 years, with an average age of 22 years. Patients with a history of bilateral wrist injuries should undergo a careful bilateral evaluation to assess the presence of this lesion and treat it appropriately, often through surgical means, to prevent the development of nonunion. It is essential to approach any conclusions with caution due to the small cohort size. Conservative treatment seems to lead to better results than surgery; however, just a few patients treated conservatively were included in this review. Timely and appropriate management strategies are paramount in minimizing the risk of complications and facilitating optimal recovery in cases of bilateral scaphoid fractures. Given the increased complexity and potential for adverse outcomes in bilateral injuries, an adequate approach is imperative to mitigate long-term sequelae. Interdisciplinary collaboration among orthopedic surgeons, radiologists, and rehabilitation specialists is essential to ensure the comprehensive evaluation and management of bilateral wrist injuries. By leveraging collective expertise and resources, healthcare teams can minimize diagnostic oversights and provide timely, effective interventions tailored to each patient's unique needs. The socioeconomic ramifications of bilateral scaphoid fractures extend beyond individual patient outcomes, encompassing healthcare resource utilization, work-related disability, and financial burdens. Proactive preventive measures and evidence-based treatment strategies are essential not only for optimizing patient care but also for mitigating the broader societal impact of these debilitating injuries.

The literature on this topic is still limited. Our study aims to highlight the possibility that these fractures may happen more frequently due to the increasingly common occurrence of high-energy traumas, emphasizing the importance for medical professionals to systematically assess both wrists for diagnosis and prevent the potential oversight of bilateral injuries. Continued research efforts are essential to expand our understanding of these fractures, including their prevalence, associated risk factors, and optimal treatment approaches. By building upon the existing body of literature, healthcare professionals can enhance diagnostic accuracy and refine treatment protocols to improve patient care.

Author Contributions: Conceptualization, L.D. and L.L.; methodology, L.L.; software, L.D. and M.F.; validation, L.D., L.L., M.F. and L.C.; formal analysis, L.D.; investigation, L.D. and M.S.G.; resources, L.D.; data curation, C.D.C.; writing—original draft preparation, L.D. and L.L.; writing—review and editing, L.D., L.L. and L.C.; visualization, M.S.G. and M.F.; supervision, L.C.; project administration, L.L. All authors have read and agreed to the published version of the manuscript.

Funding: This research did not receive external funding.

Institutional Review Board Statement: Not applicable.

Informed Consent Statement: Not applicable.

Data Availability Statement: Not applicable.

Conflicts of Interest: The authors declare no conflicts of interest.

References

1. Ghargozloo, D.; Tomarchio, A.; Ballerini, M.; Chitoni, G. Traumatic Bilateral Scaphoid Fractures. *Med. Glas. Off. Publ. Med. Assoc. Zenica-Doboj Cant. Bosnia Herzeg.* **2021**, *18*, 222–225. [CrossRef]
2. Krasin, E.; Goldwirth, M.; Gold, A.; Goodwin, D.R. Review of the Current Methods in the Diagnosis and Treatment of Scaphoid Fractures. *Postgrad. Med. J.* **2001**, *77*, 235–237. [CrossRef] [PubMed]
3. Van Tassel, D.C.; Owens, B.D.; Wolf, J.M. Incidence Estimates and Demographics of Scaphoid Fracture in the U.S. Population. *J. Hand Surg.* **2010**, *35*, 1242–1245. [CrossRef]
4. Winston, M.J.; Weiland, A.J. Scaphoid fractures in the athlete. *Curr. Rev. Musculoskelet. Med.* **2017**, *10*, 38–44. [CrossRef] [PubMed]
5. Hayat, Z.; Varacallo, M. Scaphoid Wrist Fracture. In *StatPearls*; StatPearls Publishing: Treasure Island, FL, USA, 2023.
6. Sabbagh, M.D.; Morsy, M.; Moran, S.L. Diagnosis and Management of Acute Scaphoid Fractures. *Hand Clin.* **2019**, *35*, 259–269. [CrossRef]
7. Lebowitz, C.; Massaglia, J.; Hoffman, C.; Lucenti, L.; Dheer, S.; Rivlin, M.; Beredjiklian, P.K. The Accuracy of 3D Printed Carpal Bones Generated from Cadaveric Specimens. *Arch. Bone Jt. Surg.* **2021**, *9*, 432–438. [CrossRef]
8. Kattan, A.E.; Almarghoub, M.A.; Alnujaim, N.H.; Barasain, O.; AlLababidi, N.H. Bilateral Scaphoid Fractures: Case Report and Literature Review. *Int. J. Surg. Case Rep.* **2020**, *66*, 412–415. [CrossRef] [PubMed]
9. Moher, D.; Liberati, A.; Tetzlaff, J.; Altman, D.G.; Group, T.P. Preferred Reporting Items for Systematic Reviews and Meta-Analyses: The PRISMA Statement. *PLOS Med.* **2009**, *6*, e1000097. [CrossRef]
10. Murad, M.H.; Sultan, S.; Haffar, S.; Bazerbachi, F. Methodological quality and synthesis of case series and case reports. *BMJ Evid. Based Med.* **2018**, *23*, 60–63. [CrossRef]
11. Stother, I.G. A Report of 3 Cases of Simultaneous Colles' and Scaphoid Fractures. *Injury* **1976**, *7*, 185–188. [CrossRef]
12. Kay, R.M.; Kuschner, S.H. Bilateral Proximal Radial and Scaphoid Fractures in a Child. *J. Hand Surg.* **1999**, *24*, 255–257. [CrossRef] [PubMed]
13. Kaneko, K.; Miyazaki, H.; Yamaguchi, T.; Yanagihara, Y.; Kurosawa, H. Bilateral Transcapholunate Dislocation. *Chir. Main* **2000**, *19*, 263–268. [CrossRef] [PubMed]
14. Ozkan, K.; Ugutmen, E.; Unay, K.; Poyanli, O.; Guven, M.; Eren, A. Fractures of the Bilateral Distal Radius and Scaphoid: A Case Report. *J. Med. Case Rep.* **2008**, *2*, 93. [CrossRef] [PubMed]
15. Yinusa, W.; Adetan, O.; Odatuwa-Omagbemi, D.O.; Eyo, M.U. Bilateral Simultaneous Fracture of the Carpal Scaphoid Successfully Treated with Conservative Cast Immobilisation: A Case Report. *West Afr. J. Med.* **2010**, *29*, 425–428. [CrossRef] [PubMed]
16. Yildirim, C.; Unuvar, F.; Keklikci, K.; Demirtas, M. Bilateral Dorsal Trans-Scaphoid Perilunate Fracture-Dislocation: A Case Report. *Int. J. Surg. Case Rep.* **2014**, *5*, 226–230. [CrossRef] [PubMed]
17. Saglam, F.; Gulabi, D.; Baysal, Ö.; Bekler, H.İ.; Tasdemir, Z.; Elmali, N. Chronic Wrist Pain in a Goalkeeper; Bilateral Scaphoid Stress Fracture: A Case Report. *Int. J. Surg. Case Rep.* **2015**, *7C*, 20–22. [CrossRef] [PubMed]
18. Mohamed Haflah, N.H.; Mat Nor, N.F.; Abdullah, S.; Sapuan, J. Bilateral Scaphoid Stress Fracture in a Platform Diver Presenting with Unilateral Symptoms. *Singapore Med. J.* **2014**, *55*, e159–e161. [CrossRef] [PubMed]
19. Virani, S.R.; Wajekar, S.; Mohan, H.; Dahapute, A.A. A Unique Case of Bilateral Trans-Scaphoid Perilunate Dislocation with Dislocation of Lunate into the Forearm. *J. Clin. Orthop. Trauma* **2016**, *7*, 110–114. [CrossRef] [PubMed]
20. Afshar, A.; Tabrizi, A. Bilateral Scaphocapitate Fracture Syndrome: A Case Report with Long-Term Follow up. *J. Hand Surg.* **2019**, *24*, 243–246. [CrossRef]
21. Meraghni, N.; Lahlali, M.A.; Mohand-Oussaid, A.; Derradji, M.; Kara, Z. Bilateral Combined Fractures of the Scaphoid and Distal Radius: A Case Report. *J. Orthop. Case Rep.* **2022**, *12*, 1–4. [CrossRef]
22. Johnson, N.A.; Kurien, T.; Horton, T.C. Recurrence of Scaphoid Stress Fracture: A Case Report. *J. Wrist Surg.* **2023**, *12*, 265. [CrossRef] [PubMed]
23. Korkoman, A.J.; Fayi, K.A.; Al-Turki, N.; Ramadhan, H.; Alkhathami, A.M. Bilateral Greater Arc Perilunate Injury: A Case Report. *Int. J. Surg. Case Rep.* **2023**, *113*, 109070. [CrossRef] [PubMed]
24. Reigstad, O.; Holm-Glad, T.; Korslund, J.; Thorkildsen, R.; Røkkum, M. Bilateral Carpal Pilon-Type Fractures Due to Clenched Fist Trauma: A Case Report. *Ann. Jt.* **2023**, *8*. [CrossRef] [PubMed]
25. Pidemunt, G.; Torres-Claramunt, R.; Ginés, A.; de Zabala, S.; Cebamanos, J. Bilateral Stress Fracture of the Carpal Scaphoid: Report in a Child and Review of the Literature. *Clin. J. Sport Med. Off. J. Can. Acad. Sport Med.* **2012**, *22*, 511–513. [CrossRef]
26. Testa, G.; Lucenti, L.; D'Amato, S.; Sorrentino, M.; Cosentino, P.; Vescio, A.; Pavone, V. Comparison between Vascular and Non-Vascular Bone Grafting in Scaphoid Nonunion: A Systematic Review. *J. Clin. Med.* **2022**, *11*, 3402. [CrossRef] [PubMed]
27. Lucenti, L.; Lutsky, K.F.; Jones, C.; Kazarian, E.; Fletcher, D.; Beredjiklian, P.K. Antegrade Versus Retrograde Technique for Fixation of Scaphoid Waist Fractures: A Comparison of Screw Placement. *J. Wrist Surg.* **2020**, *9*, 34–38. [CrossRef] [PubMed]
28. Cooney, W.P.; Dobyns, J.H.; Linscheid, R.L. Fractures of the Scaphoid: A Rational Approach to Management. *Clin. Orthop.* **1980**, *149*, 90–97. [CrossRef]
29. Dias, J.J.; Singh, H.P. Displaced Fracture of the Waist of the Scaphoid. *J. Bone Jt. Surg. Br.* **2011**, *93*, 1433–1439. [CrossRef] [PubMed]
30. Almigdad, A.; Al-Zoubi, A.; Mustafa, A.; Al-Qasaimeh, M.; Azzam, E.; Mestarihi, S.; Khair, Y.; Almanasier, G. A Review of Scaphoid Fracture, Treatment Outcomes, and Consequences. *Int. Orthop.* **2024**, *48*, 529–536. [CrossRef]

31. Jørgsholm, P.; Ossowski, D.; Thomsen, N.; Björkman, A. Epidemiology of Scaphoid Fractures and Non-Unions: A Systematic Review. *Handchir. Mikrochir. Plast. Chir.* **2020**, *52*, 374–381. [CrossRef]
32. Wells, M.E.; Nicholson, T.C.; Macias, R.A.; Nesti, L.J.; Dunn, J.C. Incidence of Scaphoid Fractures and Associated Injuries at US Trauma Centers. *J. Wrist Surg.* **2021**, *10*, 123–128. [CrossRef] [PubMed]
33. Duckworth, A.D.; Jenkins, P.J.; Aitken, S.A.; Clement, N.D.; Court-Brown, C.M.; McQueen, M.M. Scaphoid Fracture Epidemiology. *J. Trauma Acute Care Surg.* **2012**, *72*, E41. [CrossRef] [PubMed]
34. Suh, N.; Grewal, R. Controversies and Best Practices for Acute Scaphoid Fracture Management. *J. Hand Surg. Eur. Vol.* **2018**, *43*, 4–12. [CrossRef] [PubMed]
35. Garcia, R.M.; Leversedge, F.J.; Aldridge, J.M.; Richard, M.J.; Ruch, D.S. Scaphoid Nonunions Treated with 2 Headless Compression Screws and Bone Grafting. *J. Hand Surg.* **2014**, *39*, 1301–1307. [CrossRef]
36. DiPrinzio, E.V.; Dieterich, J.D.; Walsh, A.L.; Warburton, A.J.; Chang, A.L.; Hausman, M.R.; Kim, J.M. Two Parallel Headless Compression Screws for Scaphoid Fractures: Radiographic Analysis and Preliminary Outcome. *Hand* **2023**, *18*, 1267–1274. [CrossRef] [PubMed]
37. Vigni, G.E.; Cassata, G.; Caldarella, G.; Cirincione, R.; Licciardi, M.; Miceli, G.C.; Puleio, R.; D'Itri, L.; Lo Coco, R.; Camarda, L.; et al. Improved Bone Regeneration Using Biodegradable Polybutylene Succinate Artificial Scaffold in a Rabbit Model. *J. Funct. Biomater.* **2022**, *14*, 22. [CrossRef] [PubMed]
38. Cicero, L.; Licciardi, M.; Cirincione, R.; Puleio, R.; Giammona, G.; Giglia, G.; Sardo, P.; Edoardo Vigni, G.; Cioffi, A.; Sanfilippo, A.; et al. Polybutylene Succinate Artificial Scaffold for Peripheral Nerve Regeneration. *J. Biomed. Mater. Res. B Appl. Biomater.* **2022**, *110*, 125–134. [CrossRef] [PubMed]
39. Cicero, L.; Puleio, R.; Cassata, G.; Cirincione, R.; Camarda, L.; Caracappa, D.; D'Itri, L.; Licciardi, M.; Vigni, G.E. Peripheral Nerve Regeneration at 1 Year: Biodegradable Polybutylene Succinate Artificial Scaffold vs. Conventional Epineurial Sutures. *Polymers* **2023**, *15*, 3398. [CrossRef] [PubMed]
40. Gray, R.R.L.; Halpern, A.L.; King, S.R.; Anderson, J.E. Scaphoid Fracture and Nonunion: New Directions. *J. Hand Surg. Eur. Vol.* **2023**, *48*, 4S–10S. [CrossRef]
41. Liao, J.C.Y.; Chong, A.K.S. Pediatric Hand and Wrist Fractures. *Clin. Plast. Surg.* **2019**, *46*, 425–436. [CrossRef] [PubMed]
42. Hughes, J.L.; Bomar, J.D.; Wallace, C.D.; Edmonds, E.W. Factors influencing surgical decision-making in pediatric and adolescent scaphoid fractures. *J. Pediatr. Orthop. B* **2022**, *31*, 591–596. [CrossRef] [PubMed]
43. Lackey, J.T.; Seiler, P.L.; Lee, B.R.; Sinclair, M.K. Clinically Significant Treatment Delay in Pediatric Scaphoid Fractures. *J. Hand Surg. Am.* **2024**, *49*, 108–113. [CrossRef] [PubMed]

Journal of *Personalized Medicine*

MDPI

Article

Regional versus General Anesthesia in Postoperative Pain Management after Distal Radius Fracture Surgery: Meta-Analysis of Randomized Controlled Trials

Young Hak Roh *, Seong Gyun Park and Seung Hyun Lee

Department of Orthopaedic Surgery, Ewha Womans University Medical Center, Ewha Womans University College of Medicine, 1071 Anyangcheon-ro, Yangcheon-gu, Seoul 07985, Republic of Korea; silbsk15@naver.com (S.G.P.); priestman@naver.com (S.H.L.)
* Correspondence: rohyh@ewha.ac.kr; Tel.: +82-2-2650-2639; Fax: +82-2-2642-0349

Abstract: Distal radius fractures are the most prevalent upper extremity fractures, posing a significant public health concern. Recent studies comparing regional and general anesthesia for postoperative pain management after these fractures have yielded conflicting results. This meta-analysis aimed to compare the effectiveness of regional and general anesthesia concerning postoperative pain management and opioid consumption following distal radius fracture surgery. A comprehensive search was conducted in PubMed, Cochrane Library, and EMBASE databases to identify relevant randomized controlled trials. Four randomized trials involving 248 participants were included in the analysis. A pooled analysis revealed that regional anesthesia led to significantly reduced postoperative pain scores at 2 h compared to general anesthesia (SMD −2.03; 95% CI −2.88–−1.17). However, no significant differences in pain scores were observed between the two anesthesia types after 12 h post-surgery. Regional anesthesia was associated with lower total opioid consumption (SMD −0.76; 95% CI −1.25–−0.26) and fewer occurrences of nausea and vomiting compared to the general anesthesia. Nonetheless, opioid consumption on the first day post-discharge was significantly higher in the regional anesthesia group (SMD 0.83; 95% CI 0.47–1.20). The analgesic superiority of regional anesthesia is confined to the early postoperative hours with overall lower opioid use but a notable increase in opioid consumption on the first day post-discharge, potentially attributable to rebound pain.

Keywords: distal radius fracture; regional anesthesia; general anesthesia; postoperative pain; meta-analysis; randomized controlled trials

Citation: Roh, Y.H.; Park, S.G.; Lee, S.H. Regional versus General Anesthesia in Postoperative Pain Management after Distal Radius Fracture Surgery: Meta-Analysis of Randomized Controlled Trials. *J. Pers. Med.* **2023**, *13*, 1543. https://doi.org/10.3390/jpm13111543

Academic Editors: Markus Gesslein and Eduard Kurz

Received: 23 September 2023
Revised: 17 October 2023
Accepted: 25 October 2023
Published: 27 October 2023

1. Introduction

Distal radius fractures are the most frequently encountered fractures in the upper extremities and pose a significant public health concern, especially among the elderly [1]. Numerous studies have documented the rising prevalence of distal radius fractures as the active elderly population continues to grow [2]. These fractures make a substantial contribution to the overall morbidity in the elderly population and impose a significant socioeconomic burden [3]. Surgical fixation of distal radius fractures allows early initiation of motion exercises after surgery, leading to a faster resumption of daily activities than nonsurgical treatment methods [4,5]. However, postoperative pain can act as a deterrent to the early initiation of exercise, potentially reducing patient satisfaction and leading to suboptimal clinical outcomes owing to delayed rehabilitation [6].

Surgical fixation of distal radial fractures can be performed under general (GA) or regional anesthesia (RA). Regional anesthetic nerve blocks have a positive impact on various aspects of the surgical process, such as reduced acute postoperative pain, decreased opioid consumption, and improved patient satisfaction [7]. However, several studies have emphasized the considerable risk of "rebound pain" once the effects of RA wear off, with an incidence rate of approximately 40–50% reported following peripheral nerve block procedures [8–10]. A previous comprehensive meta-analysis investigating the use of various types of anesthesia for the treatment of distal radius fractures did not find substantial evidence favoring any particular method [11].

Randomized controlled trials comparing RA with GA for postoperative pain management after distal radius fractures have reported conflicting findings. Two early randomized controlled trials [12,13] reported no overall differences in analgesia after distal radial fracture fixation. Both studies found that RA was associated with rebound pain and worse pain scores than GA at approximately 12 h after surgery. However, more recently published randomized trials [14,15] reported contradictory results, with some indicating that RA provides better pain relief and higher patient satisfaction than GA for distal radius fracture fixation.

Although some reviews have been published on this topic, the studies included mixed levels of evidence and lacked quantitative analysis [11,16]. Therefore, we conducted a meta-analysis with evidence based solely on randomized controlled trials. This meta-analysis aimed to identify the effects of anesthesia on postoperative pain, opioid consumption, and occurrence of nausea and vomiting after surgery for distal radius fracture.

2. Materials and Methods

2.1. Study Strategy

This meta-analysis was conducted according to the Preferred Reporting Items for Systematic Review and Meta-Analysis (PRISMA) statement [17]. We did not register or publish a prior protocol of this study (see Supplementary Material). Search was conducted between April 2023 and July 2023.

Pubmed, Cochrane Central Register of Controlled Trials, and EMBASE database were searched for articles with language restricted to English. Initially, we did not restrict the research type or year of publication to avoid missing relevant articles. Search algorithms were developed for each database (without a set time limit). A combination of search terms was employed, encompassing the title, abstract, Medical Subject Headings (MeSH), Emtree, and keywords. The search terms included "wrist fracture[MeSH]", "distal radius fracture[tiab]", "anesthesia[MeSH]", "regional anesthesia[tiab]", "nerve block[tiab]", "brachial plexus[tiab]" and "general anesthesia[tiab].

2.2. Study Selection

Two authors (YHR, SGP) independently conducted the initial database search and screened the titles and abstracts for eligibility. The full manuscript was then reviewed for inclusion and exclusion criteria. Discrepancies were resolved by consensus among the reviewers. We also performed manual searches of the references in the relevant articles and previous systemic reviews to avoid missing eligible articles.

We used the Population, Intervention, Comparator, Outcome and Study design (PICOS) framework to formulate eligibility criteria [17]. Studies were included if they met the following selection criteria: participants, patients with distal radius fractures treated surgically; intervention, regional anesthesia; comparison, general anesthesia; outcome, postoperative pain; and design, randomized controlled trials. The exclusion criteria were mixed patient populations with indistinguishable datasets, publications that did not provide concrete data, patient populations under 18 years of age, and papers written in a language other than English.

2.3. Quality Assessment

The methodological quality of the studies was evaluated using the Cochrane Collaboration's Risk of Bias Tools [18]. These criteria were used to assess the potential bias level for each of the following components: random sequence generation (selection bias), allocation concealment (selection bias), double-blinding (performance or detection bias), selective data reporting (reporting bias), and missing data (attrition bias). Two reviewers (YHR and

SGP) independently assessed the methodological quality of the studies and unresolved disagreements were resolved by consensus through consultation with a third reviewer.

2.4. Data Extraction and Outcome Measure

Data extraction and outcome measurement were conducted independently by two authors (SGP and SHL). Any discrepancies that emerged were subject to discussion, and final decisions were reached through mutual agreement among the evaluators. We employed a pre-designed data spreadsheet for data extraction. It included key information such as author names, publication year, study design, sample size, patient demographics (age, sex, affected side, etc.), anesthesia methods, primary outcomes, and complications.

The following outcomes were investigated: postoperative pain levels, opioid consumption (in equivalent units), and occurrences of nausea and vomiting.

2.5. Statistical Analysis

The primary outcomes of this meta-analysis included postoperative pain levels and opioid consumption, with the secondary outcome being the incidence of nausea and vomiting following distal radius fracture surgery. We conducted a pooled analysis to compare multiple clinical outcome measures between the groups depending on the availability of the data. For all comparisons, standardized mean differences (SMD) with 95% confidence intervals (CI) were used for continuous variables, while odds ratios (OR) with 95% CI were calculated for binary variables. When medians with interquartile ranges were reported without means with standard deviations (SD), SDs were calculated using the quantile estimation approach [19]. Between-study heterogeneity was determined using the I^2 statistics. Values > 25%, >50%, and >70% were regarded as low, moderate, and high heterogeneity, respectively. Random effects models for meta-analysis were applied in the presence of moderate to high heterogeneity ($I^2 > 50\%$); otherwise, fixed effects models were used. The inverse variance weighting method was employed to derive pooled estimates of SMD or OR, along with their respective 95% CI. The results of the meta-analysis were considered statistically significant at $p < 0.05$. Subgroup analysis was only conducted for postoperative VAS pain at 24 h to explore potential explanations for the observed heterogeneity, given the limited number of articles available. A sensitivity analysis was conducted by omitting single eligible studies to evaluate the influence of each study in primary outcomes. A funnel plot was originally planned to detect publication bias but was excluded from the analysis owing to its limited interpretability caused by the small number of included studies. Statistical analyses were performed using the Comprehensive Meta-analysis Software version 3 (Biostat, Englewood, NJ, USA) and Review Manager (version 5.3; Cochrane Collaboration, Copenhagen, Denmark).

3. Results

3.1. Characteristics and Methodological Quality of the Included Study

The initial search yielded 412 articles, 4 of which met the inclusion and exclusion criteria and underwent qualitative analysis (Figure 1). All four studies included in this analysis were randomized controlled trials published in English between 2016 and 2022. Table 1 summarizes the key characteristics of these studies. Among them, two studies utilized the infraclavicular block approach, whereas one study employed the supraclavicular block approach and another study used the axillary block approach. The duration of acute postoperative pain assessment varied from 24 to 72 h across the studies.

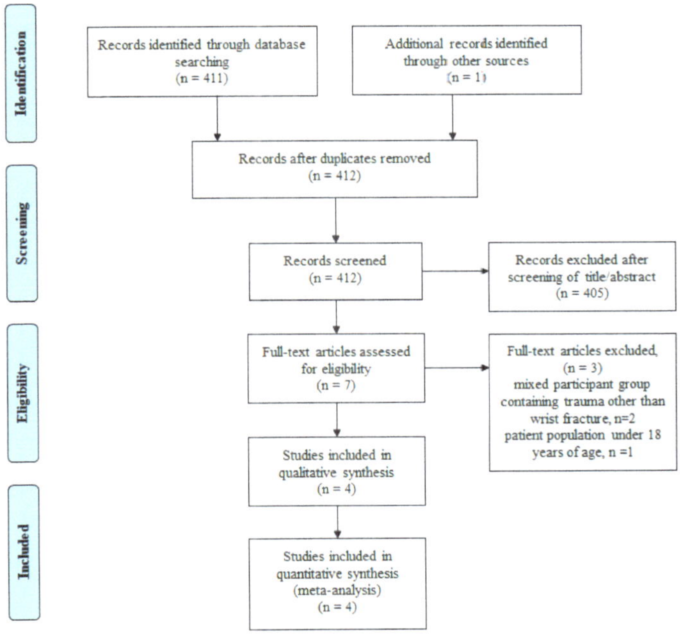

Figure 1. Preferred Reporting Items for Systematic Review and Meta-Analysis (PRISMA) flow diagram of literature selection.

Table 1. Characteristics of the included studies. RA = regional anesthesia, GA = general anesthesia, RCT = randomized controlled trial, AO/OTA = Arbeitsgemeinschaft für Osteosynthesefragen/Orthopaedic Trauma Association.

Study	Study Design	Sample Size		Mean Age (yr)		Female (%)		Injury to Right/Dominant Hand (%)		AO/OTA Fracture Type	Postoperative Pain Evaluations
		RA	GA	RA	GA	RA	GA	RA	GA		
Galos et al. (2016) [12]	RCT	18	18	54.4	54.9	50	66.6	Not reported		Type A, $n = 9$ Type B, $n = 2$ Type C, $n = 23$	2, 4, 6, 12, 24, 48, and 72 h
Rundgren et al. (2019) [13]	RCT	44	44	51.2	53.2	75	77	34	43	Type A, $n = 35$ Type B, $n = 15$ Type C, $n = 38$	0, 2, 24, 48, and 72 h, 2 weeks, and 6 months
Wong et al. (2020) [14]	RCT	26	26	59.2	58.9	69.2	73.1	42.3	42.3	Not reported	0, 1, 2, 24, and 48 h, 3 and 6 moths
Nho et al. (2022) [15]	RCT	36	36	69.5	71.2	100	100	61.1	38.9	Type A, $n = 4$ Type B, $n = 0$ Type C, $n = 68$	2, 4, 6, 12, and 24 h

Figure 2 summarizes the risk-of-bias results of the studies. Except for the absence of blinding of patients regarding the anesthesia technique and the absence of blinding of the research team responsible for collecting postoperative outcome data, the overall methodological quality of the studies was acceptable.

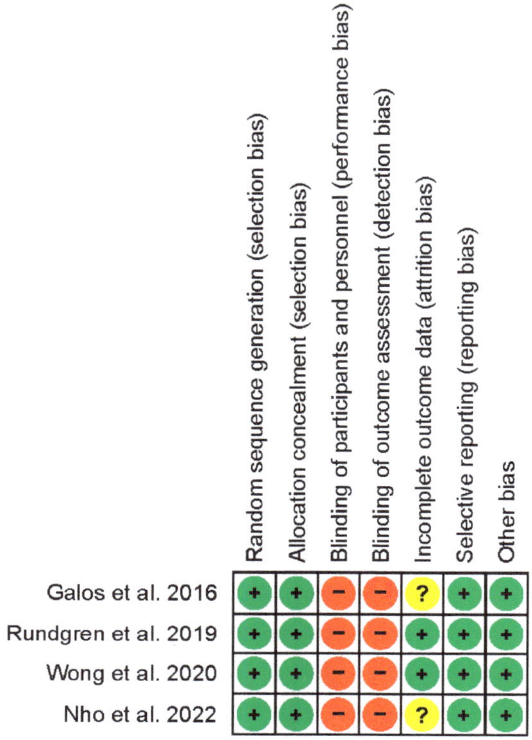

Figure 2. Methodological quality of the included randomized controlled trials. The risks of bias items were assessed and categorized as "low risk" (+), "unclear risk" (?), or "high risk" (−). Galos et al., 2016 [12]; Rundgren et al., 2019 [13]; Wong et al., 2020 [14]; Nho et al., 2022 [15].

3.2. Outcome

3.2.1. Postoperative VAS Pain Score at 2, 12, 24, and 48 h

The postoperative pain score through visual analogue scale (VAS) at 2 h was assessed in the four studies, which involved 247 wrists. All four studies consistently showed that the RA group had lower pain scores than the GA group at 2 h. The meta-analysis demonstrated significantly lower VAS scores in the RA group than the GA group. The standardized mean difference (SMD) was calculated as −2.03, with a 95% confidence interval (CI) ranging from −2.88 to −1.17 (Figure 3A). The results of the sensitivity analysis did not differ from those of the original studies.

Two studies involving 108 wrists examined VAS pain scores at 12 h. One study demonstrated significantly higher pain VAS scores in the RA group than the GA group at 12 h, whereas another study found no significant difference in pain scores between the groups. A meta-analysis revealed that the pooled effect size was not significant (Figure 3B; SMD 0.47; 95% CI −1.32–2.25; I^2 = 94%).

The VAS pain score at 24 h was assessed in four studies involving 248 wrists. Two studies reported higher VAS pain scores in the RA group compared than in the GA group, whereas one study reported higher pain scores in the GA group. The results of the meta-analysis indicated that the pooled effect size was not significant with high heterogeneity (Figure 3C; SMD 0.05; CI at 95% −0.66–0.76; I^2 = 86%). The heterogeneity was resolved when the investigations were pooled within the subgroup on outpatient surgical settings (Table 2). The sensitivity analysis did not reveal any significant differences in the outcome.

Table 2. Results of subgroup analysis for postoperative pain at 24 h according to surgical settings for regional anesthesia.

Subgroup	No. Studies	SMD; 95% CI	Heterogeneity
Outpatient surgical settings	2	0.65; 0.23–1.06	$I^2 = 0\%$, $p = 0.74$
Inpatient surgical setting	2	−0.53; −1.36–0.30	$I^2 = 80\%$, $p = 0.025$
RA with infraclavicular approach	2	−0.13; −1.8–1.6	$I^2 = 93\%$ $p = 0.01$
RA with supraclavicular approach	1	0.59; 0.07–1.11	NA
RA with axillary approach	1	−0.13; −0.59–0.34	NA

RA = regional anesthesia, SMD = standardized mean differences, CI = confidence intervals, NA = not available.

The pain score at 48 h was assessed in three studies with a total of 176 wrists. One study reported higher pain VAS scores in the GA group than in the RA group, whereas the other two investigations did not find any significant differences between the groups. The results of the meta-analysis showed no significant difference in the VAS pain scores between the two groups at 48 h (Figure 3D; SMD −0.64; CI at 95% −1.58–0.31; $I^2 = 88\%$). The sensitivity analysis revealed no change in the result of the original analysis.

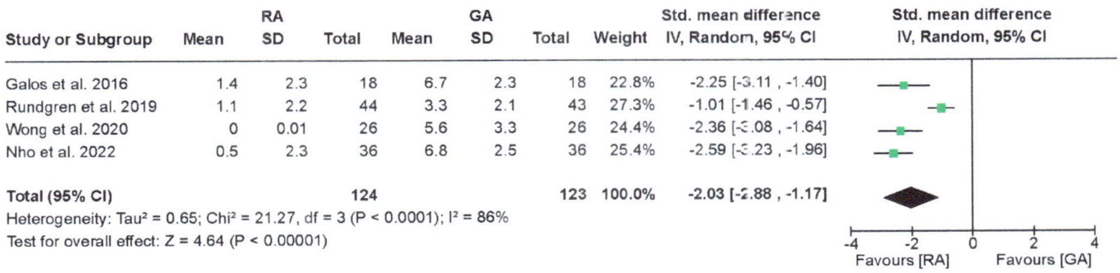

(A)

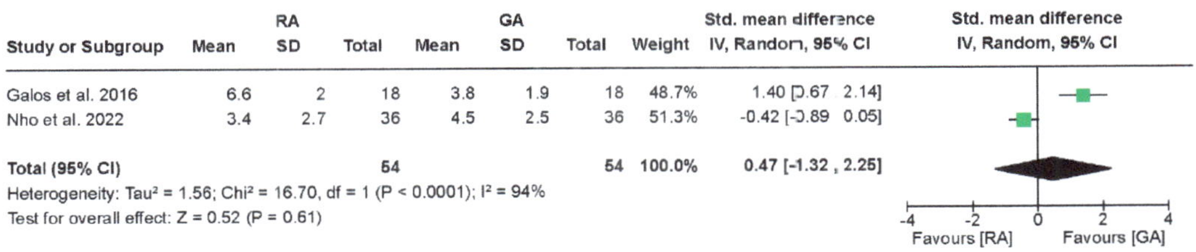

(B)

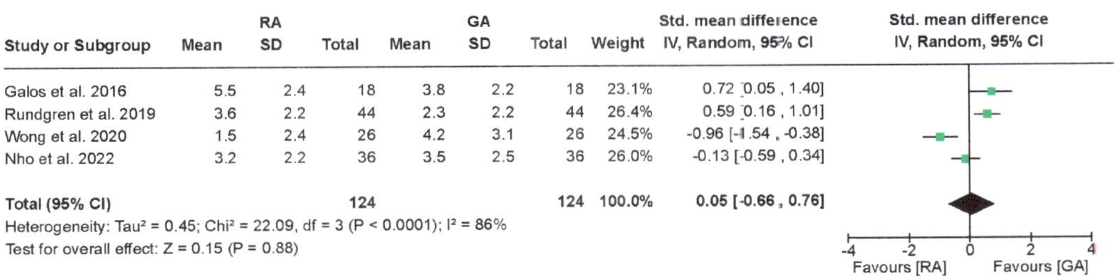

(C)

Figure 3. *Cont.*

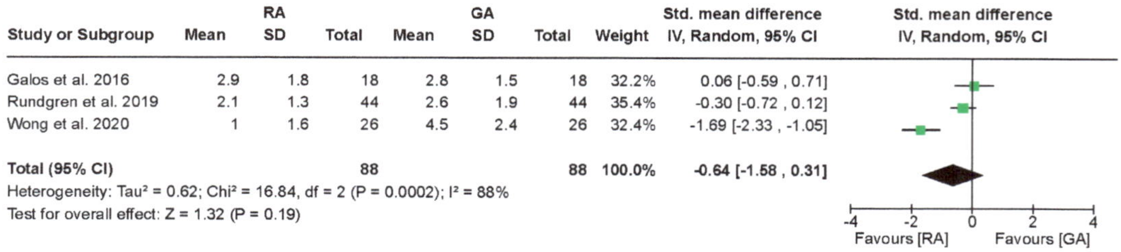

(D)

Figure 3. Forest plot of comparison in acute postoperative pain after regional or general anesthesia for distal radius fracture surgeries: (**A**) at 2 h, (**B**) at 12 h, (**C**) at 24 h, and (**D**) at 48 h. RA = regional anesthesia, GA = general anesthesia, SD = standard deviation, IV = inverse variance, CI = confidence interval. Galos et al., 2016 [12]; Rundgren et al., 2019 [13]; Wong et al., 2020 [14]; Nho et al., 2022 [15].

3.2.2. Amount of Opioid Consumption

The total amount of opioid consumption after distal radius fracture surgery was reported in three studies that included 132 wrists. Among these studies, two reported a higher total opioid consumption in the GA group than in the RA group, whereas the remaining study did not find a significant difference in opioid consumption between the two groups. The results of the meta-analysis demonstrated that the GA group had significantly greater total opioid consumption than the RA group (Figure 4A, SMD −0.76; CI at 95% −1.25–−0.26; I^2 = 58%).

Three studies consistently reported higher opioid consumption in the post-anesthesia care unit (PACU) or before discharge in the GA group than in the RA group. The meta-analysis indicated a significant increase in opioid consumption in the RA group compared to the GA group during this period (Figure 4B, SMD −0.89; CI at 95% −1.12–−0.58; I^2 = 0%).

Opioid consumption after postoperative discharge, specifically on the first day, was evaluated in two investigations focusing on ambulatory surgery. Both studies reported higher opioid use in the RA group than in the GA group. The results of the meta-analysis indicated that the patients in the RA group had significantly greater opioid use than those in the GA group (Figure 4C, SMD 0.83; CI at 95% 0.47–1.20; I^2 = 0%).

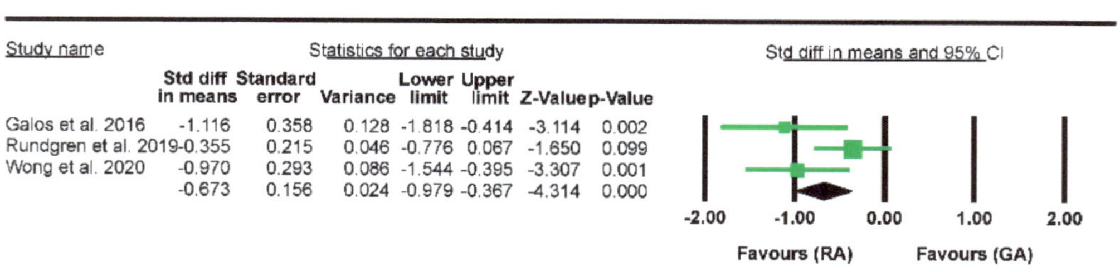

(A)

Figure 4. *Cont.*

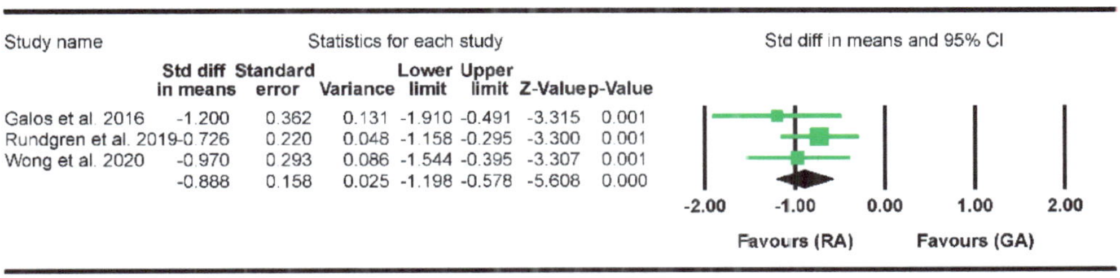

(**B**)

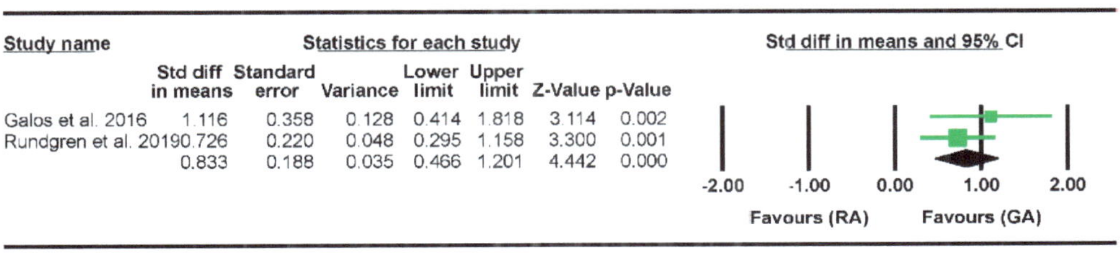

(**C**)

Figure 4. Forest plot of comparison in postoperative opioid consumption after regional or general anesthesia for distal radius fracture surgeries: (**A**) total, (**B**) in PACU, and (**C**) on the first day after discharge. RA = regional anesthesia, GA = general anesthesia, CI = confidence interval. Galos et al., 2016 [12]; Rundgren et al., 2019 [13]; Wong et al., 2020 [14].

3.2.3. Postoperative Nausea and Vomiting

Postoperative nausea and vomiting were assessed in two studies involving 140 wrists. Both studies consistently revealed a higher incidence of postoperative nausea and vomiting in the GA group than in the RA group. The results of the meta-analysis indicated a lower risk of postoperative nausea (Figure 5A; OR, 0.06, 95% CI, 0.01 to 0.35, $I^2 = 0\%$) and vomiting (Figure 5B; OR, 0.17, 95% CI, 0.04 to 0.69, $I^2 = 0\%$) in patients who underwent RA compared to those who received GA.

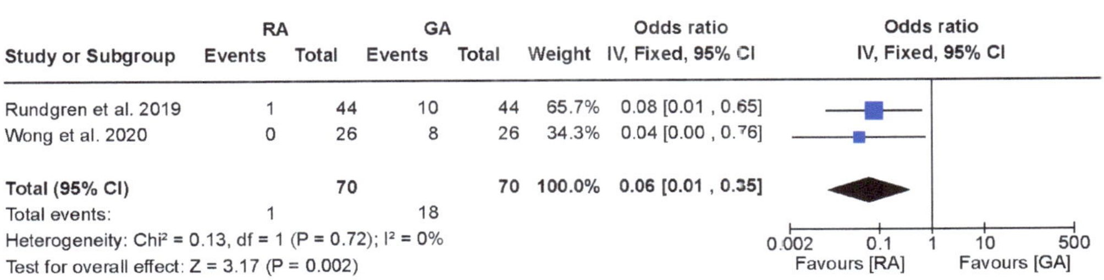

(**A**)

Figure 5. *Cont.*

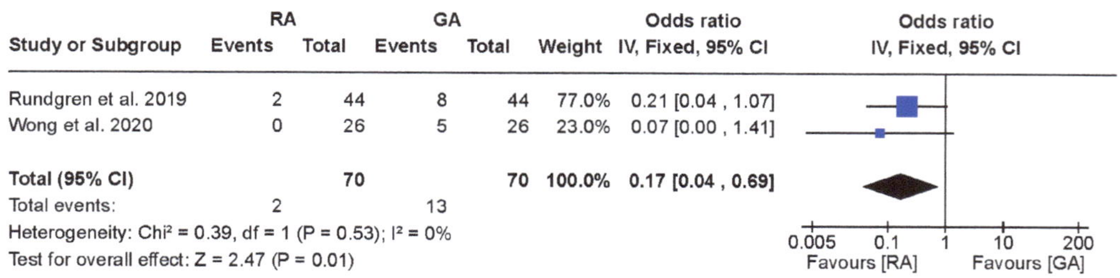

(B)

Figure 5. Forest plot of comparison in postoperative nausea and vomiting after regional or general anesthesia for distal radius fracture surgeries: (**A**) nausea and (**B**) vomiting. RA = regional anesthesia, GA = general anesthesia, SD = standard deviation, SD = standard deviation, IV = inverse variance, CI = confidence interval. Rundgren et al., 2019 [13]; Wong et al., 2020 [14].

4. Discussion

Prior investigations comparing RA and GA for postoperative pain management have yielded conflicting results, posing a significant challenge to clinicians in deciding which mode of anesthesia to employ in order to enhance post-operative recovery and reduce oral opioid use. In this meta-analysis, we identified four randomized controlled trials that directly compared RA and GA for postoperative pain management after distal radius fracture surgery. This analysis aimed to provide insights into the optimal choice of anesthesia for managing postoperative pain in this specific surgical context. Furthermore, it provides insight into rebound pain following RA, an aspect often overlooked in the context of distal radius fracture fixation. Our meta-analysis revealed significant differences in the early postoperative pain and opioid consumption based on the anesthesia method. Specifically, the GA group had significantly higher pain scores at 2 h compared to the RA group; however, opioid consumption after postoperative discharge on the first day was significantly increased in the RA group compared to in the GA group. The total opioid consumption was greater in the GA group than in the RA group. However, no significant differences were observed between the two groups in the VAS pain scores after 12 h.

RA was associated with decreased postoperative pain scores at 2 h compared to GA. However, the association diminished after the early postoperative period, and no significant differences in pain scores were observed between the GA and RA group after 12 h. The early postoperative pain patterns were significantly influenced by the anesthesia method. Patients with RA consistently exhibited a rapid increase in pain scores on the first operative day once the effects of regional anesthesia wore off. Some patients experienced an abrupt increase in pain, necessitating more pain medications. In contrast, patients with GA showed a continuous decrease in pain during the early postoperative period. Patients who received RA for ambulatory wrist fracture surgery reported a higher rate of unplanned healthcare resource utilization caused by post-discharge pain than those undergoing GA [20]. The heterogeneity in pain scores within the RA group between 12 and 24 h may have been influenced by the incidence or extent of rebound pain. Interestingly, two studies [12,13] that reported high pain scores at 12 or 24 h were conducted in outpatient surgery settings, whereas investigations [14,15] that did not show higher pain scores during this time were conducted in inpatient surgery settings. It is likely that rebound pain was more efficiently controlled in inpatient surgical settings, as previously demonstrated by the lower opioid consumption rates in shoulder surgeries performed in inpatient surgical settings than in those performed in outpatient surgical settings [21]. Patients with distal radius fractures are frequently admitted to hospital for a few days in some Asian countries, where admission costs are relatively low and patients tend to have longer hospital stays [6]. The management of rebound pain, especially in ambulatory surgical settings, is an important consideration.

Opioids continue to have a significant role for postoperative pain management, but their use in the postoperative period may elevate the risk of adverse effects, including nausea and vomiting, respiratory depression, sedation, pruritus, urinary retention, and sleep disturbances [6]. Total opioid consumption was lower in the RA group, and there were fewer instances of nausea and vomiting compared to the GA group. However, opioid consumption on the first day after discharge was significantly higher in the RA group than that in the GA group. As previously mentioned, this phenomenon can be attributed to rebound pain [8]. Rebound pain has been previously demonstrated in various upper extremity surgeries [6,22], although not reported in every investigation [14]. The results are consistent with previous findings showing that RA reduces opioid demand during the inpatient stage, but not during the outpatient period following extremity fracture surgeries [23,24]. Controversy also exists regarding the total opioid consumption after GA and RA in fracture surgeries. Several recent retrospective investigations specifically examining upper- and lower-extremity fractures have shown that RA was associated with increased inpatient and outpatient opioid demands after adjusting for baseline patient and treatment characteristics [22,25,26]. These findings contradict traditional expectations regarding the opioid-sparing effects of RA [27,28]. The observed heterogeneity may have resulted from various factors such as different types of injuries and surgeries, variation in RA technique, adjuvant medications used, and perioperative multimodal analgesia techniques employed. These factors may influence the incidence and severity of rebound pain, consequently affecting opioid consumption after RA for fracture surgeries.

The higher incidence of nausea and vomiting in the GA group could be attributed to the differences in total opioid consumption. Postoperative nausea and vomiting are known to be influenced by various factors associated with the patient, surgery, and anesthesia [29]. Previous studies demonstrated that postoperative nausea and vomiting are strongly influenced by postoperative opioid use in a dose-dependent manner [30]. This effect appears to persist as long as opioids are used for pain management in the postoperative period [31]. A substantial number of patients undergoing ambulatory hand surgery experience distressing levels of postoperative nausea and vomiting, and dissatisfaction with ambulatory hand surgery correlated with moderate as well as high levels of postoperative nausea and vomiting [32].

The main limitation of this meta-analysis is the small number of included studies. The inclusion criteria were limited to randomized controlled trials, resulting in a limited pool of studies eligible for the meta-analysis. However, it was chosen to ensure a high level of evidence and facilitate high-grade recommendations. Additionally, heterogeneity in outcome reporting was observed among the included studies. Not all of the included studies provided data on opioid consumption, and the timing of follow-up assessments varied across studies. Further research with larger sample sizes and diverse study settings is warranted to provide a more comprehensive understanding of this topic. In addition, treatment-related variables, including the time to surgery and initial closed reductions before surgery, and injury characteristics, such as open or closed fractures, could potentially influence postoperative pain level after distal radius fracture [33,34], but these potentially confounding variables were not addressed in this meta-analysis due to limitations in the available information from the included studies. Another limitation of this meta-analysis was the variability in the postoperative pain management protocols among the included trials. Although these variations reflect actual clinical practice (real-world settings), these introduced a potential source of heterogeneity into the analysis. To compensate for this inherent variability, our study examined postoperative opioid consumption and the incidence of nausea and vomiting as additional outcome measures.

5. Conclusions

The use of RA for distal radial fracture surgery was associated with reduced postoperative pain scores during the early postoperative period only. However, after 12 h, no significant differences in the VAS pain scores were observed between the GA and RA groups. Total opioid consumption was higher in the GA group than in the RA group. However, opioid consumption on the first day after discharge was significantly higher in the RA group, potentially because of rebound pain. These findings provide valuable insights for the selection of an appropriate anesthetic method based on patient characteristics and specific trauma surgeries.

Supplementary Materials: The following supporting information can be downloaded at: https://www.mdpi.com/article/10.3390/jpm13111543/s1.

Author Contributions: Conceptualization, Y.H.R., S.G.P. and S.H.L.; methodology, Y.H.R. and S.G.P.; software, S.G.P. and S.H.L.; validation, Y.H.R. and S.H.L.; formal analysis, Y.H.R. and S.H.L.; investigation, Y.H.R., S.H.L. and S.G.P.; resources, S.H.L.; data curation, S.H.L. and S.G.P.; writing—original draft preparation, Y.H.R.; writing—review and editing, all the authors; visualization, Y.H.R. and S.H.L.; supervision, Y.H.R. All authors have read and agreed to the published version of the manuscript.

Funding: This research received no external funding.

Institutional Review Board Statement: Not applicable.

Informed Consent Statement: Not applicable.

Data Availability Statement: The data presented in this study are available upon request to the corresponding author.

Conflicts of Interest: The authors declare no conflict of interest.

References

1. Chung, K.C.; Shauver, M.J.; Birkmeyer, J.D. Trends in the United States in the treatment of distal radial fractures in the elderly. *J. Bone Jt. Surg. Am.* **2009**, *91*, 1868–1873. [CrossRef] [PubMed]
2. Nellans, K.W.; Kowalski, E.; Chung, K.C. The epidemiology of distal radius fractures. *Hand Clin.* **2012**, *28*, 113–125. [CrossRef] [PubMed]
3. Shauver, M.J.; Clapham, P.J.; Chung, K.C. An economic analysis of outcomes and complications of treating distal radius fractures in the elderly. *J. Hand Surg. Am.* **2011**, *36*, 1912–1918.e3. [CrossRef] [PubMed]
4. Ring, D.; Jupiter, J.B. Treatment of osteoporotic distal radius fractures. *Osteoporos. Int.* **2005**, *16* (Suppl. 2), S80–S84. [CrossRef]
5. Mellstrand-Navarro, C.; Pettersson, H.J.; Tornqvist, H.; Ponzer, S. The operative treatment of fractures of the distal radius is increasing: Results from a nationwide Swedish study. *Bone Jt. J.* **2014**, *96*, 963–969. [CrossRef]
6. Chung, M.S.; Roh, Y.H.; Baek, G.H.; Lee, Y.H.; Rhee, S.H.; Gong, H.S. Evaluation of early postoperative pain and the effectiveness of perifracture site injections following volar plating for distal radius fractures. *J. Hand Surg. Am.* **2010**, *35*, 1787–1794. [CrossRef]
7. Joshi, G.; Gandhi, K.; Shah, N.; Gadsden, J.; Corman, S.L. Peripheral nerve blocks in the management of postoperative pain: Challenges and opportunities. *J. Clin. Anesth.* **2016**, *35*, 524–529. [CrossRef]
8. Barry, G.S.; Bailey, J.G.; Sardinha, J.; Brousseau, P.; Uppal, V. Factors associated with rebound pain after peripheral nerve block for ambulatory surgery. *Br. J. Anaesth.* **2021**, *126*, 862–871. [CrossRef]
9. Williams, B.A.; Bottegal, M.T.; Kentor, M.L.; Irrgang, J.J.; Williams, J.P. Rebound pain scores as a function of femoral nerve block duration after anterior cruciate ligament reconstruction: Retrospective analysis of a prospective, randomized clinical trial. *Reg. Anesth. Pain Med.* **2007**, *32*, 186–192. [CrossRef]
10. Dada, O.; Gonzalez Zacarias, A.; Ongaigui, C.; Echeverria-Villalobos, M.; Kushelev, M.; Bergese, S.D.; Moran, K. Does Rebound Pain after Peripheral Nerve Block for Orthopedic Surgery Impact Postoperative Analgesia and Opioid Consumption? A Narrative Review. *Int. J. Environ. Res. Public. Health* **2019**, *16*, 3257. [CrossRef]
11. Sellbrandt, I.; Brattwall, M.; Warren Stomberg, M.; Jildenstal, P.; Jakobsson, J.G. Anaesthesia for open wrist fracture surgery in adults/elderly. *F1000Research* **2017**, *6*, 1996. [CrossRef]
12. Galos, D.K.; Taormina, D.P.; Crespo, A.; Ding, D.Y.; Sapienza, A.; Jain, S.; Tejwani, N.C. Does Brachial Plexus Blockade Result in Improved Pain Scores After Distal Radius Fracture Fixation? A Randomized Trial. *Clin. Orthop. Relat. Res.* **2016**, *474*, 1247–1254. [CrossRef] [PubMed]
13. Rundgren, J.; Mellstrand Navarro, C.; Ponzer, S.; Regberg, A.; Serenius, S.; Enocson, A. Regional or General Anesthesia in the Surgical Treatment of Distal Radial Fractures: A Randomized Clinical Trial. *J. Bone Jt. Surg. Am.* **2019**, *101*, 1168–1176. [CrossRef] [PubMed]
14. Wong, S.S.; Chan, W.S.; Fang, C.; Chan, C.W.; Lau, T.W.; Leung, F.; Cheung, C.W. Infraclavicular nerve block reduces postoperative pain after distal radial fracture fixation: A randomized controlled trial. *BMC Anesthesiol.* **2020**, *20*, 130. [CrossRef] [PubMed]

15. Nho, J.H.; Jang, B.W.; An, C.Y.; Yoo, J.H.; Song, S.; Cho, H.B.; Kim, S.H.; Kim, S.I.; Jung, K.J.; Kim, B. General versus Brachial Plexus Block Anesthesia in Pain Management after Internal Fixation in Patients with Distal Radius Fracture: A Randomized Controlled Trial. *Int. J. Environ. Res. Public. Health* **2022**, *19*, 9155. [CrossRef]
16. Heroux, J.; Bessette, P.O.; Belley-Cote, E.; Lamarche, D.; Echave, P.; Loignon, M.C.; Patenaude, N.; Baillargeon, J.P.; D'Aragon, F. Functional recovery with peripheral nerve block versus general anesthesia for upper limb surgery: A systematic review. *BMC Anesthesiol.* **2023**, *23*, 91. [CrossRef]
17. Liberati, A.; Altman, D.G.; Tetzlaff, J.; Mulrow, C.; Gotzsche, P.C.; Ioannidis, J.P.; Clarke, M.; Devereaux, P.J.; Kleijnen, J.; Moher, D. The PRISMA statement for reporting systematic reviews and meta-analyses of studies that evaluate health care interventions: Explanation and elaboration. *J. Clin. Epidemiol.* **2009**, *62*, e1–e34. [CrossRef]
18. Higgins, J.P.; Altman, D.G.; Gotzsche, P.C.; Juni, P.; Moher, D.; Oxman, A.D.; Savovic, J.; Schulz, K.F.; Weeks, L.; Sterne, J.A.; et al. The Cochrane Collaboration's tool for assessing risk of bias in randomised trials. *BMJ* **2011**, *343*, d5928. [CrossRef]
19. Wan, X.; Wang, W.; Liu, J.; Tong, T. Estimating the sample mean and standard deviation from the sample size, median, range and/or interquartile range. *BMC Med. Res. Methodol.* **2014**, *14*, 135. [CrossRef]
20. Sunderland, S.; Yarnold, C.H.; Head, S.J.; Osborn, J.A.; Purssell, A.; Peel, J.K.; Schwarz, S.K. Regional Versus General Anesthesia and the Incidence of Unplanned Health Care Resource Utilization for Postoperative Pain After Wrist Fracture Surgery: Results From a Retrospective Quality Improvement Project. *Reg. Anesth. Pain Med.* **2016**, *41*, 22–27. [CrossRef]
21. Chan, J.J.; Cirino, C.M.; Vargas, L.; Poeran, J.; Zubizarreta, N.; Mazumdar, M.; Galatz, L.M.; Cagle, P.J. Peripheral nerve block use in inpatient and outpatient shoulder arthroplasty: A population-based study evaluating utilization and outcomes. *Reg. Anesth. Pain Med.* **2020**, *45*, 818–825. [CrossRef]
22. Cunningham, D.J.; LaRose, M.A.; Zhang, G.X.; Paniagua, A.R.; Klifto, C.S.; Gage, M.J. Beware the rebound effect: Regional anesthesia increases opioid utilization after humerus fracture surgery. *Shoulder Elbow* **2022**, *14*, 648–656. [CrossRef] [PubMed]
23. Cunningham, D.J.; Paniagua, A.; DeLaura, I.; Zhang, G.; Kim, B.; Kim, J.; Lee, T.; LaRose, M.; Adams, S.; Gage, M.J. Regional Anesthesia Decreases Inpatient But Not Outpatient Opioid Demand in Ankle and Distal Tibia Fracture Surgery. *Foot Ankle Spec.* **2022**, 19386400221088453. [CrossRef]
24. Cunningham, D.; LaRose, M.; Kinamon, T.; MacAlpine, E.; Au, S.; Paniagua, A.; Klifto, C.; Gage, M.J. The impact of regional anesthesia on opioid demand in distal radius fracture surgery. *J. Plast. Surg. Hand Surg.* **2023**, *57*, 299–307. [CrossRef] [PubMed]
25. Cunningham, D.J.; Paniaugua, A.R.; LaRose, M.A.; DeLaura, I.F.; Blatter, M.K.; Gage, M.J. Regional anesthesia does not decrease inpatient or outpatient opioid demand in distal femur fracture surgery. *Arch Orthop. Trauma. Surg.* **2022**, *142*, 1873–1883. [CrossRef] [PubMed]
26. Cunningham, D.J.; Robinette, J.P.; Paniagua, A.R.; LaRose, M.A.; Blatter, M.; Gage, M.J. Regional anesthesia does not decrease opioid demand in pelvis and acetabulum fracture surgery. *Eur. J. Orthop. Surg. Traumatol.* **2022**, *32*, 1357–1370. [CrossRef]
27. Cunningham, D.J.; LaRose, M.; Zhang, G.; Patel, P.; Paniagua, A.; Gadsden, J.; Gage, M.J. Regional Anesthesia Associated With Decreased Inpatient and Outpatient Opioid Demand in Tibial Plateau Fracture Surgery. *Anesth. Analg.* **2022**, *134*, 1072–1081. [CrossRef]
28. Hadzic, A.; Arliss, J.; Kerimoglu, B.; Karaca, P.E.; Yufa, M.; Claudio, R.E.; Vloka, J.D.; Rosenquist, R.; Santos, A.C.; Thys, D.M. A comparison of infraclavicular nerve block versus general anesthesia for hand and wrist day-case surgeries. *Anesthesiology* **2004**, *101*, 127–132. [CrossRef]
29. Gan, T.J. Risk factors for postoperative nausea and vomiting. *Anesth. Analg.* **2006**, *102*, 1884–1898. [CrossRef]
30. Roberts, G.W.; Bekker, T.B.; Carlsen, H.H.; Moffatt, C.H.; Slattery, P.J.; McClure, A.F. Postoperative nausea and vomiting are strongly influenced by postoperative opioid use in a dose-related manner. *Anesth. Analg.* **2005**, *101*, 1343–1348. [CrossRef]
31. Apfel, C.C.; Philip, B.K.; Cakmakkaya, O.S.; Shilling, A.; Shi, Y.Y.; Leslie, J.B.; Allard, M.; Turan, A.; Windle, P.; Odom-Forren, J.; et al. Who is at risk for postdischarge nausea and vomiting after ambulatory surgery? *Anesthesiology* **2012**, *117*, 475–486. [CrossRef] [PubMed]
32. Roh, Y.H.; Gong, H.S.; Kim, J.H.; Nam, K.P.; Lee, Y.H.; Baek, G.H. Factors associated with postoperative nausea and vomiting in patients undergoing an ambulatory hand surgery. *Clin. Orthop. Surg.* **2014**, *6*, 273–278. [CrossRef] [PubMed]
33. Khan, S.; Persitz, J.; Shrouder-Henry, J.; Khan, M.; Chan, A.; Paul, R. Effect of Time-To-Surgery on Distal Radius Fracture Outcomes: A Systematic Review. *J. Hand Surg. Am.* **2023**, *48*, 435–443. [CrossRef] [PubMed]
34. Low, S.; Papay, M.; Eingartner, C. Pain Perception following Initial Closed Reduction in the Preoperative Care of Unstable, Dorsally Displaced Distal Radius Fractures. *J. Hand Microsurg.* **2019**, *11*, 111–116.

Journal of *Personalized Medicine*

MDPI

Article

Head–Shaft Angle Influences Isometric Shoulder Strength Levels after Intramedullary Nailing of Proximal Humerus Fractures: A Pilot Study

Sebastian Grimme [1,2], Hermann Josef Bail [1], Johannes Rüther [1], Michael Millrose [1,3], Roland Biber [4], Markus Gesslein [1] and Maximilian Willauschus [1,*]

1 Department of Orthopedics and Traumatology, Paracelsus Medical University, General Hospital Nuremberg, Breslauer Straße 201, 90471 Nuremberg, Germany; sebastian.grimme@usb.ch (S.G.); hermann-josef.bail@klinikum-nuernberg.de (H.J.B.); johannes.ruether@klinikum-nuernberg.de (J.R.); michael.millrose@klinikum-gap.de (M.M.); markus.gesslein@klinikum-nuernberg.de (M.G.)

2 Department of Orthopedics and Traumatology, University Hospital Basel, 4031 Basel, Switzerland

3 Department of Trauma Surgery and Sports Medicine, Garmisch-Partenkirchen Medical Centre, 82467 Garmisch-Partenkirchen, Germany

4 Department of Traumatology, Clinic Dr. Erler gGmbH, 90429 Nuremberg, Germany; r.biber@erler-klinik.de

* Correspondence: maximilian.willauschus@klinikum-nuernberg.de

Abstract: Background: Proximal humerus fractures are common fractures of the elderly population which can lead to long-term compromise of a patient's shoulder function. Closed reduction and internal fixation with intramedullary nailing is a well-established surgical technique yielding good outcomes, as perceived by patients, obtained via Patient-Reported Outcome Measures, and objectified by clinical shoulder testing. Apart from conventional range-of-motion testing and clinical shoulder tests, strength testing of the shoulder is a yet-neglected but meaningful and standardizable outcome parameter. In this study, isometric shoulder strength is evaluated in relation to fracture morphology/postoperative reduction quality as well as with patient-reported outcomes. Methods: 25 patients (mean age 73.2 ± 10.5 years) underwent isometrics strength-testing of the shoulder joint in the scapular plane (abduction) as well as in the sagittal plane (flexion) as well as hand-grip strength-testing at 4.5 ± 1.88 years follow-up. Pre- and postoperative radiographs were analysed. Patients completed ASES and CMS questionnaires. Results: Patients exhibited a decrease in abduction and flexion force (−24.47% and −25.30%, respectively, $p < 0.001$) using the contralateral, uninjured arm as reference. Abduction force tended to be decreased in three- and four-part fractures. Patient satisfaction correlated negatively with the relatively reduced force of the affected arm. Varus-angulated humeral heads produced significantly lower abduction force output than valgus- or physiologic angulation ($p = 0.014$), whereas flexion force was unaffected ($p = 0.468$). The anatomical reduction had no influence on shoulder strength. Conclusions: Proximal humerus fractures may cause a significant reduction in shoulder function, both reported by patients and objectified by shoulder strength testing. Varus head angulation demonstrated the greatest loss of shoulder strength and should be avoided to ensure proper functioning. Further, strength testing seems a valuable outcome parameter for a thorough shoulder examination with easy obtainability.

Keywords: proximal humerus fracture; intramedullary nailing; head–shaft angle; head–neck angle; isometric strength; PROMs

Citation: Grimme, S.; Bail, H.J.; Rüther, J.; Millrose, M.; Biber, R.; Gesslein, M.; Willauschus, M. Head–Shaft Angle Influences Isometric Shoulder Strength Levels after Intramedullary Nailing of Proximal Humerus Fractures: A Pilot Study. *J. Pers. Med.* **2024**, *14*, 907. https://doi.org/10.3390/jpm14090907

Academic Editor: Jih-Yang Ko

Received: 21 July 2024
Revised: 14 August 2024
Accepted: 22 August 2024
Published: 27 August 2024

1. Introduction

Proximal humerus fractures (PHFs) are among the most common fractures affecting the elderly population [1–3]. Despite being an incisive event threatening one's autonomy, the treatment of PHFs poses an increasing burden onto health-care expenditure and care availability. Against the background of a variety of surgical treatment options for displaced fractures of the surgical neck, intramedullary nailing (IMN) is a widely used technique yielding good to excellent outcomes for patients [4–7]. Hereby, a nail is inserted into the proximal medullary canal via the humeral head and fixated by interlocking screws to stabilize the repositioned fracture parts. A relatively small incision, soft tissue and nerve sparing, early postoperative physiotherapeutic exercise and a comparably short operation time represent the major advantages [8–10] Several studies and meta-analyses have investigated outcomes based on subjective patient-centred or objective clinical measures. Established risk factors include insufficient anatomical reduction, four-part fractures, varus head–shaft angles (<130°) and the occurrence of complications [11,12]. Until now, strength measurements were commonly utilized for Constant–Murley score calculation, despite being an objective and easily measurable objective variable worthwhile addressing independently. While being linked with function, muscle strength is also a protective factor for falls, mobility, and independence [13,14]. Since it is reasonable that deficiencies in flexion and abduction force may impair activities of daily life, its magnitude and associations with other outcome parameters remain unclear. As previously described in the literature, isometric abduction force in the scapular plane and flexion force in the sagittal plane are well-established benchmarks for shoulder strength, the joint comprised by PHFs. Currently, there is limited knowledge regarding the factors that influence isometric strength following proximal humeral fractures. Additionally, there is a lack of published data concerning changes in shoulder strength after IMN.

Furthermore, since hand grip strength is a reliable indicator for frailty and adverse health outcomes, its assessment for potential declines following IMN gives valuable information concerning the long-term outcomes and compromises patients face [15].

The primary aim of this study was to evaluate the changes in shoulder strength and the according influence of anatomical reduction, fracture morphology and complications. The secondary aim was to investigate strength as related to Patient-Reported Outcome Measures (PROMs) following intramedullary nailing of proximal humerus fractures.

2. Materials and Methods

For this follow-up examination study, a total of 479 patients who underwent treatment with the Targon PH+ intramedullary nail (150 mm, Aesculap, B. Braun) for proximal humeral fractures (PHFs) at a single level-one trauma centre between 2014 and 2021 were considered. The study was approved by the institutional review committee (Paracelsus Medical University at the Nuremberg Hospital, Number IRB-2021-025).

Subsequently, exclusion criteria, as listed below, were applied, resulting in 260 patients who met the eligibility criteria for inclusion.

Exclusion criteria are comprised of patients unable to consent due to dementia and other neurologic diseases, concomitant shoulder injuries, insufficient clinical data or inadequate radiologic imaging, metastatic fractures, open fractures, additional shaft fractures, additional shoulder comorbidity (e.g., rotator cuff injuries, labral lesions), death in the follow-up period, conversion to shoulder prosthesis in the course of treatment, implantation of a nail length exceeding 150 mm and shoulder pathology at the contralateral arm. The inclusion criteria for this study encompassed patients with isolated proximal humeral fractures, unilateral injury, a minimum adequate radiological follow-up of more than one year, and provision of consent to participate. All patients who participated in the study provided written informed consent.

Thereafter, a cohort of 31 patients was successfully contacted and had provided informed consent to undergo a clinical examination and strength testing at our institution's outpatient clinic. Two patients were excluded due to underlying conditions preventing load-bearing activities and four were excluded due to the implantation of a reverse shoulder prosthesis.

Patient's contact information and general details were obtained from the clinic's database and hospital information system, specifically by SAP (Wallendorf, Baden-

Württemberg, Germany). Data analysis was performed using the statistical software IBM SPSS Statistics for Windows (version 28, 1.0.0.1406, IBM Corp., Armonk, NY, USA).

2.1. Surgical Technique

In every case, the decision for IMN was taken by a board of experienced orthopaedic and traumatological experts.

Prior to surgery, patients were positioned in beach chair position on a radiolucent table with a standard armrest. For implantation of the Targon PH+ nail, a deltoid split approach at the anterolateral margin of the acromion was performed. After a longitudinal transection of the clavipectoral fascia and the subacromial bursa, the head fragments were reduced via the joystick technique. For this procedure, a wire was used to adequately reduce the fragments.

After longitudinal splitting of the supraspinatus tendon, a guide pin at the apex of the humeral head was positioned after fluoroscopic control in two planes, and the nail was inserted 3 to 4 mm below the cartilage level. Then, a minimum of three to four locking head fixation screws were inserted into the head fragments, depending on the fracture pattern, the degree of instability, as well as bone quality.

A single or duplicate distal interlocking screw was placed through the nail percutaneously. Each procedure was performed by a board-certified trauma surgeon.

An intra- and postoperative x-ray examination was obtained from all patients to verify both the reduction and correct positioning of the implant.

The affected arm was immobilized in a sling bandage for one day to control postoperative pain and swelling. On the second postoperative day, passive mobilization by trained physiotherapists was started. An active-assisted shoulder movement below the pain threshold was then initiated as soon as wound healing was deemed satisfactory. After six weeks and radiological consolidation, weight-bearing was allowed. Every patient received postoperative standardized anterior-posterior and y-view radiographic imaging for correct fracture reposition and screw position. A radiographic follow-up after 12 months or longer was obtained from all patients.

2.2. Radiologic Fracture Assessment

The pre- and postoperative radiographic imaging was assessed using Picture Archiving and Communication System software (Version: Ashvins 2017 Build 1637), which involved the analysis of plain radiographic images or computed tomography (CT) scans. Fractures were classified based on the number of fractured parts and according to the Neer classification system [16]. The analysis also focused on assessing the postoperative anatomical reposition of the fractures and the pre- and postoperative head–shaft angle (HSA) (Figure 1).

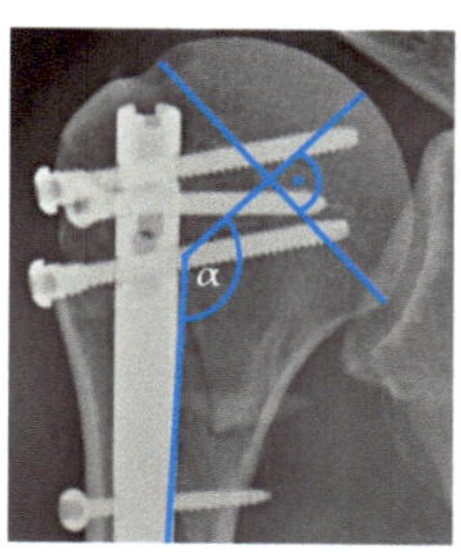

Figure 1. Head–shaft angle.

The follow-up radiographs were carefully evaluated by two experienced trauma surgeons to identify any documented complications. Anatomical reduction was considered when certain criteria had been met: the head showed neither varus angulation (<130°)

nor valgus angulation (>140°), the y-view showed no anterior or posterior tilt of the head exceeding 20°, the dislocation of the greater or lesser tuberosity was less than 3 mm, and the neck–shaft dislocation was less than 5 mm. Early postoperative complications and revisions were evaluated during the hospital stay of the patients. These included reviews of postoperative X-rays and medical records. Later, complications and revisions were recorded from hospital charts during visits at our outpatient clinic or from readmission to any hospital and finally, at the point of follow-up. All complications were categorised according to the Clavien–Dindo classification and thereupon subdivided into major (Clavien–Dindo grade III and IV) and minor complications (Clavien–Dindo grade I and II) [17]. Every secondary surgery, besides elective implant removal, was counted as a revision. Implant failure was defined as the removal of the nail and/or conversion to a different osteosynthesis or the secondary implantation of a reverse shoulder prosthesis in the instance of unsatisfactory healing of the fracture site.

2.3. Examination of Patient-Related Outcome Measures

All patients signed the informed consent form and were reassured to feel capable of performing strength measurements. The validated German versions of the America Shoulder and Elbow Surgeons score (ASES) [18], the patient-reported part of Constant–Murley score (CMS), as well as patient's satisfaction (very good, good, satisfactory, poor) were obtained. Other than ASES, besides Activities of Daily Life (ADL), CMS includes a physician-rated section including range of motion (ROM) (goniometer) and isometric strength measurements, thus being a "hybrid PROM". After completing a baseline assessment for shoulder stability and ROM, a short warm-up routine was introduced.

The examination was performed with the patient in a standing position. Evaluation of pre-injury shoulder function was presumed by the function of the contralateral, i.e., unaffected arm.

Bilateral isometric strength of both shoulder abduction in the scapular plane (kgs) and shoulder flexion (kgs) was assessed using the handheld CE-certified and calibrated electronic spring balance VOLTCRAFT HS-100 by Conrad® (Hirschau, Bavaria, Germany) (see Figure 2a). To improve standardization, the handle length was adjusted according to the patient's height. Beginning with the unaffected arm, patients were instructed to pull a handle with their straight arm parallel to the ground with maximum effort without provoking pain (see Figure 2b). Importantly, patients were instructed and observed not to augment their strength by contralateral torso inclination. Each, flexion force and abduction force were examined twice in an interval of one minute. The highest achieved orthogonal force was noted and interpreted as a measure of muscular strength (kgs).

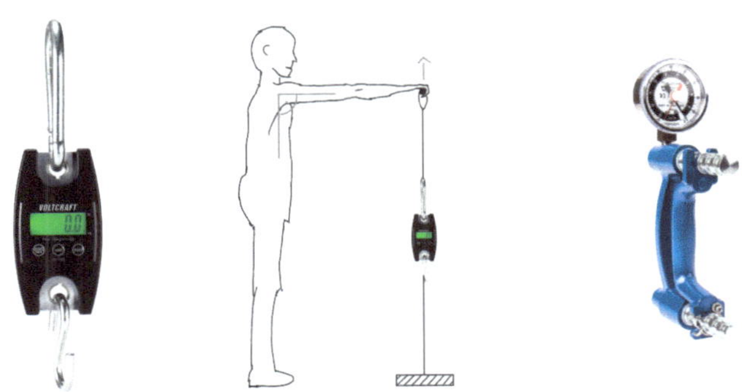

(a) (b) (c)

Figure 2. (**a**) VOLTCRAFT® CE-certified electronic spring balance, with kind permission of Conrad Electronic SE; (**b**) Test set-up for flexion force measurement with the handheld handle attached to the spring balance; (**c**) BASELINE® hand dynamometer with kind permission of Sport-Tec GmbH (Pirmasens, Germany).

2.4. Isometric Strength Measurements as a Proxy for Muscular Capability

Isometric strength measurements, although dependent on muscle length and position, show a high correlation with isokinetic measurements while being a more feasible option, since isokinetic measurement devices are rarely available in the clinical setting due to high acquisition costs. Both methods show clinical relevance and have been established for more than thirty years [19,20]. Further, isometric strength production has shown to be higher than isokinetic force production at any given joint's angle [20].

Measuring isometric strength in the 90° abduction or flexion position after a proximal humeral fracture is crucial for assessing recovery, particularly focusing on the deltoid muscle, which is known to play a key role in shoulder abduction from 15° to 90°. At this angle, the deltoid's lateral head is most active, making it an ideal position to evaluate its strength, especially considering the muscle's significance following a fracture.

This position also offers joint stability, with the head of the humerus securely positioned within the glenoid fossa. This reduces the risk of subluxation or dislocation during the assessment. Furthermore, since daily activities often require arm movement at or above shoulder height, measuring strength at 90° provides valuable insights into functional recovery.

By choosing the 90° abduction or flexion position for measuring isometric strength, we also ensure that the assessment includes the scapulothoracic joint's contribution to shoulder movement. In this position, the coordinated action of the trapezius and serratus anterior muscle is necessary to stabilize the scapula and allow smooth arm elevation. Assessing strength in this position helps evaluate not only the deltoid's function but also the integrity of the scapulothoracic joint and the effectiveness of its muscle coordination, which are critical for overall shoulder function and stability post-fracture.

Bilateral hand grip strength was assessed with the calibrated BASELINE® hydraulic hand dynamometer. After adjusting the handle grip width to hand size, subjects were instructed to compress the handle in a smooth motion in a 90° flexed elbow, whereby the measures were repeated three times to select the maximal score (kgs) (see Figure 2c).

2.5. Statistical Analysis

Because of not only sex-related but inter-individual strength differences of all subjects, it is requisite to introduce relative strength variables. Relative abduction force (AF_{rel}), relative flexion force (FF_{rel}) and relative hand grip force (HG_{rel}) depict a quotient of the affected arm's absolute force and the unaffected arm's absolute force, resulting in a percentage allowing for comparability and statistical analysis. Further, relative strength values of a patient compared to their healthy, also presumably pre-injury arm, naturally show a higher clinical relevance than absolute values.

$$\text{AFrel}(\%) = \frac{\text{Absolute flexion force(kgs) of the affected arm}}{\text{Absolute flexion force(kgs) of the unaffected arm}}$$

Statistical analysis was performed using IMB SPSS Statistics for Mac (Version 29.0.2.0, IBM Corp., Armonk, NY, USA). All variables were tested for normal distribution. For force differences between the affected and unaffected arm, analysis via Mann–Whitney U and for non-parametrical comparisons of >3 groups (varus, valgus, physiologic) Kruskal–Wallis analysis was conducted for significance. For paired samples, our data were non-parametric, which suggested the use of the Wilcoxon test. Spearman's Rho correlation was calculated for non-parametric data (relative strength and all PROMS reported p-values are two-tailed, with a well-established alpha level of <0.05 considered statistically significant. Unless otherwise stated, descriptive results are demonstrated as mean ± standard deviation (SD) and range (r).

3. Results

3.1. Patients' Characteristics

A total of 25 patients, 19 female and 6 males, were enrolled with a mean age of 73.2 ± 10.5 years and a mean follow-up time of 4.5 ± 1.9 years.

Table 1 displays demographic data of the patient collective, including fracture morphology and reduction quality, as well as complications and PROMs.

Table 1. Summary of descriptive statistics for metric and nominal data. Metric data are reported as mean ± standard deviation and range, nominal data as total number *n* and associated percentages (%).

	Total	Female	Male
Gender	*n* = 25	*n* = 19	*n* = 6
Age	73.2 ± 10.5 (SD) r: 55–88	74.63 ± 10.84 (SD) r: 55–88	68.50 ± 8.55 (SD) r: 56–81
≥65 years	18 (72.0%)	14 (73.69%)	4 (66.7%)
<65 years	7 (28.0%)	5 (26.31%)	2 (33.3%)
Follow up (years)	4.50 ± 1.88 (SD) r: 1.9–7.6	4.45 ± 1.75 (SD) r: 2.0–7.6	4.68 ± 2.42 (SD) r: 1.9–7.6
Injured side			
right	*n* = 9 (36%)	*n* = 6	*n* = 3
left	*n* = 16 (64%)	*n* = 13	*n* = 3
dominant	*n* = 8 (32%)	*n* = 7	*n* = 1
non-dominant	*n* = 17 (68%)	*n* = 12	*n* = 5
Fracture parts			
two-part	*n* = 6 (24%)	*n* = 2	*n* = 4
three-part	*n* = 16 (64%)	*n* = 14	*n* = 2
four-part	*n* = 3 (12%)	*n* = 3	*n* = 0
Head–shaft angle postoperatively	135.8 ± 7.9° (SD); r: 124.8–151.8°		
Varus < 130°	*n* = 5 (20%)	*n* = 4	*n* = 1
physiologic	*n* = 14 (56%)	*n* = 9	*n* = 5
valgus > 140°	*n* = 6 (24%)	*n* = 6	*n* = 0
Complications according to Clavien–Dindo			
I or II i.e., "minor"	*n* = 1 (4%)	*n* = 1	*n* = 0
III i.e., "major"	*n* = 5 (20%)	*n* = 4	*n* = 1
Implant failure	*n* = 4 (16%)	*n* = 3	*n* = 1
PROMs			
ASES Score	85.40 ± 16.6 (SD) r: 31.7–100	81.84 ± 17.55 (SD) r: 31.66–10.00	96.67 ± 5.35 (SD) r: 88.32–100.00
CMS affected arm	69.72 ± 13.13 (SD) r: 39.0–92.0	68.05 ± 10.60 (SD) r: 40.00–78.00	75.00 ± 19.46 (SD) r: 39.00–92.00
CMS unaffected arm	77.68 ± 11.10 (SD) r: 40.0–98.0	76.53 ± 5.73 (SD) r: 58.00–84.00	81.33 ± 21.27 (SD) r: 40.00–98.00
CMS_{rel}	89.67 ± 10% (SD) r: 57.14–101.3%	88.75 ± 11.18% r: 57.14–101.30%	92.56 ± 4.15% r: 85.19–97.50%

The patient collective had various medical preconditions. In total, fourteen patients had previously been diagnosed with osteoporosis, six patients had cardiac preconditions, three with atrial fibrillation and 1 with a history of myocardial infarction. Three patients had renal disease and pulmonary conditions. Two patients had an ongoing malignancy. Two patients had neurologic conditions. There was one patient with a history of stroke and diabetes mellitus.

Postoperative physiotherapy was prescribed to every patient for outpatient rehabilitation. In total, 24 of 25 patients attended physiotherapy. Thereof, 10 patients received ≤12 h, 10 patients up to 100 h and 4 patients even more hours of physiotherapeutic training. Additionally, 23 patients did additional exercises themselves, with 76% (*n* = 19) including strengthening exercises for the shoulder girdle. No significant relationship between duration of physiotherapy (≤12 h, <100 h, >100 h) and shoulder function was observed (flexion ROM: $p = 0.481$; abduction ROM: $p = 0.798$; AF_{rel}: $p = 0.541$, FF_{rel}: $p = 0.865$). On average, mean physical activity for one week did not differ at the point of follow-up (3:37 h pre-injury vs. 3:27 h at follow-up) ($p = 0.121$).

Table 2 reports the relative strength values in isometric force for AF_{rel}, FF_{rel} and HG_{rel}.

Table 2. Absolute and relative force measurements of the affected and unaffected arm as well as Constant–Murley scores (CMS).

	Affected Arm Force	Unaffected Arm Force	Relative Force Difference	*p*-Value
Absolute isometric abduction force (N ± SD) Relative isometric abduction force (range)	2.90 ± 1.95 kg 0.64–8.07 kg	3.86 ± 2.54 kg 1.36–10.70 kg	−24.47% (AFrel)	$p < 0.001$
Absolute isometric flexion force (N ± SD) Absolute isometric flexion force (range)	3.13 ± 1.81 kg 1.09–8.26 kg	4.21 ± 2,27 kg 1.63–10.7 kg	−25.30% (FFrel)	$p < 0.001$
Absolute isometric hand grip force (N ± SD) Relative isometric hand grip force (range)	25.64 ± 13.5 kg 12.0–67.0 kg	27.6 ± 12.4 kg 13.0–60.0 kg	−8.30% (HGrel)	$p = 0.003$
CMS relative ± SD Range	69.72 ± 13.1 39.0–92.0	77.68 ± 11.1 40.0–98.0	−10.25%	$p < 0.001$

Performing an independent-samples Mann–Whitney U test hypothesising existing differences of relative strength measures between sexes did not show significant differences of relative strength in all three measures (AF_{rel}: $p = 0.366$, FF_{rel}: $p = 0.828$, HG_{rel}: $p = 0.328$)

Significant lower force outputs of the affected arm were found compared to the unaffected arm concerning flexion and abduction forces. Relative differences in flexion force and abduction force ($p = 0.72$) as well as absolute values ($p = 0.21$) did not differ significantly, indicating a likewise reduction of both shoulder forces (abduction and flexion) after injury.

Additionally, no difference in strength measurements was obtained with regards to injury of the dominant or non-dominant side; hereby, AF_{rel} and FF_{rel} as well as HG_{rel} were identical in both groups, after injury of the dominant or non-dominant side ($p = 0.194$, $p = 0.124$, $p = 0.374$).

3.2. *Influence of Head–Shaft Angle on Strength Outcome*

When categorizing the data concerning HSA (varus, physiologic and varus), the subclassification significantly influences force output during isometric strength testing. Patients in the varus group showed only 50.41 ± 5.13% of isometric abduction force compared to their unaffected arm, compared to 81.65 ± 28.10% and 90.65 ± 25.18% in the physiologic and valgus groups, respectively ($p = 0.468$). This significant decrease in abduction force for varus-angulated humeral heads is a meaningful finding ($p = 0.014$) (see Figure 3). Flexion force appeared unaffected by head-angulation ($p = 0.522$).

The valgus group had the lowest restrictions in ROM (abduction: −23.00 ± 9.75° for varus, −22.14 ± 22.59° for physiologic, −10 ± 8.94° for valgus; flexion: −17.00 ± 10.95° for varus, −19.29 ± 15.67° for physiologic, −11.67 ± 15.71° for valgus). However, these differences were not significant (abduction: $p = 0.751$; flexion: $p = 0.939$).

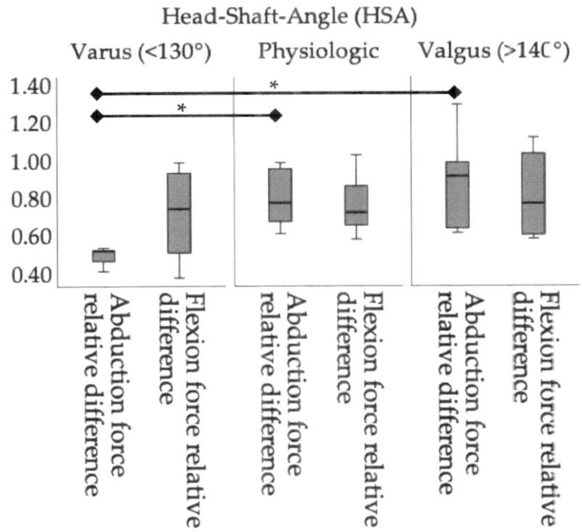

Figure 3. Relative strength differences according to the head–shaft angle group. * = $p < 0.05$.

3.3. Hand Grip Strength

Statistical analysis yielded a significant negative difference between the hand grip strength of the affected arm to the unaffected arm ($p = 0.009$). However, since most patients have injured their non-dominant, presumably weaker arm. Further hypothesis testing yielded no statistical significance correcting for this confounder. In dominant-side injuries, the mean hand grip force of the affected and unaffected arm was 26.88 kg and 26.5 kg, respectively ($p = 0.833$). Contrarily, for non-dominant side injuries, the affected arm showed statistically significant negative differences of the affected arm (25.01 kg and 28.12 kg, $p = 0.001$).

3.4. Influence of Fragment Count on Strength Outcome

Analysing two-, three-, or four-part fractures yielded different relative strength outputs in abduction force, although missing statistical significance: In two-part fractures of the surgical neck, AF_{rel} is 82.17%, which further decreases to three-part fractures (AF_{rel}: 76.22%) and two four-part fractures (AF_{rel}: 75.53%) ($p = 0.939$). For FF_{rel}, this trend is not consistent (two-part FFrel: 76.67%; three-part FF_{rel}: 75.00%; four-part FF_{rel}: 82.63%) ($p = 0.878$). A statistically significant decrease of CMS_{rel} from 96.1% in two-part to 89.4% in three-part to 78.3% in four-part PHFs ($p = 0.01$) may be attributable to decreased general function rather than significant ramifications of strength.

3.5. Relation of PROMs and Shoulder Strength

A comparison of independent variables yielded significant differences of FF_{rel} and patient-reported satisfaction with the surgery ($p = 0.046$). For $AF_{rel,}$ no significant difference was found ($p = 0.087$). To illustrate, patients who rated the surgery with 1 "very good" ($n = 11$) exhibited on average an AF_{rel} of 90.83%, whereas patients who rated a grade 2 "good" ($n = 12$) produced only an AF_{rel} of 68.46%. For FF_{rel}, the same trend was observed, where patients who rated "very good" and "good" achieved a FF_{rel} of 86.98% and 69.65%, respectively. Only three patients rated the result as "satisfactory" and demonstrated an AF_{rel} of 59.17% and a FF_{rel} of 57.67%. Regarding ASES, a significant positive correlation with AF_{rel} of the affected arm ($\rho = 0.48$; $p = 0.02$) and hand grip force of the affected arm ($\rho = 0.41$, $p = 0.04$) was demonstrated. In the same sense, a missing correlation of the unaffected arm with ASES reaffirms these findings.

3.6. Time to Follow-Up and Shoulder Function

Although displaying heterogenic times to follow up ranging from 1.9 years to 7.6 years, thus variance in possible rehabilitation improvements, shoulder function did not differ accordingly. By Kruskal–Wallis analysis for independent samples with defined groups by years to follow up no statistically significant difference was examined in the tested parameters (flexion ROM: $p = 0.118$, abduction ROM: $p = 0.439$; AF_{rel}: $p = 0.741$; FF_{rel}: $p = 0.809$). Therefore, a sufficient and statistically equal level of rehabilitation between the patients was assumed.

4. Discussion

Firstly, our investigation revealed a significant reduction of strength in the two most important shoulder movements, i.e., flexion and abduction, following years after surgery. Standardized strength measurements, i.e., with a handheld static dynamometer, seem valuable for the investigation of shoulder joint function after surgery. Secondly, patients with varus-angulated head–shaft angles have produced significantly less abduction force compared to physiologic or even valgus angulation, possibly due to leverage-arm deteriorations and early subacromial impingement. Thirdly, we have identified a significant reflection of the strength parameters in patients' subjective satisfaction with the surgery and PROMs, which undermines the clinical relevance of strength testing.

In our measurements, force in both degrees of freedom has shown to be reduced by about a quarter compared to the healthy, unaffected arm. To our knowledge, this is the first publication of raw isometric strength data after intramedullary nailing of proximal humerus fractures, and it is a challenge to contextualize these values. Despite the vast number of studies and meta-analyses comparing IMN with locking plates or conservative treatment for proximal humerus fractures regarding reduction quality, complications or PROMs, there is no comprehensive juxtaposition of long-term strength outcomes [8,21]. Standing alone, the randomized controlled trial by Zhu et al. (2011) with 51 subjects concluded a significantly higher relative abduction strength in the locking plate group compared to IMN (77.4% compared to 64%) one year postoperatively ($p = 0.032$) [10]. In their three-year postoperative follow-up, there were no differences ($p = 0.106$), contradicting our results. Unfortunately, the authors did not specify the methodology of the measurements and whether isometric or isokinetic data were involved.

Additionally, Verdano et al. (2014) have analysed isokinetic flexion–extension and internal–external rotation strength in patients with PHFs treated with open reduction and internal fixation (ORIF), i.e., locking plates and reported no statistically significant difference in flexion–extension peak torque but a notable trend [13]. Meanwhile, their CMS values differed between shoulders: their explanation infers patient's subjective perception of worse function or residual pain creates these differences, which is a statement our data cannot reinforce. Another possible explanation for the discrepancies between CMS and Verdano's strength values could be the absence of an evaluation of abduction strength, which appears to play a critical role following proximal humeral fractures and is evaluated in the CMS, potentially accounting for a maximum of 25% of points.

Further, although our strength measurements show significant differences between sides, they may still be underestimated due to the exclusion of patients with presumably below-average shoulder function: patients not reaching testing positions and patients treated with RSA as revision surgery following IMN.

The substantial influence of the head–shaft angle on isometric abduction force stands as the pivotal discovery. To our knowledge, up until now there exist no studies analysing the influence of the head–shaft angle on clinical shoulder strength.

A possible reason for presumably worse abduction force in varus-angulated humeral heads could be a superomedial shift of the greater tuberosity and, therefore, a reduced capacity of the pre-shortened supraspinatus muscle to produce force. Although it is commonly believed that its contribution to shoulder abduction ceases after the first 20 to 40°, there are studies showing a contribution during abduction in the frontal plane far beyond that [22]. Also, this finding was not observed to the same degree for flexion force. Reasoning, this movement in the sagittal plane is mainly generated by the anterior fibres of the deltoid and pectoralis major muscle, which are anatomically not impacted by HSA differences, which are constituted in the frontal plane.

Another factor can be the increased risk of early subacromial impingement during abduction due to the varus-angulated head, possibly impeding maximum strength output in the pain-provoking position where the subacromial space is most narrow. It is important to add that these observations are unlikely attributable to the training effect or any methodological errors since flexion force was tested before the abduction force. In the future, there may be room for appropriate biomechanical studies, if our findings can be reproduced.

Adjunctively, we decided to report the well-established dynamometer grip force measurements as a prognostic geriatric indicator for frailty and overall body strength for patients who obtained PHFs. Our results illustrate significantly reduced relative hand grip forces on the affected arm even years after the injury (27.03 kgs ± 12.16 -> 24.60 kgs ± 13.14), which may increase the patient's geriatric decline. Additionally, the effect of PHFs on hand grip force has not been described in previous literature, and there is a known link of grip strength to frailty and sarcopenia, two common geriatric syndromes [23].

Interestingly, we have illustrated that patients with small abduction and flexion force differences between the injured and healthy arms report consistently better overall satisfaction. Similarly, ASES, which is a validated PROMs measure for shoulder fractures, correlates with shoulder strength, indicating clear relevance of the variable "strength" on patient-reported outcomes. Following these investigations, a naturally arising question would be whether these measurable strength reductions have a practical significance in the patient's life and whether the constraint is seriously contributing to premature loss of independence in the elderly. Notably, patient satisfaction with the procedure was closely tied to strength levels, as evidenced by cohorts reporting lower satisfaction when FF_{rel} and AF_{rel} were more diminished.

Another breaking point for outcome research is the consideration of postsurgical rehabilitation and patient's lifestyle. All except one patient included in this study have completed the minimum required 12 h of physiotherapy, while many have received considerably more. No difference in ROM or strength outcomes was observed, regarding the amount of physiotherapeutic training. Here, one must acknowledge the typical "chicken or the egg" causality dilemma, where more physiotherapy following prolonged functional progression might have benefitted the patient in the end.

The present results of this pilot study should be evaluated considering specific limitations. First, this study included a small number of subjects, which is a considerable limitation, especially regarding the analysis of subgroups. For future endeavours, it would be of great interest to challenge our results in a larger patient collective. Secondly, considering the mid- to long-term follow-up time, subjects were not standardized for their different rehabilitation regime or physiotherapy, which may be a confounding variable. Finally, presuming strength reductions following surgery is based on the assumption that the contralateral arm at the point of follow-up depicts the identical (theoretical) strength of the affected arm if the injury (PHF) did not occur. Since most patients fractured their non-dominant arm, the pre-injury shoulder strength may have been lower than the contralateral shoulder beforehand.

5. Conclusions

Proximal humerus fractures cause significant strength loss in the affected shoulder, particularly in the elderly, regardless of fracture complexity or anatomical reduction. Our findings indicate that surgical reduction leading to a varus head–shaft angle should be avoided due to the associated reduction in abduction force. Isometric strength measurements are valuable clinical indicators that should be included in further research with more participants and clinical shoulder scoring systems to enhance the accuracy of evaluations.

Incorporating these measurements provides comprehensive insights into muscle function, improving the validity of research findings and patient care in shoulder-related conditions.

Author Contributions: Conceptualization, M.W., H.J.B. and S.G.; methodology, M.W. and S.G.; software, M.W. and S.G.; validation, M.G., M.W. and H.J.B.; formal analysis, M.G., M.W., J.R., M.M., R.B. and H.J.B.; investigation, M.W. and S.G. resources, H.J.B., S.G. and M.W.; data curation, S.G. and M.W.; writing—original draft preparation, S.G. and M.W.; writing—review and editing, M.G., M.W., J.R., M.M., R.B. and H.J.B.; visualization, S.G. and M.W.; supervision, M.G. and H.J.B.; project administration, M.W.; funding acquisition, M.W. All authors have read and agreed to the published version of the manuscript.

Funding: This research received no external funding.

Institutional Review Board Statement: This study was conducted in conformity with the Declaration of Helsinki and authorized by the Institutional Review Board of Paracelsus Medical University Nuremberg. (Protocol code IRB-2021-025 and date of approval: 21 December 2021).

Informed Consent Statement: Written and informed consent was obtained from all subjects involved in the study.

Data Availability Statement: Data are contained within the article.

Conflicts of Interest: Author Roland Biber was employed by the company Clinic Dr. Erler gGmbH. The company had no role in the design, execution, interpretation, or writing of the study. The remaining authors declare that the research was conducted in the absence of any commercial or financial relationships that could be construed as a potential conflict of interest.

References

1. Court-Brown, C.M.; Bugler, K.E.; Clement, N.D.; Duckworth, A.D.; McQueen, M.M. The epidemiology of open fractures in adults. A 15-year review. *Injury* **2012**, *43*, 891–897. [CrossRef] [PubMed]
2. Rupp, M.; Walter, N.; Pfeifer, C.; Lang, S.; Kerschbaum, M.; Krutsch, W.; Baumann, F.; Alt, V. The Incidence of Fractures Among the Adult Population of Germany–an Analysis From 2009 through 2019. *Dtsch. Ärzteblatt Int.* **2021**, *118*, 665–669. [CrossRef]
3. Nowak, L.L.; Dehghan, N.; McKee, M.D.; Schemitsch, E.H. Plate fixation for management of humerus fractures. *Injury* **2018**, *49* (Suppl. S1), S33–S38. [CrossRef] [PubMed]
4. Lopiz, Y.; Garríguez-Pérez, D.; Martínez-Illán, M.; García-Fernández, C.; Marco, F. Third-generation intramedullary nailing for displaced proximal humeral fractures in the elderly: Quality of life, clinical results, and complications. *Arch. Orthop. Trauma Surg.* **2022**, *142*, 227–238. [CrossRef]
5. Greenberg, A.; Rosinsky, P.J.; Gafni, N.; Kosashvili, Y.; Kaban, A. Proximal humeral nail for treatment of 3- and 4-part proximal humerus fractures in the elderly population: Effective and safe in experienced hands. *Eur. J. Orthop. Surg. Traumatol.* **2021**, *31*, 769–777. [CrossRef]
6. Boileau, P.; d'Ollonne, T.; Bessière, C.; Wilson, A.; Clavert, P.; Hatzidakis, A.M.; Chelli, M. Displaced humeral surgical neck fractures: Classification and results of third-generation percutaneous intramedullary nailing. *J. Shoulder Elb. Surg.* **2019**, *28*, 276–287. [CrossRef]
7. Paniagua Collado, M.A.; Willauschus, M.; Rüther, J.; Lahner, M.; Bail, H.J.; Geßlein, M. Intramedullary nailing as a safe and efficient treatment option for all types of displaced proximal humeral fractures in geriatric patients: A retrospective evaluation of 49 patients treated with the new Targon PH+ nail. *Technol. Health Care* **2021**, *29*, 771–780. [CrossRef]
8. Shi, X.; Liu, H.; Xing, R.; Mei, W.; Zhang, L.; Ding, L.; Huang, Z.; Wang, P. Effect of intramedullary nail and locking plate in the treatment of proximal humerus fracture: An update systematic review and meta-analysis. *J. Orthop. Surg. Res.* **2019**, *14*, 285. [CrossRef]
9. Krivohlávek, M.; Lukás, R.; Taller, S.; Srám, J. Use of angle-stable implants for proximal humeral fractures: Prospective study. *Acta Chir. Orthop. Traumatol. Cech.* **2008**, *75*, 212–220. [CrossRef]
10. Zhu, Y.; Lu, Y.; Shen, J.; Zhang, J.; Jiang, C. Locking intramedullary nails and locking plates in the treatment of two-part proximal humeral surgical neck fractures: A prospective randomized trial with a minimum of three years of follow-up. *J. Bone Jt. Surg. Am.* **2011**, *93*, 159–168. [CrossRef]
11. Tepass, A.; Rolauffs, B.; Weise, K.; Bahrs, S.D.; Dietz, K.; Bahrs, C. Complication rates and outcomes stratified by treatment modalities in proximal humeral fractures: A systematic literature review from 1970–2009. *Patient Saf. Surg.* **2013**, *7*, 34. [CrossRef]
12. Willauschus, M.; Schram, L.; Millrose, M.; Rüther, J.; Loose, K.; Bail, H.J.; Geßlein, M. Specific Radiologic Risk Factors for Implant Failure and Osteonecrosis of the Humeral Head after Interlocking Nailing with the Targon PH. *J. Clin. Med.* **2022**, *11*, 2523. [CrossRef] [PubMed]
13. Verdano, M.A.; Baudi, P.; Pellegrini, A.; Lunini, E.; Aliani, D.; Scita, G.; Schiavi, P.; Costantino, C.; Ceccarelli, F. Strength recovery after proximal humeral fractures treated with locking plate. *Musculoskelet. Surg.* **2014**, *98* (Suppl. S1), 61–69. [CrossRef] [PubMed]

14. Costantino, C.; Verdano, M.A.; Jacopetti, M.; Romiti, D.; Lunini, E.; Pellegrini, A.; Ceccarelli, F. Isokinetic strength test and functional outcomes in proximal humeral fractures treated with a locking plate. *J. Orthop. Sci.* **2014**, *19*, 776–785. [CrossRef] [PubMed]
15. Mehmet, H.; Yang, A.W.H.; Robinson, S.R. Measurement of hand grip strength in the elderly: A scoping review with recommendations. *J. Bodyw. Mov. Ther.* **2020**, *24*, 235–243. [CrossRef] [PubMed]
16. Carofino, B.C.; Leopold, S.S. Classifications in brief: The Neer classification for proximal humerus fractures. *Clin. Orthop. Relat. Res.* **2013**, *471*, 39–43. [CrossRef]
17. Dindo, D.; Demartines, N.; Clavien, P.A. Classification of surgical complications: A new proposal with evaluation in a cohort of 6336 patients and results of a survey. *Ann. Surg.* **2004**, *240*, 205–213. [CrossRef]
18. Goldhahn, J.; Angst, F.; Drerup, S.; Pap, G.; Simmen, B.R.; Mannion, A.F. Lessons learned during the cross-cultural adaptation of the American Shoulder and Elbow Surgeons shoulder form into German. *J. Shoulder Elb. Surg.* **2008**, *17*, 248–254. [CrossRef]
19. Reed, R.L.; Den Hartog, R.; Yochum, K.; Pearlmutter, L.; Ruttinger, A.C.; Mooradian, A.D. A comparison of hand-held isometric strength measurement with isokinetic muscle strength measurement in the elderly. *J. Am. Geriatr. Soc.* **1993**, *41*, 53–56. [CrossRef]
20. Kuhlman, J.R.; Iannotti, J.P.; Kelly, M.J.; Riegler, F.X.; Gevaert, M.L.; Ergin, T.M. Isokinetic and isometric measurement of strength of external rotation and abduction of the shoulder. *J. Bone Jt. Surg. Am.* **1992**, *74*, 1320–1333. [CrossRef]
21. Lange, M.; Brandt, D.; Mittlmeier, T.; Gradl, G. Proximal humeral fractures: Non-operative treatment versus intramedullary nailing in 2-, 3- and 4-part fractures. *Injury* **2016**, *47* (Suppl. S7), S14–S19. [CrossRef] [PubMed]
22. de Witte, P.B.; Werner, S.; ter Braak, L.M.; Veeger, H.E.; Nelissen, R.G.; de Groot, J.H. The Supraspinatus and the Deltoid—Not just two arm elevators. *Hum. Mov. Sci.* **2014**, *33*, 273–283. [CrossRef] [PubMed]
23. Sousa-Santos, A.R.; Amaral, T.F. Differences in handgrip strength protocols to identify sarcopenia and frailty—A systematic review. *BMC Geriatr.* **2017**, *17*, 238. [CrossRef] [PubMed]

Journal of *Personalized Medicine*

MDPI

Article

Risk Factors and Clinical Outcomes after Antegrade Intramedullary Nailing in Proximal Humeral Fractures: Insights and Implications for Patient Satisfaction

Maximilian Willauschus [1,*], Sebastian Grimme [1], Kim Loose [1], Johannes Rüther [1], Michael Millrose [1,2], Roland Biber [3], Markus Gesslein [1] and Hermann Josef Bail [1]

1 Department of Orthopedics and Traumatology, General Hospital Nuremberg, Paracelsus Medical University, Breslauer Straße 201, 90471 Nuremberg, Germany; johannes.ruether@klinikum-nuernberg.de (J.R.); hermann-josef.bail@klinikum-nuernberg.de (H.J.B.)

2 Department of Trauma Surgery and Sports Medicine, Garmisch-Partenkirchen Medical Centre, 82467 Garmisch-Partenkirchen, Germany

3 Department of Traumatology, Clinic Dr. Erler gGmbH, 90429 Nuremberg, Germany; r.biber@erler-klinik.de

* Correspondence: maximilian.willauschus@klinikum-nuernberg.de; Tel.: +49-911-3982600

Abstract: Background: Proximal humeral fractures (PHFs) are common injuries that can lead to significant functional impairment. This retrospective cohort study aimed to evaluate the clinical outcomes and complications associated with the use of the Targon PH+ (Fa. Aesculap, Germany) intramedullary nail for the treatment of PHFs. Methods: A subgroup consisting of 70 patients with a mean follow-up of 4.91 years out of 479 patients who underwent treatment with the Targon PH+ intramedullary nail for PHFs at a single center between 2014 and 2021 were included. Patient-reported outcome measures (PROMs) and health-related quality of life (HRQoL) were assessed using validated German versions of the Disabilities of the Arm, Shoulder, and Hand (DASH) questionnaire, American Shoulder and Elbow Surgeons (ASES) score, Oxford Shoulder Score (OSS), and EuroQol 5-Dimension 5-Level (EQ-5D-5L). Radiographic assessment was performed using pre- and postoperative imaging. Results: Among the 70 patients of the subgroup who completed follow-up, 21.4% experienced complications, including major complications in 15.7% of cases, all of which were revised (revision rate of 15.7%). Anatomical reduction was achieved in 48,5% of cases. The mean DASH, ASES, and OSS scores were 25.4 ± 22.0, 76.2 ± 21.1, and 38.8 ± 10.3, respectively. Significant correlations were observed among the PROMs, indicating their convergent validity. Additionally, a significant correlation of all used PROMs and patient well-being (HRQoL) was observed. Severe complications and revisions were associated with significantly lower ASES scores (−11.1%, $p = 0.013$). There was a tendency for PROM scores to slightly decline with increasing fracture complexity, although this trend did not reach statistical significance. Our findings indicate that patients over the age of 65 years tend to exhibit lower scores in PROMs and HRQoL measures. Conclusion: The use of the Targon PH+ intramedullary nail for the treatment of PHFs resulted in satisfactory clinical outcomes and acceptable complication and revision rates. The PROMs and HRQoL measures indicated varying levels of disability and symptoms, with major complications, revision surgery, and age negatively impacting shoulder function after midterm follow-up.

Keywords: proximal humeral fractures; intramedullary nail; complications; patient-reported outcome measures; health-related quality of life

check for updates

Citation: Willauschus, M.; Grimme, S.; Loose, K.; Rüther, J.; Millrose, M.; Biber, R.; Gesslein, M.; Bail, H.J. Risk Factors and Clinical Outcomes after Antegrade Intramedullary Nailing in Proximal Humeral Fractures: Insights and Implications for Patient Satisfaction. *J. Pers. Med.* **2023**, *13*, 1224. https://doi.org/10.3390/jpm13081224

Academic Editor: Jih-Yang Ko

Received: 12 July 2023
Revised: 26 July 2023
Accepted: 28 July 2023
Published: 1 August 2023

1. Introduction

Proximal humerus fractures (PHFs) rank among the most prevalent fractures in the elderly population (>65 years), with a notably higher incidence in females, approximately twice that of males [1].

The higher prevalence of fractures in females can be attributed to the increased likelihood of developing osteoporosis in postmenopausal women. Reduced bone stability in this population leads to an elevated susceptibility to fractures, even with low-energy traumas [2].

Despite the high incidence of these fractures, there remains a significant lack of consensus regarding the appropriate treatment options for various types of PHF.

Against the background of a variety of surgical treatment options for displaced fractures of the surgical neck of the humeral head, intramedullary nailing (IMN) is a widely used technique yielding good to excellent outcomes for patients [3–6]. Antegrade interlocking nail osteosynthesis is a surgical technique used in the management of proximal humeral fractures. It involves the insertion of an intramedullary nail through performing a deltoid splitting approach and using locking head fixation screws, providing stable fixation and alignment of the fractured bone segments, which facilitates early mobilization and promotes successful fracture healing.

The evolution of antegrade proximal humerus nailing can be categorized into three generations. The first generation's minimally invasive approach led to fixation failure due to inadequate stabilization. The second generation improved stability but faced issues with proximal screw loosening because of non-angular stable proximal screw designs. The third generation introduced more secure locking mechanisms, offering fixed angular stable constructs. Designs such as Stryker T2, Synthes Proximal Humeral Nail, and the Targon PH+ featured enhancements in screw design and insertion options. These advances hold promise for better patient outcomes and reduced complications in treating displaced proximal humerus fractures [7].

The current concept of straight antegrade interlocking nails poses some biomechanical benefits. They have a lower lever arm than the lateral locking plate, which may counteract varus displacing forces. Furthermore, compared to the bent proximal humeral nail, the straight design offers several advantages: Medialization of the entry point in shoulder surgery is crucial as it helps preserve the supraspinatus tendon footprint [8]. Furthermore, when the nail is medialized within the humeral head, it maintains a position which effectively enhances resistance against varus forces. Additionally, the medial entry point proves to be valuable in preventing accidental entry into the fracture zone, particularly in cases of proximal humeral fractures with a fractured greater tuberosity. Lastly, the use of a straight nail with a proximal tip efficiently anchors the densest subchondral zone, offering increased stability and protection against varus forces [9,10].

A small incision, soft tissue and nerve sparing technique, immediate postoperative joint mobilization, and a comparably short operation time represent the major advantages [11–13].

Until today, there have been several studies and meta-analyses investigating outcomes with the help of subjective (Patient-Reported Outcome Measures (PROMs), Health-Related Quality of Life (HRQoL)) and objective (radiographic assessment) parameters. Hereby, there are established risk factors including insufficient anatomical reduction, four-part fractures, varus head–shaft angle (<130°) and the occurrence of complications [14,15]. A common critique concerning many existing publications is a short follow-up time of only six to 24 months, disregarding significant improvements of shoulder function several years after surgery and the possibility of detecting long-term complications [4,16]. This goes along with a potential overestimation of the effect of surgery on PHFs on quality of life due to recent hospitalization and concomitant burdens such as pain and the rehabilitation process. As most patients would agree, the long-term outcome is a far more important measure.

Given adequate responsiveness to clinical change and the accuracy of these subjective measures, PROMs can help to save resources and manpower necessary for physical examinations [17,18].

The main aim of this study was to evaluate the mid-term patient-reported outcome measures and quality of life after antegrade nailing of proximal humeral fractures. The secondary study goal was to assess individual risk factors for differences in outcome measures.

2. Materials and Methods

2.1. Selection and Evaluation of Eligible Patients and Inclusion Criteria

In this retrospective cohort study, a total of 479 patients who underwent treatment with the Targon PH+ intramedullary nail (150 mm, Aesculap, B. Braun, Tuttlingen, Germany) for proximal humeral fractures at a single center (Klinikum Nuremberg Sued, Department of Orthopedic Surgery and Traumatology) between 2014 and 2021 were reviewed. Subsequently, exclusion criteria were applied, resulting in 260 patients who met the eligibility criteria for inclusion in this investigation. Among this group, a subset of 70 patients were successfully followed up using patient-reported outcome measures (PROMs).

The inclusion criteria for this study encompassed patients with isolated proximal humeral fractures, a minimum adequate radiological follow-up of more than 1 year, and provision of consent to participate. Exclusion criteria comprised patients with dementia, insufficient clinical data or inadequate radiologic imaging, pathological fractures, open fractures, additional shaft fractures, and the use of a nail length exceeding 150 mm. All patients who participated in the study provided written informed consent.

Patient contact information and general details were obtained from the clinic's database and hospital information system.

For EQ-5D-5L, we used the official German version 1.1 (translation certificate available) [19].

For PROMs, the following validated German versions were used:

- German DASH: Germann G. et al. (2003) [20].
- German ASES: Goldhahn J. et al. (2008) [21].
- German OSS: Huber W. et al. (2004) [22].

Parts of this dataset have been already published in a previous study about radiologic risk factors of proximal humeral nailing [15] and have been used for radiographic analysis.

2.2. Surgical Technique and Postoperative Care

In every case, the decision for IMN was taken by a board of experienced orthopedic trauma surgeons.

Prior to surgery, patients were positioned in a beach chair position on a radiolucent table with a standard armrest. For implantation of the Targon PH+ nail, a deltoid splitting approach at the anterolateral margin of the acromion was performed. After a longitudinal transection of the clavipectoral fascia and the subacromial bursa, the head fragments were reduced by direct or indirect manipulation, depending on the fracture pattern. For this procedure, a wire was used as a joystick for securing the fragments.

After longitudinal splitting of the supraspinatus tendon, a guide pin at the apex of the humeral head was positioned in both planes, and the nail was inserted 3 to 4 mm below the cartilage level. Then, a minimum of three to four locking head fixation screws were inserted into the head fragments, depending on the fracture pattern, the degree of instability, as well as bone quality.

To secure displaced fragments to the interlocking screws, a "rope-over-bitt" procedure was carried out in cases of severely displaced tuberosities (>1 cm). A single or duplicate distal interlocking screw was placed through the nail percutaneously. Board-certified trauma surgeons performed all operations.

An intra- and postoperative X-ray examination was obtained from all patients to verify both reduction and correct positioning of the implant.

The affected extremity was immobilized in an arm sling for one day to control postoperative pain. On the second postoperative day, passive mobilization by trained physiotherapists was started. The arm sling was removed as soon as the patient tolerated the pain. Active-assisted shoulder movement below the pain threshold was then initiated as soon as wound healing was deemed satisfactory. After six weeks and radiological consolidation, weight-bearing was allowed.

Every patient received postoperative standardized anterior–posterior and y-view radiographic imaging for correct fracture reduction and implant position after passive mobilization. Radiographic follow-up after 12 months or longer was obtained from all patients.

2.3. Radiologic Fracture Assessment

The pre- and postoperative radiographic imaging was assessed using PACS software, which involved the analysis of plain radiographic images or computed tomography (CT) scans. Fractures were classified based on the number of fractured parts and according to the Neer classification system [23]. The analysis also focused on assessing the quality of postoperative anatomical reduction of the fracture and the pre- as well as postoperative head–shaft angle.

The follow-up radiographs were carefully evaluated by two experienced orthopedic trauma surgeons to identify any documented complications. An anatomical reduction was considered present when certain criteria were met: the head showed neither varus angulation (<130°) nor valgus angulation (>140°), the y-view showed no anterior or posterior tilt of the head exceeding 20°, the dislocation of the greater or lesser tuberosity was less than 3 mm, and the neck-shaft dislocation was less than 5 mm.

2.4. Evaluation of Complications and Revision

Early postoperative complications and revisions were evaluated during the hospital stay of the patients. These included reviews of postoperative X-rays and medical records. Late complications and revisions were obtained from hospital charts during visits at our outpatient clinic or if the patient was re-admitted to the hospital, and finally, at the point of follow-up. All complications were organized according to the Clavien–Dindo classification and then subdivided into major (Clavien–Dindo grades III and IV) and minor complications (Clavien–Dindo grades I and II) [24]. Every subsequent surgery of the proximal humeral head apart from elective implant removal was counted as a revision. Implant failure was defined as removal of the nail and conversion to a different osteosynthesis or the secondary implantation of a reverse shoulder prosthesis.

2.5. PROMs (DASH, OSS, ASES) and HRQoL (EQ-5D-5L, EuroQol VAS)

In total, 70 subjects agreed to participate in the study by answering EQ-5D-5L, DASH, ASES, OSS, and the additional study questionnaire via a telephone interview.

2.6. Statistical Analysis

All data were obtained and analyzed retrospectively. Statistical analysis was performed using IBM SPSS Statistics for Windows (version 28, 1.0.0.1406, IBM Corp., Armonk, NY, USA). Parametric data were analyzed using t-test and ANOVA for multiple samples, nonparametric data via Mann–Whitney U and Kruskal–Wallis analysis. Pearson's P correlation coefficient was calculated for parametric data and Spearman's Rho correlation was calculated for PROMs (nonparametric data). Nominal data were analyzed using cross tables and tested with chi-square test and Fisher's exact test.

All reported *p*-values are two-tailed, with an alpha level <0.05 considered statistically significant. Unless otherwise stated, descriptive results are demonstrated as mean ± standard deviation and range.

3. Results

3.1. Descriptive Statistics

A total of 70 patients, thereof 47 female and 23 males, were enrolled with a mean age of 74.5 ± 11.6 (29–97) years and mean follow-up time of 4.9 ± 1.9 (1.6–7.8) years (Table 1).

Table 1. Summary of Descriptive Statistics for Metric and Nominal Data. Metric data are reported with mean ± SD and range, nominal data with total number *n* and associated percentage (%).

	Mean ± SD	**Range**
Age	74.5 ± 11.6	(29–97)
• Number of patients > 65 yrs.	58	-
• Number of patients < 65 yrs.	12	-
Follow-up time	4.9 ± 1.9	(1.6–7.8)
	n	%
Sex		
Female	47	67.1%
Male	23	32.9%
Injured side		
Right	34	48.6%
Left	36	51.4%
Fracture parts		
Two-part	18	25.7%
Three-part	38	54.3%
Four-part	14	20.0%
Anatomical reduction	45	48.5%
Head–shaft-angle postoperatively		
	Mean ± SD	Range
Absolute	136.0° ± 9.9	(109.7–157.6)
	n	%
Varus < 130°	13	19.1%
Physiologic	33	48.5%
Valgus > 140°	22	32.4%

Table 2 presents the classification of complications observed in the patient cases. A total of 15 patients with complications were recorded, accounting for 21.4% of the cases. Among these, 11 were classified as major complications (15.7%), while 4 were categorized as minor complications (5.7%). The revision rate was 15.7%, not counting three additional patients opting for elective implant removal. Moreover, we analyzed the incidence of complications in patients over and under 65 years. Surprisingly, patients over the age of 65 accounted for a higher rate of complications and revisions, but these differences did not reach statistical significance. The elevated overall complication rate in the study was primarily driven by a significantly increased rate of minor complications in patients under 65 years.

Table 2. Classification of Complication rate (CR)per Patient Case and Comparison between >65 and <65 years.

Complications	*n*	%	**CR in >65 vs. <65 Years in% *****
Total	15	21.4%	15.4% vs. 41.7%
Major *	11	15.7%	13.8% vs. 25.0%
Minor **	4	5.7%	1.7% vs. 20.0%
Revision surgery	11	15.7%	13.8% vs. 25.0%

* Clavien–Dindo Grades III and IV; ** Calvien-Dindo Grades I and II; *** complication rates in the subgroups >65 and <65 years.

Table 3 provides a breakdown of the major and minor complications experienced by the patients. Multiple patients experienced more than one complication.

Table 3. Number of Complications and Associated Revisions. Some patients experienced ≥1 complications.

Complications	n	Related Revisions
Nail loosening	1	1
Cut-out	1	1
Loss of reduction	3	2
Humeral head necrosis	3	3
Prolonged pain	3	3
Joint stiffness	3	1
Prolonged consolidation	1	0
Keloid scar formation	2	0

3.2. Fracture Morphology and Radiographic Analysis

Upon evaluation of all 70 patients, anatomical reduction was achieved in 64 3% of cases. Postoperatively, the integrity of the medial hinge was achieved in 78.6% of patients compared to 18.8% preoperatively.

Prior to surgery, 11.9% of fractures exhibited a physiologic head–shaft angle, while 41.8% had a valgus angle and 46.3% had a varus angle. After surgery, 48.5% of patients demonstrated a physiologic head angulation (130°–140°), 32.4% had a valgus angle (>140°), and 19.1% had a varus angle (<130°). Furthermore, the mean absolute head–shaft angle was measured as 131.8 ± 27.9 (range: 68.60°–217.20°) upon admission and 136.0 ± 9.9 (range: 109.7°–157.6°) after surgery.

3.3. PROMs and HRQoL Measures

Following IMN during the follow-up period, patients achieved a mean DASH score of 25.4 ± 22.1 (range: 0.00–81.00), indicating the level of disability and symptoms related to upper extremity function. The mean ASES score was 76.2 ± 21.1 (range: 26.6–100.0), representing the level of pain and shoulder function. The mean OSS score was 38.81 ± 10.3 (range: 9.0–54.0), reflecting the level of shoulder-related disability.

According to the DASH interpretation, 45.7% of patients achieved excellent results, indicating minimal disability and symptoms. Additionally, 27.1% of patients achieved satisfactory results, while another 27.1% of patients achieved poor results, indicating varying degrees of disability and symptoms (see Table 4).

Table 4. Overall means of PROMs and HRQoL; SD: standard deviation; n = 70.

	Mean ± SD	Range
DASH Score	25.4 ± 22.1	81.0–25.4
ASES Score	76.2 ± 21.1	100.0–76.2
OSS	38.8 ± 10.3	54.0–9.0
EQ-5D-5L index value	0.79 ± 0.23	1.00–0.05
EuroQol VAS	68.8 ± 17.9	35.0–100.0

In order to examine the association between PROMs, a correlation coefficient was computed. The analysis demonstrated a significant and statistically robust correlation among the evaluated PROMs.

Importantly, this correlation remained consistent when comparing each PROM with EQ-5D-5L (see Table 5).

Table 5. Correlation Matrix of DASH, ASES, OSS, and EQ-5D-5L Scores.

		DASH	ASES	OSS	EQ-5D-5L
DASH Score	Pearson correlation Sig. (2-tailed)	-	−0.812 * 0.000	−0.902 * 0.000	−0.823 * 0.000
ASES Score	Pearson correlation Sig. (2-tailed)	−0.813 * 0.000	-	0.852 * 0.000	0.541 * 0.000
OSS	Pearson correlation Sig. (2-tailed)	−0.902 * 0.000	0.852 * 0.000	-	0.722 * 0.000
EQ-5D-5L	Pearson correlation Sig. (2-tailed)	−0.823 * 0.000	0.541 * 0.000	0.722 * 0.000	-

* The correlation is highly significant at the level of 0.01 (2-tailed).

3.4. *Influence of Complications and Revisions on PROMs and HRQoL*

Patients sustaining major complications experienced significantly lower ASES (−11.1%: 87.3 ± 16.2 > 77.7 ± 17.9, $p = 0.013$) results compared to minor and no complication development.

However, scores of DASH (−12.8%), OSS (+2.9%), EQ-5D-5L index (+6.3%) and EuroQol VAS (−7.5%) did show a significant correlation (DASH: $p = 0.070$; OSS: $p = 0.123$; EQ-5D-5L index: $p = 0.602$; EuroQol VAS: $p = 0.885$).

When examining the influence of revision surgery, again, only ASES scores reduced significantly from 78.3 ± 21.2 to 65.1 ± 17.2 (−16.8%) ($p = 0.031$). Hereby, the increase in DASH (+14.5%) and decrease in OSS (−4.9%) was apparent, yet not significant ($p = 0.283$, $p = 0.274$).

As with complications, HRQoL measures seem to have no relation at all with revision surgery requirement, as EQ-5D-5L index and EuroQol VAS paradoxically showed a non-significant increase (+4.1%, $p = 0.948$ and +3.7%, $p = 0.631$).

3.5. *Influence of Fracture Complexity*

When observing major complications, patients with four-part fractures sustain more than twice as many complications as statistically expected (2.08, $p = 0.054$), and 41.7% of severe complications occurred in four-part fractures. Thus, the four-part complication rate is 35.7% compared to two- and three-part fractures, where only 12.5% of patients sustained a severe complication.

Analogously, the risk for revision surgery increased 2.3-fold (233.43%) in the four-part fracture group, which is a significant finding ($p = 0.036$).

Furthermore, PROMs and HRQoL measures show clear trends towards deterioration in scores with advancing fragment count, especially when considering four-part fractures (see Table 6). Mean DASH score increased with the increase in the number of fragments from two to four. ASES, OSS, and EQ-5D-5L follow the same trend but means reduced with fracture count (see Figure 1).

Table 6. PROMs and HRQoL subdivided into fragment count. Displayed via mean ± SD (range).

Outcome Measure	Two-Part n = 18	Three-Part n = 38	Four-Part n = 14
DASH	20.9 ± 21.4 (0.0–75.0)	25.1 ± 23.7 (0.0–81.0)	32.3 ± 17.1 (1.7– 61.2)
ASES	81.1 ± 19.9 (35.01–100.00)	76.4 ± 22.5 (26.66–100)	69.5 ± 18.0 (41.65–55.01)
OSS	41.0 ± 10.0 (10.0–48.0)	39.0 ± 11.0 (9.0–54.0)	36.0 ± 8.0 (25.0–51.0)
EQ-5D-5L index	0.830 ± 0.221 (0.053–1.00)	0.802 ± 0.222 (0.116–1.00)	0.699 ± 0.278 (0.077–1.00)
EuroQol VAS	66.0 ± 18.0 (40.0–100.0)	70.0 ± 19.0 (35.0–100.0)	70.0 ± 17.0 (50.0–100.0)

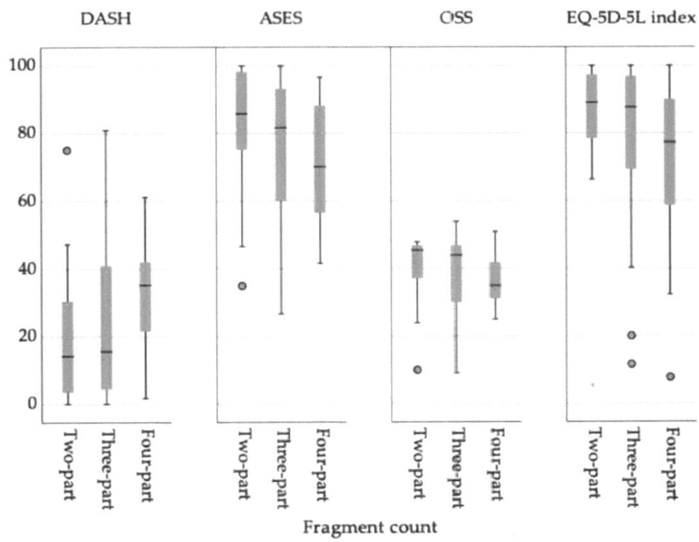

Figure 1. PROMs and HRQoL scores according to fragment count: y-axis displays absolute score values, and x-axis display sub-groups.

3.6. Age Factor

When comparing the two age groups, namely those over 65 years (n = 58) and those under (n = 12), in terms of PROMs and quality of life, we observed significant differences in all tested parameters. Patients over 65 years displayed notably reduced scores in DASH (28.6 vs. 9.9; p-value = 0.008), ASES (73.8 vs. 88.1; p-value = 0.006), OSS (37.6 vs. 44.9; p-value = 0.018), and EQ-5D-5L (0.76; p-value = 0.019).

While these differences appear to be intuitive, it is crucial to acknowledge a potential confounding factor, namely the high incidence of 4-part fractures in the subgroup of >65 years (n = 14) when compared to no 4-part fractures in the <65 years subgroup.

4. Discussion

This clinical investigation demonstrates favorable clinical and mid-term results, along with a good overall quality of life, following the use of antegrade nailing for proximal humeral fractures. The study's outcomes align with the existing scientific literature. Several factors were identified as potential risks leading to a decline in shoulder function, reflected in lower scores in PROMs.

The complexity of the fracture showed a noticeable yet statistically non-significant impact on all applied PROMs as the fragment count increased. Major complications and revision surgery also demonstrated unfavorable effects, particularly evident in reduced ASES scores.

A noteworthy correlation was observed between patients' quality of life and objective measurements obtained from the employed PROMs. However, the aforementioned factors (fracture complexity, major complications, and revisions) did not significantly influence the overall quality of life.

4.1. PROMs and HRQoL

Overall, the clinical outcome measured by PROMs after interlocking nailing of proximal humeral fractures can be considered to be good. This also applies for quality of life in our collective. In this, our data stand in line with the current literature.

Wong et al. reported slightly better results considering ASES (means: 84.4 vs. 76.2) in a large metanalysis with 448 included fractures, on average 22.6 months after intramedullary nailing [25]. Notably, the data by Wong had been collected from a 10-year-younger collective (mean age: 64.3 years), which might have skewed the outcome parameters.

Until now, few studies have evaluated quality of life after intramedullary nailing of PHF. HRQoL with the EQ-5D-5L index/EurQol VAS patients of this study averaged 0.788 and 68.78, respectively. Compared to recent data published by Lopiz et al., these results can be viewed as very satisfactory. One must acknowledge that these data were generated out of an older population with a mean age of 82.1 years, recording an EQ-5D-5L index of 0.40 and EuroQolVAS 64.2 [3]. Yet, HRQoL values after IMN are lower than in a reference German population over 65 years (.84 $\pm$ 0.22 and 73.2 $\pm$ 18.5) [26].

Maurer et al. showed high correlations between PROMs and CMS (Constant–Murley Score) as well as strong to moderate correlation between PROMs, CMS, and HRQoL, which is also depicted in our results. In this sense, shoulder integrity plays an important role in the quality of life of patients and is a valid tool in patients after PHFs, as proposed by Olerud et al. [27]. This assertion is substantiated by the outcomes obtained from our findings.

4.2. Complications and Revisions

To date, only a limited number of studies have delved adequately into exploring the impact of complications and revisions on the clinical outcomes following antegrade nailing for PHFs. Notably, the influence of major complications and revisions on most of the PROMs appears to be minimal. This intriguing finding has the potential to empower surgeons when making decisions regarding revision surgery for surgically manageable complications, given that a substantial proportion of patients with major complications underwent successful revisions.

However, it is essential to highlight that the ASES assessment demonstrated the ability to identify a clinical decline in patients who encountered major postoperative complications and required revisions. These patients exhibited significantly poorer scores in comparison to those with either no complications or minor ones.

One plausible explanation for our findings may be linked to the ASES assessment's particular focus on self-reported pain, which could strongly associate with postoperative complications and revision rates. Unlike the DASH questionnaire, where pain-related inquiries comprise only 6.67% (two questions), and the OSS, where they account for 25.00% (three questions), the ASES assessment incorporates a Pain Visual Analog Scale (VAS) that contributes to 50% of its overall scoring.

In support of our observations, Südkamp et al. reported similar results in their study involving 463 patients, wherein the CMS revealed significantly lower scores among patients with intraoperative complications, and non-significant lower scores in those with postoperative complications [28]. In contrast to our study, complications were not subclassified, which could have reduced the effect size in their data.

Zhu et al. assessed the functional outcome after surgical management of proximal fractures (using plating and nailing) in 57 patients using ASES. In contrast to our data, no significant influence of complications on ASES was established [13].

4.3. Fracture Complexity

It has previously been well published that four-part fractures pose a risk of complications and revisions [15,29]. These findings confirm our recent data with more than twice as many complications in four-part fractures compared to two- and three-part fractures. Furthermore, significantly more revisions were found.

The dataset in this study showed non-significant trends, suggesting that higher fragment counts were associated with inferior scores in PROMs and HRQol. Greenberg et al. conducted a study on the same interlocking nail (Targon PH+) and found a significant reduction in ASES scores for Neer 4 fractures compared to Neer 3 fractures [4].

Tsitsilonis et al. supported these findings in their work, when they evaluated a restricted sample of 43 patients using the CMS. Their investigation revealed significantly unfavorable outcomes for individuals with four-part fractures [30].

Interestingly, Kloub et al. conducted an analysis on a cohort of 137 patients diagnosed with three- and four-part fractures. Their findings indicated a marginal and non-significant disparity between the two groups in terms of CMSrel (Constant–Murley Score relative values) [31].

4.4. Age

Based on our data, age appears to be a critical factor influencing the clinical outcome following PHF. To assess the impact of age on PROMs and HRQol, we stratified our cohort into two subgroups: patients aged >65 and those aged <65 years. Remarkably, we identified significant differences in the measured PROMs and HRQol between these two age groups.

It has been noted that these results are very likely to be influenced by the high incidence of four-part fractures in patient over 65 years. On the contrary, a notable disparity in the incidence of complications and subsequent revisions became evident when comparing different age groups. Specifically, the younger age group exhibited a higher frequency of complications. It is worth noting that minor complications were particularly pronounced in patients under 65 years of age, which could be attributed to their potentially greater expectations and consequently, a higher tendency to report even minor incidents.

Our findings diverge from those reported by Dey Hazra et al., who reported no significant difference in clinical outcomes measured by PROMs when using the same age cutoff. Notably, Dey Hazra et al. focused predominantly on patients treated with plate osteosynthesis. Additionally, their inclusion of a substantial number of four-part fractures in the age group below 65 years might have impacted their results differently compared to ours [32].

In another study by Hättich et al., the influence of age on PROMs after antegrade nailing was investigated in a small sample size of 42. However, their data did not reveal a significant influence of age when using logistic regression analysis [29].

4.5. Limitations

Undoubtedly, this study possesses inherent limitations that warrant acknowledgment. Notably, a selection bias should be recognized due to a relatively high loss to follow-up. Patients at both extremes of the clinical spectrum demonstrate higher adherence to participation in clinical studies, with individuals who are either highly satisfied or dissatisfied showing a greater inclination to respond [33].

Moreover, the rigorous exclusion criteria, lengthy follow-up period, and the advanced mean age of the eligible pool of 260 patients resulted in the enrollment of only 70 individuals. Thus, future investigations should strive to encompass larger sample sizes.

Furthermore, it is important to note that this study was conducted solely at a single center characterized by a high level of expertise in proximal humeral nailing. To ascertain the generalizability of our findings, it is imperative to obtain data from multi-center investigations.

Nevertheless, despite these limitations, we firmly believe that this study provides valuable insights for clinicians and their patients.

5. Conclusions

This study provides valuable insights into proximal humeral fractures, offering a comprehensive understanding of multiple aspects, including risk factors, clinical outcomes, and treatment considerations related to nailing.

The clinical outcome measured by PROMs after interlocking nailing of proximal humeral fractures is considered good, consistent with the current literature. Notably, factors such as revision surgery, fracture complexity, and age appeared as influential elements, negatively impacting patient-reported outcome measure scores and well-being.

The study emphasized the significant correlation between PROMs and overall well-being, underscoring the critical role of shoulder integrity in health-related quality of life.

The current body of research on the impact of complications and revisions on clinical outcomes following antegrade nailing for PHFs remains limited. Nevertheless, our findings suggest that major complications and revisions seem to have minimal influence on most of the PROMs, potentially empowering surgeons to consider revision surgery for surgically manageable complications. However, it is crucial to acknowledge that the ASES assessment, with its focus on self-reported pain, was able to detect a decline in patients experiencing major postoperative complications and revisions, as evidenced by significantly lower scores compared to those with no or minor complications. This underscores the importance of comprehensive PROMs, which may help improve our understanding of patients' experiences and guide surgical decision making in PHF management.

Fracture complexity appears to play a substantial role in the risk of complications and revisions, as indicated by the previous literature. Our recent data reinforces these findings, revealing a significant increase in complications in four-part fractures compared to two- and three-part fractures, along with a higher rate of revisions. Moreover, our dataset suggests non-significant trends, hinting that higher fragment counts might be associated with inferior scores in PROMs and HRQol. Although many studies have indicated unfavorable outcomes for patients with four-part fractures, the literature presents contrasting findings.

Despite this, it is important to consider the consistent evidence pointing to poorer outcomes associated with four-part fractures. As a result, there is a compelling need to explore alternative treatment approaches to address the challenges of four-part fractures effectively.

Although our data indicated an influence of age on PROMs and HRQoL, further investigation is required to fully understand the role of age in PHF considering the potential impact of confounding factors.

Despite limitations such as selection bias and limited sample size, this study provides valuable insights for clinicians and patients, calling for larger, multi-center investigations to enhance generalizability and deepen our understanding of proximal humeral fractures.

Author Contributions: Conceptualization, M.W., H.J.B. and S.G.; methodology, M.W. and S.G.; software, M.W. and S.G.; validation, M.G., M.W. and H.J.B.; formal analysis, M.G., M.W., K.L., J.R., M.M., R.B. and H.J.B.; investigation, M.W. and S.G. resources, H.J.B., S.G. and M.W.; data curation, S.G. and M.W.; writing—original draft preparation, S.G. and M.W.; writing—review and editing, M.G., M.W., K.L., J.R., M.M., R.B. and H.J.B.; visualization, S.G. and M.W.; supervision, M.G. and H.J.B.; project administration, M.W.; funding acquisition, M.W. All authors have read and agreed to the published version of the manuscript.

Funding: This research received no external funding.

Institutional Review Board Statement: This study was obtained in conformity with the Declaration of Helsinki and authorized by the Institutional Review Board of Paracelsus Medical University Nuremberg. (Protocol code IRB-2021-025 and date of approval: 21 December 2021).

Informed Consent Statement: Written informed consent has been obtained from the patient to publish this paper.

Data Availability Statement: Not applicable.

Conflicts of Interest: The authors declare no conflict of interest.

References

1. Court-Brown, C.M.; Bugler, K.E.; Clement, N.D.; Duckworth, A.D.; McQueen, M.M. The epidemiology of open fractures in adults. A 15-year review. *Injury* **2012**, *43*, 891–897. [CrossRef]
2. Passaretti, D.; Candela, V.; Sessa, P.; Gumina, S. Epidemiology of proximal humeral fractures: A detailed survey of 711 patients in a metropolitan area. *J. Shoulder Elbow Surg.* **2017**, *26*, 2117–2124. [CrossRef]
3. Lopiz, Y.; Garríguez-Pérez, D.; Martínez-Illán, M.; García-Fernández, C.; Marco, F. Third-generation intramedullary nailing for displaced proximal humeral fractures in the elderly: Quality of life, clinical results, and complications. *Arch Orthop. Trauma Surg.* **2022**, *142*, 227–238. [CrossRef]
4. Greenberg, A.; Rosinsky, P.J.; Gafni, N.; Kosashvili, Y.; Kaban, A. Proximal humeral nail for treatment of 3- and 4-part proximal humerus fractures in the elderly population: Effective and safe in experienced hands. *Eur. J. Orthop. Surg. Traumatol.* **2021**, *31*, 769–777. [CrossRef]

5. Boileau, P.; d'Ollonne, T.; Bessière, C.; Wilson, A.; Clavert, P.; Hatzidakis, A.M.; Chelli, M. Displaced humeral surgical neck fractures: Classification and results of third-generation percutaneous intramedullary nailing. *J. Shoulder Elbow Surg.* **2019**, *28*, 276–287. [CrossRef]
6. Paniagua Collado, M.A.; Willauschus, M.; Rüther, J.; Lahner, M.; Bail, H.J.; Geßlein, M. Intramedullary nailing as a safe and efficient treatment option for all types of displaced proximal humeral fractures in geriatric patients: A retrospective evaluation of 49 patients treated with the new Targon PH+ nail. *Technol. Health Care* **2021**, *29*, 771–780. [CrossRef]
7. Dilisio, M.F.; Nowinski, R.J.; Hatzidakis, A.M.; Fehringer, E.V. Intramedullary nailing of the proximal humerus: Evolution, technique, and results. *J. Shoulder Elbow Surg.* **2016**, *25*, e130–e138. [CrossRef]
8. Euler, S.A.; Hengg, C.; Kolp, D.; Wambacher, M.; Kralinger, F. Lack of fifth anchoring point and violation of the insertion of the rotator cuff during antegrade humeral nailing: Pitfalls in straight antegrade humeral nailing. *Bone Jt. J.* **2014**, *96-B*, 249–253. [CrossRef]
9. Euler, S.A.; Petri, M.; Venderley, M.B.; Dornan, G.J.; Schmoelz, W.; Turnbull, T.L.; Plecko, M.; Kralinger, F.S.; Millett, P.J. Biomechanical evaluation of straight antegrade nailing in proximal humeral fractures: The rationale of the "proximal anchoring point". *Int. Orthop.* **2017**, *41*, 1715–1721. [CrossRef]
10. Hepp, P.; Josten, C. Biology and Biomechanics in Osteosynthesis of Proximal Humerus Fractures. *Eur. J. Trauma Emerg. Surg.* **2007**, *33*, 337–344. [CrossRef]
11. Shi, X.; Liu, H.; Xing, R.; Mei, W.; Zhang, L.; Ding, L.; Huang, Z.; Wang, P. Effect of intramedullary nail and locking plate in the treatment of proximal humerus fracture: An update systematic review and meta-analysis. *J. Orthop. Surg. Res.* **2019**, *14*, 285. [CrossRef] [PubMed]
12. Krivohlávek, M.; Lukás, R.; Taller, S.; Srám, J. Use of angle-stable implants for proximal humeral fractures: Prospective study. *Acta Chir. Orthop. Traumatol. Cech.* **2008**, *75*, 212–220. [CrossRef] [PubMed]
13. Zhu, Y.; Lu, Y.; Shen, J.; Zhang, J.; Jiang, C. Locking intramedullary nails and locking plates in the treatment of two-part proximal humeral surgical neck fractures: A prospective randomized trial with a minimum of three years of follow-up. *J. Bone Jt. Surg. Am.* **2011**, *93*, 159–168. [CrossRef] [PubMed]
14. Tepass, A.; Rolauffs, B.; Weise, K.; Bahrs, S.D.; Dietz, K.; Bahrs, C. Complication rates and outcomes stratified by treatment modalities in proximal humeral fractures: A systematic literature review from 1970-2009. *Patient Saf. Surg.* **2013**, *7*, 34. [CrossRef]
15. Willauschus, M.; Schram, L.; Millrose, M.; Rüther, J.; Loose, K.; Bail, H.J.; Geßlein, M. Specific Radiologic Risk Factors for Implant Failure and Osteonecrosis of the Humeral Head after Interlocking Nailing with the Targon PH. *J. Clin. Med.* **2022**, *11*, 2523. [CrossRef]
16. Zirngibl, B.; Biber, R.; Bail, H.J. Humeral head necrosis after proximal humeral nailing: What are the reasons for bad outcomes? *Injury* **2016**, *47* (Suppl. 7), S10–S13. [CrossRef]
17. Jessen, M.; Lappen, S.; Schneider, M. „Patient-reported outcome measures" (PROMs) in der Schulter- und Ellenbogenchirurgie. *Obere Extrem.* **2021**, *16*, 215–220. [CrossRef]
18. Maurer, E.; Bahrs, C.; Kühle, L.; Ziegler, P.; Gonser, C.; Stollhof, L.E. Comparability of Patient-reported Outcome Measures and Clinical Assessment Tools for Shoulder Function in Patients with Proximal Humeral Fracture. *Z. Orthop. Unfall.* **2021**, *159*, 638–648. [CrossRef]
19. Ludwig, K.; Graf von der Schulenburg, J.M.; Greiner, W. German Value Set for the EQ-5D-5L. *Pharmacoeconomics* **2018**, *36*, 663–674. [CrossRef]
20. Germann, G.; Harth, A.; Wind, G.; Demir, E. Standardisation and validation of the German version 2.0 of the Disability of Arm, Shoulder, Hand (DASH) questionnaire. *Unfallchirurg* **2003**, *106*, 13–19. [CrossRef]
21. Goldhahn, J.; Angst, F.; Drerup, S.; Pap, G.; Simmen, B.R.; Mannion, A.F. Lessons learned during the cross-cultural adaptation of the American Shoulder and Elbow Surgeons shoulder form into German. *J. Shoulder Elbow Surg.* **2008**, *17*, 248–254. [CrossRef] [PubMed]
22. Huber, W.; Hofstaetter, J.G.; Hanslik-Schnabel, B.; Posch, M.; Wurnig, C. The German version of the Oxford Shoulder Score--cross-cultural adaptation and validation. *Arch Orthop. Trauma Surg.* **2004**, *124*, 531–536. [CrossRef] [PubMed]
23. Carofino, B.C.; Leopold, S.S. Classifications in brief: The Neer classification for proximal humerus fractures. *Clin. Orthop. Relat. Res.* **2013**, *471*, 39–43. [CrossRef] [PubMed]
24. Dindo, D.; Demartines, N.; Clavien, P.A. Classification of surgical complications: A new proposal with evaluation in a cohort of 6336 patients and results of a survey. *Ann. Surg.* **2004**, *240*, 205–213. [CrossRef]
25. Wong, J.; Newman, J.M.; Gruson, K.I. Outcomes of intramedullary nailing for acute proximal humerus fractures: A systematic review. *J. Orthop. Traumatol.* **2016**, *17*, 113–122. [CrossRef]
26. Marten, O.; Greiner, W. EQ-5D-5L reference values for the German general elderly population. *Health Qual. Life Outcomes* **2021**, *19*, 76. [CrossRef]
27. Olerud, P.; Tidermark, J.; Ponzer, S.; Ahrengart, L.; Bergström, G. Responsiveness of the EQ-5D in patients with proximal humeral fractures. *J. Shoulder Elbow Surg.* **2011**, *20*, 1200–1206. [CrossRef]
28. Südkamp, N.P.; Audigé, L.; Lambert, S.; Hertel, R.; Konrad, G. Path analysis of factors for functional outcome at one year in 463 proximal humeral fractures. *J. Shoulder Elbow Surg.* **2011**, *20*, 1207–1216. [CrossRef]

29. Hättich, A.; Harloff, T.J.; Sari, H.; Schlickewei, C.; Cramer, C.; Strahl, A.; Frosch, K.H.; Mader, K.; Klatte, T.O. Influence of Fracture Reduction on the Functional Outcome after Intramedullary Nail Osteosynthesis in Proximal Humerus Fractures. *J. Clin. Med.* **2022**, *11*, 6861. [CrossRef]
30. Tsitsilonis, S.; Schaser, K.D.; Kiefer, H.; Wichlas, F. The treatment of the proximal humeral fracture with the use of the PHN Nailing System: The importance of reduction. *Acta Chir. Orthop. Traumatol. Cech.* **2013**, *80*, 250–255. [CrossRef]
31. Kloub, M.; Holub, K.; Polakova, S. Nailing of three- and four-part fractures of the humeral head—Long-term results. *Injury* **2014**, *45* (Suppl. 1), S29–S37. [CrossRef] [PubMed]
32. Dey Hazra, R.O.; Illner, J.; Szewczyk, K.; Warnhoff, M.; Ellwein, A.; Blach, R.M.; Lill, H.; Jensen, G. Age-Independent Clinical Outcome in Proximal Humeral Fractures: 2-Year Results Using the Example of a Precontoured Locking Plate. *J. Clin. Med.* **2022**, *11*, 408. [CrossRef] [PubMed]
33. Mazor, K.M.; Clauser, B.E.; Field, T.; Yood, R.A.; Gurwitz, J.H. A demonstration of the impact of response bias on the results of patient satisfaction surveys. *Health Serv. Res.* **2002**, *37*, 1403–1417. [CrossRef] [PubMed]

Journal of *Personalized Medicine*

MDPI

Article

Mid- to Long-Term Survival of Geriatric Patients with Primary Septic Arthritis of the Shoulder: A Retrospective Study over a Period of 20 Years

Johannes Rüther [1,*], Lars Taubert [1], Kim Loose [1], Maximilian Willauschus [1], Sandeep Silawal [2], Michael Millrose [1,3], Hermann Josef Bail [1] and Markus Geßlein [1]

1 Department of Orthopedics and Traumatology, Paracelsus Medical University, General Hospital Nuremberg, Breslauer Straße 201, 90471 Nuremberg, Germany
2 Institute of Anatomy and Cell Biology, Paracelsus Medical University, General Hospital Nuremberg, Prof. Ernst Nathan Str. 1, 90419 Nuremberg, Germany
3 Department of Trauma Surgery and Sports Medicine, Garmisch-Partenkirchen Medical Centre, 82467 Garmisch-Partenkirchen, Germany
* Correspondence: johannes.ruether@klinikum-nuernberg.de; Tel.: +49-911-3982600

Abstract: Septic arthritis of the shoulder is an urgent medical emergency that often occurs in elderly patients and is associated with high morbidity and mortality. Retrospectively, 56 patients aged ≥60 years, treated for primary septic monoarthritis of the shoulder at a maximum care hospital between 1 July 2001, and 30 July 2022, were included in this study. The primary aim of the study was analyzing survival rates and different bacteria in these patients. For statistical analysis, Kaplan–Meier curves were used for survival probability and the log-rank test was used to compare a survival probability of 5 years. The mean patient age was 78.7 years and a mean follow-up time of 3011.8 days. The mean survival of the entire study population was 920.3 days or 2.5 years. Significantly impaired 5-year survival was found only with increasing age and higher American Society of Anesthesiologists (ASA) physical status (PS) classification scores. Eight different types of bacteria were detected in the synovial fluid cultures. A total of 42 of 48 overall pathogens was Gram-positive and 6 were Gram-negative bacteria. *Staphylococcus aureus* was identified as the most frequent variant. We conclude that the mean survival is significantly shortened within the first 5 years with increasing age and ASA PS classification.

Keywords: septic arthritis; geriatric patients; shoulder; risk factors; infection

check for updates

Citation: Rüther, J.; Taubert, L.; Loose, K.; Willauschus, M.; Silawal, S.; Millrose, M.; Bail, H.J.; Geßlein, M. Mid- to Long-Term Survival of Geriatric Patients with Primary Septic Arthritis of the Shoulder: A Retrospective Study over a Period of 20 Years. *J. Pers. Med.* **2023**, *13*, 1030. https://doi.org/10.3390/jpm13071030

Academic Editor: Jih-Yang Ko

Received: 30 April 2023
Revised: 5 June 2023
Accepted: 20 June 2023
Published: 22 June 2023

1. Introduction

Despite current medicine, septic arthritis of the shoulder is a potentially life-threatening emergency that frequently occurs in the elderly [1–3]. Due to the disease's high morbidity and mortality, a rapid diagnosis with initiation of adequate therapy is essential [2,3]. Mostly, Gram-positive bacteria enter the affected joint through surgery or hematogenous spread [2,4,5].

The incidence of septic arthritis varies significantly in different regions of the world. It varies from 2–10 cases per 100.000 population. Moreover, the incidence seems to be rising when we examine the literature. The incidence of primary septic arthritis has been rising in recent decades. Factors influencing the prevalence are an aging population, an increase in invasive joint procedures and more comorbidities of the population. A growing prevalence of antibiotic resistant bacteria leads to an even more challenging treatment of septic arthritis of the shoulder. Studies on the epidemiology of septic arthritis of the shoulder are rare, but the incidence seems to be around 4 to 6 per 100.000 per persons and year. Septic arthritis of the shoulder varies from about 3 to 5% of these cases. The incidence varies significantly in different countries. This could be due to differences in diagnostic criteria, reporting systems and reporting mechanisms. Moreover, the incidence has varied in recent decades [4,6–8]. For instance, in the United Kingdom the incidence increased from 5.5/100.000 in 1998 to 7.8/100.000 in 2013 with 4.3% affecting the shoulder joint. In Taiwan, the incidence increased from 9.8/100.000 in 1998 to 13.3/100.000 [9,10]. Most of the infections of primary septic arthritis affect one joint (80–90%). In 10 to 20%, more than one joint, typically two or three joints, is affected. Septic arthritis affecting many joints arises typically in multimorbid patients [6,11–13]. In about half of the cases, the knee joint is affected by septic arthritis [12,14,15]. The hips (15 to 25%) and the shoulders (5 to 10% of all cases) are rare locations for primary septic arthritis [14,16,17]. The ilosacral, sternoclavicular and symphysis pubica joint are locations for atypical septic arthritis [8,18,19]. Allthough there are some studies on the incidence of septic arthritis of the shoulder, there are still scarce data about primary septic arthritis of the shoulder.

The main risk factors are age, immunosuppression, previous surgery, intra-articular joint injections, a preexisting joint disease such as osteoarthritis, rheumatoid arthritis, gout, or pseudogout; diabetes mellitus; medical prosthesis, especially joint prosthesis; or skin infections [14,20,21].

Treatment of septic arthritis requires both rapid use of effective antimicrobial agents and surgical debridement of the affected joint, with priority given to offloading the joint [4,6,22,23]. Broad-spectrum antibiotics are usually administered immediately after the joint aspiration [23,24]. Arthroscopic or open-joint debridement are gold-standard treatments for septic arthritis [6,25].

Most commonly, Gram-positive skin bacteria, especially *Staphylococcus aureus* affect septic arthritis [2,7,22]. Methicillin-resistant *Staph aureus* is the causing agent in about a quarter of the cases and seems to play a key role in primary septic arthritis too [26,27]. Biofilm-forming organisms are frequently involved, as they can absent themselves from this state and cause hematogenous spread, which is frequently seen in patients with medical implants [28]. Gram-negative agents are the causing agent in 10 to 20 % of primary septic arthritis cases [16,29–31]. This is why every patient should undergo a microbial probe to isolate the bacteria and receive the best antibiotic treatment. Cefuroxim is the first-line treatment for patients with septic arthritis. This antibiotic has a good effect on common bacteria and can be applied intravenously to reach a high level in the blood of the patient.

The American Society of Anesthesiologists (ASA) physical status classification system is a tool with the purpose of assessing a patient's medical comorbidities prior to anesthesia. It is classified preoperatively and provides a quick overview of the patient's medical condition [32–34]. The classification system was not developed to predict perioperative risks; however, many subsequent publications have shown that ASA classification strongly correlates with outcome and has the potential ability to predict intraoperative and postoperative outcomes, such as intraoperative blood loss or postoperative infection rates [32–34].

The primary aim of this study was to analyze the overall survival of older adults with acute primary septic monoarthritis of the shoulder after surgical treatment. The secondary aim was to assess the causing pathogens for primary septic arthritis of the shoulder.

2. Materials and Methods

Patients' data were evaluated using medical records from a specialized hospital over a period from 1 January 2001 to 31 July 2022. A total of 106 patients was admitted to the emergency department for primary septic arthritis of the shoulder. Of these, 56 patients were enrolled in the study.

The inclusion criteria for the study consisted of an age ≥60 years and a primary septic monoarthritis of the shoulder affecting the glenohumeral joint with fulfillment of at least one of the Newman criteria for septic arthritis [35].

Patients with unavailable survival data and incomplete medical records or follow-up were excluded (8 patients excluded). Patients with a periprosthetic joint infection (12 patients excluded), recent surgery or trauma (<6 months) (19 patients excluded), open skin wounds on the shoulder or general wound treatment were also excluded (1 patient excluded). Likewise, a diagnosis of rheumatoid arthritis, acute gout, or crystal arthropathies led to an exclusion (8 patients excluded). Patients with infections of joints other than the glenohumeral joint were also excluded (2 patients excluded).

Diagnostic measures included: hospital admission, clinical examination, blood examination, joint radiography, joint lavage with microbiological sampling during surgery, empiric antibiotic treatment after surgery, and targeted antibiotic therapy according to the results of microbiological examination and antibiograms.

Factors recorded included: Age in years, sex, joints involved, date of hospital admission, clinical examination methods, blood parameters (including leukocyte count, CRP, procalcitonin and blood cultures), ASA-PS classification scores at the time of surgery, microbiological blood culture analysis, time to follow-up in days and date of death.

The American Society of Anesthesiologists (ASA) physical status classification system is a tool with the purpose of assessing a patient's medical comorbidities prior to anesthesia. It is classified preoperatively and provides a quick overview of the patient's medical condition [32–34]. It is divided into six groups: ASA I: a normal healthy patient; ASA II: a patient with mild systemic disease; ASA III: a patient with severe systemic disease that is not life-threatening; ASA IV: a patient with a severe systemic disease that is a constant threat of life; ASA V: a moribund patient who is not expected to survive without an operation; ASA VI: a patient declared brain dead [32,33].

Joint aspirates were performed during the operation procedure by an orthopedic surgeon using a standard aseptic technique and then sent for examination for aerobic/anaerobic growth. Synovial fluid was inoculated onto aerobic chocolate, sheep blood agar, and anaerobic sheep blood agar plates and then incubated aerobically at 37 °C with 5% CO_2 and anaerobically at 37 °C for 7 days. The remaining liquid was inoculated into a thioglycolate broth for sample enrichment and then incubated for 14 days. Bacteria were identified by matrix-assisted laser desorption/ionization time-of-flight mass spectrometry (MALDI-TOF MS). Antibiotic susceptibility testing was performed according to the European Committee for Antimicrobial Susceptibility Testing (EUCAST).

All patients underwent arthroscopic surgery of the shoulder with drainage and debridement 24 h after diagnosis as an emergency treatment. Arthroscopy of the shoulder was performed in a beach-chair positiong. A standardized procedure was performed in every patient using the posterior, anterior and lateral portal [36]. A 70° arthroscopy was used for a better visualisation. Firstly, the subacromial space was examined before going through the rotator cuff into the glenohumeral joint.

Antibiotic treatment with cefuroxim 1.5 g intravenously began immediately after a diagnosis of septic arthritis as an emergency measure and ended when patients either deceased or gained normal CRP (normal reference value < 0.5 mg/dL), a normal leukocyte count (normal reference range 4.5–10 $\times$ 10^3 cells/µL), and no fever for at least 48 h (<38.5 degrees Celsius, auricular). The antibiotic agent was changed if there was a positive microbiological culture indicating any resistance to cefuroxim. The antibiotic agent was administered for at least 7 days intravenously.

The probability of patient survival over a five-year period was estimated using Kaplan–Meier survival curves, separately for each of the following groups: older adults with septic arthritis with and without medical implants, with positive and negative synovial microbiological cultures, with or without osteoarthritis and grouped according to their preoperative ASA-PS classification scores (ASA II, ASA III, or ASA IV). These survival probabilities were compared by using the log-rank test.

The follow-up was determined from subsequent in-clinic medical records of following hospitalizations, public obituaries from newspapers or internet portals.

The study was approved by a university research ethics committee (IRB-2021-031, 9 November 2021). All data were collected and analyzed anonymously.

Microsoft Excel™ 2016 and IBM SPSS Statistics version 28 were used for statistical analysis. The p value was considered significant at values of $p \leq 0.05$.

Univariate analysis was performed and is stated as mean, median, standard deviation, and confidence interval unless otherwise noted.

3. Results

A total of 56 patients (37 males and 19 females) with a mean age of 78.7 years (range: 60–96 years) was included in the study. The mean follow-up time was 3011.8 days (8.25 years). The mean survival time of the entire study population was 920.3 days (2.52 years). The 1-year survival rate was 66.1%, and the 5-year survival rate was 34.4%.

Increasing age and higher ASA PS classification scores were significantly associated with impaired 5-year survival ($p < 0.05$). No significant differences in survival were found between patients with different types of surgical debridement and antibiotic regimens.

Eight different types of bacteria were detected in the synovial fluid cultures of 48 patients. *Staphylococcus aureus* was the most common pathogen, found in 20 patients (41.7%). Other Gram-positive bacteria included *Streptococcus* species (10 patients, 20.8%), *Enterococcus* species (6 patients, 12.5%), and coagulase-negative staphylococci (6 patients, 12.5%). Gram-negative bacteria, such as *Escherichia coli*, were less common (6 patients, 12.5%). No bacteria were identified in 8 patients (14.3%).

Detailed patient characteristics are demonstrated in Table 1.

Table 1. *n* = Number of patients; % = percentage; ASA PS Classification = American Society of Anesthesiologists Physical Status classification.

		Gender		Total
		Male	*Female*	
Number of patients [*n*. (%)]		37 (66.1)	19 (33.9)	56 (100)
Age in years [Mean. Range]		77 (61–95)	82 (64–100)	78.7 (61–100)
Affected joint	Left [*n* (%)]	13 (54.2)	11 (45.8)	24 (42.9)
	Right [*n* (%)]	24 (75)	8 (25)	32 (57.1)
ASA PS Classification	ASA I [*n* (%)]	0 (0)	0 (0)	0 (0)
	ASA II [*n* (%)]	2 (66.7)	1 (33.3)	3 (5.4)
	ASA III [*n* (%)]	22 (56.4)	17 (43.6)	39 (69.6)
	ASA IV [*n* (%)]	13 (92.9)	1 (7.1)	14 (25)
Number of Operations [*n* (%)]	1 [*n* (%)]	15 (57.7)	11 (42.3)	26 (46.4)
	2 [*n* (%)]	13 (65)	7 (35)	20 (35.7)
	3 [*n* (%)]	8 (88.9)	1 (11.1)	9 (16.1)
	4 [*n* (%)]	1 (100)	0 (0)	1 (1.8)
Positive synovial fluid culture [*n* (%)]		32 (66.7)	16 (33.3)	48 (85.7)
Implants [*n* (%)]		10 (55.6)	8 (44.4)	18 (32.1)
Osteoarthritis of the shoulder [*n* (%)]		16 (53.3)	14 (46.7)	30 (53.6)
Diabetes mellitus [*n* (%)]		10 (58.8)	7 (41.2)	17 (30.4)

3.1. Survival

The mean survival of the entire study population was 920.3 days, 2.52 years, (SD ± 158.4 days; 95% CI: 609.8–1230.7) and a median of 427 days with a SD ± 126.6 days.

The mean follow-up was 3011.8 days or 8.25 years (SD ± 2040.1 days; minimum: 323 days, maximum 7620 days). The Kaplan–Meier Survival analysis is shown in Figure 1. Overall, the successful follow-up rate was 89.3%. A total of 12 (21.4%) patients achieved a 5-year survival after their diagnosis of septic arthritis in the shoulder joint, whereas 44 (78.6%) patients did not survive 5 years. A total of 29 (21.4%) patients did not survive the first year after their diagnosis, and 27 (51.8%) patients were alive 1 year after diagnosis.

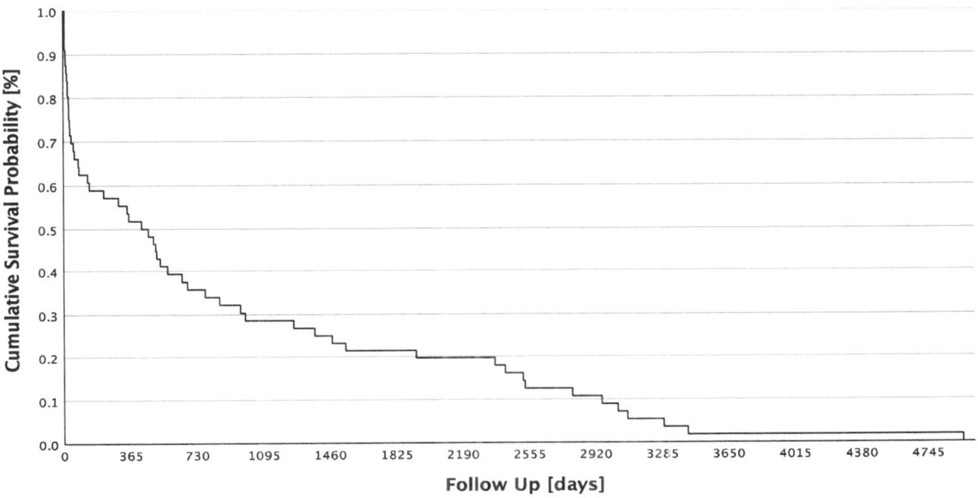

Figure 1. Cumulative Survival Probability of all patients.

3.2. Microbial Spectrum

Eight different pathogens were detected in the analyzed probes. Of the eight organisms identified, five (62.5%) were Gram-positive and three (37.5%) were Gram-negative bacteria. Altogther, 42 (75%) of the patients were detected with Gram-positive bacteria, six (10.7%) with Gram-negative bacteria, and eight (14.3%) void of any microbial detection. More detailed information can be seen in Table 2.

No significant difference in 5-year survival was observed between a positive and a negative synovial culture detection. The log-rank test calculated a p-value of 0.911 for 5-year survival between patients with a positive and a negative synovial detection, making the result statistically not significant. There was a total of eight (14.3%) patients without a pathogen detection in the synovial cultures. This group had a mean survival of 1034.8 days, 2.84 years, (SD ± 393.9 days; 95% CI: 262.7–1806.8). In 48 (85.7%) patients with bacterial septic arthritis, the median follow-up was 901.2 days, 2.47 years, (SD ± 174.2 days; 95% CI: 559.8–1242.6) (Figure 2). The 5-year survival rate without pathogen detection in synovial culture was 25% and with positive pathogen detection in synovial culture was 20.8%.

Table 2. Distribution of bacteria in septic arthritis detected in synovial fluid cultures. *MRSA* = *Methicillin-resistant Staphylococcus aureus*.

Pathogen	n (%)
Gram-positive	
Staphylococcus aureus	31 (55.4)
MRSA	3 (5.4)
Staphylococcus epidermidis	3 (5.4)
Propionibacterium acnes	3 (5.4)
Enterococcus faecalis	2 (3.6)
Total Gram-positive	42 (75)
Gram-negative	
Escherichia coli	4 (7.1)
Pseudomonas aeruginosa	1 (1.8)
Citrobacter freundii	1 (1.8)
Total Gram-negative	6 (10.7)
Negative synovial fluid culture, 8 (14.3)	
Total	56 (100)

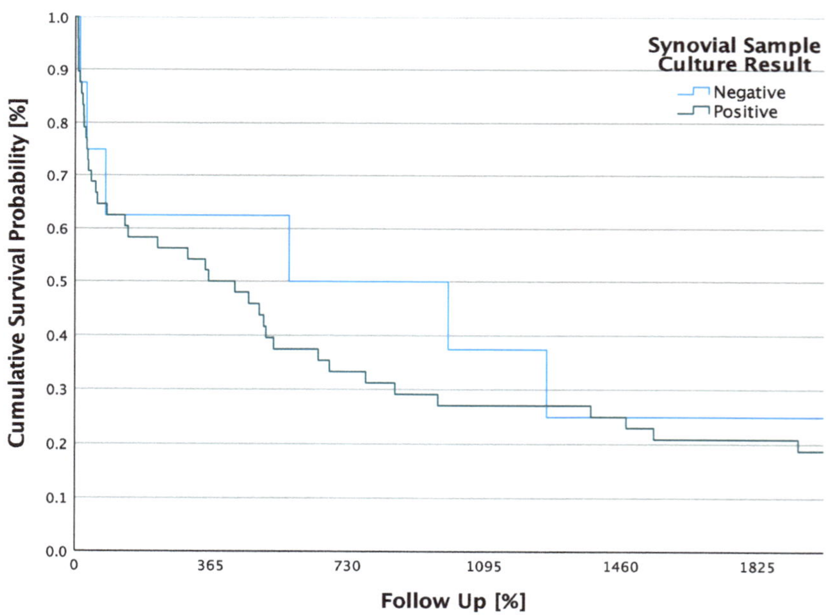

Figure 2. Cumulative 5-year survival probabilities of patients categorized as positive and negative synovial ample culture.

3.3. ASA PS Classification

Patients with an ASA II classification had a mean survival of 2631.7 days, 7.21 years, (SD ± 1302.9 days; 95% CI: 78–5185.3). Patients classified as ASA PS III had a mean survival of 1048.9 days, 2.87 years, (SD ± 182.2 days; 95% CI: 691.7–1406.1). Furthermore, the mean survival for patients in the ASA PS IV group was 195.2 days, 0.53 years, (SD ± 76.6 days; 95% CI: 45–345.4). A significant difference in 5-year survival was found between the different ASA PS classifications in ascending order with a p-value < 0.001. The 5-year survival rate in the ASA PS II group was 66.7%, in the ASA PS III group it was 25.6%, and in the ASA PS IV group it was 0%. Kaplan–Meier survival analysis is shown in Figure 3.

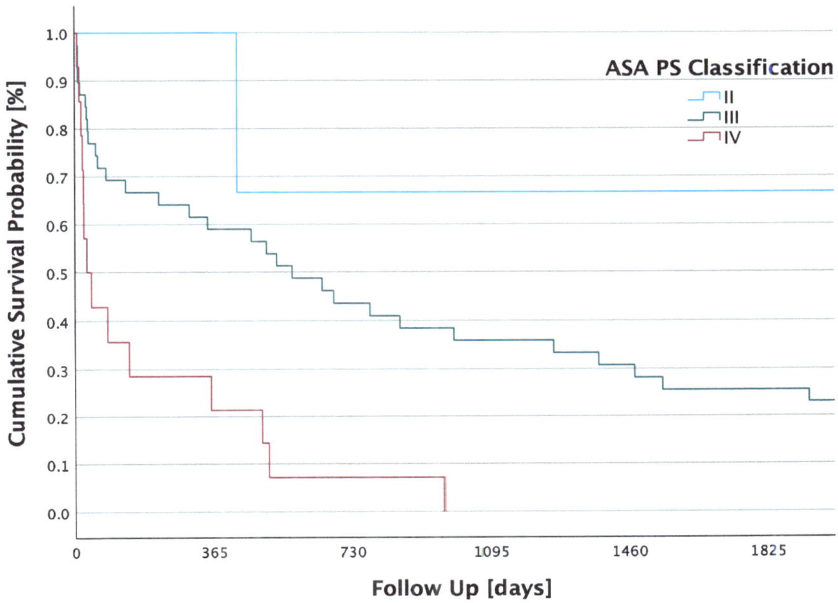

Figure 3. Cumulative 5-year survival probabilities of patients categorized as ASA PS classification II–IV.

3.4. Number of Operations

No significant difference in 5-year survival was found between the numbers of operations. The log-rank test calculated a p-value of 0.561 for the different number of surgeries, making the result statistically not significant. There was a total of 26 (46.4%) patients with surgery. These had a median survival of 1004.4 days, 2.75 years, (SD ± 257.3 days; 95% CI: 500.1–1508.7). Some 20 (35.7%) patients had two operations for therapy for their septic arthritis. In this group, the median follow-up was 846.4 days, 2.32 years, (SD ± 227.2 days; 95% CI: 401.2–1291.6). Furthermore, 9 (16.1%) patients had three operations. There, the median follow-up time was 940.3 days, 2.58 years, (SD ± 433.7 days; 95% CI: 90.3–1790.3). In 1 (1.8%) patient with four operations, the follow-up was 31 days, 0.09 years, (SD ± 0 days).

The 5-year survival rate in the one-surgery group was 26.9%, in the two-surgery group it was 15%, in the three-surgery group it was 22.2% and in the four-surgery group it was 0% (Figure 4).

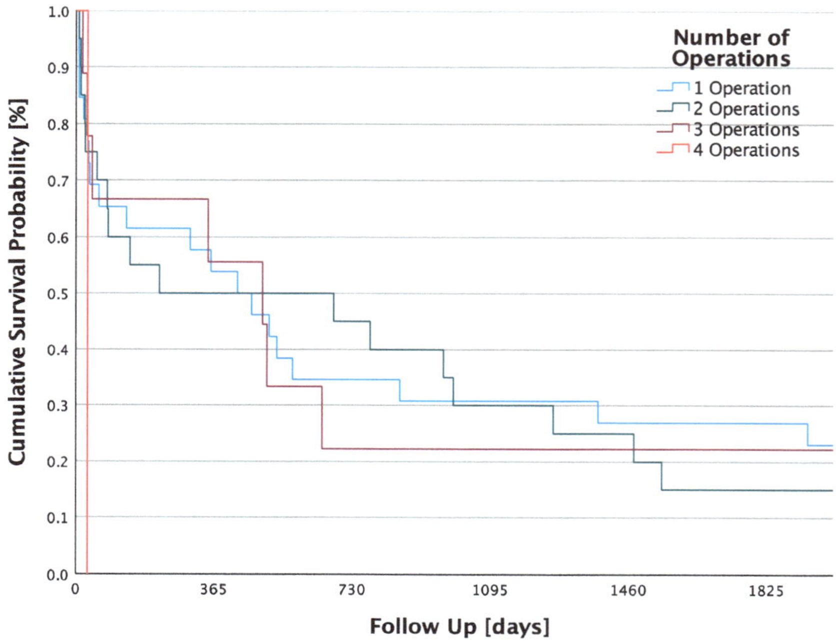

Figure 4. Cumulative 5-year survival probabilities of patients categorized as number of operations.

4. Discussion

The main finding of this study is a very low survival rate of elderly patients receiving a diagnosis of primary septic arthritis of the shoulder. The most common causes are Gram-positive pathogens, with the most frequent being *Staphylococcus aureus*. Survival of elderly patients was limited with increasing ASA PS classification and with rising age.

Due to the scarce data about mid-to long term survival on septic arthitis of the shoulder, it is hard to compare survival data. The 1-year mortality was used, as this is most often used in studies of survival data. In this study, 1-year mortality was 48.2%. In patients with pre-existing osteoarthrosis of the shoulder, the 1-year mortality after a bacterial septic arthritis was 43.3%. Comparable studies of septic arthritis of the knee report values of 23.3% [37]. A national cohort study from Taiwan from 2020 stated that the overall 1-year survival for patients with native septic arthritis of the shoulder was 80.1%. Patients from the age of 18 years or older were included in this study. The different age structure makes a comparison difficult. This study also states that a shoulder infection and an age over 65 years is a negative predictive factor for patients with native septic arthritis [10].

The reason for differences in mortality rates could be based on the anatomic characteristics of the shoulder. The shoulder joint is separated into the glenohumeral joint and the subacromial space, which are separated by the rotator cuff. Infection can spread from one compartment to another. This favours a spreading of the infection even after operation and therefore it is more difficult to eradicate the infection. The clinical signs for a septic arthritis of the shoulder are just seen at a progredient state. This could be another reason for a higher mortality rate. Furthermore, the distribution of ASA classification between this study and the study of the septic arthritis of the knee varies significantly.

Staphylococcus aureus is the primary causative and most important pathogen. *Staphylococcus aureus* but additionally other bacteria such as *Staphylococcus epidermidis* or *Pseudomonas aeruginosa* form biofilms on implants. These biofilms are responsible for the persistence of implant infections and are a source of secondary systemic dissemination of bacteria [28,38]. This hematogenous dissemination of bacteria from biofilms is a potential cause of infection in septic shoulder arthritis. Furthermore, patients who have a medical implant have other preexisting conditions, making them more susceptible to infection and, theoretically, may have higher mortality.

This study revealed a significant difference in 5-year survival between the different ASA PS classifications. Thus, patients with an ASA PS II classification survived significantly longer than those with an ASA PS IV classification, who had the shortest survival. Several studies concluded that patients with higher ASA PS classifications had more frequent medical complications or mortality after surgery than those with lower ASA PS classifications [39–42]. Therefore, ASA PS classification has been reported to be a valuable prognostic variable for both postoperative medical complications and mortality [41,43]. Because patients classified as ASA PS IV suffer from severe systemic diseases and are operated only in emergencies, such as septic arthritis, operative and postoperative risks in these patients are very high. The increased mortality rate with a higher ASA PS classification was expected because these patients suffer from diseases which, even without septic arthritis, pose a permanent threat to life.

It is also remarkable that 26 (46.4%) patients received only one surgical treatment. In the literature, a number of required operations from one to four is recommended. It is suggested that the surgical treatment be repeated as often as required until the CRP value and the fever normalize [44]. Since this study retrospectively covered a period of more than 20 years, it must be taken into consideration that guidelines and literature recommendations differed over that time.

5. Limitations

The predicated survival rate for primary septic arthritis of the shoulder in this study is limited because of a missing control group with morbidity-related matched patients. In this way, the high mortality could also be based on the morbidity these patients had before suffering from primary septic arthritis. In future, there need to be studies investigating survival rates implementing a matched control group to announce the decline in survival rates in patients.

Limitations in terms of broader applicability, scientific accuracy, and interpretation of results, such as *p*-values and confidence intervals, arise from the single-center and retrospective study method [45]. Despite these limitations, this study design was chosen because primary septic shoulder arthritis, being a medical emergency, is a rare disease.

There is a a greater risk of the accumulation of errors because the data are not actively collected, but only taken from documents, such as medical records or findings. Thus, the collection of the data set, even if performed with the greatest possible care, is dependent on the available documents and data, and their quality and completeness. Moreover, each patient record was documented by different individuals and the time difference between individual patient records may be significant, resulting in important differences even within records. In addition, chart selection bias may occur because the data were collected and recorded without the intention of a study [46].

Actual survival or death were the only outcome included in the data. Underlying diseases, medications, and functional outcomes (with the exception of diabetes mellitus and osteoarthritis) were not included in this study. Individual direct causes of death also remain unknown as there were no further investigations after death, severely limiting the follow-up. Additionally, confounding factors are likely to exist in a geriatric population with various comorbidities that may not have been studied in detail.

In addition, this study considers a period of more than 20 years. During this time, diagnostic procedures and therapies changed. Furthermore, a standardized procedure for the detection and treatment of septic arthritis was introduced at a later time. Thus, there may be differences in the selection and accuracy of diagnostic procedures, and there may be a performance bias in therapy, as different physicians with different approaches and different levels of experience performed the operations.

The strength of this study is in the fact that the risk factors affecting mid-term survival of patients with septic arthritis were observed in conjunction with each other rather than individually and that follow-up was possible for a mid-term period in the majority of patients.

Age is a significant risk factor in osteoarthritis. This study did not investigate the presence of osteoarthritis in other joints [14,47]. Patients diagnosed with other types of osteoarthritis such as gout or pseudogout were excluded from the study. However, the presence of crystals in the synovial fluid does not exclude possible infection. Therefore, we excluded potentially eligible patients who were misdiagnosed with a primary diagnosis of crystal arthropathy and secondary septic arthritis.

Another limiting factor that must be considered is selection bias, as our follow-up rate of 89.3% may introduce survival bias because we do not have information on survival or death of those patients who were not followed up.

6. Conclusions

The median survival indicates that septic shoulder arthritis remains a condition with a high mortality, even if treated in a timely manner and adequately. Significantly decreased mid- and long-term survival was found with both increasing age and increasing ASA PS classification at the time of initial surgery. The ASA-PS classification can further help to predict the outcome and therefore to select the most appropriate therapy in order to enhance the chances of survival. The main causative agent for primary septic arthritis is the Gram-positive *Staphylococcus aureus*.

Author Contributions: Conceptualization, M.G. and J.R.; methodology, M.G.; software, J.R. and L.T.; validation, H.J.B., M.G. and J.R.; formal analysis, J.R., L.T. and M.W.; investigation, J.R. and L.T.; resources, H.J.B.; data curation, L.T. and J.R.; writing—original draft preparation, J.R. and M.G.; writing—review and editing, J.R., L.T., K.L., M.W., M.M., S.S., H.J.B. and M.G.; visualization, J.R. and L.T.; supervision, M.G. and H.J.B.; project administration, M.G. All authors have read and agreed to the published version of the manuscript.

Funding: This research received no external funding.

Institutional Review Board Statement: The Institutional Review Board of Paracelsus Medical University stated that this study was conducted according to the guidelines of the Declaration of Helsinki. They approved this study (protocol code IRB-2021-031. at 9 November 2021).

Informed Consent Statement: Informed consent was not applicable due to the anonymized data generation and retrospective analysis. This is consistent with the guidelines of the Declaration of Helsinki.

Data Availability Statement: The data originate from a surgical and traumatological maximum care unit at Klinikum Nürnberg. The data can be provided anonymously in a separate file upon request. A public dataset was not used to obtain the data presented.

Conflicts of Interest: The authors declare no conflict of interest.

References

1. Margaretten, M.E.; Kohlwes, J.; Moore, D.; Bent, S. Does this adult patient have septic arthritis? *JAMA* **2007**, *297*, 1478–1488. [CrossRef]
2. Ross, J.J. Septic Arthritis of Native Joints. *Infect. Dis. Clin. North Am.* **2017**, *31*, 203–218. [CrossRef]
3. Horowitz, D.L.; Katzap, E.; Horowitz, S.; Barilla-LaBarca, M.L. Approach to septic arthritis. *Am. Fam. Physician.* **2011**, *84*, 653–660. [PubMed]
4. Shirtliff, M.E.; Mader, J.T. Acute septic arthritis. *Clin. Microbiol. Rev.* **2002**, *15*, 527–544. [CrossRef]
5. Smith, J.W.; Chalupa, P.; Shabaz Hasan, M. Infectious arthritis: Clinical features, laboratory findings and treatment. *Clin. Microbiol. Infect.* **2006**, *12*, 309–314. [CrossRef]
6. Mathews, C.J.; Weston, V.C.; Jones, A.; Field, M.; Coakley, G. Bacterial septic arthritis in adults. *Lancet.* **2010**, *375*, 846–855. [CrossRef]
7. Kaandorp, C.J.; Krijnen, P.; Moens, H.J.; Habbema, J.D.; van Schaardenburg, D. The outcome of bacterial arthritis: A prospective community-based study. *Arthritis Rheum.* **1997**, *40*, 884–892. [CrossRef]

8. Goldenberg, D.L. Septic arthritis. *Lancet* **1998**, *351*, 197–202. [CrossRef]
9. Rutherford, A.I.; Subesinghe, S.; Bharucha, T.; Ibrahim, F.; Kleymann, A.; Galloway, J.B. A population study of the reported incidence of native joint septic arthritis in the United Kingdom between 1998 and 2013. *Rheumatology* **2016**, *55*, 2176–2180. [CrossRef]
10. Huang, Y.C.; Ho, C.H.; Lin, Y.J.; Chen, H.-J.; Liu, S.-Y.; Wang, C.-L.; Lin, C.-H.; Wang, J.-J.; Chien, C.-C. Site-specific mortality in native joint septic arthritis: A national population study. *Rheumatology* **2020**, *59*, 3826–3833. [CrossRef]
11. Cesari, M.; Prince, M.; Thiyagarajan, J.A.; De Carvalho, I.A.; Bernabei, R.; Chan, P.; Gutierrez-Robledo, L.M.; Michel, J.-P.; Morley, J.E.; Ong, P.; et al. Frailty: An Emerging Public Health Priority. *J. Am. Med. Dir. Assoc.* **2016**, *17*, 188–192. [CrossRef] [PubMed]
12. Long, B.; Koyfman, A.; Gottlieb, M. Evaluation and Management of Septic Arthritis and its Mimics in the Emergency Department. *West J. Emerg. Med.* **2019**, *20*, 331–341. [CrossRef] [PubMed]
13. Shaffer, M.F.; Bennett, G.A. The passage of type III rabbit virulent pneumococci from the vascular system into joints and certain other body cavities. *J. Exp. Med.* **1939**, *70*, 293–302. [CrossRef] [PubMed]
14. Kaandorp, C.J.; Van Schaardenburg, D.; Krijnen, P.; Habbema, J.D.; van de Laar, M.A. Risk factors for septic arthritis in patients with joint disease. A prospective study. *Arthritis Rheum.* **1995**, *38*, 1819–1825. [CrossRef]
15. Kaandorp, C.J.; Dinant, H.J.; van de Laar, M.A.; Moens, H.J.; Prins, A.P.; Dijkmans, B.A. Incidence and sources of native and prosthetic joint infection: A community based prospective survey. *Ann. Rheum. Dis.* **1997**, *56*, 470–475. [CrossRef]
16. Aiello, A.; Farzaneh, F.; Candore, G.; Caruso, C.; Davinelli, S.; Gambino, C.M.; Ligotti, M.E.; Zareian, N.; Accardi, G. Immunosenescence and Its Hallmarks: How to Oppose Aging Strategically? A Review of Potential Options for Therapeutic Intervention. *Front. Immunol.* **2019**, *10*, 2247. [CrossRef] [PubMed]
17. Santoro, A.; Bientinesi, E.; Monti, D. Immunosenescence and inflammaging in the aging process: Age-related diseases or longevity? *Ageing. Res. Rev.* **2021**, *71*, 101422. [CrossRef]
18. Smith, I.D.M.; Milto, K.M.; Doherty, C.J.; Amyes, S.G.B.; Simpson, A.H.R.W.; Hall, A.C. A potential key role for alpha-haemolysin of Staphylococcus aureus in mediating chondrocyte death in septic arthritis. *Bone Joint Res.* **2018**, *7*, 457–467. [CrossRef] [PubMed]
19. Dubost, J.J.; Fis, I.; Denis, P.; Lopitaux, R.; Soubrier, M.; Ristori, J.M.; Bussiere, J.-L.; Sirot, J.; Sauvezie, B. Polyarticular septic arthritis. *Medicine* **1993**, *72*, 296–310. [CrossRef]
20. García-Arias, M.; Balsa, A.; Mola, E.M. Septic arthritis. *Best Pract. Res. Clin. Rheumatol.* **2011**, *25*, 407–421. [CrossRef]
21. Morgan, D.S.; Fisher, D.; Merianos, A.; Currie, B.J. An 18 year clinical review of septic arthritis from tropical Australia. *Epidemiol. Infect.* **1996**, *117*, 423–428. [CrossRef] [PubMed]
22. Goldenberg, D.L.; Reed, J.I. Bacterial Arthritis. *N. Engl. J. Med.* **1985**, *312*, 764–771. [CrossRef] [PubMed]
23. Hamed, K.A.; Tam, J.Y.; Prober, C.G. Pharmacokinetic optimisation of the treatment of septic arthritis. *Clin. Pharmacokinet.* **1996**, *31*, 156–163. [CrossRef]
24. Stengel, D.; Bauwens, K.; Sehouli, J.; Ekkernkamp, A.; Porzsolt, F. Systematic review and meta-analysis of antibiotic therapy for bone and joint infections. *Lancet Infect. Dis.* **2001**, *1*, 175–188. [CrossRef]
25. Johns, B.P.; Loewenthal, M.R.; Dewar, D.C. Open Compared with Arthroscopic Treatment of Acute Septic Arthritis of the Native Knee. *J. Bone Joint Surg. Am.* **2017**, *99*, 499–505. [CrossRef]
26. Abdel, M.P.; Perry, K.I.; Morrey, M.E.; Steinmann, S.P.; Sperling, J.W.; Cass, J.R. Arthroscopic management of native shoulder septic arthritis. *J. Shoulder Elb. Surg.* **2013**, *22*, 418–421. [CrossRef] [PubMed]
27. Jeon, I.H.; Choi, C.H.; Seo, J.S.; Seo, K.J.; Ko, S.H.; Park, J.Y. Arthroscopic management of septic arthritis of the shoulder joint. *J. Bone Joint Surg. Am.* **2006**, *88*, 1802–1806. [CrossRef]
28. Otto, M. Staphylococcal Biofilms. *Microbiol. Spectr.* **2018**, *6*, 4–6. [CrossRef]
29. Carpenter, C.R.; Schuur, J.D.; Everett, W.W.; Pines, J.M. Evidence-based diagnostics: Adult septic arthritis. *Acad. Emerg. Med.* **2011**, *18*, 781–796. [CrossRef]
30. Rhee, S.M.; Lee, D.K.; Park, J.S.; Rhee, Y.G. The Prognostic Value of a Novel Magnetic Resonance Imaging-Based Classification for Septic Arthritis of the Shoulder. *J. Bone Joint Surg. Am.* **2020**, *102*, 1321–1328. [CrossRef]
31. Ross, J.J.; Davidson, L. Methicillin-resistant Staphylococcus aureus septic arthritis: An emerging clinical syndrome. *Rheumatology* **2005**, *44*, 1197–1198. [CrossRef]
32. Doyle, D.J.; Goyal, A.; Garmon, E.H. American Society of Anesthesiologists Classification. In *StatPearls*; StatPearls Publishing: Tampa, FL, USA, 2022. Available online: http://www.ncbi.nlm.nih.gov/books/NBK441940/ (accessed on 6 July 2022).
33. Horvath, B.; Kloesel, B.; Todd, M.M.; Cole, D.J.; Prielipp, R.C. The Evolution, Current Value, and Future of the American Society of Anesthesiologists Physical Status Classification System. *Anesthesiology* **2021**, *135*, 904–919. [CrossRef]
34. Mayhew, D.; Mendonca, V.; Murthy, B.V.S. A review of ASA physical status—Historical perspectives and modern developments. *Anaesthesia* **2019**, *74*, 373–379. [CrossRef]
35. Newman, J.H. Review of septic arthritis throughout the antibiotic era. *Ann. Rheum. Dis.* **1976**, *35*, 198–205. [CrossRef]
36. Paxton, E.S.; Backus, J.; Keener, J.; Brophy, R.H. Shoulder arthroscopy: Basic principles of positioning, anesthesia, and portal anatomy. *J. Am. Acad. Orthop. Surg.* **2013**, *21*, 332–342. [CrossRef]
37. Haag, N.P.; Geßlein, M.; Millrose, M.; Ziegler, R.; Willauschus, M.; Steinmann, J.; Bail, H.J.; Rüther, J. Short- and Mid-Term Survival of Geriatric Patients with Septic Arthritis of the Knee and the Impact of Risk Factors on Survival. *J. Clin. Med.* **2022**, *11*, 755. [CrossRef]

38. Arciola, C.R.; Campoccia, D.; Montanaro, L. Implant infections: Adhesion, biofilm formation and immune evasion. *Nat. Rev. Microbiol.* **2018**, *16*, 397–409. [CrossRef]
39. Foley, C.; Kendall, M.C.; Apruzzese, P.; De Oliveira, G.S. American Society of Anesthesiologists Physical Status Classification as a reliable predictor of postoperative medical complications and mortality following ambulatory surgery: An analysis of 2,089,830 ACS-NSQIP outpatient cases. *BMC Surg.* **2021**, *21*, 253. [CrossRef]
40. Hackett, N.J.; De Oliveira, G.S.; Jain, U.K.; Kim, J.Y.S. ASA class is a reliable independent predictor of medical complications and mortality following surgery. *Int. J. Surg.* **2015**, *18*, 184–190. [CrossRef]
41. Li, G.; Walco, J.P.; Mueller, D.A.; Wanderer, J.P.; Freundlich, R.E. Reliability of the ASA Physical Status Classification System in Predicting Surgical Morbidity: A Retrospective Analysis. *J. Med. Syst.* **2021**, *45*, 83. [CrossRef]
42. Wolters, U.; Wolf, T.; Stützer, H.; Schröder, T. ASA classification and perioperative variables as predictors of postoperative outcome. *Br. J. Anaesth.* **1996**, *77*, 217–222. [CrossRef]
43. Sedgwick, P. Retrospective cohort studies: Advantages and disadvantages. *BMJ* **2014**, *348*, g1072. [CrossRef]
44. Gottlieb, M.; Holladay, D.; Rice, M. Current Approach to the Evaluation and Management of Septic Arthritis. *Pediatr. Emerg. Care* **2019**, *35*, 509–513. [CrossRef]
45. Bellomo, R.; Warrillow, S.J.; Reade, M.C. Why we should be wary of single-center trials. *Crit. Care Med.* **2009**, *37*, 3114–3119. [CrossRef]
46. Talari, K.; Goyal, M. Retrospective Studies—Utility and Caveats. *J. R. Coll. Physicians Edinb.* **2020**, *50*, 398–402. [CrossRef]
47. Weston, V.C.; Jones, A.C.; Bradbury, N.; Fawthrop, F.; Doherty, M. Clinical features and outcome of septic arthritis in a single UK Health District 1982–1991. *Ann. Rheum. Dis.* **1999**, *58*, 214–219. [CrossRef]

Journal of *Personalized Medicine*

MDPI

Article

The Association between Hip Joint Morphology and Posterior Wall Fracture: Analysis of Radiologic Parameters in Computed Tomography

Han Soul Kim [1], Ki Uk Mun [1] and Chul-Ho Kim [2,*]

[1] Department of Orthopedic Surgery, Gachon University Gil Medical Center, Incheon 21556, Republic of Korea
[2] Department of Orthopaedic Surgery, Asan Medical Centre, University of Ulsan College of Medicine, Seoul 05505, Republic of Korea
* Correspondence: oschulhokim@gmail.com; Tel.: +82-10-3302-1449

Abstract: Although numerous radiologic parameters of abnormal hip joint morphology are utilized in practice, studies on the relation of these parameters to acetabular fractures are limited. This study hypothesized that certain morphological features of hip joints are associated with acetabular posterior wall (PW) fracture patterns and aimed to identify morphological characteristics predictive of acetabular PW fracture. The records of 107 consecutive patients, who were diagnosed with acetabular fractures in a level I trauma center from August 2017 to April 2021, were initially reviewed. After excluding patients who lacked proper radiographic evaluation and had previous surgery or concomitant injury on the ipsilateral lower limb, a total of 99 patients were analyzed to investigate the morphological characteristics of the hip joint, measured in computed tomography, associated with acetabular posterior wall fracture. We included patient demographics, acetabular index (AI), sharp angle, acetabular depth-to-width ratio (AD/WR), center-edge angle (CEA), head–neck offset ratio (HNOR), acetabular head index (AHI), anterior acetabular sector angle (AASA), posterior acetabular sector angle (PASA), and acetabular version angle (AVA) in the univariate and multivariate analyses. The injury mechanism ($p = 0.001$) and AD/WR ($p = 0.021$) were predictors of PW fracture in the univariate analysis. In the multivariable analysis, injury mechanism ($p = 0.011$), AI (coefficient B = 0.320; Exp (B) = 1.377; $p = 0.017$), and AD/WR (coefficient B = 33.047; Exp (B) = 2.250×10^{14}; $p = 0.028$) were significant predictors of PW fracture. This study highlights the importance of morphological factors, such as a larger AI and AD/WR, that may influence joint stress distribution, resulting in acetabular PW fracture. Understanding these pathomechanisms may protect the hip joint and prevent future injuries through the early identification and treatment of pathological conditions.

Keywords: fracture; acetabulum; hip joint; diagnostic imaging; computed tomography

check for updates

Citation: Kim, H.S.; Mun, K.U.; Kim, C.-H. The Association between Hip Joint Morphology and Posterior Wall Fracture: Analysis of Radiologic Parameters in Computed Tomography. *J. Pers. Med.* **2023**, *13*, 1406. https://doi.org/10.3390/jpm13091406

Academic Editor: Markus Gesslein

Received: 27 August 2023
Revised: 11 September 2023
Accepted: 18 September 2023
Published: 20 September 2023

1. Introduction

Acetabular posterior wall (PW) fractures account for approximately one third of all acetabular fractures [1–3]. Acetabular PW fractures usually result from high-energy trauma in young patients and cause poor outcomes, thereby leaving them with functional and financial burdens many years after the index trauma [4–7]. Of patients with acetabular fractures, 85% are accompanied by dislocation, which may lead to catastrophic sequelae, such as trauma-related femoral head osteonecrosis or recurrent dislocation [5,8]. Moreover, a recent increase in elderly patients with acetabular fractures and an aging population makes the treatment of acetabular PW fracture even more challenging [9–11].

Identifying the fracture pattern using Letournel's classification system is critical to achieving a successful treatment outcome of acetabular fractures [12–15]. The specific fracture patterns are correlated with the injury mechanism, the vector of injury force application, innominate bone anatomy, and its mechanical properties [12,13]. Acetabular PW fractures mostly result from a "dashboard" injury, during which a force is transmitted to the

PW of the flexed hip joint through the femoral axis [16]. Understanding the anatomy and pathomechanism of a certain disease leads to its prevention, treatment, and improvements in treatment outcome. The dysplastic hip has been related to femoroacetabular impingement (FAI) and hip joint osteoarthritis [17–19]. However, the relation between abnormal hip joint morphology and acetabular fractures has not been identified despite the high incidence and clinical significance of acetabular PW fractures.

Although numerous radiologic parameters of abnormal hip joint morphology are utilized in practice, studies on the relation of these parameters to acetabular fractures are limited. This study hypothesized that certain morphological features of the hip joint are associated with acetabular PW fracture patterns and aimed to identify morphological characteristics predictive of acetabular PW fracture.

2. Materials and Methods

2.1. Patient Selection

The records of 107 consecutive patients, who were diagnosed with acetabular fracture in a level I trauma center from August 2017 to April 2021, were reviewed. The study was approved by our Institutional Review Board with a waiver for the need to provide written informed consent. Data collection was performed following the relevant guidelines and regulations of the committee.

All acetabular fractures, diagnosed using three-dimensional computed tomography (3D-CT) images, were included in the study to avoid misinterpretation of fracture classification. Patients who (i) lacked 3D-CT evaluation, (ii) had a combined fracture in the ipsilateral lower limb, and (iii) underwent ipsilateral hip joint or pelvic bone surgery prior to injury were excluded from the study. A total of 99 patients were included in the study (Figure 1).

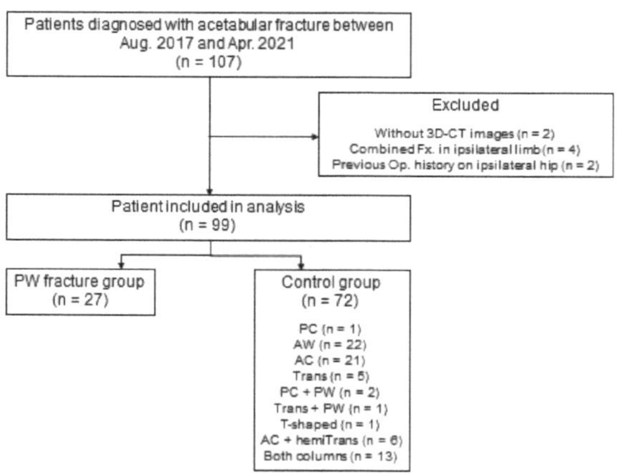

Figure 1. The flowchart of the patient selection process—3D-CT: three-dimensional computed tomography; Fx: fracture; Op: operation; PW: posterial wall; PC: posterior column; AW: anterior wall; AC: anterior column; Trans: transverse; hemiTrans: hemitransverse.

2.2. Fracture Classification and Measurement of Morphological Features

The pelvic anterior–posterior, inlet, outlet, and lateral views were gathered and examined to classify acetabular fractures according to Letournel's method [15]. Matched 3D-CT images were double-checked for precise acetabular fracture pattern identification. Patients were divided into two groups for analyses: the case group with PW fractures and the control group with other acetabular fractures.

The following nine radiologic parameters were collected to evaluate the morphological features of the acetabulum: (1) acetabular index (AI), (2) sharp angle, (3) acetabular depth-to-width ratio (AD/WR), (4) center-edge angle (CEA), (5) head–neck offset ratio (HNOR), (6) acetabular head index (AHI), (7) anterior acetabular sector angle (AASA), (8) posterior acetabular sector angle (PASA), and (9) acetabular version angle (AVA) The AI, sharp angle, AD/WR, CEA, and AHI were measured in the coronal plane of the hip center in 3D-CT [20]. The AASA, PASA, HNOR, and AVA were measured on the axial plane of the hip center [20–22]. All radiologic parameters were first measured on the injured hip, when possible. Measurements were performed on the contralateral hip if fracture displacement was >3 mm. All measurements were performed by two orthopedic surgeons who specialized in hip surgeries and were then averaged for final analysis. The measurement details of radiologic parameters are described in Figure 2.

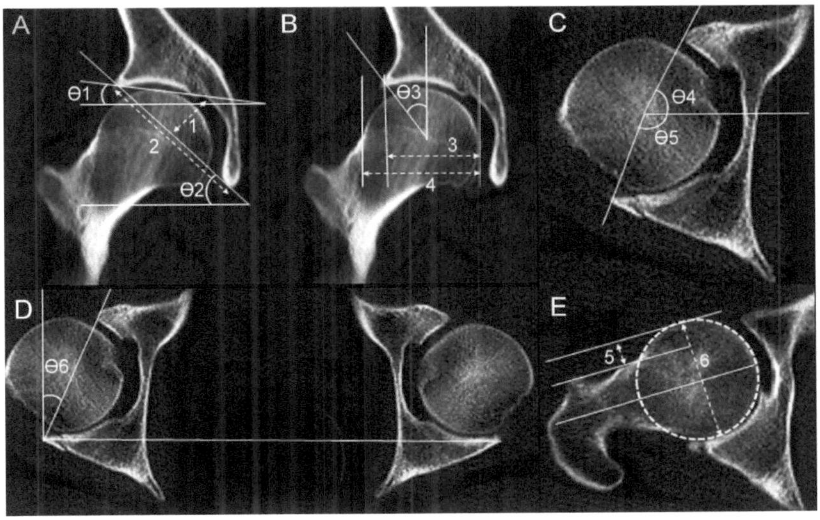

Figure 2. The measurements of radiologic hip joint parameters are shown in a 55-year-old male patient with a right acetabular posterior wall fracture. (**A**) θ1: acetabular index; θ2: sharp angle; (1/2): acetabular depth-to-width ratio. (**B**) θ3: center-edge angle; (3/4): acetabular head index. (**C**) θ4: anterior acetabular sector angle; θ5: posterior acetabular sector angle. (**D**) θ6: acetabular version angle. (**E**) (5/6): head–neck offset ratio.

2.3. Statistical Analysis

Patient sex, age, injury mechanism, and radiologic parameters were compared between the control and case groups. The independent *t*-test or Mann–Whitney U test was used for continuous variables, whereas the chi-square test or Fisher's exact test was used to evaluate categorical variables, after verifying the assumption of normal data distribution.

For the primary outcome of this study, univariate and multivariate binary logistic regression analyses were conducted to find predictors of isolated acetabular PW fracture. Univariate and multivariable binary logistic regression analyses were performed on acetabular fractures with and without PW involvement. The acetabular fractures with PW involvement included isolated PW fracture, posterior wall and column (PW + PC) fracture, and transverse and PW (Transverse + PW) fracture. All parameters which the authors presumed to be clinically associated with the incidence of PW fracture or its involvement were included in the final analyses despite the statistical insignificance in the univariate analysis.

All statistical analyses were performed using PASW Statistics version 18.0 (IBM Corp., Armonk, NY, USA). A p value of <0.05 was considered significant.

3. Results

3.1. Patient Demographics

Of the 99 patients with acetabular fracture, 27 showed isolated acetabular PW fracture. The most common type of acetabular fracture was PW (27 of 99, 27.3%) and isolated anterior wall fracture (22 of 99, 22.2%), followed by anterior column fracture (21 of 99, 21.2%). Two cases of PW + PC fracture and one case of Transverse + PW fracture were identified. The majority of the included patients were male (81 of 99, 81.8%), and the mean age was 50.1 ± 17.0 (16–71) years. The most common cause of acetabular fracture was falling from a height in the overall patient group (33 of 99, 33.3%) as well as in the control group (32 of 72, 44.4%). However, the most common cause of acetabular PW fracture was a motor vehicle accident (16 of 27, 59.3%). The radiologic parameters that characterize hip joint morphology and other details are shown in Tables 1 and 2.

Table 1. Patient demographics and injury mechanism.

	Total (n = 99)	PW (n = 27)	Control (n = 72)	*p*-Value
Male sex	81 (81.8%)	24 (88.9%)	57 (79.2%)	0.383
Age (year)	50.1 ± 17.0 (16–71)	45.3 ± 17.0 (16–71)	51.9 ± 17.6 (13–90)	0.095
Injury mechanism				<0.001
Slip down	6 (6.1%)	0 (0%)	6 (8.3%)	
Fall from height	33 (33.3%)	1 (3.7%)	32 (44.4%)	
MVA: driver	21 (21.2%)	16 (59.3%)	5 (6.9%)	
MVA: passenger	7 (7.1%)	2 (7.4%)	5 (6.9%)	
Pedestrian	11 (11.1%)	1 (3.7%)	10 (13.9%)	
Bicycle	5 (5.1%)	1 (3.7%)	4 (5.6%)	
Motorcycle	11 (11.1%)	5 (18.5%)	6 (8.3%)	
Crushing injury	5 (5.1%)	1 (3.7%)	4 (5.6%)	

PW: posterior wall fracture group; MVA: motor vehicle accident.

Table 2. Radiologic parameters of hip joint morphology.

Radiologic Parameters	Total (n = 99)	PW (n = 27)	Control (n = 72)	*p*-Value
AI (°)	8.67 ± 4.33 (0.73–20.51)	8.88 ± 4.71 (0.73–20.51)	8.59 ± 4.21 (0.82–20.32)	0.770
Sharp angle (°)	39.57 ± 4.19 (25.25–49.08)	39.76 ± 4.07 (30.71–49.08)	39.50 ± 4.25 (25.25–46.92)	0.788
AD/WR	0.27 ± 0.03 (0.21–0.41)	0.29 ± 0.04 (0.24–0.41)	0.27 ± 0.03 (0.21–0.34)	0.047
CEA (°)	32.02 ± 6.71 (14.47–48.23)	32.29 ± 6.50 (18.92–43.54)	31.91 ± 6.83 (14.47–48.23)	0.805
AHI	0.81 ± 0.07 (0.65–1.17)	0.82 ± 0.09 (0.69–1.17)	0.81 ± 0.06 (0.65–0.95)	0.797
AASA (°)	61.46 ± 7.93 (28.40–77.78)	61.63 ± 8.48 (47.73–77.24)	61.40 ± 7.77 (28.40–77.78)	0.900
PASA (°)	91.86 ± 8.04 (74.74–116.74)	92.60 ± 6.95 (79.21–108.76)	91.59 ± 8.44 (74.74–116.74)	0.580
HNOR	0.20 ± 0.06 (0.09–0.34)	0.20 ± 0.06 (0.11–0.30)	0.19 ± 0.06 (0.09–0.34)	0.847
AVA (°)	15.79 ± 5.30 (3.79–29.07)	16.28 ± 5.18 (5.75–25.53)	15.61 ± 5.37 (3.79–29.07)	0.575

PW: posterior wall fracture group; AI: acetabular index; AD/WR: acetabular depth-to-width ratio; CEA: center-edge angle; AHI: acetabular head index; AASA: anterior acetabular sector angle; PASA: posterior acetabular sector angle; HNOR: head–neck offset ratio; AVA: acetabular version angle.

3.2. The Radiologic Risk Factor Analysis for Isolated PW Fracture

The univariate and multivariable logistic regression analyses results for potential risk factors associated with isolated acetabular PW fracture are presented in Table 3. The univariate analysis revealed a significant association of injury mechanism (p = 0.001) and AD/WR (p = 0.021) with isolated PW fracture. With multivariable analysis, injury mechanism (p = 0.011), acetabular index (coefficient B = 0.320; Exp (B) = 1.377; p = 0.017), and acetabular depth-to-width ratio (coefficient B = 33.047; Exp (B) = 2.250×10^{14}; p = 0.028) were significant predictors of isolated PW fracture (Table 3).

Table 3. Results of univariate and multivariable binary logistic regression analyses for predictors of isolated posterior wall fracture.

Characteristic	Univariate Analyses			Multivariable Analyses		
	B	Exp (B)	p-Value	B	Exp (B)	p-Value
Sex (Male)	−0.744	0.475	0.272	−0.917	0.400	0.444
Age (Year)	−0.022	0.978	0.097	−0.046	0.103	0.955
Injury mechanism	N/A	N/A	0.001	N/A	N/A	0.011
AI (°)	0.015	1.016	0.767	0.320	1.377	0.017
Sharp angle (°)	0.015	1.015	0.785	−0.216	0.806	0.073
AD/WR	15.878	7.868×10^6	0.021	33.047	2.25×10^{14}	0.028
CEA (°)	0.008	1.008	0.803	0.050	1.051	0.661
AHI	0.800	2.224	0.795	−10.080	<0.001	0.405
AASA (°)	0.004	1.004	0.899	0.141	1.151	0.168
PASA (°)	0.016	1.016	0.576	0.051	1.053	0.527
HNOR	0.756	2.130	0.846	−1.032	0.356	0.876
AVA (°)	0.024	1.025	0.571	0.326	0.057	0.057

AI: acetabular index; AD/WR: acetabular depth-to-width ratio; CEA: center-edge angle; AHI: acetabular head index; AASA: anterior acetabular sector angle; PASA: posterior acetabular sector angle; HNOR: head–neck offset ratio; AVA: acetabular version angle.

3.3. The Radiologic Risk Factor Analysis for Acetabular Fractures with PW Involvement

The univariate and multivariable logistic regression analyses results for potential risk factors associated with acetabular fractures with PW involvement are presented in Table 4. The univariate analysis revealed a significant association of injury mechanism (p < 0.001) and AD/WR (p = 0.036) with acetabular fractures involving the PW component. With multivariable analysis, injury mechanism (p = 0.006) and acetabular index (coefficient B = 0.255; Exp (B) = 1.291; p = 0.035) were significantly associated with acetabular fractures involving the PW component (Table 4).

Table 4. Results of univariate and multivariable logistic regression analyses for predictors of posterior wall fracture involvement.

Characteristic	Univariate Analyses			Multivariable Analyses		
	B	Exp(B)	p-Value	B	Exp(B)	p-Value
Sex (Male)	−0.916	0.400	0.175	−1.286	0.276	0.261
Age (Year)	−0.022	0.978	0.092	−0.046	0.955	0.104
Injury mechanism	N/A	N/A	<0.001	N/A	N/A	0.006
AI (°)	0.013	1.014	0.790	0.255	1.291	0.035
Sharp angle (°)	0.042	1.043	0.434	−0.108	0.897	0.356
AD/WR	13.815	9.99×10^5	0.036	22.645	6.836×10^9	0.012
CEA (°)	0.012	1.012	0.726	0.066	1.069	0.519
AHI	0.872	2.392	0.770	−6.012	0.002	0.581
AASA (°)	0.007	1.007	0.794	0.041	1.041	0.675
PASA (°)	0.019	1.019	0.491	0.094	1.099	0.287
HNOR	0.316	1.372	0.933	−2.079	0.125	0.742
AVA (°)	0.011	1.011	0.789	0.154	1.167	0.339

AI: acetabular index; AD/WR: acetabular depth-to-width ratio; CEA: center-edge angle; AHI: acetabular head index; AASA: anterior acetabular sector angle; PASA: posterior acetabular sector angle; HNOR: head–neck offset ratio; AVA: acetabular version angle.

4. Discussion

This study primarily revealed that a larger AI and AD/WR are radiologic risk factors for acetabular PW fracture and a larger AI is the sole radiologic risk factor for any acetabular fractures with PW involvement.

Acetabular PW fractures are the most common type of acetabular fractures. An epidemiologic study in 2014 revealed that acetabular PW fractures accounted for 32% in the United States and 30% in China [3]. A review article by Kelly et al. revealed that the most common fracture types were associated with both column fractures (22.3%) and PW fractures (20.9%), followed by transverse + PW fractures (16.3%) [2]. In this study, PW fractures accounted for 27.3% of all acetabular fractures, which was comparable to the aforementioned studies. We observed a noticeable change in the mean patient age from previously published reports. The updated systematic review on acetabular fractures noted that the mean patient age has risen from 38.6 to 45.2 years compared to the previous systematic review published in 2004 [2,23]. Our results suggest that the mean patient age may have risen even higher, to 50.1, which could be attributed to the high percentage of patients aged over 60 years (30/99, 30.3%). Several authors have called for more attention to the increasing incidence of acetabular fractures among geriatric patients because these patients with increased comorbidities and osteoporosis pose a greater challenge to surgeons [9–11,24].

Multiple studies have linked hip joint morphology to FAI and osteoarthritis [17–19,25]. Regarding traumatic hip joint injury, few studies have identified proximal femoral geometry, such as neck-shaft angle and femoral anteversion, as a predisposing factor for proximal femoral fractures [26–28]. Two groups of scholars have indicated that hip dysplasia and FAI play a role in posterior hip dislocation [29,30]. Traditionally, FAI occurs in combination with morphological factors and activity level; that is, pathological repetitive contact accumulates stress concentration in certain hip joint areas until the symptom presents itself [31]. This study aimed to identify the morphological factors that may predispose the hip joint to posterior wall fractures when combined with a traumatic event. To the best of our knowledge, only one group has attempted to correlate acetabular morphology with acetabular fractures, although no relevant acetabular geometry was identified [24]. Our study identified larger AI and AD/WR as risk factors for acetabular PW fractures, and a larger AI was the sole risk factor for acetabular fractures containing PW fragments. These results suggest that less anterosuperior femoral head coverage and a deeper hip socket are predisposing factors for acetabular PW fractures.

The AI is one of the most widely used radiologic parameters to evaluate acetabular coverage [32,33]. The larger AI in our study indicates less anterior wall and more dysplastic hip coverage. A reduced load transfer area causes increased contact pressure, leading to accumulated labrum microdamage and eventual tearing, in a biomechanically unstable joint of a dysplastic hip [31]. Nearly 90% of dysplastic hips accompany labral tear [34]. An intact labrum provides equal stress distribution across the hip joint; however, a torn labrum results in increased stress concentration to the articular cartilage, which may dramatically increase under higher stress conditions, such as high-energy falls or motor vehicle accidents, compared to normal weight-bearing. Moreover, dysplastic hips are known to have cam deformities in 42% of cases, which cause impingements during hip flexion in the anterosuperior quadrant, femoral head posterior subluxation, and subsequently increased contact pressure at the acetabular PW [28]. We propose that less anterosuperior acetabular coverage in a dysplastic hip leads to increased stress concentration on the posterior compart of the hip joint, predisposing the acetabulum to fracture under trauma.

The AD/WR is also one of the representative radiologic parameters of acetabular depth [35,36]. The larger the AD/WR, the deeper the hip socket. We found that a larger AD/WR is a significant predictor of acetabular PW fracture. In deep socket or pincer deformities, femoral head overcoverage limits the range of hip motion, and greater force is transmitted to the acetabular rim as the femoral neck impinges [25]. We believe that high-energy force applied to a deeper hip joint exposes the acetabular rim to a greater risk of fragmentation as the femoral neck abuts the rim. The injury mechanism is one of the well-known predictors of the fracture pattern of the acetabulum. Our results showed that injury from motor vehicle accidents as a driver accounted for nearly 60% of all injury mechanisms in acetabular PW fractures. When we performed logistic regression analyses to eliminate the compounding effect of injury mechanism as a risk factor, the AI and AD/WR were still significantly associated with acetabular PW fracture. Therefore, we believe this study highlights the importance of morphological factors, mainly a larger AI and AD/WR, that may influence joint stress distribution, resulting in acetabular posterior wall fracture.

The current study has several limitations. First, it is retrospective in nature and the number of cases documented is relatively small. However, considering that acetabular fractures are rare injuries in heterogenous patient groups compared to other orthopedic trauma, and considering that this is the first study to successfully identify the relationship between acetabular geometry and acetabular fracture patterns, our study has shown great value in understanding the morphological risk factors of acetabular PW fractures. Second, although we attempted to measure radiologic parameters on the injured side, 69 cases were measured on the contralateral side, and gathering side-specific information was hindered by large fracture gaps and deformities. However, as Gebre et al. suggested, both sides showed a high correlation, indicating that the contralateral side is valid for radiologic parameter measurements [24]. In addition, the indicators of acetabular overcoverage or pincer deformities cannot solely be represented by the AI; that is, a CEA of >40°, an alpha angle of >55°, and an AVA of <15° may all indicate overcoverage or pincer deformities [37]. Similarly, parameters of acetabular dysplasia include not only the AD/WR but also a CEA of <20°, a sharp angle of >45°, an AI of >14°, an AHI of <75%, an AASA of <50°, and a PASA of <90° [20]. We analyzed a wide variety of radiologic parameters for hip joint morphology; however, we were able to identify only the AI and AD/WR as predictors of acetabular PW fracture. High-energy trauma and the relative position of the femoral head against the acetabulum are the two well-known predictors of acetabular fracture patterns. This study suggests that underlying pathologic conditions, such as impingement and instability, can predispose an innominate bone to acetabular PW fracture, which brings debilitating consequences to patient function and quality of life. In the era of ageing populations and a growing incidence of fragility acetabular fractures, treating such pathologic conditions before repetitive microdamage accumulates in the hip joint of geriatric patients may prevent future fractures.

5. Conclusions

This study highlights the importance of morphological factors, such as a larger AI and AD/WR, that may influence joint stress distribution, resulting in acetabular posterior wall fracture. Understanding these pathomechanisms may protect the hip joint and prevent future injuries through the early identification and treatment of pathological conditions.

Author Contributions: Methodology, data curation, validation, writing—original draft, writing—review and editing, H.S.K.; methodology, data curation, writing—original draft, K.U.M.; formal analysis, resources, supervision, investigation, conceptualization, validation, writing—review and editing, C.-H.K. All authors have read and agreed to the published version of the manuscript.

Funding: This work was supported by the Korea Evaluation Institute of Industrial Technology grant funded by the Ministry of Trade, Industry and Energy (MOTIE, Republic of Korea) (No. 20016543, Development of titanium alloy wire rod with roundness of 50 μm or less for biomedical and dental/orthopedic implant application technology).

Institutional Review Board Statement: Institutional Review Board (IRB) approval was obtained from the Gachon University Gil Medical Center Institutional Review Board (IRB NO. GAIRB2022-156) before the initiation of this study.

Informed Consent Statement: The informed consent was waived for the human protocol for this investigation and all investigations were conducted in conformity with ethical principles of research.

Data Availability Statement: The datasets generated and analyzed during the current study are not publicly available since they contain potentially identificatory information for each patient, but they are available from the corresponding author on reasonable request.

Acknowledgments: We thank Kyu Ri Kim for her dedicated assistance with illustrating.

Conflicts of Interest: The authors declare no conflict of interest.

References

1. Rickman, M.; Bircher, M.D. Acetabular fractures in the 21st century. *Trauma* **2008**, *10*, 149–173. [CrossRef]
2. Kelly, J.; Ladurner, A.; Rickman, M. Surgical management of acetabular fractures—A contemporary literature review. *Injury* **2020**, *51*, 2267–2277. [CrossRef] [PubMed]
3. Mauffrey, C.; Hao, J.; Cuellar, D.O.; Herbert, B.; Chen, X.; Liu, B.; Zhang, Y.; Smith, W. The Epidemiology and Injury Patterns of Acetabular Fractures: Are the USA and China Comparable? *Clin. Orthop. Relat. Res.* **2014**, *472*, 3332–3337. [CrossRef]
4. Moed, B.R.; Kregor, P.J.; Reilly, M.C.; Stover, M.D.; Vrahas, M.S. Current management of posterior wall fractures of the acetabulum. *Instr. Course Lect.* **2015**, *64*, 139–159. [PubMed]
5. Moed, B.R.; WillsonCarr, S.E.; Watson, J.T. Results of operative treatment of fractures of the posterior wall of the acetabulum. *J. Bone Jt. Surg. Am.* **2002**, *84*, 752–758. [CrossRef]
6. Rommens, P.M.; Giménez, M.V.; Hessmann, M. Posterior wall fractures of the acetabulum: Characteristics, management, prognosis. *Acta Chir. Belg.* **2001**, *101*, 287–293. [CrossRef]
7. Saterbak, A.M.; Marsh, J.L.; Nepola, J.V.; Brandser, E.A.; Turbett, T. Clinical failure after posterior wall acetabular fractures: The influence of initial fracture patterns. *J. Orthop. Trauma* **2000**, *14*, 230–237. [CrossRef]
8. Milenkovic, S.; Mitkovic, M.; Saveski, J.; Micic, I.; Stojiljkovic, P.; Stanojkovic, M.; Mitkovic, M.; Stamenic, S.; Spalevic, M. Avascular necrosis of the femoral head in the patients with posterior wall acetabular fractures associated with dislocations of the hip. *Acta Chir. Iugosl.* **2013**, *60*, 65–69. [CrossRef]
9. Vipulendran, K.; Kelly, J.; Rickman, M.; Chesser, T. Current concepts: Managing acetabular fractures in the elderly population. *Eur. J. Orthop. Surg. Traumatol.* **2021**, *31*, 807–816. [CrossRef]
10. Park, K.C.; Oh, C.W.; Kim, J.W.; Oh, H.K.; Shon, H.C.; Kim, J.J.; Kim, J.W. Acetabular fractures in elderly. *J. Orthop. Sci.* **2021**, *28*, 376–379. [CrossRef]
11. Butterwick, D.; Papp, S.; Gofton, W.; Liew, A.; Beaulé, P.E. Acetabular fractures in the elderly: Evaluation and management. *J. Bone Jt. Surg. Am.* **2015**, *97*, 758–768. [CrossRef] [PubMed]
12. Alton, T.B.; Gee, A.O. Classifications in Brief: Letournel Classification for Acetabular Fractures. *Clin. Orthop. Relat. Res.* **2014**, *472*, 35–38. [CrossRef] [PubMed]
13. Ly, T.V.; Stover, M.D.; Sims, S.H.; Reilly, M.C. The Use of an Algorithm for Classifying Acetabular Fractures: A Role for Resident Education? *Clin. Orthop. Relat. Res.* **2011**, *469*, 2371–2376. [CrossRef]
14. Cimerman, M.; Kristan, A.; Jug, M.; Tomaževič, M. Fractures of the acetabulum: From yestertday to tomorrow. *Int. Orthop.* **2021**, *45*, 1057–1064. [CrossRef]
15. Letournel, E. Acetabulum fractures: Classification and management. *J. Orthop. Trauma* **2019**, *33*, S1–S2. [CrossRef]
16. Scheinfeld, M.H.; Dym, A.A.; Spektor, M.; Avery, L.L.; Dym, R.J.; Amanatullah, D.F. Acetabular Fractures: What Radiologists Should Know and How 3D CT Can Aid Classification. *Radiographics* **2015**, *35*, 555–577. [CrossRef] [PubMed]
17. van Bosse, H.; Wedge, J.H.; Babyn, P. How are dysplastic hips different? A three-dimensional CT study. *Clin. Orthop. Relat. Res.* **2015**, *473*, 1712–1723. [CrossRef]
18. Tannast, M.; Hanke, M.S.; Zheng, G.; Steppacher, S.D.; Siebenrock, K.A. What are the radiographic reference values for acetabular under- and overcoverage? *Clin. Orthop. Relat. Res.* **2015**, *473*, 1234–1246. [CrossRef]
19. Ganz, R.; Leunig, M.; Leunig-Ganz, K.; Harris, W.H. The etiology of osteoarthritis of the hip: An integrated mechanical concept. *Clin. Orthop. Relat. Res.* **2008**, *466*, 264–272. [CrossRef]
20. Mimura, T.; Mori, K.; Kitagawa, M.; Ueki, M.; Furuya, Y.; Kawasaki, T.; Imai, S. Multiplanar evaluation of radiological findings associated with acetabular dysplasia and investigation of its prevalence in an Asian population: A CT-based study. *BMC Musculoskelet. Disord.* **2017**, *18*, 50. [CrossRef]
21. Kim, J.; Choi, J.A.; Lee, E.; Lee, K.R. Prevalence of Imaging Features on CT Thought to Be Associated With Femoroacetabular Impingement: A Retrospective Analysis of 473 Asymptomatic Adult Hip Joints. *AJR Am. J. Roentgenol.* **2015**, *205*, W100–W105. [CrossRef]
22. Malhotra, R.; Kannan, A.; Kancherla, R.; Khatri, D.; Kumar, V. Femoral head-neck offset in the Indian population: A CT based study. *Indian J. Orthop.* **2012**, *46*, 212–215. [CrossRef] [PubMed]
23. Giannoudis, P.V.; Grotz, M.R.; Papakostidis, C.; Dinopoulos, H. Operative treatment of displaced fractures of the acetabulum. A meta-analysis. *J. Bone Jt. Surg. Br.* **2005**, *87*, 2–9. [CrossRef]
24. Gebre, R.K.; Hirvasniemi, J.; Lantto, I.; Saarakkala, S.; Leppilahti, J.; Jämsä, T. Structural risk factors for low-energy acetabular fractures. *Bone* **2019**, *127*, 334–342. [CrossRef] [PubMed]
25. Beck, M.; Kalhor, M.; Leunig, M.; Ganz, R. Hip morphology influences the pattern of damage to the acetabular cartilage: Femoroacetabular impingement as a cause of early osteoarthritis of the hip. *J. Bone Jt. Surg. Br.* **2005**, *87*, 1012–1018. [CrossRef]

26. Pulkkinen, P.; Eckstein, F.; Lochmüller, E.M.; Kuhn, V.; Jämsä, T. Association of geometric factors and failure load level with the distribution of cervical vs. trochanteric hip fractures. *J. Bone Min. Res.* **2006**, *21*, 895–901. [CrossRef] [PubMed]
27. Frost, A.; Pavlou, G.; Richards, P.J.; Belcher, J.; Jasani, V. Influence of acetabular and femoral version on fractures of the femoral neck. *Clin. Orthop. Relat. Res.* **2010**, *468*, 2224–2229. [CrossRef] [PubMed]
28. Wells, J.; Nepple, J.J.; Crook, K.; Ross, J.R.; Bedi, A.; Schoenecker, P.; Clohisy, J.C. Femoral Morphology in the Dysplastic Hip: Three-dimensional Characterizations with CT. *Clin. Orthop. Relat. Res.* **2017**, *475*, 1045–1054. [CrossRef] [PubMed]
29. Steppacher, S.D.; Albers, C.E.; Siebenrock, K.A.; Tannast, M.; Ganz, R. Femoroacetabular impingement predisposes to traumatic posterior hip dislocation. *Clin. Orthop. Relat. Res.* **2013**, *471*, 1937–1943. [CrossRef] [PubMed]
30. Krych, A.J.; Thompson, M.; Larson, C.M.; Byrd, J.W.; Kelly, B.T. Is posterior hip instability associated with cam and pincer deformity? *Clin. Orthop. Relat. Res.* **2012**, *470*, 3390–3397. [CrossRef]
31. Henak, C.R.; Abraham, C.L.; Anderson, A.E.; Maas, S.A.; Ellis, B.J.; Peters, C.L.; Weiss, J.A. Patient-specific analysis of cartilage and labrum mechanics in human hips with acetabular dysplasia. *Osteoarthr. Cartil.* **2014**, *22*, 210–217. [CrossRef] [PubMed]
32. Shin, C.H.; Yang, E.; Lim, C.; Yoo, W.J.; Choi, I.H.; Cho, T.-J. Which Acetabular Landmarks are the Most Useful for Measuring the Acetabular Index and Center-edge Angle in Developmental Dysplasia of the Hip? A Comparison of Two Methods. *Clin. Orthop. Relat. Res.* **2020**, *478*, 2120–2131. [CrossRef] [PubMed]
33. Long, H.T.; Deng, Z.H.; Zou, M.; Lin, Z.Y.; Zhu, J.X.; Zhu, Y. Effects of the acetabular fracture index and other factors of posterior wall acetabular fracture on functional outcome. *J. Int. Med. Res.* **2017**, *45*, 1394–1405. [CrossRef] [PubMed]
34. Hartig-Andreasen, C.; Søballe, K.; Troelsen, A. The role of the acetabular labrum in hip dysplasia. A literature overview. *Acta Orthop.* **2013**, *84*, 60–64. [CrossRef] [PubMed]
35. Khaliq, S.; Qamar, A.; Khokhar, S.K.; Naseer, H. Morphometric study of acetabular depth and coverage and their clinical importance. *Pak. J. Med. Sci.* **2022**, *38*, 2227–2233. [CrossRef] [PubMed]
36. Novais, E.N.; Pan, Z.; Autruong, P.T.; Meyers, M.L.; Chang, F.M. Normal Percentile Reference Curves and Correlation of Acetabular Index and Acetabular Depth Ratio in Children. *J. Pediatr. Orthop.* **2018**, *38*, 163–169. [CrossRef]
37. Albers, C.E.; Wambeek, N.; Hanke, M.S.; Schmaranzer, F.; Prosser, G.H.; Yates, P.J. Imaging of femoroacetabular impingement-current concepts. *J. Hip Preserv. Surg.* **2016**, *3*, 245–261. [CrossRef]

Journal of *Personalized Medicine*

MDPI

Article

What Is the Best Treatment Choice for Concomitant Ipsilateral Femoral Neck and Intertrochanteric Fracture? A Retrospective Comparative Analysis of 115 Consecutive Patients

Han Soul Kim [1], Dong Keun Lee [1], Ki Uk Mun [1], Dou Hyun Moon [1] and Chul-Ho Kim [2,*]

1 Department of Orthopedic Surgery, Gachon University Gil Medical Center, Namdong-gu, Incheon 21556, Republic of Korea

2 Department of Orthopedic Surgery, Chung-Ang University Hospital, Chung-Ang University College of Medicine, Seoul 06973, Republic of Korea

* Correspondence: oschulhokim@cau.ac.kr; Tel.: +82-2-6299-1435

Abstract: Purpose: Although a concomitant ipsilateral femoral neck and intertrochanteric fracture has been considered to be a rare type of injury, its incidence has been increasing, especially among elderly hip fracture patients. However, there is limited evidence on the optimal treatment option. This study surveys surgical outcomes of different implants in order to assist in selecting the best possible implant for a combined femoral neck and intertrochanteric fracture. Methods: The postoperative complications after the treatment of a concomitant ipsilateral femoral neck and intertrochanteric fracture via cephalomedullary nail (CMN), dynamic hip screw (DHS), and hip arthroplasty groups were analyzed by retrospectively reviewing the electronic medical records of 115 consecutive patients. Results: The patient demographics and perioperative details showed no significant discrepancies amongst different surgical groups, except for the operative time; a CMN had the shortest mean operative time (standard deviation) of 85.6 min (31.1), followed by 94.7 min (22.3) during a DHS, and 107.3 min (37.2) during an HR ($p = 0.021$). Of the 84 osteosynthesis patients, 77 (91.7%) achieved a fracture union. Only one (3.2%) of the 31 HR cases had a dislocation. The sub-analysis of the different osteosynthesis methods showed a higher incidence of excessive sliding and the nonunion of the fracture fragment in the DHS group than that in the CMN group ($p = 0.004$ and $p = 0.022$, respectively). The different surgical methods did not significantly vary in other outcome variables, such as the re-operation rate, mortality, and hip function. Conclusions: For the surgical treatment of combined femoral neck and trochanteric fractures, osteosynthesis did not differ significantly from an HR in terms of the overall postoperative complications, reoperation and mortality rate, and hip function, however, the risk of nonunion and more mechanical complications should be considered when choosing a DHS. Our suggestion for the treatment of a femoral neck and ipsilateral trochanteric fracture is that a surgeon should choose wisely between an HR and a CMN depending on the patient's age, the displacement of the femoral neck, and one's expertise.

Keywords: femoral neck fracture; intertrochanteric fracture; cephalomedullary nail; dynamic hip screw; arthroplasty

Citation: Kim, H.S.; Lee, D.K.; Mun, K.U.; Moon, D.H.; Kim, C.-H. What Is the Best Treatment Choice for Concomitant Ipsilateral Femoral Neck and Intertrochanteric Fracture? A Retrospective Comparative Analysis of 115 Consecutive Patients. *J. Pers. Med.* **2022**, *12*, 1908. https://doi.org/10.3390/jpm12111908

Academic Editors: Markus Gesslein and Eduard Kurz

Received: 9 October 2022
Accepted: 13 November 2022
Published: 16 November 2022

1. Introduction

The incidence of a proximal femoral facture (PFF) is rigorously rising along with the socioeconomic burden due to the global ageing population [1]. A femoral neck fracture (FNF) and intertrochanteric fracture (ITF) are the two most common types of PFF. The clinical significance and interests in PFFs are remarkable due to detrimental postoperative complications, mortality, and a functional decline in polymorbid patients [1]. PFFs are broadly categorized into intracapsular and extracapsular fractures based on pathoanatomy and physiology for therapeutic guidance [2]. Since extracapsular fractures have an undisrupted blood supply, the preservation of the femoral head via osteosynthesis is the gold standard, however, the disruption of the terminal vascularization of the femoral head in intracapsular fractures results in a higher risk of nonunion and avascular necrosis, requiring a surgeon with more discretion in decision-making between osteosynthesis and a hip replacement (HR) [2].

A fracture such as a concomitant ipsilateral femoral neck and intertrochanteric fracture (CIFNITF), which involves both intracapsular and extracapsular fracture components, is difficult to categorize and raises a dilemma of management [3–29]. Traditionally, the CIFNITF has been considered an extremely rare type of trauma that displays a bimodal distribution in age and respective injury mechanisms [5,10,20,23]. However, recent literatures suggest otherwise [13,26,27,29]. In line with the recent literatures, the authors believe that the incidence of the CIFNITF is revving up from the expansion of an elderly population and the implementation of personal mobility vehicles. Regardless, the CIFNITF poses management challenges regarding the optimal implants due to the lack of a classification system and an evidence-based consensus. This study surveys the surgical outcomes of different implants in order to assist in selecting the best possible treatment option for the CIFNITF.

2. Materials and Methods

2.1. Patient Selection

After receiving approval by the institutional review board, we reviewed 2098 patients who underwent surgeries under the diagnosis of FNF and/or ITF between January 2010 and December 2019 in our institution. The inclusion criteria were (1) a diagnosis of CIFNITF confirmed by 3D computed tomography (CT) images, (2) a minimum follow-up period of 12 months postoperatively, and (3) patients aged >18 years. A CIFNITF was diagnosed when (1) more than 50% of the intertrochanteric fracture line was included in the intracapsular pocket at the cross-lined axial and coronal CT view where the femoral head is the largest, or (2) the additional separate fracture center is identified in the transcervical or subcapital region. The presence of a capsular sign was also screened to avoid any misdiagnosis [30]. All diagnoses and surgeries were conducted by two experienced senior orthopedic surgeons who specialized in hip and pelvic trauma for over 20 years. A total of 115 patients were included in the study (Figure 1).

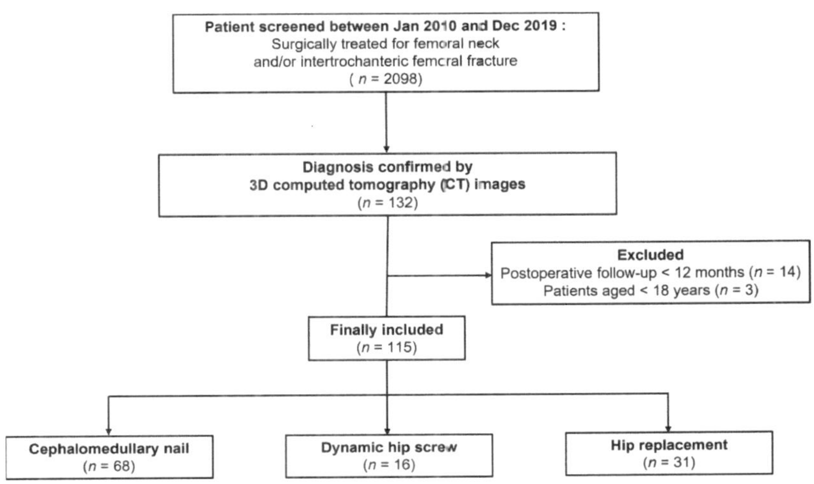

Figure 1. Flow chart of the included patients.

2.2. Surgical Technique

We categorized the included patients into three groups according to the following surgical methods: (1) a cephalomedullary nail (CMN), (2) a dynamic hip screw (DHS), and

(3) an HR. The compression hip nail (TDM, Seongnam, Republic of Korea) and proximal femoral nail (DePuy Synthes, Solothurn, Switzerland) were used for the CMN, and the DHS system (Jeil Medical, Seoul, Republic of Korea) for the DHS surgery. For the DHS surgery, a trochanteric stabilizing plate (TSP) was applied in all cases. In all osteosynthesis surgeries, we achieved an extramedullary or neutral reduction in either the anteroposterior (AP) or lateral view of a simple radiograph and an acceptable lag screw position [31,32].

For the HR surgery, we performed a total hip arthroplasty using the Bencox® Cup System (Corentec, Seoul, Republic of Korea) with the Bencox® stem (Corentec, Seoul, Republic of Korea) for patients aged <70 years and bipolar hemiarthroplasty surgery using the C2 stem (Lima-LTO, Udine, Italy) or Bencox® stem (Corentec, Seoul, Republic of Korea) for the femoral component in patients aged >70 years. We applied trochanter reinforcement procedures with either a tension band wiring or trochanteric grip plate, only when an unstable abductor or trochanteric fragment was noticed intraoperatively. Figure 2 shows example cases of each treatment modality.

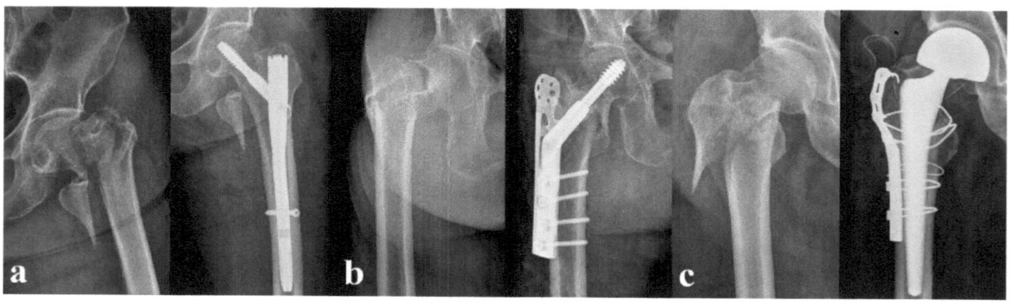

Figure 2. Pre- and postoperative plain radiographs of different treatments. (**a**) A 72-year-old female who underwent cephalomedullary nailing. (**b**) An 88-year-old female who underwent dynamic hip screw fixation. (**c**) A 73-year-old male who underwent bipolar hemiarthroplasty.

2.3. Data Collection

The demographic data and perioperative surgical variables, including the patient's age, sex, T-score of the bone mineral density (BMD), operative time, type of anesthesia during surgery, and the duration of their hospital stay were collected. The postoperative complications were analyzed for the outcome measure. In osteosynthesis surgeries, the incidence of osteonecrosis of the femoral head (ONFH), cut-out of lag screw, cut-through of lag screw, excessive sliding, and the nonunion rate were investigated. An excessive sliding was defined as a lag screw sliding > 15 mm [33]. We defined a nonunion, as no radiologic evidence of a callus or bridge formation, at least 15 weeks after surgery either AP or in the lateral view of simple radiographs [34].

In an HR, we investigated the incidence of a dislocation and implant loosening during the follow-up period. As overall complications in all surgeries, we investigated the incidence of a heterotopic ossification, venous thromboembolism (VTE), and deep infection. We also studied the revision rate in both osteosynthesis and HR surgeries and determined the mortality rate during the hospitalization period and within 12 months of the postoperative period. For the outcome measurement, we collected the data for the Harris hip score (HHS). All of the radiological parameters were measured by two forementioned orthopedic surgeons, and any disagreements were resolved through discussion between the investigators.

2.4. Statistical Analysis

The baseline characteristics were summarized using standard descriptive statistics. The continuous variables were described as the mean (standard deviation) or median (range). A Kruskal–Wallis test was utilized for the comparison of three types of surgeries, and the independent T-test or Mann–Whitney U test were applied for direct one-to-one comparisons of each type of surgery (e.g., CMN vs. DHS). The categorical data were evaluated using the chi-square test or Fisher's exact test, as applicable, and were summarized as the absolute frequency and percentages. All statistical analyses were performed using PASW Statistics version 18.0 (IBM Corp., Armonk, NY, USA). A statistical significance was defined as a p-value < 0.05. Additionally, we evaluated the trends of a CIFNIF incidence in our hospital.

3. Results

3.1. Patient Demographics and Perioperative Details

The overall incidence of CIFNITF was 6.1% (n = 115). The median (range) age of the included patients was 79 (40 to 95) years, and no significant differences were found among the three groups. Seventy-three of the 115 patients were female (63.5%), and females were predominant in all three groups. The median (range) T-score of BMD was −3.3 (−7.2 to 0.8). The CMN had the shortest mean operative time (standard deviation) of 85.6 min (31.1), followed by 94.7 min (22.3) during the DHS, and 107.3 min (37.2) during the HR (p = 0.021). When compared to the different types of surgery, significant differences in the operative time were noted between the CMN and HR (p = 0.003), and between the DHS and HR (p = 0.025). The patient demographics and perioperative details such as the types of anesthesia and the length of the hospital stay did not vary significantly. The basic demographic and perioperative details are elaborated in Table 1.

Table 1. Patient demographics and perioperative characteristics.

	All (n = 115)	CMN (n = 68)	DHS (n = 16)	HR (n = 31)	p Value			
					CMN vs. DHS	CMN vs. HR	DHS vs. HR	Overall
Age	79 (40 to 95)	47 (40 to 90)	79.5 (42 to 94)	81 (56 to 95)	0.669	0.125	0.574	0.338
Female	73 (63.5%)	39 (57.4%)	9 (56.3%)	25 (80.6%)	0.936	0.025	0.096	0.075
BMD (T-score)	−3.3 (−7.2 to 0.8)	−3.2 (−4.9 to 0.8)	−3.9 (−4.6 to −3.1)	−3.9 (−7.2 to −0.5)	0.112	0.197	0.881	0.077
Op time (min)	92.7 (33.0)	85.6 (31.1)	94.7 (22.3)	107.3 (37.2)	0.174	0.003	0.025	0.021
Anesthesia [a]	53 (46.1%)	36 (52.9%)	6 (37.5%)	11 (35.5%)	0.266	0.107	0.892	0.236
LOS (day)	14 (3 to 125)	14 (3 to 119)	16 (6 to 55)	15 (7 to 125)	0.249	0.280	0.991	0.250

Values are expressed as median (range), mean (SD), or n (%). [a] Type of anesthesia expressed in number of general anesthesia; SD: standard deviation; BMD: bone mineral density; CMN: cephalomedullary nail; DHS: dynamic hip screw; HR: hip replacement; LOS: length of stay; Op: operation.

3.2. Postoperative Complications, Mortality Rates and Outcome

We compared the incidence of ONFH, lag screw cut-out, cut-through, excessive sliding, and nonunion of the fracture between the CMN and DHS groups. Seven of the 68 patients (10.3%) in the CMN and 2 of 16 (12.5%) in the DHS were diagnosed with post-traumatic ONFH, with no significant difference between the two groups. There were two cases of cut-out, one case of cut-through in the CMN, and none in the DHS (p = 1.000 on both the cut-out and cut-through). The incidence of excessive sliding was significantly higher in the DHS than that in the CMN (25% and 1.5%, respectively [p = 0.004]). Four of 16 (25%) patients in the DHS and 3 of 68 (4.4%) patients in the CMN had a nonunion of fracture, and the difference was significant (p = 0.022). In the HR, there was one case of dislocation, and no implant loosening was observed. The overall complication rates of heterotopic ossification, VTE, deep infection, and revision surgery displayed no significant differences among the different surgical options. The overall in-hospital mortality rate was 5 of 115 (4.3%), and the one-year mortality rate was 13.9% (16 of 115). The HHS could only be collected from 24 patients in the CMN, 4 in the DHS, and 28 in the HR. The median (range) HHSs were 89.5 points (65 to 95), 78.5 points (65 to 91) in CMN, 91.0 points (85 to 95) in the DHS, and

88.5 points (85 to 95) in the HR, respectively ($p = 0.819$). The details of the surgical outcomes are listed in Table 2.

Table 2. Comparison of postoperative outcomes among different treatment modalities.

	All Patients (n = 115)	CMN (n = 68)	DHS (n = 16)	Arthroplasty (n = 31)	p Value
Osteosynthesis					
ONFH		7 (10.3%)	2 (12.5%)	N/A	0.679
Cut-out		2 (2.9%)	0 (0%)	N/A	1.000
Cut-through		1 (1.5%)	0 (0%)	N/A	1.000
Excessive sliding		1 (1.5%)	4 (25.0%)	N/A	0.004
Nonunion		3 (4.4%)	4 (25.0%)	N/A	0.022
Arthroplasty					
Dislocation		N/A	N/A	1 (3.2%)	N/A
Implant loosening		N/A	N/A	0 (0%)	N/A
Total					
HO	5 (4.3%)	4 (5.9%)	0 (0%)	1 (3.2%)	1.000
VTE	4 (3.5%)	4 (5.9%)	0 (0%)	0 (0%)	0.459
Infection	6 (5.2%)	2 (2.9%)	1 (6.3%)	3 (9.7%)	0.269
Revision surgery	6 (5.2%)	4 (5.9%)	1 (6.3%)	1 (3.2%)	1.000
In-hospital mortality	5 (4.3%)	3 (4.4%)	0 (0%)	2 (6.5%)	0.674
1-year mortality	16 (13.9%)	8 (11.8%)	1 (6.3%)	7 (22.6%)	0.293
HHS at final f/u	84.2 (65–95)	73.4 (65–91)	89.5 (85–95)	87.2 (85–95)	0.819

Values are expressed as median (range), mean (SD), or n (%). CMN: cephalomedullary nail; DHS: dynamic hip screw; HR: hip replacement; N/A: not available; f/u; follow-up; HHS: Harris hip score; HO: heterotopic ossification; ONFH: osteonecrosis of femoral head; VTE: venous thromboembolism.

3.3. Trend in the Incidence of CIFNITF

The annual trends in the incidence of CIFNITFs and PFFs were investigated. The bar graph in Figure 3 indicates that the incidence of CIFNITFs randomly varied from 4.1% to 7.0%, while the incidence of PFFs showed increasing trend over time.

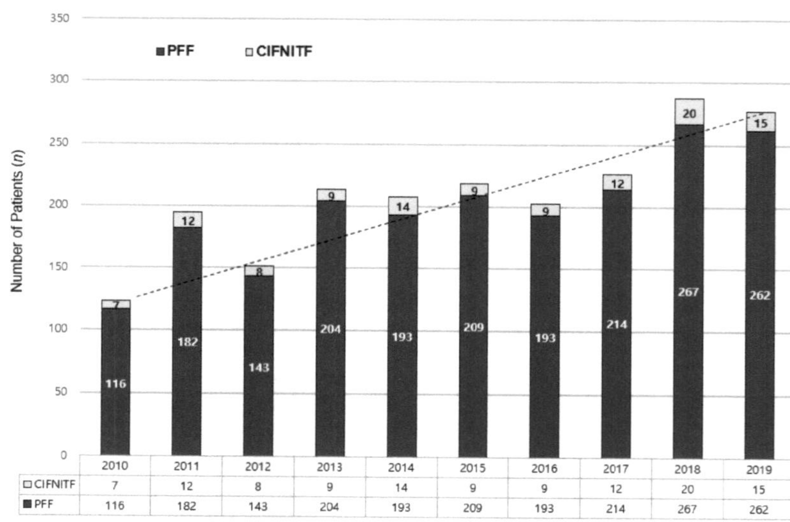

	2010	2011	2012	2013	2014	2015	2016	2017	2018	2019
CIFNITF	7	12	8	9	14	9	9	12	20	15
PFF	116	182	143	204	193	209	193	214	267	262

Figure 3. Annual trends of concomitant ipsilateral femoral neck and intertrochanter fracture. (PFF: proximal femoral fracture; CIFNITF: concomitant ipsilateral femoral neck and intertrochanter fracture).

4. Discussion

The CIFNITF is a challenging type of injury that requires proper surgical planning. However, the underestimation of the incidence of this injury has misguided surgeons in diagnosis and in reaching a consensus in both the proper diagnosis and the optimal treatment modality due to the lack of evidence in the literature. To our knowledge, our work is the first large retrospective study that analyzed the postoperative outcomes of different treatment options for a CIFNITF. Our main finding is that the CMN and DHS did not differ significantly from the HR in terms of the postoperative complications, revision rate, in-hospital and one-year mortality rate, and hip function.

We performed a literature review on a CIFNITF and found 24 case reports and 4 small retrospective studies with 128 patients in total (Supplementary Table S1) [3–29]. The age distribution and injury mechanism showed bimodal characteristics similar to our study, excluding the four retrospective studies that reported a median age with an interquartile range; the median (range) of the age in all cases was 74 (26 to 97) years. The majority of the injury mechanism (n = 104) was low-energy trauma, such as a simple fall. The sex ratio was also in concordance with our study as there were 81 females (63.3%). The studies reported before 2011 mostly used plain radiography for the diagnosis. Two studies made diagnosis intraoperatively by fluoroscopy and one study did not realize the injury until several months after the surgery, which suggests that plain radiography may not be sufficient to diagnose a CIFNITF [4,6–8,12–15,17,18,21,25]. The studies after 2011 employed CT as a final diagnostic modality, and the emergence of larger case series since then indicates that a CT-based diagnosis has enabled surgeons to fully assess and correctly diagnose concomitant injuries.

However, even with a CT-based diagnosis, most studies varied greatly in the description of the fracture pattern and in the implant choices due to the lack of a classification system. Yoo et al. categorized a CIFNITF into three-part and four-part fractures, but could not relate the category with treatment choice since they performed a CMN in all cases [29]. Tong et al. suggested a new classification system which incorporated different morphological features of a CIFNITF [26]. However, after reviewing the fracture patterns and treatment results in the literatures and in our study, we came to the conclusion that the most common fracture configuration is the reverse oblique ITF (AO Foundation/Orthopaedic Trauma Association (AO/OTA) fracture classification 31A3) with an additional fracture center at the intracapsular region (122/128 cases in the literatures and 109/115 in our study) [2]. The second most common was the segmental pattern in which a separate intracapsular fracture center ran parallel to the extracapsular fracture line (5/128 in the literature and 4 in our study). The least common was a continuous fracture line of the FNF that extended like a beak down to the lower end of the intertrochanteric ridge or below the lesser trochanteric region (one in the literature and two in our study). Based on the common morphological patterns reported in the literature and in our study, we simplified the CIFNITF classification in Figure 4.

In the literature review, the utilized implants are as follows: 14 HRs, 2 screws, 2 dynamic condylar screws, 10 plate and screw constructs, 45 DHSs, and 55 CMNs (Supplementary Table S1). Only one complication of a dislocation was reported after an HR. There were 7 nonunion or mechanical complications out of 55 CMNs and 13 nonunion, mechanical complication, or ONFH out of 45 DHSs. Similarly, we found a higher rate of excessive sliding and nonunion in the DHS than in the CMN. First, the presence of an intracapsular fracture center makes the fixation of these unstable fractures extremely challenging. The high percentage of geriatric patients with osteoporosis also makes it very challenging to operate on the pathological bone [2]. This is clinically significant since the incidence of complex type of PFFs such as CIFNITF will increase along with the rising of geriatric fragility fractures [1]. A correct diagnosis and adequate treatment are key to achieving a better outcome and minimizing the risk of complications.

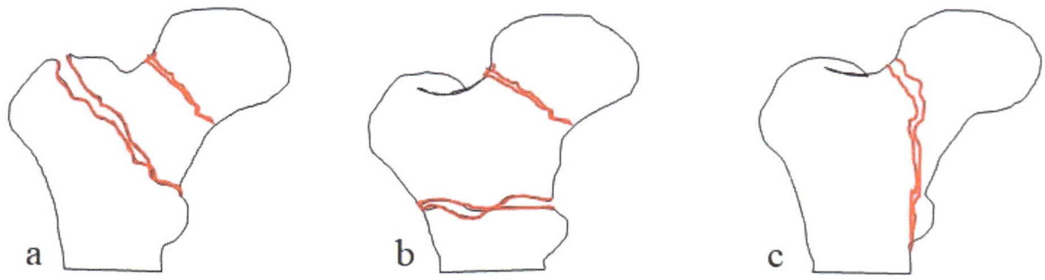

Figure 4. Classification of concomitant ipsilateral femoral neck and intertrochanter fracture. (**a**) Type I: a separate intracapsular fracture center ran parallel to extracapsular fracture line. (**b**) Reverse oblique intertrochanteric fracture with an additional fracture center at intracapsular region. (**c**) A continuous fracture line of femoral neck fracture that extended like a beak down to the lower end of intertrochanteric ridge or below lesser trochanteric region.

Unfortunately, since evidence on the treatment of a CIFNITF is weak, we suggest a provisional management algorithm based on the most appropriate surgical option of each fracture entity (Figure 5). In general, a non-displaced FNF and those of a biologically young age are candidates for a DHS, while a hip replacement (HR) is the gold standard for displaced fractures and biologically old patients [2,35–37]. Simultaneously, the most feasible fixation method for unstable ITF was weighed [2,38,39]. Although a CMN, DHS, and HR can all be considered for unstable ITF, a CMN should be the treatment of choice for younger patients, while an HR should be the gold standard for older patients when the injury is complicated by an FNF. Based on the literature review and our study results, a DHS seems to carry more burdens in terms of mechanical complications and nonunion. To date, there have been controversies regarding whether a CMN is superior to DHS in the treatment of intertrochanteric fractures [40–42]; however, in cases of unstable intertrochanteric fractures, there is a consensus that a DHS surgery is challenging because of the loss of medial buttress and a lateral wall fracture [43,44]. In fact, Kyle et al. has suggested that surgeons refrain from using DHS and consider an HR even for younger patients for managing a CIFNITF [13]. Thus, we recommend that surgeons should carefully select a DHS over a CMN when fixing a CIFNITF, considering the potential risk factors of a high nonunion and excessive displacement rate in DHS than in CMN.

As for the mortality rate in the current study, the overall in-patient mortality rate was 4.3% (5 of 115), and the overall mortality rate at the one-year postoperative follow-up was 13.9% (16 of 115). There were no significant differences among the different surgery groups; the mortality rate ranged from 0 to 6.5% during the inpatient period and from 6.3 to 22.6% during the one-year follow-up period. These data are comparable with the recent data collected from the national registries of several Asian countries from 2013 to 2017 with a review of 10 articles, which showed an inpatient mortality rate ranging from 3% to 7% and a mean 1-year mortality of 17.89% [45].

We initially hypothesized that the incidence of a combined fracture would increase over time, suspecting that this could be affected by an increased life expectancy or medication history of osteoporosis, however, the percentage of combined fractures randomly varied from 4.1% to 7.0% between 2010 and 2019, with the lowest in 2015 and the highest in 2018. Although the overall incidence of an FNF and ITF increased, there was no evidence of any trends in the incidence of a combined fracture (Figure 3). Nevertheless, the incidence of the combined fracture, which is higher than previous literatures, revealed by a preoperative CT analysis suggests that some of the combined fractures may have been either neglected or overlooked.

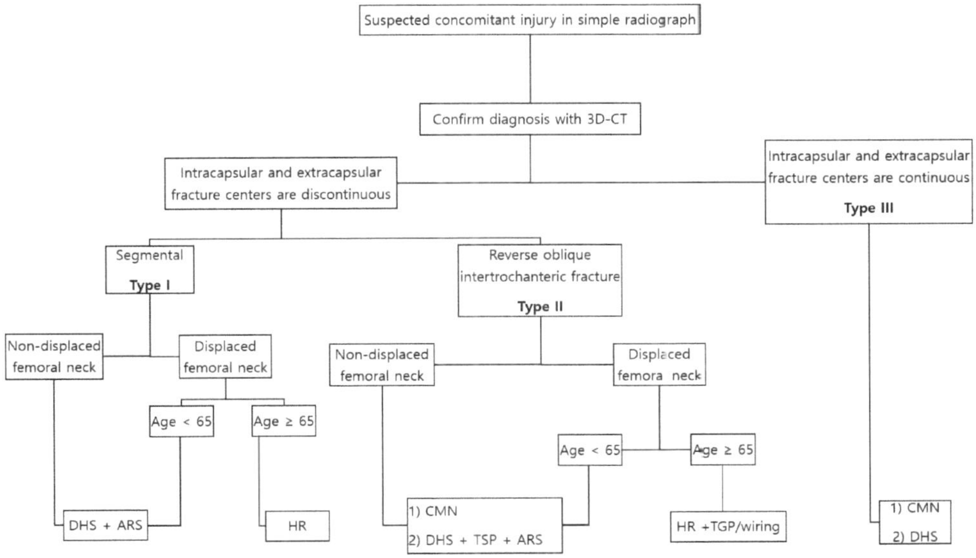

Figure 5. Algorithm for the management of concomitant ipsilateral femoral neck and intertrochanter fracture. ARS: anti-rotation screw; TGP: trochanteric grip plate.

The present study has several limitations. First, the study is retrospective and a single center cohort in nature. However, although CIFNITFs are common and most surgeons have at least some experiences with them, there have been limited large studies on this fracture type. Therefore, this study is significant in that it presents the largest sample size to date, along with a surgical outcome analysis based on the different treatment options, thereby providing a rational guide to selecting the treatment modality for this combined fracture. Second, we could not fully compare the patient-reported outcome measures as only part of the data could be extracted because of the retrospective nature of the study, and the study population consisted primarily of old age and severely injured patients. We believe a further large prospective study with more detailed outcome variables will provide surgeons with more rational guidance to approaching treatment of CIFNITFs.

5. Conclusions

For the surgical treatment of CIFNITFs, our results indicated that osteosynthesis did not differ significantly from an HR in terms of the overall postoperative complications, reoperation and mortality rate, and hip function. However, the risk of nonunion and more mechanical complications should be considered when choosing a DHS. Our suggestion for the treatment of a CIFNITF is that a surgeon should choose wisely between an HR and a CMN depending on the patient's age, displacement of the femoral neck, and one's expertise.

Supplementary Materials: The following supporting information can be downloaded at: https://www.mdpi.com/article/10.3390/jpm12111908/s1, Table S1: Summary of literatures regarding concomitant ipsilateral femoral neck and intertrochanter fracture.

Author Contributions: D.H.M. was responsible for conceptualization, methodology, validation and supervision. H.S.K. was in charge of formal analysis, validation, writing (original draft). D.K.L. and K.U.M. took charge in data curation. C.-H.K. was responsible for formal analysis and writing (original draft). All others took part in writing (review & editing). All authors have read and agreed to the published version of the manuscript.

Funding: This research received no external funding.

Institutional Review Board Statement: Institutional Review Board (IRB) approval was obtained from the Gachon University Gil Medical Center Institutional Review Board (GBIRB2021-036) prior to the initiation of this study.

Informed Consent Statement: Patient consent was waived owing to the retrospective nature of this study.

Data Availability Statement: The data presented in this study are available on request from the corresponding author. The data are not publicly available due to conditions of the ethics committee of our university.

Conflicts of Interest: The authors declare no conflict of interest.

References

1. Chang, S.M.; Hou, Z.Y.; Hu, S.J.; Du, S.C. Intertrochanteric Femur Fracture Treatment in Asia: What We Know and What the World Can Learn. *Orthop. Clin. N. Am.* **2020**, *51*, 189–205. [CrossRef] [PubMed]
2. Fischer, H.; Maleitzke, T.; Eder, C.; Ahmad, S.; Stöckle, U.; Braun, K.F. Management of proximal femur fractures in the elderly: Current concepts and treatment options. *Eur. J. Med. Res.* **2021**, *26*, 86. [CrossRef] [PubMed]
3. Ahmed, A.Z.; Shaban, A.M.; Ilyas, I. Ipsilateral femoral neck and complex pertrochanteric fracture of the femur. *J. Orthop. Trauma Surg. Relat. Res.* **2020**, *13*, 33–36.
4. An, H.S.; Wojcieszek, J.M.; Cooke, R.F.; Limbird, R.; Jackson, W.T. Simultaneous ipsilateral intertrochanteric and subcapital fracture of the hip. A case report. *Orthopedics* **1989**, *12*, 721–723. [CrossRef] [PubMed]
5. Biesemans, S. Segmental Fractures of the Neck of Femur: Fix or Replace? *Arthroplast. Today* **2021**, *8*, 247–252. [CrossRef] [PubMed]
6. Butt, M.F.; Dhar, S.A.; Hussain, A.; Gani Nu Kangoo, K.A.; Farooq, M. Femoral Neck Fracture with Ipsilateral Trochanteric Fracture: Is there room for Osteosynthesis? *Internet J. Orthop. Surg.* **2006**, *5*. [CrossRef]
7. Cohen, I.; Rzetelny, V. Simultaneous ipsilateral pertrochanteric and subcapital fractures. *Orthopedics* **1999**, *22*, 535–536.
8. Dhar, S.A.; Mir, M.R.; Butt, M.F.; Farooq, M.; Ali, M.F. Osteosynthesis for a T-shaped fracture of the femoral neck and trochanter: A case report. *J. Orthop. Surg.* **2008**, *16*, 257–259. [CrossRef]
9. Isaacs, C.; Lawrence, B. Concomitant ipsilateral intertrochanteric and subcapital fracture of the hip. *J. Orthop. Trauma* **1993**, *7*, 146–148. [CrossRef]
10. Khan, W.; Williams, R.; Hopwood, S.; Agarwal, S. Combined Intracapsular And Extracapsular Neck Of Femur Fractures Case Series, Literature Review And Management Recommendations. *Open Orthop. J.* **2017**, *11*, 600–608. [CrossRef]
11. Kothari, S.; Prakash, A. Segmental femoral neck fracture in an elderly patient: A rare case report. *Indian J. Orthop. Surg.* **2019**, *5*, 162–164. [CrossRef]
12. Kumar, R.; Khan, R.; Moholkar, K.; Smyth, H.; Borton, D. A rare combination fracture of the neck of femur. *Eur. J. Orthop. Surg. Traumatol.* **2001**, *11*, 59–61. [CrossRef]
13. Kyle, R.F.; Ellis, T.J.; Templeman, D.C. Surgical Treatment of Intertrochanteric Hip Fractures With Associated Femoral Neck Fractures Using a Sliding Hip Screw. *J. Orthop. Trauma* **2005**, *19*, 1–4. [CrossRef]
14. Lakshmanan, P.; Peehal, J.P. Management of an unusual intra- and extra-capsular subcapital femoral neck fracture. *Acta Orthop. Belg.* **2005**, *71*, 622–625. [PubMed]
15. Loupasis, G.; Ntagiopoulos, P.G.; Asimakopoulos, A. Concomitant ipsilateral subcapital and intertrochanteric fractures of the femur: A case report. *J. Med. Case Rep.* **2010**, *4*, 363. [CrossRef]
16. Neogi, D.S.; Ajay Kumar, K.V.; Trikha, V.; Yadav, C.S. Ipsilateral femoral neck and trochanter fracture. *Indian J. Orthop.* **2011**, *45*, 82–86. [CrossRef]
17. Pemberton, D.J.; Kriebich, D.N.; Moran, C.G. Segmental fracture of the neck of the femur. *Injury* **1989**, *20*, 306–307. [CrossRef]
18. Perry, D.C.; Scott, S.J. Concomitant ipsilateral intracapsular and extracapsular femoral neck fracture: A case report. *J. Med. Case Rep.* **2008**, *2*, 68. [CrossRef]
19. Poulter, R.; Ashworth, M.J. Concomitant ipsilateral subcapital and intertrochanteric fractures of the femur. *Inj. Extra* **2007**, *38*, 88–89. [CrossRef]
20. Saleeb, H.; Kanvinde, R.; Rahman, T. Literature review and case report: Current concepts for concomitant intra and extracapsular fractures of neck of femur in elderly patients. *Trauma Case Rep.* **2017**, *8*, 24–31. [CrossRef]
21. Sayegh, F.E.; Karataglis, D.; Trapotsis, S.J.; Christoforides, J.E.; Pournaras, J.D. Concomitant Ipsilateral Pertrochanteric and Subcapital Fracture of the Proximal Femur. *Eur. J. Trauma* **2005**, *31*, 64–67. [CrossRef]
22. Su, C.; Liu, Y.; Wu, P.; Yuan, J.; Lang, J.; Wu, C.; Zhang, Y.; Chen, L.; Chen, L. Ipsilateral femoral neck and intertrochanteric fractures with posterior dislocation of the hip: A report of two cases. *J. Orthop. Surg.* **2020**, *28*, 2309499019900449. [CrossRef] [PubMed]
23. Tahir, M.; Lakkol, S.; Naique, S. Segmental neck of femur fractures: A unique case report of an ipsilateral subcapital, greater trochanteric and intertrochanteric fracture and proposed management algorithm. *Int. J. Surg. Case Rep.* **2014**, *5*, 277–281. [CrossRef]
24. Taniguchi, D.; Fujiwara, H.; Kobashi, H.; Mori, G.; Yoshida, T.; Oda, R.; Kubo, T. Successful treatment of concomitant ipsilateral intracapsular and extracapsular hip fractures. *Orthopedics* **2013**, *36*, 837–839. [CrossRef]

25. Taylor, P.R.; Hepple, S.; Stanley, D. Combination subcapital and intertrochanteric fractures of the femoral neck. *Injury* **1996**, *27*, 68–71. [CrossRef]
26. Tong, D.K.; Zhou, D.X.; Zhong, X.X.; Zhou, Z.B.; Rui, B.Y.; Yue, F.; Peng, Z.D.; Guang, S.X.; Fang, Z.; Yan, G.; et al. Concomitant Ipsilateral Intracapsular and Extracapsular Hip Fractures Classification and Treatment A Multi-Hospital Experience. *Clin. Surg.* **2021**, *5*, 1–7.
27. Videla-Ces, M.; Sales-Perez, J.M.; Giros-Torres, J.; Sanchez-Naves, R.; Videla, S. A retrospective cohort study of concomitant ipsilateral extra-capsular and intra-capsular fractures of the proximal femur. Are they casual findings or an undervalued reality? *Injury* **2017**, *48*, 1558–1564. [CrossRef] [PubMed]
28. Vluggen, T.P.M.J.; van Vugt, R.; Boonen, B.; Keulen, M.H.F. Management of two-level proximal femoral fractures. *BMJ Case Rep.* **2021**, *14*, e240684. [CrossRef]
29. Yoo, H.; Cho, Y.; Hwang, S. Outcomes of Combined Neck and Trochanter Fractures of the Femur Treated with Cephallomedullary Nail in Elderly. *Hip Pelvis* **2019**, *31*, 200–205. [CrossRef]
30. Park, Y.C.; Um, K.S.; Hong, S.P.; Oh, C.W.; Kim, S.; Yang, K.H. Preoperative "Computed tomography capsular sign" for the detection of occult ipsilateral femoral neck fractures associated with femoral shaft fractures. *Injury* **2020**, *51*, 1051–1056. [CrossRef]
31. Ito, J.; Takakubo, Y.; Sasaki, K.; Sasaki, J.; Owashi, K.; Takagi, M. Prevention of excessive postoperative sliding of the short femoral nail in femoral trochanteric fractures. *Arch. Orthop. Trauma Surg.* **2015**, *135*, 651–657. [CrossRef] [PubMed]
32. John, B.; Sharma, A.; Mahajan, A.; Pandey, R. Tip-apex distance and other predictors of outcome in cephalomedullary nailing of unstable trochanteric fractures. *J. Clin. Orthop. Trauma* **2019**, *10*, S88–S94. [CrossRef] [PubMed]
33. Cho, M.R.; Lee, J.H.; Kwon, J.B.; Do, J.S.; Chae, S.B.; Choi, W.-K. The Effect of Positive Medial Cortical Support in Reduction of Pertrochanteric Fractures with Posteromedial Wall Defect Using a Dynamic Hip Screw. *Clin. Orthop. Surg.* **2018**, *10*, 292–298. [CrossRef] [PubMed]
34. Dhammi, I.; Jain, A.; Singh, A.; Rehan Ul, H.; Mishra, P.; Jain, S. Primary nonunion of intertrochanteric fractures of femur: An analysis of results of valgization and bone grafting. *Indian J. Orthop.* **2011**, *45*, 514–519. [CrossRef] [PubMed]
35. Chan, D.S. Femoral Neck Fractures in Young Patients: State of the Art. *J. Orthop. Trauma* **2019**, *33* (Suppl. S1), S7–S11. [CrossRef] [PubMed]
36. Slobogean, G.P.; Sprague, S.A.; Scott, T.; McKee, M.; Bhandari, M. Management of young femoral neck fractures: Is there a consensus? *Injury* **2015**, *46*, 435–440. [CrossRef]
37. Ma, H.H.; Chou, T.A.; Tsai, S.W.; Chen, C.F.; Wu, P.K.; Chen, W.M. Outcomes of internal fixation versus hemiarthroplasty for elderly patients with an undisplaced femoral neck fracture: A systematic review and meta-analysis. *J. Orthop. Surg. Res.* **2019**, *14*, 320. [CrossRef]
38. Cheng, Y.X.; Sheng, X. Optimal surgical methods to treat intertrochanteric fracture: A Bayesian network meta-analysis based on 36 randomized controlled trials. *J. Orthop. Surg. Res.* **2020**, *15*, 402. [CrossRef]
39. Hongku, N.; Woratanarat, P.; Nitiwarangkul, L.; Rattanasiri, S.; Thakkinstian, A. Fracture fixation versus hemiarthroplasty for unstable intertrochanteric fractures in elderly patients: A systematic review and network meta-analysis of randomized controlled trials. *Orthop. Traumatol. Surg. Res.* **2022**, *108*, 102838. [CrossRef]
40. Huang, X.; Leung, F.; Xiang, Z.; Tan, P.Y.; Yang, J.; Wei, D.Q.; Yu, X. Proximal Femoral Nail versus Dynamic Hip Screw Fixation for Trochanteric Fractures: A Meta-Analysis of Randomized Controlled Trials. *Sci. World J.* **2013**, *2013*, 805805. [CrossRef]
41. Parker, M.J.; Bowers, T.R.; Pryor, G.A. Sliding hip screw versus the Targon PF nail in the treatment of trochanteric fractures of the hip. *J. Bone Jt. Surg. Br.* **2012**, *94-B*, 391–397. [CrossRef] [PubMed]
42. Radford, P.J.; Needoff, M.; Webb, J.K. A prospective randomised comparison of the dynamic hip screw and the gamma locking nail. *J. Bone Jt. Surg. Br.* **1993**, *75-B*, 789–793. [CrossRef] [PubMed]
43. Palm, H.; Jacobsen, S.; Sonne-Holm, S.; Gebuhr, P.; Hip Fracture Study, G. Integrity of the lateral femoral wall in intertrochanteric hip fractures: An important predictor of a reoperation. *J. Bone Jt. Surg. Am.* **2007**, *89*, 470–475.
44. Kyriakopoulos, G.; Panagopoulos, A.; Pasiou, E.; Kourkoulis, S.K.; Diamantakos, I.; Anastopoulos, G.; Tserpes, K.; Tatani, I.; Lakoumentas, J.; Megas, P. Optimizing fixation methods for stable and unstable intertrochanteric hip fractures treated with sliding hip screw or cephalomedullary nailing: A comparative biomechanical and finite element analysis study. *Injury* **2022**, *in press*. [CrossRef]
45. Downey, C.; Kelly, M.; Quinlan, J.F. Changing trends in the mortality rate at 1-year post hip fracture—A systematic review. *World J. Orthop.* **2019**, *10*, 166–175. [CrossRef]

Journal of *Personalized Medicine*

MDPI

Review

Bosworth Fractures of the Ankle: A Systematic Literature Review

Ludovico Lucenti, Gianluca Testa, Chiara Nocera, Annalisa Culmone, Eleonora Dell'Agli and Vito Pavone *

Department of General Surgery and Medical Surgical Specialties, Section of Orthopaedics and Traumatology, University Hospital Policlinico "Rodolico-San Marco", University of Catania, 95123 Catania, Italy; ludovico.lucenti@gmail.com (L.L.)
* Correspondence: vitopavone@hotmail.com

Abstract: Bosworth lesions are fracture-dislocations of the ankle and are characterized by entrapment of the proximal segment of the fibula behind the posterior tubercle of the distal tibia. Treatment is challenging, mainly due to failure of a closed reduction. The aim of this study was to review the literature concerning this type of injury. A total of 103 patients with Bosworth fractures were included in the study. The analyzed studies yielded a total of 103 cases, of which 68% (n = 70) were male and 32% (n = 33) were female. Bosworth fractures are mainly due to accidental trauma (58.2%), sports-related injuries (18.4%), and traffic accidents (18.4%). More than 76% of the patients presented a Danis–Weber B fracture, 8.7% a type C fracture, and only 0.97% presented a type A fracture. In 92.2% of the patients, the attempted closed reduction was unsuccessful. A definitive treatment with open reduction and internal fixation (ORIF) was used in 96 patients (93.2%). The most frequent complication was post-traumatic arthritis (10.7%). Bosworth fractures are challenging. The available literature lacks adequate information about this fracture, and an approved standardized algorithm for treating such fractures is not available.

Keywords: Bosworth; ankle; axilla sign; fracture-dislocation

check for updates

Citation: Lucenti, L.; Testa, G.; Nocera, C.; Culmone, A.; Dell'Agli, E.; Pavone, V. Bosworth Fractures of the Ankle: A Systematic Literature Review. *J. Pers. Med.* **2023**, *13*, 713. https://doi.org/10.3390/jpm13050713

Academic Editors: Markus Gesslein and Eduard Kurz

Received: 28 March 2023
Revised: 21 April 2023
Accepted: 22 April 2023
Published: 23 April 2023

1. Introduction

In 1947, Bosworth described for the first time several cases of fracture-dislocation of the ankle, which is characterized by entrapment of the proximal segment of the fibula behind the posterior tubercle of the distal tibia [1].

This lesion is typically produced by supination and external rotation injury patterns [2–5]. A medial or lateral malleolus fracture may be involved.

A radiographic evaluation is mandatory for diagnosing this type of fracture. In the AP view, an overlap of distal tibia and fibula may be seen, while in the lateral view, a posterior subluxation of the talus and syndesmosis dissociation is usually present. On the CT scan, clearer visualization of the injury is possible, and the fibula can be seen to be displaced behind the posterior edge of the fibular notch (incisura tibiae) and locked between the distal tibia and the displaced posterior malleolus fragment.

Treatment is very challenging and is mainly due to failure of the initial reduction. To treat this complicated injury, open emergency surgery is recommended because a successful closed reduction of Bosworth fracture-dislocation is rare due to soft tissue interposition between the displaced proximal fragment of the fibula and the posterior tubercle of the tibia [6,7].

The aim of this study was to review the most recent literature concerning this type of injury.

2. Literature Search

2.1. Literature Search Strategy

This systematic review of the current literature was performed according to the Preferred Reporting Items for Systematic Reviews and Meta-Analyses (PRISMA) guidelines [8]. We searched the PubMed, Cochrane Library, and Web of Science databases using the keyword "Bosworth fracture" to identify published case reports or case series evaluating the clinical and radiological characteristics of this fracture, treatment, and short-, medium-, and long-term functional outcomes of patients with Bosworth fractures. The search was restricted to case series, case reports, and prospective and retrospective analyses published in the last 20 years from 1 July 2002 to 1 July 2022. In addition, the reference lists of all possible articles were analyzed to further identify studies potentially suitable for our research, and the articles were evaluated according to the inclusion and exclusion criteria.

2.2. Selection Criteria

The studies were selected based on several inclusion criteria: (1) studies written in English language; (2) studies of any level of evidence; (3) studies published after July 1, 2002; (4) studies reporting clinical or preclinical results; and (5) studies concerning the clinical and radiological characteristics, treatment, and functional outcomes in patients with Bosworth fractures. Exclusion criteria consisted of several parameters: (1) inaccessible studies; (2) studies dealing with other topics; and/or (3) studies with poor scientific methodology or without an accessible abstract. Any duplicates were also excluded.

2.3. Data Extraction and Criteria Appraisal

Two investigators independently extracted all data from article text, tables, and figures. All data were organized in a specific order: (1) lead author; (2) year of publication; (3) number of patients; (4) participant characteristics; (5) mechanism of injury; (6) classification of fracture; (7) clinical features of the fracture; (8) type and timing of treatment; (9) complications; (10) short-, mid-, and long-term outcomes; (11) time of follow-up; and (12) results. Disagreements between the two reviewers were resolved by consensus. Any remaining controversies were discussed with two senior investigators, who evaluated the quality of the studies. The PRISMA flow chart used for the selection and screening of studies is shown in Figure 1.

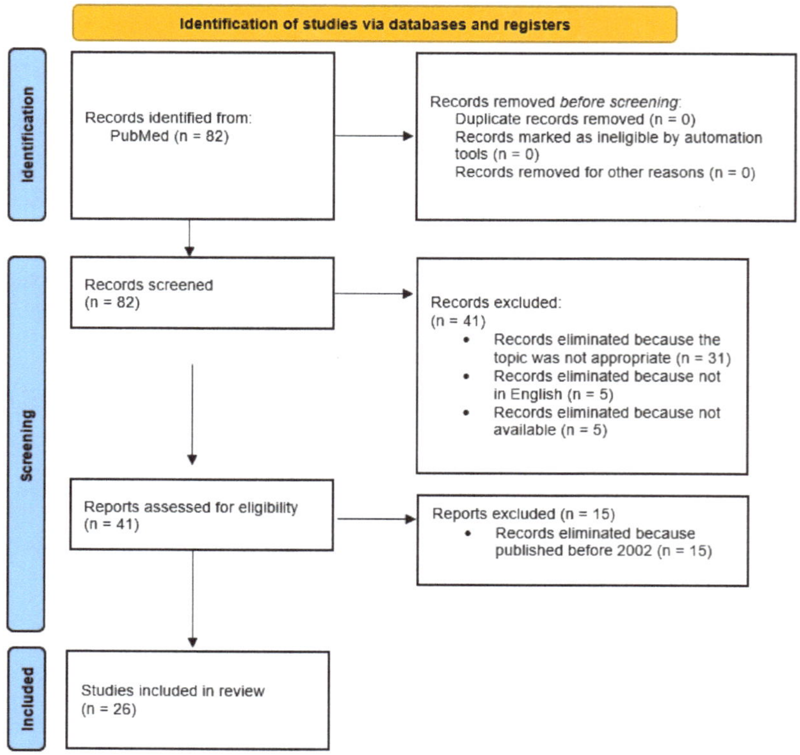

Figure 1. The Preferred Reporting Items for Systematic Reviews and Meta-Analyses (PRISMA) flow chart used for study selection and screening.

3. Studies

3.1. Study Selection

The initial search of PubMed and Web of Science databases produced 82 results, and 82 articles were examined after duplicate exclusion. Screening of abstracts and titles eliminated 56 articles because they did not meet the inclusion criteria: (1) 31 articles were eliminated from the analysis because the topic was not appropriate, (2) 15 because they were published before 2002, (3) five because they were not in English, and (4) five because they were not available. A total of 26 papers were considered eligible and included in our review. Figure 1 illustrates the study selection process. A total of 103 patients with Bosworth fractures were included in the studies.

3.2. Participants

The 26 analyzed studies described sample sizes ranging from a minimum of one patient to a maximum of 51 patients for a total of 103 cases, of which 68% (n = 70) were male, and 32% (n = 33) were female. The mean age was 38.8 years with the youngest patient being 19 years old and the oldest 72 years old. In 47.6% of cases (n = 49), the left was affected, and in 52.4% (n = 54) the right side was (Table 1).

Only the study by Fan et al. [9] reported the body mass indices (BMIs) of patients (BMI = 23 Kg/m^2), while Won et al. [10] reported the mean body weight (67 kg). Regarding the traumatic mechanism that generated the Bosworth fracture (Table 2), in 58.2% of the patients, the injury was due to accidental trauma (n = 60).

Table 1. Main parameters for each study.

AUTHOR	N° PATIENTS	ARTICLE	SEX
Arora 2021	1	Ankle Arthrodesis Using Retrograde Nail in Case of Bosworth Fracture with Ankle Dislocation: A Rare Case Report	Female
Bartonicek 2007	6	Bosworth-type fibular entrapment injuries of the ankle: the Bosworth lesion. A report of 6 cases and literature review	(I) Female (II) Male (III) Male (IV) Female (V) Female (VI) Male
Bartonicek 2022	1	Bosworth fracture complicated by unrecognized compartment syndrome: a case report and review of the literature	Male
Cappuccio 2017	1	An uncommon case of irreducible ankle fracture-dislocation: the "Bosworth-like" tibio-fibular fracture	Male
Cho 2019	15	Prognostic factors for intermediate-term clinical outcomes following Bosworth fractures of the ankle joint	11 male; 4 female
Delasotta 2013	1	Surgical management of the posterior fibula fracture dislocation: case report	Male
Ellanti 2013	1	Acutely irreducible ankle fracture dislocation: a report of a Bosworth fracture and its management	Male
Fan 2020	1	A Novel Technique for a Successful Closed Reduction of a Bosworth Fracture-Dislocation of the Ankle	Male
Foldager 2018	2	Bosworth fracture dislocation of the ankle: - Two case reports with perioperative illustration	(I) Male (II) Male
Han 2021	1	Bosworth-type fibular entrapment fracture of the ankle without dislocation: a rare case report and a review of the literature	Male
He 2020	1	Ankle arthroscopy-assisted closed reduction in Bosworth fracture dislocation	Female
Ji 2022	1	Case Report: Bosworth Fracture-Dislocation managed by Closed Reduction and Conservative Treatment	Male
Khan 2008	1	A constant radiological sign in Bosworth's fractures: "the Axilla sign"	Female
Lu 2016	1	A radiological sign (which we are calling the 'tongues of flame' sign) in irreducible trimalleolar fractures of the ankle	Female
Lui 2008	4	Ankle stiffness after Bosworth fracture dislocation of the ankle	(I) Male (II) Male (III) Female (IV) Male
Maertin-Somoza 2020	1	Bosworth fracture. An atypical case of irreducible ankle fracture-dislocation	Female
Moerenhout 2019	1	Association of Bosworth, Pilon, and Open Talus Fractures: A Very Unusual Ankle Trauma	Male
Peterson 2015	1	An Unusual Ankle Injury: The Bosworth-Pilon Fracture	Male
Ren 2019	2	Rare variants of Bosworth fracture-dislocation: Bosworth fracture-dislocation with medial malleolus adduction type fracture	2 Male
Schepers 2012	1	An irreducible ankle fracture dislocation: the Bosworth injury	Female
Wenxian 2019	1	Rare Bosworth Fracture-Dislocation Variant of an Irreducible Distal Fibula Dislocation of the Ankle Without Fibula Fracture	Male
Williams 2018	1	Bosworth Dislocation without Associated Fracture	Male
Won 2019	51	Improved functional outcome after early reduction in Bosworth fracture-dislocation	32 male; 19 female
Wright 2012	1	A contemporary approach to the management of a Bosworth injury	Male
Yang 2014	4	Assessment of Bosworth-type fracture by external oblique radiographs	4 Male
Yeoh 2013	1	Bosworth fracture-dislocation of the ankle: a case report	Male

Table 2. Mechanism of injury.

Mechanism of Injury	Percentage	N° of Patients
Accidental trauma	58.2%	60
Sport injury	18.4%	19
Traffic accident	18.4%	19
Working injury	0.97%	1
No data	3.9%	4

In most of the studies, the Danis–Weber classification was used to characterize the fracture. About three-quarters (76.7%) of the patients (n = 79) presented with a type B fracture, 8.7% (n = 9) with a type C fracture, and only one patient (0.97%) with a type A fracture (Table 3).

Table 3. Main characteristics of the trauma.

AUTHOR	INJURY TYPE (WEBER)	Axilla Sign	CT SCAN	CLOSED REDUCTION
Arora	Gustilo 2b	No data	No data	Unsuccessful
Bartonicek	Weber B	No data	Yes	Successful (2nd attempt)
Bartonicek	(I) Weber C (II) Weber C (III) Weber C (IV) Weber B (V) Weber B (VI) Weber B	No data	(I) No data (II) No data (III) No data (IV) Yes (V) no data (VI) No data	(I) Unsuccessful (II) Unsuccessful (III) Unsuccessful (III) Unsuccessful (IV) Unsuccessful (V) Unsuccessful (VI) Unsuccessful
Cappuccio	Weber C	No data	Yes	Unsuccessful
Cho	13 Weber B; 2 Weber C	No data	Yes	14 Unsuccessful; 1 successful
Delasotta	Weber B	Yes	Yes	unsuccessful
Ellanti	Weber C	No data	no data	unsuccessful
Fan	No data	No data	No	Successful (2nd attempt)
Foldager	No data	No data	(I) no data (II) no data	(I) Unsuccessful (II) unsuccessful
Han	Weber B	No data	Yes	Not attempted
He	No data	No data	Yes	Unsuccessful
Ji	Weber B	Yes	Yes	Unsuccessful
Khan	No data	Yes	No data	unsuccessful
Lu	Weber B	No data	Yes	unsuccessful
Lui	(I) No data (II)Weber B (III) Weber B (IV) No data	No data	(I) Yes (II) No data (III) No data (IV) no data	(I) Unsuccessful (II) Unsuccessful (III) Unsuccessful
Maertin-Somoza	No data	No data	No	Unsuccessful
Moerenhout	Gustilo 3B	No data	Yes	unsuccessful
Peterson	The initial radiographs showed a bimalleolar ankle fracture dislocation with a comminuted fracture of the tibial plafond	Yes	Yes	unsuccessful
Ren	Weber B	No data	Yes	Unsuccessful
Schepers	Weber B	Yes	No data	unsuccessful
Wenxian	No data	No data	Yes	Unsuccessful
Williams	No data	No data	No	Unsuccessful
Won	1 Weber A; 48 Weber B; 2 Weber C	No data	Yes	Unsuccessful
Wright	No data	No data	Yes	unsuccessful
Yang	4 Weber B	Yes	Yes	
Yeoh	Weber B	No data	No	Unsuccessful

Two studies (1.9%) used the Gustilo–Anderson classification because they were open fractures: (1) the case report described by Arora et al. [11] presented a type 2B fracture and

(2) the study by Moerenhout et al. [12] described a type 3B fracture. In 11.6% of the patients (n = 12), no data were specified. Radiographically, the "axilla sign" could be seen in these kinds of fractures. This sign was described in only nine patients (8.7%) [4,13–17] while for the rest of the patients, no data were available. A pre-operative computed tomography (CT) scan was performed in 82.5% of patients (n = 85). In four patients (3.9%), the diagnosis was only based on the X-ray study, while in 13.6% (n = 14) the use of X-rays was not reported.

3.3. Treatment

From the studies included in our review, it can be seen that a wide variety of treatments have been used for Bosworth fracture (Table 4).

Table 4. Main characteristics of the treatments.

AUTHOR	DEFINITIVE TREATMENT	Cast after Surgery	Syndesmotic Screw Removed (weeks/months)	HARDWARE Removed	USE OF EXTERNAL FIXATION	USE OF TRACTION	USE OF CAST
Arora	ORIF	6 weeks			Yes	No	Yes
Bartonicek	ORIF	6 weeks	2 months	Not reported	No	No	Yes
Bartonicek	(I) ORIF (II) cast(III) ORIF (IV) ORIF (V) ORIF (VI) ORIF	(I) 5 weeks (II)–(III) 6 weeks (IV) 6 weeks (V) 6 weeks (VI) 6 weeks	(I) 8 weeks after surgery (II)–(III) suprasyndesmotic screws were removed after 2 months (IV) suprasyndesmotic screw was removed 4 months after (V) patient declined removal of the suprasyndesmotic screw (VI) -	(I) Yes, after 2 years (IV) Yes, after 11 month	No	No	Yes
Cappuccio	ORIF	4 weeks			No	Yes	Yes
Cho		8 weeks	12–16 weeks after surgery		No	No	Yes
Delasotta	ORIF	6 weeks	4 months after surgery		No	No	Yes
Ellanti	ORIF	6 weeks	3 months after surgery		No	No	Yes
Fan	ORIF	6 weeks	No data	Not reported	No	No	Yes
Foldager	ORIF	(I) 6 weeks (II) 6 weeks	(I) no data (II) no data		No	No	Yes
Han	ORIF	6 weeks	No syndesmotic screw	Not reported	No	No	Yes
He	Ankle arthroscopy and ORIF	4 weeks	3 months after surgery		No	No	Yes
Ji	cast	9 weeks			No	No	Yes
Khan	ORIF	6 weeks	6 weeks after surgery		No	No	Yes
Lu	ORIF	6 weeks			No	No	Yes
Lui	Arthroscopy and ORIF	(I) 6 weeks (II) no cast (III) no cast (IV) no cast	(I) 12 weeks after surgery (II) 12 weeks after surgery (III)12 weeks after surgery (IV)12 weeks after surgery		No	No	Yes

Table 4. *Cont.*

AUTHOR	DEFINITIVE TREATMENT	Cast after Surgery	Syndesmotic Screw Removed (weeks/months)	HARDWARE Removed	USE OF EXTERNAL FIXATION	USE OF TRACTION	USE OF CAST
Maertin-Somoza	ORIF	10 days	No	Not reported	No	No	Yes
Moerenhout	ORIF	12 weeks			No	No	Yes
Peterson	ORIF	6 weeks	No data		Yes	No	Yes
Ren	ORIF	2 weeks	8 weeks	Not reported	No	No	Yes
Schepers	ORIF	6 weeks	Syndesmotic screws were not removed in the absence of complaints	yes, after 9 months (because of shoe wear problems caused by the implant)	No	No	Yes
Wenxian	ORIF	6 weeks	6 weeks after surgery	Yes	Yes	No	Yes
Williams	ORIF	No data	4 months	4 months	No	No	Yes
Won	ORIF	no data	No data	no data	No data	No data	No data
Wright	ORIF	6 weeks			No	No	Yes
Yang	ORIF	No data	No data	Not reported	No data	No data	No data
Yeoh	ORIF	No data	16 weeks	Not reported	No	No	No data

Almost all of the selected studies describe an attempt at closed reduction prior to definitive treatment. In 92.2% of the patients (n = 95), the attempt was unsuccessful, and in only three patients (2.9%) was it successful. Han et al. [18] did not attempt a closed reduction, while Yang et al. [4] did not report the use of this procedure in their study. Surgical treatment was used in all the studies examined, except two in which a non-surgical treatment was chosen with cast for nine weeks [13,19]. Over half (55.3%) of the patients (n = 57) involved in the studies were treated within a day of undergoing trauma, while only 23% (n = 24) were treated after 12 hours. With respect to 12 patients (11.6%), the number of days between injury and surgical treatment was not specified. In total, only five patients (4.8%) were treated with an arthroscopically assisted technique; four patients (3.9%) were treated with plate and screws at the fibula and a syndesmotic screw [20] and one patient (0.97%) with two quadricortical lag screws transversely fixed the syndesmosis [21]. Ninety-six patients (93.2%) were treated with open reduction and internal fixation (ORIF). In 39 patients (37.9%), the ORIF was constructed using a plate and was screwed to the fibula, and in 31 of these (30.1%), a syndesmosis screw was associated. The syndesmosis screws were removed between 12 and 16 weeks in 25 patients, at six weeks in two patients, and at eight weeks in five patients. In two cases, the syndesmosis screw was not removed. The ORIF procedure with screws was only used in four patients (3.9%) [19,22,23]. Moerenhout et al. [12] used K-wires and two partially threaded screws for reducing and fixing the Bosworth fracture in the presented case report. The use of an intramedullary retrograde nail with proximal and distal screws was described by Arora et al. [11]. Won et al. [10] did not specify the type of surgical treatment. After surgical treatment, a cast was placed on 100 patients (97%). Of these, in 18.4% (n = 19), the cast was removed after six weeks, in 14.6% (n = 15) after eight weeks, in only 5.8% (n = 6) was the cast was removed before six weeks, and in 1.9% (n = 2) after two weeks. In 58 patients (56.3%), these data were not specified. Three studies reported use of an external fixator as first intervention for better stability while waiting for swelling and soft tissue edema to subside or because open fractures were present [11,14,23]. Only Cappuccio et al. placed a trans-calcaneal traction to put the limb in traction awaiting definitive surgical treatment [5].

3.4. Follow-Up, Post-Operative Complications, and Outcomes

The average follow-up was 15.87 months. Nine patients 8.7% underwent a follow-up of <6 months, 56.3% (n = 58) between 6 and 12 months, 22.3% (n = 23) between 12 and 24 months, and only nine patients (8.7%) more than 24 months (Table 5).

Table 5. Complications after Bosworth fractures.

AUTHOR	Complication/Sequelae	Type of Complication	OUTCOME	FOLLOW UP
Arora	No data	No data	No data	No data
Bartonicek	Compartment syndrome	A fixed fexion contracture of the great toe was observed without the possibility of passive correction, with ulcerations on the plantar and medial aspect of the great toe and below the first metatarsal head	Unsuccessful	3 years
Bartonicek	(I) No (II) Yes (III) No (IV) Yes (V) No (VI) No	(I) The patient had no subjective complaints (II) severe posttraumatic osteoarthritis of the ankle and an arthrodesis was performed (III) the patient had no subjective complaints (IV) pain (V) no (VI) no	(I) Successful (II) unsuccessful (III) successful (IV) successful (V) successful (VI) successful	(I) 8 years (II) 2 years (III) 5 years (IV) 2.5 years (V) 3.5 years (VI) 3 years
Cappuccio	No data	No data	Successful	1 year
Cho	2 patients with joint stiffness in dorsiflexion and plantarflexion	Joint stiffness	Successful	2 years
Delasotta	No		Successful	6 months; 8 months
Ellanti	No	No complication	Successful	3 months
Fan	No	No	No data	No data
Foldager	(I) No (II) No	(I) No complication (II) No complication	(I) Successful (II) successful	(I) Lost follow up (II) 3 months
Han	No	No	Successful	2 years
He	No		Successful	6 months
Ji	No	No complication	Successful	9 months; 2.5 years
Khan	No		Successful	3 months
Lu	No		Successful	3 months

Table 5. *Cont.*

AUTHOR	Complication/Sequelae	Type of Complication	OUTCOME	FOLLOW UP
Lui	(I) Yes (II) Yes (III) Yes (IV) No	(I) Painful stiffness and using stairs, plantarflexed limitation (open anterior ankle capsular release and extensor tendon adhesiolysis at the level of superior extensor retinaculum was performed. Posterior ankle capsulectomy was also performed through posterior ankle endoscopy) (II) painful stiffness (arthroscopic ankle capsular release was performed) (III) painful stiffness (arthroscopic ankle capsular release was performed)	(I) Unsuccessful (II) unsuccessful (III) unsucessful (IV) successful	(I) 5 months (II) 3 months (III) 3 months (IV) No data
Maertin-Somoza	No	No	Successful	18 months
Moerenhout	No acute complication, but after 1 year he had an invalidating partial necrosis of the talus and pilon	No acute complication, but after 1 year he had an invalidating partial necrosis of the talus and pilon	Unsuccessful	14 months
Peterson	At 18 months, the patient was struggling owing to pain in the ankle that developed clinical and radiologic post-traumatic arthritis	At 18 months, the patient was struggling owing to pain in the ankle that developed clinical and radiologic post-traumatic arthritis	Successful	1.5 years
Ren	No	1 compartment syndrome followed by fasciotomy the day after surgery	Successful	21 months
Schepers	Problems with wearing shoes caused by the implant	Problems eith wearing shoes caused by the implant	Successful	1 years
Wenxian	No	No	Successful	22 months
Williams	No	No	Successful	6 months
Won	Yes	8 wound complication; 2 compartment syndrome; 9 osteoarthritis; 5 mal-union	Successful	6 weeks; 3–6–12 months
Wright	No		Successful	3 months;
Yang	No data	No data	No data	7–9–12–34 months
Yeoh	No	No	Successful	3 months

A patient was lost during follow-up, while no data were presented for the remaining three patients (2.9%). Regarding post-operative complications (Table 6), only 34.9% of the patients (n = 36) analyzed in the studies were affected.

Table 6. Complications after Bosworth fractures.

COMPLICATIONS	Percentage	N of Patients
Post-traumatic arthritis	10.7%	11
Wound complications	7.7%	8
Painful joint stiffness	5.8%	6
Malunion	4.8%	5
Compartment syndrome	3.9%	4
Fixed flexion contracture of great toe	0.97%	1
Necrosis of talus and pilon	0.97%	1
Total	34.9%	36

Post-traumatic arthritis of the ankle was the most frequently occurring complication in patients (n = 11, 10.7%) [10,14,19] followed by wound complications (n = 8) and malunion (n = 5) at 7.7% and 4.8%, respectively [10]. Painful joint stiffness in dorsiflexion and plantarflexion was found in six patients (5.8%) [7,19,20]. Joint stiffness and plantarflexed limitation were treated by Cho et al. with open or arthroscopic ankle capsular release [7]. Four patients (3.9%) presented a compartment syndrome followed by fasciotomy the day after surgery [2,10,24]. Bartonicek et al. observed a case of a fixed flexion contracture of the big toe without the possibility of passive correction with ulcerations on the plantar and medial aspect of the great toe and below the first metatarsal head. This toe underwent corrective fusion of the interphalangeal joint of the hallux [24]. Moerenhout et al. described a patient with an invalidating partial necrosis of the talus and pilon one year after surgery that required an ankle–hindfoot arthrodesis 14 months after the accident [12]. A case of shoe wear problems caused by the implant used in the surgical treatment was also described by Schepers et al. [15]. At the end of follow-up, 85.4% of patients (n = 88) who were treated with ORIF showed excellent clinical and radiographic outcomes. The patients presented a good recovery of plantar and dorsal flexion, no lameness, and no functional limitations for the performance of daily life. Two patients (1.94%) treated with ORIF achieved poor results. Among patients treated arthroscopically (4.85%; n = 5), 60% (n = 3) obtained unsuccessful outcomes, and 40% (n = 2) regained ankle function. Instead, cast treatment produced excellent results in one out of two patients. Functional outcomes at the end of follow-up were not reported in six patients (5.8%).

4. Discussion

In 1947, David M. Bosworth [1] reported five cases of particular fracture-dislocation of the ankle, in which the proximal fragment of the fibula was incarcerated behind the posterior tibial tubercle. Since that time, this rare ankle fracture pattern has been called "Bosworth fracture" (an example is shown in Figure 2).

According to our analyses, the Bosworth fracture is mostly observed in men with a mean age of 38.8 years old.

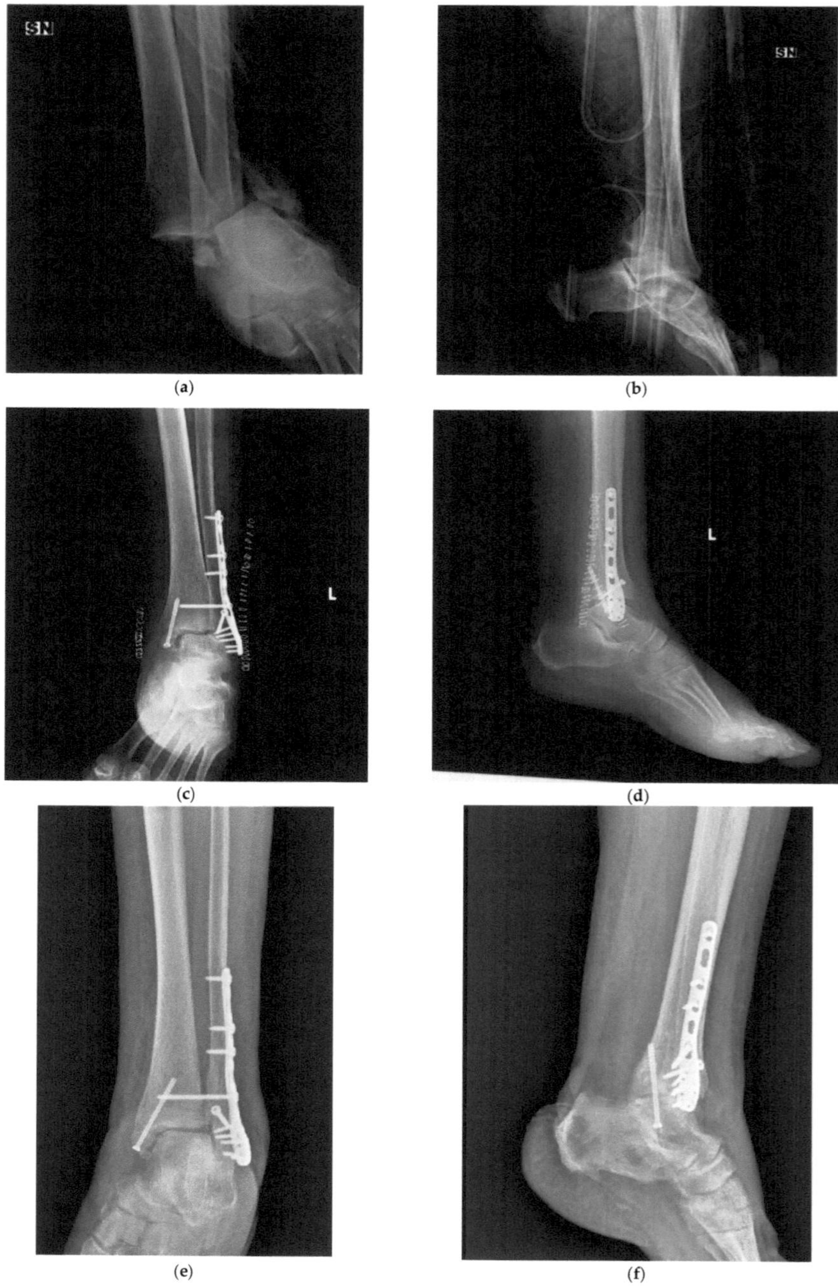

Figure 2. X-ray of a Bosworth fracture (**a**) anterio–posterior (AP) pre-operative view, (**b**) lateral pre-operative view, (**c**,**d**) post-operative X-rays, and (**e**,**f**) the 4-month follow-up.

Perry et al. performed a cadaveric experiment to identify the trauma mechanism involved in Bosworth fracture [25]. It turned out that the principal mechanism of injury was due to an external rotation force during the supination of the foot (supination-external rotation (SER) fracture according to the Lauge–Hansen classification and associated with Danis–Weber B or C fracture) [15,26]. According to our analysis, a Danis–Weber classification type B was present in 88.76% of the patients, type C in 10.11%, and type A in 1.12%. They described seven stage injury patterns: (1) stage 1: the fibula moves out of the notch after the rupture of the anterior tibiofibular ligament; (2) stage 2: rupture of the posterior tibiofibular ligament; (3) stage 3: rupture of the anteromedial part of the capsule; (4) stage 4: tear of the interosseous membrane; (5) stage 5: the fibula locks posteriorly behind the tibia; (6) stage 6: fracture of the fibula as a result of rotation of the talus; and (7) stage 7: fracture of the medial malleolus or involvement of deltoid ligament. Moerenhout et al. [12] proposed the addition of stage 8 in which a Bosworth fracture is associated with a talus fracture. Regarding the traumatic mechanism from the analysis, it was found that in 58.2%, the fracture was due to an accidental trauma (including falling from a height and falling down stairs), 18.4% due to sports-related injuries, 13.4% due to traffic incidents, 0.97% due to work-related injuries, and 3.9% due to other unspecified causes (Table 1).

In 2007, Bartonicek et al. highlighted age as a factor that can influence the fracture pattern [19]. In a person with skeletal immaturity with open physes, the result of this pattern of fracture is dislocation of the fibula with an epiphysiolysis of the distal tibia. In adults with closed physes, the dislocation of the fibula occurs without fracture. In middle-aged or older people, it is common to see a fibula fracture due to reduced bone elasticity.

The Bosworth fracture represents both a diagnostic and therapeutic challenge for orthopedic surgeons. Usually, in the emergency department, this fracture has a high rate of delayed or missed diagnosis after examination of the initial X-ray images. However, particular characteristics on standard AP, lateral, and mortise radiographs are helpful for making a diagnosis [27]:

(1) AP view: overlap of the proximal fibular fragment and the distal tibia;
(2) Lateral view: posterior displacement of the fibula and posterior subluxation of the talus;
(3) Mortise view: widened medial joint space.

In 2008, Khan and Borton described a new radiographic sign frequently present on an AP mortise view of patients with Bosworth fracture [16]. This sign was defined as the "Axilla sign", which means cortical radiodensity of the medial tibial plafond occurs because the tibia is locked in internal rotation after dislocation of the fibula occurs. The presence of an Axilla sign should alert the orthopedic surgeon to look for a Bosworth fracture. Moreover, Yang et al. suggested adding an external oblique ankle radiograph to view the position of the fibular shaft compared to the talus [4]. This Axilla sign was described in only 8.7% of all the patients considered in our study.

CT scans and three-dimensional (3D) reconstructions are not routinely used, but they are recommended for ankle fracture-dislocation after failed closed reduction. A CT scan yields additional information about the fracture pattern and interposition of soft tissues, which produces a difficult reduction. Moreover, CT images are useful for pre-operative planning. Our study showed that pre-operative CT scans were obtained in 82.5% of patients.

As mentioned before, treatment of this fracture represents a challenge. Although most ankle fracture-dislocations can be reduced using a closed reduction, in the case of a Bosworth fracture, a successful closed reduction is rare. Closed reduction in this fracture type is usually not possible due to interposition of soft tissues or entrapment of fracture fragments. As Bartonicek's study reports, a successful closed reduction in Bosworth fractures is rare [6]. In our analysis, the data are similar to Bartobieck as only three out of 103 patients (2.9%) benefitted from a successful closed reduction. Several attempts at performing a closed reduction are not recommended because they can cause further soft tissue damage. Han et al. [18] did not attempt a closed reduction in their case report but rather directly performed an open reduction.

While waiting for surgery, the ankle should be immobilized with a low knee cast and should be elevated. Typically, early open reduction and internal fixation are required to prevent adverse events such as skin necrosis, compartment syndrome [28,29], neurovascular injury, joint stiffness [20], avascular necrosis of the talus [30], and osteoarthritis that can lead to poor functional outcomes.

The ORIF was constructed using a plate and screwing it to the fibula (93.3% of cases). In 30.1% of these cases, a syndesmosis screw was associated. During surgery, it is recommended to test the stability of the syndesmosis after the fixation of fibula. In cases of instability of the syndesmosis, internal fixation is necessary [31]. Lui et al. suggest arthroscopy as a tool to guide the anatomical reduction of syndesmosis and to evaluate associated intra-articular pathology [20]. According to our study, an arthroscopically assisted technique was used in 4.8% of all cases.

In general, the goal of intra-articular fracture management is to achieve anatomical reduction to reduce the risk of developing osteoarthrosis with associated poor functional outcomes [32].

Our study showed that post-traumatic arthritis of the ankle occurred in 10.7% of patients [10,14,19] followed by wound complications (7.7%) and malunion (4.8%) [10]. Painful joint stiffness in dorsiflexion and plantarflexion was found in 5.8% of all cases.

Fournier et al. [33] did not find a correlation between a good reduction and better functional outcomes. Furthermore, in Moerenhout's study, even if an anatomical reduction and a good alignment of the ankle with hindfoot were obtained, osteoarthritis development was observed [12].

However, anatomical reduction is also necessary to find an optimal position for the arthrodesis a second time.

5. Conclusions

Bosworth fractures are lesser known and understood lesions of the ankle and are sometimes misdiagnosed or overlooked. X-ray is usually sufficient to make a diagnosis, but a CT scan is mandatory to fully understand the fracture pattern. This type of injury is challenging. It should be treated without delay and often needs an ORIF, which is characterized by a good recovery, no complications, and a good level of patient satisfaction, to yield better outcomes. The available literature lacks information and only reports a few case reports that are mainly focused on the possible available treatment options and potential complications. Establishment of a nationwide registry addressing this type of fracture could be an optimal instrument to help gain a better understanding of this injury and to design a standard way to treat the lesion. Further studies are necessary to create a clear and validated diagnostic–therapeutic algorithm.

Author Contributions: Conceptualization, L.L. and G.T.; methodology, C.N.; software, E.D.; validation, A.C., C.N. and E.D.; formal analysis, G.T.; investigation, A.C.; resources, L.L.; data curation, G.T.; writing—original draft preparation, E.D. and V.P.; writing—review and editing, L.L., V.P. and G.T.; visualization, A.C.; supervision, V.P.; project administration, G.T. All authors have read and agreed to the published version of the manuscript.

Funding: This research received no external funding.

Institutional Review Board Statement: Not applicable.

Informed Consent Statement: Not applicable.

Data Availability Statement: Not applicable.

Conflicts of Interest: The authors declare no conflict of interest.

References

1. Bosworth, D.M. Fracture-Dislocation of the Ankle with Fixed Displacement of the Fibula behind the Tibia. *J. Bone Jt. Surg. Am.* **1947**, *29*, 130–135.
2. Ren, W.; Hu, Y.-C.; Lu, J.-K. Rare Variants of Bosworth Fracture-Dislocation: Bosworth Fracture-Dislocation with Medial Malleolus Adduction Type Fracture. *Chin. J. Traumatol.* **2019**, *22*, 120–124. [CrossRef] [PubMed]
3. Martin-Somoza, F.J.; Picazo, D.R.; Cabezuelo, J.A.M.; González, A.V. Bosworth Fracture. An Atypical Case of Irreducible Ankle Fracture-Dislocation. *Trauma Case Rep.* **2020**, *28*, 100322. [CrossRef]
4. Yang, K.H.; Won, Y.; Lim, J.-R.; Kang, D.-H. Assessment of Bosworth-Type Fracture by External Oblique Radiographs. *Am. J. Emerg. Med.* **2014**, *32*, 1387–1390. [CrossRef]
5. Cappuccio, M.; Leonetti, D.; Di Matteo, B.; Tigani, D. An Uncommon Case of Irreducible Ankle Fracture-Dislocation: The "Bosworth-like" Tibio-Fibular Fracture. *Foot Ankle Surg.* **2017**, *23*, e1–e4. [CrossRef] [PubMed]
6. Bartoníček, J.; Rammelt, S.; Kostlivý, K. Bosworth Fracture: A Report of Two Atypical Cases and Literature Review of 108 Cases. *Fuß Sprunggelenk* **2017**, *15*, 126–137. [CrossRef]
7. Cho, B.-K.; Choi, S.-M.; Shin, Y.-D. Prognostic Factors for Intermediate-Term Clinical Outcomes Following Bosworth Fractures of the Ankle Joint. *Foot Ankle Surg.* **2019**, *25*, 601–607. [CrossRef]
8. Moher, D.; Liberati, A.; Tetzlaff, J.; Altman, D.G. The PRISMA Group Preferred Reporting Items for Systematic Reviews and Meta-Analyses: The PRISMA Statement. *PLoS Med.* **2009**, *6*, e1000097. [CrossRef]
9. Fan, J.; Michelin, R.M.; Jenkins, R.; Hwang, M.; French, M. A Novel Technique for a Successful Closed Reduction of a Bosworth Fracture-Dislocation of the Ankle. *Cureus* **2020**, *12*, e6632. [CrossRef]
10. Won, Y.; Lee, G.-S.; Hwang, J.-M.; Park, I.; Song, J.-H.; Kang, C.; Hwang, D.-S. Improved Functional Outcome after Early Reduction in Bosworth Fracture-Dislocation. *Foot Ankle Surg.* **2019**, *25*, 798–803. [CrossRef] [PubMed]
11. Arora, A.; Saidapur, S.K.; Uppin, R.B.; Bose, S. Ankle Arthrodesis Using Retrograde Nail in Case of Bosworth Fracture with Ankle Dislocation: A Rare Case Report. *J. Orthop. Case Rep.* **2021**, *11*, 11–15.
12. Moerenhout, K.; Gkagkalis, G.; Baalbaki, R.; Crevoisier, X. Association of Bosworth, Pilon, and Open Talus Fractures: A Very Unusual Ankle Trauma. *Case Rep. Orthop.* **2019**, *2019*, 6316137. [CrossRef] [PubMed]
13. Ji, W.-B.; Xu, Y.-F.; Lu, Z. Case Report: Bosworth Fracture-Dislocation Managed by Closed Reduction and Conservative Treatment. *Front. Surg.* **2022**, *8*, 788575. [CrossRef] [PubMed]
14. Peterson, N.D.; Shah, F.; Narayan, B. An Unusual Ankle Injury: The Bosworth-Pilon Fracture. *J. Foot Ankle Surg.* **2015**, *54*, 751–753. [CrossRef]
15. Schepers, T.; Hagenaars, T.; Den Hartog, D. An Irreducible Ankle Fracture Dislocation: The Bosworth Injury. *J. Foot Ankle Surg.* **2012**, *51*, 501–503. [CrossRef] [PubMed]
16. Khan, F.; Borton, D. A Constant Radiological Sign in Bosworth's Fractures: "The Axilla Sign". *Foot Ankle Int.* **2008**, *29*, 55–57. [CrossRef]
17. Delasotta, L.A.; Hansen, R.H.; Sands, A.K. Surgical Management of the Posterior Fibula Fracture Dislocation: Case Report. *Foot Ankle Int.* **2013**, *34*, 1443–1446. [CrossRef]
18. Han, S.-J.; Kim, J.-H.; Yang, D.-B.; Kim, B.-S.; Ok, H.-S. Bosworth-Type Fibular Entrapment Fracture of the Ankle without Dislocation: A Rare Case Report and a Review of the Literature. *Ann. Transl. Med.* **2021**, *9*, 178. [CrossRef]
19. Bartoníček, J.; Frič, V.; Svatoš, F.; Luňáček, L. Bosworth-Type Fibular Entrapment Injuries of the Ankle-The Bosworth Lesion: A Report of 6 Cases and Literature Review. *J. Orthop. Trauma* **2007**, *21*, 710–717. [CrossRef] [PubMed]
20. Lui, T.H.; Chan, K.B.; Kong, C.C.; Ngai, W.K. Ankle Stiffness after Bosworth Fracture Dislocation of the Ankle. *Arch. Orthop. Trauma Surg.* **2008**, *128*, 49–53. [CrossRef]
21. He, S.; Zhu, J. Ankle Arthroscopy–Assisted Closed Reduction in Bosworth Fracture Dislocation. *SAGE Open Med. Case Rep.* **2020**, *8*, 2050313X2097452. [CrossRef] [PubMed]
22. Williams, A.D.; Blue, M.; Douthit, C.; Caroom, C. Bosworth Dislocation without Associated Fracture. *Case Rep. Orthop.* **2018**, *2018*, 7284643. [CrossRef]
23. Wenxian, P.; Mehta, K.V. Rare Bosworth Fracture-Dislocation Variant of an Irreducible Distal Fibula Dislocation of the Ankle without Fibula Fracture. *Foot Ankle Orthop.* **2019**, *4*, 247301141882090. [CrossRef]
24. Bartoníček, J.; Rammelt, S.; Kostlivý, K. Bosworth Fracture Complicated by Unrecognized Compartment Syndrome: A Case Report and Review of the Literature. *Arch. Orthop. Trauma Surg.* **2022**, *142*, 1435–1441. [CrossRef]
25. Perry, C.R.; Rice, S.; Rao, A.; Burdge, R. Posterior Fracture-Dislocation of the Distal Part of the Fibula. Mechanism and Staging of Injury. *J. Bone Jt. Surg. Am.* **1983**, *65*, 1149–1157. [CrossRef]
26. Wright, S.E.; Legg, A.; Davies, M.B. A Contemporary Approach to the Management of a Bosworth Injury. *Injury* **2012**, *43*, 252–253. [CrossRef] [PubMed]
27. Fahey, J.J.; Schlenker, L.T.; Stauffer, R.C. Fracture Dislocation of the Ankle with Fixed Displacement of the Fibula behind the Tibia. *Am. J. Roentgenol. Radium Ther. Nucl. Med.* **1956**, *76*, 1102–1113. [PubMed]
28. Beekman, R.; Watson, J.T. Bosworth Fracture-Dislocation and Resultant Compartment Syndrome. A Case Report. *J. Bone Jt. Surg. Am.* **2003**, *85*, 2211–2214. [CrossRef]
29. Szalay, M.D.; Roberts, J.B. Compartment Syndrome after Bosworth Fracture-Dislocation of the Ankle: A Case Report. *J. Orthop. Trauma* **2001**, *15*, 301–303. [CrossRef]

30. Hockenbury, R.T.; Friermood, T.G. Dislocation of the Distal Tibiofibular Joint. *J. Orthop. Trauma* **1992**, *6*, 120–124. [PubMed]
31. Foldager, C.B.; Barckman, J.; Robertsen, K.; Borris, L.C. Bosworth Fracture Dislocation of the Ankle: Two Case Reports with Perioperative Illustration. *Trauma Case Rep.* **2018**, *17*, 39–42. [CrossRef] [PubMed]
32. Pollak, A.N.; Mccarthy, M.L.; Bess, R.S.; Agel, J.; Swiontkowski, M.F. Outcomes after treatment of high-energy tibial plafond fractures. *J. Bone Jt. Surg.* **2003**, *85*, 1893–1900. [CrossRef] [PubMed]
33. Fournier, A.; Barba, N.; Steiger, V.; Lourdais, A.; Frin, J.-M.; Williams, T.; Falaise, V.; Pineau, V.; Salle de Chou, E.; Noailles, T.; et al. Total Talar Fracture—Long-Term Results of Internal Fixation of Talar Fractures. A Multicentric Study of 114 Cases. *Orthop. Traumatol. Surg. Res.* **2012**, *98*, S48–S55. [CrossRef] [PubMed]

Journal of *Personalized Medicine*

MDPI

Article

Augmentative Plating versus Exchange Intramedullary Nailing for the Treatment of Aseptic Non-Unions of the Femoral Shaft—A Biomechanical Study in a Sawbone^{TM} Model

Matthias Georg Walcher [1,2,*], Robert E. Day [3], Markus Gesslein [2], Hermann Josef Bail [2] and Markus S. Kuster [4]

1 OC Würzburg, Oeggstr. 3, 97070 Würzburg, Germany
2 Department of Orthopedics and Traumatology, Paracelsus Medical University, Breslauer Straße 201, 90471 Nuremberg, Germany
3 Health Technology Management Unit, Royal Perth Hospital, University of Western Australia, 197 Wellington Street, Perth 6000, Australia
4 Perth Orthopaedic Sports Medicine Centre, The University of Western Australia, 31 Outram Street, Perth 6005, Australia
* Correspondence: matthias.walcher@web.de

Abstract: Background: Non-unions after intramedullary nailing of femoral shaft fractures are a significant problem. Treatment options such as augmenting with plates or exchange nailing have been proposed. The ideal treatment remains controversial. Methods: Augmentative plating using a 4.5 mm LCP or a 3.2 mm LCP leaving the nail in situ was tested biomechanically and compared to exchange intramedullary nailing in a Sawbone^{TM} model of a femoral shaft non-union. Results: The difference of fracture gap motion in axial testing was small. In rotational testing, the exchange nail allowed for the largest amount of motion. The 4.5 mm augmentative plate was the most stable construct in all loading conditions. Conclusions: Augmentative plating using a 4.5 mm LCP plate while leaving the nail in situ is biomechanically superior to exchange intramedullary nailing. A small fragment 3.2 mm LCP is undersized and does not reduce fracture motion sufficiently in a femoral shaft non-union.

Keywords: non-union long bone; biomechanics; additional plate; nail in situ; non-union revision

check for updates

Citation: Walcher, M.G.; Day, R.E.; Gesslein, M.; Bail, H.J.; Kuster, M.S. Augmentative Plating versus Exchange Intramedullary Nailing for the Treatment of Aseptic Non-Unions of the Femoral Shaft—A Biomechanical Study in a Sawbone^{TM} Model. *J. Pers. Med.* **2023**, *13*, 650. https://doi.org/10.3390/jpm13040650

Academic Editor: Jih-Yang Ko

Received: 15 March 2023
Revised: 6 April 2023
Accepted: 8 April 2023
Published: 10 April 2023

1. Article Summary

Article focus:

- The main revision options for aseptic non-unions in femoral shaft fractures after intramedullary nailing without a bony defect are revision intramedullary nailing and augmentative plating.
- Revision intramedullary nailing and augmentative plating using plates of different sizes were evaluated biomechanically in a Sawbone^{TM} model of an aseptic non-union of the femoral shaft.

Key messages:

- Augmentative plating using a 4.5 mm LCP leaving the nail in situ is more stable than exchange intramedullary nailing.
- Using a 3.2 mm LCP does not reduce fracture motion sufficiently.

Strengths and limitations of this study:

- Strengths: biomechanically based treatment recommendations for the treatment of aseptic femoral shaft non-unions after intramedullary nailing are provided.
- Limitations: no cadaver model was used, but a sawbone model was used. This is less realistic than cadaver bones but has the advantage of uniformity. Only two different plates were used for evaluation.

2. Introduction

Non-union after intramedullary nailing of fractures of diaphyseal long bones is a significant complication, leading to a long period of incapacity, implant failure, revision surgery and, in the worst case, even to the loss of a limb [1]. It is crucial to manage non-unions effectively to ensure the best chances of bone healing [1,2], especially as there is a relevant percentage of femoral non-unions which does not heal without more additional procedures, or does not heal at all [3]. The traditional approach to non-unions of long bones after intramedullary nailing without osseous defect or infection is a removal of the nail, reaming of the medullary canal, and stabilization, typically with a larger diameter nail [1,2,4]. Recently, the concept of leaving the nail in situ and adding stability with an additional plate fixation has gained popularity with promising results in the literature [5–14]. The concept behind this is that, for hypertrophic non-unions, the stability of fracture fixation is insufficient. Therefore, adding extra stability and ideally compression of the non-union is necessary to achieve bony union. Exchange nailing still seems to be the most popular amongst these treatment options. The idea of augmenting an osteosynthetic construct and adding extra stability by doing so has also been described in other fields of orthopedic trauma, like trochanteric femur fractures, where cement augmentation has been found to be safe and effective in osteoporotic fractures [15]. It needs to be stated that a lack of mechanical stability is not the only reason for a non-union. In femur fractures, the same reasons, like in any other fracture, can cause a non-union, namely too much motion at the fracture site, avascularity, bony gapping, and infection [16]. This explains the different shapes of non-unions—hypertrophic, atrophic, and oligotrophic. Bony gapping needs to be addressed with a proper operative technique, potentially in combination with a bone graft. Infectious non-unions typically need to be treated with thorough surgical debridement, rigid fixation often using external fixators, and a tailored antibiotic treatment. Avascularity is a risk factor for the non-union of femoral fractures. However, it is important to note that the effect of anatomic reduction and stability of fracture fixation can often outweigh this, allowing for a bony healing [16]. Therefore, proper osteosynthesis is in the center of the treatment of non-unions after fractures of every origin—there is basically no fracture healing without stability. To the best of our knowledge, no biomechanical investigation has been published about the effects of an augmentative plate. The aim of the present study was to evaluate the biomechanical rationale behind the concept of leaving the nail in situ while adding an additional lateral plate stabilization. The stability of additional plates of different sizes leaving the original nail in situ was compared to an exchange intramedullary nail with a larger diameter. Clinically, there are indications that the combination of an intramedullary nail and an augmentative plate offers excellent stability. The group around Passias recommends using a nail in combination with a lateral locking plate in distal femur fractures in patients that are at a higher risk for non-union or who require extra robust fixation in primary fracture care [17].

3. Materials and Methods

Fifteen large fourth generation composite femur models (Sawbones^{TM}, Pacific Research Labs, Vashon, WA, USA) with a diaphyseal outer diameter of 32 mm were used. Composite femora were used, as they have greater uniformity than cadaver bone and have been thoroughly evaluated as bone analogues [18]. To approximate a model of an aseptic non-union of the femoral shaft with the nail loose in the medullary canal, the composite femora were reamed to an inner diameter of 16 mm. A transverse fracture was created 130 mm proximal to the distal femoral end in the transition zone between the distal and the middle third of the femur with a saw. The fractures were reduced and fixed with an intramedullary nail. Three groups were tested: An exchange nail with 14 mm diameter (big nail), then a 3.2 mm augmentative plate (3.2 plate) and a 4.5 mm augmentative plate (4.5 plate), both with the 11 mm nail in situ. Five femora were used in each group.

In the exchange intramedullary nail group, five femora were fixed with a 14 mm femoral nail (AFN, Synthes^{TM}, Westchester, PA, USA) (Figure 1A). It is a hollow femoral nail design. Four regular locking bolts were used. Distally one dynamic and one static locking bolt were inserted, and proximally the regular locking option was chosen with two locking bolts from lateral to medial. The 14 mm diameter nail had a tight fit in the medullary canal, which recreated the clinical situation of a tight exchange nail.

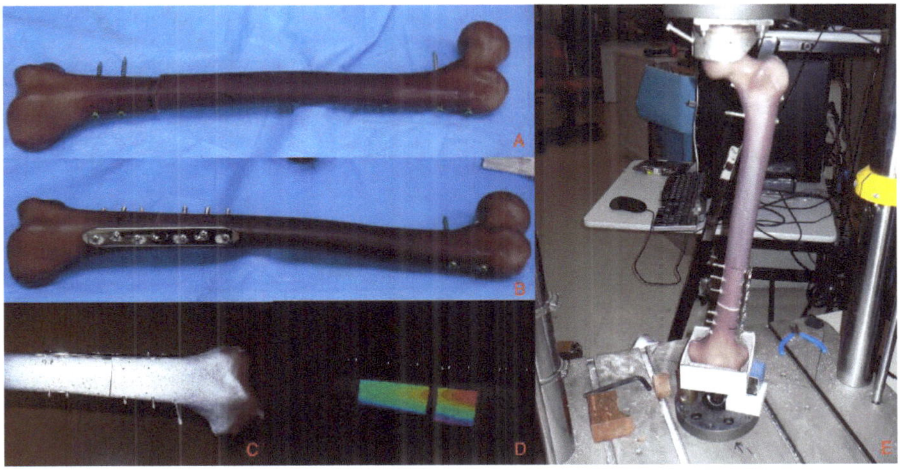

Figure 1. Test setting. (**A**) Preparation of the sawbone femora, with the nail in situ and an artificial fracture applied. (**B**) An additional plate is mounted. (**C**) The anterior surface of the bone is sprayed with a speckle pattern. (**D**) The system allows full-field 3D-surface strain and motion detection down to 200 µstrain. (**E**) Sawbone femora are potted in PMMA distally and held by a universal joint that constrained translation and axial torsion but allow rotation in the other two axes. The femoral head is held in a custom fixture rigidly attached to the load cell. The axis of the bone is aligned in 7° valgus to the rotational axis of the testing machine to recreate a loading situation as close to physiology as possible.

In the augmentative plate group, an 11 mm femoral nail (AFN, Synthes™, Westchester, PA, USA) was inserted to represent a non-union with an initially tight nail that had become loose. No distal locking bolts were inserted to simulate removing the distal locking bolts to dynamize the nonunion, which is typically part of the augmentative plating procedure. Following that, a 9-hole 3.2 stainless steel LCP plate (Synthes™, Westchester, PA, USA) was applied to five femora on the lateral side, and a 7-hole 4.5 steel LCP plate (Synthes™, Westchester, PA, USA) was applied to another five femora, again on the lateral side, to match the clinical situation of an augmentative plate, inserted from the lateral side. Both groups used three non-angular stable screws on either side of the fracture (Figure 1B). The screws were drilled and inserted eccentrically to create compression at the fracture site. The screws were inserted as non-angular-stable. The dynamic compression holes of the plate were used in a typical AO-way, not only to add stabilization, but to achieve compression at the fracture site. Ideally, the screws would be positioned in different directions—some anterior and some posterior of the nail—to position the material "around the nail". Aiming to recreate a model as close to the clinical situation as possible, we inserted the screws where they could be positioned safely bicortically, be it anterior or posterior of the nail. This resembles the situation in operation theatre, where there is typically only a very narrow corridor for screw and plate placement around the nail, not allowing readily to decide to position the screws anterior or posterior to the nail, in particular in atrophic non-unions with little vital bone available.

A total of 15 artificial composite femora, one 3 2 LCP, one 4.5 LCP, one 11 mm, and one 14 mm AFN, were used for testing.

For image correlation strain measurement, the anterior surface of the bone was sprayed with a speckle pattern (Figure 1C). The system allowed full-field 3D-surface strain and motion detection down to 200 μstrain in this set-up (VIC-3D™, Limess, Correlated Solutions) (Figure 1D). The construct was mounted to a servohydraulic biaxial testing machine (Instron 8874) with a 25 kN/100 Nm load cell. The Sawbone femora were potted in PMMA distally and held by a universal joint that constrained translation and axial torsion but allowed rotation in the other two axes. Proximally, the femoral head was held in a custom fixture that was rigidly attached to the load cell. The axis of the bone was aligned in 7° valgus to the rotational axis of the testing machine to recreate a loading situation that is as close to physiological loading as possible [19]. Of course, the model remained crude, as the forces of the surrounding muscles and soft tissue were not included (Figure 1E).

Axial loading and torsional loading were applied for 26 cycles and the mean peak strain was calculated on the bone surface in the region of the artificial fracture on the anterior surface of the femur and averaged over 20 cycles. The first three and the last three cycles were not included in the calculations to avoid the starting and stopping ramps of the test. The regions on the anterior femur immediately above and below the fracture were analyzed, as this would allow for the best assessment of the stability of the fracture fixation.

The testing started with a ramp to 10 N of axial compression over 3 s. Following that, 26 sinusoidal cycles of axial compression from 10 N to 870 N at 0.25 Hz were applied. This was followed by a ramp over 3 s to 20 N axial preload and 0 Nm torsion to ensure consistent starting conditions. Sinusoidal cyclic rotation was then applied, ranging from −6 Nm to 6 Nm at 0.25 Hz for 26 cycles. Again, for analysis, the first three and the last three cycles were discarded as the system took time to come to a full load and to a stop. The number of cycles was determined from pilot studies to give stable average measurements. The orientation and neck offset were chosen to match the study of Bergmann [19]. The loading patterns were chosen according to the study of Schneider et al., who calculated the forces acting in a femur during healing after fracture [20]. They found forces in the axial direction up to 300 N in 250 N partial weightbearing, and up to 900 N in single leg stance. The torsional moment they described was up to 6 Nm in 250 N partial weightbearing, and up to 10 Nm during torsion testing [20]. We aimed to recreate the lower range of physiological loading conditions. The constructs were compared separately for each loading condition using one-way ANOVA. The Q-Q plot was checked before testing to confirm that the data were normally distributed. The construct had a significant effect on both the axial and torsion loading results, so Tukey's honestly significant difference test was applied to look for the differences between the constructs. The level of significance was set at 5% for all tests. All analyses were performed using R version 3.1.2 running on OSX 10.9.5.

4. Results

Relative motion between the two fragments in three different directions were measured to judge the rigidity of the constructs. Figure 2 and Table 1 display the results. Us is fracture motion in the proximal/distal direction, Vs is in the medial/lateral direction, and Ws is in the anterior/posterior direction. The total interfragmentary motion D is the vector sum of the motions in the different directions.

Table 1. Mean total interfragmentary motion in mm by construct and loading directions. The big nail stands for exchange medullary nail, the 3.2 plate stands for small augmentative plate, and the 4.5 plate stands for large augmentative plate.

Construct	Axial Loading	Torsional Loading
3.2 plate	0.46	1.33
4.5 plate	0.16	0.43
big nail	0.26	2.42

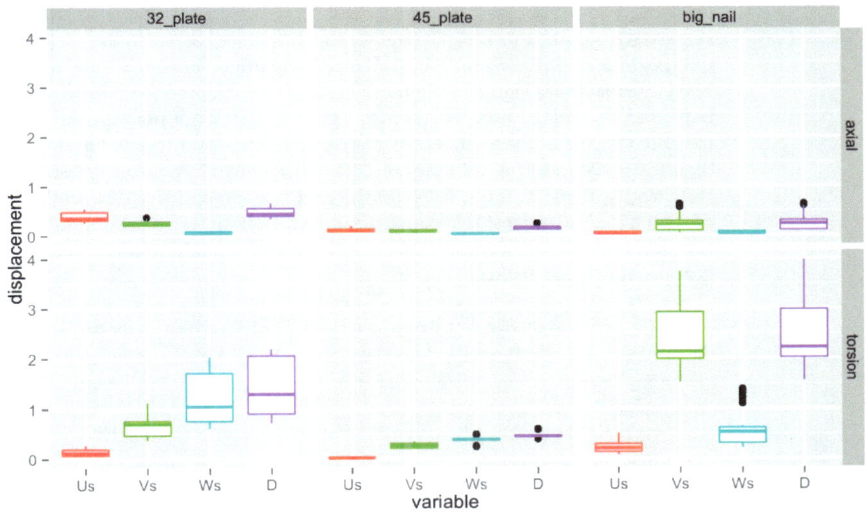

Figure 2. Us is fracture motion in the proximal/distal direction, Vs is in the medial/lateral direction, and Ws is in the anterior/posterior direction. The total interfragmentary motion D is the vector sum of the motions in the different directions. The 3.2 plate stands for the small additional plate, the 4.5 plate stands for the larger additional plate, and the big nail stands for the exchange nail. The different colors have been added to allow for a clearer determination of the information. In every column the middle horizontal line is the median. The box goes from the 25th percentile to the 75th percentile and the black dots mark the extreme values.

4.1. Axial Loading

Analyzing the total fracture gap motion, the large additional plate was the most rigid, followed by the revision intramedullary nail and the small plate, which had the largest motion. The differences between the treatment groups were all significant (see Table 2).

Table 2. Differences between constructs in axial loading. The big nail stands for exchange medullary nail, the 3.2 plate stands for small additional plates, and the 4.5 plate stands for large additional plates.

	Difference	99.9% CI	p adj
big nail versus 4.5 plate	0.0992	0.0847–0.114	<0.01
3.2 plate versus big nail	0.2063	0.1917–0.221	<0.01
3.2 plate versus 4.5 plate	0.3055	0.2910–0.320	<0.01

In the compression/distraction direction (Us), both the group with the revision intramedullary nail (big nail) and the large plate (4.5 plate) had much smaller motion at the fracture gap than the small plate (3.2 plate). The axial motion of the small plate group was significantly higher ($p < 0.05$) than either of the other groups. In the medial/lateral direction (Vs), the large additional plate group showed significantly less fracture motion ($p < 0.05$) than the other constructs, which both displayed comparable amounts of movement. In the anterior/posterior direction (Ws) all treatment groups had comparable small motion, without significant differences.

4.2. Torsional Loading

Comparing the sums of movement in the fracture gap, the large plate group had the smallest motion, followed by the small plate group, while the exchange nail was the least rigid construct in torsion. The differences between the treatment groups were all statistically significant. Table 3 displays the differences between the constructs in torsion. In the proximal/distal direction (Us), all three constructs showed a small amplitude of motion. The large plate was significantly more stable ($p < 0.05$) than the two other constructs, which both had a similar range of fracture motion. The large plate group was also the most robust construct in the medial/lateral direction (Vs), followed by the small plate group and then the revision intramedullary rod. The differences between the treatment groups were statistically significant ($p < 0.05$). In the anterior/posterior direction (Ws) the large plate proved to be the most stable construct with the smallest fracture motion, followed by the revision intramedullary nail and the small plate. The differences between the treatment groups were all statistically significant.

Table 3. Differences between constructs in torsion. The big nail stands for the exchange medullary nail, the 3.2 plate stands for the small additional plate, and the 4.5 plate stands for the large additional plate.

	Difference	99.9% CI	*p* adj
big nail versus4.5 plate	0.891	0.825–0.957	<0.01
3.2 plate versus big nail	1.094	1.028–1.160	<0.01
3.2 plate versus 4.5 plate	1.985	1.919–2.051	<0.01

Figure 3 shows the overall displacement by construct for both axial and torsion loading. In summary, all three osteosynthetic constructs showed sufficient stability under axial loading conditions with little motion in the fracture gap. Motion during torsional loading was larger than in axial compression for all constructs. In compression the difference between the constructs was generally small, with the exchange nail being superior to the 3.2 mm plate by 0.2 mm and marginally worse than the 4.5 mm plate (about 0.1 mm). A much larger motion was found in torsional loading, with the exchange nail having the largest motion. This was nearly 2 mm more than the 4.5 mm plate and over 1 mm more than the 3.2 mm plate. All of the differences in motion between the different constructs were statistically significant. The 4.5 mm plate construct had the least interfragmentary motion in all cases.

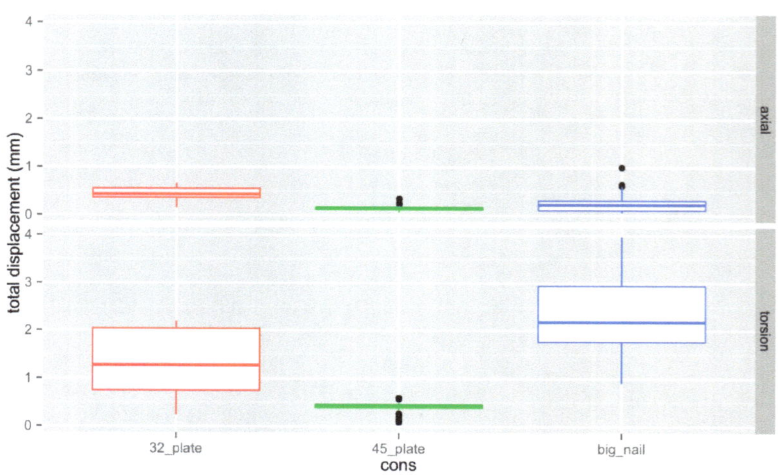

Figure 3. Overall displacement by construct for both axial and torsion loading. The different colors have been added to allow for a clearer determination of the information. In every column the middle horizontal line is the median. The box goes from the 25th percentile to the 75th percentile and the black dots mark the extreme values.

5. Discussion

There are two main options in orthopedic literature for the treatment of aseptic femoral non-unions without bony defect after intramedullary nailing which are discussed in the orthopedic literature: augmentative plating or exchange nailing. Exchange nailing is usually combined with reaming of the medullary canal and conversion to a larger diameter intramedullary nail [2].

Brinker summarized the results of the current literature for femoral non-unions after intramedullary nailing [2]. He identified several papers on augmentative plating whilst leaving the nail in situ [5–7,10–14]. These publications reported the results of a total of 147 cases, with a healing rate of 100%. The group around Zhang could confirm these excellent results with a 100% healing rate in 39 patients with a non-infectious non-union after intramedullary nail fixation for long bone fractures in the lower extremity that were treated with augmentative plating [21]. The authors pointed out the easy operative technique with a short operation time. The majority of the authors used a 4.5 mm DCP, as was done in the present biomechanical testing.

Uliana and coworkers performed a multicenter study, including 22 cases treated with plate augmentation for a femoral shaft aseptic non-union of a femur fracture with an intramedullary nail retained in situ [22]. These patients were followed for an average of 23.5 months. Among these 22 patients were 10 patients with a vascular non-union. Nonetheless, bony healing could be achieved in 19 patients, and all patients were doing well clinically. Just recently, Perisano published a literature review looking into plate augmentation in aseptic femoral shaft non-union after intramedullary nailing [23]. He included 24 studies with a total of 502 patients. Overall, 200 had an atrophic non-union, 123 had a hypertrophic non-union, and in 179 patients the type of non-union was not mentioned. The plate used the most was a DCP (dynamic compression plate). In 98% of the cases, osseous union could be achieved. Functional recovery was good, and the rate of complications was low [23].

There were eight publications investigating exchange intramedullary nailing for aseptic femoral shaft non-unions [2]. A total of 266 cases were reported with a healing rate of 89% [10,24–30].

Most of these authors did over-ream the medullary canal by 1–2 mm in the revision procedure and used a revision nail that was at least 1 mm wider than the original nail. Of particular interest is the work of Yang et al. [29]. They evaluated isthmic and non-isthmic non-unions separately. In the isthmic non-unions, 87% healed with exchange intramedullary nailing, while in the non-isthmic non-unions only 50% healed. This is in agreement with our biomechanical results—that rotational stability is an important issue—which is more prevalent in the wide, non-isthmic portion of the femur. Lai and his group looked retrospectively into 96 patients undergoing either augmentative plating or exchanging reamed nailing [31]. The overall union rate was 70.8%, even with the exclusion of hypertrophic non-unions. The union rate was significantly higher in the augmentative plate group, and the operating time was significantly shorter. Looking into the location of the non-union, the union rate was comparable for isthmic non-unions, but was significantly smaller for the augmentative plates in non-isthmic non-unions. These results were confirmed by Ru, who looked into the results of 190 non-unions of the femur after intramedullary nailing. They were treated either with exchange reamed intramedullary nailing or augmentative plating [32]. Augmentative plating could obtain a higher union rate and a shorter time to union. The intraoperative blood loss, like the complication rate, was higher for the exchange nail group.

Nonetheless, Kim and Coworkers could achieve a 100% healing rate in their patient collective of infraisthmal femoral non-unions [4]. They treated 18 patients with a non-union of a femur shaft fracture in the infraisthmic portion, not only with an exchange intramedullary nail, but enhanced the stability of their construct by adding a Poller screw and additional interlocking screws. These results confirm that exchange nailing on its own is just biomechanically not stable enough in the non-isthmic portion of the femur.

Park and his colleagues treated 17 patients with a hypertrophic non-union of the long bones with an augmentative locking compression plate [33]. They did no fracture site exposure, implant removal, or bone grafting. Still, all cases achieved osseous healing, proving the mechanical stability of augmentative plating. Sancheti combined exchange nailing with augmentative plating in 70 patients with femoral diaphyseal non-unions to profit from the benefits of both treatment options. Overall, 46 cases had a hypertrophic non-union and 24 had an atrophic non-union. All patients received autologous bone grafting and raising of periosteal flaps. The healing rate was 100% at an average time of 16.77 weeks.

It is of particular interest that the concept of augmentative plating also worked in atrophic nun-unions. In atrophic non-unions, a disturbed vascularity of the fracture regions must be assumed. The endosteal blood supply is already disturbed by the initial nailing, so plating could be potentially detrimental by inhibiting the periosteal blood supply. Obviously, the extra stability outweighs the potential side effects on vascularity [16]. These positive effects of augmentative plating in atrophic non-unions could also be observed in other studies [22,23,31].

Kontakis proposed using a primary nail/augmentative plate construct, even in primary fracture cases, and in fractures of the distal femur in geriatric, obese, or osteoporotic patients, to allow for immediate weight-bearing and range of motion exercises after fracture fixation [34].

In summary, both augmentative plating leaving the nail in situ and exchange intramedullary nailing are valid options to treat aseptic femoral non-unions. The sparse literature shows a higher success rate for augmentative plating in aseptic non-unions of the femoral shaft after intramedullary nailing, especially in the non-isthmic portion of the femur. These findings are confirmed by several recent meta-analyses comparing these two treatment options, which describe a higher rate of union, a shorter healing time, less intraoperative blood loss, and fewer complications for augmentative plating [35–37]. Ru could establish that revision with an exchange, reamed intramedullary nail instead of augmentative plating is a risk factor for the development of another non-union [38]. Our results support these findings by supplying a biomechanical explanation. In biomechanical testing conditions, augmentative plating, in particular with a 4.5 mm LCP, is clearly more stable than exchange nailing in every tested loading condition. In the axial compression testing the differences between the different constructs were small, while torsional loading produced larger motion in the fracture gap. While the exchange intramedullary nail was stronger than the small augmentative plate in axial compression, the exchange intramedullary nail was clearly the least stable in rotational testing. In summary, using a 4.5 mm LCP as an augmentative plate leaving the nail in situ in aseptic femoral shaft non-unions is biomechanically superior in both torsion and axial compression compared to an exchange nail construct. Using a 3.2 mm LCP gives larger fracture motions and seems to be an undersized construct.

6. Strengths

The study presents biomechanically based recommendations for the treatment of aseptic femoral non-unions after intramedullary nail stabilization.

These results can help orthopedic surgeons to reduce postoperative complications and reach an ideal rate of unions after revision surgery.

7. Limitations

We did not use cadaveric femora but used a sawbone model. This is a less realistic model, but the uniformity of the specimens allows for comparison between constructs without introducing the large variability of cadaveric bone qualities. We only used a 3.2 LCP and a 4.5 LCP. No other plates of different specifications and lengths were used for comparison, and no angular stable locking plates were used.

8. Conclusions

In aseptic femoral shaft non-unions, a 4.5 mm LCP as an augmentative plate leaving the nail in situ is biomechanically superior in terms of stability in both torsion and axial compression to an exchange nail construct using a bigger nail. A 3.2 mm LCP seems to produce an undersized construct.

Author Contributions: Conceptualization: M.G.W. and R.E.D. and M.S.K.; Methodology: M.G.W. and R.E.D. and M.S.K.; Software: R.E.D. and M.G.W.; Validation: M.G.W. and R.E.D. and M.S.K. and M.G. and H.J.B.; Formal Analysis: M.G.W. and R.E.D. and M.S.K. and M.G. and H.J.B.; Investigation: M.G.W. and R.E.D. and M.S.K.; Resources: M.G.W. and R.E.D. and M.S.K. and M.G. and H.J.B.; Data curation: R.E.D. and M.G.W. and M.S.K.; Writing—original draft preparation: M.G.W. and R.E.D. and M.S.K. and M.G. and H.J.B.; Writing—review and Editing: M.G.W. and R.E.D. and M.S.K. and M.G. and H.J.B.; Visualisation: M.G.W. and R.E.D. and M.S.K. and M.G. and H.J.B.; Supervision: M.S.K. and R.E.D. and H.J.B.; Project Administration: M.G.W. and R.E.D. and M.S.K. and M.G. and H.J.B.; Funding Acquisition: M.G.W. and M.S.K. All authors have read and agreed to the published version of the manuscript.

Funding: The 4.5 plate and the screws were supplied by Synthes™ PtyLtd. The sawbone femora and the 3.2 plate were supplied by the department of Orthopaedics, Royal Perth Hospital, University of Western Australia.

Institutional Review Board Statement: Not applicable.

Informed Consent Statement: Not applicable.

Data Availability Statement: The data is available at the Health Technology Management Unit, Royal Perth Hospital, University of Western Australia.

Conflicts of Interest: The authors declare no conflict of interest.

References

1. Garnavos, C. Treatment of aseptic non-union after intramedullary nailing without removal of the nail. *Injury* **2017**, *48* (Suppl. S1), S76–S81. [CrossRef] [PubMed]
2. Brinker, M.R.; O'Connor, D.P. Management of Aseptic Tibial and Femoral Diaphyseal Nonunions without Bony Defects. *Orthop. Clin. N. Am.* **2016**, *47*, 67–75. [CrossRef] [PubMed]
3. Wiss, D.A.; Garlich, J.; Hashmi, S.; Neustein, A. Risk Factors for Development of a Recalcitrant Femoral Nonunion: A Single Surgeon Experience in 122 Patients. *J. Orthop. Trauma* **2021**, *35*, 619–625. [CrossRef] [PubMed]
4. Kim, J.W.; Yoon, Y.C.; Oh, C.W.; Han, S.B.; Sim, J.A.; Oh, J.K. Exchange nailing with enhanced distal fixation is effective for the treatment of infraisthmal femoral nonunions. *Arch. Orthop. Trauma Surg.* **2018**, *138*, 27–34. [CrossRef]
5. Choi, Y.S.; Kim, K.S. Plate augmentation leaving the nail in situ and bone grafting for non-union of femoral shaft fractures. *Int. Orthop.* **2005**, *29*, 287–290. [CrossRef]
6. Birjandinejad, A.; Ebrahimzadeh, M.H.; Ahmadzadeh-Chabock, H. Augmentation plate fixation for the treatment of femoral and tibial nonunion after intramedullary nailing. *Orthopedics* **2009**, *32*, 409. [CrossRef]
7. Chen, C.M.; Su, Y.P.; Hung, S.H.; Lin, C.L.; Chiu, F.Y. Dynamic compression plate and cancellous bone graft for aseptic nonunion after intramedullary nailing of femoral fracture. *Orthopedics* **2010**, *33*, 393. [CrossRef]
8. Ateschrang, A.; Karavalakis, G.; Gonser, C.; Liener, U.; Freude, T.; Stockle, U.; Walcher, M.; Zieker, D. Exchange reamed nailing compared to augmentation compression plating leaving the inserted nail in situ in the treatment of aseptic tibial non-union: A two-centre study. *Wien. Klin. Wochenschr.* **2013**, *125*, 244–253. [CrossRef]
9. Ateschrang, A.; Albrecht, D.; Stockle, U.; Weise, K.; Stuby, F.; Zieker, D. High success rate for augmentation compression plating leaving the nail in situ for aseptic diaphyseal tibial nonunions. *J. Orthop. Trauma* **2013**, *27*, 145–149. [CrossRef]
10. Park, J.; Kim, S.G.; Yoon, H.K.; Yang, K.H. The treatment of nonisthmal femoral shaft nonunions with im nail exchange versus augmentation plating. *J. Orthop. Trauma* **2010**, *24*, 89–94. [CrossRef]
11. Gao, K.D.; Huang, J.H.; Tao, J.; Li, F.; Gao, W.; Li, H.Q.; Wang, Q.G. Management of femoral diaphyseal nonunion after nailing with augmentative locked plating and bone graft. *Orthop. Surg.* **2011**, *3*, 83–87. [CrossRef] [PubMed]
12. Hakeos, W.M.; Richards, J.E.; Obremskey, W.T. Plate fixation of femoral nonunions over an intramedullary nail with autogenous bone grafting. *J. Orthop. Trauma* **2011**, *25*, 84–89. [CrossRef] [PubMed]
13. Said, G.Z.; Said, H.G.; el-Sharkawi, M.M. Failed intramedullary nailing of femur: Open reduction and plate augmentation with the nail in situ. *Int. Orthop.* **2011**, *35*, 1089–1092. [CrossRef] [PubMed]
14. Lin, C.-J.; Chiang, C.-C.; Wu, P.-K.; Chen, C.-F.; Huang, C.-K.; Su, A.W.; Chen, W.-M.; Liu, C.-L.; Chen, T.-H. Effectiveness of plate augmentation for femoral shaft nonunion after nailing. *J. Chin. Med. Assoc.* **2012**, *75*, 396–401. [CrossRef] [PubMed]

15. Stramazzo, L.; Ratano, S.; Monachino, F.; Pavan, D.; Rovere, G.; Camarda, L. Cement augmentation for trochanteric fracture in elderly: A systematic review. *J. Clin. Orthop. Trauma* **2021**, *15*, 65–70. [CrossRef]
16. DeRogatis, M.J.; Kanakamedala, A.C.; Egol, K.A. Management of Subtrochanteric Femoral Fracture Nonunions. *JBJS Rev.* **2020**, *8*, e1900143. [CrossRef]
17. Passias, B.J.; Emmer, T.C.; Sullivan, B.D.; Gupta, A.; Myers, D.; Skura, B.W.; Taylor, B.C. Treatment of Distal Femur Fractures with a Combined Nail-Plate Construct: Techniques and Outcomes. *J. Long Term Eff. Med. Implants* **2021**, *31*, 15–26. [CrossRef]
18. Cristofolini, L.; Viceconti, M.; Cappello, A.; Toni, A. Mechanical validation of whole bone composite femur models. *J. Biomech.* **1996**, *29*, 525–535. [CrossRef]
19. Bergmann, G.; Deuretzbacher, G.; Heller, M.; Graichen, F.; Rohlmann, A.; Strauss, J.; Duda, G.N. Hip contact forces and gait patterns from routine activities. *J. Biomech.* **2001**, *34*, 859–871. [CrossRef]
20. Schneider, E.; Michel, M.C.; Genge, M.; Zuber, K.; Ganz, R.; Perren, S.M. Loads acting in an intramedullary nail during fracture healing in the human femur. *J. Biomech.* **2001**, *34*, 849–857. [CrossRef]
21. Zhang, H.J.; Wang, S.; Dong, Y.H.; Zheng, W.D.; Sun, Z.; Zheng, J. Successful management of lower limb nonunion using locking plates and bone graft with retention of intramedullary nail. *Medicine* **2019**, *98*, e14949. [CrossRef] [PubMed]
22. Uliana, C.S.; Bidolegui, F.; Kojima, K.; Giordano, V. Augmentation plating leaving the nail in situ is an excellent option for treating femoral shaft nonunion after IM nailing: A multicentre study. *Eur. J. Trauma Emerg. Surg. Off. Publ. Eur. Trauma Soc.* **2020**, *47*, 1895–1901. [CrossRef] [PubMed]
23. Perisano, C.; Cianni, L.; Polichetti, C.; Cannella, A.; Mosca, M.; Caravelli, S.; Maccauro, G.; Greco, T. Plate Augmentation in Aseptic Femoral Shaft Nonunion after Intramedullary Nailing: A Literature Review. *Bioengineering* **2022**, *9*, 560. [CrossRef] [PubMed]
24. Wu, C.C. Exchange nailing for aseptic nonunion of femoral shaft: A retrospective cohort study for effect of reaming size. *J. Trauma* **2007**, *63*, 859–865. [CrossRef]
25. Oh, J.K.; Bae, J.H.; Oh, C.W.; Biswal, S.; Hur, C.R. Treatment of femoral and tibial diaphyseal nonunions using reamed intramedullary nailing without bone graft. *Injury* **2008**, *39*, 952–959. [CrossRef]
26. Shroeder, J.E.; Mosheiff, R.; Khoury, A.; Liebergall, M.; Weil, Y.A. The outcome of closed, intramedullary exchange nailing with reamed insertion in the treatment of femoral shaft nonunions. *J. Orthop. Trauma* **2009**, *23*, 653–657. [CrossRef]
27. Gao, K.D.; Huang, J.H.; Li, F.; Wang, Q.G.; Li, H.Q.; Tao, J.; Wang, J.D.; Wu, X.M.; Wu, X.F.; Zhou, Z.H.; et al. Treatment of aseptic diaphyseal nonunion of the lower extremities with exchange intramedullary nailing and blocking screws without open bone graft. *Orthop. Surg.* **2009**, *1*, 264–268. [CrossRef]
28. Naeem-ur-Razaq, M.; Qasim, M.; Sultan, S. Exchange nailing for non-union of femoral shaft fractures. *J. Ayub Med. Coll. Abbottabad* **2010**, *22*, 106–109.
29. Yang, K.H.; Kim, J.R.; Park, J. Nonisthmal femoral shaft nonunion as a risk factor for exchange nailing failure. *J. Trauma Acute Care Surg.* **2012**, *72*, E60–E64. [CrossRef]
30. Swanson, E.A.; Garrard, E.C.; Bernstein, D.T.; O'Connor, D.P.; Brinker, M.R. Results of a systematic approach to exchange nailing for the treatment of aseptic femoral nonunions. *J. Orthop. Trauma* **2015**, *29*, 21–27. [CrossRef]
31. Lai, P.J.; Hsu, Y.H.; Chou, Y.C.; Yeh, W.L.; Ueng, S.W.N.; Yu, Y.H. Augmentative antirotational plating provided a significantly higher union rate than exchanging reamed nailing in treatment for femoral shaft aseptic atrophic nonunion—Retrospective cohort study. *BMC Musculoskelet. Disord.* **2019**, *20*, 127. [CrossRef] [PubMed]
32. Ru, J.; Xu, H.; Kang, W.; Chang, H.; Niu, Y.; Zhao, J. Augmentative compression plating versus exchanging reamed nailing for nonunion of femoral shaft fracture after intramedullary nailing: A retrospective cohort study. *Acta Orthop. Belg.* **2016**, *82*, 249–257. [PubMed]
33. Park, K.C.; Oh, C.W.; Kim, J.W.; Park, K.H.; Oh, J.K.; Park, I.H.; Kyung, H.S.; Heo, J. Minimally invasive plate augmentation in the treatment of long-bone non-unions. *Arch. Orthop. Trauma Surg.* **2017**, *137*, 1523–1528. [CrossRef] [PubMed]
34. Kontakis, M.G.; Giannoudis, P.V. Nail plate combination in fractures of the distal femur in the elderly: A new paradigm for optimum fixation and early mobilization? *Injury* **2023**, *54*, 288–291. [CrossRef] [PubMed]
35. Jin, Y.F.; Xu, H.C.; Shen, Z.H.; Pan, X.K.; Xie, H. Comparing Augmentative Plating and Exchange Nailing for the Treatment of Nonunion of Femoral Shaft Fracture after Intramedullary Nailing: A Meta-analysis. *Orthop. Surg.* **2020**, *12*, 50–57. [CrossRef] [PubMed]
36. Luo, H.; Su, Y.; Ding, L.; Xiao, H.; Wu, M.; Xue, F. Exchange nailing versus augmentative plating in the treatment of femoral shaft nonunion after intramedullary nailing: A meta-analysis. *EFORT Open Rev.* **2019**, *4*, 513–518. [CrossRef]
37. Medlock, G.; Stevenson, I.M.; Johnstone, A.J. Uniting the un-united: Should established non-unions of femoral shaft fractures initially treated with IM nails be treated by plate augmentation instead of exchange IM nailing? A systematic review. *Strateg. Trauma Limb Reconstr.* **2018**, *13*, 119–128. [CrossRef]
38. Ru, J.Y.; Chen, L.X.; Hu, F.Y.; Shi, D.; Xu, R.; Du, J.W.; Niu, Y.F. Factors associated with development of re-nonunion after primary revision in femoral shaft nonunion subsequent to failed intramedullary nailing. *J. Orthop. Surg. Res.* **2018**, *13*, 180. [CrossRef]

Journal of
Personalized Medicine

MDPI

Technical Note

Early Reduction of the Posterior Column: A Surgical Technique in AO/OTA C3 Tibial Pilon Fractures

Yanchun Gao, Hongyi Zhu, Yanjie Guo and Xingang Yu *

Department of Orthopedic Surgery, Shanghai Jiao Tong University Affiliated Sixth People's Hospital, Shanghai Jiao Tong University, Shanghai 200233, China
* Correspondence: xingangyu_sjtu@163.com

Abstract: Staged treatment for pilon fractures is widely accepted. It remains to be discussed how to reduce and fix posterior column fractures while avoiding clinical complications. We provided a staged treatment protocol with detailed surgical techniques for closed AO Foundation/Orthopaedic Trauma Association (AO/OTA) C3 tibial pilon fractures with fibular fractures. In the first stage, the internal fixation of the fibula and distal tibial posterior column is accompanied by an external fixator. After swelling, the medial and anterior columns were fixed via the posteromedial approach in the second stage. We advocate early reduction and fixation of the posterior column and lateral column. The right timing of surgery can ensure well-reduced articular surface and alignment while minimizing soft tissue complications.

Keywords: trauma; staged treatment; tibial pilon fractures

check for updates

Citation: Gao, Y.; Zhu, H.; Guo, Y.; Yu, X. Early Reduction of the Posterior Column: A Surgical Technique in AO/OTA C3 Tibial Pilon Fractures. *J. Pers. Med.* **2023**, *13*, 551. https://doi.org/10.3390/jpm13030551

Academic Editors: Markus Gesslein and Eduard Kurz

Received: 22 February 2023
Revised: 8 March 2023
Accepted: 13 March 2023
Published: 20 March 2023

1. Background

Pilon fractures, accounting for 4–10% of all tibial fractures, are a severe challenge to surgeons due to poor prognosis [1,2]. According to AO Foundation/Orthopaedic Trauma Association (AO/OTA) classification, a C3-type pilon fracture is defined as a comminuted fracture of the articular surface of the distal tibia involving the metaphysis and diaphysis. It is often accompanied by severe soft tissue injury [3]. Fibula fractures accompany about 75–90% of C3 pilon fractures. Due to axial violence, comminuted fractures often occur on the articular surface and metaphysis of the distal tibia, often accompanied by damage to the surrounding soft tissue.

The purpose of surgical treatment is to restore ankle function by fracture fixation, which acquires a high-quality reduction of articular surfaces and restoration of tibial alignment. The reduction of comminuted articular fractures places high demands on the surgeon's surgical skills. In addition, complex pilon fractures suggest high-energy injury, accompanied by increased complications such as infection, marginal necrosis, or poor wound healing [4,5].

Staged treatment for AO/OTA C3 pilon fractures has long been widely accepted [6–10]. However, the details of the surgery remain controversial, including whether to use early partial plate fixation, the choice of surgical time, choice of surgical approach, and repositioning techniques. Some surgeons advocate early fixation of the fibula for pilon fractures with fibula fractures. In contrast, others believe that the open internal fixation should be performed after the complete soft tissue detumescence [11]. Longer immobilization times provide good soft tissue detumescence, but bone resorption and osteoporosis increase the difficulty of surgical reduction [10,12]. The single anterior incision has effectively exposed and fixed the distal tibia fragment. Still, the inability to expose and reduce the posterior malleolus fragment has always been an unavoidable disadvantage [13,14]. The combined anterior and posterior approaches can effectively reduce and fix the posterior column. However, it was pointed out that excessive soft tissue dissection and double incision increased the risk of the flap and soft tissue [15,16].

This study used a staged treatment of external fixation combined with partial internal fixation to treat AO/OTA C3 pilon fractures with fibular fractures. We have gained new insights into the early fixation of the posterior column and shared surgical techniques that will contribute to clinical outcomes.

2. Methods

2.1. Preoperative Preparation

The tibiofibular X-ray, including knee and ankle joints and CT scan of the ankle joint with three-dimensional reconstruction, were completed in the emergency room to evaluate fracture line distribution, fracture displacement, and comminution. Calcaneal traction was given under local anesthesia. Ice compress and dehydration treatment were shown to reduce edema. The preoperative preparation time of the first surgery ranged from 2 to 5 days after injury (mean 2.7 days).

2.2. Surgical Technique

2.2.1. Stage I

In the stage I surgery, the patient underwent a posterolateral approach between the distal fibula and the Achilles tendon in a lateral position. The superficial peroneal and sural nerves were protected, and the fibula fracture was exposed. Soft tissues and blood clots are removed from the fibular fracture site. With the help of repositioning forceps, the fibula was anatomically reduced and fixed with an anatomical locking plate lateral to the fibula. Restoring fibula length and alignment is critical during the surgery for comminuted fibula fractures [17–19].

Exposure of the distal tibia and metaphyseal segments is very important. The conjunctions of the metaphyseal cortex determined the anatomic reduction of the posterior column. Kirschner wires temporarily fixed the posterior fragment. Subsequently, a T-type or straight locking plate was placed close to the posterior column, and short screws were used in the distal of the plates to fix the posterior fragments. Plates provided an effective buttress, and short screws were used to improve the posterior malleolus. When excessively long posterior screws break through the anterior edge of the posterior fragment, the screws will prevent the reduction during secondary surgery. Therefore, preoperative CT transverse images are measured to estimate the appropriate length of the screws in the Volkmann fragment (Figure 1). The wound was closed, and negative pressure drainage was left postoperatively. Then, the lateral calcaneus and proximal medial tibia were drilled, two 5 mm Shanz pins were inserted, respectively, and an external fixator was installed across the ankle joint. The patients were routinely treated with anti-inflammatory and detumescent therapy after the operation.

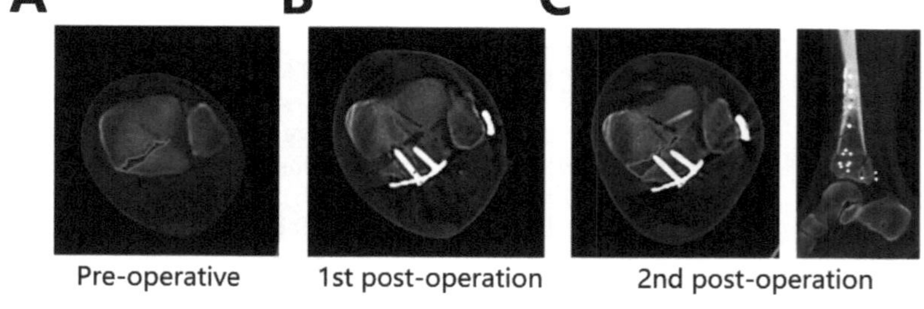

Figure 1. Inappropriate length of the screws impedes Volkmann fragment repositioning. (**A**) The CT scan shows the morphology of the preoperative pilon fracture bone fragments. (**B**) The posterior screws should be shorter than the size of the posterior ankle fragment. (**C**) CT scan of the ankle after the second surgery showed that an excessively long posterior screw impeded the reduction of the anterolateral bone fragments during the second surgery.

2.2.2. Stage II

After the soft tissue swelling had subsided entirely and skin wrinkles showed, the second phase of the operation was performed.

The external fixator was removed in a supine position three days before the second surgery. In stage II surgery, the procedure is performed using a medial incision in which the posterior edge of the tibia is past the medial ankle and extends forward (Figure 2A). The tarsal tunnel is not incised unless necessary and is protected throughout the procedure. The full-thickness skin flap was lifted over the bone, and the great saphenous vein was protected (Figure 2B). The medial fragment was lifted distally along the original fracture line, and then the medial and anterior columns were shown clearly (Figure 2C). The posterior column was reduced and fixed at the first surgery, while the mid-articular surface and anterior column were exposed under direct vision. The posterior column was used as a reference to reduce the collapsed middle articular surface and anterior column fragments. The articular surface was reduced under direct vision and temporarily fixed with point-type reduction forceps. The superior edge of the talus can also be used as a reference line for repositioning (Figure 2D). The large allogeneic bone or iliac bone was filled in the metaphyseal defect site. One to two screws were implanted from front to back above the articular surface under direct vision (Figure 2E). Subsequently, the medial column fragment was reduced and fixed with Kirschner wire, fixed by a medial anatomic plate of the distal tibia after the fracture reduction. Intraoperative fluoroscopy showed a well-reduced fracture with a flat joint surface (Figure 2F,G). The wound was flushed and closed, and drainage was left in place.

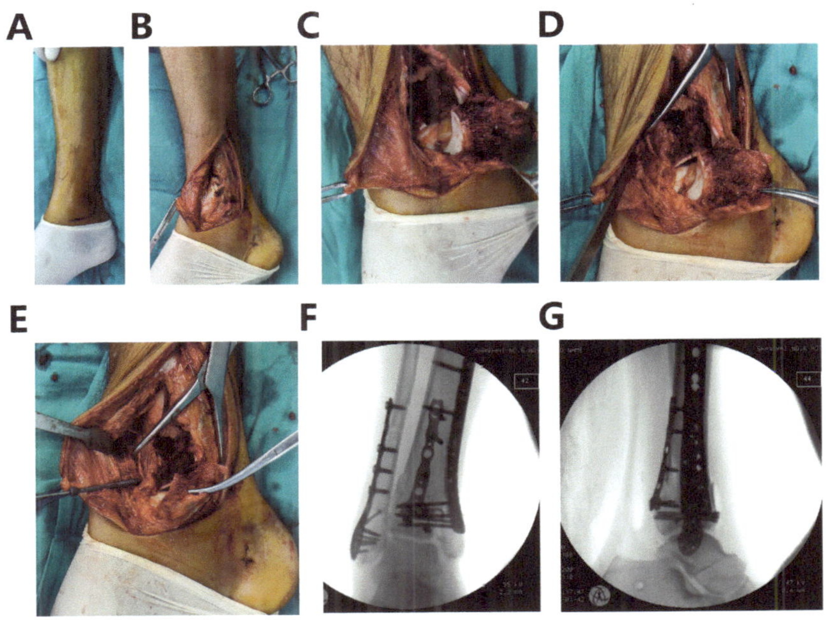

Figure 2. Intraoperative fragments and joint surface reduction procedure and intraoperative fluoroscopy. (**A**) A medial incision. (**B**) Full-thickness skin flap was lifted over the bone (**C**) The medial fragment was lifted distally along the original fracture line, and the medial and anterior columns were exposed. (**D**) The articular surface was reduced under direct vision and temporarily fixed with point-type reduction forceps. (**E**) Screws were implanted from front to back above the articular surface under direct vision. (**F**) Intraoperative fluoroscopy of anteroposterior view. (**G**) Intraoperative fluoroscopy of lateral view.

2.3. Postoperative Management

Patients were encouraged to initiate toe and ankle flexion and dorsiflexion exercises on the second day after surgery without external fixators or braces. Negative pressure suction was removed two days postoperatively. Weight-bearing walking was prohibited until radiology confirmed initial fracture healing. Partial weight-bearing walking was started gradually, 8–12 weeks postoperatively.

3. Results

From October 2016 to September 2019, 29 patients with closed complex pilon fractures combined with fibular fractures were treated in our hospital, with an average follow-up time of 25.9 months (ranging from 12 to 66 months). The average age was 54.1 years. According to Burwell–Charnley X-ray standards, the anatomical reduction was found in 21 patients, a good reduction in six patients, and a fair reduction in two patients. Postoperative imaging follow-up showed that most of the fractures healed within 2–3 months after surgery, with an average healing time of 3.38 months (ranging from 2 to 6 months). The postoperative function was assessed according to the AOFAS Ankle-Hindfoot scale. It showed that the average score was 82.4 points (range 62–94 points).

Figure 3 demonstrates a 64 year old male with an AO/OTAC3 pilon fracture who achieved good fracture reduction following our staged treatment protocol. We performed his first surgery two days after the fracture. He underwent a second surgery 19 days later. According to Burwell–Charnley's fracture reduction quality assessment, an anatomic reduction was achieved. The AOFAS score for the assessment of ankle function was 94, with a grade of Excellent.

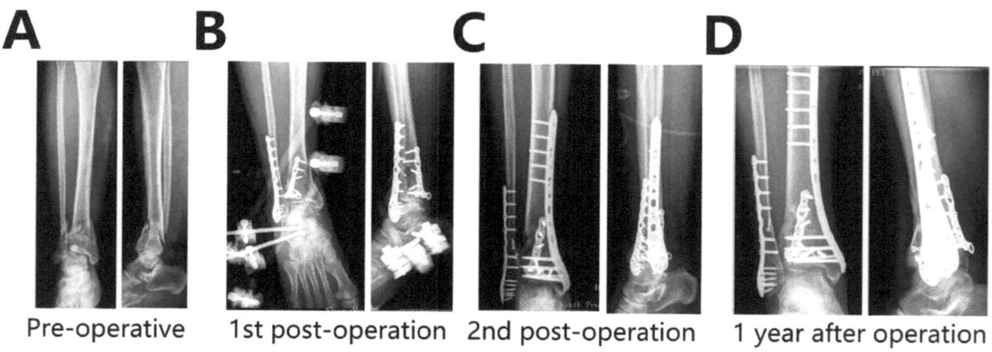

Figure 3. (**A**) Pre-operative X-ray. (**B**) The x-rays show the fixation of the posterior tibial column and fibula by a locking plate in the first surgery, combined with external fixator. (**C**) The x-ray shows the fixation of the tibia with a medial locking plate during the second surgery. (**D**) The x-ray shows the healing of the patient's fracture one year after surgery.

Figure 4 shows a CT scan of a 58 year old male before and after the first and second surgery. We can see the process of fragment reduction during the process of this staged treatment.

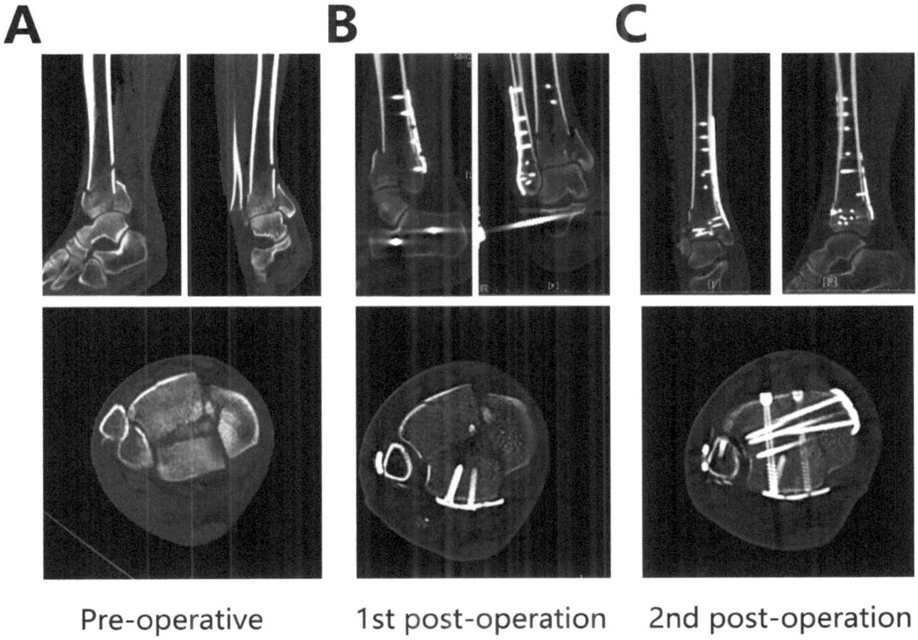

Figure 4. CT images of the patient (**A**) after the fracture (**B**) after the first surgery (**C**) after the second surgery.

4. Discussions

For AO/OTA C3 tibial pilon fractures with fibula fractures, the strategy of staged treatment has been widely accepted by most surgeons, significantly reducing the risk of soft tissue complications [20,21]. Nonetheless, the long preoperative waiting time may lead to bone loss, scarring, soft tissue contractures, increased difficulty in reducing fractures, and other complications. For the staged treatment of pilon fractures, selecting the time of the second surgery is a difficult part of the treatment decision. An excellent treatment should achieve a good reduction while avoiding soft-tissue complications, which improves fracture healing and ankle joint function.

Many researchers recommend using a temporary trans-ankle external fixator for approximately 1 to 4 weeks, followed by an open reduction and internal fixation after the surrounding soft tissues are repaired [22,23]. Despite careful soft-tissue management, the risk of wound complications was still relatively high compared with other orthopedic trauma surgeries [11,24]. Swelling reacts with a cascade of biological and biomechanical changes affecting the local circulation, inflammation, and immune reaction [25]. Ischemic necrosis, caused by the high tension of the flap or compression of displaced fracture fragments, leads to prolonged treatment or poor healing. Premature timing of the surgery often leads to difficult wound closure, poor wound healing, and muscle necrosis. In addition, scar hyperplasia, bone resorption, and callus formation can cause the loss of anatomical signs, increasing difficulties in the second surgery. Soft tissue contracture also makes it challenging for fracture reduction. These conditions suggest an early requirement for fracture reduction.

In addition to soft tissue issues, achieving anatomic reduction and fixation of the articular surface is still the key to treatment. Some studies have shown that the quality of fracture reduction may be significantly correlated with long-term functional prognosis [26,27]. The present study determined that fibular fixation was pivotal in treating pilon fractures, particularly those with severe metaphyseal comminution and joint involvement. Fibular fixation facilitated tibial reduction and helped maintain the appropriate length and alignment of the tibia [28]. Early reduction and fixation of the fibula facilitate the restoration of limb length, which provides lateral column support to the ankle fracture and contributes to the second-stage reduction of the tibial fracture. In addition, Ketz et al. proposed fibular reduction. Reduction and fixation of the fibula could restore alignment and rotation of the lateral column, simultaneously reducing the posterolateral (Volkmann) and anterolateral (Chaput) by extending the anterior and posterior tibiofibular ligaments [19]. The malreduction of the fibula will affect the reduction of the distal tibial fragments in secondary surgery. Therefore, the operator should pursue the anatomic reduction of the fibula in the first surgery to restore the length and force line of the ankle.

We want to clarify that, throughout the procedure (both the first and second stages), the articular surface is repositioned based on the principle of "back to the forward." In the first stage, the posterior column is reduced to provide the anatomical landmarks for the second stage. Based on this, the articular surfaces of the middle and anterior columns are repositioned one by one. If the fragments of the articular surface are not reduced in the first stage, the second-stage reduction becomes very difficult [13]. Ketz (2012) and Sanders (2021) reported that the loss of joint reduction was less than 2 mm in one-stage direct reduction posterior column fractures. These studies demonstrate that early removal of the posterior tibial column is reliable for restoring articular surface flattening [19,29]. For posterior column fractures with metaphyseal comminution, a pre-contorted T-shaped or straight locking plate can effectively fix the fracture, especially for shear fractures of the posterior malleolus. For posterior column fractures, the reduction of the metaphysis should precede the reduction of the posterior malleolus, which turns a C-type fracture into a B-type fracture, while reducing the limb alignment also provides a good reference mark for reducing the anterior and medial columns in the second-stage operation. We recommend an intraoperative assessment to prevent poor posterior column reduction and to apply sufficient pressure on the articular surface to ensure absolute stability of the fracture fixation.

In the treatment of complex pilon fractures, different surgical approaches have their advantages [30–32]. For AO/OTA C3 pilon fractures, single-incision surgery is insufficient for the reduction and fixation of the posterior column, and double incision shows its irreplaceable superiority in the reduction of the articular surface [8,14,16,33,34]. For dual incisions, insufficient flap width is an essential factor leading to skin necrosis, especially in patients with preoperative soft tissue contusion [35]. Even after adequate swelling and the appearance of skin wrinkling, the combined anterior–posterior approach in the treatment of pilon fractures inevitably faces the risk of difficulty in wound closure, flap necrosis, or infection [35,36]. Delayed wound closure may increase the risk of wound infection and decrease joint function [37]. In this study, staged treatment provides sufficient time to restore blood supply to the dissected tissue and skin after fixation of the posterior column by a posterolateral approach.

Many surgeons use an anterior approach to expose the anterior and part of the mid-articular surface of the tibiotalar joint [38]. However, when the anterior incision is used to reduce the fracture of the medial and anterior column, the exposure and fixation of the medial column are limited [14,39]. In our second-stage surgery, the posterior column was used as a reference to reduce the collapsed articular surface. The mid-articular surface and anterior column were exposed and fixed by a medial approach under direct vision [34]. Moreover, the medial incision ensures a sufficient flap width, which is more conducive to good medial soft tissue coverage. This may reduce the difficulty of wound closure and the possibility of skin necrosis.

Our research believes that staging surgery can ensure the healing of soft tissue. The reduction and fixation of the fibula and the posterior column of the distant tibia can effectively resist valgus stress. In contrast, the distal tibia medial locking plate supports to resist varus pressure. The anterior column was fixed by cannulated screws, effectively avoiding irritation of the nerve, tendons, and blood vessels from locking plates. Hollow screws and distal parallel screws of the medial tibia plate formed a cross-vertical fixing system. The stable fixation of the distal tibial fragments favors patients' good function through early exercises.

In recent studies, several researchers have suggested that internal fixation of fibular fractures is considered to lack statistically significant evidence in pilon fracture treatment. Despite the controversy, it helped augment external fixation and provide additional stability [40]. In 2021, a study of 76 patients by Hong C. C. et al. noted that patients without fibula fixation have more wound complications (44% vs. 25.9%, $p = 0.108$). A larger sample size may lead to statistically significant conclusions [41]. We believe that, with a large sample size cohort study, the necessity of internal fixation of fibular fractures will be reconsidered. One of the focuses of this surgical technique is to reduce wound complications. We believe that subsequent studies will prove our contention and provide guiding recommendations in the treatment of pilon fractures.

5. Conclusions

This is a technique note of the staged treatment of AO/OTA C3 type pilon fracture with fibula fractures. Early reduction and fixation of the lateral and posterior columns could provide a well-repositioned articular surface and alignment. The second-stage posteromedial incision gives a good operative view, while reducing the medial and anterior lateral columns. It ensures early postoperative functional exercise and fewer risks of complications.

Author Contributions: X.Y. and Y.G. (Yanchun Gao) designed the study. Y.G. (Yanchun Gao) wrote the article. H.Z. and X.Y. reviewed the report. Y.G. (Yanjie Guo) collected and analyzed the data. Y.G. (Yanchun Gao) and H.Z. contributed equally to this work; Y.G. (Yanjie Guo) and X.Y. contributed equally to this work. All authors have read and agreed to the published version of the manuscript.

Funding: This work was funded by the Interdisciplinary Program of Shanghai Jiao Tong University (YG2021QN98).

Institutional Review Board Statement: The Medical Ethics Committee approved this study of Shanghai Jiao Tong University Affiliated Sixth People's Hospital. Informed consent was obtained from all individual participants included in the study. All procedures performed in studies involving human participants were by the ethical standards of the institutional research committee and with the 1964 Helsinki Declaration and its later amendments or comparable ethical standards.

Informed Consent Statement: Not applicable.

Data Availability Statement: The final dataset will be available from the corresponding author.

Conflicts of Interest: The authors declare that they have no competing interests.

References

1. Zelle, B.A.; Dang, K.H.; Ornell, S.S. High-energy tibial pilon fractures: An instructional review. *Int. Orthop.* **2019**, *43*, 1939–1950. [CrossRef]
2. Kottmeier, S.A.; Madison, R.D.; Divaris, N. Pilon Fracture: Preventing Complications. *J. Am. Acad. Orthop. Surg.* **2018**, *26*, 640–651. [CrossRef]
3. Marsh, J.L.; Slongo, T.F.; Agel, J.; Broderick, J.S.; Creevey, W.; DeCoster, T.A.; Prokuski, L.; Sirkin, M.S.; Ziran, B.; Henley, B.; et al. Fracture and dislocation classification compendium—2007: Orthopaedic Trauma Association classification, database and outcomes committee. *J. Orthop. Trauma* **2007**, *21*, S1–S6. [CrossRef]
4. Chaudhry, Y.P.; Papadelis, E.; Hayes, H.; Stahel, P.F.; Hasenboehler, E.A. Fusion versus fixation in complex comminuted C3-type tibial pilon fractures: A systematic review. *Patient Saf. Surg.* **2021**, *15*, 35. [CrossRef] [PubMed]
5. Mabvuure, N.T.; Pinto-Lopes, R.; Sierakowski, A. Management of intraarticular proximal interphalangeal joint fracture-dislocations and pilon fractures with the Ligamentotaxor(R) device. *Arch. Orthop. Trauma Surg.* **2020**, *140*, 1133–1141. [CrossRef] [PubMed]
6. Patterson, M.J.; Cole, J.D. Two-staged delayed open reduction and internal fixation of severe pilon fractures. *J. Orthop. Trauma* **1999**, *13*, 85–91. [CrossRef]

7. Sirkin, M.; Sanders, R.; DiPasquale, T.; Herscovici, D., Jr. A staged protocol for soft tissue management in the treatment of complex pilon fractures. *J. Orthop. Trauma* **1999**, *13*, 78–84. [CrossRef]
8. Chen, H.; Cui, X.; Ma, B.; Rui, Y.; Li, H. Staged procedure protocol based on the four-column concept in the treatment of AO/OTA type 43-C3.3 pilon fractures. *J. Int. Med. Res.* **2019**, *47*, 2045–2055. [CrossRef] [PubMed]
9. Potter, J.M.; van der Vliet, Q.M.J.; Esposito, J.G.; McTague, M.F.; Weaver, M.; Heng, M. Is the proximity of external fixator pins to eventual definitive fixation implants related to the risk of deep infection in the staged management of tibial pilon fractures? *Injury* **2019**, *50*, 2103–2107. [CrossRef] [PubMed]
10. Deivaraju, C.; Vlasak, R.; Sadasivan, K. Staged treatment of pilon fractures. *J. Orthop.* **2015**, *12* (Suppl. S1), S1–S6. [CrossRef]
11. Halvachizadeh, S.; Klingebiel, F.K.L.; Pfeifer, R.; Gosteli, M.; Schuerle, S.; Cinelli, P.; Zelle, B.A.; Pape, H.C. The local soft tissue status and the prediction of local complications following fractures of the ankle region. *Injury* **2022**, *53*, 1789–1795. [CrossRef]
12. Olson, J.J.; Anand, K.; von Keudell, A.; Esposito, J.G.; Rodriguez, E.K.; Smith, R.M.; Weaver, M.J. Judicious Use of Early Fixation of Closed, Complete Articular Pilon Fractures Is Not Associated With an Increased Risk of Deep Infection or Wound Complications. *J. Orthop. Trauma* **2021**, *35*, 300–307. [CrossRef] [PubMed]
13. Guan, J.; Huang, M.; Wang, Q.; Chen, Y.; Wang, L. Treatment of AO/OTA 43-C3 Pilon Fracture: Be Aware of Posterior Column Malreduction. *Biomed. Res. Int.* **2019**, *2019*, 4265782. [CrossRef]
14. Malige, A.; Yeazell, S.; Nwachuku, C. Surgical fixation of pilon injuries: A comparison of the anterolateral and posterolateral approach. *Arch. Orthop. Trauma Surg.* **2019**, *139*, 1179–1185. [CrossRef] [PubMed]
15. Zhao, H.Y. Delayed combined approach with three plates internal fixation for Pilon fracture with fibula fracture. *Zhongguo Gu Shang* **2020**, *33*, 257–260. [CrossRef]
16. Huang, M.; Wang, Q.; Guan, J.; Liu, K.; Chen, Y.; Wang, L. Tips and Tricks in surgical reduction of the posterior column of AO/OTA C3 pilon fractures. *BMC Musculoskelet. Disord.* **2022**, *23*, 2. [CrossRef] [PubMed]
17. Dunbar, R.P.; Barei, D.P.; Kubiak, E.N.; Nork, S.E.; Henley, M.B. Early limited internal fixation of diaphyseal extensions in select pilon fractures: Upgrading AO/OTA type C fractures to AO/OTA type B. *J. Orthop. Trauma* **2008**, *22*, 426–429. [CrossRef] [PubMed]
18. Barei, D.P.; Nork, S.E.; Bellabarba, C.; Sangeorzan, B.J. Is the absence of an ipsilateral fibular fracture predictive of increased radiographic tibial pilon fracture severity? *J. Orthop. Trauma* **2006**, *20*, 6–10. [CrossRef]
19. Ketz, J.; Sanders, R. Staged posterior tibial plating for the treatment of Orthopaedic Trauma Association 43C2 and 43C3 tibial pilon fractures. *J. Orthop. Trauma* **2012**, *26*, 341–347. [CrossRef] [PubMed]
20. Davidovitch, R.I.; Elkhechen, R.J.; Romo, S.; Walsh, M.; Egol, K.A. Open reduction with internal fixation versus limited internal fixation and external fixation for high grade pilon fractures (OTA type 43C). *Foot Ankle Int.* **2011**, *32*, 955–961. [CrossRef] [PubMed]
21. Wang, Z.; Qu, W.; Liu, T.; Zhou, Z.; Zhao, Z.; Wang, D.; Cheng, L. A Two-Stage Protocol With Vacuum Sealing Drainage for the Treatment of Type C Pilon Fractures. *J. Foot Ankle Surg.* **2016**, *55*, 1117–1120. [CrossRef]
22. Cui, X.; Chen, H.; Rui, Y.; Niu, Y.; Li, H. Two-stage open reduction and internal fixation versus limited internal fixation combined with external fixation: A meta-analysis of postoperative complications in patients with severe Pilon fractures. *J. Int. Med. Res.* **2018**, *46*, 2525–2536. [CrossRef]
23. Nielsen, P.J.; Grossman, L.S.; Siebler, J.C.; Lyden, E.R.; Reed, L.K.; Mormino, M.A. Is It Safe to Prep the External Fixator In Situ During Second-Stage Pilon Surgical Treatment? *J. Orthop. Trauma* **2018**, *32*, e102–e105. [CrossRef]
24. Dumbre Patil Shivajirao, S.; Jadhav, N.; Dumbre Patil Sampat, V. Management of a Rare Case of Closed Pilon Fracture With Fibula Displaced in the Foot. *J. Foot Ankle Surg.* **2022**, *61*, e34–e69. [CrossRef] [PubMed]
25. Halvachizadeh, S.; Pfeifer, R.; Kalbas, Y.; Schuerle, S.; Cinelli, P.; Pape, H.C. Assessment of alternative techniques to quantify the effect of injury on soft tissue in closed ankle and pilon fractures. *PLoS ONE* **2022**, *17*, e0268359. [CrossRef] [PubMed]
26. Sousa, A.; Carvalho, J.; Amorim, J.; Rodrigues-Pinto, R. Incidence and clinical results of tendinous injuries in calcaneus and pilon fractures. *Arch. Orthop. Trauma Surg.* **2022**, *143*, 359–363. [CrossRef]
27. Middleton, S.D.; Guy, P.; Roffey, D.M.; Broekhuyse, H.M.; O'Brien, P.J.; Lefaivre, K.A. Long term trajectory of recovery following pilon fracture fixation. *J. Orthop. Trauma* **2021**, *36*, e250–e254. [CrossRef] [PubMed]
28. Huang, C.W.; Wu, W.T.; Yu, T.C.; Chen, I.H.; Wang, J.H.; Yeh, K.T. Retrograde Intramedullary Kirschner Wire Fixation as an Alternative for Treating Distal Fibular Shaft Fractures Combined with Distal Tibial Pilon Fractures. *J. Pers. Med.* **2022**, *12*, 1124. [CrossRef]
29. Sanders, D. Acute Arthrodesis of Ipsilateral Pilon and Talus Fracture: A Case Report. *JBJS Case Connect.* **2021**, *11*, e20.00175. [CrossRef]
30. Assal, M.; Ray, A.; Fasel, J.H.; Stern, R. A modified posteromedial approach combined with extensile anterior for the treatment of complex tibial pilon fractures (AO/OTA 43-C). *J. Orthop. Trauma* **2014**, *28*, e138–e145. [CrossRef]
31. Zhao, Y.; Wu, J.; Wei, S.; Xu, F.; Kong, C.; Zhi, X.; Huang, M.; Cai, X. Surgical approach strategies for open reduction internal fixation of closed complex tibial Pilon fractures based on axial CT scans. *J. Orthop. Surg. Res.* **2020**, *15*, 283. [CrossRef]
32. Labronici, P.J.; Junior, A.F.M.; da Silva, A.A.M.; da Silva, P.M.G.; Serra, M.; da Silva, A.F.; Pires, R.E.S.; Damas Dos Santos, A. CT mapping for complex tibial pilon fractures: Understanding the injury pattern and its relation to the approach choice. *Injury* **2021**, *52* (Suppl. S3), S70–S76. [CrossRef]
33. Jiang, Z.; Zhang, C.; Qin, J.J.; Wang, G.D.; Wang, H.S. Posterior pilon fracture treated by opening the fibula fracture gap. *J. Orthop. Surg. Res.* **2022**, *17*, 214. [CrossRef] [PubMed]

34. Campbell, S.T.; DeBaun, M.R.; Kleweno, C.P.; Nork, S.E. Simultaneous Posterolateral and Posteromedial Approaches for Fractures of the Entire Posterior Tibial Plafond: A Safe Technique for Effective Reduction and Fixation. *J. Orthop. Trauma* **2022**, *36*, 49–53. [CrossRef] [PubMed]
35. Chan, D.S.; Balthrop, P.M.; White, B.; Glassman, D.; Sanders, R.W. Does a Staged Posterior Approach Have a Negative Effect on OTA 43C Fracture Outcomes? *J. Orthop. Trauma* **2017**, *31*, 90–94. [CrossRef] [PubMed]
36. Leone, V.J.; Ruland, R.T.; Meinhard, B.P. The management of the soft tissues in pilon fractures. *Clin. Orthop. Relat. Res.* **1993**, *292*, 315–320. [CrossRef]
37. Singh, N.; Bluman, E.; Starnes, B.; Andersen, C. Dynamic wound closure for decompressive leg fasciotomy wounds. *Am. Surg.* **2008**, *74*, 217–220. [CrossRef] [PubMed]
38. Olszewski, N.; Tornetta, P., 3rd. Open Reduction and Internal Fixation of a Partial Articular (OTA-43B) Pilon Fracture Through a Direct Anterior Approach. *J. Orthop. Trauma* **2022**, *36* (Suppl. S3), S25–S26. [CrossRef] [PubMed]
39. Gulbrandsen, T.R.; Hulick, R.M.; Polk, A.J.; Weldy, J.M.; Howell, K.L.; Spitler, C.A.; Crist, B.D. Does surgical approach affect sagittal plane alignment and pilon fracture outcomes? *Injury* **2020**, *51*, 750–758. [CrossRef]
40. Kurylo, J.C.; Datta, N.; Iskander, K.N.; Tornetta, P., 3rd. Does the Fibula Need to be Fixed in Complex Pilon Fractures? *J. Orthop. Trauma* **2015**, *29*, 424–427. [CrossRef]
41. Hong, C.C.; Tan, S.H.S.; Saha, S.; Pearce, C.J. Fibula fixation in the treatment of tibial pilon fractures—Is it really necessary? *Foot Ankle Surg.* **2022**, *28*, 891–897. [CrossRef] [PubMed]

Journal of *Personalized Medicine*

MDPI

Article

The Impact of Anatomical Predisposition and Mechanism of Trauma on Dislocation of the Patella: A Retrospective Analysis of 104 Cases

Ilona Schubert [1,*], Patrick Morris [1], Jörg Dickschas [1,2] and Peter C. Strohm [1,3]

1 Department of Orthopedics and Trauma Surgery, Sozialstiftung Bamberg, 96049 Bamberg, Germany
2 Department of Orthopedic and Trauma Surgery, Friedrich-Alexander-University of Erlangen, 91054 Erlangen, Germany
3 Department of Orthopedic and Trauma Surgery, Albert-Ludwig-University of Freiburg, 79085 Freiburg im Breisgau, Germany
* Correspondence: ilona.schubert@sozialstiftung-bamberg.de

Abstract: Background: The aim of this study was to determine whether traumatic dislocation of the patella is provoked by the presence of predisposing factors and examine the role of the mechanism of injury. Methods: Cases diagnosed with dislocation of the patella and covered by the workers' compensation program were identified and classified as traumatic based on insurance regulations. We examined predisposing factors (e.g., frontal axis, torsional deviation, trochlear dysplasia, patella alta) in case groups based on age at dislocation and trauma mechanism. Retrospective cohort study, level of evidence III. Results: Our sample size comprised 104 cases, consisting of 54 children and 50 adults. The most common mechanism of injury in children and adults was rotational trauma. Only 20% of the children and 21% of the adults exhibited no relevant predisposing factors. Group specifically, falls accounted for the highest number of cases exhibiting none of the defined anatomical predisposing factors. Children are more frequently affected by predisposition-related dislocations than adults. Conclusion: The proportion of predispositions is high. A fall, direct impact, or rotational trauma can be viewed as an adequate mechanism of trauma. For successful treatment, it is paramount to analyze the exact mechanism of the trauma and address any underlying predispositions.

Keywords: dislocation of the patella; predisposition; patellofemoral joint; patellofemoral dysbalance

Citation: Schubert, I.; Morris, P.; Dickschas, J.; Strohm, P.C. The Impact of Anatomical Predisposition and Mechanism of Trauma on Dislocation of the Patella: A Retrospective Analysis of 104 Cases. *J. Pers. Med.* **2023**, *13*, 84. https://doi.org/10.3390/jpm13010084

Academic Editors: Markus Gesslein and Eduard Kurz

Received: 3 December 2022
Revised: 22 December 2022
Accepted: 26 December 2022
Published: 29 December 2022

1. Introduction

The instability of the patellofemoral joint has gained great attention from clinicians and researchers alike, as new revelations in biomechanics and treatment options surface. We often distinguish traumatic dislocation of the patella from habitual dislocation. For a traumatic dislocation to occur, there has to have been an adequate traumatic event, such as a direct medial impact to the patella. Habitual dislocations of the patella arise during the physiological motion of the knee joint and commonly occur repeatedly. This is due to inherent joint instability or secondary joint instability acquired through not sufficiently treated post-traumatic injuries. The patellofemoral joint is guided and correctly aligned by the interaction of active, passive, and static stabilizers. In the complete extension of the knee joint, the medial ligamentary complex acts as a passive stabilizer. At around 30° knee flexion, the patella seats itself in the trochlear groove and is supported, statically, by the osseous structures. During higher degrees of flexion, the quadriceps muscles act as active stabilizers. Anatomical abnormalities categorized as predispositions are those that promote a deviation from the optimum force vectors, thereby resulting in a lateralization of the patella or incongruence of the patella and trochlea. Known predisposing factors are valgus deformity, increased internal rotation of the femur, increased external rotation of the tibia, elevated TTTG-distance (tibial-tuberosity to trochlear groove distance), high-riding patella, or dysplasia of the trochlea or patella itself. Muscular hypotrophy and laxity of the ligamentary complex can also act as a predisposing factor. The evaluation of predisposing factors helps us to identify patients at risk for dislocation events as well as derive definitive treatment options in order to prevent the development of recurring dislocations and subsequent joint damage [1–10].

Patients who exhibit a high degree of predisposing factors have been shown to often suffer their first dislocation event as young adolescents [6,9,11]. However, the presence of a predisposition does not lead to a dislocation in every case. Depending on the severity of the predisposition, it can be difficult to distinguish whether the cause of a dislocation event is of traumatic nature or can be attributed to a predisposition. Therefore, the mechanism of trauma should be included in the process of patient assessment. A certain treatment regimen is then chosen based on the resulting injuries incurred by the dislocation event and the underlying risk factors for a dislocation to reoccur.

In the event of a first-time dislocation and lack of traumatic structural damage (e.g., flake fracture), non-surgical treatment options are usually chosen [6,12–15]. Surgical treatment is often indicated in the case of recurring dislocations of the patella after a thorough workup of predisposing factors has been performed. If the mechanism of trauma is inadequate and a relevant predisposition is present, surgical therapy should be considered even after a first-time dislocation event [6,14,16,17].

2. Aim of the Study

The aim of this study was to answer the question of whether the analysis of trauma mechanisms and certain predisposing factors in patellar dislocations can provide information about the main cause of a dislocation event. In analyzing the cause of patellar dislocations, we also wanted to examine the extent of differences between adult and pediatric patellar dislocations.

3. Methods

We included all patients who were treated in our clinic from January 2000 until April 2020 that were recorded with a diagnosis of "dislocation of the patella" on the standard form F1000 of the German statutory accident insurance. The cases that were not dislocations, after all, were excluded.

For data acquisition purposes we also utilized available digital imaging (X-ray, MRI, torsional-CT scans), as well as treatment documentation (out-patient documentation, discharge papers, surgical protocol).

Patients were grouped according to age with a cut-off at 18 years (smaller or equal to 18 and older than 18 years).

We evaluated general descriptive demographical aspects (age at trauma, gender), general information concerning the accident (mechanism of trauma, date of the traumatic event), and information concerning treatment (in-patient, total time of hospital stay, type of therapy, surgical technique chosen, time passed between trauma and surgical treatment). The data were obtained from the available form F1000 or the treatment documentation. For ease of systematical evaluation, we grouped the circumstances of the accidents in the subgroups direct impact, insufficient trauma, fall, rotational trauma, and unknown mechanism of trauma.

Furthermore, we evaluated anatomical aspects. The affected side, first-time dislocation or recurring, frontal axis, femoral and tibial torsion, TTTG value, height of the patella, shape of the trochlea according to Dejour [18], and lateral trochlear inclination angle. Frontal axis measurements were performed on full-length standing AP radiographs, and torsional angles were measured in CT scans according to the method of Waidelich [19]. Patella height was determined via the Insall-Salvati-Index [20] and Caton-Dechamps-Index [21]. Taking into consideration both methods of measurement, the cases were categorized as normal, normal to high-riding, normal to low-lying, high-riding, and low-lying. The shape of the trochlea was either classified as "normal" or categorized according to the Dejour subtypes A-D [18] by means of axial MRI scans. Additionally, we measured the lateral trochlear inclination, i.e., the resulting angle by drawing a tangent line along the lateral trochlear facet and the posterior condylar tangential line, also by means of axial MRI scans.

To analyze the relevance of predisposing factors, we determined the proportion of "knee-healthy" patients in each category with respect to the mechanism of trauma. Those labeled as "knee-healthy" exhibited measurement values that lie outside of the range of those patients who, in our clinic, are advised to undergo surgical treatment after having suffered a dislocated patella. Subjects were ruled out from the "knee-healthy" category if they exhibited a valgus deformity of greater or equal to 4 degrees, a femoral torsion of less or equal to -34 degrees, a tibial torsion of greater than or equal to 45 degrees, trochlear dysplasia Type D according to Dejour [18], a TTTG value of greater or equal to 16 mm or a high-riding patella. We excluded those subjects from our study group whose frontal axis, torsional values, and TTTG value could not be determined.

The descriptive data were recorded and analyzed with Microsoft Excel 365®. Descriptive analysis was performed primarily for both age groups, as well as in trauma category-associated subgroups. When numerical data was recorded, we specified the median and range [smallest value, largest value].

4. Results

Within the population of 18 years of age and younger, we were able to identify 54 events of a dislocated patella on 51 knees of 50 patients. Within the group older than 18 years of age we identified 50 events of a dislocated patella on 50 knees in 49 patients.

Due to inconsistent digital documentation practices, 11 cases from the adult group were derived from the years 2000 until 2010, and 39 cases from the years 2011 until 2020. In the pediatric subgroup 16 cases were derived from the years 2003 until 2010, and 38 cases from more recent years.

4.1. Adults

4.1.1. Demographics

The population consisted of 36 males and 14 females. The median age at dislocation was 28 [19; 48] years. Cases consisted of 25 dislocations on the left knee and 25 dislocations of the patella on the right knee.

4.1.2. Geometric Anatomical Data

Regarding the frontal axis, 20% presented a valgus configuration of 3 degrees [0°; 10°]. A similar number of patients (22%) presented with a varus configuration of 1 degree [0°; 4°]. In 22% of cases, we witnessed a zero-degree frontal axis, whereas no definitive conclusion on the frontal axis was reached in 36% of cases.

We found the median femoral torsion to be −24 degrees [−45°; −1°] and the tibial torsion to be 37 degrees [19°; 51°]. The original publication by Waidelich [19] defined the physiological range in patients older than 18 years of age as -20.4 ± 9.0 degrees femoral torsion and 33.1 ± 8.0 degrees tibial torsion. Therefore, our study group's median values are within the physiological range. We saw an increased tibial torsion (threshold value −24°) in 24% of cases and an increased tibial torsion (threshold value 35°) in 28% of cases.

The median TTTG value was 18 mm [8 mm; 35 mm] and was greater than 15 mm in 54% of cases. Surgical treatment is commonly recommended after a dislocation event and a present TTTG value of greater than 15 mm [14,22].

Examining subjects for a high-riding patella, we calculated a median Insall-Salvati-Index of 1.2 [0.7; 1.5] and median Caton-Dechamps-Index of 1.2 [0.8; 1.5]. A physiological range for the Insall-Salvati-Index is defined as 0.8–1.2 [20], and for the Caton-Dechamps-Index 0.6–1.2 [21]. We found 16% to be categorized as a normal to high-riding patella and 28% as high-riding. One case was found to be normal to low-lying, whereas no cases were low-lying and 46% were defined to be normal.

The trochlear shape was defined as physiological in 42% of cases, while 22% of cases could not be classified due to a lack of data. A morphological deformity (type B 56%, type

C 33%, type D 11%) was displayed in 36% of cases Lateral trochlear inclination serves as a tool to quantify the slope of the trochlea and is also seen as a predisposing factor for the dislocation of the patella. The pathological threshold value is specified at 11 degrees [22,23]. The median in the adult group was 16 degrees [7°; 24°]. Values of less than 11 degrees were shown in 16% of cases.

4.1.3. Effects of Trauma

The adult group consisted of 33 first-time and 17 recurrent dislocations. Injury to the medial patellofemoral ligament (MPFL) resulted in a complete rupture in 72%, and partial rupture or bony avulsion in 14% of cases. Due to a lack of data, no conclusion on MPFL injury could be drawn in 14% of cases. Flake fractures were present in 34% of cases. Bone bruise of the lateral femoral condyle was seen in 84% of cases. In 16%, no statement could be made regarding the aforementioned injury, while the absence of this pathognomonic phenomenon was not described in any case.

4.1.4. Treatment

In 32% of cases, non-surgical treatment was initiated, of which 10% ultimately ended up undergoing surgical treatment. Surgical treatment was described in a total of 62% of cases. In 16%, it was not possible to assign a treatment option due to a lack of data. Of the first-time dislocations, 52% were treated surgically and 33% non-surgically (no assignment possible in 15% of cases); of the recurrent dislocations, 53% were treated surgically and 29% non-surgically (no assignment possible in 18% of cases). The median duration from the time the injury was incurred to surgical treatment was 12 [1; 1100] days. In subgroup analysis, the median time to surgery, including those who primarily received non-surgical treatment, was eight [1; 137] days for primary dislocations and 24 [1; 1100] days for recurrent dislocations. The median duration of in-patient stay was three [0; 20] days, and 16 cases were treated as outpatients.

Regarding the chosen surgical treatment, reconstruction of the medial patellofemoral ligament was the most common (55%). In the case of osseous deformity correction, the following were described: 19% of cases received an osteotomy of the tibial tuberosity, a femoral or tibial osteotomy was performed in 10%, and a trochleoplasty was not documented in any case. Reattachment of a flake fracture or a bony avulsion of the MPFL was performed in 19% of cases.

4.2. *Children and Adolescents*

4.2.1. Demographics

The population consisted of 23 females and 31 males. The median age was 14 [9; 18] years. Cases consisted of 26 dislocations on the left and 28 dislocations of the patella on the right knee.

4.2.2. Geometric Anatomical Data

Regarding the frontal axis, 39% of cases showed a valgus configuration of 3 degrees [2°; 8°]. 12% of cases presented with a median varus configuration of 2 degrees [1°; 4°], 12% of cases had a straight leg axis and in 37% no conclusive statement could be made regarding the frontal axis.

We found the median femoral torsion to be −26 degrees [−51°; −12°] and median tibial torsion to be 32 degrees [17°; 48°]. We saw an increased femoral internal torsion (threshold value −24°) in 25% of cases and increased tibial external torsion (threshold value 35°) in 18% of cases. The median TTTG value was 17 mm [8 mm; 24 mm] and exceeded 15 mm in 49% of cases.

Examining patellar height, the Insall-Salvati index yielded a value of 1.3 [0.8; 2], and the Caton-Dechamps index, a value of 1.3 [0.5; 1.7]. These indices demonstrate good reliability for children despite ongoing growth [24,25]. Again, taking into consideration both methods of measurement, we found 22% to be categorized as normal to high-riding patella and 41% as high-riding. Only one case was found to be normal to low-lying and no case was low-lying. A normal height was shown in 24% of cases.

The trochlea showed a normal configuration in 20% of cases, in 23% no trochlea shape could be determined due to a lack of data. In 57% of cases a pathological trochlea configuration was found (type B 38%, type C 41%, type D 21%). The median lateral trochlear inclination angle was 12 degrees [4°; 21°], and 33% of cases showed values less than 11 degrees.

4.2.3. Effects of Trauma

The adolescent group consisted of 42 first-time dislocations and 12 recurrent dislocations. Complete rupture of the medial patellofemoral ligament occurred in 67% of cases and partial damage or elongation in 20%. A flake fracture or bony avulsion was described in 41% of cases. Bone bruise of the lateral femoral condyle was seen in 76% of cases. In 15%, no statement could be made regarding the aforementioned injury due to lack of data, while the absence of this pathognomonic phenomenon was described in 9%.

4.2.4. Treatment

In the pediatric group, 26% of the patella dislocations were treated non-surgically and 69% were treated surgically, in 5% no allocation to primary treatment groups could be made due to a lack of data. Considering first-time dislocation events, surgery was performed in 67% of cases, and non-surgical treatment was chosen in 29% (no assignment to treatment groups possible in 4% of cases); of the recurrent dislocations, 75% of cases were treated surgically and 17% of cases were treated non-surgically (no assignment possible in 8%). The median duration from the time the injury was incurred to surgical treatment was 12 [1; 214] days, in the subgroup of first-time dislocations, it was only eight [1; 34] days, while in the subgroup of recurring dislocations it was 15 [5; 214] days. The median duration of in-patient stay was three [0; 21] days, of which 12 cases only required outpatient treatment and thus accounted for a stay of zero days.

Regarding the surgical treatment options, addressing the medial ligamentary complex was the most common in 65% of cases. In 30% of cases among older adolescents, an autologous augmentation of the MPFL was performed. In the case of osseous deformity correction, the following were described: osteotomy of the tibial tuberosity in 22%, growth control as temporary hemiepiphysiodesis in 16%, reattachment of flake fractures or bony avulsions in 24%. Trochleoplasty was performed in one case of a 16-year-old male patient and recommended in one other case.

4.3. *Categories of Trauma Mechanisms*

When considering different mechanisms of trauma that cause an event of a dislocated patella, some seem more plausible than others. In order to draw conclusions on the extent to which "traumatic" patellar dislocations are triggered by predisposing factors, the raw data was broken down by accident category and the proportion of "healthy knees" in each category was determined (see Table 1, and Figure 1). Within the entire adult cohort, the proportion of "knee-healthy" patients was 21%; within the pediatric cohort, it was 20%. If the cutoff TTTG value is raised from ≥16 mm to ≥19 mm, the ratio increases to 28% in the adult group and to 24% in the pediatric group.

Table 1. The proportion of different predisposing factors and "healthy knees" in the individual trauma categories and age groups.

	Rotational Trauma		Direct Impact		Fall		Insufficient Trauma	
	Pediatric Group	Adult Group	Pediatric Group	Adult Group	Pediatric Group	Adult Group	Pediatric Group	Adult Group
1-9 median age (years)	14 [9; 18]	28 [19; 45]	14 [11; 17]	23 [19; 30]	14 [10; 18]	44 [32; 48]	17 [13; 18]	26 [20; 40]
1-9 proportion of valgus deviation (%)	32	23	33	40	57	/	29	10
1-9 median tigh torsion (°)	−29 [−44; −12]	−28 [−45; −1]	−27 [−51; −22]	−24 [−31; −20]	−22 [−25; −19]	/	−24 [−46; −13]	−21 [−33; −18]
1-9 median lower leg torsion (°)	30 [23; 48]	35 [28; 40]	36 [17; 46]	38 [27; 51]	23 [17; 38]	/	33 [28; 40]	39 [19; 42]
1-9 TTTG value (mm)	17 [9; 24]	18 [10; 27]	17 [8; 21]	16 [11; 35]	14 [8; 24]	19 [17; 22]	16 [10; 23]	18 [8; 22]
1-9 median lateral trochlear inclination (°)	14 [8; 21]	16 [7; 20]	13 [9; 18]	19 [17; 24]	17 [4; 20]	11 [9; 15]	9 [7; 16]	18 [9; 24]
1-9 ratio of trochlea dysplasia	50	38	78	0	43	75	50	20
1-9 patella height: proportion of normal to high-riding and high-riding (%)	64	42	89	60	43	25	57	70
1-9 proportion of "healthy knees" (%)	16	25	13	20	60	33	5	13

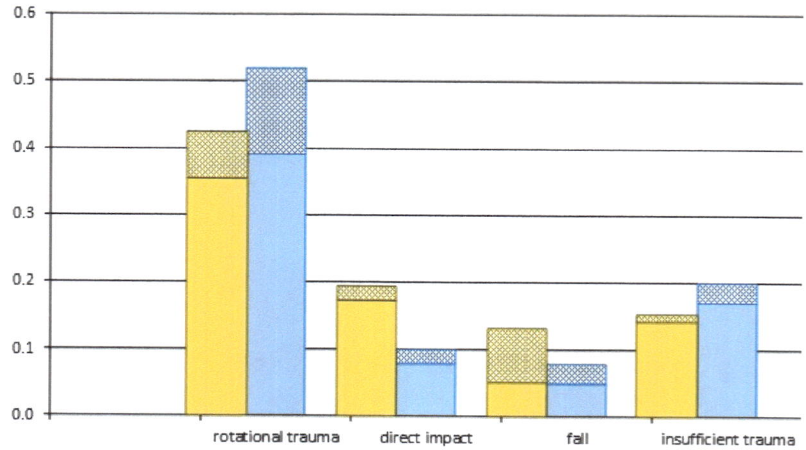

Figure 1. The distribution in terms of trauma categories: Beige columns represent the pediatric study population, blue columns the adult population. In each column, the proportion of "healthy knees" is hatched.

4.3.1. Rotational Trauma

Rotational trauma was the most common trauma entity, accounting for 42% in the pediatric population (first-time dislocation in 78%; recurrent dislocation in 22%) and 52% in the adult cohort (first-time dislocation in 73%; recurrent dislocation in 27%). The analysis for predisposing factors in this trauma category is shown in Table 1.

4.3.2. Direct Impact

Direct impact to the patella caused a dislocation event in 10% within the adult cohort and in 19% within the pediatric cohort (first-time dislocation in 80% and recurrent dislocation in 20% of cases in both age groups). In this trauma category, 20% of adults and 13% of children were "knee-healthy". In the adult population, the trochlear shape was defined as normal in 80%, while no assignment regarding the trochlea shape was possible in 20%. See Table 1.

4.3.3. Fall

13% of children and 8% of adults were assigned to this category. Falls resulted in a first-time dislocation event in 100% of both age groups. There were a significant number of cases within the adult group where a numerical value for the frontal axis and torsional values could not be determined due to a lack of digital imaging studies and/or lack of documentation. Therefore, we excluded the calculation of median torsional values as well as the percentage of valgus deformities in order to prevent bias. See Table 1.

4.3.4. Insufficient Trauma

15% of pediatric dislocations and 20% of adult dislocations were attributed to this category. This trauma category showed the highest rate of recurrent dislocations (50% of cases within the children cohort and 90% of cases within the adult cohort). The ratio of patellofemoral abnormalities was high and the proportion of "healthy knees" was lowest, at 13% in adult cases and 5% in pediatric cases. See Table 1.

5. Discussion

Rates for recurrent events after a first-time dislocation event vary among different authors [6,17]. The risk of suffering another dislocation is higher after having suffered a prior dislocation, especially when the first-time dislocation occurred at a young age [5,14,26]. The goal of successful treatment, on the one hand, is the adequate treatment of the acute trauma effects such as osteochondral damage, on the other hand, the (re-)establishment of stable joint integrity. The consequences of recurrent dislocations are far-reaching, ranging from functional limitations attributed to instability, persistent pain, and reduced stress capacity of the joint, to morphological damage such as cartilage-bone pathologies with increased risk of degenerative joint disease and reduced quality of life [6,12,27].

Traumatic first-time dislocations are usually treated non-surgically unless relevant osteochondral flakes are present [6,12–15]. However, when is the cause of a dislocation event considered traumatic? If adequate trauma cannot be assumed, the cause must be sought in potential physical abnormalities. A concept of treatment can only be causally determined after a thorough evaluation of all potential underlying factors influencing the dislocation. Germany's statutory accident insurance will cover an injury, including those with pre-existing conditions, if a certain traumatic event can, beyond a reasonable doubt, be legally accepted as the cause. Only 21% of adult knees and 20% of children's knees in our cohort showed an absence of relevant pathological deviations. Overall, our study population of traumatic dislocations showed a large proportion of predisposing factors.

A valgus configuration was a frequent predisposing factor with 20% in the adult cohort and almost twice as many in the pediatric cohort. The median values, however, did not exceed our correction limit of four degrees. There are currently no clear recommendations on the degree of axis deviation at which axis correction should be performed for patella maltracking [28], but frontal axis deviation seems to only play a minor role in traumatic dislocations.

Median torsional values also only deviated moderately from the norm for both cohorts. Considering the above-mentioned threshold values, however, clear pathologies were found in individual cases. Both groups had a similar number of cases that exceeded the threshold value for femoral torsion, while the adult population had a higher number of cases above the threshold for tibial torsion than the pediatric cohort. Thus, torsional deformities in childhood seem to play only a minor role [29]. In adults, a torsional osteotomy is recommended with a symptomatic deviation of approximately 10 degrees or more [4,28]. Dickschas et al. [4] showed that about 12% of patella instabilities are caused by torsional deformities.

With a median TTTG value of 17 mm within the pediatric and 18 mm within the adult cohort, both were slightly elevated, while both groups showed a TTTG value above the correctional limit of 15 mm [14,22] in approximately half of the cases. The risk factor "high-riding patella" also showed increased relevance, with an incidence of just under 30% in adult cases and just over 40% in pediatric cases. Osteotomies of the tibial tuberosity were performed less frequently than would be expected based on the proportion of underlying pathology.

Consistent with the increased number of cases that demonstrated trochlear dysplasia within the pediatric cohort (57%), the lateral trochlear inclination angle was less than 11 degrees in approximately one-third of cases. Unexpectedly, the adult cohort exhibited some form of trochlear abnormality in 36% of cases In severe cases of trochlea dysplasia, a surgical re-shaping (trochleoplasty) may be indicated. Trochleoplasty may influence other predisposing factors, as patellar tilt and patellar height are influenced by a dysplastic trochlear geometry, and reconfiguration of the trochlea influences the TTTG value [28]. Open growth plates are usually considered a contraindication for this procedure, in which case, one should resort to surgical procedures addressing the balance of associated soft-tissue structures.

Schmeling et al. [28] describe a classification system by naming five types of patellofemoral instability, which may assist clinicians in choosing an appropriate treatment regimen. Type 1 describes a traumatic patellar dislocation without maltracking or instability with a low risk of re-dislocation. Type 2 describes patellar instability without maltracking with a high risk of re-dislocation. Type 3 is defined as patellar instability with maltracking. Type 4 describes high-grade instability with loss of patellar guidance due to high-grade trochlear dysplasia, while type 5 is characterized by maltracking without instability. Types 1 and 2 can be easily interpreted and understood as traumatic events. However, the present study population demonstrated an increased incidence of predisposing factors hinting at pre-existing patellar maltracking, making it a necessity to evaluate the mechanism of trauma.

Mechanisms of trauma such as stepping out of a vehicle or standing up from a seated position correspond to an everyday movement and are not viewed as adequate mechanisms of trauma for a patella dislocation to occur and were therefore grouped within the category "insufficient trauma". Accordingly, this subgroup showed a low rate of "healthy knees" in both age groups. The children's cohort in particular was characterized by a high proportion of trochlear dysplasia and a high-riding patella. The question arises why the proportion of this trauma category is also high in the adult cohort, as dislocations, given a relevant predisposition, would be expected to have also occurred at a younger age. This can be explained by the high proportion of recurrent dislocations within the adult cohort within the "inadequate trauma" category. Thus, these are not to be considered as traumatic. According to the allocation to type 3 or 4 of patellofemoral instability [28], one should aim for a causal surgical therapy.

Direct impact to the medial aspect of the knee, on the other hand, can plausibly cause a traumatic dislocation [8,30]. This theory is supported by the low proportion of recurrent dislocations and the correspondingly high proportion of first-time dislocations, as well as by the rather low proportion of underlying predisposing factors in the adult group. Comparatively, a pathological trochlear shape, patellar height, and lateral trochlear inclination angle could be witnessed more often in the pediatric cohort. There is also a lower proportion of "healthy knees" in the pediatric age group. This shows that clinicians should not be misled by the plausibility of certain mechanisms of trauma and dismiss other variables, especially in children. It remains important to consider underlying predisposing factors before initiating a certain treatment option.

The pediatric subgroup in the trauma category "rotational trauma" also showed an increase in pathological values regarding predisposing factors compared to the adult subgroup. Within the latter, however, the older age of the injured policyholders was noticeable, further supporting the trauma thesis.

With 100% first-time dislocations and the highest proportion of "healthy knees" in both age groups (60% in the pediatric cohort, 33% in the adult cohort), the trauma category "fall" represents a relevant and major cause for traumatic dislocation of the patella.

Ultimately, the surgical treatment chosen should be based on the presence of relevant predisposing factors and the resulting injuries incurred due to the dislocation event itself. In accordance with the recommendation for type 1, i.e., non-surgical treatment as long as there is no relevant osteochondral damage [28], non-surgical treatment was initiated in 22% of adult cases and 26% of pediatric cases, corresponding to the proportions of "healthy knees". In type 2, surgical treatment addressing the medial soft tissue is recommended [28], while MPFL reconstruction has been shown to be superior to duplicating the retinaculum in adults [31]. In children, MPFL reconstruction is still considered critical due to the insertion point being located at the distal femoral growth plate [6]. Although there are studies that show a reduced risk of re-dislocation after surgical stabilization after having suffered a traumatic first-time dislocation with subsequent instability, long-term satisfaction in clinical scores is not significantly higher compared to non-surgical treatment [13,15,16]. In the present study cohort, surgically addressing the medial ligamentary complex was the most common procedure performed in children and adults, in line with the above-mentioned recommendation.

Contrary to one's assumption on the treatment of mere traumatic dislocations of the patella, our study population underwent correctional osteotomies by means of tibial tuberosity osteotomy, torsional and frontal axis corrections, or trochleaplasty, as recommended for maltracking instability (types 3 and 4 [28]). Within these cases, a sufficient mechanism of trauma is deemed questionable.

6. Conclusions

Dislocations of the Patella are less often of merely traumatic nature, as they are triggered by geometric malformities of the knee. Children are affected by predisposition-related dislocations more frequently than adults. Depending on the mechanism of trauma, the proposition of non-surgical treatment after a first-time dislocation event without concomitant osteochondral injury should be abandoned. Falls are an adequate mechanism of trauma. In cases of direct impact and rotational trauma, relevant trochlear dysplasia should be ruled out before trauma adequacy is determined. Especially in the cases of the insufficient mechanism of trauma, an analysis of the known predisposing factors should be performed before the initiation of a causal therapy, even in the case of a first-time dislocation event. In some cases, a definitive causal therapy can only be performed after the growth plates have closed.

Author Contributions: Conceptualization: P.C.S., J.D. and I.S.; methodology: P.C.S. and I.S.; validation: P.C.S., J.D. and I.S.; formal analysis: I.S.; investigation: I.S.; writing—original draft preparation: I.S.; writing—review and editing: P.C.S., J.D. and P.M.; visualization: I.S.; supervision: P.C.S. and J.D.; project administration: P.C.S. and I.S. All authors have read and agreed to the published version of the manuscript.

Funding: This research received no external funding.

Institutional Review Board Statement: No ethic committee votum was needed for this study.

Informed Consent Statement: Not required.

Data Availability Statement: Data is contained within the article.

Conflicts of Interest: All authors have no conflict of interest to disclose.

References

1. Chotel, F.; Bérard, J.; Raux, S. Patellar instability in children and adolescents. *Orthop. Traumatol. Surg. Res.* **2014**, *100*, S125–S137. [CrossRef] [PubMed]
2. Dejour, H.; Walch, G.; Nove-Josserand, L.; Guier, C. Factors of patellar instability: An anatomic radiographic study. *Knee Surg. Sports Traumatol. Arthrosc.* **1994**, *2*, 19–26. [CrossRef] [PubMed]
3. Dickschas, J.; Ferner, F.; Lutter, C.; Gelse, K.; Harrer, J.; Strecker, W. Patellofemoral dysbalance and genua valga: Outcome after femoral varisation osteotomies. *Arch. Orthop. Trauma Surg.* **2018**, *138*, 19–25. [CrossRef] [PubMed]
4. Dickschas, J.; Harrer, J.; Pfefferkorn, R.; Strecker, W. Operative treatment of patellofemoral maltracking with torsional osteotomy. *Arch. Orthop. Trauma Surg.* **2021**, *132*, 289–298. [CrossRef]
5. Frings, J.; Balcarek, P.; Tscholl, P.; Liebensteiner, M.; Dirisamer, F.; Koenen, P.; AGA Knee Patellofemoral Committee. Conservative versus surgical treatment for primary patellar dislocation. *Dtsch. Arztebl. Int.* **2020**, *117*, 279–286. [CrossRef]
6. Frosch, S.; Balcarek, P.; Walde, T.A.; Schüttrumpf, J.P.; Wachowski, M.M.; Ferleman, K.G.; Stürmer, K.M.; Frosch, K.H. Die Therapie der Patellaluxation: Eine systematische Literaturanalyse. *Z. Orthop. Unfallchirurgie* **2011**, *149*, 630–645. [CrossRef]
7. Jaquith, B.P.; Parikh, S.N. Predictors of recurrent patellar instability in children and adolescents after first-time dislocation. *J. Pediatr. Orthop.* **2017**, *37*, 484–490. [CrossRef]
8. Ludolph, E. Der Unfallmann. Springer Berlin Heidelberg. *Kapitel* **2013**, *14*, 393–407.
9. Obermeyer, C.; Hoffmann, D.B.; Wachowski, M.M. Patellaluxation im Kindes- und Jugendalter-Aktuelle Entwicklung bei Diagnostik und Therapie. *Orthopäde* **2019**, *48*, 868–876. [CrossRef]
10. Oestern, S.; Varoga, D.; Lippross, S.; Kaschwich, M.; Finn, J.; Buddrus, B.; Seekamp, A. Patellaluxation. *Unfallchirurg* **2011**, *114*, 345–359. [CrossRef]
11. Strich, R. Neue Aspekte in der Begutachtung der Patellaluxation. *Trauma Berufskrankh.* **2019**, *21*, 199–208. [CrossRef]
12. Ateschrang, A.; Freude, T.; Grünwald, L.; Schäffler, A.; Stöckle, U.; Schröter, S. Patellaluxation: Diagnostik- und Behandlungsalgorithmus unter Berücksichtigung der Torsion Patella Dislocation: An Algorithm for Diagnostic and Treatment Considering the Rotation. *Z. Orthop Unfallchirurgie* **2014**, *152*, 59–67. [CrossRef]
13. Buchner, M.; Baudendistel, B.; Sabo, D.; Schmitt, H. Acute traumatic primary patellar dislocation: Long-term results comparing conservative and surgical treatment. *Clin. J. Sport Med.* **2005**, *15*, 62–66. [CrossRef] [PubMed]
14. Jagodzinski, M.; Niemeyer, P.; Zeichen, J.; Balcarek, P. S1-Leitlinie 012/024: Patellaluxation. Leitlinienkommission Dtsch. Ges. Unfallchirurgie AWMF Online, 2nd ed. Available online: https://register.awmf.org/assets/guidelines/012-024l_S2e_Patellaluxation_2021-11_01.pdf (accessed on 25 October 2021).
15. Sillanpää, P.J.; Mattila, V.M.; Mäenpää, H.; Kiuru, M.; Visuri, T.; Pihlajamäki, H. Treatment with and without initial stabilizing surgery for primary traumatic patellar dislocation. A prospective randomized study *JBJS* **2009**, *91*, 263–273. [CrossRef]
16. Longo, U.G.; Ciuffreda, M.; Locher, J.; Berton, A.; Salvatore, G.; Denaro, V. Treatment of Primary Acute Patellar Dislocation: Systematic Review and Quantitative Synthesis of the Literature. *Clin. J. Sport Med.* **2017**, *27*, 511–523. [CrossRef] [PubMed]
17. Nikku, R.; Nietosvaara, Y.; Aalto, K.; Kallio, P.E. Operative treatment of primary patellar dislocation does not improve medium-term outcome: A 7-year follow-up report and risk analysis of 127 randomized patients. *Acta Orthop.* **2005**, *76*, 699–704. [CrossRef]
18. Dejour, H.; Walch, G.; Neyret, P.; Adeleine, P. Dysplasia of the femoral trochlea. *Rev. Chir. Orthop. Reparatrice Appar. Mot.* **1990**, *76*, 45–54.

19. Waidelich, H.-A.; Strecker, W.; Schneider, E. Computertomographische Torsionswinkel und Längenmessung an der unteren Extremität. Methodik, Normalwerte und Strahlenbelastung. *Fortschr. Röntgenstr.* **1992**, *157*, 245–251. [CrossRef]
20. Insall, J.; Salvati, E. Patella position in the normal knee joint. *Radiology* **1971**, *101*, 101–104. [CrossRef]
21. Caton, J.; Deschamps, G.; Chambat, P.; Lerat, J.L.; Dejour, H. Patella infera: Apropos 128 cases. *Rev. Chir. Orthop. Reparatrice Appar. Mot* **1982**, *68*, 317–325.
22. Diederichs, G.; Issever, A.S.; Scheffler, S. MR imaging of patellar instability: Injury patterns and assessment of risk factors. *Radiographics* **2010**, *30*, 961–981. [CrossRef] [PubMed]
23. Carrillon, Y.; Abidi, H.; Dejour, D.; Fantino, O.; Moyen, B.; Tran-Minh, V.A. Patellar instability: Assessment on MR images by measuring the lateral trochlear inclination—Initial experience. *Radiology* **2000**, *216*, 582–585. [CrossRef] [PubMed]
24. Balcarek, P.; Frosch, K.H. Die Patellaluxation im Kindes- und Jugendalter. *Arthroskopie* **2012**, *4*, 266–274. [CrossRef]
25. Thévenin-Lemoine, C.; Ferrand, M.; Courvoisier, A.; Damsin, J.-P.; le Pointe, H.D.; Vialle, R. Is the Caton-Deschamps index a valuable ratio to investigate patellar height in children? *JBJS* **2011**, *93*, e35. [CrossRef] [PubMed]
26. Fithian, D.C.; Paxton, E.W.; Cohen, A.B. Indications in the treatment of patellar instability. *J. Knee Surg.* **2004**, *17*, 47–56. [CrossRef] [PubMed]
27. Mäenpää, H.; Lehto, M.U. Patellofemoral osteoarthritis after patellar dislocation. *Clin. Orthop. Relat. Res.* **1997**, *339*, 156–162.
28. Schmeling, A.; Frings, J.; Akoto, R.; Frosch, K.H. Patellaluxation. *Unfallchirurg* **2020**, *123*, 969–983. [CrossRef]
29. Staheli, L.T. Torsion-treatment indications. *Clin. Orthop. Relat Res.* **1989**, *247*, 61–66. [CrossRef]
30. Ludolph, E. Patellaluxation: Anlage–habituelle Luxation–Begutachtung–Begutachtung für die Gesetzliche Unfallversicherung. *Trauma Berufskrankh.* **2002**, *4* (Suppl. 3), S206–S210. [CrossRef]
31. Ma, L.-F.; Wang, F.; Chen, B.-C.; Wang, C.H.; Zhou, J.W.; Wang, H.Y. Medial retinaculum plasty versus medial patellofemoral ligament reconstruction for recurrent patellar instability in adults: A randomized controlled trial. *Arthroscopy* **2013**, *29*, 891–897. [CrossRef]

Journal of *Personalized Medicine*

MDPI

Article

Comparative Analysis of Frailty Scores for Predicting Adverse Outcomes in Hip Fracture Patients: Insights from the United States National Inpatient Sample

Maximilian Peter Forssten [1,2,*], Yang Cao [3], Ahmad Mohammad Ismail [2], Lakshika Tennakoon [4], David A. Spain [4] and Shahin Mohseni [2,5]

1 Department of Orthopedic Surgery, Faculty of Medicine and Health, Orebro University, Orebro 701 82, Sweden; mohsenishahin@yahoo.com
2 School of Medical Sciences, Orebro University, 701 82 Orebro, Sweden; amzenik@gmail.com
3 Clinical Epidemiology and Biostatistics, School of Medical Sciences, Faculty of Medicine and Health, Orebro University, 701 82 Orebro, Sweden; yang.cao@oru.se
4 Department of Surgery, Section of Trauma and Acute Care Surgery, Stanford University School of Medicine, Stanford, CA 94305, USA; lakshika@stanford.edu (L.T.); dspain@stanford.edu (D.A.S.)
5 Division of Trauma, Critical Care & Acute Care Surgery, Department of Surgery, Sheikh Shakhbout Medical City, Abu Dhabi P.O. Box 11001, United Arab Emirates
* Correspondence: maximilian.forssten@oru.se; Tel.: +47-58-520-7702

Abstract: The aim of the current investigation was to compare the ability of several frailty scores to predict adverse outcomes in hip fracture patients. All adult patients (18 years or older) who suffered a hip fracture due to a fall and underwent surgical fixation were extracted from the 2019 National Inpatient Sample (NIS) Database. A combination of logistic regression and bootstrapping was used to compare the predictive ability of the Orthopedic Frailty Score (OFS), the Nottingham Hip Fracture Score (NHFS), the 11-factor modified Frailty Index (11-mFI) and 5-factor (5-mFI) modified Frailty Index, as well as the Johns Hopkins Frailty Indicator. A total of 227,850 patients were extracted from the NIS. In the prediction of in-hospital mortality and failure-to-rescue (FTR), the OFS surpassed all other frailty measures, approaching an acceptable predictive ability for mortality [AUC (95% CI): 0.69 (0.67–0.72)] and achieving an acceptable predictive ability for FTR [AUC (95% CI): 0.70 (0.67–0.72)]. The NHFS demonstrated the highest predictive ability for predicting any complication [AUC (95% CI): 0.62 (0.62–0.63)]. The 11-mFI exhibited the highest predictive ability for cardiovascular complications [AUC (95% CI): 0.66 (0.64–0.67)] and the NHFS achieved the highest predictive ability for delirium [AUC (95% CI): 0.69 (0.68–0.70)]. No score succeeded in effectively predicting venous thromboembolism or infections. In summary, the investigated frailty scores were most effective in predicting in-hospital mortality and failure-to-rescue; however, they struggled to predict complications.

Keywords: hip fracture; frailty; mortality; morbidity; prediction; logistic regression; Orthopedic Frailty Score; Nottingham Hip Fracture Score; Modified Frailty Index; Johns Hopkins Frailty Indicator

Citation: Forssten, M.P.; Cao, Y.; Mohammad Ismail, A.; Tennakoon, L.; Spain, D.A.; Mohseni, S. Comparative Analysis of Frailty Scores for Predicting Adverse Outcomes in Hip Fracture Patients: Insights from the United States National Inpatient Sample. *J. Pers. Med.* **2024**, *14*, 621. https://doi.org/10.3390/jpm14060621

Academic Editor: Markus Gesslein

Received: 6 May 2024
Revised: 4 June 2024
Accepted: 5 June 2024
Published: 10 June 2024

1. Introduction

Hip fractures represent a significant and growing public health concern, which predominantly afflicts the elderly population [1]. Projections indicate a substantial rise in the annual number of hip fractures, which are expected to reach 4.5 million globally by the year 2050 [2]. This is of particular concern given that the hip fracture rates in the United States are among the highest in the world [3]. With one-year postoperative mortality rates reaching 27% [4], and even exceeding 50% in certain demographic cohorts [5], this underscores the critical importance of addressing hip fractures comprehensively and urgently to mitigate their profound impact on individuals and healthcare systems alike.

This high mortality rate in hip fracture patients stems largely from the high prevalence of frailty within this population. Studies reveal stark contrasts, with crude 30-day and 90-day mortality rates among frail individuals reaching up to ten times higher than those among non-frail hip fracture patients [6]. Furthermore, frailty has been found to be the single most important predictor of in-hospital mortality in hip fracture patients [7]. Complicating matters, however, is the proliferation of published frailty scores, leading to uncertainty regarding their use and effectiveness. This equipoise underscores the need to identify which frailty scores perform optimally. Consequently, the current study aims to compare the ability of several frailty scores to predict adverse outcomes in hip fracture patients.

2. Materials and Methods

The data for the present study were sourced from the 2019 United States National Inpatient Sample (NIS). Managed by the Agency for Healthcare Research and Quality, the NIS constitutes the largest all-payer inpatient database in the United States, drawing from a 20% sample of the nation's hospitalizations. The NIS is sampled from the State Inpatient Databases, including all inpatient data that are currently contributed to the Healthcare Cost and Utilization Project. Utilizing validated sampling algorithms integrated with discharge and survey weights, the NIS ensures accurate national estimates for a substantial 97% of all inpatient hospitalizations in the country [8]. The current investigation adhered to the transparent reporting of a multivariable prediction model for individual prognosis or diagnosis (TRIPOD) guidelines.

All individuals aged 18 years or older who had undergone emergency hip fracture surgery following a traumatic fall were included in this study. Surgical procedures involved internal fixation methods, including open reduction internal fixation or intramedullary nailing, as well as arthroplasty, both total hip arthroplasty and hemiarthroplasty. To reduce heterogeneity in the dataset, patients with head, vascular, or truncal injuries were excluded. Only 0.4% of the patients (N = 925) were missing data required for calculating the frailty scores; consequently, these patients were excluded. The identification of the previously listed variables relied on the International Classification of Diseases 10th Revision (ICD-10) codes recorded in the NIS [9]. The data retrieved included patient demographics (age, sex, and race/ethnicity), comorbidities, fracture morphology, surgical management, origin of admission, and functional ability, as well as discharge disposition and in-hospital complications [10].

2.1. Calculating Frailty Scores

2.1.1. Orthopedic Frailty Score

The Orthopedic Frailty Score (OFS) was developed for the explicit purpose of evaluating frailty and predicting short-term mortality in individuals with hip fractures [11,12]. The OFS was calculated based on the presence of five dichotomous variables: congestive heart failure, a history of malignancy (excluding non-invasive skin cancer), institutionalization, dependence on assistance for activities of daily living (indicating a non-independent functional status), and an age of 85 years or older. Each of these variables was ascribed one point, resulting in a maximum possible score of 5 [11].

2.1.2. Nottingham Hip Fracture Score

The Nottingham Hip Fracture Score (NHFS) was predominantly calculated according to the original study by Maxwell et al. [13]. Patients received 3 points if their age fell within the range of 66 to 85 years and 4 points if they were 86 years or older. Additionally, patients received 1 point if they were male, institutionalized, had a history of malignancy, or presented with two or more comorbidities. In the absence of a mini-mental test score, patients received 1 point if they had dementia, aligning with the methodology employed by previous investigations [11,14–17]. Furthermore, as admission hemoglobin is not recorded in the NIS, patients also received 1 point if they were diagnosed with any deficiency anemia or chronic blood loss anemia at the time of admission [18]. Subsequently, the maximum possible score was 10.

2.1.3. 11-Factor Modified Frailty Index

The 11-factor modified Frailty Index (11-mFI) was determined based on the presence of non-independent functional status, hypertension requiring medication, chronic obstructive pulmonary disease (COPD), impaired sensorium, diabetes mellitus, previous myocardial infarction, congestive heart failure, stroke with neurologic deficit, transient ischemic attack or stroke without neurologic deficit, angina pectoris, and peripheral vascular disease [19,20]. However, pneumonia was not included in the calculation as it constituted one of the outcomes in the current analysis. Each variable present increased the score by 1, resulting in a maximum potential score of 11.

2.1.4. 5-Factor Modified Frailty Index

The 5-factor modified Frailty Index (5-mFI) was determined based on the presence of COPD, diabetes mellitus, hypertension requiring medication, congestive heart failure, and non-independent functional status [21]. Each variable present increased the score by 1, resulting in a maximum potential score of 5.

2.1.5. Johns Hopkins Frailty Indicator

As per the Johns Hopkins Frailty Indicator, patients were categorized as either non-frail or frail, contingent upon the identification of at least one diagnosis that defines frailty. These conditions included fall from wheelchair, fall on stairs or steps, abnormality of gait, difficulty in walking, inadequate material resources, inadequate housing, lack of housing, fecal incontinence, feeding difficulties and mismanagement, abnormal loss of weight and underweight, continuous urinary leakage, incontinence without sensory awareness, decubitus ulcer, profound visual impairment, senile dementia with delirium, senile dementia with delusional or depressive features, nutritional marasmus, or other severe protein-calorie malnutrition [22,23].

2.2. *Statistical Analysis*

Summary statistics were employed in order to describe the dataset. Age was summarized as a median and interquartile range, while the remaining categorical variables were presented as counts and percentages. The adverse outcomes investigated included in-hospital mortality, any complication, cardiovascular complications, delirium, venous thromboembolism, infection, and failure-to-rescue (FTR).

Any complication encompassed heart failure, cardiac arrest, cardiac tamponade, ventricular tachycardia, ventricular fibrillation, stroke, acute kidney injury, delirium, deep vein thrombosis, pulmonary embolism, embolism due to orthopedic device, urinary tract infection, pneumonia, wound infection, implant infection, sepsis, and decubitus ulcers. Cardiovascular complications were defined as heart failure, cardiac arrest, cardiac tamponade, ventricular tachycardia, ventricular fibrillation, or stroke. Venous thromboembolism comprised deep vein thrombosis, pulmonary embolism, or embolism due to orthopedic device. Infections included urinary tract infections, pneumonia, wound infections, implant infection, and sepsis. FTR was defined as in-hospital mortality subsequent to a complication.

To evaluate the predictive ability of all the included frailty scores for each outcome, LR models were fitted to the dataset with the outcome as the response variable and the frailty score as the predictor. The AUC (area under the receiver operating characteristic curve), as well as the sensitivity and specificity that maximized Youden's index (sensitivity + specificity − 1), was calculated for each model, along with their respective 95% confidence intervals (CI). The CIs were estimated using 1000 bootstrap replicates; *p*-values for the differences in the AUCs were also estimated using 1000 bootstrap replicates as follows:

1. The AUCs for two frailty scores were calculated: AUC_1 and AUC_2.
2. The observed difference in the AUCs was calculated: $D_{observed} = AUC_1 - AUC_2$.
3. 1000 bootstrap replicates of the dataset were generated.
4. The difference in the AUCs for each replicate was calculated: $D_1 \ldots D_{1000}$.

5. The standard error of the differences in the AUCs was calculated as the standard deviation across all replicates: $\sigma_{\overline{D}} = \sqrt{\frac{\sum_{i=1}^{1000}(D_i-\overline{D})^2}{1000}}$.
6. The z-value was determined by dividing the difference in the AUCs by the standard error: $Z = D_{observed}/\sigma_{\overline{D}}$.
7. A two-sided *p*-value was determined based on the proportion of a normally distributed population with a z-value > Z or < −Z (where Z is positive).

These analyses were also repeated with a subgroup of patients consisting of those who were 50 years or older. Statistical significance was defined as a two-sided *p*-value < 0.05. All *p*-values were adjusted for multiple comparisons using Bonferroni correction. All analyses were performed in the software R 4.2.2 using the tidyverse, survey, parallel, and WeightedROC packages [24].

3. Results

A total of 227,795 estimated hip fracture patients were included in the current investigation. The median age was 81, with the majority of patients being female (68.8%) and White (84.4%). Most patients had either suffered a cervical (47.3%) or pertrochanteric (46.8%) hip fracture, with the most common method of fixation being internal fixation (63.4%) (Table 1). The most common comorbidities consisted of ischemic heart disease (25.2%), fluid or electrolyte disorders (24.9%), hypothyroidism (17.1%), congestive heart failure (16.6%), and end-stage renal disease (15.5%) (Table 2). In total, 1.3% (N = 2910) died in the hospital. In 1.1% (N = 2585) of cases, this was following a complication. A total of 35.2% (N = 80,135) of patients suffered one of the included complications (Table 3). The distribution of all outcomes across the levels of each score is presented in Supplemental Table S1.

Table 1. Demographics of hip fracture patients.

	All Patients (N = 227,795)
Age, median [IQR]	81 [72.0–88.0]
Sex, n (%)	
Male	71,050 (31.2)
Female	156,745 (68.8)
Race/Ethnicity, n (%)	
White	192,245 (84.4)
Black	9230 (4.1)
Hispanic	12,235 (5.4)
Asian or Pacific Islander	3950 (1.7)
American Indian	975 (0.4)
Other	4405 (1.9)
Missing	4755 (2.1)
Type of fracture, n (%)	
Cervical	107,640 (47.3)
Basicervical	3770 (1.7)
Pertrochanteric	106,625 (46.8)
Subtrochanteric	9655 (4.2)
Missing	105 (0.0)
Type of surgery, n (%)	
Internal fixation	144,425 (63.4)
Arthroplasty	83,370 (36.6)
OFS, n (%)	
0	108,580 (47.7)
1	87,215 (38.3)
2	28,315 (12.4)
3	3420 (1.5)
4	265 (0.1)

Table 1. *Cont.*

	All Patients (N = 227,795)
NHFS, n (%)	
0	5525 (2.4)
1	12,940 (5.7)
2	7825 (3.4)
3	22,885 (10.0)
4	56,285 (24.7)
5	62,540 (27.5)
6	41,895 (18.4)
7	15,010 (6.6)
8	2655 (1.2)
9	230 (0.1)
10	5 (0.0)
11-mFI, n (%)	
0	28,390 (12.5)
1	65,070 (28.6)
2	68,060 (29.9)
3	41,230 (18.1)
4	18,345 (8.1)
5	5585 (2.5)
6	1025 (0.4)
7	85 (0.0)
8	5 (0.0)
5-mFI, n (%)	
0	46,430 (20.4)
1	98,260 (43.1)
2	61,855 (27.2)
3	18,215 (8.0)
4	2870 (1.3)
5	165 (0.1)
Johns Hopkins Frailty Indicator, n (%)	
0	164,200 (72.1)
1	63,595 (27.9)

OFS, Orthopedic Frailty Score; NHFS, Nottingham Hip Fracture Score; 11-mFI, 11-factor modified Frailty Index; 5-mFI, 5-factor modified Frailty Index.

Table 2. Comorbidities in hip fracture patients.

	All Patients (N = 227,795)
Ischemic heart disease, n (%)	57,360 (25.2)
Congestive heart failure, n (%)	37,900 (16.6)
Cerebrovascular disease, n (%)	14,280 (6.3)
Peripheral vascular disease, n (%)	10,610 (4.7)
Diabetes mellitus without complications, n (%)	20,315 (8.9)
Diabetes mellitus with complications, n (%)	27,270 (12.0)
End stage renal disease, n (%)	35,385 (15.5)
Liver disease, n (%)	4500 (2.0)
Deficiency anemia, n (%)	32,780 (14.6)
Chronic blood loss anemia, n (%)	2545 (1.1)
Coagulopathy, n (%)	12,395 (5.4)
Obesity, n (%)	10,075 (4.4)
Alcohol abuse, n (%)	7650 (3.4)
Drug abuse, n (%)	3010 (1.3)
Psychosis, n (%)	5805 (2.5)

Table 2. *Cont.*

	All Patients (N = 227,795)
Depression, n (%)	25,275 (11.1)
Current smoker, n (%)	1455 (0.6)
COPD, n (%)	42,035 (18.5)
Dementia, n (%)	58,170 (25.5)
Impaired sensorium, n (%)	66,585 (29.2)
Non-independent functional status, n (%)	10,795 (4.7)
Institutionalized, n (%)	12,405 (5.4)
Osteoporosis, n (%)	34,790 (15.3)
Osteoporosis with fracture, n (%)	2335 (1.0)
Fluid or electrolyte disorder, n (%)	56,650 (24.9)
Connective tissue disease, n (%)	7365 (3.2)
Hypothyroidism, n (%)	39,010 (17.1)
Local cancer, n (%)	4090 (1.8)
Metastatic cancer, n (%)	2465 (1.1)
Leukemia, n (%)	1405 (0.6)
Lymphoma, n (%)	1620 (0.7)

COPD, Chronic obstructive pulmonary disease.

Table 3. Crude outcomes in hip fracture patients.

	All Patients (N = 227,795)
In-hospital mortality, n (%)	2910 (1.3)
Complication, n (%)	80,135 (35.2)
Any cardiac complication	3660 (1.6)
Heart failure	505 (0.2)
Cardiac arrest	185 (0.1)
Cardiac tamponade	20 (0.0)
Ventricular tachycardia	2285 (1.0)
Ventricular fibrillation	190 (0.1)
Stroke	2160 (0.9)
Acute kidney injury	34,355 (15.1)
Delirium	8970 (3.9)
Deep vein thrombosis	1545 (0.7)
Pulmonary embolism	1240 (0.5)
Embolism due to orthopedic device	10 (0.0)
Urinary tract infection	26,680 (11.7)
Pneumonia	21,590 (9.5)
Wound infection	160 (0.1)
Implant infection	30 (0.0)
Sepsis	3995 (1.8)
Decubitus ulcer	3955 (1.7)
Failure-to-rescue, n (%)	2615 (1.1)

In the prediction of both in-hospital mortality and FTR, the OFS surpassed all other frailty measures, approaching an acceptable predictive ability for mortality [AUC (95% CI): 0.69 (0.67–0.72)] and achieving an acceptable predictive ability for FTR [AUC (95% CI): 0.70 (0.67–0.72)] [25]. All scores struggled to predict complications; however, the NHFS demonstrated the highest predictive ability [AUC (95% CI): 0.62 (0.62–0.63)]. On the other hand, the 11-mFI demonstrated the highest predictive ability for cardiovascular complications [AUC (95% CI): 0.66 (0.64–0.67)] and the NHFS achieved the highest predictive ability for delirium [AUC (95% CI): 0.69 (0.68–0.70)]. No score succeeded in effectively predicting venous thromboembolism or infections (Table 4). The results remained virtually unchanged in the subgroup analysis composed of patients who were 50 years or older (Supplemental Table S2).

Table 4. Predictive ability of frailty scores for adverse outcomes in hip fracture patients.

Outcome	AUC (95% CI)	Sensitivity (95% CI)	Specificity (95% CI)	p-Value for Difference in AUCs
In-hospital mortality				
OFS	0.69 (0.67–0.72)	0.83 (0.80–0.86)	0.48 (0.48–0.49)	Reference
NHFS	0.65 (0.63–0.67)	0.76 (0.74–0.82)	0.47 (0.43–0.47)	<0.001
11-mFI	0.61 (0.59–0.64)	0.45 (0.42–0.75)	0.71 (0.41–0.72)	<0.001
5-mFI	0.57 (0.55–0.59)	0.47 (0.18–0.68)	0.64 (0.43–0.91)	<0.001
Johns Hopkins Frailty Indicator	0.56 (0.54–0.58)	0.39 (0.35–0.43)	0.72 (0.72–0.73)	<0.001
Any complication				
OFS	0.60 (0.60–0.61)	0.64 (0.63–0.65)	0.54 (0.53–0.55)	Reference
NHFS	0.62 (0.62–0.63)	0.66 (0.65–0.66)	0.53 (0.52–0.53)	<0.001
11-mFI	0.61 (0.60–0.62)	0.69 (0.69–0.70)	0.47 (0.46–0.47)	0.147
5-mFI	0.57 (0.57–0.58)	0.44 (0.43–0.45)	0.68 (0.67–0.68)	<0.001
Johns Hopkins Frailty Indicator	0.55 (0.55–0.56)	0.35 (0.34–0.36)	0.76 (0.75–0.76)	<0.001
Cardiovascular complication				
OFS	0.60 (0.59–0.62)	0.69 (0.66–0.72)	0.48 (0.48–0.49)	Reference
NHFS	0.60 (0.58–0.61)	0.72 (0.62–0.74)	0.43 (0.43–0.54)	1.00
11-mFI	0.66 (0.64–0.67)	0.53 (0.50–0.56)	0.72 (0.71–0.72)	<0.001
5-mFI	0.56 (0.54–0.57)	0.46 (0.17–0.48)	0.64 (0.63–0.91)	<0.001
Johns Hopkins Frailty Indicator	0.53 (0.51–0.54)	0.33 (0.30–0.36)	0.72 (0.72–0.73)	<0.001
Delirium				
OFS	0.61 (0.60–0.63)	0.72 (0.70–0.74)	0.48 (0.48–0.49)	Reference
NHFS	0.69 (0.68–0.70)	0.81 (0.79–0.83)	0.47 (0.47–0.48)	<0.001
11-mFI	0.59 (0.58–0.61)	0.74 (0.72–0.76)	0.42 (0.41–0.42)	0.021
5-mFI	0.51 (0.51–0.53)	0.74 (0.29–0.77)	0.29 (0.27–0.72)	<0.001
Johns Hopkins Frailty Indicator	0.58 (0.57–0.59)	0.44 (0.41–0.46)	0.73 (0.72–0.73)	<0.001

Table 4. *Cont.*

Outcome	AUC (95% CI)	Sensitivity (95% CI)	Specificity (95% CI)	*p*-Value for Difference in AUCs
Venous thromboembolism				
OFS	0.54 (0.52–0.57)	0.60 (0.19–0.64)	0.48 (0.47–0.86)	Reference
NHFS	0.54 (0.53–0.57)	0.43 (0.33–0.79)	0.62 (0.27–0.72)	1.00
11-mFI	0.53 (0.52–0.56)	0.46 (0.34–0.75)	0.59 (0.32–0.71)	1.00
5-mFI	0.54 (0.53–0.57)	0.35 (0.25–0.66)	0.72 (0.43–0.80)	1.00
Johns Hopkins Frailty Indicator	0.54 (0.52–0.56)	0.35 (0.31–0.39)	0.72 (0.72–0.73)	1.00
Infection				
OFS	0.58 (0.58–0.59)	0.63 (0.62–0.64)	0.50 (0.50–0.51)	Reference
NHFS	0.58 (0.57–0.58)	0.63 (0.62–0.86)	0.49 (0.24–0.49)	0.764
11-mFI	0.58 (0.58–0.59)	0.69 (0.39–0.70)	0.44 (0.43–0.73)	1.00
5-mFI	0.56 (0.55–0.56)	0.44 (0.43–0.45)	0.65 (0.65–0.66)	<0.001
Johns Hopkins Frailty Indicator	0.54 (0.54–0.55)	0.34 (0.33–0.35)	0.74 (0.73–0.74)	<0.001
Failure-to-rescue				
OFS	0.70 (0.67–0.72)	0.83 (0.80–0.87)	0.48 (0.48–0.49)	Reference
NHFS	0.65 (0.63–0.67)	0.77 (0.75–0.83)	0.47 (0.43–0.47)	<0.001
11-mFI	0.61 (0.59–0.64)	0.46 (0.42–0.75)	0.71 (0.41–0.72)	<0.001
5-mFI	0.57 (0.55–0.60)	0.48 (0.19–0.70)	0.64 (0.43–0.91)	<0.001
Johns Hopkins Frailty Indicator	0.56 (0.54–0.58)	0.39 (0.35–0.43)	0.72 (0.72–0.73)	<0.001

AUC, area under the receiver operating characteristic curve; OFS, Orthopedic Frailty Score; NHFS, Nottingham Hip Fracture Score; 11-mFI, 11-factor modified Frailty Index; 5-mFI, 5-factor modified Frailty Index.

4. Discussion

By leveraging the NIS, which samples 20% of the nation's hospitalizations and accurately estimates 97% of all inpatient hospitalizations in the United States, a nationwide analysis could be conducted comparing the predictive ability of a range of frailty scores. Among the adverse outcomes investigated, these scores functioned best when predicting mortality and failure-to-rescue. Regarding these particular outcomes, the OFS demonstrated better performance compared to all other scores, with the NHFS following in second place. However, all scores comparatively struggled to accurately predict in-hospital complications, with some exceptions for specific types of complications. Notably, the NHFS stood out in its ability to predict delirium, while the 11-mFI outperformed all other scores when predicting cardiovascular complications. Of all the scores, only the OFS and NHFS were able to demonstrate an acceptable or close-to-acceptable predictive ability for any of the adverse outcomes [25].

These findings are by and large in line with prior studies. In the original study that developed and validated the OFS [AUC (95% CI): 0.77 (0.74–0.80)], it was found to surpass the 5-mFI [AUC (95% CI): 0.69 (0.66–0.73)] in predicting mortality, which is consistent with the present investigation. Furthermore, the OFS in the current study displays essentially the same predictive ability for mortality as it did in the national dataset used for developing the OFS. However, in the original study, the NHFS [AUC (95% CI): 0.76 (0.73–0.79)] and the OFS [AUC (95% CI): 0.77 (0.74–0.80)] demonstrated comparable predictive performance. This difference could be due to the lack of an admission hemoglobin or the fact that the current investigation focused on in-hospital mortality rather than 30-day postoperative mortality [11]. An analysis utilizing the NIS to explore the relationship between dementia and frailty also found that the OFS was the most important predictor of in-hospital mortality among the 52 variables analyzed, providing further support for its robust predictive capability in comparison to other scoring systems [7]. In a study conducted by Thorne et al., the AUC for in-hospital mortality was reported as 0.69 (95% CI: 0.64–0.74) for the NHFS, slightly higher than the NHFS's performance and on the same level as the OFS observed in the current investigation [26]. A separate analysis conducted by Lisk et al. reported an AUC for in-hospital mortality of 0.674 (95% CI: 0.584–0.764) for the NHFS; however, the wide confidence interval makes it difficult to judge if this indicates a true difference compared to the NHFS and the OFS in the current study [27]. Finally, when investigating the predictive performance of the 5-mFI, Ondeck et al. calculated an AUC for in-hospital mortality of 0.641 (95% CI: 0.624–0.659), which was higher than the current analysis. Nevertheless, similar to the current findings, the 5-mFI struggled to predict complications, with an AUC of 0.586 (95% CI: 0.581–0.592) for all adverse events [28].

As the current analyses were limited to the variables available in the NIS, there are many frailty scores that could not be included in the current analysis but warrant acknowledgment. In the same investigation by Thorne et al. that examined the NHFS, they also found that the Clinical Frailty Scale (CFS) yielded an AUC of 0.63 (95% CI: 0.57–0.69) for in-patient mortality, indicating a lower predictive ability compared to both the NHFS and the OFS [26]. In another study by Choi et al., the hip-multidimensional frailty score (Hip-MFS) exhibited a significantly better ability to predict postoperative complications with an AUC of 0.679 (95% CI: 0.613–0.745), surpassing all scores encompassed in the current study [29]. Additionally, the Hospital Frailty Risk Score (HFRS), composed of over 100 weighted variables, has demonstrated associations with heightened risks of in-hospital mortality, complications, prolonged length of stay, and increased total hospital costs [30,31]. Similarly, the FRAIL scale has been linked to elevated risks of postoperative complications and extended length of stay in geriatric fracture patients [32]. Employing a modified version of the Fried Frailty Index, Kistler et al. found that it was associated with an increased risk of complications [33]. Meanwhile, the Frailty Index, comprising over 50 variables, was associated with an elevated mortality rate and longer hospital stays [34].

Despite the wide range of frailty scores available, there remains a clinical equipoise when selecting which to apply in practice. Many feature an extensive variable list, which may hinder their ability to be used bedside [30,31,34,35]. Additionally, some also require patient cooperation [32,33,35], which may be challenging in a patient population where approximately 20% suffer from dementia [36]. Moreover, the need for blood tests post-admission further complicates the use of others [13,35]. Finally, some scores are inherently more subjective, relying on clinicians' judgment and observation of functional abilities in conjunction with overall health [37]. This in turn results in greater interobserver variation, especially among inexperienced users [38–41]. Despite the intuitive advantages of complexity, the current findings suggest that more intricate indices are not inherently superior. Notably, the OFS, among the simplest of the included scores, emerges as particularly adept at predicting mortality-related outcomes. Furthermore, the mFI-5, which requires fewer than half as many variables as the mFI-11, generally exhibits only marginal differences compared to the mFI-11, with the prediction of cardiovascular complications being the only exception. Additionally, research indicates that orthopedic surgeons tend to favor simpler and more straightforward scoring systems for clinical application [42]; this is an important point to consider when selecting a tool to transition from research settings to clinical practice.

This study was strengthened by the use of the NIS, which constitutes the largest all-payer inpatient database in the United States [8]. Leveraging this extensive dataset enabled the investigation of a wide array of outcomes through the inclusion of over 227,000 estimated hip fracture patients. However, a notable limitation was the absence of admission hemoglobin data, necessitating reliance on the diagnosis of anemia, which could potentially have reduced the predictive ability of the NHFS. Nonetheless, despite this alteration, the NHFS still outperformed all other scores in predicting delirium by a significant margin. Furthermore, our analyses were constrained by the variables available in the NIS, preventing the inclusion of other pertinent outcomes such as readmission, quality of life, and long-term mortality. Additionally, the risk of non-differential misclassification remains a concern when using a retrospective database, as this adds statistical noise to the data that reduces the overall predictive ability of all scores.

5. Conclusions

The investigated frailty scores were most effective in predicting in-hospital mortality and failure-to-rescue; however, they struggled to predict complications. The findings suggest that increasing complexity in frailty scores may result in diminishing returns concerning predictive ability, as evidenced by the Orthopedic Frailty Score, which, despite its simplicity in calculation, either outperforms or performs comparably to the other scores across most outcomes. Future studies should investigate how these scores can be incorporated into routine preoperative evaluations to guide care through preoperative optimization and resource allocation.

Supplementary Materials: The following supporting information can be downloaded at https://www.mdpi.com/article/10.3390/jpm14060621/s1: Table S1: Distribution of outcomes for each frailty score; Table S2: Predictive ability of frailty scores for adverse outcomes in hip fracture patients who are $\geq$50 years old.

Author Contributions: Conceptualization, M.P.F., Y.C., A.M.I., L.T., D.A.S. and S.M.; methodology, M.P.F., Y.C. and L.T.; software, M.P.F. and Y.C.; validation, Y.C.; formal analysis, M.P.F.; investigation, M.P.F., A.M.I. and S.M.; resources, D.A.S. and S.M.; data curation, M.P.F. and L.T.; writing—original draft preparation, M.P.F.; writing—review and editing, M.P.F., Y.C., A.M.I., L.T., D.A.S. and S.M.; supervision, Y.C., L.T., D.A.S. and S.M. All authors have read and agreed to the published version of the manuscript.

Funding: This research received no external funding.

Institutional Review Board Statement: The study was conducted in accordance with the Declaration of Helsinki and approved by the Institutional Review Board of Stanford University Hospital and the Swedish Ethical Review Authority (ref: 2022-03107-02).

Informed Consent Statement: Not applicable.

Data Availability Statement: Data is available upon reasonable request.

Conflicts of Interest: The authors declare no conflicts of interest.

References

1. Handoll, H.H.; Cameron, I.D.; Mak, J.C.; Panagoda, C.E.; Finnegan, T.P. Multidisciplinary Rehabilitation for Older People with Hip Fractures. *Cochrane Database Syst. Rev.* **2021**, *11*, CD007125. [CrossRef] [PubMed]
2. Gullberg, B.; Johnell, O.; Kanis, J.A. World-Wide Projections for Hip Fracture. *Osteoporos. Int. J. Establ. Result Coop. Eur. Found. Osteoporos. Natl. Osteoporos. Found. USA* **1997**, *7*, 407–413. [CrossRef] [PubMed]
3. Dhanwal, D.K.; Dennison, E.M.; Harvey, N.C.; Cooper, C. Epidemiology of Hip Fracture: Worldwide Geographic Variation. *Indian J. Orthop.* **2011**, *45*, 15–22. [CrossRef] [PubMed]
4. Gundel, O.; Thygesen, L.C.; Gögenur, I.; Ekeloef, S. Postoperative Mortality after a Hip Fracture over a 15-Year Period in Denmark: A National Register Study. *Acta Orthop.* **2020**, *91*, 58–62. [CrossRef] [PubMed]
5. Forssten, M.P.; Ismail, A.M.; Borg, T.; Ahl, R.; Wretenberg, P.; Cao, Y.; Peden, C.J.; Mohseni, S. Postoperative Mortality in Hip Fracture Patients Stratified by the Revised Cardiac Risk Index: A Swedish Nationwide Retrospective Cohort Study. *Trauma Surg. Acute Care Open* **2021**, *6*, e000778. [CrossRef] [PubMed]
6. Forssten, M.P.; Mohammad Ismail, A.; Ioannidis, I.; Wretenberg, P.; Borg, T.; Cao, Y.; Ribeiro, M.A.F.; Mohseni, S. The Mortality Burden of Frailty in Hip Fracture Patients: A Nationwide Retrospective Study of Cause-Specific Mortality. *Eur. J. Trauma Emerg. Surg. Off. Publ. Eur. Trauma Soc.* **2022**, *49*, 1467–1475. [CrossRef] [PubMed]
7. Ioannidis, I.; Forssten, M.P.; Mohammad Ismail, A.; Cao, Y.; Tennakoon, L.; Spain, D.A.; Mohseni, S. The Relationship and Predictive Value of Dementia and Frailty for Mortality in Patients with Surgically Managed Hip Fractures. *Eur. J. Trauma Emerg. Surg. Off. Publ. Eur. Trauma Soc.* **2023**, *50*, 339–345. [CrossRef] [PubMed]
8. Agency for Healthcare Research and Quality HCUP-US Methods Series. Available online: https://www.hcup-us.ahrq.gov/reports/methods/methods.jsp (accessed on 28 January 2023).
9. Tran, Z.; Hsiue, P.P.; Pan, C.; Verma, A.; Rahimtoola, R.; Stavrakis, A.; Lee, C.; Benharash, P. Impact of Delayed Intervention on Clinical Outcomes Following Traumatic Hip Fracture in the Elderly: A National Analysis. *J. Orthop.* **2021**, *27*, 74–78. [CrossRef] [PubMed]
10. Madrigal, J.; Tran, Z.; Hadaya, J.; Sanaiha, Y.; Benharash, P. Impact of Chronic Lymphocytic Leukemia on Outcomes and Readmissions After Cardiac Operations. *Ann. Thorac. Surg.* **2022**, *114*, 152–159. [CrossRef]
11. Forssten, M.P.; Cao, Y.; Trivedi, D.J.; Ekestubbe, L.; Borg, T.; Bass, G.A.; Mohammad Ismail, A.; Mohseni, S. Developing and Validating a Scoring System for Measuring Frailty in Patients with Hip Fracture: A Novel Model for Predicting Short-Term Postoperative Mortality. *Trauma Surg. Acute Care Open* **2022**, *7*, e000962. [CrossRef] [PubMed]
12. Forssten, M.P.; Cao, Y.; Mohammad Ismail, A.; Ioannidis, I.; Tennakoon, L.; Spain, D.A.; Mohseni, S. Validation of the Orthopedic Frailty Score for Measuring Frailty in Hip Fracture Patients: A Cohort Study Based on the United States National Inpatient Sample. *Eur. J. Trauma Emerg. Surg.* **2023**, *49*, 2155–2163. [CrossRef] [PubMed]
13. Maxwell, M.J.; Moran, C.G.; Moppett, I.K. Development and Validation of a Preoperative Scoring System to Predict 30 Day Mortality in Patients Undergoing Hip Fracture Surgery. *Br. J. Anaesth.* **2008**, *101*, 511–517. [CrossRef] [PubMed]
14. Boddaert, J.; Na, N.; Le Manach, Y.; Raux, M.; Cohen-Bittan, J.; Vallet, H.; Meziere, A.; Khiami, F.; Riou, B. Prediction of Postoperative Mortality in Elderly Patients with Hip Fracture: Are Specific and Geriatric Scores Better than General Scores? *Br. J. Anaesth.* **2017**, *118*, 952–954. [CrossRef] [PubMed]
15. de Jong, L.; Mal Klem, T.; Kuijper, T.M.; Roukema, G.R. Validation of the Nottingham Hip Fracture Score (NHFS) to Predict 30-Day Mortality in Patients with an Intracapsular Hip Fracture. *Orthop. Traumatol. Surg. Res. OTSR* **2019**, *105*, 485–489. [CrossRef] [PubMed]
16. Nijmeijer, W.S.; Folbert, E.C.; Vermeer, M.; Slaets, J.P.; Hegeman, J.H. Prediction of Early Mortality Following Hip Fracture Surgery in Frail Elderly: The Almelo Hip Fracture Score (AHFS). *Injury* **2016**, *47*, 2138–2143. [CrossRef] [PubMed]
17. van de Ree, C.L.; Gosens, T.; van der Veen, A.H.; Oosterbos, C.J.; Heymans, M.W.; de Jongh, M.A. Development and Validation of the Brabant Hip Fracture Score for 30-Day and 1-Year Mortality. *Hip Int. J. Clin. Exp. Res. Hip Pathol.* **2020**, *30*, 354–362. [CrossRef] [PubMed]
18. Agency for Healthcare Research and Quality Elixhauser Comorbidity Software Refined for ICD-10-CM. Available online: https://www.hcup-us.ahrq.gov/toolssoftware/comorbidityicd10/comorbidity_icd10.jsp (accessed on 2 February 2023).
19. Subramaniam, A.; Ueno, R.; Tiruvoipati, R.; Darvall, J.; Srikanth, V.; Bailey, M.; Pilcher, D.; Bellomo, R. Defining ICD-10 Surrogate Variables to Estimate the Modified Frailty Index: A Delphi-Based Approach. *BMC Geriatr.* **2022**, *22*, 422. [CrossRef] [PubMed]
20. Velanovich, V.; Antoine, H.; Swartz, A.; Peters, D.; Rubinfeld, I. Accumulating Deficits Model of Frailty and Postoperative Mortality and Morbidity: Its Application to a National Database. *J. Surg. Res.* **2013**, *183*, 104–110. [CrossRef] [PubMed]
21. Subramaniam, S.; Aalberg, J.J.; Soriano, R.P.; Divino, C.M. New 5-Factor Modified Frailty Index Using American College of Surgeons NSQIP Data. *J. Am. Coll. Surg.* **2018**, *226*, 173–181. [CrossRef] [PubMed]
22. Nieman, C.L.; Pitman, K.T.; Tufaro, A.P.; Eisele, D.W.; Frick, K.D.; Gourin, C.G. The Effect of Frailty on Short-Term Outcomes after Head and Neck Cancer Surgery. *Laryngoscope* **2018**, *128*, 102–110. [CrossRef] [PubMed]

23. ACG System Version 11.0 Technical Reference Guide. Available online: https://www.hopkinsacg.org/document/acg-system-version-11-technical-reference-guide/ (accessed on 14 May 2023).
24. R Development Core Team R: A Language and Environment for Statistical Computing. *J. Data Anal. Inf. Process.* **2008**, *3*.
25. Hosmer, D.W., Jr.; Lemeshow, S.; Sturdivant, R.X. *Applied Logistic Regression*; John Wiley & Sons: Hoboken, NJ, USA, 2013; ISBN 978-0-470-58247-3.
26. Thorne, G.; Hodgson, L. Performance of the Nottingham Hip Fracture Score and Clinical Frailty Scale as Predictors of Short and Long-Term Outcomes: A Dual-Centre 3-Year Observational Study of Hip Fracture Patients. *J. Bone Miner. Metab.* **2021**, *39*, 494–500. [CrossRef] [PubMed]
27. Lisk, R.; Yeong, K.; Fluck, D.; Fry, C.H.; Han, T.S. The Ability of the Nottingham Hip Fracture Score to Predict Mobility, Length of Stay and Mortality in Hospital, and Discharge Destination in Patients Admitted with a Hip Fracture. *Calcif. Tissue Int.* **2020**, *107*, 319–326. [CrossRef] [PubMed]
28. Ondeck, N.T.; Bovonratwet, P.; Ibe, I.K.; Bohl, D.D.; McLynn, R.P.; Cui, J.J.; Baumgaertner, M.R.; Grauer, J.N. Discriminative Ability for Adverse Outcomes After Surgical Management of Hip Fractures: A Comparison of the Charlson Comorbidity Index, Elixhauser Comorbidity Measure, and Modified Frailty Index. *J. Orthop. Trauma* **2018**, *32*, 231. [CrossRef] [PubMed]
29. Choi, J.-Y.; Kim, J.-K.; Kim, K.; Lee, Y.-K.; Koo, K.-H.; Kim, C.-H. How Does the Multidimensional Frailty Score Compare with Grip Strength for Predicting Outcomes after Hip Fracture Surgery in Older Patients? A Retrospective Cohort Study. *BMC Geriatr.* **2021**, *21*, 234. [CrossRef] [PubMed]
30. Shimizu, A.; Maeda, K.; Fujishima, I.; Kayashita, J.; Mori, N.; Okada, K.; Uno, C.; Shimizu, M.; Momosaki, R. Hospital Frailty Risk Score Predicts Adverse Events in Older Patients with Hip Fractures after Surgery: Analysis of a Nationwide Inpatient Database in Japan. *Arch. Gerontol. Geriatr.* **2022**, *98*, 104552. [CrossRef] [PubMed]
31. Kwak, M.J.; Digbeu, B.D.; des Bordes, J.; Rianon, N. The Association of Frailty with Clinical and Economic Outcomes among Hospitalized Older Adults with Hip Fracture Surgery. *Osteoporos. Int.* **2022**, *33*, 1477–1484. [CrossRef] [PubMed]
32. Gleason, L.; Benton, E.; Alvarez-Nebreda, M.; Weaver, M.; Harris, M.; Javedan, H. FRAIL Questionnaire Screening Tool and Short-Term Outcomes in Geriatric Fracture Patients. *J. Am. Med. Dir. Assoc.* **2017**, *18*, 1082–1086. [CrossRef] [PubMed]
33. Kistler, E.A.; Nicholas, J.A.; Kates, S.L.; Friedman, S.M. Frailty and Short-Term Outcomes in Patients with Hip Fracture. *Geriatr. Orthop. Surg. Rehabil.* **2015**, *6*, 209–214. [CrossRef] [PubMed]
34. Krishnan, M.; Beck, S.; Havelock, W.; Eeles, E.; Hubbard, R.E.; Johansen, A. Predicting Outcome after Hip Fracture: Using a Frailty Index to Integrate Comprehensive Geriatric Assessment Results. *Age Ageing* **2014**, *43*, 122–126. [CrossRef] [PubMed]
35. Joseph, B.; Pandit, V.; Zangbar, B.; Kulvatunyou, N.; Tang, A.; O'Keeffe, T.; Green, D.J.; Vercruysse, G.; Fain, M.J.; Friese, R.S.; et al. Validating Trauma-Specific Frailty Index for Geriatric Trauma Patients: A Prospective Analysis. *J. Am. Coll. Surg.* **2014**, *219*, 10–17. [CrossRef] [PubMed]
36. Ioannidis, I.; Mohammad Ismail, A.; Forssten, M.P.; Ahl, R.; Cao, Y.; Borg, T.; Mohseni, S. The Mortality Burden in Patients with Hip Fractures and Dementia. *Eur. J. Trauma Emerg. Surg.* **2021**, *48*, 2919–2925. [CrossRef] [PubMed]
37. Mendiratta, P.; Schoo, C.; Latif, R. Clinical Frailty Scale. In *StatPearls*; StatPearls Publishing: Treasure Island, FL, USA, 2023.
38. Pugh, R.J.; Battle, C.E.; Thorpe, C.; Lynch, C.; Williams, J.P.; Campbell, A.; Subbe, C.P.; Whitaker, R.; Szakmany, T.; Clegg, A.P.; et al. Reliability of Frailty Assessment in the Critically Ill: A Multicentre Prospective Observational Study. *Anaesthesia* **2019**, *74*, 758–764. [CrossRef] [PubMed]
39. Young, R.L.; Smithard, D.G. The Clinical Frailty Scale: Do Staff Agree? *Geriatrics* **2020**, *5*, 40. [CrossRef] [PubMed]
40. Shrier, W.; Dewar, C.; Parrella, P.; Hunt, D.; Hodgson, L.E. Agreement and Predictive Value of the Rockwood Clinical Frailty Scale at Emergency Department Triage. *Emerg. Med. J. EMJ* **2021**, *38*, 868–873. [CrossRef] [PubMed]
41. Theou, O.; Pérez-Zepeda, M.U.; van der Valk, A.M.; Searle, S.D.; Howlett, S.E.; Rockwood, K. A Classification Tree to Assist with Routine Scoring of the Clinical Frailty Scale. *Age Ageing* **2021**, *50*, 1406–1411. [CrossRef] [PubMed]
42. Archibald, M.M.; Lawless, M.; Gill, T.K.; Chehade, M.J. Orthopaedic Surgeons' Perceptions of Frailty and Frailty Screening. *BMC Geriatr.* **2020**, *20*, 17. [CrossRef] [PubMed]

Journal of Personalized Medicine

MDPI

Article

The Rising Problem of Hip Fractures in Geriatric Patients—Analysis of Surgical Influences on the Outcome

Julian Krickl [1], Till Ittermann [2], Andreas Thannheimer [1], Wolfgang Schmidt [1], Maximilian Willauschus [3], Johannes Ruether [3], Kim Loose [3], Markus Gesslein [3] and Michael Millrose [1,3,*]

[1] Department of Trauma Surgery and Sports Medicine, Garmisch-Partenkirchen Medical Centre, 82467 Garmisch-Partenkirchen, Germany; julian.krickl@klinikum-gap.de (J.K.); andreas.thannheimer@klinikum-gap.de (A.T.); wolfgang.schmidt@klinikum-gap.de (W.S.)
[2] Institute for Community Medicine, SHIP/Clinical-Epidemiological Research, University of Greifswald, 17489 Greifswald, Germany; till.ittermann@uni-greifswald.de
[3] Department of Orthopedics and Traumatology, Paracelsus Medical University, 90471 Nuremberg, Germany; maximilian.willauschus@klinikum-nuernberg.de (M.W.); johannes.ruether@stud.pmu.ac.at (J.R.); loose.kim@gmail.com (K.L.); markus.gesslein@klinikum-nuernberg.de (M.G.)
* Correspondence: michael.millrose@klinikum-gap.de; Tel.: +49-8821-771220

Abstract: Background: Hip fractures in geriatric patients often have a poor outcome in terms of mortality, mobility as well as independence. Different surgical influence factors are known that improve the outcome. Methods: In this observational cohort study, 281 patients of a geriatric trauma unit were analyzed prospectively. Demographic factors, as well as data regarding the trauma mechanism and perioperative treatment, were recorded. The nutritional status was also analyzed. The follow-up was set to 120 days. Results: The key conclusion of this study is that a high ASA classification, the use of anticoagulatory medicine and malnutrition are significantly associated with higher mortality together with worse independence ($p < 0.05$). There is no significant difference in outcome concerning the time to surgery within the first 24 h. Conclusions: Malnutrition seems to be an important risk factor for an adverse outcome of geriatric patients and therefore warrants a focus in multidisciplinary treatment. The risk factor ASA cannot be improved during the pre-surgery phase, but requires intensified care by a multidisciplinary team specialized in orthogeriatrics.

Keywords: hip fractures; geriatric trauma; orthogeriatrics; frailty; malnutrition

Citation: Krickl, J.; Ittermann, T.; Thannheimer, A.; Schmidt, W.; Willauschus, M.; Ruether, J.; Loose, K.; Gesslein, M.; Millrose, M. The Rising Problem of Hip Fractures in Geriatric Patients—Analysis of Surgical Influences on the Outcome. *J. Pers. Med.* **2023**, *13*, 1271. https://doi.org/10.3390/jpm13081271

Academic Editor: Jih-Yang Ko

Received: 21 July 2023
Revised: 13 August 2023
Accepted: 16 August 2023
Published: 17 August 2023

1. Introduction

In a population that is continuously growing older due to the demographic change marked by low birth rates and a rising life expectancy, hip fractures (HF) are among the most critical acute health risk factors in elderly patients. Since more than 90% of hip fractures are accredited to elderly patients over 65 years of age, geriatric trauma surgeons are confronted with preexisting multimorbidity, polymedication and other general geriatric challenges [1]. In Germany in 2021, the incidence of HF in patients 65 years and older is nearly six-fold compared to non-geriatric patients with an additional increased risk of death of up to nineteen-fold [2,3]. The increased mortality risk persists for up to ten years [4]. In a recent prospective study and systematic review, it is clearly depicted that HFs in geriatric patients often cause a considerable deterioration of the functional capacity for both the basic and instrumental activities of daily living. This leads to a significant decline in self-care, daily activities and mobility [5,6]. These fractures also increase the overall mortality of this group of elderly patients [7].

The treatment of HF in elderly patients is most frequently either by total hip arthroplasty or osteosynthesis, and conservative treatment is the exception [8]. Regarding the treatment of HF in geriatric patients, international guidelines, for example, by the American association of orthopedic surgeons, recommend surgery within 24–48 h [9]. These recommendations are based on population-based cohort studies and systematic reviews which show an increased risk of 30-day mortality as well as a poorer outcome [10,11]. Common complications in geriatric patients with HF are infections (systemic or surgery related), perioperative anemia, cognitive impairment (dementia and delirium), metabolic disorders and often subsequent deterioration in mobility and independence [12,13].

Considering this in combination with the estimated global increase in HF from 1.26 million in 1990 to up to 4.5 million in 2050, there is a need to further optimize the treatment and risk evaluation of geriatric patients, since those fractures and their consequences will have an increasing impact on health systems and societies and, consequently, future medical and social costs [14]. Knowing the major risk factors for higher morbidity and mortality might enable physicians to screen patients who are in need of extended diagnostics and special perioperative treatment [15,16].

In previous studies, it could be shown that elderly patients with HF who suffer from malnutrition have increased mortality within the first 12 months and functional dependence over a longer period with the need for more assisted living arrangements, such as nursing homes [17,18].

In this present study, we aim to show the effects of surgical factors, such as time interval to surgery, implant and invasiveness, as well as potential perioperative risk factors, including anemia, dementia or delirium, on mortality and functional outcome. We additionally analyze the impact of preoperative anticoagulation on the outcome and time to surgery and whether malnutrition, assessed by the nutritional risk score (NRS), exerts a higher influence on the outcome than surgical factors.

A secondary analysis of different surgically interesting questions was conducted: whether there is a correlation between the time to surgery and prescribed anticoagulants; whether more invasive surgical techniques require more blood transfusions during the inpatient stay; and whether malnourished geriatric patients show a longer time before surgery due to the need for the preoperative improvement of different blood levels.

2. Materials and Methods

This observational cohort study according to the STROBE statement was conducted from January 2018 to December 2020 at a regional care hospital with a department for geriatric trauma surgery [19]. During the entire study period, 1005 elderly patients were admitted to the department with a trauma diagnosis. All geriatric patients with a proximal femoral fracture, regardless of the traumatic or pathological cause, were invited to participate in this study. The traumatic genesis was mostly falls, either because of external reasons, such as slipping or stumbling, or internal reasons, such as syncope or the like. Out of the 1005 admitted patients, all 305 patients with HF were prospectively included in this study. Twenty-four study participants were lost during the follow-up and 281 patients (77 male and 204 female) could be analyzed. Since all patients complying with the inclusion and exclusion criteria could be included, which defined our sample group, the selection bias was minimal.

The inclusion criteria of the study were a minimum age of 65 years and an osteosynthesis of the proximal femur or joint replacement of the hip, regardless of the implant used and whether it was done in an open or closed way. The exclusion criteria of this study were the lack of consent or ability to consent (either themselves or their legal custodians) to the study. Data sets which lack three or more items, regardless of type, were also excluded to minimize the information bias.

The demographics and nutritional status of a part of the study population have already been analyzed and published [18]. The population of this study has been complemented up until December 2020 and focuses on a completely different data set.

This study was conducted in accordance with the guidelines of the declaration of Helsinki. Analysis and publication were permitted by the Institutional Review Board of Paracelsus Medical University (protocol code IRB-2022-018; approved on 23 November 2022). Written informed consent was obtained from every participant or their respective legal custodian.

At inclusion, demographic parameters such as age, sex, preoperative living situation as well as walking ability were recorded. The obtained medical information included the type of fracture classified on a preoperative x-ray, American Society of Anesthesiologists (ASA) classification and prescribed anticoagulatory medication. The time between admission and surgery was logged and categorized into four different categories: <6 h, meaning that the geriatric patient was diagnosed in the emergency department and brought directly to the operation theatre; categories 2 and 3 were 6–12 h and 12–24 h, respectively, because of additional preoperative diagnostics or the lack of OR capacity; and the fourth category of >24 h because of different influence factors. Also, the operation procedure was obtained. Additional to the surgical procedure itself, the invasiveness in terms of closed versus open procedures were differentiated. The outcome parameters were death during the inpatient stay and after 120 days and an alteration in the patient's living situation and self-dependency, as well as the ability to walk with or without the use of aiding tools. Those changes were acquired at follow-up 120 days after admission.

The preoperative hemoglobin level was routinely obtained. Also, the minimal hemoglobin level of the routine blood works during the inpatient stay of the study participant was analyzed. The need for blood transfusion and the amount of packed red blood cells (pRBCs) as well as the need for antibiotic therapy and whether it was necessary because of surgery-related complications or other infections were recorded. The presence of dementia was assessed either anamnestically or with a mini-mental state examination [20]. Also, the occurrence of postoperative delirium was included in the analysis.

To assess the nutritional status of the included geriatric patients, the body mass index (BMI) and nutritional risk score were analyzed. The nutritional risk score, a simple and well-validated tool which is one of the most often used screening tools in hospitals worldwide, was recorded during the inpatient stay by nutritionists [21]. A total score of $\geq$3 points means that the patient is at risk of malnutrition or already malnourished and, therefore, a nutritional therapy might be necessary.

The data were acquired prospectively and analyzed retrospectively. The statistical analysis was completed by using STATA 17.0 (Stata Corporation, College Station, TX, USA). Features of the study population are reported and stratified by sex as the mean and standard deviation for continuous variables and as absolute numbers and percentages for categorical variables. Associations of potential risk factors with inpatient death, mortality after 120 days, worsened mobility after surgery and worsened living situation postoperatively were analyzed by logistic regression models according to the Wald Chi-squared test. Those models were adjusted for age and sex. The results are reported as odds ratio, 95% confidence interval and p-value. A $p <$ 0.05 was considered as statistically significant. The time between admission and surgery was categorized into quarters of the first 24 h. The invasiveness and implant were categorized into procedures with bigger skin incisions and exposure as "open" versus percutaneous procedures as "closed" in order to detect the influence of additional iatrogenic trauma.

3. Results

During this prospective observational cohort study, 281 geriatric patients were included and analyzed. They sustained hip or proximal femur fractures and were admitted and treated at the geriatric trauma surgery department of a regional trauma center. Of the 281 study patients included, 204 were female with a mean age of 84.5 $\pm$ 7.2 years. The 77 men had a mean age of 83.53 $\pm$ 5.92 years. The measurements of the geriatric patients can be found in the supplementary material (Table S1). The BMIs of the included female and male patients were 23.69 $\pm$ 4.06 and 24.33 $\pm$ 3.4, respectively. Concerning the perioperative risk depicted in the ASA classification, 3 were categorized as healthy patients with ASA 1, 50 had mild systemic disease (ASA 2), 203 had severe systemic diseases (ASA 3) and 25 were constantly threatened by death because of their systemic disease-ASA 4.

During the phase of inpatient treatment, 25 patients died, while at the follow-up that occurred 120 days after surgery, the number increased to 51 patients. The details of the accessory outcome parameters (mobility and independence at follow-up) are shown in Table 1.

Table 1. Details of the additional outcome parameters: mobility and independence at follow-up.

Mobility	Independent	Cane/Crutch	Rollator	Only Indoor	Bedridden
Preadmission [n = 281]	97	31	66	76	11
Follow-up 120d [n = 230]	36	30	79	58	27
Independence	**Own home**	**Nursing home**	**Hospital**		
Preadmission [n = 281]	221	60			
Follow-up 120d [n = 230]	177	51	2		

The level of mobility of the 230 patients available at follow-up was improved or stayed the same in comparison to their preadmission level in 114 patients, while 116 patients suffered from a worse mobility level, namely those who needed additional walking support or were only mobile indoors and needed a wheelchair for outside.

With respect to the independence of the living situation, 6 patients could live in a more self-reliant manner, 201 lived as before and 23 lost their ability to live independently and were cared for in nursing homes or in a hospital.

The data of the analyzed surgical risk factors and malnutrition are depicted in Table 2 in detail. The included patients of this study groups solely used phenprocoumon as an oral anticoagulant (OAC) and Apixaban, Edoxaban or Rivaroxaban as a novel oral anticoagulant (NOAC). The prescribed platelet aggregation inhibitors were either acetylsalicylic acid or Clopidogrel.

Table 2. Details of the different analyzed surgical risk factors.

	Female (n = 204)	**Male (n = 77)**	**Total (n = 281)**
Anticoagulant			
None	104 (51.0%)	28 (36.4%)	132 (47.0%)
OAC	13 (6.4%)	8 (10.4%)	21 (7.5%)
PAI	46 (22.5%)	17 (22.1%)	63 (22.4%)
NOAC	37 (18.1%)	15 (19.5%)	52 (18.5%)
combination	4 (2.0%)	9 (11.7%)	13 (4.6%)
Any anticoagulant use	96 (47.1%)	40 (51.9%)	136 (48.4%)
Time between admission and surgery	16.1 (29.0)	15.0 (21.9)	15.8 (27.2)
<6 h	78 (38.2%)	25 (32.5%)	103 (36.7%)
6 to 12 h	31 (15.2%)	18 (23.4%)	49 (17.4%)
12 to 24 h	63 (30.9%)	23 (29.9%)	86 (30.6%)
>24 h	32 (15.7%)	11 (14.3%)	43 (15.3%)
Invasiveness			
Closed	96 (47.1%)	42 (54.5%)	138 (49.1%)
Open	108 (52.9%)	35 (45.5%)	143 (50.9%)
Implant			
DHS or nail (CRIF)	103 (50.5%)	46 (59.7%)	149 (53.0%)
Nail (ORIF) or THR	88 (43.1%)	30 (39.0%)	118 (42.0%)
Osteosynthesis	13 (6.4%)	1 (1.3%)	14 (5.0%)
Hemoglobin preoperative [g/dL]	12.5 (1.7)	12.4 (2.0)	12.5 (1.8)
Hemoglobin minimal [g/dL]	8.2 (1.4)	8.4 (1.6)	8.3 (1.5)
Transfusion	80 (39.2%)	27 (35.1%)	107 (38.1%)
Amount of transfused pRBC	1.9 (1.0)	1.9 (1.2)	1.9 (1.0)
Antibiotic	82 (40.2%)	29 (37.7%)	111 (39.5%)
Dementia	126 (62.1%)	38 (50.0%)	164 (58.8%)
Delirium	24 (11.8%)	12 (15.8%)	36 (12.9%)
NRS			
<3	53 (40.2%)	26 (49.1%)	79 (42.7%)
≥3	79 (59.8%)	27 (50.9%)	106 (57.3%)

OAC: oral anticoagulants; PAI: platelet aggregation inhibitors; NOAC: novel oral anticoagulants; DHS: dynamic hip screw; nail (CRIF): closed reduction and internal fixation with trochanteric nail; nail (ORIF): open reduction and internal fixation with trochanteric nail; THR: total hip replacement; NRS: nutritional risk score; categorial data are shown as absolute numbers and percentages.

The results of the regression analysis are shown in Table 3. The analysis shows that the preoperative prescription of an anticoagulant increases mortality during the inpatient stay. OACs especially show a highly elevated risk of death during the clinical stay as well as during the follow-up, whereas PAIs do not seem to have an influence on mortality. As expected, anticoagulation has no effect on mobility and independence. The data show that severely ill geriatric patients with ASA 4, and to a lesser degree, also ASA 3, significantly worsens the outcome with respect to mortality as well as independence. The surgical technique and implant used had a significant influence on mobility. The open surgical technique used in 64% of patients with THR improved mobility at the follow-up occurring 120 days after surgery. A preoperative anemia is associated with higher mortality and worsened mobility at follow-up. The minimal hemoglobin level during the whole inpatient stay had no significant effect on the outcome parameters. In concordance, the amount of transfused pRBCs shows a similar correlation with mortality at follow-up, similar to the preoperative hemoglobin level. The nutritional status of the geriatric patient, as reflected by a lower BMI and higher NRS, shows significantly higher mortality during inpatient stay and at follow-up, as well as worsened independence.

Table 3. Associations of surgical risk factors with mortality during inpatient stay or at follow-up and worsened mobility or domicile (n = 281).

	Mortality Inpatient OR (95%-CI)	Mortality at 120 d OR (95%-CI)	Worsened Mobility OR (95%-CI)	Worsened Domicile OR (95%-CI)
Any anticoagulant use	3.16 (1.23; 8.15) *	1.84 (0.97; 3.49)	1.17 (0.72; 1.90)	1.24 (0.72; 2.14)
OAC vs. none	9.28 (2.29; 37.6) *	3.00 (1.01; 8.90) *	1.40 (0.52; 3.75)	1.82 (0.67; 4.98)
PAI vs. none	1.76 (0.44; 7.04)	1.59 (0.69; 3.65)	1.30 (0.69; 2.42)	1.00 (0.49; 2.07)
NOAC vs. none	4.95 (1.46; 16.7) *	1.99 (0.86; 4.63)	0.99 (0.52; 1.92)	1.41 (0.68; 2.94)
ASA risk classification				
3 vs. (1,2)	3.16 (0.40; 25.2)	10.0 (1.33; 75.6) *	0.95 (0.50; 1.78)	2.45 (0.97; 6.17)
4 vs. (1,2)	15.7 (1.73; 142) *	29.2 (3.39; 251) *	1.26 (0.45; 3.54)	5.14 (1.57; 16.9) *
Time between admission and surgery				
6 to 12 h vs. <6 h	1.74 (0.57; 5.27)	1.27 (0.53; 3.00)	1.10 (0.53; 2.29)	1.02 (0.47; 2.22)
12 to 24 h vs. <6 h	0.95 (0.30; 2.95)	1.04 (0.48; 2.26)	0.61 (0.34; 1.10)	1.02 (0.53; 1.98)
>24 h vs. <6 h	0.75 (0.18; 3.21)	0.72 (0.26; 2.05)	0.82 (0.39; 1.72)	0.59 (0.24; 1.48)
Invasiveness Open vs. closed	2.07 (0.84; 5.11)	1.77 (0.93; 3.39)	0.57 (0.35; 0.93) *	1.22 (0.70; 2.12)
Implant (THR, nail ORIF) vs. (DHS, nail CRIF)	1.26 (0.52; 3.04)	1.40 (0.73; 2.70)	0.52 (0.32: 0.87) *	1.02 (0.58; 1.80)
Body mass index [kg/m^2]	0.93 (0.81; 1.06)	0.90 (0.82; 0.99) *	1.00 (0.94; 1.06)	0.94 (0.87; 1.02)
Hemoglobin preoperative [g/dL]	0.88 (0.70; 1.11)	0.79 (0.66; 0.94) *	0.84 (0.72; 0.97) *	0.84 (0.72; 0.98) *
Hemoglobin minimal [g/dL]	0.72 (0.52; 1.01)	0.86 (0.69; 1.08)	0.88 (0.75; 1.04)	0.89 (0.73; 1.08)
Transfusion	1.32 (0.55; 3.15)	1.91 (1.01; 3.62) *	1.47 (0.88; 2.43)	1.61 (0.92; 2.81)
Amount of transfused pRBCs	1.28 (0.68; 2.42)	1.77 (1.11; 2.83) *	1.37 (0.88; 2.15)	1.57 (1.01; 2.43) *
Antibiotics use	0.83 (0.34; 2.03)	1.34 (0.71; 2.52)	1.21 (0.74; 1.98)	1.43 (0.82; 2.50)
Dementia	0.65 (0.24; 1.76)	0.92 (0.43; 1.98)	0.95 (0.54; 1.68)	1.15 (0.59; 2.25)
Delirium	0.16 (0.02; 1.38)	0.92 (0.32; 2.64)	1.11 (0.48; 2.56)	1.08 (0.42; 2.74)
Nutritional risk score ≥3 vs. <3	6.69 (1.39; 32.1) *	3.30 (1.30; 8.36)*	1.44 (0.78; 2.66)	2.68 (1.25; 5.76) *

Logistic regression models adjusted for age and sex, OR: odds ratio: CI: confidence interval; statistical significance level * $p < 0.05$.; green: significantly lower risk for adverse event; red: significantly higher risk for adverse event. OAC: oral anticoagulants; PAI: platelet aggregation inhibitors; NOAC: novel oral anticoagulants; DHS: dynamic hip screw; nail (CRIF): closed reduction and internal fixation with trochanteric nail; nail (ORIF): open reduction and internal fixation with trochanteric nail; THR: total hip replacement.

The secondary analysis showed that there was no significant influence of a prescribed anticoagulation medication and the time between admission and surgery ($p = 0.76$ OAC; $p = 0.350$ PAI; $p = 0.559$ NOAC). Also, there was no significant correlation between the invasiveness of the surgical technique and the need for pRBCs ($p = 0.704$). The malnutrition of the geriatric patient had no significant effect on the time to surgery after admission ($p = 0.153$).

4. Discussion

The results of this study show different statistically significant risk factors for a worse outcome after HF in geriatric patients. Interestingly, the highest influence could be established for pre-trauma conditions, namely the use of anticoagulant medication, morbidity because of underlying health conditions (depicted in the ASA classification), as well as malnutrition. All the above-mentioned influence factors are typically present in a growing elderly patient population; therefore, they should move into the focus of geriatric trauma surgeons.

In our study, we could show that pre-surgery anemia is correlated with higher mortality and a worsened outcome in terms of mobility and independence. This is in concordance with recently published studies [22,23]. Also, elderly patients are much more likely to receive an anticoagulatory medication, which as an influence factor itself, increases mortality during inpatient stay and might also increase the severity of blood loss from the fracture, both during surgery and postoperatively. In a systematic review, it could be shown that pre-surgery anemia is an independent risk factor for surgery-related infections [24]. Hansen et al. suggested to treat the preexisting anemia with intravenous iron therapy during the inpatient stay, but could not show an effect on surgery-related infections [25]. A Cochrane systematic review from 2015 regarding the transfusion of pRBCs for patients undergoing HF surgery could show no beneficiary effect of a liberal transfusion strategy in terms of mortality [26]. Based on the results of our study, this conclusion might need reconsideration since the mortality, mobility and independence of geriatric patients are all affected. Nevertheless, our results showed, significantly, that the need for a transfusion and the amount of pRBCs during the inpatient stay increases mortality at follow-up, while the minimal hemoglobin level only showed trends, which is in concordance with a recently published study by Garg et al. [27]. Since we apply a restrictive transfusion strategy in our geriatric trauma unit, the patients most likely showed clinical signs of anemia before they were transfused, which is probably because of the comorbidities reflected in their ASA classification.

As one would expect, the data of this prospective study showed that a higher ASA risk classification of the geriatric patient dramatically increases the rate of death during the inpatient stay and after 120 days at follow-up. This is comparable to the results of Haugn et al. and Quach et al., who could show that multimorbidity, depicted by a higher ASA risk score, might lead to higher mortality 30 days and 1 year after HF, respectively [28,29]. Cher et al. even postulated that comorbidities are the dominant influence factor of mortality after HF in the geriatric patient [30]. These findings might lead to the impression that severely ill geriatric patients with HF should be treated conservatively to spare them the perioperative risks. Different systematic reviews and meta-analysis clearly show that the nonoperative treatment of those patients results in a poor prognosis with higher complication rates and mortality [8,31]. Therefore, the surgical treatment of HF in geriatric patients is strongly recommended in international guiding principles [9].

There is clear evidence in the literature that a short time to surgery has a beneficial effect on short- and long-term mortality [32,33]. An early surgery is also associated with a reduced pRBC transfusion rate [34]. Those studies define an early surgery by time to surgery of less than 48 h. The results of our study are in contrast since we were not able to show an influence of time to surgery with statistical significance. There might be different reasons. In a recent study by Fenwick et al., the authors reported that short-term mortality did not increase during the first 48 h of inpatient stay [35]. Since we categorized the time to surgery only during the first 24 h, the differences in between might be too subtle to be detected within the included patient collective. Another aspect is that a perioperative optimization of the patient, for example, in terms of electrolyte imbalances, can improve the outcome [36,37]. Since most of the preoperative interventions are completed in the emergency department, they most likely take place during the first 6-12 h with a developing effect on the geriatric patient within 24 h. Therefore, the mortality of those geriatric patients in need of optimization might decrease during the first 24 h. There is the need for an evaluation of the specific preoperative optimization within our study collective in order to substantiate this attempt of explanation.

Our study showed that prescribed oral anticoagulatory medicine, especially vitamin k antagonists, increases the risk for mortality during the inpatient stay and follow-up. This is in concordance with a prospective cohort study by Blanco et al. in 2021, who analyzed the influence of preoperative clinical variables on 30-day mortality [38]. One reason could be that particularly patients who suffer from chronic cardiac insufficiency or have a previous diagnosis of arrhythmia or previous history of ischemic heart failure usually take anticoagulant treatment. This special group of elderly patients with cardiovascular disease are at higher risk of dying, which Cha et al. demonstrated in 2019 [39]. In the studies by Blanco et al. as well as Rutenberg et al. in 2018, the use of the vitamin K antagonist was associated with a longer time to surgery, which both authors offered as a reason for increased mortality [38,40]. Interestingly, we could not show any statistically significant association between anticoagulatory medicine and time between admission and surgery. This might be due to the short time intervals defined in our study.

In accordance with our study on the effect of the malnutrition of elderly patients on mortality and mobility, we could show the same effect in this patient collective [18] Recent systematic reviews find a high prevalence of malnutrition amongst elderly patients and support the beneficial role of oral nutrition supplements [41,42]. Gonul et al. showed in their recent study that malnutrition increases the odds of mortality 30 days after surgery by 4.2 times, which is in concordance with our findings of an increased risk of dying during the inpatient stay of 6.7 and at 120 days after surgery by 3.3 [43]. A Cochrane review produced evidence that a multidisciplinary rehabilitation, including the optimization of nutrition, resulted in fewer cases of poor outcome in terms of mortality and independence [44]. Since malnutrition is a risk factor which is accessible for treatment during inpatient stay and rehabilitation treatment, this should be paid attention to by responsible physicians.

One of the strengths of our study is the prospective study design of 281 consecutive patients treated at a geriatric trauma department for HF. During the inclusion period, no essential changes in therapy were made. Also, the study population shows a high percentage of patients (85%) who were surgically treated within the first 24 h, which enables this study to further analyze the early time span of therapy, while most of the published studies differentiate between less or more than 48 h [32,33]. Additionally, concerning the important risk factor of malnutrition, the NRS in our study was obtained by nutritionists. Benoit et al. were able to show in a clinical study that the assessment of the nutritional status by surgeons was significantly worse than by nutritionists [45]. Therefore, the nutritional data acquired in our study represent the real status of the included patient more accurately.

One limitation of this study is the follow-up of only 120 days. Peeters et al. showed in a systematic review that the outcome of elderly patients after HF can improve during the first 6 months; therefore, the reported outcomes in terms of mortality or mobility and to a lesser degree, independence, might be worse after 6 months or longer of rehabilitation [5]. Confounding diagnoses likely to be present in geriatric patients, such as chronic obstructive pulmonary disease, cardiovascular disease or diabetes, were not analyzed and could have an effect on the outcome after the HF of the elderly patient. The blood iron level and the type of pre-surgery anemia should have been obtained in order to differentiate perioperative blood loss from iron-deficiency anemia.

5. Conclusions

Geriatric patients with HF are at high risk of experiencing an adverse outcome. Besides risk factors that cannot be altered pre-surgically by interdisciplinary treatment, such as the health status of the patients depicted in the ASA classification, malnutrition seems to be an important influence factor on the outcome, which should be treated during the inpatient stay and rehabilitation. Since prescribed anticoagulatory medication also shows a correlation to higher mortality, the perioperative therapy in terms of antidotes or supplementation should gain the appropriate attention.

Supplementary Materials: The following supporting information can be downloaded at: https://www.mdpi.com/article/10.3390/jpm13081271/s1, Table S1: Data of included patients.

Author Contributions: Conceptualization, M.M. and M.G.; methodology, M.M.; software, T.I.; validation, J.K., J.R. and M.W.; formal analysis, T.I.; investigation, J.K.; resources, A.T.; data curation, J.K. and W.S.; writing—original draft preparation, M.M. and J.K.; writing—review and editing, K.L. and M.G.; supervision, A.T. and W.S.; project administration, M.M. All authors have read and agreed to the published version of the manuscript.

Funding: This research received no external funding.

Institutional Review Board Statement: The study was conducted in accordance with the Declaration of Helsinki and approved by the Institutional Review Board of Paracelsus Medical University (protocol code IRB-2022-018 on 23 November 2022).

Informed Consent Statement: Informed consent was obtained from all subjects through their legal custodians involved in the study.

Data Availability Statement: The data of the study population are provided in the Supplementary Material Table S1. The complete data of the regression analysis depicted in Table 3 are available in the supplementary material complete regression analysis results (Table 3).

Acknowledgments: The authors thank Stefanie Karpik for her support in preparing and correcting the manuscript with respect to spelling and grammar.

Conflicts of Interest: The authors declare no conflict of interest.

References

1. Carpintero, P.; Caeiro, J.R.; Carpintero, R.; Morales, A.; Silva, S.; Mesa, M. Complications of hip fractures: A review. *World J. Orthop.* **2014**, *5*, 402–411. [CrossRef] [PubMed]
2. Federal Health Monitoring System. Diagnostic Data of the Hospitals by Place of Treatment for ICD S70-S79. 2021. Available online: https://www.gbe-bund.de/gbe/!pkg_olap_tables.prc_set_page?p_uid=gast&p_aid=9545559&p_sprache=D&p_help=2&p_indnr=702&p_ansnr=91237172&p_version=12&D.001=1000001&D.002=34340&D.003=1000004&D.972=1000619&D.100=10101 (accessed on 13 July 2023).
3. Federal Health Monitoring System. Death Data for ICD S70-S79. 2020. Available online: https://www.gbe-bund.de/gbe/!pkg_olap_tables.prc_set_page?p_uid=gast&p_aid=72189493&p_sprache=D&p_help=2&p_indnr=6&p_ansnr=66475306&p_version=7&D.000=3743&D.001=1000001&D.002=1000002&D.003=1000004&D.004=1000006&D.011=44302 (accessed on 13 July 2023).
4. Haentjens, P.; Magaziner, J.; Colón-Emeric, C.S.; Vanderschueren, D.; Milisen, K.; Velkeniers, B.; Boonen, S. Meta-analysis: Excess mortality after hip fracture among older women and men. *Ann. Intern. Med.* **2010**, *152*, 380–390. [CrossRef] [PubMed]
5. Peeters, C.M.; Visser, E.; Van de Ree, C.L.; Gosens, T.; Den Oudsten, B.L.; De Vries, J. Quality of life after hip fracture in the elderly: A systematic literature review. *Injury* **2016**, *47*, 1369–1382. [CrossRef] [PubMed]
6. Amarilla-Donoso, F.J.; López-Espuela, F.; Roncero-Martín, R.; Leal-Hernandez, O.; Puerto-Parejo, L.M.; Aliaga-Vera, I.; Toribio-Felipe, R.; Lavado-García, J.M. Quality of life in elderly people after a hip fracture: A prospective study. *Health Qual. Life Outcomes* **2020**, *18*, 71. [CrossRef] [PubMed]
7. Garre-Fivelsdal, T.E.; Gjertsen, J.E.; Dybvik, E.; Bakken, M.S. A standardized clinical pathway for hip fracture patients is associated with reduced mortality: Data from the Norwegian Hip Fracture Register. *Eur. Geriatr. Med.* **2023**, *14*, 557–564. [CrossRef]
8. Loggers, S.A.I.; Van Lieshout, E.M.M.; Joosse, P.; Verhofstad, M.H.J.; Willems, H.C. Prognosis of nonoperative treatment in elderly patients with a hip fracture: A systematic review and meta-analysis. *Injury* **2020**, *51*, 12407–12413. [CrossRef]
9. O'Connor, M.I.; Switzer, J.A. AAOS Clinical Practice Guideline Summary: Management of Hip Fractures in Older Adults. *J. Am. Acad. Orthop. Surg.* **2022**, *30*, e1291–e1296. [CrossRef]

10. Pincus, D.; Ravi, B.; Wasserstein, D.; Huang, A.; Paterson, J.M.; Nathens, A.B.; Kreder, H.J.; Jenkinson, R.J.; Wodchis, W.P. Association between Wait Time and 30-Day Mortality in Adults Undergoing Hip Fracture Surgery. *JAMA* **2017**, *318*, 1994–2003. [CrossRef]
11. Klestil, T.; Röder, C.; Stotter, C.; Winkler, B.; Nehrer, S.; Lutz, M.; Klerings, I.; Wagner, G.; Gartlehner, G.; Nussbaumer-Streit, B. Impact of timing of surgery in elderly hip fracture patients: A systematic review and meta-analysis. *Sci. Rep.* **2018**, *8*, 13933. [CrossRef]
12. Song, Y.; Wu, Z.; Huo, H.; Zhao, P. The Impact of Frailty on Adverse Outcomes in Geriatric Hip Fracture Patients: A Systematic Review and Meta-Analysis. *Front. Public Health* **2022**, *10*, 890652. [CrossRef]
13. Greenstein, A.S.; Gorczyca, J.T. Orthopedic Surgery and the Geriatric Patient. *Clin. Geriatr. Med.* **2019**, *35*, 65–92. [CrossRef]
14. Veronese, N.; Maggi, S. Epidemiology and social costs of hip fracture. *Injury* **2018**, *49*, 1458–1460. [CrossRef] [PubMed]
15. Zuelzer, D.A.; Weaver, D.; Zuelzer, A.P.; Hessel, E.A. Current Strategies in Medical Management of the Geriatric Hip Fracture Patient. *J. Am. Acad. Orthop. Surg.* **2023**, *31*, 620–626. [CrossRef] [PubMed]
16. Tarazona-Santabalbina, F.J.; Ojeda-Thies, C.; Figueroa Rodríguez, J.; Cassinello-Ogea, C.; Caeiro, J.R. Orthogeriatric Management: Improvements in Outcomes during Hospital Admission due to Hip Fracture. *Int. J. Environ. Res. Public Health* **2021**, *18*, 3049. [CrossRef]
17. Foo, M.X.E.; Wong, G.J.Y.; Lew, C.C.H. A systematic review of the malnutrition prevalence in hospitalized hip fracture patients and its associated outcomes. *JPEN J. Parenter. Enter. Nutr.* **2021**, *45*, 1141–1152. [CrossRef] [PubMed]
18. Millrose, M.; Schmidt, W.; Krickl, J.; Ittermann, T.; Ruether, J.; Bail, H.J.; Gesslein, M. Influence of Malnutrition on Outcome after Hip Fractures in Older Patients. *J. Pers. Med.* **2023**, *13*, 109. [CrossRef]
19. Von Elm, E.; Altman, D.G.; Egger, M.; Pocock, S.J.; Gøtzsche, P.C.; Vandenbroucke, J.P.; STROBE Initiative. The Strengthening the Reporting of Observational Studies in Epidemiology (STROBE) statement: Guidelines for reporting observational studies. *PLoS Med.* **2007**, *4*, e296. [CrossRef]
20. Folstein, M.F.; Folstein, S.E.; McHugh, P.R. "Mini-mental state": A practical method for grading the cognitive state of patients for the clinician. *J. Psychiatr. Res.* **1975**, *12*, 189–198. [CrossRef]
21. Kondrup, J.; Rasmussen, H.H.; Hamberg, O.; Stanga, Z.; Ad Hoc ESPEN Working Group. Nutritional risk screening (NRS 2002): A new method based on an analysis of controlled clinical trials. *Clin. Nutr.* **2003**, *22*, 321–336. [CrossRef]
22. Yombi, J.C.; Putineanu, D.C.; Cornu, O.; Lavand'homme, P.; Cornette, P.; Castanares-Zapatero, D. Low haemoglobin at admission is associated with mortality after hip fractures in elderly patients. *Bone Jt. J.* **2019**, *101*, 1122–1128. [CrossRef]
23. Daugaard, C.; Pedersen, A.B.; Kristensen, N.R.; Johnsen, S.P. Preoperative antithrombotic therapy and risk of blood transfusion and mortality following hip fracture surgery: A Danish nationwide cohort study. *Osteoporos. Int.* **2019**, *30*, 583–591. [CrossRef] [PubMed]
24. Potter, L.J.; Doleman, B.; Moppett, I.K. A systematic review of pre-operative anaemia and blood transfusion in patients with fractured hips. *Anaesthesia* **2015**, *70*, 483–500. [CrossRef] [PubMed]
25. Hansen, L.T.; Riis, J.; Kragholm, K.H.; Larsen, L.K.; Cavallius, C.; Mørch, M.M.; Clemmensen, S.Z.; Krogager, M.L.; Melgaard, D. Impact of postoperative intravenous iron therapy on postoperative infections in older patients with severe anaemia after hip fracture surgery. *BMC Geriatr.* **2023**, *23*, 95. [CrossRef] [PubMed]
26. Brunskill, S.J.; Millette, S.L.; Shokoohi, A.; Pulford, E.C.; Doree, C.; Murphy, M.F.; Stanworth, S. Red blood cell transfusion for people undergoing hip fracture surgery. *Cochrane Database Syst. Rev.* **2015**, *4*, CD009699. [CrossRef]
27. Garg, V.; Kandhari, V.K.; Nasim, O.; Joshi, Y. Effect of Peri-Operative Blood Transfusion on Short and Long-Term Mortality Rates in Elderly Patients with Neck of Femur Fractures: A Retrospective Study. *Cureus* **2023**, *15*, e38825. [CrossRef] [PubMed]
28. Haugan, K.; Klaksvik, J.; Foss, O.A. 30-day mortality in patients after hip fracture surgery: A comparison of the Charlson Comorbidity Index score and ASA score used in two prediction models. *Injury* **2021**, *52*, 2379–2383. [CrossRef]
29. Quach, L.H.; Jayamaha, S.; Whitehouse, S.L.; Crawford, R.; Pulle, C.R.; Bell, J.J. Comparison of the Charlson Comorbidity Index with the ASA score for predicting 12-month mortality in acute hip fracture. *Injury* **2020**, *51*, 1004–1010. [CrossRef]
30. Cher, E.W.L.; Allen, J.C.; Howe, T.S.; Koh, J.S.B. Comorbidity as the dominant predictor of mortality after hip fracture surgeries. *Osteoporos. Int.* **2019**, *30*, 2477–2483. [CrossRef]
31. Kim, S.J.; Park, H.S.; Lee, D.W. Outcome of nonoperative treatment for hip fractures in elderly patients: A systematic review of recent literature. *J. Orthop. Surg.* **2020**, *28*, 2309499020936848. [CrossRef]
32. Beaupre, L.A.; Khong, H.; Smith, C.; Kang, S.; Evens, L.; Jaiswal, P.K.; Powell, J.N. The impact of time to surgery after hip fracture on mortality at 30- and 90-days: Does a single benchmark apply to all? *Injury* **2019**, *50*, 950–955. [CrossRef]
33. Leer-Salvesen, S.; Engesæter, L.B.; Dybvik, E.; Furnes, O.; Kristensen, T.B.; Gjertsen, J.E. Does time from fracture to surgery affect mortality and intraoperative medical complications for hip fracture patients? An observational study of 73,557 patients reported to the Norwegian Hip Fracture Register. *Bone Jt. J.* **2019**, *101*, 1129–1137. [CrossRef]
34. Su, S.; Zhang, Y.; Wang, R.; Zhou, R.; Chen, Z.; Zhou, F. Early surgery within 48 h was associated with reduced perioperative blood loss and red blood cell transfusion requirements in older patients with hip fracture: A retrospective study. *Eur. Geriatr. Med.* 2023; *ahead of print*. [CrossRef]
35. Fenwick, A.; Pfann, M.; Antonovska, I.; Mayr, J.; Lisitano, L.; Nuber, S.; Förch, S.; Mayr, E. Early surgery? In-house mortality after proximal femoral fractures does not increase for surgery up to 48 h after admission. *Aging Clin. Exp. Res.* **2023**, *35*, 1231–1239. [CrossRef] [PubMed]

36. Brink, O. Hip fracture clearance: How much optimisation is necessary? *Injury* **2020**, *51*, S111–S117. [CrossRef] [PubMed]
37. Collinge, C.A.; McWilliam-Ross, K.; Beltran, M.J.; Weaver, T. Measures of clinical outcome before, during, and after implementation of a comprehensive geriatric hip fracture program: Is there a learning curve? *J. Orthop. Trauma* **2013**, *27*, 672–676. [CrossRef]
38. Blanco, J.F.; da Casa, C.; Pablos-Hernández, C.; González-Ramírez, A.; Julián-Enríquez, J.M.; Díaz-Álvarez, A. 30-day mortality after hip fracture surgery: Influence of postoperative factors. *PLoS ONE* **2021**, *16*, e0246963. [CrossRef]
39. Cha, Y.H.; Lee, Y.K.; Koo, K.H.; Wi, C.; Lee, K.H. Difference in Mortality Rate by Type of Anticoagulant in Elderly Patients with Cardiovascular Disease after Hip Fractures. *Clin. Orthop. Surg.* **2019**, *11*, 15–20. [CrossRef]
40. Frenkel Rutenberg, T.; Velkes, S.; Vitenberg, M.; Leader, A.; Halavy, Y.; Raanani, P.; Yassin, M.; Spectre, G. Morbidity and mortality after fragility hip fracture surgery in patients receiving vitamin K antagonists and direct oral anticoagulants. *Thromb. Res.* **2018**, *166*, 106–112. [CrossRef]
41. Lai, W.Y.; Chiu, Y.C.; Lu, K.C.; Huang, I.T.; Tsai, P.S.; Huang, C.J. Beneficial effects of preoperative oral nutrition supplements on postoperative outcomes in geriatric hip fracture patients: A PRISMA-compliant systematic review and meta-analysis of randomized controlled studies. *Medicine* **2021**, *100*, e27755. [CrossRef] [PubMed]
42. Liu, N.; Lv, L.; Jiao, J.; Zhang, Y.; Zuo, X.L. Association between nutritional indices and mortality after hip fracture: A systematic review and meta-analysis. *Eur. Rev. Med. Pharmacol. Sci.* **2023**, *27*, 2297–2304. [PubMed]
43. Gonul, R.; Tasar, P.T.; Tuncer, K.; Karasahin, O.; Binici, D.N.; Sevinc, C.; Turgut, M.; Sahin, S. Mortality-Related Risk Factors in Geriatric Patients with Hip Fracture. *Ann. Geriatr. Med. Res.* **2023**, *27*, 126–133. [CrossRef]
44. Handoll, H.H.; Cameron, I.D.; Mak, J.C.; Panagoda, C.E.; Finnegan, T.P. Multidisciplinary rehabilitation for older people with hip fractures. *Cochrane Database Syst. Rev.* **2021**, *11*, CD007125. [CrossRef] [PubMed]
45. Benoit, M.; Grass, F.; Demartines, N.; Coti-Bertrand, P.; Schäfer, M.; Hübner, M. Use of the nutritional risk score by surgeons and nutritionists. *Clin. Nutr.* **2016**, *35*, 230–233. [CrossRef] [PubMed]

Journal of *Personalized Medicine*

MDPI

Article

Influence of Malnutrition on Outcome after Hip Fractures in Older Patients

Michael Millrose [1,2,*], Wolfgang Schmidt [1], Julian Krickl [1], Till Ittermann [3], Johannes Ruether [2], Hermann-Josef Bail [2] and Markus Gesslein [2]

1 Department of Trauma Surgery and Sports Medicine, Garmisch-Partenkirchen Medical Centre, 82467 Garmisch-Partenkirchen, Germany
2 Department of Orthopedics and Traumatology, Paracelsus Medical University, 90471 Nuremberg, Germany
3 Institute for Community Medicine, SHIP/Clinical-Epidemiological Research, University of Greifswald, 17489 Greifswald, Germany
* Correspondence: michael.millrose@klinikum-gap.de; Tel.: +49-8821-771220

Abstract: Background: Malnutrition might lead to a worse outcome in hip fractures of older patients. The purpose of this study is to analyze different indicators of malnutrition that lead to worse outcomes. Methods: 252 patients of a geriatric trauma unit were analyzed prospectively. Different demographic, as well as data on the trauma and whether osteoporosis prophylaxis or anticoagulation was present, were recorded. The nutritional status with respect to laboratory parameters as well as nutritional risk score was also analyzed. Results: The main finding of this study is that a poor nutritional status is statistically significantly associated with higher mortality as well as worse independence ($p < 0.05$). A postoperatively prescribed osteoporosis prophylaxis is protective of mortality and independence ($p < 0.05$). Conclusions: Malnutrition of geriatric patients increases the risk for death, worse mobility, and independence after hip fractures. Osteoporosis prophylaxis prescribed during an inpatient stay enables patients to retain their independence. The nutritional status of geriatric patients with hip fractures should be obtained and provisions made.

Keywords: malnutrition; hip fractures; geriatric trauma; sarcopenia

Citation: Millrose, M.; Schmidt, W.; Krickl, J.; Ittermann, T.; Ruether, J.; Bail, H.-J.; Gesslein, M. Influence of Malnutrition on Outcome after Hip Fractures in Older Patients. *J. Pers. Med.* **2023**, *13*, 109. https://doi.org/10.3390/jpm13010109

Academic Editor: Jih-Yang Ko

Received: 22 December 2022
Revised: 31 December 2022
Accepted: 1 January 2023
Published: 3 January 2023

1. Introduction

Among older hospitalized people, malnutrition or risk of malnutrition shows a high prevalence—as high as 40% [1]. Malnutrition is defined by the World Health Organization as "deficiencies, excesses or imbalances in a person's intake of energy and/or nutrients" [2]. In different studies it could be shown that malnutrition increases hospitalization, length of hospital stays as well as the overall mortality [3,4]. It also contributes to muscle wasting and, therefore, reduced muscular power and mobility [5].

One of the most important acute health risk factors in older ages are hip fractures (HF); their incidence increases significantly with age in the German population over 65 years of age by nearly six-fold with an additional increased risk of dying of almost nineteen-fold [6,7]. It is well known that HFs in geriatric patients often cause severe functional impairment with increased disability, reduction of quality of life, and by that, consequently loss of autonomy. They also increase the mortality of this special group [8].

In a systematic review of Foo and colleagues they could provide evidence that malnutrition in patients with a hip fracture is significantly associated with increased mortality after 12 months as well as functional dependence over a longer time period with a trend to more supported living arrangements, nursing homes, etc. [1]. Therefore, HFs in malnourished older people pose an especially great threat to the individual patient as well as the health care system itself due to high costs and the risk of overburdening the nursing homes.

The overall aim would be that by addressing the malnutrition of an old patient with a HF, the clinical outcome might be improved. The malnutrition can be assessed by clinical as well as laboratory parameters. Even subclinical deficits of nutrients can have an impact on the outcome if supplemented. Additionally, the connection between malnutrition and the development of osteoporosis is well described [9]. Due to the demographic changes, the population gets older and, therefore, the prevalence of HF in this special group will continuously grow [10].

In this study we wanted to present the results of mortality, functional outcome, and independence of our geriatric trauma unit and analyze different risk factors in terms of the trauma itself and especially different indicators for malnutrition leading to a worse outcome.

2. Materials and Methods

This prospective study was conducted from January 2018 to September 2020 at a regional care hospital with a department for geriatric trauma surgery. During the entire study period, 916 patients were admitted with a trauma diagnosis. All patients suffering from a proximal femoral fracture, regardless of traumatic or pathological genesis, were asked to participate in this study. The traumatic etiology consisted mostly of falls, either because of an external reason such as slipping/tripping or internal reasons such as syncope. Out of the 916 admitted patients, 269 patients could be included in the study. Seventeen study participants were lost during the follow-up and 252 patients (69 male and 183 female) could be assessed in the analysis.

The inclusion criteria of the study were a minimum age of 65 years and a performed osteosynthesis of the femur or joint replacement of the hip, regardless of the implant used. Exclusion criteria of this study were lack of consent or ability to consent to the study. Data sets which lack 3 or more items were also discarded.

This study was conducted according to the guidelines of the declaration of Helsinki and the analysis and publication of the data was approved by the Institutional Review Board of Paracelsus Medical University (protocol code IRB-2022-018; approved on 23 November 2022). Written informed consent was obtained from every participant or their respective legal custodian.

At inclusion, the demographic parameters age, sex, American Society of Anesthesiologists (ASA) classification and living situation, the type of fracture, preoperative osteoporosis therapy as well as anticoagulation and preoperative walking ability were recorded. The time between admission and operation as well as the operation procedure were obtained. In addition to the surgery procedure itself, closed versus open procedures were differentiated. The outcome parameters were mortality during the acute inpatient stay as well as after 120 days and a change in the patient's location; the ability to walk was recorded during follow-up 120 days after admission.

For assessing the nutritional status of the included patients, the body mass index (BMI), serum albumin level, adjusted calcium as well as levels of vitamin B12, vitamin D and folic acid in serum were analyzed. The lower threshold value for those blood parameters were albumin 3.2 g/dL, calcium 4.4 mval/L, vitamin B12 197 pg/mL, vitamin D 20 ng/mL and folic acid 4.60 ng/mL, which were the lower margins for the standard range of the respective parameters provided by our laboratory. The nutritional risk score during the inpatient stay was recorded by nutritionists [11].

The data were obtained prospectively and analyzed retrospectively. The statistical analysis was performed using STATA 17.0 (Stata Corporation, College Station, TX, USA). Characteristics of the study population are reported and stratified by sex as the mean and standard deviation for continuous variables and as absolute numbers and percentages for categorical variables. Associations of potential risk factors with inner-clinic death, death after 120 days, worsened mobility after surgery and worsened domicile after surgery were analyzed by logistic regression models adjusted for age and sex. Results are reported as odds ratio, 95% confidence interval and p-value. A $p < 0.05$ was considered as statistically significant.

3. Results

During this prospective study 252 patients could be analyzed. They suffered from a hip (proximal femur) fracture and were treated at the geriatric trauma surgery department. The study population consisted of 183 females with a mean age of 84.38 ± 7.25 years and 69 men with a mean age of 83.78 ± 5.91 years. The mean BMI of all patients was 23.79 ± 3.86 with 19 underweight, 144 healthy weight, 69 overweight and 20 obese (Table 1). Of the analyzed 252 patients, 3 were categorized ASA 1, 46 ASA 2, 181 ASA and 22 ASA 4.

Table 1. Demographics of study population.

	Female [n = 183]	**Male [n = 69]**
Age [years]	84.38 (±7.25)	83.78 (±5.91)
Height [m]	1.63 (±0.07)	1.76 (±0.07)
Weight [kg]	62.79 (±11.45)	75.51 (±11.07)
BMI	23.56 (±4.02)	24.39 (±3.33)

All data is presented as mean with standard deviation.

During inpatient care 22 patients died; during the post-discharge follow-up period of 120 days after surgery the number doubled to 44 patients. The details of the two outcome parameters (mobility and independence) are depicted in Table 2.

Table 2. Details of the two outcome parameters: mobility and independence.

Mobility	**Independent**	**Cane/Crutch**	**Rollator**	**Only Indoor**	**Bedridden**
Preadmission [n = 252]	91	27	58	68	8
Follow-up 120d [n = 208]	33	28	72	53	22
Independence	Own home	Nursing home	hospital		
Preadmission [n = 252]	198	54			
Follow-up 120d [n = 208]	158	48	2		

With respect to the mobility of the 208 patients analyzed at follow-up, 18 were able to improve their mobility, 85 stayed at the same mobility level as before the fracture, while 105 patients achieved a worse mobility level, meaning that they needed additional walking support.

Concerning the independence of the living situation, 6 patients were living more self-reliant, 180 lived as before and 26 were newly nursing home depended.

The risk factors which were included into the analysis are shown in Table 3 in detail. The time between admission and surgery was recorded and showed that 90 patients were operated on within 6 h, 44 within 12 h and 83 within 24 h. Only 35 had a time span of over 24 h.

Table 3. Details of risk factors included in the statistical analysis.

Anticogulation [n = 252]	None 122	OAC 18	PAI 57	NOAC 44	combination 11
Osteoporosis prophylaxis	None	Vit. D + calcium	Specific	combination	
Preadmission [n = 252]	195	52	1	4	
Discharge [n = 230]	12	172	1	47	
Implant [n = 252]	DHS 13	Nail (CRIF) 121	Nail (ORIF) 23	THR 83	ORIF 12

OAC: oral anticoagulants; PAI: platelet aggregation inhibitors; NOAC: novel oral anticoagulants; specific osteoporosis prophylaxis: bisphosphonates, estrogen agonist/antagonist, etc.; DHS: dynamic hip screw; nail (CRIF): closed reduction and internal fixation with trochanteric nail; nail (ORIF): open reduction and internal fixation with trochanteric nail; THR: total hip replacement; ORIF: open reduction and internal fixation

The NRS was obtained by nutritionists in 158 patients with a median of 3. The serum albumin level showed a mean of 2.76 ± 0.45 g/dL (n = 227) with 211 patients below the normal range of 3.5–5.2 g/dL. The total calcium was 4.52 ± 0.25 mval/L (n = 252) with only 9 patients below the threshold of 4.4 mval/L. The mean vitamin B12 level was 469.40 ± 418.25 pg/mL (n = 235) with 47 patients below and 37 patients above the normal range of 191–771 pg/mL; the mean vitamin D level was 20.81 ± 20.14 ng/mL (n = 236) with 136 patient showing levels below the threshold of 20 ng/mL. Folic acid showed a mean of 7.68 ± 6.19 ng/mL (n = 230) with 89 patients below 4.6 ng/mL (Table S1).

During the inpatient stay at the geriatric trauma surgery department, 43 patients received oral nutritional supplements.

The results of the regression analysis are shown in Table 4. They imply that the use of anticoagulants before the trauma is statistically significantly correlated with a higher inner clinic mortality ($p < 0.05$). The prescription of osteoporosis prophylaxis at discharge is associated with a lower mortality as well as better or similar independence at follow-up ($p < 0.05$ respectively). A high perioperative risk (ASA 4) elevates the risk of death at follow-up significantly ($p < 0.05$). A poor nutritional status of the geriatric patients in terms of NRS is associated with a higher mortality ($p < 0.05$) and worse independence ($p < 0.05$). All blood parameters show a trend that levels below the lower threshold value increase the mortality and worsen the mobility and independence.

Table 4. Results of regression analysis of risk factors.

	Death Inner Clinic OR (95%-CI)	Death after 120 Days OR (95%-CI)	Worsened Mobility OR (95%-CI)	Worsened Independence OR (95%-CI)
Anticoagulant use (preadmission)	3.29 (1.12; 9.69) *	1.78 (0.87; 3.65)	1.05 (0.63; 1.77)	1.10 (0.60; 2.00)
Osteoporosis medication (preadmission)	1.57 (0.58; 4.23)	1.29 (0.60; 2.79)	0.90 (0.48; 1.67)	1.02 (0.51; 2.03)
Osteoporosis medication (at discharge)	0.10 (0.04; 0.29) *	0.19 (0.08; 0.43) *	0.75(0.33; 1.69)	0.37 (0.16; 0.82) *
ASA risk classification				
3 vs. (1,2)	2.74 (0.33; 23.0)	7.13 (0.91;55.6)	0.89 (0.44; 1.77)	1.73 (0.65; 4.57)
4 vs. (1,2)	9.27 (0.95; 91.0)	18.2 (2.01; 164) *	0.93 (0.31; 2.78)	3.25 (0.92; 11.4)
Body mass index	0.90 (0.78; 1.04)	0.89 (0.80; 0.99) *	0.99 (0.93; 1.06)	0.94 (0.87; 1.03)
Nutritional risk score	2.17 (1.37; 3.43) *	1.68 (1.16; 2.41) *	1.10 (0.80; 1.53)	1.59 (1.13; 2.22) *
Low albumin	3.19 (0.39; 26.0)	1.93 (0.53; 7.10)	2.21 (0.70; 7.00)	1.55 (0.57; 4.17)
Low calcium	1.55 (0.57; 4.17)	2.33 (1,12; 4.85) *	1.67 (0.92; 3.04)	1.53 (0.80; 2.91)
Low vitamin B12	1.32 (0.39; 4.50)	0.88 (0.34; 2.23)	1.18 (0.61; 2.30)	1.46 (0.70; 3.05)
Low vitamin D	1.41 (0.46; 4.29)	0.92 (0.43; 1.97)	1.26 (0.73; 2.17)	0.84 (0.44; 1.58)
Low folic acid	1.36 (0.47; 3.96)	2.10 (0.97; 4.51)	1.15 (0.65; 2.04)	1.83 (0.96; 3.46)
Dietary supplement	0.23 (0.03; 1.79)	1.38 (0.58; 3.29)	1.11 (0.56; 2.21)	1.21 (0.57; 2.57)

Logistic regression models adjusted for age, sex and degree of care. Lower threshold value: albumin 3.2 g/dL; calcium 4.4 mval/L; Vit. B12 197 pg/mL; vitamin D. 20 ng/mL; and folic acid 4.60 ng/mL. OR: odds ratio; CI: confidence interval; statistical significance level * $p < 0.05$.

4. Discussion

The most important finding of this study is the influence of malnutrition on outcomes after a HF. Results showed that a higher NRS (i.e., poor nutritional status) is associated with a higher mortality during the in-patient stay as well as at the follow-up 120 days after the trauma. Additionally, the independence (significant) and the mobility (on trend) of the older patient are worse with malnutrition. The prescription of osteoporosis prophylaxis shows a lower mortality and a better level of independence.

HFs in fragile and/or elderly patients causes a serious risk for their health and their independence. In the present study, it could be shown that nearly two thirds of the included patients worsened in their mobility—mostly to rollator mobility. Additionally, the number of patients that were bedridden more than doubled. The number of patients in nursing homes increased by 26, which poses a great deterioration in their independence. The independence, with respect to activities of daily life, degrades also, since there are different tasks that one is not able to perform with a rollator. Our results are in concordance with the systematic literature review by Peeters et al. [8]. They showed that HFs had a negative impact on the overall health status as well as health-related quality of life of elderly patients

(>65 years old), especially the health status in terms of physical, psychological and social aspects was seriously affected. They showed that the literature has evidence that the pre-fracture condition with respect to physical functioning or nutritional status seems to have a negative impact on the outcome. They concluded that optimizing nutritional intake and physical rehabilitation should be recommended after surgery; however, there were only few studies that examined its effectiveness [8].

A recent systematic review by Foo, Wong and Lew from 2021 analyzed the prevalence of malnutrition and its associated outcomes after a HF in a geriatric population [1]. They showed that the prevalence of malnutrition in those patients ranged from 4.0–39.4%. In our study we found that in 158 patients, most of them were at least at risk for malnutrition. They showed that malnourished older patients were significantly at risk for functional dependence, which is in concordance with our results since nearly 50% achieved a worse mobility level at the 120-day follow-up. From their results, they raised the question whether nutritional support would improve the clinical outcome. Our results of the association of malnutrition with higher mortality, as well as worse independence are in concordance with their results. We were not able to prove that a dietary supplementation improves the outcome. This could be due to the lack of consequences in prescribing these, which in our geriatrics department is a medical duty.

The systematic review by Malafarina et al. from 2018 showed that mortality was inversely associated with albumin levels [12]. They found in the literature a relative risk of dying of 1.52 with hypalbuminemia. These findings are shown in trend by our results with an odds ratio of 1.93 of death at follow-up with lower albumin levels. This systematic review also confirmed other findings of our study, such as the association of malnutrition with higher mortality and worse independence.

A clinical study form 2015 in Germany showed that malnutrition has a negative impact on the regaining of their previous Barthel index for the activities of daily living or their preadmission mobility level [13]. The nutritional status of the included patients was analyzed by the Mini Nutritional Assessment (MNA). The MNA in contrast to the NRS used in this study additionally includes measurements of anthropometric data, which might show a clearer picture of the overall status of the geriatric patient since it includes data on the muscle mass and strength respective to their decline (sarcopenia). We have used the NRS in combination with sarcopenia screening, which mostly confirmed the malnourishment with probable sarcopenia (see Table S1). A total of 45% of patients of the study by Goisser et al. were at risk of malnutrition or were malnourished. This shows the same tendency as our study, whereby nearly half of all geriatric patients are not well-nourished. They concluded that a worse preadmission nutritional status is associated with a worse functional outcome. This agrees with our results.

It is well published in the literature that malnutrition of the elderly patient is playing an important part in developing osteoporosis [14,15]. Besides pharmacological agents, nutritional support as well as fall prevention strategies and exercise can be utilized to treat osteoporosis and prevent further osteoporotic fractures, such as hip or vertebral fractures. The sufficient treatment of malnutrition and the normalization of the related blood parameter can improve the outcome of elderly patients with osteoporosis and should be routinely implemented during an in-patient stay after an osteoporosis-defining fracture. In our study we could show that more than half of the study population suffered from a deficient vitamin D level. At discharge, nearly all of the treated patients were supplemented with vitamin D and calcium to compensate this deficiency after an osteoporotic fracture.

One of the strengths of our study is that the nutritional status of our patients was routinely supported by specific parameters, for example, albumin. Cabrerizo et al. stated that albumin is a good marker of nutritional status [16]. Additionally, the NRS was obtained in our study by nutritionists. In a clinical study by Benoit et al. from 2015, it could be shown that the assessment of the nutritional status by surgeons was significantly worse than by nutritionist. Therefore, the data obtained in our study more likely represents the actual status of the included patient.

A limitation of our study is the follow-up period of only 120 days. Peeters et al. showed in their systematic review that the health status of the geriatric patient after a HF improved during the first 6 months; therefore, the reported outcomes in terms of mobility and to a lesser degree independence might be worse after 6 months or even longer of rehabilitation. Another limitation of our study is the only sparsely provided dietary supplement. A randomized controlled study with or without dietary supplement to examine the outcome after proximal femur fracture in the elderly patient is currently planned. Additionally, confounding diagnoses, such as chronic obstructive pulmonary disease or diabetes, were not recorded and might have an influence on the nutritional status and outcome after a HF.

Further high quality randomized controlled studies are needed to provide evidence that dietary supplementation improves the malnutrition of geriatric patients and, therefore, lowers the risk of mortality.

5. Conclusions

Geriatric patients with malnutrition, shown by the nutritional risk score as well as blood test parameters, have a higher risk of dying during the first 120 days after surgical treatment of a HF. Even though this study could not provide evidence that dietary supplements would improve the outcome, taking appropriate steps to improve the nutritional status of the older patient should be considered. The prescription of osteoporosis prophylaxis retains independence and should be routinely suggested.

Supplementary Materials: The following supporting information can be downloaded at https://www.mdpi.com/article/10.3390/jpm13010109/s1: Table S1: Data of included patients.

Author Contributions: Conceptualization, M.M. and M.G.; methodology, M.M. and J.R.; software, W.S.; validation, M.M., W.S. and J.K.; formal analysis, T.I.; investigation, M.M., J.K. and T.I.; data curation, W.S.; writing—original draft preparation, M.M.; writing—review and editing, J.K., J.R., H.-J.B., M.G.; visualization, M.M., M.G.; supervision, H.-J.B.; project administration, M.G.; All authors have read and agreed to the published version of the manuscript.

Funding: This research received no external funding.

Institutional Review Board Statement: The study was conducted in accordance with the Declaration of Helsinki and approved by the Institutional Review Board of Paracelsus Medical University (protocol code IRB-2022-018 on 23 November 2022).

Informed Consent Statement: Informed consent was obtained from all subjects involved in the study.

Data Availability Statement: The data of our study population is provided in the supplementary material Table S1.

Acknowledgments: The authors would like to thank Stefanie Karpik for her support in preparing and correcting the manuscript with respect to spelling and grammar.

Conflicts of Interest: The authors declare no conflict of interest.

References

1. Foo, M.X.E.; Wong, G.J.Y.; Lew, C.C.H. A systematic review of the malnutrition prevalence in hospitalized hip fracture patients and its associated outcomes. *JPEN J. Parenter. Enter. Nutr.* **2021**, *45*, 1141–1152. [CrossRef] [PubMed]
2. World Health Organization. Malnutrition. Available online: https://www.who.int/health-topics/malnutrition#tab=tab_1 (accessed on 19 October 2022).
3. Söderström, L.; Rosenblad, A.; Adolfsson, E.T.; Saletti, A.; Bergkvist, L. Nutritional status predicts preterm death in older people: A prospective cohort study. *Clin. Nutr.* **2014**, *33*, 354–359. [CrossRef] [PubMed]
4. Lew, C.C.H.; Wong, G.J.Y.; Cheung, K.P.; Chua, A.P.; Chong, M.F.F.; Miller, M. Association between Malnutrition and 28-Day Mortality and Intensive Care Length-of-Stay in the Critically ill: A Prospective Cohort Study. *Nutrients* **2017**, *10*, 10. [CrossRef] [PubMed]
5. Avenell, A.; Smith, T.O.; Curtain, J.P.; Mak, J.C.; Myint, P.K. Nutritional supplementation for hip fracture aftercare in older people. *Cochrane Database Syst. Rev.* **2016**, *11*, CD001880. [CrossRef] [PubMed]

6. Federal Health Monitoring System. Diagnostic Data of the Hospitals by Place of Treatment for ICD S70-S79. 2020. Available online: https://www.gbe-bund.de/gbe/!pkg_olap_tables.prc_set_page?p_uid=gast&p_aid=64985748&p_spracsp=D&p_help=2&p_indnr=702&p_ansnr=63284549&p_version=6&D.001=1000001&D.002=1337&D.003=1000004&D.972=1000619&D.100=10101 (accessed on 19 October 2022).
7. Federal Health Monitoring System. Death Data for ICD S70-S79. 2020. Available online: https://www.gbe-bund.de/gbe/!pkg_olap_tables.prc_set_orientation?p_uid=gast&p_aid=64985748&p_sprache=D&p_help=2&p_indnr=6&p_ansnr=92146693&p_version=3&D.000=3&D.000=3&D.002=1&D.003=3&D.004=3&D.946=2&D.011=3&D.100=1 (accessed on 19 October 2022).
8. Peeters, C.M.; Visser, E.; Van de Ree, C.L.; Gosens, T.; Den Oudsten, B.L.; De Vries, J. Quality of life after hip fracture in the elderly: A systematic literature review. *Injury* **2016**, *47*, 1369–1382. [CrossRef] [PubMed]
9. Scaturro, D.; Vitagliani, F.; Terrana, P.; Tomasello, S.; Camarda, L.; Letizia Mauro, G. Does the association of therapeutic exercise and supplementation with sucrosomial magnesium improve posture and balance and prevent the risk of new falls? *Aging Clin. Exp. Res.* **2022**, *34*, 545–553. [CrossRef]
10. Veronese, N.; Maggi, S. Epidemiology and social costs of hip fracture. *Injury* **2018**, *49*, 1458–1460. [CrossRef]
11. Kondrup, J.; Rasmussen, H.H.; Hamberg, O.; Stanga, Z.; Ad Hoc ESPEN Working Group. Nutritional risk screening (NRS 2002): A new method based on an analysis of controlled clinical trials. *Clin. Nutr.* **2003**, *22*, 321–336. [CrossRef]
12. Malafarina, V.; Reginster, J.Y.; Cabrerizo, S.; Bruyère, O.; Kanis, J.A.; Martinez, J.A.; Zulet, M.A. Nutritional Status and Nutritional Treatment Are Related to Outcomes and Mortality in Older Adults with Hip Fracture. *Nutrients* **2018**, *30*, 555. [CrossRef]
13. Goisser, S.; Schrader, E.; Singler, K.; Bertsch, T.; Gefeller, O.; Biber, R.; Bail, H.J.; Sieber, C.C.; Volkert, D. Malnutrition According to Mini Nutritional Assessment Is Associated with Severe Functional Impairment in Geriatric Patients Before and up to 6 Months After Hip Fracture. *J. Am. Med. Dir. Assoc.* **2015**, *6*, 661–667. [CrossRef]
14. Gennari, C. Calcium and vitamin D nutrition and bone disease of the elderly. *Public Health Nutr.* **2001**, *4*, 547–559. [CrossRef]
15. Chen, L.R.; Hou, P.H.; Chen, K.H. Nutritional Support and Physical Modalities for People with Osteoporosis: Current Opinion. *Nutrients* **2019**, *11*, 2848. [CrossRef]
16. Cabrerizo, S.; Cuadras, D.; Gomez-Busto, F.; Artaza-Artabe, I.; Marín-Ciancas, F.; Malafarina, V. Serum albumin and health in older people: Review and meta analysis. *Maturitas* **2015**, *81*, 17–27. [CrossRef] [PubMed]

Journal of *Personalized Medicine*

MDPI

Article

In Vitro Investigation of Pulsed Electromagnetic Field Stimulation (PEMF) with MAGCELL® ARTHRO on the Regulatory Expression of Soluble and Membrane-Bound Complement Factors and Inflammatory Cytokines in Immortalized Synovial Fibroblasts

Sandeep Silawal [1,*], Markus Gesslein [2], Maximilian Willauschus [2] and Gundula Schulze-Tanzil [1]

1 Institute of Anatomy and Cell Biology, Paracelsus Medical University, Nuremberg and Salzburg, General Hospital Nuremberg, Prof. Ernst Nathan Str. 1, 90419 Nuremberg, Germany; gundula.schulze@pmu.ac.at

2 Department of Orthopedics and Traumatology, Paracelsus Medical University, Nuremberg and Salzburg, General Hospital Nuremberg, Breslauer Str. 201, 90471 Nuremberg, Germany; markus.gesslein@klinikum-nuernberg.de (M.G.); mp-willauschus@outlook.com (M.W.)

* Correspondence: sandeep.silawal@pmu.ac.at

Abstract: Pulsed electromagnetic field stimulation (PEMF) is gaining more attention as a non-invasive arthritis treatment. In our study, immortalized synovial fibroblasts (K4IM) derived from a non-arthritic donor were exposed to MAGCELL® ARTHRO, a PEMF device, with 105 mT intensity, 8 Hz frequency, and 2 × 2.5 min sessions conducted thrice with a 1 h interval, to understand the underlying mechanism in regard to the complement system. Additionally, tumor necrosis factor (TNFα, 10 ng/mL) pre-treatment prior to PEMF stimulation, as well as 3-day versus 6-day stimulation, were compared. Gene expression of C4b binding protein-alpha and -beta (C4BPα, C4BPβ), complement factor (CF)-H, CFI, CD55, CD59, Interleukin (IL-6) and TNFα was analyzed. Immunofluorescence staining of CD55, CD59, and Ki67 was conducted. Results showed the absence of C4BPα gene expression, but C4BPβ was present. One and three days of PEMF stimulation caused no significant changes. However, after six days, there was a significant increase in CD55, CFH, and CD59 gene expression, indicating cytoprotective effects. Conversely, IL-6 gene expression increased after six days of stimulation and even after a single session in TNFα pre-stimulated cells, indicating a pro-inflammatory effect. PEMF's ambivalent, i.e., enhancing complement regulatory proteins and pro-inflammatory cytokines, highlights its complexity at the molecular level.

Keywords: cytokines; complement regulatory proteins; osteoarthritis; PEMF; synovial fibroblasts; synovitis

check for updates

Citation: Silawal, S.; Gesslein, M.; Willauschus, M.; Schulze-Tanzil, G. In Vitro Investigation of Pulsed Electromagnetic Field Stimulation (PEMF) with MAGCELL® ARTHRO on the Regulatory Expression of Soluble and Membrane-Bound Complement Factors and Inflammatory Cytokines in Immortalized Synovial Fibroblasts. *J. Pers. Med.* **2024**, *14*, 701. https://doi.org/10.3390/jpm14070701

Academic Editor: Jih-Yang Ko

Received: 24 May 2024
Revised: 24 June 2024
Accepted: 26 June 2024
Published: 29 June 2024

1. Introduction

Osteoarthritis (OA) stands as the most prevalent degenerative joint disorder globally and affects more than 500 million population worldwide [1,2]. While age is a primary factor in its occurrence, factors such as obesity, mechanical misalignment, joint instability, and genetic predisposition also significantly contribute to OA [3]. Consequently, the increasing prevalence of OA within our aging population, coupled with rising comorbidities, may substantially augment the future economic and social burdens on society. A profound comprehension of OA pathogenesis is crucial for managing the processes of cartilage depletion that lead to joint destruction.

The contemporary understanding of OA pathogenesis emphasizes its involvement across multiple tissues [4]. Hence, a comprehensive investigation into each tissue component—cartilage, subchondral bone, and synovium is necessary [5]. Synovial inflammation of varying degrees, alongside hypertrophy of the joint capsule, emerges as a principal pathological feature in OA-afflicted joints [5]. In addition to the degradation of the articular cartilage, various other pathologies such as thickening of the subchondral bone, osteophyte formation, degeneration of ligaments and menisci, inflammation of the synovium and thickening of the joint capsule are observed in the OA pathogenesis [5,6].

OA therapy poses significant clinical challenges [7]. The presence of comorbidities, the need for chronic pain management, limited functionality due to steady disease progression, and the often less optimistic outlook for improvement in patients all necessitate extensive clinical and psychological support for patients [7,8]. Current conservative pharmacological management strategies aim to alleviate pain using topical or oral nonsteroidal anti-inflammatory drugs (NSAID) and opioids for intense pain, alongside other measures such as capsaicin, antidepressants such as Duloxetine, and intra-articular anti-inflammatory agents such as glucocorticoids or hyaluronic acid. Maqbool et al., 2021 have briefly discussed the clinical prospects and management of OA [9]. Consequently, as conservative therapies falter, advanced OA joints often necessitate surgical intervention for joint replacement to restore function and alleviate pain [9]. Given the limitations and potential side effects of prolonged NSAID use, there is a pressing need for alternative approaches to OA therapy.

One promising adjunctive therapy is pulsed electromagnetic field (PEMF) therapy, which has been viewed as a potential candidate for alleviating joint pain and improving function [10–13]. However, critical reviews also suggest mixed results regarding its efficacy in treating joint pain and stiffness, though it may enhance physical function compared to placebo [14,15]. The MAGCELL® ARTHRO device, transmitting a sinusoidal waveform of strong PEMF at 105 mT with a frequency of 8 Hz, has garnered attention for its reported pain reduction and increased mobility in OA patients. A prospective, placebo-controlled, double-blind study involving 57 knee OA patients demonstrated significant reductions in pain, joint stiffness, and improved functionality with the use of this device [16]. Similarly, Bagnato et al., in a double-blind, placebo-controlled, randomized clinical trial, showed that PEMF therapy improved pain and dysfunction in patients with knee OA [12].

Despite promising clinical outcomes, the molecular mechanisms underlying PEMF therapy in joint tissues still need to be studied in detail. The complement system, a key component of the immune system, plays a crucial role in orchestrating inflammatory responses locally and systemically [4,17]. The complement system is typically activated through various pathways: the classical pathway, initiated by C1q binding directly to pathogen surfaces or antibody–antigen complexes; the lectin pathway, triggered by lectin binding to mannose-containing carbohydrates on pathogens; and the alternative pathway, initiated by spontaneous hydrolysis of complement protein C3 (Supplementary Materials) [17]. Zymogen precursors are cleaved into fragments, forming active enzymes that further split subsequent zymogens in the complement pathway, ultimately leading to cell lysis via the membrane attack complex [17]. During complement activation, anaphylatoxins such as C4a, C3a, and C5a are generated through the cleavage of proteins C4, C3, and C5, respectively, playing active roles in the inflammatory process [17]. Soluble and membrane-bound complement regulatory proteins (CrP) help regulate hyperinflammation, mitigating the detrimental effects of complement activity [18].

Complement factor I (CFI), predominantly synthesized in the liver, is an 88 kDa plasma protein consisting of two disulfide-linked chains: the α-chain (50 kDa) and the β-chain (38 kDa). Acting as a serine protease, CFI cleaves the α-chains of C4b and C3b with the assistance of cofactors such as complement factor H (CFH) and C4b binding protein (C4BP) [19]. CFI concentration increases during inflammation, with normal serum or plasma concentrations ranging between 19 and 64 μg/mL [19]. C4BP, primarily produced by hepatocytes, consists of seven thin, elongated subunits (α-chain, 70 kDa) linked to a central body (β-chain, 45 kDa), functioning as a cofactor for CFI [20]. With approximately 150 μg/mL plasma concentration, C4BP functions as mentioned before as a cofactor for CFI, which undergoes complement regulatory function interfering with the assembly of the C3- and C5-convertases, as well as contributing to the disassembly of the C3-convertase [20]. CFH

(43 kDa) is an abundant serum glycoprotein produced primarily in the hepatocytes [21]. The serum concentration can vary widely between 116 and 562 μg/mL as summarized in a review paper [22]. CFH plays a crucial role in accelerating the disassembly of the alternative pathway C3 convertase (C3bBb) and acts as a cofactor for CFI-mediated cleavage of C3b [22]. CD55 (43 kDa) and CD59 (18–20 kDa) are membrane-bound CrP that regulate complement activation at different cascade levels. CD55 accelerates C3 convertase degradation at the intermediate level, while CD59 inhibits membrane attack complex assembly [17].

Some research has described the anti-inflammatory properties of PEMF stimulation at the molecular level, such as decreasing IL1-ß or TNFα [23,24]. This study aimed to further investigate the anti-inflammatory properties of PEMF treatment, specifically its direct influence on complement regulation in synovial fibroblasts, which has never been studied before.

2. Materials and Methods

2.1. In Vitro Culture of K4IM

The SV40 T-antigen immortalized synovial fibroblast cell line, K4IM, was generously provided by Dr. Christian Kaps. These cells were cultured in a growth medium supplemented with 10% fetal bovine serum (FBS) (PAN-Biotech, Aidenbach, Germany), which also served as Ctrl Medium 1. The growth medium was composed of Dulbecco's MEM/Ham's F-12 (1:1) supplemented with 25 mg/mL ascorbic acid, 50 IU/mL streptomycin, 50 IU/mL penicillin, 2.5 μg/mL amphotericin B, and essential amino acids (all products from Carl Roth GmbH, Karlsruhe, Germany). For cell passaging, 0.05% trypsin/1.0 mM EDTA (Carl Roth GmbH) was utilized before seeding the cells for stimulation.

2.2. PEMF Stimulation of the Synovial Fibroblasts

Study 1: K4IM synovial fibroblasts were seeded at a density of 10,000/cm^2 either in T25 culture flasks (Sarstedt, Nümbrecht, Germany) or on glass coverslips, depending on the experiment, and cultured in Ctrl Medium 1 (3 mL/flask in T25 flasks). After 48 h, the supernatant was aspirated, and the cells were washed with PBS before being supplied with 1% FBS-supplemented growth medium (Ctrl Medium 2) (2 mL/flask). Following a 1 h incubation, the supernatant was replaced by fresh Ctrl Medium 2, and the stimulation commenced (Figure 1, Table 1).

Ctrl **PEMF (ON)** **PEMF (OFF)** **Sham**

Figure 1. Experimental setting for Study 1 (Table 1—Day 3). K4IM synovial fibroblasts in Ctrl Medium 2 stimulated with PEMF (ON), PEMF (OFF) or sham and non-stimulated group as control (Ctrl). Each stimulation session comprised an application time of 2 × 2.5 min 3 times with an interval of 1 h between each stimulation.

Table 1. Stimulation groups and schedule for Study 1. Ctrl Medium 1 = 10% fetal bovine serum supplemented growth medium. Ctrl Medium 2 = 1% fetal bovine serum supplemented growth medium.

	Group 1	Group 2	Group 3	Group 4
Day 1	Ctrl Medium 1	Ctrl Medium 1	Ctrl Medium 1	Ctrl Medium 1
Day 2	Ctrl Medium 1	Ctrl Medium 1	Ctrl Medium 1	Ctrl Medium 1
Day 3	Ctrl Medium 2	PEMF (ON)	PEMF (OFF)	Sham

The stimulation involved exposing the K4IM cells in the flasks to the anterior surface of the MAGCELL® ARTHRO apparatus (Catalog #AC-1815404, Physiomed Elektromedizin AG, Schnaittach, Germany) for 5 min. As a test, metal powder, distributed homogeneously on the floor of the T25 cell culture flask was treated with the apparatus. The activation of the apparatus showed that the electromagnetic influence could be seen up to the periphery of the flask (Supplementary Materials). As described in a 2015 clinical study, the MAGCELL® ARTHRO device comprises rotating magnets [16]. It utilizes time-varying magnetic fields with a flux density of 105 mT, providing current densities exceeding 10 mA/cm^2. During stimulation, the side of the culture flask with adherent cells (inferior side/floor) was positioned against the anterior surface of the apparatus. The protocol for a single-day stimulation session involved an application time of 2 $\times$ 2.5 min for 3 sessions with a 1 h interval between each stimulation. In the aforementioned clinical study, the placebo group received a similar device with an identical exterior, but non-magnetic rotating segments [16]. In our study, we referred to the stimulation group as sham. Five independent experiments (n = 5) were performed in Study 1.

Study 2: Similar to Study 1, K4IM synovial fibroblasts were seeded and cultured for the experiment in Ctrl Medium 1 (2 mL/flask in T25 flasks). After 24 h, the supernatant was aspirated, and the cells were washed with PBS. Group 1 and Group 2 received Ctrl Medium 1, while Group 3 and Group 4 were pre-stimulated with recombinant TNFα (Peprotech, Princeton, NJ, USA) at a final concentration of 10 ng/mL diluted in Ctrl Medium 1 (2 mL/flask). On day 3, following a 1 h starvation period with Ctrl Medium 2, the medium was replaced (2 mL/flask), and PEMF stimulation commenced, as described previously, with 3 sessions spaced 1 h apart for the stimulation groups (see Table 2). Five independent experiments (n = 5) were performed in Study 2.

Table 2. Stimulation groups and schedule for Study 2. Ctrl Medium 1 = 10% fetal bovine serum supplemented growth medium. Ctrl Medium 2 = 1% fetal bovine serum supplemented growth medium.

	Group 1	Group 2	Group 3	Group 4
Day 1	Ctrl Medium 1	Ctrl Medium 1	Ctrl Medium 1	Ctrl Medium 1
Day 2	Ctrl Medium 1	Ctrl Medium 1	TNFα	TNFα
Day 3	Ctrl Medium 2	PEMF (ON)	Ctrl Medium 2	PEMF (ON)

Study 3: K4IM synovial fibroblasts were seeded at a density of 5000/cm^2 and cultured in Ctrl Medium 1 (2 mL/flask in T25 flasks) for the experiment. After 24 h, the supernatant was aspirated, and the cells were washed with PBS before being supplied with Ctrl Medium 2 (2 mL/flask). Following a 1 h starvation period, the medium was replaced (2 mL/flask), and PEMF stimulation commenced, as previously described, with 3 sessions spaced 1 h apart for the stimulation groups. Groups 1–2 underwent stimulation for 3 days, while groups 3–4 were stimulated for a total of 6 days (see Table 3). Starting from day 3, 50% of the medium was replaced with fresh medium daily. Four independent experiments (n = 4) were performed in Study 3.

Table 3. Stimulation groups and schedule for Study 3. Ctrl Medium 1 = 10% fetal bovine serum supplemented growth medium. Ctrl Medium 2 = 1% fetal bovine serum supplemented growth medium.

	Group 1	Group 2	Group 3	Group 4
Day 1	Ctrl Medium 1	Ctrl Medium 1	Ctrl Medium 1	Ctrl Medium 1
Day 2	Ctrl Medium 2	PEMF (ON)	Ctrl Medium 2	PEMF (ON)
Day 3	Ctrl Medium 2	PEMF (ON)	Ctrl Medium 2	PEMF (ON)
Day 4	Ctrl Medium 2	PEMF (ON)	Ctrl Medium 2	PEMF (ON)
Day 5	-	-	Ctrl Medium 2	PEMF (ON)
Day 6	-	-	Ctrl Medium 2	PEMF (ON)
Day 7	-	-	Ctrl Medium 2	PEMF (ON)

2.3. *Gene Expression Analysis*

2.3.1. RNA Isolation and cDNA Synthesis

Synovial fibroblasts were seeded as described in Section 2.2. Post-stimulation, these cells were washed with PBS. Lysis of the cells was achieved using a 1:100 solution of β-Mercaptoethanol (Sigma-Aldrich, St. Louis, MO, USA) in RNeasy Lysis Buffer (Qiagen, Hilden, Germany). Total RNA extraction was performed using the RNeasy Mini Kit (Qiagen) following the manufacturer's protocol [25]. The RNA quantity was assessed using the NanoDropTM 2000 Spectrophotometer (Thermo Fischer Scientific, Erlangen, Germany). Subsequently, cDNA was transcribed using the QuantiTect Transcription Kit (Qiagen) with Mastercycler® (Eppendorf, Hamburg, Germany) according to the manufacturer's instructions.

2.3.2. qPCR

For real-time polymerase chain reaction (qPCR) analyses, glyceraldehyde-3-phosphate dehydrogenase (GAPDH) served as the reference gene, and specific primers from TaqMan® Gene Expression Assays (Applied Biosystems, Foster City, CA, USA), listed in Table 4, were utilized. Each well contained 20 ng of cDNA template (1 µL), TaqMan® Gene Expression Master Mix (Applied Biosystems), and water (9 µL) to perform the TaqMan® Gene Expression Assay [25]. Standard TaqMan® analysis protocol as provided by the manufacturer was followed, conducted in the Applied Biosystems StepOnePlus™ Real-Time PCR System. Relative gene expression levels were determined using the delta–delta Ct method, normalized against the reference gene [26].

Table 4. Oligonucleotides used for qPCR analysis.

Primer	Company	Sequence	Assay ID	Amplicon Length (bp)
GAPDH	ABI	*	Hs99999905_m1	122
CD55	ABI	*	Hs00167090_m1	62
CD59	ABI	*	Hs00174141_m1	70
C4bpα	ABI	*	Hs00426339_m1	105
C4bpβ	ABI	*	Hs01103672_m1	74
CFH	ABI	*	Hs00962373_m1	72
CFI	ABI	*	Hs00989715_m1	75
TNFα	ABI	*	Hs00174128_m1	80
IL-6	ABI	*	Hs00174131_m1	95

*: Sequence not provided by the company. ABI (Applied Biosystems, Foster City, CA, USA).

2.4. *Protein Expression Analysis*

Immunofluorescence Labeling of the Synovial Fibroblasts

Following PEMF stimulation, K4IM cells were washed with Tris-buffered saline (TBS) (Medicago AB, Uppsala, Sweden) and fixated with a 4% paraformaldehyde solution (PFA)

(Morphisto, Frankfurt, Germany) for 15 min. The staining protocol has been detailed in our prior publication [27]. Primary antibodies utilized are listed in Table 5. Cell nuclei were counterstained using 4′,6-diamidino-2-phenylindole (DAPI) (Roche Diagnostics GmbH, Basel, Switzerland), while the cytoskeleton was stained with Phalloidin Alexa Fluor 633 (ThermoFisher Scientific). Images of immunolabeled cells were captured using confocal laser scanning microscopy (SPE-II, Leica Microsystems, Weztlar, Germany). Total protein intensity per cell (quantified using DAPI) was measured in each group for analysis.

Table 5. Antibodies and dyes used.

Specificity and Species	Company	Catalog Number	Stock Concentration	Used Dilution
Goat anti-human CD55	R&D systems, Minneapolis, MN, USA	AF2009	200 µg/mL	1:50
Mouse anti-human CD59	Bio-Rad, Bio-Rad, Feldkirchen, Germany	MCA1054	1 mg/mL	1:50
Mouse anti-human Ki67	Chemicon International Inc., Temecula, CA, USA	MAB4190	1 mg/mL	1:50
Donkey anti-mouse-Alexa Fluor 488	Invitrogen, Waltham, MA, USA	A21202	2 mg/mL	1:200
Donkey anti-goat-cyanine (Cy)3	Dianova, Hamburg, Germany	705-165-147	1.5 mg/mL	1:200
Donkey anti-mouse-Cy3	Dianova, Hamburg, Germany	715-166-150	1.5 mg/mL	1:200

2.5. Statistical Analysis

Statistical analyses were conducted using GraphPad Prism (Version 8.1.4) from GraphPad Software, San Diego, CA, USA. The Shapiro–Wilk normality test was performed, and normalized data were presented as mean ± standard deviation (mean ± SD). Significance between experimental groups was determined at $p < 0.05$ using one-way ANOVA analysis with Tukey's multiple comparisons test. The Grubbs test was utilized to identify and exclude outliers from the statistical analysis.

3. Results

3.1. Gene Expression

C4BPα gene expression could not be detected in the immortalized synovial fibroblasts. However, the C4BPβ gene was expressed in these cells.

Study 1: In the provided setup (Table 1, Figure 1), synovial fibroblasts were exposed to MAGCELL® ARTHRO stimulation for a single day. The stimulation group PEMF(ON) was compared with several controls: a non-stimulation group (Ctrl), a group with the apparatus turned off PEMF(OFF), and a sham group (Figure 1). However, the stimulation did not exert any significant influence on the gene expression of the analyzed proteins (Figure 2).

Study 2: In this inflammation model (Table 2), synovial fibroblasts were pre-stimulated with TNFα for 24 h before PEMF stimulation PEMF(ON). Similar to Study 1, a single treatment session without TNFα pre-stimulation did not alter the gene expression of analyzed proteins compared to the Ctrl/Ctrl group. TNFα stimulation alone, i.e., TNFα/Ctrl significantly elevated the gene expression of CFH and IL-6 in K4IM compared to the Ctrl/Ctrl group. However, TNFα pre-stimulated cells, when subjected to PEMF, i.e., TNFα/PEMF(ON), exhibited increased gene expression of IL-6. Both Ctrl/Ctrl and Ctrl/PEMF(ON) consistently displayed significantly lower gene expression in analyzed complement components, except for CFI, compared to TNFα/Ctrl and TNFα/PEMF(ON) groups, respectively (Figure 3).

Study 3: In this experimental setup (Table 3), the PEMF stimulation duration was extended to examine its long-term effects compared to previous studies. Thus, synovial fibroblasts were stimulated with PEMF(ON) for 3 days and 6 days, with respective non-stimulated groups serving as controls. After 3 days of stimulation, no significant changes were observed in the gene expression of the analyzed proteins compared to the non-stimulated control. However, there was a notable trend toward increased gene expression in all proteins. After 6 days of stimulation, there was a marked increase in gene expression. Specifically, CD55, CFH, and IL-6 gene expression showed significant elevation in the PEMF(ON) stimulated group compared to the non-stimulated control (Ctrl). Furthermore, the non-stimulated group (Ctrl) displayed significantly lower gene expression of CFI and IL-6 after 3 days compared to the same group after 6 days. Interestingly, the stimulated group PEMF(ON) after 3 days also exhibited lower expression of these genes compared to the same group after 6 days (Figure 4).

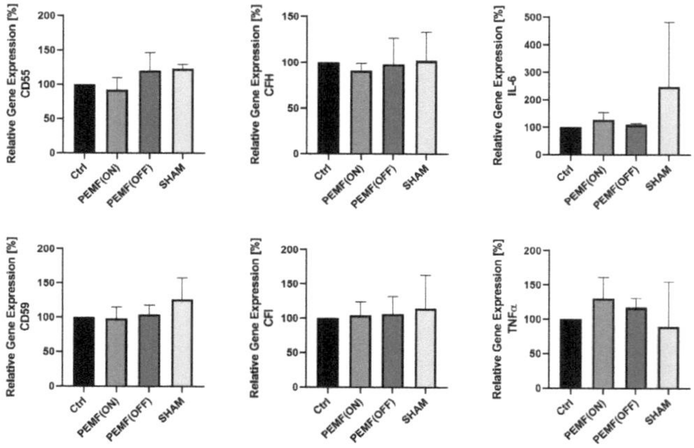

Figure 2. Graphic representation of relative gene expression of complement factors (CD55, CD59, CFH and CFI) and cytokines (IL-6 and TNFα) in synovial fibroblast cell line K4IM after PEMF stimulation [PEMF(ON)] compared to various controls. A one-day stimulation protocol was used. Ctrl = non-stimulated group, [PEMF(OFF)] = apparatus turned off, sham = apparatus turned on without electromagnetic impulse. n = 5 independent experiments. Mean with standard deviation (SD). Non-stimulated group as control has been normalized to 100. Repeated measures one-way ANOVA using Tukey's multiple comparisons.

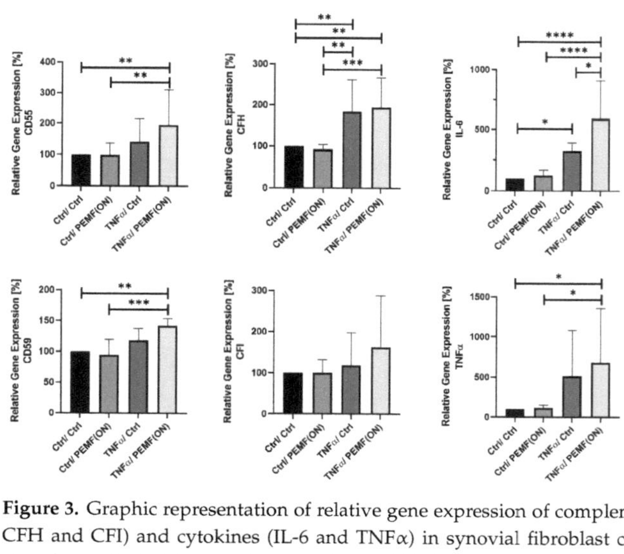

Figure 3. Graphic representation of relative gene expression of complement factors (CD55, CD59, CFH and CFI) and cytokines (IL-6 and TNFα) in synovial fibroblast cell line K4IM after PEMF stimulation [PEMF(ON)] with and without 24 h TNFα pre-stimulation (10 ng/mL) compared to their respective control (ctrl). A one-day stimulation protocol was used. Ctrl = non-stimulation. n = 5 independent experiments. Mean with standard deviation (SD). Non-stimulated group without TNFα pre-stimulation (Ctrl/Ctrl) has been normalized to 100. Repeated measures one-way ANOVA using Tukey's multiple comparisons (*). * = $p \leq 0.05$, ** = $p \leq 0.01$, *** = $p \leq 0.001$, **** = $p \leq 0.0001$.

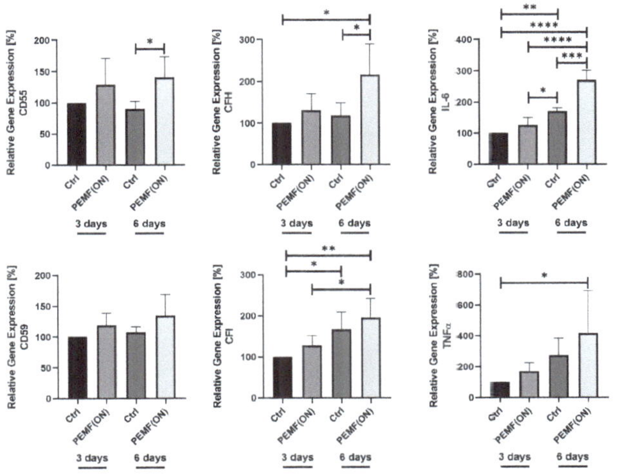

Figure 4. Graphic representation of relative gene expression of complement factors (CD55, CD59, CFH and CFI) and cytokines (IL-6 and TNFα) in synovial fibroblast cell line K4IM after PEMF stimulation [PEMF(ON)] for 3 days and 6 days compared to their respective control (ctrl). Ctrl = non-stimulation. n = 4 independent experiments. Mean with standard deviation (SD). Non-stimulated group (3 days) has been normalized to 100. Repeated measures one-way ANOVA using Tukey's multiple comparisons (*) * = $p \leq 0.05$, ** = $p \leq 0.01$, *** = $p \leq 0.001$, **** = $p \leq 0.0001$.

3.2. Protein Expression

In the analyzed synovial fibroblasts, both CD55 and CD59 protein expressions were observed. However, there were no significant changes noted after 3 days or 6 days of PEMF stimulation compared to their respective controls (Figures 5 and 6). Interestingly, PEMF(ON) stimulation did show a tendency toward increased CD55 protein expression after 3 days (Figure 5), and a similar non-significant trend was seen for CD59 protein expression after 6 days of PEMF(ON) stimulation (Figure 6).

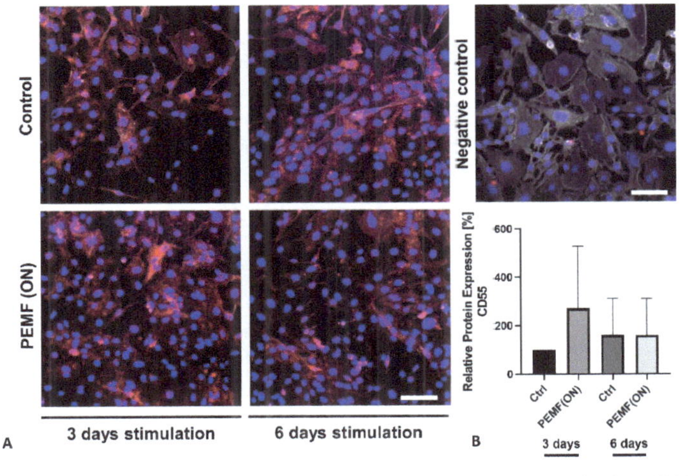

Figure 5. (**A**) Representative images of K4IM cell line after 3 days and 6 days PEMF(ON) stimulation compared to their non-stimulated counterparts, respectively (200× magnification), immunolabeled with a CD55 specific antibody and negative control of the staining. Red (Cy3) = CD55, blue (DAPI) = cell nuclei,

gray (Phalloidin Alexa Fluor 633) = F-Actin cytoskeleton. Scale bar = 100 μm. (**B**) Graphic representation of relative CD55 protein fluorescence intensity, n = 4 independent experiments, mean with standard deviation (SD). Control (3 days) has been normalized to 100. Repeated measures one-way ANOVA using Tukey's multiple comparisons.

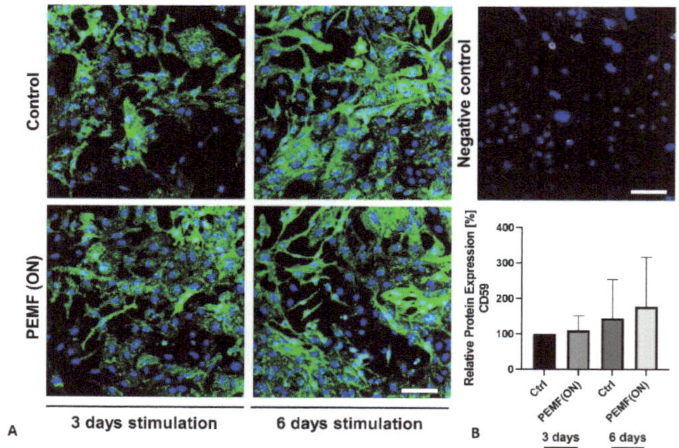

Figure 6. (**A**) Representative images of K4IM cell line after 3 days and 6 days PEMF(ON) stimulation compared to their non-stimulated counterparts, respectively (200× magnification), immunolabeled with a CD59 specific antibody and negative control of the staining. Green (Alexa Fluor 488) = CD59, blue (DAPI) = cell nuclei, gray (Phalloidin Alexa Fluor 633) = F-Actin cytoskeleton. Scale bar = 100 μm. (**B**) Graphic representation of relative CD59 protein fluorescence intensity, n = 4 independent experiments, mean with standard deviation (SD). Control (3 days) has been normalized to 100. Repeated measures one-way ANOVA using Tukey's multiple comparisons.

Also, Ki67 staining was performed (Figure 7). However, no significant differences were detected among the groups, although there was a slight inclination toward higher proliferative activity in the 6-day experimental groups. Additionally, staining with phalloidin to visualize the F-actin cytoskeleton revealed no notable differences in response to the treatment across the groups.

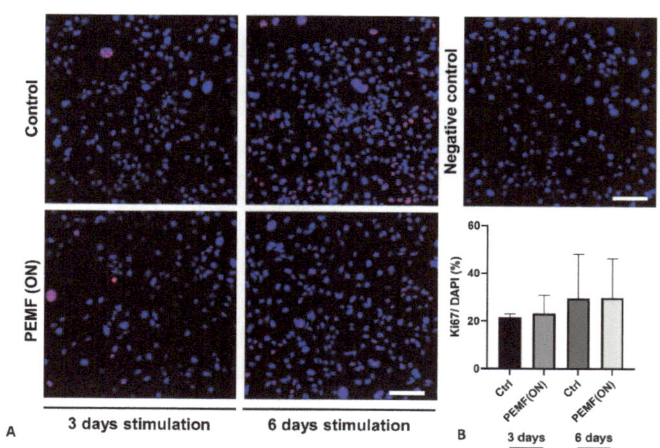

Figure 7. (**A**) Representative images of K4IM cell line after 3 days and 6 days PEMF(ON) stimulation compared to their non-stimulated counterparts, respectively (200× magnification), immunolabeled with

Ki67 specific antibody and negative control of the staining. Red (Cy3) = Ki67, blue (DAPI) = cell nuclei. Scale bar = 100 µm. (**B**) Graphic representation of relative CD55 protein fluorescence intensity, n = 3 independent experiments, mean with standard deviation (SD). Control (3 days) has been normalized to 100. Repeated measures one-way ANOVA using Tukey's multiple comparisons.

4. Discussion

In our study, utilizing K4IM cells derived from a non-arthritic donor, we observed the expression of the C4BPβ gene, while the expression of C4BPα was not detected. Similarly, Criado-García et al., (1999) found that fibroblast-like cells isolated from human ovaries expressed C4BPβ but not C4BPα at both, the gene and protein levels [28]. Another study examining arthritic joints revealed the presence of the C4BPβ peptide in the cytoplasm of lining cells and in most subintimal fibroblast-like cells within the tissue [29]. Immunohistochemistry signals and protein levels of C4BPβ were higher in synovial effusions from rheumatoid arthritis joints compared to osteoarthritic joints in that study. While C4BPα was not detected inside the cells, it was found in the interstitial areas of the subintimal region in many synovial tissues of patients with rheumatoid arthritis.

Study 1: This study aimed to investigate the impact of PEMF(ON) on the gene expression of K4IM cells compared to various control groups. In addition to non-stimulated cells (Ctrl), both, PEMF(OFF) and sham groups were included. The sham group involved cell stimulation using a non-electromagnetic apparatus, serving as a control in our experimental setup. We found that none of the groups exhibited any significant changes after a single-day session of PEMF stimulation. These findings suggest that brief exposure of K4IM cells to PEMF has minimal influence and does not yield statistically significant results in terms of gene expression. Therefore, a single session of PEMF stimulation is less likely to have any significant impact in practical use.

Study 2: In an inflammation model, cells were pre-stimulated with TNFα for 24 h before a single PEMF stimulation session. Comparing TNFα pre-stimulation of K4IM cells with no PEMF stimulation to the control group, higher gene expression levels were observed for all analyzed proteins, with statistically significant results noted for CFH and IL-6 expression. Previous research demonstrated that 24 h stimulation of K4IM cells with TNFα led to an approximately 40-fold increase in TNFα gene expression [27]. Additionally, in the same study, IL-6 gene expression was nearly 8-fold higher in the non-stimulated synovial fibroblast cell line HSE, derived from arthritic joints, compared to K4IM cells [27]. A single PEMF treatment in TNFα pre-stimulated cells also resulted in almost a 2-fold increase in IL-6 gene expression compared to TNFα pre-treated cells without PEMF stimulation. This effect could imply that either a single session of PEMF stimulation exacerbates inflammation in the inflammation model, or the increase in IL-6 gene expression needs a new interpretation. This heightened IL-6 gene expression in the inflammation model highlights the multifaceted role of IL-6 in the immune system. Despite its classification as a pro-inflammatory cytokine, recent years have seen recognition of the anti-inflammatory actions of IL-6 [30]. The increase in IL-6 gene expression could potentially contribute to the clinical improvement of patients, as demonstrated by Wuschech et al., (2015) in a prospective, placebo-controlled, double-blind study. However, it is essential to acknowledge that the simplified cell culture model used in our study may not fully reflect the systemic in vivo anti-inflammatory effect induced by PEMF in patients [16]. Moreover, future research should explore longer time frames to understand the long-term effects of PEMF stimulation in the inflammation model.

Study 3: Orfei et al. conducted a pilot study using a rat tendinopathy model induced by injecting type I collagenase into the Achilles tendon, followed by exposure to PEMF stimulation [31]. The authors observed that longer PEMF stimulation resulted in improved collagen fiber organization, reduced cell density, vascularity, and fat deposition, and restoration of physiological cell morphology compared to untreated tendons. However, similar to the 1-day stimulation in studies 1 and 2, a 3-day stimulation period did not yield significant changes in gene expression compared to the non-stimulated control group.

Nonetheless, there was a noticeable trend toward increased gene expression in the treatment groups as the stimulation period extended.

After 6 days of PEMF(ON) stimulation compared to the control group, significantly higher expression of CD55, CFH, and IL-6 genes was observed. This trend was also reflected in semi-quantitative analysis of immunofluorescence-stained images, where a stronger CD55 signal was detected in the PEMF-stimulated group compared to the non-stimulated group. Karpus et al., (2015) found abundant CD55 expression in fibroblast-like synovial fibroblasts of the intimal lining layer, associated with linearly oriented reticular fibers, which were resistant to phospholipase C treatment [32]. This suggests a protective role of CD55 in synovial tissue against immune complex-mediated arthritis. Consistent with our previous research, CD55 gene expression was nearly 4-fold lower in arthritis-derived HSE synovial fibroblasts compared to K4IM cells, further highlighting the protective role of PEMF stimulation [27]. CFH and its variant, Factor H-like protein 1, may also play a protective role against complement damage in the synovium [33]. PEMF stimulation in K4IM cells induced nearly a 2-fold increase in CFH gene expression, suggesting enhanced regulation of complement activation in synovial fibroblasts.

In terms of cytokine gene expression, Zou et al., (2017) demonstrated that IL-1β and TNFα levels secreted by nucleus pulposus cells into the culture medium were significantly reduced following PEMF exposure in an intensity-dependent manner [24]. Similarly, another study reported a reduction in IL-1β and TNFα secretion in fibroblast-like cells derived from mononuclear peripheral blood on days 14 and 21 of culture, along with an induction of IL-10 on day 21 of culture [23]. However, this study also found that PEMF did not inhibit the production of the cytokines IL-6 and IL-8. Our study revealed a significant increase in IL-6 gene expression after 6 days of PEMF stimulation compared to the non-stimulated group. This effect was observed even after a single session of stimulation, but only when the cells were pre-stimulated with TNFα. Interestingly, the group stimulated with PEMF for 6 days showed higher expression of CFI and IL-6 genes compared to the same group stimulated for only 3 days. This change in gene expression of CFI and IL-6 was also observed comparing the control groups of the two time periods, suggesting that cell proliferation or cell density may influence CFI and IL-6 gene expression, but not the expression of the other analyzed complement genes CD55, CD59, and CFH.

Regarding cellular response, PEMF stimulation has been shown to increase cell proliferation in osteoblasts and chondrocytes, cell migration as well as enhance extracellular matrix production [34–37]. However, in our study, we did not observe a significant increase in proliferative activity in K4IM synovial fibroblasts following PEMF stimulation for 3 days or 6 days. Even a longer stimulation time period should be analyzed in the future.

5. Limitations of the Study

The PEMF stimulation duration in a single session, the number of stimulation days, the stimulation intensity, and various other biological and non-biological factors could have varying impacts at the molecular level of the cells. One of the limitations of our study is the use of an immortalized cell line cultured under monolayer conditions. We recommend that future studies should be performed using different primary cell types freshly isolated from joint tissues at various time points, in a three-dimensional or co-culture model, to gain a more comprehensive understanding of PEMF stimulation effects on the entire joint. Although novel data were acquired through our study, incorporating other quantitative and semi-quantitative assays such as Western blots will be necessary to gain a broader picture of the response to PEMF stimulation in the future.

Our immune system is very complex because it requires a balanced hemostasis of soluble as well as membrane-bound pro- and anti-inflammatory components, such as cytokines and complement factors. The focus of this study was limited to two soluble complement factors (CFH and CFI), two membrane-bound complement factors (CD55 and CD59), and two pro-inflammatory cytokines (TNFα and IL-6). The overall complexity of the interplay between numerous complement factors and immune regulatory cytokines (such as IL-1, IL-33, IL-37, IL-10, etc.) needs further investigation. In addition to the production of these components, the effects of PEMF stimulation on phenotype, cell proliferation, and migration should be studied thoroughly in future research. This is also a limitation of our study.

6. Conclusions

Previous clinical studies have demonstrated the efficacy of PEMF therapy in alleviating osteoarthritic symptoms. However, both short-term (single-day) and three-day sessions of PEMF stimulation of K4IM synovial fibroblasts failed to induce significant changes in the gene expression of complement regulatory proteins (CD55, CD59, CFH, CFI) and cytokines (IL-6, TNFα) compared to six-day stimulation. This implies the importance of longer PEMF stimulation in practical applications. Nevertheless, a notable increase in the gene expression of CD55, CFH, and CD59 suggests that PEMF stimulation may confer cytoprotective and anti-inflammatory properties. Conversely, the upregulation of IL-6 gene expression indicates a pro-inflammatory response to PEMF stimulation. This dualistic effect underscores the complexity of PEMF's impact on cells at the molecular level. Further research is necessary to deepen our understanding of PEMF's influence on complement activity and cytokine regulation.

Supplementary Materials: The following supporting information can be downloaded at: https://www.mdpi.com/article/10.3390/jpm14070701/s1.

Author Contributions: Conceptualization, S.S., M.G. and G.S.-T.; methodology, S.S. and G.S.-T.; software, S.S., M.G. and G.S.-T.; validation, S.S., M.G. and G.S.-T.; formal analysis, S.S.; investigation, S.S. and M.W.; resources, M.G. and G.S.-T.; data curation, S.S.; writing—original draft preparation, S.S.; writing—review and editing, S.S., M.G., M.W. and G.S.-T.; visualization, S.S. and M.W.; supervision, G.S.-T.; project administration, S.S. and M.G.; funding acquisition, M.G. All authors have read and agreed to the published version of the manuscript.

Funding: Dr. Markus Gesslein received the research grant for this project from Nürnberger Versicherung.

Institutional Review Board Statement: Not applicable.

Informed Consent Statement: Not applicable.

Data Availability Statement: Supporting data can be obtained from the authors on request.

Acknowledgments: We gratefully acknowledge Physiomed, Schnaittach, for generously providing the MAGCELL® ARTHRO apparatus used in this study. The language optimization of the manuscript was conducted using OpenAI's ChatGPT version 3.5. The optimization was manually reviewed.

Conflicts of Interest: The authors declare no conflicts of interest.

References

1. Yao, Q.; Wu, X.; Tao, C.; Gong, W.; Chen, M.; Qu, M.; Zhong, Y.; He, T.; Chen, S.; Xiao, G. Osteoarthritis: Pathogenic signaling pathways and therapeutic targets. *Signal Transduct. Target. Ther.* **2023**, *8*, 56. [CrossRef] [PubMed]
2. Long, H.; Liu, Q.; Yin, H.; Wang, K.; Diao, N.; Zhang, Y.; Lin, J.; Guo, A. Prevalence Trends of Site-Specific Osteoarthritis From 1990 to 2019: Findings from the Global Burden of Disease Study 2019. *Arthritis Rheumatol.* **2022**, *74*, 1172–1183. [CrossRef]
3. Chen, D.; Shen, J.; Zhao, W.; Wang, T.; Han, L.; Hamilton, J.L.; Im, H.J. Osteoarthritis: Toward a comprehensive understanding of pathological mechanism. *Bone Res.* **2017**, *5*, 16044. [CrossRef] [PubMed]
4. Assirelli, E.; Pulsatelli, L.; Dolzani, P.; Mariani, E.; Lisignoli, G.; Addimanda, O.; Meliconi, R. Complement Expression and Activation in Osteoarthritis Joint Compartments. *Front. Immunol.* **2020**, *11*, 535010. [CrossRef] [PubMed]
5. Mathiessen, A.; Conaghan, P.G. Synovitis in osteoarthritis: Current understanding with therapeutic implications. *Arthritis Res. Ther.* **2017**, *19*, 18. [CrossRef] [PubMed]
6. Loeser, R.F.; Goldring, S.R.; Scanzello, C.R.; Goldring, M.B. Osteoarthritis: A disease of the joint as an organ. *Arthritis Rheum.* **2012**, *64*, 1697–1707. [CrossRef] [PubMed]
7. Sethi, V.; Anand, C.; Della Pasqua, O. Clinical Assessment of Osteoarthritis Pain: Contemporary Scenario, Challenges, and Future Perspectives. *Pain Ther.* **2024**, *13*, 391–408. [CrossRef] [PubMed]
8. Walter, N.; Hinterberger, T.; Szymski, D.; Alt, V.; Rupp, M. Psychological comorbidities in osteoarthritis in Germany. *Sci. Rep.* **2023**, *13*, 2905. [CrossRef] [PubMed]

9. Maqbool, M.; Fekadu, G.; Jiang, X.; Bekele, F.; Tolossa, T.; Turi, E.; Fetensa, G.; Fanta, K. An up to date on clinical prospects and management of osteoarthritis. *Ann. Med. Surg.* **2021**, *72*, 103077. [CrossRef]
10. Markovic, L.; Wagner, B.; Crevenna, R. Effects of pulsed electromagnetic field therapy on outcomes associated with osteoarthritis. *Wien. Klin. Wochenschr.* **2022**, *134*, 425–433. [CrossRef]
11. Wang, T.; Xie, W.; Ye, W.; He, C. Effects of electromagnetic fields on osteoarthritis. *Biomed. Pharmacother.* **2019**, *118*, 109282. [CrossRef] [PubMed]
12. Bagnato, G.L.; Miceli, G.; Marino, N.; Sciortino, D.; Bagnato, G.F. Pulsed electromagnetic fields in knee osteoarthritis: A double blind, placebo-controlled, randomized clinical trial. *Rheumatology* **2016**, *55*, 755–762. [CrossRef] [PubMed]
13. Hu, H.; Yang, W.; Zeng, Q.; Chen, W.; Zhu, Y.; Liu, W.; Wang, S.; Wang, B.; Shao, Z.; Zhang, Y. Promising application of Pulsed Electromagnetic Fields (PEMFs) in musculoskeletal disorders. *Biomed. Pharmacother.* **2020**, *131*, 110767. [CrossRef] [PubMed]
14. Chen, L.; Duan, X.; Xing, F.; Liu, G.; Gong, M.; Li, L.; Chen, R.; Xiang, Z. Effects of pulsed electromagnetic field therapy on pain, stiffness and physical function in patients with knee osteoarthritis: A systematic review and meta-analysis of randomized controlled trials. *J. Rehabil. Med.* **2019**, *51*, 821–827. [CrossRef] [PubMed]
15. McCarthy, C.J.; Callaghan, M.J.; Oldham, J.A. Pulsed electromagnetic energy treatment offers no clinical benefit in reducing the pain of knee osteoarthritis: A systematic review. *BMC Musculoskelet. Disord.* **2006**, *7*, 51. [CrossRef] [PubMed]
16. Wuschech, H.; von Hehn, U.; Mikus, E.; Funk, R.H. Effects of PEMF on patients with osteoarthritis: Results of a prospective, placebo-controlled, double-blind study. *Bioelectromagnetics* **2015**, *36*, 576–585. [CrossRef] [PubMed]
17. Janeway, C.A., Jr.; Travers, P.; Walport, M.; Shlomchik, M.J. *Immunobiology: The Immune System in Health and Disease*, 5th ed.; Garland Science: New York, NY, USA, 2001.
18. Markiewski, M.M.; Lambris, J.D. The role of complement in inflammatory diseases from behind the scenes into the spotlight. *Am. J. Pathol.* **2007**, *171*, 715–727. [CrossRef] [PubMed]
19. Hallam, T.M.; Sharp, S.J.; Andreadi, A.; Kavanagh, D. Complement factor I: Regulatory nexus, driver of immunopathology, and therapeutic. *Immunobiology* **2023**, *228*, 152410. [CrossRef] [PubMed]
20. Härdig, Y.; Hillarp, A.; Dahlbäck, B. The amino-terminal module of the C4b-binding protein alpha-chain is crucial for C4b binding and factor I-cofactor function. *Biochem. J.* **1997**, *323 Pt 2*, 469–475. [CrossRef]
21. Schwaeble, W.; Zwirner, J.; Schulz, T.F.; Linke, R.P.; Dierich, M.P.; Weiss, E.H. Human complement factor H: Expression of an additional truncated gene product of 43 kDa in human liver. *Eur. J. Immunol.* **1987**, *17*, 1485–1489. [CrossRef]
22. Lucientes-Continente, L.; Márquez-Tirado, B.; Goicoechea de Jorge, E. The Factor H protein family: The switchers of the complement alternative pathway. *Immunol. Rev.* **2023**, *313*, 25–45. [CrossRef] [PubMed]
23. Gómez-Ochoa, I.; Gómez-Ochoa, P.; Gómez-Casal, F.; Cativiela, E.; Larrad-Mur, L. Pulsed electromagnetic fields decrease proinflammatory cytokine secretion (IL-1β and TNF-α) on human fibroblast-like cell culture. *Rheumatol. Int.* **2011**, *31*, 1283–1289. [CrossRef] [PubMed]
24. Zou, J.; Chen, Y.; Qian, J.; Yang, H. Effect of a low-frequency pulsed electromagnetic field on expression and secretion of IL-1β and TNF-α in nucleus pulposus cells. *J. Int. Med. Res.* **2017**, *45*, 462–470. [CrossRef] [PubMed]
25. Silawal, S.; Wagner, M.; Roth, D.; Bertsch, T.; Schwarz, S.; Willauschus, M.; Gesslein, M.; Triebel, J.; Schulze-Tanzil, G. Complement Proteins C5/C5a, Cathepsin D and Prolactin in Chondrocytes: A Possible Crosstalk in the Pathogenesis of Osteoarthritis. *Cells* **2022**, *11*, 1134. [CrossRef] [PubMed]
26. Pfaffl, M.W. Quantification strategies in real-time PCR. *AZ Quant. PCR* **2004**, *1*, 89–113.
27. Franke, V.; Meyer, S.; Schulze-Tanzil, G.G.; Braun, T.; Kokozidou, M.; Fischlein, T.; Silawal, S. Complement Regulation in Immortalized Fibroblast-like Synoviocytes and Primary Human Endothelial Cells in Response to SARS-CoV-2 Nucleocapsid Protein and Pro-Inflammatory Cytokine TNFα. *Life* **2022**, *12*, 1527. [CrossRef] [PubMed]
28. Criado-García, O.; Fernaud-Espinosa, I.; Bovolenta, P.; de la Cuesta, R.S.; de Córdoba, S.R. Expression of the ß-chain of the complement regulator C4b-binding protein in human ovary. *Eur. J. Cell Biol.* **1999**, *78*, 657–664. [CrossRef] [PubMed]
29. Sánchez-Pernaute, O.; Esparza-Gordillo, J.; Largo, R.; Calvo, E.; Alvarez-Soria, M.A.; Marcos, M.E.; Herrero-Beaumont, G.; de Córdoba, S.R. Expression of the peptide C4b-binding protein β in the arthritic joint. *Ann. Rheum. Dis.* **2006**, *65*, 1279–1285. [CrossRef] [PubMed]
30. Borsini, A.; Di Benedetto, M.G.; Giacobbe, J.; Pariante, C.M. Pro- and anti-inflammatory properties of interleukin (IL6) in vitro: Relevance for major depression and for human hippocampal neurogenesis. *Int. J. Neuropsychopharmacol.* **2020**, *23*, 738–750. [CrossRef]
31. Orfei, C.P.; Lovati, A.B.; Lugano, G.; Viganò, M.; Bottagisio, M.; D'Arrigo, D.; Sansone, V.; Setti, S.; Girolamo, L.d. Pulsed electromagnetic fields improve the healing process of Achilles tendinopathy. *Bone Joint Res.* **2020**, *9*, 613–622. [CrossRef]
32. Karpus, O.N.; Kiener, H.P.; Niederreiter, B.; Yilmaz-Elis, A.S.; van der Kaa, J.; Ramaglia, V.; Arens, R.; Smolen, J.S.; Botto, M.; Tak, P.P.; et al. CD55 deposited on synovial collagen fibers protects from immune complex-mediated arthritis. *Arthritis Res. Ther.* **2015**, *17*, 6. [CrossRef] [PubMed]
33. Friese, M.A.; Manuelian, T.; Junnikkala, S.; Hellwage, J.; Meri, S.; Peter, H.H.; Gordon, D.L.; Eibel, H.; Zipfel, P.F. Release of endogenous anti-inflammatory complement regulators FHL-1 and factor H protects synovial fibroblasts during rheumatoid arthritis. *Clin. Exp. Immunol.* **2003**, *132*, 485–495. [CrossRef] [PubMed]

34. Sakhrani, N.; Stefani, R.M.; Setti, S.; Cadossi, R.; Ateshian, G.A.; Hung, C.T. Pulsed Electromagnetic Field Therapy and Direct Current Electric Field Modulation Promote the Migration of Fibroblast-like Synoviocytes to Accelerate Cartilage Repair In Vitro. *Appl. Sci.* **2022**, *12*, 12406. [CrossRef] [PubMed]
35. Varani, K.; Vincenzi, F.; Pasquini, S.; Blo, I.; Salati, S.; Cadossi, M.; De Mattei, M. Pulsed Electromagnetic Field Stimulation in Osteogenesis and Chondrogenesis: Signaling Pathways and Therapeutic Implications. *Int. J. Mol. Sci.* **2021**, *22*, 809. [CrossRef]
36. Stefani, R.M.; Barbosa, S.; Tan, A.R.; Setti, S.; Stoker, A.M.; Ateshian, G.A.; Cadossi, R.; Vunjak-Novakovic, G.; Aaron, R.K.; Cook, J.L.; et al. Pulsed electromagnetic fields promote repair of focal articular cartilage defects with engineered osteochondral constructs. *Biotechnol. Bioeng.* **2020**, *117*, 1584–1596. [CrossRef]
37. De Mattei, M.; Caruso, A.; Pezzetti, F.; Pellati, A.; Stabellini, G.; Sollazzo, V.; Traina, G.C. Effects of pulsed electromagnetic fields on human articular chondrocyte proliferation. *Connect Tissue Res.* **2001**, *42*, 269–279. [CrossRef]

Journal of *Personalized Medicine*

MDPI

Systematic Review

Minced Cartilage Is a One-Step Cartilage Repair Procedure for Small Defects in the Knee—A Systematic-Review and Meta-Analysis

Andreas Frodl [1,*], Markus Siegel [1], Andreas Fuchs [1], Ferdinand C. Wagner [1], Hagen Schmal [2], Kaywan Izadpanah [1] and Tayfun Yilmaz [1]

1 Department of Orthopedics and Traumatology, Freiburg University Hospital, 79106 Freiburg, Germany
2 Department of Orthopedic Surgery, University Hospital Odense, Sdr. Boulevard 29, 5000 Odense, Denmark
* Correspondence: andreas.frodl@uniklinik-freiburg.de

Abstract: Purpose: Approximately 60% of patients undergoing arthroscopy of the knee present with chondral defects. If left untreated, osteochondral lesions can trigger an early onset of osteoarthritis. Many cartilage repair techniques are mainly differentiated in techniques aiming for bone marrow stimulation, or cell-based methods. Cartilage repair can also be categorized in one- and two-stage procedures. Some two-stage procedures come with a high cost for scaffolds, extensive cell-processing, strict regulatory requirements, and limited logistical availability. Minced cartilage, however, is a one-stage procedure delivering promising results in short term follow-up, as noted in recent investigations. However, there is no available literature summarizing or synthesizing clinical data. The purpose of this study was to analyze and synthesize data from the latest literature in a meta-analysis of outcomes after the minced cartilage procedure and to compare its effectiveness to standard repair techniques. Methods: We conducted a systematic review searching the Cochrane, PubMed, and Ovid databases. Inclusion criteria were the modified Coleman methodology Score (mCMS) >60, cartilaginous knee-joint defects, and adult patients. Patient age < 18 years, biomechanical and animal studies were excluded. Relevant articles were reviewed independently by referring to title and abstract. In a systematic review, we compared three studies and 52 patients with a total of 63 lesions. Results: Analysis of Knee Injury and Osteoarthritis Outcome Score (KOOS) sub scores at 12 and 24 months showed a significant score increase in every sub score. Highest mean difference was seen in KOOS sport, lowest in KOOS symptoms (12 month: KOOS sport (Mean difference: 35.35 [28.16, 42.53]; $p < 0.0001$), lowest in KOOS symptoms (Mean difference: 20.12 [15.43, 24.80]; $p < 0.0001$)). A comparison of International Knee Documentation Committee (IKDC) scores visualized a significant score increase for both time points too ((12 month: pooled total mean: 73.00 ± 14.65; Mean difference: 34.33 [26.84, 41.82]; $p < 0.00001$) (24 month: pooled total mean: 77.64 ± 14.46; mean difference: 35.20 [39.49, 40.92]; $p < 0.00001$)). Conclusion: Due to no need for separate cell-processing, and thanks to being a one-step procedure, minced cartilage is a promising method for cartilage repair in small defect sizes (mean 2.77 cm^2, range 1.3–4.7 cm^2). However, the most recent evidence is scarce, and takes only results two years post-surgery into account. Summarized, minced cartilage presents nearly equal short-term improvement of clinical scores (IKDC, KOOS) compared to standard cartilage repair techniques.

Keywords: cartilage defects; cartilage repair; minced cartilage

check for updates

Citation: Frodl, A.; Siegel, M.; Fuchs, A.; Wagner, F.C.; Schmal, H.; Izadpanah, K.; Yilmaz, T. Minced Cartilage Is a One-Step Cartilage Repair Procedure for Small Defects in the Knee—A Systematic-Review and Meta-Analysis. *J. Pers. Med.* **2022**, *12*, 1923. https://doi.org/10.3390/jpm12111923

Academic Editors: Eduard Kurz and Jih-Yang Ko

Received: 14 October 2022
Accepted: 15 November 2022
Published: 18 November 2022

1. Introduction

Articular cartilage is highly complex tissue that undergoes metabolic changes due to trauma, degeneration, or the deprivation of nutrients through osteochondrosis dissecans (OD) or osteonecrosis [1–6]. These factors hinder tissue maintenance and repair mechanisms, leading to loss of cartilage-surface area and ultimately osteoarthrosis. However, there are treatment options to repair and restore articular cartilage surgically. Current interventions can be categorized either as techniques aiming to stimulate bone marrow, or as cell-based cartilage repair. Timewise, differentiation is possible in one- and two-stage procedures depending on whether one or two interventions are needed to integrate cartilaginous material and defect filing [4,7].

Standard procedures include matrix-assisted chondrocyte implantation (MACI) or the osteochondral autograft transfer system (OATS) and osteochondral allograft transplantation (OCA) [3,8]. Current one-stage procedures, e.g., microfracture and OATS, are applicable with good clinical results for smaller defects. Larger defects must often be handled with two-stage procedures (i.e., MACI) [3].

These cell-based procedures are associated with expensive scaffolds and the demanding conditions required for in vitro cell proliferation. Considering cost-effectiveness, a one-stage procedure not requiring demanding in vitro cell proliferation to induce hyaline or hyaline-like repair tissue, would seem more suitable. Minced cartilage implantation is just such a type of cartilage repair; it entails the implantation of autologous cartilage chips [9,10]. Recent investigations have shown promising short-term follow-up results. However, there is a paucity of data on long-term outcomes and comparisons to similar cartilage repair techniques [11]. The purpose of this study was to analyze the most recent literature reporting on outcomes after the minced-cartilage procedure, comparing its effectiveness to established cartilage repair techniques. Hence, minced cartilage is suggested as a reliable alternative in cartilage repair.

2. Methods

This descriptive systematic review was conducted according to the Preferred Reporting Items for Systematic Reviews and Meta-analyses checklist guidelines (PRISMA) and submitted to PROSPERO [12,13].

From January 2021 to October 2021, a database search was done independently by the authors (A.F. and T.Y.). MEDLINE, PubMed, Embase, Web of Science, and the Cochrane Library were searched for relevant studies reporting clinical outcome after minced cartilage therapy in patients with cartilage defects of the knee.

2.1. Search Strategy

#	Query	Results
1	exp Knee Joint/or exp Cartilage, Articular/or exp Cartilage/or exp Orthopedic Procedures/	424,841
2	exp Rehabilitation/mt, su, td [Methods, Surgery, Trends]	70,703
3	exp Cartilage/or exp Hyaline Cartilage/or exp Cartilage, Articular/	87,137
4	exp Treatment Outcome/	1,102,955
5	2 or 4	1,156,746
6	1 and 5	82,188
7	3 and 6	6059
8	exp Arthroscopy/ae, mt, rh, st, td [Adverse Effects, Methods, Rehabilitation, Standards, Trends]	10,593
9	7 and 8	572
10	minced cartilage.mp.	17
11	minced.mp.	2446
12	cartilage.mp.	98,087
13	11 and 12	65
14	10 or 13	65
15	9 or 14	637

2.2. Eligibility

Our inclusion criteria were: studies between 2000 and 2020 and a minimum follow-up of 12 months. Only publications written in German or English were included. A minimum patient age of 18 years was set to enable comparisons between fully grown adults undergoing the minced cartilage procedure. An evaluation of clinical scores (International Knee Documentation Committee IKDC, Knee Injury and Osteoarthritis Outcome Score KOOS or Lysholm) was mandatory at 12 and 24 months post-surgery. Our exclusion criteria were: an overall modified Coleman Methodology Score (mCMS) <60, follow-up rate <80%, cadaver or biomechanical studies, and animal studies.

Titles and abstracts were screened independently by the reviewers for relevance according to the aforementioned inclusion and exclusion criteria. The full text was obtained to assess the study's relevance if no abstract was available. We cross-referenced the references within included articles if they had been missed by our search algorithm to make sure we had not overlooked any suitable studies. Appropriate publications were then independently analyzed for the mCMS and risk of bias with ROBINS-I tool [12–15].

2.3. Primary Outcome Criteria

Patient demographics, number of patients, clinical scores (IKDC, KOOS), follow-up period, and surgical technique were extracted by us authors. The hypothesis was, concerning clinical scores (KOOS, IKDC), that the minced cartilage procedure is comparable to those of commonly used cartilage repair techniques.

2.4. Statistics

To analyze the collected data, Microsoft Excel (ver. 16.16, Redmond, Wahington, Microsoft Corp., 2019) and Revman5 (Version 5.3. Copenhagen: The Nordic Cochrane Centre, The Cochrane Collaboration, 2014) were used. Given means and standard deviations were pooled using continuous statistic evaluation protocol of RevMan5 and calculated in a forest plot as mean differences for each point of time.

3. Results

3.1. Study Selection

Our literature search and study selection procedure is depicted in Figure 1. A total of 637 papers were identified by our search algorithm. Moreover, one paper was added from the reference list. These papers were scanned, and any duplicates or topic-unrelated articles excluded. After analyzing the eligibility criteria, three of the five studies were included in our analysis. To acquire the original, complete data set, we contacted the corresponding authors of the papers to be included. Nevertheless, access to complete data was granted in just one case, which hindered our qualitative analysis.

We identified two prospective trails and one randomized controlled trial containing a total 52 patients with a total of 61 cartilage defects at the knee for inclusion in our review.

The number of patients included in these studies ranged from 7 to 25 with a mean age of 30.6 $\pm$ 5.4 years.

3.2. Risk of Bias Assessment

All included studies possessed evidence-level III or II. Reporting and detection biases are considerable due to the lack of randomization and blinding in two of the three studies. Surgical techniques were reported in detail in every study, minimizing the risk of operational bias even in cases in which several surgeons were operating. To calculate the risk of underlying bias, all included studies were analyzed with the ROBINS-I tool.

Our results for the risk of bias assessment and mCMS scores are shown in Tables 1 and 2.

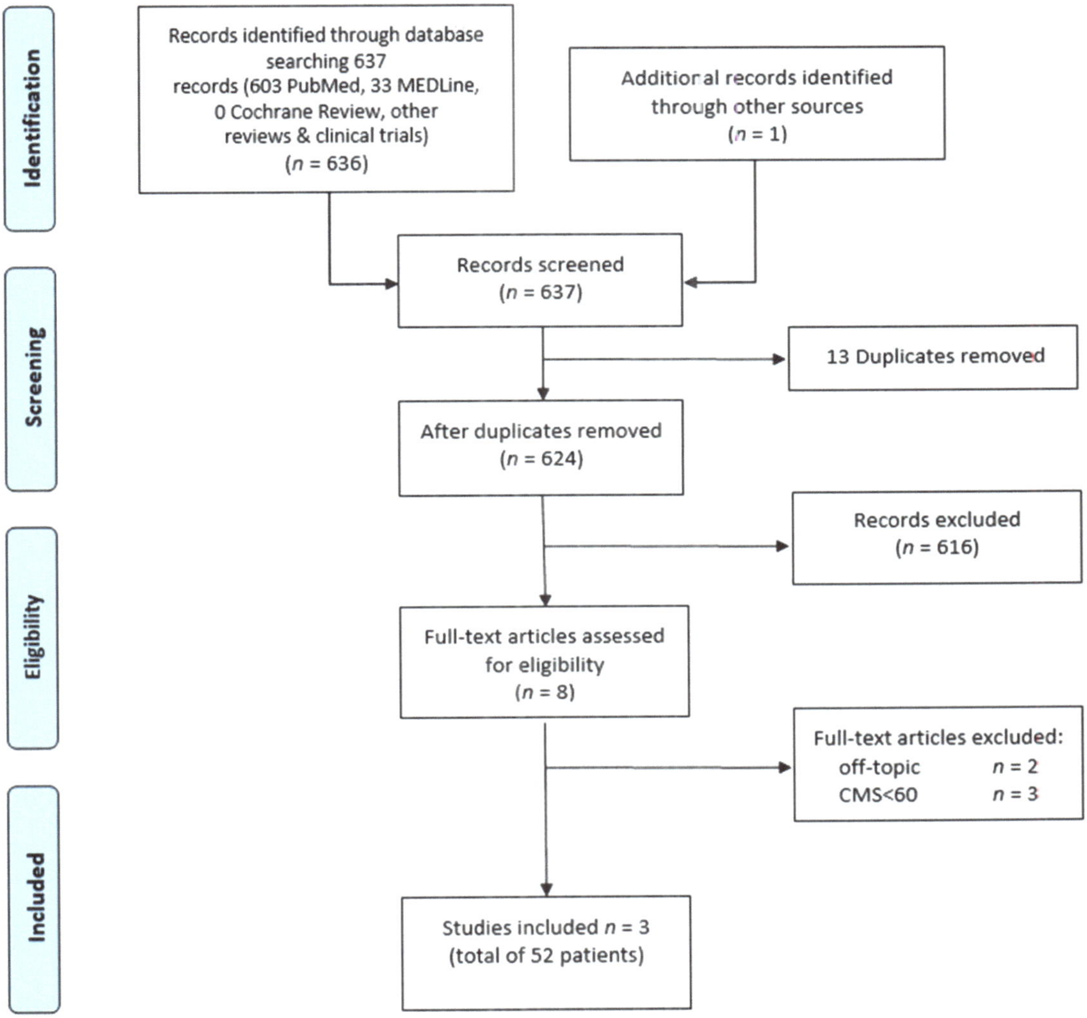

Figure 1. Overview of study selection process according to PRISMA guidelines.

Table 1. Risk of bias assessment with ROBINS-I tool implicating moderate risk of bias in included studies.

Risk of Bias Preintervention and at-Intervaention Domains					**Risk of Bias Post-Intervention Domains**			
study	Bias due to confounding	Bias due to selection of participants into study	Bias in classification of intervention	Bias due to deviation from intended intervention	Bias due to missing data	Bias in measurement of outcome	Bias in the selection of reported outcome	Overall assessment of bias
Christensen	Low	Low	Moderate	Low	moderate	moderate	Low	moderate
Farr	Low	Low	Low	low	moderate	moderate	low	moderate
Cole	low	low	low	low	low	low	low	low
Buckwalter	serious	moderate	Low	low	moderate	moderate	low	serious
Massen	serious	serious	low	Low	low	moderate	low	serious

Key: low risk of bias | moderate risk of bias | serious risk of bias

Table 2. Detailed overview of individual study mCMS scores.

Modified Coleman Methology Score					
	Christensen et al.	**Cole et al.**	**Farr et al.**	**Buckwalter et al.**	**Massen et al.**
PART A Study size	0	0	0	0	0
Mean follow up	0	0	0	0	0
Percent of Patients with follow up	3	5	5	0	5
Numbers of interventions per group	10	10	10	10	10
Type of study	10	15	10	10	0
Diagnostic certainly	5	5	5	5	5
Description of surgical technique	5	5	5	5	5
Definition of postoperativ rehabilitation	5	5	5	5	0
PART B Outcome criteria	10	10	10	8	7
Procedure of assaying outcomes	11	11	11	11	11
Description of subject selection Prozess	5	10	5	5	5
	64	76	66	59	48

3.3. *Clinical Outcome*

In all three studies, the KOOS and IKDC score was used to evaluate clinical outcome. All studies included time points of 0 and 12 months. Farr et al. and Cole et al. also conducted clinical evaluations at 24 months (Table 3). In each study minced cartilage was combined with fibrin glue, which accounts for similar surgical methods and comparable results in respect of treatment methods.

Table 3. Demographic data from the selected studies including surgical procedure.

Study	Number of Patients	Age (y)	Male-Female	Defect Location	Defect Size (cm2)	Defect Type	Operation Method
Christensen	8 Patients	32 +/− 7	female 3 male 5	Femur condyle 7 Trochlea 1	3.1 (1.5-4.7)	Osteochondritis dissecans	Minced cartilage with fibrin glue, additional bine reconstrution
Cole	20 Patients 24 Lesions	32.7 +/− 8.8	female 6 male 14	Femur condyle 14 Trochlea 10	2.75 +/− 1.5	ICRS grade III 20 ICRS grade IV 4	Minced cartilage with fibrin glue and CAIS Scaffold
Farr	25 Patients 29 Lesions	37 +/− 11.1	femlae 7 male 18	Femur condyle 18 Trochlea 11	2.7 +/− 0.8	ICRS grade III 23 ICRS grade IV 6	Juvenile minced cartilage allograft fixed with fibrin glue

CAIS: Cartilage Autograft Implantation System.

3.3.1. Clinical Scores at 12-Month Post-Surgery

Overall, 105 patients were included in this subgroup analysis. Mean differences for each KOOS sub score have been calculated and displayed in a forest plot. A significant score increase was seen in every sub score. The highest mean difference between pre- and post-surgery score was seen in KOOS sport (Mean difference: 35.35 [28.16, 42.53]; $p < 0.0001$), lowest in KOOS symptoms (Mean difference: 20.12 [15.43, 24.80]; $p < 0.0001$). A detailed analysis of the 12-month score evaluation is presented in Figure 2.

IKDC scores have been reported by Christensen et al. and Cole et al. at 12-month control. Mean difference compared to pre-surgery analysis was significantly higher than pre-surgery (pooled total mean: 73.00 ± 14.65; Mean difference: 34.33 [26.84, 41.82]; $p < 0.00001$) (Figure 3).

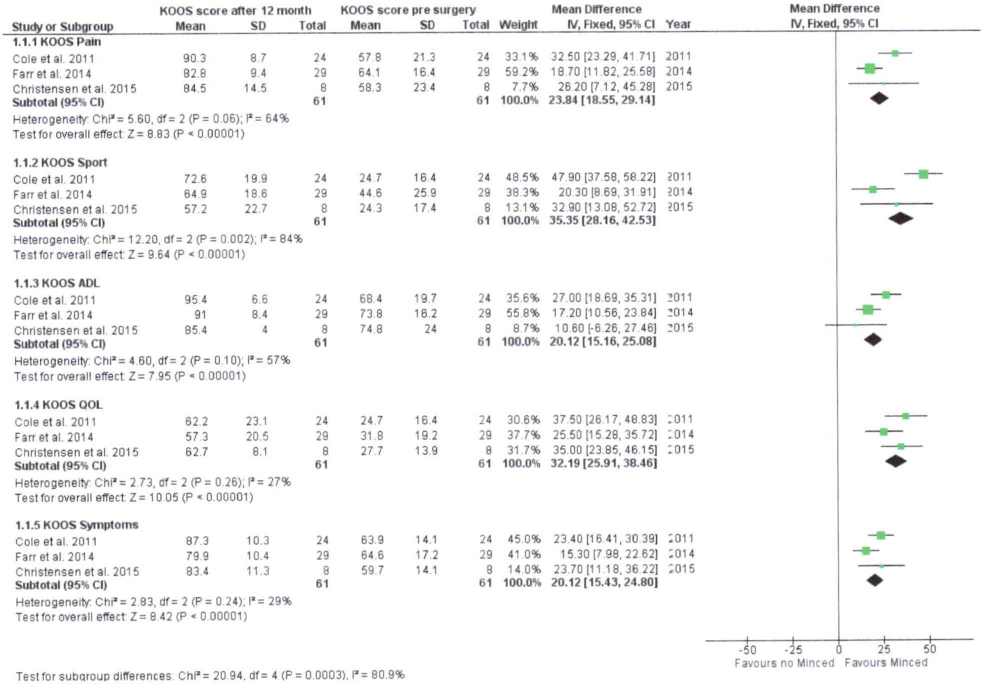

Figure 2. Forest plot estimating mean differences for 12-month time interval for each KOOS score subgroup.

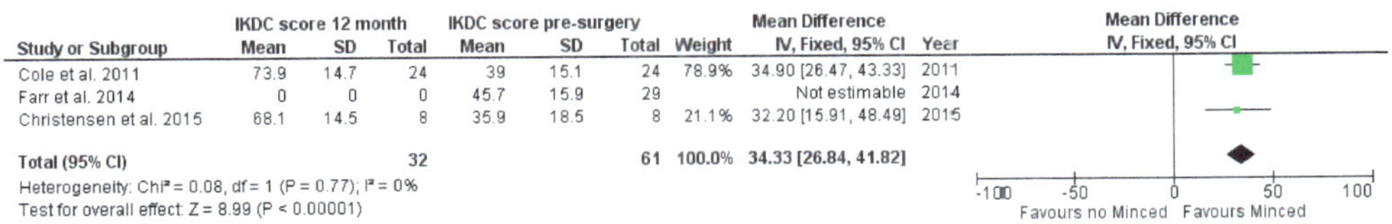

Figure 3. Forest plot estimating mean differences for 12-month time interval for IKDC score.

3.3.2. Clinical Scores at 24-Month Post-Surgery

Clinical scores for the 24-month time interval have only been reported by Cole et al. and Farr et al. Christensen et al. only reported 12-month results.

A total of 97 patients were thus included for analysis. A significant increase in every sub score was observed.

The KOOS sport subgroup showed the highest increase after 24 months (mean difference: 35.86 [27.21, 44.51]; $p < 0.00001$). Similar to 12 month results, the lowest increase of scores was seen in KOOS symptoms (mean difference: 19.29 [13.11, 25.47], $p < 0.00001$) (Figure 4).

The IKDC at 24 months was not reported by Christensen et al. A significant increase was seen compared to pre-surgery evaluation (pooled total mean: 77.64 ± 14.46; mean difference: 35.85 [20.07, 51.63]; $p < 0.00001$) and is depicted in Figure 5.

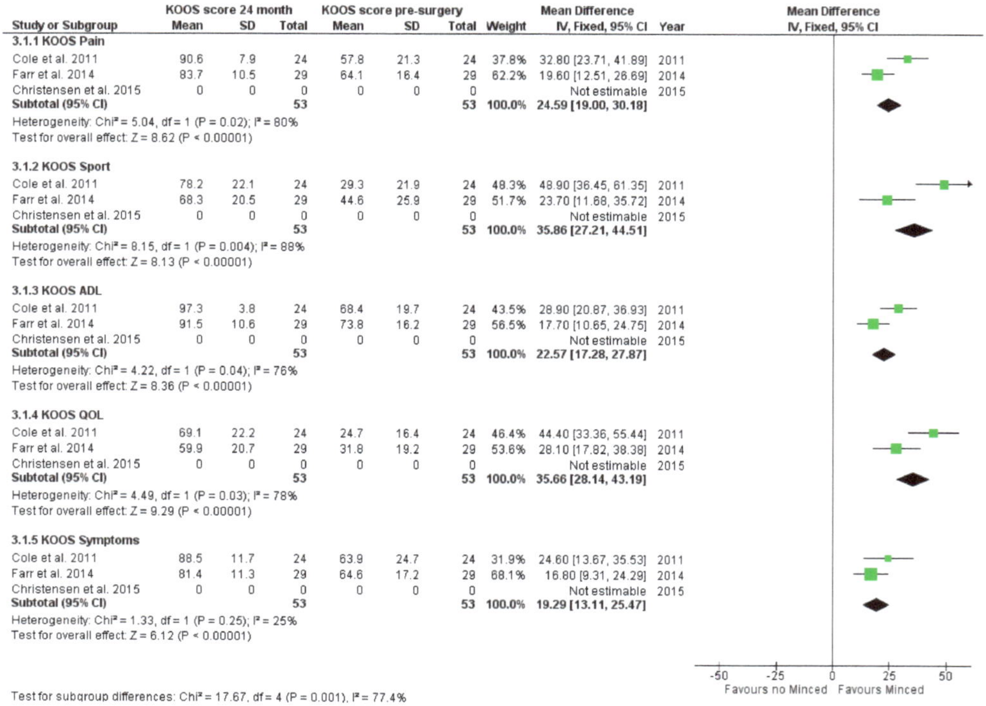

Figure 4. Forest plot estimating mean differences for 24-month time interval for each KOOS score subgroup.

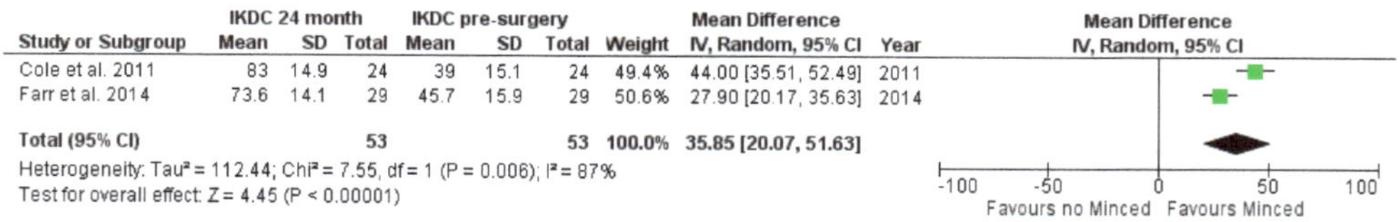

Figure 5. Forest plot estimating mean differences for 24-month time interval for IKDC score.

4. Discussion

As depicted in our overview of how we selected studies, there is very little recent data on the minced cartilage procedure and its long-term results. The studies we included represent the only evidence originating from 24-month follow-ups with satisfactory methodology. Mean defect size ranged between 1.3 and 4.7 cm^2. For inter-patient comparability, the KOOS and IKDC scores were calculated pre-surgery, and at 12- and 24-month after the minced cartilage procedure. Summarizing the outcomes of this meta-analysis, our study's key finding is that following minced cartilage procedure clinical scores (KOOS and IKDC) increased significantly at 12 and 24 months compared to pre-surgery measurements.

The latest surgical procedures for cartilage repair can be categorized as one-step or two-step procedures [16]. However, treatment options are limited to the knee's defect size [17].

The microfracture approach aims to stimulate bone marrow growth by multiple drilling into the subchondral bone, a method by which mesenchymal stem cells and growth factors are emitted from the drilling holes, filling the defect zone with fibrocartilage [18,19]. Compared to hyaline cartilage, which is mainly composed of type II collagen, such fibrocartilage is composed of a mixture of fibrous tissue and cartilage tissue containing type I and type II collagen [20]. Because of the tissue's fibrosity, fibrocartilage is both less flexible and tougher than hyaline cartilage, resulting in mechanical drawbacks compared to hyaline cartilage [3]. This is why microfracture is only suitable for smaller defects (measuring < 3 cm^2) [21]. However, the outcomes after 2–5 years after microfracture reveal alleviated pain and higher functional knee scores (KOOS, IKDC, Lysholm). In a 10-year follow-up, Gobbi et al. reported a significant improvement in two years postoperative IKDC and Lysholm scores in patients with defect sizes of 4.01 $\pm$ 0.27 cm^2 (Lysholm pre-surgery: 45.4 $\pm$ 3.5; two years: 90.4 $\pm$ 1.8; five years: 84.7 $\pm$ 3.6; 10 years: 77.2 $\pm$ 3.5.; IKDC pre-surgery 46.7 $\pm$ 2.9; two years: 82.7 $\pm$ 3.2; five years: 79.0 $\pm$ 4.1; 10 years: 71.5 $\pm$ 4.0). Nevertheless, significantly lower scores were documented between two years and 10 years post-surgery [21]. Orth et al. reported highly similar results in their systematic review of 1870 patients [22].

The microfracture approach is limited to small defects. Defect healing with fibrocartilage instead of hyaline cartilage impairs the knee's ability to shift the axial loading-force onto surrounding cartilage tissue [20]. Compared to this, the minced cartilage procedure showed slightly lower IKDC scores at 24 months post-surgery (pooled total mean: 77.64 $\pm$ 14.46). However, due to small cohort size, the standard deviation seen in our results is more affected by extreme outliers and is thus broader than in the work of Gobbi et al.

The long-term progress of scores remains unclear due to (so far) missing data. However, a smaller score-decrease can be alleged due to higher probabilities of generating hyaline cartilage compared to microfracture.

There are treatment options for larger, full-thickness cartilage repair with ACI/MACI, AMIC or OATS. Defect sizes in the literature range from 3 cm^2 to a maximum 22 cm^2, although the latter was a very rare case [23–25].

Autologous chondrocyte implantation (ACI) is a two-stage method. In the first stage, a small amount of cartilage tissue with viable chondrocytes is harvested from areas not involved in weight-bearing. The gathered tissue is then sent to a laboratory that amplifies and cultures these chondrocytes [26,27]. After cell amplification, the chondrocytes are surgically implanted in the osteochondral defect. The chondrocytes incorporated in scaffolds (hydrogel or decellularized type I collagen) are a modified version of the former ACI, the matrix-assisted ACI (MACI) [28–30].

Migliorini et al. described functional improvement after MACI in their systematic reviews [24,31]. They observed a significant improvement in IKDC and Lysholm scores. Improvement was reported over a mean 44.3-month follow-up period (IDKC pre-surgery: 38.9 $\pm$ 9.0; last follow-up: 72.1 $\pm$ 7.9; Lysholm pre-surgery: 50.1 $\pm$ 7.0; last follow-up: 82.0 $\pm$ 8.9).

However, postoperative outcomes after the ACI and MACI procedures strongly depend on the subchondral bone's integrity too [16]. There are concerns about the dedifferentiation of chondrocytes during the in vitro amplification process, a factor that would compromise the later implant's efficiency [3,32]. The main disadvantages of the ACI and MACI procedures are the high scaffold cost and demanding cell-processing conditions, both of which limit the widespread use of this approach [16,28,31]. In a randomized, prospective trial, Barie et al. compared MACI with ACI plus periosteum. Clinical scores were taken 12 and 24 months post-surgery as well as 8–11 years post-surgery. IKDC scores were 69.1 $\pm$ 23.6 at 12 month and 65.1 $\pm$ 29.4 at 24-month [33]. In direct comparison, the short-term scores of IKDC of our analysis demonstrated higher scores at the same point of time.

For defects exceeding 2 cm^2 but between 2 and 8 cm^2, autologous matrix-induced chondrogenesis (AMIC) and OATS are well-established single-step procedures [23,25,33]. Kim et al. compared the clinical and radiological outcomes of patients who underwent microfracture and AMIC. Their two-year follow up analysis demonstrated a significantly higher mean rise in IKDC and Lysholm scores following AMIC than in patients who underwent microfracture [34]. In a meta-analysis Steinwachs et al. demonstrated significant IKDC improvement in five studies after 24 months compared to baseline score. Results ranged from 65.4 to 88.0. With a pooled mean score of 77.64 ± 14.46, the IKDC scores for minced cartilage are within the same range.

Similar to the improvement and adaptation of cell application in ACI and later MACI, the minced-cartilage procedure has been optimized as well [11]. Harvesting cartilage tissue is performed using loose intra-articular cartilage fragments. However, suitable alternative options are cell harvesting from non-weightbearing sides or from defect zone edges during preparation. The latest data on tissue quality reveal topographical differences in collagen and aggrecan expression across typical cartilage-harvest sites in the knee: the lowest expression was reported from typical cartilage biopsy regions, highest collagen 1 and expression 2 at the patellofemoral joint [35]. This supports the findings of Aurich et al., who reported superior redifferentiation potential of cartilage harvested from the edge of the defect zone [36–38].

The fixation of fragmented cartilage onto the defect zone was first reported as an open, non-arthroscopic procedure in conjunction with using fibrin glue, which remains a widely used fixation method [9,37–41]. To establish healthy cartilage boarders surrounding the cartilage defect zone, the area is sharply circumcised with a scalpel and finally curettaged. Cartilage chips drawn from either the intercondylar notch or defect boarder zones is then fragmented and filled into the defect. Some authors report the pre-application of fibrin glue before and application of cartilage fragments [11]. Depending on the size and site of the defect, e.g., condylar or retropatellar, the procedure can even be fulfilled arthroscopically.

To optimize the mechanical stability of transferred minced cartilage and its adherence to the defect zone, closing the defect with a collagen membrane or specialized scaffold (CAIS scaffold: absorbable copolymer foam of 35% polycaprolactone and 65% polyglycolic acid, reinforced with polydioxanone mesh) can enhance initial stability [4,11,42,43]. The use of platelet rich plasma (PRP) in combination with autologous thrombin as an autologous sealant is currently discussed. However, the use of PRP in cartilage repair shows controversial results [11].

A comparison of inter-method results after 24 months showed improved IKDC scores in all the reported procedures. However, concerning the quality of mechanical cartilage, ACI/MACI and the minced-cartilage procedure enable the formation of hyaline-like cartilage. Fibrocartilage is more likely to form after microfracture formation: over the short term, this may alleviate symptoms and pain, but because of its rigidity, the knee's condition worsens over the long term [5,6,21]. Measured against ACI/MACI data, IKDC's improvement nearly equals that of the minced cartilage procedure. However, the defects associated with minced cartilage were smaller

Limitations

There are limitations to this study inherent in the type and small number of publications we included and in our search algorithm. Our search strategy followed an English search algorithm. Potentially suitable publications in other languages were not considered. The risk of publication bias is imminent because only published articles were included. To minimize this kind of bias, the CochraneLibrary® was scanned for clinical trials, but we detected no relevant findings, as the results of several ongoing trials have not been published yet.

All of the publications we included entail the risk for selection, detection, and reporting bias. To exclude methodologically inadequate studies, we focused on bias-assessment as conducted by ROBINS-I and mCMS. There was no critical risk of bias in any included study.

5. Conclusions

Minced cartilage is a promising method for repairing small cartilage defects as it is a single-step procedure and the production of transplant-capable material is less complex. At 12 and 24 months, clinical scores showed nearly same results as seen in commonly used cartilage repair techniques, e.g., ACI, MACI microfracture, or OATS. We thus believe it to be a very promising alternative to standard ACI/MACI, especially if these procedures are not available. There is, however, little long-term data to back up this claim at the moment, as we only have access to outcomes two years post-surgery and only a small number of studies and patients. Further research should comprise larger cohorts and the evaluation of long-term results to effectively compare results over more than five years.

Author Contributions: Conceptualization, methodology, validation, formal analysis, investigation, resources, data curation, writing—original draft preparation, writing—review and editing, visualization: A.F. (Andreas Frodl) and T.Y.; writing—review and editing, A.F. (Andreas Fuchs), T.Y., F.C.W., M.S., H.S. and K.I. All authors have read and agreed to the published version of the manuscript.

Funding: This research received no external funding. The APC was funded by Open Access Publication Fund of the University of Freiburg grant number [OA-22-476].

Institutional Review Board Statement: Not applicable.

Informed Consent Statement: Not applicable.

Data Availability Statement: All data are within the manuscript, not applicable.

Acknowledgments: We acknowledge support by the Open Access Publication Fund of the University of Freiburg.

Conflicts of Interest: The authors declare no conflict of interest.

References

1. Hjelle, K.; Solheim, E.; Strand, T.; Muri, R.; Brittberg, M. Articular cartilage defects in 1,000 knee arthroscopies. *Arthrosc. J. Arthrosc. Relat. Surg. Off. Publ. Arthrosc. Assoc. N. Am. Int. Arthrosc. Assoc.* **2002**, *18*, 730–734. [CrossRef] [PubMed]
2. Strauss, E.J.; Fonseca, L.E.; Shah, M.R.; Yorum, T. Management of focal cartilage defects in the knee—Is ACI the answer? *Bull. NYU Hosp. Jt. Dis.* **2011**, *69*, 63–72. [PubMed]
3. Chimutengwende-Gordon, M.; Donaldson, J.; Bentley, G. Current solutions for the treatment of chronic articular cartilage defects in the knee. *EFORT Open Rev.* **2020**, *5*, 156–163. [CrossRef] [PubMed]
4. Salzmann, G.M.; Calek, A.-K.; Preiss, S. Second-Generation Autologous Minced Cartilage Repair Technique. *Arthrosc. Tech.* **2017**, *6*, e127–e131. [CrossRef]
5. Carballo, C.B.; Nakagawa, Y.; Sekiya, I.; Rodeo, S.A. Basic Science of Articular Cartilage. *Clin. Sports Med.* **2017**, *36*, 413–425. [CrossRef]
6. Ulrich-Vinther, M.; Maloney, M.D.; Schwarz, E.M.; Rosier, R.; O'Keefe, R.J. Articular cartilage biology. *J. Am. Acad. Orthop. Surg.* **2003**, *11*, 421–430. [CrossRef]
7. Borrelli, J.; Olson, S.A.; Godbout, C.; Schemitsch, E.H.; Stannard, J.P.; Giannoudis, P.V. Understanding Articular Cartilage Injury and Potential Treatments. *J. Orthop. Trauma.* **2019**, *33* (Suppl. S6), S6–S12. [CrossRef]
8. Simon, T.M.; Jackson, D.W. Articular Cartilage: Injury Pathways and Treatment Options. *Sports Med. Arthrosc. Rev.* **2018**, *26*, 31–39. [CrossRef]
9. Tsuyuguchi, Y.; Nakasa, T.; Ishikawa, M.; Miyaki, S.; Matsushita, R.; Kanemitsu, M.; Adachi, N. The Benefit of Minced Cartilage Over Isolated Chondrocytes in Atelocollagen Gel on Chondrocyte Proliferation and Migration. *Cartilage* **2021**, *12*, 93–101. [CrossRef]
10. Evuarherhe, A., Jr.; Condron, N.B.; Knapik, D.M.; Haunschild, E.D.; Gilat, R.; Huddleston, H.P.; Kaiser, J.T.; Parvaresh, K.C.; Wagner, K.R.; Chubinskaya, S.; et al. Effect of Mechanical Mincing on Minimally Manipulated Articular Cartilage for Surgical Transplantation. *Am. J. Sports Med.* **2022**, *50*, 2515–2525, Erratum in *Am. J. Sports Med.* **2022**, *26*, 3635465221131042. [CrossRef]
11. Salzmann, G.M.; Ossendorff, R.; Gilat, R.; Cole, B.J. Autologous Minced Cartilage Implantation for Treatment of Chondral and Osteochondral Lesions in the Knee Joint: An Overview. *Cartilage* **2020**, *13*, 1947603520942952. [CrossRef] [PubMed]
12. Liberati, A.; Altman, D.G.; Tetzlaff, J.; Mulrow, C.; Gøtzsche, P.C.; Ioannidis, J.P.; Clarke, M.; Devereaux, P.J.; Kleijnen, J.; Moher, D. The PRISMA statement for reporting systematic reviews and meta-analyses of studies that evaluate healthcare interventions: Explanation and elaboration. *BMJ* **2009**, *21*, 339. Available online: https://www.bmj.com/content/339/bmj.b2700 (accessed on 15 July 2022). [CrossRef] [PubMed]
13. Moher, D.; Liberati, A.; Tetzlaff, J.; Altman, D.G.; Group, T.P. Preferred Reporting Items for Systematic Reviews and Meta-Analyses: The PRISMA Statement. *PLoS Med.* **2009**, *6*, e1000097. [CrossRef] [PubMed]
14. Sterne, J.A.; Hernán, M.A.; Reeves, B.C.; Savović, J.; Berkman, N.D.; Viswanathan, M.; Henry, D.; Altman, D.G.; Ansari, M.T.; Boutron, I.; et al. ROBINS-I: A tool for assessing risk of bias in non-randomised studies of interventions. *BMJ* **2016**, *12*, 355. Available online: https://www.bmj.com/content/355/bmj.i4919 (accessed on 15 July 2022). [CrossRef] [PubMed]
15. Coleman, B.D.; Khan, K.M.; Maffulli, N.; Cook, J.L.; Wark, J.D. Studies of surgical outcome after patellar tendinopathy: Clinical significance of methodological deficiencies and guidelines for future studies. *Vic. Inst. Sport Tendon. Study Group. Scand. J. Med. Sci. Sports.* **2000**, *10*, 2–11. [CrossRef] [PubMed]

16. Du, D.; Hsu, P.; Zhu, Z.; Zhang, C. Current surgical options and innovation for repairing articular cartilage defects in the femoral head. *J. Orthop. Transl.* **2020**, *21*, 122–128. [CrossRef] [PubMed]
17. Medvedeva, E.V.; Grebenik, E.A.; Gornostaeva, S.N.; Telpuhov, V.I.; Lychagin, A.V.; Timashev, P.S.; Chagin, A.S. Repair of Damaged Articular Cartilage: Current Approaches and Future Directions. *Int. J. Mol. Sci.* **2018**, *19*, 2366. [CrossRef] [PubMed]
18. Hurst, J.M.; Steadman, J.R.; O'Brien, L.; Rodkey, W.G.; Briggs, K.K. Rehabilitation following microfracture for chondral injury in the knee. *Clin. Sports Med.* **2010**, *29*, 257–265. [CrossRef] [PubMed]
19. Steadman, J.R.; Miller, B.S.; Karas, S.G.; Schlegel, T.F.; Briggs, K.K.; Hawkins, R.J. The microfracture technique in the treatment of full-thickness chondral lesions of the knee in National Football League players. *J. Knee Surg.* **2003**, *16*, 83–86.
20. Pollard, T.D.; Earnshaw, W.C.; Lippincott-Schwartz, J.; Johnson, G.T. (Eds.) Chapter 32—Connective Tissues. In *Cell Biology*, 3rd ed.; Elsevier: Amsterdam, The Netherlands, 2017; pp. 555–570. Available online: https://www.sciencedirect.com/science/article/pii/B9780323341264000323 (accessed on 15 July 2022).
21. Gobbi, A.; Karnatzikos, G.; Kumar, A. Long-term results after microfracture treatment for full-thickness knee chondral lesions in athletes. *Knee Surg. Sports Traumatol. Arthrosc. Off. J. ESSKA* **2014**, *22*, 1986–1996. [CrossRef]
22. Orth, P.; Gao, L.; Madry, H. Microfracture for cartilage repair in the knee: A systematic review of the contemporary literature. *Knee Surg. Sports Traumatol. Arthrosc.* **2020**, *28*, 670–706. [CrossRef] [PubMed]
23. Steinwachs, M.R.; Gille, J.; Volz, M.; Anders, S.; Jakob, R.; De Girolamo, L.; Volpi, P.; Schiavone-Panni, A.; Scheffler, S.; Reiss, E.; et al. Systematic Review and Meta-Analysis of the Clinical Evidence on the Use of Autologous Matrix-Induced Chondrogenesis in the Knee. *Cartilage* **2021**, *13*, 42S–56S. [CrossRef] [PubMed]
24. Migliorini, F.; Eschweiler, J.; Schenker, H.; Baroncini, A.; Tingart, M.; Maffulli, N. Surgical management of focal chondral defects of the knee: A Bayesian network meta-analysis. *J. Orthop. Surg.* **2021**, *16*, 543. [CrossRef] [PubMed]
25. Kaiser, N.; Jakob, R.P.; Pagenstert, G.; Tannast, M.; Petek, D. Stable clinical long term results after AMIC in the aligned knee. *Arch. Orthop. Trauma Surg.* **2021**, *141*, 1845–1854. [CrossRef] [PubMed]
26. Oshima, Y.; Iizawa, N.; Takai, S.; Majima, T. Maturation process of regenerated tissues after single-stage simultaneous autologous particulated cartilage implantation and open wedge high tibial osteotomy for articular cartilage defects with medial osteoarthritis of bilateral knees: A case report. *BMC Musculoskelet. Disord.* **2021**, *22*, 502. [CrossRef] [PubMed]
27. Wasiak, J.; Clar, C.; Villanueva, E. Autologous cartilage implantation for full thickness articular cartilage defects of the knee. *Cochrane Database Syst. Rev.* **2006**, *3*, CD003323.
28. Gooding, C.R.; Bartlett, W.; Bentley, G.; Skinner, J.A.; Carrington, R.; Flanagan, A. A prospective, ranomised study comparing two techniques of autologous chondrocyte implantation for osteochondral defects in the knee: Periosteum covered versus type I/III collagen covered. *Knee* **2006**, *13*, 203–210. [CrossRef]
29. Mistry, H.; Connock, M.; Pink, J.; Shyangdan, D.; Clar, C.; Royle, P.; Court, R.; Biant, L.C.; Metcalfe, A.; Waugh, N. Autologous chondrocyte implantation in the knee: Systematic review and economic evaluation. *Health Technol. Assess. Winch. Engl.* **2017**, *21*, 1–294. [CrossRef]
30. Zeifang, F.; Oberle, D.; Nierhoff, C.; Richter, W.; Moradi, B.; Schmitt, H. Autologous Chondrocyte Implantation Using the Original Periosteum-Cover Technique versus Matrix-Associated Autologous Chondrocyte Implantation: A Randomized Clinical Trial. *Am. J. Sports Med.* **2010**, *38*, 924–933. [CrossRef]
31. Migliorini, F.; Eschweiler, J.; Spiezia, F.; van de Wall, B.J.M.; Knobe, M.; Tingart, M.; Maffulli, N. Arthroscopy versus mini-arthrotomy approach for matrix-induced autologous chondrocyte implantation in the knee: A systematic review. *J. Orthop. Traumatol. Off. J. Ital. Soc. Orthop. Traumatol.* **2021**, *22*, 23. [CrossRef]
32. Jones, K.J.; Cash, B.M. Matrix-Induced Autologous Chondrocyte Implantation With Autologous Bone Grafting for Osteochondral Lesions of the Femoral Trochlea. *Arthrosc. Tech.* **2019**, *8*, e259–e266. [CrossRef] [PubMed]
33. Barié, A.; Kruck, P.; Sorbi, R.; Rehnitz, C.; Oberle, D.; Walker, T.; Zeifang, F.; Moradi, B. Prospective Long-term Follow-up of Autologous Chondrocyte Implantation With Periosteum Versus Matrix-Associated Autologous Chondrocyte Implantation: A Randomized Clinical Trial. *Am. J. Sports Med.* **2020**, *48*, 2230–2241. [CrossRef] [PubMed]
34. Valtanen, R.S.; Arshi, A.; Kelley, B.V.; Fabricant, P.D.; Jones, K.J. Articular Cartilage Repair of the Pediatric and Adolescent Knee with Regard to Minimal Clinically Important Difference: A Systematic Review. *Cartilage* **2020**, *11*, 9–18. [CrossRef]
35. Kim, J.-H.; Heo, J.-W.; Lee, D.-H. Clinical and Radiological Outcomes After Autologous Matrix-Induced Chondrogenesis Versus Microfracture of the Knee: A Systematic Review and Meta-analysis With a Minimum 2-Year Follow-up. *Orthop. J. Sports Med.* **2020**, *8*, 2325967120959280. [CrossRef] [PubMed]
36. Salzmann, G.M.; Buchberger, M.S.; Stoddart, M.J.; Grad, S.; Milz, S.; Niemyer, P.; Sudkamp, N.P.; Imhoff, A.B.; Alini, M. Varying Regional Topology Within Knee Articular Chondrocytes Under Simulated In Vivo Conditions. *Tissue Eng. Part A* **2011**, *17*, 451–461. [CrossRef] [PubMed]
37. Aurich, M.; Hofmann, G.O.; Best, N.; Rolauffs, B. Induced Redifferentiation of Human Chondrocytes from Articular Cartilage Lesion in Alginate Bead Culture After Monolayer Dedifferentiation: An Alternative Cell Source for Cell-Based Therapies? *Tissue Eng. Part A* **2018**, *24*, 275–286. [CrossRef] [PubMed]
38. Aurich, M.; Hofmann, G.O.; Rolauffs, B. Tissue engineering-relevant characteristics of ex vivo and monolayer-expanded chondrocytes from the notch versus trochlea of human knee joints. *Int. Orthop.* **2017**, *41*, 2327–2335. [CrossRef]

39. Lauer, J.C.; Selig, M.; Hart, M.L.; Kurz, B.; Rolauffs, B. Articular Chondrocyte Phenotype Regulation through the Cytoskeleton and the Signaling Processes That Originate from or Converge on the Cytoskeleton: Towards a Novel Understanding of the Intersection between Actin Dynamics and Chondrogenic Function. *Int. J. Mol. Sci.* **2021**, *22*, 3279. [CrossRef]
40. Albrecht, F.H. Closure of joint cartilage defects using cartilage fragments and fibrin glue. *Fortschr. Med.* **1983**, *101*, 1650–1652.
41. Christensen, B.B.; Olesen, M.L.; Hede, K.T.C.; Bergholt, N.L.; Foldager, C.B.; Lind, M. Particulated Cartilage for Chondral and Osteochondral Repair: A Review. *Cartilage* **2021**, *13*, 1047S–1057S. [CrossRef]
42. Christensen, B.B.; Foldager, C.B.; Jensen, J.; Lind, M. Autologous Dual-Tissue Transplantation for Osteochondral Repair: Early Clinical and Radiological Results. *Cartilage* **2015**, *6*, 166–173. [CrossRef] [PubMed]
43. Farr, J.; Tabet, S.K.; Margerrison, E.; Cole, B.J. Clinical, Radiographic, and Histological Outcomes After Cartilage Repair With Particulated Juvenile Articular Cartilage: A 2-Year Prospective Study. *Am. J. Sports Med.* **2014**, *42*, 1417–1425. [CrossRef] [PubMed]

Journal of *Personalized Medicine*

Article

Standing Steadiness and Asymmetry after High Tibial Osteotomy Surgery: A 2 Year Follow-Up Study

Eduard Kurz [1,*], Kay Brehme [2], Thomas Bartels [2], Martin Pyschik [2], Manuel Jenz [1], Wiebke Kadler [1], Karl-Stefan Delank [1] and René Schwesig [1]

[1] Department of Orthopedic and Trauma Surgery, Faculty of Medicine, Martin-Luther University Halle-Wittenberg, 06120 Halle (Saale), Germany
[2] Sports Clinic Halle, Center of Joint Surgery, 06108 Halle (Saale), Germany
* Correspondence: eduard.kurz@uk-halle.de; Tel.: +49-345-557-4897

Abstract: (1) Background: Knee osteoarthritis (OA) is a serious orthopedic problem. In this context, the high tibial osteotomy (HTO) is an established surgical procedure to reduce the load and degeneration of the affected compartment. The aim of this investigation was to judge standing steadiness and asymmetry, pain intensity and quality of life among patients who underwent HTO surgery. (2) Methods: Twenty-five male patients with medial tibiofemoral OA finished this 2 year follow-up study. Standing balance was captured using force plates with four uniaxial sensors before, 6 weeks after, 1 year after, and 2 years after HTO surgery. The percentage weight (PW) under the foot at one side, the stability (ST) index and the weight distribution (WD) index were the main outcomes. Comparisons were conducted using repeated measures analyses of variance. (3) Results: Over time, the PW under the foot at the HTO side increased on average ($p < 0.001$). In terms of standing steadiness, the average ST remained similar over the time points ($p = 0.71$). The WD index was affected by time ($p = 0.003$). (4) Conclusions: In order to judge short-term effects, the PW is recommended, whereas long-term effects can be identified either through the PW or the WD index.

Keywords: knee osteoarthrosis; surgery; posturography; pain; quality of life; long term effects

check for updates

Citation: Kurz, E.; Brehme, K.; Bartels, T.; Pyschik, M.; Jenz, M.; Kadler, W.; Delank, K.-S.; Schwesig, R. Standing Steadiness and Asymmetry after High Tibial Osteotomy Surgery: A 2 Year Follow-Up Study. *J. Pers. Med.* **2022**, *12*, 1594. https://doi.org/10.3390/jpm12101594

Academic Editor: Jih-Yang Ko

Received: 31 August 2022
Accepted: 24 September 2022
Published: 27 September 2022

Publisher's Note: MDPI stays neutral with regard to jurisdictional claims in published maps and institutional affiliations.

1. Introduction

Shifts in the load axis of the lower extremities in the sense of a varus or valgus position and the associated increased unicompartmental pressure load can contribute to the development and progression of early knee osteoarthritis (OA) with chronic pain. In addition to surgical interventions such as total or partial joint replacement surgeries, osteotomies close to the knee joint in particular offer a causal treatment approach for patients with frontal plane malalignment of the lower limb [1]. The aim of high tibial osteotomy (HTO) surgery is to provide relief to the overloaded knee compartment and restore the physiological limb alignment. Patients with an excessive genu varum suffer from on average a two-fold increased risk of knee osteoarthritis, accompanying an approximately three-fold increased disease progression compared to a sample with a normal lower limb alignment [2]. It is well known that patients with medial knee osteoarthritis suffer from functional limitations in everyday life due to stress-related pain, depending on the severity of the symptoms. As a result of a leg alignment surgery, the majority have significantly benefited from an increase in quality of life [3,4].

So far, there are only a few studies that deal with the postoperative effects of HTO surgery on postural control [5,6]. Hunt et al. [5] observed 1 year postoperatively that single-leg steadiness in patients with knee osteoarthritis is not significantly different from that prior to HTO surgery. Although not significant, reported effects were up to 0.33 and thus above trivial. Standing on one leg is challenging for patients and older participants [5,7,8]. Therefore, only 10 s trials were conducted. Possibly, longer standing duration times would have increased the effects measured. For assessments of standing steadiness, segments of at least 30 seconds' duration are recommended [9]. Posturographic systems with more than one force plate in particular overcome these limitations [10,11]. Bipedal assessments enable longer conduction times, while multiple force plates make it possible to estimate standing asymmetry (weight distribution).

In contrast, long-term follow-ups employing any kind of posturography in order to judge standing balance and weight distribution are not yet available. Thus, the main intentions of this study were to evaluate the influence of a surgical varus leg axis correction in the short term (6 weeks postoperative) and long term (1 and 2 years postoperative) on standing balance characteristics, perceived pain and health-related quality of life. The main aim of this study was to estimate possible long-term effects of the HTO surgical intervention on patients' bipedal postural control against the preoperative outcomes. We hypothesized that in the majority of patients, the HTO treatment would lead to an enhanced load proportion (percentage weight) below the foot on the affected side.

2. Materials and Methods

In this nonrandomized interventional study design, repeated measurements were performed on participants in a rehabilitation research laboratory after providing detailed explanations of the risks of the investigation and obtaining written informed consent from each participant. The study protocol of this time-series design was approved by the ethical committee of Martin-Luther-University Halle-Wittenberg (approval number: 2018-66). This is a follow-up study on the patients examined in the work of Brehme et al. [6].

Adult patients diagnosed with medial compartment knee OA verified by MRI were included. All included patients had moderate knee OA (Kellgren–Lawrence grade 3). The inclusion criteria were described in detail elsewhere [6]. Patients with knee ligament instability and contralateral knee OA requiring treatment were excluded, as were those with primary patellofemoral OA, bi-compartmental OA, tibial or femoral osteonecrosis and those with a history of an inflammatory rheumatic disease. Further, any patient who was unable to perform the posturography measurements owing to limited motion of the lower limb were excluded. Initially, 38 patients were screened for eligibility before HTO surgery (Figure 1). All measurements were performed at the same time of day and in a quiet room to minimize any disruptions during testing. Due to low sample size and to avoid distortion effects, female patients were excluded. Finally, 25 male patients were examined before, 6 weeks after, 1 year after and 2 years after HTO surgery.

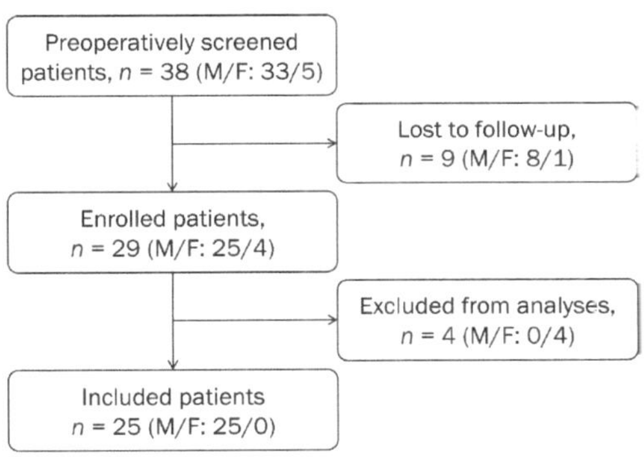

Figure 1. Flow chart of the study population.

The surgical procedure used consisted of a bi-plane medial-based open-wedge HTO, including a distal release of the superficial medial collateral ligament fibers. The aim was to shift the weight-bearing line laterally, with the post-operative mechanical axis running laterally through the tibial plateau at 62% of its entire width measured from the medial side [12]. Using standing whole leg radiographs, the amount of correction needed was determined using the Miniaci method [13]. The LOQTEQ Osteotomy Plate system (aap Implants Inc. Dover, DE, USA) was used for fixation. Only one experienced surgeon (K.B., 30 to 40 HTO surgeries yearly) performed the examination as well as surgery procedures to ensure the highest level of observation equality and to avoid any surgeon bias.

The vertical component of ground reaction force fluctuations under the left and right feet were obtained during bipedal standing (trial duration: 32 s, sampling rate: 32 Hz) using side-by-side force plates each divided into forefoot and heel components (IBS, neurodata, Vienna, Austria). This system provides a comprehensive and sufficient reference database of asymptomatic subjects ($n = 1724$) stratified by age and sex [14]. This reference database was used to select asymptomatic controls for the matched sample (selection criteria: age, sex and body height). Thus, recruitment of a control group was not necessary. The outcomes provided by the IBS system incorporating signals of four uniaxial sensors are reliable [15] and have been extensively validated [11,16,17]. The percentage weight (PW) represents pressure load under the foot at one side (sum of both sides = 100%) and contains averaged values from the forefoot and heel force plate sensors. The degree of standing steadiness was quantified in terms of the stability (ST) index with higher values indicating a more unsteady position. Standing asymmetry was quantified as the weight distribution (WD) index with higher values representing a more asymmetric standing position. Both measures represent complex outcome indices from two forefoot and two heel components. Their calculations are described elsewhere [11].

Pain intensity was derived from a single 100 mm visual analogue scale (VAS) pain measurement. The minimal clinically important difference (MCID) on VAS was shown to be 25 mm in patients with knee osteoarthritis independent of the type of treatment [18,19]. Health-related quality of life was examined using the German version of the 36-item Short Form SF-36 health survey questionnaire [20,21]. The results of the 36 questions combined into 8 different subscales were transformed into physical component summary (PCS) and mental component summary (MCS) scores. McHorney et al. [22] reported intraclass correlation coefficients from 0.65 to 0.94 across scales and patient groups. For the SF-36 health survey questionnaire, an MCID of at least 10 points was established in patients after total knee replacement [23]. Three points on the PCS or MCS scores are recommended [24].

Data are presented as mean with standard deviation (SD) and as minimum and maximum or 95% confidence interval (CI) values. The normal distribution of data was examined visually and verified using the Shapiro–Wilk test. Absolute and relative change scores were examined using the method purposed by Kaiser [25]. Percentage weight (PW) under feet before HTO surgery was compared using Student's paired t test with Cohen's d. Linear relationships between different interval scaled outcome measures were proven using Pearson's product moment correlation. Baseline comparisons of demographic characteristics, standing steadiness and asymmetry between groups were verified using Student's unpaired t tests with Cohen's d. To assess treatment effects on PW, ST (standing steadiness), WD (standing asymmetry), perceived pain and health-related quality of life, direct comparisons were conducted using separate repeated measures analyses of variance (ANOVA) with time as the within-subject factor. Practical relevance was estimated calculating partial eta-squared (η_p^2) with values ≥ 0.01, ≥ 0.06 and ≥ 0.14 indicating small, moderate or large effects, respectively. Further, indirect within-group comparisons were conducted via pre-post changes.

3. Results

Orthopedic comorbidities (ankle instability, meniscal lesions, lumbar disc disease, neck implants) were present in nine (36%) patients. Twelve (48%) patients suffered from arterial hypertension. In 12 (48%) of the patients, the operated leg was their dominant (e.g., leg used to kick a ball) lower limb. The average wedge height was 10.8 mm (SD 2.2, 7.0–15.0). The magnitude of correction was not related to patients' body mass ($r(23) = 0.28$, $p = 0.17$) or height ($r(23) = 0.17$, $p = 0.41$). The demographic characteristics of the study participants are listed in Table 1. Patients' body mass remained at comparable levels ($p > 0.18$) during the course of the 2 year follow-up (6 weeks: 98.6 (14.3) kg, 1 year: 98.8 (14.9) kg, 2 years: 99.7 (14.9) kg).

Table 1. Baseline demographic characteristics of the study participants. Values presented as mean (SD, minimum and maximum).

	HTO (n = 25)	Controls (n = 25)	p Value
Age [years]	56.6 (5.0, 43–62)	54.9 (7.1, 40–64)	0.35
Body mass [kg]	98.7 (14.1, 77.8–129.1)	87.1 (12.9, 65.5–129.4)	<0.01
Body height [m]	1.79 (0.06, 1.70–1.96)	1.78 (0.07, 1.66–1.95)	0.77
BMI [1] [kg/m^2]	30.8 (3.7, 24.0–39.6)	27.4 (3.7, 20.5–39.5)	<0.01

[1] Body mass index. p values are based on Student's unpaired t tests.

Lower percentage weight (PW) was placed on the plates at the affected side before the HTO surgery ($t(24) = -3.48$, $p = 0.002$, $d = -0.70$). Controls' percentage foot pressures did not differ between sides ($t(24) = -0.41$, $p = 0.67$, $d = -0.08$). Patients' PW under the foot of the affected leg was neither associated with their body mass ($r(23) = 0.21$, $p = 0.32$) nor with their reported pain intensity ($r(23) = -0.07$, $p = 0.74$). At baseline, patients showed large ($t(48) = 5.26$, $p < 0.001$, $d = 1.5$) and moderate ($t(48) = 2.20$, $p < 0.05$, $d = 0.62$) differences from controls regarding standing steadiness and asymmetry, respectively.

Figure 2 depicts baseline comparisons of the stability index and the weight distribution index between the patients selected for HTO surgery and age and stature matched controls. Over time, the PW under the foot at the HTO side increased on average ($F(3, 72) = 20.87$, $p < 0.001$, $\eta_p^2 = 0.47$). On average, patients put higher PW under the foot at the operated side. In terms of standing steadiness, the average ST remained similar over the time points ($F(3, 72) = 0.46$, $p = 0.71$, $\eta_p^2 = 0.02$).

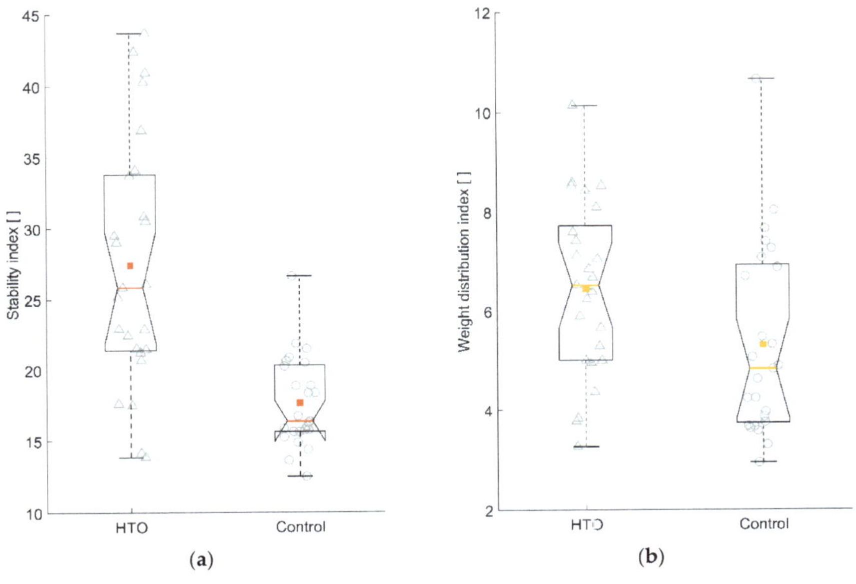

Figure 2. Box-whisker plots on the baseline (**a**) stability index and (**b**) weight distribution index data between the patients selected for high tibial osteotomy (HTO) surgery and matched controls with individual data points. The square symbol in the box denotes the mean of the group data. Higher values indicate greater standing unsteadiness and asymmetry, respectively. △, HTO individual data; ○, Control individual data.

The time factor revealed large effects on standing asymmetry ($F(3, 72) = 5.07$, $p = 0.003$, $\eta_p^2 = 0.17$). Figure 3 depicts time changes according to baseline measurements.

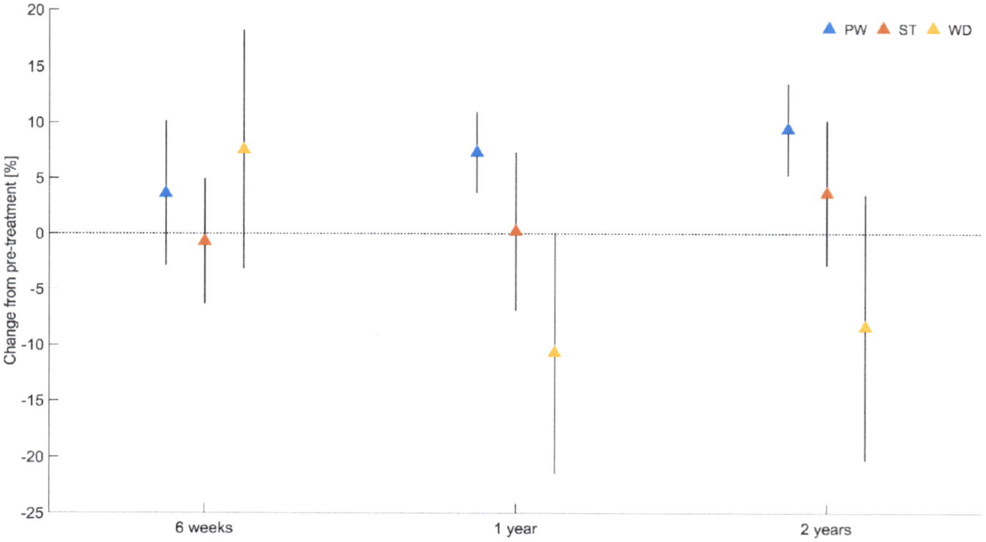

Figure 3. Time changes of percentage weight (PW), standing steadiness (ST) and standing asymmetry (WD), according to baseline measurements. Values are mean with 95% confidence intervals.

There was a large effect of time on the reported pain intensity (Table 2). However, considerably large standard deviations were seen 1 and 2 years after surgery. The repeated measures ANOVA revealed significant effects of time on the PCS and MCS. While the reported physical health increased, mental health decreased 2 years after HTO surgery.

Table 2. Baseline and follow-up results of pain intensity and health-related quality of life outcomes of the patients ($n = 25$). Values presented as mean (SD, minimum and maximum).

	Baseline	Six Weeks	One Year	Two Years	*p* Value	η_p^2 Value
VAS [1] (0–100)	49.5 (21.2) 6–100	31.8 (15.0) 4–56	20.4 (23.0) 0–72	21.0 (24.2) 0–86	<0.001	0.35
PCS [2] (0–100)	31.5 (9.0) 18–51	29.7 (8.0) 16–46	45.0 (9.9) 16–57	44.8 (12.2) 12–59	<0.001	0.56
MCS [3] (0–100)	56.8 (10.9) 34–70	55.6 (9.7) 30–68	55.1 (7.3) 36–63	51.6 (10.0) 29–65	0.01	0.14

[1] Visual analogue scale: Higher values indicate greater pain intensity. [2] Physical component summary: Higher values indicate better physical health. [3] Mental component summary: Higher values indicate better mental health. *p* values are based on repeated measures ANOVA's.

The improvements recorded for pain intensity were below the MCID of 25 mm at 1 and at 2 years after HTO surgery (Table 3). At 6 weeks after treatment, it is unclear whether the reduction in pain was clinically meaningful. The PCS clearly exceeded the MCID threshold of 3 at 1 and 2 years after surgery. Thus, the improvement in physical health was clinically important. By contrast, the MCS deteriorated, reaching the critical threshold at 2 years after HTO surgery.

Table 3. Mean absolute changes of pain intensity and health-related quality of life outcomes of the patients (n = 25) according to baseline measurements.

	6 Weeks	1 Year	2 Years
VAS [1]	−17.7	−29.1	−28.5
PCS [2]	−1.7	13.6	13.3
MCS [3]	−1.2	−1.8	−5.2

[1] Visual analogue scale: Lower values indicate greater pain reduction. [2] Physical component summary of the SF-36: Higher values indicate greater improvement in physical health. [3] Mental component summary of the SF-36: Lower values indicate greater decline in mental health.

4. Discussion

This prospective, longitudinal cohort study aimed at evaluating the influence of a surgical varus leg axis correction in the short term (6 weeks postoperative) and long term (1 and 2 years postoperative) on standing balance characteristics, perceived pain and health-related quality of life. Before HTO surgery, patients placed lower percentage weight on the force plates at the affected side and stood more unsteadily and asymmetrically as compared with asymptomatic controls. After surgery, the percentage weight under the foot at the HTO side increased on average, accompanied with reduced standing asymmetry one year after treatment. Standing steadiness, however, did not change over time.

In order to quantify standing steadiness, force plates are frequently used to record fluctuations of the center of foot pressure during upright standing. The stability index indicates the amount of force fluctuations among the four plates of the IBS [11] and thus how steady participants were able to stand during the trial. After HTO surgery, standing steadiness results were comparable to those before the intervention. Most probably, the postural task is not challenging enough and thus results in different solution strategies with high variability. Only 11 (44%) of the patients exceeded the values of the asymptomatic control subjects. Although patients had radiographically confirmed knee OA, the majority of them stood as steady as the controls. Further, comparisons of within-group differences (see Appendix A, Figure A2) confirm that changes occur with fairly equal distribution in both directions. However, during a more challenging postural task, Hunt et al. [5] failed to prove changes in standing steadiness 1 year after HTO surgery. In patients with knee OA, Hassan et al. [26] assessed bipedal standing steadiness with closed eyes before and after local anesthetic injections. Although reported pain was reduced considerably, postural steadiness remained unaltered. These findings suggest that the effects, either short-term or long-term, of mechanical leg axis corrections or pain reduction interventions on standing steadiness are negligible.

Weight distribution outcomes seem to be more sensitive to changes occurring due to disease progression or surgical interventions [10,27,28] (see Appendix A, Figures A1 and A3). Standing asymmetry was quantified by two different outcomes: the percentage weight placed on the plates at the affected side and the weight distribution index. Both outcomes differ according to the signals they contain. The percentage weight contains ground reaction force fluctuations under one foot (right or left) and thus from two force plate sensors. The weight distribution index is a complex outcome including signals from four force plate sensors (two at each side). Consequently, the weight distribution index contains medio-lateral as well as anterior–posterior information. This complex outcome increased in patients 6 weeks after HTO surgery, indicating that the standing pattern was more altered as a result of the surgical intervention. After the first year, the patients stood more evenly distributed between the four force plates, resulting in a lowered weight distribution index. The majority of asymptomatic participants evenly distribute their weight during side-by-side standing [10,28]. Interestingly, patients differ in their strategies to load the affected side. A long-term follow-up (14.7 months) in patients who underwent total knee arthroplasty but without extension limitations revealed more percentage weight put on the operated leg [27]. Patients with present knee extension deficits placed more weight at their contralateral side. The majority of patients after total hip arthroplasty tended to load the non-operated more than the operated side [10]. Amputees, in contrast, showed lower percentage weight at their prosthetic limb [28].

Patients' reported outcomes, especially the results for pain intensity, showed high variability 1 or 2 years after HTO surgery. This implies that long-term effects may rely on individual conditions. Two investigations [29,30] reported on 15 and 23 HTO patients 2 years postoperatively who were on average 6 and 11 years younger than our participants, respectively. Both studies found a quite comparable reduction of pain intensity. However, in the investigation with the on average younger patients, five (22%) went on to total knee arthroplasty [30]. Clinically meaningful reductions in pain intensity 1 and 2 years after HTO surgery go hand in hand with enhancements in physical but not mental health. This is in line with the reported magnitudes of change in different orthopedic conditions [31]. Independent of orthopedic condition, the greatest improvements were found for the physical dimensions, while only small to moderate improvements were observed for the mental and social dimensions. In a retrospective study on a cohort of comparable age, Bastard et al. [32] reported higher preoperative physical and mental subscores, both of which significantly improved 1 year postoperatively. Saier et al. [33] followed 64 patients 2 years after HTO surgery. They found inferior MCS scores to be influenced by psychopathological comorbidities examined preoperatively. The deterioration of mental health 2 years after HTO surgery in our participants, however, cannot be explained by the outcomes reported in this study and needs to be further investigated. Clinicians who aim to quantify standing balance in HTO patients are encouraged to use multi-plate posturographic systems. The percentage weight placed at one side is easy to interpret and can be used early after HTO surgery to judge rehabilitation progress.

This study comprises some limitations that need to be addressed: First, the data of the control group are from a reference database. Consequently, no longitudinal effects can be calculated. Second, a control group of patients who did not receive surgical treatment could not be included in this study. Third, obesity is a typical feature of HTO patients, in contrast to the selected asymptomatic participants. Therefore, a significant difference regarding body mass is inevitable. Finally, only data of male patients were analyzed in this investigation. In future studies, one may consider a time-series design with at least two pretest measurements to establish a baseline. Reducing variation existing in outcomes before the intervention can help researchers to estimate treatment effects more precisely. Moreover, patients' baseline measurements were taken up to 82 days preoperatively, which may have impacted the results. Changes in the selected outcomes may be due to the HTO treatment. However, the measured changes caused by HTO surgery cannot be isolated from naturally occurring outcome changes. Different experiences as well as responses following the HTO surgery and post intervention phases may particularly explain the results. Further studies need to assess how many of HTO-treated patients will get a total joint replacement of the knee or hip joints and which predictive signs are to be considered.

5. Conclusions

The findings of this study indicate that patients who have undergone HTO surgery have reduced their pain and increased their reported physical health. These changes are clinically important 1 year after the surgical intervention. The standing steadiness of the HTO patients remains fairly unaltered over the follow-up period. Standing asymmetry, in contrast, changed considerably after HTO surgery. Patients increased their percentage weight under the foot at the operated side. The weight distribution index, in contrast, increased 6 weeks after surgery and decreased 1 and 2 years after the surgical intervention. In order to judge short-term effects, the percentage weight is recommended, whereas long-term effects can be identified either through the percentage weight or the weight distribution index.

Author Contributions: Conceptualization, R.S., K.B , T.B. and M.J.; methodology, R.S., K.B. and E.K.; software, E.K.; validation, E.K., M.J. and R.S.; formal analysis, E.K. and R.S.; investigation, K.B., M.J., W.K. and R.S.; resources, T.B. and K.-S.D.; data curation, E.K. and R.S.; writing—original draft preparation, E.K. and R.S.; writing—review and editing, E.K., R.S. and K.-S.D.; visualization, E.K.; supervision, M.P.; project administration, T.B., K.B. and R.S. All authors have read and agreed to the published version of the manuscript.

Funding: This research received no external funding.

Institutional Review Board Statement: The study was conducted in accordance with the Declaration of Helsinki and approved by the Institutional Review Board of Martin-Luther-University Halle-Wittenberg (approval number: 2018-66, date of approval: 9 July 2018).

Informed Consent Statement: Informed consent was obtained from all subjects involved in the study.

Data Availability Statement: The datasets used and analyzed during the current study are available from the corresponding author on reasonable request.

Acknowledgments: The authors are grateful to all the patients who participated after HTO surgery and made this study possible and to Tara Russell for language assistance.

Conflicts of Interest: The authors declare no conflict of interest.

Appendix A

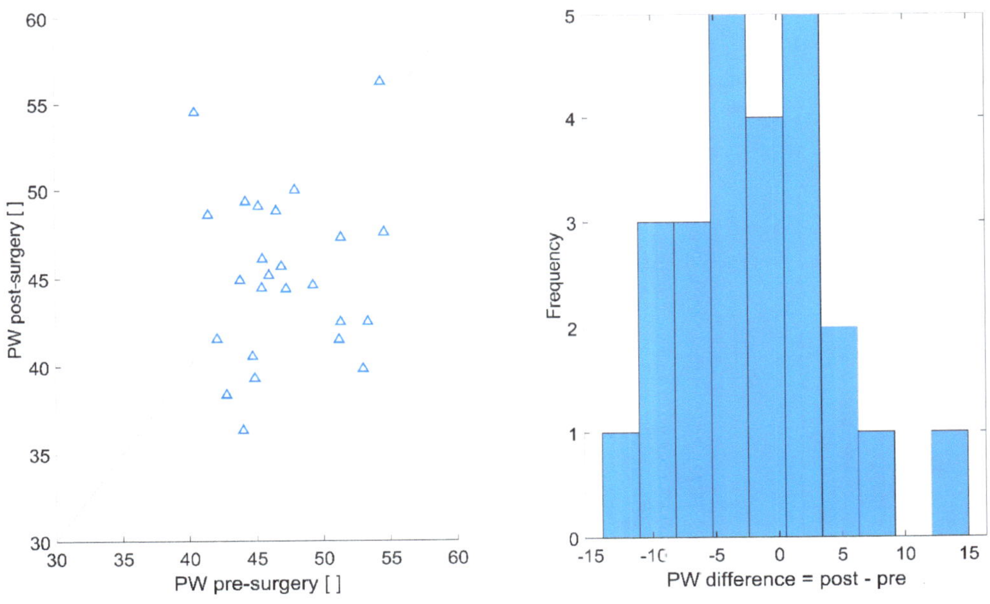

Figure A1. A scatterplot showing percentage weight (PW) data before and after (6 weeks) HTO surgery. The solid line is the one-to-one line. If there was no change between the two tests, the data point would fall on the one-to-one line. Points above this line indicate higher scores following the surgical intervention. △, HTO individual data.

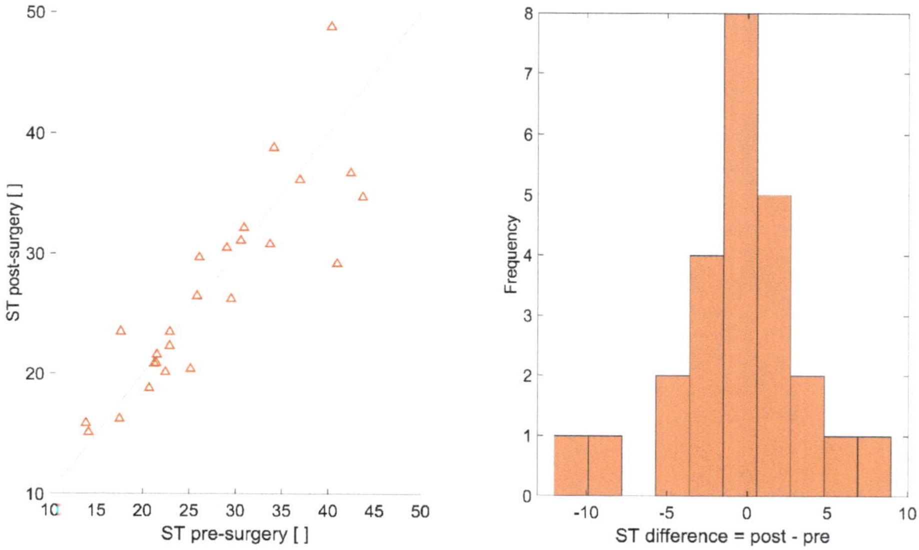

Figure A2. A scatterplot showing stability index (ST) data before and after (6 weeks) HTO surgery. The solid line is the one-to-one line. If there was no change between the two tests, the data point would fall on the one-to-one line. Points above this line indicate higher scores following the surgical intervention. △, HTO individual data.

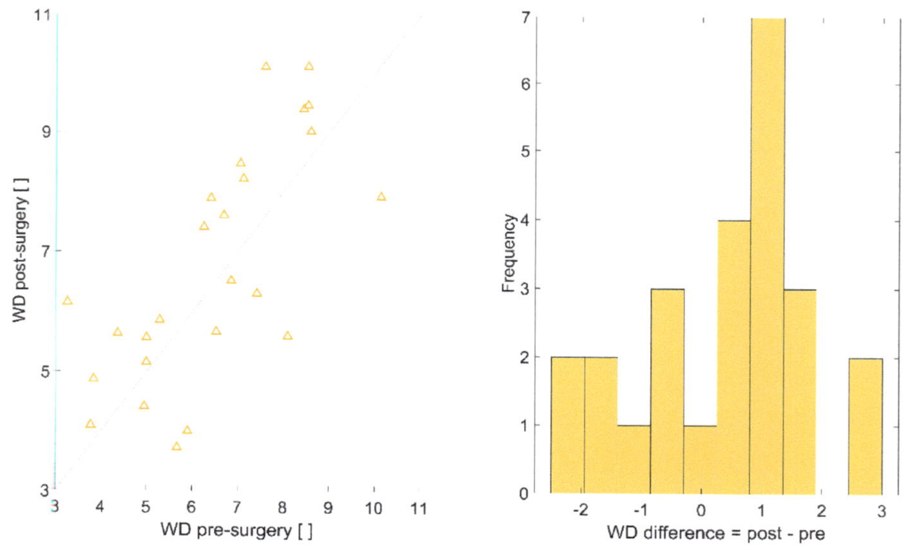

Figure A3. A scatterplot showing weight distribution (WD) index data before and after (6 weeks) HTO surgery. The solid line is the one-to-one line. If there was no change between the two tests, the data point would fall on the one-to-one line. Points above this line indicate higher scores following the surgical intervention. △, HTO individual data.

References

1. Martinek, V.; Imhoff, A.B. Tibiale Umstellungsosteotomie am Kniegelenk—Renaissance eines klassischen Verfahrens? *Dtsch. Z. Sportmed.* **2003**, *54*, 229–232.
2. Brouwer, G.M.; van Tol, A.W.; Bergink, A.P.; Belo, J.N.; Bernsen, R.M.D.; Reijman, M.; Pols, H.A.P.; Bierma-Zeinstra, S.M.A. Association between valgus and varus alignment and the development and progression of radiographic osteoarthritis of the knee. *Arthritis Rheum.* **2007**, *56*, 1204–1211. [CrossRef] [PubMed]
3. Ihle, C.; Ateschrang, A.; Grünwald, L.; Stöckle, U.; Saier, T.; Schröter, S. Health-related quality of life and clinical outcomes following medial open wedge high tibial osteotomy: A prospective study. *BMC Musculoskelet. Disord.* **2016**, *17*, 215. [CrossRef]
4. Herbst, M.; Kuwashima, U.; Ahrend, M.D.; Gueorguiev, B.G.; Schröter, S.; Ihle, C. Health-related quality of life—An underestimated factor to evaluate the treatment success after open wedge HTO surgery: Prospective 6-years follow-up. *Z. Orthop. Unfall.* **2021**, *159*, 288–297. [CrossRef]
5. Hunt, M.A.; Birmingham, T.B.; Jones, I.C.; Vandervoort, A.A.; Giffin, J.R. Effect of tibial re-alignment surgery on single leg standing balance in patients with knee osteoarthritis. *Clin. Biomech.* **2009**, *24*, 693–696 [CrossRef] [PubMed]
6. Brehme, K.; Bartels, T.; Pyschik, M.; Jenz, M.; Delank, K.S.; Laudner, K.G.; Schwesig, R. Postural stability and regulation before and after high tibial osteotomy and rehabilitation. *Appl. Sci.* **2020**, *10*, 6517. [CrossRef]
7. MacRae, P.G.; Lacourse, M.; Moldavon, R. Physical performance measures that predict faller status in community-dwelling older adults. *J. Orthop. Sports Phys. Ther.* **1992**, *16*, 123–128. [CrossRef]
8. Stensdotter, A.K.; Bjerke, J.; Djupsjobacka, M. Postural sway in single-limb and bilateral quiet standing after unilateral total knee arthroplasty. *Gait Posture* **2015**, *41*, 769–773. [CrossRef]
9. Le Clair, K.; Riach, C. Postural stability measures: What to measure and for how long. *Clin. Biomech.* **1996**, *11*, 176–178. [CrossRef]
10. Talis, V.L.; Grishin, A.A.; Solopova, I.A.; Oskanyan, T.L.; Belenky, V.E.; Ivanenko, Y.P. Asymmetric leg loading during sit-to-stand, walking and quiet standing in patients after unilateral total hip replacement surgery. *Clin. Biomech.* **2008**, *23*, 424–433. [CrossRef]
11. Reinhardt, L.; Heilmann, F.; Teicher, M.; Lauenroth, A.; Delank, K.S.; Schwesig, R.; Wollny, R.; Kurz, E. Comparison of posturographic outcomes between two different devices. *J. Biomech.* **2019**, *86*, 218–224. [CrossRef] [PubMed]
12. van der Woude, J.A.D.; Wiegant, K.; van Heerwaarden, R.J.; Spruijt, S.; van Roermund, P.M.; Custers, R.J.H.; Mastbergen, S.C.; Lafeber, F.P.J.G. Knee joint distraction compared with high tibial osteotomy: A randomized controlled trial. *Knee Surg. Sports Traumatol. Arthrosc.* **2017**, *25*, 876–886. [CrossRef] [PubMed]
13. Miniaci, A.; Ballmer, F.T.; Ballmer, P.M.; Jakob, R.P. Proximal tibial osteotomy. A new fixation device. *Clin. Orthop. Relat. Res.* **1989**, *246*, 250–259. [CrossRef]
14. Schwesig, R.; Fischer, D.; Kluttig, A. Are there changes in postural regulation across the lifespan? *Somatosens. Mot. Res.* **2013**, *30*, 167–174. [CrossRef] [PubMed]
15. Schwesig, R.; Becker, S.; Fischer, D. Intraobserver reliability of posturography in healthy subjects. *Somatosens. Mot. Res.* **2014**, *31*, 16–22. [CrossRef] [PubMed]
16. Friedrich, M.; Grein, H.J.; Wicher, C.; Schuetze, J.; Mueller, A.; Lauenroth, A.; Hottenrott, K.; Schwesig, R. Influence of pathologic and simulated visual dysfunctions on the postural system. *Exp. Brain Res.* **2008**, *186*, 305–314. [CrossRef]
17. Schwesig, R.; Becker, S.; Lauenroth, A.; Kluttig, A.; Leuchte, S.; Esperer, H.D. Neue posturographische Methode zur Differentialdiagnostik von Patienten mit Morbus Parkinson und Kleinhirnstörungen. *Biomed. Tech.* **2009**, *54*, 347–356. [CrossRef]
18. Farrar, J.T.; Young, J.P., Jr.; LaMoreaux, L.; Werth, J.L.; Poole, R.M. Clinical importance of changes in chronic pain intensity measured on an 11-point numerical pain rating scale. *Pain* **2001**, *94*, 149–158. [CrossRef]
19. Stauffer, M.E.; Taylor, S.D.; Watson, D.J.; Peloso, P.M.; Morrison, A. Definition of nonresponse to analgesic treatment of arthritic pain: An analytical literature review of the smallest detectable difference, the minimal detectable change, and the minimal clinically important difference on the pain visual analog scale. *Int. J. Inflam.* **2011**, *2011*, 231926. [CrossRef]
20. Ware, J.E., Jr.; Sherbourne, C.D. The MOS 36-item short-form health survey (sf-36). I. Conceptual framework and item selection. *Med. Care* **1992**, *30*, 473–483. [CrossRef]
21. Bullinger, M.; Kirchberger, I.; Ware, J. Der deutsche SF-36 health survey Übersetzung und psychometrische Testung eines krankheitsübergreifenden Instruments zur Erfassung der gesundheitsbezogenen Lebensqualität. *Z. Gesundh.* **1995**, *3*. 21–36. [CrossRef]
22. McHorney, C.A.; Ware, J.E., Jr.; Lu, J.F.; Sherbourne, C.D. The MOS 36-item short-form health survey (SF-36): III. Tests of data quality, scaling assumptions, and reliability across diverse patient groups. *Med. Care* **1994**, *32*, 40–66. [CrossRef] [PubMed]
23. Escobar, A.; Quintana, J.M.; Bilbao, A.; Arostegui, I.; Lafuente, I.; Vidaurreta, I. Responsiveness and clinically important differences for the WOMAC and SF-36 after total knee replacement. *Osteoarthr. Cartil.* **2007**, *15*, 273–280. [CrossRef] [PubMed]
24. Frendl, D.M.; Ware, J.E., Jr. Patient-reported functional health and well-being outcomes with drug therapy: A systematic review of randomized trials using the SF-36 health survey. *Med. Care* **2014**, *52*, 439–445. [CrossRef]
25. Kaiser, L. Adjusting for baseline: Change or percentage change? *Stat. Med.* **1989**, *8*, 1183–1190. [CrossRef]
26. Hassan, B.S.; Doherty, S.A.; Mockett, S.; Doherty, M. Effect of pain reduction on postural sway, proprioception, and quadriceps strength in subjects with knee osteoarthritis. *Ann. Rheum. Dis.* **2002**, *61*, 422–428. [CrossRef]
27. Harato, K.; Nagura, T.; Matsumoto, H.; Otani, T.; Toyama, Y.; Suda, Y. Extension limitation in standing affects weight-bearing asymmetry after unilateral total knee arthroplasty. *J. Arthroplast.* **2010**, *25*, 225–229. [CrossRef]

28. Summers, G.D.; Morrison, J.D.; Cochrane, G.M. Foot loading characteristics of amputees and normal subjects. *Prosthet. Orthot. Int.* **1987**, *11*, 33–39. [CrossRef]
29. Ghinelli, D.; Parma, A.; Baldassarri, M.; Olivieri, A.; Mosca, M.; Pagliazzi, G.; Buda, R. High tibial osteotomy for the treatment of medial osteoarthritis of the knee with new ibalance system: 2 years of follow-up. *Eur. J. Orthop. Surg. Traumatol.* **2016**, *26*, 523–535. [CrossRef]
30. Thompson, K.A.; Darden, C.N.; Katsman, A.; Alaia, M.J.; Strauss, E.J.; Jazrawi, L.M. Short-term clinical outcomes of high tibial osteotomy with the ibalance hto system. *Bull. Hosp. Jt. Dis.* **2019**, *77*, 256–262.
31. Busija, L.; Osborne, R.H.; Nilsdotter, A.; Buchbinder, R.; Roos, E.M. Magnitude and meaningfulness of change in sf-36 scores in four types of orthopedic surgery. *Health Qual. Life Outcomes* **2008**, *6*, 55. [CrossRef] [PubMed]
32. Bastard, C.; Mirouse, G.; Potage, D.; Silbert, H.; Roubineau, F.; Hernigou, P.; Flouzat-Lachaniette, C.H. Return to sports and quality of life after high tibial osteotomy in patients under 60 years of age. *Orthop. Traumatol. Surg. Res.* **2017**, *103*, 1189–1191. [CrossRef] [PubMed]
33. Saier, T.; Minzlaff, P.; Feucht, M.J.; Lämmle, L.; Burghoff, M.; Ihle, C.; Imhoff, A.B.; Hinterwimmer, S. Health-related quality of life after open-wedge high tibial osteotomy. *Knee Surg. Sports Traumatol. Arthrosc.* **2017**, *25*, 934–942. [CrossRef] [PubMed]

Journal of Personalized Medicine

MDPI

Article

Prediction of the Hamstring Graft Size for ACL Reconstruction Using Different Axial Layers in Preoperative MRI

Moritz Florian Mayr [1,*], Markus Siegel [1], Elham Taghizadeh [2], Peter Obid [1], Hagen Schmal [1,3] and Kaywan Izadpanah [1]

1 Department of Orthopedic Surgery and Traumatology, Freiburg University Hospital, Albert Ludwigs University Freiburg, Hugstetter Straße 55, 79106 Freiburg, Germany; markus.siegel@uniklinik-freiburg.de (M.S.); peter.obid@uniklinik-freiburg.de (P.O.); hagen.schmal@freenet.de (H.S.); kaywan.izadpanah@uniklinik-freiburg.de (K.I.)

2 Fraunhofer Institute for Digital Medicine MEVIS, Max-von-Laue-Str. 2, 28359 Bremen, Germany; elham.taghizadeh@mevis.fraunhofer.de

3 Department of Orthopedic Surgery, University Hospital Odense, Sdr. Boulevard 29, 5000 Odense, Denmark

* Correspondence: moritz.mayr@uniklinik-freiburg.de

Abstract: Background: In ACL reconstruction, it is desirable to assess preoperatively whether a sufficient graft diameter can be achieved with the planned tendon graft. The present study investigated the effect of the location of the cross-sectional area (CSA) measurement of the hamstring tendons in preoperative MRI on the correlation of the CSA with the intraoperative graft diameter. In addition, we analyzed whether the measurement results of examiners with different skill levels were comparable. Methods: A total of 32 subjects undergoing a single bundle ACL reconstruction using an autologous ipsilateral quadrupled hamstring graft (STGT) were included. The CSA of the semitendinosus and gracilis tendon was determined in preoperative MRI on six defined levels by three examiners. The intraclass correlation coefficient between the measurements of these observers was determined. The correlation between the sum of the CSA of both tendons (CSA STGT) and the graft diameter was investigated. Results: The interrater reliability was excellent on most of the investigated levels. A significant correlation between CSA STGT and the graft diameter was seen on all levels. The strongest correlation was found on the level 10 mm above the joint line. Conclusions: The measurement of the CSA STGT in the preoperative MRI 10 mm above the joint line enabled a good assessment of the achievable graft diameter in ACL reconstruction, independent of the examiners' training level.

Keywords: ACL surgery; knee surgery; sports injuries; arthroscopy; individual surgery; hamstring; MRI; digital imaging

Citation: Mayr, M.F.; Siegel, M.; Taghizadeh, E.; Obid, P.; Schmal, H.; Izadpanah, K. Prediction of the Hamstring Graft Size for ACL Reconstruction Using Different Axial Layers in Preoperative MRI. *J. Pers. Med.* **2024**, *14*, 582. https://doi.org/10.3390/jpm14060582

Academic Editor: Gunther Maderbacher

Received: 21 April 2024
Revised: 20 May 2024
Accepted: 27 May 2024
Published: 29 May 2024

1. Introduction

Rupture of the ACL is a common injury in individuals doing sports, either professional or recreational. Due to an increase in physically active patients, the numbers are even on the rise. The autologous hamstring tendons are frequently used grafts for ACL reconstruction with low donor-site morbidity compared to other autografts [1,2]. Recent studies have shown a significant higher risk of rerupture if hamstring autografts with a diameter less than 8 mm are used [3–5]. However, the semitendinosus and gracilis tendons show a considerable interindividual variance [6–8]. Therefore, preoperative prediction of the hamstring graft size would be of considerable importance to offer an individualized surgical procedure for each patient. There are several studies that have investigated the correlation of radiological quantifiable parameters and the intraoperative diameter of a hamstring graft. Magnetic resonance imaging (MRI) measurements of the hamstring tendons' cross-sectional area (CSA) show promising results. In the existing studies on this topic, different methods were used regarding the localization of the CSA measurement with divergent results. Often, only the data of a few selected MR scanners have been used [9–16]. This is in contrast with everyday clinical practice, where patients bring along their MRI data from different devices with varying quality into the orthopedic outpatient department, and the MR images are often evaluated by examiners with different skill levels. We hypothesize that the results of observers with different levels of expertise are comparable and that there is an effect of the location of the measurement on the correlation of the CSA with the intraoperative graft diameter. The aim of the present study was to investigate these two hypotheses.

2. Materials and Methods

The present retrospective observational study included 32 patients who underwent a single bundle ACL reconstruction using an autologous ipsilateral quadrupled hamstring graft (semitendinosus tendon and gracilis tendon each doubled). The patients were identified from the electronic database of the clinical center retrospectively according to the ICD-code for ACL-rupture and only a single surgeon. The study was conducted in accordance with the Declaration of Helsinki. Approval was obtained from the institutional review board (447/19). The study was registered in the German Trials Register (DRKS 00031502). Inclusion criteria for patients consisted of an age between 14 and 60 years, ACL reconstruction with an autologous ipsilateral quadrupled hamstring graft (STGT), complete preoperative MRI dataset of the injured knee joint, and an intraoperative documented graft diameter. Patients with an age under 14 years and older than 60 years, previous ipsilateral ACL reconstruction, acute or chronic ipsilateral hamstring injury as well as missing or corrupted digital MRI data were excluded. A total of 121 patients with ACL rupture were identified. In seventy subjects, the ACL reconstruction was performed by a single expert knee surgeon and they were screened. Seventeen of these records had to be excluded due to a rerupture after ACL-reconstruction, six records because another graft-type was used, three records due to an age under 14 years or over 60 years, and twelve records because of missing or corrupted MRI sequences, resulting in thirty-two full records for further analysis (Figure 1).

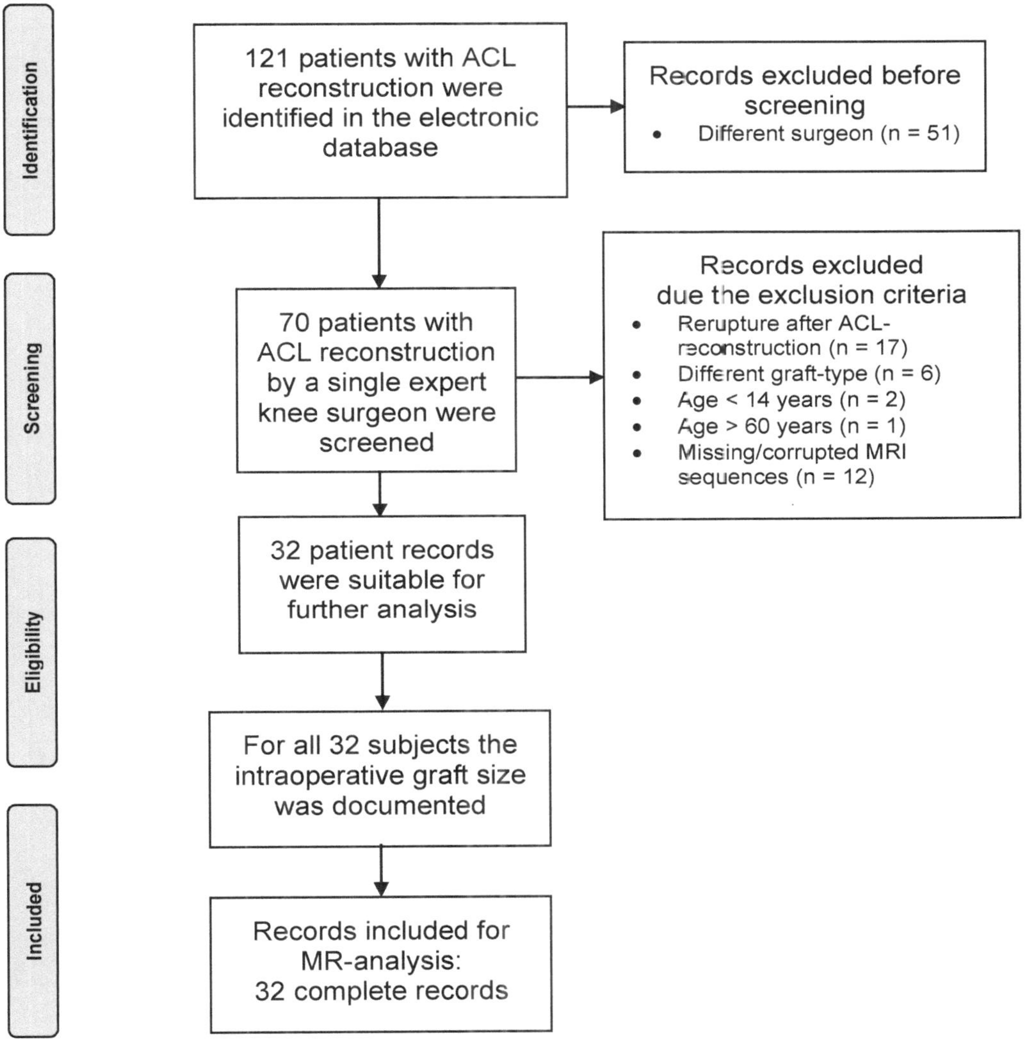

Figure 1. Flowchart of the selection process.

2.1. Participants

Mean age was 27.8 ± 11.4 years, and 62.5% of the subjects were male. The anthropometric data of the subject pool are summarized in Table 1.

Table 1. Distribution of the anthropometric data of 32 subjects included in the study.

	Mean	SD	Minimum	Maximum
Height (cm)	174.1	±10.9	150.0	192.0
Weight (kg)	74.8	±10.5	52.0	92.0
BMI (kg/cm^2)	24.7	±3.1	18.8	32.9

2.2. MRI Measurements

The complete preoperative MRI dataset was extracted from the digital database of the clinical center. Both in-house and external MRI scans were used. The image data therefore came from a variety of MR scanners. For CSA measurement, the image processing program Horos (Version 3.2.1, Nimble Co LLC, Annapolis, MD, USA) was used. First, the joint line (JL) was determined in the sagittal plane. Afterward, axial layers 10 mm and 20 mm below and above the joint line as well as the axial layer at the level of the widest intercondylar dimension of the epicondyles were selected (Figure 2). Therefore, the following six axial slices were used for the CSA measurement:

- Level of the widest intercondylar dimension (WID);
- 20 mm over joint line (20 mm);
- 10 mm over joint line (10 mm);
- Joint line level (JL);
- 10 under joint line (−10 mm);
- 20 mm under joint line (−20 mm).

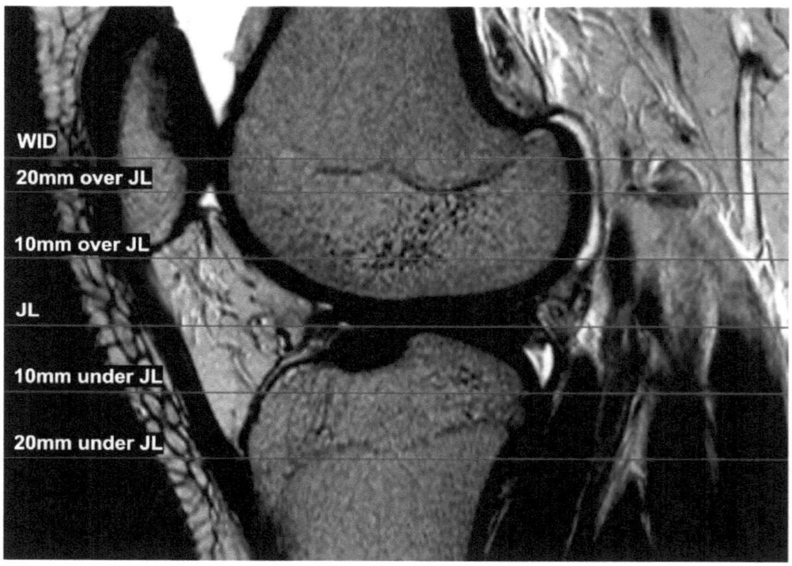

Figure 2. Sagittal presentation of the different levels where the CSA was measured in the axial layers.

The CSA of the semitendinosus tendon (CSA ST) and gracilis tendon (CSA GT) were each determined by three observers (senior physician, consultant, and medical student) on all six axial slices using the "closed polygon" function in the Horos program (Figure 3). All three observers were trained in detail about the study protocol before the manual tendon sequencing. In 27 subjects, axial T2 sequences were used after determining the different heights in the sagittal plane. Because of the variability of the available MRI datasets, five measurements were performed in axial T1 sequences. For the outline of the two tendons, the images were magnified as much as needed. Afterward, the CSA of the semitendinosus and gracilis tendon on the individual levels were summed up (CSA STGT).

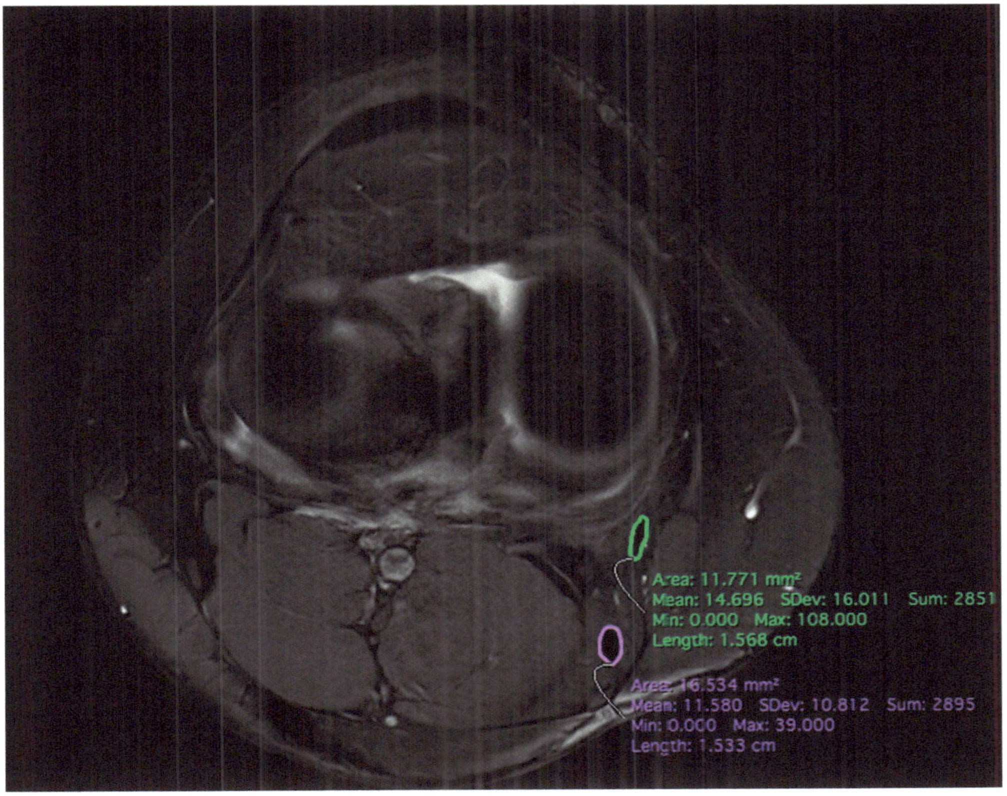

Figure 3. Measurement of the cross-sectional area of the ST (purple) and GT (green) on the level of the joint line using manually drawn closed-polygon function in Horos.

2.3. Intraoperative Measurements

Initially, the semitendinosus and gracilis tendons were harvested using a tendon stripper to detach them on the level of the musculotendinous transition. The fibers of the semitendinosus tendon that were pulling toward the medial head of the gastrocnemius muscle were detached prior to this. After the removal of the remaining muscle tissue, the two tendons were then each bundled twice. The resulting four-stranded graft (STGT graft) was then sewn together with a baseball-stitch at both ends over a length of 3 cm each. The diameter of this STGT graft was measured using a graft sizing block (Arthrex Graft Sizing Block AR-1886, Arthrex, Munich, Germany) with holes of 0.5-mm increments. This was determined by the smallest hole through which the graft could pass at its widest diameter with a tight fit but free passage.

2.4. Statistical Analysis

For the statistical analysis, IBM SPSS (Statistics for Macintosh, Version 26.0, IBM Corp. Armonk, NY, USA: IBM Corp.) was used. The intraclass correlation coefficient (ICC) was calculated using Cronbach's alpha to investigate the interrater reliability between the measurements of the three observers. In order to identify a possible correlation between the intraoperative graft diameter and the CSA on the different levels, the Pearson correlation coefficient was calculated. A p-value < 0.05 was considered significant. The Kolmogorov–Smirnov test was used to check whether the data were normally distributed.

3. Results

The complete datasets of 32 patients with preoperative MRI sequencing and intraoperatively documented graft size could be analyzed.

3.1. Interrater Reliability

The ICC was categorized as described by Koo et al. [17]. In general, the interrater reliability between the MRI measurements of the three observers was good (ICC > 0.75) to excellent (ICC > 0.9), depending on the investigated level (Table 2). For further correlation of the CSA STGT with the intraoperative four-stranded STGT graft diameter, the measurements of the doctoral student were used.

Table 2. Interrater reliability [1].

Level	CSA ST	CSA GT	CSA STGT
20 mm	0.95	0.94	0.95
10 mm	0.96	0.93	0.96
JL	0.84	0.95	0.92
−10 mm	0.95	0.91	0.95
−20 mm	0.89	0.60	0.85
WID	0.92	0.87	0.94

[1] Values are expressed as the intraclass correlation coefficient for the three investigators using Cronbach's alpha.

3.2. MRI CSA STGT

The mean CSA STGT on different levels is displayed in Table 3. At the level 20 mm under the JL, the CSA STGT could not be determined in three subjects during incomplete mapping.

Table 3. Cross-sectional area STGT [1].

Level	Mean	SD	Minimum	Maximum
20 mm	19.58	±4.26	13.12	28.74
10 mm	19.92	±4.33	13.40	28.64
JL	20.97	±5.26	10.76	31.85
−10 mm	18.84	±4.62	7.15	28.37
−20 mm	16.60	±5.30	6.95	27.85
WID	20.45	±4.32	13.38	30.09

[1] Values are expressed as the sum of the cross-sectional areas of ST and GT (STGT) in mm^2.

3.3. Correlation CSA STGT with Intraoperative Graft Diameter

The mean intraoperative four-stranded STGT graft diameter was 8.1 mm ± 0.5 mm (range: 7 mm to 9 mm). The strongest correlation between CSA STGT and the intraoperative STGT graft was found on the level 10 mm above the JL. Additionally, at the level of the joint line and 10 mm under the JL, a strong correlation was observed (Table 4).

Table 4. Correlation [1].

Level	Correlation Coefficient r	*p*-Value
20 mm	0.367	0.019 *
10 mm	0.489	0.002 *
JL	0.438	0.006 *
−10 mm	0.454	0.005 *
−20 mm	0.314	0.049 *
WID	0.410	0.010 *

[1] Values are expressed as the Pearson correlation coefficient r between the intraoperatively determined four-stranded graft diameter and cross-sectional area STGT with the level of significance *p*. Significant values are marked with *.

4. Discussion

The main finding of the present study is that there was a good correlation of the preoperatively measured CSA with the intraoperatively measured quadruple STGT graft, which was strongest at a level of 10 mm above the joint line. Furthermore, there was a negligible interrater variability between examiners with different levels of expertise.

The preoperative evaluation of whether a sufficient graft diameter can be achieved during ACL reconstruction using the hamstring tendons enables the orthopedic surgeon to better plan the surgical procedure. High interrater reliability was shown between the individual examiners regardless of medical experience. Thus, the level of training of the individual examiner appears to have very little influence on the correct segmentation of the two tendons in MRI. In practice, therefore, the measurement of the CSA does not necessarily have to be performed by an experienced senior physician but can be carried out by a trained medical assistant. In previous studies, the CSA was only measured at single levels assessed by using various diagnostic methods to investigate a possible correlation with the intraoperative graft diameter [14,15]. In most cases, the CSA was measured at the level of the WID or the widest point of the medial femoral epicondyle [16]. In comparison with the CSA values of previous studies, the results of the present study showed reproducible results at the corresponding levels [10,18–20]. In the study by Beyzadeoglu et al., the investigators used the joint line as the reference level for CSA determination [21]. This also showed comparable results to the present study. However, the degree of correlation between the CSA of the semitendinosus and gracilis tendon with the intraoperative graft diameter showed a very wide range depending on the particular study. Erquicia et al. determined the CSA of the two tendons at the level of the widest point of the medial femoral epicondyle using a standardized MRI protocol and always the same 1.5 Tesla scanner for each MRI. Using 4× magnification, there was a very strong correlation of $r = 0.86$, $p < 0.001$ of the summarized CSA STGT with the intraoperative graft diameter [10]. Corey et al. determined the CSA at two levels in MRI scans from different institutions without a standardized examination protocol and magnification. At the level at which the examiner considered the two tendons to be largest and roundest (Zakko-method), there was a weak correlation with the four-stranded graft diameter of $r = 0.388$. Using the level of the WID, the correlation increased to $r = 0.427$ [19,22]. Hollnagel et al. measured the CSA of the two tendons in images from two MRI devices (1.5 and 3 Tesla) depending on availability. The CSA was examined under maximum magnification at the level of the widest point of the medial femoral condyle and at the joint line. The correlation of the CSA of the single tendons (ST or GT), the sum of the CSA of both tendons, and the mean CSA of both tendons with the intraoperative four-stranded graft diameter was investigated. The strongest correlation was found using the 1.5 Tesla MRI scanner and the sum of the CSA of both tendons (CSA STGT) on the level of the joint line $r = 0.703$ [23]. The different focuses of the individual studies show which variables can have an influence on the correlation. Thus, in summary, the use of a single MR scanner with standardized sequences and analytical method appears to have a positive effect on the degree of correlation between the CSA of the hamstring tendons with the intraoperative graft diameter. In the present study, images from different MR scanners were included to determine the CSA. This approach realistically represents everyday clinical practice. The innovative character of this study is the measurement of the CSA at several defined levels based on the joint line and the WID. The aim was to determine the level at which the CSA of the two tendons correlated most strongly with the intraoperative graft diameter to generate a recommendation for the best patient individual surgical approach. The results of the present study showed the strongest correlation at the level of the joint line as well as 10 mm above and below it. Increasing the distance from the joint line in the cranial or caudal direction showed a decreasing correlation, which may be due to anatomical conditions. Distally, the tendons changed their shape from round to flat in the area of attachment to the pes anserinus superficialis. This makes it difficult to differentiate the tendon tissue from the surrounding structures on the MRI and makes the measurements more susceptible to segmentation errors. The interpretation of our results suggests that the CSA at the level of the WID, as used in many previous studies, does not seem to be a good predictive parameter to determine the graft diameter. Rather, in clinical practice, the focus should be on the CSA in close vicinity of the joint line to estimate the intraoperative graft diameter.

Limitations

This study has several limitations. The study population was limited to 32 patients and thus could show divergent results for larger study cohorts and should be considered exploratory. MRI datasets from many different scanner systems were analyzed, which in turn showed quantitative and qualitative variability. Nevertheless, we consider this methodology adequate due to its representation of everyday clinical practice. To generate a robust way to determine intraoperative graft size, we focused on the preoperative CSA of the graft tendons in the present study. Tendon length was not determined in the present setting. This can also help to estimate preoperatively whether a single graft is sufficient or if a STGT is needed.

5. Conclusions

The measurement of the CSA STGT in the preoperative MRI enabled a good assessment of the intraoperative achievable graft diameter in ACL reconstruction, independent of the examiners' training level. The present study suggests that the measurement should be taken in close proximity to the joint line, ideally 10 mm above it, to allow for a reliable estimation of the graft diameter. Patients in whom the adequate graft diameter of a single tendon graft cannot already be evaluated preoperatively should be informed about harvesting both ST and GT.

Author Contributions: Conceptualization, M.F.M. and K.I.; Methodology, M.F.M. and K.I.; Software, E.T.; Validation, M.F.M., K.I. and H.S.; Formal analysis, M.F.M.; Investigation, M.F.M. and K.I.; Resources, M.F.M.; Data curation, M.F.M.; Writing—original draft preparation, M.F.M.; Writing—review and editing, M.S., P.O. and H.S.; Visualization, M.F.M.; Supervision, K.I.; Project administration, K.I. All authors have read and agreed to the published version of the manuscript.

Funding: This research received no external funding.

Institutional Review Board Statement: The study was conducted in accordance with the Declaration of Helsinki, and approved by the Ethics Committee of University of Freiburg (protocol code: 447/19, date of approval: 22 October 2019).

Informed Consent Statement: Patient consent was waived due to the retrospective character of the data acquisition. The data obtained were anonymized directly in the follow-up and only then evaluated further. It is highly probable that a large proportion of patients who would be suitable for inclusion in the study on the basis of the inclusion and exclusion criteria can no longer be contacted by telephone, email, or post at the current time, for example, because they have moved or possibly incomplete/incorrect contact details are stored in the electronic database. Taking into account the benefit/risk ratio, it is therefore justified to waive the declaration of consent for study participation by the subjects, as the data were collected exclusively retrospectively and anonymously, and the undeliverability of the declaration of consent leads to a disproportionately high loss of data, which jeopardizes the scientific benefit of the study.

Data Availability Statement: All data are within the manuscript.

Acknowledgments: We acknowledge support by the Open Access Publication Fund of the University of Freiburg.

Conflicts of Interest: The authors declare no conflicts of interest.

References

1. Domnick, C.; Garcia, P.; Raschke, M.J.; Glasbrenner, J.; Lodde, G.; Fink, C.; Herbort, M. Trends and Incidences of Ligament-Surgeries and Osteotomies of the Knee: An Analysis of German Inpatient Records 2005–2013. *Arch. Orthop. Trauma Surg.* **2017**, *137*, 989–995. [CrossRef] [PubMed]
2. Weiler, A.; Scheffler, S.; Höher, J. Graft selection for primary anterior cruciate ligament reconstruction. *Orthopäde* **2014**, *31*, 731–740. [CrossRef] [PubMed]
3. Conte, E.J.; Hyatt, A.E.; Gatt, C.J.; Dhawan, A. Hamstring Autograft Size Can Be Predicted and Is a Potential Risk Factor for Anterior Cruciate Ligament Reconstruction Failure. *Arthrosc. J. Arthrosc. Relat. Surg.* **2014**, *30*, 882–890. [CrossRef] [PubMed]
4. Magnussen, R.A.; Lawrence, J.T.R.; West, R.L.; Toth, A.P.; Taylor, D.C.; Garrett, W.E. Graft Size and Patient Age Are Predictors of Early Revision after Anterior Cruciate Ligament Reconstruction with Hamstring Autograft. *Arthrosc. J. Arthrosc. Relat. Surg.* **2012**, *28*, 526–531. [CrossRef] [PubMed]
5. Mariscalco, M.W.; Flanigan, D.C.; Mitchell, J.; Pedroza, A.D.; Jones, M.H.; Andrish, J.T.; Parker, R.D.; Kaeding, C.C.; Magnussen, R.A. The Influence of Hamstring Autograft Size on Patient-Reported Outcomes and Risk of Revision after Anterior Cruciate Ligament Reconstruction: A Multicenter Orthopaedic Outcomes Network (MOON) Cohort Study. *Arthroscopy* **2013**, *29*, 1948–1953. [CrossRef] [PubMed]
6. Noyes, F.R.; Butler, D.L.; Grood, E.S.; Zernicke, R.F.; Hefzy, M.S. Biomechanical Analysis of Human Ligament Grafts Used in Knee-Ligament Repairs and Reconstructions. *J. Bone Jt. Surg. Am.* **1984**, *66*, 344–352. [CrossRef]
7. Pichler, W.; Tesch, N.P.; Schwantzer, G.; Fronhöfer, G.; Boldin, C.; Hausleitner, L.; Grechenig, W. Differences in Length and Cross-Section of Semitendinosus and Gracilis Tendons and Their Effect on Anterior Cruciate Ligament Reconstruction: A Cadaver Study. *J. Bone Jt. Surg. Br.* **2008**, *90*, 516–519. [CrossRef] [PubMed]
8. Tohyama, H.; Beynnon, B.D.; Johnson, R.J.; Nichols, C.E.; Renström, P.A. Morphometry of the Semitendinosus and Gracilis Tendons with Application to Anterior Cruciate Ligament Reconstruction. *Knee Surg. Sports Traumatol. Arthrosc.* **1993**, *1*, 143–147. [CrossRef]
9. An, V.; Scholes, C.; Mhaskar, V.A.; Parker, D.A.; Fritsch, B.A. Regression Modelling Combining MRI Measurements and Patient Anthropometry for Patient Screening and Prediction of Graft Diameter in Hamstring Autograft Arthroscopic ACL Reconstruction. *Asia-Pac. J. Sports Med. Arthrosc. Rehabil. Technol.* **2017**, *8*, 24–31. [CrossRef] [PubMed]
10. Erquicia, J.I.; Gelber, P.E.; Doreste, J.L.; Pelfort, X.; Abat, F.; Monllau, J.C. How to Improve the Prediction of Quadrupled Semitendinosus and Gracilis Autograft Sizes with Magnetic Resonance Imaging and Ultrasonography. *Am. J. Sports Med.* **2013**, *41*, 1857–1863. [CrossRef]
11. Leiter, J.; Elkurbo, M.; McRae, S.; Chiu, J.; Froese, W.; MacDonald, P. Using Pre-Operative MRI to Predict Intraoperative Hamstring Graft Size for Anterior Cruciate Ligament Reconstruction. *Knee Surg. Sports Traumatol. Arthrosc.* **2016**, *25*, 229–235. [CrossRef] [PubMed]
12. Offerhaus, C.; Albers, M.; Nagai, K.; Arner, J.W.; Höher, J.; Musahl, V.; Fu, F.H. Individualized Anterior Cruciate Ligament Graft Matching: In Vivo Comparison of Cross-Sectional Areas of Hamstring, Patellar, and Quadriceps Tendon Grafts and ACL Insertion Area. *Am. J. Sports Med.* **2018**, *46*, 2646–2652. [CrossRef] [PubMed]
13. Serino, J.; Murray, R.; Argintar, E.H. Use of Magnetic Resonance Imaging to Predict Quadrupled Semitendinosus Graft Diameter in All-Inside Anterior Cruciate Ligament Reconstruction. *Orthopedics* **2017**, *40*, e617–e622. [CrossRef] [PubMed]
14. Liau, Z.Q.G.; Ng, M.S.P.; Low, S.S.E.; Chin, B.Z.; Hui, J.H.P.; Kagda, F.H.Y. A Novel Practical Method to Predict Anterior Cruciate Ligament Hamstring Graft Size Using Preoperative MRI. *Knee Surg. Relat. Res.* **2024**, *36*, 17. [CrossRef] [PubMed]
15. Bhamare, D.S.; Sirasala, S.; Jivrajani, P.; Nair, A.; Taori, S. Preoperative MRI Assessment of Hamstring Tendons to Predict the Quadruple Hamstring Graft Diameter in Anterior Cruciate Ligament Reconstruction. *Cureus* **2022**, *14*, e21753. [CrossRef] [PubMed]
16. Pérez-Mozas, M.; Payo-Ollero, J.; Montiel-Terrón, V.; Valentí-Nin, J.R.; Valentí-Azcárate, A. Predicción Preoperatoria Del Diámetro de La Plastia Autóloga de Isquiotibiales En Reconstrucción de Ligamento Cruzado Anterior. *Rev. Española Cirugía Ortopédica Y Traumatol.* **2020**, *64*, 310–317. [CrossRef] [PubMed]
17. Koo, T.K.; Li, M.Y. A Guideline of Selecting and Reporting Intraclass Correlation Coefficients for Reliability Research. *J. Chiropr. Med.* **2016**, *15*, 155–163. [CrossRef] [PubMed]
18. Bickel, B.A.; Fowler, T.T.; Mowbray, J.G.; Adler, B.; Klingele, K.; Phillips, G. Preoperative Magnetic Resonance Imaging Cross-Sectional Area for the Measurement of Hamstring Autograft Diameter for Reconstruction of the Adolescent Anterior Cruciate Ligament. *Arthrosc. J. Arthrosc. Relat. Surg.* **2008**, *24*, 1336–1341. [CrossRef]
19. Corey, S.; Mueller, T.; Hartness, C.; Prasad, B.M. Correlation of Intra-Operative Hamstring Autograft Size with Pre-Operative Anthropometric and MRI Measurements. *J. Orthop.* **2018**, *15*, 988–991. [CrossRef]
20. Grawe, B.M.; Williams, P.N.; Burge, A.; Voigt, M.; Altchek, D.W.; Hannafin, J.A.; Allen, A.A. Anterior Cruciate Ligament Reconstruction with Autologous Hamstring: Can Preoperative Magnetic Resonance Imaging Accurately Predict Graft Diameter? *Orthop. J. Sports Med.* **2016**, *4*, 2325967116646360. [CrossRef]
21. Beyzadeoglu, T.; Akgun, U.; Tasdelen, N.; Karahan, M. Prediction of Semitendinosus and Gracilis Autograft Sizes for ACL Reconstruction. *Knee Surg. Sports Traumatol. Arthrosc.* **2012**, *20*, 1293–1297. [CrossRef] [PubMed]

22. Zakko, P.; Eck, C.F.; Guenther, D.; Irrgang, J.J.; Fu, F.H. Can We Predict the Size of Frequently Used Autografts in ACL Reconstruction? *Knee Surg. Sports Traumatol. Arthrosc.* **2015**, *25*, 3704–3710. [CrossRef] [PubMed]
23. Hollnagel, K.; Johnson, B.M.; Whitmer, K.K.; Hanna, A.; Miller, T.K. Prediction of Autograft Hamstring Size for Anterior Cruciate Ligament Reconstruction Using MRI. *Clin. Orthop. Relat. Res.* **2019**, *477*, 2677–2684. [CrossRef] [PubMed]

Journal of Personalized Medicine

Article

MRI Characteristics and Alterations in Patellar Height in Patients with Patellar Tendinopathy—A Retrospective Study

Kim Loose [1,*], Sophie Pennekamp [1], Wolfgang Hitzl [2,3,4], Maximilian Willauschus [1], Johannes Rüther [1], Sandeep Silawal [5], Philipp Schuster [1,6], Hermann Josef Bail [1], Michael Millrose [1,7] and Markus Geßlein [1,8]

1 Department of Orthopedics and Traumatology, Paracelsus Medical University, Breslauer Straße 201, 90471 Nuremberg, Germany; sophie.pennekamp@stud.pmu.ac.at (S.P.); maximilian.willauschus@klinikum-nuernberg.de (M.W.); johannes.ruether@klinikum-nuernberg.de (J.R.); philipp_schuster@gmx.de (P.S.); hermann-josef.bail@klinikum-nuernberg.de (H.J.B.); m.millrose@icloud.com (M.M.); markus.gesslein@klinikum-nuernberg.de (M.G.)
2 Research and Innovation Management (RIM), Biostatistics and Publication of Clinical Trial Studies, Paracelsus Medical University, 5020 Salzburg, Austria; wolfgang.hitzl@pmu.ac.at
3 Department of Ophthalmology and Optometry, Paracelsus Medical University Salzburg, 5020 Salzburg, Austria
4 Research Program Experimental Ophthalmology and Glaucoma Research, Paracelsus Medical University, 5020 Salzburg, Austria
5 Institute of Anatomy and Cell Biology, Paracelsus Medical University, Prof. Ernst Nathan Str. 1, 90419 Nuremberg, Germany; sandeep.silawal@pmu.ac.at
6 Center for Sports Orthopedics and Special Joint Surgery, RKH Orthopedic Hospital Markgröningen, 71706 Markgröningen, Germany
7 Department of Trauma Surgery and Sports Medicine, Garmisch-Partenkirchen Medical Centre, 82467 Garmisch-Partenkirchen, Germany
8 GOTS (German-Austrian-Swiss Society for Orthopaedic Traumatologic Sports Medicine), Muscle and tendon Committee, 07743 Jena, Germany
* Correspondence: kim.loose@klinikum-nuernberg.de; Tel.: +49-911-3982600

Abstract: (1) Background: Patellar tendinopathy (PT) is an overuse condition of the knee extensor mechanism characterized by ventral knee pain at the lower pole of the patella and limited functionality. (2) Methods: In this retrospective study, a group of patients with PT ($n = 41$) was compared with a control group ($n = 50$) in terms of patient-related data and magnetic resonance imaging (MRI) characteristics. (3) Results: Patellar height was higher in the PT patient group and there was a significant difference in Caton–Deschamps index (CD) compared to the control group ($p = 0.021$). Patients with PT showed a lower patella–patellar tendon angle (PPTA) ($p = 0.011$). The patellar tendon thickness (PTT) in the proximal (PTTprox), middle (PTTmid) and distal (PTTdistal) part of the tendon was significantly thickened ($p < 0.001$). Increased signal intensity in MRI was detected in symptomatic tendons over 6 months compared to a duration of less than 6 months ($p = 0.025$). A significant relationship between the PTTprox and an increased signal intensity was observed ($p < 0.001$). (4) Conclusions: Patients with PT showed a significant difference in the patellar height and PPTA. With persistence of symptoms over 6 months, MRI seems suitable to detect the morphologic tendon changes and further identify patients suitable for surgical procedures.

Keywords: patellar tendinopathy; jumper's knee; magnetic resonance imaging; patellar tendon thickness

check for updates

Citation: Loose, K.; Pennekamp, S.; Hitzl, W.; Willauschus, M.; Rüther, J.; Silawal, S.; Schuster, P.; Bail, H.J.; Millrose, M.; Geßlein, M. MRI Characteristics and Alterations in Patellar Height in Patients with Patellar Tendinopathy—A Retrospective Study. *J. Pers. Med.* **2023**, *13*, 698. https://doi.org/10.3390/jpm13040698

Academic Editor: Jih-Yang Ko

Received: 29 March 2023
Revised: 11 April 2023
Accepted: 12 April 2023
Published: 21 April 2023

1. Introduction

Patellar tendinopathy (PT) is a common orthopedic overuse condition of the knee extensor mechanism characterized by anterior knee pain [1]. Repetitive eccentric quadriceps loading in sports such as volleyball, track and field, basketball, long distance running and skiing can lead to PT, also called jumper's knee [1,2]. This overuse condition occurs more often in males and from adolescence through to the fourth decade of life. Up to 45% of elite jumping athletes will have PT at any point during their career [3,4]. A prospective study reported that 53% of athletes with PT were forced to quit competitive sport [5]. PT can also cause limitations in non-athletes [6].

Microtrauma of the tendon may lead to the degeneration of individual fibers due to tension in the tendon, which can accumulate over time and lead to chronic tendinopathy [7]. It is noticeable that the tendon does not recover from this degeneration/microtears. The exact pathogenesis of chronic tendinopathy remains to be definitively elucidated but is likely to develop independent of inflammation [8]. Histology shows intratendinous collagen degeneration, hypercellularity, neovascularization and, occasionally, calcifications and local necrosis [9]. PT is a clinical diagnosis; patients report progressive activity-related pain at the distal patellar pole [4]. Activities of daily living may be affected by climbing the stairs, squatting and prolonged sitting [4].

In the jumper's knee classification (JKC) according to Blazina et al., PT develops in four stages: stage 1 is pain after activity without functional impairment, stage 2 is pain during and after activity but still with satisfactory performance, stage 3 is characterized by persistent pain during and after activity and difficult performance and stage 4 is tendon rupture [10]. Ultrasound and magnetic resonance imaging are suitable for diagnosing PT [1]. Conservative treatment is the initial treatment method for PT and includes pain modulation, stress management, stress progression and functional strengthening [6]. Surgical therapy is used if six months of extensive conservative management is unsuccessful [11]. Good results have been achieved with arthroscopic and open surgical procedures in refractory cases of PT [8]. Refractory response to treatment and long-term symptoms of PT are reported in several studies, making targeted prevention and adapted treatment of the sports injury all the more important [1,12].

In addition to extrinsic risk factors for PT, such as physical activity and training volume, there are known intrinsic risk factors such as patellar height, patellar thickness and increased patellar signal intensity [4]. Some of these patellofemoral instability measurements have been investigated in a few studies: in 1986, Kujala et al. compared athletes with patellar chondropathy ($n = 20$), jumper's knee ($n = 20$) and a control group of symptom-free athletes ($n = 20$) and showed an accumulation of patella alta associated with jumper's knee ($p < 0.05$) [13]. This correlation was confirmed by Tscholl et al. in their case–control study [14]. Patellar height affects the pattern of the articulation of the patella with the femur during flexion and has an inconsistent effect on the biomechanics of the patellar tendon [15], which again can lead to tendinopathy.

In a study of elite university athletes with PT symptoms ($n = 16$) compared to athletes with PT ($n = 49$), Nishida et al. found a patellar tendon thickness of more than 7 mm on MRI to be significant [16].

In physiologic conditions, the patellar tendon appears on MRI as a homogeneous strip with low signal intensity and can be clearly distinguished from the infrapatellar fat pad. In patellar tendinopathy, there is a widening of the patellar tendon and a heterogenous increase in signal intensity [11].

Another discussed aspect in the etiology of PT is patellofemoral instability and trochlear dysplasia. Trochlear dysplasia results in a loss of bone stability of the patellofemoral joint [17], altering the distribution of force to the patellar tendon. In their study, Ivengar et al. found hypertrophy of the medial portion of the patellar tendon in patients with patellofemoral joint instability and trochlear dysplasia. Fifty MRIs of the knee joint with trochlear dysplasia and fifty MRIs with normal patellofemoral joint morphology were compared [18].

The understanding of PT by analyzing different measurements on MRI is crucial for the improvement of treatment. Regarding the MRI measurements, there has so far been limited data with concrete thresholds and any large potential.

The primary aim of the study was to evaluate if typical patellofemoral instability measurements on MRI correlate with the presence of PT to better guide the prevention and treatment of a protracted injury. The study further investigated whether there is a correlation of PT with symptoms over 6 months and the radiological image.

2. Materials and Methods

2.1. Study Population

From a total of 80 patients with PT treated in the period from March 2012 to September 2021 at a maximum care hospital, 41 patients were included in the study for a retrospective analysis (Figure 1). The control group consisted of 50 patients who underwent surgery for anterior cruciate ligament rupture without any prior symptoms of anterior knee pain at the same maximum care hospital from May 2018 to April 2022. The control group was chosen as all cases had standard MRIs and usually represent a comparable age group.

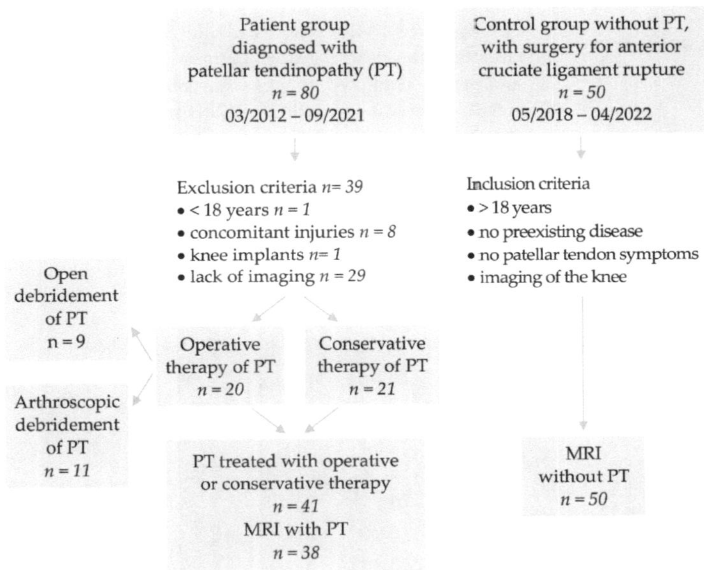

Figure 1. Study flowchart.

Inclusion criteria were a minimum patient age of 18 years and surgical (arthroscopic debridement or open debridement) or conservative treatment of PT. The patients considered in the study were affected only unilaterally and not bilaterally. Exclusion criteria were concomitant injuries that affected the ipsilateral knee, patients with existing knee implants or prior surgery on the knee, as well as lack of imaging. The control group included patients 18 years of age and older, without preexisting disease or patellar tendon symptoms, and with imaging of the knee.

2.2. Diagnosis and Treatment

Patient history, clinical examination with ultrasound, X-rays- and/or MRI -examination followed by an adequate diagnosis of all patients were undertaken by specialists at the orthopedic trauma surgery outpatient clinic. PT was treated either conservatively or surgically if symptoms persisted for more than six months. Conservative therapy was the treatment of choice in early stages and at the point of initial diagnosis. It consisted of physiotherapy with manual therapy and, in sports therapy, eccentric training of the quadriceps femoris muscle. In cases of persisting symptoms after conservative therapy, surgery was performed by an experienced orthopedic surgeon. The applied surgical techniques were either arthroscopic debridement or open debridement of the patellar tendon. Postoperatively, the affected leg was put in an orthosis with the leg fully extended. All patients were mobilized with partial weight bearing using crutches on the second day after surgery under instructions from a competent physiotherapist. Full loading of the leg was allowed after 7 days. Patients were instructed to wear the knee straightening brace with a ROM 0-0-90 for 2 weeks. Physiotherapy was allowed after one week. Full knee flexion and active knee extension were allowed after 6 weeks.

2.3. Patient-Related Data

Patient-related data such as gender assigned at birth, age at diagnosis in years, body mass index (BMI), sports, stage of the disease classified according to the JKC and duration of complaints under or over 6 months were extracted from hospital medical records.

2.4. Evaluation of Preoperative MRI

Preoperative MRI was available for evaluation in 38 patients with PT and in 50 patients in the control group. Images were acquired using 1.5 or 3.0 Tesla scanners. T1-, T2- and proton-weighted images were considered. The sequences used were SE (spin-echo), SE fs (spin-echo fat saturation), TSE (turbo spin-echo) and STIR (short-tau inversion recovery). Sagittal and axial cross-sectional images were evaluated. Images were evaluated retrospectively on an approved PACS workstation (Ashvins, MedicalCommunications, Heidelberg, Germany).

2.5. Measurement Methods in MRI

The Insall–Salvati index (IS) (B/A), Caton–Deschamps index (CD) (D/C) and the Blackburne–Peel index (BP) (F/E) were applied to determine patellar height (Figure 2).

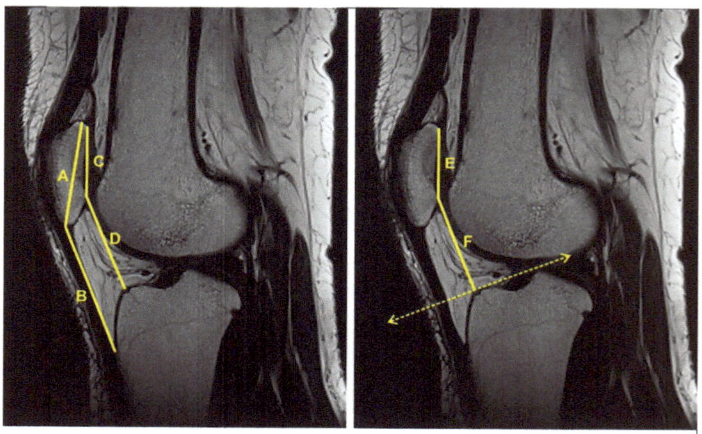

(a) (b)

Figure 2. (**a**) Insall–Salvati index (IS) (B/A): patellar tendon length from the lower pole of the patella to the tibial tuberosity (B) to the diagonal length of the patella (A). Caton–Deschamps Index (CD) (D/C): distance from the inferior edge of the articular surface of the patella to the anterior edge of the tibial plateau (D) to the articular surface of the patella (C). (**b**) Blackburne–Peel index (BP) (F/E): distance between the tibial plateau line and the inferior point of the patellar articular surface (F) to the length of the patellar articular surface (E).

In the axial plane, the patella tilt angle (PTA) was measured in the sectional view with the greatest spread of the patella from medial to lateral using the Dejour method (Figure 3a). To describe the trochlea and possible trochlear dysplasia, the lateral trochlear inclination angle (LTI) and the sulcus angle (SA) were obtained in the axial plane. Both measurements were performed in the image with the greatest distance between the intercondylar fossa and the facies patellaris femoris. The LTI describes the angle between the facies patellaris femoris lateralis and posterior condylar tangential line. The SA is formed from the angle between the lateral and medial facies patellaris femoris (Figure 3b).

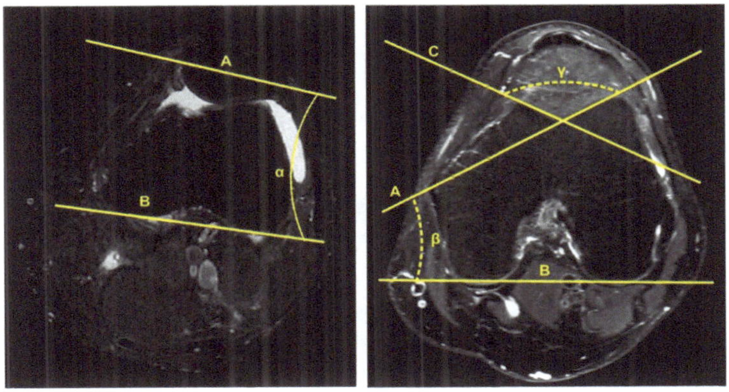

(a) (b)

Figure 3. (**a**) Patella tilt angle (PTA) = α: angle between a straight line through the transverse axis of the patella (A) and a tangent along the posterior femoral condyles (B). (**b**) Lateral trochlear inclination angle (LTI) = β: angle between the tangent along the facies patellaris femoris lateralis (A) and the posterior condylar tangential line (B). Sulcus angle (SA) = γ: angle between the lateral facies patellaris femoris (A) and the medial facies patellaris femoris (C).

Patellar tendon thickness (PTT) was measured in the sagittal plane on the MRI image with the greatest distance between the base patella and apex patella. Tendon thickness was measured proximally at the origin of the tendon (PTTprox), distally at the insertion at the tibial tuberosity (PTTdistal) and at the midpoint of these two points (PTTmid) (Figure 4a). In the same image, the presence of signal intensity was also checked (Figure 4b). The patella–patellar tendon angle (PPTA) measures the angle between the line drawn from the base of the patella to the apex (A) and the line drawn between the apex of the patella to the insertion of the patellar tendon on the tibial tuberosity (B) (Figure 4c).

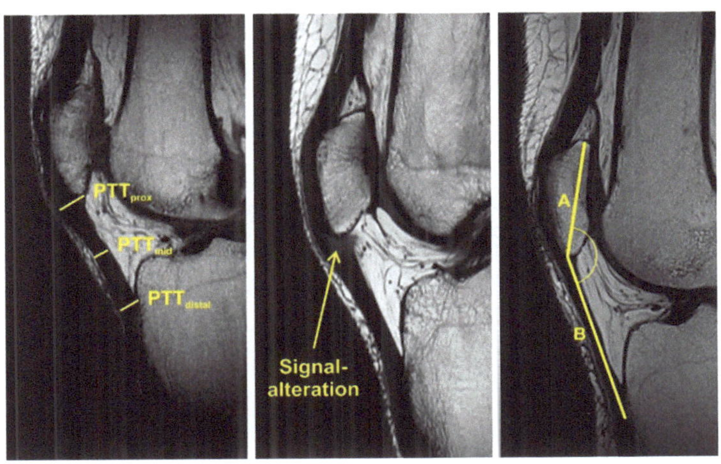

(a) (b) (c)

Figure 4. (**a**) Patellar tendon thickness (PTT): shown here on MRI, a thickened patellar tendon with PT in the proximal (PTTprox), middle (PTTmid) and distal segments (PTTdistal). (**b**) Signal intensity (alteration) in PT at the origin of the patellar tendon. (**c**) Patella–patellar tendon angle (PPTA): angle between the line drawn from the base of the patella to the apex (A) and the line drawn between the apex of the patella to the insertion of the patellar tendon on the tibial tuberosity (B).

2.6. Statistical Analysis

All data were obtained and analyzed retrospectively and the statistical analysis was conducted with IBM SPSS Statistics version 28.0 (version 28, Armonk, NY, USA). For metric parameters, Levene's test of equality of variance and bootstrap test with 4000 samples and 95% confidence interval were applied. For nominal variables, Pearson's chi-square test or Fisher's exact test were used. In addition, the least significant difference test (LSD test) was applied to represent significant differences. The p-values are two-sided and considered statistically significant at an alpha level of <0.05. Results are presented as mean ± standard deviation and range with associated minimum and maximum.

The study was approved by the institutional review committee (Paracelsus Medical University at the Nuremberg Hospital, Number IRB-2022-007) following national legal guidelines.

3. Results

A total of 91 patients were considered, of which 41 had PT (patient group) and 50 were without PT (control group).

3.1. Demographic Data

On average, patients with PT were 31.5 years old and had a BMI of 25.6 (Table 1). A total of 31.7% of the patient group was female and 68.3% was male. There was no significant difference comparing the genders of both groups ($p = 0.976$). PT was localized on the left side in 43.9% and on the right side in 56.1%, thus the probability of PT was equal on both sides ($p = 0.295$). There was equality of means in all listed patient-related data and no significant difference between patient and control group ($p > 0.05$). Thus, the two groups were well comparable (Table 1).

Table 1. Patient-related data.

	Patients with PT	Control Group	p-Value
Number of patients (n)	41	50	>0.05
Gender, female (n)	13	16	0.976
Gender, male (n)	28	34	0.976
Age, y (mean ± SD)	31.5 ± 11.9	30.9 ± 8.1	0.784
BMI [1] (mean ± SD)	25.6 ± 5.6	25.9 ± 4.8	0.826

[1] BMI: body mass index.

A total of 78.0% of patients with PT (32 of 41) participated in popular or professional sports, and 22.0% (9 of 41) did not participate in any sports. Soccer (22.0%) was the most popular sport. Among others, a variety of ball sports were present (39.1%) as well as running (12.2%) and martial arts (9.7%).

There was a significant correlation (Pearson's chi-squared test, $p < 0.001$) between the stage of the disease, classified according to the JKC, and the duration of symptoms under or over 6 months. The higher the stage of PT, the longer the symptoms persisted in patients. Thirty-three of the forty-one patients with PT showed symptoms for more than 6 months, and eight patients less than 6 months. All patients suffered from functional impairment.

3.2. Biomechanical Differences between Patient and Control Group in MRI

3.2.1. Patellar Height

For the indices of patellar height, the MRI evaluation showed a significant difference in the Caton–Deschamps (CD) between the patients and control group, a borderline significant finding in the Blackburne–Peel (BP), and no significant difference in the Insall–Salvati (IS) indices (Table 2).

Table 2. Comparison of patellar height on MRI.

	Patients with PT $n = 38$	Patients without PT $n = 50$	p-Value
IS [1] (mean ± SD (range))	1.21 ± 0.19 (0.78–1.57)	1.17 ± 0.14 (0.93–1.57)	0.244
CD [2] (mean ± SD (range))	1.22 ± 0.19 (0.82–1.77)	1.13 ± 0.15 (0.86–1.51)	0.021
BP [3] (mean ± SD (range))	1.14 ± 0.18 (0.77–1.46)	1.07 ± 0.16 (0.75–1.47)	0.053

[1] IS = Insall–Salvati index; [2] CD = Caton–Deschamps index; [3] BP = Blackburne–Peel index.

3.2.2. Patella–Patellar Tendon Angle (PPTA) and Patellar-Tilt Angle (PTA)

PPTA was found to be significantly lower in the patient group with PT at 142.8° ± 4.8° (vs. 145.7° ± 5.5°) ($p = 0.011$). No significant difference was found in PTA between the patient group (6.6° ± 4.3°) and the control group (7.8° ± 4.9°) ($p = 0.214$).

3.2.3. Trochlear Dysplasia

There were no significant differences in trochlear dysplasia measured in sulcus angle (SA) and lateral trochlear inclination angle (LTI) between the patient group and control group ($p > 0.05$).

3.2.4. Patellar Tendon Thickness (PTT) and Signal Intensity

The patellar tendon was widened in all tendon segments in the patient group and was significantly thicker than in the control group ($p < 0.001$) (Table 3). In the proximal portion (PTTprox), the mean value in PT was 6.83 mm, almost 2 mm higher than that of the control group. It was the largest mean value for PTTprox in the patient group compared to the other tendon sections. The mean tendon thickness was 5.34 mm, and the smallest mean value measured in PT as well as the same pattern seen in the control group was 3.73 mm. PTTdistal was the largest measured mean value in the control group with 4.98 mm.

Table 3. Comparison of patellar tendon thickness (PTT) and signal intensity in the MRI of patient versus control group.

	Patients with PT $n = 38$	Patients without PT $n = 50$	p-Value
PTTprox [1], mm (mean ± SD (range))	6.83 ± 2.36 (3.67–13.59)	4.76 ± 0.97 (2.84–7.02)	<0.001 *
PTTmid [2], mm (mean ± SD (range))	5.34 ± 1.54 (3.30–9.80)	3.73 ± 0.66 (2.22–5.25)	<0.001 *
PTTdistal [3], mm (mean ± SD (range))	6.15 ± 1.61 (3.64–9.26)	4.98 ± 1.07 (3.29–7.30)	<0.001 *
Signal intensity, yes/no	30/8	5/45	<0.001 *

[1] PTTprox = proximal patellar tendon thickness; [2] PTTmid = middle patellar tendon thickness; [3] PTTdistal = distal patellar tendon thickness; * significant.

The increased signal intensity occurred in 78.9% of the cases with PT (30 of the 38 patients). This was significantly higher ($p < 0.001$) than the control group (10%, 5 of the 50 patients). Patients with surgical therapy of PT had significantly more increased signal intensity of PT compared with patients without surgical therapy ($p = 0.017$) (Table 4).

Table 4. Comparison of patellar tendon thickness (PTT) and signal intensity in the MRI of patients with and without surgical therapy of PT.

	Patients with Surgical Therapy of PT *n* = 19	Patients without Surgical Therapy of PT *n* = 19	*p*-Value
PTTprox [1], mm (mean ± SD (range))	7.61 ± 2.89 (3.81–13.59)	6.05 ± 1.37 (3.67–9.13)	0.118
PTTmid [2], mm (mean ± SD (range))	5.80 ± 1.76 (3.35–9.80)	4.89 ± 1.16 (3.30–8.57)	0.129
PTTdistal [3], mm (mean ± SD (range))	6.14 ± 1.76 (3.64–9.00)	6.16 ± 1.50 (3.73–9.26)	0.861
Signal intensity, yes/no	18/1	12/7	0.017 *

[1] PTTprox = proximal patellar tendon thickness; [2] PTTmid = middle patellar tendon thickness; [3] PTTdistal = distal patellar tendon thickness; * significant.

3.3. Correlation of Stage of Disease and Radiological Image

Increased signal intensity in chronic PT (>6 months) occurred in 87.1% of patients (n = 27) compared with 42.86% of patients with PT lasting less than 6 months (n = 3) (p = 0.025). In PT of less than 6 months, no increased signal intensity was present in 57.14% (n = 4). The patellar height correlated with higher PTTprox, however, not significantly. There was a significant correlation between PTTprox and the presence of increased signal intensity ($p < 0.001$). In the patients with PT, increased signal intensity was present in 30 of 38 cases with a PTTprox of 7.30 ± 2.39 mm. The 8 of 38 cases without increased signal intensity showed a significantly lower PTTprox of 5.09 ± 1.20 mm.

4. Discussion

The main findings of this study were that the patellar height was significantly higher, the patella–patellar tendon angle (PPTA) was significantly lower and the patellar tendon thickness (PTT) was significantly thicker in the patients group compared to the control group. Furthermore, the risk of an increased signal intensity of patellar tendon at symptoms > 6 months was higher than with symptoms < 6 months.

Patella alta is considered a risk factor of PT in the literature [19]. In this study, the indices of patellar height in MRI evaluation showed a significant difference in the Caton–Deschamps (CD) between the patients and control group and a borderline significant finding in the Blackburne–Peel (BP). Thus, this study is in agreement with the study by Tscholl et al., in which patients with PT had higher patellar scores in CD (1.17 ± 0.17) [14]. In contrast to other measured indices, the IS index (Table 2) does not show a high significant difference between the patient and control groups. This could be explained by the fact that the original measure was evaluated on plain lateral X-ray images and there might be a certain loss of resolution for the exact measurement in the MRI T1 imaging [20]. This study confirms the hypothesis of a higher patella as an intrinsic risk factor for PT. Thus, especially in athletes from sports such as volleyball, soccer, or basketball and either asymptomatic or in the early stages of anterior knee pain, the measurement of patellar height may prove useful to prevent the risk of chronic PT and the potential functional consequences, up to and including withdrawal from sport. Patella alta is caused in part to a shortening of the quadriceps femoris muscle due to muscle tension, which underscores the importance of eccentric training and physiotherapy as therapeutic options for PT.

In this study, PPTA was found to be significantly lower in the patient group with PT than in the control group. Comparable studies to PPTA in PT were not found in the literature research. Published work considers PPTA in the context of other patellofemoral conditions such as medial patellar plica syndrome, chondromalacia patellae and infrapatellar fat pad syndrome. The angle was found to be lower in patients with anterior knee pain caused by one of these above stated diagnoses than in the control groups [21]. A biomechanical reason for the lower PPTA could be that the infrapatellar fat pad undergoes a fibrosis during nonphysiological loading [21].

In this study, all tendon segments of the PTT had significant thickening in comparison to the control group. A widened patellar tendon, especially in the proximal portion of the tendon, is considered a typical characteristic of PT [16,22]. Various thresholds for patellar tendon thickness PTT have been described in the literature [16,22,23]. El-Khoury et al. found that the mean patellar tendon thickness of the healthy participants (n = 10) was 3.7 mm, whereas that of the patients (n = 50) to have been 10.9 mm [23]. In this study, the mean value of the control group for PTTprox was 4.76 ± 0.97 (2.84–7.02). The results of Nishida et al. showed that PTT thicker than 7.0 mm has clinical significance on both US and MRI [16]. In this study, patients with PT had a PTTprox of 6.83 ± 2.36 (3.67–13.59). Nishida et al. described in their work the PTTmid as 4.6 ± 0.9 mm (p = 0.005), while in this study, it was 5.34 ± 1.54 mm, which coincided with the observation that in this tendon section, there was also a significantly larger value for the PT. The trend of a significantly thickened tendon was also seen in the PTTdistal in this study, but this contradicts the observations of Nishida et al. who did not show any significant differences in the distal portion (p = 0.67) [16]. It would be beneficial to measure the patellar tendon thickness divided into PTTprox, PTTmid and PTTdistal by MRI in further studies to generate clear cut-off values.

In their review, Dan et al. examined the relationship between biomechanics of the knee extensor mechanism and its relationship to PT. They claim that so far, no intrinsic morphological risk factors of the patella for patellar tendinopathy can be identified. Biomechanically, the patella should be considered as a lever in the sagittal plane and this should be considered as a basis for identifying morphological intrinsic risk factors [15]. In a subsequent study by Dan et al., based on measurements from lateral radiographs, the patellar tendon in patients with patellar tendinopathy (n = 52) had smaller lever and moment arms than in the control group (n = 53). What is striking in this study is that in the various measurements of patellofemoral instability mentioned above, only the sagittal measurements yielded significant results. The measurements in the axial planes remained without significance. The previously mentioned information supports the thesis of Dan et al. in that the patella should be seen as a lever in the sagittal plane and is also related to the transmission of force to the tibia.

In this study, the risk of an increased signal intensity of PT at symptoms > 6 months was higher than with symptoms < 6 months. It has been described in the literature that MRI shows an increased T2 signal intensity within the patellar tendon in degenerative changes [11]. An increased signal intensity in T1 and T2 at the patellar tendon origin at the lower pole of the patella in this study was observed in 30 of the 38 cases from the patient group. Additionally, there was a significant correlation between PTTprox and the presence of increased signal intensity. The thicker the patellar tendon, the more often there is increased signal intensity, indicating that there are probably also different stages of PT. These could, in addition to the JKC according to Elazina et al., represent a way of classifying the overuse condition, which requires further research. In contrast to JKC, this would be a more objective and reproducible method to assess the severity of PT and to plan a further therapeutic approach.

The results of this study indicate that if conservative treatment fails, a radiological examination by MRI should be performed at the latest after 6 months of persistent symptoms. Furthermore, surgical therapy can be considered at this stage since morphological changes in the tendon are already present by then.

Surgical procedures for PT can generally be performed open or arthroscopically based on the surgeon's preference. Tendon necrosis is mostly identified on MRI centrally within the tendon in the proximal to mid-substance parts, requiring careful surgical debridement. In those cases, the authors favor an open longitudinal surgical approach to ensure complete excision of the affected tissue and secure tendon sheet closure with resorbable sutures. During surgery, special attention should be paid to the resection of a bony "patellar nose" which can occur on the distal patellar pole and which often causes bony edema and impingement to the proximal patellar tendon and is visible on MRI as a bone bruise on the patellar side with visible adjacent inflammatory tendon changes on the T2 weighted images.

As the study highlights the role of patella alta, the possibility of patella lowering tuberosity osteotomy in cases of chronic PT should be further discussed. Besides lowering the patella height, additional elevation of the tuberosity could also increase the PPTA and improve the surgical outcome. However, as the exact role of an elevated PPTA in this context remains unclear, further studies on the clinical implications are necessary.

Therefore, physiotherapy remains the main focus and should aim on altering individual risk factors such as patellar height. Targeted training of the posterior thigh and gluteal muscles, coordination training for the thigh extensor and stretching of the anterior thigh muscles are viable options [24].

Limitations

The study had several limitations. It was a retrospective study and the results were dependent on the quality and quantity of patient data collected. It was a single-center study, which calls into question the generalizability of the results. The case number of 41 patients with PT was small; for example, some patients had to be excluded from the collective due to lack of imaging. Many studies on PT investigated professional athletes with high jump loads, whereas in this study, a wide range of different sports and mainly amateur athletes were represented. In the surgically treated patients, it must be noted that there are often several weeks to months between the radiological examination of the knee and the operation, which means that the extent of the disease may not have been adequately represented in the MRI.

Compared to other studies, the control group is an aspect worth emphasizing, because with an equality of means in demographic data, the two groups were comparable. Various measurements were performed on the patellar tendon, patella, trochlea and patellofemoral joint. This broad view can be cited as a strength of the study, as can the fact that conservatively and surgically treated patients were considered, thus representing a wide range of injury stages.

5. Conclusions

Patients with PT showed a significant increase in the patellar height, tendon thickness and PPTA. Conservative therapy should focus on the early stages of improvement of those individual risk factors. With a persistence of symptoms of over 6 months, MRI seems suitable to detect the further morphologic tendon changes as a thickening or partial necrosis and further identify patients suitable for surgical procedures.

Author Contributions: Conceptualization, M.G. and S.P.; methodology, M.G. and S.P.; software, S.P. and W.H.; validation, W.H., S.P., K.L. and M.G.; formal analysis, W.H., K.L. and M.G.; investigation, S.P. and K.L.; data curation, S.P., K.L. and M.G.; writing—original draft preparation, K.L. and M.G.; writing—review and editing, K.L., S.P., M.W., J.R., S.S., P.S., H.J.B., M.M. and M.G.; visualization, K.L.; supervision, M.G.; project administration, K.L. and S.P. All authors have read and agreed to the published version of the manuscript.

Funding: This research received no external funding.

Institutional Review Board Statement: The Institutional Review Board of Paracelsus Medical University stated that this study was conducted according to the guidelines of the Declaration of Helsinki. They approved this study (protocol code IRB-2022-007) on 31 March 2022.

Informed Consent Statement: Informed consent was obtained from all subjects involved in the study.

Data Availability Statement: Not applicable.

Acknowledgments: The authors would like to thank Lucy Ryan for correcting the manuscript with respect to spelling and grammar.

Conflicts of Interest: The authors declare no conflict of interest.

References

1. Fornaciari, P.; Kabelitz, M.; Fucentese, S.F. Jumper's Knee. *Praxis* **2018**, *107*, 513–519 [CrossRef]
2. Khan, K.M.; Mavulli, N.; Coleman, B.D.; Cook, J.L.; Taunton, J.E. Patellar Tendinopathy: Some Aspects of Basic Science and Clinical Management Tendon Injuries Account for a Substantial Proportion of Overuse Injuries in Sports. *Br. J. Sport. Med.* **1998**, *32*, 346–355. [CrossRef]
3. Lian, Ø.B.; Engebretsen, L.; Bahr, R. Prevalence of Jumper's Knee among Elite Athletes from Different Sports: A Cross-Sectional Study. *Am. J. Sport. Med.* **2005**, *33*, 561–567. [CrossRef]
4. Rudavsky, A.; Cook, J. Physiotherapy Management of Patellar Tendinopathy (Jumper's Knee). *J. Physiother.* **2014**, *60*, 122–129. [CrossRef]
5. Kettunen, J.A.; Kvist, M.; Alanen, E.; Kujala, U.M. Long-Term Prognosis for Jumper's Knee in Male Athletes. A Prospective Follow-up Study. *Am. J. Sport. Med.* **2002**, *30*, 689–692. [CrossRef]
6. Muaidi, Q.I. Rehabilitation of Patellar Tendinopathy. *J. Musculoskelet. Neuronal Interact.* **2020**, *20*, 535–540. [PubMed]
7. Rees, J.D.; Maffulli, N.; Cook, J. Management of Tendinopathy. *Am. J. Sport. Med.* **2009**, *37*, 1855–1867. [CrossRef]
8. Schwartz, A.; Watson, J.N.; Hutchinson, M.R. Patellar Tendinopathy. *Sport. Health* **2015**, *7*, 415–420. [CrossRef]
9. Roels, J.; Martens, M.; Mulier, J.C.; Burssens, A. Patellar Tendinitis (Jumper's Knee). *Am. J. Sport. Med.* **1978**, *6*, 362–368. [CrossRef]
10. Blazina, M.E.; Kerlan, R.K.; Jobe, F.W.; Carter, V.S.; Carlson, G.J. Jumper's Knee. *Orthop. Clin. N. Am.* **1973**, *4*, 665–678. [CrossRef]
11. Rosen, A.B.; Wellsandt, E.; Nicola, M.; Tao, M.A. Clinical Management of Patellar Tendinopathy. *J. Athl. Train.* **2022**, *57*, 621–631. [CrossRef]
12. Christian, R.A.; Rossy, W.H.; Sherman, O.H. Patellar Tendinopathy—Recent Developments toward Treatment. *Bull. Hosp. Jt. Dis.* **2014**, *72*, 217–224.
13. Kujala, U.M.; Osterman, K.; Kvist, M.; Aalto, T.; Friberg, O. Factors Predisposing to Patellar Chondropathy and Patellar Apicitis in Athletes. *Int. Orthop.* **1986**, *10*, 195–200. [CrossRef]
14. Tscholl, P.M.; Biedert, R.M.; Wanivenhaus, F.; Fucentese, S.F. Patellar Tendinopathy with Intratendinous Alteration on MRI May Be Related to Patellofemoral Dysplasia. *Scand. J. Med. Sci. Sport.* **2018**, *28*, 1443–1450. [CrossRef]
15. Dan, M.; Parr, W.; Broe, D.; Cross, M.; Walsh, W.R. Biomechanics of the Knee Extensor Mechanism and Its Relationship to Patella Tendinopathy: A Review. *J. Orthop. Res.* **2018**, *36*, 3105–3112. [CrossRef] [PubMed]
16. Nishida, Y.; Nishino, T.; Tanaka, K.; Onishi, S.; Kanamori, A.; Yamazaki, M. An Objective Measure of Patellar Tendon Thickness Based on Ultrasonography and Mri in University Athletes. *J. Clin. Med.* **2021**, *10*, 4092. [CrossRef]
17. Wolfe, S.; Varacallo, M.; Thomas, J.D.; Carroll, J.J.; Kahwaji, C.I. *Patellar Instability*; NIH: Treasure Island, FL, USA, 2022.
18. Iyengar, K.P.; Kho, J.; Azzopardi, C.A.; Haleem, S.; Ezegbe, F.; Botchu, R. Alteration of Patellar Tendon Morphology in Patellofemoral Instability (Trochlear Dysplasia). *J. Clin. Orthop. Trauma* **2022**, *26*, 101786. [CrossRef]
19. Johnson, D.P.; Wakeley, C.J.; Watt, I. Magnetic Resonance Imaging of Patellar Tendonitis. *J. Bone Jt. Surg. Br.* **1996**, *78*, 452–457. [CrossRef]
20. Verhulst, F.V.; van Sambeeck, J.D.P.; Olthuis, G.S.; van der Ree, J.; Koëter, S. Patellar Height Measurements: Insall-Salvati Ratio Is Most Reliable Method. *Knee Surg. Sport. Traumatol. Arthrosc.* **2020**, *28*, 869–875. [CrossRef] [PubMed]
21. Kim, T.; Kim, J.K.; Lee, H.S.; Kim, D.K. Patella-Patellar Tendon Angle in Relation to the Medial Patellar Plica Syndrome, Chondromalacia Patella, and Infrapatellar Fat Pad Syndrome. *PLoS ONE* **2022**, *17*, e0265331. [CrossRef]
22. Yu, J.S.; Popp, J.E.; Kaeding, C.C.; Lucas, J. Correlation of MR Imaging and Pathologic Findings in Athletes Undergoing Surgery for Chronic Patellar Tendinitis. *AJR Am. J. Roentgenol.* **1995**, *165*, 115–118. [CrossRef] [PubMed]
23. El-Khoury, G.Y.; Wira, R.L.; Berbaum, K.S.; Pope, T.L.J.; Monu, J.U. MR Imaging of Patellar Tendinitis. *Radiology* **1992**, *184*, 849–854. [CrossRef] [PubMed]
24. Van Linschoten, R.; van Middelkoop, M.; Berger, M.Y.; Heintjes, E.M.; Verhaar, J.A.N.; Willemsen, S.P.; Koes, B.W.; Bierma-Zeinstra, S.M. Supervised Exercise Therapy versus Usual Care for Patellofemoral Pain Syndrome: An Open Label Randomised Controlled Trial. *BMJ* **2009**, *339*, b4074. [CrossRef] [PubMed]

Journal of *Personalized Medicine*

MDPI

Article

Quadriceps Tendon Ruptures in Middle-Aged to Older Patients: A Retrospective Study on the Preoperative MRI Injury Patterns and Mid-Term Patient-Reported Outcome Measures

Kim Loose [1,*], Justus Rudolph [1], Martin Schlösser [2], Maximilian Willauschus [1], Johannes Rüther [1], Philipp Schuster [1,3], Hermann Josef Bail [1], Michael Millrose [1,4] and Markus Geßlein [1]

1 Department of Orthopedics and Traumatology, Paracelsus Medical University, Breslauer Straße 201, 90471 Nuremberg, Germany
2 Department of Radiology, Paracelsus Medical University, Breslauer Straße 201, 90471 Nuremberg, Germany
3 RKH Orthopedic Hospital Markgröningen, Center for Sports Orthopedics and Special Joint Surgery, 71706 Markgröningen, Germany
4 Department of Trauma Surgery and Sports Medicine, Garmisch-Partenkirchen Medical Centre, 82467 Garmisch-Partenkirchen, Germany
* Correspondence: kim.loose@klinikum-nuernberg.de; Tel.: +49-911-3982600

Abstract: (1) Quadriceps tendon rupture (QTR) is a rare pathology, usually occurring in elderly patients with comorbidities, requiring surgical therapy. The aim of this study was to analyze rupture patterns and concomitant injuries using preoperative magnetic resonance imaging (MRI) and to evaluate patient-reported outcome measures. (2) In this retrospective cross-sectional study, 113 patients with QTR were screened and rupture patterns/concomitant injuries (n = 33) were analyzed via MRI. Clinical outcome was assessed in 45 patients using the International Knee Documentation (IKDC) and Lysholm score with a mean follow-up of 7.2 (±5.0) years. (3) The evaluation of preoperative MRIs showed multiple ruptures of subtendons in 67% with concomitant knee injuries in 45%. The most common associated pathology detected using MRI was pre-existing tendinosis (31.2%). Surgical refixation demonstrated good results with a mean post-operative IKDC score of 73.1 (±14.1) and mean Lysholm score of 84.2 (±16.1). Patient characteristics and individual radiologic rupture patterns did not significantly affect the clinical outcome of patients. (4) Acute QTRs are complex injuries with common involvement of multiple subtendons. MRI imaging can be useful for achieving an accurate diagnosis as pre-existing tendinosis as well as concomitant injuries are common, and might be useful for providing an individual surgical strategy and improving outcomes.

Keywords: knee surgery; tendon rupture; knee extensor mechanism injuries; magnetic resonance imaging

check for updates

Citation: Loose, K.; Rudolph, J.; Schlösser, M.; Willauschus, M.; Rüther, J.; Schuster, P.; Bail, H.J.; Millrose, M.; Geßlein, M. Quadriceps Tendon Ruptures in Middle-Aged to Older Patients: A Retrospective Study on the Preoperative MRI Injury Patterns and Mid-Term Patient-Reported Outcome Measures. *J. Pers. Med.* **2023**, *13*, 364. https://doi.org/10.3390/jpm13020364

Academic Editor: Jih-Yang Ko

Received: 20 January 2023
Revised: 13 February 2023
Accepted: 15 February 2023
Published: 18 February 2023

1. Introduction

Quadriceps tendon rupture (QTR) is an uncommon injury with an incidence of 1.37 per 100,000 [1]. The mechanism of a rupture can be direct or indirect. The most common cause is a simple fall, but it is also seen because of car accidents or sporting activities. Spontaneous or even bilateral ruptures can be found as well in selected cases. The rupture of a healthy quadriceps tendon is rare, but QTR may sometimes occur in younger patients below the age of 40 years due to forceful eccentric muscle contractions during strenuous activities.

Most ruptures occur in men between 50 and 60 years old with certain comorbidities affecting tendon collagen structure [2]. Tendon injury patterns can vary within a range from partial ruptures of quadriceps' subtendons to complete ruptures of all tendons forming the quadriceps tendon. While the majority of partial ruptures can be treated nonoperatively, complete ruptures usually require surgical therapy to avoid severe functional deficits [3,4].

In addition to older patient age and male gender, known patients' risk factors for QTR include rheumatoid arthritis, systemic lupus erythematosus, diabetes mellitus, long-term corticosteroid therapy, fluoroquinolones treatment, chronic kidney disease, second or tertiary hyperparathyroidism, gouty arthritis and peripheral vascular disease [5]. Pre-existing tendinosis is common in these patients, but its exact development is not fully understood, although it may play an important role regarding surgical refixation [6]. Fatty tendon degeneration and collagen alterations due to chronic inflammation are frequently described [7]. Furthermore, data on concomitant knee injuries are also scarce [2]. As QTR is often linked to falls due to sudden loss of motor control, concomitant injuries are likely to occur. Especially when flexion and rotational movements are combined during trauma, additional intraarticular injuries such as meniscal injuries or rupture of the anterior or posterior crucial ligament may occur. Resulting persistent instability and pain can cause inferior individual outcomes. Due to the demographic development, an increasing incidence in QTR can be expected. Therefore, the improvement of diagnostic accuracy as well as further investigation of different surgical techniques seems mandatory to advance an individual therapeutic approach. In a review by Arnold et al., it was noted that the additional information provided by magnetic resonance imaging (MRI) helps in visualizing the local anatomy and may be of benefit, although the incidence of concomitant injury remains relatively low [8].

The understanding of precise quadriceps tendon anatomy is crucial for successful surgery as surgeons should be aware of the trilaminar structure combining rectus femoris (RF) as a superficial layer, an intermediate layer formed by the vastus medialis (VM) and vastus lateralis (VL) muscles and a deep layer formed by the vastus intermedius (VI) muscle [9,10].

During surgery, the identification of the exact rupture pattern can be demanding and misinterpretation or insufficient debridement of tendinosis within the stump can lead to inadequate fixation resulting in a loss of function or re-rupture. Therefore, knowledge of rupture patterns and tendon quality seems essential for careful pre-surgical planning. Regarding exact rupture patterns in QTR, there have been few studies so far and, besides clinical examination, sonography is still the common method of choice for initial diagnosis [11]. However, the sensitivity and specificity of sonography in the assessment of quadriceps injuries can vary and it is known to be operator-dependent. In contrast, MRI is more standardized and can consistently and accurately visualize a ruptured tendon [8,12,13]. The recent literature has suggested that MRI is the gold-standard test for diagnosing QTR, with reported sensitivity, specificity, and a positive predictive value of 1.0 [8,12]. Falkowski et al. evaluated 53 QTRs via magnetic resonance imaging (MRI) in 2021 and found 183 subtendon ruptures affecting different parts of the quadriceps tendon (RF, VM, VL, VI) mainly localized near the patella [13]. Furthermore, MRI can be useful for describing tendon quality at the rupture site as well as bony spurs at the insertion and, therefore, for planning optimal surgical therapy [13]. A systematic review by Oliva et al. examined the literature on QTR and demonstrated results of chronic and complex QTR. They defined complex ruptures as re-ruptures (9%), chronic (16.5%) or linked to total knee arthroplasty (74.5%). They concluded that macroscopic tendon quality, rather than time since trauma, was critical in the choice of the surgical technique [14]. As MRI sequences have the capability to demonstrate individual tendon quality at the rupture site in each individual patient, it can be assumed that surgical therapy might benefit from this additional information.

Good clinical outcomes after QTR have been reported previously in several studies. Patients returned to their pre-injury activities despite advanced age and regardless of surgical technique used for fixation [15,16]. However, only few studies have focused on a middle-aged to older population [2,16]. As QTR represents a severe injury in the elderly resulting in a loss of mobility, quality of life and associated co-morbidity, further studies are needed in this special study population.

The primary aim of this study was to analyze rupture patterns and concomitant injuries in preoperative MRIs. The secondary aim of the study was to evaluate whether rupture patterns or patient-related characteristics have an impact on patient-reported outcome measures (PROMS) after surgical refixation.

2. Materials and Methods

2.1. Study Population

From a total of 254 patients with acute QTR treated with surgical refixation in the period from January 2000 to August 2022 at a maximum care hospital, 113 patients were included in the study for a retrospective analysis (Figure 1).

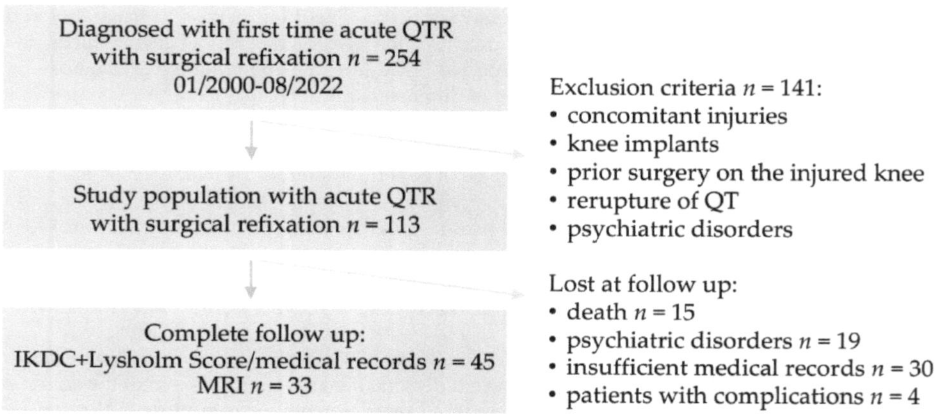

Figure 1. Study flowchart.

Inclusion criteria were a minimum patient age of 40 years and surgical treatment of QTR. Exclusion criteria were concomitant injuries that affected the ipsi- or contralateral lower extremity, psychiatric disorders (including dementia), patients with complications, patients with existing knee implants or prior surgery on the knee, as well as re-rupture of the QT. Of the 113 patients eligible for the study, 15 were dead, 19 patients suffered from psychiatric disorders (including dementia) at the time of evaluation, and 4 patients had complications of treatment. Furthermore, 30 patients had to be excluded due to insufficient medical records.

2.2. Diagnosis and Surgical Treatment

Diagnosis in all patients was performed at the emergency department including history, clinical examination, plain X-rays, and ultrasound or MRI examination. The applied surgical techniques were transosseous tendon refixation using patellar drill-holes or refixation using suture anchors. In all cases, surgery was performed by an experienced orthopedic surgeon using a vertical incision starting at the proximal patella pole and extending it proximally. Tendon rupture patterns were explored and refixation was performed on all identified ruptured quadriceps' subtendons after debridement of tendon stumps and creating a bony rim at the proximal patella pole. Postoperatively, the affected leg was put in an orthosis with the leg fully extended. All patients were mobilized with crutches or a walker on the second day after surgery under instructions from a competent physiotherapist. Patients were instructed to wear the knee straightening brace for 6 weeks. Physiotherapy was allowed after one week with passive ROM 0-0-30 and an increase in flexion for 30 degrees every two weeks. Full knee flexion and active knee extension were allowed after 6 weeks.

2.3. Evaluation of Preoperative MRI

Preoperative MRI was available for evaluation with 33 patients. Images were acquired using 1.5 or 3.0 Tesla scanners. The standard MR protocol consisted of sagittal proton density (PD) fat-saturated and sagittal PD non-fat-saturated images, coronal PD fat-saturated images, axial T2 fat-saturated images and axial T1 non-fat-saturated images.

Images were evaluated retrospectively by an experienced musculoskeletal consultant radiologist on an approved PACS workstation (Ashvins, MedicalCommunications, Heidelberg, Germany). For each quadriceps tendon rupture, the radiologist analyzed the injured subtendons of the quadriceps tendon as follows: RF, VM, VL and VI, as described in previous studies [13], and whether it was partial or complete. Furthermore, the exact location of rupture (myotendinous junction, intratendinous area and near the patella pole) was evaluated. In the case of complete rupture, the retraction distance was determined in millimeters (mm). Other concomitant injuries such as meniscal lesions, cruciate ligament ruptures and collateral ligament injuries, as well as signs of degenerative changes in the collagenous tendon structure such as tendinosis, were also evaluated.

2.4. Patient Related Data and PROMS

Patient-related data such as gender assigned at birth, age at surgery in years, length of hospital stay before and after surgery in days, American Society of Anesthesiologists Physical Status Classification System (ASA PS classification I-VI), body mass index (BMI) and prescribed medication were extracted from hospital medical records.

After hospital discharge, the clinical outcome was assessed by an experienced orthopedic surgeon in the outpatient clinic using the International Knee Documentation Committee (IKDC) and the Lysholm score [17,18], with a minimum follow-up of 1 year.

2.5. Statistical Analysis

All data were obtained and analyzed retrospectively. The statistical analysis was performed using Python (version 3.9.7. with SciPy 1.8.1, Python Software Foundation in Wilmington Delaware, USA). All p-values were statistically significant at p-value < 0.05. Normal distribution was tested using the Shapiro–Wilk test. For normally distributed data, the Pearson correlation coefficient was used. For non-normally distributed and normal-scaled data, the Spearman test was used. The Kruskal–Wallis H-test and Mann–Whitney U test were utilized depending on the group size. Unless otherwise stated, descriptive results are demonstrated as mean $\pm$ standard deviation and range.

The study was approved by the institutional review committee (Paracelsus Medical University at the Nuremberg Hospital, Number IRB-2022-007) following national legal guidelines.

3. Results

Of 113 patients with acute, first-time QTR included in the study, 97 (85.8%) were male and 16 (14.2%) were female. Complete follow-up including IKDC and Lysholm scores was obtained in 45 patients (39.8%) with a mean follow-up of 7.2 ($\pm$5.0) years.

3.1. Demographic Data and Patients' Characteristics

The characteristics of the study population including ASA PS classification, number of regularly prescribed long-term medication, time-to-surgery and time spent in hospital are demonstrated in Table 1.

3.2. Quadriceps Tendon Rupture Patterns

The majority of QTRs were localized near the patella pole (71.0%) at the bony junction between the quadriceps tendon and the patella (without bony avulsion). A small proportion of ruptures was found at the myotendinous junction (12.5%) and only few ruptures were considered to be intratendinous (9.30%). In two cases, ruptures could be found in two different localizations in the same patient. One rupture could not be precisely assigned using MRI. All identified rupture patterns are demonstrated in Figure 2.

Table 1. Demographic data and patients' related characteristics.

	All Patients (n = 113)	Female (n = 16)	Male (n = 97)
Age, y (mean ± SD)	61.2 ± 16.2	67.7 ± 17.0	60.1 ± 15.9
BMI [1] (mean ± SD)	32.5 ± 7.5	32.3 ± 7.7	32.5 ± 7.5
ASA classification [2] (n, %)			
ASA I	19 (16.8)	2 (12.5)	17 (17.5)
ASA II	57 (50.4)	7 (43.7)	50 (51.5)
ASA III	36 (31.8)	7 (43.7)	29 (29.8)
Time-to-surgery [3], d (mean ± SD)	7.6 ± 7.1	7.5 ± 7.8	7.6 ± 7.0
Hospital stay [4], d (mean ± SD)	4.9 ± 4.2	7.4 ± 4.9	4.5 ± 3.9
Number of medications, n (mean ± SD)	3.3 ± 2.9	4.1 ± 2.5	3.2 ± 3.0

[1] BMI: body mass index; [2] ASA: American Society of Anesthesiologists Physical Status Classification System; [3] time: days from rupture to operation; [4] time: days from operation to hospital discharge.

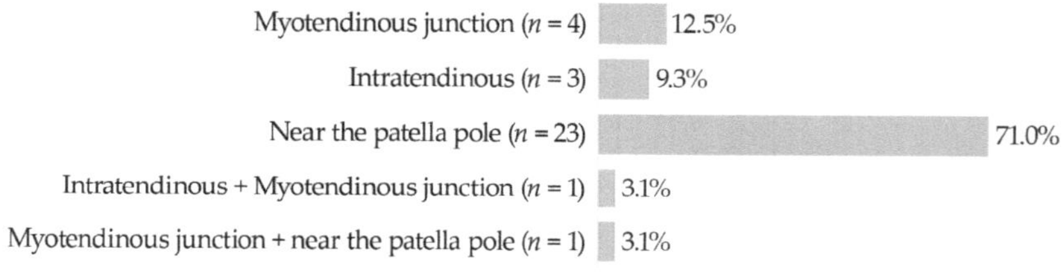

Figure 2. Rupture site based on MRI evaluation.

Altogether, the 33 evaluated MRI images showed 70 ruptures of subtendons (RF, VM, VL and VI). While 62.9% were complete ruptures, 37.1% partial ruptures could be identified (Figure 3). Complete rupture in the rectus femoris (30.0%) was the most common (Figure 4a), followed by complete rupture of the vastus medialis (15.7%). When combining the affected layers of the quadriceps tendon with the location, complete distal rupture of the RF subtendon was most common. Most patients (67%) showed multiple ruptures of subtendons compared with single tendon ruptures (33%).

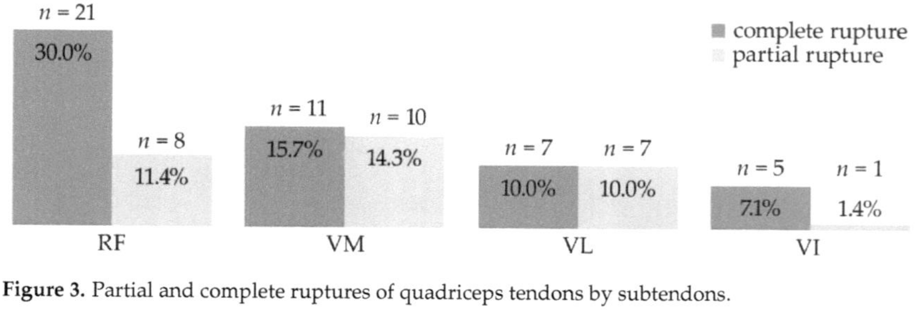

Figure 3. Partial and complete ruptures of quadriceps tendons by subtendons.

For complete QTR, the mean retraction distance was 16.5 mm (range 4.0–36.0 mm). The most common associated pathology with QTR detected via MRI was pre-existing tendinosis (31.2%). This was more frequent in complete ruptures (33.3%) than in partial ruptures (22.2%).

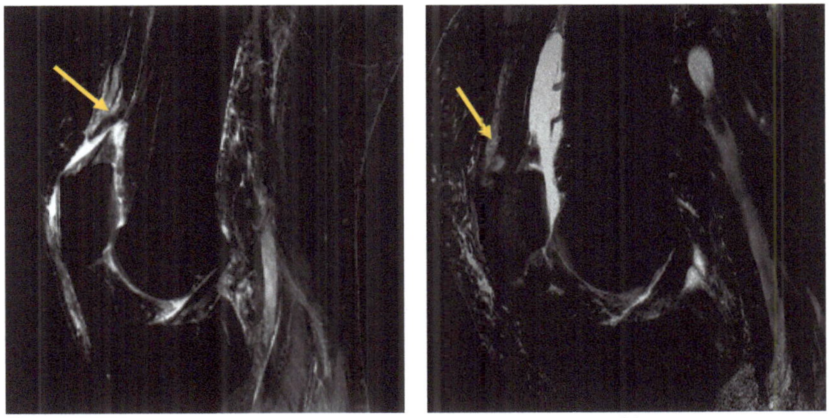

(**a**) (**b**)

Figure 4. (**a**) T2-weighted sagittal MRI demonstrating complete QTR of the rectus femoris. (**b**) T2-weighted sagittal MRI demonstrating partial QTR of the rectus femoris subtendon.

3.3. Concomitant Injuries Found in Patients with Quadriceps Tendon Ruptures

The evaluation of MRI images showed concomitant injuries with QTR in 15 patients (45.0%, Figure 5). Meniscal lesions were identified at the medial meniscus in one patient (3.1%) and at the lateral meniscus in four patients (12.5%). Partial collateral ligament lesions (34.4%) and partial lesions of the anterior and posterior cruciate ligaments (28.1%) were also detected.

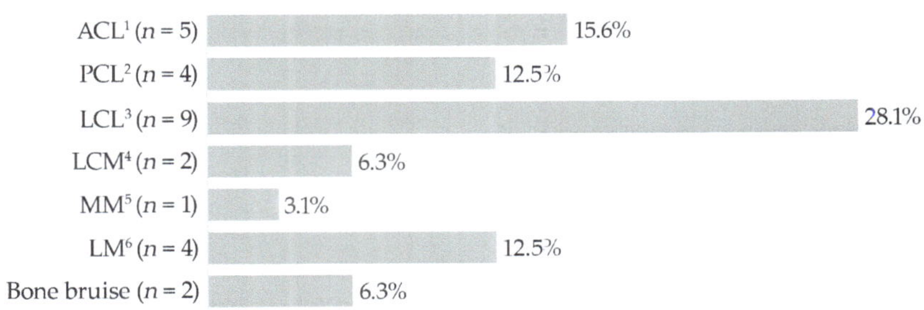

Figure 5. Concomitant injuries related to 33 patients via MRI after acute QTR. [1] Anterior cruciate ligament, [2] posterior cruciate ligament, [3] lateral collateral ligament, [4] medial collateral ligament, [5] medial meniscus and [6] lateral meniscus.

3.4. Evaluation of PROMS after Refixation of Quadriceps Tendon Ruptures

As shown in Table 2, the overall patient-related outcome measured with the IKDC score (73.1 $\pm$ 14.1) and the Lysholm score (mean 84.2 $\pm$ 16.1) was good. The investigated patients' characteristics did not show a significant correlation with clinical outcome (Table 3). Regarding the surgical technique, transosseous refixation (48%), tendon suture (27%) and the combination of transosseous refixation and intratendinous suture (25%) showed no influence on the measured outcome. Furthermore, the evaluation of rupture patterns also yielded no significant effect on PROMS after refixation of QTR with the described techniques. Re-ruptures of the QT after surgical therapy during follow-up occurred in 11 patients (9.7%). None of these patients had preoperative MRI. The percentage of patients who suffered from arterial hypertension as a cardiovascular disease was 72%, and 27% suffered from diabetes mellitus type II and obesity.

Table 2. Follow-up and patient-reported outcome measure scores.

	All Patients (*n* = 45)	Female (*n* = 3)	Male (*n* = 42)
Follow-up in years (mean, range)	7.2, 1.0–17.1	8.4, 5.2–14.3	7.1, 1.0–17.1
IKDC score (mean, range)	73.1, 38.0–87.0	56.3, 38.0–84.0	74.3, 47.0–87.0
Lysholm score (mean, range)	84.2, 39.0–100.0	62.7, 39.0–99.0	85.9, 50.0–100.0
Re-ruptures (*n*)	11	-	11

Table 3. Correlation of patient characteristics and patient-related outcome.

Patients' Characteristics for QTR	IKDC r-Value	IKDC *p*-Value	Lysholm r-Value	Lysholm *p*-Value
Age [1]	−0.262	0.083	−0.274	0.069
BMI [2]	−0.349	0.058	−0.304	0.102
ASA [3]	1.467	0.480	1.339	0.512
Time-to-surgery [1]	0.219	0.168	0.223	0.162
Time-after-surgery [2]	−0.102	0.520	−0.026	0.872
Number of medications [2]	−0.128	0.455	−0.122	0.477

[1] Pearson correlation coefficient; [2] Spearman correlation coefficient; [3] Kruskal–Wallis H-test.

4. Discussion

The main finding of this study is that most patients showed a complex rupture pattern with the involvement of different subtendons via MRI. Concomitant knee injuries were common, as well as pre-existing tendinosis. Patient-related characteristics and individual radiologic rupture patterns did not significantly affect the clinical outcome of patients having surgical treatment for QTR. Re-ruptures were more frequently found in male patients without prior MRI diagnostics.

Most patients in this study showed multiple ruptures of the quadriceps´ subtendons compared to only single-tendon ruptures. When compared to previous studies, Falkowski et al. found 183 subtendon ruptures (RF, VM, VL and VI) via 52 MRIs when two radiologists independently evaluated the ruptures [13]. Our study is in concordance with the finding that the superficial quadriceps tendon layer (RF) is more frequently affected than the middle and deep layers [13]. Furthermore, complete ruptures were more often found than partial ruptures on the evaluated MRI images. One must keep in mind that only surgically treated QTRs were included in the study, while conservative therapy may have been used for partial ruptures.

The most common rupture location found was 1–2 cm above the insertion at the patella [19]. Most QTRs in this study occurred close to the patella, which is also in line with other studies describing 84.8–93.6% of ruptures as being close to the patella [13]. A possible reason for this common site of rupture was described by Petersen et al. in 1999, as there is an approximately 30 $\times$ 15 mm oval vessel-free area about 10 mm above the patella on the quadriceps tendon [19]. In this study, rupture location and retraction distance did not significantly affect the patient-related outcome. Interestingly, the re-rupture rate was found to be 11% in the study population compared to studies including younger patients with reported re-rupture rates of 2% [2]. Oliva et al. also described 9% re-ruptures in a systematic review on complex QTR [14]. This finding may emphasize the need to identify previously described risk factors which can be addressed during surgery such as osteophytes near the patella pole or pre-existing tendinosis. Therefore, proper rim preparation and debridement of the proximal patellar pole and tendon stumps before refixation may be crucial to assure proper blood supply and provide proper conditions for tendon healing. As recent research found that women (61%) are more likely to have tendinopathy compared with men (34%), surgeons should pay special attention to tendon debridement in older female patients [12]. The reported results in this study indicate that MRI diagnostics can be useful for better identifying those individual factors and providing the best possible surgical planning as pre-existing tendinosis was frequently found in the examined, completely ruptured subtendons. For this reason, MRI diagnosis in QTR was considered to be the gold-standard diagnostic examination in the recent literature as the rate of clinical misdiagnosis ranges from 10% to 50% when compared with other radiological procedures such as ultrasonography [8]. Missing or underestimating the degenerative tendon changes during surgery may contribute to re-rupture. Therefore, when tendinosis is present in MRI, an end to end suture alone cannot be recommended in primary repairs in these cases. Nevertheless, the role of tendinosis as risk factor for QTR is not fully understood in older patients. While tendinopathy usually occurs in younger patients due to chronic eccentric tendon overload, there is limited literature that focuses directly on quadriceps tendinopathy mainly caused by microvascular damage due to systemic or local co-morbidities and medication interfering with tendon remodeling [6,20]. Pope et al. reported medical comorbidities such as renal disease and diabetes mellitus as causes of the pathologic degeneration of tendons [21]. There is evidence that the hypovascularity of tendons has been associated with progressive degeneration and subsequent rupture [22]. However, on the other hand, histological examinations conducted on chronic tendinopathy samples found many large blood vessels which were described as an 'angiofibroblastic' response [23].

There is a paucity of data regarding concomitant injuries in QTR. In the study by McKinney et al., 10% of patients investigated with QTR had concomitant intraarticular knee injuries, especially in high-energy trauma [24]. The anterior cruciate ligament, the medial meniscus, and the lateral meniscus were most often involved [24]. The analysis of rupture patterns in this study showed that QTR can be complex and highly variable. Since a high load of energy is needed to rupture the QT, it seems possible that concomitant injuries may occur. As eccentric quadriceps contraction is a typical injury mechanism for QTR, the found partial ACL/PCL injuries in this study may be considered as acute concomitant injuries due to the hyperextension of the knee joint. Concerning the significant number of meniscal lesions in this study, it remains unclear as to whether these were pre-existing degenerative lesions or acute injuries. Nevertheless, it is interesting that meniscal lesions were only found in combination with complete ruptures of the QTR in this study. As concomitant injuries showed no significant effect on clinical outcome after QTR, further studies are needed to evaluate whether the additional treatment of concomitant injuries might improve long-term outcomes.

Regarding patient-related characteristics, no influence on the clinical outcome was found in this study. In accordance with this study, Garner et al. also described a mean age of 61 years for patients with QTR [25]. Obesity has also been reported in the literature as a risk factor for QTR [26]. Demonstrated patients' characteristics in this study indicate worse, even though not statistically significant, clinical outcomes for those with obesity. Advanced age and obesity have already been associated in the literature with an increased risk of rupture due to fatty degenerative changes, loss of collagen and tendon calcification [8]. In these conditions, MRI can provide additional information about tendon quality and muscular atrophy [14]. This detailed information can be helpful for surgeons to choose the best possible strategies for tendon debridement and refixation in an older and overweight population.

Similar to this study, Rao et al. found time-to-surgery did not significantly affect clinical outcome in QTR [27]. Their study included 38 patients and compared the clinical outcome when surgery was performed within 7 days or 14 days [27], as compared to studies demonstrating inferior results after 3 weeks [25]. According to Arnold et al., delayed surgical intervention may lead to regression and atrophy of the quadriceps muscle. Outcomes of surgical repair or reconstruction may not be predictable for patients with chronic QTR and substantial fatty muscle infiltration [8]. Therefore, in cases of chronic tears, MRI can be a valuable tool in the assessment of fatty muscular atrophy and tendon retraction. However, further studies are needed to assess the role of muscular atrophy found via MRI on outcomes in these patients.

In the study by Boudissa et al., the Lysholm score including 50 knees after QTR averaged 93.7 [28]. Rao et al. studied 38 patients under 40 years of age with QTR and came up with a mean Lysholm score of 85.4 [27]. The outcome results for the evaluated middle-aged to older population in this study were similar with a Lysholm score of 84.2 (±16.1). Strikingly, there is a good clinical outcome in several studies after surgical treatment of QTR, although it has been shown to be a complex injury occurring in middle-aged to older patients with considerable risk factors [2]. Concerning the long-term outcome, recent studies have shown that up to 97% of working-age patients were able to return to work without reassignment after QTR. The authors cite direct, accurate diagnosis and surgical therapy with intensive rehabilitation as reasons for the good clinical outcomes [28]. Rao et al. reported a 63% return-to-play rate in 19 athletes under 40 years of age who had competed to varying degrees [27].

There was no difference in IKDC or Lysholm score between the different surgical techniques in this study population. Similar results have been previously described by Plesser et al. in a smaller group of patients comparing transosseous refixation to suture anchors [15].

Limitations

The study has several limitations. Firstly, it is a retrospective study with several important limitations in comparison to a prospective study setting. Investigators depend on the availability and accuracy of the medical records. Therefore, the study population is subject to selection bias because of the excluded cases due to insufficient medical records. The involvement of different healthcare professionals in patient care may also contribute to less accuracy and consistency in the measurement of risk factors and outcome(s) throughout the study than that achieved with a prospective cohort study.

Secondly, it was conducted in a single center without a control group. There was no independently, randomized image review which could have introduced a reporting bias. As the study period of 20 years was quite long, the images were acquired with two different scanners over time which might have contributed to a bias in image interpretation. The number of retrospectively obtained IKDC and Lysholm scores is small compared to the larger total sample, and no preoperative PROMS were available. However, it has to be considered that primary QTR is a rare injury and this study examined a relatively large number of patients compared with previous investigations. No strength or function tests were performed to evaluate an objective clinical outcome. Nevertheless, this study covers a long follow-up period with standardized MRI imaging as well as surgical procedures.

5. Conclusions

Acute QTRs are complex injuries with common involvement of multiple subtendons. MRI imaging can be useful for achieving an accurate diagnosis since pre-existing tendinosis as well as concomitant injuries are common. Despite good clinical outcomes after refixation, these factors should be considered prior to surgery to prevent re-rupture. Preoperative MRI evaluation might also be useful in providing an individual surgical strategy and to improve outcomes.

Author Contributions: Conceptualization, M.G. and K.L.; methodology, M.G. and K.L.; software, K.L. and J.R. (Justus Rudolph); validation, M.G. and K.L.; formal analysis, M.G. and H.J.B.; investigation, J.R. (Justus Rudolph) and K.L.; data curation, J.R. (Justus Rudolph), K.L., M.G. and M.M.; writing—original draft preparation, K.L., M.G. and P.S.; writing—review and editing, K.L., J.R. (Justus Rudolph), M.S., M.W., J.R. (Johannes Rüther), P.S., H.J.B., M.M. and M.G.; visualization, K.L. and J.R. (Justus Rudolph); supervision, M.G.; project administration, K.L. All authors have read and agreed to the published version of the manuscript.

Funding: This research received no external funding.

Institutional Review Board Statement: The Institutional Review Board of Paracelsus Medical University stated that this study was conducted according to the guidelines of the Declaration of Helsinki. They approved this study (protocol code IRB-2022-0C7) on 31 March 2022.

Informed Consent Statement: Informed consent was obtained from all subjects involved in the study.

Data Availability Statement: Not applicable.

Conflicts of Interest: The authors declare no conflict of interest.

References

1. Deopujari, S.; Kiel, J. Knee Extensor Mechanism Injuries. *Trauma. Inj. Knee* **2013**, 77–85. [CrossRef]
2. Ciriello, V.; Gudipati, S.; Tosounidis, T.; Soucacos, P.N.; Giannoudis, P.V. Clinical Outcomes after Repair of Quadriceps Tendon Rupture: A Systematic Review. *Injury* **2012**, *43*, 1931–1938. [CrossRef]
3. Clayton, R.A.E.; Court-Brown, C.M. The Epidemiology of Musculoskeletal Tendinous and Ligamentous Injuries. *Injury* **2008**, *39*, 1338–1344. [CrossRef]
4. Ochman, S.; Langer, M.; Petersen, W.; Meffert, R.H.; Tillmann, B.; Raschke, M.J. Quadrizepssehnenruptur: Diagnostik Und Therapie Einer Seltenen Verletzung. *Unfallchirurg* **2005**, *108*, 436–444. [CrossRef] [PubMed]
5. Grecomoro, G.; Camarda, L.; Martorana, U. Simultaneous Chronic Rupture of Quadriceps Tendon and Contra-Lateral Patellar Tendon in a Patient Affected by Tertiary Hyperparatiroidism. *J. Orthop. Traumatol.* **2008**, *9*, 159–162. [CrossRef]
6. King, D.; Yakubek, G.; Chughtai, M.; Khlopas, A.; Saluan, P.; Mont, M.A.; Genin, J. Quadriceps Tendinopathy: A Review-Part 1: Epidemiology and Diagnosis. *Ann. Transl. Med.* **2019**, *7*, 71. [CrossRef] [PubMed]
7. Ilan, D.I.; Tejwani, N.; Keschner, M.; Leibman, M. Quadriceps Tendon Rupture. *J. Am. Acad. Orthop. Surg.* **2003**, *11*, 192–200. [CrossRef] [PubMed]
8. Arnold, E.P.; Sedgewick, J.A.; Wortman, R.J.; Stamm, M.A.; Mulcahey, M.K. Acute Quadriceps Tendon Rupture: Presentation, Diagnosis, and Management. *JBJS Rev.* **2022**, *10*, e21. [CrossRef]
9. Zeiss, J.; Saddemi, S.; Ebraheim, N. MR Imaging of Quadrizeps Tendon: Normal Layered Configuration and Its Impartance in Cases of Tendon Rupture. *Am. J. Roentgenol.* **1992**, *159*, 1031–1034. [CrossRef]
10. Waligora, A.C.; Johanson, N.A.; Hirsch, B.E. Clinical Anatomy of the Quadriceps Femoris and Extensor Apparatus of the Knee. *Clin. Orthop. Relat. Res.* **2009**, *467*, 3297–3306. [CrossRef]
11. Foley, R.; Fessell, D.; Yablon, C.; Nadig, J.; Brandon, C.; Jacobson, J. Sonography of Traumatic Quadriceps Tendon Tears with Surgical Correlation. *J. Ultrasound. Med.* **2015**, *34*, 805–810. [CrossRef]
12. Elattar, O.; McBeth, Z.; Curry, E.J.; Parisien, R.L.; Galvin, J.W.; Li, X. Management of Chronic Quadriceps Tendon Rupture: A Critical Analysis Review. *JBJS Rev.* **2021**, *9*, e20. [CrossRef] [PubMed]
13. Falkowski, A.L.; Jacobson, J.A.; Hirschmann, M.T.; Kalia, V. MR Imaging of the Quadriceps Femoris Tendon: Distal Tear Characterization and Clinical Significance of Rupture Types. *Eur. Radiol.* **2021**, *31*, 7674–7683. [CrossRef]
14. Oliva, F.; Marsilio, E.; Migliorini, F.; Maffulli, N. Complex Ruptures of the Quadriceps Tendon: A Systematic Review of Surgical Procedures and Outcomes. *J. Orthop. Surg. Res.* **2021**, *16*, 1–19. [CrossRef]
15. Hasenoehrl, T.; Palma, S.; Keilani, M.; Reschl, M.; Vekszler, G.; Ambrozy, C.; Plesser, S.; Hajdu, S.; Crevenna, R.; Widhalm, H.K. Gait Analysis and Body Composition after Treatment of Quadriceps Tendon Ruptures Showed Equal Results Independent of Suture Anchor or Transosseus Repair Technique Used: A Pilot Study. *Disabil. Rehabil.* **2020**, *42*, 3833–3837. [CrossRef] [PubMed]
16. Ellanti, P.; Moriarty, A.; Nagle, M.; McCarthy, T. Outcomes after Quadriceps Tendon Repair in Patients over 80 Years of Age. *Muscles Ligaments Tendons J.* **2016**, *6*, 224–227. [CrossRef] [PubMed]
17. Irrgang, J.J.; Anderson, A.F.; Boland, A.L.; Harner, C.D.; Kurosaka, M.; Neyret, P.; Richmond, J.C.; Shelborne, K.D. Development and Validation of the International Knee Documentation Committee Subjective Knee Form. *Am. J. Sport. Med.* **2001**, *29*, 600–613. [CrossRef] [PubMed]
18. Barber-Westin, S.D.; Noyes, F.R. Chapter 46—Rating of Athletic and Daily Functional Activities after Knee Injuries and Operative Procedures. In *Noyes' Knee Disorders: Surgery, Rehabilitation, Clinical Outcomes*; Noyes, F.R., Barber-Westin, S.D., Eds.; W.B. Saunders: Philadelphia, PA, USA, 2010; pp. e48–e56. ISBN 978-1-4160-5474-0.
19. Petersen, W.; Stein, V.; Tillmann, B. Blutgefäßversorgung Der Quadrizepssehne. *Unfallchirurg* **1999**, *102*, 543–547. [CrossRef]
20. Sendzik, J.; Lode, H.; Stahlmann, R. Quinolone-Induced Arthropathy: An Update Focusing on New Mechanistic and Clinical Data. *Int. J. Antimicrob. Agents* **2009**, *33*, 194–200. [CrossRef]
21. Petersen, W.; Bierke, S.; Häner, M. Quadrizepssehnenruptur. *Orthopade* **2021**, *50*, 959–967. [CrossRef]
22. Frey, C.; Shereff, M.; Greenidge, N. Vascularity of the Posterior Tibial Tendon. *J. Bone Jt. Surg. Am.* **1990**, *72*, 884–888. [CrossRef]
23. Yu, J.S.; Popp, J.E.; Kaeding, C.C.; Lucas, J. Correlation of MR Imaging and Pathologic Findings in Athletes Undergoing Surgery for Chronic Patellar Tendinitis. *AJR Am. J. Roentgenol.* **1995**, *165*, 115–118. [CrossRef]
24. McKinney, B.; Cherney, S.; Penna, J. Intra-Articular Knee Injuries in Patients with Knee Extensor Mechanism Ruptures. *Knee Surg. Sport. Traumatol. Arthrosc.* **2008**, *16*, 633–638. [CrossRef]
25. Garner, M.R.; Gausden, E.; Berkes, M.B.; Nguyen, J.T.; Lorich, D.G. Extensor Mechanism Injuries of the Knee: Demographic Characteristics and Comorbidities from a Review of 726 Patient Records. *JBJS* **2015**, *97*, 1592–1596. [CrossRef]

26. Tandogan, R.N.; Terzi, E.; Gomez-Barrena, E.; Violante, B.; Kayaalp, A. Extensor Mechanism Ruptures. *EFORT Open Rev.* **2022**, *7*, 384–395. [CrossRef] [PubMed]
27. Rao, S.; Johnson, E.E.; D'Amore, T.; Szeto, S.; Otlans, P.; Cohen, S.B. Outcomes After Repair of Quadriceps Tendon Rupture in Patients Aged 40 Years and Younger. *Orthop. J. Sport. Med.* **2022**, *10*, 1–7. [CrossRef] [PubMed]
28. Boudissa, M.; Roudet, A.; Rubens-Duval, B.; Chaussard, C.; Saragaglia, D. Acute Quadriceps Tendon Ruptures: A Series of 50knees with an Average Follow-up of More than 6 Years. *Orthop. Traumatol. Surg. Res.* **2014**, *100*, 217–220. [CrossRef] [PubMed]

Journal of *Personalized Medicine*

MDPI

Article

Finite Element Analysis of Various Osteotomies Used in the Treatment of Developmental Hip Dysplasia in Children

Zsuzsanna Incze-Bartha [1,†], Sandor Incze-Bartha [2,†], Zsuzsánna Simon-Szabó [3], Andrei Marian Feier [4], Vlad Vunvulea [1,*], Alin Ioan Nechifor-Boila [1], Ylenia Pastorello [1] and Lorand Denes [1]

1 Department of Anatomy, "George Emil Palade" University of Medicine, Pharmacy, Science and Technology of Targu Mures, 540139 Targu Mures, Romania; ylenia.pastorello@umfst.ro (Y.P.); lorand.denes@umfst.ro (L.D.)
2 Department of Orthopedics and Traumatology, "Fogolyan Kristof" County Hospital Sfantu Gheorghe, 520064 Covasna, Romania
3 Department of Pathophysiology, "George Emil Palade" University of Medicine, Pharmacy, Science and Technology of Targu Mures, 540139 Targu Mures, Romania
4 Department of Orthopaedics and Traumatology, "George Emil Palade" University of Medicine, Pharmacy, Science and Technology of Targu Mures, 540142 Targu Mures, Romania
* Correspondence: vlad.vunvulea@umfst.ro
† These authors have contributed equally to this work.

Abstract: Late-discovered developmental hip dysplasia deformities often necessitate complex surgical treatments and meticulous preoperative planning. The selection of osteotomies is contingent upon the patient's age and the specific structural deformity of the hip. In our anatomical hip model, derived from the data of a 12-year-old patient, we performed virtual osteotomies that are commonly recommended for such cases. We precisely constructed geometric models for various osteotomies, including the Dega, Pemberton, Tönnis, Ganz, Chiari pelvic, and Pauwels femoral osteotomies. We employed Autodesk Inventor for the finite element analysis of the hip joint and the corrective osteotomies. In comparing one-stage osteotomies, we noted that the Dega and Ganz pelvic osteotomies, especially when combined with the Pauwels femoral osteotomy, yielded the most favorable outcomes. These combinations led to enhanced femoral head coverage and reduced intra-articular pressure. Furthermore, we calculated the femoral head-to-acetabulum volume ratio for both the Dega and Pauwels osteotomies. The encouraging results we obtained advocate for the integration of finite element analysis in virtual osteotomies of the pelvis and femur as a preoperative tool in the management of developmental hip dysplasia.

Keywords: developmental hip dysplasia; children; finite element; geometric model; hip joint; hip osteotomies; femur osteotomies; pediatric hip joint

Citation: Incze-Bartha, Z.; Incze-Bartha, S.; Simon-Szabó, Z.; Feier, A.M.; Vunvulea, V.; Nechifor-Boila, A.I.; Pastorello, Y.; Denes, L. Finite Element Analysis of Various Osteotomies Used in the Treatment of Developmental Hip Dysplasia in Children. *J. Pers. Med.* **2024**, *14*, 189. https://doi.org/10.3390/jpm14020189

Academic Editor: Günther Maderbacher

Received: 27 December 2023
Revised: 21 January 2024
Accepted: 3 February 2024
Published: 8 February 2024

1. Introduction

Pelvic osteotomies, whether combined with open reduction or not, as well as femoral osteotomies, are surgical interventions indicated and utilized for hip joint deformities in pediatric orthopedics. The primary objective of these procedures is to enhance femoral head coverage, augment the contact surface, and promote the growth and development of the joint in a normative direction. The selection of a specific procedure is contingent upon the type and degree of deformity, as well as the presence of arthrosis [1].

In recent years, corrective osteotomies employed in the treatment of developmental hip dysplasia have been classified into two principal categories: reconstructive and salvage osteotomies. Reconstructive osteotomies aim to restore normal hip anatomy and are subdivided into three groups: acetabuloplasty with distal mobilization of the acetabular roof (Pemberton, Albee–Lance, Dega, San Diego), redirectional osteotomies involving the displacement of pelvic fragments (Salter, Le Coeur, Hopf, Steel, Sutherland–Greenfield, Kalamchi, Tönnis–Trousseau, PemberSal), and pelvic osteotomies with the reorientation of the acetabulum (Blavier, Wagner, Eppright, Ninomiya, Hsieh, Ganz–Bernese). Salvage osteotomies, recommended post-bone maturity, aim to postpone arthroplasty and provide enhanced coverage of the acetabular component during arthroplasty (Chiari, Wilson, Staheli) [2–5].

For the correction of coxa valga in hip dysplasia, the most frequently utilized osteotomies are in the intertrochanteric area, specifically the Pauwels varus derotational osteotomies. Depending on the clinical requirements, the femoral osteotomy can be executed at the trochanteric or subtrochanteric level.

Complex surgical procedures that encompass pelvic osteotomies, femoral osteotomies, and open reduction are termed one-stage operations. In the majority of hip dysplasia cases, one-stage surgery is necessitated [6–9].

The indication for various osteotomies is determined by the patient's age and the extent of the deformity. In pediatric patients, the presence of growth plates dictates the application of specific types of pelvic and femoral osteotomies. Amidst the numerous indications and techniques for osteotomies in the correction of hip dysplasia, we have moved beyond the traditionally reported literature recommendations that categorize osteotomies based on dysplasia morphology. Instead, we have endeavored to identify the most efficient method through finite element analysis [8].

2. Materials and Methods

For the finite element analysis of the hip and corrective osteotomies, we utilized CT images from a 12-year-old female patient. This patient, with normal psychomotor development and a body weight of 32 kg, did not undergo screening for congenital hip dysplasia after birth. At 12 years of age, the patient presented with a painless limping gait, a positive Trendelenburg sign on the left, increased lordosis, a positive Galeazzi sign on the left, and a limitation of left hip abduction to 30 degrees. The pelvic radiographic findings included: an interruption of the Shenton–Menard line, an acetabular index of 35 degrees, and a Wieberg angle of 3 degrees, indicating developmental dysplasia of the left acetabulum with a subluxated valgus femur. The recommended treatment was a surgical one-stage operation involving both pelvis and femur osteotomies. Before surgery, the patient underwent a CT scan of the affected hip.

2.1. The Chosen Osteotomies

From the multitude of osteotomies employed in the treatment of developmental hip dysplasia, we have chosen to model the most frequently used ones, specifically focusing on osteotomies at the pelvic and femoral levels.

Prior to the closure of the Y cartilage, when bone tissue remains elastic (ages 2 to 12 years), reformation osteotomies or acetabuloplasties from pelvic reconstructive osteotomies are employed in the treatment of developmental hip dysplasia. These procedures are particularly recommended for acetabular defects in the upper or anterior region. The advantage of acetabuloplasties lies in their ability to restore the normal acetabular angle and cover the femoral head with hyaline cartilage. The most utilized techniques in this context are the Pemberton and Dega osteotomies.

Other types of osteotomies that avoid the growth cartilage but still ensure femoral head coverage are redirectional osteotomies. Redirectional osteotomies are indicated in cases of severe dysplasia where acetabuloplasty is insufficient to achieve desirable outcomes. These osteotomies can be single, double, or triple pelvic procedures. The most used redirectional osteotomies are the Salter and the Tönnis–Trousseau ones.

If the Y cartilage is closed or close to closure, pelvic osteotomies with acetabular reorientation are used to cover the acetabular head. These osteotomies involve releasing the acetabulum along a curved or polygonal osteotomy line and rotating it in the desired direction. Approximately 20 techniques are described in this context. In this group, we find the technically challenging Ganz osteotomy, which is nowadays the most frequently used.

In adult patients with slightly painful hips due to developmental dysplasia, salvage-type pelvic osteotomies can offer periodic pain relief. These salvage osteotomies also create a stable bony surface for the acetabular component of total hip arthroplasty. The described techniques differ from each other. In the Wilson and Staheli technique, a bony shelf is formed above the femoral head. One utilizes the outer acetabular roof cortex to create a lateral support for the femoral head, while the other, under the direct tendon of the rectus femoris muscle, forms an anterolateral acetabular shelf. On the other hand, the Chiari osteotomy mobilizes the whole distal pelvic fragment to cover the femoral head. We have modeled the Chiari one because, among this type of osteotomy, it is the most frequently used.

The proximal femur can also present axial and derotational deformities in developmental hip dysplasia. Usually, the femoral neck and trochanteric region are in valgus and rotated anteriorly. Femoral osteotomies are employed to correct deformities in the proximal part of the femur. The most frequently used technique is the intertrochanteric varus derotation osteotomy of Pauwels (1976). The virtual operations were executed in accordance with the operative techniques delineated in orthopedic surgical treatises [10,11].

2.2. Geometric Model Generation and Editing the Osteotomies

For finite element analysis, the models can either be patient-specific or modeled. In our study, we utilized patient volumetric data obtained from computed tomography (CT) scans conducted preoperatively. The CT data were derived from 28 slices with a resolution of 0.613281 × 0.613281 mm/pixel in the XY plane and a slice thickness of 5 mm, presented in 8-bit format.

The CT slices, stored in DICOM format, were processed using MicroDicom 0.8.8 PACS DICOM and converted to JPEG format. Subsequently, the images were imported into ImageJ (Wayne Rasband, 1997) version 1.54c. The Canny edge detector algorithm was applied to delineate the cortical and trabecular bone structures. The 8-bit images were further simplified into binary form, mapping entry points for surface construction. Using non-uniform rational B-spline (NURBS) curves and a proprietary algorithm, we minimized the distortion by reducing the number of control points.

Our in-house software facilitated the overlay of the established outlines, and with the assistance of Autodesk Inventor, we generated surface areas. The resulting three-dimensional model represented the cortical bone, trabecular bone, cartilage, and the contact nonlinearities of the hip joint as separate entities. In the dysplastic hip model, the articular components retained their original positions imported from the CT scans [12].

In the Pemberton–Zanoli technique (1965), we initiated the pericapsular osteotomy between the anterior superior and inferior iliac spines, subsequently curving it posteriorly towards the Y cartilage. On the three-dimensional model, the fragment was mobilized separately in both the anterior and lateral directions [10,13].

In the Dega (1969) pelvic osteotomy model, the osteotomy incision commenced 1.5 cm above the acetabular rim, tracing along the acetabulum over the Y cartilage. Subsequently, the liberated fragment was repositioned anteriorly and laterally. We present the process of creating the Pemberton, Dega, and Tönnis pelvic osteotomies in Figure 1 [5,7,10].

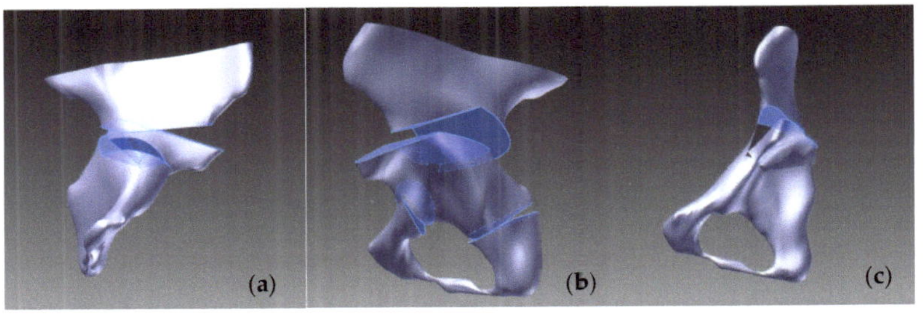

Figure 1. The process of creating (**a**) Pemberton, (**b**) Dega, and (**c**) Tönnis pelvic osteotomies.

From the redirectional oteotomies, we modeled the Tönnis–Trousseau triple osteotomy (1993), wherein the first osteotomy plane commenced 1.5 cm above the acetabular margin, directed towards the sciatic notch. The second plane was in the ischium, positioned 1 cm from the inferior acetabular margin, and the third plane was in the pubis, 1 cm from the inferior acetabular margin, inclined at 65° on the x-axis [14].

Re-orienting pelvic osteotomies involves releasing the acetabulum along a curved or polygonal osteotomy line and rotating it in the desired direction, typically indicated after the closure of the growth cartilage. Approximately 20 techniques are described in this context. The Ganz technique (1988) incorporates five osteotomy planes: the first plane starts under the anterior inferior iliac spine, directed towards the sacroiliac joint, terminating 3 cm short of reaching it. The second plane is situated in the posterior area of the acetabulum, ending before the inferior edge of the pelvis. The third plane is located in the ischium, intersecting with the fourth plane in the inferior extra-acetabular area. The fifth plane is positioned near the transverse cup on the pubis. The mobilized acetabulum is then rotated anterolaterally around the x and z axes [7,15–17].

In the case of the Chiari osteotomy (1950), the osteotomy incision was initiated just above the acetabular margin, taking a slightly oblique direction upwards at a 6° angle towards the sciatic notch. Subsequently, we mobilized the ilium laterally by 1.5 cm [7].

For the femoral Pawels osteotomy, we performed a horizontal incision in the intertrochanteric region, after which the separated fragments were repositioned at 120° on the x-axis and 15° on the z-axis. The process of creating these osteotomies is presented in Figure 2.

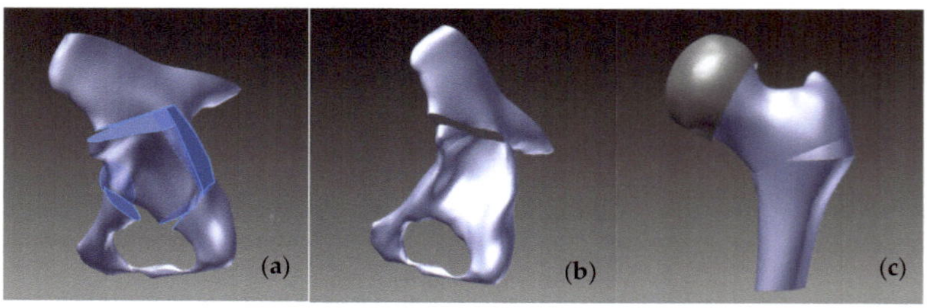

Figure 2. The process of creating (**a**) Ganz, (**b**) Chiari pelvic, and (**c**) Pauwels femoral osteotomies.

In each osteotomy, the osteotomized area was reconstructed using a material identical to bone tissue. Subsequently, we modeled the cartilage in the newly formed acetabulum and femur.

2.3. Computational Modell and Overall Assembly

To develop the mathematical model, we assembled the component parts of the model, which, until this stage of the modeling process, were processed separately as individual geometric models. In constructing the geometric model of the normal and displaced pelvis, we were careful not to alter the spatial position of any component, ensuring that both the pelvis and femur remained in their original positions as imported from the CT scans. However, in the hip osteotomy models for the left dysplastic hip, our objective was to maximize the coverage of the femoral head, necessitating modifications to the position of the femur around all axes. As a baseline reference, we performed an "open reduction" of the hip subluxation.

We also executed a rendering where the pelvis was not modified, but the osteotomized (varized) femoral component was positioned in the non-osteotomized dysplastic acetabulum. For the Pemberton osteotomy, the osteotomized femur was displaced 2 mm in the x-axis, 1.5 mm in the y-axis, and 2 mm in the z-axis. In the case of the Dega osteotomy, the displacement was 2 mm in the x-axis, 2.5 mm in the y-axis, and 2.1 mm in the z-axis. For the Tönnis osteotomy, the adjustments were 2.5 mm in the x-axis, 3 mm in the y-axis, and

2.7 mm in the z-axis. For the Ganz osteotomy, the femur was displaced by 2.4 mm in the x-axis, 2 mm in the y-axis, and 2.3 mm in the z-axis. Lastly, for the Chiari osteotomy, the osteotomized femur was repositioned by 1.5 mm in the x-axis, 1.1 mm in the y-axis, and 0.7 mm in the z-axis.

2.4. Mathematical Model: Mechanical Properties of Tissues. Boundary Conditions, and Hexagonal Mesh

For the mathematical model, we adopted linear elastic and isotropic material properties characteristic of immature bone tissue. In pediatric patients, the properties of the bone material differ from those in adults, with lower values for Young's modulus, density, yield stress, and strain. The values for immature bone material properties were sourced from the literature and considered to be an average representation. For the model validation, we utilized adult bone material properties, which were also adopted from the literature [18–21].

The overall assembly of the one-stage operations for the Pauwels reduction, Pemberton–Pauwels, Dega–Pauwels, Tönnis–Pauwels, Ganz–Pauwels, and Chiari–Pauwels osteotomies is presented in Figure 3.

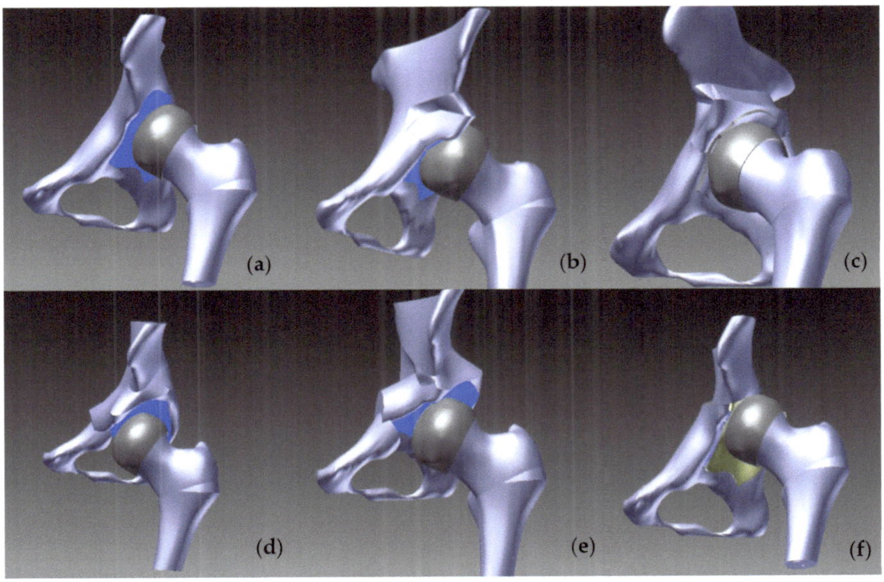

Figure 3. Overall assembly of one-stage operations: (**a**) Pauwels reduction, (**b**) Pemberton–Pauwels, (**c**) Dega–Pauwels, (**d**) Tönnis–Pauwels, (**e**) Ganz–Pauwels, and (**f**) Chiari–Pauwels osteotomies.

In our model, we differentiated between cortical and cancellous bone tissues in terms of material properties, as highlighted in Table 1. The interface between these areas was considered a fixed contact, devoid of mobility. The articular surfaces were modeled as nonlinear frictionless contact surfaces. Our study incorporated boundary conditions simulating orthostasis on a single leg. The sacroiliac joints were immobilized in the transverse plane, permitting pelvic movement only in the sagittal plane. The distal end of the femur was fixed in all directions.

Table 1. Bone material property values and boundary conditions used.

Element	Material	Boundary Conditions	Young's Modulus	Density g/cm^3	Yield Tensile Strength	Yield Compressive Strength
Cortical bone	Solid	Fixed	11.22 GPa	1.05	35 MPa	62 MPa
Cancellous bone	Solid	Fixed	5.33 MPa	0.6	7 MPa	15 MPa
Cartilage	Solid	Frictionless	0.6 MPa	1.1	1 MPa	1.5 MPa

A vertically directed force of 320 N was applied to the constructed models, centered on the middle of the force-transmitting body of the hemipelvis. This force magnitude was calculated based on the weight of the patient under examination. The balancing force of the abductor muscles was applied along the trajectory of the gluteus medius muscle [22–24].

Using Autodesk® Inventor™ 2009 (Autodesk, San Francisco, CA, USA), we generated hexagonal elements on the previously constructed geometric models, as presented in Figure 4. The research conducted by Ramos and Sim demonstrated that the accuracy of finite element models is not significantly dependent on the shape of the finite elements utilized. Hexagonal elements offer greater stability during the analysis process. The quantity of elements incorporated in our constructed model was sufficiently large to ensure a high degree of precision in the pelvic model [25,26].

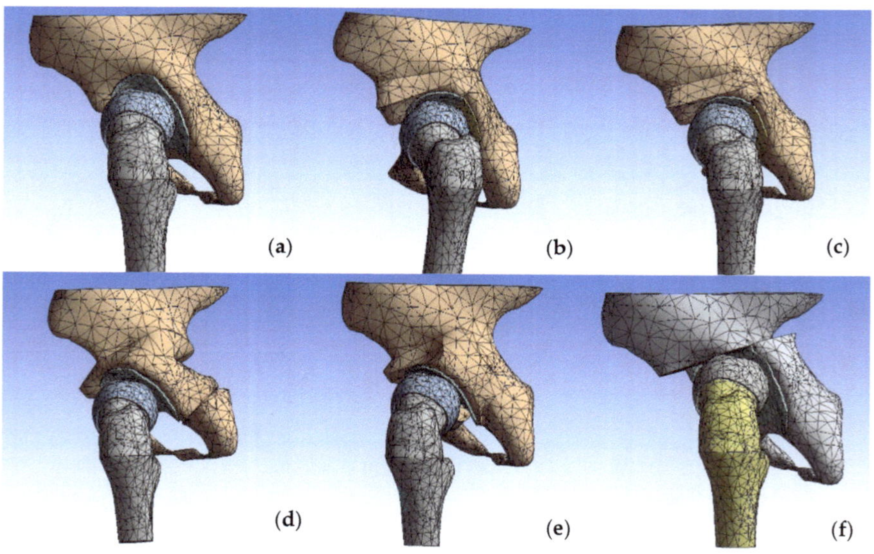

Figure 4. Hexagonal mesh in the: (**a**) left displaced hip with reduced femoral with Pauwels, one-stage operation: (**b**) Pemberton–Pauwels, (**c**) Dega–Pauwels, (**d**) Tönnis–Pauwels, (**e**) Ganz–Pauwels, (**f**) Chiari–Pauwels.

We meticulously generated each component separately—including the pelvic cortical bone, cancellous bone, and cartilage, as well as the femoral cortical bone, cancellous bone, and cartilage—for each specific model. These models encompassed the non-osteotomized pelvis with Pauwels and reduction, as well as one-stage osteotomies such as Pemberton, Dega, Ganz, Tönnis, and Chiari, combined with the varus-derotation Pauwels osteotomy.

3. Results

To comprehend the occurrence of coxarthrosis and the effectiveness of osteotomies, it is imperative to understand the biomechanics of the dysplastic hip, including the distribution of pressure zones and the magnitude of pressure and stress zones in the joint. We utilized the normal hip model as a benchmark for comparison.

The von Mises stress, built based on the invariant of the stress tensor that describes stress in 3D, is useful as it reduces the complex state of stress down to a single number. To obtain a structural analysis of the osteotomized pelvis, we compared the stress distribution across the structure of the normal and operated hips. Our structures were not tested until yield/failure, as we were interested in the response in the stress of the structure to the same load (320 N).

In the normal hip model, the distribution of the von Mises stress demonstrated a pathway originating from the sacroiliac joint, traversing through the femoral head, and extending into Adams' arch on the femoral neck, with values ranging from 8.5 to 2.9 MPa. The highest stress values, reaching 8.5 MPa, were observed at the points of constraint of the hemipelvis with the force-transmitting body, the femoral neck, along the trajectory of Adams' arch, and in the kidney-shaped anterolateral acetabular roof. Corresponding impressions on the femoral head exhibited lower stress values of approximately 2.9 MPa. We present the visual representation of the distribution of the von Mises stress in Figures 5–10.

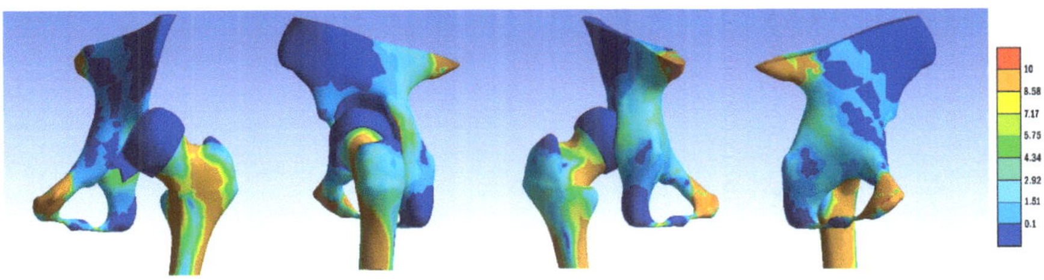

Figure 5. von Mises stress distribution in the untreated pelvis after femoral head reposition.

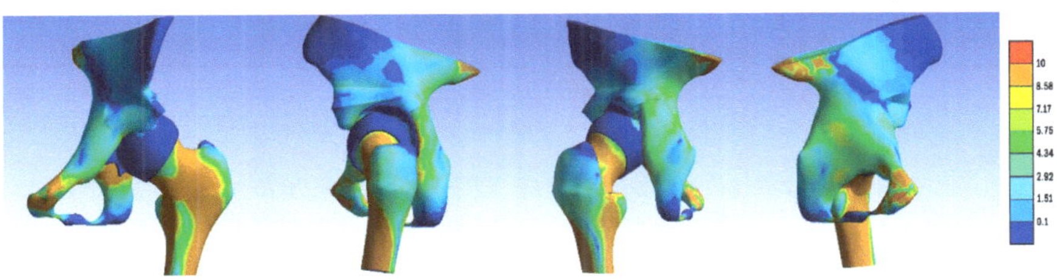

Figure 6. von Mises stress distribution after femoral head reduction, pelvic osteotomy with Pemberton technique, and Pauwels varus-derotational femur osteotomy.

We then compared the von Mises stress zones of other models, including the dysplastic pelvis, open reduction, and one-stage osteotomies (Pemberton, Dega, Ganz, Tönnis, Chiari, Pauwels) with the standard model.

The dysplastic hip model revealed a different stress distribution compared to the normal hip. The stress values in the iliac bone increased in the area of contact with the subluxated femoral head. The stress pathway over Adams' arch was narrower, while the intertrochanteric area exhibited a broader stress path. Concurrently, the stress zone expanded in size, with values approximating 8.5 MPa.

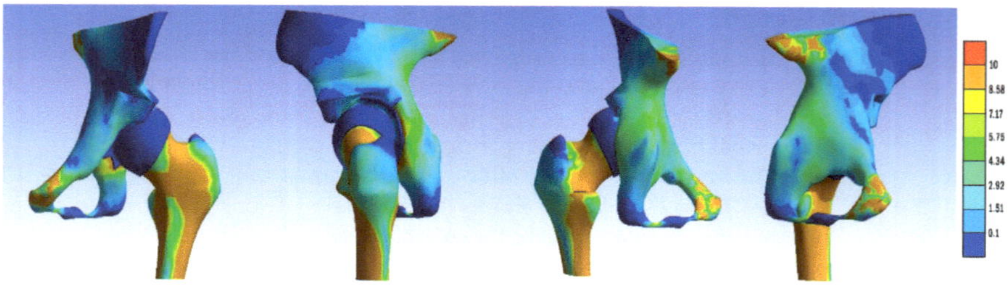

Figure 7. von Mises stress distribution in the treated hip: femoral head reduction, pelvic osteotomy with Dega technique, and Pauwels varus-derotational femur osteotomy.

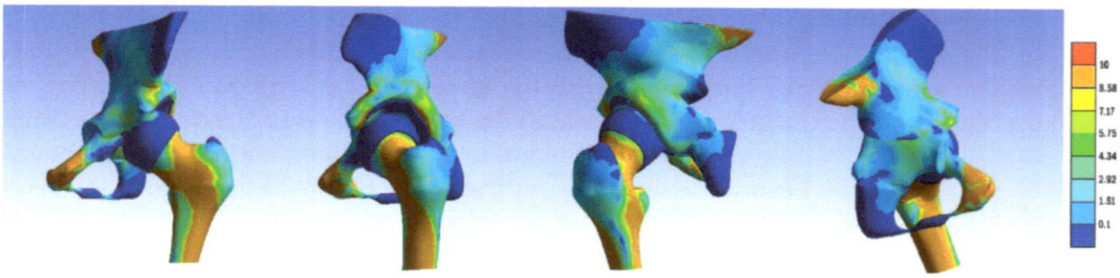

Figure 8. von Mises stress distribution in the treated hip: femoral head reduction, Tönnis pelvic osteotomy, and Pauwels varus-derotational femur osteotomy.

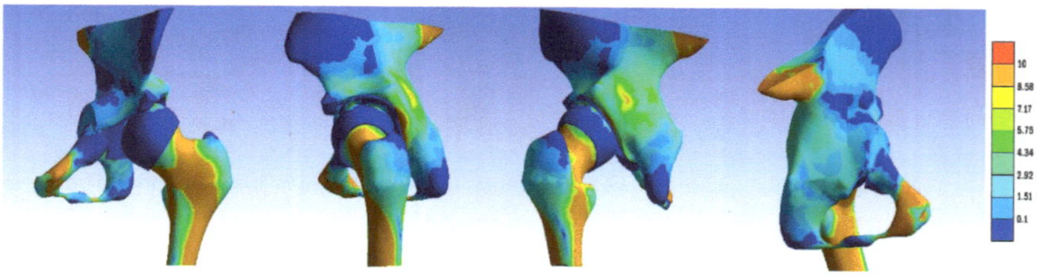

Figure 9. von Mises stress distribution in the treated hip: femoral head reduction, Ganz periacetabular osteotomy, Pauwels varus-derotational femur osteotomy.

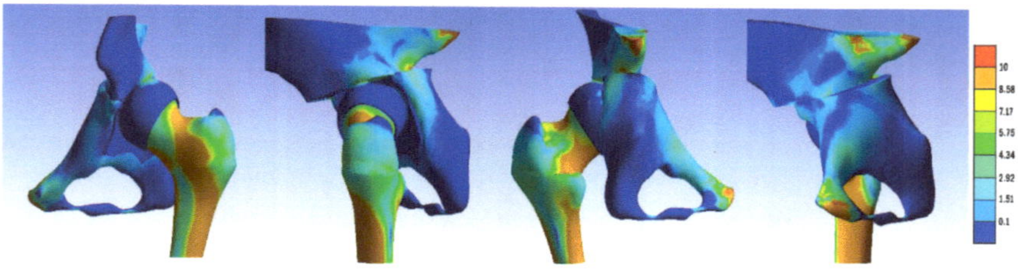

Figure 10. von Mises stress distribution in the treated hip: femoral head reduction, Chiari pelvis osteotomy, Pauwels varus-derotational femur osteotomy.

In all the pelvic osteotomy sites, except for the Chiari technique, there was an increase in the von Mises stress values. The highest stress was recorded at the fourth osteotomy site on the ilium in the Ganz technique (8.5 MPa). In the Dega technique, the pelvic surface of the iliac bone exhibited von Mises stress values of 7.2 MPa. For the Chiari osteotomy, the stress was localized only in the contact area with the femoral head.

The contact surface area was defined as sum of the surface area of the units where the contact pressure was higher than the threshold (0.5 MPa). The calculated maximum pressure was significantly lower than the pressure that would damage the cartilage; however, comparing the results based solely on maximum pressures was not practical.

In the models representing dysplasia and one-stage surgery, we calculated the contact surface area of the acetabulum and compared these values with those observed in the normal hip. During the quantitative comparison of the contact area size, the values that most closely approximated those of the normal hip were observed in the cases of one-stage surgery, specifically the Dega, Tönnis, Pemberton, Ganz, and Tönnis techniques, as visible in Table 2.

Table 2. Contact area for the various types of osteotomies.

	Contact Area in the Acetabulum (mm^2)
Normal hip	471.672
Dysplastic hip	205.272
Open reposition	207.576
Pemberton	355.975
Dega	408.619
Ganz	289.123
Tönnis	365.144
Chiari	188.394

Upon measuring the acetabular volume, we discovered that the femur/acetabulum volume ratio decreased following the Dega varus-reduction femur osteotomy. Conversely, this ratio increased in the cases of the Pemberton and Chiari varus-reduction femur osteotomies, and the facts are represented in Table 3.

Table 3. Volumes of the acetabulum, femoral head, and femur/acetabulum ratio in normal osteotomies simulations.

	Acetabulum (mm^3)	Femur (mm^3)	Femur/Acetabulum Volume Ratio
Normal	12,095.928	28,360.439	0.42
Dysplastic	8393.275	22,325	0.49
Pemberton	10,942.342	22,325	0.49
Dega	10,302.328	22,325	0.46
Chiari	11,428.190	22,325	0.51

To enhance the visualization of intra-articular contact pressures, we conducted separate comparisons of the pressure distributions on the femoral head and within the acetabulum across the constructed models, as highlighted in Figures 11 and 12. Notably, in contrast to the dysplasia and Chiari osteotomy models, the stress values decreased, yet the contact zone did not shift towards a normal configuration.

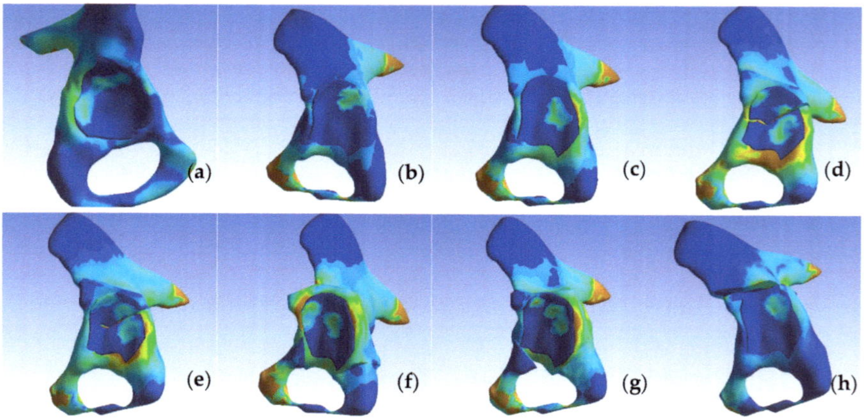

Figure 11. Contact pressure distribution in the acetabulum, non-visualized femoral component: (**a**) normal, (**b**) dysplastic, (**c**) open repositioning, (**d**) repositioning + Pemberton + Pauwels osteotomy, (**e**) repositioning + Dega + Pauwels osteotomy, (**f**) repositioning + Tönnis +Pauwels osteotomy, (**g**) repositioning + Ganz +Pauwels osteotomy, and (**h**) repositioning + Chiari + Pauwels osteotomy.

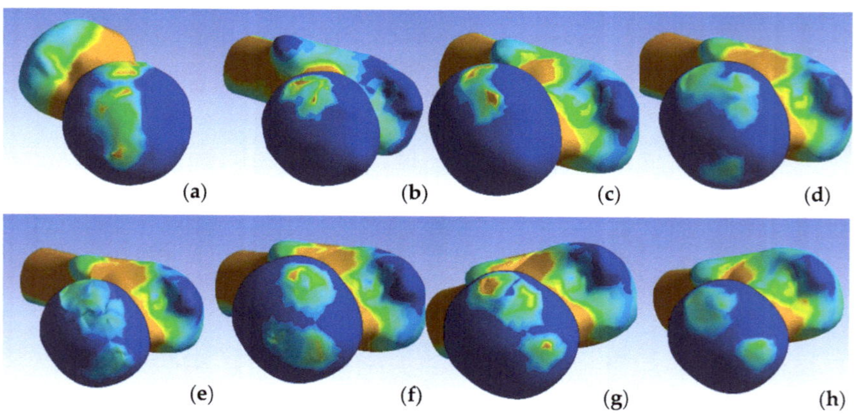

Figure 12. Contact pressure distribution in the femoral head, non-visualized acetabular component: (**a**) normal, (**b**) dysplastic, (**c**) open repositioning, (**d**) repositioning + Pemberton + Pauwels osteotomy, (**e**) repositioning + Dega + Pauwels osteotomy, (**f**) repositioning + Tönnis +Pauwels osteotomy, (**g**) repositioning + Ganz +Pauwels osteotomy, and (**h**) repositioning + Chiari + Pauwels osteotomy.

4. Discussion

Comprehending the biomechanics of the normal, dysplastic, and osteotomized hip is crucial for the effective treatment of developmental hip dysplasia. Given that in vivo measurements are impractical in pediatric patients, the analysis of the hip joint using the finite element method provides an opportunity for immediate understanding. In clinical practice, it can take months or even years to observe the outcomes from the treatment of dysplasia.

Radiographic evaluation has been widely employed to assess the effectiveness of hip osteotomies for developmental hip dysplasia. The modification in radiological parameters serves as the measure of effectiveness. Numerous studies employing this method have been published, providing insights into the changes in hip morphology during treatment. However, these studies predominantly offer information about the hip architecture in only one or two planes [6,27–29].

The advent and increased utilization of computer tomography (CT) have facilitated the three-dimensional visualization of the hip joint. Initially, a significant disadvantage of CT scanning was the requirement for sedation during the examination of children, coupled with the high dose of radiation involved. However, advancements in CT technology have led to enhanced efficiency, resulting in shorter examination times and reduced radiation exposure, thereby rendering this diagnostic tool more practical and safer for use [30].

Over time, this diagnostic tool has become the most widely used paraclinical examination of the congenital dysplastic hip. Tsumura (2005), Klaue (1988), and Tallroth (2006) used CT scans for planning pelvic corrective osteotomies in adults [31–33].

With advancements in software and the computer industry, finite element method (FEM) analysis has become increasingly prevalent in orthopedics. This analytical method provides a comprehensive understanding of hip biomechanics, with the added benefit of enabling repeated analyses at any point. Finite element method analysis is particularly effective in quantifying the biomechanical differences between dysplastic and normal hip morphologies.

One of the pioneering studies utilizing computer analysis in this field was conducted by Michaeli et al. in 1998. This study involved a comparison of hip intra-articular pressures calculated through computer analysis with those measured from cadavers. The study concluded that the computerized method accurately replicates both the magnitude and location of intra-articular pressures, thereby validating its utility in planning corrective osteotomies [34].

The majority of studies employing finite element method (FEM) analysis have concentrated on the normal or dysplastic hip in adults. The most thoroughly investigated therapeutic intervention in this context is the implantation of hip endoprostheses. Research on the dysplastic hip has primarily focused on its biomechanics, the validation of mathematical models, and the biomechanical alterations following Ganz, Dial, and Chiari periacetabular osteotomies [24,35,36].

To our knowledge, there are no studies that have examined the effectiveness of various types of corrective osteotomies in treating developmental hip dysplasia in children using finite element method (FEM) analysis. Additionally, there are relatively few FEM studies focusing on the pediatric hip.

Piszczatowski et al. conducted an evaluation of hip biomechanics in children with spastic paraparesis and adductor spasticity using the finite element method. Their study incorporated the increased muscular tone around the hip. The model they used was based on medical images of a 14-year-old patient with spastic paraparesis. The findings of their analysis indicated a medial dislocation of the load-bearing areas in both the femur and acetabulum, attributed to the traction of the adductor muscles [37].

Park et al. developed models using the finite element method based on medical imaging for hips afflicted with femoral head epiphysiolysis and Legg–Calvé–Perthes disease. They executed a shelf pelvic osteotomy on the model for Legg–Calvé–Perthes disease and a femoral osteotomy on the model with femoral head epiphysiolysis. In their finite element analysis of a normal child's hip, weighing 46.7 kg, the intra-articular pressure was found to be 1.87 MPa. In their models with pelvic osteotomies for Legg–Calvé–Perthes disease, the von Mises stress decreased, and the contact area in the hip joint increased postoperatively. Following the femoral osteotomy for Legg–Calvé–Perthes disease, there was a slight increase in the hip's contact area, but its distribution remained abnormal. The contact pressure decreased, while the contact force increased marginally [38].

Kim et al. developed finite element models for dysplastic hips using CT data from two patients with dysplastic left hips, one weighing 26 kg and the other 33 kg. They created the solid models using 3D software—Doctor 4.0 for contour detection, and Rapidform software for converting to IGES format. The tetrahedral element mesh necessary for the finite element analysis was generated using the Femap version 8.2 software. The articular cartilage was modeled based on MRI data. They simulated shelf surgery on the dysplastic pelvis in these models. Contact analysis was conducted using the ABAQUS/Standard 6.3 program. Postoperatively, in the first model, the maximum contact pressure decreased from 2.3 MPa to 1.6 MPa, and in the second model, it decreased from 6.0 MPa to 4.5 MPa. The contact area in these models increased by 22% and 23%, respectively [39].

In our model validation analysis, we preserved the geometry of the normal hip and assigned adult bone material properties for the mathematical analysis. We applied a force of 800 N, equivalent to the weight of an adult, to the sacroiliac joint, using the orthostatic position on one leg as a reference. Our results were then compared with in vivo measurements conducted by Bergmann et al. in 2010. Bergmann and colleagues measured forces on the femoral head surface at approximately 2.3 times the body weight. In our model, forces equivalent to 2.5 times the body weight were observed on the femoral head. Comparing our normal adult hip model with the one created by Zhao et al., we noted identical force distributions in both joint components of the normal hip [24].

In our finite element analysis, the von Mises stress distribution on the dysplastic hip model revealed a lateralized path compared to that of the normal hip. In the dysplastic hip, the von Mises stress pathway originated from the sacroiliac joint, directed towards the lesser trochanter through the point of contact of the femoral head with the acetabulum, but with increased stress extending towards the greater trochanter. This path became more medial following pelvic and femoral osteotomies, yet it did not fully align with the position observed in the normal hip.

For our analysis, we employed linear elastic material values characteristic of pediatric bone tissue for each joint component. The von Mises stress values increased after each femoral osteotomy due to the distalization of the femoral head. However, the intra-articular contact area showed significant improvement in cases of reconstructive osteotomies. In the Chiari salvage osteotomy, the contact surface with the posterosuperior area of the acetabulum was maintained, and an additional superior contact area with the iliac bone was established. The intra-articular pressure distribution most closely resembling the normal hip was observed following the Dega and Ganz osteotomies, where the contact area increased by 1.4 and 1.9 times, respectively.

Armiger et al. conducted discrete-element analysis of the dysplastic hip in a group of 12 adult patients with a 10-year postoperative follow-up. They created reconstructions from CT images following a Ganz osteotomy. For their discrete analysis, they utilized triangular elements, applied pressure through the center of the femoral head, and based their calculations on the values of a single elastic element. In their findings, the maximum pressure decreased by an average ratio of 1.7, and the contact area increased by an average ratio of 1.4 [35].

In our analysis, the intra-articular pressures were reduced from 5.7 MPa to values between 3.2 and 4.2 MPa for most procedures, with the exception of the Tönnis osteotomy, where they increased to 7.4 MPa over an area of 50.1 mm^2.

Through the measurement of the hip joint volume following the periacetabular osteotomies, we established that these types of osteotomies generally do not decrease the acetabular volume. The only exception was observed in the Dega technique, where the femur/acetabulum volume ratio decreased from 0.49 to 0.46 [40].

5. Conclusions

In our analysis of virtual treatment for developmental hip dysplasia in a 12-year-old patient weighing 32 kg, the most effective method was identified as the Dega osteotomy combined with femoral head reduction and the Pauwels varus-derotational femur osteotomy. It is plausible that in another patient with a different morphology of hip dysplasia, a different one-stage operation could yield postoperative results closer to normal.

The application of finite element method analysis for examining dysplastic and osteotomized hips offers significant advantages for preoperative planning and preparation. The potential integration of this preparation with intraoperative computer navigation could enable the replication of theoretical results during surgery.

Author Contributions: Conceptualization: Z.I.-B., S.I.-B. and L.D.; data curation: Z.S.-S.; formal analysis: A.I.N.-B. and Y.P.; investigation: Z.I.-B. and L.D.; methodology: S.I.-B. and A.M.F.; project administration: Z.I.-B. and S.I.-B.; resources: S.I.-B.; software: A.M.F. and V.V.; supervision: Z.I.-B., S.I.-B. and L.D.; validation: Z.I.-B., S.I.-B., Z.S.-S., A.M.F., V.V., A.I.N.-B., Y.P. and L.D.; visualization: Z.S.-S. and V.V.; writing—original draft: Z.I.-B., S.I.-B. and L.D.; writing—review and editing: Z.I.-B. and V.V. All authors have read and agreed to the published version of the manuscript.

Funding: This research received no external funding.

Institutional Review Board Statement: This study was conducted in accordance with the Declaration of Helsinki and approved by the Ethics Committee of the "Fogolyan Kristof" County Hospital Sfantu Gheorghe, Covasna, with the registration number 33/09.11.2023, approved on 25.09.2023.

Informed Consent Statement: Patient consent was waived due to approval of the Ethics Committee.

Data Availability Statement: Data are contained within the article.

Conflicts of Interest: The authors declare no conflicts of interest.

References

1. Eren, A.; Ugutmen, E. Osteotomies for acetabular dysplasia in adults and adolescents. *Acta Orthop. Traumatol. Turc.* **2007**, *41* (Suppl. 1), 74–79.
2. Millis, M.B.; Murphy, S.B.; Poss, R. Osteotomies about the hip for the prevention and treatment of osteoarthrosis. *Instr. Course Lect.* **1996**, *45*, 209–226. [CrossRef]
3. Carlioz, H. Pelvic osteotomies in children and adolescents. *Acta Orthop. Belg.* **2000**, *66*, 321–328.
4. Pogliacomi, F.; De Filippo, M.; Costantino, C.; Wallensten, R.; Soncini, G. The value of pelvic and femoral osteotomies in hip surgery. *Acta Biomed.* **2007**, *78*, 60–70.
5. Sharma, A.; Joshi, G. Master Techniques in Orthopaedic Surgery. *Med. J. Armed Forces India* **2005**, *61*, 91. [CrossRef]
6. El-Tayeby, H.M. One-stage hip reconstruction in late neglected developmental dysplasia of the hip presenting in children above 8 years of age. *J. Child. Orthop.* **2009**, *3*, 11. [CrossRef] [PubMed]
7. Sharma, A. Campbell's Operative Orthopaedics. *Med. J. Armed Forces India* **2004**, *60*, 91. [CrossRef]
8. Sankar, W.N.; Tang, E.Y.; Moseley, C.F. Predictors of the need for femoral shortening osteotomy during open treatment of developmental dislocation of the hip. *J. Pediatr. Orthop.* **2009**, *29*, 868–871. [CrossRef] [PubMed]
9. Subasi, M.; Arslan, H.; Cebesoy, O.; Buyukbebeci, O.; Kapukaya, A. Outcome in Unilateral or Bilateral DDH Treated with One-stage Combined Procedure. *Clin. Orthop. Relat. Res.* **2008**, *466*, 830. [CrossRef] [PubMed]
10. Jäger, M.; Westhoff, B.; Zilkens, C.; Weimann-Stahlschmidt, K.; Krauspe, R. Indikation und ergebnisse hüftnaher osteotomien bei dysplasie. *Orthopade* **2008**, *37*, 556–576. [CrossRef] [PubMed]
11. Sales de Gauzy, J. Pelvic reorientation osteotomies and acetabuloplasties in children. Surgical technique. *Orthop. Traumatol. Surg. Res.* **2010**, *96*, 793–799. [CrossRef] [PubMed]
12. Incze-Bartha, Z.; Incze-Bartha, S.; Simon Szabó, Z.; Feier, A.M.; Vunvulea, V.; Nechifor-Boilă, I.A.; Pastorello, Y.; Szasz, D.; Dénes, L. Finite Element Analysis of Normal and Dysplastic Hip Joints in Children. *J. Pers. Med.* **2023**, *13*, 1593. [CrossRef] [PubMed]
13. Mitchell, C.S.; Parisi, M.T. Pediatric acetabuloplasty procedures: Radiologic evaluation. *AJR Am. J. Roentgenol.* **1998**, *170*, 49–54. [CrossRef] [PubMed]
14. Farsetti, P.; Caterini, R.; De Maio, F.; Potenza, V.; Efremov, K.; Ippolito, E. Tönnis triple pelvic osteotomy for the management of late residual acetabular dysplasia: Mid-term to long-term follow-up study of 54 patients. *J. Pediatr. Orthop. B* **2019**, *28*, 202–206. [CrossRef] [PubMed]
15. Kralj, M.; Mavčič, B.; Antolič, V.; Iglič, A.; Kralj-Iglič, V. The Bernese periacetabular osteotomy: Clinical, radiographic and mechanical 7–15-year follow-up of 26 hips. *Acta Orthop.* **2005**, *76*, 833–840. [CrossRef] [PubMed]
16. Sucato, D.J. Treatment of late dysplasia with Ganz osteotomy. *Orthop. Clin. N. Am.* **2006**, *37*, 161–171. [CrossRef]
17. McKinley, T.O. The Bernese periacetabular osteotomy for treatment of adult hip dysplasia. *Skeletal Radiol.* **2010**, *39*, 1057–1059. [CrossRef]
18. Rho, J.Y.; Ashman, R.B.; Turner, C.H. Young's modulus of trabecular and cortical bone material: Ultrasonic and microtensile measurements. *J. Biomech.* **1993**, *26*, 111–119. [CrossRef]
19. Shahar, R.; Zaslansky, P.; Barak, M.; Friesem, A.A.; Currey, J.D.; Weiner, S. Anisotropic Poisson's ratio and compression modulus of cortical bone determined by speckle interferometry. *J. Biomech.* **2007**, *40*, 252–264. [CrossRef]
20. Dalstra, M.; Huiskes, R.; van Erning, L. Development and Validation of a Three-Dimensional Finite Element Model of the Pelvic Bone. *J. Biomech. Eng.* **1995**, *117*, 272–278. [CrossRef]
21. Öhman, C.; Baleani, M.; Pani, C.; Taddei, F.; Alberghini, M.; Viceconti, M.; Manfrini, M. Compressive behaviour of child and adult cortical bone. *Bone* **2011**, *49*, 769–776. [CrossRef] [PubMed]
22. Wei, H.-W.; Sun, S.-S.; Jao, S.-H.E.; Yeh, C.-R.; Cheng, C.-K. The influence of mechanical properties of subchondral plate, femoral head and neck on dynamic stress distribution of the articular cartilage. *Med. Eng. Phys.* **2005**, *27*, 295–304. [CrossRef]
23. Phillips, A.T.M.; Pankaj, P.; Howie, C.R.; Usmani, A.S.; Simpson, A.H.R.W. Finite element modelling of the pelvis: Inclusion of muscular and ligamentous boundary conditions. *Med. Eng. Phys.* **2007**, *29*, 739–748. [CrossRef] [PubMed]

24. Zhao, X.; Chosa, E.; Totoribe, K.; Deng, G. Effect of periacetabular osteotomy for acetabular dysplasia clarified by three-dimensional finite element analysis. *J. Orthop. Sci.* **2010**, *15*, 632–640. [CrossRef]
25. Ramos, A.; Simões, J.A. Tetrahedral versus hexahedral finite elements in numerical modelling of the proximal femur. *Med. Eng. Phys.* **2006**, *28*, 916–924. [CrossRef]
26. Kim, J.-E.; Li, Z.; Ito, Y.; Huber, C.D.; Shih, A.M.; Eberhardt, A.W.; Yang, K.H.; King, A.I.; Soni, B.K. Finite element model development of a child pelvis with optimization-based material identification. *J. Biomech.* **2009**, *42*, 2191–2195. [CrossRef] [PubMed]
27. Takakuwa, M.; Matsuno, T.; Gotoh, E.; Ando, M.; Funakoshi, M. Long-term results of triple osteotomy of the pelvis. *J. Orthop. Traumatol.* **2004**, *5*, 15–19. [CrossRef]
28. Armand, M.; Lepistö, J.; Tallroth, K.; Elias, J.; Chao, E. Outcome of periacetabular osteotomy: Joint contact pressure calculation using standing, A.P radiographs, 12 patients followed for average 2 years. *Acta Orthop.* **2005**, *76*, 303–313. [CrossRef]
29. Mellerowicz, H.H.; Matussek, J.; Baum, C. Long-term results of Salter and Chiari hip osteotomies in developmental hip dysplasia. A survey of over 10 years follow-up with a new hip evaluation score. *Arch. Orthop. Trauma Surg.* **1998**, *117*, 222–227. [CrossRef]
30. Fukuda, K.; Saitoh, M.; Tan, A.; Tanaka, S. Helical CT imaging of the hip joints. *Jpn. J. Rheumatol.* **1998**, *8*, 13–25. [CrossRef]
31. Tsumura, H.; Kaku, N.; Ikeda, S.; Torisu, T. A computer simulation of rotational acetabular osteotomy for dysplastic hip joint: Does the optimal transposition of the acetabular fragment exist? *J. Orthop. Sci.* **2005**, *10*, 145–151. [CrossRef] [PubMed]
32. Tallroth, K.; Lepistö, J. Computed tomography measurement of acetabular dimensions: Normal values for correction of dysplasia. *Acta Orthop.* **2006**, *77*, 598–602. [CrossRef]
33. Klaue, K.; Wallin, A.; Ganz, R. CT evaluation of coverage and congruency of the hip prior to osteotomy. *Clin. Orthop. Relat. Res.* **1988**, *232*, 15–25. [CrossRef]
34. Michaeli, D.A.; Murphy, S.B.; Hipp, J.A. Comparison of predicted and measured contact pressures in normal and dysplastic hips. *Med. Eng. Phys.* **1997**, *19*, 180–186. [CrossRef]
35. Armiger, R.S.; Armand, M.; Tallroth, K.; Lepistö, J.; Mears, S.C. Three-dimensional mechanical evaluation of joint contact pressure in 12 periacetabular osteotomy patients with 10-year follow-up. *Acta Orthop.* **2009**, *80*, 155. [CrossRef]
36. Bachtar, F.; Chen, X.; Hisada, T. Finite element contact analysis of the hip joint. *Med. Biol. Eng. Comput.* **2006**, *44*, 643–651. [CrossRef]
37. Piszczatowski, S. Analysis of the stress and strain in hip joint of the children with adductors spasticity due to cerebral palsy. *Acta Bioeng. Biomech.* **2008**, *10*, 51–56.
38. Park, W.M.; Kim, Y.H.; Kim, K.; Oh, T.Y. Non-destructive biomechanical analysis to evaluate surgical planning for hip joint diseases. *Int. J. Precis. Eng. Manuf.* **2009**, *10*, 127–131. [CrossRef]
39. Kim, Y.H.; Park, W.M.; Kim, K.; Yoo, W.J.; Cho, T.J.; Choi, I.H. Planning of Shelf Operation in Dysplastic Hip by CT and MRI Based Finite Element Contact Analysis. *JSME Int. J. Ser. C* **2005**, *48*, 586–591. [CrossRef]
40. Kavanagh, E.C.; Read, P.; Carty, F.; Zoga, A.C.; Parvizi, J.; Morrison, W.B. Three-dimensional magnetic resonance imaging analysis of hip morphology in the assessment of femoral acetabular impingement. *Clin. Radiol.* **2011**, *66*, 742–747. [CrossRef] [PubMed]

Journal of
Personalized
Medicine

MDPI

Communication

Freehand Technique for Pedicle Screw Placement during Surgery for Adolescent Idiopathic Scoliosis Is Associated with Less Ionizing Radiation Compared to Intraoperative Navigation

Peter Obid [1,*,†], Sebastian Zahnreich [2,†], Andreas Frodl [1], Tamim Rahim [3], Thomas Niemeyer [3] and Moritz Mayr [1]

1 Department of Orthopaedics and Traumatology, University Medical Center Freiburg, 79106 Freiburg, Germany; andreas.frodl@uniklinik-freiburg.de (A.F.); moritz.mayr@uniklinik-freiburg.de (M.M.)
2 Department of Radiation Oncology and Radiation Therapy, Mainz University Hospital, 55131 Mainz, Germany; zahnreic@uni-mainz.de
3 Spine and Scoliosis Center, Asklepios Klinik Wiesbaden, 65197 Wiesbaden, Germany; t.rahim@asklepios.com (T.R.); t.niemeyer@asklepios.com (T.N.)
* Correspondence: p.obid@gmx.de
† These authors contributed equally to this work.

Abstract: Purpose: We aim to compare radiation exposure and implant-related complications of the freehand (FH) technique versus intraoperative image-guided navigation (IN) for pedicle screw placement in adolescent idiopathic scoliosis (AIS) and estimate associated lifetime attributable cancer risks. Methods: A retrospective analysis of prospectively collected data from 40 consecutive AIS patients treated with pedicle screw instrumentation using the FH technique was performed. The dose area product (DAP) and effective dose (ED) were calculated. Screw-related complications were analysed, and the age- and gender-specific lifetime attributable cancer risks were estimated. The results were compared to previously published data on IN used during surgery for AIS. Results: There were no implant-related complications in our cohort. Implant density was 86.6%. The mean Cobb angle of the main curve was 75.2° (SD ± 17.7) preoperatively and 27.7° (SD ± 10.8) postoperatively. The mean ED of our cohort and published data for the FH technique was significantly lower compared to published data on the IN technique ($p < 0.001$). The risk for radiogenic cancer for our FH technique AIS cohort was 0.0014% for male patients and 0.0029% for female patients. Corresponding risks for IN were significantly higher ($p < 0.001$), ranging from 0.0071 to 0.124% and from 0.0144 to 0.253% for male and female patients, respectively. Conclusion: The routine use of intraoperative navigation in AIS surgery does not necessarily reduce implant-related complications but may increase radiation exposure to the patient.

Keywords: intraoperative navigation; freehand technique; pedicle screws; adolescent idiopathic scoliosis; radiation exposure; cancer risk; complications; safety

check for updates

Citation: Obid, P.; Zahnreich, S.; Frodl, A.; Rahim, T.; Niemeyer, T.; Mayr, M. Freehand Technique for Pedicle Screw Placement during Surgery for Adolescent Idiopathic Scoliosis Is Associated with Less Ionizing Radiation Compared to Intraoperative Navigation. *J. Pers. Med.* **2024**, *14*, 142. https://doi.org/10.3390/jpm14020142

Academic Editor: Weikuan Gu

Received: 23 November 2023
Revised: 19 January 2024
Accepted: 23 January 2024
Published: 27 January 2024

1. Introduction

First introduced by Suk et al. in 1995, the use of thoracic pedicle screws for posterior correction and fusion (PSF) has become the gold standard for the surgical treatment of adolescent idiopathic scoliosis (AIS) [1]. Segmental pedicle screw constructs allow the application of high correction forces and result in a good long-term outcome [2–5]. The accurate placement of pedicle screws is paramount since potential risks of misplaced pedicle screws include neurological or vascular complications [6,7]. The freehand (FH) technique for the implantation of pedicle screws has been proven to be effective and safe for the correction of scoliosis [8,9]. Since the introduction of navigation techniques in spine surgery, the implantation of pedicle screws using image-guided navigation (IN) has become more and more popular. An increased accuracy of pedicle screw placement using IN techniques has been described in several studies [10,11]. The use of IN techniques has also been introduced into scoliosis surgery [12,13]. However, all IN techniques have been shown to be associated with an increased radiation dose for the patient [14–16]. Any additional radiation dose administered during IN can have serious consequences for the patient and increase the risk of developing late sequelae, with radiation-induced malignancies being one of the most detrimental outcomes [17]. In particular, radiation exposure at an early age immensely increases the risk of radiogenic late consequences compared to exposures during adulthood. AIS patients are already exposed to additional higher levels of radiation due to repeated radiological examinations throughout their course of treatment [18–21].

Therefore, the risk–benefit assessment should be clearly in favor of a significantly higher clinical benefit for IN compared to the FH technique in AIS surgery. However, to date, no studies have compared implant-related complications and radiogenic cancer risk between the FH technique and IN during surgery in AIS patients. This study aims to provide a further understanding of the surgical and radiobiological implications of the use of IN during surgery for AIS compared to the FH technique.

2. Materials and Methods

Radiation exposure and radiogenic cancer risk: We retrospectively reviewed prospectively collected data of 40 consecutive patients (2 males and 38 females) at a single center that underwent PSF for AIS using the FH technique. Due to concerns about radiation exposure, we did not create a control group.

Patients with an etiology other than idiopathic scoliosis, older than 18 years, and a follow-up of less than two years were excluded. Basic demographic data, radiation exposure, and implant-related complications were analyzed. The patients' intraoperative radiation exposure was obtained as dose area product (DAP), representing the absorbed dose multiplied by the exposed area (Gy $\times$ cm^2). For a conservative estimation of the radiogenic cancer risk, the DAP was converted into an effective dose (ED) (mSv) according to models based on the ICRP (International Commission on Radiological Protection) publication 103 and UNSCEAR 2006 report for anterior–posterior and lateral thoracic spine projections using a conversion factor of 0.24 mSv $\times$ Gy^{-1} $\times$ cm^{-2} [21,22]. For the calculation of the ED from a dose length product (DLP), a conversion factor for lumbar spine examinations of 0.169 mSv $\times$ mGy^{-1} $\times$ cm^{-1} was applied according to Lee et al. [23]. The total lifetime cancer risk was determined according to Wall et al. using specific risk coefficients for cancer incidence per unit ED (% per Sv) as a function of age at exposure and sex [22]. For a comprehensive comparison of the radiation exposure and associated radiogenic cancer risk between the FH technique and IN for the treatment of AIS, we reviewed and included all the relevant literature, providing necessary radiation dose information from which we calculated the estimated lifetime attributable cancer risk. For this purpose, conversion factors for anterior–posterior and lateral thoracic spine examinations were applied for the age group of 10–19 years (male: 6.4, female: 13) or the age group of 20–29 years (male: 5.2, female: 11). We performed a comprehensive comparison of the determined ED and radiogenic cancer risk from our patient cohort with previously published studies on the use of the FH technique or IN (see Table 1). A literature search was conducted on Pubmed. The search was based on "(idiopathic scoliosis) AND (navigation)" (All Fields). Other keywords utilized were "scoliosis", "intraoperative navigation", "complications", "radiation exposure", "effective dose", and "cancer risk".

The purpose of the present study is to compare radiation exposure and associated cancer risk of IN and FH techniques for the surgical treatment of AIS. However, since most studies were found to embrace a very heterogeneous patient collective, we decided to include studies dealing with etiologies other than idiopathic scoliosis. Non-comparative studies that were otherwise deemed relevant were also included.

Freehand technique: The patient was in prone position on a radiolucent operating table (Maquet Getinge, Rastatt, Germany). After a standard posterior midline approach, entry points for pedicle screws were identified according to anatomical landmarks. Different pedicle probes were used to access the pedicles. All pedicles were tapped prior to the insertion of the pedicle screws, and the bony margins of the screw channels were checked for integrity. After the insertion of the first pedicle screw, the level was checked using fluoroscopy. The remaining pedicle screws were placed, and after placement of all pedicle screws, the implant position was checked using fluoroscopy. In the case of severe malrotation or dysplastic pedicles, additional fluoroscopies might have been necessary. After rods were inserted and deformity correction was performed, a final fluoroscopy was used to check the achieved deformity correction. During fluoroscopy, the surgical staff positioned themselves behind a mobile X-ray protective shield (WD306, Mavig GmbH, Munich, Germany). A case example is shown in Figures 1 and 2.

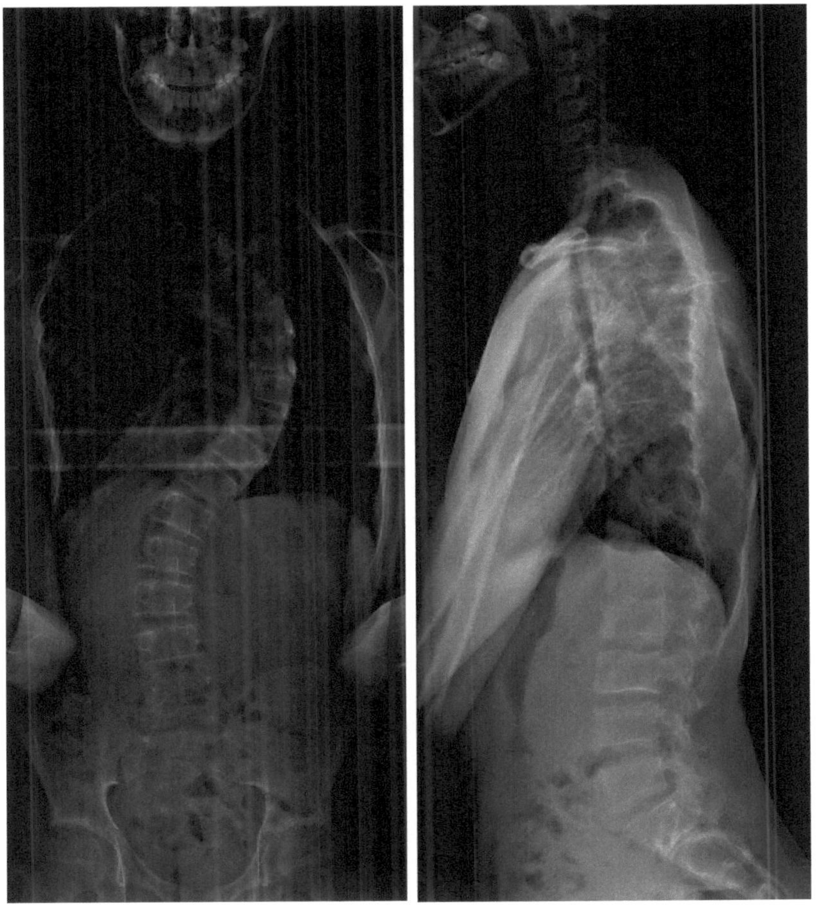

Figure 1. Preoperative biplanar radiograph of a 15-year-old patient with 62° thoracic scoliosis (Lenke I).

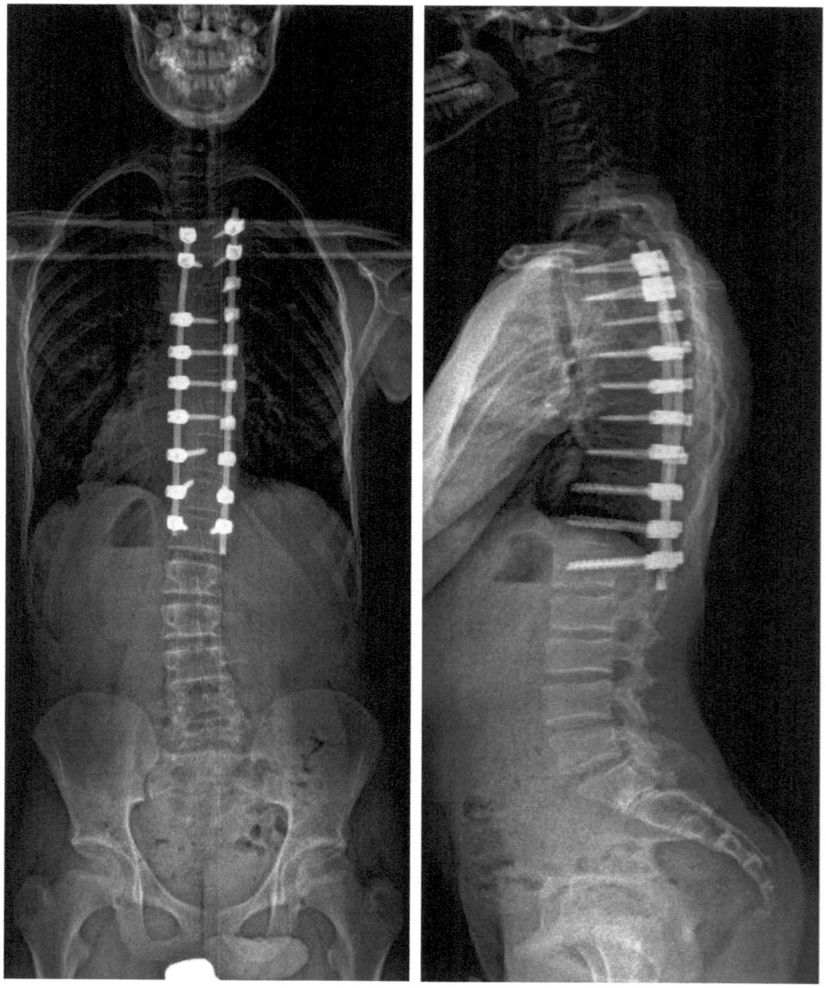

Figure 2. Biplanar radiograph of the same patient after posterior instrumented fusion using freehand technique. The intraoperative DAP was 638.2 mGy/cm^2.

3. Results

Own data on basic demographics and radiation exposure: The mean age at surgery in our cohort was 15.2 years (SD ± 1.3, range: 13–18). The mean BMI was 21.6 kg/m^2 (SD ± 4.8). The mean follow-up was 27.4 months (range: 24–33). A total of 626 pedicle screws were implanted, corresponding to an implant density of 86.6%. In three cases, hooks were used at the upper instrumented vertebra. On average, 8.8 segments (SD ± 2.3) were fused. The mean estimated blood loss was 420 cc (SD ± 292.8) corresponding to 11.4% of estimated blood volume, and the mean surgery time was 227 min (SD ± 58.1). There were no implant-related complications. The mean Cobb angle of the main curve was 75.2° (SD ± 17.7) preoperatively and 27.7° (SD ± 10.8) postoperatively.

Table 1 provides an overview of basic demographic data, radiation exposure, and estimated lifetime attributable cancer risks of previously published studies in comparison to our cohort.

Table 1. FH, freehand; IN, intraoperative navigation. Values shown in italics for other studies were calculated for this study. ** $p < 0.01$; *** $p < 0.001$.

Reference (Publication Date)	Method	No. of Patients	Mean/Median Age (Years)	Mean BMI (kg/m^2)	Mean DAP (mGy*cm^2)	Mean DLP (mGy*cm)	Mean ED (mSv)	Estimated Lifetime Attributable Cancer Risk (%)
Dabaghi-Richerand et al. (2016) [24]	FH	44	pediatric	n/a	n/a	n/a	0.34 (±0.36)	male: 0.0022 female: 0.0044
	IN	37		n/a	n/a	n/a	1.48 (±1.66) **	male: 0.0095 female: 0.019
O'Donnell et al. (2014) [25]	FH	43 (total)	14.2 (10–19)	n/a	n/a	n/a	0.189 (±0.167)	male: 0.0012 female: 0.0025
	IN			n/a	n/a	n/a	7.29–9.72 (navigation)	male: 0.0467–0.0622 female: 0.0948–0.126
							14.58–19.44 (navigation + confirmation)	male: 0.0933–0.124 female: 0.190–0.253
Su et al. *(2017) [26]*	FH	14	13 (11–18)	n/a	n/a	n/a	0.27 (±0.2)	male: 0.0017 female: 0.0035
	IN	14	14 (12–18)	n/a	n/a	n/a	1.11 (±0.3) ***	male: 0.0071 female: 0.0144
Urbanski et al. (2018) [27]	FH	22	24 (12–48)	n/a	n/a	391 (±53)	6.61 (±0.90)	male: 0.0344 female: 0.0727
	IN	27	20 (11–45)	n/a	n/a	1071 (±447) ***	18.1 (±7.6) ***	male: 0.0941 female: 0.199
Berlin et al. *(2020) [28]*	FH	73	21.0 (±9.7)	21.5 (± 4.3)	n/a	n/a	0.17 (±0.1)	male: 0.0009 female: 0.0019
Kapoor et al. *(2021) [29]*	IN	4	14.3 (±1.3)	n/a	n/a	107 (±44)	2.25 (±0.8)	male: 0.0144 female: 0.0293
Own data (2021)	FH	40	15.2 (±1.3)	20.7 (±4.8)	936 (±527.8)	n/a	0.225 (±0.13)	male: 0.0014 female: 0.0029

Radiation exposure and radiogenic cancer risk: In our patient cohort, the mean DAP during surgery for AIS using the FH technique was 936.2 mGy*cm^2 (SD ± 527.8), corresponding to an average ED of 0.225 mSv (SD ± 0.13). Table 1 provides an overview of available studies, comparing radiation dosage between surgeries for AIS with the FH technique or IN. Our average ED was within the range reported by studies using the FH technique, which showed an average ED of 1.52 mSv (SD ± 2.85). The high standard deviation is based on a strongly increased ED of 6.61 mSv from the study by Urbanski et al. [24], which was based on DLPs from 3D O-arm scans. For the studies by Dabaghi-Richerand et al. [24], Su et al. [26], O'Donnell et al. [25], and Berlin et al. [28], the mean ED is only 0.242 mSv (SD ± 0.078) and is thus more comparable to our data.

Concerning the radiation burden between surgeries for AIS with the FH technique or IN, the data show a highly significant ($p < 0.001$) increment for the mean DLP in one study (Urbanski et al.) and the mean ED available in three studies (Dabaghi-Richerand et al., Su et al., and Urbanski et al.) [24,26,27]. On average, the ED for IN was 3.7 (SD ± 0.87) times higher than for the FH technique. A deviating and up to a 51-fold increase in ED using cone-beam computed tomography IN (7.29–9.72 mSv) compared to the fluoroscopically assisted FH technique (0.189, SD ± 0.167 mSv) was reported by O'Donnell et al. [25].

The estimated lifetime attributable cancer risk was calculated from the available radiation dose data for our cohort and all the relevant literature (Table 1). According to our data, the mean ED of 0.225 mSv of the FH technique resulted in a radiogenic cancer risk of 0.0014% for males and 0.0029% for females.

These estimates are within the range of calculated data from literature values of radiation doses with an average lifetime attributable cancer risk of 0.0009–0.0344% for male patients and 0.0019–0.0727% for female patients. The respective risk projections for IN based on the relevant literature data showed significantly elevated lifetime attributable cancer risks of 0.0071–0.124% ($p = 0.023$) for male patients and 0.0144–0.253% ($p = 0.031$) for female patients, as shown in Table 1.

4. Discussion

The use of IN for the surgical treatment of AIS may improve the accuracy of pedicle screw placement compared to the FH technique, but it also increases radiation exposure of the patient and thus the risk for adverse late effects, including radiation-associated malignancies. Attempts have been made to reduce radiation exposure during IN by optimizing the imaging protocol, but the ED could not be reduced to a level comparable to the FH technique [30]. Our statement refers to the comparison of the IN versus the FH technique, while Su et al. compared the reduction in the effective dose by three different protocols for IN. Although the pediatric protocol of the IN measure by Su et al. [26] has reduced the effective dose to 0.65 mSv and thus to 1/4 of the average annual background dose of 2.4 mSv, this is still considerably higher compared to the vast majority of studies using the FH technique (see Table 1).

The question arises whether an improved—but clinically irrelevant—accuracy of pedicle screws using IN justifies a higher radiation burden for AIS patients [31,32]. In a recent meta-analysis, Chan et al. concluded that there is "moderate evidence of decreased breaches with CT-based IN compared with FH methods. Complication rates remain unknown due to the low complication rates from small sample sizes" [32]. Another meta-analysis by Baldwin et al. could show that pedicle screw placement is more accurate using IN compared to the FH technique. They also analyzed four studies that compared radiation exposure and found that "three out of four studies showed a significant difference in radiation exposure during CT navigation which favored freehand" technique. The fourth study showed "no difference in radiation in patients with smaller curves but showed significantly lower radiation using CT navigation when the starting curve was 74.0 degrees or higher". However, ED just tended to be higher in IN compared to FH without reaching significance [33]. A clinically relevant reduction in complications using IN compared to the FH technique could not be identified.

We could identify only one study that showed a higher return to the operating room for the FH technique as compared to IN [34]. However, only 49% of the patients in this study had AIS; the remainder had neuromuscular scoliosis, tumor, congenital scoliosis, or other diagnoses. There is a paucity of literature directly comparing the FH technique and IN for AIS in this regard. The previously mentioned meta-analysis by Chan et al. also found that "High-risk breach rates do not have a significant difference between FH technique and image guidance methods" [32]. So far, the only studies available are those that compared radiation doses (as the DAP, DLP, or ED) for AIS patients between the FH technique or IN. A direct comparison of radiogenic cancer risk estimates as a function of age and gender has not been conducted yet for this particular patient population.

Therefore, we estimated the attributable lifetime cancer risk for our and all available studies of interest with the appropriate age- and gender-related parameters. The attributable lifetime risk for radiation-induced cancer increases with a decrease in age at exposure and can vary by a factor of more than 9 between pediatric or adolescent and elderly patients. Adolescents have the highest risks compared to children, adults, or seniors, although there is a gender-dependent difference [18,35]. For radiation exposure to the chest region, the risk of radiogenic cancer is significantly increased for women compared to men, and this is mainly due to the high risk of breast cancer [22].

The ED that we calculated for the study by Urbanski et al. [27] based on the DLP and the conversion factor for lumbar spine examinations provided by Lee et al. [23] resulted in an elevated ED of 6.61 mSv, which is within the range of the ED of a chest CT. Kapoor et al.

used a comparable conversion factor to determine the ED from the DLP [35]. The mean ED of IN techniques from the literature spanned a wide range from 1.11 to 18.1 mSV and was significantly elevated compared to the mean ED of the FH technique.

The highest value of 18.1 mSv was also obtained for the study by Urbanski et al. [27], which was consistent with the ED from the study by O'Donnell et al. [25] with a maximum of 19.44 mSv for IN, including an intraoperative confirmation scan for implant positioning, and it was again consistent with the conversion factor of Kapoor et al. [29].

The gender- and age-specific risks for radiogenic cancer after PSF for AIS determined in this work differed significantly between the FH technique and IN. The corresponding average lifetime attributable cancer risk was 0.0070 (SD ± 0.013) for male patients and 0.0199 (SD ± 0.029) for female patients using the FH technique and 0.056 (SD ± 0.045, $p = 0.023$) for male patients and 0.116 (SD ± 0.092) for female patients using IN.

So far, only Perisinakis et al. aimed to determine the radiogenic risk for fatal cancer after scoliosis surgery using an anthropomorphic phantom [36]. However, the authors did not use the FH technique or IN but a fluoroscopically assisted technique for pedicle screw placement, which resulted in an average ED of 2.92 mSv for a mean of 4.8 pedicle screws. Because of the different techniques used for pedicle screw instrumentation, the study is not comparable. In our cohort, using the FH technique, the mean ED was 0.225 mSv (SD ± 0.13) for the whole surgery. The high cumulative number of radiological exposures during the treatment course of AIS further increases the risk for radiation-associated late sequelae. O'Donnell et al. demonstrated a doubling of the ED and the associated radiogenic cancer risk caused by a confirmation scan after PSF using IN [25]. Furthermore, to provide sufficient navigation data for the whole surgery, more than one scan might be necessary. In their study on pedicle perforation rates using cone-beam navigation, Oba et al. needed an average of 2.0 (SD ± 0.5, range: 1–3) scans to obtain the necessary navigation data [13]. Since the current model for estimating radiation-associated cancer risk is based on a linear dose–response relationship with no threshold, each additional scan doubles the risk of developing cancer.

The frequent diagnostic exposures of AIS patients to ionizing radiation are associated with an elevated incidence of various tumor entities, such as breast and endometrial cancer [16,19,20,37]. Law et al. evaluated the cumulative ED and the related cancer risk for scoliosis patients undergoing repetitive full spine radiography during their diagnosis and follow-up [35]. For patients exposed annually at an age between 5 and 30 years using digital radiography systems for full spine posteroanterior radiography, the authors calculated a cumulative effective dose of 15 mSv and a corresponding cumulative lifetime attributable cancer risk of 0.1–0.2%, with a significantly higher risk for women. In a Danish cohort of 215 consecutive AIS patients with a mean long-term follow-up of 24.5 years, the overall cancer rate was as high as 4.3%, with a relative risk of 4.8 for developing cancer compared to the normal age-matched Danish population [16].

We did not find any implant-related complications in our cohort. So far, one study has shown a reduction in pedicle screw misplacement using IN during surgery for AIS compared to the FH technique [33]. However, blood loss and the duration of surgery did not change significantly when using IN. Therefore, the use of IN would not have provided a clinical benefit for this group of AIS patients, but the use of IN increases radiation exposure and potentially the lifetime radiogenic cancer risk compared to the FH technique. This should be considered when using IN during surgery for AIS.

Limitations: Since a clinical benefit for using IN in AIS surgeries has not been shown yet, we did not include a control group in which IN has been used due to concerns about excessive radiation exposure. Therefore, we used previously published data on radiation exposure from IN to calculate and compare lifetime attributable cancer risk. Also, our mean follow-up of 27.4 months is relatively short.

5. Conclusions

IN technologies are of great benefit to the patient (and the surgeon). They may help reduce blood loss and tissue trauma in minimally invasive surgeries and help reduce the risk of relevant implant malposition in complex anatomic conditions (e.g., severe deformities or revision cases). However, for a "standard" or "straight-forward" AIS case, the use of intraoperative image-guided navigation should be considered carefully since its application significantly increases the lifetime attributable cancer risk, especially considering that AIS patients are already exposed to an increased radiation dose preoperatively and during follow-up. The use of IN may be limited to scenarios or anatomical areas that pose the most difficulty in pedicle screw placement, such as cases of congenital deformities, revision surgeries, or severely dysplastic pedicles.

Author Contributions: Conceptualization, P.O. and S.Z.; methodology, P.O., S.Z., A.F. and T.R.; validation, P.O., S.Z., A.F., T.R., T.N. and M.M.; formal analysis, A.F., T.R. and M.M.; writing—original draft preparation, P.O., S.Z. and A.F.; writing—review and editing, T.R., T.N. and M.M.; visualization, S.Z.; supervision, P.O., S.Z. and M.M. All authors have read and agreed to the published version of the manuscript.

Funding: This research received no external funding.

Institutional Review Board Statement: The ethical approval is waived due to its' a retrospective study.

Informed Consent Statement: Informed consent was obtained from all subjects involved in the study. All subjects gave written informed consent to be included in the department's registry. This includes the analysis and publication of clinical and radiological data.

Data Availability Statement: The data presented in this study are available on request from the corresponding author. The data are not publicly available due to data protection.

Conflicts of Interest: The authors declare no conflict of interest.

References

1. Suk, S.L.; Lee, C.K.; Kim, W.J.; Chung, Y.J.; Park, Y.B. Segmental pedicle screw fixation in the treatment of thoracic idiopathic scoliosis. *Spine* **1995**, *20*, 1399–1405. [CrossRef] [PubMed]
2. Aghdasi, B.; Bachmann, K.R.; Clark, D.; Koldenhoven, R.; Sultan, M.; George, J.; Singla, A.; Abel, M.F. Patient-reported Outcomes Following Surgical Intervention for Adolescent Idiopathic Scoliosis: A Systematic Review and Meta-Analysis. *Clin. Spine Surg.* **2020**, *33*, 24–34. [CrossRef] [PubMed]
3. Louer, C., Jr.; Yaszay, B.; Cross, M.; Bartley, C.E.; Bastrom, T.P.; Shah, S.A.; Lonner, B.; Cahill, P.J.; Samdani, A.; Upasani, V.V.; et al. Ten-year outcomes of selective fusions for adolescent idiopathic scoliosis. *J. Bone Jt. Surg—Am.* **2019**, *101*, 761–770. [CrossRef] [PubMed]
4. Newton, P.O.; Ohashi, M.; Bastrom, T.P.; Bartley, C.E.; Yaszay, B.; Marks, M.C.; Betz, R.; Lenke, L.G.; Clements, D. Prospective 10-year follow-up assessment of spinal fusions for thoracic AIS: Radiographic and clinical outcomes. *Spine Deform.* **2020**, *8*, 57–66. [CrossRef] [PubMed]
5. Lonstein, J.E. Selective Thoracic Fusion for Adolescent Idiopathic Scoliosis: Long-Term Radiographic and Functional Outcomes. *Spine Deform.* **2018**, *6*, 669–675. [CrossRef]
6. Mac-Thiong, J.M.; Parent, S.; Poitras, B.; Joncas, J.; Hubert, L. Neurological outcome and management of pedicle screws misplaced totally within the spinal canal. *Spine* **2013**, *38*, 229–237. [CrossRef] [PubMed]
7. Di Silvestre, M.; Parisini, P.; Lolli, F.; Bakaloudis, G. Complications of thoracic pedicle screws in scoliosis treatment. *Spine* **2007**, *32*, 1655–1661. [CrossRef] [PubMed]
8. Kim, Y.J.; Lenke, L.G.; Bridwell, K.H.; Cho, Y.S.; Riew, K.D. Free Hand Pedicle Screw Placement in the Thoracic Spine: Is it Safe. *Spine* **2004**, *29*, 333–342. [CrossRef]
9. Lehman, R.A., Jr.; Lenke, L.G.; Keeler, K.A.; Kim, Y.J.; Cheh, G. Computed tomography evaluation of pedicle screws placed in the pediatric deformed spine over an 8-year period. *Spine* **2007**, *32*, 2679–2684. [CrossRef]
10. Gelalis, I.D.; Paschos, N.K.; Pakos, E.E.; Politis, A.N.; Arnaoutoglou, C.M.; Karageorgos, A.C.; Ploumis, A.; Xenakis, T.A. Accuracy of pedicle screw placement: A systematic review of prospective in vivo studies comparing free hand, fluoroscopy guidance and navigation techniques. *Eur. Spine J.* **2012**, *21*, 247–255. [CrossRef]
11. Sun, J.; Wu, D.; Wang, Q.; Wei, Y.; Yuan, F. Pedicle Screw Insertion: Is O-Arm–Based Navigation Superior to the Conventional Freehand Technique? A Systematic Review and Meta-Analysis. *World Neurosurg.* **2020**, *144*, e87–e99. [CrossRef] [PubMed]

12. Uehara, M.; Takahashi, J.; Ikegami, S.; Kuraishi, S.; Shimizu, M.; Futatsugi, T.; Oba, H.; Kato, H. Are pedicle screw perforation rates influenced by distance from the reference frame in multilevel registration using a computed tomography-based navigation system in the setting of scoliosis? *Spine J.* **2017**, *17*, 499–504. [CrossRef] [PubMed]
13. Oba, H.; Ebata, S.; Takahashi, J.; Koyama, K.; Uehara, M.; Kato, H.; Haro, H.; Ohba, T. Pedicle Perforation While Inserting Screws Using O-arm Navigation During Surgery for Adolescent Idiopathic Scoliosis. *Spine* **2018**, *43*, 1463–1468. [CrossRef] [PubMed]
14. Li, C.; Li, H.; Su, J.; Wang, Z.; Li, D.; Tian, Y.; Yuan, S.; Wang, L.; Liu, X. Comparison of the Accuracy of Pedicle Screw Placement Using a Fluoroscopy-Assisted Free-Hand Technique with Robotic-Assisted Navigation Using an O-Arm or 3D C-Arm in Scoliosis Surgery. *Global Spine J.* **2022**, 21925682221143076. [CrossRef] [PubMed]
15. Nelson, E.M.; Monazzam, S.M.; Kim, K.D.; Seibert, J.A.; Klineberg, E.O. Intraoperative fluoroscopy, portable X-ray, and CT: Patient and operating room personnel radiation exposure in spinal surgery. *Spine J.* **2014**, *14*, 2985–2991. [CrossRef] [PubMed]
16. Simony, A.; Hansen, E.J.; Christensen, S.B.; Carreon, L.Y.; Andersen, M.O. Incidence of cancer in adolescent idiopathic scoliosis patients treated 25 years previously. *Eur. Spine J.* **2016**, *25*, 3366–3370. [CrossRef] [PubMed]
17. Mathews, J.D.; Forsythe, A.V.; Brady, Z.; Butler, M.W.; Goergen, S.K.; Byrnes, G.B.; Giles, G.G.; Wallace, A.B.; Anderson, P.R.; Guiver, T.A.; et al. Cancer risk in 680 000 people exposed to computed tomography scans in childhood or adolescence: Data linkage study of 11 million Australians. *BMJ* **2013**, *346*, f2360. [CrossRef] [PubMed]
18. Zahnreich, S.; Schmidberger, H. Childhood Cancer: Occurrence, Treatment and Risk of Second Primary Malignancies. *Cancers* **2021**, *13*, 2607. [CrossRef]
19. Hoffman, D.A.; Lonstein, J.E.; Morin, M.M.; Visscher, W.; Harris, B.S., III; Boice, J.D., Jr. Breast cancer in women with scoliosis exposed to multiple diagnostic x rays. *J. Natl. Cancer Inst.* **1989**, *81*, 1307–1312. [CrossRef]
20. Doody, M.M.; Lonstein, J.E.; Stovall, M.; Hacker, D.G.; Luckyanov, N.; Land, C.E.; US Scoliosis Cohort Study Collaborators. Breast cancer mortality after diagnostic radiography: Findings from the U.S. scoliosis cohort study. *Spine* **2000**, *25*, 2052–2063. [CrossRef]
21. Balonov, M.I.; Shrimpton, P.C. Effective dose and risks from medical x-ray procedures. *Ann. ICRP* **2012**, *41*, 129–141. [CrossRef] [PubMed]
22. Wall, B.F.; Haylock, R.; Jansen, J.T.M.; Hillier, M.C.; Hart, D.; Shrimpton, P.C. *Radiation Risks from Medical X-ray Examinations as a Function of the Age and Sex of the Patient*; Health Protection Agency: Didcot, UK, 2011.
23. Lee, S.K.; Kim, J.S.; Yoon, S.W.; Kim, J.M. Development of CT Effective Dose Conversion Factors from Clinical CT Examinations in the Republic of Korea. *Diagnostics* **2020**, *10*, 727. [CrossRef] [PubMed]
24. Dabaghi Richerand, A.; Christodoulou, E.; Li, Y.; Caird, M.S.; Jong, N.; Farley, F.A. Comparison of effective dose of radiation during pedicle screw placement using intraoperative computed tomography navigation versus fluoroscopy in children with spinal deformities. *J. Pediatr. Orthop.* **2016**, *36*, 530–533. [CrossRef] [PubMed]
25. O'Donnell, C.; Maertens, A.; Bompadre, V.; Wagner, T.A.; Krengel, W., III. Comparative radiation exposure using standard fluoroscopy versus cone-beam computed tomography for posterior instrumented fusion in adolescent idiopathic scoliosis. *Spine* **2014**, *39*, E850–E855. [CrossRef] [PubMed]
26. Su, A.W.; McIntosh, A.L.; Schueler, B.A.; Milbrandt, T.A.; Winkler, J.A.; Stans, A.A.; Larson, A.N. How Does Patient Radiation Exposure Compare with Low-dose O-arm Versus Fluoroscopy for Pedicle Screw Placement in Idiopathic Scoliosis? *J. Pediatr. Orthop.* **2017**, *37*, 171–177. [CrossRef] [PubMed]
27. Urbanski, W.; Jurasz, W.; Wolanczyk, M.; Kulej, M.; Morasiewicz, P.; Dragan, S.L.; Zaluski, R.; Miekisiak, G.; Dragan, S.F. Increased radiation but no benefits in pedicle screw accuracy with navigation versus a freehand technique in scoliosis surgery. *Clin. Orthop. Relat. Res.* **2018**, *476*, 1020–1027. [CrossRef] [PubMed]
28. Berlin, C.; Quante, M.; Thomsen, B.; Koeszegvary, M.; Platz, U.; Ivanits, D.; Halm, H. Intraoperative radiation exposure to patients in idiopathic scoliosis surgery with freehand insertion technique of pedicle screws and comparison to navigation techniques. *Eur. Spine J.* **2020**, *29*, 2036–2045. [CrossRef]
29. Kapoor, S.; O'Dowd, K.; Hilis, A.; Quraishi, N. The Nottingham radiation protocol for O-arm navigation in paediatric deformity patients: A feasibility study. *Eur. Spine J.* **2021**, *30*, 1920–1927. [CrossRef]
30. Su, A.W.; Luo, T.D.; McIntosh, A.L.; Schueler, B.A.; Winkler, J.A.; Stans, A.A.; Larson, A.N. Switching to a Pediatric Dose O-Arm Protocol in Spine Surgery Significantly Reduced Patient Radiation Exposure. *J. Pediatr. Orthop.* **2016**, *36*, 621–626. [CrossRef]
31. Chan, A.; Parent, E.; Narvacan, K.; San, C.; Lou, E. Intraoperative image guidance compared with free-hand methods in adolescent idiopathic scoliosis posterior spinal surgery: A systematic review on screw-related complications and breach rates. *Spine J.* **2017**, *17*, 1215–1229. [CrossRef]
32. Chan, A.; Parent, E.; Wong, J.; Narvacan, K.; San, C.; Lou, E. Does image guidance decrease pedicle screw—Related complications in surgical treatment of adolescent idiopathic scoliosis: A systematic review update and meta—Analysis. *Eur. Spine J.* **2020**, *29*, 694–716. [CrossRef] [PubMed]
33. Baldwin, K.D.; Kadiyala, M.; Talwar, D.; Sankar, W.N.; Flynn, J.J.M.; Anari, J.B. Does intraoperative CT navigation increase the accuracy of pedicle screw placement in pediatric spinal deformity surgery? A systematic review and meta-analysis. *Spine Deform.* **2021**, *10*, 19–29. [CrossRef] [PubMed]
34. Baky, F.J.; Milbrandt, T.; Echternacht, S.; Stans, A.A.; Shaughnessy, W.J.; Larson, A.N. Intraoperative Computed Tomography-Guided Navigation for Pediatric Spine Patients Reduced Return to Operating Room for Screw Malposition Compared With Freehand/Fluoroscopic Techniques. *Spine Deform.* **2019**, *7*, 577–581. [CrossRef] [PubMed]

35. Law, M.; Ma, W.K.; Lau, D.; Chan, E.; Yip, L.; Lam, W. Cumulative radiation exposure and associated cancer risk estimates for scoliosis patients: Impact of repetitive full spine radiography. *Eur. J. Radiol.* **2016**, *85*, 625–628. [CrossRef] [PubMed]
36. Perisinakis, K.; Theocharopoulos, N.; Damilakis, J.; Katonis, P.; Papadokostakis, G.; Hadjipavlou, A.; Gourtsoyiannis, N. Estimation of patient dose and associated radiogenic risks from fluoroscopically guided pedicle screw insertion. *Spine* **2004**, *29*, 1555–1560. [CrossRef]
37. Ronckers, C.M.; Land, C.E.; Miller, J.S.; Stovall, M.; Lonstein, J.E.; Doody, M.M. Cancer mortality among women frequently exposed to radiographic examinations for spinal disorders. *Radiat. Res.* **2010**, *174*, 83–90. [CrossRef]

Journal of
Personalized Medicine

MDPI

Article

Bony Regrowth and New Spur Formation as Possible Causes of Failed Distal Clavicle Excision—Mid-Term Outcomes after Revision Surgery in a Matched-Pair Analysis

Roman C. Ostermann [1,2,3,*], Martin Eigenschink [3,4], Philipp R. Heuberer [1,3], Paul Siegert [5], Willi Muellbacher [2], Lisa Anderl [6], Beate Schrott [7], Brenda Laky [3,7], Leo Pauzenberger [2,3] and Werner Anderl [1,3]

1 Orthopedic Department, Vienna International Shoulder and Joint Clinic, 1030 Vienna, Austria
2 2nd Orthopedic Department, Hospital of the Sacred Heart of Jesus, Baumgasse 20A, 1030 Vienna, Austria
3 Austrian Research Group for Regenerative and Orthopedic Medicine (AURROM), 1030 Vienna, Austria; ordination@ortho-eigenschink.at (M.E.)
4 AUVA Traumacenter Vienna, Meidling, 1030 Vienna, Austria
5 1st Orthopaedic Department, Orthopaedic Hospital Speising, 1030 Vienna, Austria
6 TUM School of Medicine, Technical University of Munich, 80333 München, Germany
7 Center of Clinical Research (CCR), University Clinic of Dentistry, Medical University of Vienna, 1030 Vienna, Austria
* Correspondence: roman.ostermann@me.com

Abstract: Background: Despite high rates of successful outcomes after open and arthroscopic distal clavicle excision (DCE) for symptomatic acromioclavicular joint (ACJ) degeneration, some patients present with persistent symptoms and disabilities after surgical intervention. This study aims to compare radiological, functional, and subjective outcomes of open revision surgery after failed arthroscopic DCE to primary successful arthroscopic DCE. Methods: In this retrospective case-control study, 10 patients who underwent open DCE revision were age- and gender-matched with 10 patients who did not require revision surgery after DCE. Radiographic evaluation included presence of acromioclavicular spurs and acromioclavicular joint distance. Functional and subjective outcomes were assessed using the CS, SSV, SST, VAS for pain, patient's satisfaction, ASES and quick DASH score. Results: At the latest postoperative follow-up (case: 57.3 ± 19.2 months; control: 63.5 ± 16.3 months), spur formation was detected in twice as many cases in the revision group, while acromioclavicular distance showed no significant difference. However, a significant bony regrowth was noticed in the revision group between revision surgery and latest follow-up, with a decrease of the acromioclavicular distance from 9.2 ± 1.6 mm to 5.9 ± 4.6 mm (p = 0.026) and a development of new spur formations in 30% of cases. There were no significant differences in overall CS between the revision and control group (p = 0.174) at final follow-up, but the control group scored significantly higher in the CS subgroups pain (p = 0.012) and internal rotation (p = 0.016). Mean SSV was significantly lower in the revision (65.5 ± 22.3%) compared to the control group (85.9 ± 16.4%; p = 0.031). Conclusions: Bony regrowth at the distal clavicle presenting as postoperative AC-distance narrowing and new spur formation was observed more distinctly in the revision group. Despite a slight increase in postoperative outcomes after revision surgery, subjective satisfaction and recalcitrant pain remain a concern. Level of Evidence: Therapeutic Level III, retrospective case-control study

Keywords: acromio-clavicular; AC-joint; revision; arthroscopy

check for updates

Citation: Ostermann, R.C.; Eigenschink, M.; Heuberer, P.R.; Siegert, P.; Muellbacher, W.; Anderl, L.; Schrott, B.; Laky, B.; Pauzenberger, L.; Anderl, W. Bony Regrowth and New Spur Formation as Possible Causes of Failed Distal Clavicle Excision—Mid-Term Outcomes after Revision Surgery in a Matched-Pair Analysis. *J. Pers. Med.* **2023**, *13*, 1221. https://doi.org/10.3390/jpm13081221

Academic Editors: Markus Gesslein and Eduard Kurz

Received: 10 July 2023
Revised: 26 July 2023
Accepted: 27 July 2023
Published: 1 August 2023

1. Introduction:

Distal clavicle excision (DCE) is a common surgical procedure for failed non-operative treatment of symptomatic acromioclavicular joint (ACJ) arthritis or osteolysis [1–3]. In the last decade, a transition from open to arthroscopic DCE was noticed, which might be attributed to the more invasive nature and associated complications of the open technique [4]. The increasing popularity of arthroscopic DCE was also linked to the advantage of intraoperative diagnosis and treatment of concomitant intra-articular or subacromial shoulder disorders (e.g., rotator cuff tears, long head of biceps tendon pathologies, impingement), with reported incident rates of up to 50 percent [5]. However, open distal clavicle excision has remained ingrained as a surgical approach since first described in 1941, with the advantage of better visualization of posterosuperior parts of the ACJ and the possibility of open capsular plication for stability and open denervation [6,7].

However, both approaches aim to remove the entire degenerative tissue of the affected joint, including the articular disc, degenerated articular cartilage, and adjacent bone of the lateral clavicle and medial acromion, in order to expand the intra-articular space and avoid painful contact between the articulating bones, while preserving stabilizing structures [8].

While both arthroscopic and open DCE show overall good to excellent clinical results in 76–100 percent of the cases, adequate long-term pain relief cannot be achieved in all cases, with some patients presenting with recalcitrant complaints after index surgery [1,9]. Surgical failure has been attributed to both under and over- resection, postoperative joint instability, stiffness, heterotopic ossification and spur formation, or under-diagnosed concomitant pathologies [9–12]. After a thorough reevaluation, revision surgery might be indicated [13]. To this date, there is a lack of clinical results following revision DCE in the literature.

This study aims to compare radiological, functional, and subjective outcomes of open revision surgery after failed arthroscopic DCE to primary successful arthroscopic DCE. We hypothesized that patients requiring revision surgery are more likely to present with persistent pain due to either decreased acromioclavicular distance or spur formation after primary DCE compared to patients with successful DCE.

2. Methods

2.1. Study Population

This retrospective study matched two independent cohorts for analysis. The institutional ethics committee approved this study prior to investigations. All surgical procedures were performed by fellowship-trained shoulder surgeons.

For the first cohort (group A), all patients who underwent open revision surgery within five years after primary arthroscopic DCE between January 2005 and December 2009 at a single institution were identified. Inclusion criteria were (1) male or female $\geq$ 18 years of age, (2) positive cross-body adduction test, (3) isolated ACJ complaints (persistent ACJ pain and/or clinically relevant ACJ instability), and (4) minimum follow-up of two years. All patients of this group had positive lidocaine infiltration test to the ACJ. Exclusion criteria were previous shoulder surgeries besides the index DCE surgery and secondary diagnosed concomitant glenohumeral or subacromial pathologies during revision surgery. Ten patients (three males and seven females) aged between 41 and 71 years at time of index surgery qualified for inclusion into group A.

The second cohort (group B) consisted of patients who underwent primary arthroscopic DCE between January 2005 and December 2009 at the same institution, without revision surgery. Inclusion criteria were the same as for group A. Patients with persistent ACJ complaints at final follow-up and any revision surgery were excluded. Group B was matched to group A regarding age, gender, and concomitant pathologies at index surgery.

Patient-specific parameters, including surgery-related data such as involvement of dominant shoulder, operation time (minutes), and length of hospital stay (days), were recorded for both groups.

2.2. Surgical Techniques

In the arthroscopic approach, the patient was placed in a beach-chair position. Two standard arthroscopic portals were established. After arthroscopic inspection of the glenohumeral joint for concomitant shoulder pathologies, the arthroscope was inserted into the subacromial space. The acromioclavicular joint was located by inserting a spinal needle from cranial through the joint into the subacromial space, followed by an incision of the inferior articular capsule, while carefully protecting the posterior and superior portion of the capsule through an anterior portal. A resection of the lateral clavicle and adjacent acromion with a 4 mm burr was performed. The resection was considered sufficient once the burr could easily fit between the articulating bony structures. Care was taken not to resect more than 8 mm to leave the superior joint capsule and CC ligaments attached to the lateral clavicle.

As with the arthroscopic technique, in the open approach for revision cases, the patient was positioned in a beach-chair position. A 2 to 3 cm skin incision was performed from anterior to posterior right above the AC joint. The fascia was incised perpendicular to the AC joint to expose the articular capsule. After splitting the capsule, the periosteum was elevated and the resection was carried out using an osteotome, oscillating saw, burr, or rongeur. The lateral clavicle was resected in a slight angle from superomedial to inferolateral, to parallel the normal orientation of the joint for a better congruency with the acromion. Resection length was 3–4 mm of the lateral clavicle and 2–3 mm of the adjacent acromion. By performing a flexion and adduction of the arm, the joint was examined for possible residual contact of the two articulating bones during movement. Any inferior bony spurs were additionally resected. Care was taken to perform a thorough suture of the capsule and delto-trapezial fascia, as it determines the stability of the joint postoperative. Non-absorbable sutures were used to reach watertight closure of the capsule and the delto-trapezial fascia.

2.3. Clinical and Radiographic Assessment

A comprehensive physical evaluation, including measurement of active abduction, forward flexion, and external and internal rotation using a goniometer, was conducted by two shoulder fellows (LA, BS). A cross-body adduction test for ACJ complaints was performed [3]. The Constant score (CS) [14] (0–100 points; 100 presenting the best possible outcome) was obtained as the primary clinical outcome measure. For subjective measures, visual analog scale (VAS) [15] for pain (0–10 points), Subjective Shoulder Value (SSV; 0–100%; 100% presenting the best possible outcome) [16]. Simple Shoulder Test (SST) [17], patient's satisfaction (very satisfied, satisfied, fairly satisfied, or not satisfied), American Shoulder and Elbow Surgeons (ASES) [18] score (0–100 points; 100 presenting the best possible outcome), and the Quick Disabilities of the Arm, Shoulder, and Hand (QuickDASH) [19] score (0–100 points; 100 presenting the best possible outcome) were assessed.

All patient received plain radiographic imaging, including an anteroposterior, axial and Y-view, and a Zanca view, pre- and immediately postoperative (=perioperative) as well as at the latest follow-up. Images were evaluated for acromioclavicular spur formation, and acromioclavicular joint distance (ACD; mm) was measured on Zanca views. A spur was defined as a definite bone formation inferior to the distal cortex of the lateral clavicle or the medial acromion (Figure 1). ACD and spur evaluations were carried out by two independent shoulder surgeons (PRH, LP) [20]. Additionally, postoperative computed tomographic (CT) scans (day one following surgery) of the affected AC joints were obtained for patients undergoing revision surgery (group A).

2.4. Statistical Analysis

Patient details were presented using descriptive statistics. Quantitative data were expressed as means with standard deviation or median and range. For continuous and normal distributed data, the Student's *t*-test was applied and the Mann–Whitney U test was used for ordinal or non-normally distributed data to determine differences between two independent groups. A Pearson Chi-square or Fisher's exact test was performed to analyze categorical variables. Paired *t*-tests (parametric) and Wilcoxon Signed-Rank tests (non-parametric) were performed to compare between time points within a group. All comparative tests were two-tailed and statistical significance level was set at $p < 0.05$. Statistical analyses were performed using IBM SPSS Statistics 24.0 software (IBM® Corporation, Armonk, NY, USA).

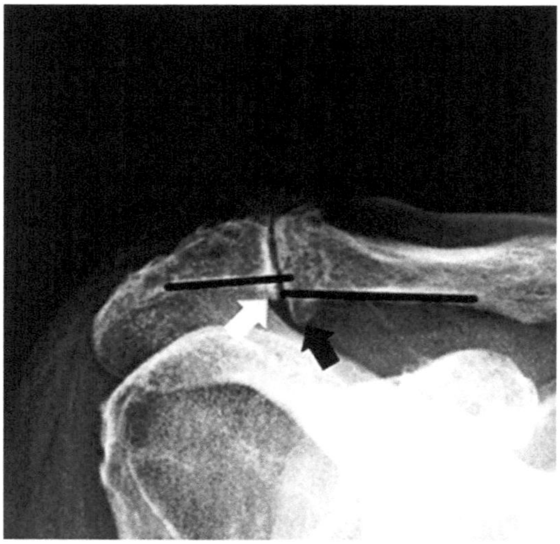

Figure 1. Zanca view radiograph of a right shoulder. The black arrow shows an inferior bony spur of the distal clavicle, the white arrow shows a small inferior bony spur of the medial acromion.

3. Results

In group A, one patient presented without any additional pathologies regarding the shoulder before the index surgery, whereas the other nine patients presented with additional shoulder-related comorbidities, including subacromial impingement ($n = 8$), rotator cuff lesions ($n = 4$), and long biceps tendon pathology ($n = 2$). All concomitant shoulder pathologies were treated surgically during index arthroscopic surgery with antero-lateral subacromial decompression, rotator cuff repair, and LHBT tenotomy, respectively. The average surgery time from incision to suture of the index surgery in this group was 48.3 ± 8.5 min, and the average length of hospital stay for the primary ACR was 2.2 ± 1.0 days. Revision surgery was indicated after a mean of 10.0 ± 8.4 months following the index surgery for persistent pain. The latest follow-up was performed 57.3 ± 19.2 months after index surgery.

In group B, one patient presented without any additional pathologies. The other nine patients presented with subacromial impingement ($n = 8$), rotator cuff lesions ($n = 4$), and long biceps tendon pathologies ($n = 2$), which were all treated surgically during the index procedure similar to group A. The average surgery time from incision to suture in group B was 35.3 ± 12.6 min, and the average length of hospital stay was 2.9 ± 1.1 days. The mean follow-up time was 63.5 ± 16.3 months.

A comparison between the age- and gender-matched cohorts showed no significant differences regarding demographic data (Table 1).

Table 1. Comparison of demographic data between both groups at the time of primary acromioclavicular joint resection.

	Group A $n = 10$	**Group B** $n = 10$	***p*-Values**
Age (years)	60.4 ± 8.9	60.8 ± 9.2	0.923
Gender (female/male)	7:3	7:3	1.000
Concomitant pathologies (yes/no)	9:1	9:1	1.000
Dominant arm involved (yes/no)	6:4	4:6	0.656

3.1. Radiographic Assessment

Before index surgery, six patients in group A showed inferior acromioclavicular spurs. According to perioperative X-rays, one day after index surgery, spurs were detected in four patients. The X-rays directly before revision surgery showed that the number of patients with spurs increased to eight. All spurs were resected during revision surgery. Thus, no spurs were detected on postoperative X-rays and CT scans immediately following revision surgery. At latest follow-up of group A, three patients presented with spurs again In group B, a decrease of spurs from six patients preoperatively to two patients directly after index surgery was noticed. The number of patients with spurs in group B increased to four at the latest follow-up. Comparing the spur formation before revision surgery in group A, which is the latest follow-up of index surgery in this group, and the latest follow-up in the control group B, twice as many patients with spurs were detected in group A (Figure 2).

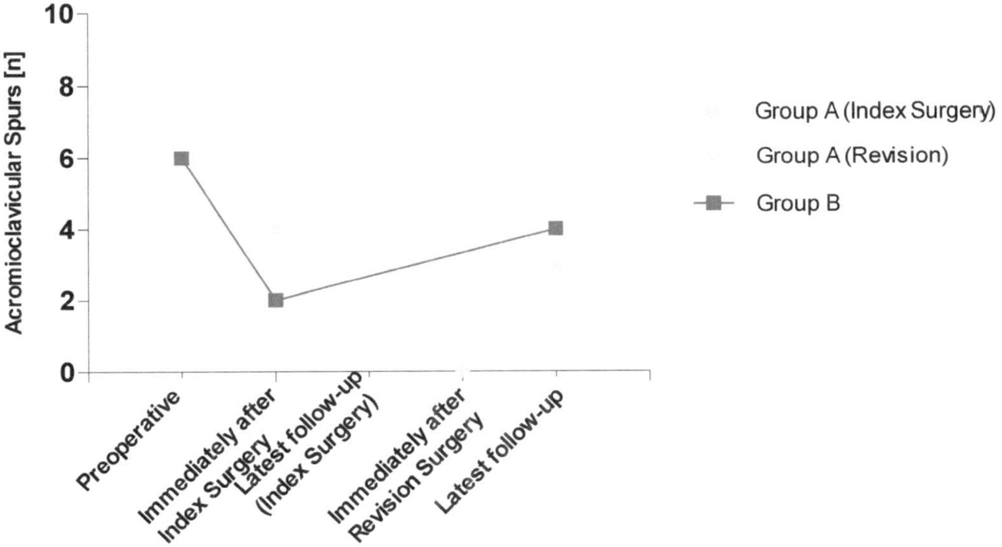

Figure 2. Detection of acromioclavicular spurs in absolute numbers (n) upon radiographic imaging. Group A (10 patients) was evaluated preoperatively, immediately after index surgery, at latest follow-up before revision surgery, immediately after revision surgery, and at latest follow-up, while group B (10 patients) was evaluated preoperatively, immediately after index surgery, and at latest follow-up.

The acromioclavicular distance in group A was 2.5 $\pm$ 2.2 mm before and 6.6 $\pm$ 2.3 mm immediately after index surgery, and decreased to 4.8 $\pm$ 3.1 mm before revision surgery. In group B, the acromioclavicular distance changed from 1.8 $\pm$ 1.6 mm before to 5.5 $\pm$ 0.9 mm immediately after surgery and remained stable with 5.7 $\pm$ 1.7 mm at latest follow-up (Figure 3).

While the decrease in acromioclavicular distance from peri- to the next postoperative follow-up regarding index surgeries of both groups was not significant (group A: $p = 0.168$; group B: $p = 0.751$), acromioclavicular distance significantly decreased from peri- (9.2 $\pm$ 1.5 mm) to the last follow-up (5.9 $\pm$ 4.6 mm; $p = 0.045$) regarding the revision surgery of group A.

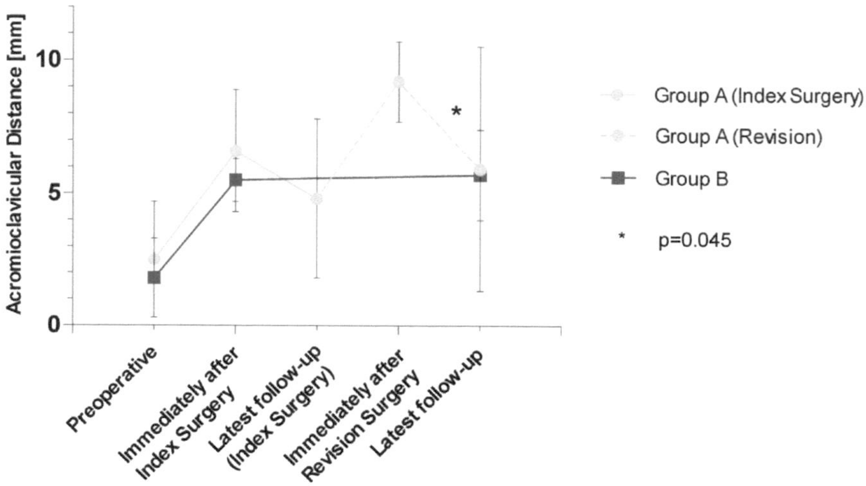

Figure 3. Acromioclavicular distance (ACD) in millimeters in both groups. Group A (10 patients) was evaluated preoperatively, immediately after index surgery, at latest follow-up before revision surgery, immediately after revision surgery, and at latest follow-up, while group B (10 patients) was evaluated preoperatively, immediately after index surgery, and at latest follow-up. Significant p-values are indicated with *.

3.2. Functional and Subjective Assessment

A comparison between group A and B of all functional and subjective outcome measures at the latest follow-up showed, with the exception of CS pain, internal rotation, and SSV, no significant differences (Table 2).

Table 2. Clinical and subjective outcome measures at latest follow-up of group A (revision) and group B (control). Respected measurements with standard deviations are indicated. Statistical differences are marked with *.

Scores	Group A $n = 10$	Group B $n = 10$	p-Values
Overall CS [points]	64.1 ± 19.3	74.9 ± 14.4	0.175
Pain	6.3 ± 6.0	13.0 ± 2.6	0.007 *
ADL	12.9 ± 5.7	17.0 ± 3.3	0.069
ROM	32.0 ± 6.0	34.6 ± 5.3	0.318
Abduction	8.4 ± 2.3	8.8 ± 2.5	0.714
Flexion	9.0 ± 1.7	9.0 ± 2.2	0.999
Internal rotation	5.0 ± 2.2	7.3 ± 1.0	0.012 *
External rotation	9.6 ± 1.3	9.6 ± 0.8	0.999
Strength	12.9 ± 6.9	10.3 ± 5.8	0.372
qDASH (points)	44.1 ± 31.3	24.3 ± 27.5	0.150
ASES (points)	56.7 ± 32.7	75.0 ± 24.3	0.172
SSV (%)	65.5 ± 22.3	85.9 ± 16.4	0.031 *
SST (points)	6.1 ± 3.6	7.9 ± 3.6	0.281
VAS (points)	4 (range 0–7)	1 (range 0–4)	0.123

Similar overall CS and CS subgroups of group A before revision surgery compared to the latest follow-up were detected (Table 3).

Table 3. Overall and subgroups of Constant score in group A before revision surgery and after revision surgery at latest follow-up.

Constant Score (Points)	Group A Before Revision	Group A After Revision (latest FU)	p-Value
Pain	5 (range, 0–15)	5 (range, 0–15)	0.478
ADL	7.5 (range, 4–20)	10.5 (range, 6–20)	0.284
ROM	24 (range, 16–38)	34 (range, 20–38)	0.185
Strength	8 (range, 4–26)	13.5 (range, 1–22)	0.575
Overall CS	51.4 ± 22.1	64.1 ± 19.3	0.316

ADL, activity of daily living; ROM, range of motion; CS, Constant Score; qDASH, Quick Disabilities of the Arm, Shoulder, and Hand score; ASES American Shoulder and Elbow Surgeons score; SSV, Subjective Shoulder Value; SST, Simple Shoulder Test; VAS, visual analog scale.

The majority of patients were very satisfied (group A: 60%; group B: 60%) or satisfied (group B: 40%) with the procedure. The remaining patients in group A were fairly satisfied (30%), while one patient was not satisfied.

4. Discussion

The most important finding of the current study was that, although a slight increase in postoperative clinical outcome could be achieved with surgical revision after failed DCE, subjective satisfaction and recalcitrant pain remain a concern. While resection length appeared to have no influence on outcomes, increased spur formation in failed DCE was noticed. Furthermore, bony regrowth at the distal clavicle resulting in AC-distance narrowing was observed and could be a possible cause for persistent pain and lower patient satisfaction following revision DCE.

To the best of our knowledge, this is the first study to report on radiological, functional, and subjective outcomes after revision distal clavicle excision. Firstly, we observed a bony regrowth at the distal clavicle representing a postoperative AC-distance narrowing and new spur formation, which could well be a possible cause for revisions. Secondly, patients with no revision after arthroscopic DCE seemed to have better internal rotation and reported slightly better subjective outcomes (CS pain, SSV, satisfaction with the procedure). Furthermore, only half of the patients seemed to be satisfied with their clinical outcomes following open revision DCE.

Even though high rates of successful outcomes have been reported for both open and arthroscopic primary DCE, treatment failures occur, and some patients remain with persistent symptoms after DCE [1,12,13]. Those patients remain challenging, and treatment options are limited. This study pursued the objective to compare clinical and subjective outcomes in patients who required revision surgery after arthroscopic DCE to patients that did not require further treatment after index DCE, to gain insight into factors for a successful treatment. Furthermore, we aimed to determine the effect of open revision surgery on the patient's satisfaction.

While the overall clinical outcome in both groups did not differ significantly, one of the main findings in our study is that the subjective outcome was significantly worse in the revision group. Our data showed that, although the active range of motion and activities of daily living were only slightly inferior to the control group, persistent ACJ pain leads to inferior subjective outcome measures. In this study, only internal rotation was impaired in the revision group, which could be attributed to pain that arises due to rotation in the ACJ during maximum glenohumeral internal rotation, rather than musculoskeletal limitation. Although we observed an overall trend of increased Constant score from before to after revision surgery, the difference was marginal and not statistically significant. Ultimately, only a little over half the patients were satisfied with open revision surgery, while the rest were not. In accordance with our findings, Kharrazi et al. found 45% of patients with ongoing symptoms and continued pain following revision DCE in their study [13].

The ideal extent of resection in DCE has been reported as a key factor for successful treatment, but is still discussed controversially [3,12,21–26]. If the resection was inadequate, meaning there is still residual bone contact after surgery, the pain is likely to remain after the procedure. On the other hand, extensive resections can lead to disruption of the acromioclavicular and coracoclavicular ligaments and, consequently, to postoperative iatrogenic instability of the joint, which is equally associated with pain [27]. When we compare the measured resection length directly following index surgeries in both groups, we did not observe significant differences. Hence, at least in this small group, we could not identify resection length as the key factor for failed DCE. Eskola et al. [22] found a correlation between over-resection and postoperative pain. In our cohort, we could not find either significant positive nor negative correlations between the perioperative acromioclavicular distance and pain measurements. Therefore, the hypothesis was not confirmed. Interestingly, though, we observed a distance decrease of around 1.8 mm between the articulating bones from directly after index surgery to before revision surgery (10.0 ± 8.4 months), and a decrease again from around 3.3 mm at latest follow-up of revision surgery compared to the postoperative CT-scans directly after revision (46.9 ± 16.1 months). This finding could either be attributed to a regrowth of bone tissue around the resected joint or a tightening of capsular structures. A bony regrowth of the lateral clavicle has been reported before and identified as a possible source of reoccurring symptoms [9,28].

Another hypothesis was that patients requiring revision surgery are more likely to present with a postoperative prominent inferior acromioclavicular spur compared to patients without failed DCE. Comparing the two groups in regard to postoperative acromioclavicular spurs, we found no difference at latest follow-up. We made the observation, though, that in the revision group the number of patients presenting with acromioclavicular spurs increased from directly postoperative after the primary DCE to preoperative before the revision surgery from four patients to eight patients. In addition, we noticed four remaining spurs directly after index surgery. In those cases, the surgeons did not perform an additional coplaning of the distal clavicle, as the spurs did not seem to interfere with underlying structures intraoperatively. Those remaining spurs, however, could as well be a possible cause for the late revision. Nevertheless, after all spurs had been resected during revision surgery, another regrowth of spurs could be observed at latest follow-up following revision surgery in two patients. This finding seems to be consistent with the statement of Levine et al. [11], that the postoperative development of spurs can cause late failure of acromioclavicular joint resection and necessitate revision surgery. Interestingly, if we compare the number of patients presenting with spurs directly before revision surgery in group A with the latest follow-up in group B, we observe twice as many spurs in the revision group. Hence, spur formation could potentially contribute to the need of late revision, which could suggest a benefit of removing any spurs during the index surgery.

Limitations

There are several limitations to this study. The current study is subject to the inherent limitations of a retrospective study design. As the number of patients undergoing revision surgery after DCE is low, the sample size of the current study is small. Although the study might be underpowered to reveal significant differences between groups, it provides novel data on a rare but clinically challenging condition. To counteract the limited cohort sizes, a matched cohort design and strictly defined inclusion criteria were employed for better comparability. Another potential bias is the difference in follow-up time. Furthermore, a limiting factor for this and other studies is the isolated evaluation of the AC joint. Pathologies affecting this joint are likely accompanied by other shoulder related comorbidities that could have an impact on the functional and subjective outcome. However, this study is one of very few studies, if not the first, reporting radiological, functional, and subjective outcomes after revision distal clavicle excision.

5. Conclusions

Bony regrowth at the distal clavicle presenting as postoperative AC-distance narrowing and new spur formation was observed more distinctly in the revision group. Resection length appeared to have no influence on outcome in our cohorts. Despite a slight increase in postoperative outcomes after index and revision surgery, subjective satisfaction and recalcitrant pain remain a concern in DCE revision surgery.

Author Contributions: R.C.O.: Conception and design of the work, revising the work. M.E.: analysis and interpretation of data, revising the work. P.R.H.: analysis and interpretation of data, revising the work. P.S.: drafting the work, analysis of data. W.M.: acquisition of data, drafting the work. L.A.: statistical analysis, revising the work. B.S.: acquisition of data, drafting the work. B.L.: statistical analysis, revising the work. L.P.: conception and design of the work, revising the work. W.A.: conception and design of the work, revising the work. All authors have read and agreed to the published version of the manuscript.

Funding: This research received no external funding.

Institutional Review Board Statement: The study was conducted in accordance with the Declaration of Helsinki, and approved by the Ethic Board of the Hospital of Sisters of Mercy, Vienna (protocol code 201209_EK03, 09/2012).

Informed Consent Statement: Informed consent was obtained from all patients.

Data Availability Statement: Not applicable.

Conflicts of Interest: W.A., R.C.O. and P.R.H. receive personal fees from Arthrex, Inc. The company had no influence on the study design, data collection, interpretation of the results, or the final manuscript. The remaining authors (M.E., P.S., W.M., L.A., B.S., B.L. and L.P.) declare that they do not have any conflict of interest pertinent to this study.

List of Abbreviations

ACD	Acromioclavicular joint distance	DCE	Distal clavicle excision
ACJ	Acromioclavicular joint	qDASH	QuickDASH score
ASES	American Shoulder and Elbow Surgeons score	SST	Simple shoulder test
CS	Constant Score	SSV	Subjective Shoulder Value
CT	Computed tomography	VAS	Visual analog scale

References

1. Rabalais, R.D.; McCarty, E. Surgical treatment of symptomatic acromioclavicular joint problems: A systematic review. *Clin. Orthop. Relat. Res.* **2007**, *455*, 30–37. [CrossRef]
2. Robertson, W.J.; Griffith, M.H.; Carroll, K.; O'Donnell, T.; Gill, T.J. Arthroscopic versus open distal clavicle excision: A comparative assessment at intermediate-term follow-up. *Am. J. Sports Med.* **2011**, *39*, 2415–2420. [CrossRef] [PubMed]
3. Mall, N.A.; Foley, E.; Chalmers, P.N.; Cole, B.J.; Romeo, A.A.; Bach, B.R., Jr. Degenerative joint disease of the acromioclavicular joint: A review. *Am. J. Sports Med.* **2013**, *41*, 2684–2692. [CrossRef] [PubMed]
4. Chronopoulos, E.; Gill, H.S.; Freehill, M.T.; Petersen, S.A.; McFarland, E.G. Complications after open distal clavicle excision. *Clin. Orthop. Relat. Res.* **2008**, *466*, 646–651. [CrossRef] [PubMed]
5. Freedman, B.A.; Javernick, M.A.; O'Brien, F.P.; Ross, A.E.; Doukas, W.C. Arthroscopic versus open distal clavicle excision: Comparative results at six months and one year from a randomized, prospective clinical trial. *J. Shoulder Elb. Surg.* **2007**, *16*, 413–418. [CrossRef]
6. Mumford, E.B. Acromioclavicular dislocation (A new operative treatment). *J. Bone Jt. Surg.* **1941**, *23*, 799–802.
7. Gurd, F.B. The Treatment of Complete Dislocation of the Outer End of the Clavicle: An Hitherto Undescribed Operation. *Ann. Surg.* **1941**, *113*, 1094–1098. [CrossRef]
8. Jerosch, J.; Steinbeck, J.; Schroder, M.; Castro, W.H. Arthroscopic resection of the acromioclavicular joint (ARAC). *Knee Surg. Sports Traumatol. Arthrosc.* **1993**, *1*, 209–215. [CrossRef] [PubMed]
9. Elhassan, B.; Ozbaydar, M.; Diller, D.; Massimini, D.; Higgins, L.D.; Warner, J.J. Open versus arthroscopic acromioclavicular joint resection: A retrospective comparison study. *Arthroscopy* **2009**, *25*, 1224–1232. [CrossRef]
10. Pensak, M.; Grumet, R.C.; Slabaugh, M.A.; Bach, B.R., Jr. Open versus arthroscopic distal clavicle resection. *Arthroscopy* **2010**, *26*, 697–704. [CrossRef]
11. Levine, W.N.; Soong, M.; Ahmad, C.S.; Blaine, T.A.; Bigliani, L.U. Arthroscopic distal clavicle resection: A comparison of bursal and direct approaches. *Arthroscopy* **2006**, *22*, 516–520. [CrossRef] [PubMed]

12. Corteen, D.P.; Teitge, R.A. Stabilization of the clavicle after distal resection: A biomechanical study. *Am. J. Sports Med.* **2005**, *33*, 61–67. [CrossRef] [PubMed]
13. Kharrazi, F.D.; Busfield, B.T.; Khorshad, D.S. Acromioclavicular joint reoperation after arthroscopic subacromial decompression with and without concomitant acromioclavicular surgery. *Arthroscopy* **2007**, *23*, 804–808. [CrossRef]
14. Constant, C.R.; Murley, A.H. A clinical method of functional assessment of the shoulder. *Clin. Orthop. Relat. Res.* **1987**, *214*, 160–164. [CrossRef]
15. Huskisson, E.C. Measurement of pain. *Lancet* **1974**, *2*, 1127–1131. [CrossRef]
16. Gilbart, M.K.; Gerber, C. Comparison of the subjective shoulder value and the Constant score. *J. Shoulder Elb. Surg.* **2007**, *16*, 717–721. [CrossRef] [PubMed]
17. Angst, F.; Schwyzer, H.K.; Aeschlimann, A.; Simmen, B.R.; Goldhahn, J. Measures of adult shoulder function: Disabilities of the Arm, Shoulder, and Hand Questionnaire (DASH) and its short version (QuickDASH), Shoulder Pain and Disability Index (SPADI), American Shoulder and Elbow Surgeons (ASES) Society standardized shoulder assessment form, Constant (Murley) Score (CS), Simple Shoulder Test (SST), Oxford Shoulder Score (OSS), Shoulder Disability Questionnaire (SDQ), and Western Ontario Shoulder Instability Index (WOSI). *Arthritis Care Res.* **2011**, *63* (Suppl. 11), S174–S188.
18. Michener, L.A.; McClure, P.W.; Sennett, B.J. American Shoulder and Elbow Surgeons Standardized Shoulder Assessment Form, patient self-report section: Reliability, validity, and responsiveness. *J. Shoulder Elb. Surg.* **2002**, *11*, 587–594. [CrossRef]
19. Hudak, P.L.; Amadio, P.C.; Bombardier, C. Development of an upper extremity outcome measure: The DASH (disabilities of the arm, shoulder and hand) [corrected]. The Upper Extremity Collaborative Group (UECG). *Am. J. Ind. Med.* **1996**, *29*, 602–608. [CrossRef]
20. Zanca, P. Shoulder pain: Involvement of the acromioclavicular joint. (Analysis of 1000 cases). *Am. J. Roentgenol. Radium Ther. Nucl. Med.* **1971**, *112*, 493–506. [CrossRef]
21. Boehm, T.D.; Kirschner, S.; Fischer, A.; Gohlke, F. The relation of the coracoclavicular ligament insertion to the acromioclavicular joint: A cadaver study of relevance to lateral clavicle resection. *Acta Orthop. Scand.* **2003**, *74*, 718–721. [CrossRef]
22. Eskola, A.; Santavirta, S.; Viljakka, H.T.; Wirta, J.; Partio, T.E.; Hoikka, V. The results of operative resection of the lateral end of the clavicle. *J. Bone Jt. Surg. Am.* **1996**, *78*, 584–587. [CrossRef] [PubMed]
23. Beitzel, K.; Sablan, N.; Chowaniec, D.M.; Obopilwe, E.; Cote, M.P.; Arciero, R.A.; Mazzocca, A.D. Sequential resection of the distal clavicle and its effects on horizontal acromioclavicular joint translation. *Am. J. Sports Med.* **2012**, *40*, 681–685. [CrossRef]
24. Branch, T.P.; Burdette, H.L.; Shahriari, A.S.; Carter, F.M., 2nd; Hutton, W.C. The role of the acromioclavicular ligaments and the effect of distal clavicle resection. *Am. J. Sports Med.* **1996**, *24*, 293–297. [CrossRef]
25. Wang, J.; Ma, J.X.; Zhu, S.W.; Jia, H.B.; Ma, X.L. Does Distal Clavicle Resection Decrease Pain or Improve Shoulder Function in Patients With Acromioclavicular Joint Arthritis and Rotator Cuff Tears? A Meta-analysis. *Clin. Orthop. Relat. Res.* **2018**, *476*, 2402–2414. [CrossRef] [PubMed]
26. Park, Y.B.; Koh, K.H.; Shon, M.S.; Park, Y.E.; Yoo, J.C. Arthroscopic distal clavicle resection in symptomatic acromioclavicular joint arthritis combined with rotator cuff tear: A prospective randomized trial. *Am. J. Sports Med.* **2015**, *43*, 985–990. [CrossRef]
27. Blazar, P.E.; Iannotti, J.P.; Williams, G.R. Anteroposterior instability of the distal clavicle after distal clavicle resection. *Clin. Orthop. Relat. Res.* **1998**, *348*, 114–120. [CrossRef]
28. Tytherleigh-Strong, G.; Gill, J.; Sforza, G.; Copeland, S.; Levy, O. Reossification and fusion across the acromioclavicular joint after arthroscopic acromioplasty and distal clavicle resection. *Arthroscopy* **2001**, *17*, E36. [CrossRef] [PubMed]

Journal of *Personalized Medicine*

MDPI

Article

Longitudinal Analysis of the ASES and Constant–Murley Scores, and the Internal Rotation/Shift and Jobe Tests Following Arthroscopic Repair of Supraspinatus Lesions

George Fieseler [1], Kevin Laudner [2,*], Jakob Cornelius [1,3], Stephan Schulze [3], Karl-Stefan Delank [3] and René Schwesig [3]

1 Clinic for Hand, Trauma and Orthopedic Surgery, Sports Medicine, Clinic Hann, Münden, 34346 Hannoversch Münden, Germany
2 Department of Health Sciences, Hybl Sports Medicine and Performance Center, University of Colorado, Colorado Springs, CO 80918, USA
3 Department of Orthopedic and Trauma Surgery, Martin-Luther-University Halle-Wittenberg, 06120 Halle, Germany
* Correspondence: klaudner@uccs.edu

Abstract: It is essential to investigate patients post-surgery using functional surveys like the American Shoulder and Elbow Surgeons Shoulder (ASES) and the Constant–Murley shoulder (CMS) scores, as well as clinical tests, such as the Internal Rotation and Shift (IRO/Shift) and Jobe tests. In this study, 51 out of an initial 87 patients underwent an arthroscopic supraspinatus repair (22 single-row, 16 double-row, 13 debridement). Testing occurred pre-surgery, and 3 and 6 months post-surgery. Both surveys showed significant improvements over time among all 87 patients, but there were no differences between groups (lesion/no lesion) ($p > 0.815$) or time × group ($p > 0.895$). The IRO/Shift test showed a stronger ability to distinguish between both groups (positive vs. negative) with respect to the ASES and CMS scores over time, but the Jobe test did not ($p > 0.100$). Improvements in the CMS scores and the Jobe test were lower following repair compared to the ASES and IRO/Shift test. Most patients returned to adequate levels of functional abilities at 6 months post-surgery. The time required to return to activities of daily living and negative clinical tests was longer for the double-row repair patients compared to the single-row and debridement groups. In conclusion, both the functional surveys and the clinical tests demonstrated improvements following surgery.

Keywords: clinical test; functional abilities; orthopedic exam; shoulder; supraspinatus; rotator cuff

Citation: Fieseler, G.; Laudner, K.; Cornelius, J.; Schulze, S.; Delank, K.-S.; Schwesig, R. Longitudinal Analysis of the ASES and Constant–Murley Scores, and the Internal Rotation/Shift and Jobe Tests Following Arthroscopic Repair of Supraspinatus Lesions. *J. Pers. Med.* **2023**, *13*, 1304. https://doi.org/10.3390/jpm13091304

Academic Editors: Markus Gesslein and Eduard Kurz

Received: 3 August 2023
Revised: 21 August 2023
Accepted: 24 August 2023
Published: 25 August 2023

1. Introduction

Rotator cuff lesions are one of the most common shoulder pathologies [1,2], with more than half of the population experiencing a lesion by the time they are 60 years old [3]. These pathologies often cause pain, decreased mobility, and decreased function during activities of daily living, resulting in the need for surgical intervention [4]. Between 1995–2009, there was a 238% increase in the number of rotator cuff repairs reported in New York alone [5]. With hundreds of thousands of these surgical interventions conducted every year [6], it is critical to understand the functional outcomes following surgery.

One of the most common repairs is to the superior rotator cuff, or more specifically, the supraspinatus tendon. After achieving the goals of pain relief and increased functional ability, the patient ultimately returns to their normal activities of daily living. Surgeons often use anchors to restore the tendon footprint, suturing the tendon directly to the greater tuberosity [7]. Two of these techniques are known as single-row repair and double-row repair. Both techniques are commonly used, but there are conflicting results regarding their benefits and functional outcomes [8–12].

Although various tests are used to determine the progression of patients following a rotator cuff repair, the American Shoulder and Elbow Surgeons Shoulder (ASES) [13–15]

and Constant–Murley shoulder (CMS) [16] scores are common patient self-evaluations used to determine functional ability. Malavolta et al. [17] previously reported that the ASES is a good prognostic tool for use at the 6-, 12-, and 24-month postoperative periods after supraspinatus repair. Similarly, the CMS has been reported to correlate with the isokinetic evaluation of the external rotation strength at the 6- and 12-month post-operative periods for rotator cuff repair [18]. In addition to functional surveys, clinical tests, such as the Jobe test [19,20] and the newly developed internal rotation and shift test (IRO/Shift test) [21,22] are used to determine the structural integrity and the functional properties of the supraspinatus tendon. Fieseler et al. [21] originally reported that the IRO/Shift test is a reliable and valid tool for assessing supraspinatus tears, demonstrating both strong intra-rater and inter-rate reliability. These authors then performed a subsequent study comparing the IRO/Shift test to the Jobe test and found that these clinical tools are comparable in the detection of supraspinatus lesions [22]. More specifically, the IRO/Shift test had a sensitivity of 96%, a specificity of 50%, and an accuracy of 77% compared to the Jobe test, which had a sensitivity of 89%, a specificity of 60%, and a similar accuracy of 77%. Unfortunately, there is little research describing the use of these questionnaires and special tests following an arthroscopic repair.

Randelli et al. [6] noted that with all the advances in rotator cuff repair techniques, it is essential for future research to investigate patient-assessed comfort and function following a surgical repair. Therefore, the primary purpose of this study was to determine the effectiveness of two popular rotator cuff patient questionnaires and two special tests following a tendon repair (double-row, single-row, debridement). Secondarily, this study used functional questionnaires and special tests to determine the effectiveness and length of healing necessary following a rotator cuff repair. More specifically, this study examined the use of the ASES, CMS, IRO/Shift test, and the Jobe test. These tools were used prior to the surgical repair of the supraspinatus tendon and again at the 3- and 6-month post-operative follow-ups to determine patient progression.

2. Materials and Methods

2.1. Subjects

Eighty-seven (36 females, 51 males; age: 53.6 ± 12.5 years; height: 1.76 ± 0.09 m; mass: 83.0 ± 15.3 kg, body mass index: 26.7 ± 3.51 kg/m^2) patients volunteered to participate in this study. The inclusion criteria included participants $\geq$ 18 years who had primarily been treated conservatively by their General Physicians and Physiotherapists between 2018–2019 to restrict their shoulder pain. The participants then completed a conservative rehabilitation program which was unsuccessful. Following the failed conservative rehabilitation, magnetic resonance imaging was used to determine a possible structural lesion, and the patient was referred to a specialized shoulder physician for consideration of surgical intervention. The exclusion criteria consisted of patients who had a history of arthritis, arthrosis, or fracture. No patients had rheumatic disease; however, information pertaining to patient history of smoking, diabetes mellitus, and thyroid disease was not collected. All subjects provided informed consent prior to any data collection. This study was approved by the Martin-Luther University Halle-Wittenberg Ethics Committee (reference number: 2018-05).

Among the 87 patients who originally participated in this study, 36 were found to not have a rotator cuff tear upon arthroscopic surgery. Therefore, 51 total patients were radiologically and subsequently arthroscopically diagnosed with a supraspinatus lesion. Among these 51 patients with confirmed supraspinatus tears (17 traumatic, 34 degenerative) requiring surgery, 22 received a single-row repair, 16 received a double-row repair, and 13 received a debridement (Figure 1). The 36 patients who did not have a confirmed rotator cuff lesion were diagnosed with Bankart lesions (4), biceps tendinitis with SLAP tears (12) or biceps tendinitis with pulley-lesions (2), bursitis with subacromial impingement (11), subcoracoid impingement (1), exophytic acromioclavicular arthrosis (4), adhesive capsulitis (1), and humeral head chondrolysis (1).

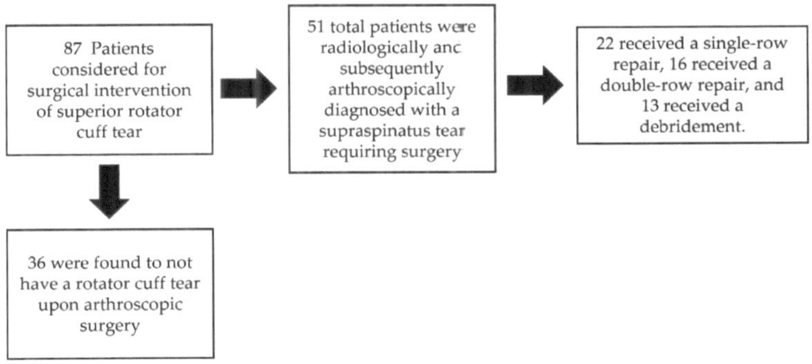

Figure 1. Participant flow diagram.

2.2. Procedures

This longitudinal study prospectively followed subjects from the time of their initial clinical evaluation with the specialized shoulder surgeon (pre-operative) to a 6-month post-operative follow-up evaluation. The initial pre-operative evaluation was performed by two physicians with extensive experience working with shoulder pathologies. During all three clinical evaluations (Exam 1: pre-surgery, Exam 2: 3-month post-operative, Exam 3: 6-month post-operative), the physicians used both the IRO/Shift test and the Jobe test to assess the function of the superior rotator cuff. Patients also completed the ASES and the CMS evaluations to determine their functional abilities.

The arthroscopic repairs consisted of tendon fixation using either a single-row or double-row suture anchor, or debridement, and were performed by the same physician who had over 20 years of shoulder surgery experience. The decision to perform either a single-row or double-row repair was made intraoperatively prior to the tendon reconstruction based on the tear size, morphology, and amount of tendon retraction. Small tears were defined when no retraction or shallow "U-shape" conditions were found. In these cases, a single-row technique with two sutures and one anchor was used. Medium-sized tears, those with minor retraction but an "L-shaped" rupture, were fixed with a double-row technique using a double suture armed with a metal anchor as the medial line, arthroscopic knots, and a bioresorbable, loaded anchor device with no suture for the lateral line of the reconstruction. Large tears with higher retraction, and deep "U-shaped" or "L-shaped" conditions were closed via a double-row technique as previously described. All anchors were manufactured by Arthrex (Arthrex Company, Naples, FL, USA). Two FiberWire® sutures were implanted and combined with bioresorbable anchors (SwiveLock® SP) for the single-row, or for the lateral line in the double-row fixations. For the medial line of the double-row fixations, two-suture loaded metal (titanium), self-cutting anchors (CorkSrew FT) were used [23]. All surgical interventions included complete reconstruction of the tendon tear regardless of the size, retraction or morphology. There were no patients treated with a salvage procedure such as a "margin convergence". Following surgery, all patients were prescribed a standard rehabilitation protocol, based on the extent of tendon damage and subsequent patient-specific physical characteristics and signs and symptoms, aimed at protecting the tendon repair, promoting healing, and ultimately returning the patient to their pre-injury level of physical abilities and function.

The IRO/Shift test required the patient to stand in a relaxed position while they actively adducted and internally rotated their shoulder, moving their involved side arm behind their back and then superiorly elevating their hand along their spine until their end ROM. At this end point, the physician then applied additional passive motion in the same direction. If the patient experienced pain, then the physician followed this test with an additional special test (e.g., O'Brien's test) to rule out labral or long head biceps tendon involvement. If the patient had pain during the adduction and internal rotation movement, but did not have pain indicating labral pathology, then the IRO/Shift test was considered positive for supraspinatus damage or load. Furthermore, none of the supraspinatus tear patients demonstrated signs of adhesive capsulitis at any point during the study. The Jobe test was also conducted standing while the physician passively elevated the patient's involved shoulder to 90° of abduction with internal rotation. In this position, the physician then applied downward pressure while the patient attempted to resist this force. A Jobe test was considered positive in the presence of pain or abnormal weakness. Both tests were conducted bilaterally.

The ASES questionnaire uses a 100-point scale to determine the patient's self-reported level of pain and ability to complete activities of daily living (each worth 50 points). Scores range from 0 (worst possible shoulder condition) to 100 (best possible shoulder condition). The ASES has a reported intra-rater reliability of 0.84–0.96 [13–15,24] and is endorsed by the American Shoulder and Elbow Surgeons Research Committee. The Constant–Murley shoulder (CMS) score uses a scale to determine the patient's level of pain and activities of daily living similar to the ASES questionnaire, but also objectively assesses the patient's ROM and strength. The pain and activities of daily living section are worth a total of 35 points, while the ROM and strength are worth 65 points, for a total of 100 points. Similar to the ASES score, more points on the CMS score indicate better shoulder function. The Constant tool has a reported intra-rater reliability of 0.80–0.96 [25], and has been endorsed by the European Society of Shoulder and Elbow Surgery [25,26].

2.3. Statistical Analysis

All statistical analyses were performed using SPSS version 28.0 for Windows (IBM, Armonk, NY, USA). A Pearson Chi^2-test was used to compare the categorical variables (IRO/Shift test and Jobe test) over the three time periods (pre-op, 3 months post-op, and 6 months post-op) and to detect any relationships. The longitudinal mean differences of the scores (ASES, CMS) were tested using a two-factor (time and tests: IRO/Shift test, Jobe test or type of surgical repair: single-row, double-row, debridement) univariate general linear model [27]. The differences between the means were considered statistically significant if the p-values were <0.05 and the partial eta-squared (η_p^2) values were >0.15 [28].

Based on the intraclass correlation coefficient (ICC) and the standard error of the mean (SEM) [29] calculation, the minimal detectable change was reported at a 95% (MDC_{95}) confidence interval [30] for ASES and CMS.

3. Results

Among the 87 patients, regardless of whether a rotator cuff tear was present or not, both the ASES and CMS scores showed significant time effects, but no significant differences between groups (lesion vs. no lesion) ($p > 0.815$) or time × group ($p > 0.895$) (Table 1).

Table 1. Analysis of ASES and CMS scores dependent on the diagnosis of a superior rotator cuff lesion.

Score	Exam	RC-Lesion (n = 51)	No RC-Lesion (n = 36)	Variance Analysis (p/η_p^2)		
				Time	Group	Time × Group
ASES	1	**45.4 ± 16.4**	44.2 ± 18.4	**<0.001/0.653**	0.828/0.001	0.913/0.001
	2	65.0 ± 16.4	65.2 ± 15.6			
	3	81.1 ± 16.3	80.3 ± 16.7			
CMS	1	41.6 ± 17.5	42.4 ± 18.5	**<0.001/0.617**	0.815/0.001	0.895/0.001
	2	61.9 ± 17.7	61.6 ± 15.6			
	3	77.5 ± 18.6	79.0 ± 21.6			

ASES = American Shoulder and Elbow Surgeons Score; CMS = Constant–Murley Shoulder Score; RC = rotator cuff; Exam 1 = pre-operative; Exam 2 = 3 months post-operative; Exam 3 = 6 months post-operative. Significant differences marked in bold.

When comparing the special tests versus the survey scores, the IRO/Shift test showed a stronger ability to distinguish between both groups (positive test vs. negative test) with respect to ASES (Table 2) and CMS (Table 3) scores over time. At follow-up exams 2 and 3, the mean differences between the positive and negative IRO/Shift tests were significant ($p < 0.001$). In contrast, the Jobe test was not able to demonstrate significant differences in positive and negative tests at any time period ($p > 0.100$). However, the number of subjects in the positive and negative test groups were vastly different between exams 1 and 3. This may have affected the variance analysis.

Table 2. Comparison of ASES scores to clinical tests.

Clinical Test		Exam 1 ASES Score	Exam 2 ASES Score	Exam 3 ASES Score
IRO/Shift	negative	50.1 ± 20.4 (n = 25)	69.7 ± 15.3 (n = 50)	82.6 ± 14.8 (n = 81)
	positive	42.8 ± 15.3 (n = 62)	58.8 ± 14.8 (n = 37)	55.8 ± 17.2 (n = 6)
	p/η_p^2	*0.072/0.04*	***0.001/0.12***	***<0.001/0.18***
Jobe	negative	45.2 ± 19.4 (n = 31)	67.8 ± 14.6 (n = 45)	82.0 ± 16.3 (n = 67)
	positive	44.8 ± 16.0 (n = 56)	62.2 ± 16.9 (n = 42)	76.6 ± 16.2 (n = 20)
	p/η_p^2	*0.902/0.00*	*0.100/0.03*	*0.193/0.02*

ASES = American Shoulder and Elbow Surgeons Score; IRO/Shift = internal rotation and shift; Exam 1 = pre-operative; Exam 2 = 3 months post-operative; Exam 3 = 6 months post-operative. Significant differences marked in bold.

Table 3. Comparison of Constant–Murley scores to clinical tests.

Clinical Test		Exam 1 CMS Score	Exam 2 CMS Score	Exam 3 CMS Score
IRO/Shift	negative	45.2 ± 20.6 (n = 25)	67.6 ± 16.2 (n = 50)	80.3 ± 16.9 (n = 81)
	positive	40.6 ± 16.5 (n = 62)	54.0 ± 14.4 (n = 37)	45.0 ± 21.8 (n = 6)
	p/η_p^2	*0.279/0.01*	***<0.001/0.16***	***<0.001/0.22***
Jobe	negative	42.1 ± 18.6 (n = 31)	64.5 ± 16.9 (n = 45)	79.3 ± 19.9 (n = 67)
	positive	41.8 ± 17.5 (n = 55)	58.9 ± 16.4 (n = 42)	72.8 ± 16.6 (n = 20)
	p/η_p^2	*0.958/0.00*	*0.125/0.03*	*0.188/0.20*

CMS = Constant–Murley Shoulder Score; IRO/Shift = internal rotation and shift; Exam 1 = pre-operative; Exam 2 = 3 months post-operative; Exam 3 = 6 months post-operative. Significant differences marked in bold.

Among the 51 patients with arthroscopically diagnosed supraspinatus tears, these patients' ASES ($p < 0.001$) and CMS ($p < 0.001$) scores showed significant improvements over time (Table 4). These improvements were similar during both time periods (exam 1 vs. exam 2, and exam 2 vs. exam 3). The average preoperative ASES and CMS scores for these patients were 45.4 ± 16.4 and 41.6 ± 17.5, respectively. However, these scores did not increase to above 70 until the 6-month post-operative follow-up, at which time the ASES and CMS scores were 81.1 ± 16.3 and 77.5 ± 18.6. Similarly, the clinical tests did not return to predominantly negative findings until the 6-month follow-up (IRO/Shift: 90%; Jobe: 65%) (Table 5). However, the proportion of positive IRO/Shift tests and Jobe tests sharply decreased during the observed periods. This reduction in positive tests was stronger for the IRO/Shift test (from 94% to 10%, an 84% difference) than for the Jobe test (from 94% to 35%, a 59% difference). This equates to a mean difference between exam 1 and exam 3 of 82% for the IRO/Shift test and 59% for the Jobe test.

When comparing the functional outcomes (ASES and CMS) of a single-row repair, double-row repair, and debridement, all surgical techniques led to significant time effects in all periods of the recovery process (Table 6). The time effects were slightly larger for single-row repair (ASES: $\eta_p^2 = 0.72$; CMS: $\eta_p^2 = 0.76$) than for double-row repair (ASES: $\eta_p^2 = 0.62$; CMS: $\eta_p^2 = 0.55$) and debridement (ASES: $\eta_p^2 = 0.65$; CMS: $\eta_p^2 = 0.73$). The largest partial time effect for the ASES scores was observed for the debridement procedure between exams 2 and 3 ($\eta_p^2 = 0.79$). Similarly, debridement also had the largest partial time effect for the CMS scores between exams 2 and 3 ($\eta_p^2 = 0.70$). Table 7 depicts the number of positive and negative clinical tests (IRO/Shift, Jobe) over time for each surgical procedure.

Table 4. Longitudinal analysis of ASES and CMS scores among patients with supraspinatus tears (n = 51).

Score	Examinations			Minimum Detectable Change (MDC)				Variance Analysis	
	Exam 1	Exam 2	Exam 3	ICC (95% CI)	SEM	MDC_{95}	Absolute Difference	(p/η_p^2)	Comparison of Adjacent Exams (η_p^2)
ASES	45.4 ± 16.4	65.0 ± 16.4	81.1 ± 16.3	0.38 (0–0.65)	16.4	43.8	1.6	**<0.001/0.66**	1 vs. 2 (0.52) 2 vs. 3 (0.58)
CMS	41.6 ± 17.5	61.9 ± 17.7	77.5 ± 18.6	0.51 (0.1–0.76)	12.6	42.8	1.2	**<0.001/0.68**	1 vs. 2 (0.61) 2 vs. 3 (0.55)

ICC = intraclass correlation coefficient; SEM = standard error of the mean; ASES = American Shoulder and Elbow Surgeons Score; CMS = Constant–Murley Shoulder Score; Exam 1 = pre-operative; Exam 2 = 3 months post-operative; Exam 3 = 6 months post-operative. Significant and clinically relevant differences are marked in bold.

Table 5. Longitudinal analysis of positive IRO/Shift and Jobe tests (n = 51).

Clinical Test	Exam 1	Exam 2	Exam 3	1 vs. 2 Chi^2/p	2 vs. 3 Chi^2/p
IRO/Shift	92%	61%	10%	2.33/0.127	3.58/0.059
Jobe	94%	71%	35%	**7.65/0.006**	**11.59/<0.001**

IRO/Shift test = Internal Rotation and Shift test; Exam 1 = pre-operative; Exam 2 = 3 months post-operative; Exam 3 = 6 months post-operative. Significant relationships marked in bold.

Table 6. Differences between single-row, double-row, and debridement repairs compared to ASES and CMS scores.

Score	Surgery	Examinations			Minimum Detectable Change (MDC)				Variance Analysis	
		Exam 1	Exam 2	Exam 3	ICC (95% CI)	SEM	MDC_{95}	Absolute Difference	(p/η_p^2)	Comparison of Adjacent Exams (η_p^2)
ASES	**Single** (n = 22)	42.1 ± 15.3	64.8 ± 17.3	81.8 ± 14.7	0.33 (0–0.66)	9.19	31.2	10.9	**<0.001/0.72**	1 vs. 2 (**0.62**) 2 vs. 3 (**0.56**)
	Double (n = 16)	46.4 ± 15.6	58.9 ± 16.4	73.8 ± 19.0	0.62 (0.07–0.86)	10.5	35.7	10.7	**<0.001/0.62**	1 vs. 2 (**0.51**) 2 vs. 3 (**0.50**)
	Debride (n = 13)	49.9 ± 19.0	73.1 ± 11.4	88.9 ± 11.9	0	14.1	47.9	2.0	**<0.001/0.65**	1 vs. 2 (**0.48**) 2 vs. 3 (**0.79**)
CMS	**Single** (n = 22)	42.1 ± 16.2	64.3 ± 15.7	78.7 ± 15.0	0.47 (0–0.77)	11.4	38.7	3.4	**<0.001/0.76**	1 vs. 2 (**0.68**) 2 vs. 3 (**0.62**)
	Double (n = 16)	38.1 ± 19.9	53.3 ± 16.9	68.8 ± 22.5	0.56 (0.05–0.83)	13.1	44.5	6.4	**<0.001/0.55**	1 vs. 2 (**0.55**) 2 vs. 3 (**0.41**)
	Debride (n = 13)	44.8 ± 17.0	68.5 ± 19.0	86.1 ± 15.3	0.38 (0–0.74)	13.5	45.8	1.0	**<0.001/0.73**	1 vs. 2 (**0.62**) 2 vs. 3 (**0.70**)

ICC = intraclass correlation coefficient; SEM = standard error of the mean; ASES = American Shoulder and Elbow Surgeons Score; CMS = Constant–Murley Shoulder Score; Single = single-row surgical repair; Double = double-row surgical repair; Exam 1 = pre-operative; Exam 2 = 3 months post-operative; Exam 3 = 6 months post-operative. Significant differences marked in bold.

Table 7. Differences between single-row (n = 22), double-row (n = 16) and debridement (n = 13) repairs compared to IRO/Shift and Jobe tests. Values are the number of positive and negative tests.

Test	Surgery	Exam 1		Exam 2		Exam 3	
		Pos/Neg	% Positive	Pos/Neg	% Positive	Pos/Neg	% Positive
IRO/Shift	Single (n = 22)	22/0	100	13/9	59	1/21	5
	Double (n = 16)	14/2	88	12/4	75	4/12	25
	Debridement (n = 13)	11/2	85	6/7	46	0/13	0

Table 7. *Cont.*

Test	Surgery	Exam 1		Exam 2		Exam 3	
		Pos/Neg	**% Positive**	**Pos/Neg**	**% Positive**	**Pos/Neg**	**% Positive**
Jobe	**Single (n = 22)**	21/1	95	16/6	73	10/12	45
	Double (n = 16)	16/0	100	13/3	81	6/10	38
	Debridement (n = 13)	11/2	85	6/7	46	2/11	15

IRO/Shift = Internal Rotation and Shift test; Single = single-row surgical repair; Double = double-row surgical repair; Exam 1 = pre-operative; Exam 2 = 3 months post-operative; Exam 3 = 6 months post-operative; Pos = positive test; Neg = negative test.

4. Discussion

With an increasing number of rotator cuff repairs performed annually [5], it is imperative for clinicians to understand the length of time necessary for patients to return to their activities of daily living and good tendon integrity. The results of this study demonstrate that arthroscopic repair of the supraspinatus requires approximately 6 months before functional abilities and the clinical integrity of the tendon return to adequate levels. Improvements in CMS scores and Jobe test results were slightly lower than ASES scores and the IRO/Shift test.

Both the ASES and CMS scores showed significant improvements over time, but did not distinguish between symptomatic patients with and without a rotator cuff tear. This suggests that the ASES and CMS are useful tools for demonstrating functional impairment of the affected shoulder, and not a specific pathology. These improvements in scores also demonstrate that patients with supraspinatus tendon tears, as well as other soft tissue pathologies, can have successful functional outcomes following an appropriate surgical intervention. In the supraspinatus tear patients, both scores increased drastically from exam 1 to exam 3 (57% ASES, 60% CMS (Table 4), reflecting the time necessary for tendon healing in conjunction with a rehabilitation program. Similar results were observed with the clinical tests, with the IRO/Shift test returning to negative in 90% of the supraspinatus tear patients at 6 months, and in 65% with the Jobe test (Table 5). Improvements in the CMS scores and the Jobe test were lower following surgery compared to the ASES and IRO/Shift test. This may be because the CMS and the Jobe test require an active abduction contraction against resistance, which may have been more difficult for patients earlier in the rehabilitation process (3 months) compared to after the tendon had more time to heal (6 months). As such, patients who present with a positive Jobe test and a low CMS score 3 months post-operatively may have deficient tendon strength as opposed to tendon failure, assuming the IRO/Shift test is negative.

Haunschild et al. [31] reported that following an arthroscopic rotator cuff repair, most patients returned to their jobs approximately 8 months post-surgery. The results of the current study support this previous work and note that the 51 patients who underwent supraspinatus repair required 6 months to return to adequate levels of function and negative clinical tests. We chose a common cutoff score of 70 to represent adequate shoulder function. At 3 months, the patients with supraspinatus repairs in this study had an ASES score of 65 and a CMS of 62. This CMS score is the same finding as a study conducted by Kukkonen et al. [32] who reported a CMS score of 62 at 3 months after rotator cuff surgery. Although this previous study did not assess CMS scores at 6 months, their 12-month score was 76, which is comparable to the current study's 6-month score of 78. Similarly, Nabergoj et al. [33] reported their patients with supraspinatus tears had an average CMS score of 80 at 12 months post-surgery. These previous findings, in combination with the results of the current study (Table 8), suggest that CMS scores may not increase much between 6 and 12 months postoperatively; however, further research is necessary to confirm this.

Table 8. Comparison of current study with previous research.

Study	Pre-Operative CMS Score	3-Month Post-Operative CMS Score	6-Month Post-Operative CMS Score	12-Month Post-Operative CMS Score	Time to Return to ADL
Haunschild et al. [31]	--------	--------	--------	--------	8 months
Kukkonen et al. [32]	53	62	--------	76	--------
Nabergoj et al. [33]	54	54	--------	80	--------
Current study	42	62	76	--------	6 months

CMS = Constant–Murley Shoulder Score; ADL = activities of daily living.

The ASES and CMS scores in the patients who had double-row repairs were slightly lower than the patients with single-row repairs and debridement (Table 6). Similarly, the number of positive IRO/Shift and Jobe tests was higher among the double-row procedures at almost every time period compared to the single-row and debridement procedures (Table 7). Only the Jobe test performed 6 months post-operatively had a higher positive rate for the single-row procedure. These findings should be considered when planning rehabilitation and return to activity protocols for single-row repair, double-row repair, and debridement patients.

There are a few limitations of this study's methodology worth noting. The main limitation of these data is the absence of objective radiological data to demonstrate tendon healing during the rehabilitation process and to objectively support the results of this study. The results of this study provided insight into the outcomes of tendon repair 3 and 6 months post-operatively, but future studies should also assess outcomes over longer periods, including over multiple years. As previously mentioned, caution should be used when interpreting the results of the comparisons between the special tests and the survey scores because the number of subjects in the positive and negative special test groups were vastly different between exams 1 and 3, which may have affected the variance analysis. Psychological distress may have also played a factor in the patients' functional outcomes. A previous systematic review investigated the association between psychosocial factors and patient-reported outcomes among rotator cuff repair patients [34]. This review found that negative psychosocial factors were associated with decreased function and increased pain both pre- and post-operatively. While the ASES may perform better psychometrically than the Patient-Reported Outcome Measurement Information System (PROMIS) [35], the impact of patient psychometric properties on tests such as the CMS, IRO/Shift test, and the Jobe test are still unclear. Finally, although all subjects with repaired tendons were prescribed the same rehabilitation protocol, not all subjects progressed through their respective protocol at the same pace, and the authors could not ensure each patient performed the protocol precisely.

5. Conclusions

The ASES and CMS scores, as well as the IRO/Shift and Jobe clinical tests showed significant improvements over time following supraspinatus repair. The main results and suggestions are:

- The IRO/Shift test showed a stronger ability to distinguish between those with positive tests compared to negative tests with respect to ASES and CMS scores.
- The average for all tools did not improve to acceptable levels until approximately 6 months post-surgery, with improvements in the CMS scores and Jobe test being slower at all time periods compared to the ASES and IRO/Shift test.
- The time required to return to activities of daily living and negative clinical tests was longer for the double-row repair patients compared to single-row and debridement.
- These results suggest that the ASES, CMS, IRO/Shift test, and Jobe test are useful, yet unique tools for determining the progress and clinical decision-making process among patients with supraspinatus repair.

Author Contributions: All authors provided substantial contributions to the conceptualization, research design, methodology, analysis of data, writing, revising, and final approval of the manuscript. All authors have read and agreed to the published version of the manuscript.

Funding: This research received no external funding.

Institutional Review Board Statement: This study was approved by the Ethics Commission of the Martin-Luther-University Halle-Wittenberg (reference number: 2018-05).

Informed Consent Statement: All subjects provided informed consent.

Data Availability Statement: Please contact the corresponding author for inquiries regarding study data.

Conflicts of Interest: The authors declare no conflict of interest.

References

1. Roquelaure, Y.; Ha, C.; Leclerc, A.; Touranchet, A.; Sauteron, M.; Melchior, M.; Imbernon, E.; Goldberg, M. Epidemiologic surveillance of upper-extremity musculoskeletal disorders in the working population. *Arthritis Rheum.* **2006**, *55*, 765–778.
2. Stevenson, J.H.; Trojian, T. Evaluation of shoulder pain. *J. Fam. Pract.* **2002**, *51*, 605–611.
3. Chakravarty, K.; Webley, M. Shoulder joint movement and its relationship to disability in the elderly. *J. Rheumatol.* **1993**, *20*, 1359–1361.
4. Sher, J.; Uribe, J.; Posada, A.; Murphy, B.; Zlatkin, M. Abnormal findings on magnetic resonance images of asymptomatic shoulders. *J. Bone Jt. Surg. Am.* **1995**, *77*, 10–15.
5. Ensor, K.L.; Kwon, Y.W.; DiBeneditto, M.R.; Zuckerman, J.D.; Rokito, A.S. The rising incidence of rotator cuff repairs. *J. Shoulder Elb. Surg.* **2013**, *22*, 1628–1632.
6. Randelli, P.; Cucchi, D.; Ragone, V.; de Girolamo, L.; Cabitza, P.; Randelli, M. History of rotator cuff surgery. *Knee Surg. Sports Traumatol. Arthrosc.* **2015**, *23*, 344–362. [PubMed]
7. Deprés-Tremblay, G.; Chevrier, A.; Snow, M.; Hurtig, M.B.; Rodeo, S.; Buschmann, M D. Rotator cuff repair: A review of surgical techniques, animal models, and new technologies under development. *J. Shoulder Elb. Surg.* **2016**, *25*, 2078–2085.
8. Dines, J.S.; Bedi, A.; ElAttrache, N.S.; Dines, D.M. Single-row versus double-row rotator cuff repair: Techniques and outcomes. *J. Am. Acad. Orthop. Surg.* **2010**, *18*, 83–93.
9. Kim, D.H.; ElAttrache, N.S.; Tibone, J.E.; Jun, B.-J.; DeLaMora, S.N.; Kvitne, R.S.; Lee, T.Q. Biomechanical comparison of a single-row versus double-row suture anchor technique for rotator cuff repair. *Am. J. Sports Med.* **2006**, *34*, 407–414.
10. Mazzocca, A.D.; Millett, P.J.; Guanche, C.A.; Santangelo, S.A.; Arciero, R.A. Arthroscopic single-row versus double-row suture anchor rotator cuff repair. *Am. J. Sports Med.* **2005**, *33*, 1861–1868.
11. Saridakis, P.; Jones, G. Outcomes of single-row and double-row arthroscopic rotator cuff repair: A systematic review. *J. Bone Jt. Surg. Am.* **2010**, *92*, 732–742.
12. Wall, L.B.; Keener, J.D.; Brophy, R.H. Double-row vs single-row rotator cuff repair: A review of the biomechanical evidence. *J. Shoulder Elb. Surg.* **2009**, *18*, 933–941.
13. Kocher, M.S.; Horan, M.P.; Briggs, K.K.; Richardson, T.R.; O'Holleran, J.; Hawkins, R.J Reliability, validity, and responsiveness of the American Shoulder and Elbow Surgeons subjective shoulder scale in patients with shoulder instability, rotator cuff disease, and glenohumeral arthritis. *J. Bone Jt. Surg. Am.* **2005**, *87*, 2006–2011.
14. Goldhahn, J.; Angst, F.; Drerup, S.; Pap, G.; Simmen, B.R.; Mannion, A.F. Lessons learned during the cross-cultural adaptation of the American Shoulder and Elbow Surgeons shoulder form into German. *J. Shoulder Elb. Surg.* **2008**, *17*, 248–254.
15. Richards, R.R.; An, K.-N.; Bigliani, L.U.; Friedman, R.J.; Gartsman, G.M.; Gristina, A.G.; Iannotti, J.P.; Mow, V.C.; Sidles, J.A.; Zuckerman, J.D. A standardized method for the assessment of shoulder function. *J. Shoulder Elb. Surg.* **1994**, *3*, 347–352. [CrossRef] [PubMed]
16. Constant, C.R.; Gerber, C.; Emery, R.J.; Søjbjerg, J.O.; Gohlke, F.; Boileau, P. A review of the Constant score: Modifications and guidelines for its use. *J. Shoulder Elb. Surg.* **2008**, *17*, 355–361. [CrossRef]
17. Malavolta, E.A.; Assunção, J.H.; Andrade-Silva, F.B.; Gracitelli, M.E.C.; Kiyomoto, H.D.; Neto, A.A.F. Prognostic Factors for Clinical Outcomes After Arthroscopic Rotator Cuff Repair. *Orthop. J. Sports Med.* **2023**, *11*, 23259671231160738.
18. Bejer, A.; Płocki, J.; Probachta, M.; Kotela, I.; Kotela, A. A Comparison Study of the Western Ontario Rotator Cuff Index, and the Constant-Murley Score with Objective Assessment of External Rotator Muscle Strength and Pain in Patients after Arthroscopic Rotator Cuff Repair. *Int. J. Environ. Res. Public Health* **2023**, *20*, 6316.
19. Jain, N.B.; Luz, J.; Higgins, L.D.; Dong, Y.; Warner, J.J.; Matzkin, E.; Katz, J.N. The Diagnostic Accuracy of Special Tests for Rotator Cuff Tear: The ROW Cohort Study. *Am. J. Phys. Med. Rehabil.* **2017**, *96*, 176–183.
20. Lädermann, A.; Meynard, T.; Denard, P.J.; Ibrahim, M.; Saffarini, M.; Collin, P. Reliable diagnosis of posterosuperior rotator cuff tears requires a combination of clinical tests. *Knee Surg. Sports Traumatol. Arthrosc.* **2021**, *29*, 2118–2133.

21. Fieseler, G.; Laudner, K.; Sendler, J.; Cornelius, J.; Schulze, S.; Lehmann, W.; Hermassi, S.; Delank, K.-S.; Schwesig, R. The internal rotation and shift-test for the detection of superior lesions of the rotator cuff: Reliability and clinical performance. *JSES Int.* **2022**, *6*, 495–499. [PubMed]
22. Fieseler, G.; Schwesig, R.; Sendler, J.; Cornelius, J.; Schulze, S.; Lehmann, W.; Hermassi, S.; Delank, K.-S.; Laudner, K. IRO/Shift Test Is Comparable to the Jobe Test for Detection of Supraspinatus Lesions. *J. Pers. Med.* **2022**, *12*, 1422. [CrossRef]
23. Piatti, M.; Gorla, M.; Alberio, F.; Omeljaniuk, R.J.; Rigamonti, L.; Gaddi, D.; Turati, M.; Bigoni, M. Comparison of all-suture anchors with metallic anchors in arthroscopic cuff repair: Structural and functional properties and clinical suitability. *J. Orthop.* **2023**, *39*, 66–69. [CrossRef] [PubMed]
24. Sallay, P.I.; Reed, L. The measurement of normative American Shoulder and Elbow Surgeons scores. *J. Shoulder Elb. Surg.* **2003**, *12*, 622–627.
25. Roy, J.S.; MacDermid, J.C.; Woodhouse, L.J. A systematic review of the psychometric properties of the Constant-Murley score. *J. Shoulder Elb. Surg.* **2010**, *19*, 157–164.
26. Yian, E.H.; Ramappa, A.J.; Arneberg, O.; Gerber, C. The Constant score in normal shoulders. *J. Shoulder Elb. Surg.* **2005**, *14*, 128–133. [CrossRef]
27. Bortz, J. Basic Principle of One-Factorial Analysis of Variance. In *Statistics for Social Scientists*; Springer: New York, NY, USA, 1999; pp. 237–249.
28. Richardson, J.T.E. Eta Squared and Partial Eta Squared as Measures of Effect Size in Educational Research. *Educ. Res. Rev.* **2011**, *6*, 135–147. [CrossRef]
29. Weir, J.P. Quantifying test-retest reliability using the intraclass correlation coefficient and the SEM. *J. Strength Cond. Res.* **2005**, *19*, 231–240.
30. Stratford, P. Getting More from the Literature: Estimating the Standard Error of Measurement from Relibility Studies. *Physiother. Can.* **2004**, *56*, 27–30. [CrossRef]
31. Haunschild, E.D.; Gilat, R.; Lavoie-Gagne, O.; Fu, M.C.; Tauro, T.; Forsythe, B.; Cole, B.J. Return to Work After Primary Rotator Cuff Repair: A Systematic Review and Meta-analysis. *Am. J. Sports Med.* **2021**, *49*, 2238–2247.
32. Kukkonen, J.; Kauko, T.; Vahlberg, T.; Joukainen, A.; Äärimaa, V. Investigating minimal clinically important difference for Constant score in patients undergoing rotator cuff surgery. *J. Shoulder Elb. Surg.* **2013**, *22*, 1650–1655.
33. Nabergoj, M.; Bagheri, N.; Bonnevialle, N.; Gallinet, D.; Barth, J.; Labattut, L.; Metais, P.; Godeneche, A.; Garret, J.; Clavert, P.; et al. Arthroscopic rotator cuff repair: Is healing enough? *Orthop. Traumatol. Surg. Res.* **2021**, *107*, 103100. [PubMed]
34. Panattoni, N.; Longo, U.G.; De Salvatore, S.; Castaneda, N.S.C.; Ambrogioni, L.R.; Piredda, M.; De Marinis, M.G.; Denaro, V. The influence of psychosocial factors on patient-reported outcome measures in rotator cuff tears pre- and post-surgery: A systematic review. *Qual. Life Res.* **2022**, *31*, 91–116. [PubMed]
35. Lavoie-Gagne, O.; Huddleston, H.P.; Fu, M.C.; Mehta, N.; Nwachukwu, B.; Forlenza, E.M.; Yanke, A.; Cole, B.J.; Verma, N.N.; Forsythe, B. PROMIS Upper Extremity underperforms psychometrically relative to American Shoulder and Elbow Surgeons score in patients undergoing primary rotator cuff repair. *J. Shoulder Elb. Surg.* **2022**, *31*, 718–725.

Journal of Personalized Medicine

Article

Individual Influence of Trochlear Dysplasia on Patellofemoral Kinematics after Isolated MPFL Reconstruction

Andreas Frodl [1,*], Thomas Lange [2], Markus Siegel [1], Hans Meine [3], Elham Taghizadeh [3], Hagen Schmal [4] and Kaywan Izadpanah [1]

1 Department of Orthopedics and Traumatology, Freiburg University Hospital, 79106 Freiburg, Germany
2 Department of Radiology, Medical Physics, Freiburg University Hospital, 79106 Freiburg, Germany
3 Fraunhofer Institute for Digital Medicine, 28359 Freiburg, Germany
4 Department of Orthopedic Surgery, University Hospital Odense, Sdr. Boulevard 29, 5000 Odense, Denmark
* Correspondence: andreas.frodl@uniklinik-freiburg.de

Abstract: Introduction: The influence of the MPFL graft in cases of patella instability with dysplastic trochlea is a controversial topic. The effect of the MPFL reconstruction as single therapy is under investigation, especially with severely dysplastic trochlea (Dejour types C and D). The purpose of this study was to evaluate the impact of trochlear dysplasia on patellar kinematics in patients suffering from low flexion patellar instability under weight-bearing conditions after isolated MPFL reconstruction. Material and Methods: Thirteen patients were included in this study, among them were eight patients with mild dysplasia (Dejour type A and B) and five patients with severe dysplasia (Dejour type C and D). By performing a knee MRI with in situ loading, patella kinematics and the patellofemoral cartilage contact area could be measured under the activation of the quadriceps musculature in knee flexion angles of 0°, 15° and 30°. To mitigate MRI motion artefacts, prospective motion correction based on optical tracking was applied. Bone and cartilage segmentation were performed semi-automatically for further data analysis. Cartilage contact area (CCA) and patella tilt were the main outcome measures for this study. Pre- and post-surgery measures were compared for each group. Results: Data visualized a trending lower patella tilt after MPFL graft installation in both groups and flexion angles of the knee. There were no significant changes in patella tilt at 0° (unloaded pre-surgery: 22.6 ± 15.2; post-surgery: 17.7 ± 14.3; $p = 0.110$) and unloaded 15° flexion (pre-surgery: 18.9 ± 12.7; post-surgery: 12.2 ± 13.0; $p = 0.052$) of the knee in patients with mild dysplasia, whereas in patients with severe dysplasia of the trochlea the results happened not to be significant in the same angles with loading of 5 kg (0° flexion pre-surgery: 34.4 ± 12.1; post-surgery: 31.2 ± 16.1; $p = 0.5$; 15° flexion pre-surgery: 33.3 ± 6.1; post-surgery: 23.4 ± 8.6; $p = 0.068$). CCA increased in every flexion angle and group, but significant increase was seen only between 0°–15° (unloaded and loaded) in mild dysplasia of the trochlea, where significant increase in Dejour type C and D group was seen with unloaded full extension of the knee (0° flexion) and 30° flexion (unloaded and loaded). Conclusion: This study proves a significant effect of the MPFL graft to cartilage contact area, as well as an improvement of the patella tilt in patients with mild dysplasia of the trochlea. Thus, the MPFL can be used as a single treatment for patient with Dejour type A and B dysplasia. However, in patients with severe dysplasia the MPFL graft alone does not significantly increase CCA.

Keywords: cartilage contact area; patella dynamics; trochlea dysplasia

check for updates

Citation: Frodl, A.; Lange, T.; Siegel, M.; Meine, H.; Taghizadeh, E.; Schmal, H.; Izadpanah, K. Individual Influence of Trochlear Dysplasia on Patellofemoral Kinematics after Isolated MPFL Reconstruction. *J. Pers. Med.* **2022**, *12*, 2049. https://doi.org/10.3390/jpm12122049

Academic Editors: Markus Gesslein and Eduard Kurz

Received: 24 November 2022
Accepted: 7 December 2022
Published: 12 December 2022

1. Introduction

Patellar instability is a frequent cause of early degenerative articular cartilage damage and developing patellofemoral osteoarthritis. Among others, trochlear dysplasia, patella alta and genu valgum are reported to predispose for patellar instability or recurrent dislocations [1]. However, patellar dislocations and instability are always accompanied by damage to the medial capsuloligamentous restrain tissue [2], consisting of mainly the medial patellofemoral ligament (MPFL), the weaker medial patellotibial ligament (MPTL), the medial patellomeniscal ligament (MPML), and the medial quadriceps tendon femoral ligament (MQTFL) [3,4]. Current studies report the MPFL as the strongest restraint against lateral translation of the patella and MPFL-plasty as a successful treatment option for patients with low flexion patellar instability [5,6]. About 50–60% of the total patellar restraining forces is provided by the MPFL between angles of 0–30° flexion [7]. The patellar tracking mechanism changes from predominant soft tissue, especially ligamentary restraints, to increasingly osseous tracking due to the engagement of the patella in the trochlear groove at about 30° of knee flexion [8].

The effects of surgical stabilization on patellar kinematics and cartilage contact mechanisms are being investigated. However, the impact of trochlear dysplasia on cartilaginous contact areas and patella kinematics after MPFL reconstruction remain unclear [9]. In order to achieve optimal post-surgery results and prevent early joint degeneration due to persistent maltracking after single MPFL reconstruction surgery, the necessity of adjunctional measures, i.e., trochlea augmentation, has to be evaluated.

Although evaluating joint kinematics and cartilage surface areas is challenging, it is greatly facilitated by fluoroscopy and Magnetic Resonance Imaging (MRI) [10,11]. Cadaveric studies primarily enabled insights into cartilage deformation under axial loading [12]. The reproducibility and quantification of MRI-guided investigations in vivo is technically difficult. Knee flexion is required to load the patellofemoral joint axially. Weight bearing under knee flexion leads to joint shaking, and consequently to motion artefacts in MRI scans. Pulley or pneumatic loading devices have been used to enable reproducible and comparable loading among a cohort of subjects [13–16]. In anticipation of motion artefacts, prospective motion correction (PMC) systems with optical tracking have successfully minimized artefacts during loaded MRI scans [16].

The purpose of this study was to evaluate the impact of trochlear dysplasia on patellar kinematics in patients suffering from low flexion patellar instability under weight-bearing conditions after isolated MPFL reconstruction. To test our hypothesis—that in patients with severe trochlea dysplasia, reconstruction of the MPFL does not sufficiently increase cartilage contact area (CCA) and patella tracking as "stand alone" treatment—we conducted pre- and post-surgery MRI scans of each patient to observe the intra-patient changes after MPFL augmentation.

2. Materials and Methods

Thirteen patients with a dysplastic trochlea and chronic patellar luxation were included in this study. Only patients with chronic patella instability, dysplasia of the trochlea, no previous surgery on the affected knee and aged between 18–40 years were included. Exclusion criteria included a history of prior patellofemoral surgery, metallic material from previous knee surgery, pregnancy, retropatellar osteoarthritis and claustrophobia. Measurements were taken pre- and post-surgery after implementation of an MPFL graft. Trochlear configurations and patellar positions were determined by measuring the patellotrochlear index, lateral trochlear inclination angle, mean osseous sulcus angle and then classified according to the Dejour classification for dysplastic trochlea in all subjects [17–20]. An overview of the patient selection process is shown in Figure 1.

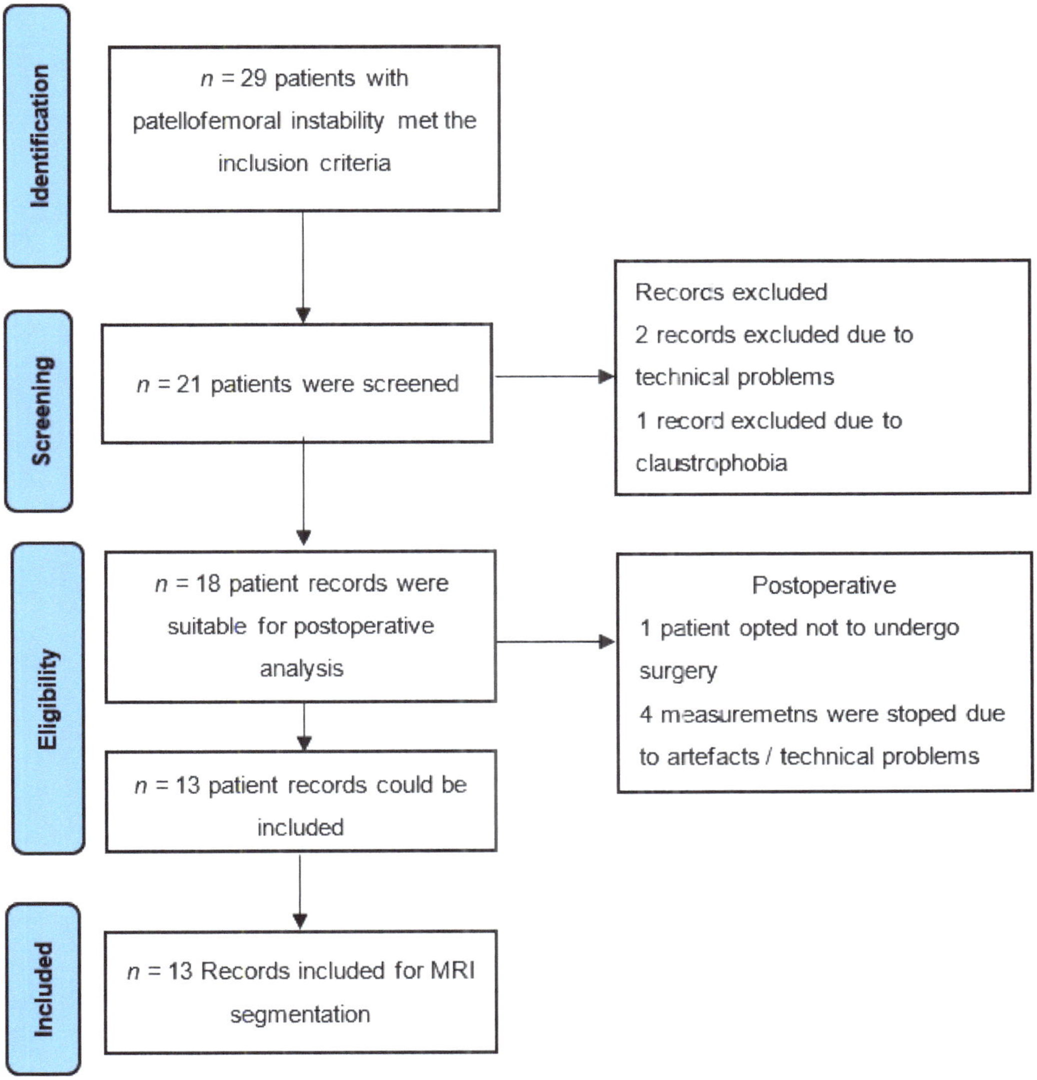

Figure 1. Patient selection process according to STROBE guideline.

The MRI measurements were performed with a Magnetom Trio 3T MRI system (Siemens Healthineers, Erlangen, Germany), using an 8-channel multipurpose coil (NORAS MRI products, Germany) for signal reception, which was attached to the thigh with a hook-and-loop fastener. The experiments were conducted with an MRI-compatible pneumatic loading device to enable accurate load adjustment in the 0–500 N range (Figure 1). A 3D turbo-spin echo (TSE) protocol with GRAPPA parallel imaging acceleration by a factor of 2 and an isotropic resolution of 0.5 mm was applied for the MRI scans. Further scan parameters were TR = 1.8 s, TE = 59 ms, receiver bandwidth = 504 Hz/Px, and scan duration = 6:20 min. For prospective motion correction, we used a moiré phase tracking (MPT) system with one in-bore camera and a single tracking marker [21]. Translational and rotational motion was optically tracked with a frame rate of 80 frames/s, enabling real-time updates of the MRI measurement volume before every excitation pulse.

The patient was tied to the scanner bed with a weight-lifting belt, and the leg to be examined was placed in the sliding carriage of the loading device (Figure 2). Measurements were performed with knee flexion angles of 0°, 15° and 30°. For accurate and reproducible angle adjustment, the knee was propped with a height-adjustable foam roll.

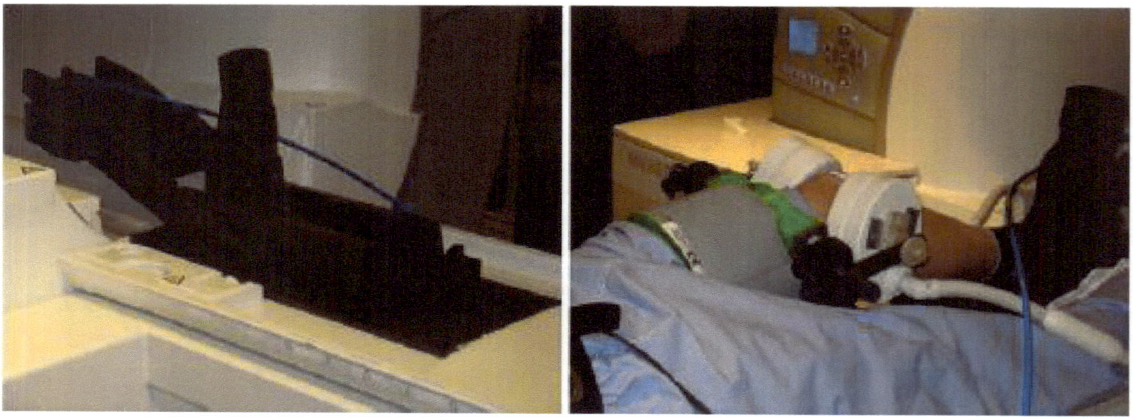

Figure 2. Experimental Setup. The knee is loaded with an MR-compatible pneumatic loading device (**left**). The multipurpose coil is attached to the thigh and the tracking marker for prospective motion correction is taped to the knee cap (**right**).

For all three flexion angles, measurements were performed first without loading and then with a load of approximately 5 kg. The MRI scans were started with a delay of 30 s after the onset of loading to allow adaptation of the cartilage to the load [11,16,22].

SATORI, the web-based application developed by Fraunhofer MEVIS was used for semi-automated segmentation of the MR images. SATORI is a customizable annotation and analysis tool based on the MeVisLab rapid prototyping environment for medical image analysis and visualization [23,24]. Bones and cartilages were semi-automatically segmented in the data acquired at knee extension position (base image). We applied inter-slice smoothing (Gaussian kernel with $\sigma = 1.25$ mm) to reduce discontinuities caused by slice-wise drawing.

The Euclidean distance between two opposing cartilage surfaces was computed. Based on these, the cartilage contact area (CCA) was defined as the cartilage surface area with an inter-cartilage distance below 1 mm. Following this definition, we achieved contiguous contact areas without holes in all our measurement data. Accurate and consistent segmentation of the cartilages were needed to compute CCAs for each flexion angle. Based on available manual segmentation masks, a convolutional neural network, known as U-Net, was trained to segment bones in all the remaining images (flexion angle > 0° and loading of 50 N) [25]. An iterative closest point algorithm on the surface meshes created from the masks was used to pre-align the femur (patella) bone in the extended knee to the flexed knee position [26]. Following this coarse alignment of the bones, a rigid image registration of the two MRI images with the normalized gradient fields (NGF) distance measure was performed to perfect the bone alignments [27]. The NGF evaluation was limited to a mask region comprising the femur (patella) bone mask and a dilated (3 mm) region around it, ensuring that the bone contours are entirely accounted for. The resulting refined transformation matrices were used to align femur and patella and the corresponding cartilages in the extension position to the corresponding bone in the flexed knee position.

For statistical analysis, our cohort was split into groups with severe (Dejour type C and D) and mild (Types A and B) dysplastic trochlea. Pre- and post-surgery data were evaluated using a Wilcoxon signed-rank test for paired data. Differences with $p < 0.05$ were considered significant.

Statistical analyses were performed using IBM SPSS Statistics 28.0 (IBM Corp., 2017. IBM *SPSS* Statistics for Windows, Armonk, NY, USA).

This study was approved by the Institutional Review Board (Freiburg University's Ethics Committee approved this study, ID 443/16) and all subjects provided written informed consent before participation. All subjects voluntarily took part in the study in accordance with the Declaration of Helsinki.

The results of all statistical tests were interpreted in an exploratory sense. No adjustments for multiple testing were made in this exploratory study.

3. Results

A total of 13 patients with trochlea dysplasia were included for data analysis. Of those, trochlea dysplasia was classified according to reported ranges in the literature [13,20]. For distributive analysis of mild and severe dysplasia, we conducted a subgroup analysis of patients with mild (Dejour types A and B; n = 8) and severe (Dejour types C + D; n = 5) trochlear dysplasia. A demographic overview is depicted in Table 1.

Table 1. Cohort demographics.

	Mild Dejour Typ A + B	**Severe Dejour Type C + D**	**Total**
Gender ratio Female:Male	4:4	4:1	8:5
Age in years	29.2 ± 9.6	23.5 ± 4.5	27.2 ± 8.4
Height in cm	*171.6 ± 8.2*	*169.8 ± 6.6*	*171.0 ± 7.4*
Weight in kg	*71.5 ± 7.5*	*70.4 ± 5.0*	*71.1 ± 6.6*
BMI	*24.4 ± 2.7*	24.5 ± 2.7	24.2 ± 2.6

The MPFL influence on patellar dynamics was evaluated via the measured differences between pre- and post-surgery patellar tilt, patellofemoral CCA and medial-to-lateral translation of the patella.

3.1. Patella Tilt

Data visualized a trending lower patella tilt after MPFL graft installation in both groups and at all flexion angles of the knee (Figures 3 and 4). Results turned out to be significant in every group except for a flexion angle of 15° with loading, as well as 0° extension with loading in case of severe dysplasia. In patients with mild trochlea dysplasia no significant change in patella tilt could be seen in full extension (0°) and 15° flexion of the knee without loading, respectively (Table 2). In patients with mild dysplasia, patella tilt was also significantly lower during 30° flexion with and without loading.

3.2. Cartilage Contact Area

The changes in cartilage contact area are depicted in Table 3. We observed a CCA increase in patients with severely and mildly dysplastic trochlea for the loaded and unloaded measurements, compared to pre-surgery measurements (Figures 5 and 6).

However, the increase in CCA at all 0°–15° flexion angles was significant in patients with mild trochlear dysplasia.

In patients with a severely dysplastic trochlea, we perceived a trending increase, but the MPFL graft only had a significant influence on the CCA in 0° flexion (unloaded) and 30° flexion (unloaded and loaded).

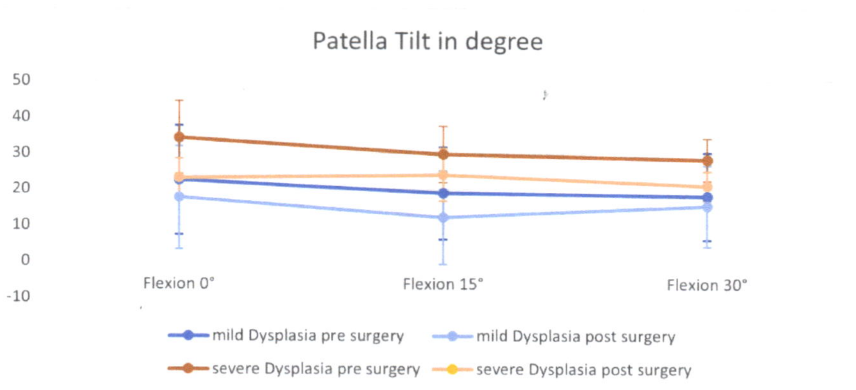

Figure 3. Patella tilt without loading before and after surgery.

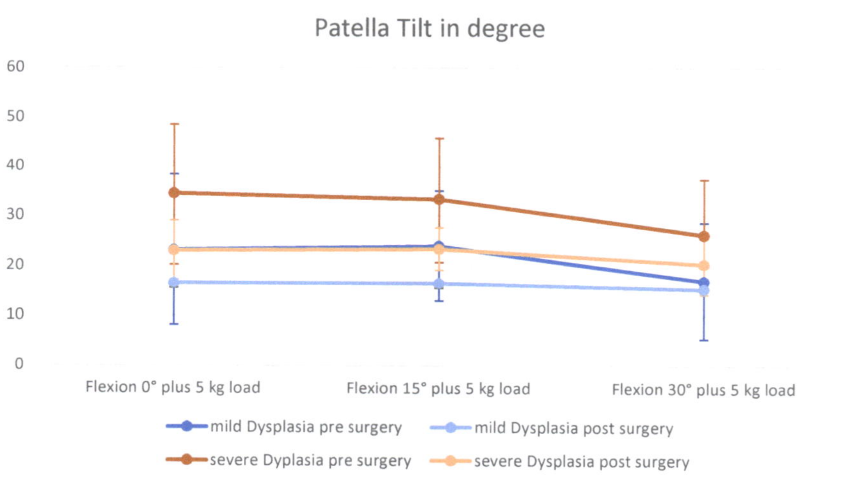

Figure 4. Patella tilt with loading before and after surgery.

Table 2. Patella tilt in degree.

Flexion Angle	*n*	Mild Pre-Operative Mean ± SD	Mild Post-Operative Mean ± SD	*p*	*n*	Severe Pre-Operative Mean ± SD	Severe Post-Operative Mean ± SD	*p*
Flexion 0°	8	22.6 ± 15.2	17.7 ± 14.3	0.110	5	34.4 ± 10.1	23.1 ± 7.6	0.043
Flexion 0° plus 5 kg load	8	23.2 ± 15.1	16.5 ± 14.1	0.051	5	34.4 ± 12.1	31.2 ± 16.1	0.500
Flexion 15°	8	18.9 ± 12.7	12.2 ± 13.0	0.052	5	29.7 ± 7.8	24.0 ± 7.4	0.043
Flexion 15° plus 5 kg load	8	24.0 ± 11.0	16.5 ± 12.5	0.044	5	33.3 ± 6.1	23.4 ± 8.6	0.068
Flexion 30°	8	18.1 ± 12.1	15.4 ± 11.3	0.041	5	28.3 ± 5.9	21.0 ± 4.0	0.043
Flexion 30° plus 5 kg load	8	17.0 ± 11.6	15.4 ± 11.2	0.039	5	26.2 ± 6.3	20.5 ± 3.1	0.043

Table 3. Cartilage contact area in mm^2.

Flexion Angle	Number of Patients	Mild Pre-Operative Mean ± SD	Mild Post-Operative Mean ± SD	*p*	Number of Patients	Severe Pre-Operative Mean ± SD	Severe Post-Operative Mean ± SD	*p*
Flexion 0°	8	74.2 ± 51.8	195.1 ± 49.9	0.007	5	56.0 ± 46.6	120.5 ± 31.1	0.043
Flexion 0° plus 5 kg load	8	76.3 ± 50.4	183.4 ± 57.7	0.011	5	84.4 ± 50.3	139.8 ± 84.5	0.225
Flexion 15°	8	114.7 ± 75.7	250.4 ± 90.0	0.007	5	131.5 ± 40.3	170.4 ± 36.5	0.345
Flexion 15° plus 5 kg load	8	104.4 ± 82.0	215.9 ± 38.2	0.009	5	89.5 ± 45.4	223.5 ± 120.4	0.068
Flexion 30°	8	339.7 ± 87.3	379.7 ± 30.2	0.128	5	148.1 ± 85.9	238.3 ± 92.5	0.043
Flexion 30° plus 5 kg load	8	336.7 ± 99.3	397.8 ± 54.8	0.056	5	187.6 ± 67.8	312.7 ± 42.2	0.043

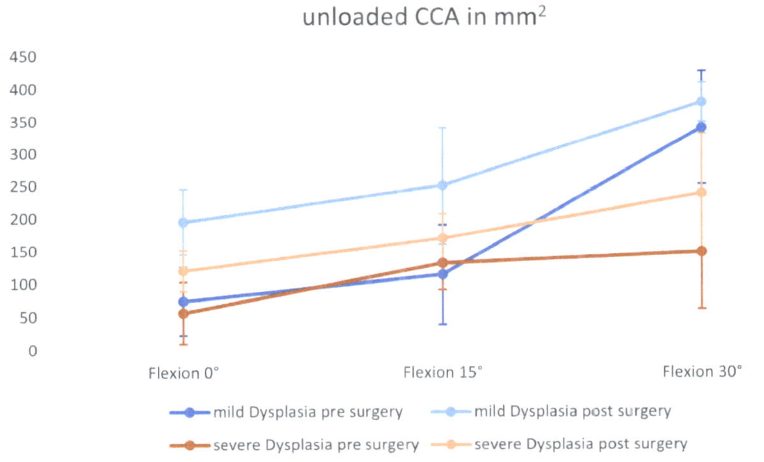

Figure 5. Changes in Cartilage Contact Area (CCA) without loading.

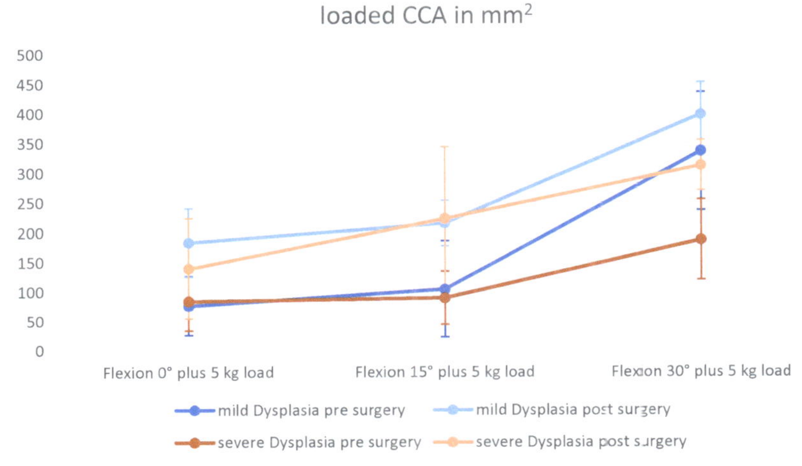

Figure 6. Changes in CCA with loading.

4. Discussion

To date, this is the first work to have quantitatively assessed patellofemoral dynamics under in situ loading. For the first time, improvement of patellar tilt and CCA could be demonstrated following medial patellofemoral ligament reconstruction for recurrent patellar instability using and A.I. algorithm based on segmented MRI in a loading device. However, in cases of dysplastic trochlea the MPFL did not increase CCA or lowered patella tilt significantly during the flexion of the knee. In an exploratory investigation under in situ loading, Clark et al. measured CCA with patella-femoral instability against healthy individuals with axial MRI sequences across different flexion angles between 0–40° of knee flexion. A continuous increase in CCA was documented over the course of flexion [28]. Similar observations were made in the present study.

Normally, to control the maltracking of the patella before and after surgery, standard imaging assessment comprises plain radiographs, CT scans or MRI scans.

Plain radiographs lack sensitivity in the detection of osteochondral lesions. In approximately 40%, small osteochondral lesions are missed [29]. CT and MRI present modalities to also account for predisposing factors associated with chronic patellar instability or patella maltracking [30,31]. However, all of the above-named imaging procedures only allow static insight into the patella-femoral joint and do not account for dynamic factors, such as quadriceps activation.

The latest literature suggests using a pulley system as a loading device [6,10,14]. Due to limited space in the scanner bore, and to enable load adjustment from the scanner control room, we decided to employ a pneumatic loading device. Motion artefacts were mitigated with camera-based prospective motion correction to optimize the image quality for robust segmentation [32,33].

The main findings of the present research were that the MPFL graft significantly increases the CCA in patients with a mildly dysplastic trochlea for flexion angles of 0°–15° (loaded and unloaded) and significantly decreases patellar tilt with quadriceps activation (loaded) at 15° knee flexion in a mildly and a severely dysplastic trochlea.

In their 2021 study, Stevens et al. reported an increase in the patellofemoral CCA when comparing patients before and after patellofemoral-stabilizing surgery using axial 5 mm slices in loaded dynamic MRI imaging. The greatest differences were shown over a range of 11–20° in active flexion [34]. The results of our study visualize a significant increase in CCA for early flexion angles (0–15°) as well but only in patients with mild dysplasia.

However, the surgical treatment of Stevens et al. comprised trochlea augmentation, tibial tubercule osteotomy and MPFL reconstruction, but in later analysis no distinction between treatment methods was found.

The patella-stabilization process is complex, involving static (joint geometry), passive (supportive retinacula) and active (quadriceps muscles) stabilizers [35]. During the first 20 degrees of knee flexion, the patella is given no bony support, but stabilization is provided by the medial and lateral retinacula—the passive stabilizers [36]. Most of the lateral retinaculum arises from the iliotibial band. Its tightness affects lateral tracking and patella tilt, primarily at around 20° flexion [35,36]. The medial retinaculum is thinner than its lateral counterpart. The MPFL is the primary restrainer to lateral patellar translation. The MPFL contributes up to 60% of the restraint to patellar lateral displacement from full extension to 20° flexion [37]. Beyond 20° flexion, the joint geometry increasingly gains influence on patellar tracking with each degree of flexion [4,38,39]. Whereas the active stabilizer loses influence at higher degrees of flexion, the impact of passive and static stabilizers rises [40,41]. In converse conclusion, the influence of a dysplastic trochlea rises during flexion, causing an alteration in the patellar tilt and reducing the cartilage contact area (CCA), as illustrated by our results. In patients with severe dysplasia, it appears that the degree of dysplasia exceeds the MFPL graft's capacity to improve the CCA.

Limitations

This study's results are limited by the small number of cases. Thus, the results should be understood as exploratory and may vary for larger study populations. As seen and faced by other studies, motion artefacts are major disadvantages of dynamic MRI imaging [22,28,34]. In anticipation we used a prospective motion correction system, which led to a lower rate of "unusable" MRI scans due to motion artefacts.

5. Conclusions

This study demonstrates a significant effect of the MPFL graft on cartilage contact area, as well as patella tilt in patients with mild dysplasia of the trochlea. Thus, the MPFL can be used as a single treatment for patients with Dejour type A and B dysplasia.

However, in patients with severe dysplasia the MPFL graft alone does not significantly increase CCA during the flexion phase of the knee. The influence of the trochlea dysplasia does not seem to be compensable with the MPFL alone. Hence, adjunctive measures, such as a surgical trochlea augmentation, seem to be needed with Dejour type C and D.

However, this exploratory study only reports results for very small cohorts. For a better understanding of in situ patella dynamics, more investigations with larger cohorts are needed.

Author Contributions: Conceptualization, K.I. and H.M.; methodology, K.I., H.M. and T.L.; software, H.M. and E.T.; validation, formal analysis and investigation A.F., T.L., M.S., K.I., E.T. and H.M.; resources, K.I., T.L. and H.M.; data curation, A.F., M.S., T.L. and E.T.; writing—original draft preparation, A.F.; writing—review and editing, A.F., T.L., M.S., K.I., E.T., H.S. and H.M.; visualization, A.F., M.S. and E.T.; supervision, K.I., H.S.; project administration, K.I., H.M.; funding acquisition, A.F. and K.I. All authors have read and agreed to the published version of the manuscript.

Funding: This research was supported by the Deutsche Forschungsgemeinschaft (Grant number: NE 2136/3-1). The APC was funded by the Open Access Publication Fund of the University of Freiburg (OA-22-540).

Institutional Review Board Statement: The study was conducted in accordance with the Declaration of Helsinki and approved by the Institutional Review Board of the University Hospital Freiburg (Nr. 91/19-210,696, date of approval 2019).

Informed Consent Statement: All patients gave written informed consent prior to participation.

Data Availability Statement: All data are available in the article.

Conflicts of Interest: The authors confirm that there is no conflict of interest that could be perceived as prejudicing the impartiality of the research reported.

References

1. Grantham, W.J.; Aman, Z.S.; Brady, A.W.; Rosenberg, S.I.; Lee Turnbull, T.; Storaci, H.W.; Dornan, G.J.; LaPrade, R.F. Medial Patellotibial Ligament Reconstruction Improves Patella Tracking When Combined with Medial Patellofemoral Reconstruction: An In Vitro Kinematic Study. *Arthrosc. J. Arthrosc. Relat. Surg. Off. Publ. Arthrosc. Assoc. N. Am. Int. Arthrosc. Assoc.* **2020**, *36*, 2501–2509. [CrossRef]
2. Huntington, L.S.; Webster, K.E.; Devitt, B.M.; Scanlon, J.P.; Feller, J.A. Factors Associated with an Increased Risk of Recurrence After a First-Time Patellar Dislocation: A Systematic Review and Meta-analysis. *Am. J. Sports Med.* **2020**, *48*, 2552–2562. [CrossRef]
3. Hinckel, B.B.; Gobbi, R.G.; Kaleka, C.C.; Camanho, G.L.; Arendt, E.A. Medial patellotibial ligament and medial patellomeniscal ligament: Anatomy, imaging, biomechanics, and clinical review. *Knee Surg. Sports Traumatol. Arthrosc. Off. J. ESSKA* **2018**, *26*, 685–696. [CrossRef]
4. LaPrade, M.D.; Kallenbach, S.L.; Aman, Z.S.; Moatshe, G.; Storaci, H.W.; Turnbull, T.L.; Arendt, E.A.; Ghala, J.; LaPrade, R.F. Biomechanical Evaluation of the Medial Stabilizers of the Patella. *Am. J. Sports Med.* **2018**, *46*, 1575–1582. [CrossRef]
5. Spang, R.; Egan, J.; Hanna, P.; Lechtig, A.; Haber, D.; DeAngelis, J.P.; Nazarian, A.; Ramappa, A.J. Comparison of Patellofemoral Kinematics and Stability After Medial Patellofemoral Ligament and Medial Quadriceps Tendon-Femoral Ligament Reconstruction. *Am. J. Sports Med.* **2020**, *48*, 2252–2259. [CrossRef]
6. Stephen, J.M.; Kittl, C.; Williams, A.; Zaffagnini, S.; Muccioli, G.M.M.; Fink, C.; Amis, A.A. Effect of Medial Patellofemoral Ligament Reconstruction Method on Patellofemoral Contact Pressures and Kinematics. *Am. J. Sports Med.* **2016**, *44*, 1186–1194. [CrossRef]

7. Desio, S.M.; Burks, R.T.; Bachus, K.N. Soft tissue restraints to lateral patellar translation in the human knee. *Am. J. Sports Med.* **1998**, *26*, 59–65. [CrossRef]
8. Imhoff, F.B.; Funke, V.; Muench, L.N.; Sauter, A.; Englmaier, M.; Woertler, K.; Imhoff, A.B.; Feucht, M.J. The complexity of bony malalignment in patellofemoral disorders: Femoral and tibial torsion, trochlear dysplasia, TT-TG distance, and frontal mechanical axis correlate with each other. *Knee Surg. Sports Traumatol. Arthrosc. Off J. ESSKA* **2020**, *28*, 897–904. [CrossRef]
9. Gold, G.E.; Besier, T.F.; Draper, C.E.; Asakawa, D.S.; Delp, S.L.; Beaupre, G.S. Weight-bearing MRI of patellofemoral joint cartilage contact area. *J. Magn. Reson. Imaging JMRI* **2004**, *20*, 526–530. [CrossRef]
10. Cotofana, S.; Eckstein, F.; Wirth, W.; Souza, R.B.; Li, X.; Wyman, B.; Graverand, M.-P.H.-L.; Link, T.; Majumdar, S. In vivo measures of cartilage deformation: Patterns in healthy and osteoarthritic female knees using 3T MR imaging. *Eur. Radiol.* **2011**, *21*, 1127–1135. [CrossRef]
11. Hosseini, A.; Van de Velde, S.K.; Kozanek, M.; Gill, T.J.; Grodzinsky, A.J.; Rubash, H.E.; Li, G. In-vivo time-dependent articular cartilage contact behavior of the tibiofemoral joint. *Osteoarthr. Cartil.* **2010**, *18*, 909–916. [CrossRef]
12. Herbort, M.; Hoser, C.; Domnick, C.; Raschke, M.J.; Lenschow, S.; Weimann, A.; Kösters, C.; Fink, C. MPFL reconstruction using a quadriceps tendon graft: Part 1: Biomechanical properties of quadriceps tendon MPFL reconstruction in comparison to the Intact MPFL. A human cadaveric study. *Knee* **2014**, *21*, 1169–1174. [CrossRef]
13. Souza, R.B.; Kumar, D.; Calixto, N.; Singh, J.; Schooler, J.; Subburaj, K.; Li, X.; Link, T.M.; Majumdar, S. Response of knee cartilage T1rho and T2 relaxation times to in vivo mechanical loading in individuals with and without knee osteoarthritis. *Osteoarthr. Cartil.* **2014**, *22*, 1367–1376. [CrossRef]
14. Shin, C.S.; Carpenter, R.D.; Majumdar, S.; Ma, C.B. Three-dimensional in vivo patellofemoral kinematics and contact area of anterior cruciate ligament-deficient and -reconstructed subjects using magnetic resonance imaging. *Arthrosc. J. Arthrosc. Relat. Surg. Off. Publ. Arthrosc. Assoc. N. Am. Int. Arthrosc. Assoc.* **2009**, *25*, 1214–1223. [CrossRef]
15. Lange, T.; Knowles, B.R.; Herbst, M.; Izadpanah, K.; Zaitsev, M. Comparative T2 and T1ρ mapping of patellofemoral cartilage under in situ mechanical loading with prospective motion correction. *J. Magn. Reason. Imaging JMRI* **2017**, *46*, 452–460. [CrossRef]
16. Lange, T.; Taghizadeh, E.; Knowles, B.R.; Südkamp, N.P.; Zaitsev, M.; Meine, H.; Izadpanah, K. Quantification of patellofemoral cartilage deformation and contact area changes in response to static loading via high-resolution MRI with prospective motion correction. *J. Magn. Reason. Imaging JMRI* **2019**, *50*, 1561–1570. [CrossRef]
17. Biedert, R.M.; Albrecht, S. The patellotrochlear index: A new index for assessing patellar height. *Knee Surg. Sports Traumatol. Arthrosc. Off. J. ESSKA* **2006**, *14*, 707–712. [CrossRef]
18. Carrillon, Y.; Abidi, H.; Dejour, D.; Fantino, O.; Moyen, B.; Tran-Minh, V.A. Patellar instability: Assessment on MR images by measuring the lateral trochlear inclination-initial experience. *Radiology* **2000**, *216*, 582–585. [CrossRef]
19. Tsavalas, N.; Katonis, P.; Karantanas, A.H. Knee joint anterior malalignment and patellofemoral osteoarthritis: An MRI study. *Eur. Radiol.* **2012**, *22*, 418–428. [CrossRef]
20. Nelitz, M.; Lippacher, S.; Reichel, H.; Dornacher, D. Evaluation of trochlear dysplasia using MRI: Correlation between the classification system of Dejour and objective parameters of trochlear dysplasia. *Knee Surg. Sports Traumatol. Arthrosc. Off. J. ESSKA* **2014**, *22*, 120–127. [CrossRef]
21. Maclaren, J.; Armstrong, B.S.R.; Barrows, R.T.; Danishad, K.A.; Ernst, T.; Foster, C.L.; Gumus, K.; Herbst, M.; Kadashevich, I.Y.; Kusik, T.P.; et al. Measurement and Correction of Microscopic Head Motion during Magnetic Resonance Imaging of the Brain. *PLoS ONE* **2012**, *7*, e48088. [CrossRef] [PubMed]
22. Lange, T.; Maclaren, J.; Herbst, M.; Lovell-Smith, C.; Izadpanah, K.; Zaitsev, M. Knee cartilage MRI with in situ mechanical loading using prospective motion correction. *Magn. Reson. Med.* **2014**, *71*, 516–523. [CrossRef] [PubMed]
23. Egger, J.; Gall, M.; Wallner, J.; Boechat, P.; Hann, A.; Li, X.; Chen, X.; Schmalstieg, D. HTC Vive MeVisLab integration via OpenVR for medical applications. *PLoS ONE* **2017**, *12*, e0173972. [CrossRef]
24. Ritter, F.; Boskamp, T.; Homeyer, A.; Laue, H.; Schwier, M.; Link, F.; Peitgen, H.-O. Medical image analysis. *IEEE Pulse* **2011**, *2*, 60–70. [CrossRef]
25. Ronneberger, O.; Fischer, P.; Brox, T. U-Net: Convolutional Networks for Biomedical Image Segmentation. In Proceedings of the Medical Image Computing and Computer-Assisted Intervention MICCAI 2015, Munich, Germany, 5–9 October 2015; Springer International Publishing: Cham, Switzerland, 2015; pp. 234–241.
26. Besl, P.J.; McKay, N.D. A method for registration of 3-D shapes. *IEEE Trans. Pattern Anal. Mach. Intell.* **1992**, *14*, 239–256. [CrossRef]
27. Haber, E.; Modersitzki, J. Intensity Gradient Based Registration and Fusion of Multi-modal Images. In Proceedings of the Medical Image Computing and Computer-Assisted Intervention—MICCAI 2006, Copenhagen, Denmark, 1–6 October 2006; Springer: Berlin, Heidelberg, Germany, 2006; pp. 726–733.
28. Clark, D.; Stevens, J.M.; Tortonese, D.; Whitehouse, M.R.; Simpson, D.; Eldridge, J. Mapping the contact area of the patellofemoral joint: The relationship between stability and joint congruence. *Bone Jt. J.* **2019**, *101-B*, 552–558. [CrossRef]
29. Jibri, Z.; Jamieson, P.; Rakhra, K.S.; Sampaio, M.L.; Dervin, G. Patellar Maltracking: An update on the Diagnosis and Treatment Strategies. Insights Imaging. Available online: https://www.ncbi.nlm.nih.gov/pmc/articles/PMC6570735/ (accessed on 10 October 2022).

30. Tscholl, P.M.; Wanivenhaus, F.; Fucentese, S.F. Conventional Radiographs and Magnetic Resonance Imaging for the Analysis of Trochlear Dysplasia: The Influence of Selected Levels on Magnetic Resonance Imaging. *Am. J. Sports Med.* **2017**, *45*, 1059–1065. [CrossRef]
31. Biedert, R.M.; Tscholl, P.M. Patella Alta: A Comprehensive Review of Current Knowledge. *Am. J. Orthop.* **2017**, *46*, 290–300.
32. Warfield, S.K.; Zou, K.H.; Wells, W.M. Validation of image segmentation by estimating rater bias and variance. *Philos. Trans. A Math. Phys. Eng. Sci.* **2008**, *366*, 2361–2375. [CrossRef]
33. Warfield, S.K.; Zou, K.H.; Wells, W.M. Simultaneous Truth and Performance Level Estimation (STAPLE): An Algorithm for the Validation of Image Segmentation. *IEEE Trans. Med. Imaging* **2004**, *23*, 903–921. [CrossRef]
34. Stevens, J.M.; Eldridge, J.D.; Tortonese, D.; Whitehouse, M.R.; Krishnan, H.; Elsiwy, Y.; Clark, D. The influence of patellofemoral stabilisation surgery on joint congruity: An MRI surface mapping study. *Eur. J. Orthop. Surg. Traumatol. Orthop. Traumatol.* **2022**, *32*, 419–425. [CrossRef]
35. Fox, A.; Wanivenhaus, F.; Rodeo, S. The Basic Science of the Patella: Structure, Composition, and Function. *J. Knee Surg.* **2012**, *25*, 127–142. [CrossRef] [PubMed]
36. McConnell, J. Rehabilitation and nonoperative treatment of patellar instability. *Sports Med. Arthrosc. Rev.* **2007**, *15*, 95–104. [CrossRef] [PubMed]
37. Mirzabeigi, E.; Jordan, C.; Gronley, J.K.; Rockowitz, N.L.; Perry, J. Isolation of the vastus medialis oblique muscle during exercise. *Am. J. Sports Med.* **1999**, *27*, 50–53. [CrossRef] [PubMed]
38. Raimondo, R.A.; Ahmad, C.S.; Blankevoort, L.; April, E.W.; Grelsamer, R.P.; Henry, J.H. Patellar stabilization: A quantitative evaluation of the vastus medialis obliquus muscle. *Orthopedics* **1998**, *21*, 791–795. [CrossRef]
39. Amis, A.A.; Firer, P.; Mountney, J.; Senavongse, W.; Thomas, N.P. Anatomy and biomechanics of the medial patellofemoral ligament. *Knee* **2003**, *10*, 215–220. [CrossRef]
40. Fulkerson, J.P.; Shea, K.P. Disorders of patellofemoral alignment. *J. Bone Jt. Surg. Am.* **1990**, *72*, 1424–1429. [CrossRef]
41. Senavongse, W.; Amis, A.A. The effects of articular, retinacular, or muscular deficiencies on patellofemoral joint stability: A biomechanical study in vitro. *J. Bone Jt. Surg. Br.* **2005**, *87*, 577–582. [CrossRef]

Journal of
Personalized Medicine

MDPI

Article

Early Complication Analysis of Dynamic Intraligamentary Stabilization versus Anterior Cruciate Ligament Reconstruction

Frank Endreß [1], Reinhard Hörner [1], Wolfgang Hauth [1], Jens Anders [1,2] and Roland Biber [1,3,*]

1 Kliniken Dr. Erler gGmbH, Kontumazgarten 4-19, 90429 Nürnberg, Germany; f.endress@erler-klinik.de (F.E.)
2 Friedrich-Alexander-University Erlangen-Nürnberg (FAU), Schloßplatz 4, 91054 Erlangen, Germany
3 Paracelsus Medical University (PMU), Prof.-Ernst-Nathan-Str. 1, 90419 Nürnberg, Germany
* Correspondence: r.biber@erler-klinik.de

Abstract: Purpose: Both dynamic intraligamentary stabilization (DIS) and reconstruction (RECO) are common treatment methods for anterior cruciate ligament (ACL) rupture. We report short term outcomes after DIS (Ligamys, Mathys, Bettlach, Switzerland) and RECO using semitendinosus tendon. We compared postoperative complications, deficits of range-of-motion (ROM), and revision rates between the two treatment options. Methods: A total of 690 patients (437 male, 253 female), after either DIS or RECO, were included. Of these, 147 patients (21%) received DIS and 543 (79%) underwent RECO. Follow-up examination focused on clinical examination, complications and revision rates. Anteroposterior instability and ROM deficits were analyzed in order to evaluate our policy of early intervention for all cases of ROM restrictions. Results: Relevant ROM restrictions occurred at a significantly higher rate after DIS than after RECO (4.8% vs. 1.3%; $p = 0.008$). Flexion was more restricted after DIS than RECO (110° vs. 124°, $p < 0.001$). Extension deficits also occurred more frequently after DIS compared to RECO (49.7% vs. 24.5%; $p < 0.001$). Total revision surgery rate was 9.1%, with patients after DIS being significantly more frequently affected (20.4% vs. 6.1%; $p < 0.001$). Conclusions: Our findings indicate a significantly higher risk for ROM restriction after DIS compared to RECO, resulting in a significantly higher revision rate.

Keywords: anterior cruciate ligament; reconstruction; dynamic intraligamentary stabilization; range-of-motion

Citation: Endreß, F.; Hörner, R.; Hauth, W.; Anders, J.; Biber, R. Early Complication Analysis of Dynamic Intraligamentary Stabilization versus Anterior Cruciate Ligament Reconstruction. *J. Pers. Med.* **2023**, *13*, 1022. https://doi.org/10.3390/jpm13071022

Academic Editor: Jih-Yang Ko

Received: 14 May 2023
Revised: 5 June 2023
Accepted: 7 June 2023
Published: 21 June 2023

1. Introduction

Knee injuries account for a high number of sports-related surgeries [1–7]. Anterior cruciate ligament (ACL) rupture is one of the most common sports injuries. The incidence varies from 36.9–78/100,000 inhabitants worldwide [4,6,8,9]. Males have a higher risk of sustaining an ACL injury than females [8–10]. Up to 65% of ACL injuries result in surgery [9]. In Germany alone, more than 40,000 anterior cruciate ligament surgeries are performed annually, demonstrating, together with other reports, the relevance of ACL ruptures in today's orthopedics [4,6,8–10].

Injuries of the ACL rarely occur in isolation. The impact of concomitant injuries, including other ligament sprains, meniscal tears, articular cartilage injuries and bone bruises, complicate the treatment and potential outcomes of ACL tears [8,11]. Reported meniscal injuries associated with acute ACL rupture range from 15% to more than 50% [11–13] and are among the most common injuries requiring additional surgery.

According to the guidelines of the German Society for Trauma Surgery (DGU) [8], conservative treatment of ACL tears is recommended only for elderly patients or patients with reduced activity requirements and no persistent symptoms of instability. However, there is no general age limit for surgical ACL reconstruction [8,14]. Elderly and physically active patients who are not satisfied with conservative treatment may also be considered for a knee stabilizing procedure provided they do not suffer from arthritic ACL-deficient knees [8,14].

Cruciate ligament surgery is subject to the constant development of anatomical and biomechanical knowledge, and consequently, of both surgical techniques and postoperative treatment. Today's gold standard for ACL surgery is ACL reconstruction (RECO), primarily developed in 1966 by H. Brückner, who proposed harvesting the middle third of the patellar tendon [15–17]. The first publication describing a technique using hamstring tendon grafts was by B. Lipscomb in 1982 [18]. In 2022, Liukkonen et.al. demonstrated that RECO is a reliable procedure with overall low historical revision rates [19]. Between 1969 and 2018, there was an overall revision rate of 3.14%, of which 2.71% related to hamstring autografts and 2.38% related to bone–patellar tendon–bone (BPTB) autografts [19].

In medical history, the procedure of ACL replacement surgery has evolved as a result of failures after primary ACL suturing [20]. The importance of repair after ACL ruptures due to the inability of spontaneous healing was published by Ivar Palmer in 1938, in which he presented a technique for primary suturing of the acutely torn anterior cruciate ligament [21]. However, the results of early repair techniques were not convincing, and overall failure and recurrence rates were very high [17,22–25]. Nevertheless, interest in preserving the anterior cruciate ligament has continued over the years. Several studies have shown that the cruciate ligaments, which have a proportion of mechanoreceptor nerve fibers in their structure (1 to 2 percent nerve fibers by volume), play an important role in proprioception of the knee joint and thus the lower extremity [20,26,27].

Anatomical and histological examinations of the knee joint have also demonstrated that the anterior cruciate ligament has healing potential [20,28–33]. The ability of the ACL to heal eventually led to the development of "dynamic intraligamentary stabilization" (DIS) of the anterior cruciate ligament, Ligamys, by Eggli et. al. and Kohl et al. [17,20,30,34–36]. In a comparative study from 2008, it was demonstrated that, in contrast to non-augmented ACL suture repair and static tape augmentation, only dynamic augmentation (Ligamys) resulted in restoration of anterior tibial translation values similar to the ACL-intact knee [37].

The biomechanical objective of Ligamys restoration is to provide temporary stabilization to the knee joint using a dynamic spring system, essentially serving as an artificial replacement for the ACL [38]. In each position, a defined spring force of 60–80 Newtons pulls the tibial head dorsally [38]. By means of this technique, it is ensured that the knee is locked in a fixed posterior translation at any degree of flexion, keeping the two ligamentous stumps in the closest possible position at all times [17]. Thereby, the reattached ACL maintains the necessary contact to be able to scar stably [17,38–40]. During the first six postoperative months, the implanted polyethylene thread "slips out" of the spring clamping mechanism in a controlled manner. The original cruciate ligament increasingly takes over the stabilization within 6 months. After this time, the implant suture no longer has tension in the spring clamping mechanism [17,30,38,40].

In our clinic, we use either the conventional technique of anterior cruciate ligament replacement with transtibial femoral drilling or the Ligamys technique as surgical treatment of anterior cruciate ligament injuries, depending on the indication. According to the current standard, indications for Ligamys treatment are fresh proximal cruciate ligament ruptures that are not older than 21 days. Contraindications are currently ACL ruptures older than 21 days, acute or chronic infections (local or systemic), arthrofibrosis, severe malposition of the knee joint, severe muscle, nerve or vascular diseases, hypersensitivity to the materials used (e.g., cobalt, chromium, nickel, etc.), insufficient bone substance and poor bone quality, which could endanger a stable anchoring of the implant. A relative contraindication is for children and adolescents with open epiphyseal gap [20,40]. According to several studies, another factor acting as a potential determinant of poor outcome after ACL repair using the DIS technique, was a high pre-injury sports activity level [41–43].

The aim of this study is to thoroughly evaluate conspicuous clinical findings and early complications during the follow-up period, in particular signs of infection, ROM deficits, incipient arthrofibrosis and the resulting revision rates after ACL rupture, comparing DIS and RECO. The knowledge gained from the study will be used in the future to develop optimized treatment strategies tailored to the individual patient, taking into account the patient's individual goals. In addition, any early complications should be detected in time and counteracted in a guideline- and evidence-based manner.

2. Patients and Methods

For this single-center prospective study, each patient treated either by ACL dynamic intraligamentary stabilization (DIS) or by ACL reconstruction (RECO) was invited for follow-up 6 weeks and 3 months postoperatively. All patients attending at least one follow-up examination were included. Data were collected from 1148 consecutive follow-up examinations from 690 patients between the years from 2012 until 2022.

Whenever DIS indication criteria were fulfilled, we recommended DIS in order to preserve the patient's natural ACL. Otherwise, we advocated for RECO as soon as all signs of inflammation had subsided and range-of-motion (ROM) had returned to normal (see Figure 1 for our treatment algorithm).

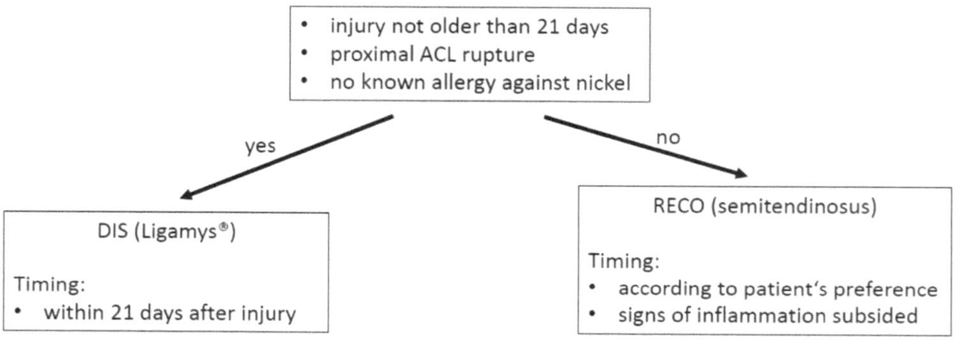

Figure 1. Treatment algorithm.

In total, 147 patients (21%) received DIS and 543 patients (79%) received ACL RECO. DIS was performed using the Ligamys method by Mathys Ltd. (Bettlach, Switzerland) according to manufacturer's instructions. For RECO, we used a transtibial technique (POSITION ACL, B. Braun Aesculap, Tuttlingen, Germany) with semitendinosus tendon as autograft [44]. The Ultrabutton Adjustable Fixation System from Smith&Nephew (London, UK) served as a suture plate for femoral implant anchoring and the Königsee Button (Allendorf, Germany) served as the suture disc for the tibial anchor. Drainages were routinely used in RECO, but not in DIS.

All surgeries, irrespective of the method used, were performed with the patient positioned in supine position and under general anesthesia. The knee joint was stored using a knee holder with, intraoperatively, 130-degree flexion ability. A tourniquet was inflated to 300–350 mm Hg. The arthroscopy portals for both surgical techniques are the same, with the exception that a larger incision is recommended for the anteromedial portal in the Ligamys treatment [40]. In both DIS and RECO, concomitant injuries were treated prior to the ACL treatment procedure. For both methods, we used devices for implant tension control. In DIS procedures, a tensiometer and a torque wrench were used to pretension the Ligamys tape to the maximum in 90° flexion. In full extension, the clamping cone was screwed in with a pretension of 60 to 80 Newtons. In RECO procedures, tibial anchorage was established using an Ultrabutton (Smith & Nephew). Fixation was achieved by knotting with a knot aid, and under tension, in the near-extension position. Residual laxity was checked and (if present) eliminated by rotating the Ultrabutton with the aid of the twister. We checked that the Ultrabutton resisted lift-off during application of at least 75 Newton meters.

Postoperative mobilization was performed during the inpatient stay according to the respective in-house post-treatment protocols. The protocols were based on the post-treatment recommendations of the manufacturers and were handed out to the patients upon discharge [45]. These included strict advice to attend continuous physiotherapy after discharge. Scheduled postoperative follow-up visits at our clinic are recommended after 6 and 12 weeks.

Our collective consisted of 437 males and 253 females (63.3% vs. 36.7%). The mean age was 32.0 years (SD 11.1 years; range 12 to 73 years). A total of 690 surgeries were performed, including 147 DIS and 543 RECO. Postoperative treatment protocols after DIS and RECO differed only slightly; both protocols were based on the post-treatment recommendations of the respective transplant manufacturers and were handed out to the patients upon discharge. Both protocols included the advice for immediate active physiotherapy without ROM restriction; however, such restrictions could have applied in cases of combined procedures (e.g., simultaneous meniscus surgery).

Scheduled postoperative follow-up visits at our clinic were offered to every ACL patient and were recommended after 6 and 12 weeks. These examinations aimed at early identification of patients with ROM deficits or incipient arthrofibrosis, which we regard as indication for early arthroscopic adhesiolysis.

Parameters evaluated during initial treatment and follow-up included:

- age and gender;
- kind of index surgery including combined procedures;
- ROM (measurement by standard goniometer);
- anteroposterior instability (measured by rolimeter) compared to the contralateral knee joint;
- clinical presence of classical inflammation signs: effusion, knee swelling and rubor;
- mean circumferences of the thigh, knee, and lower leg;
- rehabilitation activities / physiotherapy;
- revision surgery rates and procedures;
- subjective satisfaction of patient and surgeon on a scale of 1 (best) to 6 (worst).

Statistical analysis calculated both descriptive and inferential measures. Mean values and standard deviations were calculated for continuous variables, and frequency counts and percentages were obtained for the discrete variables. All confidence intervals (designated "CI") are 95% confidence intervals. The χ^2 test was used for cross tabulation analysis when all expected cell frequencies were five or greater; Fisher's exact test was used otherwise. For continuous variables, the two-sample independent *t*-test was used. Nonparametric correlations were evaluated using Spearman's ρ. All statistical analyses were performed with IBM SPSS Statistics Version 19.

3. Results

We report on the outcomes of a total of 690 cases, consisting of 147 DIS (21%) and 543 RECO (79%). Mean operation time did not relevantly differ between these two methods (55 min versus 53 min).

A total of 437 of all patients were male (63.3%). This proportion turned out to be different between the DIS and RECO subgroups: 65.7% of ACL-reconstruction patients ware male, whereas this proportion was only 54.4% in the DIS group (difference significant on χ^2 test, $p = 0.01$).

The patient ages ranged from 12 years to 73 years; the mean age was 32.0 years (SD 11.1 years). The mean patient age did not significantly differ between DIS and RECO (33.3 years versus 31.6 years; $p = 0.1$), as is shown in Figure 2. On average, female patients were slightly older than male patients (34.0 years versus 30.9 years; $p < 0.001$). Regarding the subgroups, this was true only for RECO patients (34.4 years versus 30.2 years; $p < 0.001$), whereas female DIS patients tended to be younger (32.8 years versus 33.8 years; not significant).

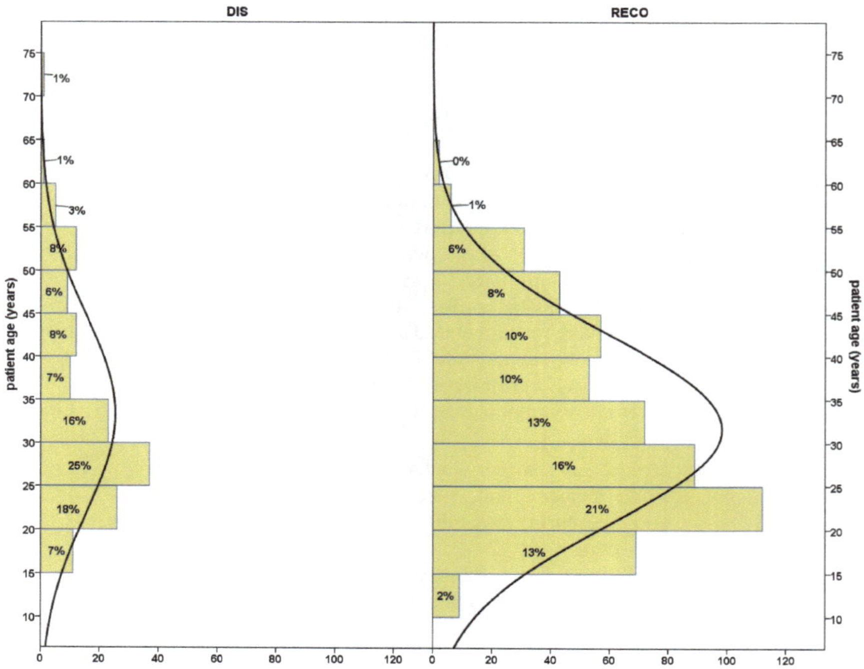

Figure 2. Age distributions of DIS and RECO patients did not differ significantly.

Index operation was combined with medial meniscus surgery in 185 cases (resection: 14.9%; suture: 11.9%), lateral meniscus surgery in 147 cases (resection: 14.4%; suture: 7.0%), medial ligament injury treatment in 14 cases (conservative: 0.9%; operative: 1.2%), cartilage surgery in 10 cases (1.5%) and PCL treatment in 4 cases (conservative: 0.1%; reconstruction: 0.4%). DIS was significantly more often accompanied by meniscal surgery than RECO (50.3% vs. 40.1%; $p = 0.026$). Details are provided in Table 1.

Table 1. Procedures combined with DIS and RECO (number, percentage in brackets).

Simultaneous Procedure	DIS	RECO	χ^2
medial meniscus (resection)	17 (11.6%)	86 (15.8%)	n.s.
medial meniscus (suture)	17 (11.6%)	65 (12.0%)	
lateral meniscus (resection)	28 (19.0%)	71 (13.1%)	$p = 0.002$
lateral meniscus (suture)	18 (12.2%)	30 (5.5%)	
any simultaneuos meniscus surgery	74 (50.3%)	218 (40.1%)	$p = 0.026$
medical collateral ligament (suture)	3 (2.0%)	5 (0.9%)	n.s.
cartilage surgery (microfracuring)	2 (1.4%)	4 (0.7%)	n.s.
cartilage surgery (flake refixation)	1 (0.7%)	0 (0%)	n.s.

Follow-Up

The mean duration until first follow-up was 53 days. Between hospital discharge and follow-up, 15 RECO patients had started or completed an inpatient rehabilitation program (2.2%). For outpatient rehabilitation programs these rates were higher (35.2% in the DIS and 41.6% in the RECO group). A total of 31.7% of DIS patients and 41.4% of RECO patients declared that they had not received any physiotherapy or rehab at all after being discharged from our hospital.

We did not detect any cases of infection or implant failure on first follow-up. When evaluating satisfaction on a subjective basis, using a scale of 1 (best) to 6 (worst), patient and surgeon perfectly agreed on a mean rating of 2.3 after RECO. DIS patients' mean rating was 2.6, as opposed to the surgeons' mean rating of 3.1.

Frequencies of major clinical findings are shown in Table 2. Active knee extension could not be fully demonstrated in 29.9% of all cases at the time of first follow-up. Although the extension deficit was mild (max. 5°) in most cases, we noticed this problem more frequently after DIS than after RECO (49.7% vs. 24.5%; $p < 0.001$). Following DIS, the problem not only seemed more frequent, but also more severe. Figure 3 shows those findings in detail.

Table 2. Outcomes on early follow-up (ø 53 days) after ACL surgery.

	DIS	RECO	Significance
ROM restriction	4.8%	1.3%	$p = 0.008$
effusion	15.8%	20.7%	n.s.
swelling	42.5%	34.3%	n.s.
rubor	0.7%	0.2%	n.s.
mean knee flexion	110°	124°	$p < 0.001$
anteroposterior instability (rolimeter) compared to contralateral	−0.14 mm	+0.02 mm	n.s.
mean circumference thigh (20 cm above knee), compared to contralateral	−2.9 cm	−3.1 cm	n.s.
mean circumference thigh (10 cm above knee), compared to contralateral	−1.3 cm	−1.6 cm	n.s.
mean circumference knee, compared to contralateral	+0.8 cm	+0.8 cm	n.s.
mean circumference lower leg (15 cm below knee), compared to contralateral	−0.9 cm	−0.8 cm	n.s.

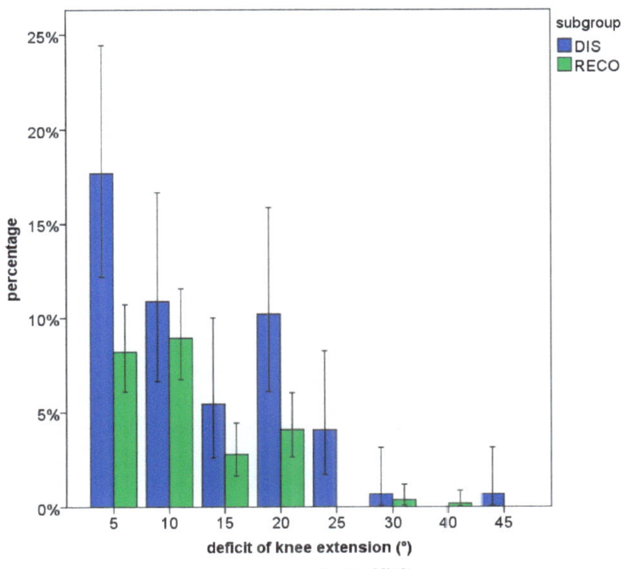

Figure 3. Proportion of patients with knee extension deficit on early follow-up.

We advised 45 patients for revision (6.5%) during the first follow-up examination. This rate was significantly higher in DIS cases than in RECO cases (17.0% vs. 3.7%; $p < 0.001$). The proportion was also higher in females than in males (10.0% versus 4.6%; $p = 0.006$).

All in all, 63 patients (9.1%) underwent at least one revision operation. In addition, this proportion was significantly higher after DIS than after RECO (20.4% vs. 6.1%; $p < 0.001$). Female patients were significantly more frequently subject to operative revision (m/f: 7.1% versus 12.6; $p = 0.015$). An association between revision rate and patient age was not evident (Spearman's $\rho = 0.01$; $p = 0.8$).

Following our strategy of early operative revision in cases of delayed regain of ROM, our most frequently applied revision procedure was arthroscopic adhesiolysis with anesthetic mobilization of the knee joint. A detailed overview of the procedures performed is given in Table 3.

Table 3. Procedures performed for early revision surgery.

Revision Procedure	DIS	RECO	χ^2
adhesiolysis/mobilization	24 (16.3%)	19 (3.5%)	$p < 0.0001$
revision for hemarthrosis (irrigation)	0 (0%)	1 (0.2%)	n.s.
revision for extracapsular hematoma	0 (0%)	3 (0.6%)	n.s.
revision for suspected infection (e.g., irrigation, synovectomy, local antibiotic treatment)	2 (1.4%)	2 (0.4%)	n.s.
revision graft	n.a.	6 (1.1%)	n.a.
conversion DIS → RECO	3 (2.0%)	n.a.	n.a.
resection of insufficient ACL	1 (0.7%)	2 (0.6%)	n.s.
Total	30 (20.4%)	33 (6.1%)	$p < 0.0001$

4. Discussion

Recently the Dutch A.C.L. repair study group published DIS results from 155 patients treated in five centers and could not demonstrate non-inferiority of DIS versus RECO [46]. These results show the need for further research on both of these methods, which may contribute to improving operative techniques, patient selection, and decision making. Our similar study based on 690 patients focuses on early complications when evaluating the results of ACL surgery, either by DIS or RECO. Our 690-patient cohort underwent surgical treatment in our hospital between 2012 and 2022. In total, 147 patients were treated with DIS and 543 patients with RECO.

As previously shown in several studies, the majority of patients are male [8–10]. However, the previously published gender difference regarding DIS was less pronounced in our collective. On average, female patients were slightly older than male patients, but in DIS cases, female patients tended to be younger, which correlates with previous studies showing that women suffer ACL ruptures at an earlier age than men [10,42,47].

Index operation was combined with medial meniscal surgery in 185 cases (resection: 14.9%; suture: 11.9%) and with lateral meniscal surgery in 147 cases (resection: 14.4%; suture: 7.0%). In the DIS group, meniscal suture, rather than resection, was the more frequently used method of treatment compared to RECO. This seems an interesting aspect worthy of further investigation. The most likely reason for this finding may be the shorter time interval between trauma and surgery in DIS patients. However, another hypothesis may be that injury mechanisms leading to proximal LCA rupture may be associated with meniscal injury more suitable for suture. Anyway, our findings are consistent with previous studies [20,34,35]. Other concomitant injuries accounted for only a small proportion and were treated according to current guidelines.

Postoperative follow-up visits at our clinic were recommended at 6 and 12 weeks. The mean duration until first follow-up was 53 days. We did not detect any cases of infection at the first follow-up. No significant differences were observed between DIS and

RECO for most postoperative clinical findings, including effusion, swelling, rubor and circumferences. Anterior-posterior instability measured with a standard rolimeter also showed no significant differences. These findings indicate that both operative trauma burden and primary stability of RECO and DIS are comparable. This is remarkable because RECO—being associated with tendon harvesting and more extensive bone canal drilling—could be expected to perform worse than DIS on early follow-up. However, postoperative inflammation and healing processes, instead, seem equal.

The fact that a significant proportion of patients (31.7% DIS and 41.4% RECO), despite explicit recommendation, did not receive any physiotherapy or rehabilitation at the time of the first follow-up is alarming. For both types of ACL surgery, early physiotherapeutic support, at least from the second week onwards, as recommended by the surgeons and the implant-manufacturers [44,45], is essential to counteract possible complications in the healing process [48]. Consistent physiotherapy is also important to avoid or correct permanent muscular asymmetries after cruciate ligament surgery. In a 2022 systematic review, it was shown that, in athletes after RECO, asymmetries in knee movements in the sagittal plane at first ground contact during the landing phase of a drop vertical jump are strong predictors of reinjury of the ACL. In addition, it was found that patients with muscular asymmetries of the lower extremity employed compensatory strategies for the hip, pelvis and trunk to compensate for imbalances.

Severe ROM restrictions (here defined as extension deficit of >10° or flexion deficit > 20°) were generally rare. However, this problem occurred significantly more frequently after DIS than after RECO (4.8% vs. 1.3%; $p = 0.008$). Mean knee flexion was more restricted after DIS (110°) compared to RECO (124°). The extension deficit of the knee joint was generally low, ranging from 1° to 5°. However, in comparison, it also occurred more frequently after DIS than after RECO, with a rate of 49.7% compared to 24.5%, respectively. Maybe this is partially explained by the significantly higher proportion of patients receiving simultaneous meniscal surgery in the DIS group (50.3% vs. 40.1%; $p = 0.03$). Other concomitant injuries, such as collateral ligament injury, accounted for only a small proportion and did not display significant differences.

We advised 45 patients (6.5%) for revision surgery at the first follow-up examination, which is part of our strategy for arthrofibrosis prevention. The rate was significantly higher after DIS. In total, 63 patients (9.1%) underwent at least one revision surgery, with DIS patients being significantly more frequently affected than RECO patients (20.4% vs. 6.1%). Female patients were significantly more often subject to operative revision (m/f: 7.1% versus 12.6%). These high rates may be attributed to our consistent attitude to advice for aggressive early revision surgery whenever recovery of ROM is delayed. Accordingly, the most frequently performed procedure was arthroscopic adhesiolysis with anesthetic mobilization of the knee joint. The percentage of arthrofibrosis or restricted ROM after DIS, however, is consistent with previous studies, according to which the percentage is widely scattered between 18% and 33% [13,35,49–51]. In three cases it was necessary to change to an anterior cruciate ligament replacement after failed Ligamys stabilization, which is around double the implant failure rate of RECO.

Evangelopulos et al. reported a combined complication rate of 78.8% that existed after DIS following midline ACL tears, of which 33.3% were due to extension deficits [49]. Recently the Dutch A.C.L. repair study group published the DIS results of 155 patients treated in five centers, which also could not demonstrate non-inferiority of DIS versus RECO [46]. They reported at least one complication in 25% of patients, comprising mainly arthrofibrosis, re-rupture and pain [46]. Most studies, including ours, have the limitation of short follow-up periods. Long-term results of our patient collective beyond the 12-week period are not known. In addition, some of the patients only showed up for one follow-up appointment. Furthermore, the examination results are influenced by the fact that a not insignificant proportion of patients had not yet received physiotherapy at the time of the follow-up appointments. The assessment of ROM deficits must therefore be viewed more critically. However, long-term study results to date following DIS tend to be ambivalent [13,34,42,52,53]. Moreover, a decision to use a particular surgical approach can neither be based solely on evidence of early complication rates nor on total complication rates. Complication rates per se do not evaluate the seriousness of the complications, which might be more or less easy to manage. Therefore, it is still important to establish a thorough indication respecting the individual patient's goals, which might include the wish to preserve the "original" cruciate ligament and preserve potential autograft tendons.

Strengths of our study are clearly the high patient numbers treated very consistently, regarding decision-making and operative technique, by only a small number of surgeons. Conversely, the single-center study design may also be regarded as a limitation. By focusing on early complication rates, aspects such as implant failure are probably underestimated. This limitation may distort a balanced comparison between DIS and RECO.

Our findings are in line with the previous studies cited above, which also demonstrate increased complication rates after DIS compared to RECO. Technical modifications may also be a strategy to overcome this drawback. The combined complication rate could be reduced to 8.7% by applying a collagen I/III membrane to the surface of the ACL. This is in line with the findings of Murray in 2007 [31] and the successful treatment of an ACL wound with a collagen and platelet-rich plasma (PRP) hydrogel, which resulted in increased filling of the wound site with repair tissue that had increased profiles of growth factors and protein expression. Therefore, the standard use of a collagen membrane for Ligamys treatment should be considered in the future to reduce the risk of arthrofibrosis or overall limitation of knee joint motion.

5. Conclusions

Comparing early outcomes, ROM deficits were significantly more frequent after DIS than after ACL reconstruction. Between 30 and 40% of all patients did not receive any physiotherapy. For both DIS and RECO, early physiotherapeutic support, on average from the second week onwards, as recommended by most surgeons and the implant-manufacturers, is essential to counteract possible complications in the healing process [45,48]. Other reasons for the higher revision rate of DIS include the necessarily earlier timing of surgery for DIS, different drainage usage, a higher proportion of accompanying meniscal surgery, and differences in the surgical methods themselves. Usage of collagen membranes or platelet-rich-plasma may contribute to future improvements in the DIS method.

A decision to use a particular surgical approach cannot be based solely on evidence of early complication rates. However, our findings are in line with several previous studies, which also demonstrated increased complication rates after DIS. Long-term study results to date following DIS tend to be ambivalent [13,34,42,52,53]. Still, it is important to establish a thorough indication respecting the individual patient's goals, which may include preserving the "original" cruciate ligament and preserving potential autograft tendons.

The need for immediate and regular physiotherapeutic co-treatment must be emphasized. In addition, we recommend routine follow-up examinations by the operating surgeon after 6 weeks, which can detect a significant proportion of patients who will profit from early arthroscopic revision (adhesiolysis or mobilization).

Author Contributions: Conceptualization, F.E. and R.B.; methodology, F.E. and R.B.; software, R.B.; validation, R.H., W.H. and J.A.; formal analysis, F.E. and R.B.; investigation, F.E.; resources, F.E. and R.B.; data curation, F.E.; writing—original draft preparation, F.E.; writing—review and editing, R.B.; visualization, F.E.; supervision, J.A. and R.B.; project administration, R.B.; funding acquisition not applicable. All authors have read and agreed to the published version of the manuscript.

Funding: This research received no external funding.

Institutional Review Board Statement: Not applicable. According to § 15 (1) Berufsordnung für die Ärzte Bayerns, no ethical review is necessary for studies not affecting the mental or physical integrity of humans and being carried out on anonymized data.

Informed Consent Statement: Informed consent was obtained from all subjects involved in the study.

Acknowledgments: The present work was performed in (partial) fulfillment of the requirements for obtaining the first author's "Dr. med." degree.

Conflicts of Interest: The authors declare that no funds, grants, or other support were received during the preparation of this manuscript. The authors have no relevant financial or non-financial interests to disclose.

References

1. Bloch, H.; Reinsberger, C.; Klein, C.; Luig, P.; Krutsch, W. High revision arthroscopy rate after ACL reconstruction in men's professional team sports. *Knee Surg. Sports Traumatol. Arthrosc.* **2023**, *31*, 142–151. [CrossRef] [PubMed]
2. Kaeding, C.C.; Léger-St-Jean, B.; Magnussen, R.A. Epidemiology and Diagnosis of Anterior Cruciate Ligament Injuries. *Clin. Sport Med.* **2017**, *36*, 1–8. [CrossRef] [PubMed]
3. Keuning, M.C.; Robben, B.J.; Brouwer, R.W.; Stevens, M.; Bulstra, S.K.; Zuurmond, R.G. Young men are at higher risk of failure after ACL hamstring reconstructions: A retrospective multivariate analysis. *BMC Musculoskelet. Disord.* **2022**, *23*, 598. [CrossRef] [PubMed]
4. Kohn, L.; Rembeck, E.; Rauch, A. Anterior cruciate ligament injury in adults: Diagnostics and treatment. *Orthopade* **2020**, *49*, 1013–1028. [CrossRef]
5. Montalvo, A.M.; Schneider, D.K.; Webster, K.E.; Yut, L.; Galloway, M.T.; Heidt, R.S., Jr.; Kaeding, C.C.; Kremcheck, T.E.; Magnussen, R.A.; Parikh, S.N.; et al. Anterior Cruciate Ligament Injury Risk in Sport: A Systematic Review and Meta-Analysis of Injury Incidence by Sex and Sport Classification. *J. Athl. Train.* **2019**, *54*, 472–482. [CrossRef]
6. Sanders, T.L.; Maradit Kremers, H.; Bryan, A.J.; Larson, D.R.; Dahm, D.L.; Levy, B.A.; Stuart, M.J.; Krych, A.J. Incidence of Anterior Cruciate Ligament Tears and Reconstruction: A 21-Year Population-Based Study. *Am. J. Sports Med.* **2016**, *44*, 1502–1507. [CrossRef]
7. Szymski, D.; Achenbach, L.; Weber, J.; Huber, L.; Memmel, C.; Kerschbaum, M.; Alt, V.; Krutsch, W. Reduced performance after return to competition in ACL injuries: An analysis on return to competition in the 'ACL registry in German Football'. *Knee Surg. Sports Traumatol. Arthrosc.* **2023**, *31*, 133–141. [CrossRef]
8. DGU. *S1-Leitlinie Vordere Kreuzbandruptur*; AWMF: Berlin, Germany, 2018.
9. Gianotti, S.M.; Marshall, S.W.; Hume, P.A.; Bunt, L. Incidence of anterior cruciate ligament injury and other knee ligament injuries: A national population-based study. *J. Sci. Med. Sport* **2009**, *12*, 622–627. [CrossRef]
10. Nordenvall, R.; Bahmanyar, S.; Adami, J.; Stenros, C.; Wredmark, T.; Felländer-Tsai, L. A population-based nationwide study of cruciate ligament injury in Sweden, 2001–2009: Incidence, treatment, and sex differences. *Am. J. Sports Med.* **2012**, *40*, 1808–1813. [CrossRef]
11. Beynnon, B.D.; Johnson, R.J.; Abate, J.A.; Fleming, B.C.; Nichols, C.E. Treatment of Anterior Cruciate Ligament Injuries, Part I. *Am. J. Sports Med.* **2005**, *33*, 1579–1602. [CrossRef]
12. Benco, M.; Tylla, A.; Stangl, R. Dynamic intraligamentary stabilization of acute anterior femoral cruciate ligament rupture: Preliminary and intermediate clinical results. *Unfallchirurg* **2019**, *122*, 706–710. [CrossRef] [PubMed]
13. Osti, M.; El Attal, R.; Doskar, W.; Höck, P.; Smekal, V. High complication rate following dynamic intraligamentary stabilization for primary repair of the anterior cruciate ligament. *Knee Surg. Sports Traumatol. Arthrosc.* **2019**, *27*, 29–36. [CrossRef] [PubMed]
14. Toanen, C.; Demey, G.; Ntagiopoulos, P.G.; Ferrua, P.; Dejour, D. Is There Any Benefit in Anterior Cruciate Ligament Reconstruction in Patients Older Than 60 Years? *Am. J. Sports Med.* **2017**, *45*, 832–837. [CrossRef] [PubMed]
15. Brückner, H. A new method for plastic surgery of cruciate ligaments. *Chirurg* **1966**, *37*, 413–414.
16. Chambat, P.; Guier, C.; Sonnery-Cottet, B.; Fayard, J.M.; Thaunat, M. The evolution of ACL reconstruction over the last fifty years. *Int. Orthop.* **2013**, *37*, 181–186. [CrossRef] [PubMed]
17. Eggli, S.; Kohlhof, H.; Zumstein, M.; Henle, P.; Hartel, M.; Evangelopoulos, D.S.; Bonel, H.; Kohl, S. Dynamic intraligamentary stabilization: Novel technique for preserving the ruptured ACL. *Knee Surg. Sports Traumatol. Arthrosc.* **2015**, *23*, 1215–1221. [CrossRef]
18. Lipscomb, A.B.; Johnston, R.K.; Snyder, R.B.; Warburton, M.J.; Gilbert, P.P. Evaluation of hamstring strength following use of semitendinosus and gracilis tendons to reconstruct the anterior cruciate ligament. *Am. J. Sports Med.* **1982**, *10*, 340–342. [CrossRef]
19. Liukkonen, R.J.; Ponkilainen, V.T.; Reito, A. Revision Rates After Primary ACL Reconstruction Performed between 1969 and 2018: A Systematic Review and Metaregression Analysis. *Orthop. J. Sports Med.* **2022**, *10*, 23259671221110191. [CrossRef]
20. Kösters, C.; Herbort, M.; Schliemann, B.; Raschke, M.J.; Lenschow, S. Dynamic intraligamentary stabilization of the anterior cruciate ligament. Operative technique and short-term clinical results. *Unfallchirurg* **2015**, *118*, 364–371. [CrossRef]
21. Palmer, I. On the injuries to the ligaments of the knee joint. A clinical study. *Acta Chir. Scand. Suppl.* **1938**, 2–282.
22. Engebretsen, L.; Benum, P.; Fasting, O.; Mølster, A.; Strand, T. A prospective, randomized study of three surgical techniques for treatment of acute ruptures of the anterior cruciate ligament. *Am. J. Sports Med.* **1990**, *18*, 585–590. [CrossRef]
23. Feagin, J.A., Jr.; Curl, W.W. Isolated tear of the anterior cruciate ligament: 5-year follow-up study. *Am. J. Sports Med.* **1976**, *4*, 95–100. [CrossRef]
24. Marshall, J.L.; Warren, R.F.; Wickiewicz, T.L. Primary surgical treatment of anterior cruciate ligament lesions. *Am. J. Sports Med.* **1982**, *10*, 103–107. [CrossRef]

25. Odensten, M.; Lysholm, J.; Gillquist, J. Suture of fresh ruptures of the anterior cruciate ligament. A 5-year follow-up. *Acta Orthop. Scand.* **1984**, *55*, 270–272. [CrossRef] [PubMed]
26. Gao, F.; Zhou, J.; He, C.; Ding, J.; Lou, Z.; Xie, Q.; Li, H.; Li, F.; Li, G. A Morphologic and Quantitative Study of Mechanoreceptors in the Remnant Stump of the Human Anterior Cruciate Ligament. *Arthroscopy* **2016**, *32*, 273–280. [CrossRef] [PubMed]
27. Lee, B.I.; Min, K.D.; Choi, H.S.; Kwon, S.W.; Chun, D.I.; Yun, E.S.; Lee, D.W.; Jin, S.Y.; Yoo, J.H. Immunohistochemical study of mechanoreceptors in the tibial remnant of the ruptured anterior cruciate ligament in human knees. *Knee Surg. Sports Traumatol. Arthrosc.* **2009**, *17*, 1095–1101. [CrossRef] [PubMed]
28. Crain, E.H.; Fithian, D.C.; Paxton, E.W.; Luetzow, W.F. Variation in anterior cruciate ligament scar pattern: Does the scar pattern affect anterior laxity in anterior cruciate ligament-deficient knees? *Arthroscopy* **2005**, *21*, 19–24. [CrossRef] [PubMed]
29. Fleming, B.C.; Carey, J.L.; Spindler, K.P.; Murray, M.M. Can suture repair of ACL transection restore normal anteroposterior laxity of the knee? An ex vivo study. *J. Orthop. Res.* **2008**, *26*, 1500–1505. [CrossRef] [PubMed]
30. Kohl, S.; Evangelopoulos, D.S.; Kohlhof, H.; Hartel, M.; Bonel, H.; Henle, P.; von Rechenberg, B.; Eggli, S. Anterior crucial ligament rupture: Self-healing through dynamic intraligamentary stabilization technique. *Knee Surg. Sports Traumatol. Arthrosc.* **2013**, *21*, 599–605. [CrossRef] [PubMed]
31. Murray, M.M.; Spindler, K.P.; Ballard, P.; Welch, T.P.; Zurakowski, D.; Nanney, L.B. Enhanced histologic repair in a central wound in the anterior cruciate ligament with a collagen-platelet-rich plasma scaffold. *J. Orthop. Res.* **2007**, *25*, 1007–1017. [CrossRef]
32. Scapinelli, R. Vascular anatomy of the human cruciate ligaments and surrounding structures. *Clin. Anat.* **1997**, *10*, 151–162. [CrossRef]
33. Steadman, J.; Cameron-Donaldson, M.; Briggs, K.; Rodkey, W. The healing response technique: A minimally invasive procedure to stimulate healing of anterior cruciate ligament injuries using the microfracture technique. *J. Knee Surg.* **2006**, *19*, 8–13. [CrossRef] [PubMed]
34. Eggli, S.; Röder, C.; Perler, G.; Henle, P. Five year results of the first ten ACL patients treated with dynamic intraligamentary stabilisation. *BMC Musculoskelet. Disord.* **2016**, *17*, 105. [CrossRef] [PubMed]
35. Kohl, S.; Evangelopoulos, D.S.; Schär, M.O.; Bieri, K.; Müller, T.; Ahmad, S.S. Dynamic intraligamentary stabilisation: Initial experience with treatment of acute ACL ruptures. *Bone Joint J.* **2016**, *98-b*, 793–798. [CrossRef] [PubMed]
36. Schliemann, B.; Herbort, M.; Lenschow, S.; Raschke, M.J.; Kösters, C. Dynamische intraligamentäre Stabilisierung frischer Rupturen des vorderen Kreuzbandes (Ligamys). *Trauma Berufskrankh.* **2015**, *18*, 254–259. [CrossRef]
37. Hoogeslag, R.A.G.; Brouwer, R.W.; Huis in 'T Veld, R.; Stephen, J.M.; Amis, A.A. Dynamic augmentation restores anterior tibial translation in ACL suture repair: A biomechanical comparison of non-, static and dynamic augmentation techniques. *Knee Surg. Sports Traumatol. Arthrosc.* **2018**, *26*, 2986–2996. [CrossRef]
38. Ihle, C.; Nakayama, H.; Gonser, C.; Stöckle, U.; Döbele, S.; Schröter, S.; Ahrend, M.-D.; Ateschrang, A. Die Naht der frischen vorderen Kreuzbandruptur mit dynamischer intraligamentärer Stabilisierung—Eine Alternative zur vorderen Kreuzbandersatzplastik bei sportlich aktiven jungen Patienten. *Z. Orthop. Unfall.* **2018**, *156*, 100–102. [CrossRef]
39. Ateschrang, A.; Ahmad, S.S.; Stöckle, U.; Schroeter, S.; Schenk, W.; Ahrend, M.D. Recovery of ACL function after dynamic intraligamentary stabilization is resultant to restoration of ACL integrity and scar tissue formation. *Knee Surg. Sports Traumatol. Arthrosc.* **2018**, *26*, 589–595. [CrossRef]
40. Mathys, A.G. *Operationstechnik Ligamys*; Mathys AG Bettlach: Bettlach, Switzerland, 2000.
41. Ahmad, S.S.; Schürholz, K.; Liechti, E.F.; Hirschmann, M.T.; Kohl, S.; Klenke, F.M. Seventy percent long-term survival of the repaired ACL after dynamic intraligamentary stabilization. *Knee Surg. Sports Traumatol. Arthrosc.* **2020**, *28*, 594–598. [CrossRef]
42. Henle, P.; Bieri, K.S.; Brand, M.; Aghayev, E.; Bettfuehr, J.; Haeberli, J.; Kess, M.; Eggli, S. Patient and surgical characteristics that affect revision risk in dynamic intraligamentary stabilization of the anterior cruciate ligament. *Knee Surg. Sports Traumatol. Arthrosc.* **2018**, *26*, 1182–1189. [CrossRef]
43. Krismer, A.M.; Gousopoulos, L.; Kohl, S.; Ateschrang, A.; Kohlhof, H.; Ahmad, S.S. Factors influencing the success of anterior cruciate ligament repair with dynamic intraligamentary stabilisation. *Knee Surg. Sports Traumatol. Arthrosc.* **2017**, *25*, 3923–3928. [CrossRef] [PubMed]
44. Operation Guide Aesculap Position ACL. The Anterior Cruciate Ligament Reconstruction System Operating Technique. B. Braun Aesculap, Tuttlingen, Germany. Available online: https://kipdf.com/aesculap-position-acl-the-anterior-cruciate-ligament-reconstruction-system-opera_5aca9a9d1723dda674abb537.html (accessed on 6 June 2023).
45. Mathys, A.G. *Ligamys Rehabilitation Leitfaden*; Mathys AG Bettlach: Bettlach, Switzerland, 2018.
46. Farid, A.; Gommers, S.A.; Sierevelt, I.N.; van Eijk, F.; van Kampen, P.M.; Haverkamp, D.; Dutch ACL Repair Study Group. Graft failure and revision rate after ACL repair with dynamic intraligamentary stabilization. One-year results of a prospective case series of 155 patients. *J. Exp. Orthop.* **2023**, *10*, 52. [CrossRef]
47. Rotllan, C.; Viscor, G. Winter Sports Injuries in Elite Female Athletes: A Narrative Review. *Int. J. Environ. Res. Public Health* **2023**, *20*, 5815. [CrossRef] [PubMed]
48. Shelbourne, K.D.; Wilckens, J.H.; Mollabashy, A.; DeCarlo, M. Arthrofibrosis in acute anterior cruciate ligament reconstruction. The effect of timing of reconstruction and rehabilitation. *Am. J. Sports Med.* **1991**, *19*, 332–336. [CrossRef] [PubMed]
49. Evangelopoulos, D.S.; Kohl, S.; Schwienbacher, S.; Gantenbein, B.; Exadaktylos, A.; Ahmad, S.S. Collagen application reduces complication rates of mid-substance ACL tears treated with dynamic intraligamentary stabilization. *Knee Surg. Sports Traumatol. Arthrosc.* **2017**, *25*, 2414–2419. [CrossRef] [PubMed]

50. Häberli, J.; Jaberg, L.; Bieri, K.; Eggli, S.; Henle, P. Reinterventions after dynamic intraligamentary stabilization in primary anterior cruciate ligament repair. *Knee* **2018**, *25*, 271–278. [CrossRef] [PubMed]
51. Meister, M.; Koch, J.; Amsler, F.; Arnold, M.P.; Hirschmann, M.T. ACL suturing using dynamic intraligamentary stabilisation showing good clinical outcome but a high reoperation rate: A retrospective independent study. *Knee Surg. Sports Traumatol. Arthrosc.* **2018**, *26*, 655–659. [CrossRef]
52. Bieri, K.S.; Scholz, S.M.; Kohl, S.; Aghayev, E.; Staub, L.P. Dynamic intraligamentary stabilization versus conventional ACL reconstruction: A matched study on return to work. *Injury* **2017**, *48*, 1243–1248. [CrossRef]
53. Kösters, C.; Glasbrenner, J.; Spickermann, L.; Kittl, C.; Domnick, C.; Herbort, M.; Raschke, M.J.; Schliemann, B. Repair with Dynamic Intraligamentary Stabilization Versus Primary Reconstruction of Acute Anterior Cruciate Ligament Tears: 2-Year Results from a Prospective Randomized Study. *Am. J. Sports Med.* **2020**, *48*, 1108–1116. [CrossRef]

Journal of *Personalized Medicine*

MDPI

Review

Management of Irreparable Posterosuperior Rotator Cuff Tears—A Current Concepts Review and Proposed Treatment Algorithm by the AGA Shoulder Committee

Jonas Pogorzelski [1,2,*], Marco-Christopher Rupp [1], Bastian Scheiderer [1], Lucca Lacheta [1], Benedikt Schliemann [3], Jakob Schanda [4], Philipp Heuberer [5], Marco Schneider [6], Michael Hackl [7], AGA Shoulder Committee—Rotator Cuff † and Olaf Lorbach [8]

1 Department of Orthopedic Sports Medicine, Technical University of Munich, Klinikum rechts der Isar, Ismaninger Str. 22, 81675 Munich, Germany
2 Private Practice OC Erlangen-Ebermannstadt, Nägelsbach Str. 25b, 91052 Erlangen, Germany
3 Herz Jesu Hospital Münster-Hiltrup, Westfalen Str. 109, 48165 Münster, Germany
4 Private Practice Dr. Schanda, Rochusgasse 17/13, 1030 Vienna, Austria
5 Private Practice OrthoCare, Kurbad Str. 14, 1100 Vienna, Austria
6 University of Witten/Herdecke, Alfred-Herrhausen-Straße 45, 58455 Witten, Germany
7 Department of Trauma-, Hand- and Elbow Surgery, University Hospital of Cologne, Kerpener Str. 62, 50937 Cologne, Germany
8 Schoen-Clinic Lorsch, Department of Shoulder Surgery and Sports Traumatology, Wilhelm-Leuschner-Straße 10, 64653 Lorsch, Germany
* Correspondence: jonas.pogorzelski@tum.de
† Collaborators/Membership of the AGA Shoulder Committee—Rotator Cuff is provided in the Acknowledgments.

check for updates

Citation: Pogorzelski, J.; Rupp, M.-C.; Scheiderer, B.; Lacheta, L.; Schliemann, B.; Schanda, J.; Heuberer, P.; Schneider, M.; Hackl, M.; AGA Shoulder Committee—Rotator Cuff; et al. Management of Irreparable Posterosuperior Rotator Cuff Tears—A Current Concepts Review and Proposed Treatment Algorithm by the AGA Shoulder Committee. *J. Pers. Med.* **2023**, *13*, 191. https://doi.org/10.3390/jpm13020191

Academic Editors: Markus Gesslein and Eduard Kurz

Received: 11 January 2023
Accepted: 19 January 2023
Published: 21 January 2023

Abstract: Posterosuperior rotator cuff tears range among the most common causes of shoulder complaints. While non-operative treatment is typically reserved for the elderly patient with low functional demands, surgical treatment is considered the gold standard for active patients. More precisely, an anatomic rotator cuff repair (RCR) is considered the most desirable treatment option and should be generally attempted during surgery. If an anatomic RCR is impossible, the adequate choice of treatment for irreparable rotator cuff tears remains a matter of debate among shoulder surgeons. Following a critical review of the contemporary literature, the authors suggest the following evidence- and experience-based treatment recommendation. In the non-functional, osteoarthritic shoulder, treatment strategies in the management of irreparable posterosuperior RCT include debridement-based procedures and reverse total shoulder arthroplasty as the treatment of choice. Joint-preserving procedures aimed at restoring glenohumeral biomechanics and function should be reserved for the non-osteoarthritic shoulder. Prior to these procedures, however, patients should be counseled about deteriorating results over time. Recent innovations such as the superior capsule reconstruction and the implantation of a subacromial spacer show promising short-term results, yet future studies with long-term follow-up are required to derive stronger recommendations.

Keywords: irreparable; rotator cuff tear; joint preservation; partial repair; superior capsular reconstruction; tendon transfer; balloon system; treatment algorithm; guideline; expert consensus

1. Introduction

In the management of tears of the posterosuperior rotator cuff, characterized by tears of the tendons of the supraspinatus and infraspinatus muscle, there exists consensus that an anatomic rotator cuff repair (RCR) is the most desirable treatment option in active patients in the absence of severe degenerative changes [1]. However, the optimal management of irreparable posterosuperior rotator cuff tears (RCTs) still remains a topic of debate among shoulder surgeons. While the definition may vary, a posterosuperior RCT is typically considered "irreparable" if a primary repair of the affected tendon to its anatomic footprint is impossible despite sufficient surgical release and the mobilization of the tendon. More specifically, the widely accepted definition by Gerber et al. [2] defines a RCT to be "irreparable" if it is impossible to achieve anatomic fixation of the torn posterosuperior tendons in <60° of glenohumeral abduction despite adequate tendon release. Factors that increase the risk of a posterosuperior RCT to be "irreparable" include decreased tendon elasticity [3], fatty infiltration [4], anterior-posterior tear size [5] and a progression to cuff arthropathy [6] that results in alterations of the glenohumeral biomechanics [7]. The term "massive" RCT—describing full thickness tears of ≥2 tendons, retraction to glenoid and/or exposure of 67% of the greater tuberosity [8]—is often incorrectly used and interchanged with the term "irreparable" and should be used separately.

If an RCR is deemed impossible prior to surgery due to an irreparable situation, treatment options such as non-operative management, the debridement of the rotator cuff with or without the tenotomy or tenodesis of the long head of the biceps tendon (LHBT), partial repair, tendon transfers (TTs), the implantation of a subacromial biodegradable spacer (SBS), superior capsular reconstruction (SCR) and the implantation of a reverse total shoulder arthroplasty (RTSA) are considered viable alternatives [2,9–14]. Specific advantages and disadvantages have been reported for each of these procedures. Consequently, the surgeon has to determine the ideal treatment approach based on patient-related, anatomic, and injury-related predictors according to an individualized treatment algorithm.

The purpose of this article was to review the current literature and to provide a current concepts overview of available contemporary treatment options for irreparable posterosuperior RCTs as well as to present an evidence- and experience-based clinically applicable treatment algorithm.

2. Non-Operative Treatment

Given a failure rate of up to 70% after RCR of chronic and massive tears [15] and acknowledging that pain relief and functional improvement do not necessarily correlate with successful structural healing of the tendon [16], non-operative treatment has been proposed as the treatment of choice for atraumatic RCT [17]. In the setting of irreparable RCTs, a special physiotherapy concept termed "anterior deltoid re-education" (ADR) has been proposed, which consists of an exercise regimen to rehabilitate the deltoid muscle to compensate for the deficient rotator cuff. This concept has been shown to result in acceptable outcomes in a group of comorbid, elderly patients in combination with subacromial injections of local anesthetics and steroids as well as an oral therapy with non-steroidal anti-inflammatory drugs [18]. More specifically, in a collective of 17 patients (mean age: 80 years), there was a statistically significant improvement in the mean Constant Score, a score combining patient subjective criteria as well as objective functional assessments [19], from 26 points prior to treatment (range, 8–41 points) to 60 points (range, 43–77 points) at a mean follow-up of nine months ($p < 0.05$). The range of motion in forward elevation improved from a mean of 40° (range 30–60°) prior to treatment to a mean of 160° (range 150–180°) [18]. However, a recent study investigating patient-reported outcomes after the same concept of ADR in 30 patients with a mean age of 74 years (range 55–89 years) has shown that only 12 patients improved by more than 20 points in the American Shoulder and Elbow Surgeons (ASES) Score, the threshold that is considered a successful non-operative treatment [20]. Persisting pain and kinematic dysfunction of the glenohumeral joint were the most common reasons for failure. In summary, careful patient selection is necessary when non-operative treatment for irreparable RCT is administered, as patients should be willing to accept discomfort and functional deficits of their shoulder in exchange for avoiding the risks of surgery. Non-operative treatment in the management of irreparable RCT is considered to yield the best outcomes in elderly patients with low functional demands or in patients for whom surgical treatment is contraindicated (Table 1).

Table 1. Advantages and disadvantages of non-operative treatment in the management of irreparable rotator cuff tears.

Advantages of Non-Operative Treatment	Disadvantages of Non-Operative Treatment
No risks related to surgery	Tear size progressing over time
	Discomfort and functional deficits may be retained
	Progression to cuff arthropathy

3. Arthroscopic Debridement, Tenotomy/Tenodesis of the Long Head of the Biceps Tendon and (Reversed) Subacromial Decompression

Irreparable RCT may also be treated by an arthroscopic debridement of the remaining rotator cuff in combination with a (reversed) subacromial decompression (SAD) and a tenotomy or tenodesis of the LHBT (Table 2) [21–23]. While lesions of the LHBT have been identified as a source of persistent pain that can be resolved with LHBT tenotomy, the purpose of arthroscopic SAD is to create a smooth, non-impinging acromiohumeral articulation by creating space, removing osteophytes and bony irregularities. A "reverse SAD", also known as tuberoplasty, has been introduced in order to protect the coracoacromial ligament—a structure at risk in regular SAD—to prevent an antero-superior escape of the humeral head, while equally ensuring a smooth acromiohumeral articulation [24,25]. Debridement combined with (reverse) SAD has been shown to result in decreased pain levels while maintaining residual rotator cuff strength (i.e., no improvement in shoulder strength). Low physical demand, sufficient residual shoulder function, older patient age and pain as the chief complaint are the main indications for arthroscopic debridement combined with (reverse) SAD and LHBT tenotomy/tenodesis. As such, a contemporary study in 19 patients (mean age 68 years) with symptomatic massive RCT reports significant improvements in functional outcomes, with an improvement in ASES Score of more than 30 points and a significant pain relief after arthroscopic debridement with SAD and the tenotomy of the LHBT after a minimum 10-year follow-up [23]. However, in this cohort 26% of patients failed and underwent RTSA during follow-up, emphasizing the importance of careful patient selection and preoperative counseling [23].

Table 2. Advantages and disadvantages of arthroscopic debridement, tenotomy/tenodesis of the long head of the biceps tendon and (reversed) subacromial decompression.

Advantages of Arthroscopic Debridement, Tenotomy/Tenodesis of the Long Head of the Biceps Tendon and (Reversed) Subacromial Decompression	Disadvantages of Arthroscopic Debridement, Tenotomy/Tenodesis of the Long Head of the Biceps Tendon and (Reversed) Subacromial Decompression
Short rehabilitation	High failure rate over time
Only minor surgical risk	Discomfort and functional deficits may be retained

4. Partial Rotator Cuff Repair

Even if a complete anatomical repair is impossible, a partial repair of an irreparable RCT is an option to restore the rotator cuff's force couple, which is created by the subscapularis muscle anteriorly and the infraspinatus and teres minor muscles posteriorly. Even in cases of a torn supraspinatus tendon, a sufficient concavity compression of the humeral head into the glenoid can be obtained by a balanced force couple and thus prevent the superior migration of the humeral head [26]. Several authors have published their short- and mid-term clinical outcomes following partial repairs of massive RCTs. Shon et al. [27] reported patient-reported outcomes after partial repairs of irreparable posterosuperior RCTs combined with or without subscapularis tendon repair in 31 patients preoperatively and at 1 and 2 years after surgery. Despite significant improvements from preoperatively to postoperatively, a slight deterioration was found from 1 to 2 years postoperatively (ASES 76 vs. 74, SST 6.6 vs. 6.1, VAS 2.1 vs. 3.2). In addition, even when considering the initial improvement, subjective patient-reported satisfaction was reported as "rather the same" or "dissatisfied" in 15 patients (48%). Teres minor fatty infiltration was identified as an independent factor affecting patient-reported satisfaction. Slightly better mid-term results were reported by Cuff et al. in 28 patients who underwent partial repairs and LHBT tenotomy at a minimum 5-year follow-up with a satisfaction rate of 75% [28]. Despite significant improvements in ASES Score (47 to 79), SST (5.7 to 9.1) and VAS for pain (6.9 to 1.9) at final follow-up compared to preoperatively, the failure rate was as high as 29%. Of note, failure was defined as an ASES score of <70 points, loss of active elevation over 90° or revision arthroplasty surgery. Summarizing these findings, initial favorable clinical results can be expected in patients whose primary complaints are pain and weakness in the setting of good preoperative shoulder function without prior osteoarthritic changes. However, patients should be counseled that clinical outcomes may deteriorate over time (Table 3).

Table 3. Advantages and disadvantages of partial rotator cuff repair.

Advantages of Partial Rotator Cuff Repair	Disadvantages of Partial Rotator Cuff Repair
Reliable short-term outcomes in patients without significant osteoarthritis	Clinical outcomes may deteriorate over time
	Long rehabilitation

5. Tendon Transfers

Tendon transfers such as the latissimus dorsi tendon transfer (LDTT) or the lower trapezius tendon transfer (LTTT) are considered viable treatment options in young patients without glenohumeral osteoarthritis (OA) but with functional deficits, especially in terms of external rotation, caused by an irreparable posterosuperior RCT (Table 4). As the coronal force couple is disrupted in this situation, higher forces are required by both the deltoid and the intact muscle-tendon units of the remaining rotator cuff to stabilize the humeral head. As these tears progress, this results in a superior migration of the humeral head and shoulder dysfunction, in particular a lack of external rotation [29].

Table 4. Advantages and disadvantages of tendon transfers.

Advantages of Tendon Transfers	Disadvantages of Tendon Transfers
Standardized surgical technique with long-term experiences with certain tendon transfers	High patient compliance necessary
Reasonable outcomes in a relatively young patient cohort	Long and complex rehabilitation
Restoration of active ROM	

5.1. *Latissimus Dorsi Tendon Transfer*

In its physiologic function, the latissimus dorsi muscle-tendon unit acts to adduct, internally rotate and extend the humerus. Its large muscle excursion makes the latissimus dorsi feasible for a muscle transfer procedure [30]. Transferring the muscle-tendon unit from its native anterior insertion at the mid-bicipital groove to posterosuperiorly to the greater tuberosity allows the tendon to close the rotator cuff defect. Furthermore, in this setting, the latissimus acts as an external rotator and a depressor of the humeral head [31], which restores the coronal force couple by taking over the biomechanical function of the posterosuperior rotator cuff tendons and improves glenohumeral function. However, the postoperative results reported after this procedure are variable [30,32–42]. Most studies evaluating clinical outcomes after LDTTs have reported a significant reduction in pain and an improvement in shoulder function. A systematic review by Namdari et al. [43] analyzed

10 studies between 1992 and 2010 to determine the expected outcomes, outcome predictors and complications of LDTTs. Patients had a frequency-weighted mean adjusted Constant Score of 45.9 preoperatively compared with 73.2 postoperatively ($p < 0.001$). Additionally, the frequency-weighted mean active forward elevation improved from 101.9° preoperatively to 137.4° postoperatively ($p < 0.001$), while the external rotation improved from 16.8° to 26.7° ($p < 0.001$). The overall reported complication rate was 9.5%, including surgical site infection, neurapraxia, hematomas, wound dehiscence and tears of the transferred tendon. Predictors of favorable outcomes included a lower degree of teres minor fatty infiltration, an LDTT as a primary procedure and the presence of an intact subscapularis tendon [43].

5.2. Combined Latissimus Dorsi and Teres Major Tendon Transfer (L'Episcopo Technique)

In 1934, L'Episcopo introduced the combined latissimus dorsi and teres major tendon transfer to regain external rotation in pediatric patients with plexus paralysis. Comparable to isolated LDTT, the transfer of both tendons allows the closing of the rotator cuff defect and allows them act as an external rotator and a humeral head depressor. Especially in patients where the teres minor muscle is already degenerated, the combined teres major and latissimus dorsi tendon transfer might be beneficial to balance the rotator cuff's compromised force couple [33]. Lichtenberg et al. [44] compared clinical outcomes between isolated LDTT and combined teres major and latissimus dorsi tendon transfer at a mean follow-up of 6 years. Significant improvements in Constant Score and active range of motion were found in 17 patients in each group. However, group comparison showed significantly better active flexion and abduction for patients undergoing isolated LDTT compared to patients undergoing combined teres major and latissimus dorsi tendon transfer. In addition, there was no progression in the degree of rotator cuff arthropathy in patients who underwent isolated LDTT [44]. Combined, these findings suggest it is preferable to rely on the isolated LDTT technique rather than a combined LDTT and teres major tendon transfer as the default technique.

5.3. Lower Trapezius Tendon Transfer (LTTT)

The theoretical advantage of the lower trapezius tendon transfer (LTTT) compared to the LDTT is the synergistic function of the lower trapezius muscle and the infraspinatus muscle, sharing a similar force vector trajectory. Accordingly, both muscles contribute to scapular retraction and glenohumeral external rotation [45]. Compared to the LDTT, the LTTT is considered to be more anatomic, which reduces the need for intensive retraining during the postoperative recovery period. However, as the lower trapezius tendon has a short excursion, an additional tendon graft such as the semitendinosus tendon is required to cover the distance between its native insertion site at the scapular spine and the footprint of the torn rotator cuff at the greater tuberosity [46]. Recently, three studies assessed clinical outcomes of an LTTT as treatment for irreparable posterosuperior RCT [46–48]. In 2016, Elhassan et al. [48] reported the outcomes of 33 patients with an average age of 53 years following LTTT prolonged by Achilles tendon allograft. At an average follow-up of 47 months, 32 patients showed significant improvements in the Subjective Shoulder Value (SSV; 54% preoperatively, 78% postoperatively; $p < 0.01$) and Disabilities of Arm, Shoulder and Hand (DASH) Score (52 $\pm$ 19 preoperatively, 18 $\pm$ 10 postoperatively; $p < 0.01$). One patient failed and required shoulder fusion subsequently. All successfully treated patients improved their ability for glenohumeral external rotation. Interestingly, patients with more than 60° of preoperative flexion had more improvement in range of motion than patients with less than 60° of glenohumeral flexion [48]. A more recent study showed similar results in 41 patients undergoing arthroscopically assisted LTTT augmented with an Achilles tendon allograft [47]. At a mean follow-up of 14 months, about 90% of the patients reported significant improvement in all outcome measures, including pain, shoulder motion and patient reported outcomes. Of note, the authors reported eight early complications with four of them being peripheral nerve symptoms due to immobilization in a custom external rotation brace for a total of 6–8 weeks postoperatively. All symptoms resolved spontaneously over a 1–3-month period after removing the brace [47]. Valenti and Werthel [46] investigated 14 patients with a mean age of 62 years undergoing LTTT with semitendinosus tendon augmentation to reconstruct irreparable posterosuperior RCTs. At a minimum 18-month follow-up, all clinical scores improved significantly with the mean Constant Score improving from 35 ± 15 to 60 ± 9 points, mean VAS for pain decreasing from 7 to 2 points and mean SSV improving from 40% to 70%. Two of the 14 patients suffered from minor early complications (hematoma and surgical site infection) requiring revision surgery [46]. Compared to the reconstruction of passive stability by the means of an SCR, the LTTT was superior in terms of functional improvement (ASES score of 84.8 ± 7.6 vs. 76.8 ± 20.3, respectively; $p = 0.045$), patient satisfaction, progression of arthritis (SCR: 22.7% vs. with LTTT: 2.8%) and graft integrity (retear rate: SCR 63.6% vs. LTTT: 8.3%) at minimum 2 years follow-up [49].

6. Superior Capsular Reconstruction (SCR)

The SCR has recently been introduced as an alternative treatment for irreparable posterosuperior RCT that is less invasive compared to tendon transfer surgery (Table 5). Biomechanically, the superior capsule of the glenohumeral joint contributes to static stabilization and is commonly disrupted when the supraspinatus and infraspinatus tendons are injured [50]. The technique of SCR in its current form was developed in Japan by Mihata et al. [51], who used a fascia lata autograft to reconstruct the superior capsule as treatment for an irreparable RCT. Today, a variety of allografts, including human acellular dermal allografts, are available [52,53]. In general, the graft prevents the superior migration of the humeral head, thereby maintaining the glenohumeral fulcrum if the force couple is still intact. Commonly accepted contraindications include high-grade glenohumeral OA (Hamada > 2), deltoid weakness, and irreparable subscapularis tears [54,55]. The main indications for an SCR include pseudoparesis (defined as active scapular plane abduction of more than 45° and less than 90°), an intact subscapularis tendon or reparable subscapularis tendon tear, and the absence of an external rotation lag sign [56,57].

Table 5. Advantages and disadvantages of superior capsular reconstruction (SCR).

Advantages of Superior Capsular Reconstruction (SCR)	Disadvantages of Superior Capsular Reconstruction (SCR)
Convincing short-term outcomes	Long-term outcomes not available
Less invasive surgery than tendon transfers	High percentage of graft retear rate
Preservation of anatomy	Limited cost-effectiveness compared to tendon transfers

Two recent systematic reviews of early postoperative clinical outcomes reported convincing clinical results in an early stage [55,58]. Clinical scores including the ASES Score improved significantly across all studies with mean postoperative scores ranging from 70–90 points [55,58]. Failure, mainly defined as graft rupture, occurred in 13–14% of cases [55,58]. This is in contrast to another recent systematic review and meta analysis that also reported satisfactory clinical and radiological results at a minimum 24 months follow-up, but with a substantially higher graft failure rate of about 23.9% [59]. Interestingly, the rupture of the graft is not always correlated with worse outcomes [54]. Consequently, SCR is believed to act as a mechanical spacer and thus improve shoulder kinematics [58,60]. This assumption, however, was disproved by a recent study that suggested that SCR may not depress the humeral head during functional abduction, and that the postoperative improvements in subjective and clinical outcomes may be affected by mechanisms other than changes in shoulder kinematics [61]. These findings are in line with a publication of Greiner et al. [62], who could not detect any differences in outcomes in the short term follow-up between the implantation of an SCR and partial infraspinatus repair in a matched-pair comparison of patients suffering from irreparable posterosuperior rotator cuff tears.

7. Subacromial Biodegradable Spacer

The SBS, also known as the balloon system, is a preshaped spacer made of a copolymer of poly-L-lactide, which biodegrades over approximately 12 months [63]. SBSs are arthroscopically inserted into the subacromial space and filled with saline solution, which subsequently depresses the humeral head. The biomechanical rationale behind performing this procedure in the setting of an irreparable RCT is to improve shoulder function by depressing the humeral head to a more central position on the glenoid, thus restoring the force couple and increasing deltoid load to improve the deltoid lever arm [11,63,64]. A recent biomechanical study confirmed these assumptions in a cadaveric model of an irreparable supraspinatus tear, showing that a balloon spacer restores intact-state glenohumeral contact pressures at most abduction angles, while also depressing the humeral head and increasing the deltoid load at time zero [11]. Moon et al. [65] summarized existing clinical results in a systematic review of seven outcome studies including 204 shoulders from 200 patients following subacromial spacer implantation due to irreparable posterosuperior RCT with Goutallier stage 3 and 4 fatty infiltration based on magnetic resonance imaging. Contraindications included cuff arthropathy stage Hamada 3 or higher, irreparable subscapularis tears, and pseudoparesis [66–68]. The mean age of patients was 68 years with a mean follow-up time of 19 months. All studies reported consistent and significant improvement in the total Constant Score or ASES Score over the duration of short-term follow-up with mean postoperative scores ranging between 60 and 70 points for the Constant Score and 70 and 80 points for the ASES Score. In a multi-center single-blinded, randomized controlled trial, Verma et al. reported comparable clinical outcomes between partial repairs, yet significantly greater forward elevation during early recovery as well as significantly shorter operating times [31]. However, another recent double-blinded randomized controlled trial comparing the arthroscopic debridement of the subacromial space with biceps tenotomy to an additional insertion of the subacromial balloon favored the group that received debridement and tenotomy only, as this group showed even slightly better outcomes after a 12-month follow-up [47]. Equally critical results were published by Garríguez-Pérez et al. at a comparable follow-up, reporting poor functional and satisfaction rates in a case series of 16 patients [9]. While a failure rate of only 3–13% had been reported at short-term follow-up [69], there are concerns regarding the effect beyond the biodegradation period of the spacer in the absence of mid- to long-term results in the literature to date [38]. This is reflected in single, low volume studies that report a preservation of a clinically relevant improvement in the postoperative outcome in only ~60% of the patients at a mid-term FU of 5 years [70]. In addition, an examination of reported adverse events such as foreign body reactions [71] and the risk of dislocations and more prospective data are needed to examine patient- and RCT-related factors predictive of the success of the procedure, providing more clarity regarding the perceived inconsistency in postoperative outcomes following the implantation of a balloon system in the contemporary literature (Table 6) [65,72,73].

Table 6. Advantages and disadvantages of the subacromial biodegradable spacer.

Advantages of the Subacromial Biodegradable Spacer	Disadvantages of the Subacromial Biodegradable Spacer
Only minor surgical risk	Long-term outcomes not available
Short rehabilitation	Biodegradation after 3 months
	Limited scientific evidence

8. Reverse Total Shoulder Arthroplasty

As surgeons have become more experienced with reverse total shoulder arthroplasty (RTSA), surgical indications have expanded over the past decade [74–76]. While irreparable RCTs with associated high-grade glenohumeral OA remain the most common indication for RTSA and show good to excellent outcomes [77], some authors suggest RTSA as a viable treatment option in patients with irreparable RCTs in the absence of glenohumeral

OA due to its predictable pain relief and return to function (Table 7) [74,75,78]. Mulieri et al. [75] reported significant improvements in patient-reported outcomes (ASES, SST, VAS) in 60 shoulders following RTSA without significant OA after a minimum 2-year follow-up. The average patient age was 71 years and overall implant survivorship was 91% at a mean follow-up of 52 months [75]. Although RTSA in patients with low-grade OA leads to substantial subjective and functional improvement without clinical deterioration beyond 10 to 15 years, it is associated with a high complication rate of up to 39%, which ultimately compromises functional outcomes [69]. Special caution should be exercised when offering RTSA to young patients (especially those under the age of 65 years) as component loosening is the main cause for failure over time [79,80]. Additionally, patients without pseudoparesis and a good forward elevation preoperatively tend to have higher dissatisfaction rates because of the possible postoperative loss of active forward elevation compared to patients with pseudoparesis and a pre-existing limited range of motion [79].

Table 7. Advantages and disadvantages of reverse total shoulder arthroplasty.

Advantages of Reverse Total Shoulder Arthroplasty	Disadvantages of Reverse Total Shoulder Arthroplasty
Predictable and very good outcomes at short- and mid-term follow-up	Limited use in patients younger than 65 years
Relatively short rehabilitation	Functional deficits related to arthroplasty design
Considerable scientific evidence	Limited load-bearing capacity

9. Discussion

The optimal management of irreparable posterosuperior RCT remains a challenge for shoulder surgeons. If possible, the anatomic reconstruction of the torn rotator cuff should be attempted. As discussed above, if anatomic reconstruction is not feasible, several treatment modalities including non-operative and operative strategies are available and should be tailored to the patient's individual preconditions and functional demand. Despite a large body of evidence on the management of irreparable posterosuperior RCT, there is still controversy about the ideal decision-making algorithm. In this current concepts review article, current evidence on surgical and non-surgical treatment strategies for irreparable posterosuperior RCT was summarized. As a result, a treatment algorithm with clinical steps and decision criteria is suggested to guide the shoulder surgeon in the decision-making for the management of irreparable posterosuperior RCTs (Figure 1).

In general, in the setting of an irreparable posterosuperior RCT, an initial non-operative treatment based on a standardized exercise program with the possibility for delayed, consecutive surgical treatment is recommended. However, purely non-operative treatment is often unsatisfactory and is mainly limited to patients not suitable for surgery or elderly patients with low functional demands. For active patients, a more selective approach is necessary as a posterosuperior RCT results in altered glenohumeral contact mechanics with the superior migration of the humeral head and accompanying force couple insufficiency. Consequently, abnormal glenohumeral contact pressure and microinstability cause rapid and early onset glenohumeral OA [7,81].

Thus, the evaluation of glenohumeral OA in patients with irreparable posterosuperior RCTs based on the classification proposed by Hamada et al. [82] should be the first step in the process of decision-making as multiple joint-preserving approaches such as tendon transfers, SCR, or spacer implantation are only suitable for patients with minimal osteoarthritic changes, typically defined as Hamada grade <2. The second step is to assess the main complaint of the patient. In patients reporting pain as their chief complaint, who only have low-grade glenohumeral OA without pseudoparesis, debridement followed by a partial rotator cuff repair to restore the force couple in combination with an LHBT tenotomy or tenodesis may be sufficient. The combination of these procedures reliably achieves pain relief and facilitates a high degree of functional compensation by a restoration of the force couple. However, as postoperative clinical outcomes may deteriorate over time, adequate patient counseling is crucial.

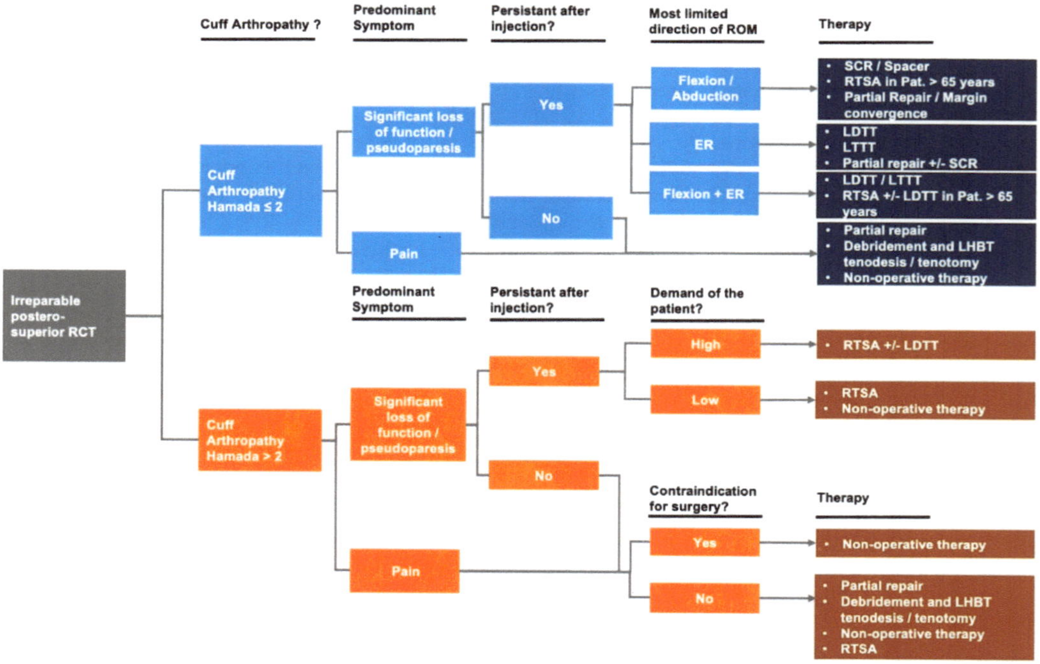

Figure 1. Treatment algorithm in the setting of irreparable posterosuperior rotator cuff tears. Abbreviations: ER, external rotation; SCR, superior capsule reconstruction; ROM, range of motion; RTSA, reverse total shoulder arthroplasty; LDTT, latissimus dorsi tendon transfer; LTTT, lower trapezius tendon transfer; LHBT, long head of the biceps tendon.

If a patient with low-grade glenohumeral OA presents with pseudoparesis or a significant loss of function as the chief complaint, the next diagnostic step is an injection of local anesthetics in patients, to determine whether the main reason for the functional deficit is primarily due to insufficient rotator cuff function or secondary to pain. In the case of a resolution of functional deficits upon injection, a partial repair combined with LBHT surgery as well as debridement are indicated. In the situation of persisting deficits following injection, it is important to determine the main limitation in the range of motion or main direction of pseudoparesis (i.e., flexion/abduction, external rotation, flexion + external rotation). In the case of deficits primarily in flexion or abduction, joint-preserving treatments such as a SCR and SBS implantation or (if possible) partial repair with margin convergence are recommended as possible treatment options, while an RTSA may also be considered in patients age > 65 years. However, although SCR and SBS implantation have become increasingly popular in recent years, as they preserve the native anatomy and do not compromise future treatment options, additional studies with long-term follow-up are required to sufficiently assess clinical outcomes, as existing early clinical outcome data are currently insufficient to provide final recommendations. In patients with deficits primarily in external rotation, an LDTT or LTTT is recommend as the treatment of choice, while a partial repair with a possible augmentation with an SCR may be a viable alternative. Of note, the clinical benefit of the combined LDTT and teres major tendon transfer (L'Episcopo technique) compared to the isolated LDTT is uncertain [44]. In patients with deficits in flexion that present with an additional positive external rotation lag sign, isolated tendon transfers as well as RTSA (in patients older than 65 years) +/− a combined LDTT are the recommended treatment options.

In patients with high-grade glenohumeral OA (i.e., Hamada > 2), only two surgical treatment options are recommended depending on the presence of pseudoparesis. A combined arthroscopic debridement, LHBT tenotomy/tenodesis, and partial rotator cuff repair is considered the first-line treatment in patients with good glenohumeral motion and no pseudoparesis, in case pain is the main symptom and/or functional deficits secondary to pain resolve upon the intra-articular injection of a local anesthetic. If severe functional deficits in any direction are present and persist after the injection of a local anesthetic in patients with high-grade glenohumeral OA, an RTSA is recommended with optional concomitant LDTT in patients with high physical demands and existing preoperative limitations of external rotation.

10. Conclusions

The management of irreparable posterosuperior RCTs remains a challenge for shoulder surgeons and treatment strategies should be tailored to the individual patient. In the non-functional, osteoarthritic shoulder, treatment strategies in the management of irreparable posterosuperior RCTs include debridement-based procedures and reverse total shoulder arthroplasty as the treatment of choice. Joint-preserving procedures aimed at restoring glenohumeral biomechanics and function should be reserved for the non-osteoarthritic shoulder. Prior to these procedures, however, patients should be counseled about deteriorating results over time. Recent innovations such as the SCR and the implantation of an SBS show promising clinical short-term results, yet current comparative studies indicate limited efficacy comparing the SCR or the in-space balloon to partial repair procedures. Taking the much higher costs for these innovative procedures into account, no fundamental recommendation can be given for either the SCR or the balloon. Future prospective comparative studies at long-term follow-up are required to delineate the sustainability of these procedures and derive stronger recommendations.

Author Contributions: J.P.; M.-C.R.; B.S. (Bastian Scheiderer); L.L.; B.S. (Benedikt Schlieman); J.S.; P.H.; M.S.; M.H.; and O.L. participated in conceptualization, J.P.; M.-C.R.; B.S. (Bastian Scheiderer); L.L.; B.S. (Benedikt Schlieman); J.S.; P.H.; O.L.; and the AGA Shoulder Committee—Rotator Cuff performed the investigation, J.P.; M.-C.R.; B.S. (Bastian Scheiderer); L.L.; B.S. (Benedikt Schlieman); J.S.; P.H.; and O.L. wrote the original draft preparation, J.P.; M.-C.R.; B.S. (Bastian Scheiderer); L.L.; B.S. (Benedikt Schlieman); J.S.; P.H.; O.L.; M.S.; M.H.; and the AGA Shoulder Committee—Rotator Cuff participated in review and editing, M.-C.R. and J.P. conducted visualization, and J.P. was involved in supervision and project administration. All authors have read and agreed to the published version of the manuscript.

Funding: This research received no external funding.

Institutional Review Board Statement: Not applicable.

Informed Consent Statement: Not applicable.

Acknowledgments: Shoulder Committee—Rotator Cuff of the AGA-Society for Arthroscopy and Joint-Surgery: Olaf Lorbach, Jonas Pogorzelski, Bastian Scheiderer, Michael Hackl, Philipp Heuberer, Barbara Wirth, Simon Euler, Ralf Müller-Rath, Benedikt Schliemann, Marco Schneider, Mathias Wellmann, Lucca Lacheta, Christian Jung, Michael Manske, and Jacob Schanda.

Conflicts of Interest: The authors declare no conflict of interest.

References

1. Pogorzelski, J.; Fritz, E.M.; Horan, M.P.; Katthagen, J.C.; Hussain, Z.B.; Godin, J.A.; Millett, P.J. Minimum Five-year Outcomes and Clinical Survivorship for Arthroscopic Transosseous-equivalent Double-row Rotator Cuff Repair. *J. Am. Acad. Orthop. Surg.* **2019**, *27*, e1093–e1101. [CrossRef] [PubMed]
2. Gerber, C.; Vinh, T.S.; Hertel, R.; Hess, C.W. Latissimus dorsi transfer for the treatment of massive tears of the rotator cuff. A preliminary report. *Clin. Orthop. Relat. Res.* **1988**, *232*, 51–61. [CrossRef]

3. Oh, J.H.; Kim, S.H.; Kang, J.Y.; Oh, C.H.; Gong, H.S. Effect of age on functional and structural outcome after rotator cuff repair. *Am. J. Sports Med.* **2010**, *38*, 672–678. [PubMed]
4. Melis, B.; Wall, B.; Walch, G. Natural history of infraspinatus fatty infiltration in rotator cuff tears. *J. Shoulder Elb. Surg.* **2010**, *19*, 757–763. [CrossRef]
5. Lee, Y.S.; Jeong, J.Y.; Park, C.D.; Kang, S.G.; Yoo, J.C. Evaluation of the Risk Factors for a Rotator Cuff Retear After Repair Surgery. *Am. J. Sports Med.* **2017**, *45*, 1755–1761. [CrossRef]
6. Zingg, P.O.; Jost, B.; Sukthankar, A.; Buhler, M.; Pfirrmann, C.W.; Gerber, C. Clinical and structural outcomes of nonoperative management of massive rotator cuff tears. *J. Bone Joint Surg. Am.* **2007**, *89*, 1928–1934. [CrossRef]
7. Muench, L.N.; Berthold, D.P.; Otto, A.; Dyrna, F.; Bell, R.; Obopilwe, E.; Cote, M.P.; Imhoff, A.B.; Mazzocca, A.D.; Beitzel, K. Increased Glenohumeral Joint Loads Due to a Supraspinatus Tear Can be Reversed with Rotator Cuff Repair: A Biomechanical Investigation. *Arthrosc. J. Arthrosc. Relat. Surg.* **2021**, *38*, 1422–1432. [CrossRef]
8. Schumaier, A.; Kovacevic, D.; Schmidt, C.; Green, A.; Rokito, A.; Jobin, C.; Yian, E.; Cuomo, F.; Koh, J.; Gilotra, M.; et al. Defining massive rotator cuff tears: A Delphi consensus study. *J. Shoulder Elb. Surg.* **2020**, *29*, 674–680. [CrossRef]
9. Burkhart, S.S. Reconciling the paradox of rotator cuff repair versus debridement: A unified biomechanical rationale for the treatment of rotator cuff tears. *Arthrosc. J. Arthrosc. Relat. Surg.* **1994**, *10*, 4–19. [CrossRef]
10. Godin, J.A.; Pogorzelski, J.; Horan, M.P.; Lacheta, L.; Hussain, Z.B.; Altintas, B.; Frangiamore, S.; Mannava, S.; Fritz, E.M.; Millett, P.J. Impact of Age and Subscapularis Tendon Reparability on Return to Recreational Sports Activities and 2-Year Outcomes After Reverse Total Shoulder Arthroplasty. *Orthop. J. Sports Med.* **2019**, *7*, 2325967119875461. [CrossRef]
11. Lobao, M.H.; Canham, R.B.; Melvani, R.T.; Abboud, J.A.; Parks, B.G.; Murthi, A.M. Biomechanics of Biodegradable Subacromial Balloon Spacer for Irreparable Superior Rotator Cuff Tears: Study of a Cadaveric Model. *J. Bone Joint Surg. Am.* **2019**, *101*, e49. [CrossRef] [PubMed]
12. Oh, J.H.; McGarry, M.H.; Jun, B.J.; Gupta, A.; Chung, K.C.; Hwang, J.; Lee, T.Q. Restoration of shoulder biomechanics according to degree of repair completion in a cadaveric model of massive rotator cuff tear: Importance of margin convergence and posterior cuff fixation. *Am. J. Sports Med.* **2012**, *40*, 2448–2453. [CrossRef] [PubMed]
13. Pogorzelski, J.; Fritz, E.M.; Godin, J.A.; Imhoff, A.B.; Millett, P.J. Nonoperative treatment of five common shoulder injuries: A critical analysis. *Obere Extrem.* **2018**, *13*, 89–97. [CrossRef] [PubMed]
14. Pogorzelski, J.; Muckenhirn, K.J.; Mitchell, J.J.; Katthagen, J.C.; Schon, J.M.; Dahl, K.D.; Hirahara, A.M.; Dines, J.S.; Adams, C.R.; Dooney, T.; et al. Biomechanical Comparison of 3 Glenoid-Side Fixation Techniques for Superior Capsular Reconstruction. *Am. J. Sports Med.* **2018**, *46*, 801–808. [CrossRef]
15. Rashid, M.S.; Cooper, C.; Cook, J.; Cooper, D.; Dakin, S.G.; Snelling, S.; Carr, A.J. Increasing age and tear size reduce rotator cuff repair healing rate at 1 year. *Acta Orthop.* **2017**, *88*, 606–611. [CrossRef]
16. Petri, M.; Ettinger, M.; Brand, S.; Stuebig, T.; Krettek, C.; Omar, M. Non-Operative Management of Rotator Cuff Tears. *Open Orthop. J.* **2016**, *10*, 349–356. [CrossRef]
17. Schemitsch, C.; Chahal, J.; Vicente, M.; Nowak, L.; Flurin, P.-H.; Heerspink, F.L.; Henry, P.; Nauth, A. Surgical repair versus conservative treatment and subacromial decompression for the treatment of rotator cuff tears: A meta-analysis of randomized trials. *Bone Joint J.* **2019**, *101*, 1100–1106. [CrossRef]
18. Levy, O.; Mullett, H.; Roberts, S.; Copeland, S. The role of anterior deltoid reeducation in patients with massive irreparable degenerative rotator cuff tears. *J. Shoulder Elb. Surg.* **2008**, *17*, 863–870. [CrossRef] [PubMed]
19. Ziegler, P.; Kühle, L.; Stöckle, U.; Wintermeyer, E.; Stollhof, L.E.; Ihle, C.; Bahrs, C. Evaluation of the Constant score: Which is the method to assess the objective strength? *BMC Musculoskelet. Disord.* **2019**, *20*, 403. [CrossRef]
20. Yian, E.H.; Sodl, J.F.; Dionysian, E.; Schneeberger, A.G. Anterior deltoid reeducation for irreparable rotator cuff tears revisited. *J. Shoulder Elb. Surg.* **2017**, *26*, 1562–1565. [CrossRef]
21. Liem, D.; Lengers, N.; Dedy, N.; Poetzl, W.; Steinbeck, J.; Marquardt, B. Arthroscopic debridement of massive irreparable rotator cuff tears. *Arthrosc. J. Arthrosc. Relat. Surg.* **2008**, *24*, 743–748. [CrossRef] [PubMed]
22. Park, J.G.; Cho, N.S.; Song, J.H.; Baek, J.H.; Rhee, Y.G. Long-term outcome of tuberoplasty for irreparable massive rotator cuff tears: Is tuberoplasty really applicable? *J. Shoulder Elb. Surg.* **2016**, *25*, 224–231. [CrossRef] [PubMed]
23. Vogler, T.; Andreou, D.; Gosheger, G.; Kurpiers, N.; Velmans, C.; Ameziane, Y.; Schneider, K.; Rickert, C.; Liem, D.; Schorn, D. Long-term outcome of arthroscopic debridement of massive irreparable rotator cuff tears. *PLoS ONE* **2020**, *15*, e0241277. [CrossRef] [PubMed]
24. Scheibel, M.; Lichtenberg, S.; Habermeyer, P. Reversed arthroscopic subacromial decompression for massive rotator cuff tears. *J. Shoulder Elb. Surg.* **2004**, *13*, 272–278. [CrossRef]
25. Verhelst, L.; Vandekerckhove, P.J.; Sergeant, G.; Liekens, K.; Van Hoonacker, P.; Berghs, B. Reversed arthroscopic subacromial decompression for symptomatic irreparable rotator cuff tears: Mid-term follow-up results in 34 shoulders. *J. Shoulder Elb. Surg.* **2010**, *19*, 601–608. [CrossRef]
26. Karduna, A.R.; Williams, G.R.; Williams, J.L.; Iannotti, J.P. Kinematics of the glenohumeral joint: Influences of muscle forces, ligamentous constraints, and articular geometry. *J. Orthop. Res.* **1996**, *14*, 986–993. [CrossRef]
27. Shon, M.S.; Koh, K.H.; Lim, T.K.; Kim, W.J.; Kim, K.C.; Yoo, J.C. Arthroscopic Partial Repair of Irreparable Rotator Cuff Tears: Preoperative Factors Associated with Outcome Deterioration Over 2 Years. *Am. J. Sports Med.* **2015**, *43*, 1965–1975. [CrossRef]

28. Cuff, D.J.; Pupello, D.R.; Santoni, B.G. Partial rotator cuff repair and biceps tenotomy for the treatment of patients with massive cuff tears and retained overhead elevation: Midterm outcomes with a minimum 5 years of follow-up. *J. Shoulder Elb. Surg.* **2016**, *25*, 1803–1809. [CrossRef]
29. Parsons, I.M.; Apreleva, M.; Fu, F.H.; Woo, S.L. The effect of rotator cuff tears on reaction forces at the glenohumeral joint. *J. Orthop. Res.* **2002**, *20*, 439–446. [CrossRef]
30. Gerber, C.; Maquieira, G.; Espinosa, N. Latissimus dorsi transfer for the treatment of irreparable rotator cuff tears. *J. Bone Joint Surg. Am.* **2006**, *88*, 113–120.
31. Henseler, J.F.; Nagels, J.; Nelissen, R.G.; de Groot, J.H. Does the latissimus dorsi tendon transfer for massive rotator cuff tears remain active postoperatively and restore active external rotation? *J. Shoulder Elb. Surg.* **2014**, *23*, 553–560. [CrossRef] [PubMed]
32. Aoki, M.; Okamura, K.; Fukushima, S.; Takahashi, T.; Ogino, T. Transfer of latissimus dorsi for irreparable rotator-cuff tears. *J. Bone Joint Surg. Br.* **1996**, *78*, 761–766. [CrossRef] [PubMed]
33. Costouros, J.G.; Espinosa, N.; Schmid, M.R.; Gerber, C. Teres minor integrity predicts outcome of latissimus dorsi tendon transfer for irreparable rotator cuff tears. *J. Shoulder Elb. Surg.* **2007**, *16*, 727–734. [CrossRef]
34. El-Azab, H.M.; Rott, O.; Irlenbusch, U. Long-term follow-up after latissimus dorsi transfer for irreparable posterosuperior rotator cuff tears. *J. Bone Joint Surg. Am.* **2015**, *97*, 462–469. [CrossRef]
35. Gerber, C. Latissimus dorsi transfer for the treatment of irreparable tears of the rotator cuff. *Clin. Orthop. Relat. Res.* **1992**, *275*, 152–160. [CrossRef]
36. Gerber, C.; Rahm, S.A.; Catanzaro, S.; Farshad, M.; Moor, B.K. Latissimus dorsi tendon transfer for treatment of irreparable posterosuperior rotator cuff tears: Long-term results at a minimum follow-up of ten years. *J. Bone Joint Surg. Am.* **2013**, *95*, 1920–1926. [CrossRef]
37. Gerhardt, C.; Lehmann, L.; Lichtenberg, S.; Magosch, P.; Habermeyer, P. Modified L'Episcopo tendon transfers for irreparable rotator cuff tears: 5-year follow-up. *Clin. Orthop. Relat. Res.* **2010**, *468*, 1572–1577. [CrossRef] [PubMed]
38. Iannotti, J.P.; Hennigan, S.; Herzog, R.; Kella, S.; Kelley, M.; Leggin, B.; Williams, G.R. Latissimus dorsi tendon transfer for irreparable posterosuperior rotator cuff tears. Factors affecting outcome. *J. Bone Joint Surg. Am.* **2006**, *88*, 342–348. [CrossRef]
39. Irlenbusch, U.; Bracht, M.; Gansen, H.K.; Lorenz, U.; Thiel, J. Latissimus dorsi transfer for irreparable rotator cuff tears: A longitudinal study. *J. Shoulder Elb. Surg.* **2008**, *17*, 527–534. [CrossRef]
40. Miniaci, A.; MacLeod, M. Transfer of the latissimus dorsi muscle after failed repair of a massive tear of the rotator cuff. A two to five-year review. *J. Bone Joint Surg. Am.* **1999**, *81*, 1120–1127. [CrossRef]
41. Moursy, M.; Forstner, R.; Koller, H.; Resch, H.; Tauber, M. Latissimus dorsi tendon transfer for irreparable rotator cuff tears: A modified technique to improve tendon transfer integrity. *J. Bone Joint Surg. Am.* **2009**, *91*, 1924–1931. [CrossRef] [PubMed]
42. Nove-Josserand, L.; Costa, P.; Liotard, J.P.; Safar, J.F.; Walch, G.; Zilber, S. Results of latissimus dorsi tendon transfer for irreparable cuff tears. *Orthop. Traumatol. Surg. Res.* **2009**, *95*, 108–113. [CrossRef] [PubMed]
43. Namdari, S.; Voleti, P.; Baldwin, K.; Glaser, D.; Huffman, G.R. Latissimus dorsi tendon transfer for irreparable rotator cuff tears: A systematic review. *J. Bone Joint Surg. Am.* **2012**, *94*, 891–898. [CrossRef]
44. Lichtenberg, S.; Magosch, P.; Habermeyer, P. Are there advantages of the combined latissimus-dorsi transfer according to L'Episcopo compared to the isolated latissimus-dorsi transfer according to Herzberg after a mean follow-up of 6 years? A matched-pair analysis. *J. Shoulder Elb. Surg.* **2012**, *21*, 1499–1507. [CrossRef] [PubMed]
45. Clark, N.J.; Elhassan, B.T. The Role of Tendon Transfers for Irreparable Rotator Cuff Tears. *Curr. Rev. Musculoskelet. Med.* **2018**, *11*, 141–149. [CrossRef]
46. Valenti, P.; Werthel, J.D. Lower trapezius transfer with semitendinosus tendon augmentation: Indication, technique, results. *Obere Extrem.* **2018**, *13*, 261–268. [CrossRef]
47. Elhassan, B.T.; Sanchez-Sotelo, J.; Wagner, E.R. Outcome of arthroscopically assisted lower trapezius transfer to reconstruct massive irreparable posterior-superior rotator cuff tears. *J. Shoulder Elb. Surg.* **2020**, *29*, 2135–2142. [CrossRef]
48. Elhassan, B.T.; Wagner, E.R.; Werthel, J.D. Outcome of lower trapezius transfer to reconstruct massive irreparable posterior-superior rotator cuff tear. *J. Shoulder Elb. Surg.* **2016**, *25*, 1346–1353. [CrossRef]
49. Baek, C.H.; Lim, C.; Kim, J.G. Superior Capsular Reconstruction Versus Lower Trapezius Transfer for Posterosuperior Irreparable Rotator Cuff Tears with High-Grade Fatty Infiltration in the Infraspinatus. *Am. J. Sports Med.* **2022**, *50*, 1938–1947. [CrossRef]
50. Petri, M.; Greenspoon, J.A.; Moulton, S.G.; Millett, P.J. Patch-Augmented Rotator Cuff Repair and Superior Capsule Reconstruction. *Open Orthop. J.* **2016**, *10*, 315–323. [CrossRef]
51. Mihata, T.; Lee, T.Q.; Watanabe, C.; Fukunishi, K.; Ohue, M.; Tsujimura, T.; Kinoshita, M. Clinical results of arthroscopic superior capsule reconstruction for irreparable rotator cuff tears. *Arthrosc. J. Arthrosc. Relat. Surg.* **2013**, *29*, 459–470. [CrossRef] [PubMed]
52. Katthagen, J.C.; Tahal, D.; Millett, P.J. Arthroscopic superior capsule reconstruction for irreparable rotator cuff tears. *Orthop. Today* **2016**.
53. Petri, M.; Greenspoon, J.A.; Millett, P.J. Arthroscopic Superior Capsule Reconstruction for Irreparable Rotator Cuff Tears. *Arthrosc. Tech.* **2015**, *4*, e751–e755. [CrossRef]
54. Lacheta, L.; Horan, M.P.; Schairer, W.W.; Goldenberg, B.T.; Dornan, G.J.; Pogorzelski, J.; Millett, P.J. Clinical and Imaging Outcomes After Arthroscopic Superior Capsule Reconstruction with Human Dermal Allograft for Irreparable Posterosuperior Rotator Cuff Tears: A Minimum 2-Year Follow-Up. *Arthrosc. J. Arthrosc. Relat. Surg.* **2020**, *36*, 1011–1019. [CrossRef] [PubMed]

55. Smith, T.J.; Gowd, A.K.; Kunkel, J.; Kaplin, L.; Hubbard, J.B.; Coates, K.E.; Graves, B.R.; Waterman, B.R. Clinical Outcomes of Superior Capsular Reconstruction for Massive, Irreparable Rotator Cuff Tears: A Systematic Review Comparing Acellular Dermal Allograft and Autograft Fascia Lata. *Arthrosc. Sport. Med. Rehabil.* **2021**, *3*, e257–e268. [CrossRef] [PubMed]
56. Ernstbrunner, L.; El Nashar, R.; Favre, P.; Bouaicha, S.; Wieser, K.; Gerber, C. Chronic Pseudoparalysis Needs to Be Distinguished from Pseudoparesis: A Structural and Biomechanical Analysis. *Am. J. Sports Med.* **2020**, *49*, 291–297. [CrossRef] [PubMed]
57. Mihata, T.; Lee, T.; Hasegawa, A.; Fukunishi, K.; Kawakami, T.; Fujisawa, Y.; Ohue, M.; Doi, M.; Neo, M. Arthroscopic Superior Capsule Reconstruction for Irreparable Rotator Cuff Tears: Comparison of Clinical Outcomes with and Without Subscapularis Tear. *Am. J. Sports Med.* **2020**, *48*, 3429–3438. [CrossRef]
58. Altintas, B.; Scheidt, M.; Kremser, V.; Boykin, R.; Bhatia, S.; Sajadi, K.R.; Mair, S.; Millett, P.J. Superior Capsule Reconstruction for Irreparable Massive Rotator Cuff Tears: Does It Make Sense? A Systematic Review of Early Clinical Evidence. *Am. J. Sports Med.* **2020**, *48*, 3365–3375. [CrossRef]
59. Werthel, J.D.; Vigan, M.; Schoch, B.; Lädermann, A.; Nourissat, G.; Conso, C. Superior capsular reconstruction—A systematic review and meta-analysis. *Orthop. Traumatol. Surg. Res.* **2021**, *107*, 103072. [CrossRef]
60. Smith, T.J.; Gowd, A.K.; Kunkel, J.; Kaplin, L.; Waterman, B.R. Superior Capsular Reconstruction Provides Sufficient Biomechanical Outcomes for Massive, Irreparable Rotator Cuff Tears: A Systematic Review. *Arthrosc. J. Arthrosc. Relat. Surg.* **2021**, *37*, 402–410. [CrossRef]
61. Kane, G.E.; LeVasseur, C.M.; Hughes, J.D.; Gabrielli, A.S.; Popchak, A.; Anderst, W.; Lin, A. Improved Outcomes Following Arthroscopic Superior Capsular Reconstruction May Not Be Associated with Changes in Shoulder Kinematics: An In Vivo Study. *Arthrosc. J. Arthrosc. Relat. Surg.* **2022**, *38*, 267–275. [CrossRef] [PubMed]
62. Greiner, S.; Kaeaeb, M.; Voss, A.; Lawton, R.; Bhide, P.; Achenbach, L. Comparison of Superior Capsular Reconstruction and Partial Infraspinatus Repair: A Matched-Pair Analysis of Irreparable Rotator Cuff Tears. *Orthop. J. Sports Med.* **2021**, *9*, 2325967120984264. [CrossRef] [PubMed]
63. Savarese, E.; Romeo, R. New solution for massive, irreparable rotator cuff tears: The subacromial "biodegradable spacer". *Arthrosc. Tech.* **2012**, *1*, e69–e74. [CrossRef] [PubMed]
64. Liu, F.; Dong, J.; Kang, Q.; Zhou, D.; Xiong, F. Subacromial balloon spacer implantation for patients with massive irreparable rotator cuff tears achieves satisfactory clinical outcomes in the short and middle of follow-up period: A meta-analysis. *Knee Surg. Sports Traumatol. Arthrosc.* **2021**, *29*, 143–153. [CrossRef]
65. Moon, A.S.; Patel, H.A.; Ithurburn, M.P.; Brabston, E.W.; Ponce, B.A.; Momaya, A.M. Subacromial Spacer Implantation for the Treatment of Massive Irreparable Rotator Cuff Tears: A Systematic Review. *Arthrosc. J. Arthrosc. Relat. Surg.* **2019**, *35*, 607–614. [CrossRef]
66. Deranlot, J.; Herisson, O.; Nourissat, G.; Zbili, D.; Werthel, J.D.; Vigan, M.; Bruchou, F. Arthroscopic Subacromial Spacer Implantation in Patients with Massive Irreparable Rotator Cuff Tears: Clinical and Radiographic Results of 39 Retrospectives Cases. *Arthrosc. J. Arthrosc. Relat. Surg.* **2017**, *33*, 1639–1644. [CrossRef]
67. Holschen, M.; Brand, F.; Agneskirchner, J.D. Subacromial spacer implantation for massive rotator cuff tears: Clinical outcome of arthroscopically treated patients. *Obere Extrem.* **2017**, *12*, 38–45. [CrossRef]
68. Piekaar, R.S.M.; Bouman, I.C.E.; van Kampen, P.M.; van Eijk, F.; Huijsmans, P.E. Early promising outcome following arthroscopic implantation of the subacromial balloon spacer for treating massive rotator cuff tear. *Musculoskelet. Surg.* **2018**, *102*, 247–255. [CrossRef]
69. Ernstbrunner, L.; Suter, A.; Catanzaro, S.; Rahm, S.; Gerber, C. Reverse Total Shoulder Arthroplasty for Massive, Irreparable Rotator Cuff Tears Before the Age of 60 Years: Long-Term Results. *J. Bone Joint Surg. Am.* **2017**, *99*, 1721–1729. [CrossRef]
70. Senekovic, V.; Poberaj, B.; Kovacic, L.; Mikek, M.; Adar, E.; Markovitz, E.; Maman, E.; Dekel, A. The biodegradable spacer as a novel treatment modality for massive rotator cuff tears: A prospective study with 5-year follow-up. *Arch. Orthop. Trauma. Surg.* **2017**, *137*, 95–103. [CrossRef]
71. Calvo, E.; Valencia, M.; Merino-Garcia, J.A.; Luengo-Alonso, G. Symptomatic foreign body reaction secondary to subacromial balloon spacer placement: A case report. *J. Shoulder Elb. Surg.* **2020**, *29*, e313–e316. [CrossRef] [PubMed]
72. Ruiz Ibán, M.A.; Lorente Moreno, R.; Ruiz Díaz, R.; Álvarez Sciamanna, R.; Paniagua Gonzalez, A.; Lorente Gómez, A.; Díaz Heredia, J. The absorbable subacromial spacer for irreparable posterosuperior cuff tears has inconsistent results. *Knee Surg. Sports Traumatol. Arthrosc.* **2018**, *26*, 3848–3854. [CrossRef] [PubMed]
73. Yamak, K.; Karahan, H.G.; Altay, T.; Kayalı, C.; Ozan, F. Is Subacromial Balloon Spacer Appropriate for Treatment of Irreparable Rotator Cuff Tears in Elderly Patients? *Ortop. Traumatol. Rehabil.* **2019**, *21*, 417–426. [CrossRef]
74. Muh, S.J.; Streit, J.J.; Wanner, J.P.; Lenarz, C.J.; Shishani, Y.; Rowland, D.Y.; Riley, C.; Nowinski, R.J.; Edwards, T.B.; Gobezie, R. Early follow-up of reverse total shoulder arthroplasty in patients sixty years of age or younger. *J. Bone Joint Surg. Am.* **2013**, *95*, 1877–1883. [CrossRef] [PubMed]
75. Mulieri, P.; Dunning, P.; Klein, S.; Pupello, D.; Frankle, M. Reverse shoulder arthroplasty for the treatment of irreparable rotator cuff tear without glenohumeral arthritis. *J. Bone Joint Surg. Am.* **2010**, *92*, 2544–2556. [CrossRef] [PubMed]
76. Petrillo, S.; Longo, U.G.; Papalia, R.; Denaro, V. Reverse shoulder arthroplasty for massive irreparable rotator cuff tears and cuff tear arthropathy: A systematic review. *Musculoskelet. Surg.* **2017**, *101*, 105–112. [CrossRef]

77. Simovitch, R.W.; Friedman, R.J.; Cheung, E.V.; Flurin, P.-H.; Wright, T.; Zuckerman, J.D.; Roche, C. Rate of Improvement in Clinical Outcomes with Anatomic and Reverse Total Shoulder Arthroplasty. *J. Bone Joint Surg. Am.* **2017**, *99*, 1801–1811. [CrossRef] [PubMed]
78. Sevivas, N.; Ferreira, N.; Andrade, R.; Moreira, P.; Portugal, R.; Alves, D.; da Silva, M.V.; Sousa, N.; Salgado, A.J.; Espregueira-Mendes, J. Reverse shoulder arthroplasty for irreparable massive rotator cuff tears A systematic review with meta-analysis and meta-regression. *J. Shoulder Elb. Surg.* **2017**, *26*, e265–e277. [CrossRef] [PubMed]
79. Boileau, P.; Gonzalez, J.F.; Chuinard, C.; Bicknell, R.; Walch, G. Reverse total shoulder arthroplasty after failed rotator cuff surgery. *J. Shoulder Elb. Surg.* **2009**, *18*, 600–606. [CrossRef] [PubMed]
80. Virk, M.S.; Nicholson, G.P.; Romeo, A.A. Irreparable Rotator Cuff Tears Without Arthritis Treated with Reverse Total Shoulder Arthroplasty. *Open Orthop. J.* **2016**, *10*, 296–308. [CrossRef]
81. Codding, J.L.; Keener, J.D. Natural History of Degenerative Rotator Cuff Tears. *Curr. Rev. Musculoskelet. Med.* **2018**, *11*, 77–85. [CrossRef] [PubMed]
82. Hamada, K.; Fukuda, H.; Mikasa, M.; Kobayashi, Y. Roentgenographic findings in massive rotator cuff tears. A long-term observation. *Clin. Orthop. Relat. Res.* **1990**, *254*, 92–96. [CrossRef]

Journal of *Personalized Medicine*

MDPI

Article

Bari Shoulder Telemedicine Examination Protocol (B-STEP): A Standard Protocol for Personalized Remote Shoulder Examination

Lorenzo Moretti [1,†], Davide Bizzoca [1,*,†], Giacomo Farì [2], Alessandro Caricato [1], Francesco Angiulli [1], Giuseppe Danilo Cassano [1], Giuseppe Solarino [1] and Biagio Moretti [1]

1 Orthopaedic and Trauma Unit, AOUC Policlinico di Bari, University of Bari "Aldo Moro", Piazza Giulio Cesare 11, 70124 Bari, Italy
2 Physical Medicine and Rehabilitation Unit, University of Bari "Aldo Moro", 70124 Bari, Italy
* Correspondence: davide.bizzoca@uniba.it
† These authors contributed equally to this work.

Abstract: The COVID-19 pandemic drastically changed many aspects of the traditional functioning of health systems all around the world. In Italy, as reported by the CIO, compared to the previous year, there was a significant reduction in 2020 in overall outpatient activities by up to 75%. These data support the need for telemedicine, which represents a current challenge and can no longer be postponed in the future. This study aims to elaborate on a possible model for remote shoulder examination based on traditional tests to improve the quality of telemedicine in orthopedic and rehabilitation. Between May 2020 and November 2020, ten orthopedic surgeons individually examined six patients with a known shoulder disorder, both in hospital and via webcam according to the previously shared protocol (B-STEP). According to the 10 observers, completing 100% of the ASES score and at least 87.5% of the Constant score is possible. Shoulder ROM and many specific tests are also reproducible via webcam, but with less sensitivity, according to the subjective opinion of observers. The B-STEP is a useful protocol for the standardization of the objective examination of the shoulder via webcam. Further studies are necessary to determine if the B-STEP protocol is useful for diagnosing pathology in unknown patients and evaluating its sensitivity and specificity for each pathology.

Keywords: telemedicine; shoulder; examination protocol

check for updates

Citation: Moretti, L.; Bizzoca, D.; Farì, G.; Caricato, A.; Angiulli, F.; Cassano, G.D.; Solarino, G.; Moretti, B. Bari Shoulder Telemedicine Examination Protocol (B-STEP): A Standard Protocol for Personalized Remote Shoulder Examination. *J. Pers. Med.* **2023**, *13*, 1159. https://doi.org/10.3390/jpm13071159

Academic Editor: Marijn Speeckaert

Received: 17 April 2023
Revised: 25 June 2023
Accepted: 18 July 2023
Published: 20 July 2023

1. Introduction

The COVID-19 pandemic drastically changed many aspects of the traditional functioning of health systems all around the world [1]. Inevitably, the closure and reconversion of entire hospitals and outpatient services led to a reduction in the assistance guaranteed in managing chronic pathologies and in the management of usual emergencies. This applied to all sectors of medicine, including those dedicated to treating musculoskeletal pathologies [2].

As reported by the CIO (Osteosynthesis Italian Club), compared to the previous year, in Italy in 2020, there was a significant reduction in overall outpatient activities by up to 75%; it registered a reduction of up to 71% in terms of trauma that required consultation, and up to 50% regarding trauma surgery, excluding fractures of the femoral neck [3].

These data support the need for telemedicine, which represents a current challenge and can no longer be postponed in the future. Great milestones in terms of time saving, easy accessibility, cost effectiveness, hospital overcrowding reduction, and health education have been achieved in the last years in remote patient care, bridging the gap between physicians and patients that the COVID-19 pandemic caused [4]. In addition, in the orthopedic and rehabilitation field, the need to exploit the new technological means the availability to prepare remote evaluations of patients is growing [5], and it requires some methods which can be standardized and reliable in objectifying data as much as possible.

The shoulder is an anatomical area that is often affected by common musculoskeletal pathologies [6], and which, moreover, lends itself well to being remotely evaluated, albeit with a series of obvious difficulties [7].

This study aims to elaborate on a possible model for remote shoulder examination based on traditional tests and new possibilities deriving from new electronic tools to improve the quality of telemedicine in orthopedics and rehabilitation.

Manuscript Organization

The following paragraphs show the steps through which we devised and developed our B-STEP protocol. The materials and methods section will explain the recruitment criteria for the patients on whom our protocol was tested. Next, the common in-patient objective examination and the various steps through which the telemedicine objective examination is conducted, describing each specific test in detail. In the Results section, we report the percentage completion data of the items performed based on the ASES score and the Constant score and the results of the reproducibility assessment of specific tests in telemedicine according to the observers.

2. Materials and Methods

We conducted an observational study between May 2020 and November 2020, at the Orthopedic and Traumatology Clinic of the Azienda Ospedaliera Universitaria Consorziale Policlinico of Bari.

We involved 10 orthopedic surgeons who are experts in the treatment of shoulder disorders; then, we asked them to elaborate on a common evaluation protocol for patients with common shoulder disease.

To study the telemedicine reproducibility of each specific test performed in a common outpatient clinical examination, we used this protocol in the follow-up of well-known patients. We recruited one patient for each of the following shoulder disorders:

1. Glenoid, humeral, and scapular fractures: a patient with proximal humerus fracture treated with plate and screws
2. Tendinitis and long head of the biceps tendon (LHB) injuries: a patient with LHB tendonitis was treated with mesotherapy
3. Rotator cuff injuries: a patient with supraspinatus (SSP) tendon lesion treated arthroscopically.
4. Acromioclavicular locations: an acromioclavicular dislocation surgically treated with a tight-rope technique
5. Glenohumeral osteoarthritis (GHO): a patient with glenohumeral osteoarthritis, not eligible for surgery, in conservative treatment
6. Shoulder instability and/or superior labrum anterior-posterior (SLAP) lesion: a patient with recurrent shoulder dislocation treated with remplissage

The recruitment criteria are shown in Table 1.

Table 1. Inclusion and exclusion criteria.

Inclusion Criteria
Over 18 years of age
Capable of expressing informed agreement
A known diagnosis
At least 6 months follow up
Availability of electronic devices with webcam (cell phones, tablets, pc)
Exclusion Criteria
The concomitance of more than shoulder pathology
Presence of other rheumatological, orthopedic, or neurological ipsilateral upper limb pathologies
Inability/unwillingness to be remotely monitored.

All patients signed informed consent at the time of recruitment. All procedures were performed by the principles of Helsinki.

Each of the orthopedists involved individually examined each of these 6 patients, both in person and via webcam according to the protocol previously shared.

At the end of the visits, we collected the impressions of the ten physicians separately about the possibilities and difficulties they encountered with the objective examination of the shoulder via webcam, compared with the traditional exam.

2.1. Outpatient Clinical Examination

First, each recruited patient was examined by each physician in a live outpatient visit. We asked the observers to use the same tests present in the B-STEP in a standardized way. (See Scheme 1).

Preliminary Phase	• Explaining the B-STEP performing • Collecting consensus agreement
Phase I: Anamnesis	• Asking for present symptoms and the next medical history • Completing ASES score
Phase II: Visual examination	• Visual examination of the affected shoulder
Phase III: Palpatory examination	• Self-shoulder palpation, guided by the Clinician
Phase IV: ROM (Constant Score application)	• Evaluation of complete ROM of the affected shoulder compared with the non-affected shoulder. • Completing Constant score
Phase V: Specific tests	• Evaluation of specific test

Scheme 1. The Bari Shoulder Telemedicine Exam Protocol (B-STEP).

After informed consensus was collected, a comprehensive examination was performed for each patient. Medical history was noted and documentation of any surgery performed was collected. The ASES-score evaluation questionnaire was submitted to each patient.

Subsequently, an objective examination was performed, visually assessing the presence of any skin alterations or anatomical profile of the pathological shoulder, compared with the non-affected shoulder.

Next, palpation of the shoulder was performed, looking for painful areas, or alterations in subcutaneous bone profiles.

In the next phase, the entire range of motion (ROM) of both shoulders was examined, evaluating any deficit in the affected shoulder. The ROM evaluation allows for completing the items of the Constant score.

In Table 2, the description, sensitivity, and specificity, as described in the literature, of each test conducted on an outpatient basis are reported.

Table 2. Sensitivity and specificity of commonly used specific tests.

Specific Test	Tested Pathology	Test Description	Criteria for Positivity	Sens/Spec in Literature
	Rotator cuff integrity			
Intra-rotation against resistance [8]	Subscapularis	Upper limb along the trunk, forearm flexed 90° in a neutral position. Intrarotation is forced against a door jamb	Test positive if evoking pain or strength deficit relative to contralateral is reported	Sens: 0.56 Spec: 0.87

Table 2. *Cont.*

Specific Test	Tested Pathology	Test Description	Criteria for Positivity	Sens/Spec in Literature
Hand behind the back [9]	Subscapularis	Shoulder intrarotation bringing the hand behind the back ("bra lacing movement")	Range of motion assessment (hand at lumbar level, hand at thoracolumbar level, . . .)	Reliability: Weighted k 0.73
Belly-press [10]	Subscapularis	The upper limb was slightly abducted and intrarotated by bringing the palm up to the navel level. The patient is asked to intrarotate the limb by pressing the hand against the abdomen	Positive test if the patient fails to keep the elbow in the frontal plane during the intrarotation movement	Sens: 0.40 Spec: 0.98
Extra-rotation against resistance [11]	Infraspinatus/teres minor	Upper limb along the trunk, forearm flexed 90° in a neutral position. The patient is asked to force extrotation against a door jamb	Test positive if evoking pain or strength deficit relative to contralateral is reported	Sens: 0.46 Spec: 0.94
Extrotation with abducted arm [11]	Infraspinatus/teres minor	Abduction of the shoulder to 90° and simultaneous external rotation.	The feeling of apprehension or pain at maximum degrees of extrotation	Sens: 0.40–0.62 Spec: 0.42–0.87
Hand behind neck [9,12]	Infraspinatus/teres minor	The patient brings the hand behind the neck	ROM amplitude is assessed: hand on the neck, hand on the ipsilateral scapula, and hand on the opposite scapula. The test is positive if it evokes pain	Reliability: Weighted k Intra-examinator: 0.80 Inter-examinator: 0.90
Jobe test [13]	SSP	Shoulder abducted to 90°, anterior flexion to approximately 45°; a supra-rotated limb. The patient is asked to further elevate the limb against known resistance	Test positive if evoking pain or strength deficit relative to contralateral is reported	Sens: 0.50–0.69 Spec: 0.62–0.87
	SLAP lesions/LHB			
Speed test [14]	SLAP lesion/long head bicep integrity	Anterior humerus flexion at 90°, upper extremity fully extended and supinated. The patient is asked to flex the elbow against resistance using a known weight	Test considered positive if it evokes pain at the level of the groove of the long head of the biceps	Sens: 0.20–0.60 Spec: 0.38–0.78
	Impingement Syndrome			
Painful ROM arch in abduction [14]	Impingement syndrome/subacromial bursitis	Abduction of the upper limb up to about 90°; possible extrotation of the limb to complete the ROM	Positive test if evokes pain	Sens: 0.56 Spec: 0.76

Table 2. *Cont.*

Specific Test	Tested Pathology	Test Description	Criteria for Positivity	Sens/Spec in Literature
	Shoulder instability			
Anterior apprehension test [15,16]		Abduction of the shoulder to 90° and simultaneous external rotation.	The feeling of apprehension or pain at maximum degrees of extrotation	Sens: 0.40–0.62 Spec: 0.42–0.87
Relocation test [17]		Supine patient; glenohumeral joint at the edge of the table. Arm in 90° abduction and fully extrotated, with the elbow flexed 90°. A direct posterior force is applied to the humeral head	The test is positive if the force application reduces pain or apprehension	Sens: 0.81 Spec: 0.92
Drawer test [17]		Supine patient; glenohumeral joint at the edge of the table. Arm at 60–80° of abduction and in neutral rotation. A force is applied to translate the humeral head anteriorly	Test positive if eliciting pain	Sens: 0.28–0.53 Spec: 0.71–0.85
Crank test [14]		Patient supine; arm elevated to 160°. An axial force is applied to the humerus while the shoulder is intra- and extra-rotated.	Test positive if eliciting pain	Sens: 0.34 Spec: 0.75
Fulcrum test [15]		Patient supine; arm abducted to 90° and elbow flexed to 90°; an axial force in compression is applied to the humerus and the shoulder is rotated.	Test positive if eliciting pain	Sens: 0.61 Spec: 0.54
Jerk test [18]		Patient seated. The arm intrarotated and elevated to 90° of abduction. Then, an abduction movement is performed on the horizontal plane and an axial force is applied to the humerus.	Test positive if eliciting pain	Sens: 0.73 Spec: 0.98

2.2. *The Bari Shoulder Telemedicine Exam Protocol (B-STEP)*

For the telemedicine examination (Scheme 1), we chose the smartphone app WhatsApp, as it is a widely used app within the population and easy to use, available for both smartphones and computers. These characteristics allowed us to avoid the use of additional apps or programs to be created "ad hoc" and the greater usability of the examination in telemedicine.

Selecting a commonly used clinical test, the Constant score, and a test for assessing the patient's quality of life for shoulder disorders, such as the ASES Score, a valid protocol was developed for the most common shoulder pathologies [19,20]. The Constant score assesses subjective factors, such as the influence of pain in daily activities and quality of life, as well as objective factors such as abduction force or shoulder range of motion in intra- and extra-rotation. The ASES score evaluates the influence of pain in shoulder disorders on quality of life and daily activities from a more subjective point of view.

In addition, we selected certain specific tests for the objective examination of the shoulder that we considered highly reproducible via webcam.

This protocol was approved by each of the ten orthopedists involved in the study.

These tests were given to the patient as an interview and as reproducible evaluation maneuvers during the objective examination.

We previously collected an informed consensus from each patient.

2.2.1. Preliminary Phase:

The patient is initially alerted to the possibility that the virtual visit may be recorded, ensuring the appropriate obscuration of the face and any sensitive reference that violates the privacy of the patient.

The sequence of movements and tests should be performed in front of the camera, about 1.5 m away; behind the patient, a bare wall with no reference to personal belongings is recommended.

Adequate space is needed for the patient to complete the full ROM of the shoulder. He/she should be free to move from clinostasis to orthostasis.

The patient is advised to blackout the camera if he/she must move within his/her home to respect his/her privacy.

The patient should wear a camisole that allows a complete visual examination of the affected shoulder.

The patient must be properly instructed in the performance of specific movements. Therefore, it may be helpful for the examiner himself to show these movements through the camera [21].

2.2.2. Phase I: Anamnesis

Ask the patient which diseases he is suffering from. If necessary, the patient is asked to indicate with one finger the site of maximal pain. In this phase, it is possible to complete the ASES score questions.

Next, medical history is collected.

2.2.3. Phase II: Visual Examination

The patient is asked to show a close-up of both shoulders to the camera to perform a visual examination as completely and accurately as possible. It is requested to show both frontally and posteriorly.

Muscle tropism is examined in comparison to the contralateral limb.

2.2.4. Phase III: Palpatory Examination

The patient should perform a self-palpation of the shoulder, and asked if they experience any swelling, muscle contractures, alterations in normal anatomy, or sites of elective pain, including the LHB tendon and the acromioclavicular joint.

2.2.5. Phase IV: ROM (Constant Score Application) [19]

In the Constant score, the first two items score the subjective assessment of pain and functional limitation in daily activities.

The remaining items score according to the degree of mobility of the shoulder, and the range of the respective ROM. Particularly, items 3 and 4 offer a qualitative and quantitative estimation of abduction.

The patient performs anterior flexion, abduction, and extension movements, either in frontal projection or with his back to the camera, comparing the ROM with the presumably healthy contralateral limb.

Each range of motion can be quantified on screen, to assign a specific Constant score (items 3, 5, 6).

- 0 pounds (0 kg)
- 1–3 pounds (0.45–1.3 kg); example: 1 to 3 packages of pasta in a bag
- 4–6 pounds (1.8–2.7 kg); example: a 2 L bottle up to 2 1.5 L bottles
- [. . .]
- >24 pounds (>10 kg) example: a bale of water with 2 L bottles

Intrarotation (item 8) can be assessed with the "hand behind the back" test; for example, women can be asked to simulate the movement to "fasten their bra", or to bring the hand behind the neck as in the act of combing their hair [22].

Extrotation (item 7) can be assessed by performing the "hand behind the neck" test with the patient placed in lateral vision, limb abducted to 90° and elbow flexed to 90°, with forearm and hand in a neutral position [23].

The patient may be asked to hold the maximum range for a few moments to capture the appropriate frame for precise range measurements using virtual goniometers.

2.2.6. Phase V: Specific Tests

The first phase of ROM assessment through the CONSTANT score, as well as the patient's compilation of the ASES score, can guide the clinician toward a diagnostic suspect. Therefore, in a normal outpatient visit, pathology-specific tests are performed. We have therefore selected some of these tests.

In the absence of direct patient contact, intra- and extra-rotation can be performed against the resistance offered by a person accompanying the patient, if present. Alternatively, the patient can perform the movements against door jambs or the wall. The test may be defined as positive if the patient reports that the movement was painful or if they feel more "weakness" than usual [23].

Subscapularis strength can be assessed with the belly press test (elbows should not move beyond the body plane). The test can be considered positive if the movement generates pain or if it reports a reduction in strength relative to the contralateral limb [24].

Empty can test (Jobe test) (SSP evaluation): shoulder abducted to 90° and in anterior flexion to approximately 45°; the upper limb is intrarotated. Normally, the patient is asked to elevate the limb further against resistance. A weight known as a standard resistance can be used [25,26].

Speed test (anterior and posterior superior cercine lesions—SLAP lesion/LHB lesions): shoulder in 90° anterior flexion. Upper limb fully extended and supinated. The patient is asked to flex the elbow against resistance, using a known weight [14,27].

To remotely evaluate shoulder instability, our observers tried to replicate the following specific tests:

Relocation test: the relocation test is a natural progression of the apprehension test and assesses relief of apprehension after manual stabilization of the shoulder. After eliciting a positive apprehension test, the examiner maintains the patient in their current position and applies a posteriorly directed force on the humeral head in an attempt to stabilize the shoulder and correct the symptoms. In a patient with anterior shoulder instability, this maneuver should bring a sub-luxed humeral head back into the correct position relative to the glenoid fossa. Resolution of guarding and apprehension suggests anterior instability and is considered a positive relocation test [19].

Apprehension test: anterior apprehension may be elicited by bringing the patient's shoulder into a position of 90° of abduction and 90° of external rotation in either the supine or upright position. A positive exam finding is the subjective feeling of impending subluxation or dislocation when in this provocative position. It is important to note that although these symptoms may be accompanied by pain, pain itself does not produce a positive test [19].

Anterior and posterior drawer test: tests are performed with the subject lying relaxed in the supine position with the shoulder slightly flexed, neutral to 30° external rotation, and

80–120° abduction. The examiner stays to the side facing the patient, stabilizing the scapula with one hand and holding the proximal humerus with the other so that the shoulder can be kept at the intended position and rotation. Then, the proximal humerus is shifted anteriorly or posteriorly whilst applying axial force. The subtle difference between the posterior to anterior laxity examination is that the patient's arm is flexed at 60–80°, slightly internally rotated, elbow flexed at 90°, and the humeral head shifts posteriorly during the maneuver [15].

Crank test: the crank test is performed with the patient in the upright or supine position. The shoulder is elevated 160° in the scapular plane, an axial load is applied by the examiner, and the humerus is internally and externally rotated. Pain elicited during this test, typically with external rotation, is a positive indication of a pathologic condition of the labrum. In addition, a click may or may not be felt that reproduces the patient's symptoms of pain or catching [16,28–31].

Fulcrum test: the fulcrum test is performed with the patient supine with one shoulder on the edge of a couch. Patients felt anxiety about dislocation when the involved shoulder is externally rotated with the shoulder at 90° of abduction and the elbow at 90° of flexion in the supine position [32].

Jerk test: while stabilizing the patient's scapula with one hand and holding the affected arm at 90° abduction and internal rotation, the examiner grasps the elbow and axially loads the humerus in a proximal direction. The arm is moved horizontally across the body. A positive result is indicated by a sudden clunk as the humeral head slides off the back of the glenoid [33].

3. Results

The reproducibility of the scores examined is shown in Table 3.

Table 3. Reproducibility of Constant score and ASES score, according to the observation of Orthopedic Surgeons, during the examination in TELEMEDICINE.

	Proximal Humeral Fracture	**LHB Tendinitis**	**SSP Lesion, Surgically Treated**	**Acromion-Clavear Luxation**	**Hemarthrosis**	**Shoulder Instability**
Orthopedic physician 1	Constant: 87.5% ASES: 100%	Constant: 100% ASES: 100%	Constant: 87.5% ASES: 100%	Constant: 100% ASES 100%	Constant: 87.5% ASES: 100%	Constant: 75% ASES: 100%
Orthopedic physician 2	Constant: 100% ASES: 100%	Constant: 100% ASES: 100%	Constant: 100% ASES: 100%	Constant: 100% ASES 100%	Constant: 100% ASES: 100%	Constant: 100% ASES: 100%
Orthopedic physician 3	Constant: 100% ASES: 100%	Constant: 87.5% ASES: 100%	Constant: 87.5% ASES: 100%	Constant: 87.5% ASES: 100%	Constant: 87.5% ASES: 100%	Constant: 87.5% ASES: 100%
Orthopedic physician 4	Constant: 87.5% ASES: 100%	Constant: 87.5% ASES: 100%	Constant: 75% ASES: 100%	Constant: 87.5% ASES: 100%	Constant: 87.5% ASES: 100%	Constant: 87.5% ASES: 100%
Orthopedic physician 5	Constant: 100% ASES: 100%	Constant: 87.5% ASES: 100%	Constant: 87.5% ASES: 100%	Constant: 100% ASES: 100%	Constant: 87.5% ASES: 100%	Constant: 87.5% ASES: 100%:
Orthopedic physician 6	Constant: 100% ASES: 100%:	Constant: 75% ASES: 100%:	Constant: 87.5% ASES: 100%:	Constant: 87.5% ASES: 100%:	Constant: 100% ASES: 100%:	Constant: 87.5% ASES: 100%:
Orthopedic physician 7	Constant: 87.5% ASES: 100%:	Constant: 87.5% ASES: 100%:	Constant: 87.5% ASES: 100%:	Constant: 87.5% ASES: 100%:	Constant: 87.5% ASES: 100%	Constant: 87.5% ASES: 100%
Orthopedic physician 8	Constant: 100% ASES: 100%:	Constant: 87.5% ASES: 100%	Constant: 100% ASES: 100%	Constant: 75% ASES: 100%	Constant: 87.5% ASES: 100%	Constant: 87.5% ASES: 100%
Orthopedic physician 9	Constant: 87.5% ASES: 100%	Constant: 87.5% ASES: 100%	Constant: 87.5% ASES: 100%	Constant: 87.5% ASES: 100%	Constant: 100% ASES: 100%	Constant: 100% ASES: 100%
Orthopedic physician 10	Constant: 87.5% ASES: 100%	Constant: 87.5% ASES: 100%	Constant: 100% ASES: 100%	Constant: 87.5% ASES: 100%	Constant: 100% ASES: 100%	Constant: 87.5% ASES: 100%

All the observers completed the ASES score questions via webcam (100% of the Ases score's item).

In 37/60 cases, the observer had not completed item 4 of the Constant score, evaluating the strength of the abduction (7/8 items completed = 87.5% of Constant score).

In 4/60 cases, the observer reported an unreliable response to item 4 and item 7 (6/8 items completed = 75% of Constant score).

They reported that in performing both items, the main problem was poor patient compliance.

Telemedicine Reproducibility of Specific Tests in the Objective Examination of the Shoulder

As shown in Table 4, there are differing opinions regarding the reproducibility of the specific tests performed in telemedicine. All orthopedic surgeons (10/10) consider the assessment of ROM reproducible, although with some limitations in the assessment of intrarotation, as evaluated above.

Table 4. Reproducibility assessment of specific tests in telemedicine according to the 10 observers.

Tab 3	ROM	Bellypress	Empty-Can Test	Speed Test	Yocum	Neer Sign	Relocation Test	Apprehension Test	Drawer Test	Crank Test	Fulcrum Test	Jerk Test
Orthopedic physician 1	Yes	Yes	Yes	Yes	No	Yes	No	Yes	No	No	No	No
Orthopedic physician 2	Yes	Yes	Yes	Yes	No	No	No	Yes	No	No	No	No
Orthopedic physician 3	Yes	No	Yes	No	Yes	No	No	Yes	No	No	No	No
Orthopedic physician 4	Yes	No	Yes	Yes	No	Yes	No	No	No	No	No	No
Orthopedic physician 5	Yes	Yes	No	Yes	No	Yes	No	Yes	No	No	No	No
Orthopedic physician 6	Yes	Yes	No	Yes	No	No	No	Yes	No	No	No	No
Orthopedic physician 7	Yes	Yes	Yes	Yes	Yes	No	No	Yes	No	No	No	No
Orthopedic physician 8	Yes	No	Yes	Yes	No	No	No	No	No	No	No	No
Orthopedic physician 9	Yes	Yes	Yes	Yes	No	No	No	Yes	No	No	No	No
Orthopedic physician 10	Yes	No	Yes	No	No	No	No	No	No	No	No	No
Reprod.:	10/10	6/10	8/10	8/10	3/10	3/10	0/10	7/10	0/10	0/10	0/10	0/10

Regarding the evaluation of the muscular-tendinous components of the rotator cuff, 8/10 observers considered the empty can test and the speed test repeatable with good reliability. A total of 6 out of 10 observers rated the belly-press test as repeatable, whereas only 3 out of 10 observers rated the Yocum test and the Neer sign conducted via webcam as reliable.

Several unfavorable opinions were obtained regarding the evaluation of tests for instability disorders. Among the various tests examined, only the apprehension test was found to be repeatable according to 7 out of 10 observers in telemedicine.

4. Discussion

The COVID-19 pandemic and the need for social distancing have created challenges in healthcare access. Telemedicine has played an important role in the delivery of medical services and is likely to be of continued importance and use even after the current pandemic.

The present study aims to depict a standardized protocol to perform a reliable remote shoulder examination. Despite a small sample size of patients evaluated, with a known diagnosis, we focused on the reproducibility of specific tests commonly used in the clinical practice of an in-person objective examination.

In line with this goal, we started with shoulder evaluation scores commonly in use in orthopedic practice, such as the Constant score and ASES score [19,20].

The ASES score, a subjective test based on the evaluation of the individual's normal daily activities and the perception of pain [34], was reproducible in all the patients examined. The Constant score was also shown to be largely reproducible during the webcam examination, and we could have a quantitative evaluation of some specific tests using everyday objects with a known weight. However, several surgeons had difficulty reproducing item 4, where the strength of abduction is quantified, and, in a few cases, there was difficulty in examining the extra rotation test.

Authors agree that most of the tests used in shoulder instability are difficult to reproduce via webcam, as these maneuvers require the in-person presence of the examiner. Such tests were excluded from our final protocol.

After performing the tests and collecting the opinions of orthopedic surgeons, we concluded that certain pathologies are more difficult to telematically assess, with likelier instability. On the other hand, many of the scores used in orthopedy can be completed with our B-STEP (e.g., Constant score, UCLA shoulder rating scale, Oxford shoulder score, ASES, DASH, and Quick DASH), allowing us to perform clinical follow-up of patients with known pathology.

The only test that could be carried out on the patient by telemedicine is the anterior apprehension test, which consists of abduction of the shoulder to 90° and simultaneous external rotation to 90°. This test has the aim of reproducing the patient's fear or feeling of a possible anterior dislocation of the shoulder, in which case the test is considered positive [15,16,19].

Given the association between joint laxity and AMBRI-type shoulder instability (atraumatic, multidirectional, bilateral), it is useful to assess the patient's ligamentous laxity according to the Beighton scale [35–38].

The most-used scale in the clinical workup of joint laxity is the Beighton nine-point scale. The scale includes only common symptoms: (1) extension of the MCF in the 5th finger beyond 90 degrees, (2) abduction of the thumb on the forearm, (3) hyperextension of the elbow by more than 10 degrees, (4) hyperextension of the knee by more than 10 degrees, and (5) contact of the palm of the hands on the floor with the lower limbs extended. One point is given for each finding. A score of $\geq$4 points is used to diagnose joint laxity [36].

The evaluation of the Beighton scale combined with medical history and signs of shoulder instability may indicate the need for further investigation, which cannot be assessed by a telemedicine visit.

Other recent studies have shown that in-person clinical examination is necessary to confirm suspected instability of the shoulder [39–44].

Our protocol appears to be in line with other methodologies used in other studies [40]. However, the use of everyday objects is promoted in our study to be able to quantify functional shoulder deficits and to standardize patient assessment and follow-up as much as possible.

5. Conclusions

The B-STEP is a useful protocol for the standardization of the objective examination of the shoulder via webcam. The purpose of this protocol is to perform the follow-up on post-treatment recovery, monitor the evolution of the pathology in patients conservatively treated, and reduce waiting lists for outpatient visits.

Further studies—blinded, or with a larger cohort of patients—are necessary to determine if the B-STEP protocol can diagnose pathology in unknown patients, evaluating its sensitivity and specificity for each pathology.

The telematics approach can also be useful in the online assessment of patient follow-up imaging, complementing the clinical assessment of the B-STEP.

Author Contributions: Conceptualization, L.M., B.M. and D.B.; methodology, L.M., G.F. and D.B.; writing—original draft preparation, F.A., A.C., G.D.C. and G.S.; writing—review and editing, L.M., G.F. and D.B.; supervision, B.M. All authors have read and agreed to the published version of the manuscript.

Funding: This research received no external funding.

Institutional Review Board Statement: Not applicable.

Informed Consent Statement: Not applicable.

Data Availability Statement: The data that support the findings of this study are available from the corresponding author upon reasonable request.

Conflicts of Interest: The authors declare no conflict of interest.

References

1. Di Renzo, L.; Gualtieri, P.; Pivari, F.; Soldati, L.; Attinà, A.; Cinelli, G.; Leggeri, C.; Caparello, G.; Barrea, L.; Scerbo, F.; et al. Eating habits and lifestyle changes during COVID-19 lockdown: An Italian survey. *J. Transl. Med.* **2020**, *18*, 229. [CrossRef] [PubMed]
2. Schwarzkopf, R.; Maher, N.A.; Slover, J.; Strauss, E.; Bosco, J.; Zuckerman, J. The Response of an Orthopedic Department and Specialty Hospital at the Epicenter of a Pandemic: The NYU Langone Health Experience. *J. Arthroplast.* **2020**, *35*, S3–S5. [CrossRef]
3. Benazzo, F.; Rossi, S.M.P.; Maniscalco, P.; Moretti, B.; Vaienti, E.; Ruggieri, P.; Massè, A.; Medici, A.; Formica, A.; Di Maggio, B.; et al. The orthopaedic and traumatology scenario during COVID-19 outbreak in Italy: Chronicles of a silent war. *Int. Orthop.* **2020**, *44*, 1453–1459. [CrossRef] [PubMed]
4. Mahajan, V.; Singh, T.; Azad, C. Using Telemedicine During the COVID-19 Pandemic. *Indian Pediatr.* **2020**, *57*, 652–657. [CrossRef]
5. Hincapié, M.A.; Gallego, J.C.; Gempeler, A.; Piñeros, J.A.; Nasner, D.; Escobar, M.F. Implementation and Usefulness of Telemedicine During the COVID-19 Pandemic: A Scoping Review. *J. Prim. Care Community Health* **2020**, *11*, 2150132720980612. [CrossRef] [PubMed]
6. Linaker, C.H.; Walker-Bone, K. Shoulder disorders and occupation. *Best Pract. Res. Clin. Rheumatol.* **2015**, *29*, 405–423. [CrossRef]
7. Pastora-Bernal, J.-M.; Martín-Valero, R.; Barón-López, F.J.; García-Gómez, O. Effectiveness of telerehabilitation programme following surgery in shoulder impingement syndrome (SIS): Study protocol for a randomized controlled non-inferiority trial. *Trials* **2017**, *18*, 82. [CrossRef]
8. Zaslav, K.R. Internal rotation resistance strength test: A new diagnostic test to differentiate intra-articular pathology from outlet (Neer) impingement syndrome in the shoulder. *J. Shoulder Elb. Surg.* **2001**, *10*, 23–27. [CrossRef]
9. Yang, J.-L.; Lin, J.-J. Reliability of function-related tests in patients with shoulder pathologies. *J. Orthop. Sports Phys. Ther.* **2006**, *36*, 572–576. [CrossRef]
10. Barth, J.R.; Burkhart, S.S.; De Beer, J.F. The bear-hug test: A new and sensitive test for diagnosing a subscapularis tear. *Arthroscopy* **2006**, *22*, 1076–1084. [CrossRef]
11. Miller, C.A.; Forrester, G.A.; Lewis, J.S. The validity of the lag signs in diagnosing full-thickness tears of rotator-cuff tears: A pre-liminary investigation. *Arch. Phys. Med. Rehabil.* **2008**, *89*, 1162–1168. [CrossRef] [PubMed]
12. Nomden, J.G.; Slagers, A.J.; Bergman, G.J.; Winters, J.C.; Kropmans, T.J.; Dijkstra, P.U. Interobserver reliability of physical examination of shoulder girdle. *Man. Ther.* **2009**, *14*, 152–159. [CrossRef] [PubMed]
13. Alqunaee, M.; Galvin, R.; Fahey, T. Diagnostic accuracy of clinical tests for subacromial impingement syndrome: A systematic review and meta-analysis. *Arch. Phys. Med. Rehabil.* **2012**, *93*, 229–236. [CrossRef]
14. Hegedus, E.J.; Goode, A.P.; Cook, C.E.; Michener, L.; Myer, C.A.; Myer, D.M.; Wright, A.A. Which physical examination tests provide clinicians with the most value when exam-ining the shoulder? Update of a systematic review with meta-analysis of individual tests. *Br. J. Sports Med.* **2012**, *46*, 964–978. [CrossRef] [PubMed]
15. Oh, J.H.; Kim, J.Y.; Kim, W.S.; Gong, H.S.; Lee, J.H. The Evaluation of various physical examinations for the diagnosis of type II superior labrum anterior and posterior lesion. *Am. J. Sports Med.* **2008**, *36*, 353–359. [CrossRef] [PubMed]
16. Guanche, C.A.; Jones, D.C. Clinical testing for tears of the glenoid labrum. *Arthrosc. J. Arthrosc. Relat. Surg.* **2003**, *19*, 517–523. [CrossRef]
17. Farber, A.J.; Castillo, R.E.; Clough, M.A.; Bahk, M.I.; McFarland, E.G. Clinical assessment of three common tests for traumatic anterior shoulder instability. *J. Bone Jt. Surg. Am.* **2006**, *88*, 1467–1474. [CrossRef]
18. Kim, S.H.; Park, J.S.; Jeong, W.K.; Shin, S.K. The Kim Test: A novel test for posteroinferior labral lesion of the shoulder—A comparison to the jerk test. *Am. J. Sports Med.* **2005**, *33*, 1188–1192. [CrossRef]
19. Constant, C.R.; Gerber, C.; Emery, R.J.; Søjbjerg, J.O.; Gohlke, F.; Boileau, P. A review of the Constant score: Modifications and guidelines for its use. *J. Shoulder Elb. Surg.* **2008**, *17*, 355–361. [CrossRef]
20. Michener, L.A.; McClure, P.W.; Sennett, B.J. American Shoulder and Elbow Surgeons Standardized Shoulder Assessment Form, patient self-report section: Reliability, validity, and responsiveness. *J. Shoulder Elb. Surg.* **2002**, *11*, 587–594. [CrossRef]
21. Tanaka, M.J.; Oh, L.S.; Martin, S.D.; Berkson, E.M. Telemedicine in the Era of COVID-19: The Virtual Orthopaedic Examination. *J. Bone Joint Surg. Am.* **2020**, *102*, e57. [CrossRef] [PubMed]

22. 22. Vicenti G, Bizzoca D, Solarino G, Moretti F, Ottaviani G, Simone F, Zavattini G, Maccagnano G, Noia G, Moretti B. The role of biophysical stimulation with pemfs in fracture healing: From bench to bedside. *J Biol Regul Homeost Agents* **2020**, *34* (Suppl. 1), 131–135.
23. Laskowski, E.R.; Johnson, S.E.; Shelerud, R.A.; Lee, J.A.; Rabatin, A.E.; Driscoll, S.W.; Moore, B.J.; Wainberg, M.C.; Terzic, C.M. The Telemedicine Musculoskeletal examination. *Mayo Clin. Proc.* **2020**, *95*, 1715–1731. [CrossRef] [PubMed]
24. Cadogan, A.; Laslett, M.; Hing, W.; McNair, P.; Williams, M. Interexaminer reliability of orthopedic special tests used in the assessment of shoulder pain. *Man Ther.* **2011**, *16*, 131–135. [CrossRef] [PubMed]
25. Michener, L.A.; Walsworth, M.K.; Doukas, W.C.; Murphy, K.P. Reliability and diagnostic accuracy of 5 physical examination tests and combination of tests for subacromial impingement. *Arch. Phys. Med. Rehabil.* **2009**, *90*, 1898–1903. [CrossRef]
26. Holtby, R.; Razmjou, H. Accuracy of the Speed's and Yergason's tests in detecting biceps pathology and SLAP lesions: Comparison with arthroscopic findings. *Arthrosc. J. Arthrosc. Relat. Surg.* **2004**, *20*, 231–236. [CrossRef]
27. Hegedus, E.J.; Goode, A.; Campbell, S.; Morin, A.; Tamaddoni, M.; Moorman, C.T., 3rd; Cook, C. Physical examination test of the shoulder: A systematic review with meta-analysis of individual tests. *Br. J. Sports Med.* **2008**, *42*, 80–92. [CrossRef]
28. Lizzio, V.A.; Meta, F.; Fidai, M.; Makhni, E.C. Clinical Evaluation and Physical Exam Findings in Patients with Anterior Shoulder Instability. *Curr. Rev. Musculoskelet. Med.* **2017**, *10*, 434–441. [CrossRef]
29. Morita, W.; Tasaki, A. Intra- and inter-observer reproducibility of shoulder laxity tests: Comparison of the drawer, modified drawer and load and shift tests. *J. Orthop. Sci.* **2018**, *23*, 57–63. [CrossRef]
30. Stetson, W.B.; Templin, K. The Crank Test, the O'Brien Test, and Routine Magnetic Resonance Imaging Scans in the Diagnosis of Labral Tears. *Am. J. Sports Med.* **2002**, *30*, 806–809. [CrossRef]
31. Liu, S.H.; Henry, M.H.; Nuccion, S.L. A prospective evaluation of a new physical examination in predicting glenoid labral tears. *Am. J. Sports Med.* **1996**, *24*, 721–725. [CrossRef] [PubMed]
32. Matsumura, N.; Oki, S.; Fukasawa, N.; Matsumoto, M.; Nakamura, M.; Nagura, T.; Yamada, Y.; Jinzaki, M. Glenohumeral translation during active external rotation with the shoulder abducted in cases with glenohumeral instability: A 4-dimensional computed tomography analysis. *J. Shoulder Elb. Surg.* **2019**, *28*, 1903–1910. [CrossRef] [PubMed]
33. Kim, S.-H.; Park, J.-C.; Park, J.-S.; Oh, I. Painful Jerk Test. *Am. J. Sport. Med.* **2004**, *32*, 1849–1855. [CrossRef]
34. Tashjian, R.Z.; Hung, M.; Keener, J.D.; Bowen, R.C.; McAllister, J.; Chen, W.; Ebersole, G.; Granger, E.K.; Chamberlain, A.M. Determining the minimal clinically important difference for the American Shoulder and Elbow Surgeons score, Simple Shoulder Test, and visual analog scale (VAS) measuring pain after shoulder arthroplasty. *J. Shoulder Elb. Surg.* **2017**, *26*, 144–148. [CrossRef] [PubMed]
35. Longo, U.G.; Rizzello, G.; Loppini, M.; Locher, J.; Buchmann, S.; Maffulli, N.; Denaro, V. Multidirectional Instability of the Shoulder: A Systematic Review. *Arthrosc. J. Arthrosc. Relat. Surg.* **2015**, *31*, 2431–2443. [CrossRef] [PubMed]
36. Murray, K.J. Hypermobility disorders in children and adolescents. *Best Pract. Res. Clin. Rheumatol.* **2006**, *20*, 329–351. [CrossRef] [PubMed]
37. Piazzolla, A.; Bizzoca, D.; Solarino, G.; Moretti, L.; Moretti, B. Vertebral fragility fractures: Clinical and radiological results of augmentation and fixation—A systematic review of randomized controlled clinical trials. *Aging Clin. Exp. Res.* **2020**, *32*, 1219–1232. [CrossRef]
38. Moretti, L.; Bizzoca, D.; Giancaspro, G.A.; Cassano, G.D.; Moretti, F.; Setti, S.; Moretti, B. Biophysical Stimulation in Athletes' Joint Degeneration: A Narrative Review. *Medicina* **2021**, *57*, 1206. [CrossRef]
39. Wang, G.; Fiedler, A.K.; Warth, R.J.; Bailey, L.; Shupe, P.G.; Gregory, J.M. Reliability and accuracy of telemedicine-based shoulder examinations. *J. Shoulder Elb. Surg.* **2022**, *31*, e369–e375. [CrossRef]
40. Gallo, R.; Applewhite, A.I.; Silvis, M.L.; Yenior, A.L.; Ton, A.N.; Ortiguera, C.J.; Pujalte, G. Shoulder Evaluation by Telephone and Video Visit: A Narrative Review. *Cureus* **2022**, *14*, e22461. [CrossRef]
41. Vicenti, G.; Bizzoca, D.; Pesare, E.; Grasso, M.; Ginestra, W.; Moretti, B. Second and Third Pandemic Waves in Apulia: How COVID-19 Affected Orthopedic and Trauma Care—A Single-Center Study. *J. Clin. Med.* **2022**, *11*, 6526. [CrossRef]
42. Moretti, L.; Vitale, E.; Bettinsoli, M.; Bizzoca, D.; Delmedico, M.; Papalia, R.; Solarino, G.; Moretti, B. The psychological and clinical impact of the COVID-19 pandemic on orthopaedic patients: An Italian gender-specific analysis. *Orthop. Rev.* **2021**, *13*, 9005. [CrossRef] [PubMed]
43. Bizzoca, D.; Vicenti, G.; Rella, C.; Simone, F.; Zavattini, G.; Tartaglia, N.; Saracino, M.; Corina, G.; Maddalena, R.; Pesce, V.; et al. Orthopedic and trauma care during COVID-19 lockdown in Apulia: What the pandemic could not change. *Minerva Ortop. Traumatol.* **2020**, *71*, 144–150. [CrossRef]
44. Ferorelli, D.; Moretti, L.; Benevento, M.; Mastrapasqua, M.; Telegrafo, M.; Solarino, B.; Dell'erba, A.; Bizzoca, D.; Moretti, B. Digital Health Care, Telemedicine, and Medicolegal Issues in Orthopedics: A Review. *Int. J. Environ. Res. Public Health* **2022**, *19*, 15653. [CrossRef] [PubMed]

Journal of
Personalized Medicine

MDPI

Opinion

Clinical Insights into the Treatment of Patellofemoral Instability with Medial Patellofemoral Ligament Reconstruction: Pearls and Pitfalls—Lessons Learned from 20 Years

Kata Papp [1,*], Bernhard M. Speth [1,2] and Carlo Camathias [3,4]

1 Department of Traumatology and Orthopaedic Surgery, Kantonsspital Aarau, CH-5000 Aarau, Switzerland
2 Department of Orthopaedic Surgery, University Children's Hospital Basel, CH-4031 Basel, Switzerland
3 Orthopädie für Kinder & Jugendliche, Praxis Zeppelin, CH-9016 St. Gallen, Switzerland
4 Faculty of Medicine, University of Basel, CH-4031 Basel, Switzerland
* Correspondence: kata.papp@ksa.ch

Abstract: Patellofemoral instability is a prevalent cause of pain and disability in young individuals engaged in athletic activities. Adolescents face a particularly notable risk of patellar dislocation, which can be attributed to rapid skeletal growth, changes in q-angle, ligamentous laxity, higher activity levels, and increased exposure to risk. Specific sports activities carry an elevated risk of patellar dislocation. Younger age and trochlear dysplasia present the highest risk factors for recurrent patellar dislocations. International guidelines recommend conservative therapy following a single patellar dislocation without osteochondral lesions but suggest surgical intervention in recurrent cases. In this study, we have compiled current scientific data on therapy recommendations, focusing on MPFL (medial patellofemoral ligament) reconstruction. We discuss patient selection, surgical indications, graft selection, location and choice of fixation, graft tensioning, and postoperative care.

Keywords: medial patellofemoral ligament reconstruction; MPFL; femoropatellar instability; young athletes

check for updates

Citation: Papp, K.; Speth, B.M.; Camathias, C. Clinical Insights into the Treatment of Patellofemoral Instability with Medial Patellofemoral Ligament Reconstruction: Pearls and Pitfalls—Lessons Learned from 20 Years. *J. Pers. Med.* **2023**, *13*, 1240. https://doi.org/10.3390/jpm13081240

Academic Editors: Markus Gesslein and Eduard Kurz

Received: 21 March 2023
Revised: 25 April 2023
Accepted: 7 August 2023
Published: 9 August 2023

1. Introduction

Patellofemoral instability is a prevalent cause of pain and disability in young individuals engaged in athletic activities. It often leads to patellar dislocations, which are more common among adolescent girls aged 10 to 17 [1]. However, recent studies indicate higher injury rates among boys, possibly due to their participation in sports such as football and wrestling [2]. Nevertheless, when comparing both genders in sports where males and females participate equally, females appear to have a higher risk of patellar dislocations. The risk of such injuries increases not only with athletic activity but also with military service before the age of 20 [3]. Certain sports, including girls' gymnastics and soccer, as well as boys' football and wrestling, carry an elevated risk of patellar dislocation [2].

According to a cohort study by Sanders et al., the overall incidence of patellar dislocation is 23.2 per 100,000 person-years [4]. This study also revealed a significantly higher incidence of patellar dislocation among adolescents than previously estimated. Therefore, the risk of patellar dislocation in this age group is notable, attributed to rapid skeletal growth, changes in the q-angle, ligamentous laxity, higher activity levels, and increased exposure to risk.

Typically, dislocations occur through a mechanism involving knee flexion and rotation, usually as a noncontact but traumatic event [5]. Contact with other players appears to increase the risk of dislocation in males, while females tend to experience femoropatellar instability without direct player contact. Mitchell et al. explain that anatomic factors play a more prominent role in females with patellar dislocations, including a higher prevalence of ligamentous laxity and changes in the q-angle, among others [2].

Various risk factors contribute to patellar dislocations, such as trochlear dysplasia, patella alta, malalignment syndromes, axial leg deformities, and neuromuscular disorders.

Among recurrent patellar dislocations, younger age and trochlear dysplasia are the highest risk factors [6]. The medial patellofemoral ligament (MPFL) is injured in approximately 87% of dislocation cases. Alongside the bone structure (static stabilizer) and the surrounding muscles (dynamic stabilizers), the MPFL is the primary passive restraint against lateralization in the initial 30° of knee flexion [7,8]. Rupture of the ligament typically occurs when it elongates to around 26 mm under a tensile force of 208 N [9].

International guidelines suggest conservative therapy after a single patellar dislocation without osteochondral lesions but recommend treatment in cases of recurrence. MPFL reconstruction is a less invasive procedure than most bony interventions, as it leaves the cartilage untouched. A systematic review of Testa et al. comparing MPFL reconstruction and trochleoplasty demonstrates significant improvement in postoperative clinical scores for both procedures [10]. Hurley et al. indicate that reconstructing the MPFL reduces rates of recurrent instability and re-dislocation postoperatively, compared to MPFL repair or nonoperative management [11]. Previous research by the same authors suggests that MPFL reconstruction can be highly effective for patients with patellofemoral instability and no or moderate risk factors [12].

Although MPFL reconstruction may require less effort than bony procedures, postoperative rehabilitation is complex and lengthy. Recurrent instability after MPFL reconstruction is a common issue [13]. Therefore, it is essential for physicians to select the most suitable operative method with minimal effort and negligible risk for patients.

The purpose of this article is to review the current literature, analyze the risks and benefits of single MPFL reconstruction in athletic youth with patellofemoral instability, and outline the key aspects of this procedure. Our aim is to provide physicians with a practical guideline for planning therapy in young individuals, as utilized in our therapy consensus.

2. Materials and Methods

We conducted a search in PubMed for recent literature that describes the outcomes of MPFL reconstruction between 2000 and 2022. In order to provide a comprehensive analysis, we also included two studies from before 2000 due to their relevance [7,14]. We did not impose any specific inclusion criteria regarding article type or level of evidence. However, we did exclude articles that focused on operations involving concurrent or previous procedures, as well as those involving patients with multi-ligamentous or osteochondral injuries.

3. Results

In recent years, both the number of publications on this topic and the frequency of this operation worldwide have increased. In this article, we aim to present the most significant findings regarding solitary MPFL reconstruction.

3.1. Patient Selection

Careful patient selection is one of the most critical factors contributing to postoperative success, as it helps determine the correct indication for surgery. In the case of patients with lateral patellar instability and recurrent dislocations without patellofemoral arthrosis, the primary selection is for MPFL reconstruction.

It is crucial to analyze the bony anatomy of each individual and plan any necessary concomitant operations. Failure to address trochlear dysplasia and neglecting the elevated TTTG (tibial tuberosity to trochlear groove) distance are the most common causes of postoperative failure [15]. Therefore, it is essential to take these factors into account. A higher degree of trochlear dysplasia, patella alta, genu valgum, and elevated femoral antetorsion are indications for bony procedures. In such cases, a bony procedure would be the preferred option, as it can effectively address these specific anatomical abnormalities and provide better outcomes.

The ideal patient for MPFL reconstruction should be younger than 30 years and free from obesity or cartilage damage. Additionally, it is worth noting that women tend to have poorer clinical outcomes [16]. According to a review by Migliorini, younger individuals who engage in athletic activities are more likely to experience re-dislocations after MPFL reconstruction [17].

3.2. Surgical Indications

Patellar instability during daily activities or sports, which has not responded to nonoperative therapy, indicates the need for operative treatment. An MRI can help confirm the diagnosis by showing a bone bruise at the typical lateral patellar facet and medial side of the lateral femoral condyle, while also ruling out any accompanying injuries [18]. In certain cases, when the MRI quality is good, it may also be possible to identify the site of the rupture.

3.3. Graft Selection

Similar to the anterior cruciate ligament (ACL), the medial patellofemoral ligament (MPFL) can be reconstructed using various options such as the quadriceps tendon, adductor magnus tendon, hamstrings, or synthetic materials. However, there is currently no consensus on the optimal graft choice [19]. Some evidence suggests that the semitendinosus tendon may yield better outcomes compared to the gracilis graft, as reported in a systematic review by Migliorini et al. [20].

A straightforward technique described by Steensen offers a convenient method for utilizing the quadriceps graft without the need for drilling a patellar tunnel, thus reducing the risks of patellar fracture or cartilage damage [21]. This technique proves beneficial for skeletally immature patients as it avoids tunneling near the growth plate, a practice we also employ in our procedures.

Achieving an appropriate graft length is crucial for a successful reconstruction, as it can impact the decision between double- or single-bundle techniques.

3.4. Location of Fixation

Another crucial factor for achieving a favorable outcome is the positioning of the tunnel and the site of fixation. The femoral attachment is considered the most critical point that affects isometry and ensures proper functioning [21]. The anatomic femoral insertion point lies between the adductor tubercle and the medial femoral epicondyle. Placing the tunnel too anteriorly and proximally is a common mistake, as it results in excessive pressure on the medial side [22].

However, it is important to note that the ligament exists as a complex with a broad attachment. Approximately 57% of the insertion is located at the patella, while 43% is found at the deep quadriceps tendon [23,24]. Tanaka also described the midpoint of this ligament complex as the point on the proximal cartilage border of the patella where the quadriceps tendon attaches. In young individuals, the ligament's origin is typically located 4.7 to 10 mm distally from the physis of the distal femur. However, this distance can vary with growth depending on the age of the patient [25]. Patients with an open physis present a special group with unique surgical challenges. The target isometric fixation point for an MPFL graft is situated just anterior to the posterior cortex of the femur and proximal to the Blumensaat line.

Regarding the patellar tunnel, it should be placed at the anatomical insertion site on the proximal two-thirds of the medial side of the patella, depending on the graft choice and technique [21]. Positioning the tunnel too anteriorly can lead to a fracture, while placing it too posteriorly can cause articular damage [26]. Therefore, obtaining an accurate lateral fluoroscopy view of the knee during the procedure is essential to determine the correct tunnel sites [27]. It is recommended to drill 3.2 mm tunnels instead of 4.5 mm to reduce the risk of fracture [28].

Techniques that are suitable for adults may not be appropriate for younger individuals due to the risk of injuring the physis and causing growth disturbances. The MPFL is closely related to the physis, making it susceptible to injury during temporary hemiepiphyseodesis of the distal femur [29]. Therefore, when choosing the location of fixation, it is important to select a distal point from the physis for a more isometric outcome compared to a proximal placement of the femoral insertion [30]. Special care should be taken with the MPFL in cases where simultaneous temporary hemiepiphyseodesis and MPFL reconstruction are planned, especially when aiming to correct the leg axis by implanting special plates on the medial femur for growth control.

3.5. Choice of Fixation

There are several methods available to address graft fixation, ranging from traditional screws to sutures. The two primary techniques for fixation are suture-based and tunnel-based approaches [31]. According to Shah et al., it is challenging to determine which approach is superior in terms of complications [31]. However, selecting a method that involves through-tunnel fixation in the femoral region is considered crucial for achieving the strongest reconstruction [9].

When it comes to fixing the patellar end, it is important to ensure minimal prominence to prevent implant-related pain and the potential need for subsequent removal. Implant pain is a common issue at the femoral insertion site and, in some cases, may require the removal of screws [26].

3.6. Graft Tensioning

Excessive tightening of the graft during surgery may lead to medial dislocations of the patella or a loss of knee motion, which could necessitate further operations [26].

In severe and uncommon cases, this restricted motion can result in a condition known as arthrofibrosis. Overconstraining the graft increases the pressure on the patellofemoral joint, causing pain and potentially leading to the development of arthrosis over time [32]. Conversely, insufficient constraint of the graft can result in recurrent dislocations.

It is crucial to consider the type of graft selected for the reconstruction. Hamstrings, for instance, tend to be stiffer and more durable than the original MPFL, requiring special care to avoid overtightening [14].

3.7. Postoperative Care

The postoperative care after MPFL reconstruction has not been standardized yet. However, it typically involves limited weight-bearing for several weeks and a gradual increase in the range of motion over a period of six weeks.

In a recent retrospective study conducted by Gulbrandsen et al., it was observed that there was a significant reduction in postoperative pain compared to the preoperative scores [33].

This finding remained consistent regardless of whether patients returned to their preoperative activity level or not [34]. It is worth noting that around 91% of patients were able to resume sports activities following isolated MPFL reconstruction. Moreover, within the same study, 67% of patients returned to the same or even higher level of activity than before their operation, with an average time of 10 months.

The reasons for this can be attributed to the fear of re-injury as well as limited knee flexion and extension strength of the operated knee. However, it should be acknowledged that the exact reasons cannot be definitively determined in every case [35].

Postoperative rehabilitation plays a crucial role in the successful return to sports activities. It is strongly recommended to follow a formal physical therapy program rather than relying solely on YouTube videos, as these videos tend to score low in terms of quality, reliability, understandability, and actionability [36].

4. Discussion

The stabilizing role of the medial patellofemoral ligament (MPFL) reaches its peak at 25–30° of flexion, showing the maximum change in strain. However, beyond 60° of flexion, the trochlea becomes the primary stabilizer [37].

Depending on the anatomical configuration, there are numerous possibilities to address the main issue of patellar instability. The key is to identify the precise factors causing instability, which often include not only a ligamentous deficiency but also skeletal anomalies such as trochlear dysplasia or patella alta, femoral/tibial rotational problems, or leg axis deformities (e.g., genu valgum). It is important to avoid relying solely on MPFL reconstruction in the presence of these anatomical changes.

Isolated MPFL reconstruction has shown reasonable patient satisfaction, reduced pain scores, and a successful return to sports on average, provided that patient selection and indications are carefully considered [38]. In carefully selected patients undergoing isolated MPFL reconstruction, the rate of recurrent instability is low, with an approximate risk of 1.9% [35]. Similar findings regarding dislocation rates were observed in a systematic review conducted by Testa et al., which compared MPFL reconstruction to trochleaplasty. Both groups showed similar dislocation rates of approximately 2% [10]. However, the review also indicated that while the patellofemoral joint was mostly stable postoperatively, most of the investigated outcome parameters did not return to normal values [10]. When discussing the operation's indication with athletic patients, it is important to emphasize that they are likely to return to their sports activities, albeit at a lower intensity level than before the operation [35].

Even the possibility of conservative treatment should be discussed with patients, particularly after a single dislocation without significant variations in the bony anatomy. Adolescents commonly experience a tear of the MPFL at the insertion site, whereas adults often have mid-substance ruptures. Vavken et al., in their systematic review, found no significant difference in clinical scores and recurrence rates between nonoperative and surgical treatments of patellar instability in children and adolescents [39].

5. Conclusions

Based on our literature research and personal experience, we recommend considering the following aspects to define the most accurate therapy:

1. For isolated MPFL reconstruction, it is preferable to consider individuals under the age of 30 without arthrosis or obesity.
2. MPFL reconstruction alone is not suitable for patients with bony anomalies such as trochlear dysplasia, increased femoral antetorsion, genu valgum, or patella alta. In such cases, bony correction should be considered to achieve successful postoperative outcomes.
3. There is no evidence supporting the necessity of surgical therapy after a single patellar dislocation. However, the psychological stress and impact on everyday activities and sports should be taken into account.
4. When harvesting the graft, it is important to ensure a sufficient length of approximately 7 cm.
5. The femoral fixation of the graft should be placed between the adductor tubercle and the medial femoral epicondyle. Intraoperative imaging from a pure lateral view can assist in locating the correct tunnel placement.
6. In patients with open physis, the location of femoral fixation should be distal to the physis to ensure optimal functionality without causing growth disturbances.
7. The patellar fixation should be positioned on the proximal medial two-thirds, and a 3.2 mm drill head should be used instead of 4.5 mm to minimize complications.
8. It is important to avoid prominent hardware in the fixation process.
9. Intraoperatively, the patellar motion should be assessed, and a translation of two or three quadrants should be allowed for, similar to the healthy contralateral side. This helps ensure proper patellar tracking. Additionally, graft fixation should be performed at approximately 30° flexion of the knee, as supported by literature references [28].
10. Postoperative care should include a tailored physical therapy program that addresses the specific needs of the patient. It is important to have open discussions with

patients about their expectations regarding returning to sports activities, ensuring that expectations are realistic and aligned with their individual recovery progress.

Author Contributions: Conceptualization, K.P.; methodology, C.C.; software, K.P.; validation, K.P., B.M.S. and C.C.; formal analysis, K.P.; investigation, K.P. and C.C.; resources, K.P.; data curation, K.P. and C.C.; writing—original draft preparation, K.P.; writing—review and editing, B.M.S. and C.C.; visualization, K.P.; supervision, C.C.; project administration, C.C.; funding acquisition, K.P. and C.C. All authors have read and agreed to the published version of the manuscript.

Funding: This research received no external funding.

Informed Consent Statement: Not applicable.

Data Availability Statement: Not applicable.

Conflicts of Interest: The authors declare no conflict of interest.

References

1. Rosinski, A.; Chakrabarti, M.; Gwosdz, J.; McGahan, P.J.; Chen, J.L. Double-Bundle Medial Patellofemoral Ligament Reconstruction with Allograft. *Arthrosc. Tech.* **2019**, *8*, e513–e520. [CrossRef] [PubMed]
2. Mitchell, J.; Magnussen, R.A.; Collins, C.L.; Currie, D.W.; Best, T.M.; Comstock, R.D.; Flanigan, D.C. Epidemiology of Patellofemoral Instability Injuries among High School Athletes in the United States. *Am. J. Sports Med.* **2015**, *43*, 1676–1682. [CrossRef] [PubMed]
3. Hsiao, M.; Owens, B.D.; Burks, R.; Sturdivant, R.X.; Cameron, K.L. Incidence of acute traumatic patellar dislocation among active-duty United States military service members. *Am. J. Sports Med.* **2010**, *38*, 1997–2004. [CrossRef] [PubMed]
4. Sanders, T.L.; Pareek, A.; Hewett, T.E.; Stuart, M.J.; Dahm, D.L.; Krych, A.J. Incidence of First-Time Lateral Patellar Dislocation: A 21-Year Population-Based Study. *Sports Health* **2018**, *10*, 146–151. [CrossRef]
5. Kramer, D.E.; Pace, J.L. Acute traumatic and sports-related osteochondral injury of the pediatric knee. *Orthop. Clin. N. Am.* **2012**, *43*, 227–236. [CrossRef]
6. Nwachukwu, B.U.; So, C.; Schairer, W.W.; Green, D.W.; Dodwell, E.R. Surgical versus conservative management of acute patellar dislocation in children and adolescents: A systematic review. *Knee Surg. Sports Traumatol. Arthrosc.* **2016**, *24*, 760–767. [CrossRef]
7. Sallay, P.I.; Poggi, J.; Speer, K.P.; Garrett, W.E. Acute dislocation of the patella. A correlative pathoanatomic study. *Am. J. Sports Med.* **1996**, *24*, 52–60. [CrossRef]
8. Felli, L.; Alessio-Mazzola, M.; Lovisolo, S.; Capello, A.G.; Formica, M.; Maffulli, N. Anatomy and biomechanics of the medial patellotibial ligament: A systematic review. *Surgeon* **2021**, *19*, e168–e174. [CrossRef]
9. Mountney, J.; Senavongse, W.; Amis, A.A.; Thomas, N.P. Tensile strength of the medial patellofemoral ligament before and after repair or reconstruction. *J. Bone Jt. Surg. Br.* **2005**, *87*, 36–40. [CrossRef]
10. Testa, E.A.; Camathias, C.; Amsler, F.; Henle, P.; Friederich, N.F.; Hirschmann, M.T. Surgical treatment of patellofemoral instability using trochleoplasty or MPFL reconstruction: A systematic review. *Knee Surg. Sports Traumatol. Arthrosc.* **2017**, *25*, 2309–2320. [CrossRef]
11. Hurley, E.T.; Colasanti, C.A.; Anil, U.; McAllister, D.; Matache, B.A.; Alaia, M.J.; Strauss, E.J.; Campbell, K.A. Management of Patellar Instability: A Network Meta-analysis of Randomized Control Trials. *Am. J. Sports Med.* **2022**, *50*, 2561–2567. [CrossRef]
12. Blanke, F.; Watermann, K.; Haenle, M.; Feitenhansl, A.; Camathias, C.; Vogt, S. Isolated Medial Patellofemoral Ligament Reconstruction Can Be an Effective Procedure in Patellofemoral Instability with Risk Factors. *J. Knee Surg.* **2020**, *33*, 992–997. [CrossRef]
13. Dall'Oca, C.; Elena, N.; Lunardelli, E.; Ulgelmo, M.; Magnan, B. MPFL reconstruction: Indications and results. *Acta Biomed.* **2020**, *91*, 128–135. [CrossRef] [PubMed]
14. Hamner, D.L.; Brown, C.H., Jr.; Steiner, M.E.; Hecker, A.T.; Hayes, W.C. Hamstring tendon grafts for reconstruction of the anterior cruciate ligament: Biomechanical evaluation of the use of multiple strands and tensioning techniques. *J. Bone Jt. Surg. Am.* **1999**, *81*, 549–557. [CrossRef] [PubMed]
15. Kita, K.; Tanaka, Y.; Toritsuka, Y.; Amano, H.; Uchida, R.; Takao, R.; Horibe, S. Factors Affecting the Outcomes of Double-Bundle Medial Patellofemoral Ligament Reconstruction for Recurrent Patellar Dislocations Evaluated by Multivariate Analysis. *Am. J. Sports Med.* **2015**, *43*, 2988–2996. [CrossRef]
16. Enderlein, D.; Nielsen, T.; Christiansen, S.E.; Faunø, P.; Lind, M. Clinical outcome after reconstruction of the medial patellofemoral ligament in patients with recurrent patella instability. *Knee Surg. Sports Traumatol. Arthrosc.* **2014**, *22*, 2458–2464. [CrossRef] [PubMed]
17. Migliorini, F.; Eschweiler, J.; Betsch, M.; Knobe, M.; Tingart, M.; Maffulli, N. Prognostic factors for isolated medial patellofemoral ligament reconstruction: A systematic review. *Surgeon* **2022**, *20*, e112–e121. [CrossRef]
18. Loeb, A.E.; Tanaka, M.J. The medial patellofemoral complex. *Curr. Rev. Musculoskelet. Med.* **2018**, *11*, 201–208. [CrossRef]

19. Stephen, J.M.; Kittl, C.; Williams, A.; Zaffagnini, S.; Marcheggiani Muccioli, G.M.; Fink, C.; Amis, A.A. Effect of Medial Patellofemoral Ligament Reconstruction Method on Patellofemoral Contact Pressures and Kinematics. *Am. J. Sports Med.* **2016**, *44*, 1186–1194. [CrossRef]
20. Migliorini, F.; Trivellas, A.; Driessen, A.; Quack, V.; Tingart, M.; Eschweiler, J. Graft choice for isolated MPFL reconstruction: Gracilis versus semitendinosus. *Eur. J. Orthop. Surg. Traumatol.* **2020**, *30*, 763–770. [CrossRef]
21. Steensen, R.N.; Dopirak, R.M.; Maurus, P.B. A simple technique for reconstruction of the medial patellofemoral ligament using a quadriceps tendon graft. *Arthroscopy* **2005**, *21*, 365–370. [CrossRef]
22. Servien, E.; Fritsch, B.; Lustig, S.; Demey, G.; Debarge, R.; Lapra, C.; Neyret, P. In vivo positioning analysis of medial patellofemoral ligament reconstruction. *Am. J. Sports Med.* **2011**, *39*, 134–139. [CrossRef] [PubMed]
23. Tanaka, M.J.; Voss, A.; Fulkerson, J.P. The Anatomic Midpoint of the Attachment of the Medial Patellofemoral Complex. *J. Bone Jt. Surg. Am.* **2016**, *98*, 1199–1205. [CrossRef] [PubMed]
24. Tanaka, M.J. Variability in the Patellar Attachment of the Medial Patellofemoral Ligament. *Arthroscopy* **2016**, *32*, 1667–1670. [CrossRef]
25. Greenrod, W.; Cox, J.; Astori, I.; Baulch, J.; Williams, J. A Magnetic Resonance Imaging Study of the Significance of the Distal Femoral Physis during Medial Patellofemoral Ligament Reconstruction. *Orthop. J. Sports Med.* **2013**, *1*, 2325967113502638. [CrossRef] [PubMed]
26. Tanaka, M.J.; Bollier, M.J.; Andrish, J.T.; Fulkerson, J.P.; Cosgarea, A.J. Complications of medial patellofemoral ligament reconstruction: Common technical errors and factors for success: AAOS exhibit selection. *J. Bone Jt. Surg. Am.* **2012**, *94*, e87. [CrossRef]
27. Redfern, J.; Kamath, G.; Burks, R. Anatomical confirmation of the use of radiographic landmarks in medial patellofemoral ligament reconstruction. *Am. J. Sports Med.* **2010**, *38*, 293–297. [CrossRef] [PubMed]
28. Smith, M.K.; Werner, B.C.; Diduch, D.R. Avoiding Complications with MPFL Reconstruction. *Curr. Rev. Musculoskelet. Med.* **2018**, *11*, 241–252. [CrossRef] [PubMed]
29. Bachmann, M.; Rutz, E.; Brunner, R.; Gaston, M.S.; Hirschmann, M.T.; Camathias, C. Temporary hemiepiphysiodesis of the distal medial femur: MPFL in danger. *Arch. Orthop. Trauma Surg.* **2014**, *134*, 1059–1064. [CrossRef]
30. Black, S.R.; Meyers, K.N.; Nguyen, J.T.; Green, D.W.; Brady, J.M.; Maher, S.A.; Shubin Stein, B.E. Comparison of Ligament Isometry and Patellofemoral Contact Pressures for Medial Patellofemoral Ligament Reconstruction Techniques in Skeletally Immature Patients. *Am. J. Sports Med.* **2020**, *48*, 3557–3565. [CrossRef]
31. Shah, J.N.; Howard, J.S.; Flanigan, D.C.; Brophy, R.H.; Carey, J.L.; Lattermann, C. A systematic review of complications and failures associated with medial patellofemoral ligament reconstruction for recurrent patellar dislocation. *Am. J. Sports Med.* **2012**, *40*, 1916–1923. [CrossRef]
32. Beck, P.; Brown, N.A.; Greis, P.E.; Burks, R.T. Patellofemoral contact pressures and lateral patellar translation after medial patellofemoral ligament reconstruction. *Am. J. Sports Med.* **2007**, *35*, 1557–1563. [CrossRef]
33. Gulbrandsen, M.T.; Hartigan, D.; Rice, R.C.; Ruckle, D.E.; Patel, K.; Chhabra, A. Do Current Stability Scores after MPFL Reconstruction Correlate with Patient Satisfaction Postoperatively? *Iowa Orthop. J.* **2022**, *42*, 187–192.
34. Meynard, P.; Malatray, M.; Sappey-Marinier, E.; Magnussen, R.A.; Bodiou, V.; Lustig, S.; Servien, E. Medial patellofemoral ligament reconstruction for recurrent patellar dislocation allows a good rate to return to sport. *Knee Surg. Sports Traumatol. Arthrosc.* **2022**, *30*, 1865–1870. [CrossRef]
35. Platt, B.N.; Bowers, L.C.; Magnuson, J.A.; Marx, S.M.; Liu, J.N.; Farr, J.; Stone, A.V. Return to Sport after Medial Patellofemoral Ligament Reconstruction: A Systematic Review and Meta-analysis. *Am. J. Sports Med.* **2022**, *50*, 282–291. [CrossRef]
36. O'Leary, B.; Saker, C.; Stamm, M.A.; Mulcahey, M.K. YouTube Videos Lack Efficacy as a Patient Education Tool for Rehabilitation and Return to Play Following Medial Patellofemoral Ligament Reconstruction. *Arthrosc. Sports Med. Rehabil.* **2022**, *4*, e1111–e1118. [CrossRef]
37. McCulloch, P.C.; Bott, A.; Ramkumar, P.N.; Suarez, A.; Isamaily, S.K.; Daylamani, D.; Noble, P.C. Strain within the native and reconstructed MPFL during knee flexion. *J. Knee Surg.* **2014**, *27*, 125–131. [CrossRef]
38. Gondolfo, R.; Emanuele, H.S.; Guerreiro, J.P.F.; Queiroz, A.O.; Danieli, M.V. Clinical results of medial patellofemoral ligament reconstruction. *Acta Ortop. Bras.* **2022**, *30*, e241172. [CrossRef]
39. Vavken, P.; Wimmer, M.D.; Camathias, C.; Quidde, J.; Valderrabano, V.; Pagenstert, G. Treating patella instability in skeletally immature patients. *Arthroscopy* **2013**, *29*, 1410–1422. [CrossRef]

Journal of Personalized Medicine

MDPI

Article

Arthrodesis of the Proximal Interphalangeal Joint of the Finger—A Biomechanical Study of Primary Stability

Michael Millrose [1,2,*], Hans Christoph Vonderlind [3], Andreas Thannheimer [1], Till Ittermann [4], Johannes Rüther [2], Maximilian Willauschus [2], Hermann-Josef Bail [2] and Andreas Eisenschenk [5]

1 Department of Trauma Surgery and Sports Medicine, Garmisch-Partenkirchen Medical Centre, 82467 Garmisch-Partenkirchen, Germany
2 Department of Orthopedics and Traumatology, Paracelsus Medical University, 90471 Nuremberg, Germany
3 Department of Trauma Surgery, HELIOS Hospital Schwerin—University Campus of MSH Medical School Hamburg, 19055 Schwerin, Germany
4 Institute for Community Medicine, SHIP/Clinical-Epidemiological Research, University of Greifswald, 17475 Greifswald, Germany
5 Department of Hand Surgery and Microsurgery, University Medicine Greifswald, 17475 Greifswald, Germany
* Correspondence: michael.millrose@klinikum-gap.de; Tel.: +49-88-2177-1220

Abstract: Background: Osteoarthritis of the proximal interphalangeal (PIP) joint of the finger often leads to global hand-function detriment. Different techniques for the arthrodesis of the proximal interphalangeal joint have been described that all lead to union in a reasonable percentage of patients and period of time. This biomechanical study aims to analyze and compare the primary stability of different techniques of arthrodesis to render postoperative immobilization unnecessary. Methods: Arthrodeses of 40° of composite cylinders were tested with different techniques in four-point bending for stability in extension as well as flexion. Results: In extension, the compression screw and the compression wires showed the highest stability—whereas in flexion, plate fusion was superior. Tension band, cerclage or compression screw fusion showed the best compromise in flexion/extension stability. Conclusions: Fusion techniques that apply compression to the fusion show superior stability. Cerclage, tension band and compression screws might be able to provide enough stability to withstand the forces exerted during unencumbered activities of daily living. Arthrodesis with plates should be limited to patients with special indications and require immobilization during consolidation.

Keywords: arthrodesis; proximal interphalangeal joint; primary stability arthrodesis; biomechanical analysis

Citation: Millrose, M.; Vonderlind, H.C.; Thannheimer, A.; Ittermann, T.; Rüther, J.; Willauschus, M.; Bail, H.-J.; Eisenschenk, A. Arthrodesis of the Proximal Interphalangeal Joint of the Finger—A Biomechanical Study of Primary Stability. *J. Pers. Med.* **2023**, *13*, 465. https://doi.org/10.3390/jpm13030465

Academic Editors: Markus Gesslein and Eduard Kurz

Received: 17 February 2023
Revised: 28 February 2023
Accepted: 1 March 2023
Published: 2 March 2023

1. Introduction

Osteoarthritis of the proximal interphalangeal (PIP) joint of the finger—either primary or secondary—can cause pain with a concomitant limited range of motion, which often leads to global hand-function deterioration [1,2]. Secondary osteoarthritis is most frequently caused by post-traumatic changes, followed by chronic instability inflammatory diseases [3]. Treatment options, beside conservative treatments such as physiotherapy and non-steroidal anti-inflammatory drugs (NSAIDs), include joint-preserving surgeries such as denervation or different arthroplasties [4]. Joint replacement options include prosthesis or arthrodesis [5]. In particular, the loss of bone stock after failed prostheses leads to the need for arthrodesis [6]. In case of a need to grip things forcefully—as, for example, for artisans—arthrodesis presents a good treatment option. The aim of all these treatments is pain reduction, leading to improved global hand function.

During recent years, research into the PIP joint arthrodesis of the fingers has been characterized by the search for improved implants and techniques with improved stability, combined with simple—as well as tissue-sparing—surgical techniques. Various surgical procedures are currently available for arthrodesis of the proximal interphalangeal joint

(PIJ); these procedures, which can achieve good functional results when indicated and used correctly, have their respective advantages and disadvantages. There are techniques which apply additional compression on the arthrodesis gap, such as tension bands or compression screws of different designs [7,8]. In particular, in trauma cases with a destroyed PIJ such as transarticular amputations, implants which can be applied quickly and induce less additional soft tissue damage are chosen. Those implants, typically k wires, do not apply any compression [9]. Plate arthrodesis is of particular importance in the case of segmental bony defects, non-unions and/or after failed endoprosthesis treatments as it can be easily combined with bone grafts, which in turn enable efficient length maintenance, and thus, less shortening of the finger [10]. A recent systematic review on the PIP joint arthrodesis showed that all these procedures lead to union in a reasonable percentage of patients and periods of time [11]. Newer published techniques, which all rely on the principles of established techniques, have not been able to prove themselves superior [8,12]. Due to the wide variety of surgical techniques for the arthrodesis of this joint, we can state that an optimal procedure has not yet been introduced.

Since almost all current techniques require immobilization after surgery of different time periods until bony fusion is reached, there are concerns that adjacent joints might suffer in mobility [13,14]. The aim of most techniques in hand surgery is the early functional treatment and normal usage of the respective hand. Previous studies have shown that most daily activities create forces of up to 20 N [15,16]. If the primary stability of the implant provided is given, one should be able to allow early functional treatment.

Therefore, this biomechanical study aims to analyze and compare the primary stability of different techniques of arthrodesis for the PIP joint of the finger to recommend techniques which qualify for further clinical studies in order to render a postoperative immobilization unnecessary.

2. Materials and Methods

The specimens were made from 42 mm-long phalangeal equivalents from short fiber-reinforced epoxy cylinders with a 10 mm outer diameter and a 1.5 mm wall in the fourth generation (Sawbone Europe AB, Malmoe, Sweden). The composite cylinders were filled with cellular rigid polyurethane-foam filling (strength 15) to mimic cancellous bone.

Each arthrodesis was performed at a standardized angle of 40°—a compromise between the recommended arthrodesis angles of the PIJ from 20–50°, depending on the finger [17]. The angle was set by cutting the distal end of each cylinder to 20°.

The tested arthrodesis techniques' respective implants were crossed K wires, crossed compression wires, intraosseous wiring, tension band wiring, compression screws and anatomical fixation, as well as locking grid plates.

The used K wires were medical stainless steel and 1.2 mm in diameter. The compression wire used had a threadless section 10 mm in length (Koenigsee, Allendorf, Germany). The arthrodesis was achieved with three turns of the trailing thread inserted into the cortical bone to ensure a standardized compression. Both wire arthrodesis techniques crossed the joint at an angle of 45° in the radio-ulnar plane; they were drilled 8 mm proximal or distal, respectively, to the arthrodesis site.

The compression screw (2.2 mm diameter, 26 mm length; Medartis, Basel, Switzerland) was inserted proximal-to-distal, perpendicular to the arthrodesis site to achieve compression. It was angulated to be fully buried into the cylinders. It consisted of two threads of a different pitch and a threadless section of 10 mm, which provided a secure and rigid cortical fixation, as well as compression of the arthrodesis. The diameter varied from 1.0 mm at the tip to 1.8 mm at the second thread. The placement of the compression wire was possible with the 50 mm-long positioning tip. The threads and main part of the compression wire followed the tip until compression was applied onto the fusion site.

Intraosseous wiring and tension band wiring were carried out with a cerclage wire of 0.8 mm and a K-wire of 1.2 mm according to well-described techniques [18,19].

Plate fixation, anatomical as well as locking grid, used a 0.8 mm low-profile plate (grid fixation plate 4 × 2), bent to accommodate the cylinder and the desired angle of 40°. To achieve fixation on both sides of the arthrodesis, 8 cortices were drilled and fixed by 1.5 mm cortical-respective locking screws (Medartis, Basel, Switzerland).

Using a universal testing machine (Zwick Z050®, Zwick GmbH & Co. KG, Ulm, Germany), tests were run in four-point bending. The extensions of the force conductor were positioned 3 mm proximally and distally from the arthrodesis site on the composite cylinders. The ends of the proximal and the distal specimen were inserted into acrylic cylinders (Figure 1). A cylinder was chosen to achieve compensatory rotational movement and a constant contact area during loading. After both composite cylinders were fused with one of the above-mentioned techniques, force was applied continuously at a rate of 100 mm/min in steps of 0.5° until a maximum of 10° flexion or extension was achieved, and the required force was recorded by a computer using testXpert III 1.1 software (Zwick GmbH & Co. KG, Ulm, Germany).

Figure 1. Four-point bending of a composite cylinder arthrodesis in 40° fused with a locking plate.

The analyzed data was the force necessary to achieve a bending of 10° either in flexion or extension. For each angle and technique, the mean extension and flexion forces over ten measurements in 0.5° steps were reported. For the angles 37.5°, 35.0°, 32.5° and 30°, differences in mean extension levels between the K wire and all the other techniques were tested using paired t-tests. For mean flexion levels, differences between the techniques for the angles 42.5°, 45.0°, 47.5° and 50° were tested. $p < 0.05$ was considered statistically significant. Furthermore, mean extension and flexion levels were plotted against all measured angles for each technique using a lowess smoother. The statistical analysis was performed using STATA 17.0 (Stata Corporation, College Station, TX, USA).

3. Results

Each technique was tested in 10° flexion or 10° extension with $n = 10$, respectively. The strength needed to bend the arthrodesis to the desired angle was measured in 0.5° steps (Tables 1 and 2).

In extension, the crossed compression wires as well as the compression screw provided the best primary stability, with a bending force necessary to achieve 10° of extension in a 40° arthrodesis of over 68 N. Both are techniques which apply compression to the fusion site. Both plates show the lowest primary stability, at 5.3 N for the fixation plate and 14.69 N for the locking plate, respectively.

Table 1. Extension of PIP joint arthrodesis with different techniques.

	K Wire $n = 10$	Cerclage $n = 10$	Compression Wire $n = 10$	Tension Band $n = 10$	Compression Screw $n = 10$	Fixation Plate $n = 10$	Locking Plate $n = 10$
39.5°	3.16 ± 0.87	5.28 ± 3.56	3.43 ± 0.67	4.18 ± 1.31	4.60 ± 1.88	2.31 ± 0.09	3.32 ± 0.58
39°	4.57 ± 1.52	8.46 ± 5.67	5.09 ± 1.70	5.74 ± 2.10	7.68 ± 3.58	2.58 ± 0.14	4.29 ± 0.76
38.5°	6.17 ± 2.01	11.68 ± 6.95	8.05 ± 3.98	7.18 ± 2.76	11.03 ± 6.20	2.82 ± 0.16	5.06 ± 1.00
38°	7.80 ± 2.64	15.03 ± 8.04	11.47 ± 6.56	8.52 ± 3.69	14.36 ± 8.80	3.03 ± 0.19	5.91 ± 1.05
37.5°	9.37 ± 3.21	18.65 ± 8.82	14.73 ± 8.21	10.28 ± 3.97	17.48 ± 9.19	3.27 ± 0.21	6.72 ± 1.06
37°	10.95 ± 3.72	22.21 ± 9.46	17.99 ± 9.75	11.95 ± 4.41	22.67 ± 13.84	3.47 ± 0.27	7.46 ± 1.04
36.5°	12.52 ± 4.21	25.56 ± 10.16	21.65 ± 11.23	13.49 ± 4.89	27.36 ± 17.44	3.68 ± 0.25	8.24 ± 1.12
36°	14.49 ± 4.75	28.80 ± 11.04	26.24 ± 12.48	15.18 ± 5.29	32.26 ± 20.00	3.87 ± 0.26	8.93 ± 1.16
35.5°	16.21 ± 5.13	31.74 ± 11.70	31.10 ± 13.45	16.78 ± 5.58	37.12 ± 20.27	4.06 ± 0.28	9.57 ± 1.24
35°	17.97 ± 5.46	34.63 ± 12.37	35.69 ± 14.42	18.41 ± 5.82	40.40 ± 17.00	4.21 ± 0.30	10.27 ± 1.19
34.5°	19.60 ± 5.79	36.99 ± 12.54	39.89 ± 15.54	19.87 ± 5.98	45.71 ± 18.63	4.34 ± 0.29	10.97 ± 1.10
34°	21.20 ± 6.13	39.38 ± 12.93	43.73 ± 16.72	21.18 ± 5.98	52.08 ± 17.90	4.47 ± 0.31	11.51 ± 1.06
33.5°	22.74 ± 6.36	40.21 ± 13.37	47.17 ± 17.78	22.54 ± 6.11	55.29 ± 19.49	4.61 ± 0.32	11.97 ± 1.00
33°	24.53 ± 6.29	42.93 ± 13.17	49.44 ± 18.26	23.76 ± 6.10	60.27 ± 20.60	4.72 ± 0.33	12.41 ± 0.92
32.5°	26.26 ± 6.20	45.11 ± 13.23	53.06 ± 17.92	24.91 ± 6.18	61.47 ± 18.24	4.84 ± 0.34	12.87 ± 0.83
32°	27.89 ± 5.87	47.10 ± 13.30	56.88 ± 16.71	25.92 ± 6.29	63.01 ± 16.00	4.97 ± 0.35	13.25 ± 0.79
31.5°	29.35 ± 5.80	49.01 ± 13.39	60.43 ± 16.28	26.71 ± 6.29	64.25 ± 14.49	5.06 ± 0.36	13.68 ± 0.67
31°	30.85 ± 5.77	50.78 ± 13.55	63.42 ± 16.28	27.80 ± 6.49	65.15 ± 13.45	5.15 ± 0.36	14.09 ± 0.58
30.5°	32.18 ± 5.81	52.51 ± 13.58	66.19 ± 16.54	28.70 ± 6.50	66.35 ± 12.68	5.24 ± 0.35	14.43 ± 0.52
30°	33.60 ± 5.43	54.11 ± 13.69	68.53 ± 16.85	29.62 ± 6.83	68.12 ± 11.07	5.33 ± 0.34	14.69 ± 0.50

Data is presented as mean ± standard deviation.

Table 2. Flexion of PIP joint arthrodesis with different techniques.

	K Wire $n = 10$	Cerclage $n = 10$	Compression Wire $n = 10$	Tension Band $n = 10$	Compression Screw $n = 10$	Fixation Plate $n = 10$	Locking Plate $n = 10$
40.5°	2.80 ± 0.27	3.39 ± 1.41	3.37 ± 0.71	3.38 ± 0.82	4.06 ± 0.97	3.37 ± 1.35	3.81 ± 0.59
41°	3.71 ± 0.21	4.53 ± 2.58	4.69 ± 1.71	4.37 ± 1.69	6.00 ± 1.82	4.45 ± 2.64	5.50 ± 1.06
41.5°	4.81 ± 0.96	6.27 ± 4.01	6.07 ± 2.74	5.84 ± 2.23	7.97 ± 3.17	7.09 ± 7.78	7.85 ± 2.49
42°	5.90 ± 1.57	8.43 ± 5.53	7.53 ± 3.69	7.35 ± 2.87	10.01 ± 4.25	10.15 ± 13.59	10.60 ± 4.75
42.5°	6.94 ± 1.93	10.84 ± 7.13	9.00 ± 4.67	8.91 ± 3.58	11.94 ± 5.15	13.34 ± 19.38	12.71 ± 4.07
43°	8.08 ± 2.36	13.28 ± 8.85	10.39 ± 5.52	10.46 ± 4.31	13.94 ± 6.01	16.73 ± 25.39	16.16 ± 5.69
43.5°	9.29 ± 2.73	15.76 ± 10.48	11.66 ± 6.20	11.95 ± 4.97	15.88 ± 6.49	20.71 ± 31.61	21.46 ± 10.50
44°	10.60 ± 3.15	18.22 ± 12.08	12.90 ± 6.80	13.28 ± 5.65	17.58 ± 6.65	24.48 ± 37.50	26.77 ± 14.80
44.5°	11.83 ± 3.59	20.86 ± 13.36	14.19 ± 7.32	14.88 ± 6.35	19.06 ± 6.65	28.06 ± 43.89	31.88 ± 18.50
45°	12.92 ± 3.91	23.54 ± 14.46	15.67 ± 7.85	16.67 ± 7.10	20.10 ± 6.19	22.19 ± 19.86	36.67 ± 21.96
45.5°	13.94 ± 4.20	26.07 ± 15.49	17.14 ± 8.47	18.29 ± 7.86	21.18 ± 5.95	27.42 ± 29.08	41.10 ± 25.05
46°	14.97 ± 4.44	28.48 ± 16.37	18.58 ± 8.99	19.97 ± 8.54	22.37 ± 5.99	33.00 ± 36.73	45.39 ± 28.24
46.5°	16.29 ± 4.54	30.72 ± 17.18	20.17 ± 9.10	21.67 ± 9.37	23.62 ± 6.18	39.51 ± 44.44	48.50 ± 30.61
47°	17.51 ± 4.79	32.88 ± 17.91	21.91 ± 9.64	23.37 ± 10.26	25.01 ± 6.45	47.12 ± 50.70	56.21 ± 34.41
47.5°	18.55 ± 5.09	34.97 ± 18.66	23.46 ± 10.27	25.21 ± 11.33	26.58 ± 6.68	52.43 ± 56.39	65.25 ± 39.34
48°	19.55 ± 5.27	36.96 ± 19.37	24.91 ± 10.92	27.12 ± 12.27	28.23 ± 6.93	61.08 ± 59.67	74.42 ± 44.06
48.5°	20.75 ± 5.17	38.85 ± 20.00	26.14 ± 11.39	29.27 ± 12.95	29.82 ± 7.24	73.03 ± 59.49	84.60 ± 46.22
49°	22.14 ± 5.21	40.76 ± 20.47	27.31 ± 11.72	31.45 ± 13.83	31.41 ± 7.56	88.06 ± 58.09	100.47 ± 49.59
49.5°	23.36 ± 5.41	42.49 ± 20.96	28.40 ± 11.97	33.48 ± 14.72	33.03 ± 7.76	103.22 ± 56.54	100.71 ± 51.63
50°	24.41 ± 5.61	44.00 ± 21.22	29.57 ± 12.17	35.41 ± 15.56	34.68 ± 7.97	118.37 ± 55.07	118.51 ± 56.84

Data is presented as mean ± standard deviation.

In flexion, the data show that both plates provided by far the highest resistance to additional bending, at 118 N each. Both techniques using wires for fusion showed the lowest primary stability, at 24.4 N for the normal k wires and 29.6 N for the compression wires.

The cerclage as well as tension band technique and the compression screw provided good results in terms of their primary stability in extension as well as in flexion (Figures 2 and 3).

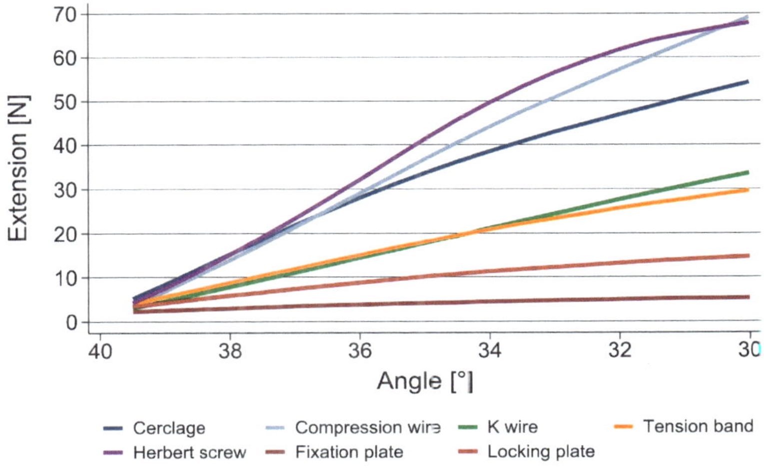

Figure 2. Stability of different fusion techniques for PIP joint arthrodeses in extension.

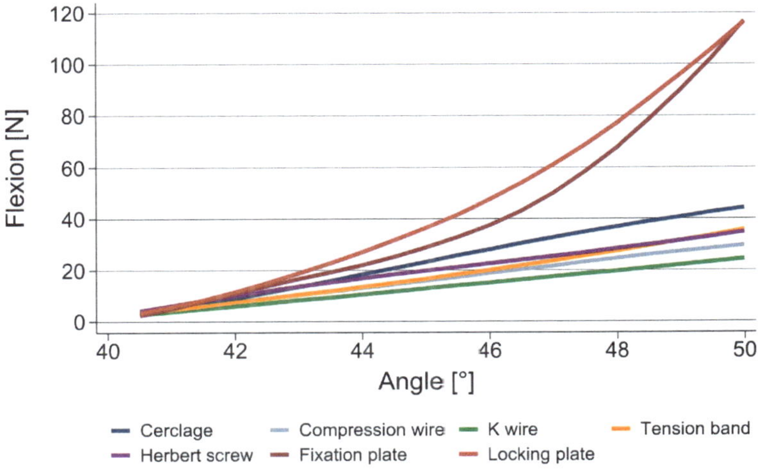

Figure 3. Stability of different fusion techniques for PIP joint arthrodeses in flexion.

The results of the statistical analysis are shown in Table 3 for extension and Table 4 for flexion. K wire fusion was the standard technique chosen to compare against, as it is the best documented technique in the literature and can be used in all situations necessitating arthrodesis.

Table 3. Differences in extension between the methods for different angles.

	37.5°	35.0°	32.5°	30.0°
Compression screw [N]	17.5; $p = 0.006$	40.4; $p = 0.002$	61.5; $p = 0.001$	68.1; $p < 0.001$
Compression wire [N]	14.7; $p = 0.142$	35.7; $p = 0.016$	53.1; $p = 0.004$	68.5; $p < 0.001$
Cerclage [N]	18.6; $p = 0.016$	34.6; $p = 0.009$	45.1; $p = 0.006$	54.1; $p = 0.003$
Tension band [N]	10.3; $p = 0.510$	18.4; $p = 0.844$	24.9; $p = 0.617$	29.6; $p = 0.145$
K wire [N]	9.4	18.0	26.3	33.6
Locking plate [N]	6.7; $p = 0.044$	10.3; $p = 0.003$	12.9; $p < 0.001$	14.7; $p < 0.001$
Fixation plate [N]	3.3; $p < 0.001$	4.2; $p < 0.001$	4.8; $p < 0.001$	5.3; $p < 0.001$

Results are reported as mean value for each method and angle; p-values are derived from a pairwise t-test in comparison to the K wire method as reference; green = statistically significantly more stable than K wire, red = statistically significantly less stable than K wire.

Table 4. Differences in flexion between the methods for different angles.

	42.5°	45.0°	47.5°	50.0°
Locking plate [N]	12.7; $p = 0.004$	36.7; $p = 0.011$	65.2; $p = 0.006$	118.5; $p < 0.001$
Fixation plate [N]	13.3; $p = 0.310$	22.2; $p = 0.147$	52.4; $p = 0.081$	118.4; $p < 0.001$
Cerclage [N]	10.8; $p = 0.153$	23.5; $p = 0.062$	35.0; $p = 0.030$	44.0; $p = 0.030$
Tension band [N]	8.9; $p = 0.217$	16.7; $p = 0.260$	25.2; $p = 0.213$	35.4; $p = 0.125$
Compression screw [N]	11.9; $p = 0.031$	20.1; $p = 0.012$	26.6; $p = 0.004$	34.7; $p < 0.001$
Compression wire [N]	9.0; $p = 0.263$	15.7; $p = 0.397$	23.5; $p = 0.232$	29.6; $p = 0.212$
K wire [N]	6.9	12.9	18.5	24.4

Results are reported as mean value for each method and angle; p-values are derived from a pairwise t-test in comparison to the K wire method as reference; green = statistically significantly more stable than K wire, red = statistically significantly less stable than K wire.

It could be shown that only the compression screw was statistically significantly more stable in both bending directions than the standard technique. The superiority of the compression wire and the cerclage in extension could also be proven.

4. Discussion

This biomechanical study of the primary stability of the proximal interphalangeal joint of the finger with different fusion techniques showed that all the used techniques beside the plates could withstand the forces which are applied during activities of daily living. The implants tested in this study—which are all standard techniques for different indications of joint fusion—demonstrated increasing resistance to bending stress, which can be at least partially attributed to the implant itself.

Although all the used implants have proven themselves to be capable of fusing the PIP joint of the finger in several clinical studies—as shown in the systematic review of Millrose et al.—some are used in special circumstances [12]. For example, plates—regardless of whether they are of a fixation or locking design—show a tendency to lead to tendon adhesions—especially when applied near or on a joint, which might impair the function of the hand [20]. In the special indication of a joint fusion which requires a corticocancellous bone transplant in order to retain the length of the finger, such after a failed prosthesis, a plate can prove itself useful, as it is able to provide a stable fixation of the joint as well as of the transplant [21]. In our study, we were able to show that fusions with plates should be immobilized for the entire consolidation time, as they are not able to adequately withstand extension forces; this is probably because of the low profile of the used plates (0.8 mm) and the fact that the implant crosses the arthrodesis site. However, one must keep in mind that in the phalanges you should not use any stronger implants, due to the above-mentioned complication.

As has been shown in biomechanical analyses of fracture fixation in the phalanges, normally, K wires are able to provide enough stability to allow for early active movement [22]. In our study, we were able to show that they could provide enough stability in extension but lack primary stability in flexion for PIP joint fusions; as such, these should also—at least in the beginning—be immobilized.

Leibovic et al., stated after a clinical study in 1994, that techniques with compression of the fusion site could provide more reliable results [23]. The primary nonunion rate was highest using Kirschner wires, intermediate using tension band wires, and lowest using Compression screws; in a systematic review on this topic, this assumption could not be proven [12]. One possible explanation was that when combining the techniques into an analysis of compression versus non-compression, more advanced techniques such as tension bands were included—which show a higher complication rate than the compression screw—and therefore no evidence could be found. This might be why there are different studies which have shown a superiority of the compression screw over other implants [23,24]. In this biomechanical study, the compression forces of the different implants were not recorded—other studies in the future should further elaborate on this topic. However, there seems to be a tendency that techniques using compression—such as the cerclage, tension band or compression screw—provide higher primary stability and therefore more reliable results.

In this study, a fusion angle of 40° was chosen. Since there are fusion angles of 20–50° that are recommended depending on the finger, this was a compromise [25]. There are several publications on the effect of the chosen arthrodesis angle on the grip strength and function of the hand. Xu et al., was able to show that the resulting strength loss, although significant, did not differ between the range of 20–40° [26]. This is in accordance with the results of Dimitrova et. al., who were also able to show a significant decrease in grip strength in all fingers beside the index [27]. However, since the functional impact of a symptomatic osteoarthritic PIP joint is higher than that of arthrodesis, fusion is still a possible treatment option. Other biomechanical studies on the arthrodesis of the proximal interphalangeal joint have used fusion angles of 20° to 40°; therefore, the chosen 40° angle seemed a reasonable compromise [28–30]. The results of these studies are in accordance with the respective techniques of this study.

In previous biomechanical studies of the stability of the PIP joint arthrodesis by Millrose et al., as well as by Vonderlind et al., in 2019, formalin-fixed human cadaver specimens were used [28,29]. Cadaver specimens have long been used in biomechanical studies, but have different disadvantages compared to those of composite bone models. They are costly and, most importantly, may vary in their structure with respect to mineralization (osteoporosis) [31]. As the purpose of this study was to firstly compare the primary stability of different implants for PIP joint arthrodesis, composite bone was used, as a lot of specimens were necessary and confounders were to be excluded as much as possible.

The forces the implant materials in this biomechanical study were able to withstand exceed most of the activities of daily life [32,33]. This is also confirmed by the results of corresponding studies, which have shown that the forces acting on the middle finger joint during everyday activities such as typing on the computer, holding a pen or playing the piano using different grip shapes range from 1.9 to 19.40 N, or that the maximum force required for flexion of the middle joints of the long fingers is 8.79 N [15,34]. Therefore, a protective immobilization of the arthrodesis might not be necessary—except for the plates—as long as the patient is capable of understanding and maintaining an early, functional, unencumbered rehabilitation. As has been shown in a systematic review of early active motion protocols after flexor tendon injury, the TAM (total active movement) of the finger is mostly preserved, which benefits the whole function of the hand [35].

Our biomechanical study has certain limitations: First of all, we did not analyze the primary stability of different fusion techniques at different angles. Since it is normally recommended to adjust the fusion angle to the respective finger—from 20° for the index up to 50° for the little finger—there might be different implants that might stabilize a lower or higher angle better, respectively. We did not analyze the resistance to torquing forces, which might be interesting—but those forces are not usually relevant. Additionally, we did not analyze the compression force applied by the implant at fusion and during application of the bending force; this might become important for clinical studies as there is a hypothesis that a compressed fusion site will heal primarily and therefore provide more reliable results [23].

A strength of this study is that it is the first biomechanical analysis which compares all the modern standard fusion techniques in such a way as to compare the primary stability of the implant in a PIP joint arthrodesis. This can be the basis of further clinical studies with and without immobilization after PIP joint fusion to better preserve the global function of the hand. By using standardized composite bone models, confounders such as low mineralization or other influences on the bone quality such as neighboring tissue lesion or injuries were excluded.

5. Conclusions

Based on the results of this biomechanical study, cerclages, tension bands and compression screws might be able to provide enough stability to withstand the forces induced by unencumbered activities of daily living and may warrant further clinical studies. Arthrodesis of the PIP joint using plates should be limited to special indications and always requires immobilization for the whole consolidation period.

Author Contributions: Conceptualization, M.M. and H.C.V.; methodology, M.M. and H.C.V.; software, T.I.; validation, M.M. and H.C.V.; formal analysis, T.I. and J.R.; investigation, M.M., H.C.V., T.I. and M.W.; data curation M.M., M.W. and A.E.; writing—original draft preparation, M.M.; writing—review and editing, A.T., H.-J.B., A.E. and J.R.; visualization, M.M. and A.T.; supervision, A.E. and H.-J.B.; project administration, A.E. All authors have read and agreed to the published version of the manuscript.

Funding: This research received no external funding.

Institutional Review Board Statement: Not applicable.

Informed Consent Statement: Not applicable.

Data Availability Statement: The data presented in this study are available on request from the corresponding author.

Acknowledgments: The authors would like to thank Stefanie Karpik for her support in preparing and correcting the manuscript with respect to spelling and grammar as well as Alexandra Amlang for her biomechanical input.

Conflicts of Interest: The authors declare no conflict of interest.

References

1. Herren, D. The proximal interphalangeal joint: Arthritis and deformity. *EFORT Open Rev.* **2019**, *4*, 254–262. [CrossRef]
2. Pellegrini, V.D.; Burton, R.I. Osteoarthritis of the proximal interphalangeal joint of the hand: Arthroplasty or fusion? *J. Hand. Surg. Am.* **1990**, *15*, 194–209. [CrossRef]
3. Jacobs, B.J.; Verbruggen, G.; Kaufmann, R.A. Proximal interphalangeal joint arthritis. *J. Hand. Surg. Am.* **2010**, *35*, 2107–2116. [CrossRef]
4. Herren, D. Current European Practice in the Treatment of Proximal Interphalangeal Joint Arthritis. *Hand. Clin.* **2017**, *33*, 489–500. [CrossRef]
5. Vitale, M.A.; Fruth, K.M.; Rizzo, M.; Moran, S.L.; Kakar, S. Prosthetic Arthroplasty Versus Arthrodesis for Osteoarthritis and Posttraumatic Arthritis of the Index Finger Proximal Interphalangeal Joint. *J. Hand. Surg. Am.* **2015**, *40*, 1937–1948. [CrossRef]
6. Beldner, S.; Polatsch, D.B. Arthrodesis of the Metacarpophalangeal and Interphalangeal Joints of the Hand: Current Concepts. *J. Am. Acad. Orthop. Surg.* **2016**, *24*, 290–297. [CrossRef]
7. Breyer, J.M.; Vergara, P.; Parra, L.; Sotelo, P.; Bifani, A.; Andrade, F. Metacarpophalangeal and interphalangeal joint arthrodesis: A comparative study between tension band and compression screw fixation. *J. Hand. Surg. Eur. Vol.* **2015**, *40*, 374–378. [CrossRef]
8. Novoa-Parra, C.D.; Montaner-Alonso, D.; Pérez-Correa, J.I.; Morales-Rodríguez, J.; Rodrigo-Pérez, J.L.; Morales-Suarez-Varela, M. Arthrodesis of the proximal interphalangeal joint of the 4th and 5th finger using an interlocking screw device to treat severe recurrence of Dupuytren's disease. *Rev. Esp. Cir. Ortop. Traumatol.* **2018**, *62*, 216–221. [CrossRef]
9. Uhiara, O.; Hussain, S.; Deshmukh, S. Plate-augmented tension band wiring for arthrodesis of the proximal interphalangeal and metacarpophalangeal joints: A case series. *J. Hand. Surg. Eur. Vol.* **2021**, *46*, 679–681. [CrossRef]

10. Tan, M.; Ho, S.W.L.; Sechachalam, S. Acute Arthrodesis of Interphalangeal Joints of the Hand in Traumatic Injuries. *J. Hand. Microsurg.* **2018**, *10*, 1–5. [CrossRef]
11. Leibovic, S.J. Internal fixation for small joint arthrodeses in the hand. The interphalangeal joints. *Hand. Clin.* **1997**, *13*, 601–613. [CrossRef] [PubMed]
12. Millrose, M.; Gesslein, M.; Ittermann, T.; Kim, S.; Vonderlind, H.C.; Ruettermann, M. Arthrodesis of the proximal interphalangeal joint of the finger—A systematic review. *EFORT Open Rev.* **2022**, *7*, 49–58. [CrossRef] [PubMed]
13. Wong, J.M. Management of stiff hand: An occupational therapy perspective. *Hand. Surg.* **2002**, *7*, 261–269. [CrossRef] [PubMed]
14. Neumeister, M.W.; Winters, J.N.; Maduakolum, E. Phalangeal and Metacarpal Fractures of the Hand: Preventing Stiffness. *Plast. Reconstr. Surg. Glob. Open* **2021**, *9*, e3871. [CrossRef] [PubMed]
15. Butz, K.D.; Merrell, G.; Nauman, E.A. A biomechanical analysis of finger joint forces and stresses developed during common daily activities. *Comput. Methods Biomech. Biomed. Engin.* **2012**, *15*, 131–140. [CrossRef] [PubMed]
16. Arauz, P.; DeChello, K.; Dagum, A.; Sisto, S.A.; Kao, I. Biomechanics and Pinch Force of the Index Finger Under Simulated Proximal Interphalangeal Arthrodesis. *J. Hand. Surg. Am.* **2017**, *42*, e1–e658. [CrossRef]
17. Vonderlind, H.C.; Eisenschenk, A.; Juergensen, I.; Kim, S.; Millrose, M. Arthrodesis of the proximal interphalangeal joint—A review. *Handchir. Mikrochir. Plast. Chir.* **2019**, *51*, 6–18.
18. Lister, G. Intraosseous wiring of the digital skeleton. *J. Hand. Surg. Am.* **1978**, *3*, 427–435. [CrossRef]
19. Hohendorff, B.; Franke, J.; Spies, C.K.; Mueller, L.P.; Ries, C. Arthrodesis of the proximal interphalangeal joint of fingers with tension band wire. *Oper. Orthop. Traumatol.* **2017**, *29*, 385–394. [CrossRef]
20. Kurzen, P.; Fusetti, C.; Bonaccio, M.; Nagy, L. Complications after plate fixation of phalangeal fractures. *J. Trauma* **2006**, *60*, 841–843. [CrossRef]
21. Büchler, U.; Aiken, M.A. Arthrodesis of the proximal interphalangeal joint by solid bone grafting and plate fixation in extensive injuries to the dorsal aspect of the finger. *J. Hand. Surg. Am.* **1988**, *13*, 589–594. [CrossRef] [PubMed]
22. Lögters, T.T.; Lee, H.H.; Gehrmann, S.; Windolf, J.; Kaufmann, R.A. Proximal Phalanx Fracture Management. *Hand* **2018**, *13*, 376–383. [CrossRef] [PubMed]
23. Leibovic, S.J.; Strickland, J.W. Arthrodesis of the proximal interphalangeal joint of the finger: Comparison of the use of the Herbert screw with other fixation methods. *J. Hand. Surg. Am.* **1994**, *19*, 181–188. [CrossRef] [PubMed]
24. Martin, L. Arthrodeses of the thumb and long finger joints. *Handchirurgie* **1981**, *13*, 221–230.
25. Murray, P.M. Treatment of the osteoarthritic hand and thumb. In *Green's Operative Hand Surgery*, 7th ed.; Wolfe, S.W., Hotchkiss, R.N., Pederson, W.C., Kozin, S.H., Cohen, M.S., Eds.; Elsevier Inc.: Philadelphia, PA, USA, 2017; pp. 345–372.
26. Xu, J.; Yong, F.C.; Lim, C.W.; Wong, K.J.H.; Teoh, L.C. The impact of various simulated arthrodesis angles of the proximal interphalangeal joint of the ring and middle finger on grip strength. *Arch. Orthop. Trauma Surg.* **2022**, *142*, 701–705. [CrossRef]
27. Dimitrova, P.; Reger, A.; Prommersberger, K.J.; van Schoonhoven, J.; Mühldorfer-Fodor, M. Effect of a single proximal interphalangeal Joint Fusion of the Index, Middle or Ring Finger on the Grip and Finger Force and Load Distribution in the Hand. *Handchir. Mikrochir. Plast. Chir.* **2022**, *54*, 409–417.
28. Millrose, M.; Zach, A.; Kim, S.; Güthoff, C.; Eisenschenk, A.; Vonderlind, H.C. Biomechanical comparison of the proximal interphalangeal joint arthrodesis using a compression wire. *Arch. Orthop. Trauma Surg.* **2019**, *139*, 577–581. [CrossRef]
29. Vonderlind, H.C.; Zach, A.; Eichenauer, F.; Kim, S.; Eisenschenk, A.; Millrose, M. Proximal interphalangeal joint arthrodesis using a compression wire: A comparative biomechanical study. *Hand. Surg. Rehabil.* **2019**, *38*, 307–311. [CrossRef]
30. Mittelmeier, W.; Lehner, S.; Gollwitzer, H.; Hauschild, M.; Werber, K.D.; Steinhauser, E. Comparing biomechanical investigations about different wiring techniques of finger joint arthrodesis. *Arch. Orthop. Trauma Surg.* **2005**, *125*, 145–152. [CrossRef]
31. Elfar, J.; Menorca, R.M.; Reed, J.D.; Stanbury, S. Composite bone models in orthopaedic surgery research and education. *J. Am. Acad. Orthop. Surg.* **2014**, *22*, 111–120.
32. Kovach, J.C.; Werner, F.W.; Palmer, A.K.; Greenkey, S.; Murphy, D.J. Biomechanical analysis of internal fixation techniques for proximal interphalangeal joint arthrodesis. *J. Hand. Surg. Am.* **1986**, *11*, 562–566. [CrossRef] [PubMed]
33. Mittelmeier, W.; Braun, C.; Hauschild, M.; Schäfer, R. Technique of finger arthrodesis by dorsal thread tension band fixation. Comparing biomechanical investigations. *Unfallchirurg* **1999**, *102*, 466–473. [CrossRef] [PubMed]
34. Lee, K.S.; Jung, M.C. Ergonomic evaluation of biomechanical hand function. *Saf. Health Work* **2015**, *6*, 9–17. [CrossRef] [PubMed]
35. Peters, S.E.; Jha, B.; Ross, M. Rehabilitation following surgery for flexor tendon injuries of the hand. *Cochrane Database Syst. Rev.* **2021**, *1*, CD012479. [CrossRef]

Journal of Personalized Medicine

Article

The Relationship between Lower Extremity Functional Performance and Balance after Anterior Cruciate Ligament Reconstruction: Results of Patients Treated with the Modified All-Inside Technique

Nizamettin Güzel [1], Ahmet Serhat Genç [1], Ali Kerim Yılmaz [2,]* and Lokman Kehribar [3]

1 Department of Orthopedics and Traumatology, Samsun Training and Research Hospital, 55090 Samsun, Türkiye
2 Departments of Recreation, Faculty of Yaşar Doğu Sport Sciences, Ondokuz Mayıs University, 55100 Samsun, Türkiye
3 Department of Orthopedics and Traumatology, Samsun University, 55090 Samsun, Türkiye
* Correspondence: akerim.yilmaz@omu.edu.tr; Tel.: +90-5424953737

Abstract: Background and Objectives: Anterior cruciate ligament (ACL) ruptures are common injuries, and ACL reconstruction (ACLR) is among the most common surgical procedures in sports surgery. Our research aims to compare the 6-month post-operative results of the modified all-inside (MAI) ACLR technique, single leg hop tests (SLHT), and Y balance tests applied in different directions on the operated and non-operated sides. Materials and Methods: A retrospective cohort of 22 male recreational athletes who underwent MAI ACLR techniques performed by the same surgeon were evaluated. The functional knee strengths of the participants on the operated and non-operated sides were evaluated with five different tests of SLHTs: single hop for distance (SH), triple hop for distance (TH), crossover triple hop for distance (CH), medial side triple hop for distance (MSTH), and medial rotation (90°) with hop for distance (MRH). Their dynamic balance was evaluated with the Y balance Test. Results: Compared to pre-operative levels, there was a significant improvement in the mean Lysholm, Tegner, and IKDC scores during the post-operative period ($p < 0.05$). There was a difference between SH, THD, CHD, MSTH, and MRH on the operated and non-operative sides ($p < 0.05$). There was no difference between Y balance scores on the operated and non-operative sides, and there were no differences between LSI scores resulting from SLHTs ($p > 0.05$). There were no significant relationships between YBT (composite scores) and SH, TH, CH, MSTH, and MRH distances in the healthy leg ($p > 0.05$), but a significant correlation with only CH in the ACL leg ($p < 0.05$). Conclusions: Our research shows that sixth-month post-operative SLHT findings were lower on the ACL side compared to the healthy side in patients tested with the MAI ACLR technique. However, when these scores are evaluated in terms of balance, it can be seen that both sides reveal similar findings. The similarity of LSIs in SLHTs applied in different directions, and balance scores of ACL and healthy sides revealed that the MAI technique is also an ACLR technique that can be used in athletes from a functional point of view.

Keywords: anterior cruciate ligament reconstruction; modified all inside; hop test; balance; return to sport

Citation: Güzel, N.; Genç, A.S.; Yılmaz, A.K.; Kehribar, L. The Relationship between Lower Extremity Functional Performance and Balance after Anterior Cruciate Ligament Reconstruction: Results of Patients Treated with the Modified All-Inside Technique. *J. Pers. Med.* **2023**, *13*, 466. https://doi.org/10.3390/jpm13030466

Academic Editors: Markus Gesslein and Eduard Kurz

Received: 6 February 2023
Revised: 27 February 2023
Accepted: 27 February 2023
Published: 2 March 2023

1. Introduction

The anterior cruciate ligament (ACL) is one of the most common orthopedic injury sites, and the mechanisms of injury have been associated with rotational movements, including sudden acceleration and deceleration, sudden changes in direction when the foot is in a fixed position, and external blows to the knee during sports [1,2]. ACL reconstruction (ACLR) involves the preparation of grafts and their restructuring through tunnels opened into the femur and tibia to restore ligament function in individuals with anterior cruciate ligament insufficiency and reduce the risk of future osteoarthritis and future degeneration of other soft tissues of the knee joint [1,3]. The most commonly used ACLR graft types today are the quadriceps tendon (QT), patellar tendon (PTG), and semitendinosus/gracilis (ST/G) tendons [4]. In addition, the so-called "all-inside" technique, with only ST and fourfold (4ST) grafting, has become widespread recently [5]. The all-inside technique has many advantages, but the modified all-inside (MAI) technique was developed by Mahirogullari et al. [6] due to the need to use a special drill while creating the socket, the problem of adjusting the socket depth, and the high cost of the all-inside technique [7,8]. In the MAI technique, the ST graft, prepared by folding the ST tendon four times, is fixed to the tibia and femur with a suspended suspension [9].

After ACLR, patients experience problems such as pain, impaired knee function, and especially quadriceps muscle weakness and atrophy [10,11]. It is known that there are proprioceptive receptors in the ACL, and these receptors form the reflex protective arc for stable muscle contractions [12,13]. When the ACL is ruptured, sensory stimuli from these mechanoreceptors in the knee to the central nervous system are absent, leading to a loss in the stabilization ability of the lower extremity, and a significant loss in proprioception occurs secondary to the instability of the knee [14]. Researchers have reported a loss of proprioceptive functions after ACL injuries [15–17]. For this reason, criteria such as functional single-leg hop tests (SLHT) and balance tests are of great importance and are widely used in decision-making for rehabilitation processes and prior to returning to sports (RTS) after ACLR [18]. Post-ACLR balance is a critical factor for sports performance and the activities of daily living [19]. Y balance test (YBT) is one of the dynamic balance tests frequently used in clinical and research settings to evaluate lower extremity function [20,21]. It is frequently used to evaluate dynamic balance disorders that may occur with lower extremity injuries, such as ACL injury and patellofemoral pain syndrome [22–25]. Lower SCT scores were associated with an increased risk of ACL disability, ACL deficiency, and ACLR [25–30].

SLHTs are widely used to evaluate athletes' functional status after ACLR, reveal limb asymmetries between the operated and non-operated sides, and follow the limb's developments [31–33]. The most significant limitation of conventionally applied SLHTs is that they mainly consist of forward movements [31,34]. Researchers have stated that multi-directional tests are important in addition to SLHTs performed in the forward direction, especially when returning to sports after injury [31,35]. Studies have also reported increased asymmetry rates in multi-directional tests compared to conventionally applied SLHTs [9,36]. For this reason, it can be said that multi-directional hop tests and traditional SLHTs after ACLR play a critical role in determining rehabilitation and return to sports time [9]. A study conducted after ACLR reported that decreased quadriceps strength on the operated side was associated with lower distances in hop tests [37]. Researchers examined limb asymmetries after ACLR and emphasized the importance of considering this factor [38,39]. These results suggest that it may be important to consider SLHTs to decide on RTS after ACLR [40].

Based on all this information, our study evaluates the results of SLHT applied both in the forward direction and in the medial and rotational directions, the resulting limb asymmetries and SCT results in patients with a 6-month post-operative ACLR history who underwent the MAI ACLR technique and compared them with the healthy sides. It was hypothesized that similar SLHT and SCT results would be obtained between the ACLR and non-operated sides of the research subjects.

2. Materials and Methods

2.1. Experimental Approach and Patients

This study used a retrospective institutional registry of patients treated for ACL tears at an academic medical center. We identified 22 active (recreational) male patients aged between 18 and 35 from a sequential case series from January 2020 to December 2021 with a diagnosis of isolated ACL tear who were treated with MAI (hamstring autograft) anterior cruciate ligament reconstruction (ACLR). Before diagnosing ACL rupture, patients presented with weakness, atrophy, pain, swelling, and slipping in the knee at the examination. Diagnosis of ACL tear was made primarily by the Lachman test and MR imaging. The number of subjects was determined by G*Power 3.1. According to the results, it would be sufficient to work with 19 patients (effect size r: 0.88, lower and upper critical p: 0.54, real power: 0.94). However, our study was completed with 22 subjects to examine the research findings. None of them was a professional sports patient, and participants consisted of those who participated in recreational activities only. Inclusion criteria were undergoing MAI (hamstring autograft) anterior cruciate ligament reconstruction (ACLR) and having no history of any other neuromuscular or musculoskeletal injury and contralateral knee surgery or injury. Exclusion criteria were non-compliance with the rehabilitation calendar, failure to complete patient-reported outcomes (PROs), SLHT, and YBTs, or exposure to various complications at the follow-up (Figure 1). The average time between the injury and the surgery was about two months. Detailed characteristics of the cohort are shown in Table 1. This study was approved by the local ethics committee with the number "SÜKAEK-2023-3/14", and all patients gave signed informed consent.

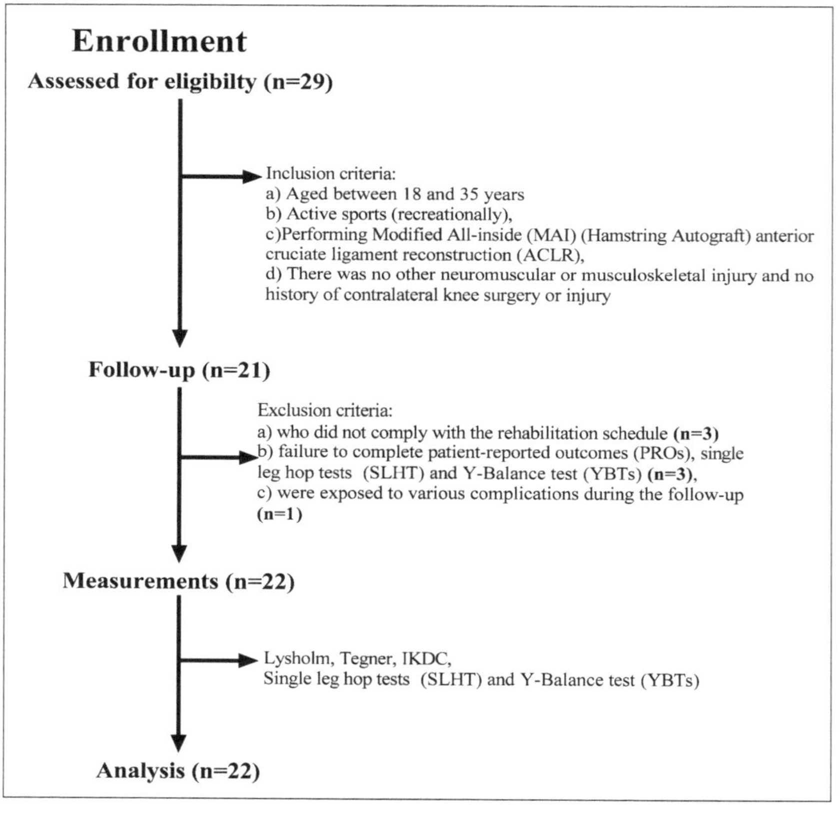

Figure 1. Flow chart diagram: description of the study population and inclusion and exclusion criteria.

Table 1. Characteristic and clinical variables of the study cohort (n = 22).

Variables	Value
Age (years)	28.32 ± 6.6
Height (cm)	176.45 ± 6.68
Weight (kg)	86 ± 12.09
BMI (kg/m^2)	27.55 ± 2.75
Time from ACLR to measurements (mon)	6.45 ± 0.67
Thigh length (cm)	
Operated side	47.27 ± 9.97
Non-operated side	47.58 ± 10.07
Lower limb length (cm)	
Operated side	90.5 ± 9.59
Non-operated side	90.73 ± 9.61
Dominant side (n (%))	
Right	17 (77.3)
Left	5 (22.7)
Surgical side (n (%))	
Right	13 (59.1)
Left	9 (40.9)

Data are presented as mean ± SD or *n* (%). ACLR, anterior cruciate ligament reconstruction; BMI, body mass index.

PROs (Lysholm, Tegner, International Knee Documentation Committee (IKDC)) (pre- and post-operative), single-leg hop test (SLHT) distances, and YBT scores obtained 6–7 months after ACLR were taken from routine clinical procedures. All reported evaluations were routinely part of the ongoing clinical procedure. The MAI ACLR technique was performed by a single surgeon specializing in soft tissue knee injuries, and data were collected from a single center. Post-surgical treatment was provided with a mandatory rehabilitation protocol as a routine clinical procedure. The rehabilitation program was the same for all patients and was followed in a controlled manner.

2.2. Surgical Treatment (Modified All-Inside Technique)

All patients underwent anatomical single-bundle ACL reconstruction using adjustable suspension fixation with a quadruple semitendinosus tendon autograft. The retrieved semitendinosus tendon (24–28 cm long) was quadrupled with adjustable loop cortical suspension fixation (Lift Loop External; Orthomed) on the tibial end and fixed loop button system (Femobutton; Orthomed) on the femoral side. The Lift Loop Outer fastening system consists of a 20 mm wide titanium button and two loops controlled by a knotless locking mechanism. The femobutton comprises a 10 mm wide titanium button and a continuous loop available in 6 lengths (15–40 mm). The anatomical femoral tunnel was carved from the anteromedial portal. First, a full bone tunnel was drilled over the guide pin using a 4.5 mm drill. The tunnel length was then measured, and a socket was opened using a router the same size as the graft, considering the 6 to 8 mm EndoButton 'flip' movement distance. A complete outside-in tibial tunnel was created at the anatomical footprint's central location. The graft was passed from the intra-articular space to the tibial and femoral tunnels, and the graft knee was stretched at 20° of flexion and flexed and extended 30 times. Graft tightness was then examined with a probe. Finally, the entire structure was re-tensioned on the tibial side, and an additional ligation was made on the adjustable suspension fixation device using a non-slip knot [6]. Figure 2 shows intraoperative fixation. Figure 3 presents a post-operative roentgenography showing the use of the materials.

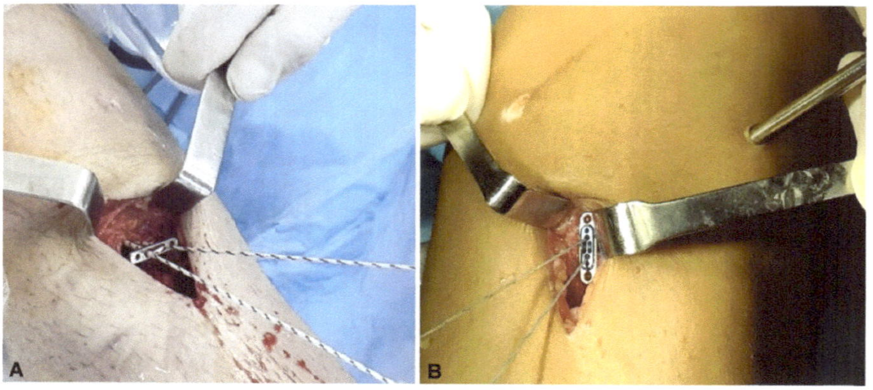

Figure 2. The graft was fixed to the tibial side with a suspended large button. (**A**) The graft was fixed to the tibial side with a Lift Loop. (**B**) The graft was fixed to the tibial side with an Ultrabutton plus Xtendobutton combination.

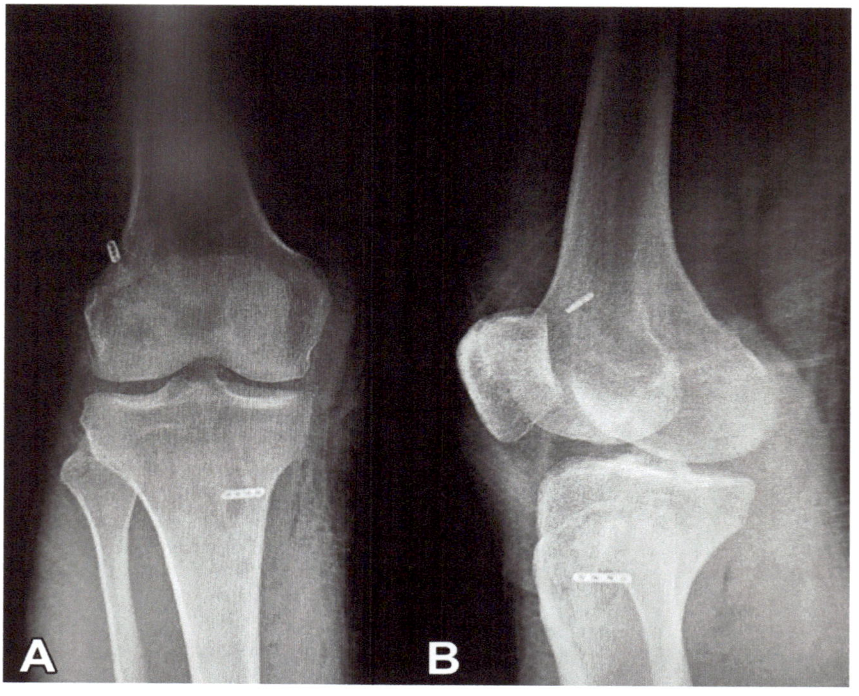

Figure 3. Post-operative roentgenography showing the use of the materials. (**A**) Post-operative roentgenography showing the use of Femobutton plus Lift Loop external. (**B**) Post-operative roentgenography shows three materials, Endobutton CL, Ultrabutton þ, and Xtendobutton in combination.

2.3. Procedures

All patients refrained from performing the high-intensity exercise the day before measurements to avoid the effects of cumulative muscle fatigue. The patients completed the YBT and SLHT functional tests consisting of anterior (ANT), posterolateral (PL), and posteromedial (PM) directions for balance and the single hop for distance (SH), triple hop for distance (TH), crossover triple hop for distance (CH), medial side triple hop for distance (MSTH), and medial rotation (90°) hop for distance (MRH) assessments. Trials were performed prior to measurements to familiarize patients with the tests fully. There were 10-min intervals between the YBT and SLHT, and all patients performed a 5-min warm-up before the tests. Figure 4 shows the setup of both tests.

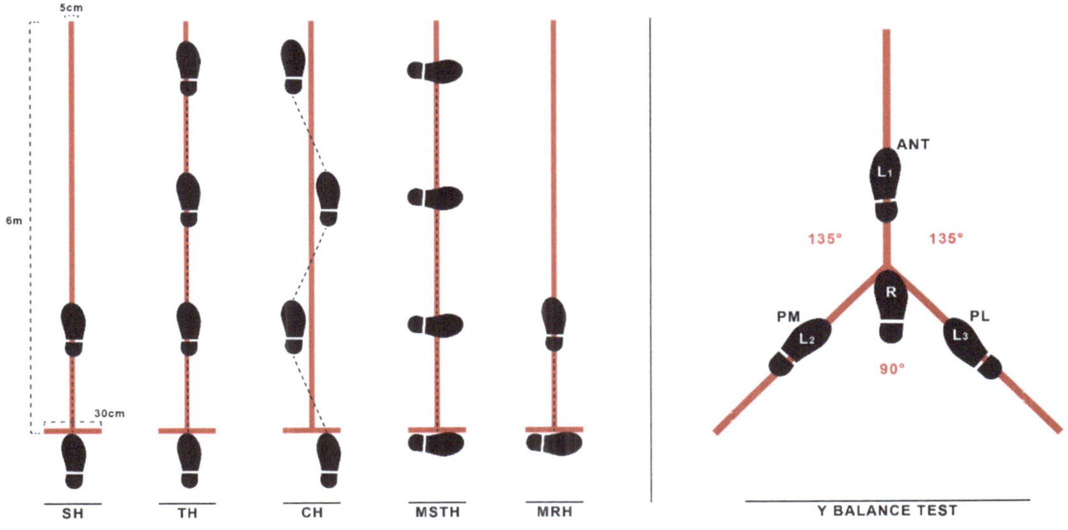

Figure 4. Demonstrations of SLHTs and Y balance tests. SH single leg hop for distance; TH triple hop for distance; CH crossover hop for distance; MSTH medial side triple hop for distance; MRH medial rotation (90°) hop for distance; ANT anterior; PL posterolateral; PM posteromedial.

2.3.1. Anthropometric Measures

Anthropometric information, such as height, weight, body mass index (BMI), thigh length, and lower limb length was collected. Thigh length was measured from the anterior superior iliac spine to the superolateral border of the patella. The lower limb length was measured bilaterally in supine position from the anterior superior iliac spine to the most distal aspect of the medial malleolus to the nearest half-centimeter [41].

2.3.2. Y Balance Test (YBT)

The YBT measures dynamic balance in the ANT, PL, and PM directions. The ground was marked in 3 different directions (⅄-shape) with a 15 cm wide tape; the angles between the anterior stripe and both posterior (posteromedial and posterolateral) stripes were 135 degrees, with 90 degrees between the two posterior stripes. Participants were instructed to place the stance foot in the marked position (zero-mark position) of the anterior, reach as far as possible in the reaching direction determined with the standing foot at the starting line, and then return the reaching foot to the starting position. ANT reach distance was measured from the toe of the stance foot in the starting position to the point reached, and PL and PM were measured from the heel of the stance foot to the point reached. The supervisor visually checked positions. Trials were canceled and repeated if the patient: 1) could not maintain his balance, and 2) failed to return the reach leg to the side of the stance leg after achieving maximal reach distance. Patients performed three trials for each direction (ANT, PL, and PM) on both legs with bare feet, and the best of the trials was recorded. Reach distances were normalized to anatomical leg length and expressed as a percentage (reach distance/limb length × 100). Composite normalized reach distance was computed for each leg as (ANT + PM + PL)/(3 × limb length) × 100 [20,27].

2.3.3. Single Leg Hop Tests (SLHT)

The SLHT included SH, TH, CH, MSTH, and MRH over a single line with maximum effort [42,43]. Kivlan et al. have demonstrated that SLHTs have good test–retest reliability in patients after ACL reconstruction [43].

The ground was marked with a 6 m long, 15 cm wide tape running perpendicular to the start and finish lines. Before starting the tests, an examiner explained and demonstrated the hop test procedures. Patients performed one practice trial (familiarizing practice) per limb, followed by three trials measured and recorded. Both limbs were tested, and subjects were not restricted in their arm movements. A 30-s rest interval was used after each attempt of the same hop tests. To minimize fatigue, a rest period of up to two minutes was given between different types of hop tests.

Patients began each trial behind a marked starting line. The best-recorded performance was the distance between the starting line and the heel of the foot where it landed at the end of the task. All hop tests were considered successful if the landings were stable. The post-jump landing was approved when it was under the participant's full control and on the tested limb. The test was repeated if the participant lost balance, touched the wall, or had additional bounces after landing [9]. The examiner visually checked this position.

2.4. Statistical Analysis

All statistical analyses were conducted using SPSS for Windows version 21 (IBM Inc., Chicago, Illinois). Descriptive data were presented as the mean, standard deviation (SD), median (Med), minimum (Min), or maximum (Max). The data were checked for normality using the Shapiro–Wilk test and were examined for kurtosis and skewness. Differences between PROs (Lysholm, IKDC, Tegner), YBT (ANT, PL, PM), and SLHT (SL, TH, CH, MSTH, MRH) distances were compared with paired samples *t*-test or Wilcoxon test. The ANOVA was also used to evaluate the limb symmetry index (LSI) differences. The Spearman's rank correlation test was used to calculate correlations between YBT (Composite score) and SLHT scores. Effect sizes were determined using the Cohen d, which defines 0.10, 0.30, and 0.50 as small, moderate, and large, respectively [44]. Significance was set at $p < 0.05$, with associated 95% confidence intervals.

3. Results

Table 1 summarizes the patient characteristics and clinical variables. There was no significant difference between pre-and post-operative in the thigh length and lower limb-length ($p > 0.05$), and pre- and post-operative Lysholm, IKDC, and Tegner scores are presented in Table 2. All PROs significantly improved after the operation (Lysholm, $p < 0.001$; IKDC, $p < 0.001$; Tegner, $p = 0.002$).

Table 2. Pre- and post-operative functional knee outcome scores (Lysholm, IKDC and Tegner).

	Pre-Operative		Post-Operative		95% CI	*p*-Value
	Mean ± SD	Med (Min-Max)	Mean ± SD	Med (Min-Max)		
Lysholm	74.09 ± 8.67	74 (58–92)	98.82 ± 2.56	100 (90–100)	−28.40 to −21.05	<0.001 [1]
IKDC	49.27 ± 8.88	48.5 (32–64)	90.73 ± 6.32	91.5 (80–100)	−45.72 to −37.18	<0.001 [2]
Tegner	6.36 ± 1.18	6 (5–9)	5.91 ± 1.31	6 (4–8)	0.229 to 0.681	0.002 [1]

[1] Wilcoxon test; [2] paired samples *t* test; IKDC, International Knee Documentation Committee; SD, standard deviation; Min, minimum; Max, maximum.

Table 3 compares anterior, posterolateral, posteromedial, and composite scores for the operated and healthy legs. Although the all-direction reach scores of the operated leg were lower than the healthy leg, it was no statistically significant differences ($p > 0.05$).

Table 3. YBT reach distances of the operated (ACL-Leg) and healthy leg (ANT, PL, and PM).

	ACL Leg Median (Min-Max)	Healthy Leg Median (Min-Max)	*p*-Value
Anterior (%)	72.54 (61.81–104.59)	74.68 (64.84–122.41)	0.123
Posterolateral (%)	67.98 (50.54–112.07)	74.57 (48.94–122.98)	0.592
Posteromedial (%)	79.88 (62.72–129.31)	82.16 (60.69–127.59)	0.884
Composite score (%)	72.96 (61.17–115.32)	76.76 (60.18–124.33)	0.291

Compares anterior, posterolateral, posteromedial, and composite scores for the operated and healthy legs. Although the all-direction reach scores of the operated leg were lower than the healthy leg, there were no statistically significant differences ($p > 0.05$). Min, minimum; max, maximum.

Significant differences were found in all tests considering the SLHT distances for the operated and healthy leg. The healthy leg had slightly higher scores compared to the operated leg in SH (114.14 ± 25.14 vs. 103.36 ± 29.21; $p = 0.006$), TH (392.05 ± 87.92 vs. 362.73 ± 88.78; $p = 0.007$), CH (340.59 ± 89.12 vs. 319.14 ± 75.53; $p = 0.038$), MSTH (310.27 ± 74.79 vs. 293.59 ± 70.16; $p = 0.049$), and MRH (106.41 ± 30.54 vs. 97.05 ± 34.04; $p = 0.030$), respectively (Figure 5).

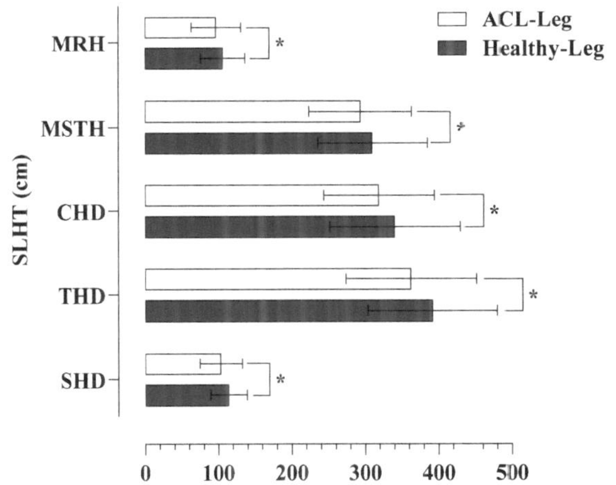

Figure 5. Differences in SLTH distances between operated (ACL-Leg) and healthy leg. * $p < 0.05$.

No statistically significant differences were found between the LSI scores ($p > 0.05$) (Table 4).

Table 4. Differences in the limb symmetry index (LSI).

Variables (%)	Mean ± SD	F	*p*-Value
SH	90.25 ± 15.06		
TH	92.73 ± 11.79		
CH	95.1 ± 14.3	0.677	0.610
MSTH	96.11 ± 15.98		
MRH	89.95 ± 20.69		

SD, standard deviation. SH, single hop for distance; TH triple hop for distance; CH crossover triple hop for distance; MSTH, medial side triple hop for distance; MRH medial rotation (90°) hop for distance.

There were no significant relationships between YBT (composite scores) and SH, TH, CH, MSTH, and MRH distances in the healthy leg ($p > 0.05$), but significant correlation with only CH in the ACL leg ($p < 0.013$; $r = -0.520$) (Figure 6).

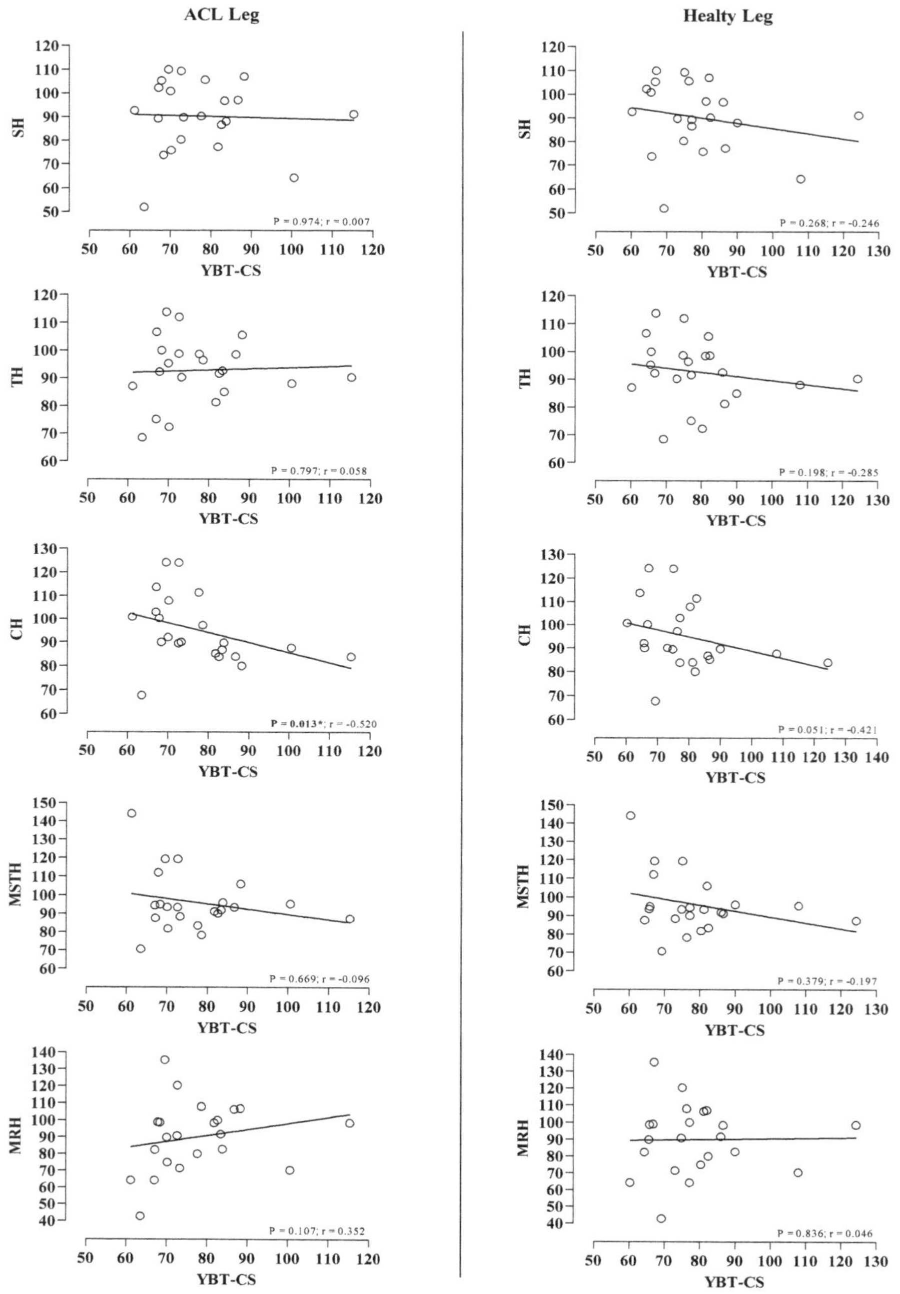

Figure 6. Correlation between YBT (composite score) and SLHT scores in operated (ACL-Leg) and healthy legs.

4. Discussion

When the major findings of our study were evaluated, the results were as follows; The MAI ACL reconstruction technique revealed an improvement between the pre-operative and post-operative findings in terms of Tegner, Lysholm, and IKDC scores, and the sixth-month post-operative findings of the patients' SH, TH, CH, MSTH, and MRH performances were lower in the ACLR sides compared to the healthy sides. When evaluated in terms of LSIs, it was determined that the resulting rates were within the normal norm ranges. Finally, when the patients were evaluated in balance scores, sixth-month post-operative findings were similar between ACLR and healthy knee.

Due to the damage to the proprioceptive receptors after ACL tears, the stimuli from the mechanoreceptors to the nervous system disappear, which may lead to a loss of balance [12–17]. In addition, the loss of muscle strength and atrophy [10,11] in the quadriceps and hamstring muscles after ACLR, especially after using hamstring autografts, led the researchers to evaluate these two critical parameters (lower extremity strength and balance). In a study evaluating balance and lower extremity function after ACLR with different graft types (PT, hamstring autograft (HT), and allograft (AG)), it was reported that the ACLR group had a decrease in anterior reach distance in balance in both the healthy and operated sides compared to the healthy group. It was reported that there was no significance in terms of balance in the operated and non-operated sides, and similar LSI rates were reported in normal norm ranges in SLHTs (SH, TH, CHD). When evaluated regarding graft types, it was reported that only the PT graft showed a decrease in PL reach distances compared to HT [45]. In their study, Bulow et al. [46] did not report any difference in ANT, PL, and PM parameters in the Y balance test on a healthy group diagnosed with an ACL tear but not undergoing reconstruction. In electromyography (EMG) studies, they stated that while reaching the ANT and PM directions during the Y balance test, the vastus medialis muscle exhibited voluntary contractions ranging from 66% to more than 100%, and in this case, the knee extensor strength is of great importance for movements that require dynamic balance [47,48]. In the literature, studies conducted on healthy subjects with an ACL injury and ACLR and healthy subjects used less reliable dynamic balance test protocols than clinical tests such as the Y balance test [23,25,49]. Researchers found different findings for ANT access in these studies between healthy controls, participants with an ACL injury, and participants with ACLR. The researchers reported the findings as follows: the performances and kinematic profiles of the subjects at ANT, PL, and PM distances were different and each test loaded the participant with different neuromuscular demands and postural control strategies [50–52]. When the results of our current research and similar studies were evaluated together, it was revealed that the graft types used in ACLR, the physical activity status of the subjects, and the anthropometric characteristics of the subjects were functional. However, different forms should be used in dynamic balance tests, such as Y-balance, which has a high margin of error. To eliminate inconsistencies in the results, the results obtained together with clinical and neuromuscular tests may be more valuable. This is especially important for a clear understanding of the results in patients with ACL tears or who have undergone ACLR in different graft types. Our current study, one of the first studies in which balance was evaluated using the MAI technique, did not reveal any difference between the operated and non-operated sides in three different directions (ANT, PL, and PM) regarding balance. Our current research revealed that the CS of the Y balance test was less than 80% of the participants, which is to be expected from recreational athletes.

Functionally, SLHTs used to determine the knee functions and RTS duration of the subjects after ACLR are tests that can be easily used to reveal the differences between the operated and non-operated sides. Researchers have reported that LSI rates calculated from SLHTs are 10%–15% in healthy limbs [53]. When the studies on patient groups who had ACLR were examined, a study was found in which SLHTs were evaluated in groups who underwent ACLR with the MAI technique. In this study, researchers found statistical significance only in the SH test in five different SLHT (SH, TH, CH, MRH, and MSTH) tests and reported that LSI rates were within normal norm ranges in all tests [9]. A study conducted with the conventional HT ACLR technique measured the SH of the subjects at the 6th and 12th months and found LSI rates of >85 and above in most of the group [54]. Another study in elite football players evaluated SH and CH strengths after post-operative ACLR and reported that LSI ratios did not reach >90, and the difference between the two limbs was not within the re-injury risk range [55]. As in balance, it is known that the ACLR side reveals very low strength rates compared to the intact side, especially between extensor forces and tests that give clinical results, such as isokinetic and functional tests such as SLHTs in knee strength [55]. The findings in our current research and in other studies suggest that the ACLR technique and rehabilitation processes applied especially for retrospective design studies result from their contribution to the healing processes. As a matter of fact, the recovery processes of sedentary and recreational athletes may vary compared to elite athletes, and the findings and literature findings support this situation. However, most researchers report that at least two SLHTs applied in different directions can reveal significant results for RTS [8,56]. When evaluated with these results, the 6-month post-operative findings obtained from five different SLHTs measured in patients in whom the MAI ACLR technique was applied in our study revealed significant differences in distances. However, the fact that the LSI ratios were in the range of <10% proves that the MAI technique is viable in terms of functional lower extremity performance. The researchers stated that although statistical differences were not observed in LSIs, there may be differences in clinical decision-making depending on the threshold values. As a matter of fact, in the study conducted, while a similarity between the limbs was found at the level of >90% in all of the subjects in traditional SLHTs, >90% similarity was found in only 68.8% of the subjects in the medial and rotational hop tests [8]. Similar findings were also emphasized in studies and review studies on healthy subjects [57,58]. Our current study observed the lowest LSI rates results, especially in the MRH test. All these results clearly show that using multi-directional SLHTs is of great importance in decision-making for post-ACLR RTS. Although the reason why multi-directional SLHTs cause high asymmetry is still debated, the researchers emphasized that limb biomechanics may vary according to the direction of jump and descent as well as dynamic postural stability, and that hip abduction and medial rotation and knee valgus movement may be restricted during descent after the multi-directional jump. [43,59–62].

Regarding ACL surgery and graft types, the "all-inside" technique has many advantages, such as using a single tendon and creating a socket instead of a tunnel on the tibial side. The disadvantages are evaluated as the need to use special burs on the tibial side, the creation of the socket, the limited margin of error in the adjustment of the socket depth, the need to place the graft from the portal and the higher cost. Considering the abovementioned disadvantages, the "MAI" technique was developed for ACLR. This technique has the advantages of the all-inside technique but does not have the disadvantages mentioned above.

Our research is the first study to evaluate the MAI technique's post-operative SLHT and balance results. In this respect, it is important to evaluate knee proprioception, stabilization, and strength together after an ACL injury. However, our study has a limitation in that it does not include tests involving fast turns and running, where ACL injuries often occur. In addition, the lack of other graft types and the absence of a control group are the main limitations of our study. In addition, one of the important limitations of our study is that the subjects did not feel any fatigue in the study. ACL injuries may occur due to voluntary or involuntary sudden movements at certain fatigue levels. It is important to plan studies by considering these limitations in future studies.

5. Conclusions

As a result, balance scores of patients who underwent MAI ACLR technique were similar on the operated and non-operated sides. In addition, although there were differences between both parties in SLHTs, this rate was not negatively reflected in LSIs. The fact that the CS scores were below 80% in the Y balance test and that the LSI scores were within the normal norm ranges reveals the outcome of the surgical technique and rehabilitation program applied, and that the non-operated sides were the least capable of the operated sides. Moreover, when supported by the literature, these findings are expected results for recreational athletes. In the future, prospective design studies on athletes, sedentary individuals, and recreational athletes, and the MAI technique's short-, medium-, and long-term findings will be evaluated. This will contribute to the literature investigating whether this technique is viable in subject groups. At the same time, clinical and biomechanical tests and functional tests will provide clear information about whether this technique harms the movement structure in patients to whom the MAI technique is applied.

Author Contributions: Conceptualization, A.S.G., N.G. and A.K.Y.; methodology, A.S.G., N.G. and A.K.Y.; writing—original draft preparation, A.S.G., N G. and A.K.Y.; writing—review and editing, N.G., A.K.Y. and L.K.; supervision, A.K.Y., A.S.G. and L.K. All authors have read and agreed to the published version of the manuscript.

Funding: This research received no external funding.

Institutional Review Board Statement: The study was conducted according to the guidelines of the Declaration of Helsinki and approved by the Clinical Research Ethics Committee at the Samsun Training and Research Hospital (Approval No: "SÜKAEK-2023-3/14").

Informed Consent Statement: Informed consent was obtained from all subjects involved in the study.

Acknowledgments: The authors thank all athletes who participated in this study.

Conflicts of Interest: The authors declare no conflict of interest.

References

1. Lehr, C.A.; Tan, C.S.; Ysseldyke, J. Alternative Schools: A Synthesis of State-Level Policy and Research. *Remedial Spec. Educ.* **2009**, *30*, 19–32. [CrossRef]
2. Wilk, K.E.; Romaniello, W.T.; Soscia, S.M.; Arrigo, C.A.; Andrews, J.R. The Relationship Between Subjective Knee Scores, Isokinetic Testing, and Functional Testing in the ACL-Reconstructed Knee. *J. Orthop. Sport. Phys. Ther.* **1994**, *20*, 60–73. [CrossRef] [PubMed]
3. Salzler, M.; Nwachukwu, B.U.; Rosas, S.; Nguyen, C.; Law, T.Y.; Eberle, T.; McCormick, F. State-of-the-Art Anterior Cruciate Ligament Tears: A Primer for Primary Care Physicians. *Phys. Sportsmed.* **2015**, *43*, 159–177. [CrossRef] [PubMed]
4. Middleton, K.K.; Hamilton, T.; Irrgang, J.J.; Karlsson, J.; Harner, C.D.; Fu, F.H. Anatomic Anterior Cruciate Ligament (ACL) Reconstruction: A Global Perspective. Part 1. *Knee Surg. Sport. Traumatol. Arthrosc.* **2014**, *22*, 1467–1482. [CrossRef] [PubMed]
5. Pasquali, M.; Plante, M.J.; Monchik, K.O.; Spenciner, D.B. A Comparison of Three Adjustable Cortical Button ACL Fixation Devices. *Knee Surg. Sport. Traumatol. Arthrosc.* **2017**, *25*, 1613–1616. [CrossRef]
6. Mahirogullari, M.; Kehribar, L.; Surucu, S.; Kayaalp, M.E.; Yilmaz, A.K.; Aydin, M. Comparative Results of Anterior Cruciate Ligament Reconstruction with Full Tibial Tunnel: Quadrupled Semitendinosus Suspensory Femoral and Tibial Fixation versus Quadrupled Semitendinosus and Gracilis Suspensory Femoral and Tibial Screw and Staple Fixation. *J. Knee Surg.* 2022; *ahead of print*.
7. Streich, N.A.; Reichenbacher, S.; Barié, A.; Buchner, M.; Schmitt, H. Long-Term Outcome of Anterior Cruciate Ligament Reconstruction with an Autologous Four-Strand Semitendinosus Tendon Autograft. *Int. Orthop.* **2013**, *37*, 279–284. [CrossRef]
8. Dingenen, B.; Truijen, J.; Bellemans, J.; Gokeler, A. Test–Retest Reliability and Discriminative Ability of Forward, Medial and Rotational Single-Leg Hop Tests. *Knee* **2019**, *26*, 978–987. [CrossRef]
9. Kehribar, L.; Yılmaz, A.K.; Karaduman, E.; Kabadayı, M.; Bostancı, Ö.; Sürücü, S.; Aydın, M.; Mahiroğulları, M. Post-Operative Results of ACL Reconstruction Techniques on Single-Leg Hop Tests in Athletes: Hamstring Autograft vs. Hamstring Grafts Fixed Using Adjustable Cortical Suspension in Both the Femur and Tibia. *Med. (B. Aires).* **2022**, *58*, 435. [CrossRef]
10. Wilk, K.E.; Arrigo, C.A. Rehabilitation Principles of the Anterior Cruciate Ligament Reconstructed Knee. *Clin. Sport. Med.* **2017**, *36*, 189–232. [CrossRef]
11. Wilk, K.E.; Macrina, L.C.; Cain, E.L.; Dugas, J.R.; Andrews, J.R. Recent Advances in the Rehabilitation of Anterior Cruciate Ligament Injuries. *J. Orthop. Sport. Phys. Ther.* **2012**, *42*, 153–171. [CrossRef]
12. Valeriani, M.; Restuccia, D.; Lazzaro, V.; Franceschi, F.; Fabbriciani, C.; Tonali, P. Clinical and Neurophysiological Abnormalities before and after Reconstruction of the Anterior Cruciate Ligament of the Knee. *Acta Neurol. Scand.* **1999**, *99*, 303–307. [CrossRef] [PubMed]
13. Reider, B.; Arcand, M.A.; Diehl, L.H.; Mroczek, K.; Abulencia, A.; Stroud, C.C.; Palm, M.; Gilbertson, J.; Staszak, P. Proprioception of the Knee before and after Anterior Cruciate Ligament Reconstruction. *Arthrosc. J. Arthrosc. Relat. Surg.* **2003**, *19*, 2–12. [CrossRef] [PubMed]

14. Hoffman, M.; Schrader, J.; Koceja, D. An Investigation of Postural Control in Post-operative Anterior Cruciate Ligament Reconstruction Patients. *J. Athl. Train.* **1999**, *34*, 130–136.
15. Barrack, R.L.; Skinner, H.B.; Buckley, S.L. Proprioception in the Anterior Cruciate Deficient Knee. *Am. J. Sport. Med.* **1989**, *17*, 1–6. [CrossRef] [PubMed]
16. Adachi, N.; Ochi, M.; Uchio, Y.; Iwasa, J.; Ryoke, K.; Kuriwaka, M. Mechanoreceptors in the Anterior Cruciate Ligament Contribute to the Joint Position Sense. *Acta Orthop. Scand.* **2002**, *73*, 330–334. [CrossRef] [PubMed]
17. Hopper, D.M.; Creagh, M.J.; Formby, P.A.; Goh, S.C.; Boyle, J.J.; Strauss, G.R. Functional Measurement of Knee Joint Position Sense after Anterior Cruciate Ligament Reconstruction11No Commercial Party Having a Direct Financial Interest in the Results of the Research Supporting This Article Has or Will Confer a Benefit upon the Author. *Arch. Phys. Med. Rehabil.* **2003**, *84*, 868–872. [CrossRef] [PubMed]
18. Clark, N.C. Functional Performance Testing Following Knee Ligament Injury. *Phys. Ther. Sport.* **2001**, *2*, 91–105. [CrossRef]
19. Gokhale, A.M.; Patel, G.R. Analysis of Variability in Tensile Ductility of a Semi-Solid Metal Cast A356 Al-Alloy. *Mater. Sci. Eng. A.* **2005**, *392*, 184–190. [CrossRef]
20. Plisky, P.J.; Gorman, P.P.; Butler, R.J.; Kiesel, K.B.; Underwood, F.B.; Elkins, B. The Reliability of an Instrumented Device for Measuring Components of the Star Excursion Balance Test. *N. Am. J. Sport. Phys. Ther.* **2009**, *4*, 92–99.
21. Shaffer, S.W.; Teyhen, D.S.; Lorenson, C.L.; Warren, R.L.; Koreerat, C.M.; Straseske, C.A.; Childs, J.D. Y-Balance Test: A Reliability Study Involving Multiple Raters. *Mil. Med.* **2013**, *178*, 1264–1270. [CrossRef]
22. Akhbari, B.; Salavati, M.; Mohammadi, F.; Safavi-Farokhi, Z. Intra- and Inter-Session Reliability of Static and Dynamic Postural Control in Participants with and without Patellofemoral Pain Syndrome. *Physiother. Canada.* **2015**, *67*, 248–253. [CrossRef] [PubMed]
23. Myers, H.; Christopherson, Z.; Butler, R.J. Relationship Between the Lower Quarter Y-Balance Test Scores and Isokinetic Strength Testing in Patients Status Post Acl Reconstruction. *Int. J. Sport. Phys. Ther.* **2018**, *13*, 152–159. [CrossRef]
24. Testerman, C.; Griend, R. Vander. Evaluation of Ankle Instability Using the Biodex Stability System. *Foot Ankle Int.* **1999**, *20*, 317–321. [CrossRef] [PubMed]
25. Herrington, L.; Hatcher, J.; Hatcher, A.; McNicholas, M. A Comparison of Star Excursion Balance Test Reach Distances between ACL Deficient Patients and Asymptomatic Controls. *Knee* **2009**, *16*, 149–152. [CrossRef]
26. Gribble, P.A.; Terada, M.; Beard, M.Q.; Kosik, K.B.; Lepley, A.S.; McCann, R.S.; Pietrosimone, B.G.; Thomas, A.C. Prediction of Lateral Ankle Sprains in Football Players Based on Clinical Tests and Body Mass Index. *Am. J. Sport. Med.* **2016**, *44*, 460–467. [CrossRef]
27. Plisky, P.J.; Rauh, M.J.; Kaminski, T.W.; Underwood, F.B. Star Excursion Balance Test as a Predictor of Lower Extremity Injury in High School Basketball Players. *J. Orthop. Sport. Phys. Ther.* **2006**, *36*, 911–919. [CrossRef]
28. Butler, R.J.; Lehr, M.E.; Fink, M.L.; Kiesel, K.B.; Plisky, P.J. Dynamic Balance Performance and Noncontact Lower Extremity Injury in College Football Players. *Sport. Health A Multidiscip Approach* **2013**, *5*, 417–422. [CrossRef]
29. Smith, C.A.; Chimera, N.J.; Warren, M. Association of Y Balance Test Reach Asymmetry and Injury in Division I Athletes. *Med. Sci. Sport. Exerc.* **2015**, *47*, 136–141. [CrossRef]
30. Delahunt, E.; Chawke, M.; Kelleher, J.; Murphy, K.; Prendiville, A.; Sweeny, L.; Patterson, M. Lower Limb Kinematics and Dynamic Postural Stability in Anterior Cruciate Ligament-Reconstructed Female Athletes. *J. Athl. Train.* **2013**, *48*, 172–185. [CrossRef]
31. Cournapeau, J.; Klouche, S.; Hardy, P. Material Costs of Anterior Cruciate Ligament Reconstruction with Hamstring Tendons by Two Different Techniques. *Orthop. Traumatol. Surg. Res.* **2013**, *99*, 196–201. [CrossRef]
32. Hegedus, E.J.; McDonough, S.; Bleakley, C.; Cook, C.E.; Baxter, G.D. Clinician-Friendly Lower Extremity Physical Performance Measures in Athletes: A Systematic Review of Measurement Properties and Correlation with Injury, Part 1. The Tests for Knee Function Including the Hop Tests. *Br. J. Sport. Med.* **2015**, *49*, 642–648. [CrossRef] [PubMed]
33. Logerstedt, D.S.; Snyder-Mackler, L.; Ritter, R.C.; Axe, M.J.; Godges, J.J. Knee Stability and Movement Coordination Impairments: Knee Ligament Sprain. *J. Orthop. Sport. Phys. Ther.* **2010**, *40*, A1–A37. [CrossRef] [PubMed]
34. Dingenen, B.; Gokeler, A. Optimization of the Return-to-Sport Paradigm after Anterior Cruciate Ligament Reconstruction: A Critical Step Back to Move Forward. *Sport. Med.* **2017**, *47*, 1487–1500. [CrossRef]
35. Harrison, J.J.; Yorgey, M.K.; Csiernik, A.J.; Vogler, J.H.; Games, K.E. Clinician-Friendly Physical Performance Tests for the Knee. *J. Athl. Train.* **2017**, *52*, 1068–1069. [CrossRef] [PubMed]
36. Gokeler, A.; Welling, W.; Zaffagnini, S.; Seil, R.; Padua, D. Development of a Test Battery to Enhance Safe Return to Sports after Anterior Cruciate Ligament Reconstruction. *Knee Surg. Sport. Traumatol. Arthrosc.* **2017**, *25*, 192–199. [CrossRef]
37. Schmitt, L.C.; Paterno, M.V.; Hewett, T.E. The Impact of Quadriceps Femoris Strength Asymmetry on Functional Performance at Return to Sport Following Anterior Cruciate Ligament Reconstruction. *J. Orthop. Sport. Phys. Ther.* **2012**, *42*, 750–759. [CrossRef]
38. Paterno, M.V.; Schmitt, L.C.; Ford, K.R.; Rauh, M.J.; Myer, G.D.; Huang, B.; Hewett, T.E. Biomechanical Measures during Landing and Postural Stability Predict Second Anterior Cruciate Ligament Injury after Anterior Cruciate Ligament Reconstruction and Return to Sport. *Am. J. Sport. Med.* **2010**, *38*, 1968–1978. [CrossRef]
39. Myer, G.D.; Schmitt, L.C.; Brent, J.L.; Ford, K.R.; Barber Foss, K.D.; Scherer, B.J.; Heidt, R.S.; Divine, J.G.; Hewett, T.E. Utilization of Modified NFL Combine Testing to Identify Functional Deficits in Athletes Following ACL Reconstruction. *J. Orthop. Sport. Phys. Ther.* **2011**, *41*, 377–387. [CrossRef]

40. Garrison, J.C.; Bothwell, J.M.; Wolf, G.; Aryal, S.; Thigpen, C.A. Y Balance Test^{TM} Anterior Reach Symmetry at Three Months Is Related to Single Leg Functional Performance at Time of Return to Sports Following Anterior Cruciate Ligament Reconstruction. *Int. J. Sport. Phys. Ther.* **2015**, *10*, 602–611.
41. Cook, G.; Plisky, P. Y Balance Test Manual. 2015. Available online: https://www.functionalmovement.com/files/Articles/660a_YBTOnlineManualv1.pdf (accessed on 20 January 2023).
42. Reid, A.; Birmingham, T.B.; Stratford, P.W.; Alcock, G.K.; Giffin, J.R. Hop Testing Provides a Reliable and Valid Outcome Measure during Rehabilitation after Anterior Cruciate Ligament Reconstruction. *Phys. Ther.* **2007**, *87*, 337–349. [CrossRef]
43. Kivlan, B.R.; Carcia, C.R.; Clemente, F.R.; Phelps, A.L.; Martin, R.L. Reliability and Validity of Functional Performance Tests in Dancers with Hip Dysfunction. *Int. J. Sport. Phys. Ther.* **2013**, *8*, 360–369.
44. Cohen, J. *Statistical Power Analysis for the Behavioral Sciences*, 2nd ed.; Laurence Erlbaum: Hillsdale, NJ, USA, 1988.
45. Clagg, S.; Paterno, M.V.; Hewett, T.E.; Schmitt, L.C. Performance on the Modified Star Excursion Balance Test at the Time of Return to Sport Following Anterior Cruciate Ligament Reconstruction. *J. Orthop. Sport. Phys. Ther.* **2015**, *45*, 444–452. [CrossRef] [PubMed]
46. Bulow, A.; Bellemare, A.; Anderson, J.E.; Leiter, J.R.S.; MacDonald, P.B.; Peeler, J.D. Lower Extremity Kinematics of the Y-Balance Test in Healthy and ACL Injured Adolescent Females. *Int. J. Sport. Phys. Ther.* **2021**, *16*, 381. [CrossRef] [PubMed]
47. Earl, J.E.; Hertel, J. Lower-Extremity Muscle Activation during the Star Excursion Balance Tests. *J. Sport Rehabil.* **2001**, *10*, 93–104. [CrossRef]
48. Norris, B.; Trudelle-Jackson, E. Hip- and Thigh-Muscle Activation During the Star Excursion Balance Test. *J. Sport Rehabil.* **2011**, *20*, 428–441. [CrossRef]
49. Zult, T.; Gokeler, A.; van Raay, J.J.A.M.; Brouwer, R.W.; Zijdewind, I.; Hortobágyi, T. An Anterior Cruciate Ligament Injury Does Not Affect the Neuromuscular Function of the Non-Injured Leg except for Dynamic Balance and Voluntary Quadriceps Activation. *Knee Surg. Sport. Traumatol. Arthrosc.* **2017**, *25*, 172–183. [CrossRef]
50. Fullam, K.; Caulfield, B.; Coughlan, G.F.; Delahunt, E. Kinematic Analysis of Selected Reach Directions of the Star Excursion Balance Test Compared With the Y-Balance Test. *J. Sport Rehabil.* **2014**, *23*, 27–35. [CrossRef]
51. Bulow, A.; Anderson, J.E.; Leiter, J.R.; MacDonald, P.B.; Peeler, J. The Modified Star Excursion Balance and Y-Balance Test Results Differ When Assessing Physically Active Healthy Adolescent Females. *Int. J. Sport. Phys. Ther.* **2019**, *14*, 192–203. [CrossRef]
52. Coughlan, G.F.; Fullam, K.; Delahunt, E.; Gissane, C.; Caulfield, B.M.; Sci, M. A Comparison Between Performance on Selected Directions of the Star Excursion Balance Test and the Y Balance Test. *J. Athl. Train.* **2012**, *47*, 366–371. [CrossRef]
53. Acar, K.; Yilmaz, A.K. Functional Dimorphism and Relationship between Different Lower Extremity Strength Tests in Young Elite Judokas. *Rev. Artes Marciales Asiáticas.* **2021**, *16*, 56–66. [CrossRef]
54. Barfod, K.W.; Feller, J.A.; Hartwig, T.; Devitt, B.M.; Webster, K.E. Knee Extensor Strength and Hop Test Performance Following Anterior Cruciate Ligament Reconstruction. *Knee* **2019**, *26*, 149–154. [CrossRef] [PubMed]
55. Herrington, L.; Ghulam, H.; Comfort, P. Quadriceps Strength and Functional Performance After Anterior Cruciate Ligament Reconstruction in Professional Soccer Players at Time of Return to Sport. *J. Strength Cond. Res.* **2021**, *35*, 769–775. [CrossRef]
56. Augustsson, J.; Thomeé, R.; Karlsson, J. Ability of a New Hop Test to Determine Functional Deficits after Anterior Cruciate Ligament Reconstruction. *Knee Surg. Sport. Traumatol. Arthrosc.* **2004**, *12*, 350–356. [CrossRef] [PubMed]
57. Hardesty, K.; Hegedus, E.J.; Ford, K.R.; Nguyen, A.-D.; Taylor, J.B. Determination of Clinically Relevant Differences in Frontal Plane Hop Tests in Women's Collegiate Basketball and Soccer Players. *Int. J. Sport. Phys. Ther.* **2017**, *12*, 182–189.
58. Abrams, G.D.; Harris, J.D.; Gupta, A.K.; McCormick, F.M.; Bush-Joseph, C.A.; Verma, N.N.; Cole, B.J.; Bach, B.R. Functional Performance Testing after Anterior Cruciate Ligament Reconstruction. *Orthop. J. Sport Med.* **2014**, *2*, 232596711351830. [CrossRef]
59. Sinsurin, K.; Vachalathiti, R.; Jalayondeja, W.; Limroongreungrat, W. Different Sagittal Angles and Moments of Lower Extremity Joints during Single-Leg Jump Landing among Various Directions in Basketball and Volleyball Athletes. *J. Phys. Ther. Sci.* **2013**, *25*, 1109–1113. [CrossRef]
60. Taylor, J.B.; Wright, A.A.; Dischiavi, S.L.; Townsend, M.A.; Marmon, A.R. Activity Demands during Multi-Directional Team Sports: A Systematic Review. *Sport. Med.* **2017**, *47*, 2533–2551. [CrossRef]
61. Liu, K.; Heise, G.D. The Effect of Jump-Landing Directions on Dynamic Stability. *J. Appl. Biomech.* **2013**, *29*, 634–638. [CrossRef]
62. Powers, C.M. The Influence of Abnormal Hip Mechanics on Knee Injury: A Biomechanical Perspective. *J. Orthop. Sport. Phys. Ther.* **2010**, *40*, 42–51. [CrossRef]

Journal of *Personalized Medicine*

MDPI

Article

Individual Concepts in Foot Surgery: A Comparison of Xenogeneic and Autologous Bone Grafts Used in Adults for Lateral Calcaneus-Lengthening Osteotomy According to Evans

Markus Siegel [1,*], Lisa Bode [1], Leonard Simon Brandenburg [2], Andreas Frodl [1], Hagen Schmal [1,3] and Jan Kühle [1]

1 Department of Orthopedic Surgery and Traumatology, Freiburg University Hospital, Albert-Ludwigs University Freiburg, Hugstetter Straße 55, 79106 Freiburg, Germany
2 Department of Oral and Maxillofacial Surgery, Freiburg University Hospital, Albert-Ludwigs University Freiburg, Hugstetter Straße 55, 79106 Freiburg, Germany
3 Department of Orthopedic Surgery, University Hospital Odense, Sdr. Boulevard 29, 5000 Odense, Denmark
* Correspondence: markus.siegel@uniklinik-freiburg.de

Abstract: Background: Xenogeneic bone grafts, when compared to autologous grafts, are supposed to provide structural benefits without donor site morbidity. To date, there have been divergent results in the use of xenogeneic grafts in foot surgery, primarily in pediatric patient cohorts. The present study examines the incorporation and maintenance of the achieved correction using autologous and xenogeneic bone grafts in adult patients with a six-month follow-up period. Material/Methods: In this retrospective study, 31 adult patients (43 feet in total) treated in our clinic by a lateral calcaneus-lengthening osteotomy, according to Evans, between 01/2006 and 12/2020 were included. The patients were assigned to study groups according to the use of xenogeneic or autologous bone grafts. The osseous incorporation following the criteria of Worth et al., correction maintenance by measuring the talo-navicular coverage angle (TNCA), the talo-first metatarsal angle (TFMA), the calcaneal pitch angle (PCA) and necessary revisions six months after surgery were extracted from the medical files retrospectively. Furthermore, the medical files were screened for the relevant comorbidities, nicotine abuse, BMI, sex and age. Results: In total, 27 autogenous (iliac crest) and 16 xenogeneic bone grafts of bovine origin were used. The evaluation of the radiographs at follow-up demonstrated that there was a mean incorporation rate of 96.3% for the autologous grafts and 57% for the patients treated with xenogeneic grafts ($p = 0.002$). Compared to the autologous group, xenogeneic grafts did not increase the loss of hindfoot alignment in the postoperative course, regardless of being incorporated or not. ΔTNCA, ΔTFMA and ΔPCA displayed no significant differences in both groups ($p = 0.45$, $p = 0.42$ and $p = 0.10$). Conclusion: Despite a significantly lower incorporation rate, the use of xenogeneic grafts was not accompanied with a greater risk of hindfoot alignment loss in the first six months after surgery. Early revision after a postoperative course of six months should not be motivated solely by the radiographic picture of incomplete osseous integration.

Keywords: calcaneus osteotomy; Evans; autologous; xenogeneic; bone graft

check for updates

Citation: Siegel, M.; Bode, L.; Brandenburg, L.S.; Frodl, A.; Schmal, H.; Kühle, J. Individual Concepts in Foot Surgery: A Comparison of Xenogeneic and Autologous Bone Grafts Used in Adults for Lateral Calcaneus-Lengthening Osteotomy According to Evans. *J. Pers. Med.* **2023**, *13*, 95. https://doi.org/10.3390/jpm13010095

Academic Editors: Markus Gesslein and Eduard Kurz

Received: 6 December 2022
Revised: 25 December 2022
Accepted: 26 December 2022
Published: 30 December 2022

1. Introduction

The symptomatic pes planovalgus, which results from stage II tibial tendon insufficiency (flexible deformity), is a foot deformity with a high prevalence, afflicting 19.5 to 26% of the world's population [1,2]. The lateral calcaneus-lengthening osteotomy, according to Evans, is a commonly performed procedure, which is conducted when conservative treatment regimens, such as foot orthotics and exercises, fail [3]. Initially, Evans performed this procedure using an autologous graft placed into an open-wedge osteotomy of the anterior calcaneal process [4]. Several authors modified his procedure [5,6].

Different grafts can be used for the lengthening of the lateral column: autologous, allogeneic, xenogeneic and synthetic grafts with or without a concomitant osteosynthesis were described. Various studies investigated the incorporation of different grafts with or without internal plate fixation, with varying results [7–12].

Allogeneic bone grafts were proven to be a good alternative to autologous bone grafts in pediatric patients, in terms of incorporation and hindfoot alignment loss [8,9]. Commonly, radiologic follow-up is performed six to twelve months after surgery. In this period, the bone graft is continuously invaded by migrating osteoblasts but not fully replaced by surrounding bone. Nevertheless, the graft was already exposed to weight-bearing mechanical loads [9].

In terms of correction maintenance and the incorporation rate, good results were also obtained in adult patients. A fundamental advantage of non-autologous grafts is the prevention of donor site morbidity [13,14]. Xenogeneic transplants have been used internationally as an alternative for reconstructive foot surgery for several years, including Evans' lengthening osteotomy [11]. The proposed benefits of xenogeneic grafts include improved structural integrity and greater mechanical stability [11,15]. The use of xenogeneic bone grafts also seems to be a good way to avoid complications in the context of donor site morbidity and to prevent high costs, at least in comparison to allogeneic grafts [3,16].

Despite these potential advantages, xenogeneic transplants are expected to become incorporated more slowly than allogeneic and autologous bone grafts [17].

Only a few studies investigated xenogeneic bone grafts regarding correction maintenance and incorporation in a lateral calcaneal-lengthening osteotomy. Mainly pediatric patients were examined in this context, yielding divergent results concerning both the incorporation rate and correction maintenance [11,18].

The purpose of this study was to investigate the rate of consolidation and the correction maintenance of autologous and xenogeneic bone grafts in adult patients with pes planovalgus, treated by calcaneus-lengthening osteotomy according to Evans.

We hypothesize that hindfoot alignment loss in the postoperative course at six months does not occur more frequently in patients treated with xenogeneic bone grafts, even if lower incorporation rates compared to autologous bone grafts were reported.

2. Materials and Methods

In this retrospective cohort study, a total of 43 feet from 31 different patients were analyzed. These patients were surgically treated by three different foot experts between 01/2006 and 12/2020 in our institution. In total, 23 right and 20 left feet were included, with 12 patients having undergone surgery on both feet in two separate operations.

The study was approved by the Institutional Review Board (the ethics committee of Freiburg University which approved this study, ID 422/18) and registered in the German Trials Register (DRKS00015478). All included patients agreed with study implementation voluntarily, in accordance with the Declaration of Helsinki.

Inclusion criteria were the diagnosis of a radiologically evident pes planovalgus with clinical symptoms (such as pain or general foot fatigue), failed conservative therapy, patient age of 18 to 75, surgery using the modified Evans' calcaneus-lengthening osteotomy with xenogeneic or autologous bone grafts and a documented postoperative course of therapy with radiographic imaging. A well-documented clinical examination and at least 1 preoperative and 2 postoperative radiographs, including the lateral and dorsoplantar view of the affected foot, were required.

Exclusion criteria were age below 18 or over 75, previous surgery of the investigated foot and other type of material used for the lengthening of the lateral column (such as allogeneic grafts).

The first postoperative radiographic imaging took place during the first two days after surgical treatment to document the postoperative result. This served as a reference point for the further postoperative course.

The primary endpoint of our study was the evaluation of the bony incorporation of the different bone grafts six months after surgery. Secondary endpoints were the postoperative maintenance of correction, the need for consecutive revision surgery and the influence of BMI, sex, age and pre-existing conditions (such as diabetes, vascular diseases and nicotine abuse) on the consolidation process.

A total of 102 records were identified and screened by searching the internal database. In accordance with the inclusion and exclusion criteria, 58 records were excluded. Consequently, 54 procedures could be considered for further analysis. Eight records had to be excluded because no bone graft had been implanted and a further three because the follow-up were not performed in our department. This left 43 procedures which could be included in the analysis. Figure 1 shows a flowchart of the selection process.

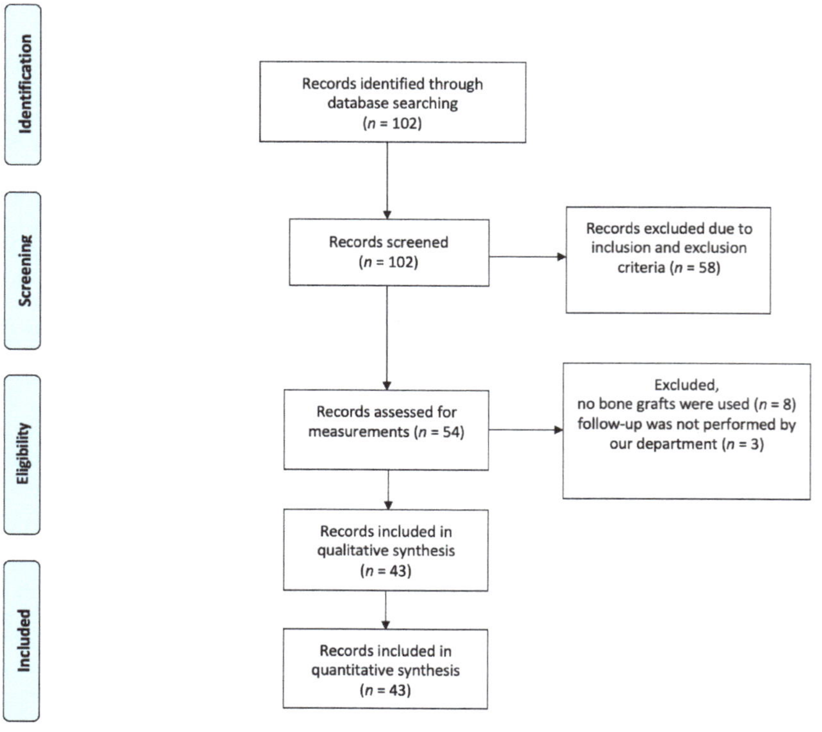

Figure 1. Flowchart of the selection process.

In our collective, 27 autologous (AG) and 16 xenogeneic (XG) bone grafts were used. The autologous grafts were taken from the iliac crest. Xenogeneic grafts were of bovine origin (Tutobone®, novomedics).

Surgical treatment with a modified lateral lengthening osteotomy according to Evans was performed by one of the surgical foot experts in our institution [6].

The procedure was performed via a lateral calcaneus approach above the peroneal tendons. In addition to the osseous lengthening of the lateral column, an Achilles tendon lengthening was performed according to Strayer, Evans or Young [19]. After surgical treatment with xenogeneic or autologous bone grafts, regular postoperative checks were performed. The majority was fixed by pressfit; in case of doubt regarding stability, plate fixation was added.

In our collective, postoperative immobilization was assured using a plaster cast and patient instruction for partial weight bearing for a total of six weeks.

As part of the six-month postoperative clinical follow-up, native weight-bearing radiological imaging was conducted. The radiologically determined incorporation of the bone graft was evaluated according to the criteria of Worth et al. (see Figure 2) by two independent, blinded investigators [20]. Grades III and IV were considered successful incorporation.

Grade 1	A lucent graft site indicating graft failure
Grade 2	An obvious lytic line at the interface of the graft to the surrounding host bone or a partially lytic graft site
Grade 3	No lytic line present, but obvious graft/host interface still to be remodeled
Grade 4	Difficulty in discerning graft site because of successful remodeling of graft trabeculi into that of the host bone

Figure 2. Adapted and reprinted grades of incorporation by Worth et al., 2005 [20].

Radiographic foot alignment evaluation was performed by measuring the talo-navicular coverage angle (TNCA), the talo-first metatarsal angle (TFMA) and the calcaneal pitch angle (PCA) in the dorsoplantar and lateral radiographic imaging, preoperatively and in the postoperative course.

For comparison, the patients were divided into 2 groups—the autologous bone graft group (AG) and the xenogeneic bone graft group (XG).

Statistical Analysis

The descriptive statistics are presented in mean values and standard deviations.

A statistical analysis of the demographic and clinical data was compared using 2-sample t-tests for normally distributed data and a Mann–Whitney rank sum test for not normally distributed data. Categorical variables were compared with Chi-square tests (e.g., gender). Differences in the incorporation rate between the different subgroups were analyzed via Fisher's exact test.

A logistic regression model was used to identify factors that increase the risk of nonunion.

Due to the retrospective design of our study, no power-analysis was performed before data collection. The calculation uses the IBM SPSS Statistics program, SPSS Statistics 24.0 (IBM Corp., New York, NY, USA). Statistical significance was defined as $p < 0.05$. All analyses were exploratory in nature. As a result, p-values and 95% confidence intervals were not corrected for multiple comparisons, and inferences drawn from them may not be reproducible.

3. Results

For the analysis, 43 complete data sets were included, of which 27 received autologous (AG) and 16 xenogeneic (XG) bone grafts.

3.1. Participants

The patients' characteristics are found in Table 1. The groups were homogenous in gender distribution and pre-existing conditions, such as vascular diseases (CAD) or diabetes. The AG patients were significantly older and had a lower BMI than the XG patients. The AG patients also showed an increased level of nicotine abuse.

3.2. Radiographic Measurements

The first postoperative radiographic imaging, which was performed on the first two postoperative days, showed a good surgical result in all the cases, with a statistically significant correction of the TMFA, TNCA and CPA in both groups ($p < 0.001$) as a starting point for the postoperative course.

Table 2 shows the radiographic angle measurements preoperatively, in the first follow-up and in the second follow-up after a mean time of 6.25 ± 1.27 months. Preoperatively, the groups differed significantly in the lateral TFMA.

Table 1. Patient characteristics including gender, age, body mass index (BMI) and health-related factors: nicotine, diabetes and coronary artery disease (CAD); * Chi-square test, # Mann–Whitney rank sum test, + Student's *t*-test.

Characteristics	All Patients	Autograft (*n* = 27)	Xenograft (*n* = 16)	*p*-Value
Gender				
Male sex (*n*)	25	16	9	0.548 *
Female sex (*n*)	18	11	7	
Age (yr)				
Mean	34.43 ± 17.25	40.1 ± 17.58	23.4 ± 9.48	<0.001 #
Range	18 to 75	20 to 75	18 to 58 years	
BMI (kg/sqm)				
Mean	25.51 ± 4.70	22.98 ± 3.9	26.99 ± 4.5	0.003 +
Range	17.6 to 36.4	17.6 to 31.0	21.19 to 36.4	
Nicotine (*n*)	9	8	1	0.038 #
Diabetes (*n*)	1	1	0	0.327 #
CAD (*n*)	2	2	0	0.161 #

Table 2. Angle measurements preoperatively and in the postoperative course of the TFMA, TNCA and CPA. * TFMA negative values equal plantar convexity, † time in months, ′ angles in degree.

Parameter	All Patients	Xenograft	Autograft	*p*-Value
Preoperative				
Mean Preop. Time (range) †	4.07 (0.1 to 16.0)	3.67 (0.2 to 11)	4.29 (0.1 to 16)	0.58
Mean TFMA * (range) ′	−13.96 (−37.7 to 14.5)	−9.94 (−32.6 to 0.8)	−16.34 (−37.7 to 14.5)	0.02
Mean TNCA (range)	18.24 (−12.3 to 46.6)	16.43 (2.1 to 27.4)	19.3 (−12.3 to 46.6)	0.32
Mean CPA (range)	16.65 (8.3 to 29.4)	15.44 (8.3 to 21.1)	17.37 (9.0 to 29.4)	0.18
1. Follow-up (FU-1)				
Mean TFMA (range)	−0.31 (−16.8 to 13.7)	−1.44 (−16.8 to 13.7)	0.34 (−11,2 to 9.3)	0.40
Mean TNCA (range)	1.57 (−20.7 to 18.4)	−2.66 (−20.7 to 9.8)	3.84 (−12.2 to 18.4)	0.04
Mean CPA (range)	24.81 (14.7 to 32.1)	23.85 (14.7 to 31.5)	25.37 (16.2 to 32.1)	0.36
2. Follow-up (FU-2)				
Mean FU Duration (range)	6.25 (4.5 to 9.5)	6.33 (4.5 to 8.0)	6.20 (4.5 to 9.5)	0.73
Mean TFMA (range)	−4.25 (−20.8 to 11.3)	−3.91 (−16.1 to 11.3)	−4.47 (−20.8 to 4.5)	0.78
Mean TNCA (range)	3.81 (−21.1 to 22.4)	0.64 (−21.2 to 13.5)	6.01 (−6.7 to 22.4)	0.06
Mean CPA (range)	20.83 (10.6 to 32.6)	19.14 (11.8 to 30.7)	21.96 (10.6 to 32.6)	0.06

In the first follow-up, the TNCA showed a significant difference between both groups. In comparison to the measurements of the second follow-up, there was no significant difference between the two groups looking at the TNCA and the CPA ($p = 0.06$). Comparing the correction maintenance as seen in Table 3, the loss of correction between the first and the second follow-ups showed no significant difference in both groups.

Table 3. Delta of the angles measured between follow-up (FU) 1 and follow-up 2.

Parameter	All Patients	Xenograft	Autograft	*p*-Value
Delta mean FU-1 to FU-2				
Delta mean TFMA (range)	−3.94 (−10.2 to 4)	−2.47 (−10.2 to 4)	−4.81 (−9.7 to 2.2)	0.42
Delta mean TNCA (range)	2.24 (−13.5 to 16.6)	3.30 (−2.8 to 16.6)	2.17 (−13.5 to 15.6)	0.45
Delta mean CPA (range)	−3.98 (−12.3 to 0.5)	−4.71 (−12.3 to 0.3)	−3.41 (−6.5 to 0.5)	0.10

3.3. Incorporation Rate

Within six months of the postoperative follow-up, the AG showed a successful incorporation of the transplanted autologous bone grafts in 26 transplants (96.3%). Only one transplant failed to demonstrate sufficient consolidation (3.7%). For the XG at a six-month follow-up, nine transplants (57%) were incorporated. Seven bone grafts (43%) showed grade I or II, according to Worth et al. [20]. Figure 3 gives an example of a successful incorporation (Figure 3A,B) and an example of a failed incorporation with a loss of correction (Figure 3C,D).

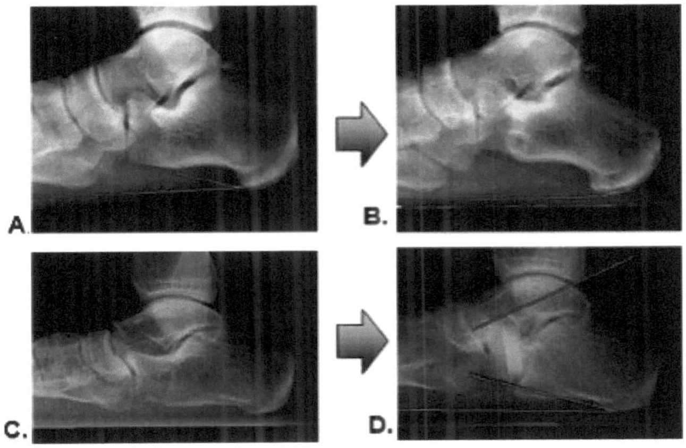

Figure 3. Radiographic example of pre- and postoperative images with successful incorporation (**A**,**B**), and loss of correction with incomplete incorporation (**C**,**D**).

When comparing the groups AG and XG in terms of the incorporation rate after six months, the autologous bone grafts were incorporated into the surrounding bone stock significantly ($p = 0.002$) more frequently than the xenogeneic bone grafts.

When comparing the hindfoot alignment of the cases with healed bone grafts (Worth III and IV) after 6 months for the XG group with grafts that had not been fully incorporated (Worth I and II), there are no significant differences in the absolute values or the FU-1 to the FU-2 delta of the respective angles (Delta TFMA: $p = 0.230$; Delta TNCA: $p = 0.412$; and Delta CPA: $p = 0.239$). Table 4 shows the radiographic angle measurements preoperatively in the first and second follow-ups for the XG, divided into cases with incorporated and non-incorporated bone grafts.

3.4. Analysis of Patient-Specific and Health-Related Factors

None of the investigated patient-specific and health-related factors were correlated significantly with the consolidation rate (age ($p = 0.107$), the physical constitution in terms of the BMI ($p = 0.536$), the gender of the patients ($p = 0.285$) or the pre-existing conditions, such as nicotine abuse ($p = 0.811$), CAD ($p = 0.999$) or diabetes mellitus ($p = 1.000$)).

Overall, within our follow-up period, one subject (6.2%) of the XG with insufficient incorporation, a significant loss of correction (TFMA−10.2 degree) and clinical symptoms (strong pain) in the postoperative course received revision surgery. Another revision surgery for the XG was conducted due to a symptomatic bony nonunion 21 weeks after surgery, without a significant loss of correction Both subjects underwent revision surgery with autologous bone grafts.

Within the first six months, we had a complication rate and a revision rate of 12.5% for the XG group, while there were neither relevant complications nor revisions in the AG group ($p = 0.16$).

Table 4. Angle measurements for the XG with incorporated bone grafts after the FU-2 and unincorporated bone grafts after the second FU-2 preoperatively, then in the postoperative course of the TFMA, TNCA and the CPA; ′ angles in degree.

Parameter	Incorporated Bone Grafts XG	Incomplete Incorporated Bone Grafts XG	p-Value
Preoperative			
Mean TFMA (range) ′	−14.80 (−32.4 to −5.7)	−16.33 (−26.6 to −5.6)	0.366
Mean TNCA (range)	14.72 (2.1 to 23.9)	18.63 (7.9 to 27.4)	0.154
Mean CPA (range)	16.13 (11.8 to 20.4)	14.54 (8.3 to 21.1)	0.234
1. Follow-up (FU-1)			
Mean TFMA (range)	5.31 (0.4 to 10.4)	−0.28 (−12.6 to 16.7)	0.117
Mean TNCA (range)	−3.60 (−20.7 to 5.1)	−1.40 (−18.4 to 9.8)	0.347
Mean CPA (range)	24.06 (14.7 to 31.2)	23.64 (17.0 to 31.5)	0.440
2. Follow-up (FU-2)			
Mean TFMA (range)	0.66 (−10.3 to 12.6)	−1.89 (−15.4 to 15.6)	0.285
Mean TNCA (range)	−0.03 (−10.9 to 9.2)	1.51 (−21.2 to 13.5)	0.384
Mean CPA (range)	18.81 (13.9 to 23.2)	19.56 (11.8 to 30.7)	0.387

4. Discussion

The main findings of the present study are that xenogeneic bone grafts have a significantly lower incorporation rate than autologous grafts six months after surgery. Nevertheless, there was no increased loss of hindfoot alignment in the XG compared to the AG. Moreover, within the XG group, there was no difference in the correction maintenance of the healed versus non-healed bone grafts. The factors sex, age, BMI, nicotine abuse, CAD and diabetes mellitus had no significant influence on the incorporation rate. Although there were two patients undergoing revision surgery in the XG and zero in the AG group, this difference did not reach significance.

The use of xenogeneic bone grafts in foot surgery has been the subject of intense discussion [11,17,18]. While the physical properties of xenogeneic grafts promise good surgical results, previous studies have shown delayed incorporation rates for xenogeneic transplants used in reconstructive foot surgery [15,17].

The results of our study show that, after a mean follow-up of 6.25 months, only about half (57%) of the xenogeneic grafts were incorporated.

Comparable to the chosen study design by Müller et al. (2016), we restricted our follow-up period from directly postoperative to 9.5 months [9]. During this time period, the transplanted graft is continuously invaded and mechanically loaded. In early postoperative follow-up care, a correction loss is closely associated with the failure or collapse of the bone graft rather than the secondary loss of correction caused by the degenerative processes of tendons or ligaments [9].

The presented incorporation rate is comparable to the results obtained by Rhodes et al. in a pediatric patient population. In this study, a prolonged incorporation rate for xenogeneic transplants of 29% after 6 months and 73% after 9 months in comparison to allogeneic bone grafts was detected. Though, their criterion for successful incorporation was Grade IV according to Worth et al. [11,20].

The poor integration of xenogeneic transplants in reconstructive foot surgery was also shown in other studies [17,18]. In 2013, Ledford et al. showed a similar incorporation rate to our findings by investigating 13 operations with xenogeneic grafts in pediatric patients with foot deformities. In their study, seven patients (53%) showed clinical symptoms of failure with a corresponding radiographically failed graft incorporation. This resulted in revision surgery in all seven cases, with allografts used for the revision [18]. However, the comparison of these studies to our results is lacking due to the pediatric patient population. In an evaluation by Shibuya et al., who investigated various reconstructive foot surgery procedures in which xenogeneic grafts were used, an incorporation rate of only 22.2% (two of nine) after six months was found for subjects treated with an Evans osteotomy [17]. In this study, an age distribution of 9 to 80 years was used for all the patients included.

For autologous bone grafts, the research has shown very good results for the incorporation rate within the first 3–6 months, with up to 100% successful incorporation [4,13,14]. The same can be seen in our population, with an incorporation rate of 96% after about six months.

In our study, a difference between the AG and XG groups showed a significantly younger average age, less tobacco use and a higher BMI for the XG group. While a younger age is considered a positive predisposing factor for faster bone healing, the opposite is true for tobacco use and a higher BMI [21]. However, there was no significant correlation between the incorporation rate and the investigated patient-specific and health-related factors extracted from the medical charts.

Despite the significantly poorer incorporation rate of xenogeneic bone grafts, there was a similar maintenance of the correction in the postoperative course of six months when considering the angles we chose for the evaluation (TFMA, TNCA and CPA) in comparing the two groups. For the evaluation of the hindfoot alignment, previous studies have shown that the TMFA is a sensitive marker for plantar pressure distribution under dynamic loads when compared to other angles [9,22]. Though the TFMA differed preoperatively in both groups, a comparable maintenance of the correction in both groups was shown in our study after surgical treatment in the six-month follow-up.

There are no significant differences between the correction maintenance of all the angles investigated between the incompletely incorporated bone grafts and the healed bone grafts within the XG group. This leads us to the assumption that, despite a slower incorporation of the xenogeneic when compared to the autologous grafts in adult Evans surgery, a loss of correction is not inevitable. Early revision after a postoperative course of six months should therefore not be motivated solely by the radiographic picture of incomplete osseous integration.

There are no statistical differences between the two groups when comparing the complication and revision rates. The possibility that delayed incorporation leads to further complications and possibly to revision surgery in the further postoperative course is certainly given but could not be ascertained in our study.

To point out the advantages of xenogeneic bone grafts, as with allogeneic or synthetic grafts, there are no donor site morbidities, such as wound infections, pain, neuralgia paresthetica or iliac wing fractures due to bone harvesting [14,23]. Though there were no reported donor site complications in this study, several studies have shown a major complication rate (between 0 and 11.5%) in harvesting autologous bone grafts from the iliac crest [16]. This suggests that the use of autologous bone grafts might be dependent on the surgical technique and experience. Using xenogeneic grafts could avoid such severe complications.

Another aspect to be considered in the choice of bone graft is the cost. As shown in several studies, another argument for xenogeneic bone grafts is that they are less expensive than allogeneic grafts [24].

To summarize the results of this study, we can state that autologous grafts appear to be superior to xenogeneic transplants in terms of the osseus incorporation within the first six months after surgery in adult patients treated by a calcaneus-lengthening osteotomy according to Evans. Despite this finding, there was no significant difference in the maintenance of correction in both groups. In our collective, there was no statistical relevance for the factors age, sex or BMI, nor for tobacco use, CAD or diabetes on the consolidation rate.

5. Limitations

The retrospective design prevented the standardization of the group size, as well as in terms of age and sex matching. Furthermore, we did not perform a power-analysis prior to the study as the group size was predefined by the retrospective design of the study. The rather short follow-up period of our study was due to our postoperative protocol, which scheduled a postoperative clinical examination and radiological control after six months in order to reveal complications (such as a loss of corrections and pseudarthrosis). One aim of this study was to reveal the causes of early failure. In this regard, the follow-up time of six months is considered appropriate for attributing complications predominantly to the osseous integration of the bone graft. The groups differed significantly in age, BMI and nicotine abuse. Nonetheless, there was no statistical influence of patient age or BMI with the incorporation rate in our study. Thus, the results of the AG and the XG groups could be compared, at least in the early postoperative course. Moreover, the AG patients were significantly older than the XG patients but displayed a significantly higher rate of consolidation after a six-month follow-up, which underlines the superiority of autologous bone grafts regarding incorporation. A larger collective might show divergent results. The study did not include patient-related outcome measures. Therefore, it cannot be evaluated whether the lacking bony integration was associated with more pain or other complaints.

6. Conclusions

Despite a significantly lower incorporation rate, xenogeneic grafts do not entail a higher risk of hindfoot alignment loss within the first six months after surgery. Thus, xenogeneic grafts prove to be an adequate alternative, without causing any donor site morbidities. Early revision after a postoperative course of six months should not be motivated solely by the radiographic picture of incomplete osseous integration.

Author Contributions: Conceptualization, M.S. and J.K.; Methodology, M.S.; Software, M.S.; Validation, M.S., J.K. and L.B.; Formal Analysis, M.S.; Investigation, M.S.; Resources, M.S.; Data Curation, M.S.; Writing—Original Draft Preparation, M.S.; Writing—Review and Editing, L.B., L.S.B., A.F., H.S. and J.K.; Visualization, M.S.; Supervision, J.K.; Project Administration, M.S.; Funding Acquisition, M.S. All authors have read and agreed to the published version of the manuscript.

Funding: This research received no external funding.

Institutional Review Board Statement: The ethics committee of Freiburg University approved this study, ID 422/18.

Informed Consent Statement: Written informed consent.

Data Availability Statement: All data are within the manuscript.

Acknowledgments: We acknowledge support by the Open Access Publication Fund of the University of Freiburg.

Conflicts of Interest: The authors declare no conflict of interest.

References

1. Deland, J.T.; Page, A.; Sung, I.-H.; O'Malley, M.J.; Inda, D.; Choung, S. Posterior Tibial Tendon Insufficiency Results at Different Stages. *HSS J.* **2006**, *2*, 157–160. [CrossRef] [PubMed]
2. Pita-Fernandez, S. Flat Foot in a Random Population and Its Impact on Quality of Life and Functionality. *J. Clin. Diagn. Res.* **2017**, *11*, LC22–LC27. [CrossRef] [PubMed]
3. Sheikh Taha, A.M.; Feldman, D.S. Painful Flexible Flatfoot. *Foot Ankle Clin.* **2015**, *20*, 693–704. [CrossRef]
4. Müller, M.A.; Frank, A.; Briel, M.; Valderrabano, V.; Vavken, P.; Entezari, V.; Mehrkens, A. Substitutes of Structural and Non-Structural Autologous Bone Grafts in Hindfoot Arthrodeses and Osteotomies: A Systematic Review. *BMC Musculoskelet. Disord.* **2013**, *14*, 59. [CrossRef]
5. Mosca, V.S. Calcaneal Lengthening for Valgus Deformity of the Hindfoot. Results in Children Who Had Severe, Symptomatic Flatfoot and Skewfoot. *J. Bone Jt. Surg. Am.* **1995**, *77*, 500–512. [CrossRef]
6. Zwipp, H.; Rammelt, S. Modifizierte Evans-Osteotomie zur operativen Behandlung des erworbenen Knick-Platt-Fußes. *Oper. Orthopädie Traumatol.* **2006**, *18*, 182–197. [CrossRef] [PubMed]
7. Hintermann, B.; Valderrabano, V.; Kundert, H.P. Lateral Column Lenghtening by Calcaneal Osteotomy Combined with Soft Tissue Reconstruction for Treatment of Severe Posterior Tibial Tendon Dysfunction. Technique and Preliminary Results. *Orthopäde* **1999**, *28*, 760–769. [CrossRef]
8. Kwak, Y.H.; Park, K.B.; Park, H.W.; Kim, H.W. Use of Allograft in Skeletally Immature Patients for Calcaneal Neck Lengthening Osteotomy. *Yonsei Med. J.* **2008**, *49*, 79. [CrossRef]
9. Müller, S.A.; Barg, A.; Vavken, P.; Valderrabano, V.; Müller, A.M. Autograft versus Sterilized Allograft for Lateral Calcaneal Lengthening Osteotomies: Comparison of 50 Patients. *Medicine* **2016**, *95*, e4343. [CrossRef]

10. Philbin, T.M.; Pokabla, C.; Berlet, G.C. Lateral Column Lengthening Using Allograft Interposition and Cervical Plate Fixation. *Foot Ankle Spec.* **2008**, *1*, 288–296. [CrossRef]
11. Rhodes, J.; Mansour, A.; Frickman, A.; Pritchard, B.; Flynn, K.; Pan, Z.; Chang, F.; Miller, N. Comparison of Allograft and Bovine Xenograft in Calcaneal Lengthening Osteotomy for Flatfoot Deformity in Cerebral Palsy. *J. Pediatr. Orthop.* **2017**, *37*, e202–e208. [CrossRef] [PubMed]
12. Templin, D.; Kerwyn, J.; Weiner, D. The Incorporation of Allogeneic and Autogenous Bone Graft in Healing of the Lateral Column Lengthening of the Calcaneus. *J. Foot Ankle Surg.* **2008**, *47*, 283–286. [CrossRef] [PubMed]
13. Dolan, C.M.; Henning, J.A.; Anderson, J.G.; Bohay, D.R.; Kornmesser, M.J.; Endres, T.J. Randomized Prospective Study Comparing Tri-Cortical Iliac Crest Autograft to Allograft in the Lateral Column Lengthening Component for Operative Correction of Adult Acquired Flatfoot Deformity. *Foot Ankle Int.* **2007**, *28*, 8–12. [CrossRef] [PubMed]
14. Grier, K.M.; Walling, A.K. The Use of Tricortical Autograft Versus Allograft in Lateral Column Lengthening for Adult Acquired Flatfoot Deformity: An Analysis of Union Rates and Complications. *Foot Ankle Int.* **2010**, *31*, 760–769. [CrossRef] [PubMed]
15. Kemper, N.; Davison, N.; Fitzpatrick, D.; Marshall, R.; Lin, A.; Mundy, K.; Cobb, R.R. Characterization of the Mechanical Properties of Bovine Cortical Bone Treated with a Novel Tissue Sterilization Process. *Cell Tissue Bank.* **2011**, *12*, 273–279. [CrossRef]
16. Pollock, R.; Alcelik, I.; Bhatia, C.; Chuter, G.; Lingutla, K.; Budithi, C.; Krishna, M. Donor Site Morbidity Following Iliac Crest Bone Harvesting for Cervical Fusion: A Comparison between Minimally Invasive and Open Techniques. *Eur. Spine J.* **2008**, *17*, 845–852. [CrossRef]
17. Shibuya, N.; Jupiter, D.C.; Clawson, L.D.; La Fontaine, J. Incorporation of Bovine-Based Structural Bone Grafts Used in Reconstructive Foot Surgery. *J. Foot Ankle Surg.* **2012**, *51*, 30–33. [CrossRef]
18. Ledford, C.K.; Nunley, J.A.; Viens, N.A.; Lark, R.K. Bovine Xenograft Failures in Pediatric Foot Reconstructive Surgery. *J. Pediatr. Orthop.* **2013**, *33*, 458–463. [CrossRef]
19. Strayer, L.M. Recession of the Gastrocnemius; An Operation to Relieve Spastic Contracture of the Calf Muscles. *J. Bone Jt. Surg. Am.* **1950**, *32*, 671–676. [CrossRef]
20. Worth, A.; Mucalo, M.; Horne, G.; Bruce, W.; Burbidge, H. The Evaluation of Processed Cancellous Bovine Bone as a Bone Graft Substitute. *Clin. Oral Implant. Res.* **2005**, *16*, 379–386. [CrossRef]
21. Gao, F.; Lv, T.-R.; Zhou, J.-C.; Qin, X.-D. Effects of Obesity on the Healing of Bone Fracture in Mice. *J. Orthop. Surg. Res.* **2018**, *13*, 145. [CrossRef] [PubMed]
22. Yalçin, N. Evaluation of the Medial Longitudinal Arch: A Comparison between the Dynamic Plantar Pressure Measurement System and Radiographic Analysis. *Acta Orthop. Traumatol. Turc.* **2010**, *44*, 241–245. [CrossRef] [PubMed]
23. Baumhauer, J.; Pinzur, M.S.; Donahue, R.; Beasley, W.; DiGiovanni, C. Site Selection and Pain Outcome After Autologous Bone Graft Harvest. *Foot Ankle Int.* **2014**, *35*, 104–107. [CrossRef] [PubMed]
24. Sonmez, M.M.; Armagan, R.; Ugurlar, M.; Eren, T. Allografts versus Equine Xenografts in Calcaneal Fracture Repair. *J. Foot Ankle Surg.* **2017**, *56*, 510–513. [CrossRef]

Journal of *Personalized Medicine*

Article

Standard Total Ankle Arthroplasty vs. Patient-Specific Instrumentation: A Comparative Study

Alberto Arceri [1], Pejman Abdi [1], Antonio Mazzotti [1,2,*], Simone Ottavio Zielli [1], Elena Artioli [1], Laura Langone [1], Federico Sgubbi [1] and Cesare Faldini [1,2]

1 1st Orthopaedics and Traumatologic Clinic, Istituto di Ricovero e Cura a Carattere Scientifico (IRCCS), Istituto Ortopedico Rizzoli, 40136 Bologna, Italy; alberto.arceri@ior.it (A.A.); pejman.abdi@ior.it (P.A.); simoneottavio.zielli@ior.it (S.O.Z.); elena.artioli@ior.it (E.A.); laura.langone@ior.it (L.L.); federico.sgubbi@ior.it (F.S.); cesare.faldini@ior.it (C.F.)

2 Department of Biomedical and Neuromotor Sciences (DIBINEM), Alma Mater Studiorum University of Bologna, 40123 Bologna, Italy

* Correspondence: antonio.mazzotti@ior.it

Abstract: Purpose: This retrospective study aims to compare surgical outcomes between two cohorts of patients who underwent total ankle arthroplasty (TAA) using either standard technique or patient-specific instrumentation (PSI). Methods: A consecutive series of patients who affected of end-staged ankle osteoarthritis were retrospectively assessed and divided into two groups based on TAA techniques: a TAA standard technique group and a TAA-using PSI group. The two groups were compared in terms of operative time, additional procedures, complications (neurovascular and wound problems, infection, loosening and osteolysis, revision and explantation rates, and perioperative fracture), clinical scores, and range of motion (ROM). Result: Fifty-one patients underwent standard TAA, while 13 patients underwent TAA with PSI. At 1-year follow-up, there were no significant differences in complication rates between the two groups ($p > 0.05$). AOFAS scores were similar, with the standard TAA group scoring 83.33 ± 7.55 and the PSI group scoring 82.92 ± 9.7 ($p = 0.870$). Likewise, the postoperative ROM did not differ significantly, with 15.12 ± 7.6 degrees for the standard TAA group and 16.05 ± 6.7 degrees for the PSI group ($p = 0.689$). However, the standard TAA group experienced significantly longer operative time (107.1 ± 22.1 min) compared to the PSI group (91.92 ± 22.9 min, $p = 0.032$). Additionally, the standard TAA group required more adjunctive procedures (29.7%) compared to the PSI group (7.7%, $p = 0.04$). Residual pain was also more frequently reported in the standard TAA group (62.7%) than in the PSI group (30.7%, $p = 0.038$). Conclusion: While both techniques resulted in comparable complication rates, clinical scores and ROM, the PSI group reported significantly shorter operative time and less residual pain, thus requiring fewer postoperative procedures.

Keywords: total ankle arthroplasty; patient-specific instrumentation; surgical outcomes; retrospective analysis; complications

check for updates

Citation: Arceri, A.; Abdi, P.; Mazzotti, A.; Zielli, S.O.; Artioli, E.; Langone, L.; Sgubbi, F.; Faldini, C. Standard Total Ankle Arthroplasty vs. Patient-Specific Instrumentation: A Comparative Study. *J. Pers. Med.* **2024**, *14*, 770. https://doi.org/10.3390/jpm14070770

Academic Editors: Markus Gesslein and Eduard Kurz

Received: 21 June 2024
Revised: 6 July 2024
Accepted: 13 July 2024
Published: 19 July 2024

1. Introduction

Total ankle arthroplasty (TAA) is an effective and increasingly viable surgical procedure for end-stage ankle osteoarthritis with an overlapping functional outcome [1–3]. In contrast to ankle fusion, TAA aims to restore motion and requires the restoration of joint anatomy and biomechanics. This speculation undoubtedly needs precision in bone cutting and prosthetic components positioning. To enhance these aspects, there have been numerous advancements in implant design and surgical techniques in recent years. One notable innovation is the introduction of patient-specific instrumentation (PSI) in TAA. This is a custom-made surgical device based on preoperative computed tomography (CT) scans and computer-generated three-dimensional (3D) models. PSI has the theoretical potential to improve accuracy in bone resection, implant positioning, and alignment and thus should result in satisfactory outcomes [4,5]. The potential advantages of using PSI also include a reduction in procedural complexity and a decrease in operative time [5], which could potentially lead to a lower incidence of complications.

Current literature is focused on the accuracy propriety of a PSI in TAA, demonstrating that a PSI system accurately and reliably assists in total ankle prosthesis implant positioning, except for the prediction of implant size [6,7]. However, limited comparative studies exist evaluating the efficacy of PSI versus standard techniques in TAA regarding operative time, additional procedures, complications (neurovascular and wound problems, infection, loosening and osteolysis, revision and explantation rates, and perioperative fracture), clinical scores, and range of motion (ROM). By examining these variables, it may be possible to better understand whether the theoretical advantages of PSI manifest in practical clinical benefits, thus offering valuable insights for orthopedic surgeons considering its adoption in TAA procedures.

Our hypothesis is that the enhanced precision of PSI, as demonstrated in various studies, should translate into improved intraoperative and postoperative clinical outcomes when compared to standard TAA techniques. We anticipate that PSI will reduce operative time, require fewer adjunctive procedures, and be associated with fewer complications while yielding superior clinical scores and ROM.

Therefore, the purpose of this study is to comprehensively compare the surgical outcomes between standard TAA and TAA performed using PSI.

2. Materials and Methods

2.1. Patient Selection

A retrospective study, using our institutional database of all patients who underwent standard TAA or TAA with PSI from January 2020 to December 2022, was conducted.

The protocol was approved by the ethical committee and all patients provided informed consent. The study was conducted in accordance with the ethical principles outlined in the Declaration of Helsinki and with the Guidelines for Good Clinical Practice. The study followed the STROBE statement and checklist for retrospective studies [8].

Specific inclusion criteria were used to select participants for our study to ensure the relevance and reliability of our findings. (1) Patients diagnosed with end-stage ankle osteoarthritis who underwent TAA were included in the study. Specifically, this included those who received either the standard TAA technique or TAA with PSI. (2) To ensure that differences in outcomes were more accurately attributed to the type of instrumentation used (standard vs. PSI) rather than to differences in surgical approach, we only included those who received either technique via an anterior surgical approach, which is routinely used at our institute. (3) Only patients who had a complete set of pre- and postoperative clinical and radiographic evaluations with a minimum follow-up of 1 year were included in the study.

Exclusion criteria were (1) revision arthroplasty or severe bone loss, (2) patients with previous ankle fusions, and (3) ankle with deformity in varus/valgus over 10 degrees and intra-articular incongruency. The 10-degree cut-off was established based on prior studies, which suggest that a deformity exceeding 10 degrees serves as an indication to perform additional bony procedures, such as osteotomies and/or adjacent joint fusions [9–11]. Furthermore, if the tibiotalar joint is incongruent, certain studies advocate balancing by medial/lateral malleolar lengthening osteotomy, lateral or medial ligament repair/reconstruction, or release, whereas if there is intra-articular congruency, ligament-balancing TAA may be feasible, usually obviating the need for additional local surgery [12–14].

These criteria were meticulously designed to create a homogeneous patient population that would allow for a robust and meaningful comparison between standard TAA and TAA with PSI.

2.2. Surgical Techniques

Two experienced foot and ankle orthopedic surgeons were responsible for performing the TAAs. One surgeon exclusively performed the TAAs using the standard technique, while the other focused exclusively on the PSI approach. Due to the nature of the procedures, the surgeons were necessarily aware of the technique they were using. However, efforts were made to ensure blinding in the evaluation process. Independent clinicians, specifically radiologists for radiographic assessments and orthopedic residents who administered clinical scores to the patients, conducted these assessments without knowledge of the surgical technique used. This approach was designed to facilitate an unbiased comparison between the outcomes of the two different surgical methods.

The standard TAA employed the FAR® mobile-bearing prosthesis (AdlerOrtho, Milan, Italy) was executed via an anterior approach. The talar and tibial resections were conducted using a conventional tibial alignment jig and cutting blocks, followed by the implantation of cementless talar and tibial components, along with an ultra-high molecular weight polyethylene (UHMWPE) meniscus.

For TAAs utilizing PSI, preoperative CT scans were used to generate a three-dimensional model of the ankle. This model facilitated the customization of cutting guides for the ankle. A virtual implantation was conducted to ensure precise component positioning, and the patient-specific cutting guides were then fabricated based on this simulation. The surgical procedure was performed using these guides through an anterior approach, preserving necessary osteophytes to maintain tactile feedback. Specifically, the distal tibial and talar resections were performed using the patient-specific guides [15,16] (Figure 1).

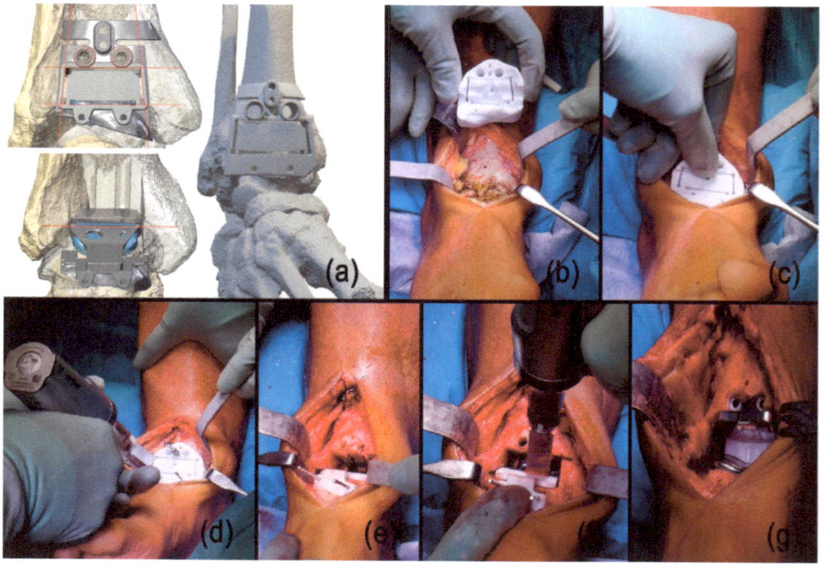

Figure 1. Preoperative CT scans were used to create a three-dimensional model of the ankle, which was used to customize cutting guides (**a**). The first cutting guide was precisely designed and manufactured to align with the anterior aspect of the joint, including the osteophytes (**b**,**c**). Using an oscillating bone saw, talar and tibial resections were carried out through the first cutting guide, which was stabilized with Kirschner wires (**d**). The second talar cutting guide was then positioned and stabilized using K-wires. This guide facilitated the posterior chamfer resection and included two holes for the preparation of the talar bone pegs (**e**,**f**). The final implant was correctly placed into the prepared bone resections (**g**).

In both the standard and PSI procedures, the ankle was immobilized in a plaster cast for three weeks, during which weight-bearing was prohibited. Following cast removal, progressive weight-bearing was permitted with the use of a walking boot, and flexion–extension exercises for the ankle were encouraged [17].

2.3. Outcomes Assessment

Patient and operative data were collected by manual chart review. Patient information included age and gender, while operative data included operative time, any subsequent postoperative additional bony procedures, revisions or explantations. Postoperative additional bony procedures include any adjunctive surgical procedures, as supramalleolar or calcaneal osteotomy, performed to correct residual deformities after TAA.

Complications, including the occurrence of neurovascular and wound problems, infection, loosening and osteolysis, revision and explantation rates, and perioperative fracture, were assessed at the clinical and radiological follow-up visits.

Residual pain was routinely assessed during the hospitalization and at follow-up visits using the Numerical Rating Scale (NRS), wherein pain intensity during daily activity is rated from 0–10, where 0 indicates no pain and 10 indicates the worst pain [18]. NRS at 1-year follow-up was collected as reference of comparison.

All patients were administered the American Orthopedic Foot and Ankle Society Ankle–Hindfoot (AOFAS-AH) Scale questionnaires both pre- and postoperatively [19]. Clinical outcomes were graded for pain (from 0 to 40 points), functional evaluation (from 0 to 50 points), and alignment (from 0 to 10 points). Thus, the overall score can range from 0 to 100 points, where 100 is a foot without pain, plantigrade, ankle–hindfoot well aligned, and no functional or gait issues. Preoperative scores were compared with ones collected at 1 year follow-up visits.

ROM was measured radiographically using an established method on postoperative weight-bearing lateral ankle radiographs, including maximal dorsiflexion, maximal plantarflexion, and total ROM [20].

2.4. Statistical Analysis

Data collection was carried out utilizing Microsoft Excel Version 365 (Microsoft Corporation, Redmond, Washington, DC, USA) for Windows 11, and statistical analysis was performed using the software Jamovi project (2022) version 2.3.

The primary parameters considered for the power analysis to determine the appropriate sample size were operative time and ROM. The alpha level was set at 0.05, with a study power of 80%. To estimate the effect size for operative time, the study by Saito et al. [21] was referenced, which reported an average operative time of 167 $\pm$ 42 min in the PSI group and 190 $\pm$ 46 min in the standard group. For the ROM effect size, data from study by Marks et al. [22] were utilized, which provided a mean postoperative ROM for standard TAA of 40.0 $\pm$ 12.3°. In the absence of a direct comparison for the PSI group, the significant average improvement in tibiotalar total ROM of 8.1° reported by Fletcher et al. [23] was used.

Continuous variables were reported as mean and standard deviation, while categorical variables were reported as percentage and frequency. Variables were analyzed for normality using the Shapiro Wilk test. Statistical analysis was performed using chi-square testing when appropriate to determine differences in categorical variables between groups, while differences in continuous variables were analyzed using Student's *t*-test.

Statistical significance was set at p value less than 0.05 per standard convention.

3. Results

3.1. Population

From a chart review, 122 patients who had undergone TAA were identified. Out of these, 64 patients fulfilled the selection criteria and were included in the study (Figure 2).

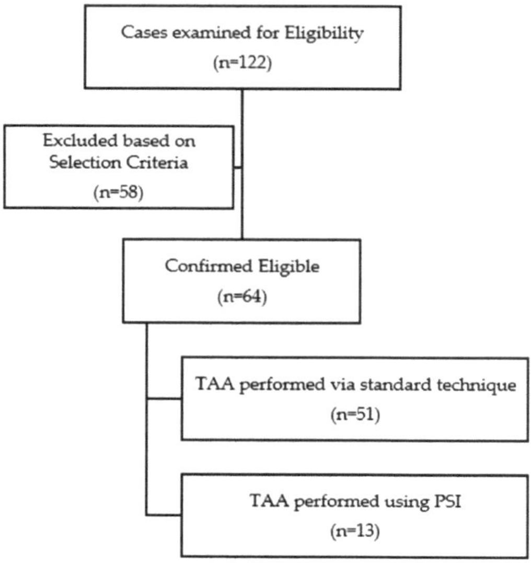

Figure 2. Flowchart of the study cohort.

The study involved 64 patients who underwent TAA. Out of these, 51 patients received standard TAA, while 13 patients underwent TAA with PSI. Their characteristics are shown in Table 1. No statistical differences were found between the two groups. The mean follow-up was 18.6 months.

Table 1. Patient's characteristics.

	Standard TAA	**PSI TAA**	***p*-Value**
Patients	51	13	
Age	59.9 ± 10.3	55.6 ± 12.1	0.344
Gender	17 M/34 F	5 M/8 F	
OA etiology	2 PRIM 1 RA 1 C 47 PT	1 PRIM 1 RA 11 PT	
OA onset	5.6 ± 4.7	6.7 ± 4.5	0.543

Abbreviations: M male; F female; PRIM primitive; RA rheumatoid arthritis; C clubfoot; PT post-traumatic; OA osteoarthritis.

3.2. TAA vs. TAA Using PSI

Comparison between the standard TAA and PSI groups revealed no statistically significant difference in complications rate, but only in operative time (Table 2). However, the standard TAA group showed a significantly higher frequency of additional procedures ($p < 0.05$) (Table 3 and Figure 3). Regarding clinical outcomes, the TAA with PSI group reported less residual pain compared to the standard TAA group ($p < 0.05$). No statistical differences in AOFAS score and ROM were found between the two groups (Table 3 and Figure 3).

Given our current sample sizes, the power analysis suggests that our study may be underpowered for detecting differences in outcomes, which may require cautious interpretation.

Table 2. Standard and PSI TAA comparison of operative time and complications.

	Standard TAA (51)	PSI (13)	*p*-Value	Cohen's d	95% CI for Cohen's d Lower	Upper
Operative time (min)	107.1 ± 22.1	91.92 ± 22.9	0.032 *	0.68	0.05	1.30
Additional procedure (%)	19 (29.7)	1 (7.7)	0.04 *			
Revision (%)	5 (9.8)	2 (15.3)	0.565			
Explantation (%)	3 (5.8)	1 (7.7)	0.81			
VNP complication (%)	2 (3.9)	2 (15.3)	0.127			
Wound problem (%)	10 (19.6)	4 (30.7)	0.385			
Infection (%)	3 (5.8)	1 (7.7)	0.81			
Fracture (%)	2 (3.9)	0 (0)	0.468			
Loosening (%)	5 (9.8)	2 (15.3)	0.565			
Residual pain (%)	32 (62.7)	4 (30.7)	0.038 *			

Abbreviations: TAA total ankle arthroplasty; PSI patient-specific instrumentation, VNP neurovascular, CI confidence intervals. * Indicates reaching statistical significance ($p < 0.05$).

Table 3. Standard and PSI TAA comparison of preoperative and postoperative AOFAS outcomes and postoperative ROM.

		Standard TAA (51)	PSI TAA (13)	*p*-Value	Cohen's d	95% CI for Cohen's d Lower	Upper
Preoperative	AOFAS tot	48.86 ± 12.92	44.69 ± 15.02	0.319	0.31	0.30	0.92
	Pain	10.98 ± 10.05	7.69 ± 10.13	0.297	0.33	−0.29	0.94
	Function	26.75 ± 9.44	25.77 ± 9.71	0.742	0.1	−0.50	0.71
	Alignment	11.14 ± 3.84	11.23 ± 3.63	0.937	−0.02	−0.63	0.58
Postoperative	AOFAS tot	83.33 ± 7.55	82.92 ± 9.7	0.87	0.51	−0.56	0.66
	Pain	24.90 ± 6.12	25.38 ± 6.6	0.804	−0.07	−0.69	0.53
	Function	46.1 ± 3.82	45.0 ± 4.53	0.376	0.27	−0.34	0.88
	Alignment	12.33 ± 3.37	12.54 ± 4.52	0.856	−0.05	−0.66	0.55
ROM postop	Total	15.12 ± 7.6	16.05 ± 6.7	0.689	−0.12	−0.73	0.48
	Dorsiflexion	6.03 ± 3.4	5.79 ± 3.5	0.822	0.07	−0.54	0.68
	Plantarflexion	9.07 ± 5.5	10.24 ± 4.4	0.48	−0.22	−0.83	0.39

Abbreviations: TAA total ankle arthroplasty; PSI patient-specific instrumentation, CI confidence intervals.

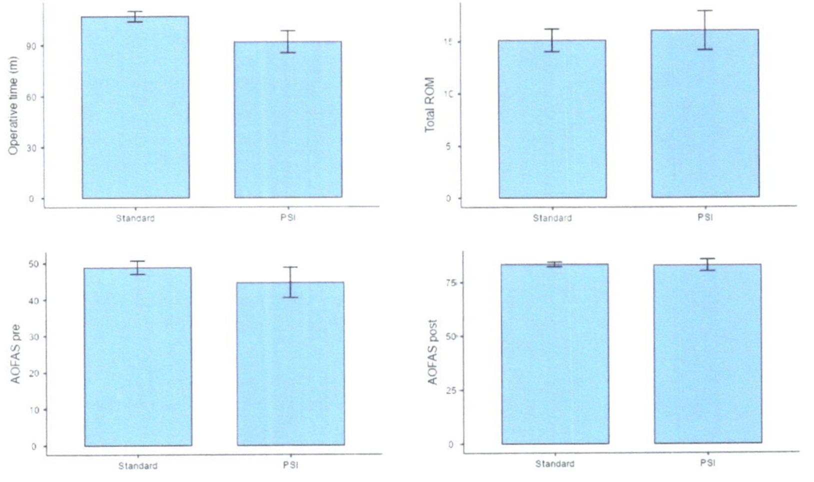

Figure 3. A bar graph showing the mean of the main outcomes for both groups, with standard error bars.

4. Discussion

The findings of this study suggest comparable overall outcomes between standard TAA and PSI in terms of operative time, complication rate, and clinical outcomes. However, the standard TAA group reported longer operative time and higher residual pain and required more postoperative surgical procedures compared to PSI group.

Significant reductions in operative time, tourniquet time, and radiation exposure in the PSI group were already reported in the majority of studies [4,5,21,24–27]. Our findings corroborate this significant difference (Table 2). However, it is important to note that this advantage may not always be as pronounced. Operative time is influenced by the surgeon's skill. Specifically, inexperienced surgeons may benefit more from PSI due to its tailored guidance, whereas experienced surgeons who are proficient in standard TAA techniques might achieve similar operative time without PSI.

To our knowledge, no studies have specifically investigated whether PSI reduces the need for concomitant or staged additional procedures to TAA. Theoretically, the use of software to plan tibial and talar cuts could improve the correction of ankle and hindfoot alignment in frontal plane deformities, when the deformity is less than 10 degrees and the tibiotalar joint remains congruent (thus not requiring additional ligamentous procedures). Implanting a prosthesis with cuts parallel to the existing deformity may leave a residual inframalleolar deformity. Attempting to correct the deformity intraoperatively with a standard TAA is challenging due to the loss of anatomical landmarks and the non-weight bearing status of the ankle, which may result in suboptimal postoperative alignment at follow-up.

Our case series demonstrated a significant reduction in the need for adjunctive procedures in the PSI group (Table 2). Therefore, PSI could play a critical role in the management of ankle deformities < 10 degrees in the frontal plane with a congruent tibiotalar joint. This approach may reduce the need for concurrent or staged corrective procedures after TAA.

The existing literature regarding complications associated with PSI TAA is limited. The studies that do report complications suggest that there is no significant difference in revision and complication rates between PSI and standard TAA [7,26,28]. However, standard TAA group reported higher residual pain compared to PSI group (Table 2). The observed results may be attributed to the precision of PSI in positioning and orientation the prosthetic components in the axial plane, which could potentially prevent component malrotation and reduce the risk of gutter impingement, a common source of residual pain post-TAA [29]. Nonetheless, it should be noted that some degree of residual pain has been reported even in the PSI group (Table 2).

In general, previous papers showed significant differences in favor of PSI on the walking/standing domain of the MOXFQ [27] and the total AOFAS score [26] at 12 months; however, in our case series, no differences in AOFAS score were found between the two groups (Table 3).

To date, comparative studies regarding ROM between standard TAA and PSI TAA have been lacking. This study represents the first attempt to compare ROM between these two groups, but no statistical differences in ROM was found (Table 3). The ROM values observed in our case series are consistent with those reported in the current literature [22,30,31]. However, when considering the prosthetic model most similar to the one used in our case series, the mean values for plantar flexion are lower than those reported in the literature [32].

Although the current results do not allow for definitive conclusions, it is crucial to recognize the important clinical implications of this type of study. If PSI demonstrates a reduction in operative time and complication rates, it could become the preferred method for performing TAA, especially in more complex cases or when performed by less experienced surgeons. Furthermore, should the investigation show that PSI leads to superior clinical scores and improved ROM, PSI could be established as a more effective technique for restoring ankle function and mobility compared to the standard approach. Overall, these findings could significantly influence surgical practices and decision-making processes, potentially leading to the broader adoption of PSI in ankle replacement surgeries and setting new benchmarks in the field.

The limitations of this study include its retrospective design and the small size of the sample, particularly for the PSI group. The limited sample size in our study is primarily attributable to the relatively low volume of this surgery and the stringent selection criteria employed. TAA and especially the PSI application are specialized procedures that are not performed in high frequencies at most medical centers. Moreover, our study applied rigorous inclusion and exclusion criteria to minimize confounding variables and enhance the reliability of our findings. The smaller sample size reflects both the rarity of the procedure and our commitment to maintaining high methodological standards. Despite the limited number of participants, our study offers valuable comparative insights into the standard TAA and PSI techniques. However, we acknowledge the potential limitations in statistical power. These factors underscore the need for continued investigation and larger-scale studies to fully elucidate the comparative benefits and risks of these approaches in TAA. To address these challenges, future studies should consider strategies such as multi-center collaboration to increase the sample size, the collection of more detailed baseline data to improve effect size estimation, and potentially, the use of advanced statistical methods that can handle small sample sizes and high variability. Additionally, ongoing data collection and analysis will help refine effect size estimates, thereby facilitating more accurate sample size calculations in subsequent research.

An additional limitation of our study is the absence of a preoperative assessment of ROM, which could be a crucial parameter that can significantly influence postoperative outcomes. Future studies should include this measure to provide a more comprehensive analysis.

Finally, to better evaluate the long-term clinical and functional outcomes, and to determine the sustained effectiveness of PSI in TAA, a follow-up period extending beyond one year is essential.

Future studies could explore computational simulations or in silico modeling [33,34] to further investigate this topic. Computational simulations could provide valuable insights into biomechanical aspects such as joint kinematics, stress distribution on implants, and functional outcomes. By simulating different surgical scenarios and patient-specific anatomies, researchers can potentially predict the impact of PSI on surgical precision, implant positioning, and overall ankle joint function. Such studies could complement clinical findings, offering a deeper understanding of how PSI influences surgical outcomes and guiding refinements in surgical techniques and implant designs.

5. Conclusions

While both techniques resulted in comparable complication rates, the PSI group reported significantly shorter operative time and less residual pain, thus requiring fewer postoperative procedures. Although the current results do not allow definitive conclusions to be drawn, these preliminary findings suggest that the PSI technique may offer potential advantages due to the improved reliability and accuracy of prosthetic implant positioning provided by the guided approach of PSI. This may be particularly beneficial in more complex cases or when performed by less experienced surgeons.

Further research efforts with larger sample sizes and rigorous study designs are warranted to provide clearer insights into the comparative efficacy between standard TAA and TAA with PSI.

Author Contributions: Conceptualization, A.A. and A.M.; methodology, P.A.; software, S.O.Z. and F.S.; formal analysis, S.O.Z. and E.A.; investigation, P.A. and A.A.; data curation, P.A. and L.L.; writing—original draft preparation, A.A.; writing—review and editing, A.M.; supervision, C.F. All authors have read and agreed to the published version of the manuscript.

Funding: This publication has been financed with the support by Ministero della Salute 5 × 1000 fund raising campaign 5xmille 2020 contributions to the IRCCS, Istituto Ortopedico Rizzoli.

Institutional Review Board Statement: Ethical approval for this study was obtained from Institutional Review Board CE AVEC 370/2022/Oss/IOR (Rizzoli Orthopaedic Institute, Bologna, Italy)—protocol number: 0008996—approval date 31 May 2022. The study was conducted according to the guidelines of the Declaration of Helsinki. All patients given a written consent to participate.

Informed Consent Statement: All the patients provided their written informed consent at admission on data collection.

Data Availability Statement: Some or all data and models that support the findings of this study are available from the corresponding author upon reasonable request.

Conflicts of Interest: The authors declare no conflict of interest.

References

1. Dalat, F.; Trouillet, F.; Fessy, M.H.; Bourdin, M.; Besse, J.-L. Comparison of quality of life following total ankle arthroplasty and ankle arthrodesis: Retrospective study of 54 cases. *Orthop. Traumatol. Surg. Res.* **2014**, *100*, 761–766. [CrossRef] [PubMed]
2. Arceri, A.; Mazzotti, A.; Zielli, S.; Bonelli, S.; Artioli, E.; Abdi, P.; Langone, L.; Faldini, C. Return to sports after total ankle arthroplasty: A systematic review and meta-analysis. *J. Orthop.* **2023**, *44*, 57–65. [CrossRef] [PubMed]
3. Cho, B.-K.; An, M.-Y.; Ahn, B.-H. Comparison of Clinical Outcomes after Total Ankle Arthroplasty between End-Stage Osteoarthritis and Rheumatoid Arthritis. *Foot Ankle Int.* **2021**, *42*, 589–597. [CrossRef] [PubMed]
4. Mazzotti, A.; Arceri, A.; Zielli, S.; Bonelli, S.; Viglione, V.; Faldini, C. Patient-specific instrumentation in total ankle arthroplasty. *World J. Orthop.* **2022**, *13*, 230–237. [CrossRef] [PubMed]
5. Wang, Q.; Zhang, N.; Guo, W.; Wang, W.; Zhang, Q. Patient-specific instrumentation (PSI) in total ankle arthroplasty: A systematic review. *Int. Orthop.* **2021**, *45*, 2445–2452. [CrossRef] [PubMed]
6. April, P.-M.; Locke, E.; Champagne, P.-H.; Angers, M.; Martinez-Gomez, A.P.; Seidel, A.; Jibri, Z.; Lalonde, K.-A.; Meulenkamp, B. Accuracy of a Patient-Specific Total Ankle Arthroplasty Instrumentation. *Foot Ankle Int.* **2023**, *44*, 1150–1157. [CrossRef] [PubMed]
7. Heisler, L.; Vach, W.; Katz, G.; Egelhof, T.; Knupp, M. Patient-Specific Instrumentation vs Standard Referencing in Total Ankle Arthroplasty: A Comparison of the Radiologic Outcome. *Foot Ankle Int.* **2022**, *43*, 741–749. [CrossRef] [PubMed]
8. von Elm, E.; Altman, D.G.; Egger, M.; Pocock, S.J.; Gøtzsche, P.C.; Vandenbroucke, J.P. STROBE Initiative the Strengthening the Reporting of Observational Studies in Epidemiology (STROBE) statement: Guidelines for reporting observational studies. *J. Clin. Epidemiol.* **2008**, *61*, 344–349. [CrossRef] [PubMed]
9. Haskell, A.; Mann, R.A. Ankle arthroplasty with preoperative coronal plane deformity: Short-term results. *Clin. Orthop. Relat. Res.* **2004**, *424*, 98–103. [CrossRef]
10. Kim, B.S.; Choi, W.J.; Kim, Y.S.; Lee, J.W. Total ankle replacement in moderate to severe varus deformity of the ankle. *J. Bone Jt. Surg. Br.* **2009**, *91*, 1183–1190. [CrossRef]
11. Trincat, S.; Kouyoumdjian, P.; Asencio, G. Total ankle arthroplasty and coronal plane deformities. *Orthop. Traumatol. Surg. Res.* **2012**, *98*, 75–84. [CrossRef]
12. Plaass, C.; Louwerens, J.W.; Claassen, L.; Ettinger, S.; Yao, D.; Lerch, M.; Stukenborg-Colsman, C.; Donken, C. Treatment concepts for pes planovalgus with concomitant changes of the ankle joint. *Orthopäde* **2020**, *49*, 991–999. [CrossRef] [PubMed]
13. What is Feasible in Total Ankle Arthroplasty? In *Total Ankle Arthroplasty*; Springer: Vienna, Austria, 2005; pp. 135–162, ISBN 978-3-211-21252-3.
14. Dodd, A.; Daniels, T.R. Total Ankle Replacement in the Presence of Talar Varus or Valgus Deformities. *Foot Ankle Clin.* **2017**, *22*, 277–300. [CrossRef]
15. Faldini, C.; Mazzotti, A.; Langone, L.; Arceri, A.; Bonelli, S.; Zielli, S.O.; Artioli, E. Custom-made total ankle arthroplasty with patient-specific instrumentation for severe bone loss conditions: A case series. *Eur. J. Orthop. Surg. Traumatol.* **2024**, *34*, 379–387. [CrossRef]
16. Faldini, C.; Mazzotti, A.; Belvedere, C.; Durastanti, G.; Panciera, A.; Geraci, G.; Leardini, A. A new ligament-compatible patient-specific 3D-printed implant and instrumentation for total ankle arthroplasty: From biomechanical studies to clinical cases. *J. Orthop. Traumatol.* **2020**, *21*, 16. [CrossRef]
17. Mazzotti, A.; Viglione, V.; Gerardi, S.; Bonelli, S.; Zielli, S.; Geraci, G.; Faldini, C. Post-operative management after total ankle arthroplasty: A systematic review of the literature. *Foot Ankle Surg.* **2022**, *28*, 535–542. [CrossRef] [PubMed]
18. Thong, I.S.K.; Jensen, M.P.; Miró, J.; Tan, G. The validity of pain intensity measures: What do the NRS, VAS, VRS, and FPS-R measure? *Scand. J. Pain.* **2018**, *18*, 99–107. [CrossRef]
19. Leigheb, M.; Janicka, P.; Andorno, S.; Marcuzzi, A.; Magnani, C.; Grassi, F. Italian translation, cultural adaptation and validation of the "American Orthopaedic Foot and Ankle Society's (AOFAS) ankle-hindfoot scale". *Acta Biomed.* **2016**, *87*, 38–45. [PubMed]
20. Coetzee, J.C.; Castro, M.D. Accurate measurement of ankle range of motion after total ankle arthroplasty. *Clin. Orthop. Relat. Res.* **2004**, *424*, 27–31. [CrossRef]
21. Saito, G.H.; Sanders, A.E.; O'Malley, M.J.; Deland, J.T.; Ellis, S.J.; Demetracopoulos, C.A. Accuracy of patient-specific instrumentation in total ankle arthroplasty: A comparative study. *Foot Ankle Surg.* **2019**, *25*, 383–389. [CrossRef]

22. Marks, R.M. Mid-Term Prospective Clinical and Radiographic Outcomes of a Modern Fixed-Bearing Total Ankle Arthroplasty. *J. Foot Ankle Surg.* **2019**, *58*, 1163–1170. [CrossRef] [PubMed]
23. Fletcher, A.N.; Campbell, J.T.; Maloney, P.J.; Cerrato, R.A.; Jeng, C.L. The Effect of Tibiotalar Joint Line Level Alterations on Tibiotalar Range of Motion Following Total Ankle Arthroplasty. *Foot Ankle Int.* **2024**, *45*, 20–29. [CrossRef] [PubMed]
24. Hamid, K.S.; Matson, A.P.; Nwachukwu, B.U.; Scott, D.J.; Mather, R.C.; DeOrio, J.K. Determining the Cost-Savings Threshold and Alignment Accuracy of Patient-Specific Instrumentation in Total Ankle Replacements. *Foot Ankle Int.* **2017**, *38*, 49–57. [CrossRef] [PubMed]
25. Adams, S.B.; Spritzer, C.E.; Hofstaetter, S.G.; Jakoi, A.M.; Pietrobon, R.; Nunley II, J.A.; Easley, M.E. Computer-assisted tibia preparation for total ankle arthroplasty: A cadaveric study. *Int. J. Med. Robot. Comput. Assist. Surg.* **2007**, *3*, 336–340. [CrossRef] [PubMed]
26. Giardini, P.; Di Benedetto, P.; Mercurio, D.; Gisonni, R.; Molinari, M.; Causero, A.; Cortese, F. Infinity ankle arthroplasty with traditional instrumentation and PSI prophecy system: Preliminary results. *Acta Biomed.* **2020**, *91*, e2020021. [CrossRef]
27. Yau, J.; Emmerson, B.; Kakwani, R.; Murty, A.N.; Townshend, D.N. Patient-Reported Outcomes in Total Ankle Arthroplasty: Patient Specific Versus Standard Instrumentation. *Foot Ankle Spec.* **2023**, *17*, 19386400231179124. [CrossRef] [PubMed]
28. Escudero, M.I.; Symes, M.; Bemenderfer, T.B.; Barahona, M.; Anderson, R.; Davis, H.; Wing, K.J.; Penner, M.J. Does Patient-Specific Instrumentation Have a Higher Rate of Early Osteolysis Than Standard Referencing Techniques in Total Ankle Arthroplasty? A Radiographic Analysis. *Foot Ankle Spec.* **2020**, *13*, 32–42. [CrossRef] [PubMed]
29. Najefi, A.-A.; Ghani, Y.; Goldberg, A. Role of Rotation in Total Ankle Replacement. *Foot Ankle Int.* **2019**, *40*, 1358–1367. [CrossRef] [PubMed]
30. Malherbe, C.; Deleu, P.-A.; Devos Bevernage, B.; Birch, I.; Maldague, P.; Gombault, V.; Putzeys, P.; Leemrijse, T. Early-Term Results of the Cadence Total Ankle Prosthesis: An European Noninventor Study. *Foot Ankle Int.* **2023**, *44*, 1–12. [CrossRef] [PubMed]
31. Ajis, A.; Henriquez, H.; Myerson, M. Postoperative range of motion trends following total ankle arthroplasty. *Foot Ankle Int.* **2013**, *34*, 645–656. [CrossRef]
32. Najefi, A.; Malhotra, K.; Chan, O.; Cullen, N.; Goldberg, A. The Bologna-Oxford ankle replacement: A case series of clinical and radiological outcomes. *Int. Orthop.* **2019**, *43*, 2333–2339. [CrossRef]
33. Tauviqirrahman, M.; Ammarullah, M.I.; Jamari, J.; Saputra, E.; Winarni, T.I.; Kurniawan, F.D.; Shiddiq, S.A.; van der Heide, E. Analysis of contact pressure in a 3D model of dual-mobility hip joint prosthesis under a gait cycle. *Sci. Rep.* **2023**, *13*, 3564. [CrossRef]
34. Hidayat, T.; Ammarullah, M.I.; Ismail, R.; Saputra, E.; Lamura, M.D.P.; N, C.K.; Bayuseno, A.P.; Jamari, J. Investigation of contact behavior on a model of the dual-mobility artificial hip joint for Asians in different inner liner thicknesses. *World J. Orthop.* **2024**, *15*, 321–336. [CrossRef]

Journal of Personalized Medicine

MDPI

Article

Determination of Relationships between Symmetry-Based, Performance-Based, and Functional Outcome Measures in Patients Undergoing Total Hip Arthroplasty

Jana Kirschner [1,*], Sven Michel [1], Roland Becker [2,3], Olaf Stiebitz [1], Hagen Hommel [1,4], Robert Schulz [5], Maciej Daszkiewicz [6], Aleksandra Królikowska [7,*] and Robert Prill [2,3,*]

1 Department of Therapy Sciences, Brandenburg University of Technology Cottbus-Senftenberg, 03046 Cottbus, Germany; sven.michel@b-tu.de (S.M.); olafstiebitz@lebenslang-lernen.eu (O.S.); h.hommel@khmol.de (H.H.)
2 Center of Orthopaedics and Traumatology, University Hospital Brandenburg a.d.H., Brandenburg Medical School Theodor Fontane, 14770 Brandenburg an der Havel, Germany; roland_becker@yahoo.de
3 Faculty of Health Sciences Brandenburg, Brandenburg Medical School Theodor Fontane, 14770 Brandenburg an der Havel, Germany
4 District Hospital Märkisch Oderland GmbH, 16269 Wriezen, Germany
5 BIH QUEST Center for Responsible Research, Berlin Institute of Health at Charité, Universitätsmedizin Berlin, 10178 Berlin, Germany; robert.schulz@bih-charite.de
6 eMKaMED Medical Center, 53110 Wroclaw, Poland; daszkiewiczmaciej98@gmail.com
7 Ergonomics and Biomedical Monitoring Laboratory, Department of Physiotherapy, Faculty of Health Sciences, Wroclaw Medical University, 50367 Wroclaw, Poland
* Correspondence: kirschne@b-tu.de (J.K.); aleksandra.krolikowska@umw.edu.pl (A.K.); robert.prill@mhb-fontane.de (R.P.)

Citation: Kirschner, J.; Michel, S.; Becker, R.; Stiebitz, O.; Hommel, H.; Schulz, R.; Daszkiewicz, M.; Królikowska, A.; Prill, R. Determination of Relationships between Symmetry-Based, Performance-Based, and Functional Outcome Measures in Patients Undergoing Total Hip Arthroplasty. *J. Pers. Med.* **2023**, *13*, 1046. https://doi.org/10.3390/jpm13071046

Academic Editor: Alexander J. Nedopil

Received: 16 May 2023
Revised: 6 June 2023
Accepted: 23 June 2023
Published: 26 June 2023

Abstract: Due to the high heterogeneity in outcome measures after total hip arthroplasty (THA), the prospective observational study investigated the relationships between symmetry-based (SBMs), performance-based (PBMs), and functional outcome measures in THA patients to determine necessary or redundant categories of tests. The study material consisted of 24 patients with end-stage hip osteoarthritis scheduled for THA. The patients were examined one day before surgery and consecutively on the 4th day, 9th day, and 10th week postoperatively using the SBMs (weight-bearing chair-rising test, measurements of the maximal isometric torque of the quadriceps muscle); the PBMs (10-m walk, timed up-and-go, and stair-climbing tests); and the functional outcome measure (Harris Hip Score). The results obtained in a given category of tests at different time points were compared, and the correlations between the tests were determined. The reliability of the outcome measures was determined. The results of tests in the studied categories statistically significantly ($p < 0.05$) improved at the 10th week postoperatively compared to preoperative results. No strong correlations were revealed between the three studied types of outcome measures in THA patients. Therefore, none of them can be considered redundant. It also means that the relevance of symmetry for a core measurement set to describe the domain function in THA patients must be further clarified.

Keywords: biomedical monitoring; hip replacement; measurement set; outcome set; reliability; performance-based measures

1. Introduction

There is a need to improve reporting standards in orthopedics [1]. Currently, there are very heterogeneous statements on outcome measures for patients with total knee (TKA) or total hip (THA) arthroplasties [2,3]. Diverse outcomes, like balance aspects or muscle strength measured under isokinetic conditions, are used to describe the recovery and performance of patients after the THA [4,5]. Also, different outcome assessment scores are largely used to assess functional capacity in patients after THA or TKA [6–8]. Whether the function of the hip joint can be comprehensively evaluated with the help of questionnaires remains debatable. One of the most commonly used questionnaires for THA patients is the Harris Hip Score (HHS) [9].

Often, there is inconstancy concerning the follow-up time points, the measurement tools, and the whole idea of how 'function' should be defined. Reliable and valid test instruments specifically for postoperative THA treatment are needed to assess therapy success objectively. In recent years, there have been intensive efforts to close these gaps. For this purpose, the Initiative for Method, Measurement, and Pain Assessment in Clinical Trials (IMMPACT) and the Outcome Measures in Rheumatology (OMERACT) try to determine core outcome set (COS) and core measurement sets to help choose assessments for the domain physical function after THA. Their tasks include refining physical function measurements, including an evaluation with PROMs and performance-related measures [10]. According to Singh et al. (2015), there is, for now, no single measurement tool to evaluate surgical outcomes after THA or TKA. The heterogeneous different examination tools make it challenging to compare postoperative results [11].

Performance-based measures (PBMs), such as sit-to-stand transfer, timed up-and-go test, or stair-climbing test, are practical assessments for the objective representation of the domain functions of the lower extremity [12–14]. The role of symmetry and the redundance of established functional assessment tools and PBMs against each other must be discussed. Therefore, the present study investigates the relationships between symmetry-based, performance-based, and functional outcome measures in THA patients to determine necessary or redundant categories of tests. It was hypothesized that the symmetry-based, performance-based, and functional outcome measures do not correspond with each other among this group of patients.

2. Materials and Methods

The prospective, observational cohort study was conducted in a highly specialized hospital for arthroplasty. It was reviewed and approved by the local medical chamber under the number S3(a)/2017. Informed consent was obtained from all subjects involved in the study. The study had a repeated-measures design.

The studied material consisted of patients with end-stage osteoarthritis (OA) of the hip joint who were scheduled for THA. There were the first 24 patients who met all inclusion criteria: being scheduled for primary THA; end-stage hip OA; age > 60 years; ability to walk more than 25 m before THA; not living in a nursing home. Exclusion criteria were present and/or previous musculoskeletal disorders other than hip OA; current injuries of the lower extremities or trunk; cardio-pulmonary diseases with reported influence on limiting performance; or any diseases affecting balance like vestibular or brain-affecting diseases.

All patients were operated on by a senior main surgeon from a certified arthroplasty center. The operation was performed via a minimally invasive anterolateral approach. All endoprostheses were implanted uncemented. There were no intraoperative or postoperative complications during the observation period. The postoperative radiological control of the position of the prosthesis did not reveal any abnormalities.

The following categories of tests were used to assess the physical function of the hip joint: (1) symmetry-based measures, including weight-bearing chair-rising tests (CRT) and measurements of the maximal isometric torque (IT) of the quadriceps muscle; (2) performance-based measures: 10-m walk test (10 MWT), timed up-and-go test (TUG), stair-climbing test (SCT); (3) functional outcome measures: HHS.

2.1. Symmetry-Based Measures

2.1.1. Weight-Bearing Chair-Rising Test

The weight-bearing CRT was carried out with simultaneous measurement of ground reaction forces using two force plates separately for involved and uninvolved limbs (Leonardo Mechanograph®GRFP STD and Leonardo Mechanography v. 4.4 Software, Novotec Medical GmbH, Pforzheim, Germany). The examined patient was instructed to stand up from the bench, take a complete stance, and sit down again. There were two repetitions of the test. The measured parameter was the between-limbs difference, expressed in kilonewtons (kN).

2.1.2. Measurement of the Maximal Isometric Strength of the Quadriceps Muscle

The measurement of the maximal isometric strength of the quadriceps muscle was performed using a hand-held dynamometer (MicroFet 2, Hogan Health Industries, Inc., West Jordan, UT, USA). The torque was measured bilaterally, starting from the uninvolved limb. The measurements were carried out in a seated position with the examined limb flexed to 90 degrees. The hand-held dynamometer was held by the examiner proximally to the ankle joint. The examined patient was instructed to perform maximal-effort contraction of the quadriceps muscle against resistance applied by the examiner. The resistance had to be large enough to make extending the knee on the examined side impossible. The measured parameter was a peak expressed in Newtons (N).

2.2. *Performance-Based Measures*

2.2.1. Ten-Meter Walk Test (10 MWT)

The 10 MWT assessed the time a patient walked a distance of 10 m. To ensure that possible starting and stopping distances did not influence the measurement results, the patient started and ended 2 m before and after the actual test distance of 10 m. If necessary, the patient could use a walking aid during the 10 MWT, but this had to be noted in the documentation. Verbal commands were used. The measurement was carried out using a photoelectric sensor and stopwatch. The measured parameter was time, expressed in seconds (s).

2.2.2. Timed Up-and-Go Test (TUG)

The TUG test assessed the time required for a patient to rise from a standardized chair with a seat height of 46 cm, walk three meters at a comfortable pace to a mark placed on the floor, turn around at the three meters mark, walk back to the starting point, and finally, return to sitting in the chair. During the TUG test, the patient could use a walking aid, if necessary, although it had to be recorded in the documentation. Verbal commands were used. The measurement was carried out using a stopwatch. The measured parameter was time, expressed in seconds (s).

2.2.3. Stair-Climbing Test (SCT)

The SCT evaluates whether and in how much time the patient can walk up and down a given number of stairs. As the 9-step SCT was found to be very sensitive to initial deterioration and improvement in the early phase of THA, a 14-step stair-climbing test was selected [15]. Preoperatively, the test was carried out with no walking aids. Postoperatively, walking support and the stair railing were used. When going up the stairs, the patient placed the uninvolved limb first. When going down the stairs, the patient placed the involved limb first. The measurement was carried out using a stopwatch. The measured parameter was time, expressed in seconds (s).

2.3. *Functional Outcome Measures*

Harris Hip Score

The HHS was collected in the German version of Fortbildungen für Orthopädische Medizin und Manuelle Therapie (FOMT). The HHS assesses patients in terms of pain (one item, 0–44 points), function (seven items, 0–47 points), absence of deformity (one item, 4 points), and range of motion (two items, 5 points), giving a total score out of 100 [9]. The so-called function domain consists of daily activities, including stair use, using public transportation, sitting and managing shoes and socks, and gait, including limp, support needed, and walking distance.

Patients were assessed at four defined time points, as presented in Table 1, precisely T0, one day before surgery; T1, on the 4th postoperative day (3.96 ± 0.95 days); T2 on the 9th postoperative day (8.62 ± 0.77 days); T3, at the 10th (9.29 ± 1.71 weeks) week postoperatively.

Table 1. The scheme of measures that were performed at four defined time points.

Category	Measure	T0	T1	T2	T3
Symmetry-based measures	Weight-bearing chair-rising test (CRT)	x	x	x	x
	Isometric torque (IT)	x	x	x	x
Performance-based measures	Ten-meter walk test (10 MWT)	x	x	x	x
	Timed up-and-go test (TUG)	x	x	x	x
	Stair-climbing test (SCT)	x	-	x	x
Functional outcome measures	Harris Hip Score (HHS)	x	-	-	x

T0, one day before surgery; T1, on the 4th postoperative day (3.96 ± 0.95 days); T2 on the 9th postoperative day (8.62 ± 0.77 days); T3, at the 10th (9.29 ± 1.71 weeks) week postoperatively.

2.4. Reliability of the Outcome Measures

The reliability of all symmetry-based and performance-based measures was established at every time point, namely T0, T1, T2, and T3. All patients in the studied group performed the given tests twice on the same day. According to the methodology described in Sections 2.1 and 2.2 of the present paper, The tests were administered by the same examiner. The reliability and validity of HHS in THA patients have already been proven; therefore, it was not included in the present study [16,17].

2.5. Statistical Analysis

Due to missing pilot values, the sample-size calculation was carried out power-based, and following a similar study on TKA [18], with the help of G-Power 3.1.9.2, assuming mean effects of 0.5, p 0.05, and a power of 80%, 24 patients were required for sufficient correlation analysis.

The anonymized raw data, which were received in the form of data protocols and completed questionnaires, were transferred to an Excel spreadsheet and then imported into "R" and evaluated with the statistics programs "R (Version 3.4.4)" and "R-Studio (Version 1.1.442)". The study assumed a comparison of the results obtained in a given category of tests at different time points and correlations between the tests at different time points.

The arithmetic mean (x) and standard deviation (±) of the studied parameters were calculated for the studied group of patients. The normal distribution of the studied parameters was determined using the Shapiro-Wilk test.

The assessment results with the use of HHS, a total score of fewer than 70 points, was considered a poor result; 70–80 points a fair result; 80–90 points a good result, and 90–100 points an excellent result [19].

According to the results of the Shapiro-Wilk test, the results of the performed measures involved in three different categories obtained at T0 were compared to results reported at T3 using parametric or nonparametric t-tests for dependent samples. The level of significance was defined as $p < 0.05$.

The linear Pearson's correlation coefficient (r) for normally distributed data and Spearman's correlation coefficient for not normally distributed data were calculated to measure any relationships between the results of the tests belonging to three different categories at all time points. The magnitudes of all of the bivariate associations were classified as negligible (0.00–0.30), low (0.31–0.50), moderate (0.51–0.70), high (0.71–0.90), and very high (0.91–1.00) [20]. Additionally, the coefficient of determination, the r-squared (r^2), was calculated to give a proportion of the variance of one variable that is predictable from the other variable. In other words, r^2 represents the percentage of data points closest to the line of best fit.

The relative reliability assessment of each test was based on the intraclass correlation coefficient (ICC) calculation according to the guidelines described by Shrout and Fleiss (1979) [21]. As the intra-rater test reliability was determined, a two-way mixed-effects model, single measurement type, and absolute agreement definition were used [22]. Values less than 0.50 indicated poor reliability; values between 0.50 and 0.75 demonstrated moderate reliability; values between 0.75 and 0.90 were determined to have good reliability; and values greater than 0.90 indicated excellent reliability [22].

3. Results

The final sample of 24 patients included 18 males and six females. The mean age of the patients was 63.5 ± 8.3 years (males 64.6 ± 7.6 years; females 60.0 ± 10.0 years), and a mean body mass exceeded 89.2 ± 15.97 kg. The mean Body Mass Index amounted to 30.0 ± 4.3 kg/m^2 (men 30.1 ± 4.6 kg/m^2; women 30.0 ± 3.90 kg/m^2).

3.1. Symmetry-Based Measures

The postoperative result of the weight-bearing CRT with simultaneous measurement of ground reaction forces statistically significantly improved ($p < 0.05$) compared to the preoperative result, as presented in Figure 1. The between-limbs difference at T3 that exceeded x = −5.47 ± 4.01 kN was reduced compared to the between-limbs difference noted at T0, amounting to x = −7.13 ± 8.42 kN.

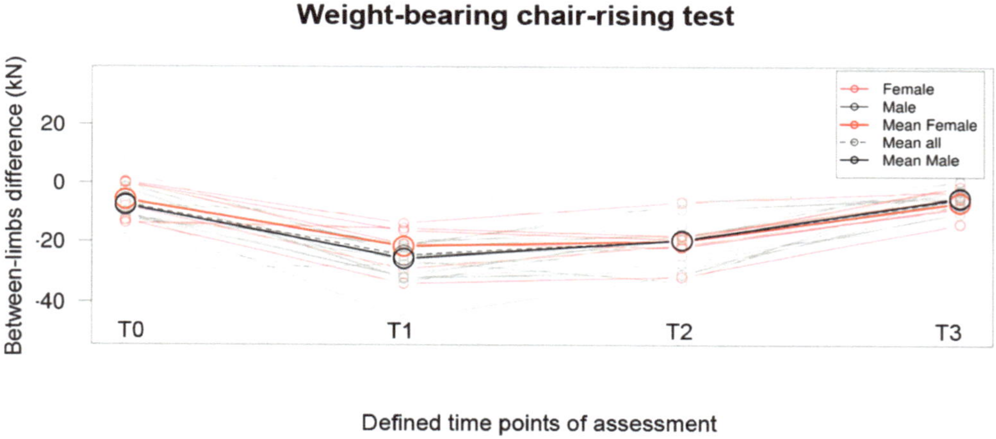

Figure 1. Comparison of the weight-bearing chair raising test with simultaneous measurement of ground reaction forces results obtained at the four defined time points, namely T0, one day before surgery; T1, on the 4th postoperative day; T2, on the 9th postoperative day; T3, at the 10th week postoperatively.

Also, the between-limb symmetry regarding maximal IT of the quadriceps muscle improved postoperatively compared to preoperative results, as presented in Figure 2. The between-limbs difference at T3, which amounted to x = −16.71 ± 44.34 N, was reduced when compared to the between-limbs difference noted at T0, amounting to x = −42.18 ± 11.52 kN.

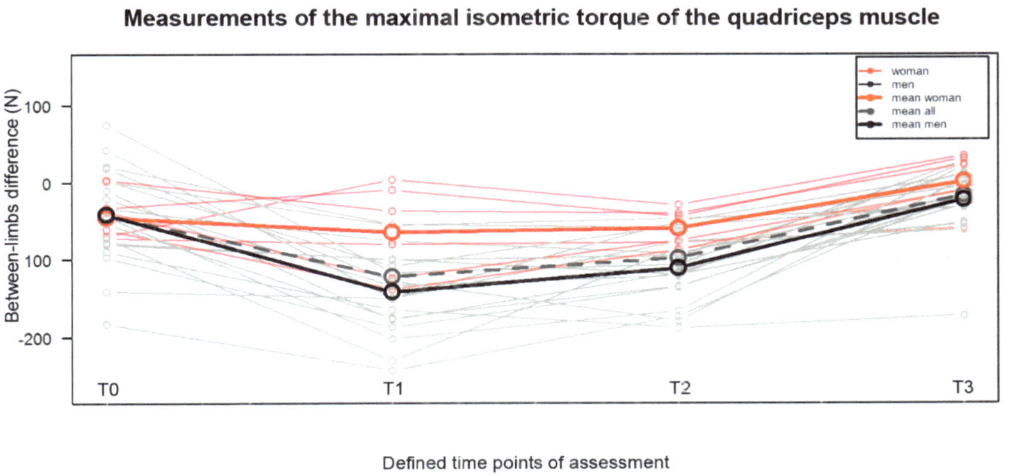

Figure 2. Comparison of the results of the measurement of the maximal isometric strength of the quadriceps muscle obtained at the four defined time points, namely T0, one day before surgery; T1, on the 4th postoperative day; T2, on the 9th postoperative day; T3, at the 10th week postoperatively.

3.2. Performance-Based Measures

The results of the 10 MWT statistically significantly ($p < 0.05$) improved at T3 (x = 7.67 ± 1.30 s) when compared to T0 (x = 9.24 ± 2.38 s), as shown in Figure 3.

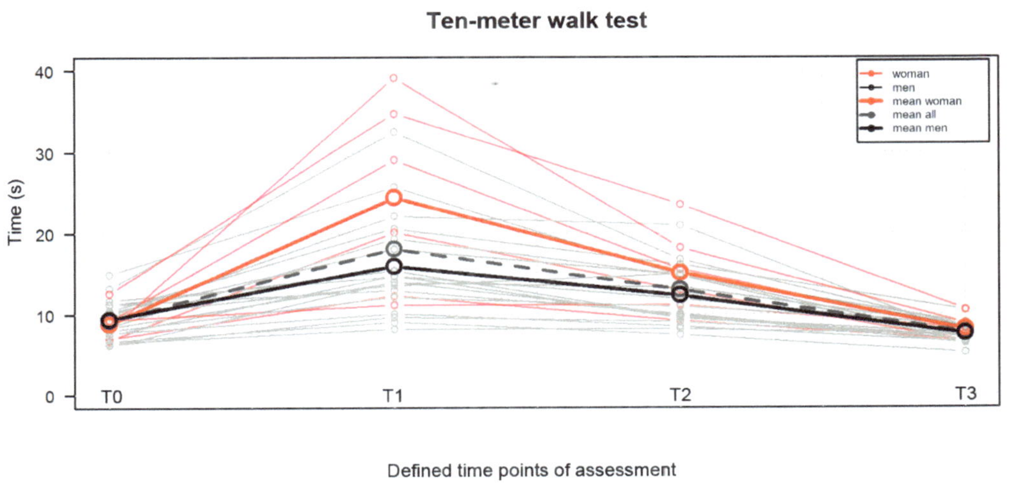

Figure 3. Comparison of the ten-meter walk test results obtained at the four defined time points, namely T0, one day before surgery; T1, on the 4th postoperative day; T2, on the 9th postoperative day; T3, at the 10th week postoperatively.

A statistically significant improvement ($p < 0.05$) was also reported for the TUG results. The time to perform the test decreased at T3, amounting to x = 8.74 ± 1.72 s, from T0, amounting to x = 11.22 ± 3.11 s. The results are presented in Figure 4.

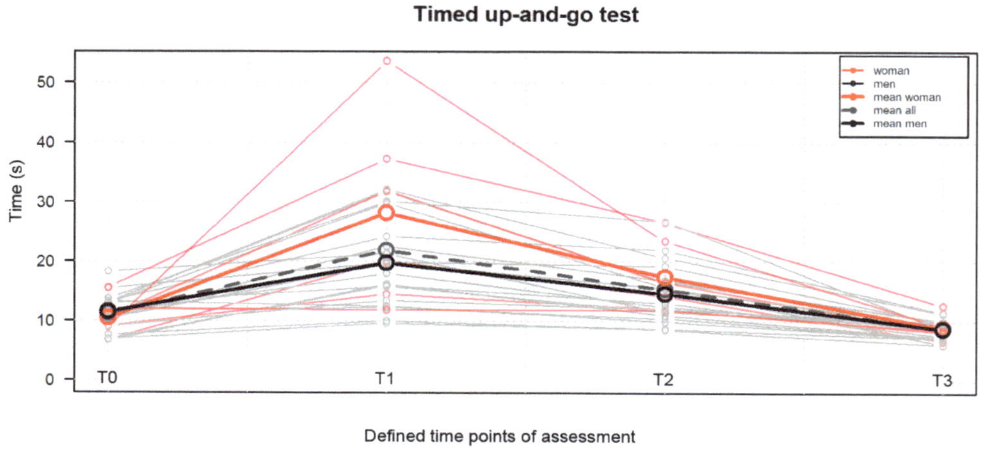

Figure 4. Comparison of the timed up-and-go test results obtained at the four defined time points, namely T0, one day before surgery; T1, on the 4th postoperative day; T2, on the 9th postoperative day; T3, at the 10th week postoperatively.

Comparable to the two above performance-based measures, the SCT results were also statistically significant ($p < 0.05$) better postoperatively than preoperatively, as presented in Figure 5. The time to perform the test noted at T3 exceeded x = 15.51 ± 3.72 s was decreased when compared to T0 when it amounted to x = 21.00 ± 8.36 s.

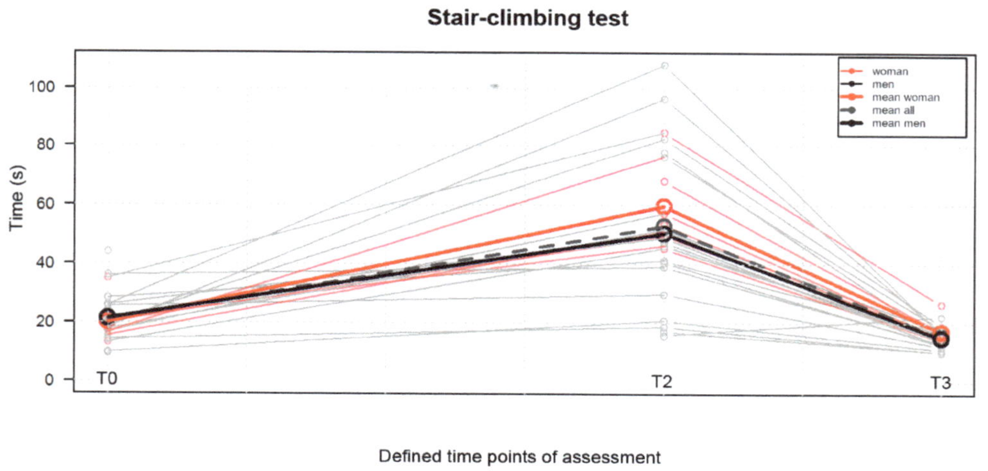

Figure 5. Comparison of the stair-climbing test results obtained at the three defined time points, namely T0, one day before surgery; T2, on the 9th postoperative day; T3, at the 10th week postoperatively.

3.3. Functional Outcome Measures

The total HHS statistically significantly ($p < 0.05$) improved from a poor result (x = 48.25 ± 12.04 points) at T0 to a good result (x = 87.54 ± 11.91 points), reported at T3. The pain domain statistically significantly ($p < 0.05$) improved from x = 12.92 ± 4.64 points at T0 to x = 39.83 ± 7.87 points at T3. A statistically significant ($p < 0.05$) improvement was also reported for the daily activities domain, which increased from x = 7.79 ± 3.19 points at T0 to x = 12.25 ± 1.82 points at T3. Also, the domain of gait statistically significantly ($p < 0.05$) improved from x = 21.12 ± 5.97 points at T0 to x = 27.12 ± 4.40 points at T3. A minor improvement was reported regarding the absence of deformity as the results obtained at T0 and T3 amounted to x = 3.21 ± 1.10 points and x = 3.92 ± 0.41 points, respectively. The domain of range of motion statistically significantly ($p < 0.05$) improved from x = 3.21 ± 0.83 points at T0 to x = 4.33 ± 0.64 points at T3.

3.4. Correlations between the Categories of Tests

Despite an enormous number of possibilities between all the studied variables at all time points, no high nor very high correlations were noted ($r < 0.70$). In Figure 6, there were correlations at the last time point.

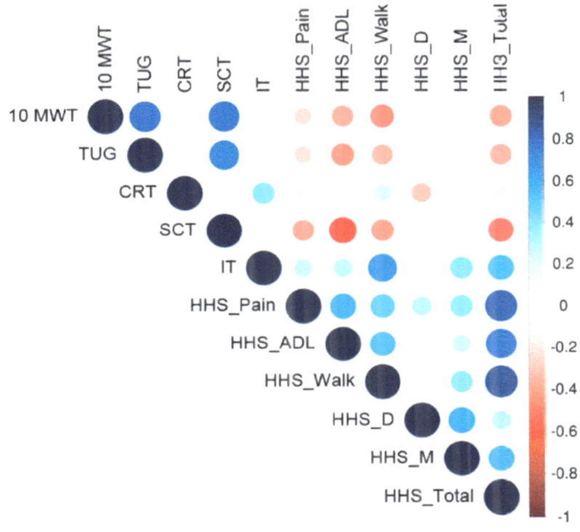

Figure 6. Correlations between the results of the tests performed at the 10th week postoperatively representing three different categories of measurements, namely symmetry-based measures: CRT, weight-bearing chair raising test with simultaneous measurement of ground reaction forces; IT, measurements of the maximal isometric torque of the quadriceps muscle; performance-based measures: 10 MWT, 10-m walk test; TUG, timed up-and-go test; SCT, stair-climbing test; functional outcome measures: HHS, Harris Hip Score, including pain (HHS_Pain), activities of daily living (HHS_ADL), gait (HHS_Walk), absence of deformity (HHS_D), and range of motion (HHS_M).

A moderate negative correlation was noted between the HHS gait domain at T0 and 10 MTW at T3 and ($r = -0.60$). The better score in the HHS gait domain preoperatively, the shorter time to cover the 10 MWT at the 10th week postoperatively. The 36% of the total variation in time to cover the 10 MWT at the 10th week postoperatively can be explained by the negative relationship between the 10 MWT at T3 and obtained HHS gait domain score at T0 ($r^2 = 0.36$). The other 64% of the total variation remains unexplained.

Also, a moderately negative correlation was recorded for the SCT at T0 and HHS activities of a daily living domain at T3 ($r = -0.65$). The SCT at T0 was also moderately negatively correlated with the HHS gait domain at T3 ($r = -0.61$). The shorter the time to cover the SCT preoperatively, the better the score in HHS activities of daily living and the HHS gait domain at the 10th week postoperatively. The 42% of the total variation in HHS activities of a daily living domain at T3 can be explained by the negative relationship between the SCT at T0 and HHS activities of a daily living domain at T3 ($r^2 = 0.42$). Similarly, 37% of the total variation in the HHS gait domain at T3 can be explained by the negative relationship between the SCT at T0 and the HHS gait domain at T3 ($r^2 = 0.37$). The other 58% and 63% consecutively remain unexplained.

Next, the SCT at T3 was moderately negatively associated with the HHS gait domain at T3 ($r = -0.60$). In the 10th week postoperatively, the shorter time to cover SCT, the better score in the HHS gait domain. The 36% of the total variation in time to cover the SCT can be explained by the negative relationship between the SCT and HHS gait domain ($r^2 = 0.36$). The other 64% of the total variation remains unexplained.

A moderately negative correlation was noted between the difference in maximal IT of the quadriceps muscle in the involved and uninvolved limbs at T0 and the HHS gait domain at T3. The r^2 ranged from 0.36 to 0.42. From 36% to 42% of the total variation in the HHS gait domain at the 10th week postoperatively can be explained by the negative relationship between the difference in maximal IT of the quadriceps muscle in the involved and uninvolved limbs at T0 and the HHS gait domain at T3. The other 64% and 58% consecutively remain unexplained.

The rest of the correlations were lower than moderate.

3.5. Reliability of the Outcome Measures

As presented in Table 2, the symmetry-based measures were characterized by excellent reliability, with the ICC ranging from 0.962 to 0.989.

Table 2. The test-retest reliability results of the symmetry-based measures at the four defined time points.

	Symmetry-Based Measures				
Measure	Studied Limb	T0	T1	T2	T3
Weight-bearing chair-rising test (CRT)	Involved	0.964	0.962	0.938	0.970
	Uninvolved	0.980	0.976	0.970	0.970
Isometric torque (IT)	Involved	0.981	0.989	0.986	0.986
	Uninvolved	0.986	0.973	0.978	0.988

The values are expressed as the intraclass correlation coefficient. T0, one day before surgery; T1, on the 4th postoperative day; T2, on the 9th postoperative day; T3, at the 10th week postoperatively.

The values of the ICC for the performance-based measures ranged from 0.902 to 0.995, indicating excellent reliability, as presented in Table 3.

Table 3. The test-retest reliability results of the performance-based measures at the four defined time points.

	Performance-Based Measures			
Measure	T0	T1	T2	T3
Ten-meter walk test (10 MWT)	0.962	0.951	0.967	0.929
Timed up-and-go test (TUG)	0.964	0.981	0.973	0.961
Stair-climbing test (SCT)	0.995	-	0.902	0.966

The values are expressed as the intraclass correlation coefficient. T0, one day before surgery; T1, on the 4th postoperative day; T2, on the 9th postoperative day; T3, at the 10th week postoperatively.

4. Discussion

The present study investigated the relationships between symmetry-based, performance-based, and functional outcome measures in THA patients. There was no strong relationship between the three studied categories of tests. Therefore, it was proven that the symmetry-based, performance-based, and functional outcome measures do not correspond with each other among this group of patients. Because of that, no categories of tests were considered redundant.

With a prevalence of 10% in people over 55 years of age, OA is the most common joint disease, causing significant pain and disability [23]. THA is undertaken to relieve pain and improve function in patients with advanced hip OA whose symptoms are impossible to manage with conservative treatment. The procedure is one of the most successful in orthopedics, although it is also one of the most cost-effective. The number of THAs performed is increasing every year [24]. The evaluation of THA patients after the intervention is critical. The hip joint is one of the most stable but, at the same time, mobile joints of the lower limbs. The function of the hip joint is based primarily on the muscular contribution to support during weight-bearing movements that result in significant loads and stress on cartilage and soft tissues [25].

Of course, it is worth keeping in mind that the outcome measures might be considered an artificial construct as the life and well-being of the patients are influenced by many other factors than a specific illness or treatment. Patients after THA should be regularly put through different measures to monitor their clinical and functional outcomes, evaluate their progress, and introduce eventual changes in postoperative physiotherapeutic procedures. The follow-ups also define the THA's final clinical and functional effects. Moreover, constant outcome measures remain reliable for detecting early signs of failure. It is vital in light of early intervention, as the extensive bone damage makes revision surgery more difficult and significantly increases treatment costs [26].

The results of the individual tests taken together always constitute a large amount of data. Therefore, for another group of examined patients, the authors proposed a strategy that decreased the number of data points to be analyzed without reducing the number of tests performed, like composite scores [27,28]. However, considering the demographics of THA patients, it seems more rational to identify the measures of the highest importance.

The study did not reveal any strong relationships between the studied outcome measures. Some correlations were noted on a moderate level; however, they should be analyzed cautiously. It must be highlighted that only 36% to 42% of the total variation can be explained by those correlations (r^2 = 0.36–0.42). That means the other 58–64% of the total variation remains unexplained.

The study has also determined the reliability of the utilized symmetry-based and performance-based measures. Reliability refers to the extent to which results of a given measure for patients whose status has not changed are reproducible. The reliability can be assessed under several conditions. Presented in the given study, test-retest represents testing over time. Retesting can be performed by different raters on the same occasion (inter-rater reliability) or by the same rater on different occasions (intra-rater reliability) [29]. It's crucial in orthopedics and physiotherapy that the outcome measures must be evaluated regarding their reliability, validity, and responsiveness [30–33]. For the present study purposes, the reliability of the symmetry-based and performance-based measures was determined separately for each defined point of assessment. It was determined that symmetry-based measures (weight-bearing chair raising test with simultaneous measurement of ground reaction forces, measurement of the maximal isometric strength of the quadriceps muscle), as well as performance-based measures (ten-meter walk test, timed up-and-go test, stair-climbing test), are characterized by excellent reliability. The reliability of the HHS was not assessed as it had already been done by other authors [16,17].

One of the limitations of the study was the unequal number of male and female patients in the study group. It has been highlighted in the literature that the relationship between sex and gender-related factors and postoperative outcome and satisfaction in patients undergoing THA is unclear, and more studies on this issue are needed [34]. Therefore, it is hard to judge how the inequality of men and women in the studied group could affect the results.

It also needs to be emphasized that in the final decision on which performance- or time-based measurement to include, a larger sample with subgroups in terms of gender, activity level, and severity of symptoms prior to surgery might be necessary. The authors support the idea of stratification in terms of those confounders for interventional studies. For the observational character of this study in terms of outcome progress, neither the exclusion of all potential confounders nor the exact power analysis seems crucial. Still, the authors agree that the low power of the study, due to a small sample size, constitutes a significant limitation.

Another limitation might be the usage of only HHS regarding functional outcomes, as it is a clinician-based questionnaire designed to be administered by a healthcare professional. In the future, it would be good to include patient-reported outcome measures, as clinician-based tools may not accurately reflect the perception of the patient following the intervention [35]. Therefore, patient-reported outcome measures constitute a gold standard in evaluating patients with musculoskeletal disorders, whose perspective and health-related quality of life are important [9]. Next to the HHS that was utilized in the present study, the most commonly used functional outcome measures primarily developed for patients after THA include the Merle d'Aubigné and Postel Hip Score [36]. However, the score is also a clinician-based one. The patient-based questionnaires also have an added value: they don't need to be administered by a trained healthcare specialist [37]. Other functional outcome measures dedicated to the hip joint are the Hip Disability and Osteoarthritis Outcome Score (HOOS) and the Oxford Hip Score (OHS) [9]. Additionally, sometimes the general status of health is assessed, commonly with the use of Short Form 36 (SF-36) [38]. However, the undoubted value of using HHS is that it is widely used in the orthopedic literature; therefore, comparing the obtained results with other studies is more manageable.

5. Conclusions

The study revealed no strong relationships between body symmetry-based, performance-based, and patient-reported outcome measures in THA patients. Therefore, none of the studied categories of tests can be considered redundant. It also means that the relevance of symmetry for a core measurement set to describe the domain 'function' in THA patients must be further clarified.

Author Contributions: Conceptualization, J.K. and R.P.; methodology, J.K. and R.P.; investigation, J.K., H.H. and R.S.; data curation, J.K. and O.S.; writing—original draft preparation, J.K., A.K. and R.P.; writing—review and editing, S.M, R.B. and M.D., visualization, R.S.; supervision, S.M. and R.B. All authors have read and agreed to the published version of the manuscript.

Funding: This research received no external funding.

Institutional Review Board Statement: The study was conducted in accordance with the Declaration of Helsinki and approved by the Ethics Committee of the Brandenburg State Medical Association (protocol code: S.3(a)/2017; date of approval: 22 February 2017).

Informed Consent Statement: Informed consent was obtained from all subjects involved in the study.

Data Availability Statement: Data is available on request from the corresponding author.

Conflicts of Interest: The authors declare no conflict of interest.

References

1. Prill, R.; Królikowska, A.; Becker, R.; Karlsson, J. Why there is a need to improve evaluation standards for clinical studies in orthopedic and sports medicine. *Knee Surg. Sport. Traumatol. Arthrosc.* **2023**, *31*, 4–5. [CrossRef]
2. Prill, R.; Singh, J.A.; Seeber, G.H.; Nielsen, S.M.; Goodman, S.; Michel, S.; Kopkow, C.; Schulz, R.; Choong, P.; Hommel, H. Patient, physiotherapist and surgeon endorsement of the core domain set for total hip and total knee replacement in Germany: A study protocol for an OMERACT initiative. *BMJ Open* **2020**, *10*, e035207. [CrossRef]
3. Adriani, M.; Becker, R.; Milano, G.; Lachowski, K.; Prill, R. High variation among clinical studies in the assessment of physical function after knee replacement: A systematic review. *Knee Surg. Sport. Traumatol. Arthrosc.* **2023**. [CrossRef]
4. Labanca, L.; Ciardulli, F.; Bonsanto, F.; Sommella, N.; Di Martino, A.; Benedetti, M.G. Balance and proprioception impairment, assessment tools, and rehabilitation training in patients with total hip arthroplasty: A systematic review. *BMC Musculoskelet. Disord.* **2021**, *22*, 1055. [CrossRef] [PubMed]
5. Cankaya, D.; Inci, F.; Karakuş, D.; Turker, H.B.; Kahve, Y.; Neyisci, C. Isokinetic performance and function are similar after total hip arthroplasty applied with a posterior or anterolateral approach: A randomised controlled trial. *Hip Int.* **2023**, *33*, 67–72. [CrossRef] [PubMed]
6. Austin, M.S.; Urbani, B.T.; Fleischman, A.N.; Fernando, N.D.; Purtill, J.J.; Hozack, W.J.; Parvizi, J.; Rothman, R.H. Formal Physical Therapy After Total Hip Arthroplasty Is Not Required: A Randomized Controlled Trial. *J. Bone Jt. Surg. Am. Vol.* **2017**, *99*, 648–655. [CrossRef] [PubMed]

7. Capone, A.; Bienati, F.; Torchia, S.; Podda, D.; Marongiu, G. Short stem total hip arthroplasty for osteonecrosis of the femoral head in patients 60 years or younger: A 3- to 10-year follow-up study. *BMC Musculoskelet. Disord.* **2017**, *18*, 301. [CrossRef]
8. Goyal, P.; Lau, A.; Naudie, D.D.; Teeter, M.G.; Lanting, B.A.; Howard, J.L. Effect of Acetabular Component Positioning on Functional Outcomes in Primary Total Hip Arthroplasty. *J. Arthroplast.* **2017**, *32*, 843–848. [CrossRef]
9. Harris, W.H. Traumatic arthritis of the hip after dislocation and acetabular fractures: Treatment by mold arthroplasty: An end-result study using a new method of result evaluation. *J. Bone Jt. Surg. Am.* **1969**, *51*, 737–755. [CrossRef]
10. Taylor, A.M.; Phillips, K.; Patel, K.V.; Turk, D.C.; Dworkin, R.H.; Beaton, D.; Clauw, D.J.; Gignac, M.A.M.; Markman, J.D.; Williams, D.A.; et al. Assessment of physical function and participation in chronic pain clinical trials: IMMPACT/OMERACT recommendations. *Pain* **2016**, *157*, 1836–1850. [CrossRef]
11. Singh, J.A.; Dohm, M.; Sprowson, A.P.; Wall, P.D.; Richards, B.L.; Gossec, L.; Hawker, G.A.; Riddle, D.L.; Buchbinder, R. Outcome Domains and Measures in Total Joint Replacement Clinical Trials: Can We Harmonize Them? An OMERACT Collaborative Initiative. *J. Rheumatol.* **2015**, *42*, 2496–2502. [CrossRef] [PubMed]
12. Okoro, T.; Whitaker, R.; Gardner, A.; Maddison, P.; Andrew, J.G.; Lemmey, A. Does an early home-based progressive resistance training program improve function following total hip replacement? Results of a randomized controlled study. *BMC Musculoskelet. Disord.* **2016**, *17*, 173. [CrossRef] [PubMed]
13. Hendrickx, C.; De Hertogh, W.; Van Daele, U.; Mertens, P.; Stassijns, G. Effect of percutaneous assisted approach on functional rehabilitation for total hip replacement compared to anterolateral approach: Study protocol for a randomized controlled trial. *Trials* **2014**, *15*, 392. [CrossRef] [PubMed]
14. Judd, D.L.; Dennis, D.A.; Thomas, A.C.; Wolfe, P.; Dayton, M.R.; Stevens-Lapsley, J.E. Muscle strength and functional recovery during the first year after THA. *Clin. Orthop. Relat. Res.* **2014**, *472*, 654–664. [CrossRef]
15. Bennell, K.; Dobson, F.; Hinman, R. Measures of physical performance assessments: Self-Paced Walk Test (SPWT), Stair Climb Test (SCT), Six-Minute Walk Test (6MWT), Chair Stand Test (CST), Timed Up & Go (TUG), Sock Test, Lift and Carry Test (LCT), and Car Task. *Arthritis Care Res.* **2011**, *63* (Suppl. 11), S350–S370. [CrossRef]
16. Söderman, P.; Malchau, H. Is the Harris hip score system useful to study the outcome of total hip replacement? *Clin. Orthop. Relat. Res.* **2001**, *384*, 189–197. [CrossRef]
17. Söderman, P.; Malchau, H.; Herberts, P. Outcome of total hip replacement: A comparison of different measurement methods. *Clin. Orthop. Relat. Res.* **2001**, *390*, 163–172. [CrossRef]
18. Prill, R.; Becker, R.; Schulz, R.; Michel, S.; Hommel, H. No correlation between symmetry-based performance measures and patient-related outcome prior to and after total knee arthroplasty. *Knee Surg. Sport. Traumatol. Arthrosc.* **2022**, *30*, 3155–3161. [CrossRef]
19. Nilsdotter, A.; Bremander, A. Measures of hip function and symptoms: Harris Hip Score (HHS), Hip Disability and Osteoarthritis Outcome Score (HOOS), Oxford Hip Score (OHS), Lequesne Index of Severity for Osteoarthritis of the Hip (LISOH), and American Academy of Orthopedic Surgeons (AAOS) Hip and Knee Questionnaire. *Arthritis Care Res.* **2011**, *63* (Suppl. 11), S200–S207. [CrossRef]
20. Mukaka, M.M. Statistics corner: A guide to appropriate use of correlation coefficient in medical research. *Malawi Med. J.* **2012**, *24*, 69–71.
21. Shrout, P.E.; Fleiss, J.L. Intraclass correlations: Uses in assessing rater reliability. *Psychol. Bull.* **1979**, *86*, 420–428. [CrossRef] [PubMed]
22. Koo, T.K.; Li, M.Y. A Guideline of Selecting and Reporting Intraclass Correlation Coefficients for Reliability Research. *J. Chiropr. Med.* **2016**, *15*, 155–163. [CrossRef] [PubMed]
23. Peat, G.; McCarney, R.; Croft, P. Knee pain and osteoarthritis in older adults: A review of community burden and current use of primary health care. *Ann. Rheum. Dis.* **2001**, *60*, 91–97. [CrossRef] [PubMed]
24. Hegde, V.; Stambough, J.B.; Levine, B.R.; Springer, B.D. Highlights of the 2022 American Joint Replacement Registry Annual Report. *Arthroplast. Today* **2023**, *21*, 101137. [CrossRef]
25. Torry, M.R.; Decker, M.J.; Cowley, J.C.; Keeley, D.; Kernozek, T.W.; Shelburne, K.B. Function of the Normal Hip. In *Hip Joint Restoration: Worldwide Advances in Arthroscopy, Arthroplasty, Osteotomy and Joint Preservation Surgery*; McCarthy, J.C., Noble, P.C., Villar, R.N., Eds.; Springer: New York, NY, USA, 2017; pp. 77–89.
26. Duwelius, P.J.; Southgate, R.D.; Crutcher, J.P., Jr.; Rollier, G.M.; Li, H.F.; Sypher, K.S.; Tompkins, G.S. Registry Data Show Complication Rates and Cost in Revision Hip Arthroplasty. *J. Arthroplast.* **2023**, *33*, S29–S33. [CrossRef]
27. Oleksy, Ł.; Królikowska, A.; Mika, A.; Kuchciak, M.; Szymczyk, D.; Rzepko, M.; Bril, G.; Prill, R.; Stolarczyk, A.; Reichert, P. A Compound Hop Index for Assessing Soccer Players' Performance. *J. Clin. Med.* **2022**, *11*, 255. [CrossRef]
28. Oleksy, Ł.; Mika, A.; Królikowska, A.; Kuchciak, M.; Stolarczyk, M.; Kielnar, R.; Racheniuk, H.; Szczegielniak, J.; Łuszczki, E.; Stolarczyk, A. Composite Score of Readiness (CSR) as Holistic Profiling of Functional Deficits in Footballers Following ACL Reconstruction. *J. Clin. Med.* **2021**, *10*, 3570. [CrossRef]
29. Mokkink, L.B.; Terwee, C.B.; Patrick, D.L.; Alonso, J.; Stratford, P.W.; Knol, D.L.; Bouter, L.M.; de Vet, H.C. The COSMIN study reached international consensus on taxonomy, terminology, and definitions of measurement properties for health-related patient-reported outcomes. *J. Clin. Epidemiol.* **2010**, *63*, 737–745. [CrossRef]
30. Schlüter, I.M.; Prill, R.; Królikowska, A.; Cruysen, C.; Becker, R. A Pilot Study on the Reliability of Ultrasound-Based Assessment of Patella Diameter and Sulcus Angle. *Diagnostics* **2022**, *12*, 3164. [CrossRef]

31. Oleksy, Ł.; Królikowska, A.; Mika, A.; Reichert, P.; Kentel, M.; Kentel, M.; Poświata, A.; Roksela, A.; Kozak, D.; Bienias, K.; et al. A Reliability of Active and Passive Knee Joint Position Sense Assessment Using the Luna EMG Rehabilitation Robot. *Int. J. Environ. Res. Public Health* **2022**, *19*, 15885. [CrossRef]
32. Królikowska, A.; Maj, A.; Dejnek, M.; Prill, R.; Skotowska-Machaj, A.; Kołcz, A. Wrist motion assessment using Microsoft Azure Kinect DK: A reliability study in healthy individuals. *Adv. Clin. Exp. Med.* **2023**, *32*, 203–209. [CrossRef] [PubMed]
33. Królikowska, A.; Mika, A.; Plaskota, B.; Daszkiewicz, M.; Kentel, M.; Kołcz, A.; Kentel, M.; Prill, R.; Diakowska, D.; Reichert, P.; et al. Reliability and Validity of the Athletic Shoulder (ASH) Test Performed Using Portable Isometric-Based Strength Training Device. *Biology* **2022**, *11*, 577. [CrossRef] [PubMed]
34. Solarino, G.; Bizzoca, D.; Moretti, A.M.; D'Apolito, R.; Moretti, B.; Zagra, L. Sex and Gender-Related Differences in the Outcome of Total Hip Arthroplasty: A Current Concepts Review. *Medicina* **2022**, *58*, 1702. [CrossRef] [PubMed]
35. Lieberman, J.R.; Dorey, F.; Shekelle, P.; Schumacher, L.; Thomas, B.J.; Kilgus, D.J.; Finerman, G.A. Differences between patients' and physicians' evaluations of outcome after total hip arthroplasty. *J. Bone Jt. Surg. Am.* **1996**, *78*, 835–838. [CrossRef]
36. D'Aubigne, R.M.; Postel, M. Functional results of hip arthroplasty with acrylic prosthesis. *J. Bone Jt. Surg. Am.* **1954**, *36*, 451–475. [CrossRef]
37. Young, S.W.; Safran, M.R. Outcome Instruments for Assessment of Hip Function. In *Hip Joint Restoration: Worldwide Advances in Arthroscopy, Arthroplasty, Osteotomy and Joint Preservation Surgery*; McCarthy, J.C., Noble, P.C., Villar, R.N., Eds.; Springer: New York, NY, USA, 2017; pp. 91–114.
38. Söderman, P.; Malchau, H. Validity and reliability of Swedish WOMAC osteoarthritis index: A self-administered disease-specific questionnaire (WOMAC) versus generic instruments (SF-36 and NHP). *Acta Orthop. Scand.* **2000**, *71*, 39–46. [CrossRef] [PubMed]

Journal of *Personalized Medicine*

MDPI

Article

Accuracy of Positioning and Risk Factors for Malpositioning Custom-Made Femoral Stems in Total Hip Arthroplasty—A Retrospective Multicenter Analysis

Philip Mark Anderson [1,*], Tizian Heinz [1], Dominik Rak [1], Jörg Arnholdt [2], Boris Michael Holzapfel [2], Silke Dorsch [1], Manuel Weißenberger [1], Rüdiger von Eisenhart-Rothe [3], Max Jaenisch [4], Max Ertl [5], Michael Wagner [6], Henning Windhagen [7], Maximilian Rudert [1] and Axel Jakuscheit [1]

1 Department of Orthopedics, König-Ludwig-Haus, University of Würzburg, Brettreichstraße 11, 97070 Wuerzburg, Germany
2 Department of Orthopedic Surgery, Klinikum Grosshadern, Ludwig-Maximilians-University of Munich, Marchioninistraße 15, 81377 Munich, Germany
3 Department of Orthopaedics and Sports Orthopaedics, Klinikum Rechts der Isar, School of Medicine, Technical University of Munich, Ismaninger Straße 22, 81675 Munich, Germany
4 Department of Orthopedics and Trauma Surgery, University Hospital Bonn, Venusberg Campus 1, 53127 Bonn, Germany
5 Department of Orthopaedic Surgery, Klinikum Rechts der Isar, Technical University of Munich, Ismaninger Straße 22, 81675 Munich, Germany; max.ertl@mri.tum.de
6 Department of Orthopaedics and Trauma Surgery, Klinikum Nürnberg Campus Süd, Paracelsus Medical University, Breslauer Straße 201, 90472 Nürnberg, Germany; michael.wagner@klinikum-nuernberg.de
7 Department of Orthopaedic Surgery, Annastift Hannover, Medical School Hannover, Borries Str. 1-6, 30625 Hannover, Germany; henning.windhagen@diakovere.de
* Correspondence: p-anderson.klh@uni-wuerzburg.de; Tel.: +49-931-8033135

check for updates

Citation: Anderson, P.M.; Heinz, T.; Rak, D.; Arnholdt, J.; Holzapfel, B.M.; Dorsch, S.; Weißenberger, M.; von Eisenhart-Rothe, R.; Jaenisch, M.; Ertl, M.; et al. Accuracy of Positioning and Risk Factors for Malpositioning Custom-Made Femoral Stems in Total Hip Arthroplasty—A Retrospective Multicenter Analysis. *J. Pers. Med.* **2023**, *13*, 1285. https://doi.org/10.3390/jpm13091285

Academic Editors: Markus Gesslein and Eduard Kurz

Received: 31 July 2023
Revised: 17 August 2023
Accepted: 20 August 2023
Published: 22 August 2023

Abstract: Total hip arthroplasty (THA) is commonly performed using off-the-shelf implants. In the case of a severe mismatch between the anatomy of the proximal femur and the geometry of the stem, the use of custom-made stems might become necessary. The goal of this study was to investigate the precision of the implantation of custom-made stems of one manufacturer (CTX stem, AQ Implants) and to determine risk factors for malpositioning. All patients receiving a custom-made CTX stem between 2014 and 2020 at six high-volume academic centers were retrospectively recruited. The achieved position of the stem, as determined by stem version, stem coronal angle, and implantation depth on radiographs, was compared to the plan. The influence of radiographic and demographic parameters on the position was investigated. The results revealed a high variability of the achieved implant position in relation to the preoperative plan. While the stem coronal angle only differed slightly from the intended position, the stem version and the implantation depth showed a high frequency and amount of deviation. Right stems showed significantly higher positions than planned. Surgeons must be aware of this potential problem when implanting custom-made stems.

Keywords: total hip arthroplasty; custom-made implant; radiographs; hips

1. Introduction

In light of the tremendous achievements made in this field during the 20th century, total hip arthroplasty was considered the "operation of the century" [1]. Today, total hip arthroplasty is one of the most frequently performed orthopedic operations worldwide, supported by strong evidence of the good results in terms of gain of quality of life and longevity of the implants, even in geriatric patients [2].

While the vast majority of these operations are performed with standard off-the-shelf implants of many different manufacturers, there are cases with strongly abnormal anatomy, making it impossible to use these standard implants mainly because of a mismatch between the shape of the implant and the proximal femur. These deformities can include alterations in all three plains leading to distortions of the version and the angularity of the proximal and diaphyseal femur [3]. Furthermore, often there are abnormalities of the width and the length of the femoral neck and the medullary cavity present, leading to a significant mismatch with standard implants, which are designed for the typical anatomy. This mismatch can lead to severe alterations in the biomechanics of the reconstructed hip, leg length discrepancies, and fractures of the femur and often prevents the stable fixation of cementless implants.

D'Antonio et al. presented a classification system of femoral abnormalities in total hip arthroplasty in 1993 [3]. This classification system summarizes six possible abnormalities: segmental bone defects, cavitary bone defects, combined bone defects, malalignment of the femur, femoral stenosis within the medullary canal, and femoral discontinuity.

The reasons for the distorted anatomy of the femur are either of innate or developmental nature or often a combination of both. The most frequent causal pathologies are hip dysplasia, previous femoral osteotomies, previous fractures of the femur, and Paget disease [4].

In these complex cases, either modular stem designs or custom-made stems have been proposed to achieve well-fixed implants and restore the physiological biomechanics of the hip [5]. Due to their extensive use in revision surgery, modular implants are usually easily available to address these complex cases. Furthermore, modular implants allow for intraoperative adjustments that might become necessary to achieve full stability and a free range of motion. However, modular implants have several potential disadvantages linked to modular junctions, e.g., corrosion, implant fracture, and release of metal ions. In particular, fractures of the neck and stem junction are a rather common devastating complication that might lead to extremely invasive revision surgeries.

To avoid such complications, custom implants might be used instead, but these implants are expensive and make preoperative computed tomography necessary [6,7]. Nevertheless, their use in cases of complex anatomy is supported by literature showing promising results with implant survival of 93–100% after 6–14 years [8]. While there is some evidence regarding the clinical results of these rarely used implants, there is a lack of evidence concerning their precise implantation, which is crucial for success. In order to close this gap of knowledge in the existing literature, this retrospective multicenter study was conducted to investigate the precision of the implantation of custom-made stems of one manufacturer (CTX stem, AQ Implants) in relation to the preoperative plan and to determine risk factors for malpositioning. Especially nowadays, where personalization is one of the most current topics in modern medicine, the use of custom-made implants is spreading as technical progress eases their availability and makes them more and more affordable. For this, it is important to gain more knowledge on the achievable results with these implants.

The investigated CTX individual endoprosthesis has been implanted since 1992. Since 1998, prosthesis manufacturing has been based on 3D data. The CTX individual endoprosthesis is made of a titanium alloy. It is proximally coated with hydroxyapatite, with the tip left uncoated. The CTX prosthesis is a short stem prosthesis, individually tailored based on computer tomography (CT) scans of the proximal femur and two X-ray images. The stem tapers distally to avoid contact with the diaphyseal cortex. The CTX prosthesis is intended to be implanted with metaphyseal press-fit with cortical bone contact desired only at the calcar, the lateral prosthetic shoulder, and the area of the trochanter minor. The goal of this fixation method is to ensure metaphyseal loading [9]. Planning and templating are 3D based and, therefore, likely to be more accurate than 2D planning [10].

2. Materials and Methods

After approval by the local ethics committee, all patients receiving a custom-made CTX stem between 2014 and 2020 at six high-volume academic centers were retrospectively recruited. Next to the implanted CTX stem, the only inclusion criterion was the availability of the pre- and postoperative records, including the preoperative CT scan with the planned position of the stem and the postoperative standardized X-rays revealing the achieved position. There were no exclusion criteria.

The protocol for the manufacturing of all CTX individual endoprostheses was based on CT scans of the proximal femur and additional slices through the knee condyles to determine the antetorsion. The physician marked the planned hip center and cup position. Based on this information, the proximal femur was 3D-reconstructed. The manufacturer used dedicated software to carry out virtual prosthetic planning and implantation into the 3D-reconstructed femur until stable cortical press-fit was achieved.

After calibrating the pre- and postoperative X-rays (pelvis AP view) using the known diameter of the reference body or the implanted femoral head, the following radiographic landmarks and lines were determined (Figure 1):

- The center of the femoral head on the affected side.
- Both teardrops.
- Greater and lesser trochanter.
- A line tangent to the inferior border of both teardrops (inter-teardrop line).
- A line parallel to the inter-teardrop line passing through the center of the femoral head on the affected side.
- A line perpendicular to the inter-teardrop line crossing the deepest point of the teardrop (vertical teardrop line).
- A horizontal line connecting the ischial tuberosities.
- A horizontal line connecting the inferior borders of the lesser trochanter on both sides.
- The anatomical axis of the proximal femur.
- The longitudinal axis of the prosthetic neck.
- The longitudinal axis of the implanted stem.

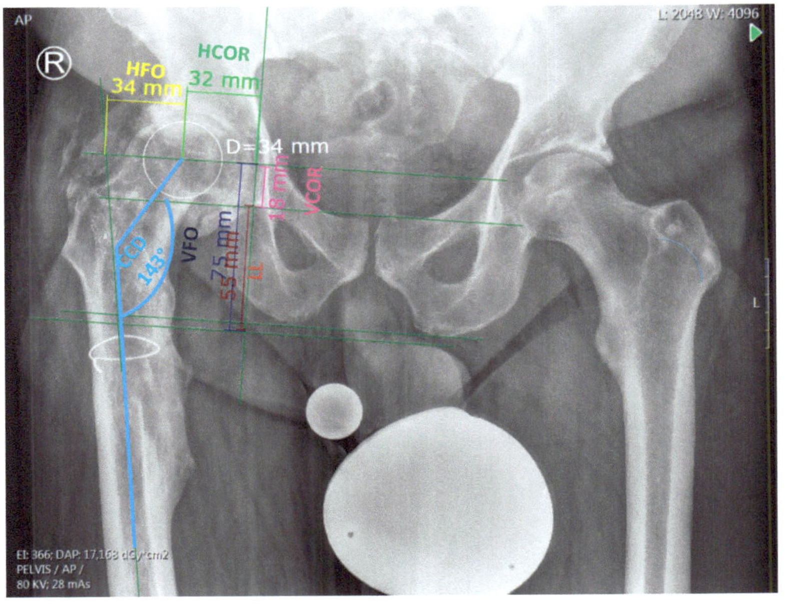

Figure 1. Preoperative measurements on AP pelvis X-ray HFO: horizontal femoral offset. VFO: vertical femoral offset. HCOR: horizontal center of rotation. VCOR: vertical center of rotation. CCD: projected neck-to-shaft angle. LL: leg length.

Using these lines and landmarks, the following biomechanical parameters were measured on the preoperative X-ray using dedicated software (mediCAD, version 6.0, Hectec GmbH, Altdorf, Germany) (Figure 1):

- Horizontal femoral offset (HFO):

  The horizontal femoral offset is defined as the distance from the center of rotation of the femoral head to the anatomical axis of the proximal femur.
- Vertical femoral offset (VFO):

  The vertical femoral offset is defined as the distance from the center of rotation of the femoral head to the lower margin of the lesser trochanter.
- Horizontal center of rotation (HCOR):

  The horizontal center of rotation is defined as the distance from the center of rotation of the femoral head to the vertical teardrop line.
- Vertical center of rotation (VCOR):

  The vertical center of rotation is defined as the distance from the center of rotation of the femoral head to the inter-teardrop line.
- Projected neck-to-shaft angle (CCD):

  The projected neck-to-shaft angle is defined as the projected angle between the anatomical axis of the femur and the axis of the femoral neck.
- Leg length discrepancy (LLD):

  The leg length discrepancy is defined as the difference in the distance between the inferior margin of the lesser trochanter and the inter-teardrop line.

The following parameters were measured on the postoperative X-ray to determine the achieved stem position (Figure 2):

- Stem version:

  The stem version was calculated by the formula presented by Weber et al. [11]:

$$\text{Stem version} = \text{arcus}\ ((\tan\ (\text{projected neck-to-shaft angle})/\tan\ (\text{real neck-shaft-angle}))$$

- Stem coronal angle:

  The stem coronal angle is defined as the angle between the longitudinal axis of the implanted stem and the anatomical axis of the proximal femur.
- Implantation depth of the stem:

  The implantation depth of the stem is defined as the distance from the lower margin of the lesser trochanter to the shoulder of the stem.

The following information concerning the preoperative CT-based planning was recorded:

- Planned stem version;
- Planned stem coronal angle;
- Planned implantation depth of the stem.

Additionally, the following patient-related information was gathered:

- Body mass index (BMI);
- Gender;
- Age at time of surgery;
- Side operated on;
- Dorr classification of the morphology of the proximal femur [11];
- Surgical approach to the hip;
- Staging of osteoarthritis by the classification of Kellgren and Lawrence [12].

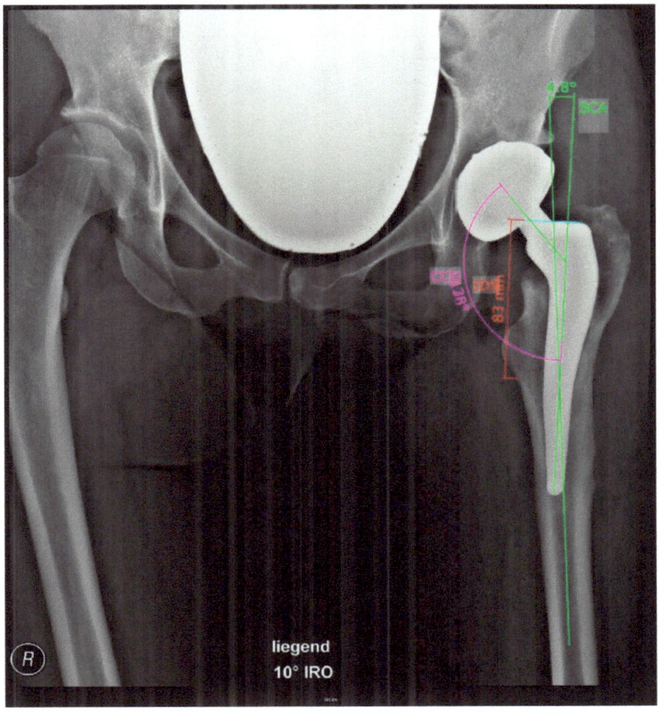

Figure 2. Postoperative measurement on AP pelvis X-ray in 10° internal rotation (IRO). Implantation depth of the stem (IDS), projected neck-to-shaft angle (CCD), and stem coronal angle (SCA).

The primary outcome was the position of the implanted stem compared to the preoperative planning. The position was defined and compared by the implantation depth of the stem, the stem version, and the stem coronal angle.

Secondary outcome parameters were the influence of the patient-related factors and the biomechanical parameters on the achieved position.

Statistical analysis was performed using SPSS Statistics version 27 (IBM Inc., Armonk, NY, USA). Categorial data are presented as rates. Numerical data are presented as mean $\pm$ standard deviation. Group comparison of normal distributed numerical data was performed by t-test. Linear regression models were used to reveal the influence of the preoperative biomechanical and patient-related factors on the achieved position of the implanted stem. The level of significance was set at $p < 0.05$.

The histograms were designed with SigmaPlot for Windows, version 13.0 (Systat Software Inc., San Jose, CA, USA).

3. Results

The complete records of 41 hips in 38 patients were available. Of these, 31 were female (76%). The mean age at the time of surgery was 50 ($\pm$14) years and the mean BMI was 27.9 ($\pm$5.4) kg/m^2. In 17 (41.5%) patients, the right side was operated on. In 15 (37%) cases, the anterior approach was chosen; in 10 (24%), the anterolateral; and in 16 (39%), the lateral. Table 1 gives an overview of the radiographic measurements on the preoperative X-rays that were performed on all 41 hips.

Table 1. Preoperative radiographic measurements.

Measured Parameters	Mean	±SD
HFO (mm)	26.4	±13.2
VFO (mm)	68.4	±19.9
HCOR (mm)	32.7	±11.0
VCOR (mm)	24.6	±9.3
CCD (°)	136	±22.4
LLD (mm)	9.4	±18.4
Kellgren and Lawrence Score	*n*	(%)
0	5	(12.2)
1	2	(4.9)
2	15	(36.6)
3	8	(19.5)
4	11	(26.8)
Dorr Classification Type	*n*	(%)
A	15	(36.6)
B	18	(43.9)
C	8	(19.5)

HFO—horizontal femoral offset; VFO—vertical femoral offset; HCOR—horizontal center of rotation; VCOR—vertical center of rotation; CCD—projected neck-to-shaft angle; LLD—leg length discrepancy.

The stems were implanted on average 4.4 mm (±2.4 mm) less deep than planned. Only 15 (36.6%) stems were implanted within ±5 mm. A total of 17 (41.5%) stems were implanted more than 5 mm less deep than planned. In 18 patients (43.9%), the implantation depth of the stem differed by 10 mm or more from the plan (Figure 3).

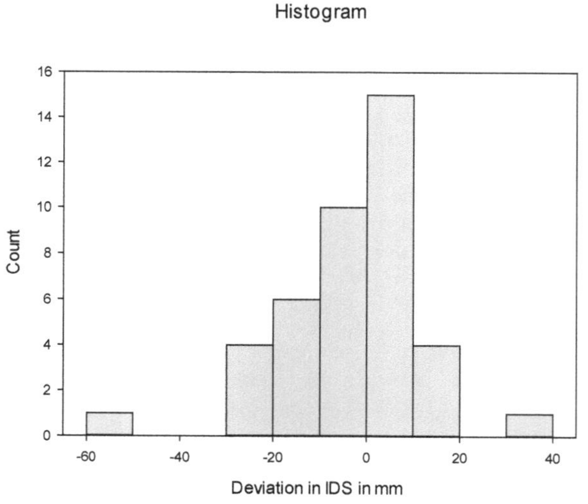

Figure 3. Deviation in postoperative stem implantation depth (IDS) from targeted IDS.

The stems were implanted on average in 15.3° (±15.6°) more antetorsion than planned. Only three (7.3%) stems were implanted within ±5°. In 29 (70.7%) patients, the stem version differed 10° or more from the plan (Figure 4).

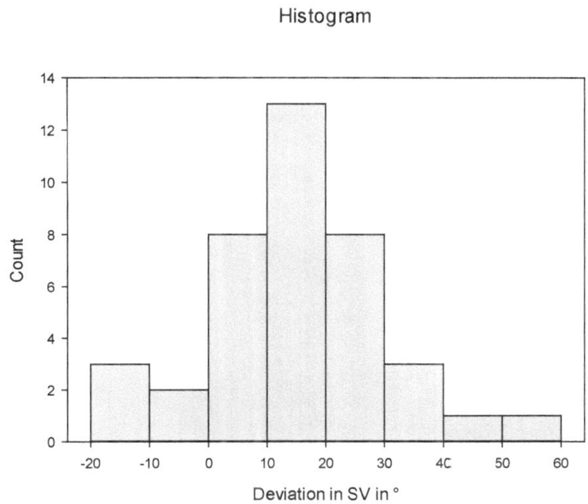

Figure 4. Deviation in the postoperative stem version (SV) from targeted SV.

The stems were implanted on average in 0.1° (±2.3°) more valgus than planned. In all stems (100%), the stem coronal angle was within ±5° (Figure 5).

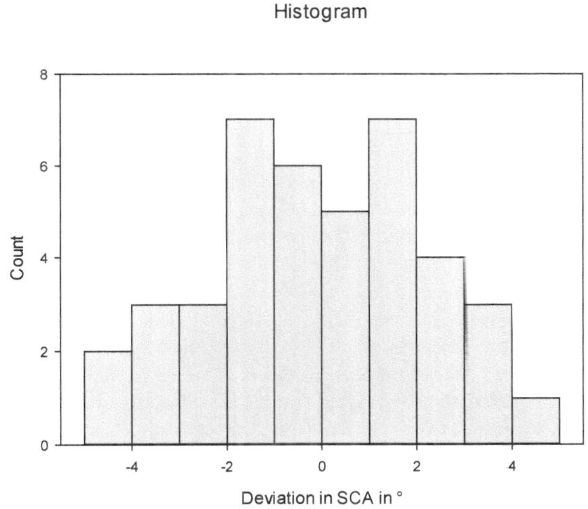

Figure 5. Deviation in postoperative stem coronal angle (SCA) from targeted SCA.

Only four (9.8%) stems were implanted within ±5° stem version, ±5 mm implantation depth, and ±5° stem coronal angle. Eight (19.8%) stems were implanted within ±10° stem version, ±10 mm implantation depth, and ±10° stem coronal angle.

None of the three parameters defining the position of the stem differed significantly between minimally invasive muscle-sparing approaches and the transgluteal lateral approach (Table 2).

Table 2. Deviations of postoperative stem positions from the targeted position under minimally invasive muscle-sparing approaches and the transgluteal lateral approach.

	Minimally Invasive Anterior and Antero-Lateral Approach (n = 25)		**Lateral Approach (n = 16)**		p
Stem version deviation in ° (mean ± SD)	18.0	±17.9	11.4	±10.9	0.198
Stem implantation depth deviation in mm (mean ±SD)	−5.3	±17.8	−3.1	±10.8	0.657
Stem coronal angle deviation in ° (mean ± SD)	0.1	±2.2	−0.4	±2.4	0.516

n—number of hips; SD—standard deviation; p—significance.

Regression analysis was performed to study the influence of the biomechanical parameters and the patient's characteristics on the achieved position.

A short distance between the inferior margin of the lesser trochanter and the inter-teardrop line was a predictor for increased antetorsion of the stem ($p = 0.03$). The right side was strongly associated with a less deep position of the stem ($p = 0.01$). In relation to the planned position, stems on the right side were implanted 12.4 mm (±19.1 mm) less deep, whereas stems on the left side were implanted 1.2 (±8.6) mm deeper than planned.

4. Discussion

The most important finding of this study is the high variability of the achieved implant position in relation to the preoperative plan. Not even 10% of the stems showed a position within a commonly accepted range of ±5° of stem version and stem coronal angle and ±5 mm of implantation depth. While the stem coronal angle only differed slightly from the intended position, the stem version and the implantation depth showed a high frequency and amount of deviation from the plan.

Especially, the fact that over 40% of the stems were implanted more than 5 mm less deep than intended raises concerns about clinical disadvantages for the patient as this will often lengthen the leg in a substantial manner. Even though lengthening of the leg is common after total hip arthroplasty, it is known to be a potential source of persistent pain and functional impairment and therefore is the most common reason for litigation against the surgeon [13]. Within the orthopedic literature, the incidence of substantial leg length discrepancy reported varies between 1% and 27%, whereby lengthening of the operated leg is far more frequent than shortening [14]. It remains uncertain where the boundaries between asymptomatic and symptomatic leg length differences lie. Some authors even postulated that there was no association between leg length discrepancy on the one side and patient satisfaction and functional outcome on the other side at all [15]. Despite these controversies, within most orthopedic communities, it seems to be a consensus that 10 mm differences are usually well tolerated by the patient. When applying this boundary, there remains 12% of patients in the investigated cohort in which the operated leg was lengthened substantially due to insufficient deep implantation of the stem.

The only predictive factor for higher stem position was the implantation in right hips, whereas left hips tended to have deeper implantation than planned. The reason for this finding remains unclear. A possible explanation might be the fact that approximately 90% of the population are right-handed and that this can be assumed to be true for orthopedic surgeons too, even though clear evidence on this is lacking [16,17]. While implanting a stem in a left hip in the supine position can be easily performed with the mallet in the right hand, implantation in a right hip makes either the use of the weaker left hand or a backhand technique necessary, both possibly preventing the surgeon from achieving the necessary amount of force. Pennington et al. published their investigation on implant position in total hip arthroplasty in left- and right-handed surgeons [18]. In contrast to the position of the acetabular component, they could not find a significant effect on the stem position. Mehta et al. published their investigation on the influence of handedness in total knee arthroplasty and were able to show that a right-handed surgeon produced less favorable results when implanting knee prostheses in left knees while changing the side of the table he stood on according to the side operated on [19]. The odds of having a poor outcome were nearly 1.4 times higher on the left side. While not proven, a relation to differences in dexterity and proprioception on the non-dominant side was assumed. In a similar manner, Molony et al. investigated the influence of handedness on malpositioning and consecutive failure of sliding hip screws in internal fixation of proximal femur fractures [20]. They reported a significantly higher risk in left hips when the surgeon was right-handed. In summary, although not well investigated, the available research points to handedness as a relevant factor in joint surgery.

Minimally invasive approaches during total hip arthroplasty are often deemed technically demanding, especially because of the limited exposition of the proximal femur. Spaans et al., for example, reported twice as long operating time, twice as much blood loss, and elevated early complication rates in the minimally invasive anterior approach compared to the postero-lateral approach without learning effect after 46 patients [21]. Martin et al. published the results of a prospective randomized controlled trial on the minimally invasive antero-lateral approach compared to a standard lateral approach [22]. They, too, found a significantly increased operating time and encountered several challenges in exposing the proximal femur. Despite these well-described technical difficulties in minimally invasive approaches to the hip, there was no significant difference in the achieved stem position between those stems implanted via the minimally invasive anterior or antero-lateral approach and those implanted via the standard lateral approach. This is in accordance with a systematic review by Den Hartog et al. showing no difference in the position of the components between the anterior approach and others [23]. Innmann et al. investigated the position of the cup and the stem between a minimally invasive anterolateral and a standard lateral approach [24]. They also could not find any influence on the stem position, whereas the risk of malpositioning the cup was higher in patients operated on via the anterolateral approach.

The observed high deviation in the stem version might lead to impingement of the prosthesis neck at the acetabular component, possibly causing reduced range of motion, instability, and increased wear, whereby the combined functional anteversion as determined by the stem version and the functional version of the acetabular component must be taken into account [25]. An increasing number of studies are highlighting the importance of spino-pelvic alignment in total hip arthroplasty, as the position of the acetabular component is highly variable throughout different movements and between different individuals with their variable spino-pelvic relationships [26]. Therefore, static parameters like the traditional "Lewinnek safe zone" are not reliable for determining if the combined positions of the stem and the cup will work together or lead to mechanical conflicts [27].

The only predictive factor for an increased antetorsion of the stem was a short distance between the inferior margin of the lesser trochanter and the inter-teardrop line on the preoperative radiograph. The reason for this remains unclear. To the best of our knowledge, there has been no published study describing this correlation before. A short distance between the inferior margin of the lesser trochanter and the inter-teardrop line refers to hips with short varus femoral necks with low neck-to-shaft angles. This anatomic variant has been identified as a risk factor for coronal malalignment in former studies [28,29]. The possible influence on the stem version was not investigated. A possible explanation might be the reduced working space in these hips leading to a limited exposition of the proximal femur. Regardless of the explanation of this finding, literature reports a high variability and bad predictability of the stem version in cementless off-the-shelf stems with a trend to increased anteversion [30–33]. Our results support the thesis that this is true for custom-made stems, too.

To our knowledge, the only other study investigating the accuracy of implantation of custom-made CTX stems was published by Rittmeister et al. in 2004 [34]. Within 107 hips, they found 31 (29%) stems implanted more than 5 mm less deep than intended—similar to our results. Within their cohort, they were able to show that dysplasia of the hip was associated with a less deep stem position. Furthermore, the position was dependent on the operating surgeon. In contrast to our findings, the side showed no effect on the stem position.

Independent of the accuracy of the achieved stem position, custom-made stems show good results throughout orthopedic literature. Jacquet et al. published their results on the use of customized stems in patients aged 50 years or less and patients suffering from developmental dysplasia of the hip. When taking aseptic loosening with consecutive revision as an endpoint, the survival rates were 96.8% and 96.81%, respectively. Good functional outcomes measured by the Harris Hip Score and the Merle D'Aubigne–Postel score were achieved in these demanding patient populations [35]. Pakos et al. published long-term results of 67 patients with congenital hip disease who had undergone total hip arthroplasty with the Symbiose custom-made femoral stem. At 10 years follow-up, the overall survival of the femoral component was 97.6%, and the survival related to aseptic loosening was 100% [36].

When interpreting the results of this study, one must be aware of its limitations, caused mainly by the retrospective design and the small number of included patients. Even though six high-volume academic centers reported their cases over a period of six years, only 41 hips could be recruited, highlighting the problem of achieving sufficient case numbers in studies on these rarely used implants. Furthermore, even though only implants of one manufacturer were included, every custom-made stem is unique by definition and therefore adds an additional possible confounder to the results, limiting the transferability to custom-made stems of other manufacturers.

During implantation, no navigation tool or robotic device was used. Current literature mostly shows the benefit of these techniques in terms of achieving the planned position of the cup and restoring the biomechanics of the hip [37–39]. Concerning the stem, there is conflicting evidence on whether the planned position can be achieved more accurately with navigation tools or robots. Koper et al. in 2019 published their multicenter randomized controlled trial on the value of navigation in hip resurfacing using the BrainLab Ci™ ASR System 1.0 (BrainLAB AG, Feldkirchen, Germany) [40]. Compared to conventional implantation, there was no significant difference in the accuracy of the placement of the femoral component. Wang et al. recently published their meta-analysis on radiological and clinical outcomes between robot-assisted and conventional total hip arthroplasty [38]. With the use of robots, the desired coronal alignment of the stem was achieved significantly better.

In light of the recent literature, it can be postulated that additional technical support by means of navigation tools or robots can reduce the risk of malpositioning the stem and—if used—might have changed the presented results significantly.

Neither the position of the acetabular component nor clinical or radiographic follow-up examinations were included in this study. Consequently, the clinical impact of the deviation in the stem version and the implantation depth of the stem cannot be determined.

Furthermore, the limitations of measuring the stem version on radiographs must be acknowledged even though the method used has shown high accuracy and high correlation compared to CT-based measurements [41].

Nevertheless, despite the mentioned limitations, this study adds important findings to the existing literature and might be of great interest to orthopedic surgeons dealing with these highly complicated cases.

5. Conclusions

Even though they are custom-made on the basis of individual anatomy, CTX stems are likely to show relevant deviation from the planned position. Especially in right hips, surgeons must be aware of the danger of implanting the stem substantially higher than planned, with possible negative effects on the leg length. Despite the preoperative plan and the custom-made nature of the implant, the achieved stem version often differs significantly from the planned position. Therefore, it might be helpful to carry out a trial reduction with trial implants in place in order to be able to adjust the position of the stem and or the cup in case of severe leg length discrepancy, impingement, or instability.

Author Contributions: M.R., S.D., J.A. and A.J. were responsible for the conceptualization, methodology, and data collection. Formal analysis was performed by P.M.A. and A.J. Writing and draft preparation was performed by P.M.A. Review and editing were performed by S.D., A.J., B.M.H., M.R., T.H., D.R., M.W. (Michael Wagner), M.W. (Manuel Weißenberger), H.W., M.E., M.J., R.v.E.-K., and J.A. All authors have read and agreed to the published version of the manuscript.

Funding: This research received no external funding.

Institutional Review Board Statement: This study was conducted in accordance with the Declaration of Helsinki and approved by the Institutional Ethics Committee of the University of Wuerzburg (protocol code 2018072701, date of approval 8 February 2018).

Informed Consent Statement: Patient consent was waived due to the retrospective analysis of formerly obtained data.

Data Availability Statement: The data presented in this study are available on request from the corresponding author. The data are not publicly available due to privacy policy.

Conflicts of Interest: The authors declare no conflict of interest.

References

1. Learmonth, I.D.; Young, C.; Rorabeck, C. The operation of the century: Total hip replacement. *Lancet* **2007**, *370*, 1508–1519. [CrossRef] [PubMed]
2. Anderson, P.M.; Vollmann, P.; Weißenberger, M.; Rudert, M. Total hip arthroplasty in geriatric patients—A single-center experience. *SICOT-J* **2022**, *8*, 12. [CrossRef] [PubMed]
3. D'Antonio, J.; McCarthy, J.C.; Bargar, W.L.; Borden, L.S.; Cappelo, W.N.; Collis, D.K.; Steinberg, M.E.; Wedge, J.H. Classification of femoral abnormalities in total hip arthroplasty. *Clin. Orthop. Relat. Res.* **1993**, *296*, 133–139. [CrossRef]
4. Mortazavi, S.M.J.; Restrepo, C.; Kim, P.J.W.; Parvizi, J.; Hozack, W.J. Cementless femoral reconstruction in patients with proximal femoral deformity. *J. Arthroplast.* **2011**, *26*, 354–359. [CrossRef]
5. McCarthy, J.C.; Bono, J.V.; O'Donnell, P.J. Custom and modular components in primary total hip replacement. *Clin. Orthop. Relat. Res. (1976–2007)* **1997**, *344*, 162–171. [CrossRef]
6. Berry, D.J. Utility of modular implants in primary total hip arthroplasty. *J. Arthroplast.* **2014**, *29*, 657–658. [CrossRef] [PubMed]
7. Lakstein, D.; Eliaz, N.; Levi, O.; Backstein, D.; Kosashvili, Y.; Safir, O.; Gross, A.E. Fracture of cementless femoral stems at the mid-stem junction in modular revision hip arthroplasty systems. *JBJS* **2011**, *93*, 57–65. [CrossRef] [PubMed]
8. Tsiampas, D.T.; Pakos, E.E.; Georgiadis, G.C.; Xenakis, T.A. Custom-made femoral implants in total hip arthroplasty due to congenital disease of the hip: A review. *HIP Int.* **2016**, *26*, 209–214. [CrossRef]
9. Yanik-Karaca, Z. Erfassung der ctx-individualschaftposition mit ebra-fca: Abgleich präoperativer planung mit postoperativem prothesensitz bei künstlichem hüftgelenksersatz. Ph.D. Thesis, Frankfurt University of Applied Sciences, Frankfurt, Germany, 2008.
10. Bishi, H.; Smith, J.B.V.; Asopa, V.; Field, R.E.; Wang, C.; Sochart, D.H. Comparison of the accuracy of 2d and 3d templating methods for planning primary total hip replacement: A systematic review and meta-analysis. *EFORT Open Rev.* **2022**, *7*, 70–83. [CrossRef]
11. Dorr, L.D.; Faugere, M.-C.; Mackel, A.M.; Gruen, T.A.; Bognar, B.; Malluche, H.H. Structural and cellular assessment of bone quality of proximal femur. *Bone* **1993**, *14*, 231–242. [CrossRef]
12. Kellgren, J.H.; Lawrence, J. Radiological assessment of osteo-arthrosis. *Ann. Rheum. Dis.* **1957**, *16*, 494. [CrossRef] [PubMed]
13. Desai, A.S.; Dramis, A.; Board, T.N. Leg length discrepancy after total hip arthroplasty: A review of literature. *Curr. Rev. Musculoskelet. Med.* **2013**, *6*, 336–341. [CrossRef] [PubMed]
14. Ranawat, C.S.; Rodriguez, J. Functional leg-length inequality following total hip arthroplasty. *J. Arthroplast.* **1997**, *12*, 359–364. [CrossRef] [PubMed]
15. White, T.O.; Dougall, T.W. Arthroplasty of the hip. *J. Bone Jt. Surg. Br. Vol.* **2002**, *84*, 335–338. [CrossRef]
16. Othman, B.; Chandra, R.; Nestel, D. Age of the leftie: The lived experience of left-handed surgeons. *ANZ J. Surg.* **2022**, *92*, 2082–2087. [CrossRef]
17. Papadatou-Pastou, M.; Ntolka, E.; Schmitz, J.; Martin, M.; Munafò, M.R.; Ocklenburg, S.; Paracchini, S. Human handedness: A meta-analysis. *Psychol. Bull.* **2020**, *146*, 481. [CrossRef]
18. Pennington, N.; Redmond, A.; Stewart, T.; Stone, M. The impact of surgeon handedness in total hip replacement. *Ann. R. Coll. Surg. Engl.* **2014**, *96*, 437–441. [CrossRef]

19. Mehta, S.; Lotke, P.A. Impact of surgeon handedness and laterality on outcomes of total knee arthroplasties: Should right-handed surgeons do only right tkas? *Am. J. Orthop.-Belle Mead* **2007**, *36*, 530.
20. Moloney, D.; Bishay, M.; Ivory, J.; Pozo, J. Failure of the sliding hip screw in the treatment of femoral neck fractures: 'Left-handed surgeons for left-sided hips'. *Injury* **1994**, *25*, SB9–SB13. [CrossRef]
21. Spaans, A.J.; Hout, J.A.A.M.v.d.; Bolder, S.B.T. High complication rate in the early experience of minimally invasive total hip arthroplasty by the direct anterior approach. *Acta Orthop.* **2012**, *83*, 342–346. [CrossRef]
22. Martin, R.; Clayson, P.E.; Troussel, S.; Fraser, B.P.; Docquier, P.-L. Anterolateral minimally invasive total hip arthroplasty: A prospective randomized controlled study with a follow-up of 1 year. *J. Arthroplast.* **2011**, *26*, 1362–1372. [CrossRef]
23. den Hartog, Y.M.; Mathijssen, N.M.C.; Vehmeijer, S.B.W. The less invasive anterior approach for total hip arthroplasty: A comparison to other approaches and an evaluation of the learning curve—A systematic review. *HIP Int.* **2016**, *26*, 105–120. [CrossRef] [PubMed]
24. Innmann, M.M.; Streit, M.R.; Kolb, J.; Heiland, J.; Parsch, D.; Aldinger, P.R.; Königshausen, M.; Gotterbarm, T.; Merle, C. Influence of surgical approach on component positioning in primary total hip arthroplasty. *BMC Musculoskelet. Disord.* **2015**, *16*, 180. [CrossRef] [PubMed]
25. O'Connor, P.B.; Thompson, M.T.; Esposito, C.I.; Poli, N.; McGree, J.; Donnelly, T.; Donnelly, W. The impact of functional combined anteversion on hip range of motion: A new optimal zone to reduce risk of impingement in total hip arthroplasty. *Bone Jt. Open* **2021**, *2*, 834–841. [CrossRef]
26. Krebs, V.E.; Hameed, D.; Mont, M.A. The evolving influence of spino-pelvic biomechanics and research on hip arthroplasty. *J. Arthroplast.* **2023**, *38*, 611–613. [CrossRef]
27. Dorr, L.D.; Callaghan, J.J. Death of the lewinnek "safe zone". *J. Arthroplast.* **2019**, *34*, 1–2. [CrossRef] [PubMed]
28. Haversath, M.; Busch, A.; Jäger, M.; Tassemeier, T.; Brandenburger, D.; Serong, S. The 'critical trochanter angle': A predictor for stem alignment in total hip arthroplasty. *J. Orthop. Surg. Res.* **2019**, *14*, 165. [CrossRef] [PubMed]
29. Free, M.D.; Barnes, I.; Hutchinson, M.; Harvie, P. Preoperative radiographs to predict component malposition in direct anterior approach total hip arthroplasty. *HIP Int.* **2023**, *33*, 207–213. [CrossRef] [PubMed]
30. Worlicek, M.; Weber, M.; Craiovan, B.; Wörner, M.; Völlner, F.; Springorum, H.R.; Grifka, J.; Renkawitz, T. Native femoral anteversion should not be used as reference in cementless total hip arthroplasty with a straight, tapered stem: A retrospective clinical study. *BMC Musculoskelet. Disord.* **2016**, *17*, 399. [CrossRef]
31. Emerson, R.H. Increased anteversion of press-fit femoral stems compared with anatomic femur. *Clin. Orthop. Relat. Res.* **2012**, *470*, 477–481. [CrossRef]
32. Marcovigi, A.; Ciampalini, L.; Perazzini, P.; Caldora, P.; Grandi, G.; Catani, F. Evaluation of native femoral neck version and final stem version variability in patients with osteoarthritis undergoing robotically implanted total hip arthroplasty. *J. Arthroplast.* **2019**, *34*, 108–115. [CrossRef]
33. Belzunce, M.A.; Henckel, J.; Di Laura, A.; Hart, A. Uncemented femoral stem orientation and position in total hip arthroplasty: A ct study. *J. Orthop. Res.* **2020**, *38*, 1486–1496. [CrossRef] [PubMed]
34. Rittmeister, M.; Eisenbeis, K.; Hanusek, S.; Yanik-Karaca, Z.; Starker, M.; Arabmotlagh, M. Bewertung der implantatposition von 107 individuellen ctx-hüftendoprothesenschäften mit ebra-fca. *Z. Orthop. Ihre Grenzgeb.* **2004**, *142*, 659–665. [CrossRef]
35. Jacquet, C.; Flecher, X.; Pioger, C.; Fabre-Aubrespy, M.; Ollivier, M.; Argenson, J.N. Long-term results of custom-made femoral stems. *Orthopade* **2020**, *49*, 408–416. [CrossRef] [PubMed]
36. Pakos, E.E.; Stafilas, K.S.; Tsovilis, A.E.; Vafiadis, J.N.; Kalos, N.K.; Xenakis, T.A. Long term outcomes of total hip arthroplasty with custom made femoral implants in patients with congenital disease of hip. *J. Arthroplast.* **2015**, *30*, 2242–2247. [CrossRef]
37. Liu, Z.; Gao, Y.; Cai, L. Imageless navigation versus traditional method in total hip arthroplasty: A meta-analysis. *Int. J. Surg.* **2015**, *21*, 122–127. [CrossRef]
38. Wang, Y.; Wang, R.; Gong, S.; Han, L.; Gong, T.; Yi, Y.; Feng, Y.; Wang, G.; Xu, W. A comparison of radiological and clinical outcomes between robotic-assisted and conventional total hip arthroplasty: A meta-analysis. *Int. J. Med. Robot. Comput. Assist. Surg.* **2023**, *19*, e2463. [CrossRef]
39. Domb, B.G.; El Bitar, Y.F.; Sadik, A.Y.; Stake, C.E.; Botser, I.B. Comparison of robotic-assisted and conventional acetabular cup placement in tha: A matched-pair controlled study. *Clin. Orthop. Relat. Res.* **2014**, *472*, 329–336. [CrossRef]
40. Koper, M.C.; Reijman, M.; van Es, E.M.; Waarsing, J.H.; Koot, H.W.J.; Keizer, S.B.; Jansen, I.; van Biezen, F.C.; Verhaar, J.A.N.; Bos, P.K. No added value for computer-assisted surgery to improve femoral component positioning and patient reported outcomes in hip resurfacing arthroplasty; a multi-center randomized controlled trial. *BMC Musculoskelet. Disord.* **2019**, *20*, 473. [CrossRef] [PubMed]
41. Weber, M.; Lechler, P.; Kunow, F.v.; Völlner, F.; Keshmiri, A.; Hapfelmeier, A.; Grifka, J.; Renkawitz, T. The validity of a novel radiological method for measuring femoral stem version on anteroposterior radiographs of the hip after total hip arthroplasty. *Bone Jt. J.* **2015**, *97*, 306–311. [CrossRef] [PubMed]

Journal of Personalized Medicine

MDPI

Review

Allograft Prosthesis Composite (APC) for Proximal Humeral Bone Loss: Outcomes and Perspectives

Pietro Gregori [1,2], Giovanni Perricone [1,2], Edoardo Franceschetti [1,2,*], Giancarlo Giurazza [1,2], Giuseppe Francesco Papalia [1,2], Pierangelo Zà [1,2] and Rocco Papalia [1,2]

1 Fondazione Policlinico Universitario Campus Bio-Medico, Via Alvaro del Portillo, 00128 Roma, Italy; pietro.gregori@unicampus.it (P.G.); giovanni.perricone@unicampus.it (G.P.); g.giurazza@unicampus.it (G.G.); g.papalia@policlinicocampus.it (G.F.P.); p.za@unicampus.it (P.Z.); r.papalia@policlinicocampus.it (R.P.)

2 Research Unit of Orthopaedic and Trauma Surgery, Department of Medicine and Surgery, Università Campus Bio-Medico di Roma, Via Alvaro del Portillo, 00128 Roma, Italy

* Correspondence: e.franceschetti@policlinicocampus.it

Abstract: (1) Background: Allograft prosthetic composite (APC) represents one of the techniques used for reconstruction in large proximal humeral bone deficits. The present systematic review aimed at summarizing the state of the art of the technique and analyzing its outcomes. (2) Methods: The PRISMA guidelines were followed to perform this systematic review. A systematic electronic search was performed using PubMed (MEDLINE), EMBASE, and the Cochrane Library databases. All the studies analyzing the rates of allograft prosthesis composite were pooled, and the data were extracted and analyzed. (3) Results: A total of 10 studies were eligible for inclusion in this systematic review for a total of 239 patients. The rate of patient satisfaction with surgery was reported in 7 studies with a mean of 86.4% $\pm$ 13.64. The mean constant score was 45.7 $\pm$ 3.51, the mean ASES score was 63.58 $\pm$ 8.37, and the mean SST was 4.6 $\pm$ 1.04. The mean revision rate observed was 10.32% $\pm$ 3.63 and the mean implant survival was 83.66% $\pm$ 14.98. (4) Conclusions: Based on the currently available data, allograft prosthesis composite represents a valuable option for the reconstruction of proximal humeral deficits. All studies analyzed showed the favorable impact of this surgical technique on clinical outcomes and patient satisfaction.

Keywords: shoulder replacement; shoulder revision surgery; allograft prosthesis composite; APC; humeral allograft; APC results

Citation: Gregori, P.; Perricone, G.; Franceschetti, E.; Giurazza, G.; Papalia, G.F.; Zà, P.; Papalia, R. Allograft Prosthesis Composite (APC) for Proximal Humeral Bone Loss: Outcomes and Perspectives. *J. Pers. Med.* **2023**, *13*, 1301. https://doi.org/10.3390/jpm13091301

Academic Editors: Markus Gesslein and Eduard Kurz

Received: 4 July 2023
Revised: 17 August 2023
Accepted: 20 August 2023
Published: 25 August 2023

1. Introduction

Proximal humeral bone loss (PHBL) may occur in cases of humeral resection for tumors, complex fractures, or following a revision of a failed shoulder arthroplasty [1–3]. Regardless of the etiology, the absence of the proximal humeral bone and the consequent lack of attachment of the capsule and the rotator cuff tendons limit the available effective alternatives for the management of this issue [4]. Traditionally, different reconstructive techniques such as arthrodesis [5], osteoarticular allografts [6], or tumor prostheses have been proposed. The main theoretical advantages of an osteoarticular allograft over a prosthesis are the supply of soft-tissue attachments with which reconstruct the remnant host soft tissues, lateralization of the pull of the deltoid, reconstitution of bone stock and improved contour of the shoulder. However, there are lots of complications associated with this type of reconstruction, such as subchondral collapse, fracture, infection, non-union, and late degenerative arthritis [7]. Although using an endoprosthesis avoids most of these complications [8], the prosthesis does not have the effective soft-tissue attachment sites of an allograft, and therefore it can be unstable, loose, or lead to fracture. Using an allograft prosthesis composite (APC) (Figure 1) combines the potential advantages of an allograft's soft-tissue tendinous and capsular attachments with the benefits of a humeral prosthesis and theoretically avoids the disadvantages of each [7]. Modern reconstructive options for

PHBL or resection include the use of a mega-tumor reverse prosthesis and reverse shoulder arthroplasty (RSA) with or without allograft augmentation [9]. Patients undergoing APC reconstructions of the proximal humerus seem to enjoy relatively good early function with respect to pain relief, active range of motion (ROM), and Musculoskeletal Tumor Society (MSTS) scores [10,11]. In particular, APC-RSA restores the lever arm of the greater tubercle and the deltoid "wrapping effect", increasing ROM and implant stability while also restoring bone stock to spread out mechanical stress on the humerus, improving implant survival [12]. However, because there are no prospective or randomized trials, it is difficult to know which approach is best in terms of functional outcome, implant survivorship, or complications. Multiple previous studies have shown the difficulty associated with treating PHBL with APC [13,14]. Significant heterogeneity exists regarding the severity of bone loss encountered during surgery. There are factors that may predispose patients to more advanced bone loss, resulting in the use of larger allografts, longer stems, and additional modes of fixation [15,16]. Management of severe proximal humeral bone loss thus remains a surgical challenge. The purpose of our systematic review is to summarize the state of the art of the APC for humeral bone loss surgical technique and analyze its outcomes.

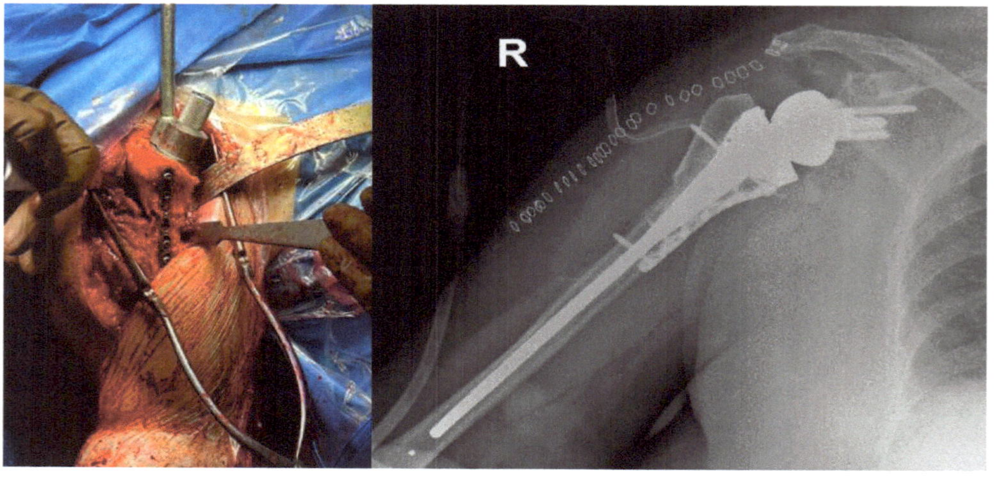

Figure 1. Intra-operative and post operative X-ray of Allograft Prosthesis Composite (APC) technique.

2. Materials and Methods

2.1. Information Sources and Search

An electronic systematic search of PubMed was carried out by two reviewers to identify eligible studies (Figure 2). Interrater reliability for study eligibility was measured using the statistic. The search was executed on 20 July 2023. Then, a manual search of the bibliography of each published study was performed, to find relevant articles that could potentially have been missed. Reviews, systematic reviews, and meta-analyses were also retrieved and read to broaden the search to include studies that might have been missed. The remaining articles were analyzed by two reviewers to exclude studies not fulfilling the eligibility criteria. The reviewers were not blinded to the authors, year, or journal of publication. Studies eligible for inclusion were categorized by study type, according to the Oxford Centre for Evidence-Based Medicine. The following categories were utilized: case reports, randomized controlled trials, case series, and cohort studies.

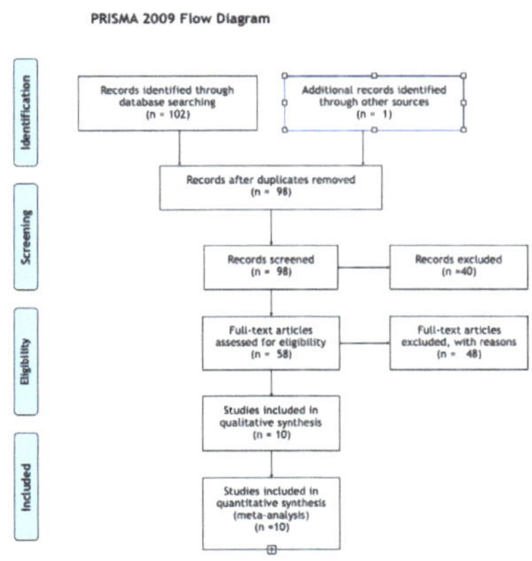

Figure 2. PRISMA flow chart of the study selection process.

2.2. Data Collection Process

Two assessors independently extracted data from the eligible studies using a predefined data extraction procedure. For each study, we extracted data concerning the epidemiological characteristics of participants (age, sex) and the assessment of results (clinical outcomes, mean follow-up, complications, radiographic evaluation, revision rate). Data were analyzed using the R software (2020: R Core Team). The primary endpoint was the clinical outcome after APC in reverse shoulder replacement surgery.

2.3. Quality of the Studies

The quality of the included studies was evaluated using the MINORS (Methodological Index for Nonrandomized Studies) score. The following domains were assessed: a clearly stated aim, inclusion of consecutive patients, prospective data collection, endpoints appropriate to the aim of the study, unbiased assessment of study endpoints, follow-up period appropriate to the aim of the study, loss to follow-up of <5%, prospective calculation of the study size, adequate control group, contemporary group, baseline group equivalence, and adequate statistical analysis (Table 1).

Table 1. Study details.

Study	Year of Publication	Study Type	Level of Evidence	Mean Follow-Up (Months)	No. of Patients (Shoulders)	Women/Men	Mean Age	MINORS
Houdek et al. [17]	2020	Prospective study	III	84	10 (10)	5:5	57 ± 18	11/16
El Beaino et al. [9]	2017	Retrospective cohort study	IV	97	21 (21)	14:7	41 (20–80)	11/16
Chacon et al. [4]	2009	Retrospective cohort study	IV	30.2	25 (25)	23:2	-	12/16
Callamand et al. [11]	2020	Retrospective cohort study	IV	30	11 (11)	7:4	51 (19–87)	10/16
Abdeen et al. [7]	2009	Retrospective cohort study	IV	60	36 (36)	16:20	23 (6–74)	12/16

Table 1. *Cont.*

Study	Year of Publication	Study Type	Level of Evidence	Mean Follow-Up (Months)	No. of Patients (Shoulders)	Women/Men	Mean Age	MINORS
Sotelo et al. [18]	2017	Retrospective cohort study	IV	64	26 (26)	16:10	62 (33–86)	12/16
Black et al. [19]	2007	Retrospective cohort study	IV	55	6 (6)	4:2	40.7 (15–73)	12/16
Cox et al. [20]	2019	Case series	IV	67.9	73 (73)	55:18	67 ± 10	12/16
Boileau et al. [8]	2020	Case series	IV	48	25 (25)	16:9	59 (18–82)	13/16
Lazerges et al. [21]	2017	Case series	IV	70.8	6 (6)	2:4	65.5 (41–79)	12/16

3. Results

A total of 23 studies were identified in the electronic search; of these, 10 were eligible for inclusion in this systematic review. No further studies were identified as relevant through the manual search. All the individual studies were retrospective. Study details are summarized in Table 1. The main indication was bone deficits after tumor resection or hemiarthroplasty/reverse shoulder arthroplasty failure. A total of 239 patients were included in the 10 studies. The mean age was 46.62 years (range: 6 to 87 years) (Table 2), with an average follow-up of 60.67 ± 19.55 months (Table 1). The average value was 11.7 on the MINORS scale.

3.1. Functional Outcome and Patient Satisfaction

The rate of patient satisfaction with surgery was reported in seven studies with a mean of 86.4 ± 13.64%. Different scores were applied to evaluate clinical outcomes. When reported, the mean constant score was 45.7 ± 3.51, the mean ASES score was 63.58 ± 8.37, and the mean SST was 4.6 ± 1.04. Regarding the range of motion of the shoulder, the mean forward elevation was 89.93° ± 11.86°, the mean external rotation was 21.1° ± 10.05°, the mean internal rotation was 4° or L4, and the mean abduction was 70.6° ± 10.09.

3.2. Revision Rate and Complications

The mean revision rate observed was 10.32 ± 3.63%, and the mean implant survival, reported in two studies, was 83.66% ± 14.98 at 10 years follow-up. The most common complications were dislocation of the implant, non-union/resorption of the graft, periprosthetic fractures, and infections. (Table 3).

3.3. Radiological Assessment

Nine studies analyzed radiological assessments through the study of post-operative X-rays. The majority of these studies analyzed the incorporation of the bone graft with a mean of 86.4% (range 84 to 96%). The mean time to union of the allograft–host junction was 7 months (range, 3 to 13 months). Another radiological complication was the scapular notching, with a mean of 24.75% (range: 0 to 48%). The remaining radiological assessments are summarized. Because of heterogeneity and sensitivity to bias when not including randomized controlled trials, we narratively reported our results. Survival and complications are reported as proportions of the included patients. (Table 3).

Table 2. Main outcomes.

Study		Boileau et al. [8]	El Beaino et al. [9]	Chacon et al. [4]	Cox et al. [20]	Callamand et al. [12]	Black et al. [19]	Lazerges et al. [21]	Abdeen et al. [7]	Houdek et al. [17]	Sotelo et al. [18]
Constant Score	Pre	21 (5–47)	-	-	-	-	-	-	-	-	
	Post	42 (13–73)	-	-	-	49	-	46.1	-	-	
Patient overall satisfaction		76%	-	76% good/excellent 20% satisfactory 4% unsatis-factory	70% good/excellent 17% satisfactory 13% unsatis-factory	100%	-	81%	-	100% (RSA + APC) 75% (RSA)	7 excellent, 10 satisfactory, 9 unsatisfac-tory
Forward elevation	Pre	50		32.7	49	-	-	-	-	-	41
	Post	90	101 (1 y follow-up) 92 (5 y follow-up)	82.4	75	105	-	97	70 (deltoid intact) 59 (partial resection) 23 (total resection)	100 (RSA + APC) 76 (RSA)	98
External rotation	Pre	0	-	9.9	-	-	-	-	-	-	11
	Post	10	-	17.6	-	23	-	11	-	34 (RSA + APC) 27 (RSA)	31
Internal rotation	Pre	2	-	Sacrum		-	-	-	-	-	
	Post	4	-	L4	-	4	-	L4	-	-	
Abduction	Pre	-	-	40.4	45	-	-	-	-		
	Post	-	-	81.4	72	-	-	57	72 (deltoid intact); 52 (partial resection); 19 (total resection)	-	
VAS	Pre	6	-	-	-	-	-	5,1	-	-	
	Post	2	-	-	-	-	-	2,3	-	-	

Table 2. *Cont.*

Study		Boileau et al. [8]	El Beaino et al. [9]	Chacon et al. [4]	Cox et al. [20]	Callamand et al. [12]	Black et al. [19]	Lazerges et al. [21]	Abdeen et al. [7]	Houdek et al. [17]	Sotelo et al. [18]
ASES	Pre	-	-	31.7	33.8	-	-	-	-	-	
	Post	-	-	69.4	51.4	-	59 (last follow-up)	-	-	72 (RSA + APC) 61 (RSA)	66.1
SST	Pre	-	-	1.4	1.3	-	-	-	-	-	
	Post	-	-	4.5	3.5	-	-	-	-	6 (RSA + APC) 4 (RSA)	4.4

Table 3. Technique, survival, and complications.

Study	Cause of Proximal Humeral Bone Loss	Type of Technique	Revisions/Reoperations (%)	Implant Survival	Complications	Radiologic Assessment
Boileau et al. [8]	2 primary after tumor resection, 6 after failed MTP, 12 after failed RSA and 5 after failed HA.	15 step osteotomy 10 cement + screws 9 associated L'episcopo procedure	8 (32%) reoperation	-	- Instability: 6 - Loosening (glenoid): 4 - Infection: 2 - Allograft fracture: 1 - Temporary radial palsy: 1	Incorporation of the host: 24 (96%) scapular notching 12 (48%) NO humeral loosening
El Beaino et al. [9]	3 benign tumor 17 malignant tumor 1 renal cancer metastasis	7 cement + plate 14 cement	10.1% revision	-	- Loosening: 3 - Local recurrence: 2 - Periprosthetic fracture: 1 - Infection: 1	- Subluxation: 12 - Delayed union: 10 - Allograft resorption: 9
Chacon et al. [4]	24 after failed hemiarthroplasty 1 after failed bipolar hemiarthroplasty	Step cut osteotomy + cables 1 required plate for stability	-	-	- Dislocation: 2 - Allograft fracture: 1 - Non-displaced fracture of acromion: 1	- NO scapular notching - 1 humeral subluxation - Non-incorporation of the graft in the absence of resorption or fragmentation: 4

Table 3. *Cont.*

Study	Cause of Proximal Humeral Bone Loss	Type of Technique	Revisions/Reoperations (%)	Implant Survival	Complications	Radiologic Assessment
Cox et al. [20]	54 after HA (43 glenoid erosion and instability 3 tuberosity non-union 2 periprosthetic fracture 1 humeral stem loosening 5 infection) 17 after RSA 1 after anatomic 1 primary APC	Step cut + cables	-	88% (5 y) 78% (10 y) 67% (10 y+)	- Allograft fractures: 8 - Dislocations: 4 - Loosening: 5 - Glenosphere dissociation: 2 - Infections: 2 - 10 pz humeral loosening	- Radiographic incorporation: 53% (metaphysis) and 84% diaphysis
Callamand et al. [11]	6 chondrosarcoma 2 osteosarcoma 1 B-cell lymphoma 1 metastatic disease	4 fixation plate 7 self-stabilized Chevron osteotomy 5 associated with L'episcopo procedure	1 revision	-	- Dislocation: 1	- Humeral allograft consolidation: 73% - NO humeral loosening - Glenoid Allograft consolidation 100% - Scapular notching: 2
Black et al. [19]	4 chondrosarcoma 1 osteosarcoma 1 metastatic thyroid cancer	Chevron cut and cement	1 reoperation—1 revision		- Shoulder pain: 1 (did not follow indications)	- Radiographic non-union: 1
Lazerges et al. [21]	2 chondrosarcoma 1 Plasmacytoma 3 metastatic disease	Cemented	-	-	-	- Allograft host–junction consolidation: 5 - Non-union: 1 - Scapular notching: 2
Abdeen et al. [7]	19 high-grade osteogenic sarcoma 8 chondrosarcoma 3 metastasis 1 Ewing sarcoma 1 giant cell tumor 1 malignant fibrous histiocytoma 1 secondary high-grade osteogenic sarcoma 1 angiosarcoma 1 plasma cell myeloma	Cemented 5 + 31 Cemented + plate	3 revisions	88% at 10 y	- Dislocation: 1 - Loosening: 2 - Wound complications: 2	- Radiographic superior migration: 5 - Persisted radiolucent line: 1

Table 3. *Cont.*

Study	Cause of Proximal Humeral Bone Loss	Type of Technique	Revisions/Reoperations (%)	Implant Survival	Complications	Radiologic Assessment
Houdek et al. [17]	-	Cement		-	-	-
Sotelo et al. [18]	5 previous trauma, 3 tumor resection,	Cement, compression plate	80% reoperation rate, 2 revisions	96%	Hematoma with deep infection 1, dislocation 1, allograft fracture 1, periprosthetic fracture (distal to APC) 1, acromial fracture 1	Delayed union 1 Graft resorption 1

4. Discussion

Revision shoulder arthroplasty has been more extensively performed in recent years for a variety of reasons, most commonly after a periprosthetic joint infection or mechanical failure. Moreover, the proximal humerus is the third most common site for osteosarcoma and the second most common site for all osseous sarcomas, most of them requiring surgery. While Ewing's sarcomas and osteosarcomas occur characteristically in young adults and teenagers, chondrosarcomas occur in older individuals [22]. In patients undergoing wide excision of the shoulder girdle, this method has been used in tumor reconstruction to provide a functional upper extremity, although most of this literature focuses on the use of an anatomic humeral prosthesis [23]. Advocates of allograft prosthetic composites in RTSA address their benefits as the reattachment of the subscapularis tendon insertion, lateralization of the pull of the deltoid, reconstitution of bone stock, and improved contour of the shoulder [4,13]. In this setting, reconstruction of the shoulder after resection of a malignant or benign, locally aggressive primary bone tumor of the proximal humerus remains challenging. This is due to the increased bending and torsional forces on the humeral component when significant bone loss is present [18]. Moreover, in order to perform an adequate resection of the tumor, the deltoid musculature and joint capsule, and occasionally the rotator cuff, axillary nerve, glenoid, or scapula, may be included. Since there are no quality randomized prospective trials and there is no consensus on the best reconstructive technique after PHBL, reviewing the literature might be useful to evaluate the use of this treatment. Multiple risks of using allograft persist, including infection from the donor graft to the host, increased nidus for the de novo infection, graft resorption, cost of the graft, increased operative complexity and time, and non-union or malunion [19–21]. This systematic review aimed to evaluate the survivorship, as well as the clinical and radiological outcomes, of the allograft prosthesis composite (APC) technique (Figure 1) for the management of proximal humeral bone loss in shoulder arthroplasty. The main finding was that this technique has shown high survival (83.66% ± 14.98) up to ten-year follow-up, associated with an important increase in average ROM compared with pre-operative evaluation and also a high rate of patient satisfaction (86.4%). Even though the quality of the published evidence is low, the use of bone allograft support for the humeral revision construct demonstrates promising results. Furthermore, the high rate of satisfaction represents today a central variable for the decision-making process of the surgeon [24,25]. A direct comparison between APC reconstruction and no-bone-loss reconstruction techniques was performed in one study only [17]. They analyzed and compared patients with proximal humeral bone loss undergoing endoprosthetic replacement or a reverse shoulder replacement with or without the use of APC. Reverse reconstructions had improved American Shoulder and Elbow Surgeons scores (65 vs. 57; $p = 0.01$) and Musculoskeletal Tumor Society 93 scores (72 vs. 63; $p < 0.001$) versus hemiarthroplasty. Moreover, reverse reconstructions had improved forward elevation (85° vs. 44°, $p < 0.001$) and external rotation (30° vs. 21°; $p < 0.001$) versus a hemiarthroplasty. Teunis et al. [26] conducted a systematic review in 2014 comparing different proximal humerus surgical reconstruction techniques to identify which offered the best functional outcome measured by the Musculoskeletal Tumor Society (MSTS) score, had the longest survival rate, and had the lowest complication rate. They found that allograft prosthesis composites and prostheses had similar functional outcomes and survival rates, and both were able to avoid the fractures that were observed with osteoarticular allografts. These results must be viewed with caution since the small sample sizes and substantial heterogeneity of the patient populations hardly demonstrate the superiority of one technique over another. Excellent outcomes have been reported in other retrospective studies focusing on APC constructs of survivorship and the related satisfaction of the patients. Sanchez-Sotelo et al. [18] found an increased elevation of 41° to 98°, a mean ASES functional score at final follow-up of 66.1, a survival rate of 96%, and an overall survival rate of 80% at 5 years. Moreover, another retrospective consecutive case series of prospectively collected data from Cox et al. [20] found an overall survival rate of 88% at 5 years, as well as a humeral-sided survival rate of 94% at 5 years, 89% at 10 years, and 75% beyond 10 years, along with an increase in elevation from 49° to 75° and an increase in the ASES score from 33.7 to 51.1 at final follow-up. A meta-analysis of the different types of bone loss or techniques of APC could not be conducted due to the limited data in the literature. Regarding infection, the use of this technique is still debated. In their study, Boileau et al. performed nine two-stage APC procedures after failed arthroplasty. They found that the nine patients with preoperative infections, operated on in two stages, were considered to be cured at the last follow-up. They also found no significant differences in complications, reoperations, or clinical outcomes when they compared patients with postoperative infection vs. those without infection [9]. Sotelo et al. found that the infection and fracture rates in their study were relatively low, even if some patients had a history of infection or a malignant tumor requiring immunosuppressive therapy. They also did not find any statistical differences in both outcomes and complications rates between the primary and revision groups [18]. Chacon et al. performed their procedure on three patients who had a previous infection; they noted that two of them had some of the same outcomes and complications rates as the non-infected patients; the last patient with a previous infection suffered implant instability, but according to the author, this was caused by severe deltoid atrophy [4]. Finally, in their study, Cox et al. performed five APC procedures after failed HA for infection and two after failed RSA for infection, and they found the same outcomes and complications rate as the other patients [20]. Our study had several limitations. First, the high heterogeneity between the studies should lead to a critical interpretation of the results. The most relevant limitation is the low evidence level of the included studies, as most are level IV studies, and the low number of studies available on this topic, as well as the difficulty of comparing functional and satisfaction outcomes, which is related to the differences in study design. Moreover, the indications for APC were heterogeneous, and the absence of prospective studies with significant groups of controls decreased the quality of the data. Finally, a lack of subgroup analysis related to the different quantities of bone loss among the patients could produce a high risk of bias. This impedes the predictive value of our results for specific patients.

5. Conclusions

This study underlies high clinical improvements and low complication rates in patients treated with allograft prosthesis composite (APC) for proximal humeral bone loss. At the current state of research and with the highest number of studies analyzed on the topic, the APC technique represents a valid option for the treatment of these challenging cases. Further research and higher-quality studies should be carried out to individuate long-term outcomes and to compare this treatment with other techniques or implants in order to define where to find the right indication.

Author Contributions: Conceptualization, P.G. and E.F.; methodology, P.G.; software, E.F.; validation, G.P., G.G. and R.P.; formal analysis, G.F.P.; investigation, P.Z.; resources, P.Z.; data curation, G.P.; writing—original draft preparation, P.G.; writing—review and editing, E.F.; visualization, R.P.; supervision, G.G.; project administration, G.G.; funding acquisition, R.P. All authors have read and agreed to the published version of the manuscript.

Funding: This research received no external funding.

Institutional Review Board Statement: Not applicable.

Informed Consent Statement: Not applicable.

Conflicts of Interest: The authors declare no conflict of interest.

References

1. Ascione, F.; Domos, P.; Guarrella, V.; Chelli, M.; Boileau, P.; Walch, G. Long-term humeral complications after Grammont-style reverse shoulder arthroplasty. *J. Shoulder Elb. Surg.* **2018**, *27*, 1065–1071. [CrossRef] [PubMed]
2. Levy, J.C.; Virani, N.; Pupello, D.; Frankle, M. Use of the reverse shoulder prosthesis for the treatment of failed hemiarthroplasty in patients with glenohumeral arthritis and rotator cuff deficiency. *J. Bone Jt. Surg.* **2007**, *89*, 189–195. [CrossRef] [PubMed]
3. Merolla, G.; Wagner, E.; Sperling, J.W.; Paladini, P.; Fabbri, E.; Porcellini, G. Revision of failed shoulder hemiarthroplasty to reverse total arthroplasty: Analysis of 157 revision implants. *J. Shoulder Elb. Surg.* **2018**, *27*, 75–81. [CrossRef] [PubMed]
4. Chacon, A.; Virani, N.; Shannon, R.; Levy, J.C.; Pupello, D.; Frankle, M. Revision Arthroplasty with Use of a Reverse Shoulder Prosthesis-Allograft Composite. *J. Bone Jt. Surg.* **2009**, *91*, 119–127. [CrossRef] [PubMed]
5. Chauhan, V.S.; Vaish, A.; Vaishya, R. Reverse shoulder arthroplasty after failed megaprosthesis for osteosarcoma of the proximal humerus: A case report and review of literature. *J. Clin. Orthop. Trauma* **2019**, *10*, 526–530. [CrossRef]
6. Abdeen, A.; Healey, J.H. Allograft-prosthesis composite reconstruction of the proximal part of the humerus: Surgical technique. *J. Bone Jt. Surg. Am.* **2010**, *92 Pt 2* (Suppl. S1), 188–196. [CrossRef]
7. Abdeen, A.; Hoang, B.H.; Athanasian, E.A.; Morris, C.D.; Boland, P.J.; Healey, J.H. Allograft-prosthesis composite reconstruction of the proximal part of the humerus: Functional outcome and survivorship. *J. Bone Jt. Surg. Am.* **2009**, *91*, 2406–2415. [CrossRef]
8. Boileau, P.; Raynier, J.-L.; Chelli, M.; Gonzalez, J.-F.; Galvin, J.W. Reverse shoulder–allograft prosthesis composite, with or without tendon transfer, for the treatment of severe proximal humeral bone loss. *J. Shoulder Elb. Surg.* **2020**, *29*, e401–e415. [CrossRef]
9. El Beaino, M.; Liu, J.; Lewis, V.O.; Lin, P.P. Do Early Results of Proximal Humeral Allograft-Prosthetic Composite Reconstructions Persist at 5-year Followup? *Clin. Orthop. Relat. Res.* **2019**, *477*, 758–765. [CrossRef]
10. Franceschetti, E.; Gregori, P.; Giurazza, G.; Papalia, G.; Caraffa, A.; Papalia, R. Short to Early-Mid Term Clinical Outcomes and Survival of Pyrocarbon Shoulder Implants: A Systematic Review and Meta-Analysis. *J. Shoulder Elb. Arthroplast.* **2023**, *7*, 24715492231152143. [CrossRef]
11. Callamand, G.; Barret, H.; Saint-Genez, F.; Bonnevialle, P.; Mansat, P.; Bonnevialle, N. Reconstruction by allograft-prosthetic composite reverse shoulder arthroplasty after proximal humerus tumor resection: Clinical and radiographic assessment at a minimum 2 years' follow-up. *Orthop. Traumatol. Surg. Res.* **2022**, *108*, 102957. [CrossRef] [PubMed]
12. van de Sande, M.A.J.; Dijkstra, P.D.S.; Taminiau, A.H.M. Proximal humerus reconstruction after tumour resection: Biological versus endoprosthetic reconstruction. *Int. Orthop.* **2011**, *35*, 1375–1380. [CrossRef] [PubMed]
13. Levy, J.; Frankle, M.; Mighell, M.; Pupello, D. The use of the reverse shoulder prosthesis for the treatment of failed hemiarthroplasty for proximal humeral fracture. *J. Bone Jt. Surg. Am.* **2007**, *89*, 292–300. [CrossRef]
14. Favard, L. Revision of total shoulder arthroplasty. *Orthop. Traumatol. Surg. Res.* **2013**, *99* (Suppl. S1), S12–S21. [CrossRef] [PubMed]
15. McLendon, P.B.; Cox, J.L.; Frankle, M.A. Large diaphyseal-incorporating allograft prosthetic composites: When, how, and why: Treatment of advanced proximal humeral bone loss. *Orthopade* **2017**, *46*, 1022–1027. [CrossRef] [PubMed]
16. Walker, M.; Brooks, J.; Willis, M.; Frankle, M. How Reverse Shoulder Arthroplasty Works. *Clin. Orthop. Relat. Res.* **2011**, *469*, 2440–2451. [CrossRef]
17. Houdek, M.T.; Bukowski, B.R.; Athey, A.G.; Elhassan, B.T.; Barlow, J.D.; Morrey, M.E.; Rose, P.S.; Wagner, E.R.; Sanchez-Sotelo, J. Comparison of reconstructive techniques following oncologic intraarticular resection of proximal humerus. *J. Surg. Oncol.* **2021**, *123*, 133–140. [CrossRef]
18. Sanchez-Sotelo, J.; Wagner, E.R.; Houdek, M.T. Allograft-Prosthetic Composite Reconstruction for Massive Proximal Humeral Bone Loss in Reverse Shoulder Arthroplasty. *JBJS Essent. Surg. Tech.* **2018**, *8*, e3. [CrossRef]
19. Black, A.W.; Szabo, R.M.; Titelman, R.M. Treatment of malignant tumors of the proximal humerus with allograft-prosthesis composite reconstruction. *J. Shoulder Elb. Surg.* **2007**, *16*, 525–533. [CrossRef]
20. Cox, J.L.; McLendon, P.B.; Christmas, K.N.; Simon, P.; Mighell, M.A.; Frankle, M.A. Clinical outcomes following reverse shoulder arthroplasty–allograft composite for revision of failed arthroplasty associated with proximal humeral bone deficiency: 2- to 15-year follow-up. *J. Shoulder Elb. Surg.* **2019**, *28*, 900–907. [CrossRef]
21. Lazerges, C.; Dagneaux, L.; DeGeorge, B.; Tardy, N.; Coulet, B.; Chammas, M. Composite reverse shoulder arthroplasty can provide good function and quality of life in cases of malignant tumour of the proximal humerus. *Int. Orthop.* **2017**, *41*, 2619–2625. [CrossRef] [PubMed]
22. Dorfman, H.D.; Czerniak, B. Bone cancers. *Cancer* **1995**, *75* (Suppl. S1), 203–210. [CrossRef] [PubMed]
23. Dudkiewicz, I.; Velkes, S.; Oran, A.; Pritsch, M.; Salai, M. Composite grafts in the treatment of osteosarcoma of the proximal humerus. *Cell Tissue Bank.* **2003**, *4*, 37–41. [CrossRef] [PubMed]
24. Franceschetti, E.; Gregori, P.; de Sanctis, E.G.; Palumbo, A.; Paciotti, M.; Maffulli, N.M.; Franceschi, F. Return to Sport After Total Shoulder Arthroplasty and Hemiarthroplasty: A Systematic Review. *Sports Med. Arthrosc. Rev.* **2023**, *31*, 27–32. [CrossRef]

25. Franceschetti, E.; de Sanctis, E.G.; Gregori, P.; Palumbo, A.; Paciotti, M.; Di Giacomo, G.; Franceschi, F. Return to sport after reverse total shoulder arthroplasty is highly frequent: A systematic review. *J. ISAKOS* **2021**, *6*, 363–366. [CrossRef]
26. Teunis, T.; Nota, S.P.F.T.; Hornicek, F.J.; Schwab, J.H.; Lozano-Calderón, S.A. Outcome After Reconstruction of the Proximal Humerus for Tumor Resection: A Systematic Review. *Clin. Orthop. Relat. Res.* **2014**, *472*, 2245–2253. [CrossRef]

Journal of *Personalized Medicine*

MDPI

Article

Effect of Surgeon Volume on Mechanical Complications after Resection Arthroplasty with Articulating Spacer

Chih-Yuan Ko [1,2], Chun-Hao Tsai [1,3], Yi-Chin Fong [1,3], Hui-Yi Chen [4], Hsien-Te Chen [1,3] and Tsung-Li Lin [1,2,3,*]

1 Department of Orthopedics, China Medical University Hospital, Taichung 40447, Taiwan; 014333@tool.caaumed.org.tw (C.-Y.K.); 007940@tool.caaumed.org.tw (C.-H.T.); 001762@tool.caaumed.org.tw (Y.-C.F.); 002326@tool.caaumed.org.tw (H.-T.C.)
2 Graduate Institute of Biomedical Sciences, China Medical University, Taichung 40402, Taiwan
3 Department of Sports Medicine, College of Health Care, China Medical University, Taichung 40604, Taiwan
4 Department of Radiology, China Medical University Hospital, Taichung 40447, Taiwan; 007396@tool.caaumed.org.tw
* Correspondence: 018144@tool.caaumed.org.tw; Tel.: +886-4-2205-2121

Abstract: Two-stage revision with an antibiotic-loaded cement articulating spacer is a standard treatment for chronic prosthetic knee infection (PKI); however, mechanical complications can occur during the spacer period. There is limited evidence on the association between surgeon volume and mechanical complications after resection arthroplasty (RA) using an articulating spacer. This study aimed to compare the rates of mechanical complications and reoperation after RA with articulating spacers by surgeons with high volumes (HV) and low volumes (LV) of RA performed and analyzed the risk factors for mechanical failure. The retrospective study investigated 203 patients treated with PKIs who underwent RA with articulating spacers and were divided according to the number of RAs performed by the surgeons: HV (≥14 RAs/year) or LV (<14 RAs/year). Rates of mechanical complications and reoperations were compared. Risk factors for mechanical complications were analyzed. Of the 203 patients, 105 and 98 were treated by two HV and six LV surgeons, respectively. The mechanical complication rate was lower in HV surgeons (3.8%) than in LV surgeons (36.7%) ($p < 0.001$). The reoperation rate for mechanical complications was lower in HV surgeons (0.9%) than in LV surgeons (24.5%) ($p < 0.001$). Additionally, 47.2% of patients required hinge knees after mechanical spacer failure. Medial proximal tibial angle < 87°, recurvatum angle > 5°, and the use of a tibial spacer without a cement stem extension were risk factors for mechanical complications. Based on these findings, we made the following three conclusions: (1) HV surgeons had a lower rate of mechanical complications and reoperation than LV surgeons; (2) mechanical complications increased the level of constraint in final revision knee arthroplasty; and (3) all surgeons should avoid tibial spacer varus malalignment and recurvatum deformity and always use a cement stem extension with a tibial spacer.

Keywords: prosthetic knee infection; high volume; low volume; resection arthroplasty; articulating spacer; mechanical complications

Citation: Ko, C.-Y.; Tsai, C.-H.; Fong, Y.-C.; Chen, H.-Y.; Chen, H.-T.; Lin, T.-L. Effect of Surgeon Volume on Mechanical Complications after Resection Arthroplasty with Articulating Spacer. *J. Pers. Med.* **2024**, *14*, 490. https://doi.org/10.3390/jpm14050490

Academic Editor: Kenneth P.H. Pritzker

Received: 23 March 2024
Revised: 20 April 2024
Accepted: 1 May 2024
Published: 3 May 2024

1. Introduction

Prosthetic knee infections (PKIs) occur in 1–2% of primary total knee arthroplasties (TKA) [1]. In addition, approximately 25% of revision TKAs are performed for PKIs [2]. Furthermore, PKIs increasingly concern arthroplasty surgeons due to their high financial burden on the healthcare system, large negative effect on patient outcomes, and high mortality rates [3].

Acute PKIs may be treated effectively with debridement, antibiotics, and implant retention procedure; however, chronic infections often require a two-stage revision [4]. Two-stage revision with antibiotic-loaded cement articulating spacers is a standard treatment for chronic PKIs, as it provides a greater range of motion and has better functional scores, high infection eradication, and easier revision approaches [5]. Despite their advantages, less than half of articulating spacers are considered optimally sized and positioned, and the rates of various mechanical complications, such as spacer fracture or migration, periprosthetic fracture, joint subluxation or dislocation, or extensor mechanism disruption, as high as 57%, have been reported [6]. These mechanical complications may lead to additional spacer exchange, compromised functional outcomes, prolonged treatment course, the need for more constrained prostheses, and decreased survivorship after reimplantation [7,8].

High-volume (HV) surgeons achieve better results and lower complication rates than low-volume (LV) surgeons in primary TKAs [9]. Similarly, HV surgeons have better outcomes and lower re-revision rates following aseptic revision TKA than LV surgeons, and they support the development of revision teams within arthroplasty centers to offer patients the best outcomes [10]. Furthermore, establishing a specialized PKI center with experienced surgeons performing HV infection procedures is recommended to improve the rate of reimplantation [11]. However, the effects of surgeon volume on mechanical complications following resection arthroplasty (RA) using articulating spacers remain unclear. To the best of our knowledge, only one study found that being an LV surgeon is a risk factor associated with dislocation and malalignment/malpositioning of spacers [8]. Understanding the outcomes of HV and LV surgeons is helpful in counseling patients and taking measures to improve the success of two-stage revisions.

Therefore, this retrospective study had the dual aims of comparing the rate of mechanical complications and reoperation after RA with articulating spacers by HV or LV surgeons and analyzing the risk factors for mechanical failure. We hypothesized that HV surgeons would have fewer mechanical complications and reoperations.

2. Material and Methods

2.1. Articulating Spacers

Posterior-stabilized (PS) cement articulating spacers were used (CADAS; EverYoung BioDimensions, Taichung, Taiwan) [12]. A PS spacer with a post-cam construct was fabricated intraoperatively using a silicone mold. There are six sizes of femoral and tibial trials and silicone molds available, with sizes increasing from 1 to 6. A depth scale was used in the tibial trials and molds to determine the adequate thickness of the tibial spacers.

2.2. Patients

The study was approved by the local institutional review board, and all study participants provided informed consent. The study population was retrospectively derived from a database and included 278 adult patients with chronic PKI, based on the Musculoskeletal Infection Society criteria [13], who underwent a two-stage revision with PS spacers between January 2015 and March 2022, with a minimum follow-up of 1 year. Those with incomplete radiographic data, PKIs after primary TKA with highly constrained prostheses (e.g., NexGen Legacy Constrained Condylar or Rotating Hinge Knees), PKIs after revision TKA, fungal or tuberculous PKIs, above-knee amputation or death during the spacer period, spacer period of more than 1 year, or permanent spacer retention were excluded from the study. Consequently, there were 203 included patients.

All 203 surgeries were performed by eight fellow-trained arthroplasty surgeons in one center in order to ensure similar approaches to the surgery. The patients were divided into two consecutive cohorts on the basis of the annualized volume of RAs performed by the surgeon, which was obtained by dividing the total number of RA performed by a given surgeon per year by the number of years in which a surgeon performed at least one RA [14]. The estimates were then inspected, and surgeon cutoff points were chosen to divide the patients into two approximately equal volume-based medians. In other words, there were approximately 100 patients in each cohort. Therefore, high volume was defined as $\geq$14 RAs/year, and low volume was defined as <14 RAs/year [8].

2.3. Surgical Technique

All antibiotic-loaded cement spacers were prepared using a 1:5 ratio of antibiotics to bone cement (CMW3; DePuy Synthes, Warsaw, IN, USA). Vancomycin and ceftazidime were routinely used [15]. All spacers were implanted using a routine medial parapatellar approach.

The femoral and tibial trials were sized against the prostheses retrieved after resection, and the corresponding molds were chosen. By applying the femoral trial to the distal femur, the knee joint was positioned at full extension and 90° flexion to determine the adequate thickness of the tibial spacer with a depth scale for gap balance and bony defects. The femoral and tibial spacers were fabricated simultaneously. The femoral spacer was first cemented to the distal femur using an additional package of antibiotic-loaded cement. The tibial spacer was cemented to the proximal tibia after curing the femoral portion. The knee was fully extended with an appropriate alignment to allow the tibial cement to set. Lateral patellar release was performed if necessary. After November 2018, the cam of the femoral spacer was augmented with a K-wire as an endoskeleton to reinforce mechanical strength, according to the study design [8].

2.4. Postoperative Protocol

Radiographic evaluations, including knee weight-bearing anteroposterior, lateral, and Merchant views, and lower-limb scanogram, were conducted 1 week after RA (Figure 1). Radiographs were obtained monthly before reimplantation or when the patient experienced knee pain, swelling, instability, or deformity. Rehabilitation was partial weight-bearing with crutches without a hinged brace. Each patient received at least 4 weeks of organism-specific intravenous antibiotics according to the recommendations of an infection consultant. Oral antibiotics were maintained until C-reactive protein levels and erythrocyte sedimentation rates decreased.

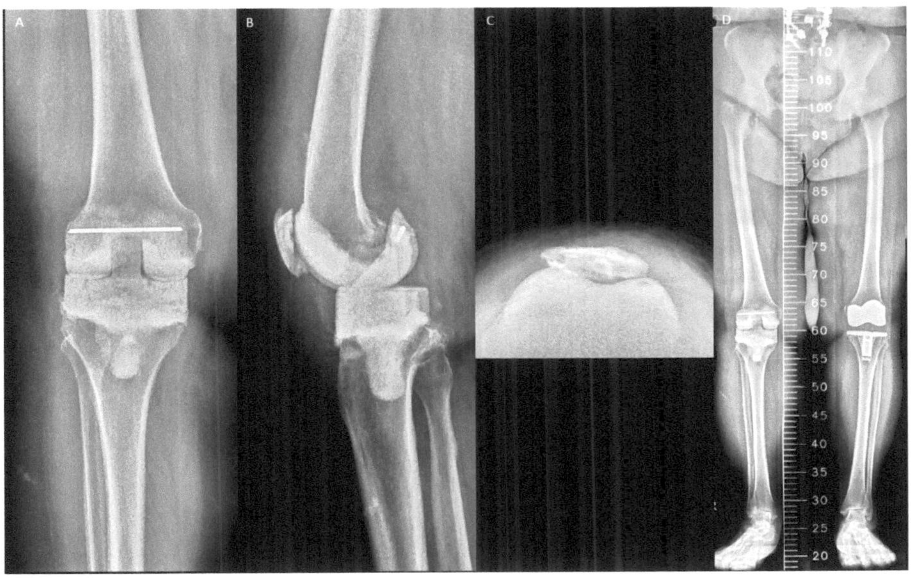

Figure 1. Radiographs of the articulating spacer obtained 1 week after resection arthroplasty: (**A**) standing anteroposterior view; (**B**) standing lateral view; (**C**) merchant view; and (**D**) scanogram of the lower limb.

The criteria for reimplantation included negative clinical signs of infection, normalized C-reactive protein and erythrocyte sedimentation rates, and negative arthrocentesis culture after a 2-week antibiotic holiday, which was defined as the period off antibiotics prior to reimplantation [16].

2.5. Evaluation

Demographic patient data, including medical comorbidities, microorganisms, spacer information, and spacer period, were extracted from medical charts. The following radiograph findings before reimplantation were recorded as mechanical complications using the INFINITT Picture Archiving and Communications System (INFINITT, Seoul, South Korea): spacer fracture, spacer migration, periprosthetic fracture, joint subluxation/dislocation, or extensor mechanism disruption [10,12,17]. Those with the progression of radiolucent lines around spacers without the above complications were excluded. Mixed complications were counted only as one major complication for each patient (e.g., spacer migration followed by joint dislocation). A musculoskeletal radiologist and two arthroplasty surgeons independently assessed and recorded all radiographic data.

Reoperations for mechanical complications, such as open reduction and partial or total spacer exchange, and extensor mechanism repair data were extracted from operation records; unexpected early reimplantation was also recorded. However, procedures for persistent infections, such as debridement, arthrotomy, and spacer exchange, were not included. Patients with simultaneous complications of mechanical and persistent infections were excluded from the reoperation count.

Spacer malalignment was defined as medial distal femoral angle $<2°$ or $>8°$ of valgus [18] (Figure 2A), medial proximal tibial angle (PTA) $<-3°$ or $>3°$ deviation from neutral [19] (Figure 2B), posterior tibial slope angle $<0°$ or $>7°$ of flexion [20] (Figure 2C), flexion contracture $>15°$ of flexion [21] (Figure 2D), and recurvatum deformity $>5°$ of extension [22] (Figure 2E); spacer malposition such as anterior femoral notching >3 mm in depth [23] (Figure 2F), and mediolateral or anteroposterior overhang of tibial spacers >3 mm [24] (Figure 2G,H). Cementing technique of spacers with or without cement stem extension was also recorded (Figure 2I,J). All the above data were measured on anteroposterior and lateral short-leg radiographs obtained 1 week after RA [25]. Lower-limb malalignment was defined as a hip–knee–ankle angle $<-10°$ (severe varus) or $>10°$ (severe valgus) on a scanogram [26] (Figure 2K). Patellar maltracking was defined as a lateral patellar tilt $>10°$ on the Merchant view [27] (Figure 2L). Data were measured separately by the same three physicians, and the mean value was calculated. Mixed outliers were also recorded for each patient.

2.6. Data Analyses

Data analyses were performed using the Statistical Package for the Social Sciences version 24.0 (IBM Corp., Armonk, NY, USA). Descriptive statistics are presented as means and 95% confidence intervals for continuous variables and as counts and percentages for categorical variables. Differences between two continuous variables were assessed using the Student's *t*-test, a statistical test used to determine if the means of two groups significantly differ from each other. Differences between two categorical variables were assessed using Fisher's exact test, a statistical test used to determine if the proportions of categories in two groups significantly differ from each other.

The reliability of mechanical complications was examined using the intraclass correlation coefficient (ICC). ICC is usually found to have a value between 0 and 1. An ICC value below 0.5 is regarded as a sign of poor reliability. On the other hand, an ICC value above 0.9 is regarded as a sign of excellent reliability. Associations between covariates and mechanical complications were determined with univariate logistic regression analysis. Covariates with statistically significant association on univariate analysis were included in a multivariable logistic regression model. Statistical significance was set at $p < 0.05$.

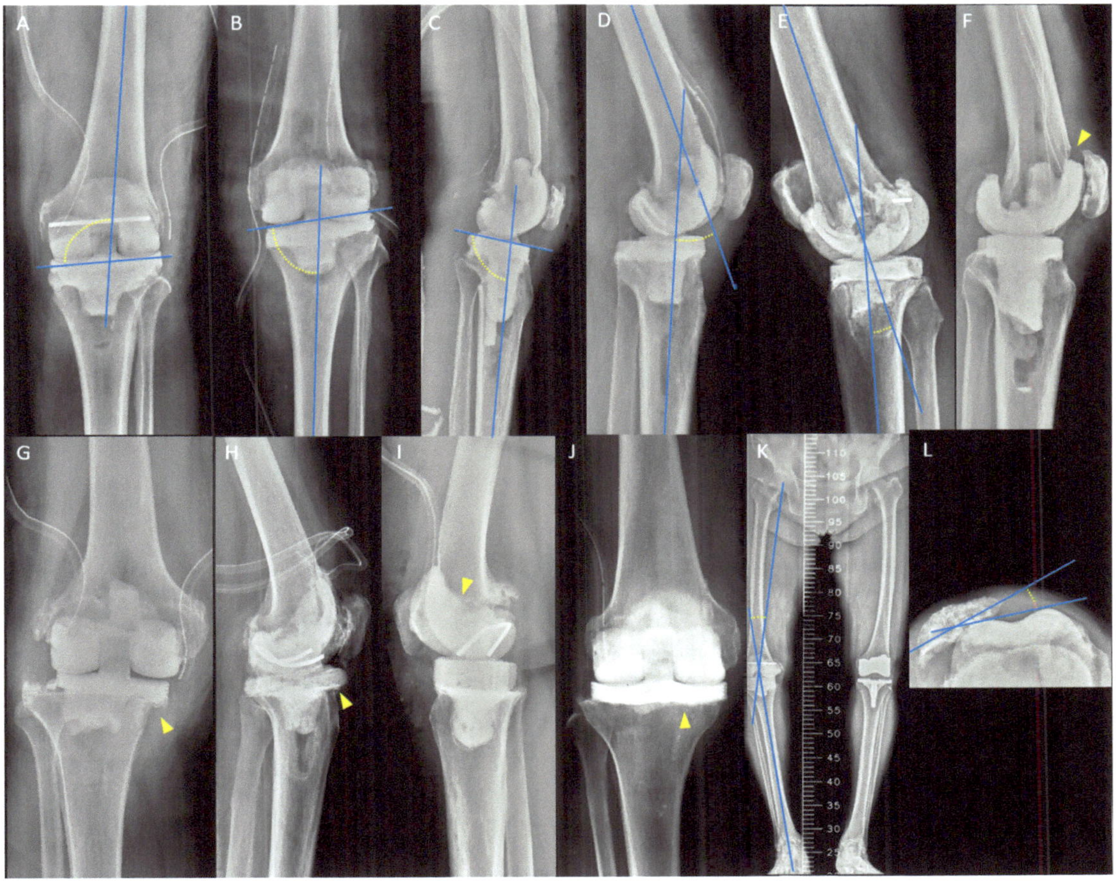

Figure 2. Malalignment and malposition of articulating spacer: (**A**) medial distal femoral angle, 100°; (**B**) medial proximal tibial angle, 83°; (**C**) posterior tibial slope angle, 96°; (**D**) flexion contracture, 26°; (**E**) recurvatum deformity, 11°; (**F**) femoral spacer notching (arrowhead); (**G**) medial overhang of tibial spacer (arrowhead); (**H**) anterior overhang of tibial spacer (arrowhead); (**I**) femoral spacer without cement stem extension (arrowhead); (**J**) tibial spacer without cement stem extension (arrowhead); (**K**) severe varus deformity with hip–knee–ankle angle, 16°; and (**L**) maltracking with lateral patellar tilt, 14°.

3. Results

Of the 203 included patients, 105 and 98 were treated by two HV and six LV surgeons, respectively. A single knee was used for each patient. The mean spacer duration was 14.3 weeks. The average follow-up time was 54.1 months (minimal and maximal were 13 and 79 months, respectively). Figure 3 presents the Strengthening the Reporting of Observational Studies in Epidemiology flowchart detailing the study design. There were no significant differences in the demographic data between patients who underwent HV or LV surgery (Table 1).

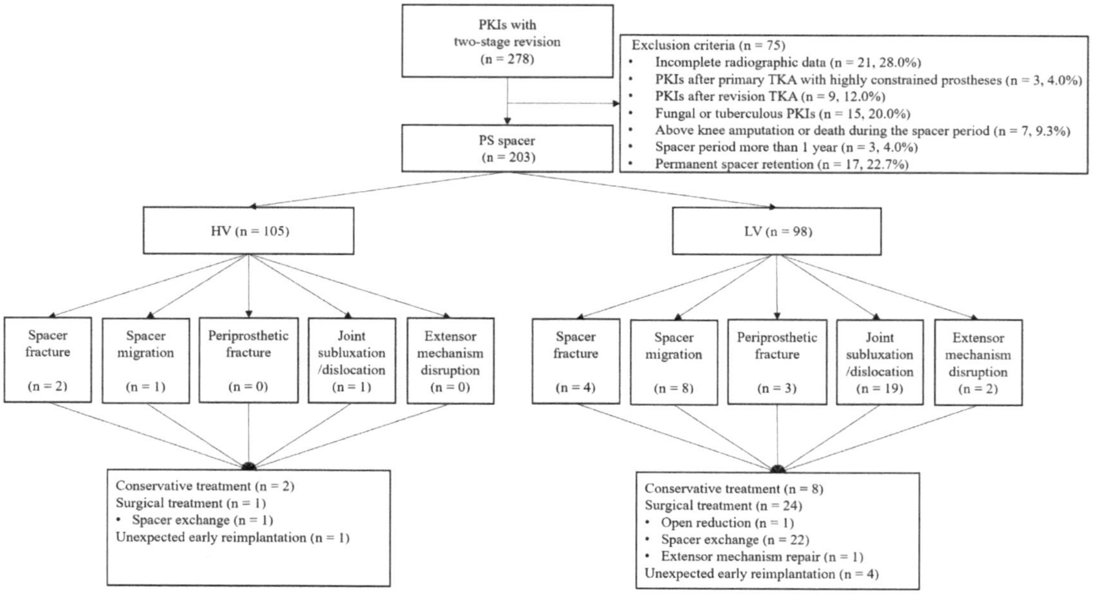

Figure 3. Strengthening the Reporting of Observational Studies in Epidemiology flowchart detailing the design of the study. PKIs, prosthetic knee infections; TKA, total knee arthroplasty; HV, high volume; LV, low volume.

Table 1. Demographic data of patients performed on by HV or LV surgeons.

Variables	Surgeon Volume		*p*-Value
	HV (n = 105)	**LV (n = 98)**	
Age (years)	67.4 (59–83)	66.1 (58–82)	0.713
Sex (female)	59 (56.2)	60 (61.2)	0.554
BMI (kg/m^2)	27.3 (21.0–34.6)	28.6 (21.7–33.2)	0.676
Right laterality	54 (51.4)	46 (46.9)	0.262
Constraint of the original prosthesis			
Cruciate-retaining	37 (35.2)	35 (35.7)	0.906
Posterior-stabilized	68 (64.8)	63 (64.3)	0.894
Smoking status			
Never-smokers	49 (46.7)	42 (42.9)	0.648
Current/ex-smokers	56 (53.3)	56 (57.1)	0.517
Insurance status			
Insured	104 (99.0)	98 (100.0)	0.930
Uninsured	1 (1.0)	0 (0.0)	0.964
Socioeconomic status			
Low	42 (40.0)	39 (39.8)	0.911
Middle	38 (36.2)	37 (37.8)	0.890
High	25 (23.8)	22 (22.4)	0.835
Charlson Comorbidity Index			
0	52 (49.5)	47 (48.0)	0.876
1–2	38 (36.2)	36 (36.7)	0.937
3+	15 (14.3)	15 (15.3)	0.882

Table 1. *Cont.*

Variables	Surgeon Volume		*p*-Value
	HV (n = 105)	LV (n = 98)	
McPherson host grade			
Uncompromised	51 (48.6)	45 (45.9)	0.704
Compromised	30 (28.5)	29 (29.6)	0.816
Substantially compromised	24 (22.9)	24 (24.5)	0.773
Microorganisms			
Staphylococcus species	35 (33.3)	33 (33.7)	0.938
Streptococcus species	20 (19.0)	18 (18.4)	0.863
Gram-positive cocci	15 (14.3)	14 (14.3)	0.999
Gram-negative species	16 (15.2)	17 (17.3)	0.803
Polymicrobial	8 (7.6)	8 (8.2)	0.830
Culture negative	11 (10.5)	8 (8.2)	0.709
Spacer information			
Femoral spacer size (No.)	3.9 (2–5)	3.8 (2–5)	0.774
Femoral spacer with endoskeleton	52 (49.5)	47 (48.0)	0.808
Tibial spacer size (No.)	4.0 (2–5)	3.9 (2–5)	0.822
Tibial spacer thickness (mm)	13.2 (10–19)	13.7 (10–21)	0.762
Spacer period (weeks)	14.6 (11–24)	14.4 (10–23)	0.825
Follow-up period (months)	54.7 (13–79)	53.6 (13–77)	0.692

Values are presented as means [95% confidence intervals] or n (%). HV, high volume; LV, low volume; BMI, body mass index; No., number.

There were 40 (19.7%) mechanical complications in 203 spacers. The ICC value for mechanical complications was 0.974 (range, 0.918 to 0.990; $p < 0.001$). The outcomes of spacer placement performed by HV or LV surgeons are presented in Table 2. The overall mechanical complication rate was lower in HV surgeons (3.8%) than in LV surgeons (36.7%) ($p < 0.001$), especially for joint subluxation or dislocation. The overall reoperation rate was lower in HV surgeons (0.9%) than in LV surgeons (24.5%) ($p < 0.001$), especially in spacer exchanges (Figure 4A–F). Among the cases that had 36 mechanical complications in LV surgeons, 66.7% (24/36) underwent reoperation, and 91.7% (22/24) of the reoperation procedures were spacer exchanges. There were 1 (0.9%) and 4 (4.1%) knees that had unexpected early reimplantation that were performed smoothly after mechanical complications by HV and LV surgeons, respectively (Figure 4G–I). Table 3 presents the details of these mechanical complications, in which all mechanisms were atraumatic and 92.5% had early failure (≤6 weeks) after spacer insertion. All knees requiring reoperation were treated with hinge braces and were free of instability before reimplantation.

Table 2. Outcomes of PKIs performed by HV and LV surgeons.

Parameters	Surgeon Volume		*p*-Value
	HV (n = 105)	LV (n = 98)	
Overall mechanical complication	4 (3.8)	36 (36.7)	**<0.001**
Spacer fracture	2 (1.9)	4 (4.1)	0.284
Spacer migration	1 (0.9)	8 (8.2)	0.051
Periprosthetic fracture	0 (0.0)	3 (3.1)	0.417
Joint subluxation/dislocation	1 (0.9)	19 (19.4)	**0.006**
Extensor mechanism disruption	0 (0.0)	2 (2.0)	0.596
Overall reoperation	1 (0.9)	24 (24.5)	**<0.001**
Open reduction	0 (0.0)	1 (1.0)	0.891
Spacer exchange	1 (0.9)	22 (22.4)	**0.003**
Extensor mechanism repair	0 (0.0)	1 (1.0)	0.891
Unexpected early reimplantation	1 (0.9)	4 (4.1)	0.110

Table 2. *Cont.*

Parameters	Surgeon Volume		*p*-Value
	HV (n = 105)	LV (n = 98)	
Constraint of reimplantation prosthesis			
Legacy constrained condylar knee	104 (99.1)	81 (82.7)	**0.007**
Rotating hinge knee	1 (0.9)	17 (17.3)	**0.008**

Values are presented as n (%). *p*-values in bold are statistically significant. PKI, prosthetic knee infection; HV, high volume; LV, low volume.

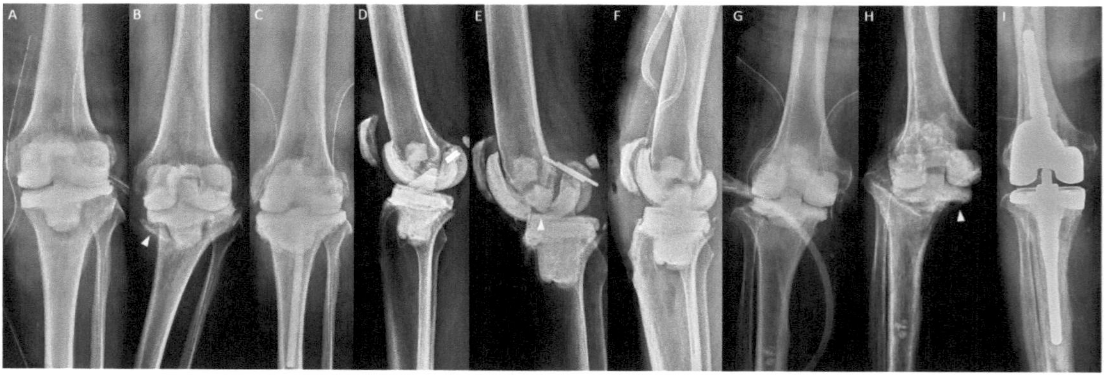

Figure 4. Examples of mechanical complication: (**A**) a 68-year-old man after resection arthroplasty (RA) with varus malalignment of tibial spacer (medial proximal tibial angle, 83°); (**B**) periprosthetic tibial fracture (arrowhead) 3 weeks after spacer insertion; (**C**) reoperation with tibial spacer exchange; (**D**) a 63-year-old man after RA with recurvatum deformity (12°); (**E**) femoral spacer fracture (arrowhead) 4 weeks after spacer insertion; (**F**) reoperation with both spacers exchange; (**G**) a 76-year-old woman after RA with tibial spacer without cement stem extension; (**H**) tibial spacer migration (arrowhead) 6 weeks after spacer insertion; and (**I**) unexpected early reimplantation smoothly.

Table 3. Details of patients with mechanical complications by HV and LV surgeons.

Patient No.	Age (Years)	Sex	Radiographic Outlier	Mechanical Complication	Timing after Spacer (Weeks)	Intervention
			HV surgeons			
1	69	F	PTA < 87°; HKA < −10°	TS migration	3	Hinge knee brace
2	64	M	TS overhang; DFA > 98°	TS fracture	4	Hinge knee brace
3	76	M	PTA > 90°; FS notching	TS fracture	5	Both spacers exchange
4	77	F	Recurvatum deformity	Joint dislocation	6	Unexpected early reimplantation
			LV surgeons			
1	73	M	PTA < 87°; DFA < 92°	TS fracture	2	Hinge knee brace
2	66	M	PTA < 87°; DFA > 98°	Joint subluxation	5	Hinge knee brace
3	82	M	PTA < 87°; FS notching	Joint subluxation	4	Hinge knee brace
4	63	F	PTA < 87°; patellar maltracking	Joint dislocation	4	Open reduction
5	66	F	PTA < 87°; HKA < −10°	Joint subluxation	3	TS exchange
6	86	F	PTA < 87°; flexion contracture	Joint dislocation	6	TS exchange
7	77	F	PTA < 87°	Joint dislocation	4	TS exchange
8	71	M	PTA < 87°; FS without cement stem extension	Joint dislocation	5	TS exchange

Table 3. *Cont.*

Patient No.	Age (Years)	Sex	Radiographic Outlier	Mechanical Complication	Timing after Spacer (Weeks)	Intervention
9	68	M	PTA < 87°	Periprosthetic tibial fracture	3	TS exchange
10	73	M	PTA < 87°; flexion contracture	Periprosthetic tibial fracture	4	Both spacers exchange
11	72	F	PTA < 87°; HKA < −10°	Both spacers fracture	6	Unexpected early reimplantation
12	68	F	Recurvatum deformity; DFA > 98°	Joint subluxation	5	Hinge knee brace
13	58	M	Recurvatum deformity	Joint subluxation	7	Hinge knee brace
14	72	M	Recurvatum deformity; TS without cement stem extension	Joint dislocation	4	Both spacers exchange
15	80	F	Recurvatum deformity; TS overhang	Joint dislocation	6	Both spacers exchange
16	66	F	Recurvatum deformity; HKA angle < −10°	Joint dislocation	7	FS exchange
17	60	M	Recurvatum deformity; DFA < 92°	Joint dislocation	4	FS exchange
18	80	F	Recurvatum deformity; patellar maltracking	Joint dislocation	8	FS exchange
19	67	F	Recurvatum deformity; FS notching	Joint dislocation	4	FS exchange
20	63	M	Recurvatum deformity	FS fracture	4	Both spacers exchange
21	60	M	Recurvatum deformity; FS without cement stem extension	Periprosthetic tibial fracture	6	Unexpected early reimplantation
22	61	M	TS without cement stem extension	Joint subluxation	4	Hinge knee brace
23	73	F	TS without cement stem extension; DFA > 98°	Joint subluxation	5	Hinge knee brace
24	61	M	TS without cement stem extension	Joint dislocation	3	Both spacers exchange
25	72	M	TS without cement stem extension; DFA < 92°	Joint dislocation	6	Both spacers exchange
26	75	F	TS without cement stem extension; patellar maltracking	TS migration	3	TS exchange
27	68	F	TS without cement stem extension	TS migration	4	TS exchange
28	69	M	TS without cement stem extension; PSA > 90°	TS migration	4	Both spacers exchange
29	73	F	TS without cement stem extension; TS overhang	TS migration	3	TS exchange
30	60	F	TS without cement stem extension; HKA angle < −10°	TS migration	5	TS exchange
31	66	M	TS without cement stem extension; PTA > 93°	TS migration	3	TS exchange
32	76	F	TS without cement stem extension	TS migration	6	Unexpected early reimplantation
33	77	F	TS overhang; DFA > 98°	Extensor mechanism disruption	3	TS exchange
34	67	M	Patellar maltracking; DFA < 92°	Extensor mechanism disruption	4	Lateral release
35	64	F	HKA < −10°; DFA < 92°	TS migration	6	Unexpected early reimplantation
36	72	F	HKA < −10°; flexion contracture	Both spacers fracture	3	Hinge knee brace

HV, high volume; LV, low volume; No., number; PTA, medial proximal tibial angle; HKA, hip–knee–ankle; TS, tibial spacer; PSA, posterior tibial slope angle; FS, femoral spacer; DFA, medial distal femoral angle.

Moreover, LV spacers were found to be associated with an increased requirement for highly constrained prostheses, such as rotating hinge knees (LV vs. HV, 17.3% vs. 0.9%, $p = 0.008$), and these knees had mechanical complications during the spacer period. Additionally, 47.2% (17/36) of patients required hinge knees after mechanical spacer failure. Table 4 presents the univariate risk factors for mechanical complications. The multivariate logistic regression analysis identified PTA < 87°, recurvatum deformity, and tibial spacer use without a cement stem extension as independent risk factors for mechanical failure (Table 5).

Table 4. Univariate analysis for risk factors associated with mechanical complications.

Variables	No Mechanical Complications (n = 163)	Mechanical Complications (n = 40)	Odds Ratio (95% CI)	*p*-Value
Age ≥ 65 years	128 (78.5)	30 (75.0)	0.91 (0.19–2.43)	0.638
Sex (female)	98 (60.1)	21 (52.5)	0.87 (0.12–1.92)	0.507
BMI ≥ 25.0 kg/m^2	96 (58.9)	28 (70.0)	4.43 (1.84–23.5)	**0.038**
Right laterality	77 (47.2)	23 (57.5)	1.99 (0.78–5.37)	0.059
Constraint of the original prosthesis				
Cruciate-retaining	57 (35.0)	15 (37.5)	1.02 (0.30–3.84)	0.710
Posterior-stabilized	106 (65.0)	25 (62.5)	0.90 (0.44–2.89)	0.569
Smoking status				
Never-smokers	73 (44.8)	18 (45.0)	1.00 (0.18–1.76)	0.996
Current/ex-smokers	90 (52.4)	22 (55.0)	1.02 (0.11–3.03)	0.834
Insurance status				
Insured	162 (99.4)	40 (100.0)	1.01 (0.33–1.68)	0.914
Uninsured	1 (0.6)	0 (0.0)	0.99 (0.22–1.37)	0.939
Socioeconomic status				
Low	64 (39.3)	17 (42.5)	1.09 (0.40–4.13)	0.456
Middle	61 (37.4)	14 (35.0)	0.93 (0.13–2.10)	0.741
High, n (%)	38 (23.3)	9 (22.5)	0.98 (0.38–2.31)	0.853
Charlson Comorbidity Index				
0	81 (49.7)	18 (45.0)	0.91 (0.51–2.27)	0.727
1–2	59 (36.2)	15 (37.5)	1.02 (0.52–2.63)	0.734
3+	23 (14.1)	7 (17.5)	1.03 (0.49–3.11)	0.515
McPherson host grade				
Uncompromised	79 (48.5)	17 (42.5)	0.89 (0.28–2.16)	0.760
Compromised	47 (28.8)	12 (30.0)	1.09 (0.35–4.61)	0.672
Substantially compromised	37 (22.7)	11 (27.5)	1.11 (0.59–5.48)	0.518
Microorganisms				
Staphylococcus species	54 (33.1)	14 (35.0)	1.07 (0.39–3.44)	0.630
Streptococcus species	31 (19.0)	7 (17.5)	0.98 (0.20–2.11)	0.832
Gram-positive cocci	24 (14.7)	5 (12.5)	0.97 (0.21–2.09)	0.814
Gram-negative species	27 (16.6)	6 (15.0)	0.98 (0.23–2.07)	0.827
Polymicrobial	13 (7.9)	3 (7.5)	1.00 (0.12–1.97)	0.964
Culture negative	14 (8.6)	5 (12.5)	1.12 (0.13–6.40)	0.633
Spacer information				
Femoral spacer size ≤ 2	76 (46.6)	23 (57.5)	3.97 (1.19–12.3)	**0.042**
Femoral spacer with endoskeleton	81 (49.7)	18 (45.0)	0.91 (0.48–2.33)	0.718
Tibial spacer size ≤ 2	73 (44.8)	21 (52.5)	3.19 (1.28–10.6)	0.051
Tibial spacer thickness ≥ 13 mm	79 (48.5)	22 (55.0)	3.04 (1.21–10.3)	0.053
Spacer malalignment and malposition				
DFA > 98°	18 (11.0)	5 (12.5)	1.02 (0.38–3.20)	0.863
DFA < 92°	11 (6.7)	5 (12.5)	3.03 (1.19–9.73)	0.056
PTA > 93°	8 (4.9)	1 (2.5)	0.90 (032–2.00)	0.778
PTA < 87°	15 (9.2)	12 (30.0)	6.76 (2.88–43.3)	**0.005**

Table 4. *Cont.*

Variables	No Mechanical Complications (n = 163)	Mechanical Complications (n = 40)	Odds Ratio (95% CI)	*p*-Value
PSA > 90°	7 (4.3)	2 (5.0)	1.01 (0.76–1.86)	0.901
PSA < 83°	2 (1.2)	0 (0.0)	0.98 (0.63–1.69)	0.806
Flexion contracture	15 (9.2)	3 (7.5)	0.97 (0.23–1.98)	0.716
Recurvatum deformity	9 (5.5)	11 (27.5)	7.31 (2.04–50.7)	**0.003**
FS notching	17 (10.4)	3 (7.5)	0.95 (0.51–1.87)	0.706
TS overhang	26 (15.9)	5 (12.5)	0.91 (0.28–1.83)	0.739
FS without cement stem extension	18 (11.0)	2 (5.0)	0.72 (0.13–1.76)	0.132
TS without cement stem extension	7 (4.3)	12 (30.0)	8.94 (2.76–62.6)	**<0.001**
HKA angle > 10°	2 (1.2)	0 (0.0)	0.98 (0.20–2.02)	0.870
HKA angle < −10°	10 (6.1)	7 (17.5)	4.11 (2.24–22.1)	**0.041**
Patellar maltracking	23 (14.1)	4 (10.0)	0.94 (1.11–2.04)	0.703
Spacer period (weeks)	14.1 (12–24)	14.9 (11–22)	1.11 (0.48–3.78)	0.529
Follow-up period (months)	54.2 (13–78)	53.9 (14–75)	0.97 (0.46–2.32)	0.631

Values are presented as n (%) and means [95% confidence intervals]. *p*-values in bold are statistically significant. BMI, body mass index; DFA, medial distal femoral angle; PTA, medial proximal tibial angle; PSA, posterior tibial slope angle; FS, femoral spacer; TS, tibial spacer; HKA, hip–knee–ankle; CI, confidence interval.

Table 5. Multivariate analysis for risk factors associated with mechanical complications.

Variables	Adjusted Odds Ratio (95% CI)	*p*-Value
BMI ≥ 25.0 kg/m^2	3.01 (1.32–12.4)	0.054
Femoral spacer size ≤ 2	2.63 (1.04–9.18)	0.127
Medial PTA < 87°	5.30 (2.11–37.6)	**0.009**
Recurvatum deformity	5.43 (1.93–41.4)	**0.007**
TS without cement stem extension	7.25 (2.24–53.5)	**0.002**
HKA angle < −10°	2.99 (1.62–11.4)	0.075

p-values in bold are statistically significant. CI, confidence interval; BMI, body mass index; PTA, proximal tibial angle; TS, tibial spacer; HKA, hip–knee–ankle.

4. Discussion

This study compared the mechanical complications in RA with articulating spacers according to surgeon volume and analyzed the risk factors. Our results showed that HV surgeons have a lower risk of overall mechanical complications and reoperations. The mechanical complications increased the level of constraint in the final revision of knee arthroplasty. The identified risk factors for mechanical failure were varus malalignment of the tibial spacer, recurvatum deformity, and tibial spacer without cement stem extension.

Tan et al. and George et al. reported that the rates of spacer mechanical complications were approximately 17 to 18% [28,29], and our rate, 19.7%, was consistent with these data. Our mechanical complications were all atraumatic, with the majority (92.5%) of spacers failing at <6 weeks, which might be related to the unstable biomechanical environment in PKI after RA [12], in addition to the weaker strength of the cement due to high-dose antibiotics [8], and the spacers were fixed in the bone ends without tight interdigitation for ease of removal with minimal bone loss [17]. Lin et al. found that a low surgical volume was a risk factor for spacer fractures with dislocation (odds ratio, 8.13) [8]. Moreover, better infection control and reimplantation rates have been reported in patients treated at specialized PKI centers [11]. Superior results have been found for the treatment at subspecialty centers of bone and joint infections [30]. Although our institution performed protocol-driven RAs with articulating spacers, the clinical results were different after analysis by surgical volume; HV surgeons had only 3.8% mechanical complications compared with 36.7% in LV surgeons. This discrepancy might be related to the learning curve or other associated risk factors, as LV surgeons only performed 4.23 RAs/year (range, 1.9 to 8.6 RAs/year), compared with HV surgeons, who treated multiple referred PKIs per month and performed many RAs annually.

Lau et al. found that 34.9% (22/63) of coronal subluxations of the articulating spacers required highly constrained knee systems during reimplantation [7]. Lin et al. reported that 70% (14/20) of dislocated spacers needed rotating hinge knees during reimplantation and proposed the concept "added insult to injury" when knee dislocations occur after RA with articulating spacers, suggesting that surgeons must strive to prevent joint dislocation during the spacer period [12]. Lin et al. also mentioned that to prevent post-cam failure of the PS spacer, an endoskeleton-reinforced femoral cam or additional hinged brace protection should be considered [8]. The finding that hinge knees were required in 47.2% of patients after mechanical spacer failure in our LV group is substantial, as this illustrates that the mechanical complications increased the level of constraint in final revision knee arthroplasty [12]. The above findings indicate that soft tissue compromise after an unstable joint event might be a predictor for highly constrained revision prostheses, which should be prepared and made available at the time of reimplantation. In two-stage revisions, Struelens et al. reported that only 43% of 154 articulating spacers were considered to be optimally sized and positioned, and mechanical complications were as high as 57% [6]. Lin et al. proposed that malalignment and malrotation of femoral and tibial spacers may lead to spacer overstress [8]. Our PTA <87° tibial spacers had a 5.3 times higher risk of mechanical complications. Medial tibial plateau bony loss or tibial bowing may lead to the varus placement of the tibial spacer [31]. Gililland et al. reported a useful technique to restore a stable bony tibial platform perpendicular to the mechanical axis by using a standard extramedullary tibial cutting jig and drop rod to avoid malalignment of the tibial spacer [32]. From our HV surgeon's experience, cement augmentation for bony defects is a promising technique if there is huge bone loss after the cut or if the joint line is too low.

Recurvatum deformity is the rarest form of instability after TKA and is difficult to correct [33]. Many studies have reported the use of rotating hinge knees as extension stops to treat post-TKA recurvatum [34]. The etiologies of pre-TKA and range of motion pre-RA in our cohort were not fully documented because of the multiple surgeons involved in the current study. However, multiple surgeries, including RA, might result in weak quadriceps and compromise of the extensor mechanism locking during the stance phase of gait, which adds to the risk of recurvatum, as hyperextension helps stabilize the knee [35]. A huge bony defect, insufficient collateral ligaments, and compromised posterior capsule during infection and after RA might result in joint laxity, even with a spacer [8,12,17]. Moreover, the post-cam mechanism of the PS spacer cannot resist the force during knee hyperextension; thus, mechanical complications occur. Although surgeons have used a depth scale within tibial trials and molds to ensure accurate thickness of the tibial spacer to fulfill the gap balance, there is still a learning curve during spacer fabrication and implantation [8]. Based on our HV surgeon's experience, careful examination of the ligament status during RA and placement of the knee with a spacer in slight flexion will avoid recurvatum complications.

In revision TKA, Oh et al. reported that metaphyseal fixation is important for ensuring the stability of the component [36]; Lee et al. recommended the use of an extension stem with a tibial canal filling ratio > 0.85 to minimize tibial component loosening [37]. Jung et al. reported that the spiked bottom of the tibial spacer provides superior stability and lowers spacer translation and tilting than a flat bottom [38]. Tsai et al. used the bottom of a tibial spacer with a cement stem extension to fill the metaphyseal defect and stably implant a spacer to minimize spacer loosening and migration [17]. Gililland et al. suggested a stemmed spacer with a cement dowel for increased antibiotic elution and added stable fixation [32]. We found that a femoral spacer without a cement stem extension was not associated with mechanical failure, suggesting a difference in bony end morphology and kinematics between the distal femur and the proximal tibia [8]. Our tibial spacers without cement stem extensions had a 7.25 times higher risk of mechanical complications, indicating that the stem plays an important role in additional stability. We strongly suggest using cement stem extensions of the tibial spacer to avoid spacer dislodgement.

This study has some limitations. This was a retrospective cohort study involving only one center. A higher-powered, prospective, controlled study with multiple centers is needed to confirm the effect of surgical volume. The physical activity levels of patients were not measured, and detailed chart records of every non-traumatic event were lacking, but there was no clear association between mechanical failure and more active patients [39]. The lack of computer tomography prevented spacer malrotation measurements, which may have caused loosening and further failure [40]. We did not evaluate the effect of spacers malfabricated by LV surgeons, which might increase the risk of mechanical complications [8].

Nevertheless, this was the first study to directly compare surgical volume in RA and radiographic measurements of articulating spacers with a large sample size and sufficient power to detect effects. Our results broadly indicate better outcomes for HV surgeons than for LV surgeons. This finding provides a further research platform for establishing a new subspecialty for prosthetic joint infection with experienced HV surgeons to improve the outcomes of two-stage revision.

5. Conclusions

This was the first retrospective study to compare the effect of surgeon volume on mechanical complications after RA with an articulating spacer in a two-stage revision of PKI. Based on the findings, we made the following three conclusions: (1) HV surgeons had a lower rate of mechanical complications and reoperations than LV surgeons; (2) mechanical complications increased the level of constraint in the final revision of knee arthroplasty; and (3) all surgeons should avoid tibial spacer varus malalignment and recurvatum deformity and always use a cement stem extension with a tibial spacer.

Author Contributions: Conceptualization, T.-L.L.; data curation, H.-Y.C.; formal analysis, Y.-C.F. and C.-H.T.; methodology, H.-T.C.; writing—original draft, C.-Y.K.; writing—review and editing, C.-Y.K., C.-H.T., Y.-C.F., H.-Y.C., H.-T.C. and T.-L.L. All authors have read and agreed to the published version of the manuscript.

Funding: The authors acknowledge receipt of a grant from the China Medical University Hospital (DMR-113-199).

Institutional Review Board Statement: The study was approved by the local institutional review board (CMUH107-REC3-130) and was performed in accordance with the ethical standards in the 1964 Declaration of Helsinki.

Informed Consent Statement: Informed consent was obtained from all subjects involved in the study.

Data Availability Statement: The authors declare that all data supporting the findings of this study are available within the article.

Acknowledgments: We thank the X-Dimension Center for Medical Research and Translation, Department of Orthopedics, Department of Radiology, Division of Infection Disease, Department of Internal Medicine, China Medical University Hospital, Taichung, Taiwan, for providing facility support services for the study.

Conflicts of Interest: The authors declare no conflicts of interest.

References

1. DeBoer, D.K. Comparison of Traditional Molded, First-Generation Premolded, and Second-Generation Premolded Antibiotic-Loaded Polymethylmethacrylate Articulating Spacers for Treatment of Chronic Prosthetic Joint Infection of the Knee. *J. Arthroplast.* **2020**, *35*, S53–S56. [CrossRef] [PubMed]
2. Magruder, M.L.; Yao, V.J.H.; Rodriguez, A.N.; Ng, M.K.; Piuzzi, N.S.; Mont, M.A. History of Diabetic Foot Ulcer is Associated With Increased Risk of Prosthetic Joint Infection and Sepsis After Total Joint Arthroplasty. *J. Arthroplast.* **2024**, *39*, 250–254. [CrossRef] [PubMed]
3. Piple, A.S.; Wang, J.C.; Kebaish, K.J.; Mills, E.S.; Oakes, D.A.; Lieberman, J.R.; Christ, A.B.; Heckmann, N.D. Does Prednisone Dose Affect Rates of Periprosthetic Joint Infection Following Primary Total Hip and Total Knee Arthroplasty? *J. Arthroplast.* **2023**, *38*, 1024–1031. [CrossRef] [PubMed]

4. Florance, J.; Chang, J.; Kelly, P.J.; Smith, D.; Bolognesi, M.; Seyler, T.; Ryan, S.P. Inferior Outcomes for Patients Transferred Between Surgical Stages for Knee Periprosthetic Joint Infection. *J. Arthroplast.* **2024**, *39*, 490–493. [CrossRef] [PubMed]
5. Nahhas, C.R.; Chalmers, P.N.; Parvizi, J.; Sporer, S.M.; Berend, K.R.; Moric, M.; Chen, A.F.; Austin, M.S.; Deirmengian, G.K.; Morris, M.J.; et al. A Randomized Trial of Static and Articulating Spacers for the Treatment of Infection Following Total Knee Arthroplasty. *J. Bone Jt. Surg. Am.* **2020**, *102*, 778–787.
6. Struelens, B.; Claes, S.; Bellemans, J. Spacer-related problems in two-stage revision knee arthroplasty. *Acta Orthop. Belg.* **2013**, *79*, 422–426. [PubMed]
7. Lau, A.C.; Howard, J.L.; Macdonald, S.J.; Teeter, M.G.; Lanting, B.A. The Effect of Subluxation of Articulating Antibiotic Spacers on Bone Defects and Degree of Constraint in Revision Knee Arthroplasty. *J. Arthroplast.* **2016**, *31*, 199–203. [CrossRef] [PubMed]
8. Lin, T.L.; Tsai, C.H.; Fong, Y.C.; Shie, M.Y.; Chen, H.Y.; Chen, Y.W. Posterior-Stabilized Antibiotic Cement Articulating Spacer with Endoskeleton-Reinforced Cam Reduces Rate of Post-Cam Mechanical Complications in Prosthetic Knee Infection: A Preliminary Study. *J. Arthroplast.* **2022**, *37*, 1180–1188.e1182. [CrossRef] [PubMed]
9. Wilson, S.; Marx, R.G.; Pan, T.J.; Lyman, S. Meaningful Thresholds for the Volume-Outcome Relationship in Total Knee Arthroplasty. *J. Bone Jt. Surg. Am.* **2016**, *98*, 1683–1690. [CrossRef]
10. Roof, M.A.; Sharan, M.; Merkow, D.; Feng, J.E.; Long, W.J.; Schwarzkopf, R.S. High-volume revision surgeons have better outcomes following revision total knee arthroplasty. *Bone Jt. J.* **2021**, *103*, 131–136. [CrossRef]
11. Fehring, T.K.; Otero, J.E.; Curtin, B.M.; Fehring, K.A.; Metcalf, R.; Rowe, T.M.; Springer, B.D. Does Treatment at a Specialized Prosthetic Joint Infection Center Improve the Rate of Reimplantation? *J. Arthroplast.* **2023**, *38*, S314–S317. [CrossRef] [PubMed]
12. Lin, T.L.; Tsai, C.H.; Fong, Y.C.; Shie, M.Y.; Chen, H.Y.; Chen, Y.W. Cruciate-Retaining vs Posterior-Stabilized Antibiotic Cement Articulating Spacers for Two-Stage Revision of Prosthetic Knee Infection: A Retrospective Cohort Study. *J. Arthroplast.* **2021**, *36*, 3750–3759.e3752. [CrossRef] [PubMed]
13. Parvizi, J.; Gehrke, T. Definition of periprosthetic joint infection. *J. Arthroplast.* **2014**, *29*, 1331. [CrossRef] [PubMed]
14. Wright, J.D.; Hershman, D.L.; Burke, W.M.; Lu, Y.-S.; Neugut, A.I.; Lewin, S.N.; Herzog, T.J. Influence of surgical volume on outcome for laparoscopic hysterectomy for endometrial cancer. *Ann. Surg. Oncol.* **2012**, *19*, 948–958. [CrossRef] [PubMed]
15. Hsu, Y.H.; Hu, C.C.; Hsieh, P.H.; Shih, H.N.; Ueng, S.W.; Chang, Y. Vancomycin and Ceftazidime in Bone Cement as a Potentially Effective Treatment for Knee Periprosthetic Joint Infection. *J. Bone Jt. Surg. Am.* **2017**, *99*, 223–231. [CrossRef] [PubMed]
16. Diaz-Ledezma, C.; Higuera, C.A.; Parvizi, J. Success after treatment of periprosthetic joint infection: A Delphi-based international multidisciplinary consensus. *Clin. Orthop. Relat. Res.* **2013**, *471*, 2374–2382. [CrossRef] [PubMed]
17. Tsai, C.-H.; Hsu, H.-C.; Chen, H.-Y.; Fong, Y.-C.; Ho, M.-W.; Chou, C.-H.; Chen, Y.-W.; Shie, M.-Y.; Lin, T.-L. A preliminary study of the novel antibiotic-loaded cement computer-aided design-articulating spacer for the treatment of periprosthetic knee infection. *J. Orthop. Surg. Res.* **2019**, *14*, 136. [CrossRef] [PubMed]
18. Ritter, M.A.; Davis, K.E.; Meding, J.B.; Pierson, J.L.; Berend, M.E.; Malinzak, R.A. The effect of alignment and BMI on failure of total knee replacement. *J. Bone Jt. Surg. Am.* **2011**, *93*, 1588–1596. [CrossRef] [PubMed]
19. Abane, L.; Anract, P.; Boisgard, S.; Descamps, S.; Courpied, J.P.; Hamadouche, M. A comparison of patient-specific and conventional instrumentation for total knee arthroplasty: A multicentre randomised controlled trial. *Bone Jt. J.* **2015**, *97*, 56–63. [CrossRef]
20. Kim, Y.H.; Park, J.W.; Kim, J.S.; Park, S.D. The relationship between the survival of total knee arthroplasty and postoperative coronal, sagittal and rotational alignment of knee prosthesis. *Int. Orthop.* **2014**, *38*, 379–385. [CrossRef]
21. Song, S.J.; Lee, H.W.; Park, C.H. Predictors of Recurrent Flexion Contracture after Total Knee Arthroplasty in Osteoarthritic Knees with Greater Than 15° Flexion Contracture. *Clin. Orthop. Surg.* **2023**, *15*, 770–780. [CrossRef] [PubMed]
22. Shagawa, M.; Maruyama, S.; Sekine, C.; Yokota, H.; Hirabayashi, R.; Hirata, A.; Yokoyama, M.; Edama, M. Comparison of anterior knee laxity, stiffness, genu recurvatum, and general joint laxity in the late follicular phase and the ovulatory phase of the menstrual cycle. *BMC Musculoskelet. Disord.* **2021**, *22*, 886. [CrossRef] [PubMed]
23. Puranik, H.G.; Mukartihal, R.; Patil, S.S.; Dhanasekaran, S.R.; Menon, V.K. Does Femoral Notching During Total Knee Arthroplasty Influence Periprosthetic Fracture. A Prospective Study. *J. Arthroplast.* **2019**, *34*, 1244–1249. [CrossRef] [PubMed]
24. Nielsen, C.S.; Nebergall, A.; Huddleston, J.; Kallemose, T.; Malchau, H.; Troelsen, A. Medial Overhang of the Tibial Component Is Associated With Higher Risk of Inferior Knee Injury and Osteoarthritis Outcome Score Pain After Knee Replacement. *J. Arthroplast.* **2018**, *33*, 1394–1398. [CrossRef]
25. Tammachote, N.; Kriengburapha, N.; Chaiwuttisak, A.; Kanitnate, S.; Boontanapibul, K. Is Regular Knee Radiograph Reliable Enough to Assess the Knee Prosthesis Position? *J. Arthroplast.* **2018**, *33*, 3038–3042. [CrossRef]
26. Thienpont, E.; Parvizi, J. A New Classification for the Varus Knee. *J. Arthroplast.* **2016**, *31*, 2156–2160. [CrossRef]
27. Kleimeyer, J.P.; McQuillan, T.J.; Arsoy, D.; Aggarwal, V.K.; Amanatullah, D.F. Agreement and Reliability of Lateral Patellar Tilt and Displacement following Total Knee Arthroplasty with Patellar Resurfacing. *J. Knee Surg.* **2021**, *34*, 717–720. [CrossRef]
28. George, J.; Miller, E.M.; Curtis, G.L.; Klika, A.K.; Barsoum, W.K.; Mont, M.A.; Higuera, C.A. Success of Two-Stage Reimplantation in Patients Requiring an Interim Spacer Exchange. *J. Arthroplast.* **2018**, *33*, S228–S232. [CrossRef]
29. Tan, T.L.; Goswami, K.; Kheir, M.M.; Xu, C.; Wang, Q.; Parvizi, J. Surgical Treatment of Chronic Periprosthetic Joint Infection: Fate of Spacer Exchanges. *J. Arthroplast.* **2019**, *34*, 2085–2090.e2081. [CrossRef]

30. Ferry, T.; Seng, P.; Mainard, D.; Jenny, J.-Y.; Laurent, F.; Senneville, E.; Grare, M. Jolivet-Gougeon, A.; Bernard, L.; Marmor, S. The CRIOAc healthcare network in France: A nationwide Health Ministry program to improve the management of bone and joint infection. *Orthop. Traumatol. Surg. Res.* **2019**, *105*, 185–190. [CrossRef]
31. Palanisami, D.; Jagdishbhai, C.P.; Manohar, M.; Ramesh, P.; Natesan, R.; Shanmuganathan, R. Improving the accuracy of tibial component placement during total knee replacement in varus knees with tibial bowing: A prospective randomised controlled study. *Knee* **2019**, *26*, 1088–1095. [CrossRef] [PubMed]
32. Gililland, J.M.; Carlson, V.R.; Fehring, K.; Springer, B.D.; Griffin, W.L.; Anderson, L.A. Balanced, Stemmed, and Augmented Articulating Total Knee Spacer Technique. *Arthroplast. Today* **2020**, *6*, 981–986. [CrossRef] [PubMed]
33. Mesnard, G.; Batailler, C.; Fary, C.; Schmidt, A.; Servien, E.; Lustig, S. Posterior-Stabilized TKA in Patients with Severe Genu Recurvatum Achieves Good Clinical and Radiological Results at 5-year Minimum Follow-Up: A Case-Controlled Study. *J. Arthroplast.* **2021**, *36*, 3154–3160. [CrossRef] [PubMed]
34. Al-Jabri, T.; Brivio, A.; Maffulli, N.; Barrett, D. Management of instability after primary total knee arthroplasty: An evidence-based review. *J. Orthop. Surg. Res.* **2021**, *16*, 729. [CrossRef] [PubMed]
35. Cottino, U.; Sculco, P.K.; Sierra, R.J.; Abdel, M.P. Instability After Total Knee Arthroplasty. *Orthop. Clin. N. Am.* **2016**, *47*, 311–316. [CrossRef] [PubMed]
36. Oh, J.H.; Scuderi, G.R. Zonal Fixation in Revision TKA: The Key Is Metaphyseal Fixation. *J. Knee Surg.* **2021**, *34*, 1402–1407. [CrossRef] [PubMed]
37. Lee, S.H.; Shih, H.N.; Chang, C.H.; Lu, T.W.; Chang, Y.H.; Lin, Y.C. Influence of extension stem length and diameter on clinical and radiographic outcomes of revision total knee arthroplasty. *BMC Musculoskelet. Disord.* **2020**, *21*, 15. [CrossRef] [PubMed]
38. Jung, K.H.; Lee, C.C.; Kim, T.H.; Han, J.W.; Park, K.B. Does spiked tibial cement spacer reduce spacer-related problems in two-stage revision total knee arthroplasty for infection? *Int. Orthop.* **2022**, *46*, 2009–2017. [CrossRef]
39. Sylvia, L.G.; Bernstein, E.E.; Hubbard, J.L.; Keating, L.; Anderson, E.J. Practical guide to measuring physical activity. *J. Acad. Nutr. Diet.* **2014**, *114*, 199–208. [CrossRef]
40. Yoshino, K.; Hagiwara, S.; Nakamura, J.; Tsukeoka, T.; Tsuneizumi, Y.; Ohtori, S. Intra- and interobserver reliability and agreement in three-dimensional computed tomography measurements of component positions after total knee arthroplasty. *Knee* **2019**, *26*, 1102–1110. [CrossRef]

Journal of *Personalized Medicine*

MDPI

Article

Postoperative Activity and Knee Function of Patients after Total Knee Arthroplasty: A Sensor-Based Monitoring Study

Sebastian Kersten [1,2], Robert Prill [1,3,*], Hassan Tarek Hakam [1], Hannes Hofmann [1], Mahmut Enes Kayaalp [1,4], Jan Reichmann [5] and Roland Becker [1,3,*]

1 Center of Orthopaedics and Traumatology, University Hospital Brandenburg/Havel, Brandenburg Medical School Theodor Fontane, 14770 Brandenburg an der Havel, Germany
2 Department of Orthopaedic Surgery, Sana Kliniken Sommerfeld, 16766 Sommerfeld, Germany
3 Faculty of Health Sciences Brandenburg, Brandenburg Medical School Theodor Fontane, 14770 Brandenburg an der Havel, Germany
4 Istanbul Kartal Research and Training Hospital, 34865 Istanbul, Turkey
5 StatConsult GmbH, 39112 Magdeburg, Germany
* Correspondence: robert.prill@mhb-fontane.de (R.P.); roland_becker@yahoo.de (R.B.)

Abstract: Inertial measurement units (IMUs) are increasingly being used to assess knee function. The aim of the study was to record patients' activity levels and to detect new parameters for knee function in the early postoperative phase after TKA. Twenty patients (n = 20) were prospectively enrolled. Two sensors were attached to the affected leg. The data were recorded from the first day after TKA until discharge. Algorithms were developed for detecting steps, range of motion, horizontal, sitting and standing postures, as well as physical therapy. The mean number of steps increased from day 1 to discharge from 117.4 (SD ± 110.5) to 858.7 (SD ± 320.1), respectively. Patients' percentage of immobilization during daytime (6 a.m. to 8 p.m.) was 91.2% on day one and still 69.9% on the last day. Patients received daily continuous passive motion therapy (CPM) for a mean of 36.4 min (SD ± 8.2). The mean angular velocity at day 1 was 12.2 degrees per second (SD ± 4.4) and increased to 28.7 (SD ± 16.4) at discharge. This study shows that IMUs monitor patients' activity postoperatively well, and a wide range of interindividual motion patterns was observed. These sensors may allow the adjustment of physical exercise programs according to the patient's individual needs.

Keywords: inertial measurement unit; TKA; remote monitoring; knee function; rehabilitation

Citation: Kersten, S.; Prill, R.; Hakam, H.T.; Hofmann, H.; Kayaalp, M.E.; Reichmann, J.; Becker, R. Postoperative Activity and Knee Function of Patients after Total Knee Arthroplasty: A Sensor-Based Monitoring Study. *J. Pers. Med.* **2023**, *13*, 1628. https://doi.org/10.3390/jpm13121628

Academic Editor: Günther Maderbacher

Received: 20 October 2023
Revised: 14 November 2023
Accepted: 20 November 2023
Published: 21 November 2023

1. Introduction

There is a growing interest in the use of sensors in orthopaedics. Sensor-assisted Total Knee Arthroplasty or Sensor-guided Arthroscopy Training are common examples [1,2].

Wearable inertial measurement units (IMUs) are commonly used for assessing lower extremity kinematics or after knee surgery [3]. They have gained popularity for monitoring gait and joint function after TKA [4,5]. These systems allow a more objective assessment of joint function offering detailed information about patients' activity and, thus, allowing the adjustment of postoperative rehabilitation programs on a more individual basis [6]. A large number of different sensor types are used, depending on the aim of the investigations with the result of a wide variety of spatiotemporal gait parameters [3]. No standardization in measurement techniques or test methodologies has been established so far [4]. Sensors with triaxial accelerometers are most frequently used in the functional assessment of patients. Other uses include either inertial sensors equipped with triaxial accelerometers, gyroscopes or magnetometers [4,6]. In addition, there is a significant variance in the sampling rate of data recordings and the application of the sensor to the patient [4]. Accelerometers allow extended monitoring of the general activity of patients due to better battery life and low power consumption. The limitation of accelerometers is the lack of data quality referred to as complex gait parameters in comparison to IMUs. Most accelerator systems were attached to the trunk, at the hip or back. Knee joint-specific parameters have not been primarily investigated so far [4]. IMUs are valid and reliable for TKA patients.

IMUs enable the acquisition of more complex gait parameters and activity patterns. Previously published studies analyzed the pre- and postoperative activity level measured by the patient's maximum knee angle, the impact load, the asymmetry in the impact load, the knee joint flexion while walking, the number of steps taken and the time spent sitting. The most commonly used measurement of patients' activity level is the number of steps taken and range of knee motion [4,7–12]. However, current IMUs are unable to clearly differentiate between certain activities of daily life (ADL), such as walking, sitting, standing, lying down, climbing stairs, getting passive motion or physical therapy [7]. Most sensor-based examinations after knee replacement refer to the functional analysis preoperatively and weeks to months postoperatively. As the research group showed in previous work, sensor-based assessment is reliable, but the functional assessment of improvement after TKA and also in hip arthroplasty is diverse, not consistent and lacks adherence to core outcome sets [6,13–16]. Most sensor-based examinations after knee replacement refer to the functional analysis preoperatively and weeks to months postoperatively. It is known that early and standardized recovery concepts after TKA and total hip arthroplasty (THA) significantly reduce the length of hospital stay and the occurrence of complications [17,18]. However, the function of the knee joint and activity of the patient in the early postoperative phase after TKA lacks objective data.

The aim of this study was to record the postoperative level of patients' activity in more detail during their hospital stay after primary TKA using a novel IMU. Therefore, the main purpose of the study is to prove the feasibility of providing complex and novel movement parameters during a hospital stay. Novel parameters were measured analyzing knee function and patient's activity after surgery better. It was hypothesized that the novel IMU would allow the characterization of different patient activities in more detail quantitatively during the early postoperative period.

2. Materials and Methods

2.1. Participants

From September 2021 to March 2022, a total of 20 patients (n = 20) were prospectively enrolled in the study.

The inclusion criterion for this study comprised patients treated with primary total knee arthroplasty. Patients with unicondylar knee arthroplasty; postoperative complications, such as bleeding, signs of infection, severe joint effusion; an allergy to plasters, cognitive deficits or multimorbidity (ASA III–IV), were excluded from the study.

Ethical approval was obtained from the Ethics Committee of the Landesärztekammer Brandenburg, Germany (ref.-number: AS 92(bB)/2018). All participants gave written consent.

2.2. Sensors

The Orthronic Smart Knee (OSK) sensor works with an inertial measurement unit, including a 3-axis accelerometer/gyroscope/magnetometer. Other integrated sensors record air pressure and temperature. The size of the sensor was 37 mm × 26 mm × 8 mm (L × W × H) and a total weight of 6 g. The sensors recorded knee motion at a sampling rate of 25 Hz and transmitted it to a tablet via Bluetooth. The sampling rate for measuring the air pressure, temperature and the number of steps was reduced to 1 Hz.

If the IMU was unable to transfer data directly via Bluetooth, measurement data were automatically saved in the integrated 512 MB memory in offline mode.

In the present study, the 290 mAh lithium polymer battery enabled measurement intervals of 24 h in total.

The reliability of the sensors was proven during development by comparison to a 10-camera motion capture system (VICON-MX-S, Vicon Motion System Ltd., Oxford, UK) with defined movements. A deviation with a mean of 5 degrees was shown [19]. Other patient-related factors that could affect the reliability and accuracy of the sensors, such as the attachment to the skin or the tilting of the sensors due to muscle movements, were also examined using the VICON® camera system. The more complex motion analysis of the sensors showed a deviation of a maximum of six degrees.

2.3. Sensor Application

The measurements started on the first postoperative day. The proximal and distal sensors were positioned on the ventral distal thigh over the quadriceps tendon and the medial tibia (Figure 1). After connecting the sensors with the mobile application via Bluetooth, the data were continuously recorded in offline mode for 24 h a day. The duration of recording the motion was up to six days, from the first postoperative day to when the patients were discharged (POD 1–6) from the hospital. The sensors were changed daily every afternoon. For documentation purposes, a picture of the patient's knee with the applied sensor was taken every time. Additionally, patients protocoled their daily activities, such as walking, going to the toilet, different exercises, physical therapy and climbing stairs, including the corresponding time in which these activities took place. After the sensors were removed, data were extracted to a tablet via Bluetooth.

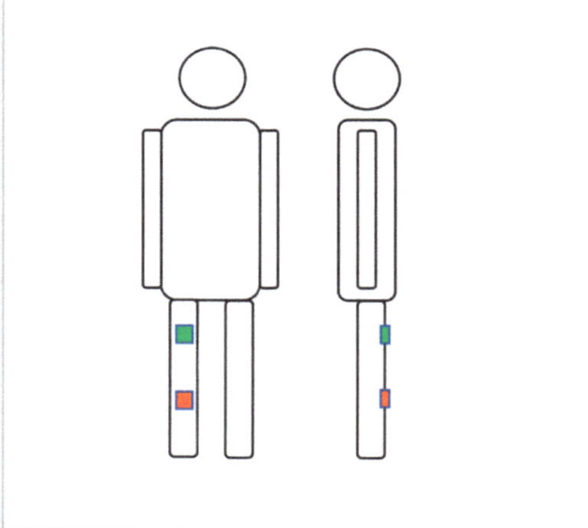

Figure 1. Sensor placement at the femur and tibia using sticky tape.

2.4. Measurements

2.4.1. Daily Distribution of Standing, Horizontal and Sitting Position

The integrated accelerators allowed the analysis of patients' positions. A standing position was characterized by the vertical orientation of the thigh, a sitting position by the horizontal orientation of the thigh and the vertical orientation of the lower leg and a horizontal position by the horizontal orientation of both the upper and the lower part of the leg (Figure 2). To ensure that sensor vibrations did not affect the recording of the posture, measurements made using the accelerometer were filtered based on average data. The size of the data window was 25 samples, which corresponds to a data collection time of one second.

The definition of a standing position was the inclination of the upper leg with more than 45 degrees with respect to the transverse plane ($\alpha > 45°$). The definition of a sitting position was the combination of an inclination of the upper leg with less than 45 degrees and an inclination of the lower leg with more than 30 degrees with respect to the transverse plane each ($\alpha < 45°$, $\beta > 30°$). The lying position was defined as a combination of an inclination of the upper leg with less than 12 degrees and an inclination of the lower leg with less than 30 degrees with respect to the transverse plane each ($\alpha < 12°$, $\beta < 30°$). The data were analyzed for daytime from 6 a.m. to 8 p.m.

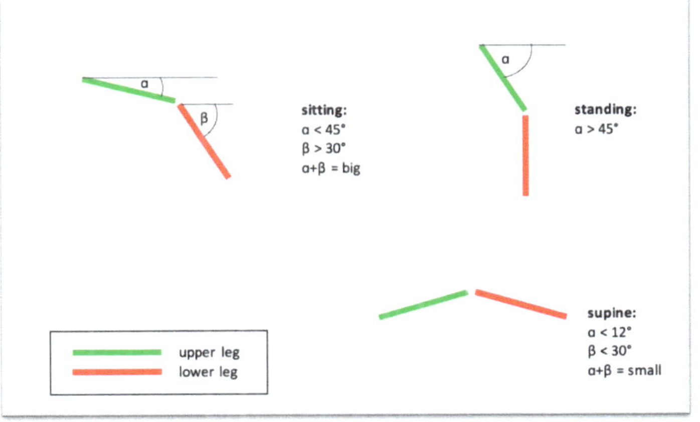

Figure 2. Algorithms for detecting different postures of the patients, such as sitting, standing and lay position by analyzing the position of the femur and tibia.

The postures were calculated at each measurement time. After that, all samples at which a standing, sitting or lying posture was detected were counted. The total time for each posture was computed relating to the number of samples meeting the described angle criteria.

2.4.2. Movements

A detected movement of the knee joint is defined as a significant change in knee angle and direction of movement (Figure 3). The threshold was set for a significant change in knee angle to 10 degrees to distinguish clearly between real movements and tremors or flickers. In addition, it was not assumed that the knee angle would return to the initial value after the movement. An example of that is the transition from a horizontal to a sitting position. A new movement was detected as soon as the knee angle moved more than 10 degrees from the last maximum and ended when a counter movement of more than 10 degrees was detected.

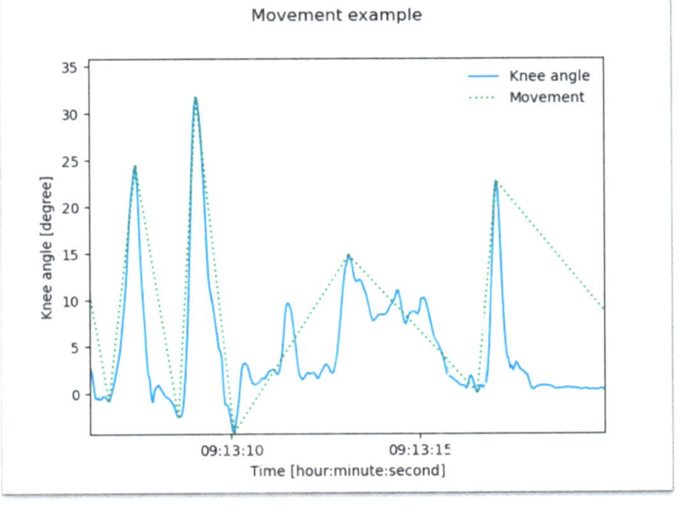

Figure 3. Algorithm to detect and discriminate real movements from trembling or flickering. A significant change in flexion angle greater than 10 degrees (threshold 10°) was counted as a movement.

2.4.3. Steps

Algorithms for counting steps, which are standard in many accelerators, are supposed to be less sensitive at the early stage after surgery because of the slow walking speed and limited weight bearing.

Patient step counting was calculated based on the detected posture and changes in the knee angle. A standing posture with a significant change of the knee angle of more than 20 degrees was counted as a step (Figure 4).

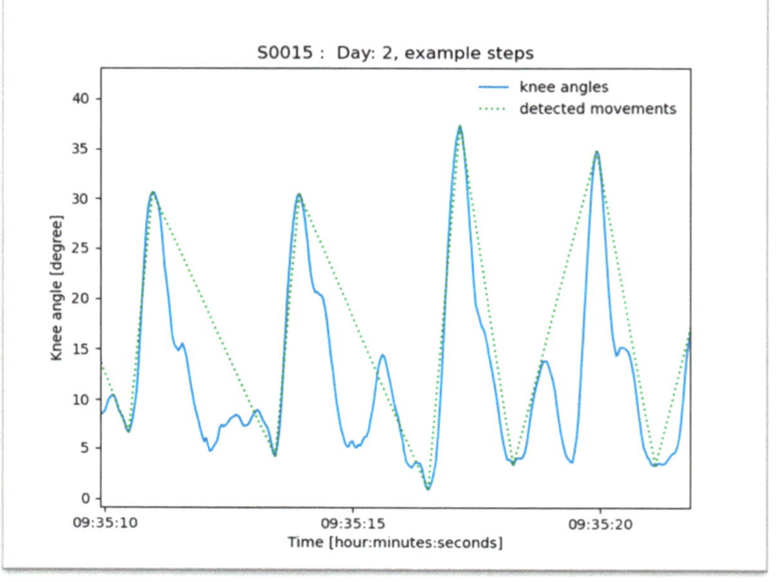

Figure 4. Algorithm to detect steps. Steps were detected when a standing posture with a significant change of the knee angle greater than 20 degrees was measured by IMUs (threshold > 20 degrees).

2.4.4. Movement over Angular Velocity

The knee angle is measured at a rate of 25 times per second by the OSK-Sensor. The angular velocity can be calculated using the first-time derivative.

$$Angular\ velocity = \frac{change\ of\ knee\ angle}{time\ interval}$$

In the present measurements, the time interval was 40 ms.

2.4.5. Continuous Passive Motion (CPM)

CPM was defined as a significant change in knee movements with an angle change of more than 10 degrees in a supine position of the patient. Furthermore, the movements were slow and lasted longer than 12 s but less than a minute. A CPM training session involved at least 10 resembling motions within a 15 min interval.

A minimum of 20 repetitions were executed during a session.

2.4.6. Stationary Cycling (MOTOmed®)

During the hospital stay, patients performed training sessions on a stationary bike (MOTOmed®, RECK-Technik GmbH & Co. KG, 88422 Betzenweiler, Germany). This exercise was performed while being seated and took less than three seconds per repetition. During practice, these repetitive motions took place in rapid succession. If at least 250 of these repetitive motions were detected without a break, the period was classified as a MOTOmed® exercise session.

2.5. Statistical Analysis

All patient and operation-related data were pseudonymized and transferred to an Excel sheet using the serial number "S00XX". The patient's names and case numbers were removed. Assigning case-related data to specific patients was thereby not possible. All continuous variables were expressed as mean and standard deviation. Figures have been prepared with Python for SPSS V26 (IBM SPSS Inc., Chicago, IL, USA).

No inductive statistics were used as no statistical hypothesis had to be proven. Analysis was performed on a visual and documentation basis.

3. Results

3.1. Demographics

A total of 20 patients participated in the study, including nine males (n = 9) and eleven females (n = 11), with a mean age of 69.6 years (SD ± 8.8) and a mean BMI of 29.9 (SD ± 5.6). The demographic data of the participants are shown in Table 1.

Table 1. Patients' demographic data.

Patient Data	Overall (n = 20)
female: male (n)	11: 9
Age (years)	69.6 (SD ± 8.8)
Heigth (in m)	1.7 (SD ± 0.1)
Weight (in kg)	84.4 (SD ± 19)
BMI (kg/m^2)	29.9 (SD ± 5.6)

All continuous variables were expressed as mean and standard deviation. Abbreviations: BMI—Body Mass Index.

3.2. Measurements of Patient Activity and Knee Function

3.2.1. Steps per Hour

The mean number of steps on the first day after surgery was 117.4 (SD ± 110.5) and increased to 858.7 steps (SD ± 320.1) at the time of discharge (POD 6). The average activity level of patient S0011 is shown in Figure 5. The activity level was reflected by the number of steps taken per hour. Step frequency increased significantly from the first postoperative day.

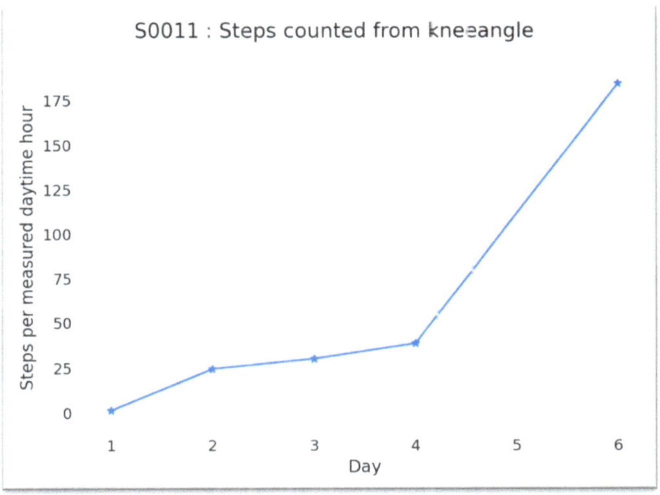

Figure 5. Steps normalized per hour. Day 1 = first day after surgery, Patient S0011.

The average number of steps performed by the patients during the entire measurement period between day 1 and day 6 was 526.2 steps per day (SD ± 431.9). The mean knee flexion angle of all patients during walking was 30.9° across all days (SD ± 6.5). The mean walking time of the patients during the entire measurement period was 16.1 min per day (SD ± 11.1).

3.2.2. Analysis of Postures during Daily Activity

There was an increase in the general activity with an increasing time of standing and sitting positions and a decrease in the time of horizontal position. Figure 6 shows the results of patient S0017 for illustrative purposes. The distribution of the position of the patients on the first day after surgery was 91.2% in a horizontal position, 6.4% in a sitting position and 2.4% in a standing position and changed to 69.9% in a horizontal position, 22.9% in a sitting position and 7.1% in a standing position at the final day.

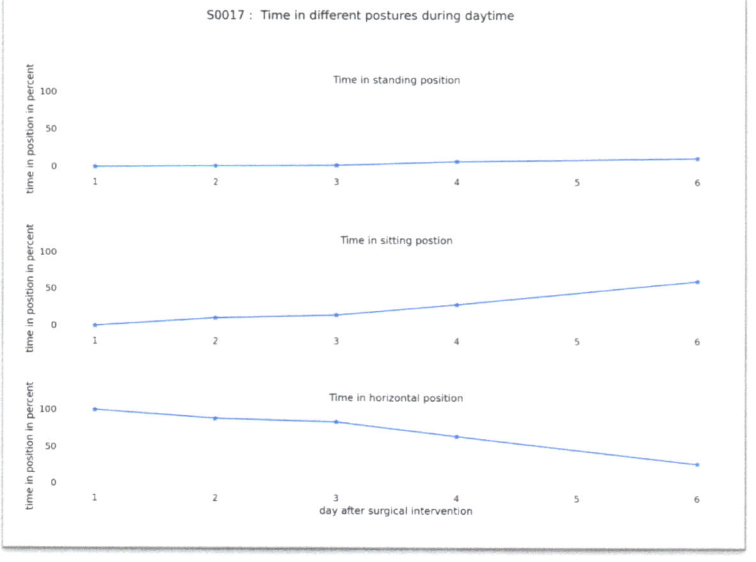

Figure 6. Time given in percentage of different postures from 6 a.m. to 8 p.m. of Patient S0017.

3.2.3. Physical Therapy, CPM-Device

CPM was performed for a mean of 36.4 min per day (SD ± 8.2). An example of a detected CPM training is shown in Figure 7. The mean knee angle achieved was 82.9 degrees (SD ± 21.7) with a repetition rate of 42.1 (SD ± 16.3) movements per application. Ninety degrees of knee flexion was achieved between the third and fourth day after surgery (Figure 8).

3.2.4. Knee Flexion over Angular Velocity

Data evaluation of angular velocities used in motion revealed a significant trend over the period of early postoperative rehabilitation. The mean velocity for a defined knee joint movement with an angle change of more than 10 degrees regardless the activity was 12.2 degree per second on day one of the measurement (SD ± 4.4). On the last day of the measurement, the mean velocity was 28.7 degrees per second (SD ± 16.4). Patients moved their affected knee joints in a small range and at low speeds in the first few days after surgery. During their hospital stay, they increased their range of motion and angular velocity, which could already be seen at the time of discharge from the hospital. This is shown in Figures 9 and 10 using patients S0011 and S0017 as an example.

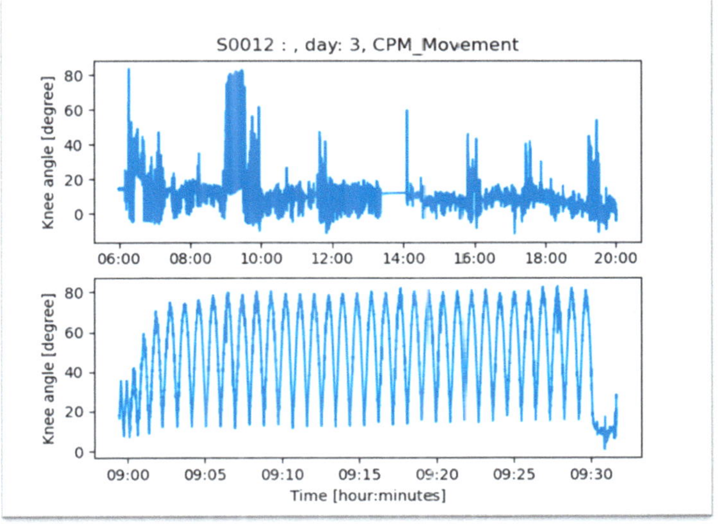

Figure 7. Example of a detected CPM training session for patient S0012 during 9.00 and 9.30 am. Despite the fact that the patients can move the knee up to 80 degrees, active movement was significantly lower throughout the day.

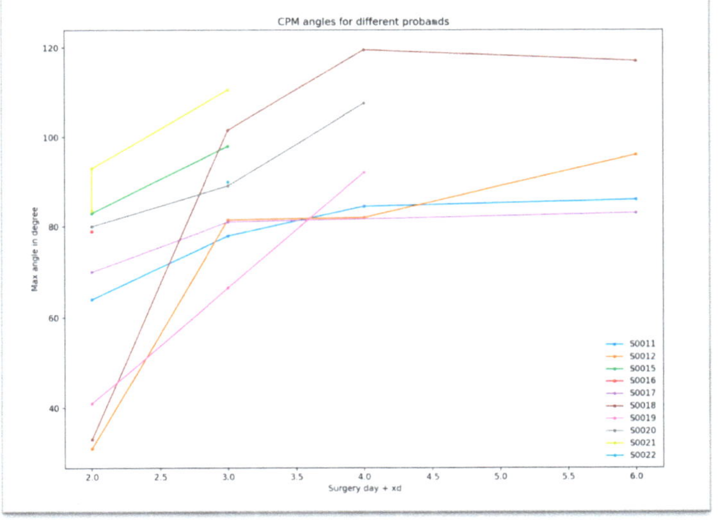

Figure 8. CPM-detected continuous passive motion angles over days.

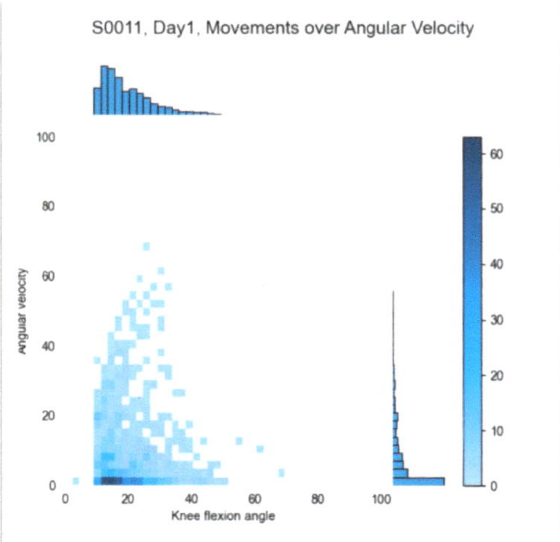

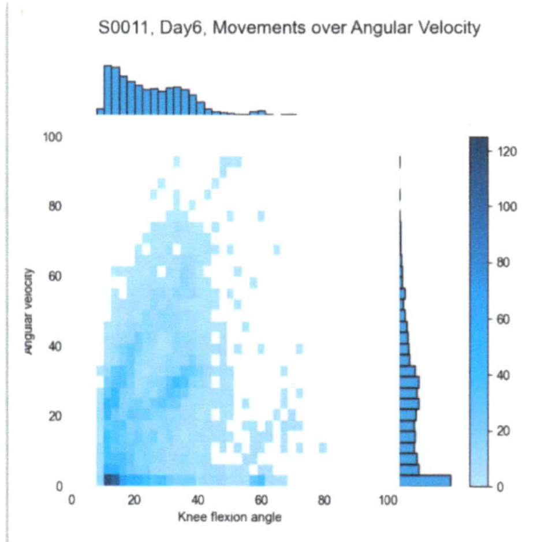

(a) (b)

Figure 9. Knee flexion over angular velocity for patient S0011: (**a**) first day after surgery and (**b**) last day of measurement, sixth postoperative day. The color of the squares shows the distribution of the flexion angle in regard to the angle velocity. The dark color means more frequent movement.

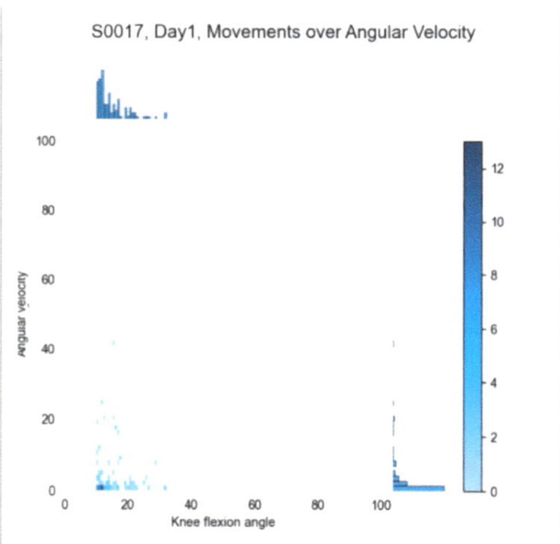

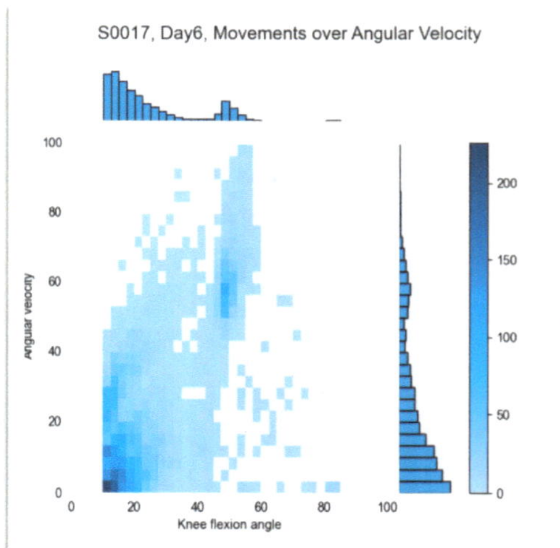

(a) (b)

Figure 10. Knee flexion over angular velocity for patient S0017: (**a**) first day after surgery and (**b**) last day of measurement, sixth postoperative day. The color of the squares shows the distribution of the flexion angle in regard to the angle velocity. The dark color means more frequent movement.

3.2.5. Analysis of Stair Climbing

Going up and down stairs is one of the patient's most important therapeutic goals for the activities of daily life. This is characterized by a rapid change in altitude detected by the air pressure sensor with simultaneous knee activity, as shown in Figure 11. Using an elevator is an alternative to climbing stairs in the early period after surgery. The change in altitude is also recognized by air pressure, but the associated knee movement is missing (Figure 12).

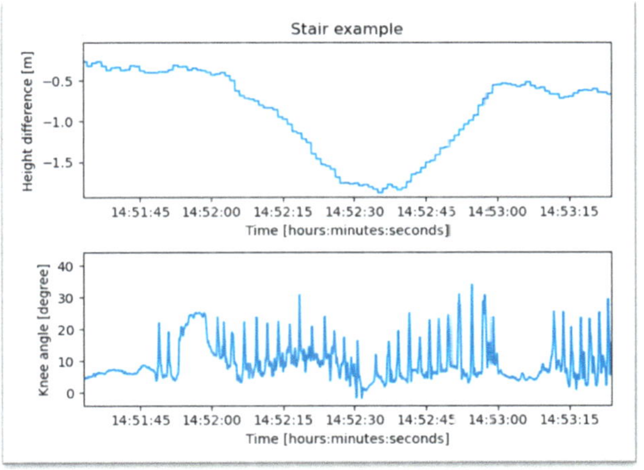

Figure 11. Example of a patient who is climbing stairs.

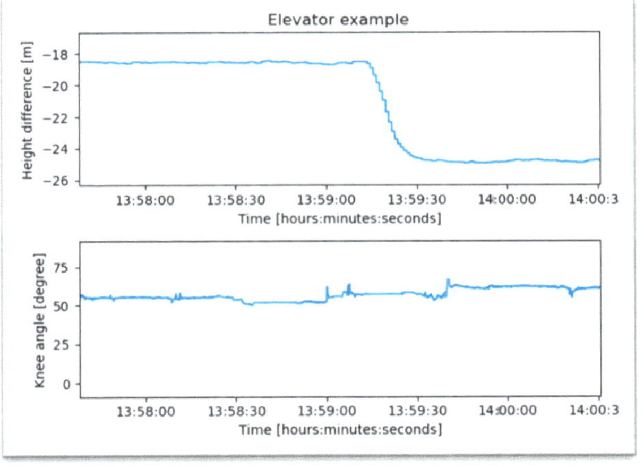

Figure 12. A patient using an elevator.

4. Discussion

The most important finding of the study is that IMUs offered detailed monitoring of different patient's activities after TKA in the early postoperative period. The hypothesis, that the novel IMU allows to characterize different patient's activities in more detail quantitatively at the early postoperative period, can be accepted. The recording of several functional parameters of the knee joint made it possible to compare the activity level of the participants with the result of major differences in the level of postoperative mobilization between patients.

In addition to recording functional parameters of the knee joint, the sensors allowed to identify various activities of the patient's daily life. Patient's therapy goals during the hospital stay and criteria of discharge can be objectified quantitatively with the help of data analysis of the IMU. The sensors could be an important tool to analyze the postoperative progress in patient mobilization and recovery in real time. Due to the more detailed assessment of patients, activity rehabilitation programs can be individualized according to patients' need.

The integration of different sensor types into the IMU allowed us to distinguish different functional and activity parameters, which could be quantified using algorithms. However, some of the parameters had to be read out manually, such as climbing stairs or using an elevator. Different methods are used in sensor-based measurements of knee patients. The review of Small et al. has shown that the type of sensors being used for gait analysis changed between 2008 and 2016. Initially, uniaxial accelerometers were used. Recent publications mostly imply triaxial IMUs [1].

A two-sensor-based measurement method was used in the current study, with one sensor placed over the anterior distal femur and the other over the proximal medial tibial surface. A similar position was reported by others [9]. However, alternatively, sensor positioning may also be in the sagittal plane at the lateral femur and tibia for instance [20]. There are also measuring systems with multiple sensors available, which are attached in different positions on the body and often fixed with multiple belts, such as in the DynaPort®-System [21,22]. It seems that none of the sensor systems allow a consistently high level of accuracy in all motion variables. It depends on which parameters are examined [23]. The usage of more than two sensors makes recording daily activity rather impossible. The key determinant of applicability seems to be the simple usage for patients.

The threshold of 20 degrees was chosen to distinguish true steps with near-normal knee function from movements rather than performed with a stiff knee joint while walking. Thus, a change of knee angle of less than 20 degrees was not detected routinely by the current algorithm. Therefore, the number of steps may be underrepresented. The threshold is still of debate and may be reduced to 15 degrees for instance. Hayashi et al. used an IMU attached with a belt around the waist and showed a significantly higher rate of counted steps per day for patients in the early postoperative phase after TKA (POD 3–10) [24]. However, the IMU attached to the waist might also have detected other motion activities than the overall step counting. The number of steps is recorded by others using a single IMU according to an algorithm, which distinguished between static and dynamic activities, counting all peaks in the vertical signal as steps and all peaks in the anteroposterior signal as a complete gait cycle (2 steps) [25].

In order to record the different postures of the patients, we used an algorithm that calculated the position of the two sensors in relation to one another. Based on previously defined angles, we differentiated between a sitting, standing or horizontal position. Lipperts et al. chose another method by using the inclination angle of a single triaxial accelerometer for distinguishing between sitting and standing positions and defined them as static activities [25].

An increase in the physical activity of the patients was observed during early rehabilitation at the hospital. While patients predominantly were lying in bed during the first few days, the activity pattern shifted to a higher percentage of standing and sitting positions. However, significant heterogeneity between the patients was noted.

Patients' motivation but also patients' preparation for surgery and the early postsurgical period showed a significant impact [17,18]. Taking the current data into consideration, it shows that early inpatient mobilization after TKA was rather insufficient. There seems to be a lack of encouragement for patients' mobilization. The results emphasize the need for advanced recovery concepts. Generally, perioperative care and integrated education programs are insufficiently focused but hold high potential for outcome improvement [26].

The movements in regard to angular velocities showed an interesting trend over the entire time of recording. Already in the first days after surgery, the patients used higher knee angles with a higher speed more frequently. An explanation might include the reduction in pain and swelling, improved coordination while walking and a wider range of motion. Muscle function may improve at the early phase as well, which requires a more detailed investigation of muscle function prior to and at the early stage after surgery.

The level of overall activity during hospital the stay was rather low. These data cannot be generalized, however, as they represent the results of a single center. These data are valuable to improve patients' care during hospitalization and can be used for quality assessment purposes.

The collected data showed that patients presented a very individual level of mobility. In some cases, the percentage between the standing and supine positions differed in an inversely proportional manner to the number of days after surgery. Some patients had a larger number of steps taken on the 2nd than on the 3rd or 4th postoperative day. Nevertheless, it should be noted that the recording period involved the weekends as well. Less activity during the weekend in general seems to be partially caused by fewer physiotherapy sessions. It also showed that the patient's mobility seems to rely mainly on daily physiotherapy compared to self-mobilization, which is partially in line with other findings [27] but supports the need for defining precise programs for home-based approaches. This also shows the strong need for digital and virtual reality rehabilitation approaches [28–30]. The importance of compliance issues is still unknown when self-mobilization is recommended. A high variance in the level of activity of patients after TKA was reported by others, but it has to be mentioned that the measurement with a watch is a limiting factor in the comparison of the data with our study [31].

The sensors are not only applicable for general monitoring of patient's daily activity but may also be used while performing specific exercises. The Sensor and accompanying System will be used in an already planned clinical trial on Compliance parameters for physical therapy interventions after Total Knee Arthroplasty with a focus on APP-based rehabilitation. It might also be used in combination with Neural Networks for decision making in primary or revision TKA [32]. They could also further inform predictive model studies on pain or length of stay [33].

Due to the ease of use and independence from the clinical environment, the IMU could also measure patients' mobilization after discharge from the hospital providing important information to the Orthopedic Surgeon and practitioner at the follow-up examination.

5. Conclusions

The objective parameter of movements over angular velocities could be a new decisive criterion for the patient's knee function. The wearable IMUs are a quick and easy solution to measure a patient's activity level and could contribute to low-threshold monitoring of postoperative knee function to help quantify improvements in both activity and knee motion following TKA. The usage of IMUs might aid in the development of home-based physical exercise programs. These programs might be adjusted to the individual needs of patients.

6. Limitation

A limiting confounder could be the visible presence of the sensors, which could have led to an increase in compliance by both the patients and the physical therapists. Furthermore, some of the participants in the study were on different wards with different therapists, who might have administered different rehabilitative exercises. On the other hand, this often reflects the reality of inpatient physiotherapeutic care. The number of steps might have been underrepresented due to our algorithm.

Additionally, the length of inpatient stays was different in some cases, which had an impact on the duration of the postoperative measurement and, thus, on the increase in the activity pattern.

Author Contributions: Conceptualization, R.B., R.P. and J.R.; methodology, R.P. and J.R.; software, J.R.; validation, S.K., H.H. and J.R.; formal analysis, S.K. and J.R.; investigation, S.K. and H.T.H.; resources, J.R. and R.B.; data curation, J.R.; writing—original draft preparation, S.K., R.P. and H.H.; writing—review and editing, R.B., H.H., H.T.H. and M.E.K.; visualization J.R.; supervision, R.P. and R.B.; funding acquisition, J.R. and R.B. All authors have read and agreed to the published version of the manuscript.

Funding: Funded by the Ministry of Science, Research and Cultural Affairs of the State of Brandenburg. Material delivery and Sensor development was financed by StatConsults GmbH. Funded by the Brandenburg Medical School (MHB) publication fund supported by the German Research Foundation (DFG).

Institutional Review Board Statement: Ethical approval was obtained from the Ethics Committee of the Landesärztekammer Brandenburg, Germany (ref.-number: AS 92(bB)/2018).

Informed Consent Statement: Informed consent was obtained from all subjects involved in the study.

Data Availability Statement: Additional Data is available on request from the corresponding author.

Conflicts of Interest: Author Jan Reichmann was employed by the company StatConsult GmbH. The remaining authors declare that the research was conducted in the absence of any commercial or financial relationships that could be construed as a potential conflict of interest.

References

1. Sava, M.-P.; Hara, H.; Alexandra, L.; Hügli, R.W.; Hirschmann, M.T. Verasense sensor-assisted total knee arthroplasty showed no difference in range of motion, reoperation rate or functional outcomes when compared to manually balanced total knee arthroplasty: A systematic review. *Knee Surg. Sports Traumatol. Arthrosc.* **2023**, *31*, 1851–1858. [CrossRef]
2. Aksoy, M.E.; Kocaoglu, B.; İzzetoglu, K.; Agrali, A.; Yoner, S.I.; Polat, M.D.; Kayaalp, M.E.; Yozgatli, T.K.; Kaya, A.; Becker, R. Assessment of learning in simulator-based arthroscopy training with the diagnostic arthroscopy skill score (DASS) and neurophysiological measures. *Knee Surg. Sports Traumatol. Arthrosc.* **2023**. *Epub ahead of print*. [CrossRef]
3. Zeng, Z.; Liu, Y.; Hu, X.; Tang, M.; Wang, L. Validity and Reliability of Inertial Measurement Units on Lower Extremity Kinematics During Running: A Systematic Review and Meta-Analysis. *Sports Med. Open* **2022**, *8*, 86. [CrossRef]
4. Small, S.R.; Bullock, G.S.; Khalid, S.; Barker, K.; Trivella, M.; Price, A.J. Current clinical utilisation of wearable motion sensors for the assessment of outcome following knee arthroplasty: A scoping review. *BMJ Open* **2019**, *9*, e033832. [CrossRef]
5. Gianzina, E.; Kalinterakis, G.; Delis, S.; Vlastos, I.; Platon Sachinis, N.; Yiannakopoulos, C.K. Evaluation of gait recovery after total knee arthroplasty using wearable inertial sensors: A systematic review. *Knee* **2023**, *41*, 190–203. [CrossRef]
6. Prill, R.; Walter, M.; Królikowska, A.; Becker, R. A Systematic Review of Diagnostic Accuracy and Clinical Applications of Wearable Movement Sensors for Knee Joint Rehabilitation. *Sensors* **2021**, *21*, 8221. [CrossRef]
7. Bolam, S.M.; Batinica, B.; Yeung, T.C.; Weaver, S.; Cantamessa, A.; Vanderboor, T.C.; Yeung, S.; Munro, J.T.; Fernandez, J.W.; Besier, T.F.; et al. Remote Patient Monitoring with Wearable Sensors Following Knee Arthroplasty. *Sensors* **2021**, *21*, 5143. [CrossRef]
8. Chapman, R.M.; Moschetti, W.E.; Van Citters, D.W. Stance and swing phase knee flexion recover at different rates following total knee arthroplasty: An inertial measurement unit study. *J. Biomech.* **2019**, *84*, 129–137. [CrossRef]
9. Chiang, C.Y.; Chen, K.H.; Liu, K.C.; Hsu, S.J.; Chan, C.T. Data Collection and Analysis Using Wearable Sensors for Monitoring Knee Range of Motion after Total Knee Arthroplasty. *Sensors* **2017**, *17*, 418. [CrossRef]
10. Daugaard, R.; Tjur, M.; Sliepen, M.; Lipperts, M.; Grimm, B.; Mechlenburg, I. Are patients with knee osteoarthritis and patients with knee joint replacement as physically active as healthy persons? *J. Orthop. Transl.* **2018**, *14*, 8–15. [CrossRef]
11. Webber, S.C.; Strachan, S.M.; Pachu, N.S. Sedentary Behavior, Cadence, and Physical Activity Outcomes after Knee Arthroplasty. *Med. Sci. Sports Exerc.* **2017**, *49*, 1057–1065. [CrossRef]
12. Christiansen, M.B.; Thoma, L.M.; Master, H.; Voinier, D.; Schmitt, L.A.; Ziegler, M.L.; LaValley, M.P.; White, D.K. Feasibility and Preliminary Outcomes of a Physical Therapist-Administered Physical Activity Intervention After Total Knee Replacement. *Arthritis Care Res.* **2020**, *72*, 661–668. [CrossRef]
13. Prill, R.; Becker, R.; Schulz, R.; Michel, S.; Hommel, H. No correlation between symmetry-based performance measures and patient-related outcome prior to and after total knee arthroplasty. *Knee Surg. Sports Traumatol. Arthrosc.* **2022**, *30*, 3155–3161. [CrossRef]
14. Adriani, M.; Becker, R.; Milano, G.; Lachowski, K.; Prill, R. High variation among clinical studies in the assessment of physical function after knee replacement: A systematic review. *Knee Surg. Sports Traumatol. Arthrosc.* **2023**, *31*, 3854–3860. [CrossRef]
15. Singh, J.A.; Murphy, S.; Bhandari, M. Assessment of the methodologic quality of medical and surgical clinical trials in patients with arthroplasty. *J. Rheumatol.* **2009**, *36*, 2642–2654. [CrossRef]
16. Singh, J.A.; Dohm, M.; Sprowson, A.P.; Wall, P.D.; Richards, B.L.; Gossec, L.; Hawker, G.A.; Riddle, D.L.; Buchbinder, R. Outcome Domains and Measures in Total Joint Replacement Clinical Trials: Can We Harmonize Them? An OMERACT Collaborative Initiative. *J. Rheumatol.* **2015**, *42*, 2496–2502. [CrossRef]

17. Kehlet, H. Fast-track hip and knee arthroplasty. *Lancet* **2013**, *381*, 1600–1602. [CrossRef]
18. Khan, S.K.; Malviya, A.; Muller, S.D.; Carluke, I.; Partington, P.F.; Emmerson, K.P.; Reed, M.R. Reduced short-term complications and mortality following Enhanced Recovery primary hip and knee arthroplasty: Results from 6000 consecutive procedures. *Acta Orthop.* **2014**, *85*, 26–31. [CrossRef]
19. Kayaalp, M.E.; Agres, A.N.; Reichmann, J.; Bashkuev, M.; Duda, G.N.; Becker, R. Validation of a Novel Device for the Knee Monitoring of Orthopaedic Patients. *Sensors* **2019**, *19*, 5193. [CrossRef]
20. Rahman, J.; Tang, Q.; Monda, M.; Miles, J.; McCarthy, I. Gait assessment as a functional outcome measure in total knee arthroplasty: A cross-sectional study. *BMC Musculoskelet. Disord.* **2015**, *16*, 66. [CrossRef]
21. Zhang, H.H.; Yan, S.H.; Fang, C.; Guo, X.Y.; Zhang, K. Clinical Evaluation and Gait Characteristics before and after Total Knee Arthroplasty Based on a Portable Gait Analyzer. *Orthop. Surg.* **2016**, *8*, 360–366. [CrossRef]
22. Van Hemert, W.L.; Senden, R.; Grimm, B.; Kester, A.D.; van der Linde, M.J.; Heyligers, I.C. Patella retention versus replacement in total knee arthroplasty; functional and clinimetric aspects. *Arch. Orthop. Trauma Surg.* **2009**, *129*, 259–265. [CrossRef]
23. Sharifi Renani, M.; Myers, C.A.; Zandie, R.; Mahoor, M.H.; Davidson, B.S.; Clary, C.W. Deep Learning in Gait Parameter Prediction for OA and TKA Patients Wearing IMU Sensors. *Sensors* **2020**, *20*, 5553. [CrossRef]
24. Hayashi, K.; Kako, M.; Suzuki, K.; Takagi, Y.; Terai, C.; Yasuda, S.; Kadono, I.; Seki, T.; Hiraiwa, H.; Ushida, T.; et al. Impact of variation in physical activity after total joint replacement. *J. Pain Res.* **2018**, *11*, 2399–2406. [CrossRef]
25. Lipperts, M.; van Laarhoven, S.; Senden, R.; Heyligers, I.; Grimm, B. Clinical validation of a body-fixed 3D accelerometer and algorithm for activity monitoring in orthopaedic patients. *J. Orthop. Transl.* **2017**, *11*, 19–29. [CrossRef]
26. Ho, C.J.; Chen, Y.T.; Wu, H.L.; Huang, H.T.; Lin, S.Y. The Effects of a Patient-Specific Integrated Education Program on Pain, Perioperative Anxiety, and Functional Recovery following Total Knee Replacement. *J. Pers. Med.* **2022**, *12*, 719. [CrossRef]
27. Rak, D.; Nedopil, A.J.; Sayre, E.C.; Masri, B.A.; Rudert, M. Postoperative Inpatient Rehabilitation Does Not Increase Knee Function after Primary Total Knee Arthroplasty. *J. Pers. Med.* **2022**, *12*, 1934. [CrossRef]
28. Lebleu, J.; Pauwels, A.; Anract, P.; Parratte, S.; Van Overschelde, P.; Van Onsem, S. Digital Rehabilitation after Knee Arthroplasty: A Multi-Center Prospective Longitudinal Cohort Study. *J. Pers. Med.* **2023**, *13*, 824. [CrossRef]
29. Gazendam, A.; Zhu, M.; Chang, Y.; Phillips, S.; Bhandari, M. Virtual reality rehabilitation following total knee arthroplasty: A systematic review and meta-analysis of randomized controlled trials. *Knee Surg. Sports Traumatol. Arthrosc.* **2022**, *30*, 2548–2555. [CrossRef]
30. LeBrun, D.G.; Martino, B.; Biehl, E.; Fisher, C.M.; Gonzalez Della Valle, A.; Ast, M.P. Telerehabilitation has similar clinical and patient-reported outcomes compared to traditional rehabilitation following total knee arthroplasty. *Knee Surg. Sports Traumatol. Arthrosc.* **2022**, *30*, 4098–4103. [CrossRef]
31. Luna, I.E.; Kehlet, H.; Wede, H.R.; Hoevsgaard, S.J.; Aasvang, E.K. Objectively measured early physical activity after total hip or knee arthroplasty. *J. Clin. Monit. Comput.* **2019**, *33*, 509–522. [CrossRef]
32. Klemt, C.; Uzosike, A.C.; Harvey, M.J.; Laurencin, S.; Habibi, Y.; Kwon, Y.M. Neural network models accurately predict discharge disposition after revision total knee arthroplasty? *Knee Surg. Sports Traumatol. Arthrosc.* **2022**, *30*, 2591–2599. [CrossRef]
33. Shemesh, S.S.; Dieterich, J.D.; Chen, D.; Sharon, R.; Bronson, M.J.; Frenkel Rutenberg, T.; Moucha, C.S. Preoperative Pain Catastrophizing and Neuropathic Pain Do Not Predict Length of Stay and Early Post-Operative Complications following Total Joint Arthroplasty. *J. Pers. Med.* **2023**, *13*, 216. [CrossRef]

Journal of *Personalized Medicine*

MDPI

Review

Total Knee Arthroplasty in Patients with Ipsilateral Hip Fusion: Technical Notes and Literature Review

Liliana Savin [1,2], Tudor Pinteala [1,2,*], Paul Botez [2], Smaranda Miu [3], Norin Forna [1,2], Dan Mihailescu [1,2], Dragos Cristian Popescu [1,4], George Enescu [1,2] and Paul Dan Sirbu [1,2]

1 Department of Orthopedics and Traumatology, Faculty of Medicine, "Grigore T. Popa" University of Medicine and Pharmacy, 700115 Iasi, Romania; liliana.savin@umfiasi.ro (L.S.); norin.forna@umfiasi.ro (N.F.); dan.mihailescu1@umfiasi.ro (D.M.); dragos.popescu@umfiasi.ro (D.C.P.); enescu_george@d.umfiasi.ro (G.E.); paul.sirbu@umfiasi.ro (P.D.S.)

2 Department Orthopedics and Traumatology, Clinical Rehabilitation Hospital, 700661 Iasi, Romania; paulbotez@yahoo.com

3 Department of Rehabilitation, Clinical Rehabilitation Hospital, 700661 Iasi, Romania; miu.smaranda@scr.ro

4 Department Orthopedics and Traumatology, Sf. Spiridon' County Emergency Hospital, 700661 Iasi, Romania

* Correspondence: tudor_pinteala@umfiasi.ro; Tel.: +40-731-550-503

Abstract: Numerous studies report the success and outcomes of the total knee arthroplasty (TKA); however, few papers present patients with knee osteoarthritis and ipsilateral hip fusion. One controversy when treating patients requiring a TKA with prior ipsilateral hip fusion is whether to first perform a total hip arthroplasty (THA) of the fused hip, followed by the ipsilateral TKA, or to proceed with the TKA without replacing the hip; studies suggest that the position of the fused hip is a key factor when making this therapeutical decision. In addition, performing a TKA in patients with an ipsilateral fused hip may require modifications to the surgical technique generated by the lack of joint mobility in the hip. We identified 12 studies encompassing 30 patients with hip fusion and ipsilateral TKA in current orthopedic literature, but only six offered insights on patient positioning on the operating table during surgery. This study aims to review the current literature on patients with knee osteoarthritis and prior ipsilateral hip fusion and to present some technical considerations when performing a TKA on a 75-year-old patient with hip ankylosis who underwent a total ipsilateral knee arthroplasty in our clinic.

Keywords: total knee arthroplasty (TKA); ipsilateral hip fusion

check for updates

Citation: Savin, L.; Pinteala, T.; Botez, P.; Miu, S.; Forna, N.; Mihailescu, D.; Popescu, D.C.; Enescu, G.; Sirbu, P.D. Total Knee Arthroplasty in Patients with Ipsilateral Hip Fusion: Technical Notes and Literature Review. *J. Pers. Med.* **2023**, *13*, 1705. https://doi.org/10.3390/jpm13121705

Academic Editor: Jih-Yang Ko

Received: 14 November 2023
Revised: 4 December 2023
Accepted: 11 December 2023
Published: 13 December 2023

1. Introduction

Hip arthrodesis was considered the surgical intervention of choice in the early 20th century, when treating patients with painful hips caused by osteoarthritis, infection, inflammatory diseases, or traumatic or congenital pathologies [1]. While the initial absence of pain in the fused hip significantly improved patient quality of life, this benefit was not without consequences, as studies have shown that patients with a fused hip started developing pain in neighboring joints such as the ipsilateral knee, spine, or contralateral hip over long periods of time (20 or more years) [1–4]. Biomechanical studies of the fused hip proved that the lack of movement in the hip joint is compensated by an increase in pelvic transverse and sagittal rotation, increased ipsilateral knee flexion, and an increased mobility of the contralateral hip [2]. In a gait analysis study, Thambyah et al. [5] mentioned that tilting the pelvis induced a pseudo-extension of the hip by increasing lumbar lordosis, whereas the ipsilateral knee presented a decrease in knee flexion moment and a decrease of 95% of the muscle extension moment (similar to a quadriceps avoidance gait).

The weakness of the quadriceps muscle, commonly seen in this group, can impair their ability to absorb shock while walking. This often leads to more pronounced symptoms associated with knee osteoarthritis and may result in changes in their walking pattern [6–8].

Therefore, hip fusion, in time, causes degenerative changes in neighboring joints such as the lumbar spine, sacroiliac joint, ipsilateral knee, and contralateral hip [1,2,4].

To date, a small number of papers exist on the management of patients with symptomatic knee osteoarthritis and prior ipsilateral hip fusion, and there is no guideline on how to best manage this situation: to perform a total knee arthroplasty (TKA) or to first convert the fused hip into a total hip arthroplasty (THA) and after to perform the ipsilateral TKA. Studies have mentioned that the position of the fused hip should be considered when deciding between these two therapeutic options. Callaghan et al. [1] clinically measured the hip position in the frontal plane by measuring the angle between the axis of the femur and a line that passes through the anterior superior iliac spine and the femoral vessels (parallel with the midline of the body). Garvin et al. [3] determined the rotational position of the fused hip by measuring the angle between the axis that passes through the second ray and a line perpendicular to the body's axis in the frontal plane. Romnes et al. [9] performed a TKA in patients with a hip arthrodesis in the "optimal" position (22.5° hip flexion, 5° hip abduction, neutral hip rotation), whereas, in his group of patients considered to have a fused hip in suboptimal position (an average of 38.8° hip flexion, 2.1° hip abduction and 3.3° hip external rotation), the author first converted the fused hip into a THA and subsequently performed the ipsilateral TKA.

In the rare cases of patients with knee osteoarthritis below a fused hip, surgeons must take into account that the conversion of a fused hip to a THA is a technically challenging procedure with a high rate (up to 54%) of complications such as nerve palsy, early prosthetic failure, heterotopic ossification, and infection [10]. In addition, the rate of hip prosthetic loosening at 10 years is clearly higher in the case of hips with surgical arthrodesis (48.3%) when compared to spontaneous ankylosis of the hip (5%) [11]. Furthermore, given the potential risks of this surgery, most authors suggest that a de-arthrodesis (converting the fused hip into a THA) is controversial if the hip has been fused in an acceptable position (the ideal position for hip fusion is considered 15°–30° of hip flexion (HF), 5°–10° of hip adduction (HA), and 0°–10° hip external rotation (HER) [12]. Therefore, patients with hip fusion (in the correct position), who suffer from end stage osteoarthritis of their ipsilateral knee, can be treated through a TKA only.

Performing a TKA on patients with a same side hip fusion poses multiple and unique challenges, both in terms of surgical technique and postoperative management as well as possible complications. Current literature offers few data regarding the performance of TKA in patients with ipsilateral hip fusion. This article aims to review the current literature on this subject and present our surgical approach of a patient with end-stage knee osteoarthritis and prior ipsilateral hip fusion.

2. Materials and Methods

In 2018, a 75-year-old female patient who underwent right hip fusion at the age of 15 due to bacillary osteoarthritis was admitted to our clinic. She complained of severe pain and a significantly decreased range of motion (ROM) in her right (ipsilateral) knee; the patient was ambulating with the aid of crutches. The clinical examination of her right knee (Figure 1) revealed a right knee flexion of 90°, a 15° flexion contracture, and no mediolateral ligamentous instability. Her fused right hip was fixed in a flexion of 15°–20°, neutral abduction/adduction, and neutral internal/external rotation; she did not report any pain/discomfort in this hip. Her leg length discrepancy was measured at 3 cm, and she compensated for it with custom insoles.

Her X-ray findings (Figure 2) were the following: stage IV Kelgreen and Lawrence right knee osteoarthritis; her right hip was ankylosed in 15° of flexion, neutral abduction/adduction, and internal/external rotation; in addition, she had a thoracolumbar kyphoscoliosis, with L2–L4 vertebral block and severe degenerative changes.

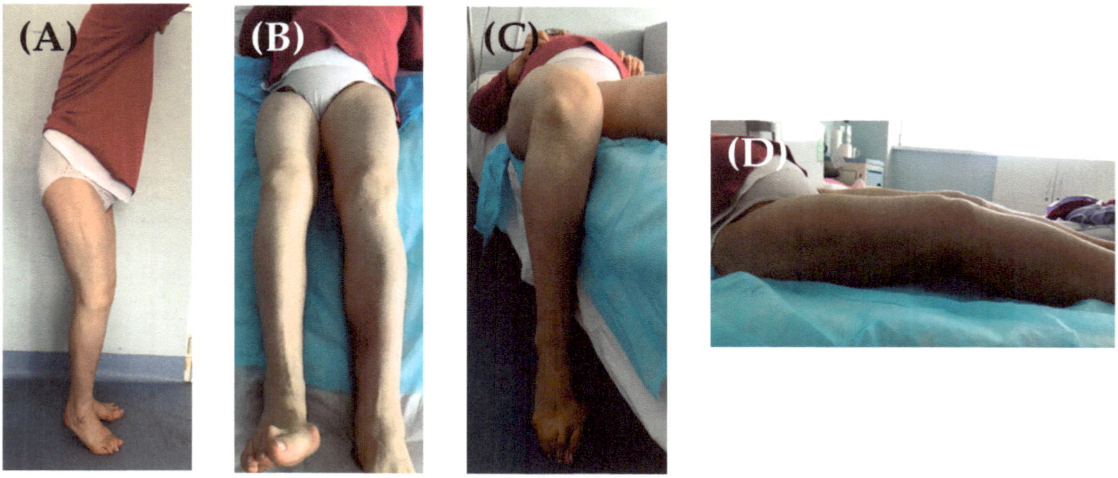

Figure 1. Right hip fusion, with the hip fused in 15° HF (**A**), neutral HA, and neutral HER (**B**). Right knee flexion of 90° (**C**) and 10° of flexion contracture (**D**).

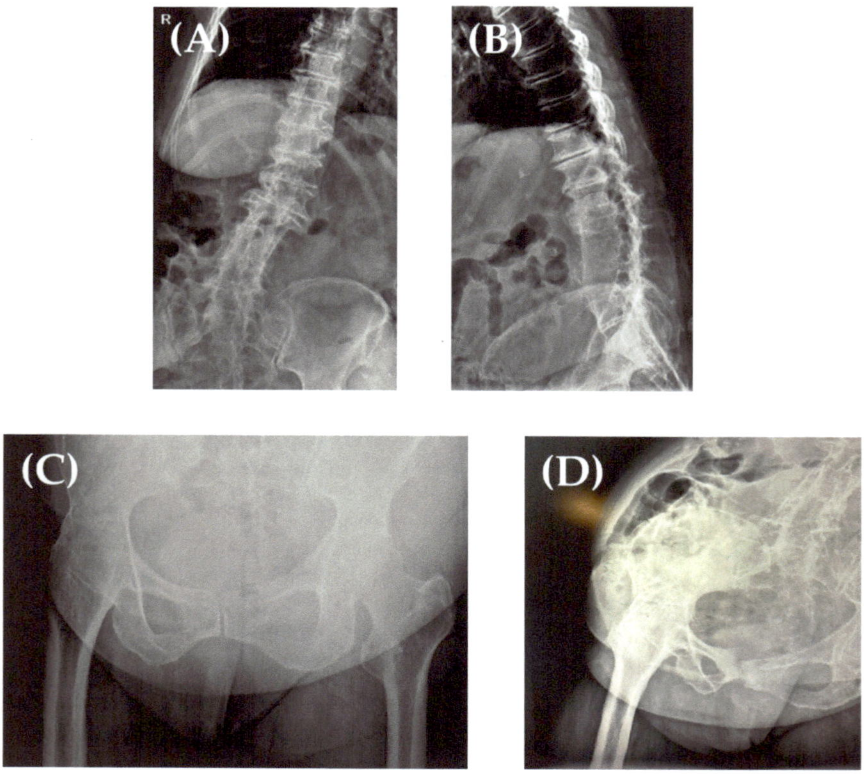

Figure 2. Thoracolumbar kyphoscoliosis, with L2–L4 vertebral block (**A**,**B**). Right hip fused in 15° of hip flexion, neutral hip abduction/adduction, and neutral hip internal/external rotation (AP view—**C**, Lateral view—**D**).

Taking all of this into account, we decided to perform a total knee arthroplasty on her right knee. Her preoperative planning long leg X-ray (Figure 3A,B) showed a grade II varus deformity of 12° (Hip-Knee-Angle—HKA of 168°), a lateral distal femoral angle (LDFA) of 89°, and a medial proximal tibial angle (MPTA) of 78° [13] with an increased posterior tibial slope and an IM angle (the angle between the femoral anatomical and mechanical axis) of 6°. Furthermore, in our clinic, we routinely perform a "seated view" X-ray which, according to our previous studies [14], helps in determining the distal femoral torsion. Using this special X-ray, where the patient sits on a radiolucent support with the knee flexed at 90°, we determine the posterior condyle angle (PCA) (and thus, the distal femoral rotation) by measuring the angle between the femoral trans-epicondylar axis (a line from the lateral epicondyle connecting it to the sulcus of the medial epicondyle) and the line connecting the posterior femoral condyles (Figure 3D); in this case, the femoral torsion was determined to be 4°.

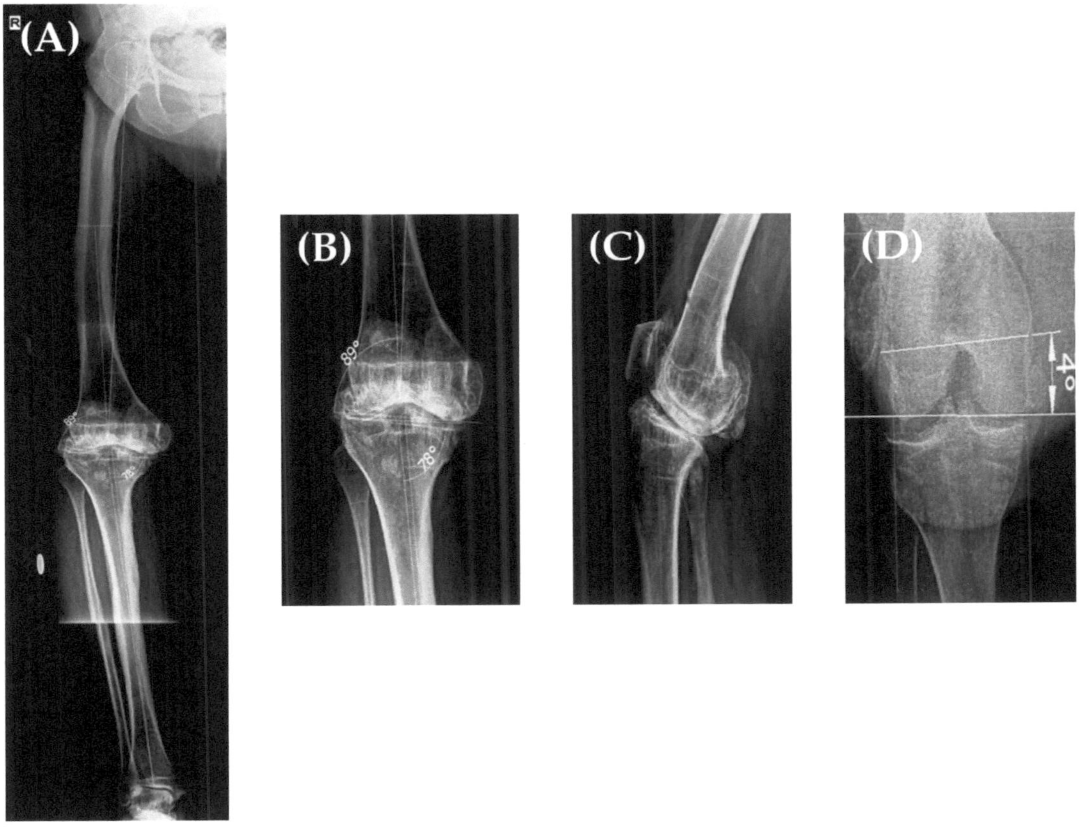

Figure 3. Long leg X-ray of the patient's right lower limb showing a HKA = 168°, LDFA = 89°, MPTA = 78° (**A**,**B**). Sagittal X-ray (performed at 30° of knee flexion)—increased tibial slope (**C**). Seated view X-ray showing a PCA (femoral torsion) of 4° (**D**).

Before surgery, we obtained written (approved by our hospital's ethics committee) and verbal consent from our patient to use her clinical history, radiographs, as well as photos from before, during, and after surgery for research/presentation purposes.

In general, when performing a TKA, the patient is supine on the surgical table, and a minimum of 90° of knee flexion is necessary. However, in the case of patients with hip fusion, an ipsilateral knee flexion of 90° is impossible to achieve without making any changes to the surgical table or patient's position on the operating table.

3. Results

In order to achieve a minimum knee flexion of 90° intraoperatively, we positioned the patient on the operating with the following modifications (compared to a standard supine position when we perform a TKA) (Figure 4): a thigh support fixed in 20° of flexion was used; we broke the ipsilateral distal half of the table, thus causing the knee to be flexed at >90°, and kept the distal arm of the table in the normal position (which provides a solid support the patient's foot). Therefore, the patient kept her thigh flexion position; when placing the knee at the level of the table break (Figure 4A,B), it could be flexed to more than 90°, and the patient's foot being placed on the distal part of the surgical table allowed for good structural support (useful during tibial impaction of the prosthetic) (Figure 4A,B). The contralateral limb was positioned in slight flexion (Figure 4C).

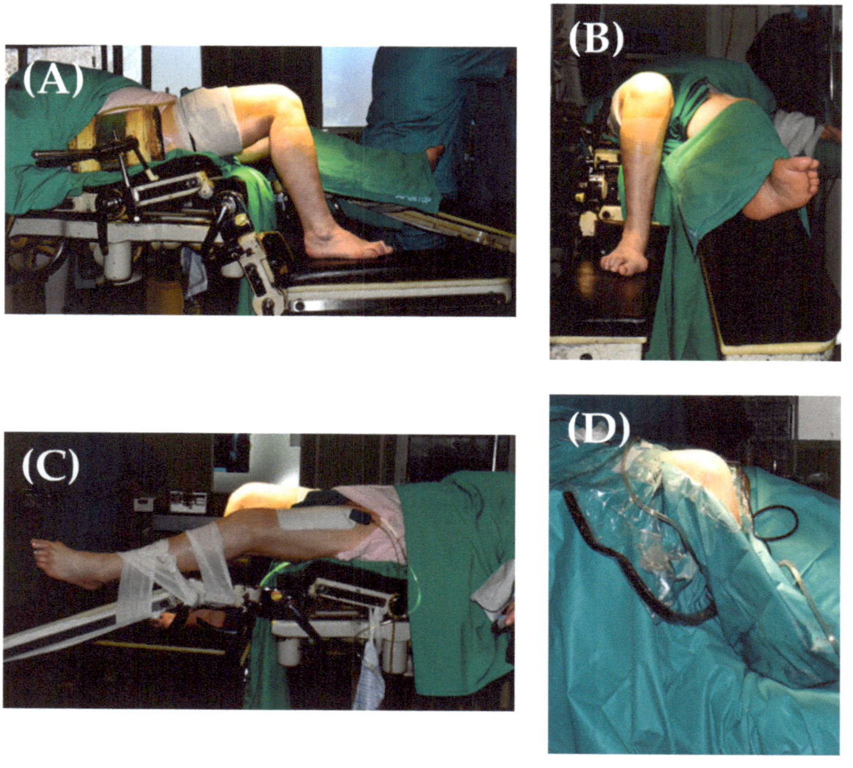

Figure 4. Patient's position on the operating table and modifications made: thigh support (**A**,**B**); breaking the distal half of or table, which allowed for knee flexion of more than 90° and keeping the distal portion of the table in the normal position (**A**,**B**); positing of the contralateral limb in slight flexion (**C**); operating knee after sterile draping (**D**).

A total knee replacement was performed using a cemented postero-stabilized (PS) prosthesis (NexGen) through a medial parapatellar approach, using the extension gap first technique [15]. This type of implant was chosen due to the fact that the patient had competent collateral ligaments, and the ideal flexion-extension gap was intraoperatively obtained using a PS insert. Having the knee flexed and the foot stable on the operating table provided optimal conditions for this surgery. Knee extension was held by a surgical assistant when needed; bone cuts were made according to the preoperative plan (Figure 5A,B).

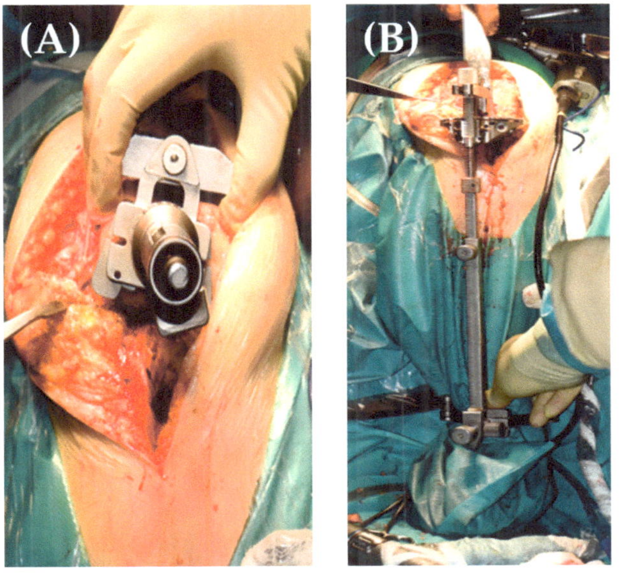

Figure 5. Medial parapatellar approach, placing the intramedullar distal cutting guide with the knee flexed at 90° (**A**), placing the extramedullar tibial cutting guide with the patient's foot stable on the surgical table (**B**).

Postoperative X-rays revealed adequate positioning and sizing of the prosthetic components and a restored alignment of the lower limb. Personalized rehabilitation was initiated 24 h after the surgery, with knee flexion being recovered at the edge of the bed, and knee extension in lateral decubitus; weight bearing was allowed on the 1st postoperative day, and the patient started walking with the aid of crutches. Subsequent periodic radiological follow-ups at 3, 6, 12 months and annually up to 5 years (Figure 6A,B) did not show any signs of loosening (Figure 3). At her 3-month follow-up, the patient had an active range of motion (ROM) of 110° knee flexion and full extension (in lateral decubitus), walking without any aiding devices, with an increase in KSS from 26 to 84 points and WOMAC increase from 40 to 77.

Searching the PubMed and Google Scholar databases for "total knee arthroplasty" and "ipsilateral hip fusion/arthrodesis/ankylosis", we found 81 articles, out of which 12 included our criteria: patients with hip fusion/ankylosis/arthrodesis and ipsilateral knee osteoarthritis treated with a total knee replacement. These 12 studies followed and described the total knee arthroplasty in a total of 30 patients with ipsilateral hip fusion (Table 1) [1,3,9,16–24]. Callaghan et al. [1] was the first to mention the possibility of performing a TKA in a patient with same side hip fusion, without giving too many details about the respective patients or surgical technique. Out of these 12 articles, only 6 described the changes made to the surgical table during the arthroplasty [18,19,21–24], and only 2 others mentioned the need for changes [3,17]. Goodman et al. [18] positioned their patients as supine, with the table in Trendelenburg, broke the table at the level of the knee, causing the knee to hang outside; during the surgery, the table was lowered and elevated periodically. Koo et al. [19] placed a sandbag under the ipsilateral buttock, tilted the table towards the operating knee, and hung the knee outside the surgical table. Tang et al. [21] suspended the knee over the distal half of the broken table with the contralateral limb in the lithotomy position. Samborski et al. [22] used multiple mattresses to raise the upper part of the body, the knee being positioned on the surgical table in the Mayo support that offered the possibility of controlled flexion. Ullan et al. [23] positioned the limb in the arthroscopy support. Ashkenazi et al. [24] positioned his patient similarly to Goodman et al. [18]: table in Trendelenburg, broke the distal half of the table, the non-operating knee secured to the table break, and the operated knee bent at the level of the table break.

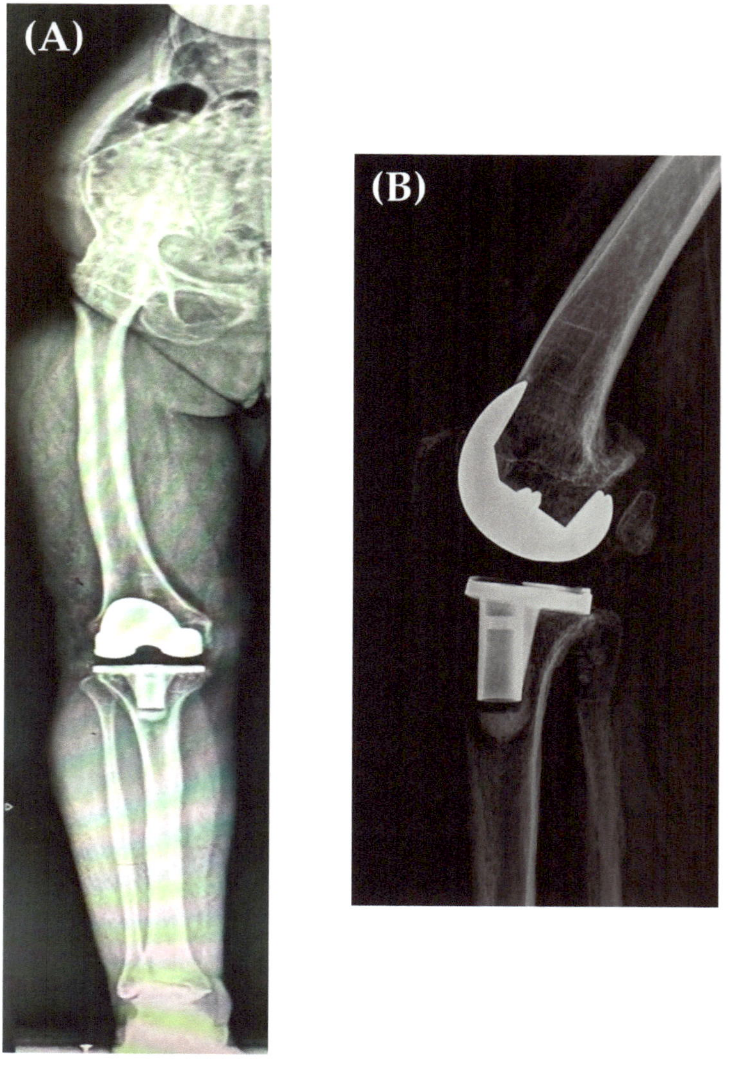

Figure 6. Five years postoperative AP long leg X-ray (**A**) and sagittal X-ray (performed at 30° of knee flexion) (**B**) showing the correct alignment of the femoral and tibial prosthetic components, neutral alignment of the right lower limb, and no component loosening.

Table 1. Findings from the reviewed literature of patients with TKA and ipsilateral hip fusion.

No. Crt.	Author	No. of Pts	Hip Fusion Position	HSS Knee Score	Results	Follow-Up Interval	Type of Prosthesis	Complications	Patient Position during Surgery
1	Callaghan (1985) [1]	2	-						-
2	Garvin (1989) [3]	9	37° HF 1° HA 8° HER	-	3 excellent 4 good 1 fair 1 poor (infect)	29–177 months	1 duo-condylar 1 guepar 3 total condylar 4 PS	7 multiple manipulations 1 Guepar prosthesis—quadriceps rupture, late infection (7 years), amputation	Bolster to elevate the hip fused
3	Romness (1991) [6]	4	20°–30° HF 0°–10° Abd 0° HER	43.5 to 72.1	3 good 1 poor	2.3–10 years	7 Kinematic 3 PCA 1 Kinematic stabilized 1 Stabo-condylar 1 Tavernetti 1 Geomedic	1 manipulation	-
4	Rittmeister (1999) [13]	3	20°–30° HF 10° HA 10°–20° HER	36 to 44	2 poor 1 fair	24–86 months	-	1 deep infection, three revision, and arthrodesis	-
5	Katsumitsu Arai (2001) [14]	1	25° HF 5° HA 0° HER		Limited ROM	1 year	PS	Fused hip converted to THA, Knee manipulation Quadriceps plasty	Semilateral position
6	Goodman (2014) [15]	2	15°–17.5° HF 5° HA 0°–10° HER	33 to 83 40 to 78	satisfactory	4.7–7.8 years	PS NexGen	-	Table in Trendelenburg Knee at the level of the break in the table Flexed the table foot to 90°
7	Koo (2015) [16]	1	30° of hip flexion Neutral hip adduction	WOMAC score 98	Excellent	6 months	PS	-	Sandbag placed under the ipsilateral buttock Table tilted about 20° towards the operated knee Hang the leg over the table
8	De la Hera (2017) [17]	2	0°–5° HF 5° HA–5° Abd 0°–5° HER	49 to 65 13 to 88		1–11 years	1 PS NexGen 1 semi-constrained LCCK	-	-
9	Tang (2019) [18]	1	15° HF 15° HA 20° HER	-	Good	6 months	-	-	Knee suspended over the broken table Contralateral leg in lithotomy position
10	Samborski (2020) [19]	1	20° HF 5° HA 20° HER	-	-	6 months	PS Triathlon with short tibial stem	Respiratory failure, acute kidney injury, hemodialysis, decubitus ulcers, foot drop	Multiple operating table mattress pads under the upper body Remove the pads under the operated leg De Mayo Surgical leg positioner

Table 1. *Cont.*

No. Crt.	Author	No. of Pts	Hip Fusion Position	HSS Knee Score	Results	Follow-Up Interval	Type of Prosthesis	Complications	Patient Position during Surgery
11	Ullan (2021) [20]	3	15°–30° HF 5°–10° HA 0°–10° HER	24 to 72 45 to 71 32 to 78	Satisfactory	1–5 years	Rotating hinge	-	Knee suspended with arthroscopy leg holder
12	Ashkenazi (2023) [21]	1	5°–10° HF 10° HER	-	-	2 weeks	Robotic assisted surgery PS	-	Table in Trendelenburg, broke the distal half of the table, and secured the contralateral leg to it. The operated knee was placed at the level of the table break (similar to Goodman et al.)

Pts = Patients; HF = Hip Flexion; HA = Hip Adduction; HER = Hip External Rotation; Abd = Abduction; PS = Postero-Stabilized; ROM = range of motion; manipulation = closed manipulation under anesthesia.

In these 12 studies, the patient follow-up period was between 6 and 177 months, and the number of patients included in each study varied from 1 to 9. The postoperative results varied both within the study [3,9] and between studies from fair to excellent, with the most common postoperative complication being the need for multiple closed manipulations due to postoperative knee stiffness [3,9,17].

Garvin et al. [3] followed nine patients, with a mean follow-up period of 7 years (interval between 29 and 177 months), reporting three with excellent results, four as good, one as fair, and one as poor, with seven patients requiring multiple manipulations due to stiffness, and one case finally necessitating an amputation due to a late infection. Rittmeister et al. [16], in a 33-year retrospective study that followed 18 patients with hip ankylosis and ipsilateral knee ostheoarthritis, of which only 3 patients benefited from TKA (follow-up interval between 24 and 86 months) without converting the hip fusion to a THA, reported two poor results and one fair, with 1 patient subsequently requiring knee arthrodesis after three failed revision surgeries. Romness et al. [9] reported three good and one poor outcome in a study that included four patients with TKA and ipsilateral hip fusion, with a follow-up between 2.3 and 10 years.

Arai et al. [17] placed their patient in a semilateral position; in addition, intraoperatively, they performed a tibial tubercle osteotomy and rectus snip in order to achieve better exposure. The authors present the case of a 50-year-old patient with rheumatoid arthritis who at the age of 36 had a spontaneous right hip fusion (25° HF, 5° HA, 0° HRE) and an ipsilateral arthritic stiff knee (ROM −22° to 30°), where they performed a TKA (Kinemax plus system); intraoperatively, the surgeons decided not to use a PS implant. The authors report immediate postop ROM of the operated knee of 0° to 90°; however, after 1-year post surgery, the patient had a limited ROM of −15° to 30°. Due to this poor result, the authors decided to convert the patient's ankylosed hip to a THR and unsuccessfully tried a closed manipulation of her knee (due to severe stiffness). Therefore, one month later, they performed a quadricepsplasty and retinacular release on the patient's right knee; during this surgery, the authors reported to have found severe adhesions and fibrous tissue surrounding the knee prosthesis (to which they attribute the poor outcome of the TKR). One year after the THA, they reported a knee ROM of −15° to 75° and a hip ROM of 0° in extension, 70° in flexion, 30° in abduction, 20° in adduction, and 5° in internal and external rotation [17]. Goodman et al. [18] decided to first perform the patella and tibial cuts followed by the femoral cuts and used NexGen implants. They published two cases: an 87-year-old with their hip fused at 17.5° HF, 5° HA, and 10° HER and a 63-year-old with their hip fused at 15° HF, 5° HA, and 0° HER. They reported good outcomes in both cases at follow-ups (4.7 years and 7.8 years postoperatively) with KSS scores of 83 (33 preop) and 73 (40 preop) in each respective case; however, their second patient had limited flexion due to scarring of their quadriceps. Samborski et al. [22] published the case of a 72-year-old patient, obese and diabetic (type II) with end-stage osteoarthritis of the knee with ipsilateral hip fusion (20° HF, 5 HA°), where they implanted a PS cemented Stryker Triathlon (short-stemmed tibia). During the surgery, the authors encountered no complications; however, after discharge, their patient was diagnosed with chronic kidney injury and required hemodialysis; due to a prolonged immobilization, the patient developed bedsore lesions (heel and dorsal sacrum). At the 2-week postop follow-up, they noted wound healing but noticed a foot drop on the operated side; at the 3-month follow-up, the patient's wounds were healed, and she was mainly ambulating in a wheelchair with a limited walking distance, aided by a foot drop orthosis and a walker; and, at 6 months after the surgery, the authors reported a passive knee flexion on 0° to 135° [22].

Other articles report satisfactory/good surgical outcomes without any complications in their respective follow-ups (Table 1) [18–21,23,24]. In 2017, de la Hera et al. [20] published two cases; their first case was a 72-year-old patient with a fused hip at 5° HF, 5° HA, and 0° HER, where they performed an ipsilateral TKA (PS NEX-GEN); the authors reported no complications after 11 years postop, with an improvement in KSS from 49 to 65 and increase from 38 to 68 in the WOMAC score. Their second case was a 59-year-old patient with a fused hip at 0° HF, 5° of hip abduction, and 5° HER, who underwent a previous tibia vara osteotomy and later a supracondylar varus osteotomy and subsequently developed end-stage knee arthritis, for which they performed a TKA (LCCK implant with femoral and tibial stems); one year postop, they reported no complications with an increase in KSS from 13 to 88 and WOMAC from 40 to 59. Koo et al. [16] presented the case of a 67-year-old patient with a hip fused at 30° HF and an active flexion of her ipsilateral arthritic knee of 70°. The patient did not suffer any postoperative complications, and the authors established a WOMAC score of 98 at the 6 months postop follow-up with a ROM of the operated knee (PS implant) of 0° to 120°. Tang et al. [21] published the case of a 64-year-old patient with a fused hip at 15° HF, 15° HA, and 20° HER with ipsilateral knee osteoarthritis and previous reduction and internal fixation of her lumbar spine. At the 6 months follow-up, the authors reported no complications, and the patient had a ROM of 0° to 100°. Ullan et al. [23] presented three cases of patients with hip arthrodesis (fused in 15° to 30° HF, 5° to 10° HA, and 0° to 10° HER) and ipsilateral symptomatic knee osteoarthritis, for which they performed a rotational hinge TKA. Their first case was a 72-year-old who had a ROM of the operated knee of −5° to 115°, no sign of complications or loosening, and an increase in the HSS score from 24 (preop) to 72 (postop) at five years postop. The second case was an 81-year-old with hip arthrodesis due to bilateral congenital dislocation who subsequently developed severe valgus knee and complete extension valgus laxity. For this case, the authors chose a constrained rotational hinge implant; at the 3 years follow-up, the authors reported no complications, with a knee ROM of 0° to 100° and an increase in KSS from 45 (preop) to 71 (postop). The third case the authors presented was a 79-year-old patient with hip arthrodesis and a lower limb shortening of 12 cm who developed severe knee osteoarthritis on the same limb (12° genu varum and 11° flexion contracture); in this case, the surgeons decided to implant a hinged TKA (due to the severe malalignment and flexion contracture). One year after surgery, the authors reported no complications, an operated knee ROM of 0° to 110°, and an increase in the KSS score from 32 (preop) to 78 points (postop).

The position of the ankylosed hip was followed in all 12 studies, varying from patient to patient, but generally remaining within acceptable limits and thus permitting a total knee replacement of the ipsilateral osteoarthritic knee. Considering the long period of time in which these studies were published (38 years), the diversified types of prostheses used for these patients is understandable. The patients in the first studies [3,4] benefited from the existing prostheses at the time: guepar, duocondylar, PCA (Porous-coated Anatomic), stabocondylar, or Geomedic; since the 2000s, PS (posterostabilized) [17–20,22,24], semi-constrained [20], or rotating hinge prostheses [23] have been used. Ashkenazi et al. [24] implanted a PS knee with the use of CAN (computer-assisted navigation); however, the authors acknowledge the limited use of navigation and/or robotic assisted surgery when performing a TKA in a patient with an ipsilateral hip arthrodesis, as the hip center of rotation cannot be determined during surgery due to the lack of hip motion. They solved this issue by registering the femur as a reflected tibia (Table 1).

4. Discussion

Nowadays, hip fusion is a rare therapeutical option (primarily due to the advancements in joint replacements) and can be considered in selected cases, such as uncontrolled joint infection or for patients considered too young to be a viable candidate for primary hip replacement [12,25]. The literature suggests that 35–70% of patients complain of back pain, 17–28% develop contralateral hip pain, and 24–57% have ipsilateral knee pain [1,26–28] after more than 20 years post hip fusion. In addition, radiographical evidence of ipsilateral knee osteoarthritis is present in over 68% of patients with hip arthrodesis, after several years have passed since surgery [1,9].

Treating patients with end-stage knee osteoarthritis and an ipsilateral fused hip provides controversies regarding which joint to replace first. Some suggest that one should first perform a de-arthrodesis of the hip (transform the fused hip into a THA) and then, after a period of time, assess the actual need of a TKA; however, Callaghan et al. [1] reported that 44% of patients have persisting knee pain after a hip de-arthrodesis is performed. Rittmeister et al. [16], in a retrospective study of 18 patients with hip fusion and ipsilateral symptomatic knee osteoarthritis, reported that the most improved HSS knee score (from 33 to 78) were the 4 out of the 18 patients who first underwent a conversion from hip fusion to THA followed by an ipsilateral TKA, whereas the 3 out of 18 patients who "just" underwent a TKA had a mild increase in their HSS knee score (from 35 to 44). Therefore, the authors concluded that hip fusion is a poor prognosis for TKA and that TKA should be performed after hip de-arthrodesis, this therapeutical approach offering the advantage of lower limb mechanical axis realignment and thus decreasing the risk of early implant failure and loosening. Romness and Morrey [9] reported similar HSS knee scores improvements (from 43 to 72 vs. 28 to 72) when they treated 16 patients with same side hip fusion and knee osteoarthritis the following way: 4 patients underwent TKA vs. 12 patients who underwent hip fusion conversion to THA followed by TKA (5.5 years mean follow-up); however, the authors stated that appropriate positioning and good alignment of the hip joints is the most important prognostic factor for TKA. Moreover, Garvin et al. [3] reported similar functional results when comparing nine patients with hip arthrodesis (in a proper position) who underwent an ipsilateral TKA to patients without hip fusion who underwent a TKA. Considering all mentioned above, Cho [29] recommends that when considering treatment options in patients with hip fusion and ipsilateral knee osteoarthritis, the surgeon should consider the position of the hip fusion and the condition of neighboring joints (contralateral hip, spine and sacroiliac joint). Furthermore, Roberts et al. [30] states that the fused hip should be in 25° or less of flexion, 0° abduction, and 5° or less of external rotation in order to obtain good results for the TKA [30].

Performing a TKA without first converting the fused hip to a THA poses unique technical challenges due to the lack of mobility in the ipsilateral fused hip, such as patient positioning during surgery, as the usual patient position on the operating table requires certain modifications to obtain adequate knee flexion in order to perform a total knee arthroplasty. Furthermore, surgeons might need some alterations to their surgical technique generated by the lack of joint mobility in the hip for example: Arai et al. performed a tibial tubercle osteotomy and rectus snip for better knee exposure [17]; Goodman et al. started with the patella and tibial cuts followed by the femoral cuts [18]; and Ashkenazi et al., using CAN, made the femoral cut as a reflected tibia without determining the hip center of rotation [24].

In general, joint resurfacing requires a profound understanding of normal anatomy and biomechanics [31], and the surgeon should take into account certain surgical technique alterations when performing this kind of surgery in particular patients [32].

In our case, the decision to perform a total knee replacement followed the recommendations of available orthopedic literature, our patient presenting a fused right hip in 15° of hip flexion, 0° hip abduction, and 0° hip external rotation. The modification we made to the surgical table allowed us to perform the TKA in safe conditions; furthermore, we feel that maintaining the distal portion of the surgical table in its normal position provides a great benefit, as it stabilizes the foot throughout the surgery, in addition to it being of great help when impacting the tibial component, and thus the surgical technique we used did not require any changes compared to other patients without hip fusion. Moreover, having the operating knee positioned as we did allowed us, during surgery, to flex the knee more than 90°; the knee was kept in extension by a surgical assistant when checking for the extension gap. We believe that these technical modifications aided in the correct implantation of the knee prosthesis, especially the tibial component. Only Samborski et al. [22] maintained a stable foot support in the distal part of the surgical table, having the patient's body elevated with the help of several mattresses.

Our patient did not present any ligamentous laxity/instability in the frontal plane, so we decided to use a PS implant without the need to augment bone defects; the patient's flexion contracture was corrected by performing a posterior capsular release and removal of the posterior osteophytes. Probably, if we were to intervene on this patient now, we would use a cemented short tibial stem in hopes of having a longer implant stability [33–35]. The postoperative evolution of our patient was favorable without any complications, and the operated knee mobility at 5 years was preserved (110° of flexion and 0° of extension—measured in the lateral decubitus).

5. Conclusions

To the best of our knowledge, in the current orthopedic literature, few studies report the management of patients with hip fusion and ipsilateral end-stage knee osteoarthritis. Currently, there are no clear guidelines for how to best manage this condition. In our opinion, patients with this particular pathology can benefit from a total knee arthroplasty with good to excellent outcomes, if the hip is fused in an optimal position. We consider that preoperative clinical and radiological planning is crucial in determining the therapeutic approach (performing a TKA and keeping the hip fused vs. converting the fused hip into a THA followed by TKA) and selecting the appropriate knee implant (taking into consideration collateral ligaments' stability and bone defect). In addition, the surgeon should consider patient positioning modifications in order to obtain more than 90° of knee flexion and good stability of the joint during surgery (despite the lack of hip mobility). Even though our study focuses on a single case, we believe our surgical table modification can assist other surgeons dealing with advanced knee osteoarthritis and ipsilateral hip fusion.

Author Contributions: Conceptualization, L.S., T.P. and P.D.S.; methodology, L.S., T.P. and G.E.; validation, L.S. and P.D.S.; investigation, T.P., S.M., N.F., D.M., D.C.P. and G.E.; data curation, L.S., T.P., P.B. and N.F.; writing—original draft preparation, L.S., T.P., S.M., D.M., N.F. and P.D.S.; writing—review and editing, L.S., T.P., D.C.P., N.F., S.M. and D.M.; supervision, P.D.S. All authors have read and agreed to the published version of the manuscript.

Funding: This research received no external funding.

Institutional Review Board Statement: The study was conducted in accordance with the Declaration of Helsinki and approved by The Ethics and Research Committee of the Clinical Rehabilitation Hospital, protocol code 61/9.11.2023.

Informed Consent Statement: Informed consent was obtained from all subjects involved in the study.

Data Availability Statement: The dataset used and/or analyzed during the current study is available from the corresponding author on reasonable request.

Conflicts of Interest: The authors declare no conflict of interest.

References

1. Callaghan, J.J.; Brand, R.A.; Pedersen, D.R. Hip Arthrodesis. A Long-Term Follow-Up. *J. Bone Jt. Surg. Am.* **1985**, *67*, 1328–1335. [CrossRef]
2. Gore, D.R.; Murray, M.P.; Sepic, S.B.; Gardner, G.M. Walking Patterns of Men with Unilateral Surgical Hip Fusion. *J. Bone Jt. Surg. Am.* **1975**, *57*, 759–765. [CrossRef]
3. Garvin, K.L.; Pellicci, P.M.; Windsor, R.E.; Conrad, E.U.; Insall, J.N.; Salvati, E.A. Contralateral Total Hip Arthroplasty or Ipsilateral Total Knee Arthroplasty in Patients Who Have a Long-Standing Fusion of the Hip. *J. Bone Jt. Surg. Am.* **1989**, *71*, 1355–1362. [CrossRef]
4. Sponseller, P.D.; McBeath, A.A.; Perpich, M. Hip Arthrodesis in Young Patients. A Long-Term Follow-up Study. *J. Bone Jt. Surg. Am.* **1984**, *66*, 853–859. [CrossRef]
5. Thambyah, A.; Hee, H.T.; Das De, S.; Lee, S.M. Gait Adaptations in Patients with Longstanding Hip Fusion. *J. Orthop. Surg.* **2003**, *11*, 154–158. [CrossRef] [PubMed]
6. Clermont, C.A.; Barden, J.M. Accelerometer-Based Determination of Gait Variability in Older Adults with Knee Osteoarthritis. *Gait Posture* **2016**, *50*, 126–130. [CrossRef] [PubMed]
7. Kierkegaard, S.; Jørgensen, P.B.; Dalgas, U.; Søballe, K.; Mechlenburg, I. Pelvic Movement Strategies and Leg Extension Power in Patients with End-Stage Medial Compartment Knee Osteoarthritis: A Cross-Sectional Study. *Arch. Orthop. Trauma. Surg.* **2015**, *135*, 1217–1226. [CrossRef]
8. Spinoso, D.H.; Bellei, N.C.; Marques, N.R.; Navega, M.T. Quadriceps Muscle Weakness Influences the Gait Pattern in Women with Knee Osteoarthritis. *Adv. Rheumatol.* **2018**, *58*, 26. [CrossRef]
9. Romness, D.W.; Morrey, B.F. Total Knee Arthroplasty in Patients with Prior Ipsilateral Hip Fusion. *J. Arthroplast.* **1992**, *7*, 63–70. [CrossRef]
10. Richards, C.J.; Duncan, C.P. Conversion of Hip Arthrodesis to Total Hip Arthroplasty: Survivorship and Clinical Outcome. *J. Arthroplast.* **2011**, *26*, 409–413. [CrossRef]
11. Strathy, G.M.; Fitzgerald, R.H., Jr. Total Hip Arthroplasty in the Ankylosed Hip. A Ten-Year Follow-Up. *J. Bone Jt. Surg. Am.* **1988**, *70*, 963–966. [CrossRef]
12. Stover, M.D.; Beaulé, P.E.; Matta, J.M.; Mast, J.W. Hip Arthrodesis: A Procedure for the New Millennium? *Clin. Orthop. Relat. Res.* **2004**, *418*, 126–133. [CrossRef] [PubMed]
13. Paley, D.; Herzenberg, J.E.; Tetsworth, K.; McKie, J.; Bhave, A. Deformity Planning for Frontal and Sagittal Plane Corrective Osteotomies. *Orthop. Clin. N. Am.* **1994**, *25*, 425–465. [CrossRef]
14. Savin, L.; Botez, P.; Mihailescu, D.; Predescu, V.; Grierosu, C. Pre-Operative Radiological Measurement of Femoral Rotation for Prosthetic Positioning in Total Knee Arthroplasty. *Int. Orthop.* **2016**, *40*, 1855–1860. [CrossRef] [PubMed]
15. Daines, B.K.; Dennis, D.A. Gap Balancing vs. Measured Resection Technique in Total Knee Arthroplasty. *Clin. Orthop. Surg.* **2014**, *6*, 1–8. [CrossRef] [PubMed]
16. Rittmeister, M.; Starker, M.; Zichner, L. Hip and Knee Replacement after Longstanding Hip Arthrodesis. *Clin. Orthop. Relat. Res.* **2000**, *371*, 136–145. [CrossRef]
17. Arai, K.; Hanyu, T.; Murai, T.; Onda, N.; Kanda, T. Total Knee Arthroplasty in a Patient with Rheumatoid Arthritis and Spontaneous Fusion of the Right Hip. *Mod. Rheumatol.* **2001**, *11*, 344–347. [CrossRef]
18. Goodman, S.B.; Huddleston, J.I., 3rd; Hur, D.; Song, S.J. Total Knee Arthroplasty in Patients with Ipsilateral Fused Hip: A Technical Note. *Clin. Orthop. Surg.* **2014**, *6*, 476–479. [CrossRef]
19. Koo, K.; Pang, K.C.; Wang, W. Total Knee Arthroplasty in a Patient with a Fused Ipsilateral Hip. *J. Orthop. Surg. Res.* **2015**, *10*, 127. [CrossRef]
20. de la Hera, B.; Rubio-Quevedo, R.; Gomez-Garcia, A.; Gomez-Rice, A. Total Knee Arthroplasty in Patients with Prior Ipsilateral Hip Arthrodesis. *Eur. J. Orthop. Surg. Traumatol.* **2018**, *28*, 521–524. [CrossRef]
21. Tang, X.; Yan, Z.; Wang, W.; Liu, T. Total Knee Arthroplasty in a Patient with Ipsilateral Hip Ankylosis: A Case Report. *Medicine* **2019**, *98*, e16889. [CrossRef] [PubMed]
22. Samborski, S.A.; Quinzi, D.; Balkissoon, R. Total Knee Arthroplasty with Prior Ipsilateral Hip Arthrodesis. *Arthroplast. Today* **2020**, *6*, 830–834. [CrossRef] [PubMed]
23. Encinas-Ullán, C.A.; Gómez-Cardero, P.; Rodríguez-Merchán, E.C. Total Knee Arthroplasty in Patients with Prior Ipsilateral Hip Fusion: A Surgical Technique. *HSS J.* **2022**, *18*, 175–181. [CrossRef] [PubMed]
24. Ashkenazi, I.; Habibi, A.; Schwarzkopf, R. Total Knee Arthroplasty in a Patient with Ipsilateral Painless Hip Arthrodesis: Technical Surgical Tips and a Novel Method for Intraoperative Navigation-Assisted Planning. *Surg. Technol. Int.* **2023**, *42*, sti42-1664. [CrossRef] [PubMed]
25. Clohisy, J.C.; Keeney, J.A.; Schoenecker, P.L. Preliminary Assessment and Treatment Guidelines for Hip Disorders in Young Adults. *Clin. Orthop. Relat. Res.* **2005**, *441*, 168–179. [CrossRef]
26. Amstutz, H.C.; Sakai, D.N. Total Joint Replacement for Ankylosed Hips. Indications, Technique, and Preliminary Results. *J. Bone Jt. Surg. Am.* **1975**, *57*, 619–625. [CrossRef]
27. Lubahn, J.D.; Evarts, C.M.; Feltner, J.B. Conversion of Ankylosed Hips to Total Hip Arthroplasty. *Clin. Orthop. Relat. Res.* **1980**, *153*, 146–152. [CrossRef]

28. Hardinge, K.; Murphy, J.C.; Frenyo, S. Conversion of Hip Fusion to Charnley Low-Friction Arthroplasty. *Clin. Orthop. Relat. Res.* **1986**, *211*, 173–179. [CrossRef]
29. Cho, W. TKA in Specific Conditions. In *Knee Joint Arthroplasty*; Cho, W., Ed.; Springer: Berlin/Heidelberg, Germany, 2014; pp. 143–194. ISBN 9783642393891.
30. Roberts, C.S.; Fetto, J.F. Functional Outcome of Hip Fusion in the Young Patient. Follow-up Study of 10 Patients. *J. Arthroplast.* **1990**, *5*, 89–96. [CrossRef]
31. Chethan, K.N.; Bhat, S.N.; Zuber, M.; Satish, S.B.; Jin, Z. Evolution of different designs and wear studies in total hip prosthesis using finite element analysis: A review. *Cogent Eng.* **2022**, *9*, 2027081.
32. Poboży, T.; Wojciech, K.; Hordowicz, M. TKA with retained hardware guided by in-traoperative ultrasonography—A case report. *BMC Surg.* **2019**, *19*, 126. [CrossRef] [PubMed]
33. Secrist, E.S.; Rowe, T.; Li, K.K.; Fehring, T.K. Variability of Sagittal Plane Proximal Tibial Morphology and Its Effect on Stem Placement in Total Knee Arthroplasty. *Arthroplast. Today* **2022**, *13*, 55–61. [CrossRef] [PubMed]
34. Filip, A.C.; Cuculici, S.A.; Cristea, S.; Filip, V.; Negrea, A.D.; Mihai, S.; Pantu, C.M. Tibial Stem Extension versus Standard Configuration in Total Knee Arthroplasty: A Biomechanical Assessment According to Bone Properties. *Medicina* **2022**, *58*, 634. [CrossRef] [PubMed]
35. Scott, C.E.H.; Biant, L.C. The Role of the Design of Tibial Components and Stems in Knee Replacement. *J. Bone Jt. Surg. Br.* **2012**, *94*, 1009–1015. [CrossRef]

Journal of *Personalized Medicine*

MDPI

Communication

The Real Post-Operative Range of Motion Differs from the Virtual Pre-Operative Planned Range of Motion in Reverse Shoulder Arthroplasty

Julien Berhouet [1,2,*], Ramy Samargandi [1,3], Luc Favard [1], Céline Turbillon [1], Adrien Jacquot [4] and Marc-Olivier Gauci [5]

1. CHRU Trousseau Service d'Orthopédie Traumatologie, Faculté de Médecine de Tours, Université de Tours, 1C Avenue de la République, 37170 Tours, France; ramy.samargandi@hotmail.com (R.S.)
2. Equipe Reconnaissance de Forme et Analyse de l'Image, Laboratoire d'Informatique Fondamentale et Appliquée de Tours EA6300, Ecole d'Ingénieurs Polytechnique Universitaire de Tours, Université de Tours, 64 Avenue Portalis, 37200 Tours, France
3. Department of Orthopedic Surgery, Faculty of Medicine, University of Jeddah, Jeddah 23218, Saudi Arabia
4. Centre for Chirurgie des Articulations et du Sport (ARTICS), 24 rue du XXIème Régiment d'Aviation, 54000 Nancy, France
5. Institut Locomoteur et du Sport, Hôpital Pasteur 2, 30 Voie Romaine, 06000 Nice, France

* Correspondence: julien.berhouet@gmail.com; Tel.: +33-247478644; Fax: +33-247475912

Abstract: Introduction: The purpose of this study was to analyze the real range of motion (RoM) measured in patients operated on for reverse shoulder arthroplasty (RSA) and compare it to the virtual RoM provided by the preoperative planning software. Hypothesis: There was a difference between virtual and real RoM, which can be explained by different factors, specifically the scapula-thoracic (ST) joint. Methods: Twenty patients with RSA were assessed at a minimum follow-up of 18 months. Passive RoM in forward elevation abduction, without and with manually locking the ST joint, and in external rotation with arm at side were recorded. The humerus, scapula, and implants were manually segmented on post-operative CTs. Post-operative bony structures were registered to preoperative bony elements. From this registration, a post-operative plan corresponding to the real post-operative implant positioning was generated and the corresponding virtual RoM analysis was recorded. On the post-operative anteroposterior X-rays and 2D-CT coronal planning view, the glenoid horizontal line angle (GH), the metaphyseal horizontal line angle (MH), and the gleno-metaphyseal angle (GMA) were measured to assess the extrinsic glenoid inclination, as well as the relative position of the humeral and glenoid components. Results: There were some significant differences between virtual and post-operative passive abduction and forward elevation, with (55° and 50°, $p < 0.0001$) or without ST joint participation (15° and 27°, $p < 0.002$). For external rotation with arm at side, there was no significant difference between planning (24° ± 26°) and post-operative clinical observation (19° ± 12°) ($p = 0.38$). For the angle measurements, the GMA was significantly higher (42.8° ± 15.2° vs. 29.1° ± 18.2°, $p < 0.0001$), and the GH angle, significantly lower on the virtual planning (85.2° ± 8.8° vs. 99.5° ± 12.5°, $p < 0.0001$), while the MH was not different ($p = 0.33$). Conclusions: The virtual RoM given by the planning software used in this study differs from the real post-operative passive RoM, except for external rotation. This can be explained by the lack of ST joint and soft tissues simulation. However, in focusing on the virtual GH participation, the simulation looks informative. Some modifications between the glenoid and humerus starting positions before running the motion analysis could be provided for making it more realistic and predictive of the RSA functional results. Level of evidence: III.

Keywords: reverse shoulder arthroplasty; range of motion; preoperative planning; scapulothoracic joint; soft tissues; motion analysis

Citation: Berhouet, J.; Samargandi, R.; Favard, L.; Turbillon, C.; Jacquot, A.; Gauci, M.-O. The Real Post-Operative Range of Motion Differs from the Virtual Pre-Operative Planned Range of Motion in Reverse Shoulder Arthroplasty. *J. Pers. Med.* **2023**, *13*, 765. https://doi.org/10.3390/jpm13050765

Academic Editor: Alexander J. Nedopil

Received: 28 March 2023
Revised: 21 April 2023
Accepted: 26 April 2023
Published: 29 April 2023

1. Introduction

Restoring range of motion (RoM) is one of the main challenges of reverse shoulder arthroplasty (RSA). Functional outcomes are closely related to bony or prosthetic impingement, especially during adduction and external rotation with elbow at side [1,2]. Scapular notching is one of the consequences of this inferior impingement [3].

Three-dimensional (3D) preoperative planning solutions have been developed to improve glenoid implant positioning and optimize RSA biomechanics [4,5]. In a computer study, Berhouet et al. [6] reported that the placement of the glenoid baseplate was more accurate and reliable when using 3D rather than two-dimensional (2D) preoperative visualization. Virtual RoM information can also be provided by the planning software. This helps to guide the surgeon in the choice and placement of the implants and provides a glimpse into the functional outcomes of the surgery [7]. Werner et al. [8] conducted a virtual study with this modelling software to analyze the effect of glenosphere design and size with the aim of reducing the incidence of scapular notching. However, one of the study's limitations is that the RoM analysis was only based on the glenohumeral joint and bony impingement, without taking into account scapulothoracic (ST) or scapulohumeral movements and soft tissues constraints.

ST participation is more important for RSA biomechanics than for a normal shoulder [9–11]. While the glenohumeral (GH) joint is 2.5 times more active than the ST joint during abduction or forward elevation in normal conditions, the participation ratio (GH/ST) decreases from 1.5 to 1 after RSA, especially over 30° of range [12]. The benefits of preoperative planning, which provides incomplete and potentially mismatched functional information relative to the real conditions, must be questioned [13]. Additionally, the clinical impact of such software can only be correctly addressed if the planned implant positioning is precisely replicated during the real procedure. In practice, this implies that accurate post-operative imaging is needed [14].

The purpose of this study was to analyze the real RoM measured in patients operated on for an RSA and compare it with the virtual RoM given by the preoperative planning software in the same implant positioning conditions (Figure 1). The hypothesis was that there was a difference between virtual and real RoM, which can be explained by different factors, specifically the ST joint.

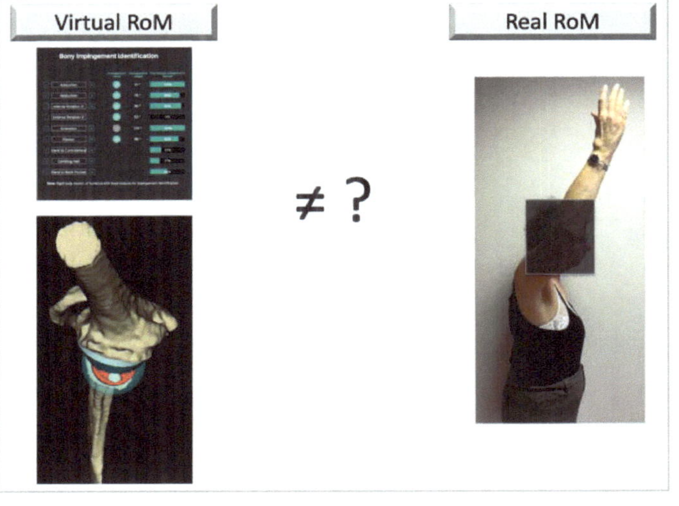

Figure 1. Picture illustrating the goal of the study: comparison of the virtual RoM to the real RoM in case of RSA implantation.

2. Materials and Methods

2.1. Ethics

The study was approved by our institutional review board (IRB: 13B-T-SHOULDER-RM).

2.2. Patients and Implants

This was a retrospective clinical and experimental study involving 20 patients (6 males, 14 females), who were operated on for RSA by two surgeons between 1 January and 31 December 2016 (IRB: 13B-T-SHOULDER-RM). The mean age was 73.8 (63–85) years. The mean BMI was 27.1 (19.5–34.4) kg/m^2. The surgical indications were cuff tear arthropathy (CTA) for 11 patients, primary osteoarthritis (OA) for 7 patients, and rheumatoid arthritis and post-instability arthritis for 2 patients.

In four patients, the AEQUALIS™ REVERSED II shoulder system was used, while the AEQUALIS™ ASCEND™ FLEX shoulder system was implanted in 16 patients (Wright Medical, Memphis, TN, USA).

Patients with missing post-operative RoM data and pre- and post-operative imaging were excluded.

2.3. D Pre-Operative Planning Software

The BluePrint® 3D Planning software (version 2.1.6, Wright Medical France, Monbonnot Saint Martin, France) provides a validated method for automatic segmentation, 3D reconstruction, and accurate 3D pre-operative glenoid (version and inclination) and humeral (inclination and posterior subluxation) measurements [15,16]. This information can then be used to precisely plan the size and position of the glenoid and humeral implants.

At the end of the planning session for the RSA procedure, the BluePrint® software runs a virtual RoM analysis in the different degrees of freedom: abduction/adduction, flexion/extension, and internal/external rotation arm at side. The maximum range of motion relative to bony or implant impingement between the humeral and scapular sides is recorded (Figure 2).

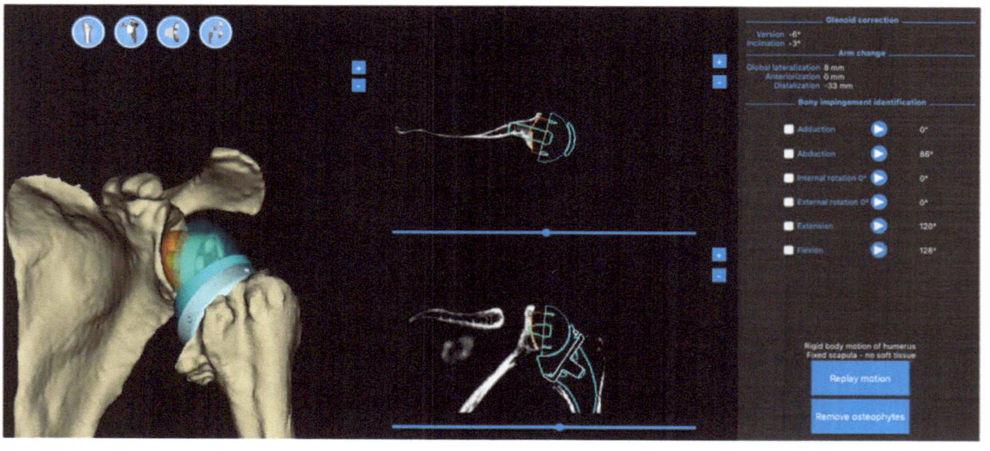

Figure 2. Screenshot of RSA planning with the positioning of the glenoid and humeral components and the virtual RoM analysis results on the right of the picture.

2.4. Study Protocol

There were seven steps in the protocol.

1. After the usual pre-operative evaluation (clinical exam, X-ray, and CT scan), the RSA procedure was performed for each patient;

2. At a minimum follow-up of 18 months, a post-operative CT scan was performed for each patient using the acquisition characteristics required by the BluePrint®3D Planning software, in addition to the routine clinical and radiological assessments, which were conducted by a single surgeon;
3. For all shoulders, the planning software's automated segmentation algorithm was used to extract the pre-operative humerus and scapula 3D models from the pre-operative CTs (Figure 3A);
4. On all the post-operative CTs, the humerus, scapula, and implants were segmented manually (Figure 3B);
5. Registration and superposition were performed between the pre- and post-operative bony structures using the PTC Creo®Version 6.0 software (Parametric Technology Corporation, Needham, MA, USA) (Figure 3C);
6. A software program that takes the registration data and generates "post-operative planning" files equivalent to the real post-operative implant positioning was developed;
7. Surgeons could open the post-operative planning file and access all BluePrint®3D Planning software measurements, in particular, the RoM analysis (Figure 3D).

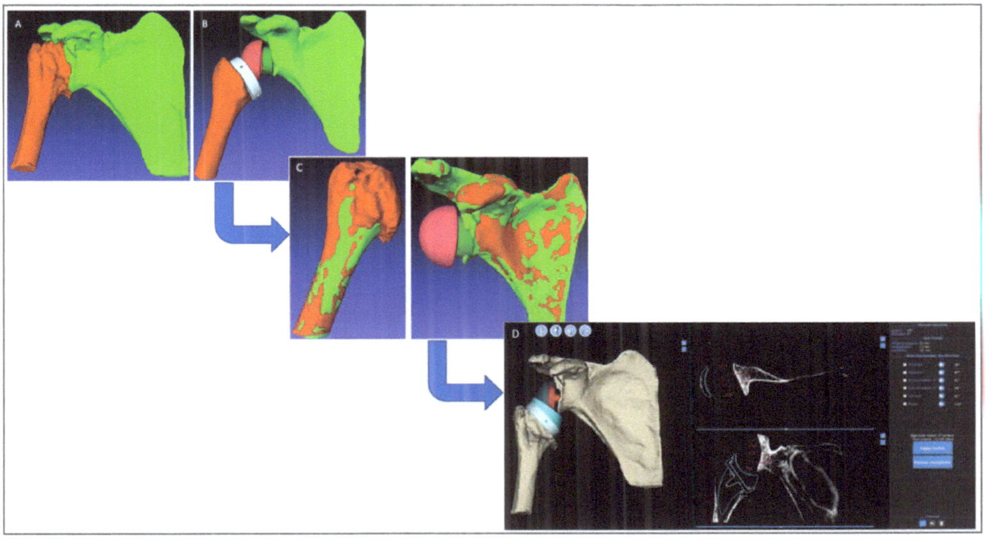

Figure 3. Study protocol. (**A**) Pre-operative automated joint segmentation for extracting humerus and scapula 3D models from the pre-operative CT scans. (**B**) Post-operative manual joint segmentation of the humerus, scapula, and implants from the post-operative CT scan. (**C**) Registration between pre-operative (green) and post-operative (orange) bony structures. (**D**) Post-operative planning with virtual RoM analysis after registration.

This protocol guaranteed that the implant choice and positioning were identical between the virtual and real conditions, making it possible to compare the RoM measurements between both conditions.

2.5. Clinical and Radiological Assessment Criteria

Passive RoM in abduction and forward elevation without and with manually locking the scapula-thoracic joint, as well as the external and internal rotation with elbow at side, were assessed clinically in each patient at a minimum follow-up of 18 months. Those measurements were compared to the virtual RoM generated by the "postoperative planning".

Three angles were measured on post-operative anteroposterior x-rays in neutral rotation, and on a planning 2D-CT view passing by the larger diameter of the sphere. The first was the glenometaphyseal angle (GMA) [17] between the prosthetic humeral metaphysis and the backside of the baseplate. The two other angles—metaphyseal horizontal line angle (MH) and glenoid horizontal line angle (GH)—corresponded to the angles between the humeral metaphysis and the horizontal line, and the glenoid component and this latter line, respectively (Figure 4).

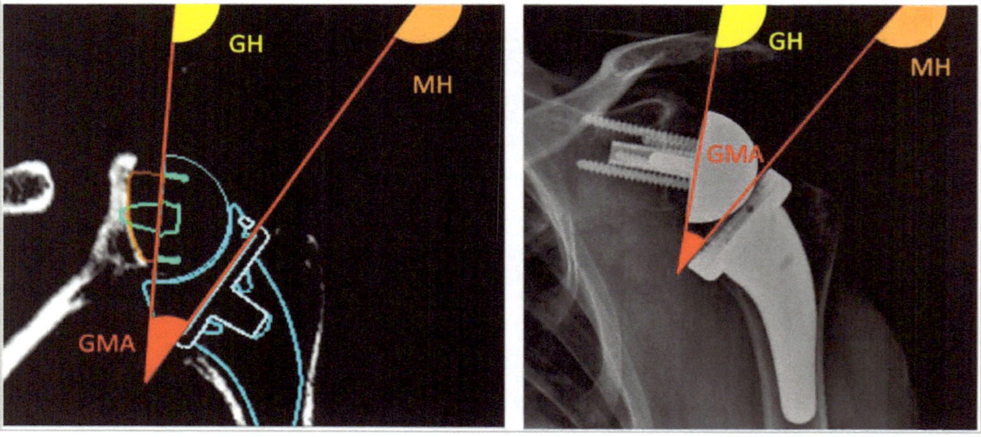

Figure 4. Angle measurements on planning 2D-CT coronal view (**left**) and post-operative anteroposterior X-ray view (**right**). Glenometaphyseal angle (GMA) in red, metaphyseal horizontal line angle (MH) in orange, and glenoid horizontal line angle (GH) in yellow.

2.6. Statistical Analysis

The statistical analysis was performed by a biostatistician using XLSTAT Life Sciences software (Addinsoft, Bordeaux, France). The two-sided Wilcoxon–Mann–Whitney test was used to compare two variables. The significance level was set at $p \leq 0.05$.

3. Results

At a mean follow up of 26.3 months (18–38), the mean post-operative passive abduction was 136° (±28°), while the virtual abduction given by the software was 81° (±13°). There was a significant difference of 55° between the two conditions ($p < 0.0001$). Passive real forward elevation was 141° (±24°), while the virtually planned motion was 91° (±24°). This 50° difference was significant ($p < 0.0001$).

When manually locking the scapula-thoracic joint, the mean post-operative passive abduction was 65° (±18°), and the passive forward elevation was 63° (±20°). The differences with the planned virtual motions were, respectively, 15° and 27° ($p < 0.002$ and $p < 0.0001$). In those conditions, the abduction ratio (GH/ST) was 0.92. The forward elevation ratio (GH/ST) was 0.81.

For the external rotation with elbow at side, there was no significant difference between the virtual motion (24° ± 26°) and the post-operative clinical observation (19° ± 12° with ST; 22° ± 19° without ST) ($p = 0.38$; 0.42). Internal rotation was not interpretable because the measurement scale differs between the virtual motion given by the software (in degrees) and the real post-operative motion assessed by the surgeon (vertebral level reached) (Table 1).

Table 1. Comparison of virtual and real post-operative RoM (GH: gleno humeral joint; ST: scapula thoracic joint). The two-sided Wilcoxon–Mann–Whitney test was used to compare two variables.

	Virtual (1)	Real GH (2)	Real GH + ST (3)	*p*-Value (1 vs. 2)	*p*-Value (1 vs. 3)
Passive Abduction (°)	80.8 ± 13.6	65.6 ± 18.2	136.7 ± 27.7	<0.0001	<0.0001
Passive Forward Elevation (°)	90.8 ± 23.1	63.3 ± 19.7	141.1 ± 23.82	<0.002	<0.0001
External rotation arm at side (°)	24.1 ± 25.6	22.1 ± 19.4	19.2 ± 12.4	0.42	0.38

For the different radiological angles, the GMA was significantly higher on the virtual planning (42.8° ± 15.2°) than on the post-operative X-ray (29.1° ± 18.2°) ($p < 0.0001$). The GH angle was significantly lower on the virtual planning (85.2° ± 8.8° vs. 99.5° ± 12.5°, $p < 0.0001$), while the MH angle was not different ($p = 0.33$) (Table 2).

Table 2. Comparison of post-operative X-rays and CT planned angles measurements. The two-sided Wilcoxon–Mann–Whitney test was used to compare two variables.

	Planning	Post-Operative	*p*-Value
Gleno-Metaphyseal Angle (°)	42.8 ± 15.2	29.1 ± 18.2	<0.0001
GH angle (°)	85.2 ± 8.8	99.5 ± 12.5	<0.0001
MH angle (°)	133.3 ± 11.1	128.6 ± 12.3	0.33

4. Discussion

Post-operative passive forward elevation and abduction were significantly different to the virtual post-operative RoM, while the features and the positioning of the components were exactly the same between planning and surgery. However, those differences were less important when the ST joint was manually neutralized during the clinical evaluation.

The ST joint was not taken into account during the RoM simulation. In our study, the ratios (GH/ST) for passive forward elevation and abduction in usual clinical exam conditions (ST joint free) are estimated at 0.82 and 0.92. Those values are close to previous values observed by different authors based on kinematic studies [10,12], where a more important participation of the ST joint is reported in case of RSA (1 < ratio (GH/ST) < 1.5) [18]. It could, thus, be considered that GH participation in passive forward elevation and abduction after RSA is quite correctly simulated by the software used in our study. The remaining differences we found between ST locked virtual and real conditions could be explained by the lack of reliability or accuracy of the measurements by the observer, as well as by the soft tissues. The results for the passive external rotation with elbow at side reinforce this interpretation. There was no significant difference between the virtually planned and the real post-operative passive external rotation. The ST joint's involvement in this motion is different and minimal [19]. In practice, this could mean that passive external rotation is limited by bony inferior scapular impingement, as well as by soft tissues with subscapularis tensioning (12 closing tendon repairs in this study) or fibrosis. Consequently, the virtual RoM analysis generated by the software looks informative for the GH participation rendering in the different passive motions (abduction and forward elevation). It looks even more realistic for external rotation elbow at side, mainly based on GH participation.

ST joint participation in shoulder RoM partially depends on glenoid inclination. Some authors reported that the more the glenoid is downward, the greater the action of the ST joint for carrying out this motion, especially for abduction and forward elevation [20,21]. For this reason, different angles were measured to investigate the differences between virtual and real conditions. The glenoid inclination analysis may be divided into extrinsic tilt, according to the position of the full scapula, and intrinsic tilt, according to the position of the glenoid joint line relative to the rest of the scapula in the coronal plane. Thanks to the study protocol, the intrinsic glenoid tilt could be considered as the same between post-operative X-rays and virtual planning. However, the GMA was different between those both conditions. This difference came from the GH angle, which assesses the extrinsic and intrinsic glenoid inclination. MH did not change because it reflected the humerus metaphysis inclination, which was not dependent of the scapular inclination. In practice, the software placed every scapula in the same position, with the supraspinatus fossa aligned to a horizontal line. The extrinsic glenoid tilt is systematically neutral in these planning conditions, while it is not the case in reality after surgery. For most of the patients, there is an inferior tilt of the scapula in resting position, which theoretically could prevent inferior impingement and scapular notching, and improve the rotations [22]. Consequently, it could be interesting to place the scapula in the same position as it was pre-operatively during the planning. It could also inform us about the GH joint's participation, as well as the ST joint's reserve to carry out certain motions. The conditions of virtual motion would, thus, be closer to the post-operative real conditions.

Other factors may explain the differences between virtual and post-operative real RoM. The humerus position during simulated abduction is not strictly aligned with the glenoid surface (Figure 5). The abduction motion is consequently not performed in the plane of the scapula, although it aims to be during the clinical exam. It may affect the range of abduction by changing the zones of bony impingement. Thus, the humerus position relative to the glenoid could be modified in the software to optimize the virtual RoM. Soft tissue interposition and tensioning may also limit the RoM before the bony impingement occurs [23]. There is currently no information on how to measure or assess this restriction in the passive motion.

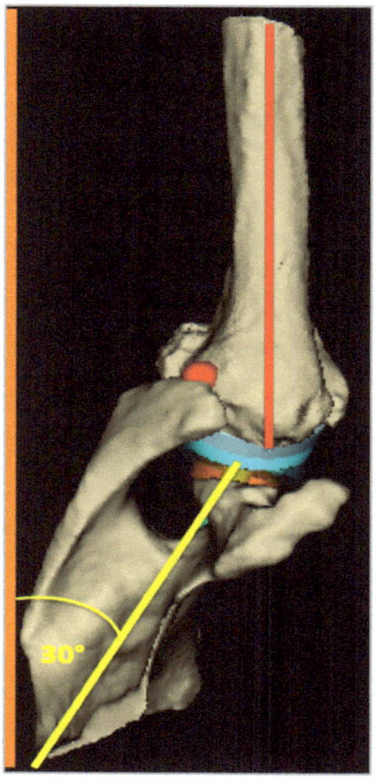

Figure 5. Screenshot of the virtual abduction given by the software (3D superior view). Note the nonaligned position of the humerus (red line) with the scapular plane (yellow line), which makes about a 30° angle with the frontal plane, while the humeral plane is parallel to the frontal plane (orange axis).

Our study has several limitations. The number of patients involved is relatively small. The method for measuring the post-operative passive mobilities with the manual neutralization of the ST joint is not so easy achieve and it is not reproducible, and consequently, not so accurate for assessing the range of motion. This could be one of the reasons, as reported above, for explaining the difference observed between virtual and real RoM results. Additionally, this clinical evaluation remains subjective and surgeon-dependent.

Nevertheless, to our knowledge, this is the first study reported aiming to address the relevance of a pre-operative planning software program with a RoM simulation. Our study protocol was rigorous. The same conditions were used for implant choice and positioning between the planning and the surgery. The passive RoM with and without the ST joint was assessed. Utilizing the limited available data, an associated radiological analysis was produced.

5. Conclusions

The virtual RoM given by the 3D planning software programs used in this study differs from the true post-operative RoM, except for the external rotation with elbow at side. This can likely be explained by the lack of simulation rendering for ST joint and soft tissues during the virtual analysis. However, the RoM simulation provided looks informative for GH participation. Some modifications, such as the starting scapula inclination, as well as its position relative to the humerus during abduction, could be inputted in the planning software to make it more realistic and predictive of the RSA functional results.

Author Contributions: J.B.: study design, data acquisition, statistical and data analysis, manuscript writing and editing. M.-O.G.: manuscript editing. F.S.: manuscript editing. C.T.: statistical analysis A.J.: manuscript editing. L.F.: data acquisition, manuscript editing. All authors reviewed and approved the final submitted version. All authors have read and agreed to the published version of the manuscript.

Funding: This research received no external funding.

Institutional Review Board Statement: Ethical approval for this study was obtained from our institutional review board (IRB: 13B-T-SHOULDER-RM).

Informed Consent Statement: Informed consent was obtained from all subjects involved in the study.

Data Availability Statement: Data supporting reported results can be available in the Department of Orthopedic Surgery of the CHRU Trousseau of Tours, France.

Conflicts of Interest: Julien Berhouet: Wright consulting; Marc-Olivier Gauci: Wright consulting; François Boux de Casson: Wright employee; Adrien Jacquot: Wright consulting; Luc Favard: Wright consulting and royalties.

References

1. Bacle, G.; Nove-Josserand, L.; Garaud, P.; Walch, G. Long-Term Outcomes of Reverse Total Shoulder Arthroplasty: A Follow-up of a Previous Study. *J. Bone Jt. Surg. Am.* **2017**, *99*, 454–461. [CrossRef]
2. Arenas-Miquelez, A.; Murphy, R.J.; Rosa, A.; Caironi, D.; Zumstein, M.A. Impact of humeral and glenoid component variations on range of motion in reverse geometry total shoulder arthroplasty: A standardized computer model study. *J. Shoulder Elb. Surg.* **2021**, *30*, 763–771. [CrossRef]
3. Mollon, B.; Mahure, S.A.; Roche, C.P.; Zuckerman, J.D. Impact of scapular notching on clinical outcomes after reverse total shoulder arthroplasty: An analysis of 476 shoulders. *J. Shoulder Elb. Surg.* **2017**, *26*, 1253–1261. [CrossRef]
4. Lädermann, A.; Denard, P.J.; Collin, P.; Zbinden, O.; Chiu, J.C.-H.; Boileau, P.; Olivier, F.; Walch, G. Effect of humeral stem and glenosphere designs on range of motion and muscle length in reverse shoulder arthroplasty. *Int. Orthop.* **2020**, *44*, 519–530. [CrossRef]
5. Werner, B.S.; Hudek, R.; Burkhart, K.J.; Gohlke, F. The influence of three-dimensional planning on decision-making in total shoulder arthroplasty. *J. Shoulder Elb. Surg.* **2017**, *26*, 1477–1483. [CrossRef] [PubMed]
6. Berhouet, J.; Gulotta, L.; Dines, D.; Craig, E.; Warren, R.; Choi, D.; Chen, X.; Kontaxis, A. Preoperative planning for accurate glenoid component positioning in reverse shoulder arthroplasty. *Orthop. Traumatol. Surg. Res.* **2017**, *103*, 407–413. [CrossRef]

7. Lädermann, A.; Tay, E.; Collin, P.; Piotton, S.; Chiu, C.-H.; Michelet, A.; Charbonnier, C. Effect of critical shoulder angle, glenoid lateralization, and humeral inclination on range of movement in reverse shoulder arthroplasty. *Bone Jt. Res.* **2019**, *8*, 378–386. [CrossRef] [PubMed]
8. Werner, B.S.; Chaoui, J.; Walch, G. Glenosphere design affects range of movement and risk of friction-type scapular impingement in reverse shoulder arthroplasty. *Bone Jt. J.* **2018**, *100-B*, 1182–1186. [CrossRef] [PubMed]
9. Blaimont, P.; Taheri, A.; Vanderhofstadt, A. Displacement of the instantaneous center of rotation of the humeral head during abduction: Implication for scapulohumeral muscular function. *Rev. Chir. Orthop. Reparatrice L'appareil Mot.* **2005**, *91*, 399–406. [CrossRef]
10. Kwon, Y.W.; Pinto, V.J.; Yoon, J.; Frankle, M.A.; Dunning, P.E.; Sheikhzadeh, A. Kinematic analysis of dynamic shoulder motion in patients with reverse total shoulder arthroplasty. *J. Shoulder Elb. Surg.* **2012**, *21*, 1184–1190. [CrossRef]
11. Terrier, A.; Scheuber, P.; Pioletti, D.P.; Farron, A. Activities of daily living with reverse prostheses: Importance of scapular compensation for functional mobility of the shoulder. *J. Shoulder Elb. Surg.* **2013**, *22*, 948–953. [CrossRef]
12. Boileau, P.; Walch, G.; Liotard, J.P. Radio-cinematographic study of active elevation of the prosthetic shoulder. *Rev. Chir. Orthop. Reparatrice L'appareil Mot.* **1992**, *78*, 355–364.
13. Moroder, P.; Urvoy, M.M.; Raiss, P.; Werthel, J.-D.; Akgün, D.; Chaoui, J.; Siegert, P. Patient Posture Affects Simulated ROM in Reverse Total Shoulder Arthroplasty: A Modeling Study Using Preoperative Planning Software. *Clin. Orthop. Relat. Res.* **2022**, *480*, 619–631. [CrossRef] [PubMed]
14. Marcoin, A.; Nerot, C.; Lestra, T.; Blasco, L.; Ferrier, A.; Siboni, R.; Ohl, X. The precision of patient-specific instrumentation guides for the positioning of the glenoid component in total reverse shoulder arthroplasty: An in vivo scanographic study. *Int. Orthop.* **2020**, *44*, 1761–1766. [CrossRef]
15. Boileau, P.; Cheval, D.; Gauci, M.O.; Holzer, N.; Chaoui, J.; Walch, G. Automated Three-Dimensional Measurement of Glenoid Version and Inclination in Arthritic Shoulders. *J. Bone Joint Surg Am.* **2018**, *100*, 57–65. [CrossRef]
16. Walch, G.; Vezeridis, P.S.; Boileau, P.; Deransart, P.; Chaoui, J. Three-dimensional planning and use of patient-specific guides improve glenoid component position: An in vitro study. *J. Shoulder Elbow Surg.* **2015**, *24*, 302–309. [CrossRef] [PubMed]
17. Falaise, V.; Levigne, C.; Favard, L. Scapular notching in reverse shoulder arthroplasties: The influence of glenometaphyseal angle. *Orthop. Traumatol. Surg. Res.* **2011**, *97* (Suppl. S6), S131–S137. [CrossRef]
18. Alta, T.D.; de Toledo, J.M.; Veeger, D.; Janssen, T.; Willems, W.J. The active and passive kinematic difference between primary reverse and total shoulder prostheses. *J. Shoulder Elb. Surg.* **2014**, *23*, 1395–1402. [CrossRef]
19. Maier, M.W.; Caspers, M.; Zeifang, F.; Dreher, T.; Klotz, M.C.; Wolf, S.I.; Kasten, P. How does reverse shoulder replacement change the range of motion in activities of daily living in patients with cuff tear arthropathy? A prospective optical 3D motion analysis study. *Arch. Orthop. Trauma. Surg.* **2014**, *134*, 1065–1071. [CrossRef] [PubMed]
20. Poppen, N.; Walker, P. Normal and abnormal motion of the shoulder. *J. Bone Jt. Surg. Am.* **1976**, *58*, 195–201. [CrossRef]
21. Poppen, N.K.; Walker, P.S. Forces at the glenohumeral joint in abduction. *Clin. Orthop. Relat. Res.* **1978**, *135*, 165–170. [CrossRef]
22. Patel, M.; Martin, J.R.; Campbell, D.H.; Fernandes, R.R.; Amini, M.H. Inferior tilt of the glenoid leads to medialization and increases impingement on the scapular neck in reverse shoulder arthroplasty. *J. Shoulder Elb. Surg.* **2021**, *30*, 1273–1281. [CrossRef] [PubMed]
23. Pitocchi, J.; Plessers, K.; Wirix-Speetjens, R.; Debeer, P.; van Lenthe, G.H.; Jonkers, I.; Pérez, M.A.; Sloten, J.V. Automated muscle elongation measurement during reverse shoulder arthroplasty planning. *J. Shoulder Elb. Surg.* **2021**, *30*, 561–571. [CrossRef] [PubMed]

Journal of
Personalized Medicine

MDPI

Article

Comparison of the Radiographic and Clinical Outcomes between the Sinus Tarsi and Extended Lateral Approaches for Intra-Articular Calcaneal Fractures: A Retrospective Study

Jui-Ting Mao [1,2,†], Chien-Ming Chen [1,2,†], Chung-Wei Lin [1,2], Hsuan-Lun Lu [3] and Chien-Chung Kuo [1,2,*]

1 Department of Orthopedics, China Medical University Hospital, Taichung 40447, Taiwan; jtmao0911@gmail.com (J.-T.M.); 023325@tool.caaumed.org.tw (C.-M.C.); visionaryshade@gmail.com (C.-W.L.)
2 Department of Orthopedics, School of Medicine, China Medical University, Taichung 40447, Taiwan
3 Department of Biomedical Engineering, Da-Yeh University, Changhua 51591, Taiwan; hllu@mail.dyu.edu.tw
* Correspondence: chienck@mail.cmu.edu.tw
† These authors contributed equally to this work.

Abstract: The aim of this study was to compare the radiological and functional outcomes of the extended lateral and sinus tarsi approaches for managing displaced intraarticular calcaneal fractures. This retrospective study involved 44 patients with displaced intra-articular calcaneal fractures. The patients were treated with either the extended lateral or sinus tarsi approach and followed up for at least a year. The radiological and clinical outcomes were compared between the approaches. The waiting time for surgery was shorter and the complication rate was lower in the sinus tarsi approach group than in the other group. There were no significant differences in the American Orthopedic Foot and Ankle Society ankle–hindfoot score, Foot Function Index, or visual analog scale score between the groups. In both groups, the radiological outcomes (Böhler angle, calcaneal width, and calcaneal height) were better postoperatively than preoperatively. The sinus tarsi approach is a safe and effective alternative to the extended lateral approach for managing displaced intraarticular calcaneal fractures. It is associated with a lower complication rate and a shorter waiting time for surgery than the extended lateral approach, with similar functional and radiological outcomes.

Keywords: extended lateral approach; sinus tarsi approach; intra-articular calcaneal fractures

Citation: Mao, J.-T.; Chen, C.-M.; Lin, C.-W.; Lu, H.-L.; Kuo, C.-C. Comparison of the Radiographic and Clinical Outcomes between the Sinus Tarsi and Extended Lateral Approaches for Intra-Articular Calcaneal Fractures: A Retrospective Study. *J. Pers. Med.* **2024**, *14*, 259. https://doi.org/10.3390/jpm14030259

Academic Editors: Norio Yamamoto and Henrik Galbo

Received: 27 January 2024
Revised: 13 February 2024
Accepted: 26 February 2024
Published: 28 February 2024

1. Introduction

Calcaneal fractures constitute approximately 60% of all tarsal fractures and are typically caused by falls from heights or motor-vehicle accidents [1–3]. Displaced intra-articular fractures account for 60–75% of all calcaneal fractures [4]. These fractures often lead to soft tissue injuries and a loss of function in the subtalar joint area [3]. However, the treatment of calcaneal fractures can be challenging for surgeons. The conservative management of these fractures is associated with a poor prognosis, with patients progressing to subtalar joint arthritis, malunion, and poor functional outcomes [5]. Therefore, open reduction and internal fixation are recommended for managing displaced intra-articular calcaneal fractures [6,7].

The extensile lateral approach (ELA) is considered the gold standard for calcaneal fractures [8–10]. However, despite adequate exposure, the ELA is associated with a high incidence of postoperative complications, such as skin edge necrosis, wound infection, and nonunion [11,12]. The sinus tarsi approach (STA), a minimally invasive technique, has recently gained popularity for reducing the complications associated with calcaneal fracture treatment. The purpose of this study was to compare the radiological and clinical functional outcomes of the ELA and STA to add to the existing literature on calcaneal fracture management.

2. Materials and Methods

2.1. Patient Population

Between January 2018 and January 2020, 44 patients with displaced intra-articular calcaneal fractures (Sanders types II and III) were treated by two experienced surgeons at a level-1 trauma center. Each of the two surgeons performed the ELA and STA separately. All patients were 18–65 years old; they had sustained injuries from falls or motor-vehicle accidents and had undergone open reduction and internal fixation with either the ELA or STA. Patients were excluded if they were heavy smokers (>20 cigarettes per day); had an open fracture, multiple fracture sites, or underlying medical comorbidities (all with Type 2 diabetes); and had been followed up for less than 1 year (Figure 1). The sex distribution did not differ significantly between the groups (Group 1: 78.3% men and 21.7% women; Group 2: 90.5% men and 9.5% women; p = 0.416). Furthermore, the two groups did not differ significantly in terms of the mean patient age (Group 1: 45.30 ± 10.00 years; Group 2: 42.86 ± 12.95 years; p = 0.581) and mean BMI (Group 1: 24.72 ± 4.20 kg/m^2; Group 2: 24.93 ± 3.84 kg/m^2; p = 0.7). The fracture types included Sanders types II and III, and their proportions did not differ significantly between the groups (Table 1).

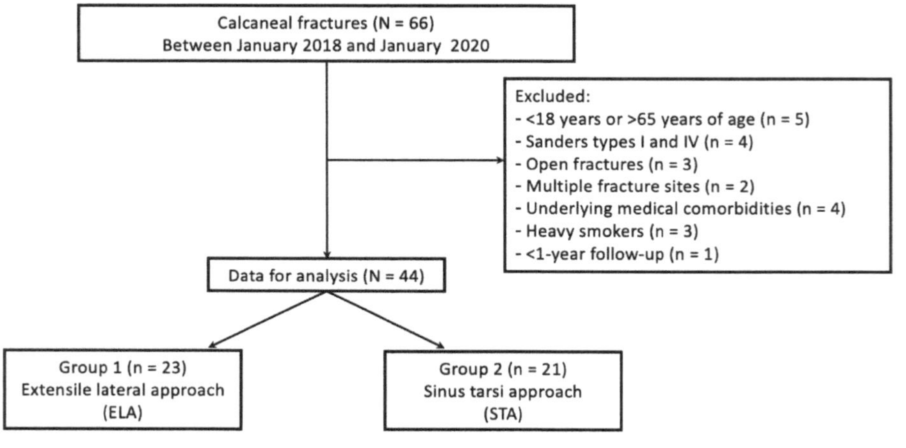

Figure 1. Participant enrollment flow chart. ELA: extensile lateral approach; STA: sinus tarsi approach.

Table 1. Patient characteristics.

	ELA	STA	*p*
Number of patients	23	21	
Mean age (years)	45.30 ± 10.00	42.86 ± 12.95	0.581
Sex			0.416
Male	18 (78.3%)	19 (90.5%)	
Female	5 (21.7%)	2 (9.5%)	
BMI (kg/m^2)	24.72 ± 4.20	24.93 ± 3.84	0.7
Etiology			0.439
Falls	21	21	
Traffic accidents	2	0	
Sanders type			0.880
Type IIA	3	4	
Type IIB	6	5	
Type IIC	4	3	
Type IIIAB	2	1	
Type IIIAC	1	2	
Type IIIBC	0	0	

BMI: body mass index; ELA: extensile lateral approach; STA: sinus tarsi approach.

2.2. *Surgical Technique*

The patients were placed in the lateral decubitus position under spinal anesthesia using a tourniquet. In both groups, all patients were administered prophylactic antibiotic therapy with first-generation cephalosporin, both preoperatively and postoperatively; the postoperative course comprised three doses spaced 8 h apart.

2.2.1. Group 1 (ELA)

The standard ELA described by Benirschke and Sangeorzan in 1993 was used to expose the subtalar joint and lateral wall [13]. After making an incision directly into the bone, a subperiosteal flap was raised and held using 1.6 mm Kirschner wires. Following the direct reduction of the fracture site, locking calcaneal plates (Acumed Combo Calcaneal plate) were used for fixation.

2.2.2. Group 2 (STA)

The STA used in this study was based on the technique described by Chiang et al. in 2021 [14]. The first incision, approximately 3–4 cm in length, was made horizontally along the tip of the lateral malleolus to a level distal to the calcaneocuboid joint. The incision allowed for adequate exposure of the subtalar joint from the calcaneocuboid joint to the posterior facet of the calcaneus. The peroneal tendons were identified and pulled inferiorly. A reduction was achieved and temporarily fixed using Kirschner wires. Another incision was made in the posterior calcaneal tuberosity. A locking plate (Acumed Combo Calcaneal plate) was used to fix the calcaneus through these two incisions (Figure 2).

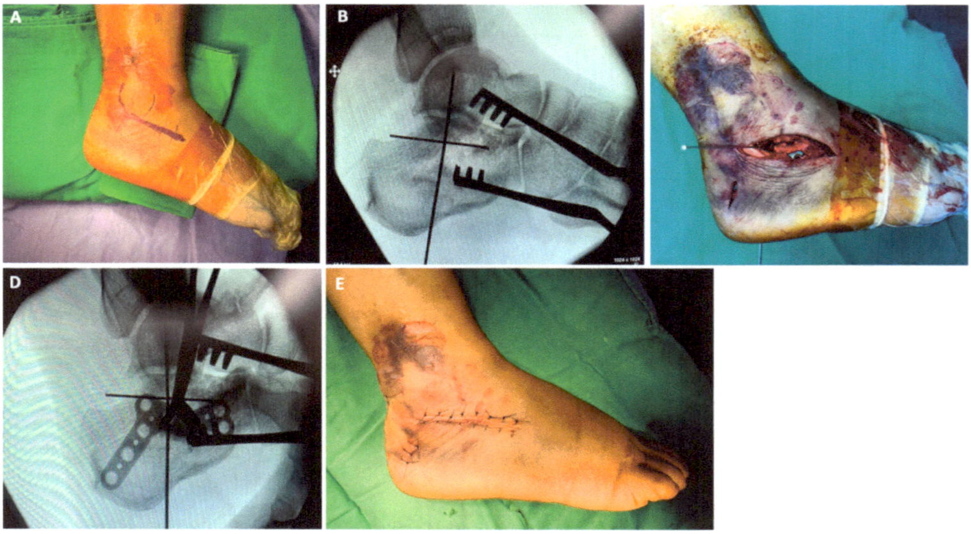

Figure 2. Two-incision STA for a right DIACF in a 37-year-old woman. (**A**) Skin incision in the STA. (**B**) Reduction and fixation with Kirschner wires. (**C**) Application of a locking plate to fix the calcaneus through two incisions. (**D**) Confirmation of adequate plate positioning under fluoroscopy. (**E**) Removal of stitches 2 weeks after operation. DIACF: displaced intra-articular calcaneal fracture; STA: sinus tarsi approach.

2.3. Postoperative Protocol

Dressings were changed regularly to monitor the wound condition, and stitches were removed 2 weeks postoperatively if there was no surgical site infection. Patients in both groups started active and passive range motion exercise regimes for the foot and ankle joints immediately after the surgery, based on their tolerance. Partial weight bearing was initiated at 8 weeks postoperatively, and full weight bearing was allowed at 12 weeks postoperatively once radiographic union was achieved. We determined radiographic union by regularly following up with X-rays to monitor fracture healing. Radiographic union was defined based on radiological findings such as callus formation and cortical bridging.

2.4. Variables, Data Sources, and Outcome Assessment

All patients were followed up for at least 1 year. Postoperatively, radiographs were obtained at the 1-month, 3-month, 6-month, and 1-year follow-ups. The data obtained from the medical records included basic demographics (age, sex, body mass index [BMI], and etiology), Sanders type, waiting time for surgery, blood loss, operation time, and complication rates. The clinical outcomes included wound complications, the American Orthopedic Foot and Ankle Society (AOFAS) ankle–hindfoot score, the Foot Function Index (FFI), and the visual analog scale (VAS) score. Radiological assessments comprised an evaluation of the reduction quality of the articular surface and measurements of the Böhler angle (measured between a line drawn from the highest point of the anterior process and posterior articular facet and another line joining the highest point of the posterior articular facet with the highest point of the calcaneal tuberosity), calcaneal width, and calcaneal height.

Complications were categorized as minor or major. Minor complications, including superficial infections, were defined as those that were treated without reoperation. Major complications were defined as deep infections that required surgical intervention. The AOFAS ankle–hindfoot score, FFI, and VAS score were self-assessed by a single resident doctor for each patient 1 year after surgery to determine their clinical outcomes. Articular surface anatomic reduction was defined as no obvious displacement in an X-ray. Nearly anatomic reduction was defined as a displacement between 0 mm and 2 mm, and poor reduction was defined as a displacement of >2 mm among the reduction qualities.

2.5. Statistical Analysis

This study was approved by the institutional ethical review board (CMUH111-REC3-132). We conducted a power analysis using G*Power to ensure that the sample size was adequate for comparing the ELA and STA. According to the G*Power analysis, when the power was >0.8, the number of cases required was 18 for each group. Therefore, our case number is sufficient for the analysis. Dichotomous data were compared between the groups using Fisher's exact test. Differences in the measured variables were determined using Mann–Whitney U tests. The treatment effects were analyzed using Wilcoxon signed-rank tests. The statistical significance was set at $p < 0.05$. Data analyses were performed using SPSS ver. 19.0 (SPSS Inc., Chicago, IL, USA).

3. Results

The waiting time for surgery was significantly shorter in Group 2 than in Group 1 (3.76 ± 1.81 days vs. 10.78 ± 2.76 days, $p < 0.05$). However, Groups 1 and 2 did not differ significantly in terms of the operation time (131.00 ± 49.34 min vs. 129.38 ± 37.49 min, $p = 0.689$), blood loss volume (43.70 ± 42.91 mL vs. 24.05 ± 16.25 mL, $p = 0.091$), and hospital stay duration (6.22 ± 4.17 days vs. 5.19 ± 1.94 days, $p = 0.867$) (Table 2).

Table 2. Perioperative period.

	ELA	STA	*p*
Waiting time for surgery (days)	10.78 ± 2.76	3.76 ± 1.81	<0.05
Operation time (min)	131.00 ± 49.34	129.38 ± 37.49	0.689
Blood loss (mL)	43.70 ± 42.91	24.05 ± 16.25	0.091
Hospital stay (days)	6.22 ± 4.17	5.19 ± 1.94	0.867

ELA: extensile lateral approach; STA: sinus tarsi approach.

The postoperative radiographic parameters (Böhler angle, calcaneal width, and calcaneal height) significantly improved compared to the corresponding preoperative values (Table 3).

Table 3. Radiographic parameters.

	Preoperative	Postoperative	*p*
ELA			
Böhler angle (°)	11.27 ± 9.01	28.10 ± 5.10	<0.001
Calcaneal width (cm)	5.14 ± 0.48	4.02 ± 0.45	<0.001
Calcaneal height (cm)	5.16 ± 0.51	5.90 ± 0.39	<0.001
STA			
Böhler angle (°)	10.75 ± 6.88	29.89 ± 5.44	<0.001
Calcaneal width (cm)	5.10 ± 0.57	4.09 ± 0.42	<0.001
Calcaneal height (cm)	5.16 ± 0.47	5.78 ± 0.42	<0.001

ELA: extensile lateral approach; STA: sinus tarsi approach

Both groups achieved a 100% union rate, with no significant differences in the union time between them (Group 1: 86.96 ± 20.32 days; Group 2: 87.14 ± 21.48 days; $p = 0.806$; Table 4). No significant differences were detected in the Böhler angle or the reduction quality of the articular surface between the groups. Furthermore, the functional scores (comprising the AOFAS ankle–hindfoot score, FFI, and VAS score) did not differ significantly between the groups at the 1-year follow-up (Table 4).

Table 4. Radiographic and clinical outcomes.

	ELA	STA	*p*
Union rate	100%	100%	
Union time (days)	86.96 ± 20.32	87.14 ± 21.48	0.806
Bohler angle (Pre) (°)	11.27 ± 9.01	10.75 ± 6.88	0.953
Bohler angle (Post) (°)	28.10 ± 5.10	29.89 ± 5.44	0.329
Reduction quality			0.228
Anatomic	12 (52.17%)	15 (71.43%)	
Nearly anatomic	11 (47.83%)	6 (28.57%)	
VAS pain score	0.74 ± 0.69	0.81 ± 0.75	0.779
FFI	18.78 ± 9.93	17.67 ± 8.14	0.906
AOFAS score	81.26 ± 6.02	81.10 ± 5.80	0.887

AOFAS: American Orthopedic Foot and Ankle Society; ELA: extensile lateral approach; FFI: Foot Function Index; Pre: preoperative; Post: postoperative; STA: sinus tarsi approach; VAS: visual analog scale.

Patients with minor complications required antibiotic treatment and wound care. For major complications, patients underwent debridement for infection control, fortunately without the need for implant removal. The overall complication rate was significantly higher in Group 1 than in Group 2 (66.67% vs. 33.52%, $p < 0.05$). Six of the eight patients with infections were from Group 1; the remaining two were from Group 2. Among the patients with infections, three in Group 1 had major complications (Table 5).

Table 5. Complications.

	ELA 23	STA 21	*p*
Overall complications	14 (60.87%)	5 (23.81%)	<0.05
Subtalar arthrosis	5 (21.74%)	3 (14.29)	
Superficial infection	6 (26.09%)	2 (9.52%)	
Deep infection	3 (13.04%)	0	

ELA: extensile lateral approach; STA: sinus tarsi approach.

4. Discussion

In this study, we aimed to retrospectively compare the short-term outcomes of displaced intraarticular calcaneal fractures treated with the ELA or STA. Our results suggested that patients who underwent the STA experienced shorter waiting times for surgery and lower complication rates. There were no significant differences in the functional and radiological outcomes between the approaches.

In 2008, Weber et al. conducted the first comparative study on the ELA and STA in 50 patients; they compared the outcomes of displaced intraarticular calcaneal fractures treated using the ELA with lateral plate fixation and the STA with cannulated screw fixation. Their study revealed a shorter operative time with the STA; however, no significant differences in the complication rate, clinical outcomes, and radiographic results were noted between the approaches [15]. In 2013, Kline et al. performed a study on 112 patients; the overall complication rates differed significantly between the ELA and STA groups (29% vs. 6%). However, the functional and radiographic outcomes were similar between the groups [16]. In both studies, screw fixation was performed in the STA group.

Kir et al. compared cannulated screw and miniplate fixation treatments using the STA in 60 patients with Sanders type II and III calcaneal fractures and reported better functional outcomes and a lower reoperation rate with miniplate fixation [17]. As plate fixation has better outcomes than other methods, several studies have used locking plate fixation via the STA approach. Xia et al. compared the outcomes of 49 and 59 patients treated with the ELA and STA, respectively; the fractures in both groups were fixed with plates. There were no significant differences in the radiographic outcomes between the groups; however, the STA group had a shorter operative time and fewer postoperative wound complications [18]. Basile et al. conducted another comparative study on 38 patients and reported similar clinical and radiographic outcomes between the STA and ELA groups; however, the STA group experienced a significantly faster surgical procedure and shorter waiting time for surgery. Although a lower wound complication rate was observed in the STA group, the difference was not significant [19]. Chiang et al. reported a modified two-incision STA with plate fixation and demonstrated its safety and effectiveness [14]. Our study is the first to compare the two-incision STA with the ELA.

Despite surgical intervention for intraarticular calcaneal fractures, subtalar arthritis may develop. Herscovici et al. reported on thirty-five patients who underwent surgery for a calcaneal fracture, after which posttraumatic subtalar arthritis developed in twelve patients [20]. The higher incidence of wound complications after calcaneal fracture treatment with the ELA is because wound healing depends on the subperiosteal flap covering the lateral wall, which is thin and vulnerable [21,22]. Surgery via the ELA is safe when the skin shows a positive wrinkle sign without pitting edema [23]. In our study, the mean waiting time for the ELA was 10.78 days. The STA causes less damage to the subperiosteal flap and requires less demanding skin conditions. The mean waiting time for surgery with the STA was 3.76 days. However, the STA has some limitations, including poor exposure of the fracture site and difficulties in accessing the fracture for reduction [16,24]. In our opinion, early intervention, standard calcaneal plate insertion via a two-incision wound, and arthroscopic-assisted reduction could overcome these limitations. Our study also had some limitations, including its retrospective design and small sample size. In addition, the follow-up period was limited to 12 months, which may not fully capture the long-term outcomes of the intervention under investigation. Future studies should address these limitations with a larger sample and longer follow-up.

5. Conclusions

In conclusion, the STA demonstrated lower rates of wound complications and shorter waiting times for surgery than the ELA while achieving similar functional and radiological outcomes in the treatment of displaced intra-articular calcaneal fractures. Therefore, we recommend using the STA to treat these fractures.

Author Contributions: All authors contributed to the study conception and design. Material preparation, data collection, and analysis were performed by H.-L.L. The first draft of the manuscript was written by J.-T.M. and C.-M.C. All authors commented on previous versions of the manuscript. All authors have read and agreed to the published version of the manuscript.

Funding: This research received no external funding.

Institutional Review Board Statement: This study was approved by the institutional ethical review board (CMUH111-REC3-132).

Informed Consent Statement: Informed consent was obtained from all subjects involved in the study.

Data Availability Statement: Further data that support the reported results are available from the corresponding author upon reasonable request.

Conflicts of Interest: The authors declare that they have no competing interests.

References

1. Hollawell, S. Wound closure technique for lateral extensile approach to intra-articular calcaneal fractures. *J. Am. Podiatr. Med. Assoc.* **2008**, *98*, 422–425. [CrossRef]
2. Molloy, A.P.; Lipscombe, S.J. Hindfoot arthrodesis for management of bone loss following calcaneus fractures and nonunions. *Foot Ankle Clin.* **2011**, *16*, 165–179. [CrossRef]
3. Yu, T.; Xiong, Y.; Kang, A.; Zhou, H.; He, W.; Zhu, H.; Yang, Y. Comparison of sinus tarsi approach and extensile lateral approach for calcaneal fractures: A systematic review of overlapping meta-analyses. *J. Orthop. Surg.* **2020**, *28*, 2309499020915282. [CrossRef]
4. Dhillon, M.S.; Bali, K.; Prabhakar, S. Controversies in calcaneus fracture management: A systematic review of the literature. *Musculoskelet. Surg.* **2011**, *95*, 171–181. [CrossRef]
5. Epstein, N.; Chandran, S.; Chou, L. Current concepts review: Intra-articular fractures of the calcaneus. *Foot Ankle Int.* **2012**, *33*, 79–86. [CrossRef]

6. Buckley, R.; Tough, S.; McCormack, R.; Pate, G.; Leighton, R.; Petrie, D.; Galpin, R. Operative compared with nonoperative treatment of displaced intra-articular calcaneal fractures: A prospective, randomized, controlled multicenter trial. *J. Bone Jt. Surg. Am.* **2002**, *84*, 1733–1744. [CrossRef] [PubMed]
7. Thordarson, D.B.; Krieger, L.E. Operative vs. nonoperative treatment of intra-articular fractures of the calcaneus: A prospective randomized trial. *Foot Ankle Int.* **1996**, *17*, 2–9. [CrossRef] [PubMed]
8. Kołodziejski, P.; Czarnocki, Ł.; Wojdasiewicz, P.; Bryłka, K.; Kuropatwa, K.; Deszczyński, J. Intraarticular fractures of calcaneus—Current concepts of treatment. *Pol. Orthop. Traumatol.* **2014**, *79*, 102–111.
9. Makki, D.; Alnajjar, H.M.; Walkay, S.; Ramkumar, U.; Watson, A.J.; Allen, P.W. Osteosynthesis of displaced intra-articular fractures of the calcaneum: A long-term review of 47 cases. *J. Bone Jt. Surg. Br.* **2010**, *92*, 693–700. [CrossRef] [PubMed]
10. Ibrahim, T.; Rowsell, M.; Rennie, W.; Brown, A.R.; Taylor, G.J.; Gregg, P.J. Displaced intra-articular calcaneal fractures: 15-year follow-up of a randomised controlled trial of conservative versus operative treatment. *Injury* **2007**, *38*, 848–855. [CrossRef] [PubMed]
11. Maskill, J.D.; Bohay, D.R.; Anderson, J.G. Calcaneus fractures: A review article. *Foot Ankle Clin.* **2005**, *10*, 463–489. [CrossRef]
12. De Groot, R.; Frima, A.J.; Schepers, T.; Roerdink, W.H. Complications following the extended lateral approach for calcaneal fractures do not influence mid- to long-term outcome. *Injury* **2013**, *44*, 1596–1600. [CrossRef]
13. Benirschke, S.K.; Sangeorzan, B.J. Extensive intraarticular fractures of the foot. Surgical management of calcaneal fractures. *Clin. Orthop. Relat. Res.* **1993**, *292*, 128–134. [CrossRef]
14. Wang, C.S.; Tzeng, Y.H.; Yang, T.C.; Lin, C.C.; Chang, M.C.; Chiang, C.C. Radiographic and clinical results of modified 2-incision sinus tarsi approach for treatment of calcaneus fracture. *Injury* **2021**, *52*, 1971–1977. [CrossRef] [PubMed]
15. Weber, M.; Lehmann, O.; Sägesser, D.; Krause, F. Limited open reduction and internal fixation of displaced intra-articular fractures of the calcaneum. *J. Bone Jt. Surg. Br.* **2008**, *90*, 1608–1616. [CrossRef] [PubMed]
16. Kline, A.J.; Anderson, R.B.; Davis, W.H.; Jones, C.P.; Cohen, B.E. Minimally invasive technique versus an extensile lateral approach for intra-articular calcaneal fractures. *Foot Ankle Int.* **2013**, *34*, 773–780. [CrossRef] [PubMed]
17. Kir, M.C.; Ayanoglu, S.; Cabuk, H.; Dedeoglu, S.S.; Imren, Y.; Karslioglu, B.; Yuce, A.; Gurbuz, H. Mini-plate fixation via sinus tarsi approach is superior to cannulated screw in intra-articular calcaneal fractures: A prospective randomized study. *J. Orthop. Surg.* **2018**, *26*, 2309499018792742. [CrossRef]
18. Xia, S.; Lu, Y.; Wang, H.; Wu, Z.; Wang, Z. Open reduction and internal fixation with conventional plate via L-shaped lateral approach versus internal fixation with percutaneous plate via a sinus tarsi approach for calcaneal fractures—A randomized controlled trial. *Int. J. Surg.* **2014**, *12*, 475–480. [CrossRef] [PubMed]
19. Basile, A.; Albo, F.; Via, A.G. Comparison between sinus tarsi approach and extensile lateral approach for treatment of closed displaced intra-articular calcaneal fractures: A multicenter prospective study. *J. Foot Ankle Surg.* **2016**, *55*, 513–521. [CrossRef]
20. Dolfi, H., Jr.; James, W.; Julia, M.S.; Roy, W.S.; Arthur, W. Operative treatment of calcaneal fractures in elderly patients. *J. Bone Jt. Surg. Am.* **2005**, *87*, 1260–1264. [CrossRef]
21. Borrelli, J., Jr.; Lashgari, C. Vascularity of the lateral calcaneal flap: A cadaveric injection study. *J. Orthop. Trauma* **1999**, *13*, 73–77. [CrossRef] [PubMed]
22. Cavadas, P.C.; Landin, L. Management of soft-tissue complications of the lateral approach for calcaneal fractures. *Plast. Reconstr. Surg.* **2007**, *120*, 459–466. [CrossRef]
23. Sanders, R. Intra-articular fractures of the calcaneus: Present state of the art. *J. Orthop. Trauma* **1992**, *6*, 252–265. [CrossRef] [PubMed]
24. Yeo, J.H.; Cho, H.J.; Lee, K.B. Comparison of two surgical approaches for displaced intra-articular calcaneal fractures: Sinus tarsi versus extensile lateral approach. *BMC Musculoskelet. Disord.* **2015**, *16*, 63. [CrossRef] [PubMed]

Journal of Personalized Medicine

Article

Preoperative Planning for Reverse Shoulder Arthroplasty: Does the Clinical Range of Motion Match the Planned 3D Humeral Displacement?

Diane Ji Yun Yoon [1,2], Guillaume-Anthony Odri [2,3], Luc Favard [1], Ramy Samargandi [1,4] and Julien Berhouet [1,5,*]

1 Service de Chirurgie Orthopédique et Traumatologique, CHRU Trousseau, Faculté de Médecine de Tours, Université de Tours, 1C Avenue de la République, 37170 Chambray-les-Tours, France
2 Inserm U1132 BIOSCAR, Université Paris Cité, 75010 Paris, France
3 Service de Chirurgie Orthopédique et Traumatologique, Centre Hospitalier Universitaire Lariboisière, 75010 Paris, France
4 Department of Orthopedic Surgery, Faculty of Medicine, University of Jeddah, Jeddah 23218, Saudi Arabia
5 Equipe Reconnaissance de Forme et Analyse de l'Image, Laboratoire d'Informatique Fondamentale et Appliquée de Tours EA6300, Ecole d'Ingénieurs Polytechnique Universitaire de Tours, Université de Tours, 64 Avenue Portalis, 37200 Tours, France
* Correspondence: julien.berhouet@gmail.com; Tel.: +33-02-4747-8644; Fax: +33-02-4747-5912

Abstract: Introduction: The functional outcome after reverse shoulder arthroplasty (RSA) is closely linked to how much the humerus shifts because of the implants. While two-dimensional (2D) angle measurements have been used to capture this shift, it can be measured in three dimensions (3D) as the arm change position (ACP). In a previous study, the ACP was measured using 3D preoperative planning software with the passive virtual shoulder range of motion obtained after RSA. The main objective of this study was to evaluate the relationship between the ACP and the actual active shoulder range of motion measured after RSA. The hypothesis was that the ACP and the active clinical range of motion are related such that the ACP is a reliable parameter to guide the preoperative planning of an RSA. The secondary objective was to assess the relationship between 2D and 3D humeral displacement measurements. Materials and methods: This prospective observational study enrolled 12 patients who underwent RSA and had a minimum follow-up of 2 years. The active range of motion in shoulder flexion, abduction, and internal and external rotation was measured. At the same time, ACP measurements were taken from a reconstructed postoperative CT scan, in addition to the radiographic measurements of humeral lateralization and distalization angles on AP views in neutral rotation. Results: The mean humeral distalization induced by RSA was 33.3 mm (±3.8 mm). A non-statistically significant increase in shoulder flexion was observed for humeral distalization beyond 38 mm ($R^2 = 0.29$, $p = 0.07$). This "threshold" effect of humeral distalization was also observed for the gains in abduction, as well as internal and external rotations, which seemed better with less than 38 mm or even 35 mm distalization. No statistical correlation was found between the 3D ACP measurements and 2D angle measurements. Conclusion: Excessive humeral distalization seems to be detrimental to joint mobility, especially shoulder flexion. Humeral lateralization and humeral anteriorization measured using the ACP seem to promote better shoulder range of motion, with no threshold effect. These findings could be evidence of tension in the soft tissues around the shoulder joint, which should be taken into consideration during preoperative planning.

Keywords: reverse shoulder arthroplasty; range of motion; preoperative planning; arm change position; motion analysis

Citation: Yoon, D.J.Y.; Odri, G.-A.; Favard, L.; Samargandi, R.; Berhouet, J. Preoperative Planning for Reverse Shoulder Arthroplasty: Does the Clinical Range of Motion Match the Planned 3D Humeral Displacement? *J. Pers. Med.* **2023**, *13*, 771. https://doi.org/10.3390/jpm13050771

Academic Editors: Markus Gesslein and Eduard Kurz

Received: 1 April 2023
Revised: 22 April 2023
Accepted: 24 April 2023
Published: 29 April 2023

1. Introduction

Reverse total shoulder arthroplasty (RSA) has shown excellent functional outcomes and therefore has become widely accepted as the treatment of choice for cuff tear arthropathy [1]. In recent years, the indications of RSA have been expanded for the treatment of various conditions, including pseudoparalysis due to massive irreparable rotator cuff tears, acute three-part or four-part proximal humerus fractures, revision arthroplasty, chronic irreducible shoulder dislocation, tumor resection, and inflammatory joint conditions such as rheumatoid arthritis [2–9]. In the last decade, numerous lines of evidence demonstrated that RSA is a safe and effective treatment in patients younger than 65 years, which offers a substantial functional improvement in pain level, range of motion, and strength [10–13]. Previous studies reported a 10-year survival of 91.0% in case of primary RSA and 80.9% in case of revision RSA for failed arthroplasty [12]. The Australian National Joint Registry reported similar results, with a 10-year survival for the diagnosis of rotator cuff arthropathy of 94.1% [14]. The incidence of primary RSA in the United States has increased from 22,835 procedures in 2011 to 62,705 procedures in 2017. The greatest increase in incidence was found in male patients and patients aged between 50 and 64 years, whereas the incidence of hemiarthroplasty has markedly decreased in recent years [15]. In France, primary shoulder arthroplasty procedures increased by 47% between 2012 and 2018 and are estimated to grow from 31% to 322% between 2018 and 2050 [16].

An RSA design was first proposed by Paul Grammont in 1985 to lower and shift the humerus medially [17,18]. Since the deltoid's lever arm increases, the shoulder's mobility in flexion and abduction (ABD) improves [3,19]; however, rotation with the elbow at the side is often deficient [20,21]. Scapular notching is another limitation of this humeral medialization [20,22]. Scapular notching was described by Sirveaux et al. and was theorized to be caused by the mechanical impingement of the medial side of the humeral polyethylene component against the inferior scapular neck [23]. Scapular notching can be avoided by the lateralization of the center of rotation either through a lateralized RSA design or bony increased-offset RSA [20,24,25].

At the same time, excessive arm lengthening can lead to neurological impairment [26,27]. Thus, it is hard to grasp how each component of humeral displacement impacts the functional outcomes after RSA.

Boutsiadis et al. [28] recently described two radiographic angle measurements: "lateralization shoulder angle" (LSA) and "distalization shoulder angle" (DSA). These evaluate humeral lateralization and distalization, respectively, induced by the RSA. The main limitation of these angles is that they are measured with anteroposterior X-ray views, in a two-dimensional (2D) frame of reference. Additionally, the quality of the images affects the measurement accuracy.

Berhouet et al. [29] recently described another parameter, called the "arm change position" (ACP), to evaluate humeral displacement after RSA but this time in three dimensions (3D). The ACP is calculated using a 3D preoperative planning software program for the placement of shoulder replacement implants. In that study, the virtual range of motion (RoM) analysis and the ACP measurement of humeral displacement after the RSA showed that lowering and shifting the humerus laterally improved shoulder mobility in flexion–extension, external rotation with the elbow at the side (ER1), and adduction (ADD). Better internal rotation with the elbow at the side (IR1) was also found when the humerus was shifted anteriorly [29]. Nevertheless, this virtual analysis was somewhat unrealistic because it did not consider how the scapulothoracic joint contributes to the shoulder joint's movements. The Scapulothoracic participation plays a more important role in RSA biomechanics than in normal shoulder biomechanics [30,31].

The main objective of this study was to determine the relationship between 3D humeral displacements based on the ACP after RSA and the true active RoM measured in a clinical setting. If these are found to be related, we hypothesize that the ACP is a reliable parameter to guide the choice and positioning of RSA implants during preoperative planning. The study's secondary objective was to determine the relationship between the humeral displacements measured in 3D using the ACP and those in 2D using the LSA and DSA.

2. Materials and Methods

2.1. Materials

2.1.1. Patients

The research protocol approval was obtained from our institutional review board (IRB: 13B-T-SHOULDER-RM). This prospective observational study enrolled 12 patients (9 women and 3 men), with a mean age of 73 ± 4.3 years (range, 63 to 81 years), who had undergone RSA surgery between 9 January 2015 and 31 August 2016 (Figure 1). RSA indications were primary osteoarthritis and cuff tear arthropathy (CTA), with six patients in each group. Their average body mass index was 28.5 ± 3 kg/m^2 (19.5–34.4). RSA was performed on the dominant side in nine cases. The shoulder undergoing RSA had never been operated on. The preoperative demographic, clinical, and imaging data of the study population are presented in Table 1.

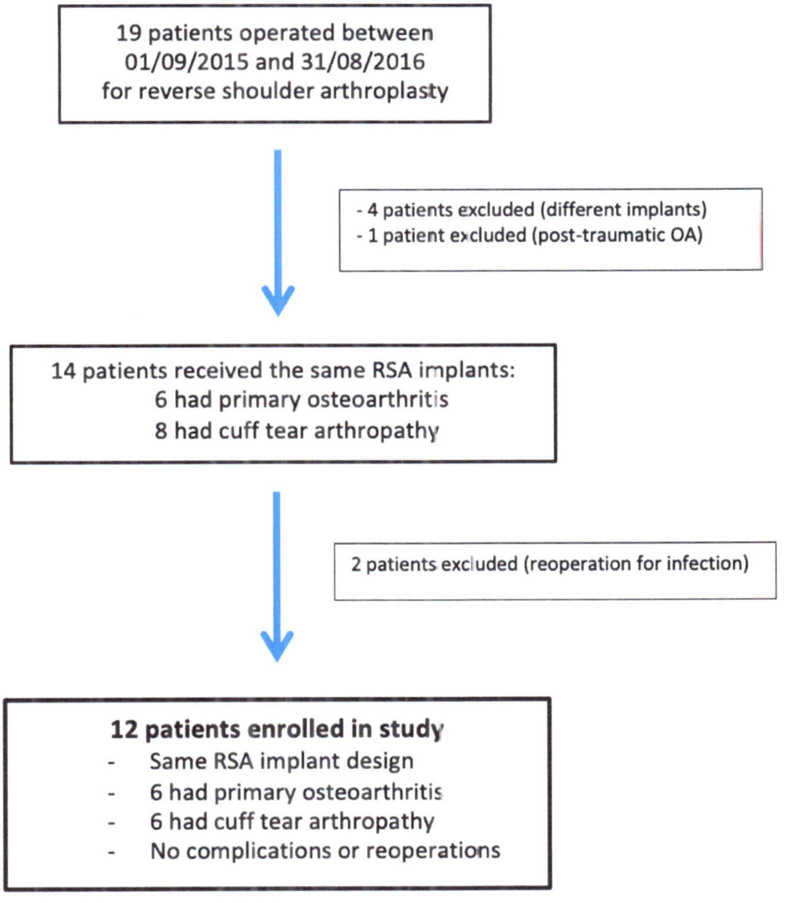

Figure 1. Flowchart of patients included in the study. RSA: reverse shoulder arthroplasty; OA: osteoarthritis.

Table 1. Preoperative demographic, clinical, and radiological data.

		Age	BMI	Flex	ABD	ER1	ER2	IR1	Constant	CSA	LSA	DSA
OA group	Patient 1	81	32.1	30	20	−10	0	Thigh = 0	8	31	92	13
	Patient 2	77	31.6	80	70	5	20	Thigh = 0	25	29	89	8
	Patient 3	74	27.4	80	60	−5	0	Thigh = 0	21	42	87	14
	Patient 4	79	34.4	90	80	0	0	Thigh = 0	23	28	82	−3
	Patient 5	79	29.4	80	80	−10	20	Lumbosacral = 4	12	37	81	3
	Patient 6	68	30.1	90	60	−10	0	Thigh = 0	21	39	100	12
Mean OA		76 ± 5 (68–81)	30.8 ± 2.4 (27.4–34.4)	75 ± 23 (30–90)	62 ± 22 (20–80)	−5 ± 6 (−10–5)	7 ± 10 (0–20)	1 ± 2 (0–4)	18 ± 7 (8–25)	34 ± 6 (28–42)	89 ± 7 (81–100)	8 ± 7 (−3–14)
CTA group	Patient 7	72	27.1	120	110	10	50	L3 = 6	32	39	125	6
	Patient 8	70	19.5	90	80	0	0	T7 = 10	38	52	41	30
	Patient 9	63	29.4	40	40	30	40	Buttocks = 2	18	39	96	6
	Patient 10	74	28.0	80	70	10	30	Buttocks = 2	24	42	55	5
	Patient 11	71	27.5	110	100	25	50	T7 = 10	34	34	105	18
	Patient 12	73	25.4	150	150	0	20	T7 = 10	47	40	105	18
Mean CTA		71 ± 4 (63–73)	26.2 ± 3.5 (19.5–29.4)	98 ± 38 (40–150)	92 ± 38 (40–150)	13 ± 13 (0–30)	32 ± 19 (0–50)	7 ± 4 (2–10)	32 ± 10 (18–47)	41 ± 6 (34–52)	88 ± 33 (41–125)	14 ± 10 (5–30)
p		0.04	0.02	0.22	0.12	0.01	0.02	0.006	0.002	0.08	0.96	0.25
Overall Mean		73 ± 4 (63–81)	28.5 ± 3 (19.5–34.4)	87 ± 31 (30–150)	77 ± 31 (20–150)	4 ± 10 (−10–30)	19 ± 16 (0–50)	4 ± 4 (0–10)	25 ± 9 (8–47)	38 ± 6 (28–52)	88 ± 24 (41–125)	11 ± 9 (−3–30)

OA: primary osteoarthritis; CTA: cuff tear arthropathy; BMI: body mass index (kg/m^2); Flex: flexion (°); ABD: abduction (°); ER1: external rotation with elbow at side (°); ER2: external rotation with 90° shoulder abduction; IR1: internal rotation with the highest level reached at the back with the elbow at the side; Constant: Constant–Murley Shoulder Outcome Score; CSA: critical shoulder angle (°); LSA: lateralization shoulder angle (°); DSA: distalization shoulder angle (°).

2.1.2. RSA Implants

A reversed shoulder prosthesis (Wright Medical France, Monbonnot Saint Martin, France) was used on all shoulders. The humeral stem was an AEQUALIS ASCEND™ FLEX system (Tornier, Bloomington, MN, USA), which has a 132.5° fixed inclination Onlay stem mounted with a 1.5 mm to 3.5 mm lateralized humeral plateau with variable indexing and a standard polyethylene insert having a fixed tilt angle of 12.5°. The final humeral configuration had a 145° inclination.

The glenosphere implant was Reversed AEQUALIS™ (Tornier, Bloomington, MN, USA). It included a 25 mm (n = 7) or 29 mm (n = 5) diameter baseplate with a short or long peg and a 36 mm (n = 9) or 42 mm (n = 3) centered sphere. A 7–10 mm thick lateralizing glenoid bone graft (BIO-RSA) was added in nine patients.

2.1.3. Software

The BluePrint® 3D Planning software (Tornier SAS France, Monbonnot Saint Martin, France) was used to select the type of implants to use in a given patient and to position them in a virtual shoulder joint. Once the implant configuration is finalized, the software simulates the passive RoM and measures humeral displacement in the frontal, sagittal, and axial planes. This is the ACP parameter (expressed in millimeters), which comprises 3 values that correspond to the superior–inferior, mediolateral, and anteroposterior humeral displacements after the virtual implantation of the RSA (Figure 2). These three values are positive when the displacement is superior, lateral, and anterior. They are negative when the displacement is inferior, medial, and posterior (Figure 3).

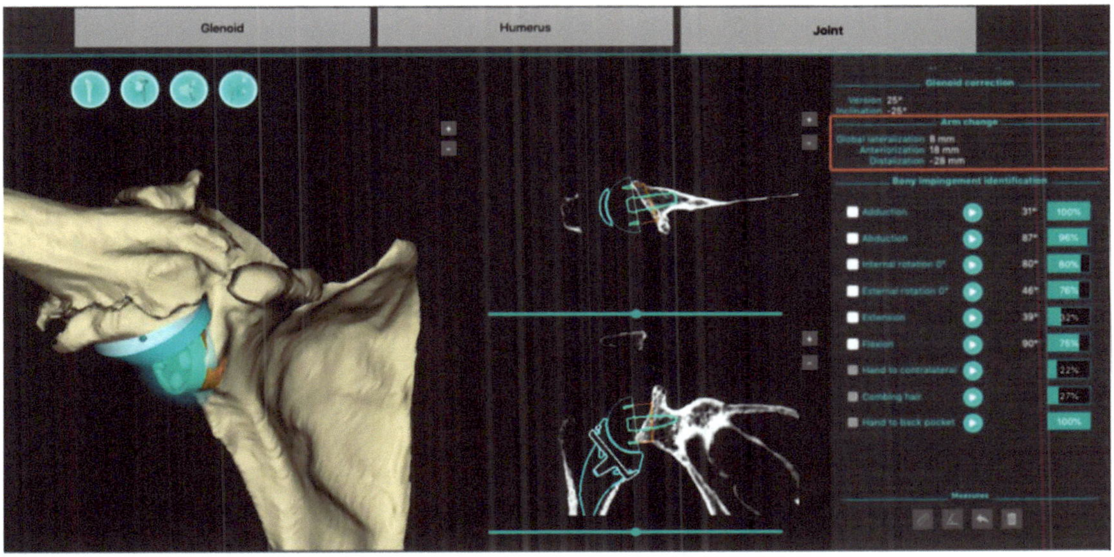

Figure 2. Screenshot of a preoperative plan for RSA from the BluePrint® 3D Planning software (Stryker, Monbonnot Saint Martin, France). In addition to a virtual analysis of the passive joint range of motion, information on humeral positioning (arm change position) is included (red frame).

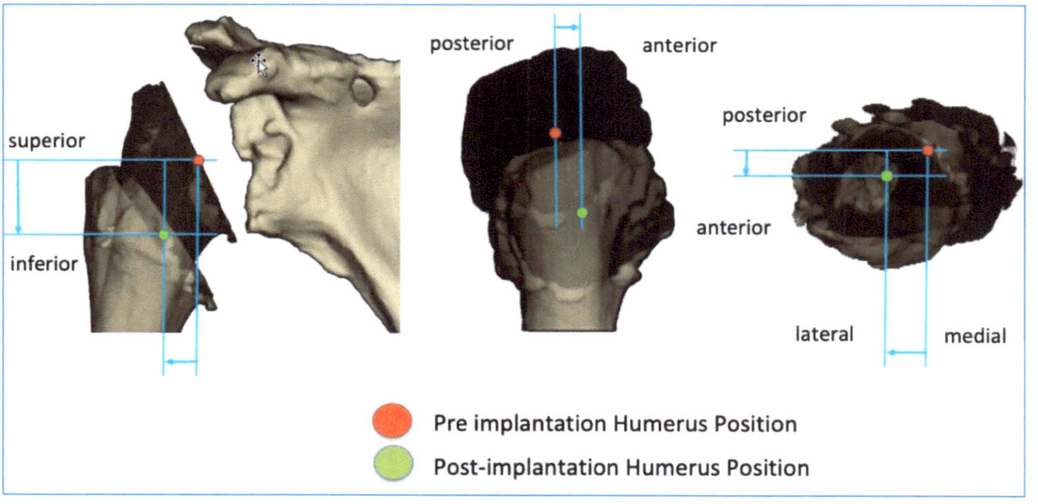

Figure 3. Illustration of how the arm change position (ACP) is measured in three dimensions.

Postoperatively, another software package (PTC Creo® version 6.0, Parametric Technology Corporation, Needham, MA, USA) was used to capture a postoperative CT scan of the patient's shoulder with the implants in place. These images were uploaded into the Blueprint software to measure the ACP postoperatively, as this corresponds to the actual implant positioning.

2.2. Methods

The preoperative imaging assessment included standard AP X-rays in three different rotations plus an axillary view and a CT scan of the shoulder for 3D planning purposes. All the patients were operated on using the deltopectoral approach for the implantation of the prosthesis, in a beach chair position and under general anesthesia combined with an interscalene block. The procedure was performed in a standard manner with no particular technical point during the procedure. The retroversion of the humeral implant was adjusted relative to the forearm axis and ranged from 10° to 30°, depending on the patient. The subscapularis was repaired when it was still present. Biceps tenodesis to the pectoralis major was carried out at the end of the procedure. The patients were discharged 48 h after surgery. Postoperative care included shoulder immobilization with an abduction pillow for 3 to 6 weeks. Physiotherapy was started immediately or at 3 weeks if BIO-RSA had been performed.

The 12 patients were reviewed at a minimum follow-up of 2 years postoperatively. Their active shoulder joint mobility was measured in degrees for flexion (F), abduction (ABD), external rotation (ER1/ER2) with the elbow at the side/with the arm abducted at 90°, and internal rotation (IR1) with the elbow at the side, as it had been during the preoperative consultation. The active internal rotation range of motion measurement was defined as the highest midline vertebral segment of the back that can be reached. This measurement was converted into a 10-point scale according to the Constant–Murley Shoulder Outcome Score guidelines [32].

A complete radiographic assessment identical to that carried out preoperatively (AP view in three rotations and axillary view) was performed; fluoroscopy was used beforehand to ensure the images were reproducible between the patients. The LSA and DSA in degrees (°) were measured on AP views in neutral rotation, as described by Boutsiadis et al. [12] (Figure 4).

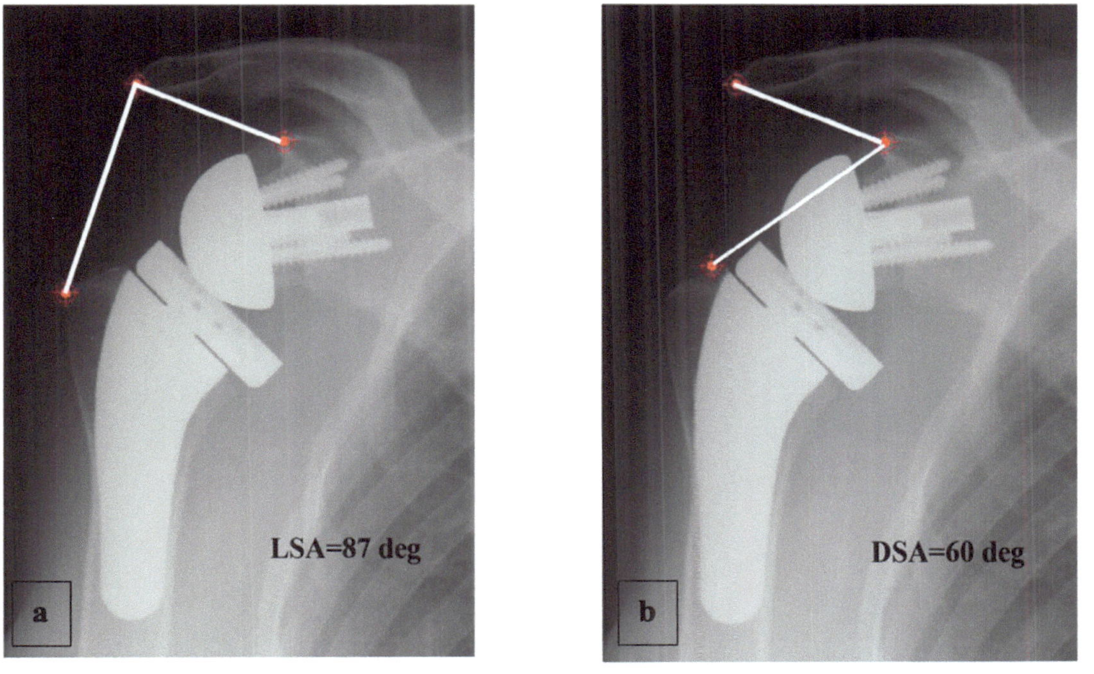

Figure 4. Example of LSA (**a**) and DSA (**b**) measured on an AP radiograph in neutral shoulder rotation, as described by Boutsiadis et al. [12].

Finally, at the same minimum follow-up of 2 years, a CT scan of the operated shoulder was carried out to measure the ACP after RSA using the BluePrint® planning software. However, the presence of the implants meant that these CT scans were not directly supported by the BluePrint® software. Several image preparation and processing steps had to be completed before the postoperative ACP could be measured in each patient (Figure 5).

These processing steps were as follows:

1. The extraction of the 3D geometry of the humerus and scapula from the preoperative CT scan;
2. The manual registration of the preoperative 3D geometry of the humerus and scapula, with RSA implants from the postoperative CT scan, using the PTC Creo® software;
3. The creation of planning files integrating the readjusted bone and implant geometries in the BluePrint® software to measure the postoperative ACP corresponding to the RSA implant configuration specific to each patient.

The data were collated in an Excel® spreadsheet and analyzed using the JMP® 11.0.0 software (SAS Institute Inc.©, Cary, NC, USA). A Shapiro–Wilk test was used to evaluate the normal distribution of the continuous quantitative variables. A Student's *t*-test was performed to compare the means between indication subgroups; linear regression was carried out for the correlations. The significance level was set at $p < 0.05$.

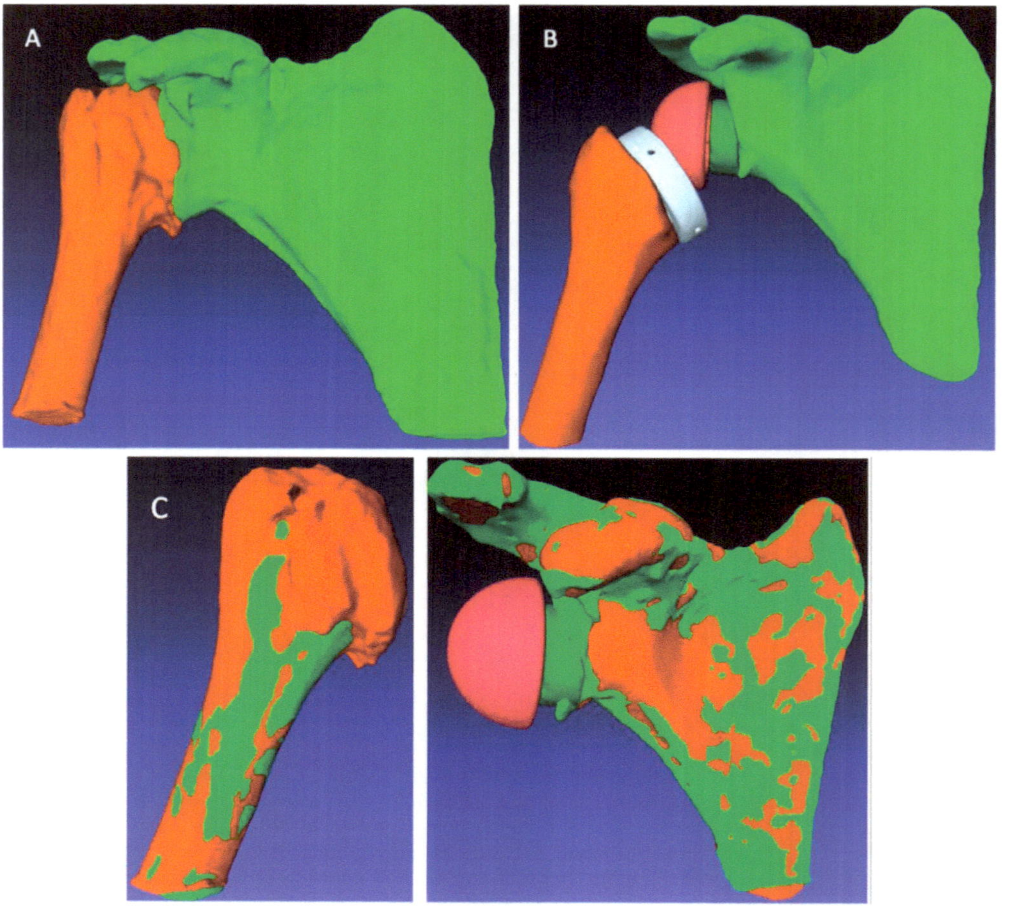

Figure 5. Image processing steps for postoperative arm change position (ACP) measurement: (**A**) automatic segmentation to extract the humeral and scapular bones from the postoperative CT scan; (**B**) manual segmentation to extract the bones and implants from the postoperative CT scan; (**C**) registration overlay of the preoperative (green) and postoperative (orange) structures to measure the ACP corresponding to the postoperative situation.

3. Results

3.1. Descriptive Data

RSA improved the active shoulder RoM in all patients. These improvements were similar for the two groups (Table 2). Internal rotation was unchanged in the CTA group, whereas it was improved in the primary osteoarthritis group relative to the preoperative measurement. The other RoM (ΔF, ΔABD, ΔER1, and ΔER2) had also further increased compared with the preoperative RoM for the primary osteoarthritis group (Table 3).

Table 2. Postoperative clinical and radiological data.

		EA	ABD	RE1	RE2	RI1	Constant	CSA	LSA	DSA	ACP Latéralisation	ACP Antériorisation	ACP Distalisation
OA group	Patient 1	80	80	0	40	Buttocks = 2	32	28	78	65	8	0	−33
	Patient 2	130	100	10	60	Lumbosacral = 4	67	19	60	34	3	6	−31
	Patient 3	110	100	0	30	Buttocks = 2	66	24	90	46	5	−1	−34
	Patient 4	140	140	20	80	Thigh = 0	67	24	85	56	9	18	−27
	Patient 5	160	160	10	60	T12 = 8	79	23	79	57	8	12	−29
	Patient 6	130	120	5	40	L3 = 6	73	30	84	45	−1	−10	−42
Mean OA		125 ± 27 (80–160)	117 ± 29 (80–160)	8 ± 8 (0–20)	52 ± 18 (30–80)	4 ± 3 (0–8)	64 ± 17 (32–79)	25 ± 4(19–30)	79 ± 10 (60–90)	51 ± 11 (34–65)	5 ± 3 (−1–9)	4 ± 8 (−10–18)	−33 ± 4 (−42– −27)
CTA group	Patient 7	110	100	10	70	L3 = 6	67	18	87	48	3	8	−38
	Patient 8	100	100	0	40	T12 = 8	60	34	65	44	8	10	−29
	Patient 9	110	90	50	20	T12 = 8	55	47	92	38	10	13	−31
	Patient 10	160	160	10	70	L3 = 6	79	30	73	53	0	11	−30
	Patient 11	130	130	30	70	T7 = 10	66	22	96	38	−1	1	−34
	Patient 12	160	160	0	60	Lumbosacral = 4	72	34	82	58	0	4	−41
Mean CTA		128 ± 26 (100–160)	123 ± 31 (90–160)	17 ± 20 (0–50)	55 ± 21 (20–70)	7 ± 2 (4–10)	69 ± 7 (60–79)	31 ± 10 (18–47)	83 ± 12 (65–96)	47 ± 8 (38–58)	3 ± 4 (−1–10)	8 ± 4 (1–13)	−34 ± 4 (−41– −29)
p		0.83	0.71	0.31	0.77	0.05	0.57	0.20	0.63	0.49	0.53	0.96	0.21
Overall mean		127 ± 27 (80–160)	120 ± 30 (80–160)	12 ± 15 (0–50)	53 ± 20 (20–80)	5 ± 3 (0–10)	66 ± 13 (32–79)	28 ± 8 (18–47)	81 ± 11 (60–96)	49 ± 10 (34–65)	4 ± 4 (−1– 10)	6 ± 6 (−10–18)	−33 ± 4 (−42– −27)

OA: primary osteoarthritis; CTA: cuff tear arthropathy; BMI: body mass index (kg/m^2); Flex: flexion (°); ABD: abduction (°); ER1: external rotation with elbow at side (°); ER2: external rotation with 90° shoulder abduction; IR1: internal rotation with the highest level reached at the back with the elbow at the side; Constant: Constant–Murley Shoulder Outcome Score; CSA: critical shoulder angle (°); LSA: lateralization shoulder angle (°); DSA: distalization shoulder angle (°); ACP: arm change position (mm).

Table 3. Gains in the clinical range of motion (Δ = postoperative measurement − preoperative measurement) and postoperative ACP.

		Δflex	ΔABD	ΔER1	ΔER2	ΔIR1	ΔConstant	ACP lat	ACP ant	ACP dist
OA group	Patient 1	50	60	10	40	2	24	8	0	−33
	Patient 2	50	30	5	40	4	42	3	6	−31
	Patient 3	30	40	5	30	2	45	5	−1	−34
	Patient 4	50	60	20	80	0	44	9	18	−27
	Patient 5	80	80	20	40	4	67	8	12	−29
	Patient 6	40	60	15	40	6	52	−1	−10	−42
Mean OA		50 ± 16 (30–80)	55 ± 18 (30–80)	13 ± 7 (5–20)	45 ± 18 (30–80)	3 ± 2 (0–6)	64 ± 17 (32–79)	5 ± 3 (−1–9)	4 ± 8 (−10–18)	−33 ± 4 (−42– −27)
CTA group	Patient 7	−10	−10	0	20	0	35	3	8	−38
	Patient 8	10	20	0	40	−2	22	8	10	−29
	Patient 9	70	50	20	−20	6		10	13	−31
	Patient 10	80	90	0	40	4	55	0	11	−30
	Patient 11	20	30	5	20	10	32	−1	1	−34
	Patient 12	10	10	0	40	−6	25	0	4	−41
Mean CTA		30 ± 36 (−10–80)	32 ± 35 (−10–90)	4 ± 8 (0–20)	23 ± 23 (−20–40)	2 ± 6 (−6–10)	69 ± 7 (60–79)	3 ± 4 (−1–10)	8 ± 4 (1–13)	−34 ± 4 (−41– −29)
p		0.25	0.17	0.08	0.10	0.70	0.17	0.53	0.96	0.21
Overall Mean		40 ± 29 (−10–80)	43 ± 29 (−10–90)	8 ± 8 (0–20)	34 ± 23 (−20–80)	3 ± 4 (−6–10)	40 ± 14 (22–55)	4 ± 4 (−1– 10)	6 ± 6 (−10–18)	−33 ± 4 (−42– −27)

OA: primary osteoarthritis; CTA: cuff tear arthropathy; Flex: flexion (°); ABD: abduction (°); ER1: external rotation elbow at side (°); ER2: external rotation with 90° shoulder abduction; IR1: internal rotation with the highest level reached at the back with the elbow at the side; Constant: Constant–Murley Shoulder Outcome Score; ACP: Arm Change Position (mm).

The mean humeral distalization induced by RSA, assessed using the ACP, was 33.3 mm (±3.8 mm). The mean lateralization was 4.3 mm (±3.7 mm). The mean anterior displacement of the humerus ("anteriorization") was 6 mm (±6 mm). There was no significant difference between the indications for these three parameters (Table 2). The mean LSA and DSA values were 80.9° (±11.1°) and 48.5° (±9.7°), with no difference between the two indications (Table 2).

3.2. Analytical Data

RoM and Measurement of 3D Humeral Displacement Using the ACP (Figure 6).

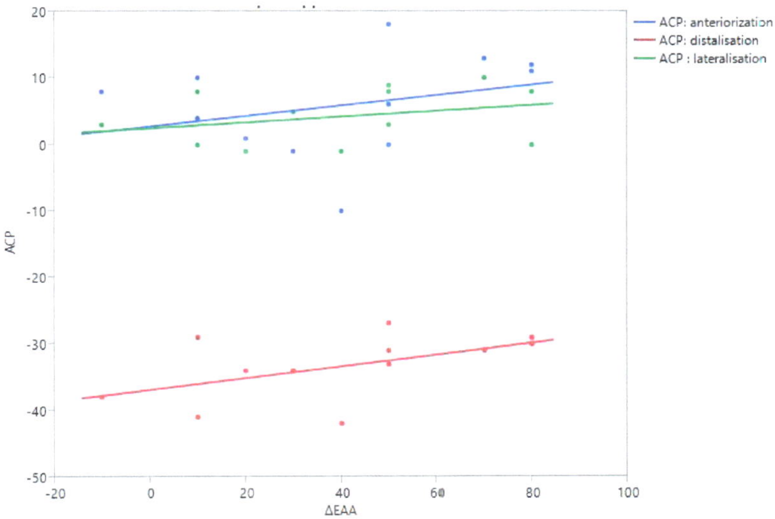

(A)

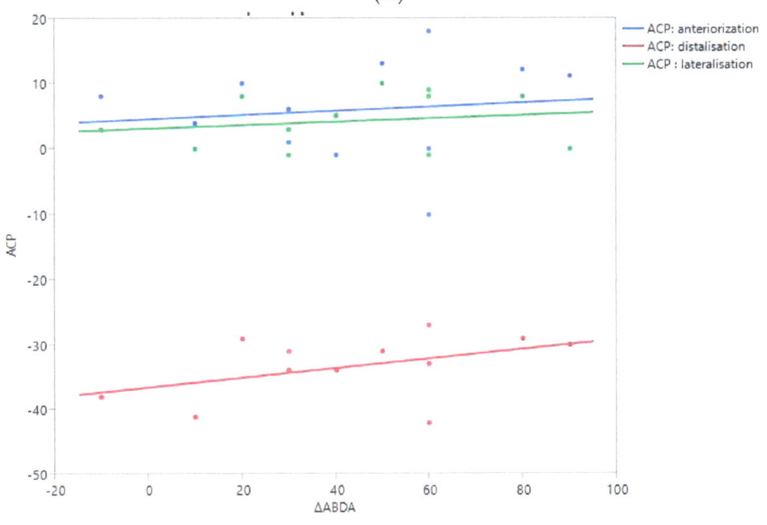

(B)

Figure 6. *Cont.*

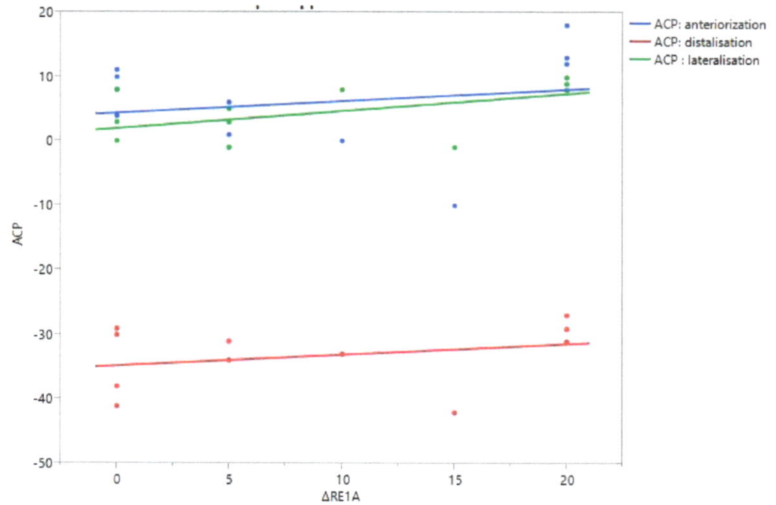

(**C**)

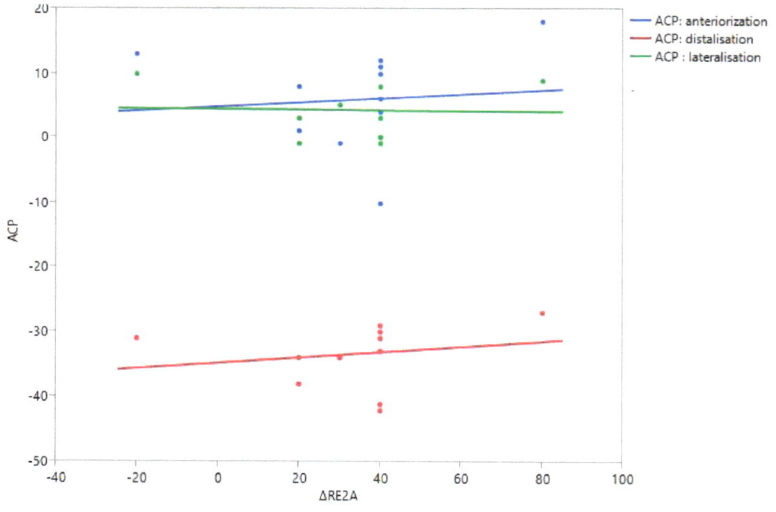

(**D**)

Figure 6. *Cont.*

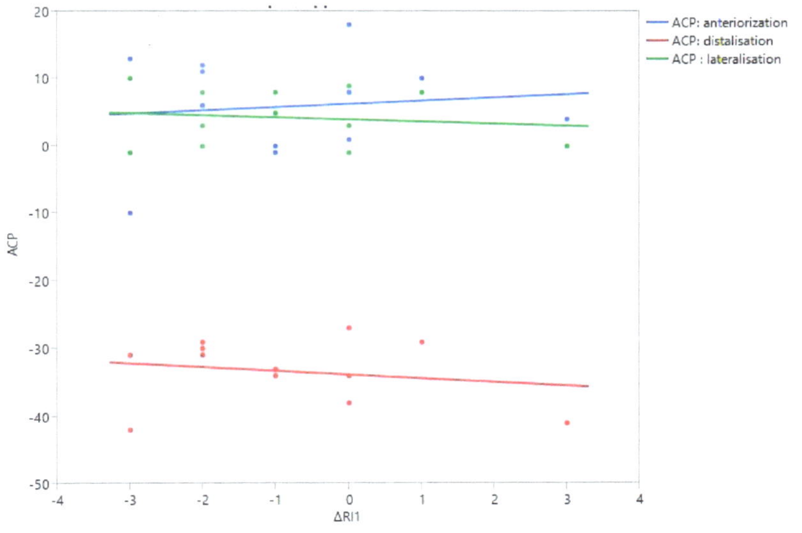

(E)

Figure 6. Shoulder motion and 3D humeral displacement according to the arm change position (ACP). (**A**) Change in active shoulder flexion amplitude (Δ F) relative to ACP. (**B**) Change in active abduction amplitude (Δ ABD) relative to ACP. (**C**) Change in active external rotation elbow at side (Δ RE1A) relative to ACP. (**D**) Change in active external rotation with 90° shoulder abduction (Δ RE2A) relative to ACP. (**E**) Change in active internal rotation elbow at side (Δ RI1) relative to ACP.

There were no statistically significant findings; however, there were two interesting trends observed:

- The first observation was the gains in flexion. Flexion tended to decrease with humeral distalization (or lowering). There was a threshold around 38 mm, which was associated with the worse RoM values; shoulder flexion was better when the resulting humerus position was less distal ($R^2 = 0.29$, $p = 0.07$). This "threshold" effect of humeral distalization was also observed for other shoulder motions, but the correlation was not as strong. The gains in ER1, ABD, ER2, and IR1 seemed better when the humerus was lowered less than 35 or 38 mm, depending on the patient.
- The second observation was the gains in ER1, which tended to improve with humeral lateralization ($R^2 = 0.29$, $p = 0.07$).

Among our other results, a positive linear relationship was found between the improvement in ABD, F, IR1, and ER2 and the anterior humeral displacement. Humeral lateralization was associated with a slightly worse range of motion in ER2 and IR1.

RoM and Measurement of 2D Humeral Displacement Using the DSA and LSA (Figure 7).

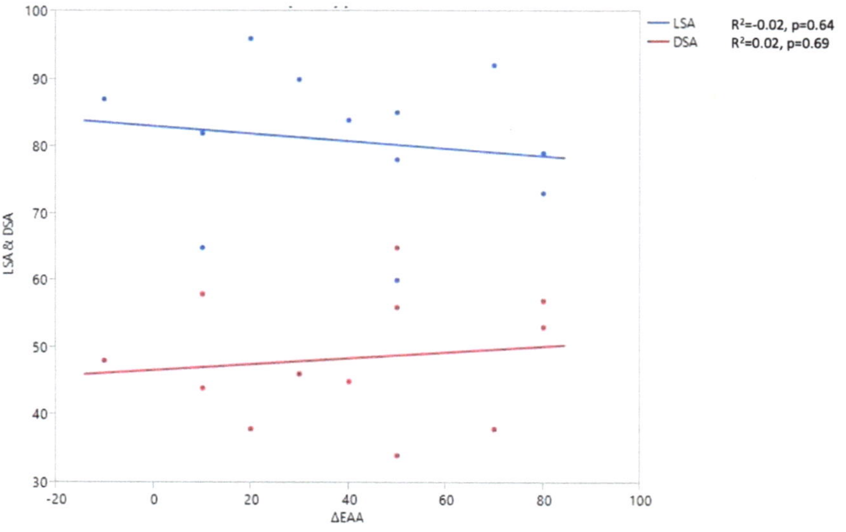

(**A**)

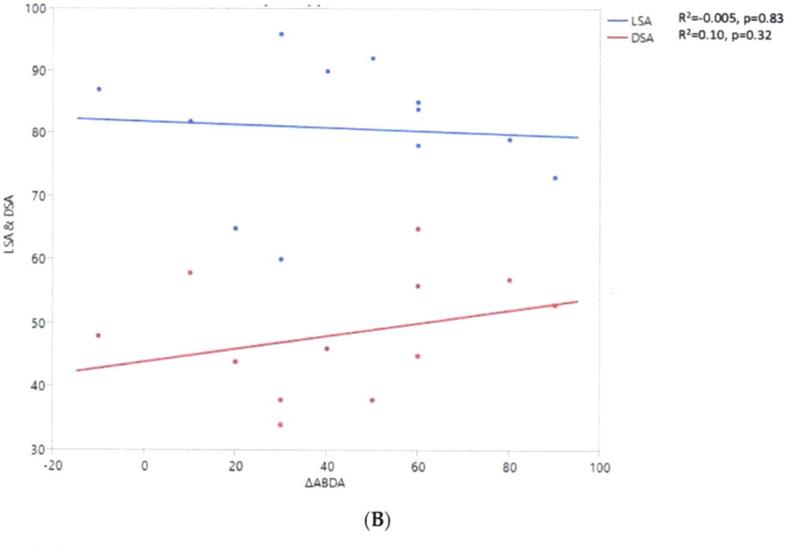

(**B**)

Figure 7. *Cont.*

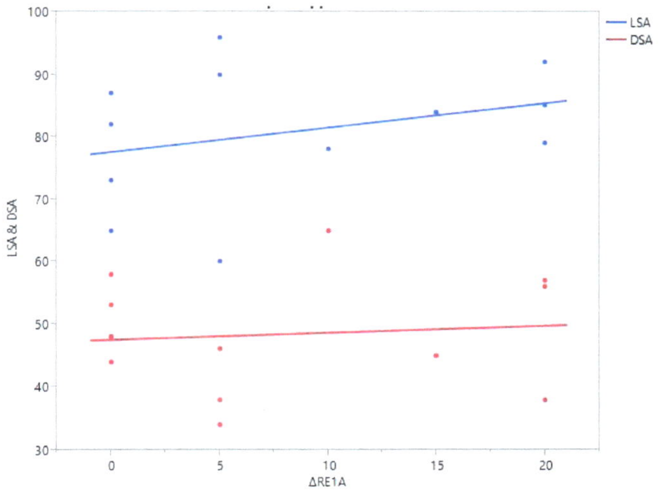

(**C**)

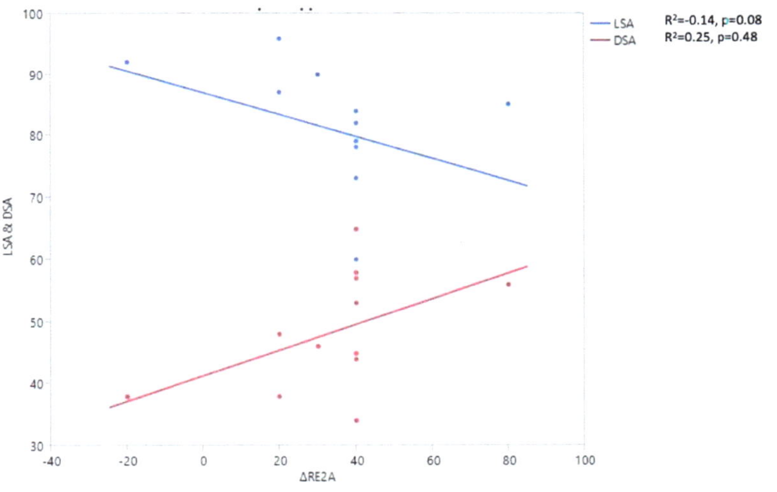

(**D**)

Figure 7. *Cont.*

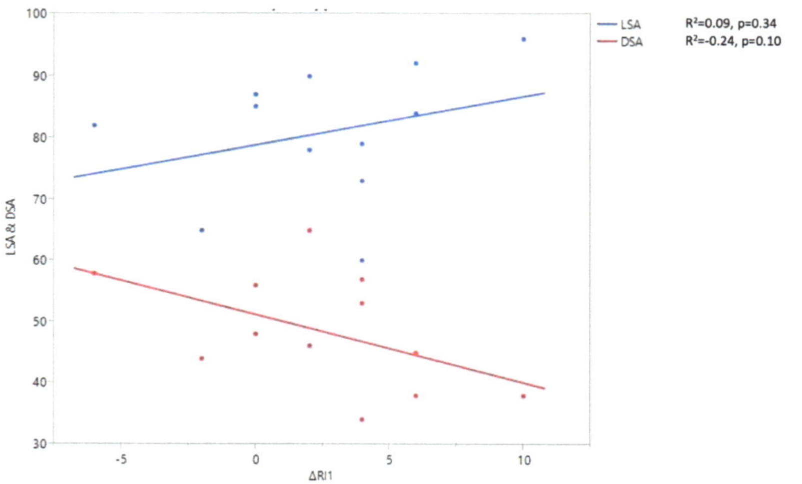

(**E**)

Figure 7. Shoulder motion and 2D humeral displacement using DSA and LSA. (**A**) Change in active shoulder flexion amplitude (Δ F) relative to DSA and LSA. (**B**) Change in active abduction (Δ ABD) relative to DSA and LSA. (**C**) Change in active external rotation elbow at side (Δ RE1A) relative to DSA and LSA. (**D**) Change in active external rotation with arm in abduction (Δ RE2A) relative to DSA and LSA. (**E**) Change in active internal rotation elbow at side (Δ RI1) relative to DSA and LSA.

A positive linear regression was observed between F, ABD, ER1, ER2, and humeral distalization assessed using the DSA. On the other hand, the IR1 decreased when the DSA increased. The greater humeral lateralization assessed using the LSA was associated with better ER1 and IR1 but worse F and ER2. However, none of these findings were statistically significant.

The 2D and 3D Humeral Displacement Measurements (Figure 8).

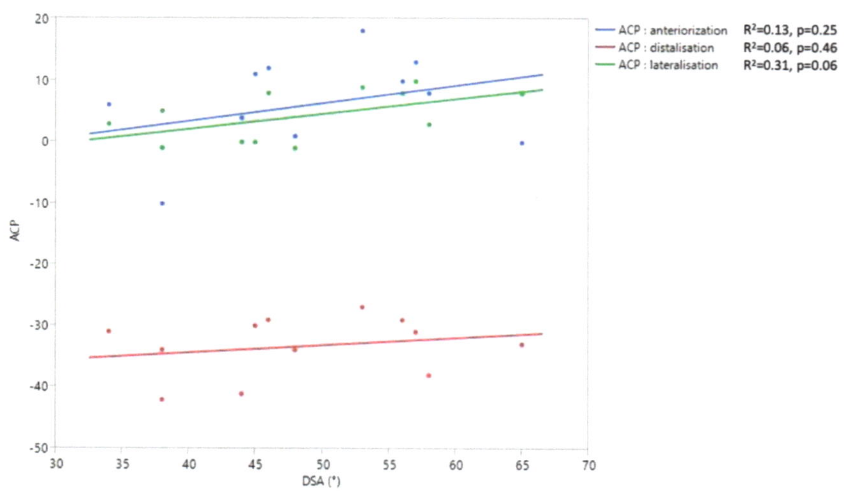

(**A**)

Figure 8. *Cont.*

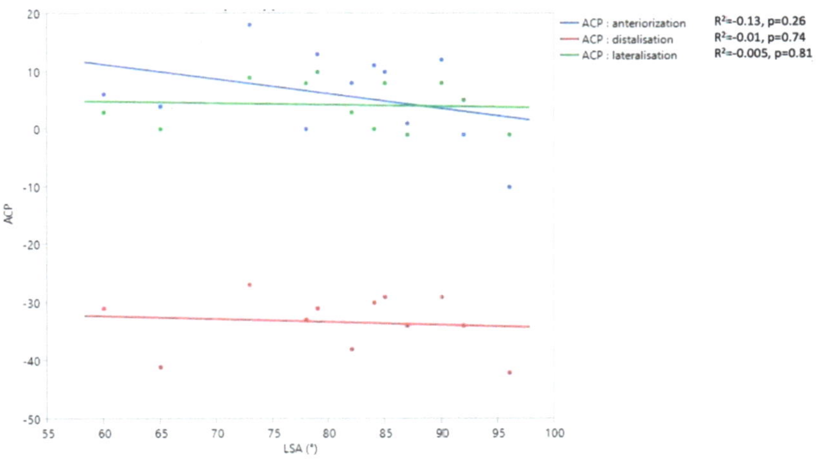

(B)

Figure 8. Relationship between the 2D DSA (**A**) and 2D LSA (**B**) measurements relative to the 3D ACP measurement.

No correlation was found between the measurements of 3D humeral displacement using the ACP and 2D displacement using the LSA and DSA.

4. Discussion

Restoring active shoulder flexion is one of the main functional objectives of RSA. In this clinical study, the main finding was that excessive humeral distalization (or lowering) after RSA implantation somewhat reduced flexion amplitude. This observation was made based on the 3D and 2D analyses of humeral displacement. The mean distalization ACP was 33.3 ± 3.8 mm, and the flexion amplitude was better for the smallest ACP values. The mean DSA was 48.5 ± 9.7°, and the flexion amplitude was also better for the smallest DSA values. Werner et al. [33] reported similar results; they found that arm lengthening ranging between 1 and 2.5 cm was associated with a better constant score. Shoulder flexion increased until humeral lowering reached a value of 25 mm. Beyond that, shoulder flexion decreased. Other previously published studies also supported this finding. Jobin et al. [34] published a prospective cohort study that included 49 patients who underwent RSA for cuff tear arthropathy. They evaluated deltoid lengthening and medialization of the center of rotation radiographically and correlated with RoM. They demonstrated that deltoid lengthening was significantly correlated with superior shoulder flexion. Lädermann et al. [27] compared 183 patients with arm lengthening and those with arm shortening after RSA; they found that postoperative shoulder flexion was significantly greater after arm lengthening, 145° versus 122°, with a mean difference of 23°. However, a lengthening threshold was not found in their study.

In our study, the more uniform distribution of ACP distalization values allowed us to identify a threshold value of around 35 mm. However, the greater dispersion of DSA values made it impossible to identify a humeral lowering threshold value for this criterion. In a recent study by Berhouet et al. [29], it was revealed that the greater the humeral distalization, the better the abduction, with no threshold effect. However, that study was a computer analysis of passive virtual glenohumeral mobility. These new findings leave us wary of the preoperative planning data provided using the dedicated software and their application during the intervention. In other words, even though lowering the humerus is theoretically beneficial for improving the action of the deltoid and therefore joint mobility, too much humeral distalization is harmful in practice, with less RoM in flexion. It becomes nonsensical to plan more than 25 or 30 mm humeral distalization for RSA. Additionally, it would be difficult to physically reduce the implanted prosthesis at the end of the procedure when the planned excessive humeral lowering is carried out.

The counterproductive threshold effect of excessive humeral distalization, measured using the ACP, was also observed for the other RoM values measured in this study: ABD, ER1, ER2, and IR1. On the other hand, no functional limitation was observed relative to humeral lateralization measured using the ACP. This tended to mainly improve the ER1 but was statistically not significant ($p = 0.07$). In the axial plane, anterior humeral displacement correlates with better RoM in different directions. No threshold effect was observed. These clinical observations are therefore comparable to those reported in a virtual planning study by Berhouet et al. [29]. We propose the following explanation: The humeral displacement induced by RSA in the anteroposterior (6 ± 6 mm) and mediolateral (4.3 ± 3.7 mm) planes is less important than that generated in the frontal plane with humeral distalization. In fact, the soft tissue loading in these different planes is probably less, not exceeding the limits of muscle elongation. We think the rotations are improved by humeral lateralization and anteriorization, as this prevents impingement with the scapular pillar and restores muscular configuration to a more favorable status for the recruitment of the remaining rotator cuff. Thus, virtual preoperative planning may be the most reliable and the least impacted by soft tissues when evaluating shoulder rotation, based on the humeral displacement in the anteroposterior and mediolateral planes [35,36].

The functional consequences of humeral displacement assessed using 2D angle measurements slightly differ from those previously reported with the ACP. Moreover, we found no correlation between the 3D ACP measurements and the 2D angle measurements. The potential deleterious effect on the shoulder flexion of too much distalization, assessed using the DSA, was not observed for other shoulder motions. Lateralization, assessed using the LSA, was favorably related to ER1, ER2, and IR1 but unfavorably to flexion, and had no impact on ABD. The wide dispersion of these different angle values for humeral displacement in our limited study population very likely explains the difficulty in interpreting these results. This may also reflect a limit on these angle measurements being carried out using a 2D image since such measurements are less precise than those in a 3D reference frame. Measuring the 3D humeral displacement via the ACP at the millimeter level allowed for the identification of a threshold value of distalization on the different shoulder motions. This was not the case with the LSA and DSA, which were less accurate. This is further evidence that a 2D parameter is not always suitable for characterizing a 3D movement or position.

The main limitation of this study is the small number of patients included. This contributes to a lack of statistical power and, thus, the inability to identify statistically significant differences. However, this study has several methodological strengths. It was prospective and based on real clinical data. Additionally, it was built around a specific imaging processing protocol for postoperative CT scans, which facilitated the precise measurement of the humeral displacement obtained after RSA implantation and its comparison with that of the preoperative humeral position. In this study, we sought to objectively evaluate the quality of virtual planning information (via the ACP) by comparing it to actual clinical observations. Further investigation with a larger number of patients and long-term follow-up could be carried out to strengthen our findings.

5. Conclusions

Beyond the 2D angular or 3D millimeter-level evaluation of humeral displacement after RSA, this study reveals how soft tissue tension impacts the functional results with this type of implant. Determining threshold values for the distalization, lateralization, or even "anteriorization" of the humerus is probably one of the first steps in understanding how to restore the humeral position in order to optimize shoulder joint mobility and prevent deleterious effects. Notably, 3D planning software programs are important elements of this process. The ACP measurement generated with the software used in this study is one of the first parameters designed to explore this issue.

Author Contributions: D.J.Y.Y.: manuscript writing and revisions; J.B.: study design, data acquisition, statistical and data analysis, manuscript writing and revisions; G.-A.O.: critical revisions; R.S.: critical revisions; L.F.: data acquisition. All authors reviewed and approved the final submitted version. All authors have read and agreed to the published version of the manuscript.

Funding: No specific grants were received from public, commercial, or non-profit organizations for this study.

Institutional Review Board Statement: Ethics approval for this study was obtained from our institutional review board (IRB: 13B-T-SHOULDER-RM).

Informed Consent Statement: Informed consent was obtained from all subjects involved in the study.

Data Availability Statement: The data used to support the findings of this study are included within the article. Raw data are available from the corresponding author.

Acknowledgments: The authors thank Joanne Archambault for providing language editing support.

Conflicts of Interest: Julien Berhouet: Wright consulting; Luc Favard: Wright consulting and royalties.

References

1. Grammont, P.M.; Baulot, E. Delta shoulder prosthesis for rotator cuff rupture. *Orthopedics* **1993**, *16*, 65–68. [CrossRef]
2. Drake, G.N.; O'Connor, D.P.; Edwards, T.B. Indications for reverse total shoulder arthroplasty in rotator cuff disease. *Clin. Orthop. Relat. Res.* **2010**, *468*, 1526–1533. [CrossRef]
3. Boileau, P.; Watkinson, D.; Hatzidakis, A.M.; Hovorka, I. Neer Award 2005: The Grammont reverse shoulder prosthesis: Results in cuff tear arthritis, fracture sequelae, and revision arthroplasty. *J. Shoulder Elb. Surg.* **2006**, *15*, 527–540. [CrossRef]
4. Petrillo, S.; Longo, U.G.; Papalia, R.; Denaro, V. Reverse shoulder arthroplasty for massive irreparable rotator cuff tears and cuff tear arthropathy: A systematic review. *Musculoskelet. Surg.* **2017**, *101*, 105–112. [CrossRef]
5. Bufquin, T.; Hersan, A.; Hubert, L.; Massin, P. Reverse shoulder arthroplasty for the treatment of three- and four-part fractures of the proximal humerus in the elderly: A prospective review of 43 cases with a short-term follow-up. *J. Bone Jt. Surg. Br.* **2007**, *89*, 516–520. [CrossRef] [PubMed]
6. Wall, B.; Nové-Josserand, L.; O'Connor, D.P.; Edwards, T.B.; Walch, G. Reverse total shoulder arthroplasty: A review of results according to etiology. *J. Bone Jt. Surg. Am.* **2007**, *89*, 1476–1485.
7. Holcomb, J.O.; Hebert, D.J.; Mighell, M.A.; Dunning, P.E.; Pupello, D.R.; Pliner, M.D.; Frankle, M. Reverse shoulder arthroplasty in patients with rheumatoid arthritis. *J. Shoulder Elb. Surg.* **2010**, *19*, 1076–1084. [CrossRef]
8. Walker, M.; Willis, M.P.; Brooks, J.P.; Pupello, D.; Mulieri, P.J.; Frankle, M.A. The use of the reverse shoulder arthroplasty for treatment of failed total shoulder arthroplasty. *J. Shoulder Elb. Surg.* **2012**, *21*, 514–522. [CrossRef] [PubMed]
9. Ryan, D. Indications for Reverse Shoulder Arthroplasty. In *Shoulder Arthro-Plasty: The Shoulder Club Guide*; Huri, G., Familiari, F., Moon, Y.L., Doral, M.N., Marcheggiani Muccioli, G.M., Eds.; Springer International Publishing: Cham, Switzerland, 2021; pp. 97–102.
10. Muh, S.J.; Streit, J.J.; Wanner, J.P.; Lenarz, C.J.; Shishani, Y.; Rowland, D.Y.; Riley, C.; Nowinski, R.J.; Edwards, T.B.; Gobezie, R. Early follow-up of reverse total shoulder arthroplasty in patients sixty years of age or younger. *J. Bone Jt. Surg. Am.* **2013**, *95*, 1877–1883. [CrossRef] [PubMed]
11. Samuelsen, B.T.; Wagner, E.R.; Houdek, M.T.; Elhassan, B.T.; Sánchez-Sotelo, J.; Cofield, R.; Sperling, J.W. Primary reverse shoulder arthroplasty in patients aged 65 years or younger. *J. Shoulder Elb. Surg.* **2017**, *26*, e13–e17. [CrossRef]
12. Chelli, M.; Cunsolo, L.L.; Gauci, M.-O.; Gonzalez, J.-F.; Domos, P.; Bronsard, N.; Boileau, P. Reverse shoulder arthroplasty in patients aged 65 years or younger: A systematic review of the literature. *JSES Open Access* **2019**, *3*, 162–167. [CrossRef] [PubMed]
13. Leathers, M.P.; Ialenti, M.N.; Feeley, B.T.; Zhang, A.L.; Ma, C.B. Do younger patients have better results after reverse total shoulder arthroplasty? *J. Shoulder Elb. Surg.* **2018**, *27*, S24–S28. [CrossRef] [PubMed]
14. Australian Orthopaedic Association National Joint Replacement Registry. *Hip, Knee & Shoulder Arthroplasty Annual Report 2019*; AOA: Adelaide, Australia, 2019.
15. Best, M.J.; Aziz, K.T.; Wilckens, J.H.; McFarland, E.G.; Srikumaran, U. Increasing incidence of primary reverse and anatomic total shoulder arthroplasty in the United States. *J. Shoulder Elb. Surg.* **2021**, *30*, 1159–1166. [CrossRef] [PubMed]
16. Villatte, G.; Erivan, R.; Barth, J.; Bonnevialle, N.; Descamps, S.; Boisgard, S. Progression and projection for shoulder surgery in France, 2012–2070: Epidemiologic study with trend and projection analysis. *Orthop. Traumatol. Surg. Res.* **2020**, *106*, 1067–1077. [CrossRef] [PubMed]
17. Baulot, E.; Sirveaux, F.; Boileau, P. Grammont's idea: The story of paul grammont's functional surgery concept and the development of the reverse principle. *Clin. Orthop. Relat. Res.* **2011**, *469*, 2425–2431. [CrossRef]
18. Grammont, P.; Trouilloud, P.; Laffay, J.; Deries, X. Étude et réalisation d'une nouvelle prothèse d'épaule. *Rhumatologie* **1987**, *39*, 407–418.
19. Berliner, J.; Regalado-Magdos, A.; Ma, C.; Feeley, B. Biomechanics of reverse total shoulder arthroplasty. *J. Shoulder Elb. Surg.* **2015**, *24*, 150–160. [CrossRef]

20. Boileau, P.; Moineau, G.; Roussanne, Y.; O'Shea, K. Bony increased-offset reversed shoulder arthroplasty minimizing scapular impingement while maximizing glenoid fixation. *Clin. Orthop. Relat. Res.* **2011**, *469*, 2558–2567. [CrossRef]
21. Lädermann, A.; Denard, P.J.; Boileau, P.; Farron, A.; Deransart, P.; Terrier, A.; Ston, J.; Walch, G. Effect of humeral stem design on humeral posi-tion and range of motion in reverse shoulder arthroplasty. *Int. Orthop.* **2015**, *39*, 2205–2213. [CrossRef]
22. Lädermann, A.; Gueorguiev, B.; Charbonnier, C.; Stimec, B.V.; Fasel, J.H.; Zderic, I.; Hagen, J.; Walch, G. Scapular Notching on Kinematic Simulated Range of Motion After Reverse Shoulder Arthroplasty Is Not the Result of Impingement in Adduction. *Medicine* **2015**, *94*, e1615. [CrossRef]
23. Sirveaux, F.; Favard, L.; Oudet, D.; Huquet, D.; Walch, G.; Molé, D. Grammont inverted total shoulder arthroplasty in the treatment of glenohumeral osteoarthritis with massive rupture of the cuff. Results of a multicentre study of 80 shoulders. *J. Bone Jt. Surg. Br.* **2004**, *86*, 388–395. [CrossRef] [PubMed]
24. Katz, D.; Valenti, P.; Kany, J.; Elkholti, K.; Werthel, J. Does lateralisation of the centre of rotation in reverse shoulder arthroplasty avoid scapular notching? Clinical and radiological review of one hundred and forty cases with forty five months of follow-up. *Int. Orthop.* **2016**, *40*, 99–108. [CrossRef]
25. Athwal, G.; MacDermid, J.; Reddy, K.; Marsh, J.; Faber, K.; Drosdowech, D. Does bony increased-offset reverse shoulder arthro-plasty decrease scapular notching? *J. Shoulder Elb. Surg.* **2015**, *24*, 468–473. [CrossRef] [PubMed]
26. Lädermann, A.; Edwards, T.B.; Walch, G. Arm lengthening after reverse shoulder arthroplasty: A review. *Int. Orthop.* **2014**, *38*, 991–1000. [CrossRef]
27. Lädermann, A.; Walch, G.; Lubbeke, A.; Drake, G.N.; Melis, B.; Bacle, G.; Collin, P.; Edwards, T.B.; Sirveaux, F. Influence of arm lengthening in reverse shoulder arthroplasty. *J. Shoulder Elb. Surg.* **2012**, *21*, 336–341. [CrossRef]
28. Boutsiadis, A.; Lenoir, H.; Denard, P.J.; Panisset, J.-C.; Brossard, P.; Delsol, P.; Guichard, F.; Barth, J. The lateralization and distalization shoulder an-gles are important determinants of clinical outcomes in reverse shoulder arthroplasty. *J. Shoulder Elb. Surg.* **2018**, *27*, 1226–1234. [CrossRef]
29. Berhouet, J.; Jacquot, A.; Walch, G.; Deransart, P.; Favard, L.; Gauci, M.-O. The Arm Change Position: Additional Information for Optimizing Range of Motion After Reverse Shoulder Arthroplasty. *Orthop. Traumatol. Surg. Res.* **2022**, *108*, 103246. [CrossRef]
30. Kwon, Y.W.; Pinto, V.J.; Yoon, J.; Frankle, M.A.; Dunning, P.E.; Sheikhzadeh, A. Kinematic analysis of dynamic shoulder motion in patients with reverse total shoulder arthroplasty. *J. Shoulder Elb. Surg.* **2012**, *21*, 1184–1190. [CrossRef]
31. Terrier, A.; Scheuber, P.; Pioletti, D.P.; Farron, A. Activities of daily living with reverse prostheses: Importance of scapular compensation for functional mobility of the shoulder. *J. Shoulder Elb. Surg.* **2013**, *22*, 948–953. [CrossRef]
32. Constant, C.R.; Murley, A.H. A clinical method of functional assessment of the shoulder. *Clin. Orthop. Relat. Res.* **1987**, *214*, 160–164. [CrossRef]
33. Werner, B.; Ascione, F.; Bugelli, G.; Walch, G. Does arm lengthening affect the functional outcome in onlay reverse shoulder arthroplasty? *J. Shoulder Elb. Surg.* **2017**, *26*, 2152–2157. [CrossRef] [PubMed]
34. Jobin, C.M.; Brown, G.D.; Bahu, M.J.; Gardner, T.R.; Bigliani, L.U.; Levine, W.N.; Ahmad, C.S. Reverse total shoulder arthroplasty for cuff tear arthropathy: The clinical effect of deltoid lengthening and center of rotation medialization. *J. Shoulder Elb. Surg.* **2012**, *21*, 1269–1277. [CrossRef] [PubMed]
35. Ferrier, A.; Blasco, L.; Marcoin, A.; De Boissieu, P.; Siboni, R.; Nérot, C.; Ohl, X. Geometric modification of the humeral position after total reverse shoulder arthroplasty: What is the optimal lowering of the humerus? *J. Shoulder Elb. Surg.* **2018**, *27*, 2207–2213. [CrossRef] [PubMed]
36. Schwartz, D.; Cottrell, B.; Teusink, M.J.; Clark, R.E.; Downes, K.L.; Tannenbaum, R.S.; Frankle, M.A. Factors that predict postoperative motion in patients treated with reverse shoulder arthroplasty. *J. Shoulder Elb. Surg.* **2014**, *23*, 1289–1295. [CrossRef] [PubMed]

Journal of Personalized Medicine

Communication

Biomechanical Comparison of Conventional Plate and the C-Nail® System for the Treatment of Displaced Intra-Articular Calcaneal Fractures: A Finite Element Analysis

Roxana Maria Pînzaru [1], Silviu Dumitru Pavăl [2,*], Mihaela Perțea [3,*], Ovidiu Alexa [1,†], Paul Dan Sîrbu [1], Alexandru Filip [1,†], Adrian Claudiu Carp [1], Liliana Savin [1], Norin Forna [1] and Bogdan Veliceasa [1]

1 Department of Orthopaedics and Traumatology, Surgical Science (II), Faculty of Medicine, "Grigore T. Popa" University of Medicine and Pharmacy, 16, University Street, 700115 Iasi, Romania
2 Department of Computer Science and Engineering, "Gheorghe Asachi" Technical University, 27, Dimitrie Mangeron, 700050 Iasi, Romania
3 Department of Plastic Surgery and Reconstructive Microsurgery, Surgical Science (I), Faculty of Medicine, "Grigore T. Popa" University of Medicine and Pharmacy, 16, University Street, 700115 Iasi, Romania
* Correspondence: silviu-dumitru.paval@academic.tuiasi.ro (S.D.P.); mihaela.pertea@umfiasi.ro (M.P.)
† These authors contributed equally to this work.

Abstract: The C-Nail® system is a novel intramedullary fixation method for displaced intra-articular calcaneal fractures. The aim of this study was to evaluate the biomechanical performance of the C-Nail® system and compare it with conventional plate fixation for the treatment of displaced intra-articular calcaneal fractures using finite element analysis. The geometry of a Sanders type-IIB fracture was constructed using the computer-aided design software Ansys SpaceClaim. The C-Nail® system (Medin, Nové Mesto n. Morave, Czech Republic) and the calcaneal locking plate (Auxein Inc., 35 Doral, Florida) and screws were designed according to the manufacturer specifications. Vertical loading of 350 N and 700 N were applied to the subtalar joint surfaces to simulate partial weight bearing and full weight bearing. Construct stiffness, total deformation, and von Mises stress were assessed. The maximum stress on the C-Nail® system was lower compared with the plate (110 MPa vs. 360 MPa). At the bone level the stress was found to have higher values in the case of the plate compared to the C-Nail® system. The study suggests that the C-Nail® system can provide sufficient stability, making it a viable option for the treatment of displaced intra-articular calcaneal fractures.

Keywords: calcaneal fracture; finite element analysis; biomechanics; interlocking nail; C-Nail; locking plate

check for updates

Citation: Pînzaru, R.M.; Pavăl, S.D.; Perțea, M.; Alexa, O.; Sîrbu, P.D.; Filip, A.; Carp, A.C.; Savin, L.; Forna, N.; Veliceasa, B. Biomechanical Comparison of Conventional Plate and the C-Nail® System for the Treatment of Displaced Intra-Articular Calcaneal Fractures: A Finite Element Analysis. *J. Pers. Med.* **2023**, *13*, 587. https://doi.org/10.3390/jpm13040587

Academic Editors: Markus Gesslein and Eduard Kurz

Received: 31 January 2023
Revised: 17 March 2023
Accepted: 21 March 2023
Published: 27 March 2023

1. Introduction

Calcaneus fractures account for 1–4% of all fractures (60% of all tarsal fractures) [1], with 60–75% of them being displaced intra-articular calcaneal fractures (DIACFs) [2,3]. Non-surgical treatment of patients with DIACFs frequently has poor functional outcomes due to secondary subtalar arthritis or malunion leading to variable periods of disability [4,5]. The management of these fractures is complex and there are different opinions about the optimal surgical method. Open reduction and internal fixation (ORIF) with plate and screws using an extended lateral approach is currently considered the gold standard treatment in DIACFs [6,7]. This approach is frequently associated with complications, such as major wound-healing problems (5.8–43%), including haematoma, wound-edge necrosis, superficial and deep wound infections, peroneal tendon problems and cutaneous nerve injury [8–10].

In order to reduce these complications and facilitate early recovery, surgeons have shifted their focus towards minimally invasive techniques, including percutaneous pinning or screw fixation, external fixator combined with limited internal fixation (EFLIF), the sinus tarsi approach, the limited posterior approach, and arthroscopic-assisted fixation [11–18]. Intramedullary nail fixation has been recently included in the minimally invasive treatment of DIACFs. The C-Nail® and Calcanail® are the most common intramedullary fixation systems used [19–21].

The aim of this study was to compare the biomechanical performance and stability of a conventional calcaneal plate in comparison to the C-Nail® system for the treatment of DIACFs using finite element analysis (FEA).

2. Materials and Methods

2.1. Geometry Reconstruction

The right foot of a female patient was scanned by computed tomography (Siemens Somatom Emotion) for the reconstruction of calcaneal bony structure. The subject was 30 years old, 175 cm tall and weighed 70 kg. The patient agreed to have the scan used in this research. The scan was taken in the transverse direction at 0.6 mm intervals and a pitch factor of 0.8.

The geometry of a Sanders type-IIB fracture was then constructed by using the computer-aided design software Ansys SpaceClaim. The original CT scan was imported into SpaceClaim, cleansed and the fracture inserted. A gap of 0.5 mm was added between the fractured elements. Three major cuts were made according to Blake et al. [22] (see Figure 1):

- from the midpoint of the anterior facet to the medial calcaneal tuberosity;
- a vertical cut near the angle of Gissane extending from the lateral cortex to the first cut;
- a vertical cut separating the posterior tuberosity from the posterior facet.

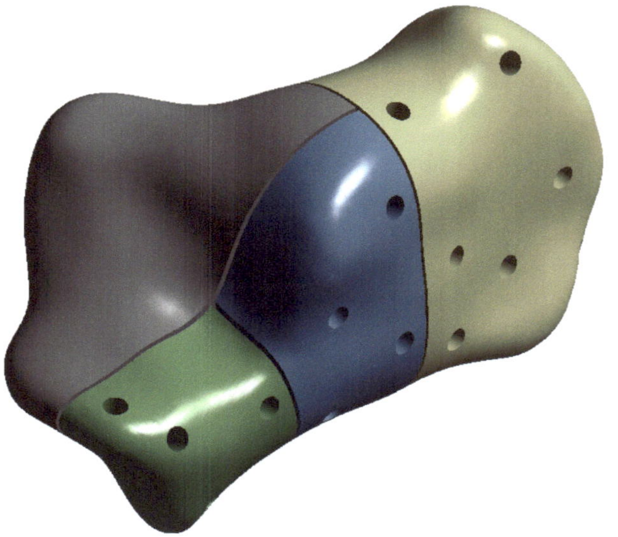

Figure 1. Sanders fracture type-IIB.

The C-Nail® implant (Medin, Nové Město n. Moravě, Czech Republic) has a 8 mm diameter and 65 mm long hollow nail which can be extended and closed by an end cap with sizes: 0 mm; 5 mm or 10 mm. Figure 2 shows the C-Nail® implant designed in SpaceClaim according to the manufacturer specifications. The screws were simplified and modelled as cylinders with 3.5 mm diameter. This implant was inserted into the fractured calcaneal bone, as previously detailed.

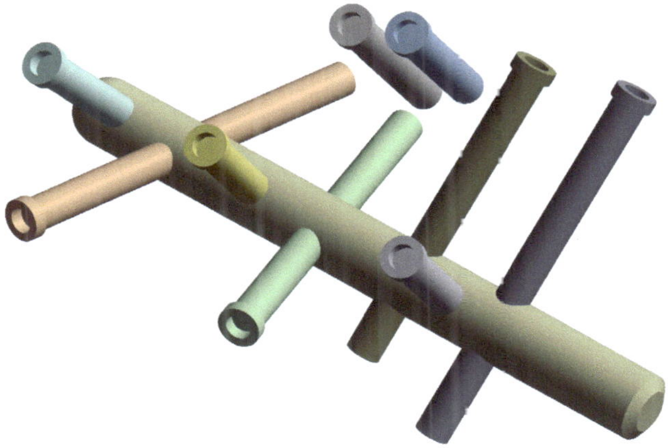

Figure 2. C-Nail® implant.

The second calcanail implant was realized via a calcaneal locking plate (Auxein Inc., Doral, FL, USA) and screws with simplified cylinders with 3.5 mm diameter (Figure 3).

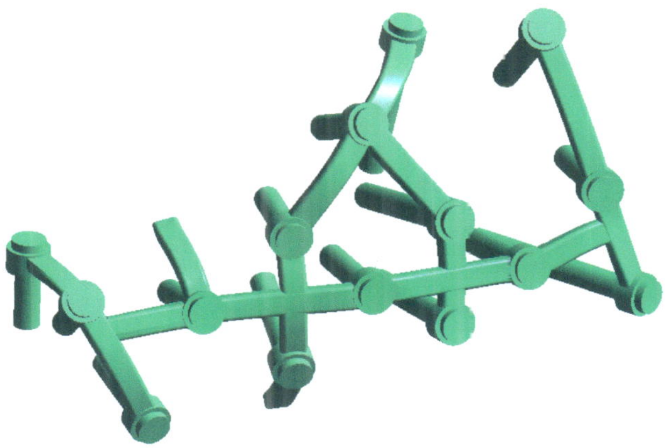

Figure 3. Calcaneal locking plate.

2.2. Material Properties and Mesh Creation

The material properties of the finite element models (bone and implants) were assigned according to the values listed in Table 1.

Table 1. Physical characteristics for the finite element models.

Material	Young's Modulus (MPa) [1]	Poisson's Ratio [2]
Calcaneal bone	1.89×10^4	0.3
Steel (implants)	2×10^5	0.3

[1] Young's modulus is a measure of a material's stiffness and is used to model the elastic behaviour of materials such as metals and bone. [2] Poisson's ratio is a measure of a material's lateral contraction and is used to model the deformation of materials in response to applied loads.

Tetrahedral finite elements were created by Ansys on the calcaneus bone and implants. The first system of calcaneus bone with the C-Nail® implant was meshed with

336,293 elements and 580,519 nodes. The second system of calcaneus bone with the locking plate was meshed with 239,988 elements and 420,096 nodes.

2.3. Boundary and Loading Conditions

The friction coefficient between the bone and implant was assigned a value of 0.3, except for the bone–screw interaction which was fully bonded. Two loads of approximately 350 N and 700 N were applied vertically to the surface of the subtalar joint (Figure 4) in order to simulate full stance and single stance standing positions, respectively. In Ansys, we defined the stress force by using absolute loading values on the X, Y and Z axes (absolute component values) with the force distributed by using a surface effect (as opposed to direct force applied at one point only). For the 700 N force, the corresponding X, Y and Z values are: -100 N, -100 N and -686 N, respectively. In the case of 350 N force, the absolute component values are set to half of the previously stated values.

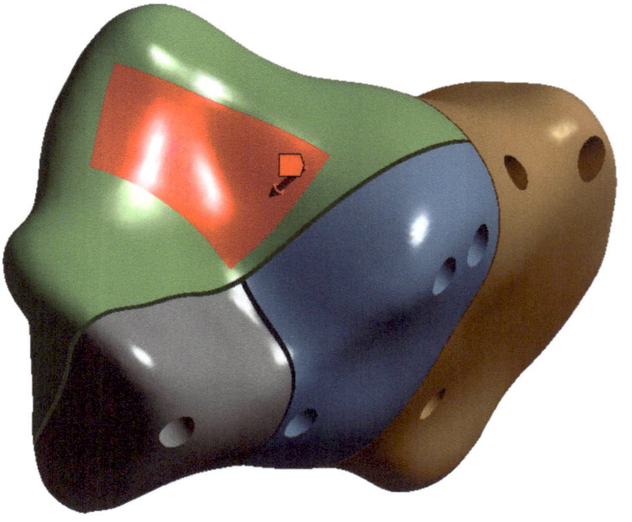

Figure 4. Force distribution and orientation.

The ends of the calcaneus bone were fully constrained, as seen in Figure 5. In finite element analysis (FEA) a fully constrained surface is a surface on which all degrees of freedom (DOF) are restricted. This means that the nodes (or points) on the surface cannot move in any direction (such as translation or rotation) in response to the applied loads or boundary conditions.

All other areas of the bone not fully constrained were set to be of the elastic support-type in Ansys. For the volume delimited by the elastic area, we set a foundation stiffness of 10^9 N/m^3.

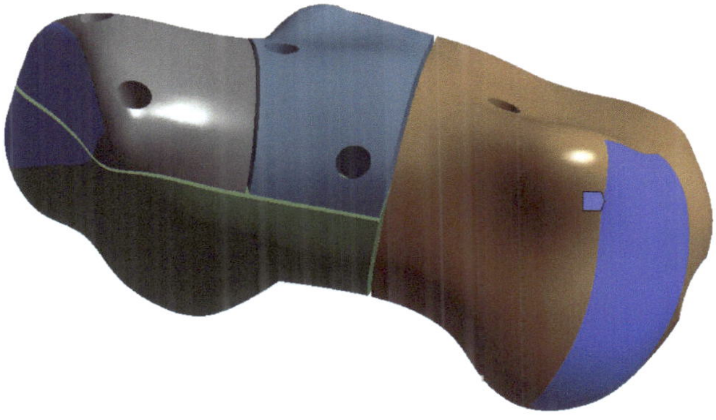

Figure 5. Fixed boundaries (coloured in dark blue).

3. Results

In this section we evaluate the stress levels and total displacement of bone and implant materials under both single-stance and full-stance loads. To achieve this, we used the von Mises stress calculation obtained from the finite element analysis process performed by Ansys. The von Mises stress is a measure of the equivalent stress in a material, taking into account the combined effects of the normal and shear stresses on a material. This is also known as the "equivalent stress" or "average stress".

The simulation results are presented in the upcoming subsections using pictures of a colour mesh overlay on the bone surface to aid visualization. The colour codes are visualized in Figures 6 and 7 where the red/warm colours represent higher values of stress and deformation while blue/cold colours represent lower values of stress and deformation. For the specific absolute maximum values measured during the simulation we used numeric representation found in the tables inserted in their respective subsections.

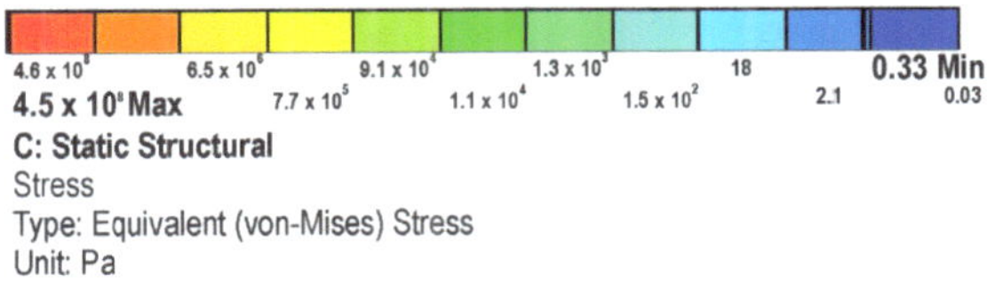

Figure 6. Colour-coded scale for deformation simulations.

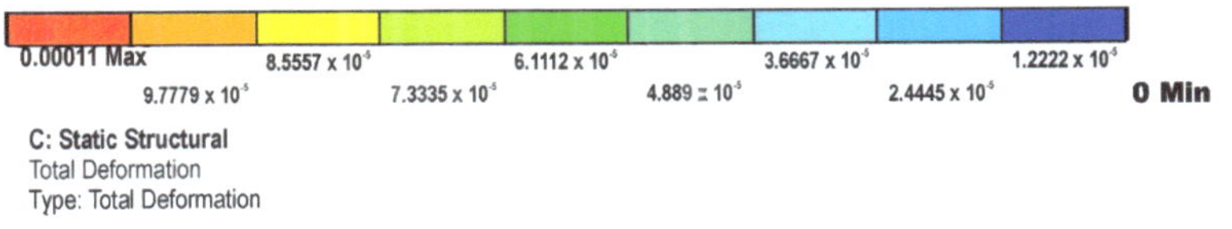

Figure 7. Colour-coded scale for stress simulations.

3.1. *Stresses on Calcaneal C-Nail® Implant*

The FEA simulations show greater stresses occur on the C-Nail® main rod at the first fracture line and junction between the nail and screw located in the anterior process, as seen in Figure 8 (red areas). The magnitude of the stresses at these locations in the case of the single-stance standing position was about 110 MPa. In the same case, the average stress on the C-Nail® structure was 1.8 MPa.

Figure 8. The von Mises stress on the C-Nail® implant structure.

3.2. Stresses on Calcaneal Plate Implant

The stress in the case of the plate implant was found to be higher with a maximum value of 360 MPa in the red areas seen in Figure 9. On the other hand, the average stress was lower, with a value of 0.1 MPa. These values were for the single-stance standing position. The simulation shows the distribution of stress to be uneven for the plate implant, with parts of the structure heavily loaded while others are minimally loaded.

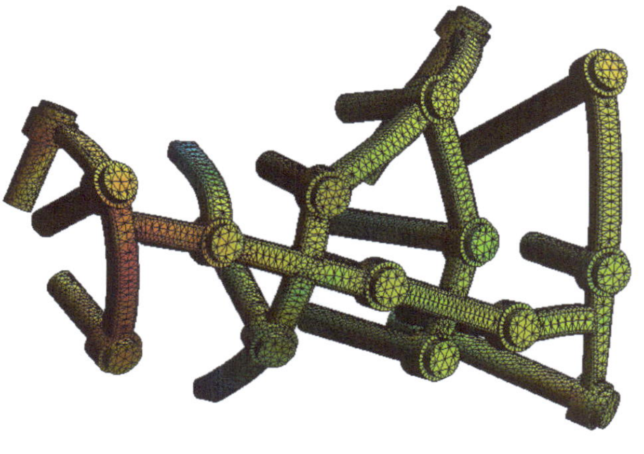

Figure 9. The von Mises stress on the plate implant structure.

3.3. Stresses on Calcaneal Bone

At the bone level the stress was found to be higher in the case of the plate implant compared to the C-Nail® implant. Figures 10 and 11 show the von Mises stress on the calcaneal bone with the C-Nail® implant and plate implant, respectively.

The C-Nail® implant maximum stress on bone was 28 MPa while the plate implant was 65 MPa. This is again for the case of single stance standing (maximum force).

Figures 12 and 13 show a different view point for the same C-Nail® implant and plate implant cases, respectively.

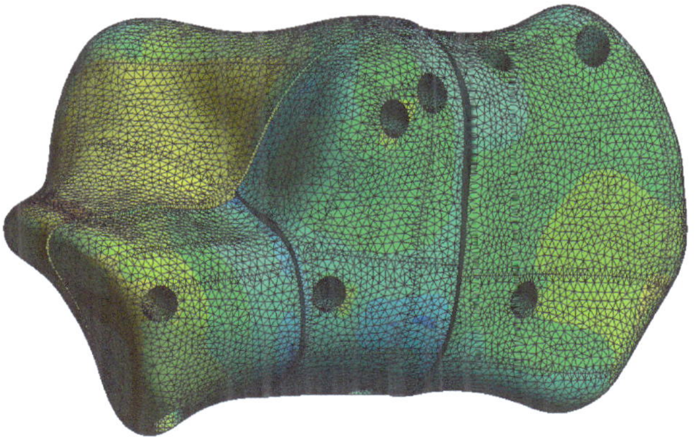

Figure 10. The von Mises stress for calcaneal bone with the C-Nail® implant (left lateral view).

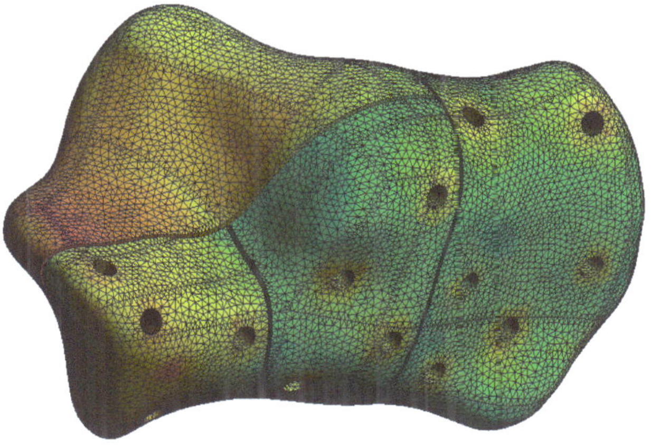

Figure 11. The von Mises stress for calcaneal bone with the plate implant (left lateral view).

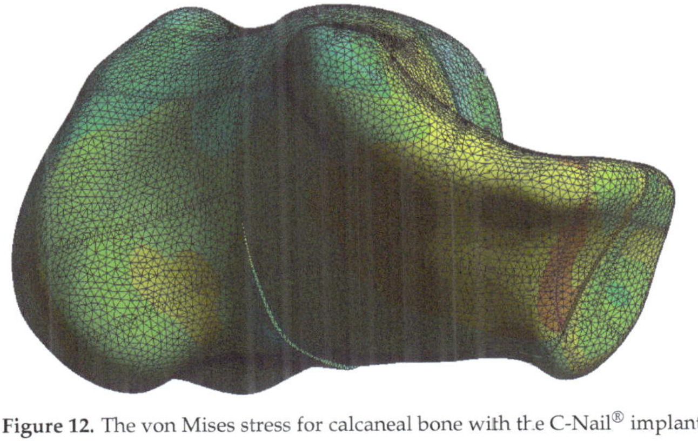

Figure 12. The von Mises stress for calcaneal bone with the C-Nail® implant (left medial view).

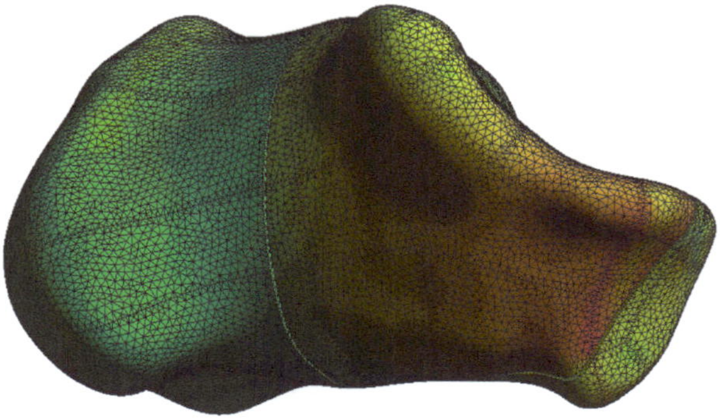

Figure 13. The von Mises stress for calcaneal bone with the plate implant (left medial view).

3.4. Displacement on Calcaneal Bone

Calcaneal bone displacement measures the movement or shifting of the calcaneus when force is applied. Lower displacement values are better as displacement usually results in pain, difficulty walking and difficulty bearing weight on the affected foot.

During our simulations we found the C-Nail® system (see Figure 14) to have displacements approximately 10 times lower than the conventional plate system (see Figure 15), under same load conditions.

Table 2 details the exact maximum stress and displacement values for single-stance and full-stance standing positions observed during the simulation.

Table 2. Maximum stress and displacement values obtained during the simulations.

	Max von Mises Stress (MPa)	Max Displacement (mm)
C-Nail® System		
Single-standing stance (700 N)	110	0.01
Full-standing stance (350 N)	56	0.007
Calcaneal Plate		
Single-standing stance (700 N)	360	1.1
Full-standing stance (350 N)	185	0.6

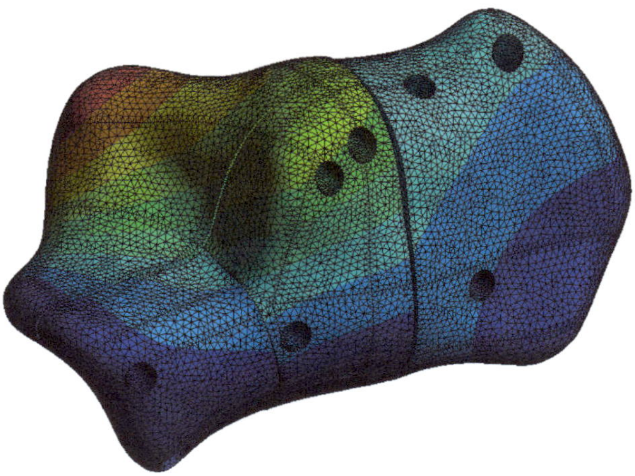

Figure 14. Calcaneal bone displacement in the case of the C-Nail® simulation.

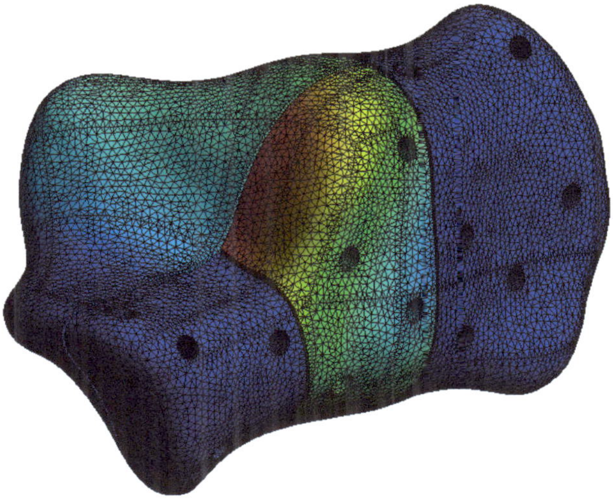

Figure 15. Calcaneal bone displacement in the case of the plate implant simulation.

4. Discussion

The use of the C-Nail® system for treating DIACFs has been recently introduced as a treatment method [23]. Restoration of the posterior calcaneal facet is obtained through a sinus tarsi approach and fixed with two lag screws in order to achieve anatomical reduction. After reducing all fragments to the articular block, the calcaneal fracture fixation is performed percutaneously with an interlocking nail, the C-Nail®, which results in the restoration of the overall shape of the calcaneus. This facilitates the rapid recovery of the patient and reduces the risks of complications associated with the extended lateral approach. In recent literature, clinical studies have shown favourable results in cases of minimally invasive approaches combined with intramedullary nail (C-Nail®) fixation [24]. With the use of minimally invasive techniques the rate of wound-healing complications decreased from 24.9% in the extended lateral approach to 4.9% in the sinus tarsi approach [24–26]. The studies of Zwipp et al. [19] and Pompach et al. [27] obtained a low incidence of wound-edge necrosis (1.9%) and deep infection (0.9%). Another study conducted by Veliceasa et al. [24] reported a low rate of complications, with wound-edge necrosis appearing in 4% of the patients, 1.3% developed a superficial infection, and no deep infections.

From a biomechanical perspective, the structural design of the C-Nail® system contributes to better axial stability, while the calcaneus conventional plate provides better lateral support. This innovative system provides angular stability and firm fixation of the bone fragments [28]. The results of this study show that the C-Nail® system improves the stability of the fixed fracture and has the potential to facilitate consolidation.

In 2016 Reinhardt et al. [29] conducted a biomechanical study on 21 cadavers, comparing a polyaxial interlocking plate, the C-Nail® system and the Calcanail®. The authors created a Sanders type-IIB fracture model and started biomechanical tests with a vertical load of 1000 N, increasing to 2500 N, and then tested the load to failure at a maximum load of 5000 N. Only one specimen in the C-Nail® group reached the maximum load to failure at 5000 N. The highest load to failure, with a mean of 2808 N (±973.6 N), was in the C-Nail® group, followed by the Rimbus plate group, with a mean of 2041 N (±603.6 N), and the Calcanail® group, with a mean of 1751 N (±736.3 N). In our finite element analysis, the vertical load applied on the subtalar joint was 350 N and the results support the C-Nail® as the calcaneal implant with a higher stress resistance and better efficiency in the treatment of DIACFs in comparison with the plate. We obtained a peak bone stress of 56 MPa in the C-Nail® system, while in the plate implant was 185 MPa.

A finite element analysis of Ni et al. [30] compared the plate fixation with the Calcanail® and a modified version of it with a 3.5 mm diameter transfixation screw for a Sanders type-IIIAB calcaneal fracture. They applied a vertical load of 700 N to the subtalar joint and compared the construct stiffness, fracture migration and maximum stress of the implant and calcaneus. In the plate construct, the maximum stress was 102.68 MPa and in the modified Calcanail® the lowest was 84.78 MPa. The peak von Mises stress was also lower in the modified Calcanail® construct [30]. Under similar loading conditions, our study supports the fact that a minimally invasive centromedullar implant has lower stress values and thus a higher overall resistance. The maximum stress on the C-Nail® system was 110 MPa, while on the conventional plate was 360 MPa.

Stress distribution is an important indicator in hardware failure and the stress magnitude is relevant to the safety of the implant under given loading conditions when validated by practical biomechanical tests. In this study, the maximal load applied vertically on the subtalar joint was 700 N. The C-Nail® system experienced the least stress, distributed between the nail and screws, with minimal effects on the model. The maximum stress in the C-Nail® system occurs on the nail, at the level of the first fracture line and the junction between the nail and screw located in the anterior process. The fixation with the locked plate revealed the highest stress, concentrated in the anterior and central branches of the plate extending across the fracture line, as well as in the plate–screw junction. Yu et al. [31] measured the stress on the screw in a finite element model of a Sanders type-IIB fracture fixed with a conventional plate and an anatomical plate and their results are similar to our analysis. The stress observed on the screws is higher on the screws close to the fracture line than those apart from it. It is important to note that the applied forces did not exceed the maximum resistance of the material from which the plate is made, stainless steel (up to 600 MPa). This may suggest that the implant could be considered safe, with a small chance of failure with loading forces equivalent to full weight bearing. These results are similar with the ones reported by Ouyang et al. [32].

To the best of our knowledge, this is the first study comparing the biomechanical performance of a conventional calcaneal plate with the C-Nail® system for DIACFs using finite element analysis (FEA).

One of the strengths of our study is the mesh creation. The calcaneus bone and locking plate were meshed with 239,988 elements and 420,096 nodes, while the calcaneus bone with the C-Nail® implant were meshed with 336,293 elements and 580,519 nodes. The number of elements and nodes is important in FEA because it impacts the accuracy of the simulation. Usually a greater number of elements and nodes provides a more detailed representation of the deformation and displacement of the model. A higher number of nodes allows a more accurate representation of the displacement field and can capture more localized behaviours. However, there are limitations imposed by the computational power as a higher number of elements will require a greater number of computations to be made.

There are still some limitations of this study that should be noted. Firstly, only a Sanders type-IIB fracture was simulated and tested with the FEA. Further research, both biomechanical and clinical, concerning other types of calcaneal fractures should be carried out in future. Secondly, the soft tissue envelope was excluded from the model and axial loads were applied only at the level of the posterior facet as in the other tests that were previously made. Nevertheless, we believe that the used model is reliable in testing the two fixation's biomechanical conduct.

5. Conclusions

The C-Nail® system provided greater stability than the conventional calcaneal plate in the treatment of DIACFs. The C-Nail® system fixation combined with the low complication rate of the sinus tarsi approach represents a viable option for the treatment of DIACFs. The results of this study may provide surgeons with useful information when choosing the optimum implant for the fixation of DIACFs.

Author Contributions: Conceptualization, R.M.P., B.V., O.A. and A.F.; data curation, R.M.P., A.C.C. and B.V.; formal analysis, S.D.P., A.F., N.F. and P.D.S.; software, S.D.P.; writing—original draft, R.M.P., M.P., A.F., A.C.C., N.F. and L.S.; writing—review and editing, P.D.S., L.S., O.A., M.P. and S.D.P.; supervision, O.A., M.P. and B.V. All authors have read and agreed to the published version of the manuscript.

Funding: This research received no external funding.

Informed Consent Statement: Informed consent was obtained from all subjects involved in the study.

Conflicts of Interest: The authors declare no conflict of interest.

References

1. Cianni, L.; Vitiello, R.; Greco, T.; Sirgiovanni, M.; Ragonesi, G.; Maccauro, G.; Perisano, C. Predictive Factors of Poor Outcome in Sanders Type III and IV Calcaneal Fractures Treated with an Open Reduction and Internal Fixation with Plate: A Medium-Term Follow-Up. *J. Clin. Med.* **2022**, *11*, 5660. [CrossRef] [PubMed]
2. Coughlin, M.; Saltzman, C.; Anderson, R. *Mann's Surgery of the Foot and Ankle*, 9th ed.; Elsevier: Amsterdam, The Netherlands, 2014.
3. Volpin, G.; Cohen, Z.; Kirshner, G.; Shtarker, H. Displaced intra-articular calcaneal fractures: Current concepts and modern management. *J. Foot Ankle Surg.* **2014**, *1*, 24–31. [CrossRef]
4. Wei, N.; Yuwen, P.; Liu, W.; Zhu, Y.; Chang, W.; Feng, C.; Chen, W. Operative versus nonoperative treatment of displaced intra-articular calcaneal fractures: A meta-analysis of current evidence base. *Medicine* **2017**, *96*, e9027. [CrossRef] [PubMed]
5. Rammelt, S.; Sangeorzan, B.J.; Swords, M.P. Calcaneal Fractures—Should We or Should We not Operate? *Indian J. Orthop.* **2018**, *52*, 220–230. [CrossRef] [PubMed]
6. Rammelt, S.; Zwipp, H. Fractures of the calcaneus: Current treatment strategies. *Acta Chir. Orthop. Traumatol. Cech.* **2014**, *81*, 177–196.
7. Razik, A.; Harris, M.; Trompeter, A. Calcaneal fractures: Where are we now? *Strat. Traum Limb Recon* **2018**, *13*, 1–11. [CrossRef]
8. Luo, X.; Li, Q.; He, S.; He, S. Operative Versus Nonoperative Treatment for Displaced Intra-Articular Calcaneal Fractures: A Meta-Analysis of Randomized Controlled Trials. *J. Foot Ankle Surg.* **2016**, *55*, 821–828. [CrossRef]
9. Backes, M.; Schepers, T.; Beerekamp, M.S.H.; Luitse, J.S.; Goslings, J.C.; Schep, N.W. Wound infections following open reduction and internal fixation of calcaneal fractures with an extended lateral approach. *Int. Orthop.* **2014**, *38*, 767–773. [CrossRef]
10. Schepers, T.; Den Hartog, D.; Vogels, L.M.; Van Lieshout, E.M. Extended lateral approach for intra-articular calcaneal fractures: An inverse relationship between surgeon experience and wound complications. *J. Foot Ankle Surg.* **2013**, *52*, 167–171. [CrossRef]
11. Caravelli, S.; Gardini, G.; Pungetti, C.; Gentile, P.; Perisano, C.; Greco, T.; Rinaldi, V.G.; Marcheggiani Muccioli, G.M.; Tigani, D.; Mosca, M. Intra-Articular Calcaneal Fractures: Comparison between Mini-Invasive Approach and Kirschner Wires vs. Extensive Approach and Dedicated Plate—A Retrospective Evaluation at Long-Term Follow-Up. *J. Clin. Med.* **2023**, *12*, 20. [CrossRef]
12. Wu, J.; Zhou, F.; Yang, L.; Tan, J. Percutaneous reduction and fixation with Kirschner wires versus open reduction internal fixation for the management of calcaneal fractures: A meta-analysis. *Sci. Rep.* **2016**, *6*, 30480. [CrossRef] [PubMed]
13. Tantavisut, S.; Phisitkul, P.; Westerlind, B.O.; Gao, Y.; Karam, M.D.; Marsh, J.L. Percutaneous Reduction and Screw Fixation of Displaced Intra-articular Fractures of the Calcaneus. *Foot Ankle Int.* **2017**, *38*, 367–374. [CrossRef] [PubMed]
14. Pan, M.; Chai, L.; Xue, F.; Ding, L.; Tang, G.; Lv, B. Comparisons of external fixator combined with limited internal fixation and open reduction and internal fixation for Sanders type 2 calcaneal fractures: Finite element analysis and clinical outcome. *Bone Jt. Res.* **2017**, *6*, 433–438. [CrossRef]
15. Yeo, J.H.; Cho, H.J.; Lee, K.B. Comparison of two surgical approaches for displaced intra-articular calcaneal fractures: Sinus tarsi versus extensile lateral approach. *BMC Musculoskelet Disord.* **2015**, *16*, 63. [CrossRef] [PubMed]
16. Park, C.H.; Lee, D.Y. Surgical treatment of Sanders type 2 calcaneal fractures using a sinus tarsi approach. *Indian J. Orthop.* **2017**, *51*, 461–467. [CrossRef]
17. Pastides, P.S.; Milnes, L.; Rosenfeld, P.F. Percutaneous arthroscopic calcaneal osteosynthesis: A minimally invasive technique for displaced intraarticular calcaneal fractures. *J. Foot Ankle Surg.* **2015**, *54*, 798–804. [CrossRef]
18. Park, C.H.; Yoon, D.H. Role of subtalar arthroscopy in operative treatment of Sanders type 2 calcaneal fractures using a sinus tarsi approach. *Foot Ankle Int.* **2018**, *39*, 443–449. [CrossRef]
19. Zwipp, H.; Paša, L.; Žilka, L.; Amlang, M.; Rammelt, S.; Pompach, M. Introduction of a New Locking Nail for Treatment of Intraarticular Calcaneal Fractures. *J. Orthop. Trauma* **2016**, *30*, e88–e92. [CrossRef]
20. Goldzak, M.; Mittlmeier, T.; Simon, P. Locked nailing for the treatment of displaced articular fractures of the calcaneus: Description of a new procedure with calcanail. *Eur. J. Orthop. Surg. Traumatol.* **2012**, *22*, 345–349. [CrossRef]
21. Bernasconi, A.; Iorio, P.; Ghani, Y.; Argyropoulos, M.; Patel, S.; Barg, A.; Smeraglia, F.; Balato, G.; Welck, M. Use of intramedullary locking nail for displaced intraarticular fractures of the calcaneus: What is the evidence? *Arch. Orthop. Trauma Surg.* **2022**, *142*, 1911–1922. [CrossRef]

22. Blake, M.H.; Owen, J.R.; Sanford, T.S.; Wayne, J.S.; Adelaar, R.S. Biomechanical evaluation of a locking and nonlocking reconstruction plate in an osteoporotic calcaneal fracture model. *Foot Ankle Int.* **2011**, *32*, 432–436. [CrossRef]
23. Allegra, P.R.; Rivera, S.; Desai, S.S.; Aiyer, A.; Kaplan, J.; Gross, C.E. Intra-articular calcaneus fractures: Current concepts review. *Foot Ankle Orthop.* **2020**, *5*, 2473011420927334. [CrossRef]
24. Veliceasa, B.; Filip, A.; Pinzaru, R.; Pertea, M.; Ciuntu, B.; Alexa, O. Treatment of Displaced Intra-articular Calcaneal Fractures With an Interlocking Nail (C-Nail). *J. Orthop. Trauma* **2020**, *34*, e414–e419. [CrossRef] [PubMed]
25. Fascione, F.; Di Mauro, M.; Guelfi, M.; Malagelada, F.; Pantalone, A.; Salini, V. Surgical treatment of displaced intraarticular calcaneal fractures by a minimally invasive technique using a locking nail: A preliminary study. *Foot Ankle Surg.* **2019**, *25*, 679–683. [CrossRef] [PubMed]
26. Nosewicz, T.L.; Dingemans, S.A.; Backes, M.; Luitse, J.S.K.; Goslings, J.C.; Schepers, T. A systematic review and meta-analysis of the sinus tarsi and extended lateral approach in the operative treatment of displaced intra-articular calcaneal fractures. *Foot Ankle Surg.* **2019**, *25*, 580–588. [CrossRef]
27. Pompach, M.; Carda, M.; Amlang, M.; Zwipp, H. Treatment of calcaneal fractures with a locking nail (C-Nail). *Oper Orthop. Traumatol.* **2016**, *28*, 218–230. [CrossRef]
28. Sejda, F.; Frydrýšek, K.; Pleva, L.; Pompach, M.; Hlinka, J.; Sadílek, M.; Murčinková, Z.; Krpec, P.; Havlíček, M.; Madeja, R.; et al. Numerical Analysis of the Calcaneal Nail C-NAIL. *Appl. Sci.* **2022**, *12*, 5265. [CrossRef]
29. Reinhardt, S.; Martin, H.; Ulmar, B.; Döbele, S.; Zwipp, H.; Rammelt, S.; Richter, M.; Pompach, M.; Mittlmeier, T. Interlocking Nailing Versus Interlocking Plating in Intra-articular Calcaneal Fractures: A Biomechanical Study. *Foot Ankle Int.* **2016**, *37*, 891–897. [CrossRef]
30. Ni, M.; Wong, D.W.; Niu, W.; Wang, Y.; Mei, J.; Zhang, M. Biomechanical comparison of modified Calcanail system with plating fixation in intra-articular calcaneal fracture: A finite element analysis. *Med. Eng. Phys.* **2019**, *70*, 55–61. [CrossRef]
31. Yu, B.; Chen, W.C.; Lee, P.Y.; Lin, K.P.; Lin, K.J.; Tsai, C.L.; Wei, H.W. Biomechanical comparison of conventional and anatomical calcaneal plates for the treatment of intraarticular calcaneal fractures—A finite element study. *Comput. Methods Biomech. Biomed. Eng.* **2016**, *19*, 1363–1370. [CrossRef]
32. Ouyang, H.; Deng, Y.; Xie, P.; Yang, Y.; Jiang, B.; Zeng, C.; Huang, W. Biomechanical comparison of conventional and optimized locking plates for the fixation of intraarticular calcaneal fractures: A finite element analysis. *Comput. Methods Biomech. Biomed. Eng.* **2017**, *20*, 1339–1349. [CrossRef] [PubMed]

Journal of
Personalized Medicine

MDPI

Article

The Differential Systemic Biological Effects between Computer Navigation and Conventional Total Knee Arthroplasty (TKA) Surgeries: A Prospective Study

Shu-Jui Kuo [1,2], Ka-Kit Siu [3], Kuan-Ting Wu [4], Jih-Yang Ko [4,*] and Feng-Sheng Wang [5]

1 School of Medicine, China Medical University, Taichung 404, Taiwan
2 Department of Orthopedic Surgery, China Medical University Hospital, Taichung 404, Taiwan
3 Department of Orthopedic Surgery, Park One International Hospital, Kaohsiung 813, Taiwan
4 Department of Orthopedic Surgery, Kaohsiung Chang Gung Memorial Hospital, Kaohsiung 833, Taiwan
5 Department of Medical Research, Kaohsiung Chang Gung Memorial Hospital, Kaohsiung 833, Taiwan
* Correspondence: kojy@cgmh.org.tw

Abstract: Distal femur reaming-free total knee arthroplasty (TKA) was reported to possess lower risk of acute myocardial infarction (AMI) or venous thromboembolism (VTE) than conventional TKA in a retrospective population-based study. We tried to offer prospective biological evidence by comparing the levels of AMI and VTE serum surrogate markers among the patients undertaking navigation and conventional TKAs to support these observations. Thirty-four participants undertaking navigation TKA and 34 patients receiving conventional TKA were recruited between February 2013 and December 2015. Blood samples were drawn from all participants before TKA, and 24 and 72 h after TKA, to assess the concentration of soluble P-selectin, matrix metalloproteinase-9 (MMP-9), C-reactive protein (CRP), and interleukin-8 (IL-8) between the participants undergoing navigation and conventional TKAs. We showed that significantly lower serum levels of soluble P-selectin 24 h after, as well as CRP 24 and 72 h after TKA could be observed in the navigation cohort. The more prominent surge of serum soluble P-selectin and CRP were perceived 24 and 72 h after TKA among the participants undergoing conventional TKA. Based upon our prospective biological evidence, the merits of navigation TKA are strengthened by lower levels of AMI and VTE serum surrogate markers.

Keywords: total knee arthroplasty (TKA); navigation TKA; soluble P-selectin; matrix metalloproteinase-9 (MMP-9); C-reactive protein (CRP); interleu-kin-8 (IL-8)

Citation: Kuo, S.-J.; Siu, K.-K.; Wu, K.-T.; Ko, J.-Y.; Wang, F.-S. The Differential Systemic Biological Effects between Computer Navigation and Conventional Total Knee Arthroplasty (TKA) Surgeries: A Prospective Study. *J. Pers. Med.* **2022**, *12*, 1835. https://doi.org/10.3390/jpm12111835

Academic Editors: Markus Gesslein and Eduard Kurz

Received: 17 October 2022
Accepted: 31 October 2022
Published: 3 November 2022

1. Introduction

Total knee arthroplasty (TKA) is an acknowledged procedure for the treatment of advanced knee osteoarthritis and could bring about high satisfaction [1]. However, TKA is not free from complications, and these complications could be potentially fatal [2,3]. The absolute 6-week risk of acute myocardial infarction (AMI) was 0.21% among TKA patients [2]. Venous thromboembolism (VTE) after TKA can manifest from a subclinical manifestation of deep vein thrombosis (DVT) to a potentially deadly pulmonary embolism (PE). The incidence of in-hospital VTE after TKA is about 0.9%~1.9% [3]. Thus, it is an important issue to mitigate the risk of AMI and/or VTE after TKA.

P-selectin belongs to the selectin family and is bound on the membranes of the Weibel–Palade bodies of endothelial cells and α-granules of platelets, and soluble P-selectin is derived from the alternative splicing [4]. Soluble P-selectin not only reflects the extent of platelet activation but also directly induces a coagulation cascade [5]. Matrix metalloproteinase-9 (MMP-9), also known as gelatinase B, is associated with both arterial atherosclerotic plaque instability and vein wall biomechanics during thrombus formation [6,7]. C-reactive protein (CRP) directly takes part in the arterial atherosclerotic process and is a known predictor of novo ruptured plaque and vascular death [8]. CRP also affects

VTE risk by inducing monocytes to express tissue factor [9]. Interleukin-8 (IL-8), a C-X-C chemokine, mediates the activation of integrin-mediated adhesion of neutrophils and triggers the adhesion of monocytes to vascular endothelium [10,11]. Elevated plasma IL-8 levels are related to an increased risk of coronary artery disease and VTE in population-based studies [10–12]. P-selectin, MMP-9, CRP, and IL-8 are all involved in the processes of both arterial and venous hemostatic processes.

Prosthetic malalignment is an important contributing factor for painful TKA prosthesis [13]. The navigation-assisted or robotic technique was initially employed to optimize the prosthetic alignment. However, under the navigation-assisted or robotic TKA, bone cutting can be executed precisely in the extra-medullary way, thus diminishing the violation of the bone marrow cavity. Diminishing bone marrow insult is an additional benefit other than the optimization of prosthetic alignment. Some authors have indicated the relationships between the violation of bone marrow cavity and the risk of AMI and VTE after TKA [14,15]. A recent population-based retrospective study suggested that the distal femur reaming-free TKA seems to have lower risk of AMI and/or VTE than the conventional TKA. However, support of these observations from prospectively designed clinical studies and biological corroboration are lacking [16].

Due to the unresolved issue, we conducted a prospective study to offer biological evidence for the observations derived from a recent retrospective population-based study. We hypothesize that the mitigation of the intramedullary reaming of the bone marrow of the distal femur might alleviate the postoperative increases of selected markers involved in both arterial and venous hemostatic processes. This study prospectively compared the selected markers between the patients undertaking navigation and conventional TKAs as the surrogate contrast of AMI and VTE. We hypothesize that differential levels of markers involved in both arterial and venous hemostatic processes could be observed after navigation and conventional TKAs.

2. Materials and Methods

Our study was approved by the Institutional Review Board of Chang Gung Memorial Hospital (IRB number: 100-0038A3, 101-0050C, 104-9154D) and registered in ClnicalTrials.gov website (registration number: NCT02206321). Written informed consents were obtained from all the participants. The patients undergoing TKA surgeries performed by the senior author (J.-Y.K.) and professor Ching-Jen Wang between February 2013 and December 2015 were all enrolled. The participants were self-separated in accordance with their clinical registration to the two senior authors. Professor Ching-Jen Wang performed conventional TKA, and the senior author (J.-Y.K.) performed navigation TKA. Both surgeons had performed more than one thousand conventional or navigation TKAs, respectively. Elderly (>90 years), those with malignancy, autoimmune disease, undergoing renal replacement therapy, active infection, previous AMI or VTE history, any kind of surgery performed less than 6 months before the index TKA, and any previous surgery on the knee receiving index TKA (such as fracture repair operation or high tibia osteotomy) were excluded from the study. All the enrolled patients submitted eligible informed consent and knew their group assignment before index TKAs, as the patients undertaking navigation TKAs had four additional tiny stab wounds for tracker-pinning.

Navigation TKAs were carried out under the guidance of the navigation system (Vector Vision, Brain Lab, Heimstetten, Germany, CT-free). Both tibia and femur bone cutting were carried out in an extramedullary way among patients assigned to the navigation group, and femoral bone cuts were led by intramedullary instruments among patients allocated to the conventional group.

Ten milliliters of venous blood attained from each participant before TKA and 24 and 72 h after surgery were collected and stored at −80 °C until analysis. Concentrations of soluble P-selectin, MMP-9, CRP, and IL-8 were quantified using ELISA kits (R & D Systems Inc., Minneapolis, MN, USA). Each patient started continuous passive range of motion exercises and quadriceps strengthening exercises 24 h after TKA, and off-bed ambulation was suggested for most patients on the same day. Oral aspirin 500 mg was given every day for VTE prophylaxis if not contraindicated. Doppler ultrasonography or venography would be arranged if DVT was clinically suspicious (e.g., positive Homan's sign, increased calf or thigh circumference, etc.), whereas, lung perfusion scan or chest CT were arranged if PE was suspected. For the patients free from DVT suspicion, rigorous physiotherapy and off-bed ambulation was arranged for the mechanical prophylaxis of VTE.

Hemoglobin (Hb) concentrations were obtained prior to and 24 h after TKA. If the patients had Hb < 8 mg/dL or Hb 8~9 mg/dL with unstable vitals, blood transfusion with packed red blood cells was ordered. Oral iron supplement was given for patients with Hb 8~9 mg/dL with stable vital signs. Clinical evaluations for DVT were performed regularly until discharge and at regular outpatient follow-ups.

The least sample size necessary for the two groups was pre-determined by the G*Power 3.1.9.2 software (http://www.gpower.hhu.de/en.html (accessed on 1 January 2013)) before the recruitment of participants. The priori power calculation employed a 2-tailed Wilcoxon signed rank test to reckon the sample size of at least 27 for each group (α level: 0.05; power: 80%; effect size: 0.8; allocation ratio: 1).

The data were shown as median (lower quartile, upper quartile). Categorical variables were compared by the Chi-square testing or Fisher's exact test as appropriate. The Mann–Whitney U test was applied to compare the between-group differences. The Friedman test was utilized for the assessment of repeated within-group measurements for continuous variables, and the Wilcoxon signed-rank test was employed for post hoc analysis. All statistics were carried out by IBM SPSS software version 21.0 (IBM Corp. Armonk, NY, USA), and the p value of < 0.05 was defined to be statistically significant.

3. Results

After excluding one patient undergoing conventional surgery without offering valid informed consent, sixty-eight eligible patients offering valid informed consent between February 2013 and December 2015 were recruited. There were five males and twenty-nine females in the navigation group (67.0 (64.0, 73.0) years old). The conventional group comprised 10 males and 24 females (66.0 (61.0, 69.0) years old). There were no differences in gender (p = 0.144), age (p = 0.300), operated side (p = 0.215), type of preoperative coronal plane deformity (p = 0.230), or body mass index (p = 0.124) between the two groups. There were no between-group differences in the proportion of subjects suffering from diabetes (p = 0.355), hypertension (p = 1.000), coronary artery disease (p = 1.000), and stroke (p = 1.000) (Table 1).

Table 1. The demographic profiles between the navigation and conventional groups.

	Navigation (n = 34)	Conventional (n = 34)	p-Value
Gender (male/female)	5/29	10/24	0.144
Age (years)	67.0 (64.0; 73.0)	66.0 (61.0; 69.0)	0.300
Side (left/right)	16/18	11/23	0.215
Deformity (valgus/varus)	9/25	5/29	0.230
Body mass index (kg/m^2)	27.10 (25.70; 29.20)	28.70 (25.70; 31.00)	0.124
Diabetes	8	5	0.355
Hypertension	23	23	1.000
Coronary artery disease	1	2	1.000
Stroke	1	0	1.000

Before TKA, serum concentrations were commensurate between the two groups for soluble P-selectin (p = 0.897), MMP-9 (p = 0.682). CRP (p = 0.263), and IL-8 (p = 0.992). The p-values for the Friedman test were 0.013, <0.001, <0.001 and <0.001 for soluble P-selectin, MMP-9, CRP, and IL-8 in the navigation group; corresponding values were <0.001, <0.001, <0.001 and <0.001 in the conventional group. The respective p-values that were less than 0.050 of the post-hoc Wilcoxon signed-rank test are provided in Tables 2–5.

Table 2. Serum soluble P-selectin concentrations (ng/mL) before TKA, then at 24 and 72 h after TKA.

	Navigation (n = 34) [A]	**Conventional (n = 34) [B]**	**# p-Value**
Baseline	20.36 (9.55; 41.83) [ab]	15.19 (11.54; 35.86) [cd]	0.897
24 h	37.46 (27.85; 62.82) [a]	68.35 (45.63; 70.97) [c]	< 0.001
72 h	43.15 (29.25; 66.93) [b]	62.26 (33.4;, 66.47) [d]	0.352
24 h-baseline	21.94 (−0.79; 36.17)	41.20 (18.57; 55.63)	0.002
72 h-baseline	19.50 (−1.19; 43.71)	43.59 (17.55; 53.44)	0.046

Mann-Whitney U test, [A]: $p = 0.013$; [B]: $p < 0.001$ by Friedman test, [a]: $p = 0.015$; [b]: $p = 0.003$; [c]: $p < 0.001$; [d]: $p < 0.001$ by post-hoc Wilcoxon signed-rank test.

Table 3. Serum metalloproteinase-9 (MMP-9) concentrations (ng/mL) before TKA, then at 24 and 72 h after TKA.

	Navigation (n = 34) [A]	**Conventional (n = 34) [B]**	**# p-Value**
Baseline	243.14 (167.04; 369.10) [ab]	217.56 (172.23; 354.21) [cd]	0.682
24 h	841.79 (583.57; 1073.78) [a]	903.34 (505.68; 1310.89) [c]	0.818
72 h	927.37 (681.93; 1430.69) [b]	1119.43 (594.61; 1846.24) [d]	0.920
24 h-baseline	520.88 (353.43; 771.57)	576.94 (279.84; 1094.34)	0.803
72 h-baseline	724.08 (423.66; 1054.85)	710.52 (357.28; 1563.52)	0.904

Mann–Whitney U test, [A]: $p < 0.001$; [B]: $p < 0.001$ by Friedman test, [a]: $p < 0.001$; [b]: $p < 0.001$; [c]: $p < 0.001$; [d]: $p < 0.001$ by post-hoc Wilcoxon signed-rank test.

Table 4. Serum C-reactive protein concentrations (ng/mL) before TKA, then at 24 and 72 h after TKA.

	Navigation (n = 34) [A]	**Conventional (n = 34) [B]**	**# p-Value**
Baseline	1282.17 (350.45; 2741.22) [ab]	1059.40 (681.85; 4560.36) [cd]	0.263
24 h	5782.79 (5530.72; 6516.73) [a]	7522.12 (7009.85; 8018.18) [c]	<0.001
72 h	5987.70 (5804.76; 7429.90) [b]	8127.07 (7630.61; 8161.80) [d]	<0.001
24 h-baseline	4260.38 (3373.61; 5334.24)	6280.45 (3053.30; 6780.72)	0.016
72 h-baseline	4098.27 (2700.19; 5332.74)	6843.20 (5039.93; 7513.91)	0.004

Mann–Whitney U test. [A]: $p < 0.001$; [B]: $p < 0.001$ by Friedman test. [a]: $p < 0.001$; [b]: $p < 0.001$; [c]: $p < 0.001$; [d]: $p < 0.001$ by post-hoc Wilcoxon signed-rank test.

Table 5. Serum interleukin-8 concentrations (pg/mL) before TKA, then at 24 and 72 h after TKA.

	Navigation (n = 34) [A]	**Conventional (n = 34) [B]**	**# p-Value**
Baseline	13.48 (9.80; 28.46) [ab]	15.08 (10.19; 32.20) [cd]	0.992
24 h	61.49 (29.23; 65.79) [a]	62.94 (32.06; 70.33) [c]	0.263
72 h	58.30 (30.35; 66.96) [b]	38.88 (33.20; 65.48) [d]	0.881
24 h-baseline	27.39 (7.93; 52.12)	28.61 (12.31; 55.20)	0.617
72 h-baseline	32.90 (2.55; 50.41)	21.45 (0.74; 52.93)	0.826

Mann–Whitney U test, [A]: $p < 0.001$; [B]: $p < 0.001$ by Friedman test, [a]: $p < 0.001$; [b]: $p < 0.001$; [c]: $p < 0.001$; [d]: $p < 0.001$ by post-hoc Wilcoxon signed-rank test.

At 24 h after TKA, the median serum soluble P-selectin (37.46 (27.85; 62.82)) in the navigation group was 45.2 % ($p < 0.001$) lower than that in the conventional group (68.35 (45.63; 70.97)). The median increment in serum soluble P-selectin was 46.7% (21.94 (−0.79; 36.17)) and 44.7% (19.50 (−1.19; 43.71)) lower in the navigation cohort versus that in the conventional group (41.20 (18.57; 55.63)) and (43.59 (17.55; 53.44)) 24 h ($p = 0.002$) and 72 h ($p = 0.046$) after TKA, respectively (Table 2) (unit: ng/mL).

The median serum CRP (5782.79 (5530.72; 6516.73)) was 23.1% lower in the navigation group than that in the conventional cohort (7522.12 (7009.85; 8018.18)) ($p < 0.001$) 24 h after TKA. At 72 h after surgery, the median serum CRP 5987.70 (5804.76; 7429.90) in the navigation group was 26.3% ($p < 0.001$) lower than that in the conventional group (8127.07 (7630.61; 8161.80)). The median increment in serum CRP was 32.2% lower ($p = 0.016$) in the navigation group (4260.38 (3373.61; 5334.24)) versus that in the conventional group (6280.45 (3053.30; 6780.72)) 24 h after TKA. At 72 h after operation, the median increment in serum CRP was 40.1% lower ($p = 0.004$) in the navigation cohort (4098.27 (2700.19; 5332.74)) versus that in the conventional group (6843.20 (5039.93; 7513.91)) (Table 4) (Unit: ng/mL).

There were no between-group differences before and 24, 74 h after TKA in terms of MMP-9 and IL-8 (Tables 2 and 5).

According to a previous population-based study, AMI risk peaked to 30.9-fold during the first 2 weeks but did not differ from controls thereafter among TKA patients [2]. The median time to the presentation of DVT and PE was 20 and 12 days after TKA, respectively, and the cumulative risk of VTE lasted for one month after TKA [17]. Based upon these previous studies, all the participants among the two cohorts were followed for 6 weeks after index TKA surgeries. None of the participants suffered from AMI or symptomatic VTE within 6 weeks after TKA.

4. Discussion

The applications of technology-assisted (e.g., navigation, robotics) TKA surgeries have recently been gaining popularity. One recent US nationwide study tried to identify elective primary TKA patients from 2015 to 2020 [16]. Among 847,496 patients enrolled, 49,317 (5.82%) and 24,460 (2.89%) underwent computer navigation (CN)-TKA and robotic assistance (RA)-TKA, respectively. CN-TKA application and RA-TKA deployment increased from 5.64% to 6.41% and 0.84% to 5.89%, respectively, from 2015 to 2020. After adjusting for confounders, CN-TKA was related to lower pulmonary embolism ($p < 0.001$), acute respiratory failure ($p = 0.015$), and periprosthetic joint infection ($p = 0.001$) risk compared to conventional TKA. RA-TKA was associated with lower myocardial infarction ($p = 0.013$), pulmonary embolism ($p = 0.001$), and deep vein thrombosis ($p < 0.001$) risk than conventional TKA. Lower opioid prescription was observed with CN-TKA and RA-TKA than with conventional TKA ($p < 0.001$) 24 h after operation. Lower postoperative opioid consumption was also seen in RA-TKA ($p < 0.001$) 24 h after surgery. In keeping with these findings, our group also showed that the risk of periprosthetic joint infection is lower in the navigation TKA cohort than in the conventional TKA cohort [18]. These data reinforce the safety of RA-TKA and CN-TKA compared to conventional TKA.

In addition to the acute complications requiring admission, the differential systemic biological effects between conventional and technology-assisted TKAs have become an issue attracting increasing attention. The previous works from us and other groups trying to demonstrate the differential levels of systemic biological factors are summarized in Table 6. The present study offered novel prospective biological evidence to support the observations that the risks of AMI and VTE are lower among patients undertaking distal femur reaming-free TKAs in a recent retrospective population-based study [16].

In our study, we tried to suggest the potential risk reduction effect for arterial and venous thromboembolic events for navigation TKAs by comparing serum soluble P-selectin, MMP-9, CRP, and IL-8 between the two TKA techniques. We demonstrated that significantly lower serum levels of soluble P-selectin 24 h after, as well as CRP 24 and 72 h after TKA could be observed in the navigation cohort. The more prominent surge of serum soluble P-selectin and CRP were perceived 24 and 72 h after TKA among the participants undergoing conventional TKA. The differential expression of serum soluble P-selectin and CRP after navigation and conventional TKA deserves attention.

Table 6. The comparison of systemic biological factors between the conventional and the technique avoiding intramedullary reaming.

	Participants	Design	Results
Kuo et al. (2015) [19]	44N/33C	Prospective cohort	Patients undergoing navigation TKAs had fewer blood loss and lower CAMs in serum and hemovac drainage. Milder post-op elevation of serum PECAM-1 ($p = 0.003$) and ICAM-1 ($p = 0.022$) from the pre-op basis was also noted.
Kuo et al. (2018) [1]	44N/53C	Prospective cohort	Serum levels of IL-6, IL-10, TNF-α and TGF-β1 were increased from baseline by smaller increments in the navigation cohort compared with the conventional cohort at 24 h and at 72 h. IL-10 levels in hemovac drainage 24 h after TKA were also significantly lower in the navigation cohort.
Siu et al. (2019) [20]	89N/85C	Prospective cohort	A decreased plasma D-dimer level and a less apparent increase in the plasma D-dimer level were seen in patients undertaking navigation TKA in contrast to patients undertaking conventional TKA 24 h after operation.
Kayani et al. (2021) [21]	30R/30C	Prospective RCT	Participants undertaking conventional TKA and robotic TKA had commensurate changes in the systemic inflammatory and localized thermal response at 6 h, 24 h, 48 h, and day 28 after surgery. Robotic TKA had apparently lower levels of IL-6 ($p < 0.001$), TNF-α ($p = 0.021$), ESR ($p = 0.001$), CRP ($p = 0.004$), lactate dehydrogenase ($p = 0.007$), and creatine kinase ($p = 0.004$) at day 7 after operation compared with conventional TKA.
Xu et al. (2022) [22]	34R/31C	Retrospective	IL-6 serum concentrations were apparently lower in the MA-TKA group on postop day 1 (11.4 (5.2, 21.0) vs. 24.6 (86.3, 170.8), $p = 0.031$). This difference in inflammatory markers was more pronounced at 3 days after the TKA because IL-6, ESR, CRP and CK values were significantly lower in the MA-TKA group 72 h after operation ($p < 0.05$).

N: navigation; C: conventional; R: robotic; CAM: cell adhesion molecule; ICAM: intercellular cell adhesion molecule; VCAM: vascular cell adhesion molecule; PECAM: platelet endothelial cellular adhesion molecule-1; IL-6: interleukin-6; IL-12: interleutkin-10; TNF-α: tumor necrosis factor-alpha; TGF-β1: transforming growth factor-beta 1; IL-10: interleukin-10; ESR: erythrocyte sedimentation rate; CRP: C-reactive protein; MA-TKA: MAKO-robotic assisted total knee arthroplasty; CK: creatine kinase.

The association between serum level of soluble P-selectin and the occurrence of AMI and VTE has been reported before. The concentration of plasma-soluble P-selectin has been shown to be positively correlated with the Gensini score, the numbers of vessels lesions, and the type of coronary artery lesion among patients with angiocardiography-documented coronary heart disease [23]. Ramacciotti et al. has shown that a combination of serum-soluble P-selectin level and Wells score could establish (cut point $\geq$ 90 ng/mL + Wells $\geq$ 2) (specificity: 96%; positive predictive value: 100%) and refute DVT diagnosis (cut point $\leq$ 60 ng/mL + Wells < 2) (sensitivity: 99%; specificity: 33%; negative predictive value: 96%) [24]. Interestingly, the median postoperative serum-soluble P-selectin levels were below and above 60 ng/mL for the navigation and conventional cohort, respectively.

Serum CRP levels also correlate with the onset of both arterial and venous thromboembolic effects as well. According to one Taiwanese study, the serum CRP levels were remarkably higher in subjects with AMI onset < 6 h than in individuals with stable angina (2.7 $\pm$ 2.3 mg/L vs. 1.4 $\pm$ 0.7 mg/L, $p < 0.001$) and in control individuals (2.7 $\pm$ 2.3 mg/L vs. 1.0 $\pm$ 0.6 mg/L, $p < 0.001$) [8]. Kunutsor et al. established a prospective study with 25-year follow up and concurrent meta-analysis to unravel the correlation between CRP level and VTE risk. In this 25-year longitudinal study, CRP was metered in blood samples at baseline from 2420 men. After age adjustment for 119 VTE patients, the regression dilution ratio for log CRP was 0.57 (95% CI: 0.51–0.64). The meta-analysis part of the study merged nine studies comprising 2225 VTE cases and 81625 participants. The fully adjusted estimated VTE risk was 1.14 (95% CI: 1.08–1.19) per standard deviation increase in log baseline CRP. The pooled risk estimate for VTE per 5 mg/L increase in CRP levels was 1.23 (95% CI: 1.09–1.38). The author thus concluded that increased CRP is related to greater VTE risk [25]. The publications mentioned above demonstrated the potential predictive capacity of soluble P-selectin and CRP for AMI and VTE events. The findings of our study cannot be extrapolated to assert that navigation TKAs lead to fewer post-TKA AMI and VTE. The relationships between the serum VTE markers and the true occurrence of AMI VTE among patients undergoing TKA necessitate a larger cohort for further validation.

Previous studies have shown that MMP-9 activity and expression level are increased during thrombus resolution in vivo [26]. MMP-9 mediates macrophage migration and macrophage-induced thrombus resolution [27,28]. Increased level of IL-8 is favorable for thrombus resolution, and impaired inflammatory response by cytokine receptor deletion or neutropenia also decreases thrombus resolution [28–30]. In our study, no between-group differences of serum MMP-9 concentration could be observed. The decoupling of the expression of thrombus-formation marker (soluble P-selectin) and thrombus-resolution marker (MMP-9, IL-8) between the navigation and conventional cohort might reflect the increased net thrombus accumulation in the conventional cohort. The phenomenon of decoupled thrombus formation and resolution and the differential VTE events between the two groups warrant further investigation.

Our study is not flawless. Firstly, we could not prospectively show the superiority of navigation over conventional TKA regarding the true AMI and VTE incidences because of the small sample size. Secondly, the patients were free to choose which senior surgeon they favored, thus jeopardizing the randomization process, in spite of the rigorous anonymization and de-identification process. Thirdly, we performed clinical physical assessment to exclude DVT only in this study. More detailed clinical assessment by the application of modern functional tools, such as baropodometry, should be performed for the participants undertaking TKA [31]. Despite these limitations, the participants did not know which technique would be performed before admission, and no participant shifted to the other group before the surgery. Baseline demographic profiles were comparable between the two groups. The strategies mentioned above might lessen the potential influence of selection bias.

5. Conclusions

Conventional TKAs contribute to higher gush of serum AMI and VTE surrogate markers. Our study demonstrates lower systemic AMI and VTE surrogate markers as emerging biochemical signatures that complement the known advantages of navigation TKAs. We offered novel prospective biological evidence to support the observed lower risk of AMI and VTE for distal femur reaming-free TKAs.

Author Contributions: Conceptualization, K.-K.S. and J.-Y.K.; methodology, S.-J.K. and K.-T.W.; validation, F.-S.W.; formal analysis, S.-J.K.; investigation, S.-J.K.; resources, J.-Y.K. and F.-S W.; data curation, F.-S.W.; writing—original draft preparation, S.-J.K.; writing—review and editing, J.-Y.K.; visualization, K.-K.S.; supervision, J.-Y.K. and F.-S.W.; project administration, J.-Y.K.; funding acquisition, S.-J.K. and J.-Y.K. All authors have read and agreed to the published version of the manuscript.

Funding: This research was funded by the Minister of Science and Technology, Taiwan (110-2314-B-182A-149-, 111-2314-B-182A-055-, and 109-2314-B-039-018-MY3); China Medical University Hospital, Taiwan (DMR-110-111, DMR-110-224, DMR-111-114, and DMR-111-230); and Kaohsiung Chang Gung Memorial Hospital, Taiwan (CRRPG8J0071~3).

Institutional Review Board Statement: The study was conducted in accordance with the Declaration of Helsinki and approved by the Institutional Review Board of Chang Gung Memorial Hospital (IRB number: 100-0038A3, 101-0050C, 104-9154D).

Informed Consent Statement: Informed consent was obtained from all subjects involved in the study.

Data Availability Statement: The data presented in this study are available on request from the corresponding author.

Acknowledgments: The author group is thankful for Ching-Jen Wang for his selfless efforts in contributing the conventional group participants just before his retirement to make our research possible.

Conflicts of Interest: The authors declare no conflict of interest.

References

1. Kuo, S.J.; Hsu, H.C.; Wang, C.J.; Siu, K.K.; Hsu, Y.H.; Ko, J.Y.; Tang, C.H. Effects of computer-assisted navigation versus conventional total knee arthroplasty on the levels of inflammation markers: A prospective study. *PLoS ONE* **2018**, *13*, e0197097. [CrossRef] [PubMed]
2. Lalmohamed, A.; Vestergaard, P.; Klop, C.; Grove, E.L.; de Boer, A.; Leufkens, H.G.; van Staa, T.P.; de Vries, F. Timing of acute myocardial infarction in patients undergoing total hip or knee replacement: A nationwide cohort study. *Arch. Intern. Med.* **2012**, *172*, 1229–1235. [CrossRef] [PubMed]
3. Keller, K.; Hobohm, L.; Barco, S.; Schmidtmann, I.; Munzel, T.; Engelhardt, M.; Eckhard, L.; Konstantinides, S.V.; Drees, P. Venous thromboembolism in patients hospitalized for knee joint replacement surgery. *Sci. Rep.* **2020**, *10*, 22440. [CrossRef]
4. Andre, P.; Hartwell, D.; Hrachovinova, I.; Saffaripour, S.; Wagner, D.D. Pro-coagulant state resulting from high levels of soluble P-selectin in blood. *Proc. Natl. Acad. Sci. USA* **2000**, *97*, 13835–13840. [CrossRef]
5. Antonopoulos, C.N.; Sfyroeras, G.S.; Kakisis, J.D.; Moulakakis, K.G.; Liapis, C.D. The role of soluble P selectin in the diagnosis of venous thromboembolism. *Thromb. Res.* **2014**, *133*, 17–24. [CrossRef] [PubMed]
6. Liu, X.Q.; Mao, Y.; Wang, B.; Lu, X.T.; Bai, W.W.; Sun, Y.Y.; Liu, Y.; Liu, H.M.; Zhang, L.; Zhao, Y.X.; et al. Specific matrix metalloproteinases play different roles in intraplaque angiogenesis and plaque instability in rabbits. *PLoS ONE* **2014**, *9*, e107851. [CrossRef]
7. Nguyen, K.P.; McGilvray, K.C.; Puttlitz, C.M.; Mukhopadhyay, S.; Chabasse, C.; Sarkar, R. Matrix Metalloproteinase 9 (MMP-9) Regulates Vein Wall Biomechanics in Murine Thrombus Resolution. *PLoS ONE* **2015**, *10*, e0139145. [CrossRef]
8. Yip, H.K.; Wu, C.J.; Chang, H.W.; Yang, C.H.; Yeh, K.H.; Chua, S.; Fu, M. Levels and values of serum high-sensitivity C-reactive protein within 6 hours after the onset of acute myocardial infarction. *Chest* **2004**, *126*, 1417–1422. [CrossRef]
9. Cermak, J.; Key, N.S.; Bach, R.R.; Balla, J.; Jacob, H.S.; Vercellotti, G.M. C-reactive protein induces human peripheral blood monocytes to synthesize tissue factor. *Blood* **1993**, *82*, 513–520. [CrossRef]
10. van Aken, B.E.; Reitsma, P.H.; Rosendaal, F.R. Interleukin 8 and venous thrombosis: Evidence for a role of inflammation in thrombosis. *Br. J. Haematol.* **2002**, *116*, 173–177. [CrossRef]
11. Kadi, F.A.; Yuniati, T.; Sribudiani, Y.; Rachmadi, D. The association of rs187238, rs19465518 and rs1946519 IL-8 polymorphisms with acute kidney injury in preterm infants. *Biomedicine* **2021**, *11*, 43–50. [CrossRef] [PubMed]
12. Boekholdt, S.M.; Peters, R.J.; Hack, C.E.; Day, N.E.; Luben, R.; Bingham, S.A.; Wareham, N.J.; Reitsma, P.H.; Khaw, K.T. IL-8 plasma concentrations and the risk of future coronary artery disease in apparently healthy men and women: The EPIC-Norfolk prospective population study. *Arterioscler. Thromb. Vasc. Biol.* **2004**, *24*, 1503–1508. [CrossRef] [PubMed]
13. Spinarelli, A.; Pesce, V.; Campagna, C.; Maccagnano, G.; Moretti, B. Painful knee prosthesis: CT scan to assess patellar angle and implant malrotation. *Muscles Ligaments Tendons J.* **2016**, *6*, 461–466. [CrossRef] [PubMed]
14. Gandhi, R.; Petruccelli, D.; Devereaux, P.J.; Adili, A.; Hubmann, M.; de Beer, J. Incidence and timing of myocardial infarction after total joint arthroplasty. *J. Arthroplast.* **2006**, *21*, 874–877. [CrossRef] [PubMed]
15. Santana, D.C.; Emara, A.K.; Orr, M.N.; Klika, A.K.; Higuera, C.A.; Krebs, V.E.; Molloy, R.M.; Piuzzi, N.S. An Update on Venous Thromboembolism Rates and Prophylaxis in Hip and Knee Arthroplasty in 2020. *Medicina* **2020**, *56*, 416. [CrossRef]
16. Wang, J.C.; Piple, A.S.; Hill, W.J.; Chen, M.S.; Gettleman, B.S.; Richardson, M.; Heckmann, N.D.; Christ, A.B. Computer-Navigated and Robotic-Assisted Total Knee Arthroplasty: Increasing in Popularity Without Increasing Complications. *J Arthroplast.* 2022. [CrossRef] [PubMed]
17. Bjornara, B.T.; Gudmundsen, T.E.; Dahl, O.E. Frequency and timing of clinical venous thromboembolism after major joint surgery. *J. Bone Jt. Surg. Br.* **2006**, *88*, 386–391. [CrossRef]
18. Chen, J.W.; Kuo, F.C.; Kuo, S.J.; Siu, K.K.; Ko, J.Y. Avoidance of intramedullary violation in computer-assisted total knee arthroplasty lowers the incidence of periprosthetic joint infection compared with conventional total knee arthroplasty: A propensity score matching analysis of 5342 cases. *Knee* **2022**, *35*, 164–174. [CrossRef]
19. Kuo, S.J.; Wang, F.S.; Wang, C.J.; Ko, J.Y.; Chen, S.H.; Siu, K.K. Effects of Computer Navigation versus Conventional Total Knee Arthroplasty on Endothelial Damage Marker Levels: A Prospective Comparative Study. *PLoS ONE* **2015**, *10*, e0126663. [CrossRef]
20. Siu, K.K.; Wu, K.T.; Ko, J.Y.; Wang, F.S.; Chou, W.Y.; Wang, C.J.; Kuo, S.J. Effects of computer-assisted navigation versus the conventional technique for total knee arthroplasty on levels of plasma thrombotic markers: A prospective study. *Biomed. Eng. Online* **2019**, *18*, 99. [CrossRef]
21. Kayani, B.; Tahmassebi, J.; Ayuob, A.; Konan, S.; Oussedik, S.; Haddad, F.S. A prospective randomized controlled trial comparing the systemic inflammatory response in conventional jig-based total knee arthroplasty versus robotic-arm assisted total knee arthroplasty. *Bone Jt. J.* **2021**, *103-B(1)*, 113–122. [CrossRef] [PubMed]
22. Xu, J.Z.; Li, L.L.; Fu, J.; Xu, C.; Zhang, G.Q.; Chai, W.; Hao, L.B.; Li, X.; Chen, J.Y. Comparison of serum inflammatory indicators and radiographic results in MAKO robotic-assisted versus conventional total knee arthroplasty for knee osteoarthritis: A retrospective study of Chinese patients. *BMC Musculoskelet Disord.* **2022**, *23*, 418. [CrossRef] [PubMed]
23. Qu, X.B.; Sun, Z.Q.; Chen, M.J.; Chen, L.L. The study of soluble P-selectin levels and it's correlation to the severity of coronary artery lesions in coronary heart disease. *Zhonghua Liu Xing Bing Xue Za Zhi* **2005**, *26*, 617–621. [PubMed]
24. Ramacciotti, E.; Blackburn, S.; Hawley, A.E.; Vandy, F.; Ballard-Lipka, N.; Stabler, C.; Baker, N.; Guire, K.E.; Rectenwald, J.E.; Henke, P.K.; et al. Evaluation of soluble P-selectin as a marker for the diagnosis of deep venous thrombosis. *Clin. Appl. Thromb. Hemost.* **2011**, *17*, 425–431. [CrossRef]

25. Kunutsor, S.K.; Seidu, S.; Blom, A.W.; Khunti, K.; Laukkanen, J.A. Serum C-reactive protein increases the risk of venous thromboembolism: A prospective study and meta-analysis of published prospective evidence. *Eur. J. Epidemiol.* **2017**, *32*, 657–667. [CrossRef]
26. Dahi, S.; Lee, J.G.; Lovett, D.H.; Sarkar, R. Differential transcriptional activation of matrix metalloproteinase-2 and membrane type-1 matrix metalloproteinase by experimental deep venous thrombosis and thrombin. *J. Vasc. Surg.* **2005**, *42*, 539–545. [CrossRef]
27. Gong, Y.; Hart, E.; Shchurin, A.; Hoover-Plow, J. Inflammatory macrophage migration requires MMP-9 activation by plasminogen in mice. *J. Clin. Investig.* **2008**, *118*, 3012–3024. [CrossRef]
28. Varma, M.R.; Varga, A.J.; Knipp, B.S.; Sukheepod, P.; Upchurch, G.R.; Kunkel, S.L.; Wakefield, T.W.; Henke, P.K. Neutropenia impairs venous thrombosis resolution in the rat. *J. Vasc. Surg.* **2003**, *38*, 1090–1098. [CrossRef]
29. Wang, Z.W.; Wang, J.J.; Zhang, J.Z.; Xue, Z.J.; Miao, J.; Li, L.; Hu, W.X. Thrombolysis of deep vein thrombosis and inhibiting chemotaxis of macrophage by MCP-1 blockage. *Eur. Rev. Med. Pharmacol. Sci.* **2017**, *21*, 1695–1701.
30. Henke, P.K.; Wakefield, T.W.; Kadell, A.M.; Linn, M.J.; Varma, M.R.; Sarkar, M.; Hawley, A.; Fowlkes, J.B.; Strieter, R.M. Interleukin-8 administration enhances venous thrombosis resolution in a rat model. *J. Surg. Res.* **2001**, *99*, 84–91. [CrossRef]
31. Notarnicola, A.; Maccagnano, G.; Fiore, A.; Spinarelli, A.; Montenegro, L.; Paoloni, M.; Pastore, F.; Tafuri, S.; Moretti, B. Baropodometry on patients after total knee arthroplasty. *Musculoskelet Surg.* **2018**, *102*, 129–137. [CrossRef] [PubMed]

Journal of *Personalized Medicine*

MDPI

Article

Development of the INpaTiEnt Rehabilitation App Compliance QuesTionnaire [INTERACT]

Hassan Tarek Hakam [1,2,3,*], Felix Mühlensiepen [2,3,4], Mikhail Salzmann [1], Jonathan Lettner [1,2,3], Roland Becker [1,2], Sebastian Kopf [1] and Robert Prill [1,2,3,*]

1 Center of Orthopaedics and Traumatology, University Hospital Brandenburg/Havel, Brandenburg Medical School, Hochstr. 29, 14770 Brandenburg an der Havel, Germany
2 Faculty of Health Science, Brandenburg Medical School, 14770 Brandenburg an der Havel, Germany
3 Center of Evidence Based Practice in Brandenburg (EBB), A JBI Affiliated Group, Brandenburg Medical School, 14770 Brandenburg an der Havel, Germany
4 Center for Health Services Research, Brandenburg Medical School, 15562 Rüdersdorf, Germany
* Correspondence: hassantarek.hakam@mhb-fontane.de (H.T.H.); robert.prill@mhb-fontane.de (R.P.)

Abstract: Background: The elements of previously designed questionnaires do not take into consideration the burdens encountered in an inpatient hospital setting. The purpose of this study is to validate elements of a non-compliance survey in an in-hospital setting and discuss aspects of compliance with telerehabilitative physiotherapy in the early postoperative period. Methods: A literature search was conducted to identify elements that prevent patients from performing their prescribed physical therapy exercises. These items were then evaluated by the expert review technique as described by Ikart. Afterwards, the synthesized survey was handed out to patients for the assessment of the quality of its items. Results: The results of the expert review technique identified some conceptual and grammatical problems. This led to the adjustment of some of the elements. The quality of the resulting questionnaire was deemed to be good, as patients were able to fully understand the concepts and answer accordingly. A statistical analysis was conducted to evaluate the responses. Conclusions: The items of this second questionnaire have proven to be reliable to assess the compliance of patients in an in-hospital setting. These items will be added to the cross-culturally adapted items of a previous questionnaire.

Keywords: compliance; physical therapy; in-hospital setting; rehabilitative care; joint replacement; expert review; telerehabilitation

Citation: Hakam, H.T.; Mühlensiepen, F.; Salzmann, M.; Lettner, J.; Becker, R.; Kopf, S.; Prill, R. Development of the INpaTiEnt Rehabilitation App Compliance QuesTionnaire [INTERACT]. *J. Pers. Med.* **2023**, *13*, 1638. https://doi.org/10.3390/jpm13121638

Academic Editors: Markus Gesslein and Eduard Kurz

Received: 21 September 2023
Accepted: 21 November 2023
Published: 24 November 2023

1. Introduction

Physical therapy has been proven to be an effective treatment modality in patients with varying musculoskeletal conditions. Patients with conditions ranging from lower back pain to postoperative rehabilitation after knee or hip replacement can benefit from physiotherapeutic rehabilitative intervention [1–3]. One factor determining the success of rehabilitative measures is the patient's compliance with those measures [4]. However, non-adherence with physiotherapeutic care seems to be an active issue in orthopaedic patients, preventing them from achieving optimal outcomes [5]. One systematic review [6] examined a multitude of studies aimed at improving a patient's adherence to physical rehabilitation. Of the examined studies, only two revealed a statistically significant improvement between the intervention group and the comparators. The interventions included a print-out of the exercises in one study [7] and the addition of cognitive behavioral training in the other [8].

The research group at the Center of Orthopaedics and Trauma Surgery is aiming to conduct a high-quality randomized controlled trial to identify the barriers to compliance with physiotherapeutic care in the early postoperative period as delivered by a mobile health application. Therefore, a non-compliance to physiotherapy survey [9] was cross-culturally adapted and validated [10]. However, the elements were deemed insufficient, as they tackled concerns of patients in private physiotherapeutic practice. Thus, a comparative study contrasting inpatient to outpatient physical therapy might facilitate the identification of some non-adherence parameters. However, a systematic review attempting to compare the effectiveness of inpatient with outpatient physiotherapeutic care found that no scientific evidence is in existence concerning the research question at hand [11].

The effectiveness of delivering musculoskeletal physical therapy by the means of telerehabilitation has been proven for a variety of conditions ranging from lower back pain to rehabilitation after knee or hip replacement [12,13]. In fact, telerehabilitation has been hypothesized to have the same effect as outpatient professionally supervised physiotherapy [14]. In light of this assumption, great implications for the healthcare system can be foreseen, as a decreased workload on physiotherapists would be a direct consequence. This would prove to be beneficial in light of the shortage affecting skilled workers in physiotherapeutic practice [15], the risk of transmitting an infection in the age of the pandemic [16], and the ability to reduce socioeconomic costs by implementing telerehabilitation in everyday practice [17]. However, the effect of mobile health applications on compliance has not yet been studied, and a systematic review attempting to evaluate these apps identified several problems with their conception [18]. The design of the application used for the previously mentioned randomized controlled trial will be the subject of another paper.

The purpose of this study is to conceive items for a questionnaire that tackles non-compliance with physical therapy in an inpatient setting using the expert review technique and to put these elements in light of an early postoperative rehabilitative app-guided physical therapy program.

2. Materials and Methods

A three-step method was used in accordance with the suggested approach of Ikart [19]. The first step aimed to identify potential non-adherence elements by conducting a literature search and their synthesis into survey items. The second step included pretesting the synthesized elements using the expert review technique. The final step consisted of standard-pretesting the resulting questionnaire using patient feedback. A total number of twenty-eight patients were surveyed.

To evaluate the mobile application-based approach to physical therapy, the German adaption of the System Usability Score was used [20,21]. At the top of each questionnaire, a self-assessment of compliance to physical therapy for the current and past rehabilitative measures was asked for.

The methods used for this study are summarized in Figure 1.

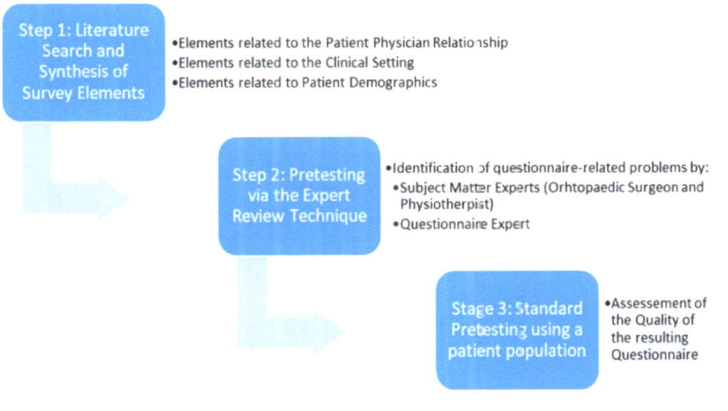

Figure 1. A summary of the steps of the expert review method as applied to the questionnaire at hand.

2.1. Step 1: Literature Search and Synthesis of Survey Elements

In his work, Geisler (1992) identifies physician-related factors that influence their patient's adherence to a given therapeutic regimen. The empathy the physician shows, his/her skills in motivating patients in taking ownership of their health issues, and the ability to give specific instructions that lead to specific results without over or underestimation were identified as key factors [22].

A cross-sectional study examined the medical staff's opinions on attendance to physical therapy [23]. The parameters deemed to be applicable for our survey were the patient's perception about the severity of the disease, the patient's perception regarding the intensity of the exercise, and the environment in which physical therapy takes place.

Finally, the patient's demographics correlate with their compliance to physical therapy. Younger patients, male patients, and patients with a higher educational background are more likely to follow health recommendations [24]. As these data are recorded in a standardized manner in any randomized controlled trial, pretesting was not deemed necessary. Another parameter that was identified in this study is the living situation of the corresponding patient. These parameters should be presented at the end of the survey since participants dealing with a questionnaire will have peak interest at the beginning. Afterwards, energy levels of the patient begin to decline, and subsequently, easier questions should be presented in the last section [19]. Patient's living with their families and receiving support were more likely to be compliant with health recommendations.

2.2. Step 2: Pretesting via the Expert Review Technique

The experts were divided into two groups: the survey and questionnaire experts and the substantive or subject-matter experts [25]. The first group was made up of two researchers who are primarily engaged in qualitative research, and the second group was made up of an orthopaedic surgeon and a physiotherapist. All involved experts had many years of academic and research experience.

The experts then evaluated the individual elements using the Questionnaire Appraisal Scheme (QAS) [26]. When an issue was encountered, the appropriate QAS element was mentioned, and an alternative to the element in question was given. The QAS questionnaire will be provided in Appendix A.

2.3. Step 3: Standard Pretesting Using a Patient Population

After the questionnaire was reviewed by experts, it was handed to twenty-eight patients for standard pretesting by checking the answers on the written survey. Multiple-choice answers allowed patients to rate the elements from one (completely disagree or false) to five (completely agree or completely right). An extra space was left blank at the end of the questionnaire for comments provided by patients. This step helps in evaluating the quality of the survey. Patients were made aware of their role and asked to provide feedback if any difficulties were encountered.

3. Results

3.1. Results of Stage 1: Literature Search and Synthesis of Survey Elements

The literature search yielded the items that are presented in the table in Appendix B. Item thirteen was eliminated from the questionnaire, as questioning the support of the prothesis might reveal insecurity towards the support it provides and hinder future execution of exercises.

3.2. Results of Stage 2: Pretesting via the Expert Review Technique

The results of the subjective matter expert's evaluation (orthopaedic surgeon and physiotherapist) as well as the evaluation of the subject-matter expert are presented as tables in Appendix C. The survey items have a designated QAS, a commentary section, and suggestions for improvement. The items are shown in Table 1. The most important modifications included the translation of scientific terms into commonly understood language.

"Prothesis" was hence replaced by "artificial joint" and "early postoperative period" was replaced by "first days after the operation".

Table 1. Represents the elements after modification using the expert review method.

Item Number	Items after Expert Review
1	Was your physician empathetic?
2	How did this affect your adherence to your prescribed physical therapy exercises?
3	Did your physician motivate you to take ownership in ameliorating your situation?
4	How did this affect your adherence to your prescribed physical therapy exercises?
5	Did your physician give you specific recommendations regarding rehabilitation for the first days of in-hospital stay after your surgery?
6	How did this affect your adherence to your prescribed physical therapy exercises?
7	Did your physician give you a realistic assessment regarding the first days after surgery?
8	How did this affect your adherence to your prescribed physical therapy exercises?
9	Did pain hinder your completion of your prescribed physical therapy exercises?
10	Were you confident that your artificial joint will support the execution of the exercises?
11	Was it unpleasant to do the exercises in front of a third party (other patients, medical staff, etc.)?
12	Did your family support you (emotionally/bodily) while performing the exercises?

3.3. Results of Stage 3: Standard Pretesting Using a Patient Population

The standard pretesting did not reveal any comprehension problems with the questionnaire after the adjustments made in stage 2. Comments provided by patients are presented in Appendix D; the most important comment being that answering the questionnaire depends on the patient's psychological status as well as the amount of perceived pain.

A simple statistical analysis of the patient's responses to the individual items are presented in Table 2.

Table 2. Results of the patient's responses to the individual items. The percentage is presented by the number of responses per total number of participants.

Item Number	1	2	3	4	5
1	0	1 (4%)	5 (18%)	4 (14%)	18 (64%)
2	0	0	8 (29%)	12 (43%)	7 (25%)
3	0	2 (7%)	3 (11%)	9 (32%)	13 (46%)
4	2 (7%)	0	4 (14%))	12 (43%)	7 (25%)
5	0	3 (11%)	4 (14%))	7 (25%)	10 (36%)
6	4 (14%))	0	8 (29%)	9 (32%)	9 (32%)
7	1 (4%)	2 (7%)	4 (14%)	10 (36%)	6 (21%)
8	7 (25%)	0	11 (39%)	10 (36%)	6 (21%)
9	8 (29%)	7 (25%)	3 (11%)	8 (29%)	2 (7%)
10	8 (29%)	3 (11%)	14 (50%)	2 (7%)	1 (4%)
11	0	2 ()	2 (7%)	13 (46%)	10 (36%)
12	1 (4%)	0	0	5 (18%)	21 (75%)

4. Discussion

The primary result of this study yielded a non-compliance to the physiotherapy questionnaire adjusted to an inpatient setting. A literature review yielded the synthesis of appropriate items for the survey, an evaluation by subject-matter and questionnaire experts resulted in the refinement of the individual elements, and a pretesting method ensured the quality assessment of the survey. When adding the results of this study to the noncompliance questionnaire provided by "Correlates of exercise compliance in physical therapy", a comprehensive evaluation of the patient's motivation to not adhere to physical therapy in a hospital setting can be evaluated. Tackling these issues might lead to an increase in the adherence to recommended therapies and hence ameliorate the outcome of the target population.

The literature search yielded thirteen questions dealing with a variety of aspects of physiotherapy. The physician-specific aspects were discussed in one paper [22] while other studies focused on the environmental aspects affecting the patient's compliance [23,24]. Intrinsic patient-related factors are discussed in "Correlates of exercise compliance in physiotherapy". Deeper psychological aspects affecting patient compliance are out of the scope of this questionnaire, and the resulting RCT will exclude patients with psychological disorders that might affect compliance. This aspect, however, is not very well researched and should be considered as a topic for future RCTs dealing with early postoperative rehabilitation. Suggested elements for this type of research can be taken over from studies dealing with non-compliance to pharmaceutical therapy [27]. Hence, this questionnaire only considers the environmental factors that are specific to a clinical setting. Issues like the psychological state and patient expectations are not accounted for in this survey, as they deal with factors that can be controlled in a hospital setting.

The results of the expert review led to the revision and the adjustment of individual elements. Items were adjusted to imply the patient's specific situation. QAS items 4 (Vague/Unclear), 7 (Undefined/Vague Term), 15 (Undefined Period), 20 (Complex Estimation), 22 (Undefined Term), and 23 (Vague Term) were used to evaluate survey questions and responses. This led to the adjustments presented in the results of stage 2. Since the patient's comprehension of the question is key in the evaluation process [28], terms like "postoperative" and "prosthesis" were simplified to "after surgery" and "artificial joint", respectively. The implementation of precise elements was also tackled, leading to the modification of the terms "early postoperative" and "PT exercises" to the first days after your operation" and to "your prescribed PT exercises", respectively. German grammatical errors were corrected by the questionnaire expert.

Pretesting on a patient population revealed no comprehension-related issues. Patients primarily used the "Suggestions and Comments" part of the questionnaire to express their views on compliance. The empathy physicians expressed towards the patients' situations, the motivation physicians provide to patients in taking ownership regarding rehabilitative measures, the presentation of a realistic image concerning the early postoperative period, and the communication of a specific rehabilitative plan all play a substantiative role that affect patients' adherence to their exercises. Familial support seemed to play the biggest role with regards to compliance with rehabilitative measures. The results of this analysis, however, are not representative for a wide patient population with varying disorders since the evaluation only took place in the confined space of the wards of the orthopaedics and traumatology department and mostly included in-hospital patients treated with reconstruction of the joints.

Generally, the expert review method ensures that multiple dimensions concerning validity are ensured. Face validity was established by searching the literature for commonly perceived problems with compliance and the review of these elements by subject-matter experts and patients [29]. Content validity was ensured by the review of content experts (physiotherapists and orthopaedic surgeons) on the relevance of the presented items and by comments supplemented by patients regarding ambiguity, clarity, and simplicity [30], as the questionnaire addressed the importance of reporting any issue regarding these measures before beginning to answer individual items. Predictive validity is difficult to assess in the realm of future compliance with physical therapy measures, as multiple factors, including the patient's psychological state, concurrent medical issues, and simple logistic issues like transportation play an immense role [31].

The before-mentioned randomized controlled trial (RCT) will evaluate the effect of an mHealth application on the compliance and the outcome of patients in need of in-hospital rehabilitation. To measure the parameters dealing with the practicality of the application, subjects will be provided with the system usability score (SUS) [32]. Since the SUS is an already validated instrument that aims to evaluate the end-user's perspectives with respect to the applied system, the research team investigating the use of the mobile rehabilitation application will be implying slightly modified terms to refer to the app. So for example, "system" will be replaced with "app" to facilitate comprehension. These elements will provide the basis on which subsequent modifications targeting the improvement of physiotherapeutic mobile applications will be made.

Finally, the importance of having an active lifestyle and the compliance to physical therapy cannot be understated. New measures to increase the adherence of patients to physiotherapy, specifically, and to an active lifestyle, more generally, must be found. At the Center of Orthopaedics and Trauma Surgery of the University Clinic of Brandenburg, new methods, such as the utilization of a mobile physiotherapeutic reminder application and the use of knee motion sensors, are currently under investigation. While patient-reported compliance to physical therapy is an important measure, the objective assessment of this parameter is of utmost importance, as patient-reported adherence to treatment seldomly reflects the objective reality of the matter.

4.1. Limitations

Limitations of this study include the pretesting conducted on patients in German. The use of this questionnaire in English-speaking countries would require the re-evaluation of step 3 using a patient population. As well, this questionnaire did not tackle psychosocial factors such as the patient's mental health or emotional well-being. This, however, should be taken with caution, as the purpose of this paper is to investigate clinical aspects of compliance in a healthcare facility. Involving the patient's psychosocial and emotional states is an important aspect that is not in the scope of the current investigation but to which specialists should attend in order to elevate the rate of compliance to the recommended interventions.

4.2. Implications for Future Research

This questionnaire can be used in the future to assess environmental factors that lead to non-compliance with physical therapy in Germany and German-speaking countries. The use of this measuring instrument in English should be proceeded by a validation process targeting the affected population, namely, patients undergoing physical rehabilitation after an injury or a surgical procedure. As the relationship between exercise compliance and better outcomes has been proven by previously mentioned studies, this questionnaire will enable orthopaedic surgeons and physiotherapists to investigate and overcome certain barriers dealing with the matter at hand.

5. Conclusions

Adherence to physiotherapeutic measures in an inpatient setting can present with some challenges. Physician- and patient-related aspects as well as environmental conditions of the hospital setting play a role in that regard. Additionally, new methods of treatment delivery such as the use of an application might play an immense role in the ever-growing strain on healthcare. A questionnaire was developed in accordance with the "expert review method" to tackle the aspects of compliance in an in-hospital setting. The questionnaire contained aspects of the patient-physician relationship and considerations to the environment in which physical therapy takes place. Additionally, elements of the SUS were added to ensure the quality of the application and its ability to improve the desired outcomes.

Author Contributions: H.T.H. is the primary author of this paper. H.T.H. and R.P. made a literature search to identify the elements for a non-compliance questionnaire in an inpatient setting. F.M. oversaw the methodological soundness of the approach and provided feedback and took the role of the questionnaire expert. R.P. and M.S. were the subject-matter experts. J.L. and H.T.H. oversaw the distribution of the pre-final version at the university clinic. R.B. and S.K. oversaw the conception of the study and provided necessary feedback for the study design. All authors have read and agreed to the published version of the manuscript.

Funding: Funded by the Brandenburg Medical School publication fund supported by the German Research Foundation and the Ministry of Science, Research and Cultural Affairs of the State of Brandenburg.

Institutional Review Board Statement: Not applicable due to the non-interventional nature of the study. Please refer to the statement below for detailed information.

Informed Consent Statement: The study was conducted in compliance with current data protection regulations and the Helsinki Declaration. All study participants were informed about the research project. Data were anonymized before analysis. This is in accordance with the German data regulation requirements DSGVO, the national regulatory law for data protection and regulation, on anonymized data.

Data Availability Statement: Additional data are available after request from the corresponding authors.

Conflicts of Interest: The authors declare no conflict of interest.

Appendix A

Figure representing the evaluation using the Questionnaire Appraisal Scheme (Stage 2):

Code problem label
Problem types:
Interviewer Difficulties
1. Inaccurate instruction
2. Complicated instruction
3. Difficult for interviewer to administer
Question Content
4. Vague/unclear question
5. Complex topic
6. Topic carried over from earlier Q
7. Undefined/vague term
Question structure
8. Transition needed
9. Unclear respond instruction
10. Question too long
11. Complex/awkward syntax
12. Erroneous assumption
13. Several questions
Referenced period
14. Period carried over from earlier Q
15. Undefined period
16. Unanchoring/rolling period
Retrieval from Memory
17. Shortage of memory cues
18. High detail require or info unavailable
19. Long recall or reference period
Judgement & Evaluation
20. Complex estimation
21. Potentially sensitive/bias
Respond terminology
22. Undefined term
23. Vague terms
Response Units
24. Responses use wrong or mismatching units
25. Unclear to respondent what the response options are
26. Multi-dimensional response set
Response structure
27. Overlapping response categories
28. Missing response categories
Grand Total

Appendix B

Synthesis of the elements after conducting a literature review (results of Stage 1):

Source	Survey Item
Geisler (1992) [22]	1. Was your physician empathetic and did he stand by your side?
	2. How did this affect your adherence to the PT exercises?
	3. Did your physician motivate you to take responsibility for your situation?
	4. How did this affect your adherence to the PT exercises?
	5. Did your physician give you specific recommendations for your early postoperative rehabilitation?
	6. How did this affect your adherence to the PT exercises?
	7. Did your doctor give you a realistic assessment regarding the early postoperative phase?
	8. How did this affect your adherence to the PT exercises?
Mbada (1970) [23]	9. Did pain hinder you from completing your PT exercises?
	10. Were you confident that your prothesis would support the execution of the exercises?
	11. Was it unpleasant to do the exercises in front of a third party (other patients, medical staff, etc.)?
	12. Were you confident that your direct surroundings (chairs, bed, tables, etc.) would support you while performing the exercises?
Groen (2012) [24]	13. Did your family support you (emotionally/physically) while performing the exercises?

Appendix C

Results of the Questionnaire Appraisal by the experts:

Subject-Matter Expert 1: Orthopaedic Surgeon			
Item Number	QAS Number	Comment	Suggestions
1	0	None	None
2	find out	Not very specific	"to your prescribed PT exercises" instead of "the PT exercises"
3	0	None	None
4	0	None	None
5	7 and 15	Some patients might not understand "early postoperative"	Instead use "in the first days after your operation"
6	0	None	None
7	7 and 15	Some patients might not understand "early postoperative"	Instead use "in the first days after your operation"
8	0	None	None
9	4	Patients receive specific programs from the physiotherapists	Use "your prescribed PT exercises instead of "your PT exercises"
10	7	Prosthesis is too scientific	Use "your artificial joint" instead of "prothesis"
11	0	None	None
12	7	Some patients might read this rapidly and understand psychological instead of physical	Use "bodily" instead of "physically"

Appendix D

Patient's comments:

Patient 1: "These factors change from day to day. In the first day after the surgery, I would have answered the questionnaire very differently".

Patient 2: "Familial support is always nice. But having a competent doctor is more important than having an empathetic one".

References

1. Lewis, A.; Morris, M.E.; Walsh, C. *Are Physiotherapy Exercises Effective in Reducing Chronic Low Back Pain?* Centre for Reviews and Dissemination: York, UK, 2008. Available online: https://www.ncbi.nlm.nih.gov/books/NBK75764/ (accessed on 1 June 2022).
2. Artz, N.; Elvers, K.T.; Lowe, C.M.; Sackley, C.; Jepson, P.; Beswick, A.D. Effectiveness of physiotherapy exercise following total knee replacement: Systematic review and meta-analysis. *BMC Musculoskelet. Disord.* **2015**, *16*, 15. [CrossRef]
3. Coulter, C.L.; Scarvell, J.M.; Neeman, T.M.; Smith, P.N. Physiotherapist-directed rehabilitation exercises in the outpatient or home setting improve strength, gait speed and cadence after elective total hip replacement: A systematic review. *J. Physiother.* **2013**, *59*, 219–226. [CrossRef]
4. Pisters, M.F.; Veenhof, C.; de Bakker, D.H.; Schellevis, F.G.; Dekker, J. Behavioural graded activity results in better exercise adherence and more physical activity than usual care in people with osteoarthritis: A cluster-randomised trial. *J. Physiother.* **2010**, *56*, 41–47. [CrossRef]
5. Cinthuja, P.; Krishnamoorthy, N.; Shivapatham, G. Effective interventions to improve long-term physiotherapy exercise adherence among patients with lower limb osteoarthritis. A systematic review. *BMC Musculoskelet. Disord.* **2022**, *23*, 147. [CrossRef]
6. McLean, S.M.; Burton, M.; Bradley, L.; Littlewood, C. Interventions for enhancing adherence with physiotherapy: A systematic review. *Man. Ther.* **2010**, *15*, 514–521. [CrossRef] [PubMed]
7. Schneiders, A.; Zusman, M.; Singer, K. Exercise therapy compliance in acute low back pain patients. *Man. Ther.* **1998**, *3*, 147–152. [CrossRef]
8. Göhner, W.; Schlicht, W. Preventing chronic back pain: Evaluation of a theory-based cognitive-behavioural training programme for patients with subacute back pain. *Patient Educ. Couns.* **2006**, *64*, 87–95. [CrossRef] [PubMed]
9. Sluijs, E.M.; Kok, G.J.; van der Zee, J. Correlates of Exercise Compliance in Physical Therapy. *Phys. Ther.* **1993**, *73*, 771–782. [CrossRef] [PubMed]
10. Beaton, D.E.; Bombardier, C.; Guillemin, F.; Ferraz, M.B. Guidelines for the Process of Cross-Cultural Adaptation of Self-Report Measures. *Spine* **2000**, *25*, 3186–3191. [CrossRef] [PubMed]
11. Klugarova, J.; Klugar, M.; Mareckova, J.; Gallo, J.; Kelnarova, Z. The effectiveness of inpatient physical therapy compared to outpatient physical therapy in older adults after total hip replacement in the post-discharge period: A systematic review. *JBI Evid. Synth.* **2016**, *14*, 174–209. [CrossRef] [PubMed]
12. Pastora-Bernal, J.M.; Martín-Valero, R.; Barón-López, F.J.; Estebanez-Pérez, M.J. Evidence of Benefit of Telerehabilitation After Orthopedic Surgery: A Systematic Review. *J. Med. Internet Res.* **2017**, *19*, e142. [CrossRef] [PubMed]
13. Suso-Martí, L.; La Touche, R.; Herranz-Gómez, A.; Angulo-Díaz-Parreño, S.; Paris-Alemany, A.; Cuenca-Martínez, F. Effectiveness of Telerehabilitation in Physical Therapist Practice: An Umbrella and Mapping Review with Meta–Meta-Analysis. *Phys. Ther.* **2021**, *101*, pzab075. [CrossRef] [PubMed]
14. Seron, P.; Oliveros, M.-J.; Gutierrez-Arias, R.; Fuentes-Aspe, R.; Torres-Castro, R.C.; Merino-Osorio, C.; Nahuelhual, P.; Inostroza, J.; Jalil, Y.; Solano, R.; et al. Effectiveness of Telerehabilitation in Physical Therapy: A Rapid Overview. *Phys. Ther.* **2021**, *101*, pzab053. [CrossRef] [PubMed]
15. Schwinger, M.; Grafe, M. Exploring barriers, advantages and potentials in realising clinical education in private physiotherapy practice settings in Germany. *OpenPhysio J.* **2020**. [CrossRef] [PubMed]
16. Werneke, M.W.; Deutscher, D.; Grigsby, D.; Tucker, A.C.; Mioduski, E.J.; Hayes, D. Telerehabilitation During the COVID-19 Pandemic in Outpatient Rehabilitation Settings: A Descriptive Study. *Phys. Ther.* **2021**, *101*, pzab110. [CrossRef]
17. Nelson, M.; Russell, T.; Crossley, K.; Bourke, M.; McPhail, S. Cost-effectiveness of telerehabilitation versus traditional care after total hip replacement: A trial-based economic evaluation. *J. Telemed. Telecare* **2019**, *27*, 359–366. [CrossRef]
18. Bahadori, S.; Wainwright, T.W.; Ahmed, O.H. Smartphone apps for total hip replacement and total knee replacement surgery patients: A systematic review. *Disabil. Rehabil.* **2018**, *42*, 983–988. [CrossRef]
19. Ikart, E.M. Survey Questionnaire Survey Pretesting Method: An Evaluation of Survey Questionnaire via Expert Reviews Technique. *Asian J. Soc. Sci. Stud.* **2019**, *4*, 1. [CrossRef]
20. Jordan, P.W.; Thomas, B.; McClelland, I.L.; Weerdmeester, B. *Usability Evaluation in Industry*; Chapman and Hall/CRC: Boca Raton, FL, USA; London, UK, 2014.
21. Berke, L.; Huenerfauth, M.; Patel, K. Design and Psychometric Evaluation of American Sign Language Translations of Usability Questionnaires. *ACM Trans. Access. Comput.* **2019**, *12*, 1–43. [CrossRef]
22. Geisler, L. *Arzt und Patient—Begegnung im Gespräch*; Peter Hoffmann/Pharma Verlag: Frankfurt, Germany, 1992. Available online: http://www.linus-geisler.de/ap/ap00_inhalt.html (accessed on 26 June 2023).
23. Mbada, C.; Ajayi, O.; Agbeja, O.B.; Mbada, K.; Awotidebe, T.; Oghumu, S. Non-Attendance for Out-patient Physiotherapy: Evaluation, Prediction and Physiotherapists' Perceptions—A Cross-Sectional Study: Semantic Scholar. 1970. Available online: https://www.semanticscholar.org/paper/Non-Attendance-for-Out-Patient-Physiotherapy%3A-and-a-Mbada-Ajayi/dd0d9b27bfd25bc7371461bc8988187f098993eb#citing-papers (accessed on 26 June 2023).
24. Groen, J.-W.; Stevens, M.; Kersten, R.F.; Reininga, I.H.; Akker-Scheek, I.v.D. After total knee arthroplasty, many people are not active enough to maintain their health and fitness: An observational study. *J. Physiother.* **2012**, *58*, 113–116. [CrossRef]
25. Fricker, R. Evaluating Survey Questions. In Proceedings of the Meeting of Naval Postgraduate School, Monterey, CA, USA, 26 June 2012.

26. Lessler, J.T. A Coding System for Appraising Questionnaires. In *Answering Questions Methodology for Determining Cognitive and Communicative Processes in Survey Research*, 1st ed.; Forsyth, B.H., Ed.; Jossey-Bass Inc.: San Francisco, CA, USA, 1996; pp. 259–291.
27. Jin, J.; Sklar, G.E.; Oh, V.M.S.; Li, S.C. Factors affecting therapeutic compliance: A review from the patient's perspective. *Ther. Clin. Risk Manag.* **2008**, *4*, 269–286. [CrossRef]
28. Gangavalli, A.K.; Malige, A.; Rehman, S.; Nwachuku, C.O. Patient Comprehension and Compliance Survey to Assess Postoperative Pain Regimens in the Orthopaedic Trauma Population. *J. Orthop. Trauma* **2017**, *31*, e190–e194. [CrossRef]
29. Thomas, S.D.; Hathaway, D.K.; Arheart, K.L. Face Validity. *West. J. Nurs. Res.* **1992**, *14*, 109–112. [CrossRef] [PubMed]
30. Yaghmaie, F. Content Validity and Its Estimation. *J. Med. Educ.* **2003**, *3*, e105015. [CrossRef]
31. Morisky, D.E.; Green, L.W.; Levine, D.M. Concurrent and Predictive Validity of a Self-reported Measure of Medication Adherence. *Med. Care* **1986**, *24*, 67–74. [CrossRef] [PubMed]
32. Hyzy, M.; Bond, R.; Mulvenna, M.; Bai, L.; Dix, A.; Leigh, S.; Hunt, S. System Usability Scale Benchmarking for Digital Health Apps: Meta-analysis. *JMIR mHealth uHealth* **2022**, *10*, e37290. [CrossRef] [PubMed]

Journal of
Personalized Medicine

MDPI

Article

Effects of Sensorimotor Training on Transversus Abdominis Activation in Chronic Low Back Pain Patients

Felix Marchand [1,2], Kevin Laudner [3], Karl-Stefan Delank [1], René Schwesig [1] and Anke Steinmetz [4,*]

1 Department of Orthopedic and Trauma Surgery, Martin-Luther-University Halle-Wittenberg, 06120 Halle, Germany; f.marchand@jhwaf.de (F.M.); stefan.delank@uk-halle.de (K.-S.D.); rene.schwesig@uk-halle.de (R.S.)
2 Department of Orthopedic and Trauma Surgery, Josephs-Hospital Warendorf, 48231 Warendorf, Germany
3 Department of Health Sciences, Hybl Sports Medicine and Performance Center, University of Colorado, Colorado Springs, CO 80918, USA; klaudner@uccs.edu
4 Department of Trauma, Reconstructive Surgery and Rehabilitation Medicine, University Medicine Greifswald, Physical and Rehabilitation Medicine, 17475 Greifswald, Germany
* Correspondence: anke.steinmetz@med.uni-greifswald.de

Abstract: (1) Background: The aim of this study was to investigate and compare the effect of sensorimotor training on transversus abdominis activation. (2) Methods: Seventy-five patients with chronic low back pain were randomly assigned to one of three groups (whole body vibration training using Galileo®, coordination training using Posturomed®, or physiotherapy (control)). Transversus abdominis activation was measured by using sonography pre- and post-intervention. Second, changes in clinical function tests and their correlation with the sonographic measurements were determined. (3) Results: All three groups showed an improvement in activation of the transversus abdominis post-intervention, with the Galileo® demonstrating the largest improvement. There were no relevant ($r > 0.5$) correlations between activation of the transversus abdominis muscle and any clinical tests. (4) Conclusions: The present study provides evidence that sensorimotor training on the Galileo® significantly improves the activation of the transversus abdominis muscle.

Keywords: Galileo®; Posturomed®; chronic low back pain; sensorimotor training; transversus abdominis activation; physiotherapy; rehabilitation

check for updates

Citation: Marchand, F.; Laudner, K.; Delank, K.-S.; Schwesig, R.; Steinmetz, A. Effects of Sensorimotor Training on Transversus Abdominis Activation in Chronic Low Back Pain Patients. *J. Pers. Med.* **2023**, *13*, 817. https://doi.org/10.3390/jpm13050817

Academic Editors: Markus Gesslein and Eduard Kurz

Received: 14 April 2023
Revised: 9 May 2023
Accepted: 10 May 2023
Published: 11 May 2023

1. Introduction

Back pain is one of the largest health problems in Germany and the entire Western world [1,2]. Current studies clearly point to the high risk of chronic back pain, with 65% of first-time patients with non-specific back pain reporting a relapse within the first year [3]. About 10% of back pain patients report significant restrictions with work and quality of life [4]. If therapy is not successful after six weeks, a multidisciplinary assessment should be conducted to reduce the risk of chronicity. This includes the determination of specified therapy techniques such as multimodal pain therapy [5]. As such, sensory-motor training is usually conducted in addition to various physiotherapeutic, psychoeducational and psychotherapeutic measures.

The aim of sensorimotor training is to increase proprioceptive input and the subsequent motor response during dynamic situations. This can be achieved, for example, through exercises performed on unstable surfaces [6]. For patients with chronic non-specific back pain, the aim is to improve the intra- and inter-muscular coordination of the superficial and deep trunk muscles. In addition to the multifidi muscles, the pelvic floor and the diaphragm, and the transversus abdominis play an important role in the dynamic segmental stabilization of the lumbar spine [7–9]. The transversus abdominis exerts traction on the thoracolumbar fascia and thus, via co-contraction with the lumbar multifidi muscles, exerts a stabilizing effect on the spine [8,10]. Of all the abdominal muscles, the transversus abdominis has the highest activity in maintaining intra-abdominal pressure [11].

Sensorimotor training also increases the motor adaptability, segmental stability and overall performance of muscles, which thereby reduces pain frequency and intensity [12]. However, despite extensive research in recent decades, no concrete recommendations for the implementation of sensorimotor training have been derived regarding the exact form, frequency and intensity for optimal training effects [6,13].

The aim of this study was to investigate the effect of sensorimotor training on the Galileo® and on the Posturomed® have on transversus abdominis activation in patients with chronic low back pain during inpatient multimodal pain or complex therapy compared to conventional sensorimotor physiotherapy. Activation of the transversus abdominis was assessed using sonography and clinical tests.

2. Materials and Methods

2.1. Design

The study was conducted as a prospective, randomized and confirmatory intervention. Patients (n = 75) with chronic low back pain were randomly divided into two intervention groups (Galileo® and Posturomed®) and a control group (conventional physiotherapy) with 25 patients per group. The participants were recruited from the patient collective of the conservative orthopedic department of the University Hospital Halle (Saale). All participants were patients with specialist indication for a two-week multimodal therapy for chronic low back pain (according to the National Treatment Guideline [5]). Participants had to be at least 18 years old and suffer from chronic low back pain that had been present for more than 12 weeks. Exclusion criteria included trauma, infection, tumor, spinal surgery history and pregnancy or breastfeeding. During the inpatient stay, the control group received six sessions of sensorimotor training using conventional physiotherapy, while the intervention groups received six units of sensorimotor training on the Galileo® or the Posturomed®. Sonographic examination of the transversus abdominis and functional clinical tests were used to determine segmental stability pre- (T0) and post-intervention (T1).

The study was conducted according to the Declaration of Helsinki in the current version between July 2015 and January 2017. Ethical clearance was granted by the Institutional Ethics Committee (reference number: 2015-84), and written informed consent was obtained from all participants. Data regarding the effect of sensorimotor training on postural control and functional status have been previously published [14].

2.2. Sonography of Transversus Abdominis

To measure activity of the transversus abdominis, an ultrasound probe was positioned in transverse alignment on the patient's right side, above the iliac crest, lateral to the anterior superior iliac spine and in the area of the mid-axillary line [15]. The sonographic images were then captured of the cross-section muscle thickness of the transversus abdominis. This measurement was taken under two conditions:

1. supine position with the abdominal muscles relaxed (50° hip flexion and 90° knee flexion) [16],
2. supine position during targeted contraction of the muscle using the "abdominal draw-in maneuver" (patient holds his/her breath after expiration and gently draws their navel inwards without moving the pelvis) [17].

By measuring muscle thickness in different starting positions, the absolute changes in thickness (cm) and the relative changes in muscle thickness (%) due to tension (abdominal draw-in maneuver) were able to be determined. The draw-in maneuver is considered a measure of the activation of the transversus abdominis. The percentage increase was then calculated using the following formula [16,18–20]:

$$\text{Activation} = \frac{(\text{muscle thickness relaxation} - \text{muscle thickness tension})}{\text{muscle thickness relaxation}}$$

As in comparable studies, functional ultrasound of the transversus abdominis muscle was chosen due to its proven validity in the assessment of segmental stability in patients with chronic low back pain [17,18,21–25]. Sonography seems to have an advantage over surface EMG, which cannot adequately differentiate between the external abdominal oblique, internal abdominal oblique and transversus abdominal muscles [20,23,26].

2.3. Clinical Tests

Clinical tests were used to obtain a clinical picture of the quality of movement control as well as activity and coordination of the deep trunk muscles. Seven clinical tests (Table 1), with good to very good reliability (ICC range: = 0.687–0.895), were completed [24]. The tests were rated as "inconspicuous" or "conspicuous" in terms of performance.

Table 1. Clinical tests with execution and evaluation criteria.

Clinical Test	Implementation	Evaluated as Conspicuous at
Inspiration in supine position	Spontaneous breathing	Thoracic high breathing
Hip flexion in supine position	Lifting both legs	Lumbar lordosis, cranialization upper body
Trunk flexion in supine position	Supine position with feet apart, upper body erect	Heels leave base, scapula does not lift off
Hip abduction in lateral position	Lift contralateral outstretch leg	Leg rotation
Vele test in standing position	Shift weight forward	No gripping function of the toes
Matthias test in standing position	Arm elevation to 90° with internal roation	Scapula winging, Retroflexion upper body
Single leg stand	Flex contralateral hip and knee	Pelvic tilt to the healthy side (Trendelenburg sign)

2.4. Intervention

Sensorimotor training on the Galileo® and the Posturomed® are considered widely used therapies in daily clinical practice.

2.4.1. Galileo® Training

A Galileo® Med M vibration platform (Novotec Medical GmbH, Pforzheim, Germany) was used for training the Galileo® group (n = 25). This therapy platform uses a side-alternating movement pattern such as a seesaw with variable amplitude and frequency. For this training, a tilting movement of the pelvis generates a movement pattern, similar to a human gait, but much more frequent. The body reacts in a compensatory manner with rhythmic muscle contractions, which are reflex controlled from a frequency of approximately 10 Hz. This activates the muscles in the legs, abdomen and back throughout the torso. The frequency of the vibration plate is continuously adjustable between 5 and 30 Hz, as is the amplitude of ±4.5 mm. Low frequencies can be used for mobilization, medium for muscle function and coordination and high to increase muscle performance and endurance [14]. A total of six training sessions were carried out by the Galileo® group under physiotherapeutic supervision by an experienced physiotherapist. The total training time per session was 15 min.

2.4.2. Posturomed® Training

The Bioswing Posturomed® 202 platform (Haider Bioswing GmbH, Pullenreuth, Germany) was used for training the Posturomed® group (n = 25). This is a sensorimotor therapy and training device with a 60 × 60 cm damped oscillating unstable surface. The surface, which is suspended from eight steel cables, enables progressively damped evasive movements. Different oscillation amplitudes (max. surface deflection mediolateral 25 mm (locked)/50 mm (unlocked); max. surface deflection anterior-posterior 25 mm (locked)/50 mm (unlocked)) can be achieved by unlocking additional ropes. Maximum surface deflection in the anterior–posterior directions (40 mm/80 mm) and oscillation frequencies (1.0 to 4.2 Hz) are enabled. This allowed for an ideal severity adjustment of the exercises [14]. A total of six training sessions were carried out by the Posturomed® group under physiotherapeutic supervision by an experienced physiotherapist. The total training time per session was 15 min.

2.5. Statistics

The sample size and number of cases per intervention group were based on comparable studies [16,22]. According to Bortz [25], an effect size of 0.7 ($\alpha = 0.05$; $1 - \beta = 0.8$) requires 25 patients per intervention group. The significance level was set at $p < 0.05$ and $\eta_p^2 \geq 0.10$. All data were analyzed using SPSS 28.0 for Windows (SPSS, Chicago, IL, USA).

Descriptively, means, standard deviations and 95% confidence intervals (95% CI) were calculated. A Chi-square test was used for comparison of gender distribution. Age and anthropometric data were compared using a univariate analysis of variance. For the evaluation of the clinical tests, the multivariate linear model was used, and effect sizes (d) were calculated according to Hartmann et al. [27]. For the evaluation of the sonographic thickness of the transversus abdominis, a post-hoc test was used in addition to the general linear model.

Correlations between the data of the clinical tests and the sonographic data were investigated by calculating the Pearson correlation coefficient.

3. Results

3.1. Participants

A total of 48 (64%) of the participants were female and 27 (36%) were male (Table 2). The Chi-square test showed an unequal gender distribution between the groups, but this was not statistically significant ($p = 0.105$). No significant differences between groups were found for age or the anthropometric parameters of height, body weight and body mass index (Table 2).

Table 2. Description of the investigated sample regarding age, sex and anthropometric parameters. Results reported as mean ± standard deviation (95% CI)

	Galileo®	Posturomed®	Control	Variance Analysis	
				p	η_p^2
Age [years]	58.3 ± 11.6 (53.8–62.9)	62.0 ± 11.8 (57.4–66.5)	59.9 ± 10.7 (55.3–64.4)	0.529	0.018
Height [m]	1.70 ± 0.10 (1.66–1.74)	1.72 ± 0.12 (1.68–1.76)	1.68 ± 0.07 (1.64–1.72)	0.343	0.029
Weight [kg]	79.6 ± 19.2 (72.9–86.4)	87.2 ± 15.9 (80.4–93.9)	81.7 ± 15.6 (75.0–88.5)	0.277	0.035
BMI [kg/m²]	27.6 ± 5.42 (25.5–29.6)	29.5 ± 4.76 (27.5–31.6)	29.1 ± 5.34 (27.0–31.1)	0.385	0.026
Sex, male:female	10:15	12:13	5:20	$p = 0.105$ Chi-Squared: 4.51	

Analysis of the functional status measured with the Oswestry Disability Index (ODI) was described elsewhere. The mean values between groups varied between 17.3 and 22.3 [14].

3.2. Sonography of Transversus Abdominis Muscle

The absolute muscle thicknesses (cm) of the transversus abdominis in the relaxed position changed minimally from T0 to T1 in all groups, while a slight increase in muscle thickness occurred from 0.30 ± 0.08 cm to 0.37 ± 0.11 cm in the Posturomed® group (Table 3). Only muscle activation for a relevant group effect was calculated ($p = 0.003$;

$\eta_p^2 = 0.147$). Relevant time and interaction effects were only detected for muscle contracted ($p < 0.001$; $\eta_p^2 = 0.225$) and muscle relaxed ($p = 0.019$; $\eta_p^2 = 0.105$).

Table 3. Descriptive comparison (mean ± standard deviation), variance analysis and calculation of the effect size η_p^2 between examination 1 and 2 depending on groups. Relevance level: $p < 0.05$ and $\eta_p^2 \geq 0.10$. Relevant differences are marked in bold.

Parameter	Patients with Chronic Low Back Pain (n = 75)						Variance Analysis					
	Examination 1			Examination 2			Group		Time		Group × Time	
	Galileo®	Posturomed®	Control	Galileo®	Posturomed®	Control	p	η_p^2	p	η_p^2	p	η_p^2
Muscle Relaxed [cm]	0.33 ± 0.12	0.30 ± 0.08	0.36 ± 0.09	0.32 ± 0.10	0.37 ± 0.11	0.37 ± 0.09	0.269	0.036	0.059	0.049	**0.019**	**0.105**
Muscle Contracted [cm]	0.49 ± 0.16	0.40 ± 0.13	0.50 ± 0.13	0.56 ± 0.17	0.55 ± 0.15	0.53 ± 0.10	0.288	0.034	**<0.001**	**0.225**	0.028	0.095
Muscle Activation [%]	0.52 ± 0.27	0.32 ± 0.22	0.43 ± 0.25	0.81 ± 0.48	0.53 ± 0.25	0.49 ± 0.29	**0.003**	**0.147**	**<0.001**	**0.265**	0.049	0.080

The absolute muscle thicknesses (cm) of the transversus abdominis in the tensed position showed a slight increase from T0 to T1 in all three groups. However, there were no significant differences between the groups.

At T0, the percentage increases in muscle thickness were 52% in the Galileo® group, 32% in the Posturomed® group and 43% in the control group. T1 showed muscle thickness increases of 81% in the Galileo® group, 53% in the Posturomed® group and 49% in the control group. After comparing the two measurement points, there was a clear improvement from 52% to 81% in the Galileo® group and the Posturomed® group (32% to 53%), while the control group improved from 43% to 49%. There were different time effects in all three groups, with the Galileo® group having a significant group effect compared to the control and Posturomed® groups (Figure 1).

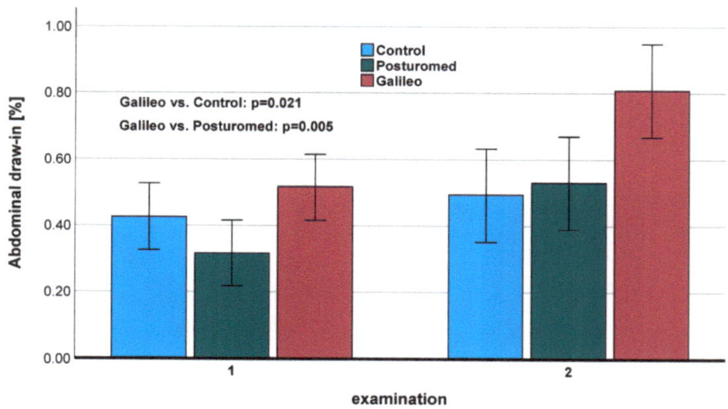

Figure 1. Targeted activation of the transversus abdominis using the abdominal draw-in maneuver. Increase in muscle thickness (%) at examination 1 and 2. The Galileo® group showed a significant advantage compared with the Posturomed® group and the control group.

3.3. Clinical Tests

Large effects were only seen during hip flexion in both the Posturomed® group (d = 1.24) and the control group (d = 1.00; Table 4). Medium effects were found in the Galileo® group during inspiration in the supine position, trunk flexion in the supine position, the Vele test and the Matthiaß test. The Posturomed® group also showed medium effects for trunk flexion in the supine position and the single-leg stand. There were no medium effects in the control group.

Table 4. Descriptive comparison (mean ± standard deviation) and effect size calculation (d) for clinical tests. Examination: exam. Relevant differences (d > 0.5) marked in bold.

Clinical Test	Galileo®			Posturomed®			Control		
	Exam 1	Exam 2	d	Exam 1	Exam 2	d	Exam 1	Exam 2	d
Inspiration in supine position	0.12 ± 0.33	0.00 ± 0.00	**0.73**	0.24 ± 0.44	0.08 ± 0.28	0.34	0.12 ± 0.33	0.06 ± 0.22	0.22
Hip flexion in supine position	0.04 ± 0.20	0.00 ± 0.00	0.40	0.68 ± 0.48	0.12 ± 0.33	**1.24**	0.18 ± 0.38	0.00 ± 0.00	**1.00**
Trunk flexion in supine position	0.24 ± 0.43	0.04 ± 0.02	**0.63**	0.60 ± 0.50	0.24 ± 0.44	**0.77**	0.50 ± 0.50	0.48 ± 0.51	0.04
Hip abduction in lateral position	0.16 ± 0.37	0.04 ± 0.02	0.42	0.20 ± 0.41	0.12 ± 0.33	0.22	0.16 ± 0.35	0.08 ± 0.28	0.25
Vele test in standing position	0.08 ± 0.28	0.00 ± 0.00	**0.57**	0.00 ± 0.00	0.00 ± 0.00	0.00	0.12 ± 0.33	0.04 ± 0.02	0.30
Matthias test in standing position	0.12 ± 0.03	0.00 ± 0.00	**0.73**	0.08 ± 0.28	0.04 ± 0.20	0.16	0.12 ± 0.03	0.04 ± 0.20	0.30
Single leg stand	0.00 ± 0.00	0.00 ± 0.00	0.00	0.08 ± 0.28	0.00 ± 0.00	**0.57**	0.04 ± 0.20	0.00 ± 0.00	0.40

No relevant correlations were found for activation of the transversus abdominis using sonography compared to any of the clinical tests (Table 5). The largest, but still not relevant relationship, was found during inspiration in a supine position (r = −0.276).

Table 5. Pearson correlation coefficients (r) between clinical tests and the parameter "sonography M. transversus abdominis" (n = 75). Relevant correlations (r > 0.5) marked in bold.

Clinical Test	Activation M. Transversus Abdominis
Inspiration in supine position	−0.276
Matthias test in standing position	0.212
Hip flexion in supine position	−0.211
Trunk flexion in supine position	−0.178
Single leg stand	0.155
Hip abduction in lateral position	0.095
Vele test in standing position	−0.060

4. Discussion

The primary aim of this study was to investigate the effects of different forms of sensorimotor training on transversus abdominis activation in patients with chronic low back pain. When comparing the entire group of subjects (n = 75) with each of the three groups (n = 25), there was a homogeneous distribution of age and anthropometric data. Therefore, a good comparison of the data can be assumed. The unequal gender distribution in favor of females reflects the epidemiological situation in Germany where more women suffer from chronic low back pain than men [5]. Furthermore, the gender-specific differences in the different groups were not statistically significant.

Some authors have limited their sonographic examination to the absolute increase in thickness (hypertrophy) of the transversus abdominis achieved during 2–3 months of interventional muscle training [28,29]. These previous results showed that the increase in thickness was associated with a reduction in pain intensity and improvement in functional mobility. More recent work shows that, in addition to the absolute increase in thickness, the change in recruitment (i.e., activation) is crucial in chronic low back pain patient outcomes [9,16,19]. The intervention period in the present study was limited to two weeks and six therapy sessions, so no relevant muscle morphological changes were expected. A significant increase in the cross-sectional area of the entire muscle as well as individual muscle fibers, which results partially from the increase in the size and number of myofibrils, can only be expected after a training period of eight to twelve weeks [30].

The average muscle thickness of the transversus abdominis among all subjects during relaxation was approximately 0.34 cm. Comparably, Rankin et al. [31] described an average muscle cross-section of the transversus abdominis of 0.46 cm. However, this can differ significantly depending on gender, underlying diseases and age. The lower average thickness in the test group of the current study compared to the normal population is not surprising. All subjects included were patients with chronic low back pain. It has been well established that hypotrophy of the deep trunk muscles and especially of the transversus abdominis can be both a cause and a consequence of chronic low back pain [8,11,16,32,33]. In addition, the high proportion of female subjects and the average age of over 60 years within the subject population may possibly explain this deviation.

The mean values of the muscle thickness during tension showed a slight increase in all three groups T0 to T1. The positive time effects in the two intervention groups (Galileo® and Posturomed®) were slightly higher than the control group but with no significant differences between groups. This indicates that both sensorimotor training on the Galileo® and Posturomed® and sensorimotor training through conventional physiotherapy can lead to improved contraction of the transversus abdominis. These results support previous research, which compared specialized trunk-stabilizing exercises with conventional physiotherapy in patients with chronic low back pain and found that muscle thickness of the transversus abdominis also served as a parameter for segmental trunk stability. Again, positive effects were seen in both groups with slight advantages among the intervention group, but with no significant group or interaction effects [26].

The results of this study indicate that even two weeks of sensorimotor training leads to an improvement in the targeted activation of the transversus abdominis in patients with chronic back pain. Sensorimotor training on the Galileo® led to a significantly greater improvement in targeted muscle activation compared to sensorimotor training using the control and Posturomed® group.

Cruz-Diaz et al. [34] used a similar study design to investigate the effect of Pilates training on transversus abdominis activation in patients with chronic low back pain and reported similar results. For this study, two different forms of Pilates training (Pilates on a mat and equipment-assisted Pilates) were compared with a control group. While no significant group and interaction effects were found for muscle thickness at rest, significant improvements were noted for the percentage increase in muscle thickness through targeted contraction (activation) in the intervention groups. In contrast to the present study, the intervention period was twelve weeks in total and the data were collected at three time periods. However, it is worth noting that the majority of the positive effects on the targeted activation of the transversus abdominis occurred between T0 and T1. This suggests that the improvement in activation of the transversus abdominis muscle is possible after a short intervention period and can occur independently of muscle morphological adaptations. Sensorimotor training primarily addresses the neuromuscular system, which controls inter- and intramuscular coordination. The improvement of muscle activation can be achieved through more effective recruitment of existing motor units and an increased frequency of nerve impulses. These neurological adaptations seem to have great potential already in the initial phases of training sessions [30]. This could explain the improved activation of the transversus abdominis despite the short intervention period.

Dong et al. [35] compared the muscle activity of different trunk muscles during the performance of motor exercises with and without the influence of whole-body vibration. These authors found that at a vibration frequency of 15 Hz, muscle activity increased by 190–247% of the multifidi muscles, erector spinae, external obliques, and rectus abdominis, but not on the transversus abdominis. However, a similar effect can also be assumed in the area of the deep abdominal muscles. This may partially explain the higher training effectiveness on the Galileo®.

Wang et al. [36] provided a review on the effect of whole-body vibration training in chronic back pain patients. Pain intensity and functional abilities were evaluated, but activity of the transversus abdominis was not included as a measure of lumbar segmental stability. Whole-body vibration therapies showed some significant improvements when compared to classical physiotherapy and conventional stability exercises, similar to the present study.

Another aim of the present study was to determine the effect of different forms of sensorimotor training on the performance of clinical tests in patients with chronic low back pain. All three groups showed a slight improvement in the performance of the clinical tests, which partially led to positive time effects, although significant interaction effects were not demonstrated. Since these seven clinical tests have not been previously validated as a measure for assessing lumbar segmental stability, the correlation coefficients between the functional clinical diagnostics and sonography were determined (r = −0.276–0.212). Unfortunately, no correlations were found in any of the clinical tests, so the interpretation of functional clinical diagnostics with regard to segmental stability in patients with chronic low back pain remains limited. Despite ongoing research, no reliable clinical tests for assessing segmental stability except the abdominal-draw-in maneuver while using a pressure pad have been established. When visually assessing movement quality and control during clinical examinations, there is already a problem of low inter- and intra-rater reliability, even among experienced healthcare professionals [37,38]. Therefore, more objective assessments of movement characteristics, such as sensor-based kinematic data, are needed. The potential here lies primarily in being able to assign underlying pathologies of the lumbar spine (e.g., segmental instability) to specific, abnormal movement patterns. This can help to identify the cause and effectiveness of treatments [39–41].

Limitations

There are some limitations of this study that need to be acknowledged. Blinding of subjects could not be implemented, as each subject's group allocation was obvious based on the subject's physiotherapist conducting the training sessions through the respective intervention.

All subjects were recruited from the patient collective of a two-week inpatient multimodal pain therapy. In addition to the respective interventions, the test persons also received pain medicine treatments, exercise therapy and psychotherapeutic measures. Therefore, co-effects in the results cannot be excluded with certainty. Furthermore, it cannot be excluded that concomitant diseases may also have had an influence on the outcome of this study.

All subjects (n = 75) were randomly allocated to the three groups (n = 25 per group), each with different forms of sensorimotor training. Two groups (Galileo® and Posturomed®) were defined as intervention groups. Since the control group also received standardized sensorimotor training under physiotherapeutic guidance, no group was disadvantaged with lesser therapeutic measures. However, this led to a lack of a true intervention-free control group. An intervention-free control group might have demonstrated clearer effects of sensorimotor training.

The inpatient setting in which the study was conducted limited the intervention period to two weeks. Furthermore, this study did not investigate whether the positive effects of the interventions are temporary or lasting. This would have required a follow-up study and potential for subject attrition.

Finally, lumbar stability was not assessed directly but via measurement of transversus abdominis activation in patients with chronic low back pain.

5. Conclusions

The results of this study demonstrate that the use of sensorimotor training in different forms can be effective in the rehabilitation of patients with chronic low back pain. More specifically, sensorimotor training on the Galileo® may lead to greater improvement in activation of the transversus abdominis compared to sensorimotor training on the Posturomed® or sensorimotor training through conventional physiotherapy (control group). This could justify a permanent establishment of sensorimotor training on the Galileo® when using multimodal pain therapy. However, the use of supportive training devices should not replace the presence of medical professionals. As described in the present study, detailed instruction of the exercises by an experienced physiotherapist was necessary. The validity of using functional clinical diagnostics to determine the effectiveness of sensorimotor training on segmental stability could not be refuted in the present study. However, clinical diagnostics are and remain an important tool for clinicians. Future research should investigate how functional clinical diagnostics could be supplemented by more objective, sensory measurement methods of movement control in the evaluation of therapeutic measures for chronic low back pain patients.

Author Contributions: Conceptualization, F.M., K.-S.D. and A.S.; Data curation, F.M., K.L. and R.S.; Formal analysis, F.M., K.L. and R.S.; Investigation, F.M.; Methodology, F.M., K.-S.D., R.S. and A.S.; Project administration, A.S.; Resources, K.-S.D. and R.S.; Supervision, A.S.; Validation, K.L. and R.S.; Visualization, F.M., K.L. and R.S.; Writing—original draft, F.M. and A.S.; Writing—review and editing, F.M., K.L., K.-S.D., R.S. and A.S. All authors have read and agreed to the published version of the manuscript.

Funding: This research received no external funding.

Institutional Review Board Statement: The study was conducted in accordance with the Declaration of Helsinki and approved by the Ethics Committee of the University Hospital Halle (Saale) (reference number: 2015-84) (9 September 2015).

Informed Consent Statement: Informed consent was obtained from all participants involved in the study.

Data Availability Statement: Not applicable.

Acknowledgments: The authors would like to thank all the staff involved in patient care at the conservative orthopedic Section 2 of the University Hospital Halle (Saale), as well as in particular the physiotherapists involved for their excellent support of this study.

Conflicts of Interest: The authors declare no conflict of interest.

References

1. Schmidt, C.; Kohlmann, T. Was Wissen Wir Über Das Symptom Rückenschmerz? *Z. Orthop.* **2005**, *143*, 292–298. [CrossRef]
2. Hartvigsen, J.; Hancock, M.; Kongsted, A.; Louw, Q.; Ferreira, M.; Genevay, S.; Hoy, D.; Karppinen, J. What Low Back Pain Is and Why We Need to Pay Attention. *Lancet* **2018**, *391*, 2356–2367. [CrossRef]
3. Itz, C.J.; Geurts, J.W.; Nelemans, M.V.K.P. Clinical Course of Non-Specific Low Back Pain: A Systematic Review of Prospective Cohort Studies Set in Primary Care. *Eur. J. Pain* **2013**, *17*, 5–15. [CrossRef] [PubMed]
4. Schmidt, C.; Raspe, H.; Pfingsten, M.; Hasenbring, M.; Basler, H.D.; Eich, W.; Kohlmann, T. Back Pain in the German Adult Population: Prevalence, Severity, and Sociodemographic Correlates in a Multiregional Survey. *Spine* **2007**, *32*, 2005–2011. [CrossRef]
5. Bundesärztekammer (BÄK), Kassenärztliche Bundesvereinigung (KBV), Arbeitsgemeinschaft Der Wissenschaftlichen Medizinischen Fachgesellschaften (AWMF). Nationale VersorgungsLeitlinie Nicht-Spezifischer Kreuzschmerz—Langfassung, 2. Auflage. Version 1. Available online: www.kreuzschmerz.versorgungsleitlinie.de (accessed on 27 March 2023).
6. McCaskey, M.; Schuster-amft, C.; Wirth, B.; Bruin, E.D. De Effects of Postural Specific Sensorimotor Training in Patients with Chronic Low Back Pain: Study Protocol for Randomised Controlled Trial. *Trials* **2015**, *16*, 571. [CrossRef]
7. Hodges, P.; Richardson, C. Inefficient Muscular Stabilization of the Lumbar Spine Associated with Low Back Pain: A Motor Control Evaluation of Transversus Abdominis. *Spine* **1996**, *21*, 2640–2650. [CrossRef] [PubMed]
8. Chen, Y.; Chai, H.; Shau, Y.; Wang, C.; Wang, S. Increased Sliding of Transverse Abdominis during Contraction after Myofascial Release in Patients with Chronic Low Back Pain. *Man. Ther.* **2016**, *23*, 69–75. [CrossRef]
9. Kim, J.; Seok, C.; Jeon, H. Abdominal Draw-in Maneuver Combined with Simulated Weight Bearing Increases Transversus Abdominis and Internal Oblique Thickness. *Physiother. Theory Pract.* **2017**, *33*, 954–958. [CrossRef] [PubMed]
10. Tesh, K.; Shaw, J.; Evans, J. The Abdominal Muscles and Vertebral Stability. *Spine* **1987**, *12*, 501–508. [CrossRef] [PubMed]
11. Cresswell, A.G.; Thorstensson, A. Changes in Intra-Abdominal Pressure, Trunk Muscle Activation and Force during Isokinetic Lifting and Lowering. *Eur. J. Appl. Physiol.* **1994**, *68*, 315–321. [CrossRef]
12. Hwang, J.; Bae, S.; Kim, G.; Kim, K. The Effects of Sensorimotor Training on Anticipatory Postural Adjustment of the Trunk in Chronic Low Back Pain Patients. *J. Phys. Ther. Sci.* **2013**, *25*, 1189–1192. [CrossRef]
13. McCaskey, M.; Schuster-amft, C.; Wirth, B.; Suica, Z.; Bruin, E. Effects of Proprioceptive Exercises on Pain and Function in Chronic Neck- and Low Back Pain Rehabilitation: A Systematic Literature Review. *BMC Musculoskelet. Disord.* **2014**, *15*, 382. [CrossRef]
14. Rüger, A.; Laudner, K.; Delank, K.-S.; Schwesig, R.; Steinmetz, A. Effects of Different Forms of Sensorimotor Training on Postural Control and Functional Status in Patients with Chronic Low Back Pain. *J. Pers. Med.* **2023**, *13*, 634. [CrossRef]

15. McMeeken, J.; Beith, I.; Newham, D.; Milligan, P.; Critchley, D. The Relationship between EMG and Change in Thickness of Transversus Abdominis. *Clin. Biomech.* **2004**, *19*, 337–342. [CrossRef] [PubMed]
16. Ferreira, P.H.; Ferreira, M.L.; Maher, C.G.; Refshauge, K.; Herbert, R.D.; Hodges, P.W.; Ferreira, P.H.; Ferreira, M.L.; Maher, C.G.; Refshauge, K.; et al. Changes in Recruitment of Transversus Abdominis Correlate with Disability in People with Chronic Low Back Pain. *Br. J. Sports Med.* **2009**, *44*, 1166–1172. [CrossRef] [PubMed]
17. Richardson, C.; Snijders, C.; Hides, J. The Relation Between the Transversus Abdominis Muscles, Sacroiliac Joint Mechanics, and Low Back Pain. *Spine* **2002**, *27*, 399–405. [CrossRef]
18. Koppenhaver, S.; Parent, E.; Teyhen, D. The Effect of Averaging Multiple Trials on Measurement Error During Ultrasound Imaging of Transversus Abdominis and Lumbar Multifidus Muscles in Individuals With Low Back Pain. *J. Orthop. Sport. Phys. Ther.* **2009**, *39*, 604–611. [CrossRef] [PubMed]
19. Mens, J.; Pool-Goudzwaard, A. The Transverse Abdominal Muscle Is Excessively Active during Active Straight Leg Raising in Pregnancy-Related Posterior Pelvic Girdle Pain: An Observational Study. *BMC Musculoskelet. Disord.* **2017**, *18*, 372. [CrossRef]
20. Shabnam, S.; Massoud, A.; Ebrahimi, E. Ultrasound Measurement of Abdominal Muscles during Clinical Isometric Endurance Tests in Women with and without Low Back Pain. *Physiother. Theory Pract.* **2019**, *35*, 130–138. [CrossRef]
21. Ghamkhar, L.; Emami, M.; Mohammad, A.; Behtash, H. Application of Rehabilitative Ultrasound in the Assessment of Low Back Pain: A Literature Review. *J. Bodyw. Mov. Ther.* **2011**, *15*, 465–477. [CrossRef]
22. Park, S.; Yu, S. The Effects of Abdominal Draw-in Maneuver and Core Exercise on Abdominal Muscle Thickness and Oswestry Disability Index in Subjects with Chronic Low Back Pain. *J. Exerc. Rehabil.* **2013**, *9*, 286–291. [CrossRef]
23. Shamsi, M.; Sarrafzadeh, J.; Jamshhidi, A. The Effect of Core Stability and General Exercise on Abdominal Muscle Thickness in Non-Specific Chronic Low Back Pain Using Ultrasound Imaging. *Physiother. Theory Pract.* **2016**, *32*, 277–283. [CrossRef] [PubMed]
24. Niemier, K.; Schmidt, S.; Engel, K.; Steinmetz, A. Functional diagnostics of mobility control, mobility stabilization and hypermobility. Reliability of clinical tests—Results of a multicenter study. *Orthopade* **2009**, *38*, 847–854. [CrossRef] [PubMed]
25. Bortz, J. *Lehrbuch Der Empirischen Forschung*; Springer: Berlin/Heidelberg, Germany, 1984; pp. 277–283.
26. Nabavi, N.; Bandpei, A. The Effect of 2 Different Exercise Programs on Pain Intensity and Muscle Dimensions in Patients With Chronic Low Back Pain: A Randomized Controlled Trial. *J. Manipulative Physiol. Ther.* **2018**, *41*, 102–110. [CrossRef] [PubMed]
27. Hartmann, A.; Herzog, T.; Drinkmann, A. Psychotherapy of Bulimia Nervosa: What Is Effective? A Meta-Analysis. *J. Psychosom. Res.* **1992**, *36*, 159–167. [CrossRef] [PubMed]
28. Mannion, A.; Pulkovski, N.; Loupas, T.; Schenk, P.; Gerber, H.; Sprott, H. Muscle Thickness Changes during Abdominal Hollowing: An Assessment of between-Day Measurement Error in Controls and Patients with Chronic Low Back Pain. *Eur. Spine J.* **2008**, *17*, 494–501. [CrossRef] [PubMed]
29. Akbari, A.; Khorashadizadeh, S.; Abdi, G. The Effect of Motor Control Exercise versus General Exercise on Lumbar Local Stabilizing Muscles Thickness: Randomized Controlled Trial of Patients with Chronic Low Back Pain. *J. Back Musculoskelet. Rehabil.* **2008**, *21*, 105–112. [CrossRef]
30. Folland, J.; Williams, A. The Adaptations to Strength Training. Morphological and Neurological Contributions to Increased Strength. *Sport. Med.* **2007**, *37*, 145–168. [CrossRef]
31. Rankin, G.; Stokes, M.; Newham, D. Abdominal Muscle Size and Symmetry in Normal Subjects. *Muscle Nerve.* **2006**, *34*, 320–326. [CrossRef]
32. Hodges, P.; Richardson, C. Delayed Postural Contraction of Transversus Abdominis in Low Back Pain Assocated with Movement of the Lower Limb. *J. Spinal. Disord.* **1998**, *11*, 46–56. [CrossRef]
33. Hodges, P.; Moseley, G. Pain and Motor Control of the Lumbopelvic Region: Effect and Possible Mechanisms. *J. Electrmyogaphy Kinisiology* **2003**, *13*, 361–370. [CrossRef] [PubMed]
34. Cruz-Diaz, D.; Bergamin, M.; Gobbo, S.; Hita-Contreras, F. Comparative Effects of 12 Weeks of Equipment Based and Mat Pilates in Patients with Chronic Low Back Pain on Pain, Function and Transversus Abdominis Activation. A Randomized Controlled Trial. *Complement. Ther. Med.* **2017**, *33*, 72–77. [CrossRef] [PubMed]
35. Dong, Y.; Wang, H.; Zhu, Y.; Chen, B.; Zheng, Y.; Liu, X.; Qiao, J.; Wang, X. Effects of Whole Body Vibration Exercise on Lumbar-Abdominal Muscles Activation for Patients with Chronic Low Back Pain. *BMC Sports Sci. Med. Rehabil.* **2020**, *12*, 78. [CrossRef]
36. Wang, W.; Wang, S.; Lin, W.; Li, X.; Andersen, L.; Wang, Y. Efficacy of Whole Body Vibration Therapy on Pain and Functional Ability in People with Non-Specific Low Back Pain: A Systematic Review. *BMC Complement. Med. Ther.* **2020**, *20*, 158. [CrossRef] [PubMed]
37. Biely, S.; Silfies, S.; Smith, S.; Hicks, G. Clinical Observation of Standing Trunk Movements: What Do the Aberrant Movement Patterns Tell Us? *J. Orthop. Sport. Phys. Ther.* **2014**, *44*, 262–272. [CrossRef] [PubMed]
38. Beange, K.; Chan, A.; Beaudette, S.; Ryan, B. Concurrent Validity of a Wearable Imu for Objective Assessments of Functionalv-movement Quality and Control of the Lumbar Spine. *J. Biomech.* **2019**, *97*, 109356. [CrossRef]
39. Spinelli, B.; Wattananon, P.; Sil, S.; Talaty, M.; Ebaugh, D. Using Kinematics and a Dynamical Systems Approach to Enhance Understanding of Clinically Observed Aberrant Movement Patterns. *Man. Ther.* **2015**, *20*, 221–226. [CrossRef

40. Bauer, C.; Rast, F.; Ernst, M.; Kool, J.; Oetiker, S.; Rissanen, S.; Suni, J.; Kankaanpää, M. Concurrent Validity and Reliability of a Novel Wireless Inertial Measurement System to Assess Trunk Movement. *J. Electromyogr. Kinesiol.* **2015**, *25*, 782–790. [CrossRef]
41. Laird, R.; Kent, P.; Keating, J. How Consistent Are Lordosis, Range of Movement and Lumbo-Pelvic Rhythm in People with and without Back Pain? *BMC Musculoskelet. Disord.* **2016**, *17*, 403. [CrossRef]

Journal of
Personalized Medicine

MDPI

Article

Effects of Different Forms of Sensorimotor Training on Postural Control and Functional Status in Patients with Chronic Low Back Pain

Alex Rüger [1,2], Kevin Laudner [3], Karl-Stefan Delank [1], René Schwesig [1] and Anke Steinmetz [4,*]

1 Department of Orthopedic and Trauma Surgery, Martin-Luther-University Halle-Wittenberg, 06120 Halle, Germany; alex.rueger@uk-halle.de (A.R.); stefan.delank@uk-halle.de (K.-S.D.)
2 Department of Ophthalmology, Martin-Luther-University Halle-Wittenberg, 06120 Halle, Germany
3 Department of Health Sciences, University of Colorado, Colorado Springs, CO 80918, USA; klaudner@uccs.edu
4 Department of Trauma, Reconstructive Surgery and Rehabilitation Medicine, University Medicine Greifswald, Physical and Rehabilitation Medicine, 17475 Greifswald, Germany
* Correspondence: anke.steinmetz@med.uni-greifswald.de

Abstract: The aim of this study was to compare three sensorimotor training forms in patients with chronic low back pain to determine their effects on the reduction of pain-related impairment and changes in posturography. Over two weeks, during the multimodal pain therapy (MMPT) period, six sessions of sensorimotor physiotherapy or training in the Galileo® or Posturomed® ($n = 25$ per group) were performed. A significant reduction in pain-related impairment after the intervention phase was shown across all groups (time effect: $p < 0.001$; $\eta_p^2 = 0.415$). There was no change in postural stability (time effect: $p = 0.666$; $\eta_p^2 = 0.003$), but there was a significant improvement in the peripheral vestibular system (time effect: $p = 0.014$; $\eta_p^2 = 0.081$). An interaction effect was calculated for the forefoot-hindfoot ratio ($p = 0.014$; $\eta_p^2 = 0.111$). Only the Posturomed® group showed an improvement in anterior-posterior weight distribution (heel load: 47% vs. 49%). These findings suggest that these forms of sensorimotor training in the context of MMPT are suitable for reducing pain-related impairment. Posturography demonstrated stimulation of a subsystem, but no improvement in postural stability.

Keywords: Galileo®; Posturomed®; chronic; low back pain; sensory; motor; physiotherapy; rehabilitation

check for updates

Citation: Rüger, A.; Laudner, K.; Delank, K.-S.; Schwesig, R.; Steinmetz, A. Effects of Different Forms of Sensorimotor Training on Postural Control and Functional Status in Patients with Chronic Low Back Pain. *J. Pers. Med.* **2023**, *13*, 634. https://doi.org/10.3390/jpm13040634

Academic Editors: Markus Gesslein and Eduard Kurz

Received: 13 March 2023
Revised: 28 March 2023
Accepted: 3 April 2023
Published: 5 April 2023

1. Introduction

According to the national guideline for non-specific back pain, physical activation of the patient is important during rehabilitation, especially with the aim of maintaining activities of daily living. The inclusion of multimodal treatment concepts in the therapy of non-specific back pain should be initiated at an early stage, especially if there is a risk of chronicity or if chronicity has already occurred [1]. Interdisciplinary multimodal assessment alone seems to improve long-term outcomes for patients with chronic low back pain [2]. Sensorimotor training is often part of this multimodal approach and covers a wide range of interactions between afferent and efferent signals [3].

The aim of sensorimotor training in patients with chronic low back pain is to improve postural stabilization and function of the deep muscles (e.g., transversus abdominis muscles and multifidus muscles) and to increase intermuscular coordination with the superficial muscles (e.g., oblique abdominis muscles or latissimus dorsi muscle) [4]. In the context of three-dimensional training, Müller et al. [5] used electromyography and posturography during coordinative training in patients with non-specific back pain and showed this group had a significant improvement in intermuscular coordination of the trunk muscles and postural stability compared to patients performing passive training. As a result, a significant reduction in back pain was also observed.

Essential for muscular stabilization of the lumbar spine and for postural control is the so-called deep stabilization system, consisting of the deep musculi (Mm.) multifidi (medial tract of the autochthonous back muscles), the musculus (M.) transversus abdominis as well as the diaphragm and pelvic floor [6]. The lumbar multifidi muscles and the deep abdominal muscles, especially the transversus abdominis, are critical for the dynamic control of the individual vertebral segments and spinal stability [7–10]. As such, neuromuscular control and function of the transversus abdominis and deep multifidi muscles are often impaired in patients with back pain [11,12] and lumbar instability [13,14].

Various training programs targeting the deep stabilization system and thus an effector of postural control have shown good therapeutic results in patients with chronic back pain [15,16]. This is especially true for patients with back pain due to lumbar instability [17]. Tsao and Hodges [18] reported that patients with chronic low back pain who performed a single session of isolated sensorimotor training of the transversus abdominis muscle modified their electromyographic activity closer to asymptomatic patients and could cause direct improvements in the feedforward activation of this muscle even in untrained individuals and during more complex movements. They suggested that the earlier feedforward activation of the transversus abdominis in the isolated training group was potentially caused by a change in the central nervous system's mode of preparation for posture and movement. Overall, patients with chronic low back pain have been shown to have reduced postural control compared to healthy individuals. This is especially true for more demanding balance tests [19,20]. It is assumed that critical joint positions are more likely to be adopted, which could subsequently cause injuries [3]. Researchers have concluded that sensorimotor training methods with the aim of improving postural control should be part of rehabilitative programs for chronic back pain [20] and a variety of sensorimotor training methods can be considered. These methods may include stabilizing and coordinating exercises within the sensorimotor system, as well as other therapeutic instruments such as Posturomed® and Galileo®.

Posturomed® is used to keep the freely swinging surface as still as possible during the exercises and to make as few compensating movements as possible [21]. The physical principle of action is that the neuromuscular system must adapt to the resulting oscillation frequencies (i.e., intercept its own movement impulses (feedforward mechanism)). The aim is to increase the performance of the postural system and consequently improve posture and balance regulation. During training with the Galileo® platform, the vibrating platform (10 Hz) causes reflex-controlled muscle contractions via the stretch reflex of the muscle spindles and their afferent receptors. These reflex contractions then help in postural stabilization. The Meissner corpuscles are responsible for the perception and processing of low-frequency vibrations, and the Vater-Pacini corpuscles are responsible for high-frequency vibrations. At the same time, vibrational stimuli also have inhibitory effects on the neuromuscular system [22]. A significant improvement in proprioceptive properties in the lumbar region [23] as well as a reduction in pain intensity and an increase in postural stability in patients with chronic back pain [24–26] have been demonstrated.

The aim of this study was to compare the effects of sensorimotor training on postural control and pain-related impairment in the context of inpatient multimodal pain therapy (MMPT) among patients with chronic non-specific low back pain, using training on the Galileo® or the Posturomed® in comparison with conventional physiotherapy (PT) focusing on sensorimotor training.

2. Materials and Methods

2.1. Participants

Notably, 75 participants were included in this prospective, randomized, confirmatory intervention study and randomly assigned to 1 of the 3 training groups. During a two-week inpatient MMPT stay, the participants were each given six sessions of sensorimotor training in Galileo®, Posturomed®, or traditional sensorimotor physiotherapy training. The participants were examined pre-intervention and postintervention using the Oswestry

Low Back Pain Questionnaire and posturography (Figure S1). The initial orthopedic examinations took place between July 2015 and January 2017 in an orthopedic ward of the university hospital in Halle (Saale). To be included, participants had to be 18 years old or older and diagnosed with chronic low back pain present for greater than 12 weeks. Exclusion criteria included a history of trauma, infection, tumor, or operation of the spine as well as pregnancy or breastfeeding. Current imaging was available for all patients, so low back pain requiring surgical intervention, e.g., radiculopathies or relevant instabilities, was also excluded. The MMPT included an individual combination of physiotherapy, manual medicine, interventional pain therapy, medical training therapy, relaxation techniques, oral pain medication, psychoeducation, and psychotherapy with at least 30 therapy units. The study was conducted according to the Declaration of Helsinki. Ethical clearance was granted by the Institutional Ethics Committee (reference number: 2015-84), and written informed consent was obtained from all participants prior to any data collection.

2.2. Oswestry Low Back Pain Questionnaire

The Oswestry *Low Back Pain Questionnaire*, also in the 2006 German translation, is an established instrument to assess the pain and functional status of patients with chronic low back pain [27,28]. Participants' pain-related limitations in various everyday activities, such as personal hygiene, sitting, sleeping, or traveling, were surveyed using the Oswestry Disability Index (ODI). This index indicates that patients with mild limitations in activities of daily living have a degree of limitation from 0 to 20%, moderate from 21 to 40%, severe from 41 to 60%, disabling from 61 to 80%, and bedridden from 81 to 100%.

2.3. Posturography

Posturography is an established method for identifying training effects in the area of balance regulation in various settings, including rehabilitation [29]. The Interactive Balance System (IBS) (neurodata GmbH, Wien, Austria) was used to measure postural stability and regulation. This assessment differentiated vertical forces in the forefoot and hindfoot area on two force measurement platforms (each with a forefoot and hindfoot plate) at an anteriorly open angle of 30° using strain gauges. As a stimulus response, body sway was determined by the force measurement platforms using different parameters (Table S1) [30,31]. Compared to an asymptomatic population, back pain patients conspicuously present with higher values in the stability indicator (ST), which suggests greater instability and a comparatively insufficient peripheral vestibular system (frequency band F2–4) [32].

2.4. Interventions

2.4.1. Sensorimotor Physiotherapy Training

For PT, a total of six training sessions were completed in the PT group (n = 25). This group performed all six sessions under physiotherapeutic supervision by an experienced physiotherapist involved in each case. The total training time per session was 15 min and included a series of exercises using both stable and unstable surfaces (Table S2).

2.4.2. Gallileo® Training

A Galileo® Med M vibration platform (Novotec Medical GmbH, Pforzheim, Germany) (Figure S2) was used for training the Galileo® group (n = 25). This therapy platform uses a side-alternating movement pattern such as a seesaw with variable amplitude and frequency. For this training, a tilting movement of the pelvis generates a movement pattern, similar to a human gait, but much more frequent. The body reacts in a compensatory manner with rhythmic muscle contractions, which are reflex controlled from a frequency of approximately 10 Hz. This activates the muscles in the legs, abdomen, and back throughout the torso. The frequency of the vibration plate is continuously adjustable between 5 and 30 Hz, as is the amplitude of ±4.5 mm. Low frequencies can be used for mobilization, medium for muscle function and coordination, and high to increase muscle performance and endurance.

A total of six training sessions were carried out by the Galileo® group under physiotherapeutic supervision by one of two experienced physiotherapists. The total training time per session was 15 min and included a warm-up/acclimatization task, followed by 6 exercises (Table S3).

2.4.3. Posturomed® Training

The Bioswing Posturomed® 202 platform (Haider Bioswing GmbH, Pullenreuth, Germany) (Figure S3) was used for training the Posturomed® group (n = 25). This is a sensorimotor therapy and training device with a 60 × 60 cm damped oscillating unstable surface. The surface, which is suspended from eight steel cables, enables progressively damped (i.e., as the deflection of the therapy surface increases, the damping/resistance to deflection increases) evasive movements. Different oscillation amplitudes (max. surface deflection mediolateral 25 mm (locked)/50 mm (unlocked); max. surface deflection anterior-posterior 25 mm (locked)/50 mm (unlocked)) can be achieved by unlocking additional ropes. Maximum surface deflection in the anterior-posterior directions (40 mm/80 mm) and oscillation frequencies (1.0 to 4.2 Hz) are enabled. This allowed for an ideal severity adjustment of the exercises.

The exercises performed by the Posturomed® group consisted of six total training sessions, which were performed under physiotherapeutic guidance. The total time per training session was 15 min and consisted of a consecutive sequence of nine sensorimotor exercises on the Posturomed® platform (Table S4). Two 30 s exercise periods and 30 s rest periods were planned for each level.

If a performance error occurred in any of the groups during a specific exercise, then an additional repetition(s) was performed until the exercise was performed safely and correctly. The progress in the exercises was decided individually based on the physiotherapists' assessment of the participant's abilities and correct execution of the elements.

2.5. Statistics

Data were analyzed using the statistical program SPSS 28.0 for Windows (SPSS, Chicago, IL, USA). Analyses of variance were performed to compare mean values or detect any relationships.

For the sample size calculation (power analysis), this study was based on the results of a comparable study design [24]. In this previous study, 2 groups with 25 participants per group suffering from chronic, non-specific back pain were compared with regard to vibration training. An effect size for the ODI of d = 0.72 was obtained in this work. The goal was to achieve a similar medium-to-high effect for each group and variable. In this regard, according to Bortz [33], with an effect size of 0.7 ($\alpha = 0.05$; $1 - \beta = 0.8$), 25 participants per group were required to obtain a relevant change.

The critical level of significance was adjusted using the Bonferroni correction. For the Bonferroni correction, a significance level (p) of 0.05 was divided by the number of total tests [10]. As such, differences between means were considered significant, if p values were <0.005 or partial eta squared (η_p^2) values were greater than 0.15 [34].

3. Results

3.1. Participants

A total of 75 participants (women: n = 48 (64%)) equally divided into 3 groups (n = 25) were included in this study. Anthropometric parameters such as age, height, weight, and body mass index (BMI) did not show any significant differences between groups (Table 1).

Table 1. Demographic and anthropometric data of the sample (n = 75). Values are given as mean ± standard deviation.

Group	Body Height [m]	Age [Years]	Body Weight [kg]	BMI [kg/m]^2
PT	1.68 ± 0.07	59.9 ± 10.7	81.7 ± 15.6	29.1 ± 5.34
Galileo®	1.70 ± 0.10	58.3 ± 11.6	79.6 ± 19.2	27.6 ± 5.42
Posturomed®	1.72 ± 0.12	62.0 ± 11.8	87.2 ± 15.9	29.5 ± 4.76
p	0.343	0.529	0.277	0.385
η_p^2	0.029	0.018	0.035	0.026

PT: conventional physiotherapy; p: significance level.

3.2. Oswestry Disability Index (ODI)

The evaluation of ODI showed a significant reduction in pain-related limitations for all groups following the two-week inpatient stay. Thus, the mean total ODI scores of the PT group decreased from 22.3 ± 2.32 to 17.1 ± 8.57 (d = 0.95). In the Galileo® group, these scores decreased from 17.3 ± 7.10 to 11.7 ± 7.72 (d = 0.76) and in the Posturomed® group, from 21.8 ± 5.81 to 15.1 ± 8.27 (d = 0.95) (Figure 1). In the ODI, this translates to a mean reduction in limitations from 45% to 34% for the PT group, from 35% to 23% for the Galileo® group, and from 44% to 30% in the Posturomed® group. From this, an improvement in the level from severe to moderate pain-related limitation can be shown for the PT and Posturomed® groups. The Galileo® group moves from the upper to the lower limit of moderate limitation. Large effect sizes for the reduction of pain-related limitations were found for all three groups, with those of the PT and Posturomed® groups exceeding those of the Galileo® group. Analysis of variance showed a strong time effect ($p < 0.001$; $\eta_p^2 = 0.470$) for the reduction of pain-related limitations from pre- to postintervention. There were no cross-sectional baseline differences between the groups ($p = 0.086$; $\eta_p^2 = 0.094$). However, there was also no significance demonstrated for the interaction effect (i.e., a difference in time trend between groups (time × group)) ($p = 0.763$; $\eta_p^2 = 0.011$) (Table 2). However, for the individual pre-/postintervention developments, the largest effect sizes were found in the range of a medium or large effect according to Cohen's d (medium effect d = 0.5–0.8, large effect d > 0.8), especially in areas that the participants could easily experience during the inpatient stay. Thus, medium to large effect sizes (d = 0.65–1.61; $p < 0.001$; $\eta_p^2 = 0.415$) were observed for pain intensity across all groups over time and likewise for impairment during standing (d = 0.64–0.75; $p < 0.001$; $\eta_p^2 = 0.338$). However, again, no significant differences were observed in the cross-section of each group or between groups over time for pain intensity ($p = 0.937$; $\eta_p^2 = 0.002$ (group); $p = 0.053$; $\eta_p^2 = 0.078$ (time × group)) or impairment on standing ($p = 0.287$; $\eta_p^2 = 0.034$ (group); $p = 1.000$; $\eta_p^2 = 0.000$ (time × group); Table 2).

3.3. Posturography

For posturography, no significant improvements in postural control were calculated in any of the three groups over the measurement period. The stability indicator as a parameter of postural stability (the larger the value, the higher the instability) showed no significant change in the PT group from 31.2 to 29.6 (d = 0.11), in the Galileo® group from 30.2 to 30.1 (d = −0.01) and in the Posturomed® group from 27.7 to 30.6 (d = −0.26) (Figure 2). No significant changes were found for the stability indicator, group effect ($p = 0.927$; $\eta_p^2 = 0.002$), time effect ($p = 0.666$; $\eta_p^2 = 0.003$), and interaction effect ($p = 0.109$; $\eta_p^2 = 0.060$; Table 3).

Table 2. Descriptive group comparison (mean ± SD) using the Oswestry Disability Index.

Parameter	Oswestry Disability Index														
	Physiotherapy Group			Galileo® Group			Posturomed® Group			Variance Analyses (ALM)					
	Pre-Score	Post-Score	d	Pre-Score	Post-Score	d	Pre-Score	Post-Score	d	Group		Time		Time × Group	
										p	η_p^2	p	η_p^2	p	η_p^2
Pain intensity	2.44 ± 0.65	1.60 ± 0.82	**1.14**	2.36 ± 0.76	1.80 ± 0.96	**0.65**	2.72 ± 0.68	1.40 ± 0.96	**1.61**	0.937	0.002	**<0.001**	**0.415**	0.053	0.078
Body care	0.92 ± 0.95	0.44 ± 0.65	**0.60**	0.72 ± 0.74	0.52 ± 0.65	0.29	0.96 ± 0.98	0.44 ± 0.65	**0.63**	0.910	0.003	**<0.001**	**0.212**	0.298	0.033
Gait	1.28 ± 1.17	0.88 ± 0.93	0.38	1.12 ± 0.83	0.76 ± 0.88	0.42	1.40 ± 1.12	1.08 ± 0.99	0.32	0.492	0.020	**0.001**	**0.141**	0.953	0.001
Sedentary lifestyle	2.28 ± 1.06	2.04 ± 1.17	0.21	2.04 ± 0.79	1.44 ± 1.04	**0.65**	2.48 ± 0.92	1.76 ± 0.97	**0.76**	0.161	0.050	**<0.001**	**0.199**	0.259	0.037
stand	2.80 ± 0.87	2.00 ± 1.26	**0.75**	2.40 ± 1.04	1.60 ± 1.29	**0.68**	2.80 ± 1.12	2.00 ± 1.38	**0.64**	0.287	0.034	**<0.001**	**0.338**	1.000	0.000
sleep	1.68 ± 1.03	1.20 ± 0.91	0.49	1.72 ± 1.02	1.40 ± 1.04	0.31	1.80 ± 0.91	1.04 ± 0.68	**0.95**	0.798	0.006	**<0.001**	**0.219**	0.297	0.033
Sexuality	2.06 ± 2.08	1.69 ± 1.92	0.19	1.56 ± 1.50	0.88 ± 1.46	0.46	2.19 ± 1.63	1.33 ± 1.62	**0.53**	0.465	0.030	**<0.001**	**0.245**	0.451	0.031
Social life	2.36 ± 1.25	1.64 ± 1.47	0.44	2.08 ± 1.22	1.20 ± 1.32	**0.69**	2.16 ± 1.14	1.24 ± 1.09	**0.82**	0.456	0.022	**<0.001**	**0.310**	0.843	0.005
Travel	2.52 ± 1.33	1.40 ± 1.29	**0.85**	1.92 ± 1.58	1.08 ± 1.12	**0.62**	2.48 ± 1.39	1.76 ± 1.62	0.48	0.165	0.049	**<0.001**	**0.276**	0.619	0.013
Lifting	2.76 ± 1.20	2.28 ± 1.49	0.36	1.96 ± 1.24	1.72 ± 1.40	0.18	2.80 ± 1.16	2.20 ± 1.53	0.45	0.084	0.066	**0.002**	**0.128**	0.545	0.017
Sum	22.3 ± 2.32	17.1 ± 8.57	0.95	17.3 ± 7.10	11.7 ± 7.72	0.76	21.8 ± 5.81	15.1 ± 8.27	0.95	0.086	0.094	**<0.001**	**0.470**	0.763	0.011

Analysis of variance (relevant differences marked in bold), calculation of effect size (η_p^2). Scores of sensorimotor interventions (physiotherapy (PT), Galileo®, and Posturomed®). Values are given as mean ± standard deviation for the single categories and the summary score. Variance analysis (group, time, interaction effects) for the two weeks of multimodal pain therapy. ALM: General linear model (p, η_p^2); effect size (d).

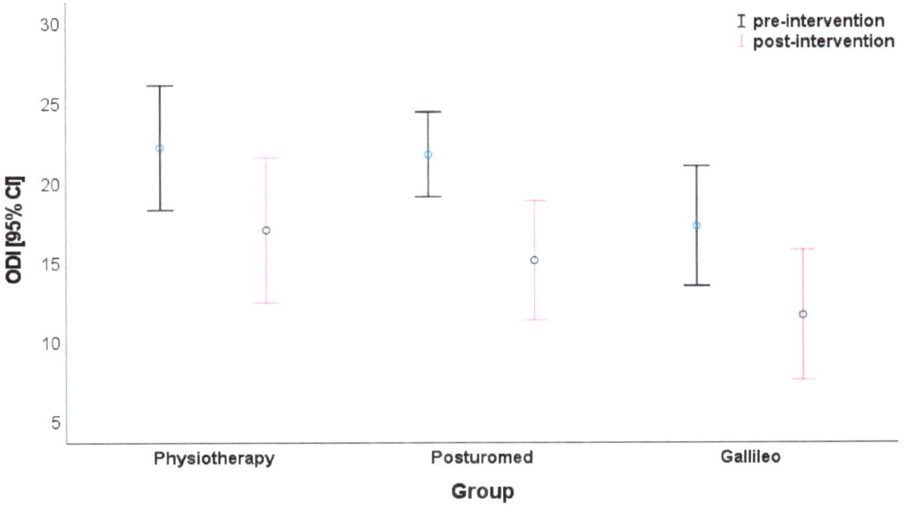

Figure 1. Comparison of the Oswestry Disability Index (ODI) depending on group and time (time effect: $p < 0.001$; $\eta_p^2 = 0.470$).

In the respective frequency ranges one to eight, there was a weak significant effect for time, but only for frequency band F2–4, which maps the peripheral vestibular system ($p = 0.014$; $\eta_p^2 = 0.081$). There was no effect for this frequency band by group ($p = 0.723$; $\eta_p^2 = 0.009$) or for group x time ($p = 0.767$; $\eta_p^2 = 0.007$). Another significant effect was found for the interaction over time for heel loading ($p = 0.014$; $\eta_p^2 = 0.111$). Here, the participants in the PT group moved from 48% to 46% (d = 0.32) towards an increased forefoot load, while those in the Posturomed® group moved from 47% to 49% (d = −0.30) towards an increased rearfoot load. The participants in the Galileo® group showed little change in terms of heel loading from 47.4% to 47.2% (d = 0.02; Table 3).

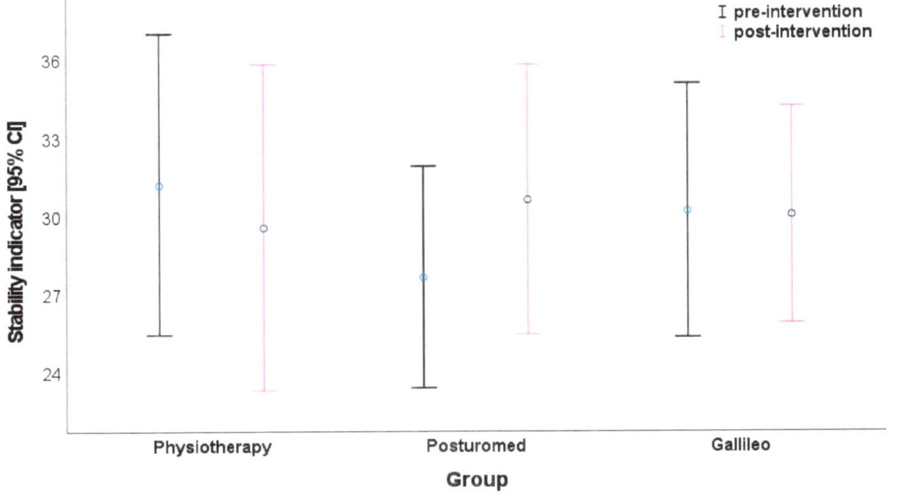

Figure 2. Comparison of stability indicator depending on group and time (time effect: $p = 0.666$; $\eta_p^2 = 0.003$).

Table 3. Descriptive group comparison (mean ± SD) using posturography.

Parameter	Posturography														
	Physiotherapy Group			Galileo® Group			Posturomed® Group			Variance Analyses (ALM)					
	Pre	Post	d	Pre	Post	d	Pre	Post	d	Group		Time		Time × Group	
										p	η_p^2	p	η_p^2	p	η_p^2
F 1	19.4 ± 9.07	18.0 ± 6.59	0.17	17.7 ± 5.59	18.1 ± 5.49	−0.08	18.4 ± 6.61	23.6 ± 25.6	−0.33	0.467	0.021	0.426	0.009	0.315	0.032
F 2–4	11.9 ± 3.99	10.3 ± 3.41	0.42	11.6 ± 3.64	10.8 ± 2.78	0.26	11.0 ± 5.15	10.1 ± 2.11	0.25	0.723	0.009	**0.014**	**0.081**	0.767	0.007
F 5–6	5.63 ± 2.37	5.08 ± 2.30	0.24	5.29 ± 1.89	5.38 ± 1.85	−0.05	4.80 ± 1.69	5.17 ± 1.92	−0.20	0.748	0.008	0.844	0.001	0.050	0.080
F 7–8	1.06 ± 0.55	1.04 ± 0.55	0.04	0.96 ± 0.44	1.02 ± 0.37	−0.15	0.92 ± 0.37	1.07 ± 0.53	−0.33	0.884	0.003	0.59	0.049	0.103	0.061
ST	31.2 ± 14.0	29.6 ± 15.2	0.11	30.2 ± 11.8	30.1 ± 10.1	−0.01	27.7 ± 10.4	30.6 ± 12.6	−0.26	0.927	0.002	0.666	0.003	0.109	0.060
WDI	6.27 ± 2.52	5.64 ± 2.34	0.26	5.67 ± 2.27	5.32 ± 2.04	0.16	5.06 ± 1.41	5.46 ± 1.75	−0.25	0.424	0.024	0.341	0.013	0.103	0.061
Synch.	672 ± 155	636 ± 145	0.24	635 ± 128	634 ± 132	0.01	670 ± 139	644 ± 120	0.20	0.781	0.007	0.117	0.034	0.551	0.016
Left	49.6 ± 6.76	50.1 ± 4.19	−0.09	51.5 ± 4.74	50.9 ± 3.88	0.13	50.5 ± 4.67	49.0 ± 5.48	0.30	0.468	0.021	0.226	0.020	0.174	0.047
Heel	47.9 ± 7.51	45.5 ± 7.20	0.32	47.4 ± 7.73	47.2 ± 8.22	0.02	46.7 ± 5.92	48.7 ± 7.35	−0.30	0.886	0.003	0.781	0.001	**0.014**	**0.111**

Analysis of variance (relevant differences marked in bold) and calculation of effect size (η_p^2) are presented. Values are given as mean ± standard deviation for the different posturographic parameters depending on sensorimotor interventions (physiotherapy (PT), Galileo®, and Posturomed®). Variance analysis (group, time, interaction effects) for the two weeks of multimodal pain therapy. ALM: General linear model (p, η_p^2); effect size (d). F 1: Frequency band 1, visual and nigrostriatal system; F 2–4: Frequency band 2–4, peripheral vestibular system; F 5–6: Frequency band 5–6, somatosensory system; F 7–8: Frequency band 7–8, cerebellar system; WDI: weight distribution index; Synch.: Synchronization; Heel: forefoot-hindfoot ratio, percentage of weight distribution forefoot vs. hindfoot with a description of heel loading; Left: load on the left side, percentage of weight distribution left vs. right with the description of left side loading.

4. Discussion

The reduction of the ODI scores within this study across all groups shows the effectiveness of the different sensorimotor training forms and, in connection with their effectiveness, that of the MMPT, for a reduction in pain-related limitations in patients with lower back pain. Under comparable baseline values, there was no significant difference in the reduction of pain-related limitations between the individual groups. These results suggest that, on the one hand, sensorimotor training using physiotherapeutic exercises is equivalent to training on the Galileo®/Posturomed® platforms. On the other hand, these devices can also be a supplement for sensorimotor training, as they have similar effects on the reduction of pain-related limitations in chronic low back pain. It should also be noted that although the individual parameters of the ODI are also shown in Table 2, these have no relevance in the evaluation, since the ODI is evaluated as an overall questionnaire and no evaluation of the individual questions took place independently of the overall result.

Several studies have already investigated the effects of sensorimotor training on the subjective, pain-related limitation of participants measured by the ODI. Significant reductions in pain-related limitations are shown in many studies [24,25,35,36]. Yang and Seo [25] compared a group using only sensorimotor physiotherapy training with a group combined with vibration training. As in our study, there was a significant reduction in the ODI values for both groups, but no significance for an effect in the development over time of each other could be demonstrated.

However, other studies did not demonstrate any positive effects of sensorimotor training in the reduction of pain-related impairment [37,38]. This is contrary to the results obtained in our study. For example, McCaskey et al [37] did not find a significant reduction in ODI scores for either the intervention or control group during a four-week training with a total of nine 45-min interventions. In comparison, this constellation resembles the exercise design of the interventions performed in this study with respect to both the sequence with supplemental physical therapy and the intervention duration on the Posturomed®, as well as the low total number of exercises (single digits). However, the baseline ODI represents a significant difference with similar anthropometric conditions in the study populations. In McCaskey et al. [37] during the pre-intervention, there was an ODI percentage limitation of 20% for participants in the intervention group and 18% for those in the control group. In contrast, baseline values in our study were 45% for the physiotherapy group, 44% for the Posturomed® group, and 35% for the Galileo® group. If the results of del Pozo-Cruz et al. [24] are also considered, where the baseline values were 27% (intervention) and 29% (control group), it is conceivable that the comparatively lower baseline values did not allow for a significant effect. This is also suggested by Wegener et al. [38] who reported no significant reduction in ODI scores, even over a longer intervention period of 18 weeks with two training sessions per week. Their baseline ODI values were 21% in the physiotherapy group (in terms of sensorimotor training) and 18% in the vibration training group

The results of this study showed that there were no significant positive effects on postural control as detected by posturography, for sensorimotor physiotherapeutic training, Galileo® training, or Posturomed® training. This is contrary to the majority of studies that demonstrated significant postural improvements following sensorimotor physiotherapeutic training and sensorimotor vibration training [5,24,26,38–44]. However, it is difficult to relate the results of these studies to the current study. This is because the measurements of postural control across the various studies were conducted using different measurement systems. Only two studies used the IBS [5,39] which was used in the current investigation. In the majority of subsequent studies, the ability to balance on a Posturomed® or gait belt was measured using electromyographic recordings as a representation of postural control [37,40–42]. Other studies have used the Biodex Balancesystem® [24], the Bretz Stabilometer® [44], the Leonardo Mechanograph® [36], or the MFT S3 Check® [38] for this same purpose. Although all of these studies did mention that their respective balance system was a valid assessment tool of the postural system. Unfortunately, these differences in instrumentation make comparing results across studies difficult.

Due to similarities in methodology, the results of this study can be directly compared with the study conducted by Schwesig et al. [39]. These authors investigated the effects of three months of sensorimotor training (15 min per week on an unstable mat) among osteoporosis patients. This group was then compared to an asymptomatic control group, which individually completed daily sensorimotor training at home according to the authors' instructions. This study reported that the postural performance, measured by the stability indicator, significantly improved by 80% of the participants in the osteoporosis group and 75% in the control group [39]. This suggests that an individual continuation of sensorimotor training at home among patients with low back pain may also provide positive results. The importance of this assumption can be concluded from the fact that Schwesig et al. [39] again showed a significant deterioration in postural performance for the three-month and two-year follow-up without sensorimotor training intervention, which was even greater in the osteoporosis group compared to the control group.

The results of this study showed that all three groups had a significant improvement in the performance of the peripheral vestibular system, without any effect between them over time. This speaks to the comparability of the three sensorimotor training methods, as already established in the development of the scores of the ODI. Lauenroth et al. [32] previously reported that this postural subsystem shows an insufficiency in back pain patients compared to the asymptomatic population. Unfortunately, there are currently no comparative data for sensorimotor training in chronic back pain patients. Nevertheless, this development shows that the postural system was at least stimulated with the completed sensorimotor trainings, even if not yet with a direct influence on the stability indicator. Thus, it can also be assumed from this result that a longer training duration and a combination of different or very varied sensorimotor training methods, in order to challenge the postural system, would be more likely to bring about an improvement in postural control as measured by the IBS. To preserve an achieved improvement in postural control, a permanent execution of the exercises seems to be indispensable. However, future research is necessary to determine which training methods or combinations with the corresponding training intensities are necessary for improvement.

A similar conclusion was reached by Jafarzadeh et al. [45] who compared sensorimotor training against the same training with transcranial direct current stimulation of the motor cortex. When compared to the current study, Jafarzadeh et al. [45]'s study had a similar duration and number of training sessions, but their low back pain patients were much younger. Similarly to the current study, Jafarzadeh et al. only found an improvement in postural control among the group that received transcranial direct current stimulation in addition to the sensorimotor training [45]. Because of the differences in patient ages between these studies, it is important for future research to determine whether the results presented by Jafarzadeh et al. can be reproduced in older patients who are more likely to suffer from chronic back pain.

Limitations

One unique feature of this current study was the inpatient setting. Since all patients were participating in an inpatient MMPT, there was potentially a higher degree of limitations in activities of daily living compared to patients in an outpatient setting. This is comparable to other sensorimotor training studies which used outpatient chronic pain patients [24,37] or persons without a history of back pain [42,43].

During the recruitment of study participants, there was no specific questioning about diseases that may have had an influence on the postural system and its subsystems [3], such as Parkinson's or cerebellar disorders [31]. Another factor that influences postural control is the degree of physical activity of the individual participants [3], which was not included in the study questionnaire.

Double blinding was not possible due to the initial fixed randomization list, since both the investigators/treatment providers and the participants could recognize the corresponding group assignment via their intervention. The participants were assigned to the next free place in the randomization list one after the other, depending on their inclusion time. If a loss-to-follow-up occurred, the vacant place in the randomization list was assigned to the next available participant included in the study, without categorizing the cause of the loss-to-follow-up or the number of dropouts per group or in total.

Three groups were compared in this study, each with an intervention of sensorimotor training using Galileo®, Posturomed®, or a physiotherapeutic training program. Consequently, there was no untrained control group, which could have led to a bias of results due to a more intensive or additive therapy of additional sensorimotor training of the intervention groups. Thus, it is likely that the observed effects with respect to the reduction of pain-related impairment by means of the CDI over the two-week intervention phase are, at least in part, effects co-induced by the MMPT. This point is particularly unremarkable since this is precisely one of the goals of MMPT.

5. Conclusions

Sensorimotor training in the context of multimodal pain therapy is suitable for reducing pain-related impairment. Posturography shows stimulation of a subsystem, but no improvement in postural stability. Sensorimotor training on the Galileo® or Posturomed® platforms is equivalent to a physiotherapeutic sensorimotor exercise program. Therefore, these devices can be a supplement for sensorimotor training, as they have similar effects on the reduction of pain-related limitations in chronic back pain.

Supplementary Materials: The following supporting information can be downloaded at https://www.mdpi.com/article/10.3390/jpm13040634/s1, Figure S1: Flow-Chart to the study. Figure S2: Galileo®. Figure S3: Posturomed®. Table S1: IBS parameters [30,31]. Table S2: Exercise program in physiotherapy. Table S3: Exercise program in Galileo®-Group. Table S4: Exercise program in Posturomed®-Group.

Author Contributions: Conceptualization, A.R., K.-S.D. and A.S.; Data curation, A.R., K.L. and R.S.; Formal analysis, A.R., K.L. and R.S.; Investigation, A.R.; Methodology, A.R., K.-S.D., R.S. and A.S.; Project administration, A.S.; Resources, K.-S.D. and R.S.; Supervision, A.S.; Validation, K.L. and R.S.; Visualization, A.R., K.L. and R.S.; Writing—original draft, A.R. and A.S.; Writing—review and editing, A.R., K.L., K.-S.D., R.S. and A.S. All authors have read and agreed to the published version of the manuscript.

Funding: This research received no external funding.

Institutional Review Board Statement: The study was conducted in accordance with the Declaration of Helsinki, and approved by the Ethics Committee of the University Hospital Halle (Saale) (reference number: 2015-84) (9 September 2015).

Informed Consent Statement: Informed consent was obtained from all participants involved in the study.

Data Availability Statement: Not applicable.

Acknowledgments: The authors would like to thank all the staff involved in patient care at the conservative orthopedic section 2 of the University Hospital Halle (Saale), as well as in particular the physiotherapists involved for their excellent support of this study.

Conflicts of Interest: The authors declare no conflict of interest.

References

1. Chenot, J.F.; Greitemann, B.; Kladny, B.; Petzke, F.; Pfingsten, M.; Schorr, S.G. Non-Specific Low Back Pain. *Dtsch. Arztebl. Int.* **2017**, *114*, 883–890. [CrossRef] [PubMed]
2. Rothman, M.G.; Ortendahl, M.; Rosenblad, A.; Johansson, A.-C. Improved quality of life, working ability, and patient satisfaction after a pretreatment multimodal assessment method in patients with mixed chronic muscular pain: A randomized-controlled study. *Clin. J. Pain* **2013**, *29*, 195–204. [CrossRef] [PubMed]
3. Taube, W. Neuronale Mechanismen der Posturalen Kontrolle und der Einfluss von Gleichgewichtstraining. *J. Neurol. Neurochir. Psychiatr.* **2012**, *14*, 55–63.
4. Otte, C.; Rašev, E. Posturale Aspekte der Schmerztherapie des Bewegungssystems. *Man. Med.* **2010**, *48*, 267–274. [CrossRef]

5. Müller, K.; Schwesig, R.; Leuchte, S.; Riede, D. Coordinative treatment and quality of life—A randomised trial of nurses with back pain. *Gesundheitswesen* **2001**, *63*, 609–618. [CrossRef] [PubMed]
6. Hodges, P.W.; Moseley, G.L. Pain and motor control of the lumbopelvic region: Effect and possible mechanisms. *J. Electromyogr. Kinesiol.* **2003**, *13*, 361–370. [CrossRef]
7. Wilke, H.J.; Wolf, S.; Claes, L.E.; Arand, M.; Wiesend, A. Stability increase of the lumbar spine with different muscle groups. A biomechanical in vitro study. *Spine* **1995**, *20*, 192–198. [CrossRef]
8. Tesh, K.M.; Dunn, J.S.; Evans, J.H. The abdominal muscles and vertebral stability. *Spine* **1987**, *12*, 501–508. [CrossRef]
9. Cresswell, A.G.; Thorstensson, A. Changes in intra-abdominal pressure, trunk muscle activation and force during isokinetic lift ing and lowering. *Eur. J. Appl. Physiol. Occup. Physiol.* **1994**, *68*, 315–321. [CrossRef]
10. Cresswell, A.G.; Oddsson, L.; Thorstensson, A. The influence of sudden perturbations on trunk muscle activity and intra-abdominal pressure while standing. *Exp. Brain Res.* **1994**, *98*, 336–341. [CrossRef]
11. Hodges, P.W.; Richardson, C.A. Inefficient muscular stabilization of the lumbar spine associated with low back pain. A motor control evaluation of transversus abdominis. *Spine* **1996**, *21*, 2640–2650. [CrossRef]
12. Hides, J.A.; Stanton, W.R.; McMahon, S.; Sims, K.; Richardson, C.A. Effect of stabilization training on multifidus muscle cross-sectional area among young elite cricketers with low back pain. *J. Orthop. Sport. Phys. Ther.* **2008**, *38*, 101–108. [CrossRef]
13. Sihvonen, T.; Partanen, J. Segmental hypermobility in lumbar spine and entrapment of dorsal rami. *Electromyogr. Clin. Neurophysiol.* **1990**, *30*, 175–180.
14. Lindgren, K.A.; Sihvonen, T.; Leino, E.; Pitkänen, M.; Manninen, H. Exercise therapy effects on functional radiographic findings and segmental electromyographic activity in lumbar spine instability. *Arch. Phys. Med. Rehabil.* **1993**, *74*, 933–939.
15. Smith, D.; Bissell, G.; Bruce-Low, S.; Wakefield, C. The effect of lumbar extension training with and without pelvic stabilization on lumbar strength and low back pain. *J. Back Musculoskelet. Rehabil.* **2011**, *24*, 241–249. [CrossRef]
16. França, F.R.; Burke, T.N.; Caffaro, R.R.; Ramos, L.A.; Marques, A.P. Effects of muscular stretching and segmental stabilization on functional disability and pain in patients with chronic low back pain: A randomized, controlled trial. *J. Manip. Physiol. Ther.* **2012**, *35*, 279–285. [CrossRef]
17. Javadian, Y.; Behtash, H.; Akbari, M.; Taghipour-Darzi, M.; Zekavat, H. The effects of stabilizing exercises on pain and disability of patients with lumbar segmental instability. *J. Back Musculoskelet. Rehabil.* **2012**, *25*, 149–155. [CrossRef]
18. Tsao, H.; Hodges, P.W. Immediate changes in feedforward postural adjustments following voluntary motor training. *Exp. Brain Res.* **2007**, *181*, 537–546. [CrossRef]
19. Tsigkanos, C.; Gaskell, L.; Smirniotou, A.; Tsigkanos, G. Static and dynamic balance deficiencies in chronic low back pain. *J. Back Musculoskelet. Rehabil.* **2016**, *29*, 887–893. [CrossRef]
20. da Silva, R.A.; Vieira, E.R.; Fernandes, K.B.P.; Andraus, R.A.; Oliveira, M.R.; Sturion, L.A.; Calderon, M.G. People with chronic low back pain have poorer balance than controls in challenging tasks. *Disabil. Rehabil.* **2018**, *40*, 1294–1300. [CrossRef]
21. Müller, O.; Günther, M.; Krauß, I.; Horstmann, T. Physical characterization of the therapeutic device Posturomed as a measuring device–presentation of a procedure to characterize balancing ability. *Biomed. Tech.* **2004**, *49*, 56–60. [CrossRef] [PubMed]
22. Bongiovanni, L.G.; Hagbarth, K.E.; Stjernberg, L. Prolonged muscle vibration reducing motor output in maximal voluntary contractions in man. *J. Physiol.* **1990**, *423*, 15–26. [CrossRef] [PubMed]
23. Fontana, T.L.; Richardson, C.A.; Stanton, W.R. The effect of weightbearing exercise with low frequency, whole body vibration on lumbosacral proprioception: A pilot study on normal subjects. *Aust. J. Physiother.* **2005**, *51*, 259–263. [CrossRef] [PubMed]
24. Del Pozo-Cruz, B.; Hernández-Mocholí, M.A.; Adsuar, J.C.; Parraca, J.A.; Muro, I.; Gusi, N. Effects of whole body vibration therapy on main outcome measures for chronic non-specific low back pain: A single-blind randomized controlled trial. *J. Rehabil. Med.* **2011**, *43*, 689–694. [CrossRef] [PubMed]
25. Yang, J.; Seo, D. The effects of whole body vibration on static balance, spinal curvature, pain, and disability of patients with low back pain. *J. Phys. Ther. Sci.* **2015**, *27*, 805–808. [CrossRef]
26. Chow, D.H.K.; Lee, T.Y.; Pope, M.H. Effects of whole body vibration on spinal proprioception in healthy individuals. *Work* **2018**, *61*, 403–411. [CrossRef]
27. Mannion, A.F.; Junge, A.; Fairbank, J.C.T.; Dvorak, J.; Grob, D. Development of a German version of the Oswestry Disability Index. Part 1: Cross-cultural adaptation, reliability, and validity. *Eur. Spine J.* **2006**, *15*, 55–65. [CrossRef]
28. Gaul, C.; Mette, E.; Schmidt, T.; Grond, S. Praxistauglichkeit einer Deutschen Version des Oswestry Low Back Pain Disability Questionnaire. Ein Fragebogen zur Beeinträchtigung durch Rückenschmerzen. *Schmerz* **2008**, *22*, 51–58. [CrossRef]
29. Schwesig, R.; Becker, S.; Fischer, D. Intraobserver reliability of posturography in healthy subjects. *Somatosens. Mot. Res.* **2014**, *31*, 16–22. [CrossRef]
30. Schwesig, R.; Lauenroth, A.; Müller, A.; Becker, S.; Hottenrott, K. Parametrisierung posturaler Subsysteme mit Posturografie. *Man. Med.* **2006**, *44*, 376–384. [CrossRef]
31. Schwesig, R.; Becker, S.; Lauenroth, A.; Kluttig, A.; Leuchte, S.; Esperer, H.D. A novel posturographic method to differentiate sway patterns of patients with Parkinson's disease from patients with cerebellar ataxia. *Biomed. Tech.* **2009**, *54*, 347–356. [CrossRef]
32. Lauenroth, A.; Reinhardt, L.; Schulze, S.; Laudner, K.G.; Delank, K.S.; Schwesig, R. Does low back pain affect the reliability of postural regulation? *Somatosens. Mot. Res.* **2019**, *36*, 116–121. [CrossRef]
33. Bortz, J. Basic principle of one-factorial analysis of variance. In *Statistics for Social Scientists*; Springer: New York, NY, USA, 1999; pp. 237–249.

34. Richardson, J.T.E. Eta squared and partial eta squared as measures of effect size in educational research. *Educ. Res. Rev.* **2011**, *6*, 135–147. [CrossRef]
35. Brooks, C.; Kennedy, S.; Marshall, P.W.M. Specific trunk and general exercise elicit similar changes in anticipatory postural adjustments in patients with chronic low back pain: A randomized controlled trial. *Spine* **2012**, *37*, E1543–E1550. [CrossRef]
36. Kaeding, T.S.; Karch, A.; Schwarz, R.; Flor, T.; Wittke, T.C.; Kück, M.; Böselt, G.; Tegtbur, U.; Stein, L. Whole-body vibration training as a workplace-based sports activity for employees with chronic low-back pain. *Scand. J. Med. Sci. Sports* **2017**, *27*, 2027–2039. [CrossRef]
37. McCaskey, M.A.; Wirth, B.; Schuster-Amft, C.; de Bruin, E.D. Postural sensorimotor training versus sham exercise in physiotherapy of patients with chronic non-specific low back pain: An exploratory randomised controlled trial. *PLoS ONE* **2018**, *13*, e0193358. [CrossRef]
38. Wegener, V.; Rarack, S.; Tiffe, T.; Grill, E.; Melcher, C.; Birkenmaier, C.; Jansson, V.; Wegener, B. Effects of Whole Body Vibration Therapy and Classic Physiotherapy on Postural Stability in People with Back Pain: A Randomized Trial. *Clin. Spine Surg.* **2019**, *32*, E214–E220. [CrossRef]
39. Schwesig, R.; Müller, K.; Becker, S.; Kreutzfeldt, A.; Hottenrott, K. Sensomotorisches Training im Alter und bei Osteoporose. *Akt. Rheumatol.* **2006**, *31*, 196–203. [CrossRef]
40. Granacher, U.; Gollhofer, A.; Strass, D. Training induced adaptations in characteristics of postural reflexes in elderly men. *Gait Posture* **2006**, *24*, 459–466. [CrossRef]
41. Korsten, K.; Mornieux, G.; Walter, N.; Gollhofer, A. Gibt es Alternativen zum Sensomotorischen Training? *Schweiz. Z. Sportmed. Sporttraumatol.* **2008**, *56*, 150–155.
42. Granacher, U.; Gruber, M.; Gollhofer, A. Auswirkungen von Sensomotorischem Training auf die Posturale Kontrolle Älterer Mäner. *Deut. Z. Sportmed.* **2009**, *60*, 387–393.
43. Giboin, L.S.; Gruber, M.; Kramer, A. Task-specificity of balance training. *Hum. Mov. Sci.* **2015**, *44*, 22–31. [CrossRef] [PubMed]
44. Mikó, I.; Szerb, I.; Szerb, A.; Poor, G. Effectiveness of balance training programme in reducing the frequency of falling in estab-lished osteoporotic women: A randomized controlled trial. *Clin. Rehabil.* **2017**. *31*, 217–224. [CrossRef] [PubMed]
45. Jafarzadeh, A.; Ehsani, F.; Yosephi, M.H.; Zoghi, M.; Jaberzadeh, S. Concurrent postural training and M1 anodal transcranial direct current stimulation improve postural impairment in patients with chronic low back pain. *J. Clin. Neurosci.* **2019**, *68*. 224–234. [CrossRef]

Journal of *Personalized Medicine*

MDPI

Article

Impact of Spino-Pelvic Parameters on the Prediction of Lumbar and Thoraco-Lumbar Segment Angles in the Supine Position

Philipp Schenk [1,*], Arija Jacobi [2], Carolin Graebsch [1], Thomas Mendel [3,4], Gunther Olaf Hofmann [3,4] and Bernhard Wilhelm Ullrich [3,4]

1 Department of Science, Research and Education, BG Klinikum Bergmannstrost Halle gGmbH, 06112 Halle, Germany
2 Department of Orthopedic and Trauma Surgery, DIAKO Ev. Diakonie-Krankenhaus gGmbH, 28239 Bremen, Germany
3 Department of Trauma and Reconstructive Surgery, BG Klinikum Bergmannstrost Halle gGmbH, 06112 Halle, Germany
4 Department of Trauma, Hand and Reconstructive Surgery, Jena University Hospital, Friedrich Schiller University Jena, 07747 Jena, Germany
* Correspondence: philipp.schenk@bergmannstrost.com; Tel.: +49-3641-132-7804

Abstract: Background: The correction of malposition according to vertebral fractures is difficult because the alignment at the time before the fracture is unclear. Therefore, we investigate whether the spinal alignment can be determined by the spino-pelvic parameters. Methods: Pelvic incidence (PI), pelvic tilt (PT), sacral slope (SS), adjacent endplate angles (EPA), age, sex, body weight, body size, BMI, and age were used to predict mono- and bisegmental EPA (mEPA, bEPA) in the supine position using linear regression models. This study was approved by the Ethics Committee of the Medical Association of Saxony-Anhalt Germany on 20 August 2020, under number 46/20. Results: Using data from 287 patients, the prediction showed R^2 from 0.092 up to 0.972. The adjacent cranial and caudal EPA showed by far the most frequently significance in the prediction of all parameters used. Anthropometric and spino-pelvic parameters showed sparse impact, which was frequently in the lower lumbar regions. On average, a very good prediction was found. For two mEPA (L3/4 $R^2 = 0.914$, L4/5 $R^2 = 0.953$) and two bEPA (L3 $R^2 = 0.899$, L4 $R^2 = 0.972$), the R^2 was >0.8. However, the predicted EPA differed for individual patients, even in these very effective prediction models—roughly around ±10° as compared to the measured EPA. Conclusions: In general, the prediction showed good to perfect results. In the supine position, the spinopelvic and anthropometric parameters show sparse impact on the prediction of mEPA or bEPA.

Keywords: spinal alignment; sagittal profile; mono-segmental EPA; bi-segmental EPA

Citation: Schenk, P.; Jacobi, A.; Graebsch, C.; Mendel, T.; Hofmann, G.O.; Ullrich, B.W. Impact of Spino-Pelvic Parameters on the Prediction of Lumbar and Thoraco-Lumbar Segment Angles in the Supine Position. *J. Pers. Med.* **2022**, *12*, 2081. https://doi.org/10.3390/jpm12122081

Academic Editors: Markus Gesslein and Eduard Kurz

Received: 1 December 2022
Accepted: 15 December 2022
Published: 17 December 2022

1. Introduction

Approximately 80% of spinal fractures are located in the lumbar or thoracic sections of the spine, of which 50% are located at the thoracolumbar junction (Th11-L2) [1,2]. Spine fractures can lead to a malalignment, and thus to kyphotic deformity of the spine. When surgical therapy is indicated, one of the surgeon's goals is to correct the fracture-related malalignment. Achieving the former spinal alignment may be important to avoid static misload and, inter alia, the consequent development of pain. The goal of restoring the physiological spinal alignment can be difficult to achieve due to missing pre-trauma radiological images of the patient's spine.

There are efforts and investigations to elucidate the sagittal profile of the spine using pelvic parameters [3–5]. The most important parameters are: pelvic incidence (PI), pelvic tilt (PT), and sacral slope (SS). These were primarily determined using X-rays in a standing position [6–8]. Correlations between spino-pelvic parameters and lumbar lordosis, as well as between age and the spino-pelvic parameters, could be found [4,9,10].

However, it is unclear whether there is a relationship between the sagittal spine profile and the pelvic parameters, even in the supine position, and whether the pelvic parameters are suitable for estimating the sagittal spine segment angles.

Therefore, the aim of this study is to investigate whether the spinal alignment, captured by the mono- and bi-segmental sagittal Cobb angles, can be determined by the spino-pelvic parameters and the Cobb angles of the adjacent segments.

2. Materials and Methods

This study is a retrospective analysis of cross-sectional data, and it was conducted in accordance with the Declaration of Helsinki and approved by the Ethics Committee of the Medical Association of Saxony-Anhalt Germany on 20 August 2020, under number 46/20. Initially, polytrauma CT scans of 1826 patients were examined and evaluated for suitability according to the inclusion and exclusion criteria. The CT images were acquired using the Aquilion PRIME (Canon medical systems Europe) in the supine position. Included in this study were patients with intact spine, pelvis and femur. Patients under 18 years old, with spine deformity or spondylolisthesis with a Meyerding grade greater than 1 [11]; spine, pelvic, or femur fracture, or previous spinal surgery, were excluded, leading to a sample size of 287 patients.

2.1. Parameters

The following spino-pelvic parameters were determined in the CT images: pelvic incidence (PI), pelvic tilt (PT), and sacral slope (SS). The spino-pelvic parameters were measured according to the specifications of Roussouly et al. and Le Huec et al. [3,4].

The PI was measured as the angle between the line perpendicular to the midpoint of the cranial endplate of the sacrum (white dotted line in Figure 1) and the line connecting the midpoint of the sacral plate to the center of one femoral head (red/white dotted line). The PT was measured between the line connecting the midpoint of the cranial endplate of the sacrum to the center of one femoral head and the coronal plane (red solid line). Thus, the PI describes the inclination of the sacrum in relation to the femur head, and the PT indicates the inclination of the sacrum in relation to the body plumb line, which is usually used in the standing position. The SS was measured as the angle between the midpoint of the cranial endplate of the sacrum and the body's transversal plane (black bold lines in Figure 1). The transversal plane stands perpendicular to the body plumb line. For a better understanding, Figure 1 shows how the measurement of the spino-pelvic parameters was carried out.

In addition to the spino-pelvic parameters, the sagittal vertebral Cobb angles between the cranial and caudal endplates were mono-segmentally (mEPA) measured for each segment from Th8/Th9 to L4/L5 and bi-segmentally (bEPA) for the segments of Th8 to L4, respectively. The mEPA is the angle between the cranial endplate of the cranial vertebra and the caudal endplate of the caudal vertebra of the segment. The bEPA is the angle between the cranial endplate of the nearest cranial vertebra and the caudal endplate of the adjacent caudal vertebra (e.g., bEPA Th12 represents the angle between the cranial endplate Th11 and the caudal endplate L1). However, the procedure for the prediction of the caudal segments of the lumbar spine L3/5, L4/5 (mEPA) and L3, L4 (bEPA) were different due to the issue that no complete mEPA or bEPA could be measured as described above. Therefore, the segment angles of L4/S1 and L5/S1 were used as caudal segments for the prediction and were measured between the cranial endplates of L4 or L5 and S1, respectively.

Lordotic angles are described by positive values and kyphotic angles by negative values.

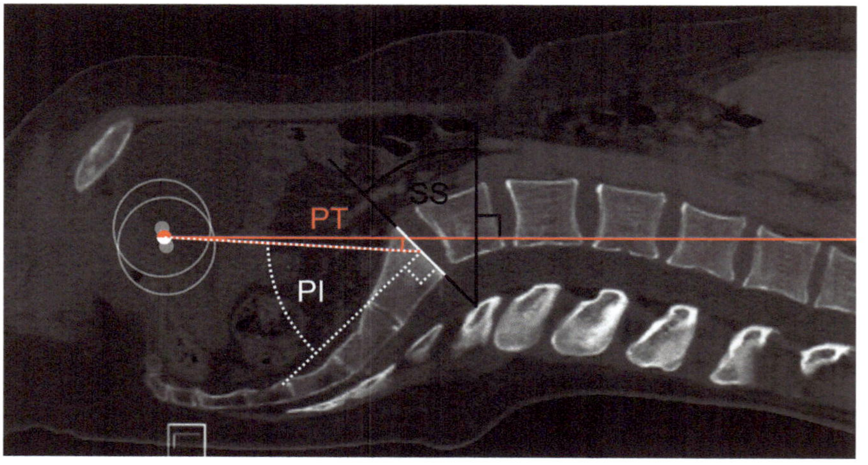

Figure 1. Measurement of the spino-pelvic parameters. The head of both femurs (grey circles) and their mean is marked with a white/red circle. A horizontal red solid line is drawn crossing the mean of the femur heads. The cranial endplate of the first sacral vertebra is highlighted with a white solid line. The pelvic tilt (PT) was measured as the angle between the red horizontal line and the white/red dotted line between the mean of the femur heads and the center of the cranial endplate of S1. The angle of the pelvic incidence (PI) was measured as the angle between the white/red dotted line between the mean of the femur heads and the center of the cranial endplate of S1, perpendicular to the midpoint of the cranial endplate of S1 (white dotted line). The sacral slope (SS) was measured as the angle between the cranial endplate of S1 perpendicular to the horizontal red solid line, highlighted with black solid lines.

2.2. Statistical Analyses

To identify differences between older and younger patients, the cohort was divided in older and younger patients, according to the median age. Differences regarding each mEPA and bEPA between the age groups were examined using a linear general model for repeated measures. The mEPA and bEPA of the segments were used for repeated measurements. The impact of sex, body weight, body size, BMI, and spino-pelvic parameters (PT, PI, and SS) on the segmental angles were adjusted by using them as covariates. Differences in spino-pelvic parameters were examined using a multivariate general linear model. Sex, body weight, body size, and BMI were used as covariates. Sex was coded with 1 for men and 2 for women.

The results of the Bonferroni-Holm post hoc pairwise comparison between the younger and older group regarding mEPA and bEPA for each segment, as well as the spino-pelvic parameters, are reported as means, standard deviation (sd), lower and upper limit (LL, UL) of 0.95 confidence interval (CI), and minima and maxima (min, max).

To predict the mono- and bisegmental angles from Th9/10 to L4/5 (mEPA) and Th10 and L4 (bEPA), linear regression models were used for each segmental level, respectively.

For the prediction of mEPA, the adjacent cranial and caudal segments were chosen in the following way. For example, if the mEPA of Th10/11 is the focus for the prediction, it can be predicted using Th8/9 and Th11/12, or using Th9/10 and Th12/L1 as the cranial and caudal segments, respectively. In other words, Th8/9 and Th11/12 can be used as the adjacent mEPA for the prediction of the mEPA of Th9/10 and Th10/11, respectively. Figure 2 shows an exemplary overview of the EPA used for the prediction. In addition to the adjacent EPS, the following variables were used in the regression model: the spino-pelvic parameters (PT, PI, and SS), age, sex, body weight, body size, and BMI.

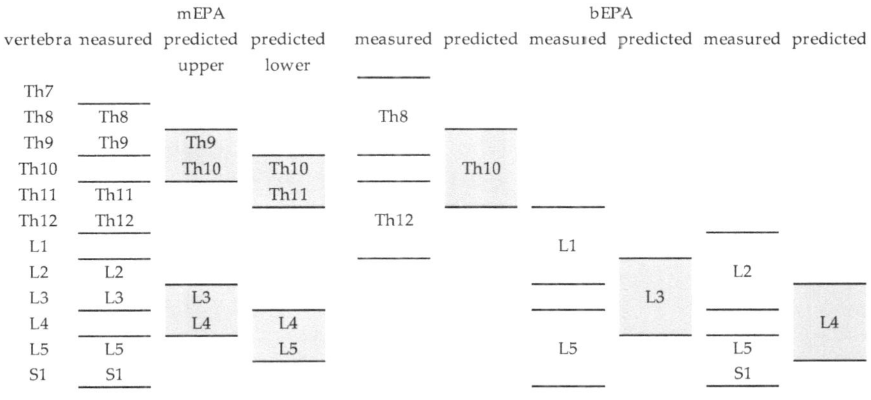

Figure 2. Schematic and exemplary representation of the measured (light grey) and the predicted (dark gray) monosegmental (mEPA) and bisegmental endplate angles (bEPA). The EPAs were measured as the enclosing angles of the cranial endplate of the cranial vertebra and the caudal endplate of the caudal vertebra. With one exception at S1 (first vertebra of the sacrum), here the cranial endplate was used as the caudal end of the EPA. The further cranial and caudal EPAs are shown as examples. The setup was chosen to predict the EPA in fractures of single vertebrae from Th10 to L4. The second example for bEPA, predicting L3, uses the bEPA of L1 and the EPA of the cranial endplates of L4 and S1. Predicting the bEPA of L4, the bEPA of L2 and the EPA of the cranial endplates of L5 and S1 were used.

In an iterative process, the dependent variables were successively reduced to such an extent that all variables ultimately used in the regression model showed significance.

Corrected R^2 was used to assess model quality of the linear regression. Model quality with an R^2 greater than or equal to 0.8 was set to define the prediction as valid. Predictions of valid EPA were subsequently compared with the measured EPA, using difference plots. Therefore, the measured EPAs were used, and the difference between the predicted and the measured EPAs was calculated. The limits of agreement were set using a 1.96-fold standard deviation of the differences between both methods. Linear regression lines were fitted to this data in order to reveal the tendency impact, depending on the values. SPSS V.28 (IBM Corp. Released 2021. IBM SPSS Statistics for Windows, Armonk, NY, USA: IBM Corp) was used for statistical analyses, with $p = 0.05$.

3. Results

In total, the data of 208 male and 79 female adults were be included in this study. Due to the median split (40 years), a group of 143 younger adults (28 ± 6 years, range: 18–39 years) and a group of 144 older adults (56 ± 10 years, range: 40–87 years) was generated.

The younger group included significantly more males (79%, male/female: 113/90) compared to the older group (66%, male/female: 95/49, $p = 0.017$).

In general, no differences could be found for mEPA between the younger and older group ($p = 0.302$). The mEPA differed significant between the segmental levels ($p < 0.001$), except between L4/L5 and L5/S1 ($p = 0.121$), Th12/L1 and Th8/9 ($p = 1.000$), Th9/10 ($p = 0.660$), Th10/11 ($p = 1.000$), between Th11/12 and Th8/9 ($p = 1.000$), and between Th10/11 and Th8/9 ($p = 1.000$). The covariates showed a significant effect for sex ($p < 0.001$), PT ($p = 0.015$), and SS ($p = 0.009$). Body weight ($p = 0.584$), body size ($p = 0.389$), BMI ($p = 0.509$), and PI ($p = 0.051$) showed no significant effect.

For the bEPA, similar results can be found for the main effects regarding age group ($p = 0.469$) and the segments ($p < 0.001$). No significant differences in bEPA could be found between Th10, Th11, and Th12 ($p > 0.535$). Sex was the only covariate which showed significance ($p = 0.001$). The other covariates showed no significance (body weight: $p = 0.483$, body size: $p = 0.414$, BMI: $p = 0.470$, PI: $p = 0.131$, PT: $p = 0.236$, SS: $p = 0.113$).

The multivariate general linear model showed no differences between age groups for PT ($p = 0.188$), PI ($p = 0.725$), and SS ($p = 0.121$). The covariates showed no significant effects on spino-pelvic parameters (sex: $p > 0.205$, body weight: $p > 0.233$, body size: $p > 0.121$, and BMI: $p > 0.282$). Descriptive data for mEPA, bEPA, the spino-pelvic parameters, and the results of the pairwise comparison of the younger and older group are given in Table 1.

Table 1. Angles of thoraco-lumbar mono- and bisegmental EPA and spino-pelvic parameters (PT: pelvic tilt, PI: pelvic incidence, SS: sacral slope) of younger and older adults. Data are given as mean and standard deviation, the lower (LL) and upper limits (UL) of the 95% confidence interval (CI), minimum and maximum, and p-value of pairwise comparison.

		Younger than 40 Years (N = 143)				40 Years and Older (N = 144)				
		Mean ± Sd	0.95 CI (LL UL)	Min	Max	Mean ± Sd	0.95 CI (LL UL)	Min	Max	*p* Value
monosegmental	Th8/9	−4.3 ± 4.3	(−5.0 −3.5)	−24	3	−5.0 ± 4.2	(−5.6 −4.3)	−15	4	0.164
	Th9/10	−2.8 ± 4.2	(−3.5 −2.1)	−18	6	−3.0 ± 3.9	(−3.7 −2.4)	−13	8	0.607
	Th10/11	−4.5 ± 3.8	(−5.1 −3.9)	−16	5	−4.2 ± 4.0	(−4.9 −3.6)	−17	7	0.510
	Th11/12	−5.3 ± 4.0	(−5.9 −4.6)	−17	5	−5.4 ± 4.7	(−6.1 −4.6)	−16	6	0.842
	Th12/L1	−4.4 ± 4.5	(−5.1 −3.6)	−16	12	−3.4 ± 4.6	(−4.1 −2.6)	−13	11	0.058
	L1/L2	−1.3 ± 4.9	(−2.1 −0.5)	−16	13	−0.3 ± 5.2	(−1.2 0.5)	−11	13	0.102
	L2/3	4.1 ± 5.1	(3.2 4.9)	−6	21	5.0 ± 4.9	(4.2 5.8)	−8	19	0.120
	L3/4	10.3 ± 5.5	(9.4 11.2)	−4	26	10.4 ± 5.6	(9.5 11.4)	−2	27	0.877
	L4/L5	22.4 ± 6.7	(21.3 23.5)	6	45	21.8 ± 6.2	(20.8 22.8)	6	38	0.389
	L5/S1	23.9 ± 6.0	(22.9 24.9)	6	50	23.7 ± 6.9	(22.6 24.9)	0	49	0.814
bisegmental	Th10	−6.0 ± 4.8	(−6.8 −5.2)	−21	5	−6.1 ± 4.9	(−6.9 −5.3)	−17	7	0.866
	Th11	−6.7 ± 4.7	(−7.5 −6.0)	−23	6	−6.6 ± 5.4	(−7.5 −5.7)	−22	5	0.856
	Th12	−6.7 ± 5.4	(−7.6 −5.8)	−21	11	−6.2 ± 5.7	(−7.2 −5.3)	−19	11	0.489
	L1	−2.3 ± 6.0	(−3.3 −1.4)	−16	16	−1.2 ± 6.3	(−2.2 −0.1)	−14	18	0.101
	L2	4.1 ± 7.0	(2.9 5.2)	−10	27	6.0 ± 7.0	(4.8 7.1)	−10	27	0.024
	L3	14.2 ± 7.8	(12.9 15.5)	−9	41	15.6 ± 7.9	(14.3 16.9)	0	41	0.144
	L4	30.7 ± 9.1	(29.2 32.2)	11	52	30.9 ± 8.2	(29.6 32.3)	8	51	0.837
	PT	8.8 ± 6.8	(7.7 9.9)	−7	29	10.0 ± 6.0	(9.0 11.0)	−4	26	0.116
	PI	50.5 ± 11.0	(48.7 52.3)	27	86	52.7 ± 10.6	(50.9 54.4)	28	86	0.092
	SS	41.9 ± 8.0	(40.6 43.3)	21	71	42.7 ± 7.7	(41.5 44)	18	73	0.390

Using the iterative process to reduce the number of variables to predict mEPAs using the linear regression with only significant variables led to the exclusion of body weight, body size, BMI, and SS. None of these showed a significant impact on explaining the variance. The coefficients and the model quality of the logistic regression are shown in Table 2. The adjacent mEPA used showed the most frequent impact on the prediction. The model quality for mEPA prediction ranged from R^2 0.092 (L2/3) up to 0.953 (L4/5). Valid model quality ($R^2 \geq 0.8$) was reached for the prediction of L3/4 (using L1/2 and L4/5 $R^2 = 0.867$, using L2/3 and L5/S1 $R^2 = 0.914$) and L4/5 ($R^2 = 0.953$). The closest prediction, which failed the validity threshold of 0.8, was L2/3, with an R^2 of 0.712.

It can be observed that when predicting the mEPA, for nearly every case, the cranial and caudal mEPA had a significant impact. Age showed significance in just two cases, and seemed to have a low impact. Only in the caudal segments of the spine (L3/4 and L4/5), sex, PT, and PI showed significant impacts.

For bEPA the body weight and BMI showed no significant impact on the prediction. The adjacent bEPA showed the most frequently significant impact. The model quality ranged from 0.294 (L1) up to 0.972 (L4). The prediction of L3 ($R^2 = 0.899$) and L4 ($R^2 = 0.962$) met the validity criteria of $R^2 \geq 0.8$. For both predictions, the adjacent caudal bEPA showed significance, while the cranial bEPA did not.

Table 2. Results of the linear regression model for the prediction of mono- and bisegmental endplate angles (mEPA, bEPA). The explained variance (R^2) and the significant coefficients (ExpB) for the parameters used are given. Sex was coded as 1 for men and 2 for women.

	EPA Used				ExpB							
	Predicted	**Cranial**	**Caudal**	**R^2**	**Cranial**	**Caudal**	**Age**	**Sex**	**Body Size**	**PT**	**PI**	**SS**
monosegmental	Th9/10	Th8/9	Th11/12	0.568	0.597 ^{A}							
	Th10/11	Th8/9	Th11/12	0.645	0.334 ^{A}	0.472 ^{A}						
	Th10/11	Th9/10	Th12/L1	0.578	0.698 ^{A}	0.345 ^{A}						
	Th11/12	Th9/10	Th12/L1	0.626	0.419 ^{A}	0.760 ^{A}						
	Th11/12	Th10/11	L1/2	0.581	0.867 ^{A}	0.232 ^{A}						
	Th12/L1	Th10/11	L1/2	0.523	0.485 ^{A}	0.613 ^{A}						
	Th12/L1	Th11/12	L2/3	0.576	0.634 ^{A}	0.229 ^{A}	−0.028 ^{B}					
	L1/2	Th11/12	L2/3	0.295	0.404 ^{A}	0.457 ^{A}						
	L1/2	Th12/L1	L3/4	0.425	0.661 ^{A}	0.181 ^{A}						
	L2/3	Th12/L1	L3/4	0.712	0.245 ^{A}	0.539 ^{A}						
	L2/3	L1/2	L4/5	0.092	0.405 ^{A}							
	L3/4	L1/2	L4/5	0.867	0.215 ^{A}	0.367 ^{A}					0.050 ^{B}	
	L3/4	L2/3	L5/S1	0.914	0.426 ^{A}	−0.295 ^{A}	−0.033 ^{B}	2.718 ^{A}		−0.393 ^{A}	0.332 ^{A}	
	L4/5	L2/3	L5/S1	0.953		−0.316 ^{A}		4.745 ^{A}		−0.729 ^{A}	0.588 ^{A}	
bisegmental	Th10	Th8	Th12	0.696	0.492 ^{A}	0.233 ^{A}						
	Th11	Th9	L1	0.548	0.698 ^{A}	0.436 ^{A}						
	Th12	Th10	L2	0.413	0.867 ^{A}	0.258 ^{A}						
	L1	Th11	L3	0.294	0.320 ^{A}	0.370 ^{A}			−0.029 ^{A}			
	L2	Th12	L4	0.611	0.335 ^{A}	0.236 ^{A}	0.054 ^{B}		−0.075 ^{A}			0.255 ^{A}
	L3	L1	L4/S1	0.899		−0.576 ^{A}		5.005 ^{A}	−0.047 ^{A}	−0.648 ^{A}	0.533 ^{B}	0.414 ^{B}
	L4	L2	L5/S1	0.972		−0.611 ^{A}		6.135 ^{A}		−1.140 ^{A}	0.933 ^{A}	

^{A}: $p < 0.001$. ^{B}: $p < 0.05$.

For L3 and L4, an excellent model quality could be found with results of 0.899 and 0.972. In contrast to the mEPA, the predicted bEPA showed significant consideration of body height and SS. For both, predicting mEPA and bEPA for L3 and L4, sex showed a significant impact. However, the spino-pelvic parameters showed significance for the prediction of mEPA in just 5 of 42 possible cases (12%) and for the prediction of bEPA, significance was shown in only 6 of 21 possible cases (29%), respectively.

A visual comparison of the predicted (with model quality $R^2 \geq 0.8$) and the measured mEPA (L3/4 and L4/5) and bEPA (L3 and L4), shown as difference plots, is given in Figure 3. For all five predictions, the mean difference is close to zero, which is considered perfect. Nonetheless, the limits of agreement span over a considerable area of roughly ±10°. The linear regression showed a significant trend for each prediction, but with varying degrees. The linear regression of bEPA L4 was very low, with an R^2 of 0.07. However, the direction of the general trend of the linear regression is comparable in each figure, showing a positive trend. Visually, the scatter clouds show that the variability of the prediction seems not to be dependent on the value of the predicted EPA.

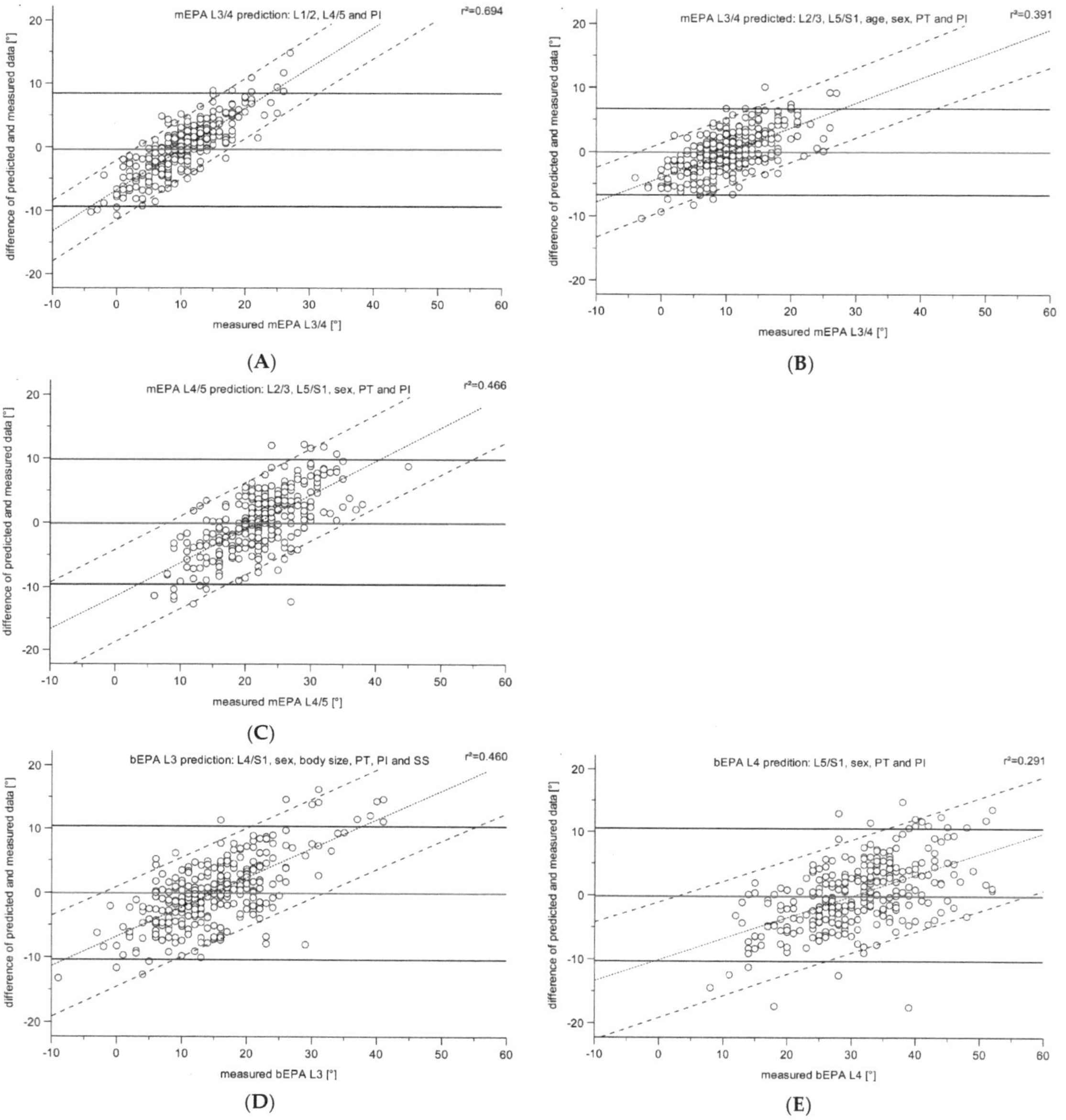

Figure 3. Difference plots of the measured mEPA ((**A**): L3/4, (**B**): L3/4, (**C**): L4/5) and bEPA ((**D**): L3, (**E**): L4) and their difference from the predicted monosegmental (mEPA) and bisegmental endplate angles (bEPA), respectively. Horizontal lines are the limits of agreement (black) and the mean difference (red); dotted and dashed lines are the regression line and their 0.95% confidence intervals. To detect the dependency of differences between measured and predicted angles, R^2 is given in the upper right corner of each plot.

4. Discussion

The aim of this study was to investigate if spinal segment angles of the thoraco-lumbar and lumbar spine can be sufficiently predicted using spino-pelvic parameters. In general, the predictions showed good to perfect results. However, this only applies to the generalization of our results. Limiting the threshold of the model, as we have done with a model quality greater than 0.8, resulted in only two or three predictions being supported, depending on whether mono- or bisegmental angles were predicted, respectively. Surprisingly, the spino-pelvic parameters had little influence on the predictions, with the only effect found in the lumbar region. The measurement of the cranial endplate of S1 is mandatory to determine spino-pelvic parameters. In other words, segments that are directly related to S1 are the most likely to be influenced by the orientation of the S1 cranial endplate, which could have a direct effect in this study. Here, the orientation of the sacrum, respectively S1, seems to have direct impact on the segment angles in the lower lumbar segments. Segments that are more cranial seem to be influenced indirectly by the pelvic alignment.

In several studies, PI was found to be the key factor in sagittal balance [8,10,12,13]. In our study, regarding the prediction of single segmental angles, the PI shows significance only in the caudal region of the lumbar spine. The studies mainly examine the influence of the spino-pelvic parameters on the general alignment, i.e., on the thoracic kyphosis and/or lumbar lordosis. The prediction of lumbar lordosis based on spino-pelvic parameters has previously been shown in two studies [8,14]. In addition, in these studies, the parameters are determined in the standing position. There is no consideration of their influence on individual the segment angles.

As for the PI, only a rare impact of the PT in the segmental angles was found. Significance was found for the last two segments of the lumbar spine (mEPA and bEPA), with negative coefficients found for PT and positive coefficients for PI, respectively. The coefficients, and thus, the actual impact on the prediction, is distinctly stronger for the PT than for the PI.

For the PI, the influence on the spinal profile, and thus, on these segments, is easy to imagine. The more the sacrum is tilted, the greater the PI, the greater the lordosis, and thus, the larger the individual segment angles. The less tilted it is, the more it is parallel to the horizontal plane in supine position.

Regarding the geometric similarity for the PT, the more the pelvis is retro verted, the higher the PI and the more pronounced the lordosis.

Our data and analyses indicate that the closer the segments are located to the sacrum, the stronger the impact of the PT and the PI on the segment angles. In the cranial segments of the spine, their influence seems to completely disappear. For the SS, we would have assumed a similar relationship, compared to the PT and PI. However, no significant impact on the prediction can be determined for the monosegmental segments.

We assume that the influence of the three spino-pelvic parameters and their effects interact and thus, cannot be reduced at a one-dimensional level. Additionally, other variables, including the position in which the parameters are determined, may interact and affect the segmental angles [15].

In our prediction, we considered body weight, height, and BMI as anthropometric. However, none of these variables were found to affect the segment angles in the supine position.

The body weight, and thus, the BMI, have an effect on the spino-pelvic alignment in the standing position [16–18]; however, their impact on segment angles was not examined here. Therefore, it is unclear whether their impact on the individual segment angles exists in the standing position, or whether the supine position is the reason for the disappearance of this impact. Of course, it may be that the effect on the individual segment angles is quite small, and that this effect only becomes apparent when considering the thoracic kyphosis or lumbar lordosis.

Age and gender show only a rare significance. These findings affect the predictions of the mEPA and bEPA in a comparable manner.

In investigating the effect of age on the spino-pelvic parameters, no differences could be found between age groups. Various authors have so far, been able to demonstrate age effects on the spino-pelvic parameters [9,12,17,19].

Different authors have been able to show differences between men and women with regard to the spino-pelvic parameters in the standing position [13,17,20]. Our study is based on data collected from patients in the supine position. We suggest that this is the main reason that we did not find comparable results. However, our surgeons assessed the extent of fracture reduction in the patient in CT images collected in the supine position.

Naturally, effects of positional changes—standing versus supine—are to be expected. Radiological diagnosis is also possible in the standing position, but this is not possible for every patient, depending on the injury. Additionally, not every hospital has the technical equipment for this type of diagnostic method.

It would be interesting to investigate the prediction of mEPA and bEPA for a population in which radiological diagnostics were performed in both a standing and a supine position in order to compare the results of both predictions. Nevertheless, the consideration of spino-pelvic parameters in fracture correction is recommended [21]. Furthermore, the spino-pelvic parameters also show an impact on the patient's quality of life [9], and therefore, should be of particular consideration.

In this respect, the correction or non-correction of fractures not only has an impact on the spino-pelvic alignment, but also has a secondary effect on the patient's quality of life.

In this study, we did not examine lumbar lordosis per se. The aim of this study was to determine valuable parameters to predict segment angles. These parameters should not have a great deal of influence if a fracture has occurred. In a fractured spine, it can be assumed that depending on the severity and location of the fracture, the thoracic kyphosis or lumbar lordosis might have changed, and it is likely unclear how the fracture has affected the other segmental angles. Differences between mEPA or bEPA in the lower lumbar region are greater than in the more cranial segment angles, respectively. This might certainly be a reason why there is a better prediction in the lumbar area, and why the influences of the adjacent segment angles are greater here. In the more cranial regions, the differences in segment angles between may be too small to differentiate between segments.

If the surgeon orientates the fracture reduction on the lumbar lordosis from S1 to the inflection point, however, in our experience, the inflection point appears to be more cranial when lying down than when in a standing position, which makes lumbar lordosis less suitable as a target parameter in the supine position.

During the surgical reduction of the fracture, the avoidance of insufficient or excessive reduction of fractures, lacking sufficient information regarding the individual physiological profile, is also in surgical focus. With precise individual adjustment of the fracture, less additional compensation may be necessary, resulting in better rehabilitation after surgery. Particularly after spondylodesis and existing malalignment, unfavorable loads on the adjacent movement segment are to be expected. Misalignments due to poor fracture position likely diminish the success of subsequent rehabilitation and can lead to pain, adjacent fracture, or other complications. The combination of nutritional supplements and rehabilitation programs can reduce the patient's pain and improve the quality of life [22]. However, spinal alignment is not the only thing that determines recovery. Even with poor posture, the health of patients can be positively influenced by training [23].

Finally, only a few of the parameters used in this study are suitable for predicting segment angles in general. Overall, the caudally and cranially adjacent segment angles seems to be sufficient parameter for use in predicting the segment angles between them. The prerequisite for this, however, is that the adjacent segments are intact. Looking closer at our results, the predictions achieve a wide range, which is shown by the relatively large limits of agreement in the difference plots. Regarding the mean values of the mEPA and bEPA, plus/minus 10 degrees is too great to provide a good individual recommendation for the scope of correction. Ultimately, further investigations must be carried out in order to predict the segment angles in the supine position, considering the spino-pelvic parameters. Questions also arise regarding the amount of malalignment for which the human body can physiology compensate, as well as the way in which this compensation changes with age.

Author Contributions: Conceptualization, P.S. and B.W.U.; Methodology, P.S. and B.W.U.; Software, P.S.; Validation, P.S. and A.J.; Formal analysis, P.S.; Investigation, P.S. and A.J.; Resources, A.J.; Data curation, P.S.; Writing—original draft, P.S., A.J., C.G., T.M. and B.W.U.; Writing—review & editing, P.S., C.G. and B.W.U.; Visualization, P.S.; Supervision, C.G., T.M., G.O.H. and B.W.U.; Project administration, C.G. and G.O.H. All authors have read and agreed to the published version of the manuscript.

Funding: This research received no external funding.

Institutional Review Board Statement: The study was conducted in accordance with the Declaration of Helsinki and approved by the Ethics Committee of the Medical Association of Saxony-Anhalt Germany on 20 August 2020, under number 46/20.

Informed Consent Statement: Informed consent was obtained from all subjects involved in the study.

Data Availability Statement: The data from all subjects involved and analyzed in this study are available from the corresponding author on reasonable request.

Conflicts of Interest: The authors declare no conflict of interest. B.W. Ullrich has received research funding from Medizintechnik Marquardt, Germany, as well as payments for lectures from B. Braun Aesculap, Germany.

References

1. Leucht, P.; Fischer, K.; Muhr, G.; Mueller, E.J. Epidemiology of traumatic spine fractures. *Injury* **2009**, *40*, 166–172. [CrossRef] [PubMed]
2. Maier, B.; Ploss, C.; Marzi, I. Verletzungen der thorakolumbalen Wirbelsäule. *Der Orthop.* **2010**, *39*, 247–255. [CrossRef] [PubMed]
3. Roussouly, P.; Gollogly, S.; Berthonnaud, E.; Dimnet, J. Classification of the normal variation in the sagittal alignment of the human lumbar spine and pelvis in the standing position. *Spine* **2005**, *30*, 346–353. [CrossRef]
4. Le Huec, J.-C.; Aunoble, S.; Philippe, L.; Nicolas, P. Pelvic parameters: Origin and significance. *Eur. Spine J.* **2011**, *20*, 564–571. [CrossRef] [PubMed]
5. Laouissat, F.; Sebaaly, A.; Gehrchen, M.; Roussouly, P. Classification of normal sagittal spine alignment: Refounding the Roussouly classification. *Eur. Spine J.* **2018**, *27*, 2002–2011. [CrossRef]
6. Stagnara, P.; DE Mauroy, J.C.; Dran, G.; Gonon, G.P.; Costanzo, G.; Dimnet, J.; Pasquet, A. Reciprocal angulation of vertebral bodies in a sagittal plane: Approach to references for the evaluation of kyphosis and lordosis. *Spine* **1982**, *7*, 335–342. [CrossRef]
7. Duval-Beaupere, G.; Schmidt, C.; Cosson, P. A Barycentremetric study of the sagittal shape of spine and pelvis: The conditions required for an economic standing position. *Ann. Biomed. Eng.* **1992**, *20*, 451–462. [CrossRef]
8. Legaye, J.; Duval-Beaupère, C.; Hecquet, J.; Marty, C. Pelvic incidence: A fundamental pelvic parameter for three-dimensional regulation of spinal sagittal curves. *Eur. Spine J.* **1998**, *7*, 99–103. [CrossRef]
9. Hasegawa, K.; Okamoto, M.; Hatsushikano, S.; Shimoda, H.; Ono, M.; Watanabe, K. Normative values of spino-pelvic sagittal alignment, balance, age, and health-related quality of life in a cohort of healthy adult subjects. *Eur. Spine J.* **2016**, *25*, 3675–3686. [CrossRef]
10. Zhou, S.; Xu, F.; Wang, W.; Zou, D.; Sun, Z.; Li, W. Age-based normal sagittal alignment in Chinese asymptomatic adults: Establishment of the relationships between pelvic incidence and other parameters. *Eur. Spine J.* **2020**, *29*, 396–404. [CrossRef]
11. Meyerding, H. Spondylolisthesis. *Surg. Gynecol. Obs.* **1932**, *54*, 371–377.
12. Legaye, J. Influence of age and sagittal balance of the spine on the value of the pelvic incidence. *Eur. Spine J.* **2014**, *23*, 1394. [CrossRef]
13. Asai, Y.; Tsutsui, S.; Oka, H.; Yoshimura, N.; Hashizume, H.; Yamada, H.; Akune, T.; Muraki, S.; Matsudaira, K.; Kawaguchi, H.; et al. Sagittal spino-pelvic alignment in adults: The Wakayama Spine Study. *PLoS ONE* **2017**, *12*, e0178697. [CrossRef]
14. Boulay, C.; Tardieu, C.; Hecquet, J.; Benaim, C.; Mouilleseaux, B.; Marty, C.; Prat-Pradal, D.; Legaye, J.; Duval-Beaupère, G.; Pélissier, J. Sagittal alignment of spine and pelvis regulated by pelvic incidence: Standard values and prediction of lordosis. *Eur. Spine J.* **2006**, *15*, 415–422. [CrossRef]
15. Park, S.-A.; Kwak, D.-S.; Cho, H.-J.; Min, D.-U. Changes of spinopelvic parameters in different positions. *Arch. Orthop. Trauma Surg.* **2017**, *137*, 1223–1232. [CrossRef]
16. Romero-Vargas, S.; Zárate-Kalfópulos, B.; Otero-Cámara, E.; Rosales-Olivarez, L.; Alpízar-Aguirre, A.; Morales-Hernández, E.; Reyes-Sánchez, A. The impact of body mass index and central obesity on the spino-pelvic parameters: A correlation study. *Eur. Spine J.* **2013**, *22*, 878–882. [CrossRef]
17. Zhu, Z.; Xu, L.; Zhu, F.; Jiang, L.; Wang, Z.; Liu, Z.; Qian, B.P.; Qiu, Y. Sagittal alignment of spine and pelvis in asymptomatic adults: Norms in Chinese populations. *Spine* **2014**, *39*, E1–E6. [CrossRef]
18. Khan, J.M.; Basques, B.A.; Kunze, K.N.; Grewal, G.; Hong, Y.S.; Pardo, C.; Louie, P.K.; Colman, M.; An, H.S. Does obesity impact lumbar sagittal alignment and clinical outcomes after a posterior lumbar spine fusion? *Eur. Spine J.* **2020**, *29*, 340–348. [CrossRef] [PubMed]

19. Zhu, X. Comparison of four methods for handing missing data in longitudinal data analysis through a simulation study. *Open J. Stat.* **2014**, *4*, 933. [CrossRef]
20. Janssen, M.M.; Drevelle, X.; Humbert, L.; Skalli, W.; Castelein, R.M. Differences in male and female spino-pelvic alignment in asymptomatic young adults: A three-dimensional analysis using upright low-dose digital biplanar X-rays. *Spine* **2009**, *34*, E826–E832. [CrossRef]
21. Vialle, R.; Levassor, N.; Rillardon, L.; Templier, A.; Skalli, W.; Guigui, P. Radiographic analysis of the sagittal alignment and balance of the spine in asymptomatic subjects. *J. Bone Joint Surg. Am.* **2005**, *87*, 260–267. [CrossRef] [PubMed]
22. Farì, G.; Santagati, D.; Pignatelli, G.; Scacco, V.; Renna, D.; Cascarano, G.; Vendola, F.; Bianchi, F.P.; Fiore, P.; Ranieri, M.; et al. Collagen peptides, in association with vitamin C, Sodium Hyaluronate, Manganese and Copper, as Part of the Rehabilitation Project in the Treatment of Chronic Low Back Pain. *Endocr. Metab. Immune Disord. Drug. Targets* **2022**, *22*, 108–115. [CrossRef] [PubMed]
23. Notarnicola, A.; Farì, G.; Maccagnano, G.; Riondino, A.; Covelli, I.; Bianchi, F.P.; Tafuri, S.; Piazzolla, A.; Moretti, B. Teenagers' perceptions of their scoliotic curves. an observational study of comparison between sports people and non- sports people. *Muscles Ligaments Tendons J.* **2019**, *9*, 225–235. [CrossRef]

Journal of *Personalized Medicine*

MDPI

Article

Effects of Different Orthoses on Neuromuscular Activity of Superficial and Deep Shoulder Muscles during Activities of Daily Living and Physiotherapeutic Exercises in Healthy Participants

Casper Grim [1,2], Christian Baumgart [3], Manuel Schlarmann [4], Thilo Hotfiel [1,5], Sasha Javanmardi [3], Natalie Hoffmann [1], Eduard Kurz [6], Jürgen Freiwald [3], Martin Engelhardt [1] and Matthias W. Hoppe [7,*]

[1] Center for Musculoskeletal Surgery Osnabrück (OZMC), Klinikum Osnabrück, Am Finkenhügel 1, 49076 Osnabrueck, Germany
[2] Institute for Health Research and Education (IGB), Faculty of Human Sciences, University of Osnabrueck, Nelson-Mandela-Straße 11, 49076 Osnabrueck, Germany
[3] Department of Movement and Training Science, University of Wuppertal, Gaußstraße 20, 42119 Wuppertal, Germany
[4] Institute of Human Movement and Sport Scienes, University of Osnabrueck, Jahnstraße 75, 49080 Osnabrueck, Germany
[5] Department of Trauma and Orthopaedic Surgery, University Hospital Erlangen, Friedrich-Alexander-Universität Erlangen-Nürnberg, Krankenhausstraße 12, 91054 Erlangen, Germany
[6] Department of Orthopedic and Trauma Surgery, Faculty of Medicine, Martin-Luther University Halle-Wittenberg, Ernst-Grube-Straße 40, 06112 Halle (Saale), Germany
[7] Movement and Training Science, Leipzig University, Jahnallee 59, 04109 Leipzig, Germany
* Correspondence: matthias.hoppe@uni-leipzig.de; Tel.: +49-341-97-31823

Citation: Grim, C.; Baumgart, C.; Schlarmann, M.; Hotfiel, T.; Javanmardi, S.; Hoffmann, N.; Kurz, E.; Freiwald, J.; Engelhardt, M.; Hoppe, M.W. Effects of Different Orthoses on Neuromuscular Activity of Superficial and Deep Shoulder Muscles during Activities of Daily Living and Physiotherapeutic Exercises in Healthy Participants. *J. Pers. Med.* **2022**, *12*, 2068. https://doi.org/10.3390/jpm12122068

Academic Editor: Jih-Yang Ko

Received: 7 November 2022
Accepted: 13 December 2022
Published: 15 December 2022

Abstract: Background: This study aimed to investigate the effects of different shoulder orthoses on the neuromuscular activity of superficial and deep shoulder muscles during activities of daily living (ADL) and physiotherapeutic exercises. Methods: Ten participants with healthy shoulders (31 ± 3 years, 23.1 ± 3.8 kg/m^2) were randomized to receive a "shoulder sling", an "abduction pillow" and a "variably adjustable orthosis" on the dominant side. With each orthosis, they completed seven ADL with and four physiotherapeutic exercises without wearing the orthoses. An electromyographic system was used to record the neuromuscular activity of three superficial (trapezius, deltoid, pectoralis major) and two deep shoulder muscles (infraspinatus, supraspinatus) using surface and intramuscular fine-wire electrodes. Results: The neuromuscular activity differs between the orthoses during ADL ($p \leq 0.045$), whereby the "variably adjustable orthosis" mostly showed the highest activation levels associated with the worst subjective wearing comfort rated on a visual analog scale. In addition, differences exist between the physiotherapeutic exercises ($p \leq 0.006$) demonstrating the highest activations of the infra- and supraspinatus muscles for assistive elevation and wipe across a table, middle for pendulum and lowest for continuous passive motion exercises. Conclusions: The neuromuscular activity of superficial and deep shoulder muscles differs between the orthoses during ADL and also between the physiotherapeutic exercises.

Keywords: conservative therapy; CPM; EMG; mobilization; musculoskeletal complains; prevention; rehabilitation; rotator cuff; tendon; upper extremity

1. Introduction

Immobilization is a fundamental part of conservative therapy and postoperative rehabilitation of various diseases and injuries of the shoulder joint and its girdle [1,2]. Over the last 20 years, orthotics has developed into a leading field of medical products [3,4]. However, the basic idea remained the same: partial immobilization, ensuring a prescribed joint position, and thereby stabilization and relief of the injured or surgically treated structures [1,5]. Additionally, ideas such as early functional treatment approaches and pathology-adapted immobilization positions were implemented over time [1]. Nowadays, it is well established to treat various acute and overuse injuries of the shoulder with different orthosis concepts, either as a primary conservative therapy option or part of the postoperative rehabilitation [1,3,6].

For a shoulder orthosis, different requirements can be defined. Predominantly, its use should lead to sufficient immobilization and reduction of neuromuscular activity to protect the injured or repaired tissue [5,7–10]. Another aspect is related to the early functional treatment, namely that it should be easy to take the orthosis on and off and allow the shoulder to be exercised by simply opening few fasteners or straps [5,9]. Additionally, it is important that sufficient mobility for activities of daily living (ADL) is retained [5]. It is self-evident that this includes not only tasks such as personal hygiene and dressing, but also independently putting on and taking off an orthosis. During such activities, it is known that the wearing comfort of an orthosis has an influence on the patient compliance [1]. In shoulder orthotics, three different retention and construction concepts can be distinguished: (i) sling-like orthosis with no abduction and mainly internally rotated position (hand on the belly), (ii) soft shoulder orthosis (abduction pillow) with an abducted position and a slightly variable rotation component and (iii) variable adjustable orthosis (more rigid construction) with an abduction and more fixed rotation position [1,6].

Postoperative rehabilitation protocols of the shoulder aim to balance between the deconditioning and stiffness of the joint that can be related to immobilization and harmful over-activity effects [6–10]. In fact, excessive neuromuscular activation of the rotator cuff muscles after surgery can contribute to a retearing of the reconstructed tendons [6,11,12]. Furthermore, an increased joint stiffness has also been shown to have adverse effects on clinical outcome and recovery [8]. Therefore, several postoperative protocols have been designed to retain joint mobility while safely loading the repaired tendons, including passive mobilization (e.g., continuous passive motion; CPM) and more active physiotherapeutic exercises (e.g., pendulum exercises) [5,6].

Since it is clear that superficial and deep shoulder muscles are simultaneously activated or coactivated during some motor tasks [13,14], it is of special interest to concurrently investigate the neuromuscular activity of both different muscle layers for a comprehensive evaluation of orthoses during ADL and physiotherapeutic exercises. Previous studies investigated either superficial or deep shoulder muscle activations during ADL or physiotherapeutic exercises including CPM via different electromyographic (EMG) approaches [15–17]. Moreover, only one previous study evaluated the effects of a not commercially available orthosis on the neuromuscular activity of shoulder muscles [18]. To the best of our knowledge, there is no study taking ADL and physiotherapeutic exercises into account by a comprehensive EMG-approach while also comparing the effects of different commercial orthoses with each other. Such knowledge may help to optimize shoulder orthoses and rehabilitation protocols.

The aim of the study was to investigate the effects of different shoulder orthoses on the neuromuscular activity of superficial and deep shoulder muscles during ADL. Additionally, the neuromuscular activity was examined during different physiotherapeutic exercises without wearing the orthoses. It was hypothesized that the neuromuscular activity of shoulder muscles differs between the orthoses during ADL and also between the physiotherapeutic exercises.

2. Materials and Methods

2.1. Participants

Five female and five male participants were recruited. The inclusion criteria were: (i) 18–35 years old, (ii) body mass index < 28.0 kg/m^2, (iii) right-handed and (iv) signed written informed consent to participate. The exclusion criteria were: (i) previous or acute injuries, complaints or misalignments at the cervical and thoracic spine or shoulder joint, (ii) cognitive or psychological impairments, (iii) intake of blood thinners and (iv) neuromuscular diseases. The participants were informed about the purposes, procedures and potential risks of the study. All procedures were pre-approved and accepted by the Ethics Committee of the University of Wuppertal (MS/BBL; 190702) and were conducted in accordance with the Declaration of Helsinki. Table 1 summarizes the anthropometric characteristics of the participants.

Table 1. Anthropometric characteristics of the participants.

Variable	Mean ± SD
Age (years)	31± 3
Body height (m)	1.72 ± 0.08
Body mass (kg)	68.9 ± 14.8
Body mass index (kg/m^2)	23.1 ± 3.8

2.2. Design

In this observational-experimental study, the data collection took place under standardized conditions in an examination room of a hospital over a 14-day period. Per each participant, the data collection lasted 180 min and was carried out on one day. After the participants had signed the written informed consent, a certified physiotherapist investigated the musculoskeletal system by routine clinical examinations for any exclusion criteria described above. Additionally, a specialist in orthopedics and trauma surgery examined the shoulder joint by a sonography device with a standardized preset. In case the participants could be included in the study, they were randomized to receive three different shoulder orthoses on the dominant right side. With each orthosis, the participants completed seven different ADL. Additionally, they performed four different physiotherapeutic exercises without wearing the orthoses. The ADL and physiotherapeutic exercises were defined due to their common occurrence and practical application as well according to previous studies [18–20]. During all movement tasks, an EMG-system was used to record the neuromuscular activity of selected superficial and deep shoulder muscles. In addition, an inertial measurement unit (IMU), a video camera and a visual analogue scale (VAS) were used to quantify the level of immobilization, beginning and end of each movement task and wearing comfort of each orthosis, respectively.

2.3. Shoulder Orthoses

Figure 1 shows the three used shoulder orthoses. These orthoses were selected due to their commercial availability, frequent application after surgical care and completely different immobilization concepts. Orthosis #1 was a "shoulder sling" that positioned the arm in internal rotation in front of the body and 0° abduction. Orthosis #2 was an "abduction pillow" with 10° internal rotation and 20° abduction; Orthosis #3 was a "variably adjustable orthosis" with neutral rotation position and 20° abduction. All orthoses were fitted according to the manufacturer's instructions and under supervision of the physiotherapist.

2.4. Activities of Daily Living

With each orthosis, the participants performed seven different ADL in the following order:

2.4.1. Stand Up-Sit Down from a Chair

The participants stood in front of a 50 cm high treatment couch, which corresponds to a seat high of many commonly used chairs in daily living. They sat down at their own pace, remained seated for 3 s and stood back up.

2.4.2. Open-Close a Water Bottle

The participants stood in front of a 90 cm high worktop on which a bottle was placed. They were asked to grab the bottle with their left and open-close the bottle with their right hand. Then, they removed both hands from the bottle and remained standing for 3 s.

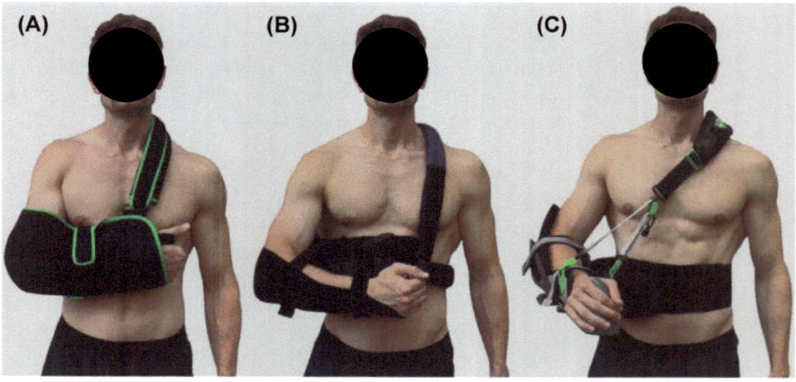

Figure 1. Applied shoulder orthoses. Note: (**A**) Orthosis #1 (shoulder sling); (**B**) Orthosis #2 (abduction pillow); (**C**) Orthosis #3 (variably adjustable orthosis).

2.4.3. Type the Alphabet on a Computer Keyboard

The participants sat on a 50 cm high chair in front of a 75 cm high desk. A standard computer keyboard was placed 12 cm from the edge of the table. Both hands were on the table. Then, they were asked to type the alphabet from A to Z using both hands. Upon completion, they took both hands off the table and remained seated for 3 s.

2.4.4. Put On-Take Off a Jacket

In a standing position, the participants took their jacket from the back of a chair immediately to their left side. The jacket was considered to be put on, when the left arm was completely passed through the jacket sleeve and the jacket was placed over the right shoulder. Then, the participants remained standing for 3 s and took the jacket back off.

2.4.5. Lie Down-Stand Up on a Couch

The participants stood in front of a 50 cm high treatment couch. The headboard was elevated by 10°. They were asked to lie down completely on their back, remain there for 3 s and stand back up.

2.4.6. Walk Slowly over a Short Distance

The participants were asked to walk a 10 m distance on level ground at a self-chosen pace, remain standing for 3 s and walk the same distance back again.

2.4.7. Put On-Take Off an Orthosis

According to the manufacturer's instructions, the participants removed the orthosis independently and placed it on a chair immediately to their left side. Then, they remained in place for 3 s and put the orthosis back on.

2.5. Physiotherapeutic Exercises

Additionally, the participants completed the four different physiotherapeutic exercises without an orthosis accordingly.

2.5.1. Wipe across a Table

The participants sat on a 50 cm high chair. Both hands lay folded on a towel placed on a 75 cm high table. The participants pushed the towel forwards to a mark placed 60 cm from the edge of the table. The left arm should guide the movement and the right shoulder should remain relaxed.

2.5.2. Pendulum

The participants stood next to a 75 cm high table and placed their left hand so far forward on the table that their upper body was bent forward by 50°. From this starting position, the right arm was allowed to hang relaxed and swing loosely back and forth 10 times. Due to the relatively short duration of the movement, the participants performed the swinging 3 times.

2.5.3. Assistive Elevation

The participants grabbed their right wrist with the left hand and moved the right arm upwards to 130° elevation of the shoulder joint. They were instructed to activate their right shoulder muscles as little as possible. Passive elevation was instructed and supervised during the measurements by a physiotherapist.

2.5.4. Continuous Passive Motion

Finally, the participants sat on a CPM-chair. The chair was set to an initial position of 25° abduction. The right arm was passively moved until 60° abduction and then back to the initial position. The exercise was set to a duration of 5 min with an angular velocity of 90° per min.

Each movement task of the ADL and physiotherapeutic exercises was performed under the supervision of the physiotherapist 10 times and was separated by a 3 s rest period with exception of the pendulum und CPM exercises.

2.6. Data Collection

The collection of the EMG-, IMU-, video- and VAS-data was as follows:

2.6.1. Electromyography

A telemetric EMG-system (NORAXON, Clinical DTS, Scottsdale, AZ, USA) was used to record the neuromuscular activity of three superficial and two deep shoulder muscles. As superficial muscles, the trapezius (pars descendens), deltoid (pars acromialis) and pectoralis major (pars sternalis) muscles were captured using surface electrodes (HEX Dual Electrodes, NORAXON, Scottsdale, AZ, USA) in a bipolar configuration. The arrangement of the electrodes and skin preparation were conducted according to established guidelines [21,22]. Additionally, deep shoulder muscles of the rotator cuff, namely the infraspinatus and supraspinatus muscles, were examined using intramuscular fine-wire electrodes. To ensure an accurate placement of the electrodes [23], an ultrasound device (M-Turbo, Fujifilm Sonosite, Amsterdam, The Netherlands) was utilized for a real-time sonography-guided percutaneous application. The applications were performed by a specialist in orthopedics and trauma surgery, who is skilled and certified according to the German Society of Ultrasound in Medicine (DEGUM). The intrusion depth was set at 40 mm and the focus was centered at 50% cross-sectional muscle diameter. A linear probe (HFL50x, Sonosite, Amsterdam, The Netherlands) with a frequency of 12 MHz within standardized musculoskeletal preset was positioned at the posterior shoulder aspect in longitudinal axis for the infraspinatus and supraspinatus muscle separately. A 75 mm needle (Chalgren Enterprises, Gilroy, CA, USA) was then aligned via in-plane technique visualizing the entire shaft and needle tip. When an accurate position was achieved, the needle was released leaving the fine-wire electrodes intramuscularly centered. The procedure was digitally recorded for each participant. All procedures were performed under sterile standards according to established recommendations [24]. Figure 2 shows the placement of the EMG-electrodes to the investigated shoulder muscles.

For each of the shoulder muscles, a maximal voluntary contraction (MVC) test was conducted to normalize the recorded EMG-data. Therefore, ramped maximal isometric muscle contractions over 5 s in duration were performed and repeated after a 60 s recovery period 3 times. The particular procedure for each muscle was as follows: The trapezius (pars descendens) muscle was activated caudally by giving resistance of the physiotherapist from above on the shoulder. Then, the participants were instructed to pull the shoulder upwards. For the pectoralis major (pars sternalis) muscle, the participants were instructed to grasp a rod with both hands a shoulder-width apart and hold it at shoulder height in front of the upper body. The physiotherapist stood behind the participants and applied resistance by pulling the rod dorsally. The participants were instructed to push the rod away from themself. For the supraspinatus and deltoideus muscles, the upper arm was in zero position in the shoulder joint with the forearm flexed by 90° to control the neutral rotational position of the shoulder joint. Then, the upper arm was tensed against resistance inducted by the physiotherapist at the distal upper arm in abduction. For the infraspinatus muscle, the upper arm was tensed in external rotation at 90° from zero position on the shoulder joint. To prevent abduction, the participants were instructed to fix the left hand of the physiotherapist to their upper body with their elbows.

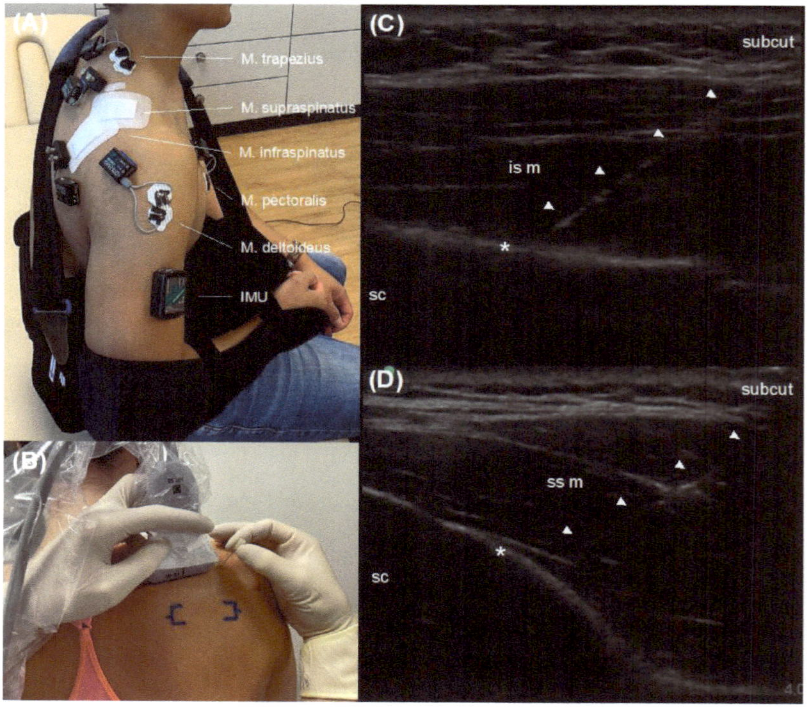

Figure 2. Placement of the electromyography electrodes on the investigated shoulder muscles. The placement of the inertial measurement unit is also shown. Note: (**A**) placement of electrodes and the inertial measurement unit (IMU); (**B**) sonography-guided application of the fine-wire electrodes; (**C**) positioning of electrode in the infraspinatus muscle; (**D**) positioning of electrode in the supraspinatus muscle. Abbreviations: subcut = subcutaneous fat tissue; is m = infraspinatus muscle; sc = scapula; ss m = supraspinatus muscle; asterisk = cortical interface; arrows = intramuscular needle position.

All EMG amplifiers and cables were fixed by double-sided adhesive tape and sterile patches (Figure 2). The EMG-data were acquired at 1500 Hz, processed using standard procedures (25 Hz Butterworth high-pass filter, 200 ms RMS) and MVC-normalized (500 ms) over all trials as mean activation per movement [25,26].

2.6.2. Inertial Measurement Unit

An IMU-device (NORAXON, myoMOTION Research Pro, Arizona, USA), collecting 3D accelerometer data at 100 Hz, was used to quantify the total amount of immobilization of the shoulder joint, due to both active muscle contraction and passive external caused movements, induced by the different orthoses. The unit was placed by double-sided adhesive tape on the middle distance of the upper arm (Figure 2). To quantify the global movement of the upper arm during the ADL, the numerical integral of the resulting acceleration was calculated and averaged over all trials [27].

2.6.3. Video Analysis

A video camera (Logitech, C920; Lausanne, Switzerland) operating at 30 Hz was used to detect the beginning and end of each movement task within the recorded EMG- and IMU-data. A lamp was placed in the recorded video for synchronizing.

2.6.4. Visual Analogue Scale

A 10 cm VAS was used to rate the wearing comfort of each orthosis. The ratings were conducted by the participants after all devices and materials have been removed from the body.

2.7. Statistical Analysis

All data were collected and proceeded by manufacturer software (NORAXON, MyoResearch 3.18, Arizona USA). The statistical analysis was performed in RStudio (vers. 2022.7.1.554) [28,29]. Due to the non-given normal distribution of several data and small sample size, non-parametric statistical tests were applied. Friedman tests were used to detect significant mean effects between the three orthoses during ADL and between the four physiotherapeutic exercises. In terms of global significance, Wilcoxon-Signed-Rank tests were used for post-hoc analyses. Corresponding effects sizes according to Kedall's W were computed and interpreted accordingly: 0.1 to <0.3, small; 0.3 to <0.5, moderate and ≥0.5, large [30]. The threshold for global statistical significance was set at $p < 0.05$, whereas Holm corrections were applied to control for type 1 errors during post-hoc testing.

3. Results

3.1. Effects of Orthoses on Neuromuscular Activity during Activities of Daily Living

Figure 3 shows the effects of the orthoses on the neuromuscular activity of the shoulder muscles during the ADL. The neuromuscular activity of shoulder muscles differs between the selected orthoses during ADL. Particularly, the variably adjustable orthosis (orthosis #3) showed the highest activation levels (Figure 3). The Friedman tests revealed several global significant effects ($p \leq 0.045$). The corresponding effect sizes ranged from moderate to large ($W \geq 0.31$). The Wilcoxon-Signed-Rank tests showed numerous significant post-hoc effects summarized in Figure 3.

Figures 4 and 5 show the effects of the orthoses on the total amount of immobilization of the shoulder joint and wearing comfort during the ADL, respectively. The Friedman tests showed several global significant effects on the immobilization ($p \leq 0.02$) and wearing comfort ($p < 0.001$). The associated effect size(s) ranged from moderate to large ($W \geq 0.39$). The Wilcoxon-Signed-Rank tests showed numerous significant post-hoc effects summarized in Figures 4 and 5.

3.2. Effects of Physiotherapeutic Exercises on Neuromuscular Activity

Figure 6 shows the effects of the physiotherapeutic exercises on the neuromuscular activity of the shoulder muscles. The Friedman tests demonstrated several global significant effects ($p \leq 0.006$) supported by moderate to large effect sizes ($W \geq 0.41$). The Wilcoxon-Signed-Rank tests indicated numerous significant post-hoc effects summarized in Figure 6.

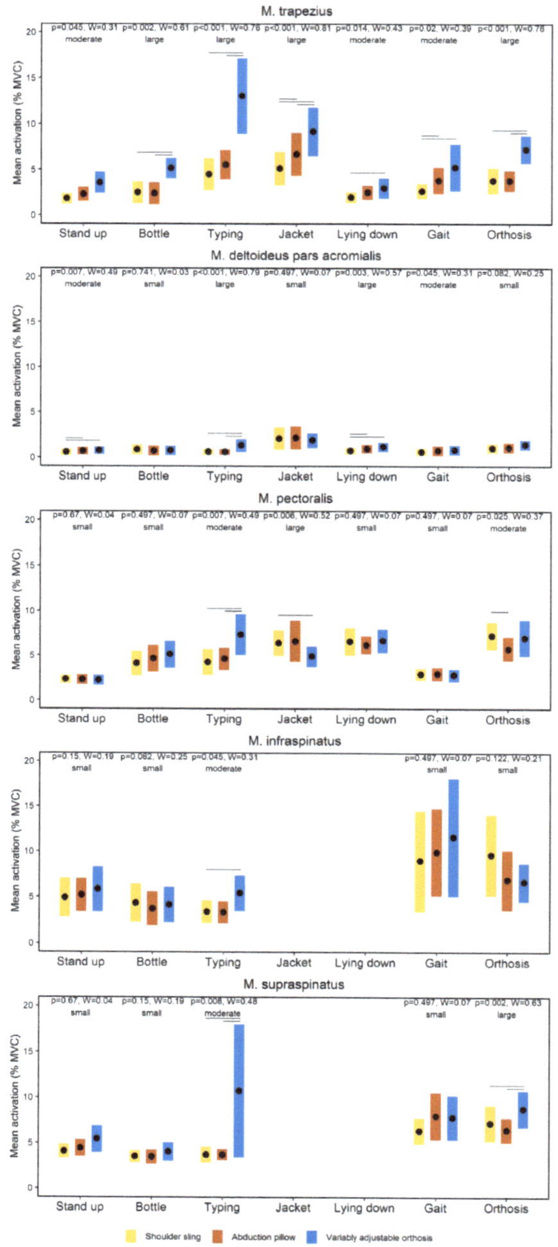

Figure 3. Effects of the orthoses on the neuromuscular activity of the shoulder muscles during the activities of daily living. Note: Diamonds and bars indicate mean activations (%MVC) and 90% confidence intervals, respectively. Horizontal lines indicate significant post-hoc effects. Data of the infraspinatus and supraspinatus muscles during two movement tasks (jacket and lying down) could not be analyzed due to movement artefacts within the fine-wire electrodes.

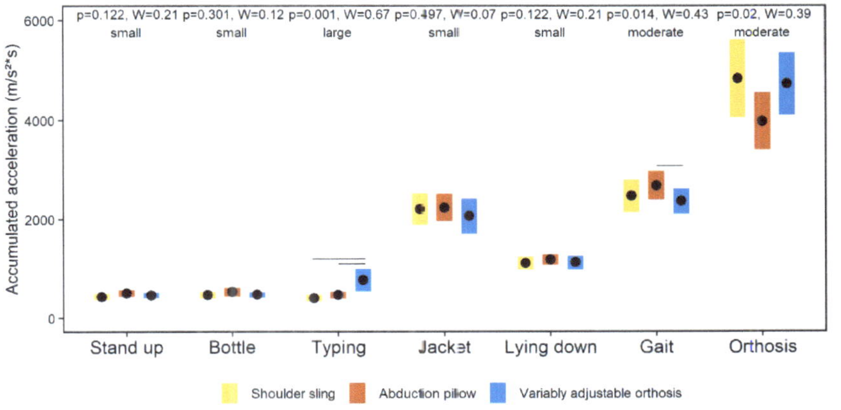

Figure 4. Effects of the orthoses on the total amount of immobilization of the shoulder joint during the activities of daily living. Note: Diamonds and bars indicate accumulated accelerations and 90% confidence intervals, respectively. Horizontal lines indicate significant post-hoc effects.

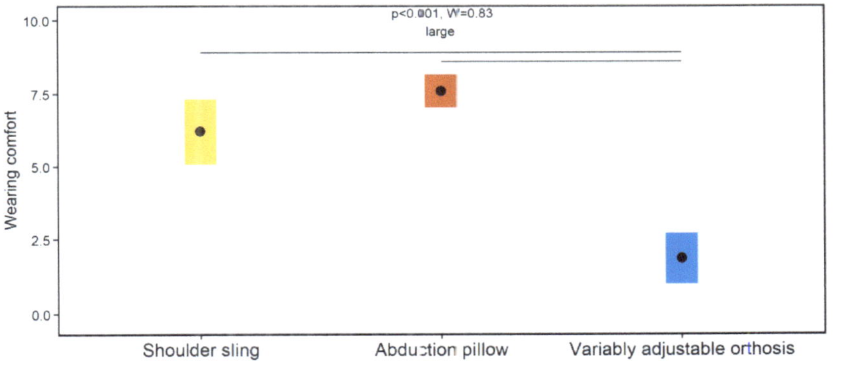

Figure 5. Effects of the orthoses on the wearing comfort during the activities of daily living. Note: Diamonds and bars indicate mean ratings and 90% confidence intervals, respectively. Horizontal lines indicate significant post-hoc effects.

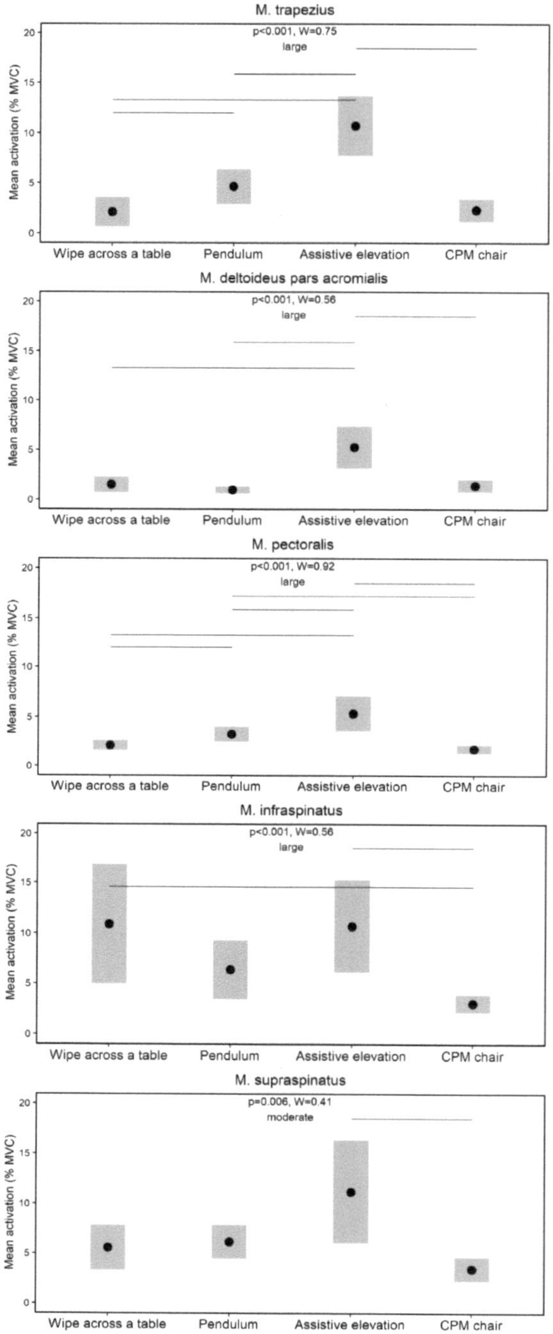

Figure 6. Effects of the physiotherapeutic exercises on the neuromuscular activity of the shoulder muscles. Note: Diamonds and bars indicate mean activations (%MVC) and 90% confidence intervals, respectively. Horizontal lines indicate significant post-hoc effects.

4. Discussion

This study investigated the effects of three different orthoses on the neuromuscular activity of superficial and deep shoulder muscles during ADL. Additionally, the neuromuscular activity was examined during four different physiotherapeutic exercises without wearing the orthoses. Our main findings were that the neuromuscular activity of shoulder muscles differ between (i) the orthoses during ADL, whereby the variably adjustable orthosis mostly showed highest activation levels; and between (ii) the physiotherapeutic exercises demonstrating highest activations for the assistive elevation and wipe across a table exercises, especially of the infra- and supraspinatus muscles.

The first main finding was that the neuromuscular activity of shoulder muscles differs between the selected orthoses during ADL. Particularly, the variably adjustable orthosis (orthosis #3) showed the highest activation levels (Figure 3). A previous study shows that neuromuscular rotator cuff activity of healthy participants did not exceed 11% MVC during ADL while wearing an orthosis [18], which is similar to the MVC-data of this study (Figure 3) and supports our methodological approach. However, compared to the previous study [18], all of our investigated shoulder orthoses are commercially available, allowing more external valid conclusions for clinical care. In our study, it was hypothesized that the different immobilization concepts of the orthoses (Figure 1) lead to different neuromuscular activations of the shoulder muscles during ADL. This could not be clearly confirmed by our results. Generally, the variable adjustable orthosis (orthosis #3) shows higher activations compared to the shoulder sling (orthosis #1) and abduction pillow (orthosis #2), whereby both latter orthoses mainly did not differ (Figure 3). One speculation for these findings may be that the neuromuscular activation of shoulder muscles is primarily not caused by the immobilization concepts of the orthoses, but rather by the perceived subjective wearing comfort [1]. This assumption is supported by both the findings of a previous study, showing that soft shoulder orthoses are perceived as more comfortable than harder ones [31] and our data. In fact, the total amount of immobilization of the shoulder joint induced by the different orthoses during ADL, as quantified by IMU-data, did overall not systematically differ (Figure 4). Instead, the variable adjustable orthosis (orthosis #3), constructed of harder material (Figure 1), showed a lower wearing comfort than both other orthoses (Figure 5), and this worst comfort was associated with the highest neuromuscular activation (Figure 3). Thus, the wearing comfort of a shoulder orthosis must be considered when interpreting its immobilization effect by measuring neuromuscular activation levels and is an essential aspect for patient compliance [1]. However, it is recommended that a neuromuscular activation of <15% MVC is required for a safe healing of the injured or surgery treated tissue during the shoulder rehabilitation [32,33]. From this perspective, all three shoulder orthoses can be considered as safe for ADL during the rehabilitation (Figure 3), requiring more research.

The second main finding was that the neuromuscular activity of shoulder muscles differs between the chosen physiotherapeutic exercises. Surprisingly, the highest activations were evident for the passive assistive elevation and wipe across a table exercises, especially of the infra- and supraspinatus muscles (Figure 6). A classification of our data with those of previous studies [15,19,33–35] is difficult due to differences in the selection and execution of exercises. However, a systematic review [32] identified several physiotherapeutic exercises that were considered to be appropriate during the early postoperative rehabilitation after surgery of the infra- and supraspinatus muscles due to their low neuromuscular activation (MVC < 15%). These included comparable wipe across a table, pendulum and assistive elevation exercises that were also examined in a comparable manner in our study (Figure 6), supporting their application during the rehabilitation and again our methodological approach. A new finding of our study was that also a commonly performed passive exercise during the early postoperative rehabilitation, namely a CPM-chair [6], was found to have a low neuromuscular activation (MVC < 15%) (Figure 6). A previous study compared the neuromuscular activity of shoulder muscles during different passive exercises only and found lowest activation levels for CMP [36], which is in line with our findings (Figure 6).

Although a clear differentiation between active and passive rehabilitation exercises is difficult, especially in patients where protective muscle co-activations can be evident [14,15], the CPM-chair always showed lowest neuromuscular activation levels, which should be considered for the progression of exercise selection during the early postoperative rehabilitation [6]. Taken overall, the investigated physiotherapeutic exercises can be considered as safe due their low neuromuscular activation (MVC < 15%), whereas the highest activations of the infra- and supraspinatus muscles were found for the assistive elevation and wipe across a table, the middle for the pendulum, and the lowest for the CPM-chair. These hierarchical outcomes can be helpful for the periodization of rehabilitation protocols after surgical shoulder treatments, for which further investigations are essential.

An interesting additional finding was that the neuromuscular variability of deep was higher than those of superficial shoulder muscles. Predominantly, this higher variability was evident during the more dynamic movement tasks, such as walk slowly and put on-take off orthosis (Figure 3) as well as wipe across table and assistive elevation (Figure 6). Previous studies also showed higher neuromuscular variations for deep than superficial shoulder muscles during similar exercises [33,36], as indicated by higher dispersion data, which is in line with our calculations (Figures 3 and 6). While the underlying reasons remain unknown, both methodological and physiological causes can be assumed. For example, and compared to non-invasive surface electrodes, potential fewer active motor units are derived by the intramuscular fine-wire approach [37], which may explain the higher variability from an EMG-methodological point of view. A physiological point can be related to potential perceived pains or discomforts inducted by the invasive approach that could lead to more spontaneous discharges of motor units [37]. One could also speculate about a less voluntary motor control of the deep compared to the superficial shoulder muscles, which has been discussed for other deep muscles such as the multifidi [38,39]. However, from an evolutionary perspective, the human shoulder muscles are extremity muscles [40,41] and have a different type of innervation compared to the multifidus muscles [42]. Thus, more research to clarify the observed higher variability of deep shoulder muscles is warranted. However, from a clinical perspective, our variability outcomes underline that an "one treatment fits all" approach is—of course—not timely in the present context. Instead, a personalized postoperative treatment approach concerning the selection of shoulder orthoses and physiotherapeutic exercises is obligatory.

While our study undoubtedly increased the knowledge on the neuromuscular activity of superficial and deep shoulder muscles during ADL while wearing different shoulder orthoses and during physiotherapeutic exercises performed without the orthoses, the main noteworthy limitation is related to our relatively young and shoulder healthy participants. Since it is clear that the shoulder biomechanics of injured or surgery treated patients is completely different to healthy participants [43], caution should be used in transferring our results into the medical care. When designing the study, we could not rule out that some settings may exceed a certain limitation of shoulder immobilization, so that patients might have been jeopardized. Based on our results and the defined measurement setup, further studies recruiting surgically treated patients should be implemented to relating data to clinical conditions. Future studies should focus on older patients with traumatic injuries, degenerative changes or generally on patients at post-surgical situations. We acknowledge that our measurements including healthy patients represent short-term effects of ADL and physiotherapeutic exercises on neuromuscular activity. A continuous application of orthosis or a long-term immobilization may lead to functional impairments and adaptions. Hence, our data are not transferable to long-term conditions. Follow-up investigations assessing changes of neuromuscular activity related to the long-term use of orthosis need to be conducted.

5. Conclusions

This study suggests that it is predominantly not the immobilization concept of shoulder orthoses, but rather the wearing comfort that supports a low neuromuscular activation of superficial and deep shoulder muscles during ADL. In regard to common physiotherapeutic exercises performed during the early postoperative rehabilitation, the highest activations of the infra- and supraspinatus muscles were found for the assistive elevation and wipe across a table, the middle for the pendulum, and the lowest for the CPM-chair. Our results could be used to guide the orthosis selection and type of physiotherapeutic exercises. We are convinced that our results can provide useful information for both orthosis designers and clinicians.

Author Contributions: Conceptualization, C.G., C.B., M.S. and M.W.H.; investigation, M.S., T.H., S.J., N.H. and M.W.H.; methodology, C.G., C.B., M.S., T.H., E.K. and M.W.H.; software, C.B.; resources, C.G., M.E. and J.F.; writing—original draft preparation, C.G. and M.W.H.; writing—review and editing, C.B., T.H., E.K., J.F. and M.E.; visualization, C.B.; supervision, C.G. and M.W.H.; project administration, M.W.H. All authors have read and agreed to the published version of the manuscript.

Funding: This research received no external funding.

Institutional Review Board Statement: The study was conducted in accordance with the Declaration of Helsinki, and approved by the Ethics Committee of the University of Wuppertal (MS/BBL; 2 February 2019).

Informed Consent Statement: Informed consent was obtained from all subjects involved in the study.

Data Availability Statement: The datasets used and analyzed during the current study are available from the corresponding author on reasonable request.

Acknowledgments: The authors are grateful to all the participations for their participation and to Joana Brochhagen for her language and formatting assistance.

Conflicts of Interest: The authors declare no conflict of interest.

References

1. Krukenberg, A.; Thierbach, A.; Thiele, K.; Scheibel, M. Aktuelle Entwicklungen und Einsatzgebiete von Schulterorthesen. *Obere Extrem.* **2016**, *11*, 10–15. [CrossRef]
2. Bucksch, S.; Hoffmann, N.; Osterkamp, N.; Yilmaz, S. *Heil- und Hilfsmittelreport 2018*, BARMER, ed.; 37 Grad Analyse & Beratung: Köln, Germany, 2018; Volume 13.
3. Greitemann, B.; Maronna, U. Technische Orthopädie: Stellenwert in einem zunehmend operativ ausgerichteten Fach. *Orthopäde* **2013**, *42*, 842–853. [CrossRef] [PubMed]
4. Markets and Markets. Orthopedic Braces & Supports Market by Product (Knee, Ankle, Spine, Shoulder, Neck, Elbow, Wrist, Facial), Category (Soft, Hard), Application (Ligament (ACL, LCL), Preventive, OA), Distribution, and Region—Global Forecast to 2024. Available online: https://www.marketsandmarkets.com/Market-Reports/orthopedic-braces-support-market-24806829.html (accessed on 28 September 2022).
5. Sgroi, T.A.; Cilenti, M. Rotator cuff repair: Post-operative rehabilitation concepts. *Curr. Rev. Musculoskelet. Med.* **2018**, *11*, 86–91. [CrossRef] [PubMed]
6. Jung, C.; Tepohl, L.; Tholen, R.; Beitzel, K.; Buchmann, S.; Gottfried, T.; Grim, C.; Mauch, B.; Krischak, G.; Ortmann, H.; et al. Rehabilitation following rotator cuff repair: A work of the Commission Rehabilitation of the German Society of Shoulder and Elbow Surgery e.V. (DVSE) in collaboration with the German Association for Physiotherapy (ZVK) e.V., the Association Physical Therapy, Association for Physical Professions (VPT) e.V. and the Section Rehabilitation-Physical Therapy of the German Society for Orthopaedics and Trauma e.V. (DGOU). *Obere Extrem.* **2018**, *13*, 45–61. [CrossRef] [PubMed]
7. Brislin, K.J.; Field, L.D.; Savoie, F.H., 3rd. Complications after arthroscopic rotator cuff repair. *Arthroscopy* **2007**, *23*, 124–128. [CrossRef] [PubMed]
8. Huberty, D.P.; Schoolfield, J.D.; Brady, P.C.; Vadala, A.P.; Arrigoni, P.; Burkhart, S.S. Incidence and treatment of postoperative stiffness following arthroscopic rotator cuff repair. *Arthroscopy* **2009**, *25*, 880–890. [CrossRef] [PubMed]
9. Peltz, C.D.; Sarver, J.J.; Dourte, L.M.; Wurgler-Hauri, C.C.; Williams, G.R.; Soslowsky, L.J. Exercise following a short immobilization period is detrimental to tendon properties and joint mechanics in a rat rotator cuff injury model. *J. Orthop. Res.* **2010**, *28*, 841–845. [CrossRef]
10. Sarver, J.J.; Peltz, C.D.; Dourte, L.M.; Reddy, S.; Williams, G.R.; Soslowsky, L.J. After rotator cuff repair, stiffness but not the loss in range of motion increased transiently for immobilized shoulders in a rat model. *J. Shoulder Elb. Surg.* **2008**, *17*, 108–113. [CrossRef]
11. Ghodadra, N.S.; Provencher, M.T.; Verma, N.N.; Wilk, K.E.; Romeo, A.A. Open, mini-open, and all-arthroscopic rotator cuff repair surgery: Indications and implications for rehabilitation. *J. Orthop. Sports Phys. Ther.* **2009**, *39*, 81–89. [CrossRef]
12. Millett, P.J.; Wilcox, R.B., 3rd; O'Holleran, J.D.; Warner, J.J. Rehabilitation of the rotator cuff: An evaluation-based approach. *J. Am. Acad. Orthop. Surg.* **2006**, *14*, 599–609. [CrossRef]

13. Gomi, H.; Osu, R. Task-dependent viscoelasticity of human multijoint arm and its spatial characteristics for interaction with environments. *J. Neurosci.* **1998**, *18*, 8965–8978. [CrossRef] [PubMed]
14. Yamazaki, Y.; Suzuki, M.; Ohkuwa, T.; Itoh, H. Coactivation in arm and shoulder muscles during voluntary fixation of a single joint. *Brain Res. Bull.* **2003**, *59*, 439–446. [CrossRef] [PubMed]
15. Cross, J.A.; deVries, J.; Mocarski, M.; Ketchum, N.C.; Compty, E.; Krimmer, M.; Fritz, J.M.; Vetter, C.S. Electromyography of the shoulder musculature during passive rehabilitation exercises. *J. Shoulder Elb. Arthroplast.* **2020**, *4*, 2471549220960044. [CrossRef]
16. Smith, J.; Dahm, D.L.; Kaufman, K.R.; Boon, A.J.; Laskowski, E.R.; Kotajarvi, B.R.; Jacofsky, D.J. Electromyographic activity in the immobilized shoulder girdle musculature during scapulothoracic exercises. *Arch. Phys. Med. Rehabil.* **2006**, *87*, 923–927. [CrossRef] [PubMed]
17. Smith, J.; Dahm, D.L.; Kotajarvi, B.R.; Boon, A.J.; Laskowski, E.R.; Jacofsky, D.J.; Kaufman, K.R. Electromyographic activity in the immobilized shoulder girdle musculature during ipsilateral kinetic chain exercises. *Arch. Phys. Med. Rehabil.* **2007**, *88*, 1377–1383. [CrossRef] [PubMed]
18. Alenabi, T.; Jackson, M.; Tetreault, P.; Begon, M. Electromyographic activity in the immobilized shoulder musculature during ipsilateral elbow, wrist, and finger movements while wearing a shoulder orthosis. *J. Shoulder Elb. Surg.* **2013**, *22*, 1400–1407. [CrossRef]
19. Gurney, A.B.; Mermier, C.; LaPlante, M.; Majumdar, A.; O'Neill, K.; Shewman, T.; Gurney, J.G. Shoulder electromyography measurements during activities of daily living and routine rehabilitation exercises. *J. Orthop. Sports Phys. Ther.* **2016**, *46*, 375–383. [CrossRef]
20. Yoon, S.H.; Lee, D.H.; Jung, M.C.; Park, Y.U.; Lim, S.Y. Electromyographic activities of the rotator cuff muscles during walking, eating, and washing. *Am. J. Phys. Med. Rehabil.* **2016**, *95*, e169–e176. [CrossRef]
21. Seniam. Available online: http://www.seniam.org/ (accessed on 28 September 2022).
22. Hermens, H.J.; Freriks, B.; Disselhorst-Klug, C.; Rau, G. Development of recommendations for SEMG sensors and sensor placement procedures. *J. Electromyogr. Kinesiol.* **2000**, *10*, 361–374. [CrossRef]
23. Menkes, D.L.; Pierce, R. Needle EMG muscle identification: A systematic approach to needle EMG examination. *Clin. Neurophysiol. Pr.* **2019**, *4*, 199–211. [CrossRef]
24. Muller, T.; Martiny, H.; Merz, E.; Doffert, J.; Wustner, M.; Lessel, W.; Heynemann, H.; Enzmann, T.; Dudwiesus, H.; Nuernberg, D.; et al. DEGUM recommendations on infection prevention in ultrasound and endoscopic ultrasound. *Ultraschall Med.* **2018**, *39*, 284–303. [CrossRef] [PubMed]
25. Berckmans, K.; Castelein, B.; Borms, D.; Palmans, T.; Parlevliet, T.; Cools, A. Analysis of scapular kinematics and muscle activity by use of fine-wire electrodes during shoulder exercises. *Am. J. Sports Med.* **2020**, *48*, 1213–1219. [CrossRef] [PubMed]
26. Werin, M.; Maenhout, A.; Palmans, T.; Borms, D.; Cools, A. Does verbal and tactile feedback change EMG activity in shoulder girdle and trunk muscles during plyometric exercise in overhead athletes with and without shoulder pain? *Phys. Ther. Sport* **2022**, *54*, 65–73. [CrossRef] [PubMed]
27. Hoppe, M.W.; Hotfiel, T.; Stuckradt, A.; Grim, C.; Ueberschar, O.; Freiwald, J.; Baumgart, C. Effects of passive, active, and mixed playing strategies on external and internal loads in female tennis players. *PLoS ONE* **2020**, *15*, e0239463. [CrossRef] [PubMed]
28. RStudio Team. RStudio: Integrated Development Environment for R. *RStudio, PBC, Boston, MA.* Available online: https://www.rstudio.com/ (accessed on 28 September 2022).
29. R Core Team. R: A Language and Environment for Statistical Computing. R Foundation for Statistical Computing, Vienna, Austria. Available online: https://www.r-project.org/ (accessed on 28 September 2022).
30. Tomczak, M.; Tomczak, E. The need to report effect size estimates revisited. An overview of some recommended measures of effect size. *Trends Sport Sci.* **2014**, *1*, 19–25.
31. Sullivan, L.G.; Bailie, R.; Weiss, N.; Miller, B.S. An evaluation of shoulder external rotation braces. *Arthrosc. J. Arthrosc. Relat. Surg.* **2007**, *23*, 129–134. [CrossRef] [PubMed]
32. Edwards, P.K.; Ebert, J.R.; Littlewood, C.; Ackland, T.; Wang, A. A systematic review of electromyography studies in normal shoulders to inform postoperative rehabilitation following rotator cuff repair. *J. Orthop. Sports Phys. Ther.* **2017**, *47*, 931–944. [CrossRef]
33. Long, J.L.; Ruberte Thiele, R.A.; Skendzel, J.G.; Jeon, J.; Hughes, R.E.; Miller, B.S.; Carpenter, J.E. Activation of the shoulder musculature during pendulum exercises and light activities. *J. Orthop. Sports Phys. Ther.* **2010**, *40*, 230–237. [CrossRef]
34. Edwards, P.K.; Kwong, P.W.H.; Ackland, T.; Wang, A.; Donnelly, C.J.; Ebert, J.R. Electromyographic evaluation of early-stage shoulder rehabilitation exercises following rotator cuff repair. *Int. J. Sports Phys. Ther.* **2021**, *16*, 1459–1469. [CrossRef]
35. Reinold, M.M.; Macrina, L.C.; Wilk, K.E.; Fleisig, G.S.; Dun, S.; Barrentine, S.W.; Ellerbusch, M.T.; Andrews, J.R. Electromyographic analysis of the supraspinatus and deltoid muscles during 3 common rehabilitation exercises. *J. Athl. Train.* **2007**, *42*, 464–469.
36. Dockery, M.L.; Wright, T.W.; LaStayo, P.C. Electromyography of the shoulder: An analysis of passive modes of exercise. *Orthopedics* **1998**, *21*, 1181–1184. [CrossRef] [PubMed]
37. Merletti, R.; Farina, D. Analysis of intramuscular electromyogram signals. *Philos. Trans. A Math. Phys. Eng. Sci.* **2009**, *367*, 357–368. [CrossRef] [PubMed]
38. Stokes, I.A.F.; Henry, S.M.; Single, R.M. Surface EMG electrodes do not accurately record from lumbar multifidus muscles. *Clin. Biomech.* **2003**, *18*, 9–13. [CrossRef] [PubMed]

39. Zhang, S.; Xu, Y.; Han, X.; Wu, W.; Tang, Y.; Wang, C. Functional and morphological changes in the deep lumbar multifidus using electromyography and ultrasound. *Sci. Rep.* **2018**, *8*, 6539. [CrossRef] [PubMed]
40. Arias-Martorell, J. The morphology and evolutionary history of the glenohumeral joint of hominoids: A review. *Ecol. Evol.* **2019**, *9*, 703–722. [CrossRef]
41. Schmidt, M.; Voges, D.; Fischer, M.S. Shoulder movements during quadrupedal locomotion in arboreal primates. *Z. Morphol. Anthropol.* **2002**, *83*, 235–242. [CrossRef]
42. Macintosh, J.E.; Valencia, F.; Bogduk, N.; Munro, R.R. The morphology of the human lumbar multifidus. *Clin. Biomech.* **1986**, *1*, 196–204. [CrossRef]
43. Rathi, S.; Taylor, N.F.; Soo, B.; Green, R.A. Glenohumeral joint translation and muscle activity in patients with symptomatic rotator cuff pathology: An ultrasonographic and electromyographic study with age-matched controls. *J. Sci. Med. Sport* **2018**, *21*, 885–889. [CrossRef]

Journal of
Personalized
Medicine

MDPI

Article

Risk and Influence Factors of Fall in Immobilization Period after Arthroscopic Interventions

Johannes Rüther [1,*], Luka Boban [2], Christoph Paus [1], Kim Loose [1], Maximilian Willauschus [1], Hermann Josef Bail [1] and Michael Millrose [1,3]

1 Department of Orthopedics and Traumatology, Paracelsus Medical University, Breslauer Straße 201, 90471 Nuremberg, Germany
2 Department of Trauma and Orthopedic Surgery, Schwarzwald-Baar Clinic, 78052 Villingen Schwenningen, Germany
3 Department of Trauma Surgery and Sports Medicine, Garmisch-Partenkirchen Medical Centre, 82467 Garmisch-Partenkirchen, Germany
* Correspondence: johannes.ruether@klinikum-nuernberg.de; Tel.: +49-911-3982600

Abstract: Knee injuries are one of the most common injuries. Falls during the immobilization period can deteriorate the postoperative outcome. The risk factors causing falls after initial injury and the question of whether a rigid orthosis serves as a protective factor remain unclear. The primary aim of the study was to record the fall rate in the first six weeks after arthroscopic intervention. The secondary aim was to assess the influences of risk factors and protective factors on these fall ratios. Different scores were examined and compared in the groups 'fall event' and 'no fall'. Data from 51 patients (39 males, 12 females) with a mean age of 31.2 years (19–57 years) were collected. A total of 20 patients suffered at least one fall event within the observation period. A total of 18 of 23 fall events happened within the first three weeks postoperatively. The Extra Short Musculoskeletal Function Assessment Questionnaire (XSMFA) showed a significant difference between the groups ($p = 0.02$). People with multiple injuries to the knee joint were more likely to suffer fall events. Conclusively, patients with limited knee functions appeared to fall more frequently within the first three weeks postoperatively. Therefore, appropriate measures should be taken to protect the postoperative outcome. Physical therapy and patient behavioural training should be practiced perioperatively in patients at risk.

Keywords: knee surgery; arthroscopy; sports injury; orthosis; postoperative treatment

Citation: Rüther, J.; Boban, L.; Paus, C.; Loose, K.; Willauschus, M.; Bail, H.J.; Millrose, M. Risk and Influence Factors of Fall in Immobilization Period after Arthroscopic Interventions. *J. Pers. Med.* **2022**, *12*, 1912. https://doi.org/10.3390/jpm12111912

Academic Editor: Eduard Kurz

Received: 17 October 2022
Accepted: 12 November 2022
Published: 16 November 2022

1. Introduction

Injuries to the lower extremities are common injuries that occur during sporting activities or in everyday life. According to Bednarski et al., the knee injury rate is highest in the 11–20 age group with equal sex distribution [1]. Meniscal injuries and injuries to the anterior cruciate ligament (ACL) account for the largest share [1–4]. After reconstruction of the aforementioned anatomical structures, the use of postoperative orthosis with a limitation of range of motion (ROM) under 90° of flexion as well as partial weight bearing is common [5,6].

Several studies observed falling after surgery of the knee. Most of these studies focus on total replacement of the knee. In a review, Di Laura Frattura et al. stated fall rates between 12 and 38 percent in the first 6 months after surgery [7]. The overall fall rate of patients without pre-existing illnesses was calculated to be about 20 percent [8]. Overall, there are only a few studies on the risk factors for a tendency to fall after surgical interventions [9–12]. An increased tendency to slip, trip or fall after surgical intervention has been seen in elderly patients [9–12]. To date, no difference in the postoperative fall frequency has been found in relation to the body mass index (BMI) or the American Society of Anesthesiology (ASA) score [11]. In older patients, the preoperative tendency to fall was also found to be a risk factor for postoperative falls [12]. The risk factors 'fear of falling' and musculoskeletal comorbidities showed a tendency towards significance [12].

Although it is widely used, there is no evidence for a protective factor of an orthosis in the literature. Several studies did not find a significant difference between patients wearing an orthosis and patients not wearing one [13–15]. The systematic review of Lowe et al. states the positive benefit on the in vivo kinematic [16].

The aim of the present study was the prospective recording of falls and the calculation of fall rates after arthroscopic interventions, in general during the immobilization phase, and the evaluation of possible risk factors as well as the recording of fall-related re-interventions.

2. Materials and Methods

This prospective study was conducted from February 2019 to March 2021 at a maximum care hospital with a department for sports orthopaedics and arthroscopic surgery. During the entire study period, 200 patients underwent arthroscopic knee repair.

The inclusion criteria of the study were a minimum age of 18 years and a performed arthroscopic reconstructive knee surgery—regardless of the anatomical structure—, which necessitated partial weight bearing of 15 kg and the usage of a rigid orthosis with a range-of-motion (ROM) limitation of full extension with restricted flexion. Exclusion criteria of this study were a lack of consent or ability to consent to the study, rejection of the after-treatment plan and the orthosis, use of orthostasis-promoting medication, injuries to the opposite side, wound infection, thrombosis, neurological diseases, psychiatric diseases and diabetes mellitus.

At inclusion into the study, the age, gender, height and BMI as well as data on injury and therapy were recorded. Postoperative pain was measured by using the NRS (numeric pain rating scale). The pain was rated from 0–10 with 10 as maximum pain and 0 without pain [17]. A score median has been calculated and used for analysis. Moreover, the ROM (range of motion) was checked every week and was analysed 6 weeks postoperatively with a comparison of the maximum range of flexion and extension. Every week, the difference of knee swelling was measured using the circumference 10 cm above the patella in comparison with the non-operated knee. The median was calculated for every participant to compare it in the groups of falls and no falls. The duration of wearing the orthosis was recorded in hours every week, and the median was compared statistically.

The patients were assessed with different assessment tools at different times of the study. Preoperatively, IKDC (International Knee Documentation Committee), the fall risk assessment STRATIFY, the XSMFA-D (Short Musculoskeletal Function Assessment) and the Tegner Activity Scale were completed. [18–21] Furthermore, IKDC was checked every week. All these scores were secondary scores, as the primary aim of the study was to evaluate the fall rates and the tendency to fall. The IKDC is a tool to measure the function of the knee and has a score range of 0–100%. A total of 100% is a normal function of the knee, and a low score is a marker for an impaired function of the knee [18]. The STRATIFY score is divided into five questions. Every question aims to find a risk of falling. The maximum score is five points, zero points implies no falling tendency, one point is a moderate falling tendency, and two points or more represent a high tendency to fall [19]. The XSMFA-D consists of many single-choice questions, which are converted into numbers. The overall score ranges from 0–100, and a higher score indicates a worse function of the knee joint [20]. The Tegner score is a score to evaluate the activity of every participant. The score ranges from 0–10, where zero is doing no activities at all and 10 is performing professional competitive sport [21].

After surgery—regardless of which anatomical reconstruction was in question—, each participant received a four-point-rigid-knee-orthosis M.4s comfort (Fa. Medi, Bayreuth) with ROM limitations adapted to the operation: meniscal suture; ACL reconstruction in combination with meniscal suture or medial patellofemoral ligament (MPFL) reconstruction—extension/flexion 0-0-60° for 3 weeks and 0-0-90° for another

3 weeks; ACL reconstruction—0-0-90° for 6 weeks. In addition, all patients were mobilized postoperatively with physiotherapeutic support on two crutches, applying a partial load of 15 kg. The physiotherapists could start lymph drainage on the second postoperative day. In the case of meniscal sutures, passive movement was allowed for the first 3 weeks, and active movement with weight bearing was allowed after 6 weeks. In the case of ACL reconstruction or MPFL reconstruction, active movement was allowed from the second postoperative day. All patients received manual lymphatic drainage, local cryotherapy, as well as range-of-motion physiotherapy during the follow-up-period.

Over a period of six weeks after discharge, the frequency of falls or near-falls (defined as losing balance and applying one's full weight to prevent a fall), the range of motion, the NRS and the length of time during which the orthosis was worn were queried in a weekly telephone interview. Possible complications after falls were evaluated immediately by means of a clinical examination on the day after the fall, sonography, X-ray and MRI.

For statistical analysis, falls and near-falls were combined, as both have an elevated risk of injury.

A post-hoc power analysis was conducted with a power level of 0.8 as well as a significance level of 0.05 for significant data, which has been fulfilled for XSMFA-D scores in the fall and no fall groups (sample size of 52).

The data were obtained prospectively and analysed retrospectively. The statistical analysis was performed using SPSS Statistics for Windows (Version 25; IBM Corp, Nuremberg, Germany). All p-values described are tested paired and are rated as statistically significant by $p < 0.05$. Data are presented as a mean ± standard deviation and range unless otherwise specified. Normal distribution was tested using the Shapiro–Wilk test. To evaluate statistically significant differences, the t-test for independent variables, the Mann–Whitney–U-test or the chi-square test was used, depending on the scale level and the extent of the normal distribution.

3. Results

Of the 200 operated patients, 60 could be included in the study. A total of 9 study participants were lost during the follow-up, and 51 patients (39 male, 12 female) could be assessed in the analysis. The participants had a mean age of 31 years (19 years–57 years). The average BMI was 25.1 (16.6–37.2). There were no differences in the number of dropouts between the two study groups, fall event and no fall event.

As mentioned above, different types of surgery were included in the study: ACL-reconstructions (43.1%), meniscal sutures (27.5%), a combination of both of the aforementioned (13.7%), MPFL reconstructions (5.9%) as well as femoral cartilage repair (9.8%).

In this study, 39.2% (20 patients) of patients had a fall event and 60.8% (31 patients) did not. There were no significant differences in fall events between men and women.

The majority (78.3%) of the falls observed occurred in the first three postoperative weeks. The fall rate was highest in the first postoperative week (30.4%) (Figure 1). Only 21.8% of all falls in the observed period occurred in the last three weeks (Figure 1).

The preoperative XSMFA-D scores showed a significant difference ($p = 0.02$) between patients with and without a fall event. The mean for the group with a fall event was 52.1 points, and the mean for the group without a fall was 45.3 points (Table 1). The other risk factors, such as BMI, age, gender, knee swelling, range of motion, NRS, wearing time, IKDC and Tegner score, showed no significant difference between the two groups (Table 1).

The evaluation of risk factors showed a significant difference between the fall and no fall group after the sixth week for the risk factors ROM, NRS and the wearing time of the orthosis. During the entire observation period, none of the patients required revision or intervention due to the fall event.

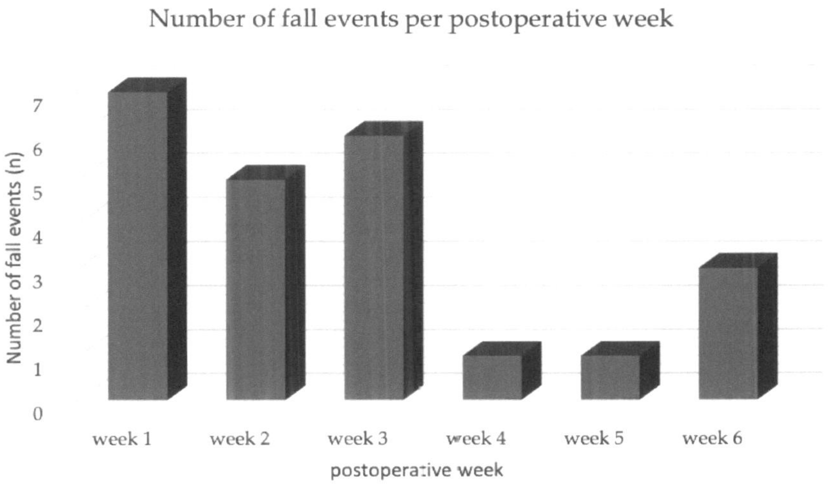

Figure 1. Number of fall events compared for each of the six weeks.

Table 1. Analyses of risk factors for postoperative fall events after arthroscopic reconstructive knee surgery.

	Total	Fall Event	No Fall	*p*-Value
Age (median ± SD, range)	31.2 ± 10.6, 19–57	31.1 ± 11 7, 19–57	31.4 ± 8.7, 20–52	0.5
BMI [1] (median ± SD, range)	25.1 ± 3.8, 16.6–37.2	24.7 ± 3.1, 16.6–30.4	25.7 ± 4.6, 20.8–37.2	0.4
NRS [2] (median ± SD, range)	3.2 ± 2.2, 0–8	3 ± 2.2, 0–8	3.4 ± 2.4, 1–8	0.6
Circumference [3] (median ± SD, range)	1.9 ± 1.4, −4–6	1.8 ± 1.3, −3.5–5	2 ± 1.5, −4–6	0.7
ROM [4] (median ± SD, range)	77.6 ± 19.4, 0–90	80.9 ± 15.9, 30–90	72.5 ± 23.5, 0–90	0.2
Wearing duration (median ± SD, range)	14 ± 6.5, 1–24	14.2 ± 6.5, 1–24	13.8 ± 6.8, 3–24	0.9
IKDC [5] (median ± SD, range)	45.7 ± 15.7, 23–79,3	49 ± 17.1, 23–79.3	40.6 ± 12.3, 23–65,5	0.1
STRATIFY (median ± SD, range)	0.7 ± 0.7, 0–2	0.7 ± 0.7, 0–2	0.8 ± 0.7, 0–2	0.7
XSMFA-D [6] (median ± SD, range)	47.9 ± 10.5, 22–71	45.3 ± 10 22–68	52.1 ± 10, 31–71	0 02
Tegner (median ± SD, range)	5.6 ±2.1; 1–9	5.5 ± 2.5, 1–9	5.8 ± 1.6, 3–9	0 95

[1] BMI: body mass index; [2] NRS: numeric pain rating scale; [3] Circumference: difference of circumference 10 cm above patella pole from operated and not operated knee; [4] ROM: range of motion; [5] IKDC: International Knee Documentation Committee; [6] XSMFA-D: Extra Short Musculoskeletal Function Assessment Questionnaire.

There was no difference between the kind of surgery in relation to the tested scores and risk factors, nor were there any differences between the surgery types and the frequency of falls.

4. Discussion

The most important finding in this prospective study is the high rate of falls (39.2%) after arthroscopic knee joint surgery within the first six postoperative weeks. In the first three weeks in particular, patients were at risk of falling. The overall condition of the musculoskeletal system as well as pain and the length of time during which the orthosis was worn were also identified as potential risk factors. Despite the high number of falls, no revisions or interventions had to be carried out.

Postural stability is an objective measurable parameter for the study of functional limitation. There are studies examining both postural stability after knee joint surgery and the relationship between postural stability and orthoses [22,23]. In the study by Gokalp et al., patients undergoing arthroscopic ACL reconstruction were assessed for fall risk using a posturographic analysis. One result was that patients with a preoperative ACL rupture are, on average, in the high-risk group for a fall. The highest risk of falling was measured in the study during the fourth postoperative week, where, on average, the patients could also be assigned to the high-risk group postoperatively. The result corresponds to our observations that the majority of falls (78.3%) occur in the first three weeks postoperatively (Figure 1). Since Gokalp et al. only collected data in the fourth, eighth and twelfth postoperative week, it might be possible to assess whether the actual peak in fall risk is found within the first three weeks [22]. One can postulate a higher risk of falls in the early postoperative phase, even after minimal invasive operations.

In the present study, one can note that wearing a knee joint orthosis tends to have a positive influence on the number of falls. There is conflicting information about the effectiveness of a knee joint orthosis after arthroscopic knee surgery. Almost all studies relate to rehabilitation after cruciate ligament surgery, while no data is available for a broader spectrum. In their study, Derouin et al. describe the muscular contribution to rotational stability in ACL injuries and after ACL reconstruction [24]. The result of the study describes a 24% higher rotational stability of the knee joint while it is in a splint. The authors have concluded that an increased rotational stability of the orthosis group is triggered by increased muscle activity. It might therefore be assumed that an orthosis promotes muscular activity and thus minimizes falls. It must be noted, however, that there are several studies that do not find any significant differences in various parameters [13,16,25]. Bordes et al. examine athletes after ACL reconstruction and describe that there are no significant differences between the individual groups [13]. There is no significant difference in clinical outcome and no significant influence of the orthosis on the short-term complication rate. A systematic review by Lowe et al. shows that there could be potential benefits of orthoses for in vivo kinematics, such as protection of the joint implant [16].

This study on falls after minimally invasive arthroscopic knee surgery further confirms that patients with an impaired knee function have an increased risk of falling, after surgery on the knee joint, within the first three weeks, even when crutches and an orthosis are used. Therefore, after the prescription of appropriate measures such as an orthosis and crutches, direct attention should be paid to adequate training in handling the aids or mobilizing the patient under the guidance of a therapist.

Further, it seems particularly important to pass on this information regarding the risk of falling to patients and to educate them accordingly. This study showed that the XSMFA-D could be a suitable questionnaire for identifying young patients at risk of falling after knee joint surgery. The introduction of a specific "fall training" could therefore also be particularly important for patients in the early rehabilitation phase. Consistent wearing of an orthosis and sufficient postoperative analgesia could have a positive effect on the fall rate. For this reason, possible problems with pain or the orthosis should be addressed intensively during follow-up treatment, especially in the first weeks. The exact influence of wearing an orthosis on the occurrence or prevention of recorded falls was not examined in this study. It can only be stated that no patients suffered injuries requiring revision or intervention during the fall with the orthosis. In this regard, studies with a larger number of cases and possibly a different study design should be developed.

A strength of this study is that the rate of falls was examined prospectively through close controls. The weekly telephone interviews and surveys provided differentiated data, based on which the fall event rates in the immobilization period could be evaluated. In addition, this study focuses particularly on risk factors for slips, trips and falls. The fall event rate is explicitly determined and not just, as in many other studies, the mere probability of a fall event. In addition to the telephone surveys, close controls also included regular clinical examinations, another strength of this study.

The limitations of this study are both the missing control group and the heterogeneity of the operations. Due to a missing control group, the orthosis cannot adequately be analysed as a protective factor. The heterogeneity of the different operations is a limitation that could affect falls differently. Nevertheless, in our opinion, the most important factor, which was also an inclusion criterion, for a fall was partial weight bearing as well as ROM-limitation after arthroscopic surgery. In addition, there was no further fall analysis. The place or the reason for falls was not analysed in this study and should be analysed in further studies. The occurrence of a recall bias is favoured by the telephone survey with regard to the fall events and the wearing time of the orthosis. Further limitations are possible comorbidities that have not been adequately considered and recorded.

5. Conclusions

There is a considerable fall rate during the immobilization period in patients with arthroscopic interventions. Risk factors seem to be the low condition of the musculoskeletal system as well as pain, while the role of wearing an orthosis in this context remains controversial. Intensified physical therapy and patient behavioural training should be recommended perioperatively to patients at risk.

Author Contributions: Conceptualization, J.R., M.M. and H.J.B.; methodology, J.R., M.M. and H.J.B.; software, M.W., L.B. and J.R.; validation, M.M., J.R.; formal analysis, C.P., K.L. and L.B.; investigation, M.M., L.B. and K.L.; data curation, C.P. and J.R.; writing—original draft preparation, L.B., M.M. and J.R.; writing—review and editing, M.M. and J.R.; visualization, C.P., L.B., K.L., M.W. and J.R.; supervision, H.J.B.; project administration, H.J.B., J.R. and M.M. All authors have read and agreed to the published version of the manuscript.

Funding: This research received no external funding.

Institutional Review Board Statement: The Institutional Review Board of Paracelsus Medical University stated that this study was conducted according to the guidelines of the Declaration of Helsinki. They approved this study (protocol code IRB-2020-005 on 22 January 2020).

Informed Consent Statement: Informed consent was obtained from all subjects involved in the study. Written informed consent has been obtained from the patient(s) to publish this paper.

Data Availability Statement: Not applicable.

Acknowledgments: The authors would like to thank Stefanie Karpik for her support in preparing and correcting the manuscript with respect to spelling and grammar.

Conflicts of Interest: The authors declare no conflict of interest.

References

1. Bednarski, P.; Piekarska, K. Traumatic Knee Injuries in 2016–2019—An Analysis of Newly Diagnosed Patients Based on NHF Data Reporting. *Ortop. Traumatol. Rehabil.* **2021**, *23*, 181–192. [CrossRef] [PubMed]
2. Nordenvall, R.; Bahmanyar, S.; Adami, J.; Stenros, C.; Wredmark, T.; Felländer-Tsai, L. A population-based nationwide study of cruciate ligament injury in Sweden, 2001–2009: Incidence, treatment, and sex differences. *Am. J. Sports Med.* **2012**, *40*, 1808–1813. [CrossRef]
3. Ferry, T.; Bergström, U.; Hedström, E.M.; Lorentzon, R.; Zeisig, E. Epidemiology of acute knee injuries seen at the Emergency Department at Umeå University Hospital, Sweden, during 15 years. *Knee Surg. Sport. Traumatol. Arthrosc.* **2014**, *22*, 1149–1155. [CrossRef]
4. Clayton, R.A.; Court-Brown, C.M. The epidemiology of musculoskeletal tendinous and ligamentous injuries. *Injury* **2008**, *39*, 1338–1344. [CrossRef] [PubMed]
5. MARS Group. Rehabilitation predictors of clinical outcome following revision ACL reconstruction in the MARS cohort. *J. Bone Jt. Surg. Am.* **2019**, *101*, 779–786. [CrossRef] [PubMed]
6. Koch, M.; Memmel, C.; Zeman, F.; Pfeifer, C.G.; Zellner, J.; Angele, P.; Weber-Spickschen, S.; Alt, V.; Krutsch, W. Early Functional Rehabilitation after Meniscus Surgery: Are Currently Used Orthopedic Rehabilitation Standards Up to Date? *Rehabil. Res. Pract.* **2020**, *2020*, 3989535. [CrossRef] [PubMed]

7. Talbot, L.A.; Musiol, R.J.; Witham, E.K.; Metter, E.J. Falls in young, middle-aged and older community dwelling adults: Perceived cause, environmental factors and injury. *BMC Public Health* **2005**, *5*, 86. [CrossRef]
8. Di Laura Frattura, G.; Filardo, G.; Giunchi, D.; Fusco, A.; Zaffagnini, S.; Candrian, C. Risk of falls in patients with knee osteoarthritis undergoing total knee arthroplasty: A systematic review and best evidence synthesis. *J. Orthop.* **2018**, *15*, 903–908. [CrossRef]
9. Verma, S.K.; Willetts, J.L.; Corns, H.L.; Marucci-Wellman, H.R.; Lombardi, D.A.; Courtney, T.K. Falls and fall-related injuries among community-dwelling adults in the United States. *PLoS ONE* **2016**, *11*, e0150939. [CrossRef]
10. Fujita, Y.; Mera, H.; Watanabe, T.; Furutani, K.; Kondo, H.O.; Wakai, T.; Kawashima, H.; Ogose, A. Significantly earlier ambulation and reduced risk of near-falls with continuous infusion nerve blocks: A retrospective pilot study of adductor canal block compared to femoral nerve block in total knee arthroplasty. *BMC Musculoskelet. Disord.* **2022**, *23*, 768. [CrossRef]
11. Pelt, C.E.; Anderson, A.W.; Anderson, M.B.; Van Dine, C.; Peters, C.L. Postoperative falls after total knee arthroplasty in patients with a femoral nerve catheter: Can we reduce the incidence? *J. Arthroplast.* **2014**, *29*, 1154–1157. [CrossRef] [PubMed]
12. Levinger, P.; Wee, E.; Margelis, S.; Menz, H.B.; Bartlett, J.R.; Bergman, N.R.; McMahon, S.; Hill, K.D. Pre-operative predictors of post-operative falls in people undergoing total hip and knee replacement surgery: A prospective study. *Arch. Orthop. Trauma Surg.* **2017**, *137*, 1025–1033. [CrossRef] [PubMed]
13. Bordes, P.; Laboute, E.; Bertolotti, A.; Dalmay, J.F.; Puig, P.; Trouve, P.; Verhaegue, E.; Joseph, P.A.; Dehail, P.; Seze, M.D. No beneficial effect of bracing after anterior cruciate ligament reconstruction in a cohort of 969 athletes followed in rehabilitation. *Ann. Phys. Rehabil. Med.* **2017**, *60*, 230–236. [CrossRef]
14. McDevitt, E.R.; Taylor, D.C.; Miller, M.D.; Gerber, J.P.; Ziemke, G.; Hinkin, D.; Uhorchak, J.M.; Arciero, R.A.; Pierre, P.S. Functional bracing after anterior cruciate ligament reconstruction: A prospective, randomized, multicenter study. *Am. J. Sports Med.* **2004**, *32*, 1887–1892. [CrossRef] [PubMed]
15. Kartus, J.; Stener, S.; Köhler, K.; Sernert, N.; Eriksson, B.I.; Karlsson, J. Is bracing after anterior cruciate ligament reconstruction necessary? *Knee Surg. Sports Traumatol. Arthrosc.* **1997**, *5*, 157–161. [CrossRef]
16. Lowe, W.R.; Warth, R.J.; Davis, E.P.; Bailey, L. Functional bracing after anterior cruciate ligament reconstruction: A systematic review. *J. Am. Acad. Orthop. Surg.* **2017**, *25*, 239–249. [CrossRef]
17. Hawker, G.A.; Mian, S.; Kendzerska, T.; French, M. Measures of Adult Pain: Visual Analog Scale for Pain (Vas Pain), Numeric Rating Scale for Pain (NRS Pain), McGill Pain Questionnaire (MPQ), Short-Form McGill Pain Questionnaire (SF-MPQ), Chronic Pain Grade Scale (Cpgs), Short Form-36 Bodily Pain Scale (SF-36 BPS), and Measure of Intermittent and Constant Osteoarthritis Pain (ICOAP). *Arthritis Care Res.* **2011**, *63*, 240–252. [CrossRef]
18. Irrgang, J.J.; Anderson, A.F.; Boland, A.L.; Harner, C.D.; Kurosaka, M.; Neyret, P.; Richmond, J.C.; Shelborne, K.D. Development and validation of the international knee documentation committee subjective knee form. *Am. J. Sport. Med.* **2001**, *29*, 600–613. [CrossRef]
19. Oliver, D.; Britton, M.; Seed, P.; Martin, F.C.; Hopper, A.H. Development and evaluation of evidence based risk assessment tool (STRATIFY) to predict which elderly inpatients will fall: Case-control and cohort studies. *BMJ* **1997**, *315*, 1049–1053. [CrossRef]
20. Wollmerstedt, N.; Kirschner, S.; Boehm, D.; Faller, H.; Koenig, A. Design and evaluation of the Extra Short Musculoskeletal Function Assessment Questionnaire XSMFA-D. *Z. Fur Orthop. Und Ihre Grenzgeb.* **2003**, *141*, 718–724. [CrossRef]
21. Tegner, Y.; Lysholm, J. Rating systems in the evaluation of knee ligament injuries. *Clin. Orthop. Relat. Res.* **1985**, *198*, 43–49. [CrossRef]
22. Gokalp, O.; Akkaya, S.; Akkaya, N.; Buker, N.; Gungor, H.R.; Ok, N.; Yorukoglu, C. Preoperative and postoperative serial assessments of postural balance and fall risk in patients with arthroscopic anterior cruciate ligament reconstruction. *J. Back Musculoskelet. Rehabil.* **2016**, *29*, 343–350. [CrossRef] [PubMed]
23. Bartels, T.; Brehme, K.; Pyschik, M.; Schulze, S.; Delank, K.-S.; Fieseler, G.; Laudner, K.G.; Hermassi, S.; Schwesig, R. Pre- and postoperative postural regulation following anterior cruciate ligament reconstruction. *J. Exerc. Rehabil.* **2018**, *14*, 143–151. [CrossRef] [PubMed]
24. Derouin, A.; Potvin, J.R. The Effect of Functional Knee Braces on Muscular Contributions to Joint Rotational Stiffness in Anterior Cruciate Ligament-Deficient and -Reconstructed Patients. *J. Appl. Biomech.* **2019**, *35*, 344–352. [CrossRef]
25. Knapik, D.M.; Singh, H.; Gursoy, S.; Trasolini, N.A.; Perry, A.K.; Chahla, J. Functional Bracing Following Anterior Cruciate Ligament Reconstruction: A Critical Analysis Review. *JBJS Rev.* **2021**, *9*, e21.00056. [CrossRef]

Journal of
Personalized Medicine

MDPI

Article

Contribution of Different Impairments to Restricted Knee Flexion during Gait in Individuals with Cerebral Palsy

Faustyna Manikowska [1,*], Sabina Brazevič [1], Marek Jóźwiak [1] and Maria K. Lebiedowska [2]

1 Department of Pediatric Orthopedics and Traumatology, Poznań University of Medical Sciences, 61-701 Poznań, Poland
2 Independent Researcher, Edmonton, AB T5R 2R2, Canada
* Correspondence: foka@interia.pl

Abstract: The coexistence of overlapping impairments modulates the knee pattern in the swing phase of walking in children with cerebral palsy (CP). The impact and contribution of each impairment to the reduction of knee range-of-motion is unknown. The aim of the study was to establish the gradation of the impact of individual coexisting impairments on the knee flexion range-of-motion. Passive range-of-motion, selective motor control, strength, and spasticity from 132 patients (Male = 76, Female = 56, age:11 ± 4 years) with spastic CP were tested with clinical tools. Knee flexion range-of-motion at terminal stance, pre-swing, and initial swing phases were assessed by gait analysis. Hypertonia ($\beta = -5.75$) and weakness ($\beta = 2.76$) of knee extensors were associated with lower range of knee flexion ($R^2 = 0.0801$, $F = 11.0963$, $p < 0.0001$). The predictive factors ($R^2 = 0.0744$, $F = 7.2135$, $p < 0.0001$) were strength ($\beta = 4.04$) and spasticity ($\beta = -2.74$) of knee extensors and strength of hip flexors ($\beta = -2.01$); in swing those were knee extensors hypertonia ($\beta = -2.55$) and passive range of flexion ($\beta = 0.16$) ($R^2 = 0.0398$, $F = 3.4010$, $p = 0.01$). Hypertonia of knee extensors has the strongest impact on knee flexion range-of-motion; secondary is the strength of knee extensors. The knee extensors strength with knee extensors hypertonia and strength of hip flexors contributes in stance. Knee extensors hypertonia with passive knee flexion range-of-motion contributes in swing.

Keywords: cerebral palsy; impairments; motor dysfunction; gait; motion analysis

Citation: Manikowska, F.; Brazevič, S.; Jóźwiak, M.; Lebiedowska, M.K. Contribution of Different Impairments to Restricted Knee Flexion during Gait in Individuals with Cerebral Palsy. *J. Pers. Med.* **2022**, *12*, 1568. https://doi.org/10.3390/jpm12101568

Academic Editors: Markus Gesslein and Eduard Kurz

Received: 24 August 2022
Accepted: 20 September 2022
Published: 23 September 2022

1. Introduction

Children with motor disorders such as cerebral palsy (CP) presented a combination of clinical signs as consequences of upper motor neuron lesions. As an effect of brain damage, positive (increased involuntary muscle activity, movement, or movement pattern; e.g., hypertonia) and negative motor signs (insufficient muscle activity or its control; e.g., weakness, impaired selective muscle control) are observed. Multiple impairments may occur simultaneously and affect functional skills such as gait in CP individuals [1,2].

The knee range of motion (ROM) during a gait cycle (GC) in healthy individuals is usually around 60 deg. There usually exist two periods when the knee is flexed during GC. The first one occurs at the beginning of GC, and the second one starts in the middle of the terminal stance (TSt; ~40% of GC) and ends with knee maximum flexion in the middle of the initial swing (ISw; ~70% of GC). The first period of knee flexion occurs in the loading response phase and serves as a shock absorber and supports stability [3–5]. The second period of knee flexion is especially important as a major factor, besides terminal swing knee extension, affecting knee ROM during gait. The second period of knee flexion consists of two distinct periods: 1. flexion in stance: from the beginning of knee flexion at TSt to the toe off, increase from full knee extension to 40° of flexion; and 2. flexion in swing: increase from 40° to 60° of flexion in initial swing phase.

The gait efficiency, foot clearance, step length, and walking velocity are usually determined by proper knee flexion ROM, its velocity, and the timing of those events [4]. The decrease in ranges and velocity and delay of the peak of knee flexion was reported in a patient with CP [5–7].

The most common impairments (mainly hypertonia, weakness, and lack of selective motor control) affect the functional abilities of patients with CP. The selected impairments at different joints were previously reported as factors contributing to deviations in gait patterns. The weakness and/or hypertonia of knee extensors, hypertonia of knee flexors, hypertonia of plantar flexors, knee extensors contraction, limited thigh advancement caused by weakness of hip flexors or insufficient toe-off, and primitive synergies are all potential factors of pathological knee flexion in the swing phase of walking [4,5,8–10].

Previous studies usually focused on a single disregarding coexisting impairment. The effect of spasticity [10–12], inappropriate electrical firing activity of rectus femoris (RF) [6,7], RF muscle maximum length [8], growth and age [13], and decrease of range-of-motion (ROM) of knee flexion were recognized as factors decreasing ROM.

In addition, the most common orthopedic treatment surgeries are based on the same assumption of the significance of isolated factors. For example, RF transfer is supposed to improve the range and/or timing and velocity of knee flexion. The results of the treatment are not consistently successful [14–17]. Campbell et al. reported that the intervention is ineffective based on their meta-analysis [18]. Inconsistency in reported effectiveness seems to be dependent on the individual's status. It was shown that this treatment is more beneficial in patients with predominant swing phase activity of RF and only those on higher functional status (GMFCS levels I & II) than those who are not independent walkers (GMFCS levels III & IV) [14,19]. It suggested that not only hypertonia, but weakness, lack of selective motor control, and contraction might affect the effectiveness of RF transfer surgery.

Thus, it seems that the coexistence of many overlapping impairments modulates the knee pattern in the swing phase of walking in patients with CP. However, the impact and contribution of each impairment to the reduction of knee ROM are unknown.

The present study aimed to establish the gradation of the impact of individual coexisting impairments on the knee flexion ROM. To achieve the aims of the study, the results on selected knee ranges of motion (three-dimensional gait analysis) and clinical tests (strength [MMT], hypertonia [HYP], selective motor control [SEL], and passive knee flexion ROM (CON) were analyzed.

2. Materials and Methods

In total, 132 patients (male, n = 76; female, n = 56; age: 11 $\pm$ 4 years) with spastic hypertonia were recruited for the study. Data from 262 limbs were analyzed; due to technical errors, data from two limbs were excluded. Data were collected from a local Gait and Motion Analysis Laboratory. To be enrolled in the study group, patients had to meet the inclusion criteria: complete data from clinical and gait analysis; no botulinum injection or surgery within six months before the assessment; be able to walk without physical assistance. The appropriate Institutional Review Board approved the study, and written consent for using the examination data was acquired from all participants. For participants under the age of 18, consent was obtained from a parent or legal guardian. Data from lower limb clinical examinations and gait analysis were investigated. Clinical examination, including passive range of motion, selective motor control, muscle strength, and muscle tone (around hip, knee, and ankle joints), were tested.

2.1. Protocols

Data from clinical examinations and gait analysis were investigated.

2.1.1. Clinical Examination

- Passive knee flexion ROM: It was tested with a handheld goniometer; the passive angular value of movement was measured, and the presence or absence of impairment was documented according to the normal values [20];

- Selective motor control (SMC): SMC was graded from 0 (no ability to perform isolated movement) to 2 (presented movement is completely isolated). The presence of impairment was defined in limbs with SMC < 2; otherwise, the limbs were defined as not impaired [21];
- Manual muscle testing (MMT): MMT was graded from 0 (no evidence of muscle contraction) to 5 (ability to produce movement against the full external resistance). The presence of strength impairment was defined in limbs with a grade less than 4 [20];
- Muscle hypertonia (HYP): HYP was assessed with: Modified Ashworth Scale [22] and Tardieu Scale [23]. The presence of hypertonia was defined in limbs with a score higher than 0.

2.1.2. Gait Analysis

An eight-camera three-dimensional gait analysis system (6 Bonita cameras and 2 Vero cameras; Vicon Motion Systems Ltd., Oxford, UK) sampling at 120 Hz was used. Reflective markers were applied according to the standard Plug-in-Gait marker placement model to each patient. Participants walked barefoot along a 10-m walkway at a self-selected speed. Pre-swing (PSw), and initial swing (ISw) phases of gait, the kinematic data were collected to measure ROM and velocity of knee flexion in the sagittal plane during terminal stance (TSt).

2.2. Data Analysis

2.2.1. Outcome Measures

The primary outcome measures were selected ranges of knee motion in the sagittal plane during terminal stance (TSt), pre-swing (PSw), and initial swing (ISw). As shown in Figure 1, five key points in sagittal knee kinematics were identified in each GC for further analysis: [K1] Knee position at initial contact; [K2] In a normal pattern of gait, it refers to the maximum knee flexion at initial stance; in knee hyperextension, it indicates a minimum of knee extension in stance; [K3] Minimum knee position during terminal stance; [K4] Knee position at toe-off; and [K5] Maximum knee flexion during swing phase. Only K3, K4, and K5 were used in the present study. The ROM was considered as one value (from TSt to ISw) K3–K5 and also was divided into two parts: flexion in stance: K3–K4 and flexion in swing: K4–K5. In addition, knee flexion velocity at toe-off was included in the analysis.

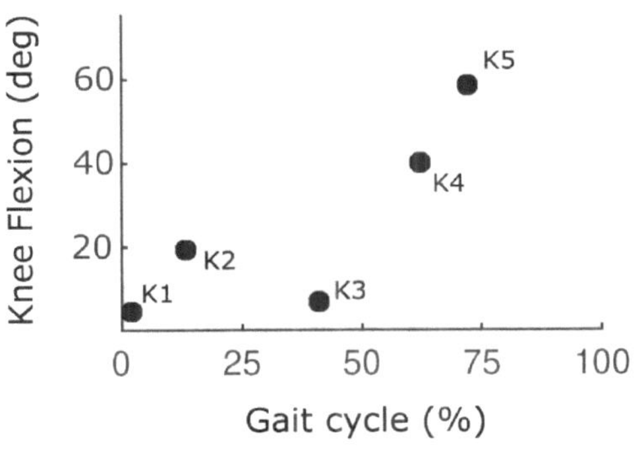

Figure 1. Knee kinematics in sagittal plane. [K1] Knee position at initial contact; [K2] maximum knee flexion at initial stance; [K3] Minimum knee position during terminal stance; [K4] Knee position at toe off; [K5] Maximum knee flexion during swing phase.

Major factors that play a potential role in the biomechanics of knee joint flexion during TSt, PSw, and ISw phases were considered in the analysis. The hypertonia (hip and knee flexors, knee extensors, and ankle plantar flexors at slow and fast movement velocity), muscle strength (hip flexors and knee extensors), selective motor control (hip flexors and knee extensors), and restricted passive range of motion (knee extensors) were included in the analysis. The potential outcome measures within each impairment are listed in Table 1.

Table 1. Impairment outcome measures (MMT, HYP, SMC, CON measures) involved in knee flexion in terminal stance, pre-swing and initial swing (K3–K5).

Impairment	Muscle Group	Outcome Measure
Weakness (MMT)	Hip flexors Knee extensors	MMT_HIP_FL MMT_KNEE_EXT
Hypertonia (HYP)	Knee extensors Knee flexors Plantar flexors	HYP_KNEE_EXT HYP_KNEE_FL HYP_PF
Lack of selectivity (SMC)	Hip flexors Knee extension	SMC_HIP_FL SMC_KNEE_EXT
Restricted range of motion (ROM)	Knee extensors	CON_KNEE_EXT

MMT: manual muscle testing; HYP: muscle hypertonia; SMC: selective motor control; ROM: range of motion; HIP_FL: hip flexors; KNEE_EXT: knee extensors; KNEE_FL: knee extensors; PF: plantar flexors; CON: contraction.

2.2.2. Statistical Analysis

A multiple linear regression analysis was used to determine an impact. A multiple linear regression model was built to investigate the simultaneous effects of each of the outcome measures within selected impairments (independent variables) on the selected knee ROMs (dependent variable). It was assumed that $n \geq 50 + 8$ k, where k is the number of explanatory variables in the model. The analysis of variance test (F test) was used to test the significance of the model. *T*-test was used to test the significance of individual parameters. The β coefficients were estimated using the classical least squares method. Statistical analysis was performed with the application of Statistica 13 (TIBCO Software Inc., Palo Alto, CA, USA), Stata (StataCorp LLC, College Station, TX, USA), and StatXact (Cytel Inc., Walthan, MA, USA) with a level of significance of $p \leq 0.05$.

3. Results

Multiple Regression Analysis (Gradation of Impact of Single Impairment When Coexisting of Multi-Impairments Were Stated)

Results of the multiple linear regression analysis showed that increased HYP_KNEE_EXT and reduced MMT_KNEE_EXT were associated with a lower range of knee flexion K5–K3. Moreover, HYP_KNEE_EXT was a twice stronger predictor than MMT_KNEE_EXT explained by the model.

Reduced range of knee flexion at initial (K5–K4) was associated with lower CON_KNEE_EXT and increased HYP_KNEE_EXT. Although two variables, SMC_HIP_FLEX and SMC_KNEE_EXT, did not show any significant association with K5–K4, removing them from the model would significantly reduce the adjusted R^2. It must be noted that HYP_KNEE_EXT is the strongest predictor in this model. It is 16 times stronger than CON_KNEE_EXT.

MMT_KNEE_EXT, MMT_HIP_FL as well as HYP_KNEE_EXT were found to be significant predictive factors in the case of the K4–K3 range. An increase in HYP_KNEE_EXT, as well as MMT_HIP_FL, reduced the range of knee flexion during terminal stance (K4–K3). On the other hand, decreased MMT_KNEE_EXT, which is a twice stronger predictor than the other two in this model (MMT_HIP_FL and HYP_KNEE_EXT), will result in reduced K4–K3.

Increased knee angular velocity at toe-off was associated with a higher SMC_HIP_FLEX score and decreased HYP_RF. Both predictors revealed similar strengths in this model.

Neither HYP_PF nor HYP_KNEE_FL or SMC_KNEE_EXT were significantly associated with knee ROMs (Table 2, Figure 2).

Table 2. Multiple linear regression analysis with selected knee ROM and angular velocity at toe-off as independent variables (n = 262).

	β	95% CI	*p*-Value
K5–K3			
HYP_KNEE_EXT	−5.75	[−8.74; −2.76]	0.0002
MMT_KNEE_EXT	2.76	[0.08; 5.44]	0.0433
Adjusted $R^2 = 0.0801$, $F = 11.0963$, $p < 0.0001$			
K4–K3			<0.0001
HYP_KNEE_EXT	−2.74	[−4.98; −0.49]	0.0170
MMT_HIP_FL	2.01	[−3.89; −0.14]	0.0353
MMT_KNEE_EXT	4.04	[1.87; 6.21]	0.0003
Adjusted $R^2 = 0.0744$, $F = 7.2135$, $p < 0.0001$			
K5–K4			
CON_KNEE_EXT	0.16	[0.01; 0.31]	0.0400
HYP_KNEE_EXT	−2.55	[−4.96; −0.14]	0.0384
SMC_HIP_FL	2.02	[1.56; −1.05]	0.1962
SMC_KNEE_EXT	−2.42	[2.03; −6.43]	0.2353
Adjusted $R^2 = 0.0398$, $F = 3.4010$, $p = 0.01$			
V			
HYP_ KNEE_EXT	−39.86	[−60.60; −19.11]	0.0002
SMC_HIP_FL	39.68	[15.27; 64.09]	0.0016
Adjusted $R^2 = 0.1152$, $F = 16.0988$, $p < 0.0001$			

β: slope of regression; CI: confidence interval at 95%; V: angular velocity at toe-off.

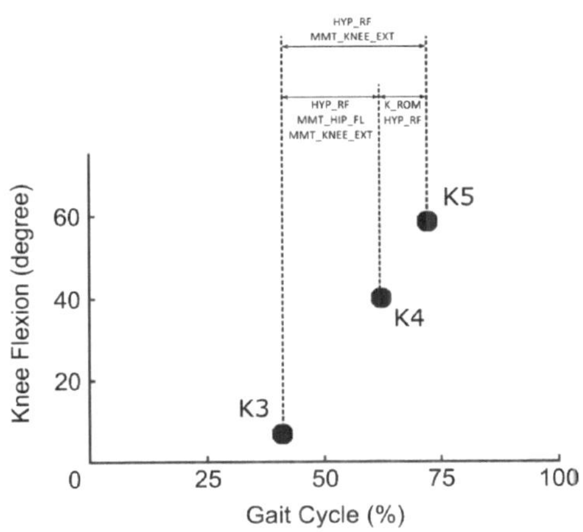

Figure 2. Contribution of impairments to knee kinematics during gait.

4. Discussion

The present study aimed to establish the gradation of the impact of individual coexisting impairments on the selected knee flexion ranges of motion in the sagittal plane.

We found that the hypertonia of the knee extensor muscles had the most significant contribution (β = −5.75) to the restriction of knee flexion ROM from late stance to early swing K5–K3 (Figure 2; Table 2). The secondary factor that contributed to the increase of the K5–K3 range of flexion in those phases was the strength of knee extensor muscles (β = 2.76) (two times weaker than the contribution of hypertonia).

The knee flexion ROM during the stance phase from the minimum in TSt to the toe-off (K4–K3) depended strongest on the knee extensors strength ($\beta = 4.04$) and around 30% less than its hypertonia ($\beta = -2.74$) and 50% less than the strength of hip flexors ($\beta = 2.01$).

The knee flexion ROM between maximum and the beginning of swing (K5–K4) depended strongest on the knee extensors hypertonia ($\beta = -2.55$) and next on the knee flexion passive ROM ($\beta = 0.16$), knee extensors hypertonia was around 20 times stronger decreasing factor than passive ROM.

The knee flexion velocity at TSt/ISw was affected by knee extensor hypertonia ($\beta = -39.86$) and SMC of hip flexor with similar power ($\beta = 39.68$). The rest of investigated impairments did not show any effects on studied periods of knee flexion applying a developed statistical model of gradation of impairments.

We investigated the second period of knee flexion from the minimum in terminal stance and maximum in the swing phases of GC. The knee motion during this period affects the entire knee ROM and velocity of walking. During the entire period of knee flexion (K5–K3) and its parts (K4–K3) and (K5–K4), we found a substantial impact of RF hypertonia, strongest in K5–K3 and K5–K4. During this period, predominantly passive knee flexion takes place [4]. The participation of RF in these periods can be explained by the chain of biomechanical and neurophysiologic events. The predominantly passive knee flexion stretches the knee extensors (including RF) that might trigger spasticity (through exaggerated stretch reflexes) and the next increase in resistance to external stretch (spastic hypertonia). Thus, our data confirmed a potent contribution of knee extensor hypertonia on the restrictions of knee ROM leading the gait deterioration in ambulatory (GMCFS group I to IV) patients with CP. The role of RF hypertonia on knee flexion ROM is still debatable [24–27]. The origin of RF over activity is not clear, and the lack of standard, objective methods of hypertonia evaluation add some confusion. Some studies confirm its contribution, and some contradict it [24–27]. It was reported that different factors might affect knee flexion in the stance than in the swing phase based on an analysis of bioelectrical activation of RF muscle during gait. It was suggested that RF activity could decrease knee ROM during the pre-swing but not at the initial swing phase [26,27]. We found that RF hypertonia was the primary factor that affected knee flexion both in the in stance and in swing phases. However, we also found that in stance phase secondary and third, respectively, were the strength of hip flexors and strength of knee extensors. In swing, the secondary knee flexion ROM. It was previously widely discussed that hypertonia coexists with muscle weakness and impaired selective motor control [10,28–32]. Muscle strength can be affected by impaired neural function due to damage to the central nervous system and by muscle properties. Activation of muscles with proper reciprocal inhibition from antagonist during voluntary movement and/or improper motor unit recruitment might also be connected to spastic hypertonia [33,34]. Reduced selective muscle control is a concern, such as the impaired ability to produce isolated activation of the muscle. This occurs due to lesions in the descending spinal path but can also be affected by reflex irradiation and hypersensitivity to non-specific stimuli (e.g., sound) [1]. Thus, it can also be related to hypertonia.

It was previously reported that diminished knee flexion might be related to low knee flexion velocity at toe-off [12,35] because of the decreased push-off power caused by insufficient plantar flexors activity [4,15,36]. Our data agree that the lower velocity at push-off, the lower range of the motion. However, the knee velocity of the toe-off is inherently linked to the ROM (as a first-time derivate of motion). Our data suggested that the velocity of flexion in toe-off depends exactly to the same extent as knee extensors hypertonia as to the ability to produce selective hip flexion movement but is not related to the plantar-flexors hypertonia. Therefore, with the application of our statistical model, we could not confirm at this time an effect of plantar flexors hypertonia on knee flexion ROM.

It is generally accepted that hip flexors play an important role in walking. Walking with self-paced velocity usually does not require the significant activity of hip flexors because pendular leg motion is driven through the forward velocity of a body. Knee flexion in the swing phase is predominantly passive as a result of passive double pendulum leg movement that is propelled by forward body motion with a potential hip flexors activity and some contribution from push-off [4,5,37]. At a slower velocity of walking, typically acquired by patients with CP, the voluntary hip flexors activation might be necessary to replace the reduced dynamic drive [4]. Our finding proposed that in the stance (K3–K4), the primary contribution came from knee extensor strength, with secondary contribution from knee extensor hypertonia and strength of hip flexors confirmed by previous studies. We found that when the leg is in full swing (K5–K4), not only knee extensor hypertonia but also the passive range of knee flexion participated in the reduction of ROM.

The transfer or lengthening of the rectus femoris muscle is the most common orthopedic procedure to increase knee flexion during the swing phase of walking [38–40]. Theoretically, it was assumed that the transfer of RF and changing its function from knee extensor to flexion should increase the knee flexion ROM [39]; however, the mechanism and benefits of the surgery were recently seriously questioned [18,41,42]. Most likely, the lengthening of the RF and reducing knee extensors force led to the improvement in flexion [42]. Our data does not support this claim—it shows the positive influence of knee extensor strength on kinematics as a secondary predictor (with knee extensor hypertonia as a major factor). It was reported that RF surgery improves gait parameters predominantly in patients of GMFCS levels I and II but not in patients of GMFCS levels III and IV [14]. The more advanced the GMFCS level the more overlapping impairments not only restricted to muscles or joints might be responsible for this [43]. In the future, it seems reasonable to analyze the significance of coexisting impairments separately for CP independent and assisted walkers.

Our study provides the first quantitative data supporting not only an impact but also the gradation of the impact of multiple impairments on knee flexion during walking. Our data also offers strong confirmation of an effect of rectus femoris muscle biomechanical and neurophysiological properties on knee flexion in terminal stance and initial swing. In general, the primary effect is apparent through the muscle spastic hypertonia and a secondary effect through weakness of knee extensors and hip flexors and impairment of SMC of hip flexors.

Limitations

The main limitation of the study was that we investigated data from all CP subjects as one study group putting aside the fact that patients with different levels of involvement presented different levels of severity in symptoms. Patients walking with aids were once weaker and more spastic, with the presence of more impairments than independent walkers [43].

Our study was based on the analysis of the presence or absence of specific impairment based on the results of clinical tests. Taking the low reliability of clinical measures into consideration, we reduced the results of the original tests into two categories. It is expected that the objective, quantitative measures within impairment could increase our understanding of the complex interactions between function and overlapping impairments.

In sum, the very strong contribution of RF hypertonia through the entire period of knee flexion and its parts was accompanied by the weaker contribution of the strength of knee extensors (K5–K3, K4–K3), hip flexors (K4–K3), passive knee range of the motion (K5–K4), and SMC of hip flexors (knee velocity at toe-off). The relationship between clinical parameters and GC is not easy to be established because of the different nature of this data, which is a result of the complex relationship between anatomical, clinical, biomechanical, and physiological factors [27,44,45]. Our data support the claim that, considering all present impairments, their severity and interaction potentially could customize the treatments to an individual's needs.

5. Conclusions

Hypertonia of knee extensors has the primary impact on knee flexion ROM. The strength of knee extensors at the terminal stance up maximum in the swing has a secondary impact. In the stance, the primary contribution came from knee extensor strength, with the secondary contribution from knee extensor hypertonia and strength of hip flexors. In the swing, the primary contribution came from knee extensor hypertonia, with the secondary contribution from passive knee flexion ROM.

Author Contributions: Conceptualization, F.M. and M.K.L.; methodology, F.M. and M.K.L.; software, S.B.; validation, F.M., S.B., M.J. and M.K.L.; formal analysis, F.M., S.B. and M.K.L.; investigation, F.M. and S.B.; resources, M.J.; data curation, F.M., S.B., M.J. and M.K.L.; writing—original draft preparation, F.M. and M.K.L.; writing—review and editing, F.M., S.B., M.J. and M.K.L.; supervision, M.J. and M.K.L.; project administration, F.M. All authors have read and agreed to the published version of the manuscript.

Funding: This research received no external funding.

Institutional Review Board Statement: The study was conducted in accordance with the Declaration of Helsinki, and approved by Research Ethics Committees of Poznań University of Medical Sciences, Poland (approval #244/20 11 March 2020).

Informed Consent Statement: Informed consent was obtained from all subjects involved in the study. Written informed consent has been obtained from the patients to publish this paper.

Data Availability Statement: The datasets generated during and/or analyzed during the current study are available from the corresponding author on reasonable request.

Acknowledgments: The authors wish to thank Brian Po-Jung Chen for his perceptive editing of the paper.

Conflicts of Interest: The authors declare no conflict of interest.

References

1. Sanger, T.D.; Chen, D.; Delgado, M.R.; Gaebler-Spira, D.; Hallett, M.; Mink, J.W. Taskforce on Childhood Motor Disorders. Definition and Classification of Negative Motor Signs in Childhood. *Pediatrics* **2006**, *118*, 2159–2167. [CrossRef] [PubMed]
2. Rosenbaum, P.; Paneth, N.; Leviton, A.; Goldstein, M.; Bax, M.; Damiano, D.; Dan, B.; Jacobsson, B. A Report: The Definition and Classification of Cerebral Palsy April 2006. *Dev. Med. Child Neurol. Suppl.* **2007**, *109*, 8–14. [PubMed]
3. Lebiedowska, M.K.; Wente, T.M.; Dufour, M. The Influence of Foot Position on Body Dynamics. *J. Biomech.* **2009**, *42*, 762–766. [CrossRef] [PubMed]
4. Babyar, S.R. Gait Analysis: Normal and Pathological Function. Perry J, Thorofare, NJ, Slack Inc, 1992. *J. Phys. Ther. Educ.* **1994**, *8*, 47–48. [CrossRef]
5. Dreher, T.; Wolf, S.I.; Maier, M.; Hagmann, S.; Vegvari, D.; Gantz, S.; Heitzmann, D.; Wenz, W.; Braatz, F. Long-Term Results after Distal Rectus Femoris Transfer as a Part of Multilevel Surgery for the Correction of Stiff-Knee Gait in Spastic Diplegic Cerebral Palsy. *J. Bone Jt. Surg. Am.* **2012**, *94*, e142. [CrossRef]
6. Goldberg, S.R.; Ounpuu, S.; Arnold, A.S.; Gage, J.R.; Delp, S.L. Kinematic and Kinetic Factors That Correlate with Improved Knee Flexion Following Treatment for Stiff-Knee Gait. *J. Biomech.* **2006**, *39*, 689–698. [CrossRef]
7. Ounpuu, S.; Muik, E.; Davis, R.B., 3rd; Gage, J.R.; DeLuca, P.A. Rectus Femoris Surgery in Children with Cerebral Palsy. Part II: A Comparison between the Effect of Transfer and Release of the Distal Rectus Femoris on Knee Motion. *J. Pediatr. Orthop.* **1993**, *13*, 331–335. [CrossRef]
8. Rha, D.-W.; Cahill-Rowley, K.; Young, J.; Torburn, L.; Stephenson, K.; Rose, J. Biomechanical and Clinical Correlates of Swing-Phase Knee Flexion in Individuals with Spastic Cerebral Palsy Who Walk with Flexed-Knee Gait. *Arch. Phys. Med. Rehabil.* **2015**, *96*, 511–517.
9. Perry, J. Distal Rectus Femoris Transfer. *Dev. Med. Child Neurol.* **1987**, *29*, 153–158. [CrossRef]
10. Lebiedowska, M.K.; Fisk, J.R. Quantitative Evaluation of Reflex and Voluntary Activity in Children with Spasticity. *Arch. Phys. Med. Rehabil.* **2003**, *84*, 828–837. [CrossRef]
11. Bar-On, L.; Molenaers, G.; Aertbeliën, E.; Monari, D.; Feys, H.; Desloovere, K. The Relation between Spasticity and Muscle Behavior during the Swing Phase of Gait in Children with Cerebral Palsy. *Res. Dev. Disabil.* **2014**, *35*, 3354–3364. [CrossRef] [PubMed]
12. Tuzson, A.E.; Granata, K.P.; Abel, M.F. Spastic Velocity Threshold Constrains Functional Performance in Cerebral Palsy. *Arch. Phys. Med. Rehabil.* **2003**, *84*, 1363–1368. [CrossRef]

13. Bell, K.J.; Ounpuu, S.; DeLuca, P.A.; Romness, M.J. Natural Progression of Gait in Children with Cerebral Palsy. *J. Pediatr. Orthop.* **2002**, *22*, 677–682. [CrossRef] [PubMed]
14. Sousa, T.C.; Nazareth, A.; Rethlefsen, S.A.; Mueske, N.M.; Wren, T.A.L.; Kay, R.M. Rectus Femoris Transfer Surgery Worsens Crouch Gait in Children with Cerebral Palsy at GMFCS Levels III and IV. *J. Pediatr. Orthop.* **2019**, *39*, 466–471. [CrossRef]
15. Kay, R.M.; Rethlefsen, S.A.; Kelly, J.P.; Wren, T.A.L. Predictive Value of the Duncan-Ely Test in Distal Rectus Femoris Transfer. *J. Pediatr. Orthop.* **2004**, *24*, 59–62. [CrossRef]
16. Metaxiotis, D.; Wolf, S.; Doederlein, L. Conversion of Biarticular to Monoarticular Muscles as a Component of Multilevel Surgery in Spastic Diplegia. *J. Bone Jt. Surg. Br.* **2004**, *86*, 102–109. [CrossRef]
17. Salazar-Torres, J.J.; Church, C.; Shields, T.; Lennon, N.; Shrader, M.W.; Sees, J.P.; Miller, F. Comparison of Surgical Outcomes for Distal Rectus Femoris Transfer and Resection Surgeries in Children With Cerebral Palsy with Stiff Knee Gait. *J. Pediatr. Orthop.* **2021**, *41*, 520–524. [CrossRef]
18. Campbell, R.; Tipping, N.; Carty, C.; Walsh, J.; Johnson, L. Orthopaedic Management of Knee Joint Impairment in Cerebral Palsy: A Systematic Review and Meta-Analysis. *Gait Posture* **2020**, *80*, 347–360. [CrossRef]
19. Miller, F.; Cardoso Dias, R.; Lipton, G.E.; Albarracin, J.P.; Dabney, K.W.; Castagno, P. The Effect of Rectus EMG Patterns on the Outcome of Rectus Femoris Transfers. *J. Pediatr. Orthop.* **1997**, *17*, 603–607. [CrossRef]
20. Hislop, H.; Avers, D.; Brown, M. *Daniels and Worthingham's Muscle Testing: Techniques of Manual Examination and Performance Testing*, 9th ed.; Elsevier Health Sciences: Amsterdam, The Netherlands, 2013
21. Trost, J.P. Physical Assessment and Observational Gait Analysis. In *The Treatment of Gait Problems in Cerebral Palsy*; Gage, J.R., Ed.; MacKeith Press: London, UK, 2004; pp. 71–89.
22. Bohannon, R.W.; Smith, M.B. Interrater Reliability of a Modified Ashworth Scale of Muscle Spasticity. *Phys. Ther.* **1987**, *67*, 206–207. [CrossRef]
23. Boyd, R.N.; Graham, H.K. Objective Measurement of Clinical Findings in the Use of Botulinum Toxin Type A for the Management of Children with Cerebral Palsy. *Eur. J. Neurol.* **2007**, *6*, S23–S35. [CrossRef]
24. Young, R.R.; Wiegner, A.W. Spasticity. *Clin. Orthop. Relat. Res.* **1987**, *219*, 50–62. [CrossRef]
25. Ross, S.A.; Engsberg, J.R. Relationships between Spasticity, Strength, Gait, and the GMFM-66 in Persons with Spastic Diplegia Cerebral Palsy. *Arch. Phys. Med. Rehabil.* **2007**, *88*, 1114–1120. [CrossRef] [PubMed]
26. Sutherland, D.H.; Santi, M.; Abel, M.F. Treatment of Stiff-Knee Gait in Cerebral Palsy: A Comparison by Gait Analysis of Distal Rectus Femoris Transfer versus Proximal Rectus Release. *J. Pediatr. Orthop.* **1990**, *10*, 433–441. [CrossRef]
27. Desloovere, K.; Molenaers, G.; Feys, H.; Huenaerts, C.; Callewaert, B.; Van de Walle, P. Do Dynamic and Static Clinical Measurements Correlate with Gait Analysis Parameters in Children with Cerebral Palsy? *Gait Posture* **2006**, *24*, 302–313. [CrossRef]
28. Sanger, T.D.; Delgado, M.R.; Gaebler-Spira, D.; Hallett, M.; Mink, J.W. Task Force on Childhood Motor Disorders. Classification and Definition of Disorders Causing Hypertonia in Childhood. *Pediatrics* **2003**, *111*, e89–e97. [CrossRef]
29. Ross, S.A.; Engsberg, J.R. Relation between Spasticity and Strength in Individuals with Spastic Diplegic Cerebral Palsy. *Dev. Med. Child Neurol.* **2002**, *44*, 148–157. [CrossRef]
30. Fowler, E.G. Concepts in Spasticity and Selective Motor Control in Children with Spastic Cerebral Palsy. *Technol. Disabil.* **2010**, *22*, 207–214. [CrossRef]
31. Fellows, S.J.; Kaus, C.; Ross, H.F.; Thilmann, A.F. Agonist and Antagonist EMG Activation during Isometric Torque Development at the Elbow in Spastic Hemiparesis. *Electroencephalogr. Clin. Neurophysiol.* **1994**, *93*, 106–112. [CrossRef]
32. Levin, M.F.; Hui-Chan, C. Ankle Spasticity Is Inversely Correlated with Antagonist Voluntary Contraction in Hemiparetic Subjects. *Electromyogr. Clin. Neurophysiol.* **1994**, *34*, 415–425.
33. Myklebust, B.M.; Gottlieb, G.L.; Penn, R.D.; Agarwal, G.C. Reciprocal Excitation of Antagonistic Muscles as a Differentiating Feature in Spasticity. *Ann. Neurol.* **1982**, *12*, 367–374. [CrossRef]
34. Grimby, L.; Hannerz, J. Recruitment Order of Motor Units on Voluntary Contraction: Changes Induced by Proprioceptive Afferent Activity. *J. Neurol. Neurosurg. Psychiatry* **1968**, *31*, 565–573. [CrossRef] [PubMed]
35. Goldberg, S.R.; Ounpuu, S.; Delp, S.L. The Importance of Swing-Phase Initial Conditions in Stiff-Knee Gait. *J. Biomech.* **2003**, *36*, 1111–1116. [CrossRef]
36. Thawrani, D.; Haumont, T.; Church, C.; Holmes, L.; Dabney, K.W.; Miller, F. Rectus Femoris Transfer Improves Stiff Knee Gait in Children With Spastic Cerebral Palsy. *Clin. Orthop. Relat. Res.* **2011**, *470*, 1303–1311. [CrossRef]
37. Piazza, S.J.; Delp, S.L. The Influence of Muscles on Knee Flexion during the Swing Phase of Gait. *J. Biomech.* **1996**, *29*, 723–733. [CrossRef]
38. Young, J.L.; Rodda, J.; Selber, P.; Rutz, E.; Graham, H.K. Management of the Knee in Spastic Diplegia: What Is the Dose? *Orthop. Clin. N. Am.* **2010**, *41*, 561–577. [CrossRef] [PubMed]
39. Ellington, M.D.; Scott, A.C.; Linton, J.; Sullivan, E.; Barnes, D. Rectus Femoris Transfer versus Rectus Intramuscular Lengthening for the Treatment of Stiff Knee Gait in Children with Cerebral Palsy. *J. Pediatr. Orthop.* **2018**, *38*, e213–e218. [CrossRef] [PubMed]
40. Josse, A.; Pons, C.; Printemps, C.; Chan-Waï-Nam, J.; Affes, H.; Brochard, S.; Thépaut, M. Rectus Femoris Surgery for the Correction of Stiff Knee Gait in Cerebral Palsy: A Systematic Review and Meta-Analysis. *Orthop. Traumatol. Surg. Res.* **2021**, *24*, 103022. [CrossRef] [PubMed]

41. Riewald, S.A.; Delp, S.L. The Action of the Rectus Femoris Muscle Following Distal Tendon Transfer: Does It Generate Knee Flexion Moment? *Dev. Med. Child Neurol.* **1997**, *39*, 99–105. [CrossRef] [PubMed]
42. Asakawa, D.S.; Blemker, S.S.; Rab, G.T.; Bagley, A.; Delp, S.L. Three-Dimensional Muscle-Tendon Geometry after Rectus Femoris Tendon Transfer. *J. Bone Jt. Surg. Am.* **2004**, *86*, 348–354. [CrossRef]
43. Dan, B.; Mayston, M.; Paneth, N.; Rosenbloom, L. (Eds.) Clinics in Developmental Medicine. In *Cerebral Palsy*; MacKeith Press: London, UK, 2014.
44. Bonnefoy-Mazure, A.; Sagawa, Y., Jr.; Pomero, V.; Lascombes, P.; De Coulon, G.; Armand, S. Are Clinical Parameters Sufficient to Model Gait Patterns in Patients with Cerebral Palsy Using a Multilinear Approach? *Comput. Methods Biomech. Biomed. Engin.* **2016**, *19*, 800–806. [CrossRef] [PubMed]
45. McMulkin, M.L.; Gulliford, J.J.; Williamson, R.V.; Ferguson, R.L. Correlation of Static to Dynamic Measures of Lower Extremity Range of Motion in Cerebral Palsy and Control Populations. *J. Pediatr. Orthop.* **2000**, *20*, 366–369. [CrossRef] [PubMed]

Journal of *Personalized Medicine*

MDPI

Article

Standardized Rehabilitation or Individual Approach?—A Retrospective Analysis of Early Rehabilitation Protocols after Isolated Posterior Cruciate Ligament Reconstruction

Clemens Memmel [1,*], Matthias Koch [2,3], Dominik Szymski [2], Lorenz Huber [2], Christian Pfeifer [2,4], Christian Knorr [1], Volker Alt [2] and Werner Krutsch [2,5]

1 Department of Pediatric Surgery and Orthopedics, Clinic St. Hedwig, Barmherzige Brueder Regensburg, KUNO Pediatric University Medical Center and FIFA Medical Center of Excellence, University Medical Center Regensburg, 93053 Regensburg, Germany
2 Department of Trauma Surgery and FIFA Medical Center of Excellence, University Medical Center Regensburg, 93053 Regensburg, Germany
3 Department of Orthopedic Surgery and Traumatology, Kantonsspital Baselland, Bruderholz, 4101 Basel, Switzerland
4 Department of Trauma Surgery, InnKlinikum Altoetting, 84503 Altoetting, Germany
5 SportDocs Franken, 90455 Nuremberg, Germany
* Correspondence: clemens.memmel@ukr.de

Abstract: (1) Background: Isolated posterior cruciate ligament (PCL) tears represent a severe type of injury. In hospitals, PCL reconstruction (PCL-R) is less frequently performed than other types of knee surgery. It is unclear whether there is consensus among surgeons on how to perform rehabilitation after PCL-R or if there are different, more individual approaches in daily routines. (2) Methods: Rehabilitation protocols and their main criteria (the progression of weight bearing and range of motion, the use of knee braces, rehabilitation training, and sports-specific training) were retrospectively analyzed after PCL-R. (3) Results: Only 33 of 120 (27.5%) analyzed institutes use rehabilitation protocols after PCL-R. The applied protocols showed vast differences between the individual rehabilitation criteria, especially with regard to the progression of weight bearing and the range of motion. The only standardized recommendations were the obligatory use of knee braces and the general restriction of weight bearing and range of motion immediately post-surgery. Therefore, because of the lack of a consensus about a standardized rehabilitation protocol after PCL-R, no recommendation can be made on one particular protocol. (4) Conclusion: There is no acknowledged standardized rehabilitation protocol after PCL-R. In clinical practice, recommendations are influenced by, i.a., surgeons' opinions and experience. The lack of scientific evidence on a particular standardized rehabilitation protocol after PCL-R suggests that rehabilitation protocols need to be tailored to the individual patient.

Keywords: posterior cruciate ligament reconstruction; rehabilitation; weight bearing; range of motion; standardized rehabilitation

Citation: Memmel, C.; Koch, M.; Szymski, D.; Huber, L.; Pfeifer, C.; Knorr, C.; Alt, V.; Krutsch, W. Standardized Rehabilitation or Individual Approach?—A Retrospective Analysis of Early Rehabilitation Protocols after Isolated Posterior Cruciate Ligament Reconstruction. *J. Pers. Med.* **2022**, *12*, 1299. https://doi.org/10.3390/jpm12081299

Academic Editor: Jih-Yang Ko

Received: 1 July 2022
Accepted: 5 August 2022
Published: 8 August 2022

1. Introduction

The posterior cruciate ligament (PCL) extends from the lateral aspect of the medial femoral condyle, adherent to the articular cartilage, down to the PCL facet of the posterior tibia, approximately 1 to 1.5 cm below the tibial plateau. Its course is intrasynovial but extraarticular [1–3]. The main function of the PCL is to limit posterior tibial translation relative to the femur, with an increasing effect the higher the flexion in the knee joint [1,4]. Other tasks, albeit to a lesser extent, include restraining internal and external as well as varus–valgus rotation [1,5]. As previously described, the biomechanical stress on the posterior cruciate ligament increases with the amount of flexion in the knee joint, when the knee joint is passively moved. During active movement, certain muscle groups play a special role, which, when activated, load or unload the posterior cruciate ligament. While activation of the gastrocnemius muscles significantly increases the tension on the posterior cruciate ligament at flexion from 40°, the quadriceps muscles relieve the posterior cruciate ligament at angles between 20 and 60° of flexion. In turn, when the hamstring muscles are activated, this increases the tension on the posterior cruciate ligament just at angles of 70 to 110 degrees of flexion [6]. These anatomical and biomechanical principles must always be present not only if one wants to perform sufficient PCL-R but also postoperative rehabilitation.

Even in patients with asymptomatic, isolated PCL injuries, abnormal knee kinematics can be found during functional activities such as stair ascent and running [7]. In general, tears of the PCL are far less frequent than tears of the anterior cruciate ligament (ACL). The incidence of complete isolated PCL ruptures is 2 per 100,000 [8]. PCL ruptures occur less often isolated than in combination with multi-ligament injuries of the knee or with cartilage defects [9–11]. In 16% of isolated PCL reconstruction (PCL-R), meniscus injury could also be detected. The majority (64%) of isolated PCL ruptures that require PCL-R result from sports injuries. In comparison, in isolated ACL reconstruction, the mechanism of injury is 89% sports-related [11]. Even though the rupture of the PCL is often well tolerated clinically, such ruptures significantly increase the biomechanical stress on the medial and retropatellar articular cartilage and, thus, the risk of medial and retropatellar gonarthrosis [8,12–14]. Studies have shown that conservative treatment of acute isolated PCL ruptures may achieve good functionality and stability of the injured knee joint [15–18]. Since the development of modern surgical techniques, however, arthroscopically assisted surgery has become increasingly sufficient even in the case of acute isolated PCL ruptures with significant posterior tibial translation or chronic PCL tears with symptoms such as problems with deceleration or incline descending and posterior tibial translation [19,20]. In addition, long-term outcome measures of PCL-deficient knees and the role of the PCL in physiological knee biomechanics have become of increasing importance, so that PCL-R has become a suitable therapy for isolated PCL injury [4,21]. Still, the ideal management of isolated PCL injuries, that is conservative therapy vs. surgery and the choice of surgical technique, is still controversially discussed [22].

The rehabilitation phase must be at least as important in the overall treatment of the patient and must be accompanied in a similar way as the surgery itself. Its purpose is not only to protect the surgical result, but also, among other things, to restore the proprioceptive, coordinative, and conditional abilities lost since the injury. During the month-long rehabilitation phase, the patient should not only be accompanied by the orthopedic surgeon, but rather by a rehabilitation team, which, in addition to the patient and surgeon, also consists of a physiotherapist and at best people from the home environment (e.g., parents and partner). There should also be regular exchanges between the patient, the orthopedic surgeon, and the physiotherapist, not only to prevent possible undesirable developments in rehabilitation, but also to document progress and thus keep the patient motivated to continue. Because of a mostly relatively young, sportive patient group with high pre-injury activity levels, the aims of surgery and the following rehabilitation phase are set high. In principle, the pre-injury activity level of the injured person as well as the demand on the quality of life, including the participation in sports after the injury, should be considered when deciding on conservative or surgical therapy. The same principle applies to the rehabilitation phase. Adequate rehabilitation is the key factor for graft healing (particularly of hamstring grafts) and for improving patient outcome. Yet, standardized post-treatment protocols are often used to guide patients through the rehabilitation process and to instruct their physical therapists on weight bearing, the range of motion (ROM), and the begin of rehabilitative training. As recently shown, post-treatment protocols vary significantly, even for well-established and frequently used surgical therapies of the knee joint, such as ACL reconstruction [23] or cartilage-reparative procedures [24].

Based on this finding, the aim of this study is to answer the question whether there exists a standardized approach or if there are different and more individual approaches among orthopedic surgeons on how to perform postsurgical rehabilitation after isolated PCL-R in daily routines.

2. Materials and Methods

Design—In a retrospective cohort study, the early rehabilitation after PCL-R was analyzed in 120 institutes for orthopedic surgery. The medical device company OPED (GmbH, Valley, Germany) developed a tabular template for creating a post-treatment protocol after knee surgery (PCL R). This template was offered to orthopedic departments and orthopedic outpatient centers in German-speaking countries for free, so that associated orthopedic surgeons may create their own individual, but for their own institution standardized, post-treatment rehabilitation protocol using the criteria described in Table 1. OPED itself was not involved in the content of the rehabilitation protocols. All protocols were collected and blinded for study purposes. The investigated protocols contain general early rehabilitation recommendations made by orthopedic surgeons that were given to the patients and their physical therapists. The investigation focused on the availability of rehabilitation protocols after PCL-R and the criteria of early rehabilitation as described in Table 1.

Table 1. Criteria of early rehabilitation.

Criteria of Early Rehabilitation	
Weight bearing	No body weight (NBW)
	Partial body weight (PBW; loading up to 20 kg)
	Half body weight (HBW; loading more than 20 kg)
	Full body weight (FBW; unlimited loading)
Range of motion (extension/flexion)	Immobilization 0°
	0–0–30°
	0–0–60°
	0–0–90°
	Free
Use of braces	Yes/No
	Recommended wearing time (weeks)
Continuous passive/active motion	No CPM/CAM recommended
	CPM recommended
	CAM recommended
Start of rehabilitation training	Weeks after surgery
Start of specific training	Weeks after surgery

Participants—For this qualitative study, we analyzed early rehabilitation protocols after knee surgery (PCL R) that are currently used in daily routines in German-speaking countries (Germany, Austria, and Switzerland). The protocols originated from 120 different orthopedic institutions, of which 63 were outpatient centers, 57 clinical centers, and 4 university medical centers.

Measures—This study focuses on the early rehabilitation phase. This phase can basically be divided into three episodes, each with its own goals. Within knee surgery, these phases are quite similar and are not necessarily PCL-specific. The timing of the phases, in turn, varies individually. The first episode includes the period immediately following surgery. It is also called the acute, inflammatory, or early protective phase. The therapeutic goals are to relieve pain, reduce swelling, and maintain or increase the mobility of the knee joint, especially the patellofemoral joint. The second episode is a transitional phase. It is characterized by a progressive increase in all dynamic modules of post-treatment. This mainly concerns the modules of loading and ROM. In addition, there is the restoration of coordinative abilities. Therefore, the start of rehab training usually falls into this phase. At the end of the second episode, full weight bearing, and the full ROM should be achieved. The third episode of early rehabilitation is the remodeling phase. Its therapeutic goal is to strengthen the important muscle groups and increase endurance with a view to mastering everyday skills. Everyday motor skills should be restored at the end of this episode. During this phase, sport-specific training usually begins. The early rehabilitation protocols were analyzed with regard to the following rehabilitation criteria (see Table 1): postoperative weight-bearing recommendations, restriction of range of motion (ROM), use of braces, recommended continuous passive/active motion (CPM/CAM), as well as the recommended start of rehabilitation training and specific training. Rehabilitation training was defined as any type of basic sports activity to restore coordination and muscle strength, such as ergometer, cycling, aqua jogging, general strength training, or crawling. Specific training included roadwork training, coordination, and proprioception training, as well as sports-specific training. Criteria for the rehabilitation progress such as the duration and time points of limitations and recommendations were documented. All categories, excluding rehabilitation or specific training, were analyzed up to three days postoperatively, seven days postoperatively, and then weekly until full weight bearing and full range of motion were allowed or until knee braces or further CPM/CAM training were no longer required.

Data analysis—Statistical analysis was performed with SPSS® (Version 25, IBM, Armonk, NY, USA). Data are presented as mean $\pm$SD or absolute and relative frequencies. Continuous data between two or more groups were compared with analysis of variance (ANOVA). Categorical data were compared using the chi-squared-test of independence. In case of significance, further post-hoc tests (according to Fisher's least significant difference) were calculated. A probability (p) value of $\leq$0.05 was considered to be significant for each test. Graphical illustrations were generated with GraphPad Prism® (Version 5.01, GraphPad Software, La Jolla, CA, USA) and Microsoft PowerPoint 2013® (Microsoft Corporation, Redmond, WA, USA).

3. Results

3.1. Algorithm of Evaluation

In total, 620 protocols from 120 different orthopedic institutions were collected. A total of 33 (5.3%) of the 620 protocols described the early rehabilitation phase after isolated PCL-R. Twenty (60.6%) of the institutions, one of them a university medical center, provided stationary health care, thirteen (39.4%) were ambulatory. None of the PCL-R rehabilitation protocols were identical.

3.2. Progression of Weight Bearing and Range of Motion

Of the 33 PCL-R protocols, 31 contained detailed information on how to proceed with weight bearing and range of motion (ROM) (see Figures 1 and 2). For the first postoperative week, recommendations ranged from 'no body weight' (n = 3; 9.7%) to 'full body weight' (n = 1; 3.2%), although most recommendations were 'partial weight bearing' (n = 20; 64.5%) or even 'half body weight' (n = 7; 22.6%). 'Full body weight' was recommended immediately after surgery in 1 (3.2%) protocol, after two weeks in 6 (19.4%) protocols, after four weeks in 4 (12.9%) protocols, after six weeks in 13 (41.9%) protocols, and, at the latest, after 12 weeks in 1 (3.2%) protocol.

In the first postoperative week, 11 (35.5%) of 31 protocols recommended 0° to 10° of flexion of the knee joint, 14 (45.2%) protocols up to 30°, and 5 (16.1%) protocols up to 60°. One (3.2%) protocol even allowed pain-adapted, but in principle unrestricted, flexion and extension from the beginning. The majority of orthopedic surgeons recommended a steady increase in ROM in steps of further 30° over the following weeks. After six weeks, the majority of protocols (n = 24; 77.4%) recommended release of the full range of motion in flexion, whereas 7 (22.6%) protocols set the limit at 90°. This limit was finally to be lifted after 9 weeks in 12.9%, after 13 weeks in 6.5%, and after 17 weeks in 3.2% of the protocols.

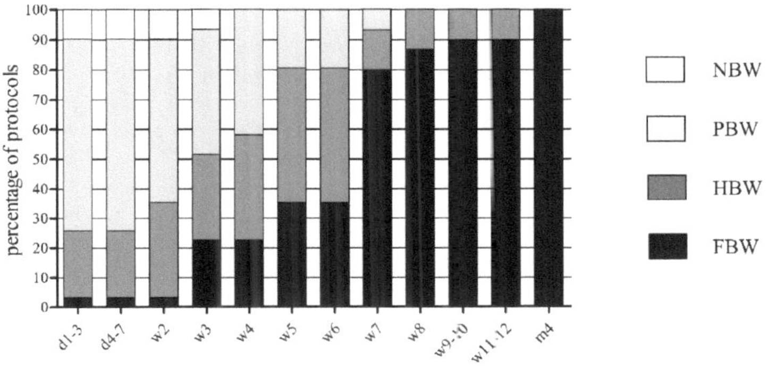

Figure 1. Early rehabilitation phase after PCL-R concerning weight bearing, expressed in percentage of protocols (%). NBW: no body weight, PBW: partial body weight, HBW: half body weight, FBW: full body weight, d: day, w: week.

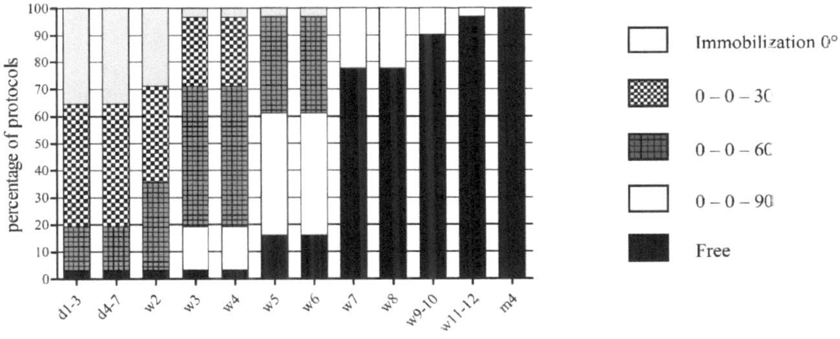

Figure 2. Early rehabilitation phase after PCL-R concerning range of motion (ROM), expressed in percentage of protocols (%). d: day, w: week.

3.3. Use of Knee Braces

According to the post-treatment protocols, the restriction of ROM in case of full restriction of flexion was mostly ensured by the use of an adjustable knee brace or a posterior tibial support brace (PTS). Thirty (90.9%) of the thirty-three included protocols provided specific information on the type of brace and their wearing period. Each protocol that included information about braces (100%) recommended the use of a brace immediately after surgery, approximately one third (n = 9; 30.0%) a PTS, and two thirds (n = 21; 70.0%) an adjustable knee brace. The mean wearing period of a knee brace after PCL-R is 12.2 ± 2.2 weeks (range: 6 to 24 weeks).

3.4. Rehabilitation and Specific Training

A total of 27 (87.1%) of the 33 protocols contained specific recommendations on rehabilitation and specific training after PCL. Rehabilitation training was recommended at the earliest after 8 weeks in four (14.8%) protocols and at the latest after 4 months in nine (14.8%). The majority of protocols (n = 14; 51.9%) recommended starting rehabilitation training between the 8th and 12th postoperative week. Two (7.4%) out of twenty-seven protocols recommended starting specific training at the earliest after 12 weeks and 10 (37.0%) after 6 months and later.

4. Discussion

The early rehabilitation protocols after PCL-R of various institutes did not correspond with regard to individual rehabilitation procedures. Some protocols contained comparable procedures, yet no standardized and more individual and personalized procedures. The main consensus found in over 90% of the protocols was on wearing a knee brace during early rehabilitation. The recommended duration of wearing a brace for an average of 12.2 weeks was almost twice as long as after isolated anterior cruciate ligament reconstruction (6.8 weeks; $p < 0.001$) [23]. This finding shows how highly orthopedic surgeons value additional posterior tibial support and guidance of the knee joint during the first weeks and months after knee surgery. If the type of brace was indicated on the rehabilitation protocol, it was to a large extent a posterior tibial support brace. The main function of this brace is to take over the primary function of the PCL, which is to limit posterior tibial translation relative to the femur. The posterior tibial support should minimize traction forces on the freshly inserted graft. Another point that can be considered consensus is the fact that well over 90% of surgeons limit weight bearing to a maximum of half body weight, as well as clearly limiting the range of motion in flexion to a maximum allowed 60° for at least the first two weeks after surgery (see Figures 1 and 2). These measures are intended to address the low biomechanical loading capacity of the inserted graft. The results clearly show that restriction of weight bearing and ROM after PCL-R is standard procedure in clinical practice.

However, the main finding of our study was the wide range of different rehabilitation protocols after PCL-R, in particular regarding the progression of weight bearing and ROM of the knee joint. Previous researchers did also not find any relevant consensus when systematically reviewing post-treatment recommendations after PCL-R [10,25,26]. The lack of both standardization and scientific evidence is principally a lack for each medical treatment procedure but opens the way for individualized rehabilitation protocols. Surgeons' experience and the activity level of the existing patient collective after PCL-R provide the basis for a personalized but not a standardized rehabilitation protocol; however, surgical techniques still differ significantly and thus play an important role in early post-surgical rehabilitation [4,20,21]. In addition to the preoperative activity level and the surgical technique used, the individual goal of rehabilitation must also be taken into account when planning the rehabilitation phase. This should already be formulated together with the patient preoperatively and also be included in the indication for surgical treatment. These individual goals can range from 'regaining the stability of the knee joint and mobility necessary for everyday life' to 'return to high-performance sports'. Accordingly, the training plans must be adapted to this, and a longer rehabilitation phase must be scheduled the higher the performance goal.

The fact that different post-treatment protocols—based on fixed postoperative time points—partly differ significantly in their respective rehabilitative criteria is not PCL-specific but also applies to other surgical therapies such as repair of the meniscus [27,28], the ACL [23,29], the medial patellofemoral ligament [30,31], or cartilage [24].

Of course, the range of different recommendations seen from the results cannot be attributed solely to the presumption that an individualized approach is preferred. In this respect, the presented results cannot be generalized. One of the limitations that must be blamed for this is the fact that the investigated protocols do not take into account the time of indication for surgical therapy (surgical therapy as primary therapy versus secondary after unsuccessful conservative therapy). Furthermore, due to the retrospective design of the descriptive analyses of rehabilitation protocols of different medical settings, individual clinical outcome or complications during the rehabilitation phase were not included in the assessment of rehabilitation protocols in the context of this study. The same applies to the lack of sociodemographic data such as sex, age, or the pre-operative activity level of the patients. Furthermore, there is no information about the surgical technique applied. However, this study shows the current clinical routine and daily practices of orthopedic surgeons and therefore the current status of postoperative care after PCL-R.

Standards in rehabilitation should obviously not be presented in the form of timeline-based protocols and schedules that refer to the time since surgery; here, a much more differentiated approach seems to be required. The first step towards a patient-centered, personalized rehabilitation standard would be to disregard the notion that 'time since surgery' is a fixed criterion for reaching the next step, such as increasing weight bearing or the release of ROM. A scheduled rehabilitation protocol may suit one patient but may proceed too fast for another with a possible negative clinical outcome. Vice versa, however, professional athletes may achieve their rehabilitation goals more quickly than recreational athletes with a lower preoperative activity level. By setting rehabilitation goals rather than time points, a patient's deficits and potentials can be adequately addressed. This way, post-surgical rehabilitation becomes more personalized and tailored to the individual patient. A similar approach has been proposed for the far more commonly performed reconstruction of the ACL [32] and meniscal repair [33]. For PCL-R, detailed rehabilitation standards already exist that do not contain any time points, but different 'phases' separated by 'goals', which have to be achieved before entering the next rehabilitation 'phase' [25]. This concept also means that rehabilitation does not end after a certain period of time, for instance, after three or six months, but best after the knee joint has regained and maintained its full functionality and the patient has returned to sports activity.

5. Conclusions

The only consensus in the currently existing rehabilitation protocols after PCL-R was on wearing a brace after knee surgery as well as the restriction of weight bearing and ROM immediately post-surgery. However, the wide range of different rehabilitation procedures, especially with regard to the progression of weight bearing and ROM, shows that a standardized post-surgical rehabilitation protocol is not available in clinical practice. Rather, individual rehabilitation seems to be the current solution in daily routine, depending on the time of indication, the surgical technique applied, and most importantly individual patient factors such as pre-operative activity level and the desired goal at the end of the rehabilitation phase.

Author Contributions: Conceptualization, C.M. and C.M.; Methodology, C.M.; Software, C.M.; Validation, L.H., M.K. and D.S.; Formal Analysis, C.K.; Investigation, C.M.; Resources, V.A. and C.P.; Data Curation, M.K.; Writing—Original Draft Preparation, C.M.; Writing—Review and Editing, W.K.; Visualization, L.H.; Supervision, V.A.; Project Administration, W.K. All authors have read and agreed to the published version of the manuscript.

Funding: This research received no external funding.

Institutional Review Board Statement: Not applicable.

Informed Consent Statement: Not applicable.

Data Availability Statement: The data presented in this study are available on request from the corresponding author. The data are not publicly available due to privacy reasons.

Conflicts of Interest: The authors declare no conflict of interest.

References

1. Amis, A.A.; Bull, A.M.J.; Gupte, C.M.; Hijazi, I.; Race, A.; Robinson, J.R. Biomechanics of the PCL and related structures: Posterolateral, posteromedial and meniscofemoral ligaments. *Knee Surgery, Sports Traumatol. Arthrosc.* **2003**, *11*, 271–281. [CrossRef] [PubMed]
2. Bowman, K.F.; Sekiya, J.K. Anatomy and Biomechanics of the Posterior Cruciate Ligament, Medial and Lateral Sides of the Knee. *Sports Med. Arthrosc. Rev.* **2010**, *18*, 222–229. [CrossRef] [PubMed]
3. Matava, M.J.; Ellis, E.; Gruber, B. Surgical Treatment of Posterior Cruciate Ligament Tears: An Evolving Technique. *J. Am. Acad. Orthop. Surg.* **2009**, *17*, 435–446. [CrossRef] [PubMed]
4. Bedi, A.; Musahl, V.; Cowan, J.B. Management of Posterior Cruciate Ligament Injuries: An Evidence-Based Review. *J. Am. Acad. Orthop. Surg.* **2016**, *24*, 277–289. [CrossRef]

5. Kennedy, N.I.; Wijdicks, C.A.; Goldsmith, M.T.; Michalski, M.P.; Devitt, B.M.; Årøen, A.; Engebretsen, L.; LaPrade, R.F. Kinematic analysis of the posterior cruciate ligament, part 1: The individual and collective function of the anterolateral and posteromedial bundles. *Am. J. Sports Med.* **2013**, *41*, 2828. [CrossRef]
6. Dürselen, L.; Claes, L.; Kiefer, H. The Influence of Muscle Forces and External Loads on Cruciate Ligament Strain. *Am. J. Sports Med.* **1995**, *23*, 129–136. [CrossRef]
7. Goyal, K.; Tashman, S.; Wang, J.H.; Li, K.; Zhang, X.; Harner, C. In Vivo Analysis of the Isolated Posterior Cruciate Ligament–Deficient Knee During Functional Activities. *Am. J. Sports Med.* **2012**, *40*, 777–785. [CrossRef]
8. Sanders, T.L.; Pareek, A.; Barrett, I.J.; Kremers, H.M.; Bryan, A.J.; Stuart, M.J.; Levy, B.A.; Krych, A.J. Incidence and long-term follow-up of isolated posterior cruciate ligament tears. *Knee Surgery, Sports Traumatol. Arthrosc.* **2017**, *25*, 3017–3023. [CrossRef]
9. Fanelli, G.C. Posterior cruciate ligament injuries in trauma patients. *Arthrosc. J. Arthrosc. Relat. Surg.* **1993**, *9*, 291–294. [CrossRef]
10. Senese, M.; Greenberg, E.; Todd Lawrence, J.; Ganley, T. Rehabilitation following isolated posterior cruciate ligament reconstruction: A literature review of published protocols. *Int. J. Sports Phys. Ther.* **2018**, *13*, 737–751. [CrossRef]
11. Zsidai, B.; Horvath, A.; Winkler, P.W.; Narup, E.; Kaarre, J.; Svantesson, E.; Musahl, V.; Hamrin Senorski, E.; Samuelsson, K. Different injury patterns exist among patients undergoing operative treatment of isolated PCL, combined PCL/ACL, and isolated ACL injuries: A study from the Swedish National Knee Ligament Registry. *Knee Surg., Sports Traumatol. Arthrosc.* **2022**. Online ahead of print. [CrossRef] [PubMed]
12. Okazaki, K.; Takayama, Y.; Osaki, K.; Matsuo, Y.; Mizu-Uchi, H.; Hamai, S.; Honda, H.; Iwamoto, Y. Subclinical cartilage degeneration in young athletes with posterior cruciate ligament injuries detected with T1ρ magnetic resonance imaging mapping. *Knee Surg. Sports Traumatol. Arthrosc.* **2014**, *23*, 3094–3100. [CrossRef] [PubMed]
13. Schüttler, K.F.; Ziring, E.; Ruchholtz, S.; Efe, T. Posterior cruciate ligament injuries. *Unfallchirurg* **2017**, *120*, 55–68. [CrossRef] [PubMed]
14. Skyhar, M.J.; Warren, R.F.; Ortiz, G.J.; Schwartz, E.; Otis, J.C. The effects of sectioning of the posterior cruciate ligament and the posterolateral complex on the articular contact pressures within the knee. *J. Bone Jt. Surg. Am* **1993**, *75*, 694–699. [CrossRef]
15. Fowler, P.J.; Messieh, S.S. Isolated posterior cruciate ligament injuries in athletes. *Am. J. Sports Med.* **1987**, *15*, 553–557. [CrossRef]
16. Jacobi, M.; Reischl, N.; Wahl, P.; Gautier, E.; Jakob, R.P. Acute isolated injury of the posterior cruciate ligament treated by a dynamic anterior drawer brace: A preliminary report. *J. Bone Jt. Surg. Br.* **2010**, *92*, 1381–1384. [CrossRef]
17. Shelbourne, K.D.; Clark, M.; Gray, T. Minimum 10-Year Follow-up of Patients After an Acute, Isolated Posterior Cruciate Ligament Injury Treated Nonoperatively. *Am. J. Sports Med.* **2013**, *41*, 1526–1533. [CrossRef]
18. Shelbourne, K.D.; Muthukaruppan, Y. Subjective results of nonoperatively treated, acute, isolated posterior cruciate ligament injuries. *Arthrosc. J. Arthrosc. Relat. Surg.* **2005**, *21*, 457–461. [CrossRef]
19. Ahn, S.; Lee, Y.S.; Song, Y.D.; Chang, C.B.; Kang, S.B.; Choi, Y.S. Does surgical reconstruction produce better stability than conservative treatment in the isolated PCL injuries? *Arch. Orthop. Trauma. Surg.* **2016**, *136*, 811–819. [CrossRef]
20. Vaquero-Picado, A.; Rodríguez-Merchán, E.C. Isolated posterior cruciate ligament tears: An update of management. *EFORT Open Rev.* **2017**, *2*, 89–96. [CrossRef]
21. Johnson, P.; Mitchell, S.M.; Görtz, S. Graft Considerations in Posterior Cruciate Ligament Reconstruction. *Curr. Rev. Musculoskelet. Med.* **2018**, *11*, 521–527. [CrossRef]
22. Montgomery, S.R.; Johnson, J.S.; McAllister, D.R.; Petrigliano, F.A. Surgical management of PCL injuries: Indications, techniques, and outcomes. *Curr. Rev. Musculoskelet. Med.* **2013**, *6*, 115–123. [CrossRef] [PubMed]
23. Memmel, C.; Krutsch, W.; Szymski, D.; Pfeifer, C.; Henssler, L.; Frankewycz, B.; Angele, P.; Alt, V.; Koch, M. Current Standards of Early Rehabilitation after Anterior Cruciate Ligament Reconstruction in German Speaking Countries—Differentiation Based on Tendon Graft and Concomitant Injuries. *Int. J. Environ. Res. Public Health* **2022**, *19*, 4060. [CrossRef] [PubMed]
24. Memmel, C.; Krutsch, W.; Koch, M.; Riedl, M.; Henssler, L.; Zeman, F.; Knorr, C.; Alt, V.; Pfeifer, C. The Conformity of Rehabilitation Protocols Used for Different Cartilage Repairs of the Knee Joint—A Review on Rehabilitation Standards in German Speaking Countries. *Appl. Sci.* **2021**, *11*, 8873. [CrossRef]
25. Pierce, C.M.; O'Brien, L.; Griffin, L.W.; LaPrade, R.F. Posterior cruciate ligament tears: Functional and postoperative rehabilitation. *Knee Surgery, Sports Traumatol. Arthrosc.* **2012**, *21*, 1071–1084. [CrossRef] [PubMed]
26. de Paula Leite Cury, R.; Kiyomoto, H.D.; Rosal, G.F.; Bryk, F.F.; de Oliveira, V.M.; de Camargo, O.P.A. Rehabilitation protocol after isolated posterior cruciate ligament reconstruction. *Rev. Bras. Ortop.* **2012**, *47*, 421–427. [CrossRef] [PubMed]
27. Koch, M.; Memmel, C.; Zeman, F.; Pfeifer, C.G.; Zellner, J.; Angele, P.; Weber-Spickschen, S.; Alt, V.; Krutsch, W. Early Functional Rehabilitation after Meniscus Surgery: Are Currently Used Orthopedic Rehabilitation Standards Up to Date? *Rehabil. Res. Pr.* **2020**, *2020*, 3989535. [CrossRef]
28. O'Donnell, K.; Freedman, K.B.; Tjoumakaris, F.P. Rehabilitation Protocols After Isolated Meniscal Repair: A Systematic Review. *Am. J. Sports Med.* **2016**, *45*, 1687–1697. [CrossRef]
29. Valle, C.; Hirschmüller, A.; Schmitt-Sody, M.; Haller, B.; Keller, M.; Schoch, W.; Hoffman, H.; Minzlaff, P. Standards in rehabilitation following anterior cruciate ligament reconstruction in the German-speaking world. *Sportverletz. Sportschaden* **2018**, *32*, 103–110.
30. Fisher, B.; Nyland, J.; Brand, E.; Curtin, B. Medial Patellofemoral Ligament Reconstruction for Recurrent Patellar Dislocation: A Systematic Review Including Rehabilitation and Return-to-Sports Efficacy. *Arthrosc. J. Arthrosc. Relat. Surg.* **2010**, *26*, 1384–1394. [CrossRef]

31. Hilber, F.; Pfeifer, C.; Memmel, C.; Zellner, J.; Angele, P.; Nerlich, M.; Kerschbaum, M.; Popp, D.; Baumann, F.; Krutsch, W. Early functional rehabilitation after patellar dislocation—What procedures are daily routine in orthopedic surgery? *Injury* **2019**, *50*, 752–757. [CrossRef] [PubMed]
32. Cavanaugh, J.T.; Powers, M. ACL Rehabilitation Progression: Where Are We Now? *Curr. Rev. Musculoskelet Med.* **2017**, *10*, 289–296. [CrossRef] [PubMed]
33. Cavanaugh, J.T.; Killian, S.E. Rehabilitation following meniscal repair. *Curr. Rev. Musculoskelet Med.* **2012**, *5*, 46–58. [CrossRef] [PubMed]

MDPI AG
Grosspeteranlage 5
4052 Basel
Switzerland
Tel.: +41 61 683 77 34

Journal of Personalized Medicine Editorial Office
E-mail: jpm@mdpi.com
www.mdpi.com/journal/jpm

www.ingramcontent.com/pod-product-compliance
Lightning Source LLC
Chambersburg PA
CBHW040853230926
51850CB00014BA/3428
9783725832842